ELEMENTS OF
PROPERTIES OF MATTER

ELEMENTS OF
PROPERTIES OF MATTER

With Typical Solved Numerical Problems

For B.Sc. Students of all Indian Universities

D.S. MATHUR

S Chand And Company Limited
(ISO 9001 Certified Company)

S Chand And Company Limited
(ISO 9001 Certified Company)
Head Office: D-92, Sector–2, Noida – 201301, U.P. (India), Ph. 91-120-4682700
Registered Office: A-27, 2nd Floor, Mohan Co-operative Industrial Estate,
New Delhi – 110 044, Phone: 011-49731800
www.**schandpublishing.com**; e-mail: **info@schandpublishing.com**

Marketing Offices:

Chennai : Ph: 23632120; chennai@schandpublishing.com
Guwahati : Ph: 2738811, 2735640; guwahati@schandpublishing.com
Hyderabad : Ph: 40186018; hyderabad@schandpublishing.com
Jalandhar : Ph: 4645630; jalandhar@schandpublishing.com
Kolkata : Ph: 23357458, 23353914; kolkata@schandpublishing.com
Lucknow : Ph: 4003633; lucknow@schandpublishing.com
Mumbai : Ph: 25000297; mumbai@schandpublishing.com
Patna : Ph: 4011400; patna@schandpublishing.com

First Edition 1949
Subsequent Editions and Reprints 1969, 72, 74, 75, 76, 78, 79, 80, 81, 82, 83, 84, 85, 87, 89, 90, 92, 93, 94, 96, 98, 99, 2000, 2001, 2003 (Twice), 2005, 2006, 2007, 2008, 2009, 2010, 2011, 2012, 2013, 2014 (Twice), 2016, 2018 (Twice), 2020 (Twice), 2021, 2022 (Thrice)

Reprint 2026

ISBN: 978-81-219-0815-3 **Product Code:** H6PTM68PHYS10ENAK92O

PRINTED IN INDIA

By Shri Shyamlal Printing Press Private Limited, Plot 20/4, Site-IV, Industrial Area Sahibabad, Ghaziabad – 201 010 and Published by S Chand And Company Limited, A-27, 2nd Floor, Mohan Co-operative Industrial Estate, New Delhi – 110 044.

PREFACE TO THE ELEVENTH EDITION

It gives us great satisfaction that the book still continues to enjoy warm reception; as it has all along been, at the hands of its users—the students and teachers alike. However, the need was felt to update the information given in the text, in the light of latest developments and strides made in the subject-matter of the book. This has been done, while thoroughly revising it, deleting obselete or old classical details and adding new ones. One chapter on "The Special Theory of Relativity" — so vital to the subject has also been added. Many of the diagrams have been redrawn and many more new ones added. The same is true of 'worked out examples' given at the end of each chapter. As usual, latest questions, set at various University Examinations, have been added, both among the solved examples and exercises.

In the light of all the above additions/modifications, etc. it is earnestly hoped that the book, in its new format, will be found to be more useful not only to the B.Sc. degree students, but also those going in for further technical education.

Constructive suggestions for the improvement of the book will be thankfully received.

February, 1992 PUBLISHERS
New Delhi.

PREFACE TO THE FIRST EDITION

A word of explanation is perhaps necessary in adding one more to the existing number of books on the mechanical properties of matter, including some, written by well-known authorities on the subject.

All such standard works, however, take for granted, and rightly so, a certain amount of preliminary knowledge on the part of the student, —an assumption, unfortunately not always true in the case of the average student, who has his own initial difficulties to be overcome, and hazy ideas to be clarified, before he can properly profit by them. It is for students, such as these, that need was felt for a book that could initiate them into the subject, almost from scratch, or *ab initio*, as it were. This volume has been attempted with this end in view, and is intended, primarily, for Pass Degree students of our Indian Universities. Particular care has been taken to make the treatment detailed, yet simple, with each step fully explained, even at the cost of quite a few extra pages, and the risk of annoying the cleverer section of students, who may feel many a step to be superfluous or unnecessary.

Further, knowing from experience now bordering over a quarter of a century, that a *mere* knowledge of theory is no guarantee that the student will be able to tackle a numerical problem with confidence or success, a number of typical problems have been solved at the end of each chapter, and their solutions, with the use of logarithms, shown on the margin to the left. Indeed, these solved problems are responsible for about half the bulk of this volume, their number being large enough to have justified a separate volume, all to themselves, but they have been purposely included here after the relevant chapters, as it was felt that the student will be able to understand them more readily and thoroughly, while the knowledge of the theory is yet fresh in his mind after a perusal of the chapter concerned. This will, it is sincerely hoped, not only stimulate the student's desire to apply his knowledge of theory to solve numerical problems and thus to test for himself how far he has grasped and imbibed the basic principles he has been taught, but will also help him to put his solutions down neatly, concisely and clearly.

The unrivalled methods of Calculus have been used throughout, and, for the benefit of those, not altogether free or familiar with its use, an appendix, on its basic principles has been added at the end, as also other appendices on the Trigonometrical relations required and the use of Logarithms.

No originality is claimed, or indeed can be claimed, in a work, confined to a precisely defined course of this nature, but an honest and humble attempt has been made to present the subject in the simplest possible manner, keeping in view the difficulties and the limitations of the average class of students.

I shall be failing in my duty if I did not unreservedly acknowledge my indebtedness to the numerous books and authors on the subject I have freely consulted during the preparation of this volume, and if I did offer my grateful thanks to Messrs Kanwar Behari Lal, Kishan Behari Lal, Satya Narain, and Dharam Pal Tandon, all my old students who have been of invaluable help to me in fairing out the manuscript, and in various other ways, and whose insistence on the completion of this volume is in main responsible for its appearance today in this form.

A word of apology is also perhaps due to the reader, who may come across some errors and misprints that inadvertently creep in despite all care, due to the great volume and rush of work which our Indian presses have to cope with in the present times.

Suggestions for the improvement of the book will be most gratefully received.

DELHI
October 15, 1949.

D.S. MATHUR

CONTENTS

CHAPTER	PAGES
Chapter I—Unit and Dimensions	**1—18**
Units—Fundamental and Derived Units—Principal Systems of Units—Dimensions—Dimensional Formulae and Equations—Uses of Dimensional Equations—Limitations of Dimensional Analysis—Worked Examples—Exercise I.	
Chapter II—Motion along a Curve—The Projectile	**19—43**
Rotation—Angular Velocity—Angular Acceleration—Couple—Work done by a Couple—Relation between Couple and Angular Acceleration—The Hodograph—Velocity in the Hodograph—Uniform Circular Motion—Centripetal Force—Centrifugal Force—Practical Applications of Centripetal and Centrifugal Forces—Other Effects and Applications of Centrifugal Force—The Projectile—Motion of a projectile in a non-resisting medium—Horizontal Range of a Projectile—Maximum Height attained by a Projectile—Angle of Projection for Maximum Range—Range on an Inclined Plane—Resultant Velocity of a Projectile at a given instant—Worked Examples—Exercise II.	
Chapter III—Moment of Inertia—Gyroscopic Motion	**44—101**
Moment of Inertia and its Physical Significance—Radius of Gyration—Expression for Moment of Inertia—Torque—General Theorems on Moment of Inertia—Calculation of the Moment of Inertia of a Body—Its Units etc.—Particular Cases of Moments of Inertia—Table of Moments of Inertia—Routh's Rule—Practical Methods for the Determination of Moments of Inertia—Angular Moment and Angular Impulse—Law of Conservation of Angular Momentum—Law of Rotation—Kinetic Energy of Rotation—Acceleration of a body rolling down an Inclined Plane—Graphical Representation of Plane Vectors—Precession—The Gyrostat—Gyroscope—The Gyrostatic Pendulum—Case of a Rolling Disc or Hoop—Gyrostatic and Gyroscopic Applications—Worked Examples—Exercise III.	
Chapter IV—Simple Harmonic Motion	**102—131**
Definition—Characteristics of a Linear *S. H. M.*—Equation of Simple Harmonic Motion—Composition of Two Simple Harmonic Motions (Graphical and Analytical Methods)—Composition of two equal circular motions in opposite directions—Energy of a Particle in simple Harmonic Motion—Average Kinetic and Potential Energies of a Particle in *S. H. M.*—Worked Examples—Exercise IV.	
Chapter V—Measurement of Mass—The Balance	**132—143**
Mass and Weight—The Common Balance—Essentials or Requisites of a Good Balance—Faults in a Balance—Determination of True Weight—Correction for Buoyancy—Worked Examples—Exercise V.	
Chapter VI—Acceleration due to Gravity	**144—199**
Acceleration due to Gravity—The Simple Pendulum—*Borda's Pendulum*—Compound Pendulum—Interchangeability of the Centres of Suspension and Oscillation—Centre of Percussion—Other points, Col-	

CHAPTER PAGES

linear with the *c.g.*, about which the time-period is the same—Conditions for Maximum and Minimum Time-periods—Bar Pendulum—Owen's modification of the bar pendulum—*Kater's Reversible Pendulum—Kater's Method of Coincidences*—Computed Time—*Bessel's Contribution*—Errors in the Compound Pendulum and their Remedies—Other Improvements due to Bessel—Conical Pendulum—Steam Engine Governor—Other methods for the determination of *g*—Variation of the value of *g*—Determination of the value of *g* at Sea—Local and Temporal Changes in the value of *g*—Gravity Survey—Geophysical Prospecting—Worked Examples—Exercise VI.

Chapter VII—Gravitation 200—248

Historical—*Kepler's Laws*—Note on Newton's deductions from Kepler's laws—Newton's Law of Gravitation—Determination of the Gravitational Constant—Density of the Earth—Qualities of Gravitation—Law of Gravitation and the Theory of Relativity—Gravitational Field—Intensity of the Field—Gravitational Potential—Potential Energy—Gravitational potential at a point distant *r* from a body of mass-*m*—Velocity of Escapo—Equipotential Surface—Potential at a point Outside and Inside a Spherical Shell—Gravitational Field Inside a Spherical Shell or a Hollow Sphere—Potential and Field Intensity due to a Solid Sphere at a point (*i*) Inside the Sphere and (*ii*) Outside the Sphere—Intensity and Potential of the Gravitational Field at a Point due to a Circular Disc—Intensity and Potential of the Gravitational field at a point due to an Infinite Plane—Inertial and Gravitational Mass—Earthquakes—Seismic Waves and Seismographs—Seismology—Seismographs—*Galitzin's Seismograph*—Determination of the Epicentre and the Focus—Modern Application of Seismology. Worked Examples—Exercise VII.

Chapter VIII—Elasticity 249—326

Introductory—Stress and Strain—*Hooks's Law*—Three types of Elasticity—Equivalence of a shear to a Compression and an Extension at right angles to each other—Shearing stress equivalent to an equal linear tensile stress and an equal compression stress at right angles to each other—Work done per Unit Volume in a Strain—Deformation of a Cube—Bulk Modulus—Modulus of Rigidity—*Young's modulus*—Relation connecting the Elastic Constants—*Poisson's Ratio*—Determination of *Young's Modulus*—Determination of *Poisson's Ratio* for Rubber—Resilience—Effect of a suddenly applied load—Twisting Couple on a Cylinder (or wire)—Variation of stress in a twisted cylinder (or wire)—strain energy in a twisted cylinder (or wire)—Alternative expression for strain energy in terms of stress—Torsional Pendulum—Determination of the Coefficient of Rigidity (η) for a Wire—Determination of Moment of Inertia with the help of a Torsional Pendulum—Bending of Beams—Bending Moment—The Cantilever—(*i*) Loaded at the free end (*ii*) Loaded uniformly—Limitations of the Simple Theory of Bending—Strongly bent beams—Transverse vibrations of a loaded cantilever—Depression of a beam supported at the ends (*i*) when the beam is loaded at the centre (*ii*) when the beam is loaded uniformly—Searle's Method for the comparison of Young's Modulus and coefficient of Rigidity for a given material. Strain energy in a bent beam—Resilience of bent beams—Columns, Pillars and Struts—Critical load for long columns (*i*) when the two ends of the column are rounded or hinged (*ii*) when the two ends of the column are fixed—(*iii*) when one end

CHAPTER PAGES

of the column is fixed and the other loaded—Elastic waves (*i*) Compressional waves—Impact—coefficient of Restitution—loss of kinetic energy on Impact—Relative masses of colliding bodies—Worked Examples—Exercise VIII.

Chapter IX—Hydrostatics **327—347**

Fluids—Liquids and Gases—Hydrostatic Pressure—Hydrostatic Pressure due to a liquid column—The Hydrostatic Paradox—A liquid transmits Pressure equally in all directions—*Pascal' Law*—Thrust on an Immersed Plane—Centre of Pressure—Change of Depth of Centre of Pressure—Particular Cases of Centre of Pressure—Change of Depth of Centre of Pressure—*Principle of Archimedes*—Equilibrium of Floating Bodies—Stability of Equilibrium—Rolling and Pitching of a Ship—Determination of Metacentric Height—Pressure due to a Compressible Fluid or a Gas—Measurement of Atmospheric Pressure—Correction of Barometric Reading—Change of Pressure with Altitude—Worked Examples—Exercise IX.

Chapter X—Jet Planes, Rockets and Satellites **348—361**

Jet propulsion—Thrust supplied by the jet—Efficiency of the jet—Effect of smaller cross-section of the jet—Rocket planes—Rocket fuel—Specific impulse—Shape of the Rocket—The Multi-stage Rocket—Take off of the rocket—Salvaging of the various stage rockets—Satellites—Conditions for a satellite to be placed in orbit—Launching of the satellite—Stability of the rocket during flight—Form of the satellite—Weight and size of the satellite—Material of the frame of the satellite—Duration of satellite's existence—Other essentials—Return of artificial satellite—Uses of an artificial satellite—Exercise X.

Chapter XI—Friction and Lubrication—Principle of Virtual Work and its Simple Applications **362—381**

Static Friction—Laws of Friction—Sliding Friction—Angle of Friction—Cone of Friction—Acceleration down an Inclined Plane—Rolling Friction—Friction and Stability—Friction, a necessity—Simple Practical Applications of Friction—Rope Machines—*(i) The Prony Brake—(ii) The Rope Brake—(iii) The Band Brake*—Mechanism of Friction—Lubricants—Principle of Virtual Work—*(i) Case of a body in equilibrium on a smooth Inclined Plane under the action of a force—(ii) Case of equilibrium of a body on a rough Inclined Plane—(iii) Case of equilibrium of a system of two or more connected bodies—(iv) Relation between Equilibrium and potential energy—(v) Tension in a Flywheel*—Worked Examples—Exercise XI.

Chapter XII—Flow of Liquids—Viscosity **382—422**

Rate of Flow of a liquid—Lines and Tubes of Flow—Energy of the Liquid—*Bernoulli's Theorem* and its important Applications—Important Applications of *Bernoulli's Equation*—Viscosity—Coefficient of Viscosity—Fugitive Elasticity—Critical Velocity—*Poiseuille' Equation* for flow of liquid through a tube—Experimental Determination of η for a liquid—*Poiseulle's Method*—Motion in a Viscous Medium—Determination of Coefficient of Viscosity of a Liquid—*Stokes' Method*—Rotation Viscometer—Variation of Viscosity of a Liquid with Temperature—Comparison of Viscosities—*Ostwald* Viscometer—Determination of Viscosity

CHAPTER PAGES

of Gases—*Rankine's Method* for the determination of the Viscosity of a Gas—Worked Examples—Exercise XII.

Chapter XIII—Diffusion and Osmosis **423—439**

Diffusion—Fick's law—Relation between Time of Diffusion and Length of Column—Experiment Measurement of Diffusivity—*Graham's Law for Diffusion of Gases*—Effusion—Transpiration and Transfusion—Osmosis and Osmotic Pressure—Laws of Osmotic Pressure—Kinetic Theory of Solutions—Osmosis and Vapour Pressure of a Solution—Osmosis and Boiling Point of a Solution—Osmosis and Freezing Point of a Solution—Determination of Percentage of Dissociation of an Electrolyte—Determination of Molecular Weight of a Substance from Elevation of Boiling point or Depression of Freezing-point of a Solution of the substance—Worked Examples—Exercise XIII.

Chapter XIV—Surface Tension—Capillarity **440—496**

Molecular Force—Molecular Range—Sphere of Influence—Surface Tension—Explanation of Surface Tension—Surface-Film and Surface Energy—Free Energy of a Surface and Surface Tension—Pressure Difference across a Liquid Surface—Drops and B ubbles—Excess Pressure inside a Liquid Drop—Excess Pressure inside a Soap Bubble—Determination of the Surface Tension of a Bubble—Work done in blowing a Bubble—Curvature, Pressure and Surface Tension—Layer of Liquid between two plates—Shape of Liquid Meniscus in a Capillary Tube—Angle of Contact—Measurement of the Angle of Contact—Rise of Liquid in a Capillary Tube—Rise of liquid in a Tube of Insufficient Length—Rise of liquid in a Conical Capillary Tube—Energy required to raise a liquid in a Capillary Tube—Rise of a liquid between two Parallel Plates—Force between Bodies Partly Immersed in a Liquid—Shape of a Liquid Drop on a Horizontal Plate—Experimental Determination of Surface Tension. (Different Methods) Surface Tension of Liquid Interfaces—Factors affecting Surface Tension—Experimental Study of the Variation Surface Tension with temperatures—Surface Tension and Vapour Pressure over a Liquid Surface—Effect on Evaporation and Condensation—Worked Examples—Exercise XIV.

Chapter XV—Gases—Kinetic Theory **497—536**

The Kinetic Theory—Introduction—Kinetic Theory of Gases—Pressure Exerted by a Perfect Gas—Value of c—Relation between c and T—Deduction of Gas Laws on the basis of the Kinetic Theory—Kinetic Energy of a Molecule—Value of the Gas Constant—*van der Waal's Equation*—Mean Free Path of a Molecule—Viscosity of Gases—Production of Low Pressure—Exhaust Pumps—Exhaust Pumps and their characteristics—Different Types of Pumps—The Common Air Pump—Rotary Oil Pumps (*Gaede* and *Hyvac* types)—Molecular Pumps—Diffusion Condensation Pumps (*Gaede* and *Waran* types)—Other methods of producing Vacua—Measurement of Low Pressures)—Manometers and Gauges—Common Mercury Manometers—The *Bourdon Gauge*—*Mcleod* Vacuum Gauge—Improved modification of *Mcleod* Gauge—The *Pirani* Resistance Gauge—Thermocouple Gauge—Ionisation Gauge—α-ray Ionisation Gauge—The *Knudsen* Gauge—Worked Examples—Exercise XV.

CHAPTER PAGES

Chapter XVI—Special Theory of Relativity **537—570**

Some common terms—particle, event, frame of reference; Limitations of Newton's Laws of Motion—Inertial frames of reference; Galilian transformation—Position, length, velocity, acceleration; Classical or Newtonion Relativity—Galilian Invariance; Non-inertial frames of reference—fictitious forces; Search for absolute frame of reference; Michelson-Morely experiment; Einstein's concept of Relativity; Special Theory of Relativity; Lorentz Transformations Length-contraction, time-dilation, simultaneity, velocity; Relativity of Mass. Relatvistic Law of Conservation of Momentum : Ultimate speed of a material particle, Equivalence of Mass and Energy—Its importance; Transformation of Relativstic Momentum and Energy; Conservation of Relativistic Energy—relation between the two; Mass, velocity, momentum and energy of a particle of zero rest mass; Worked Examples—Exercise XVI.

APPENDICES

Tables of Constants **571—574**

I.	Densities of Common Substances	571
II.	Elastic Constants	572
III.	Coefficients of Restitution	572
IV.	Coefficients of Viscosity	572
V.	Molecular Elevation of Boiling Points of Solvents	573
VI.	Molecular Depression of Freezing Point of Solvents	573
VII.	Surface Tensions of Important Liquids	573
VIII.	Molecular Constants	574

Logarithmic and Antilogarithmic Tables 575—580

CHAPTER I

UNIT AND DIMENSIONS

1.1. Unit. The physicist always seeks to reduce his physical concepts and conclusions to measurable quantities, in the spirit of Lord Kelvin's very shrewd and apt remark that *'when you can measure what you are speaking about and express it in numbers, you know something about it, but when you cannot measure it in numbers, your knowledge is meagre and unsatisfactory'*, — a remark which is at once a challenge and an inspiration to men of science to sift and clarify their ideas and notions until they became quite precise and clear-cut.

Now, measurement inevitably involves comparison with a chosen *standard* or *unit* of a similar kind ; so that, the first essential step to be taken is the selection of a suitable standard or unit in accordance with the nature of the physical quantity to be measured, and the second, to determine its value in terms of the chosen unit. In other words, to form an exact idea of the magnitude of a physical quantity, it is necessary to express *(i) the standard or unit in which the quantity is measured and (ii) the number of times the quantity contains that unit.*

Thus, for example, when we speak of a distance as being equal to 5 *miles*, we mean that the standard or unit in which it is measured is the *mile*, and that the distance in question is *five times this unit*. If we choose the *yard* (which is 1/1760 of 1 *mile*) or the *foot* (which is 1/1760× 3 of 1 *mile*) as our unit, the *same* distance will be equal to 8800 *yards* or 26,400 *feet* respectively, *i.e.*, its numerical value will be 1760 times or 1760× 3 times 5. Thus, *for a given physical quantity, the larger the unit chosen, the smaller its numerical value and viseversa*. It follows, that *the product of the numerical value of the quantity and the magnitude of the unit in which it is expressed is a constant*. Thus,

$$5 \times 1 = 8800 \times \frac{1}{1760} = 26400 \times \frac{1}{1760 \times 3}$$

Or, in general, if n_1 and n_2 be the numerical values of a given physical quantity corresponding to the units x_1 and x_2 respectively, we have $n_1 x_1 = n_2 x_2$

1.2. Fundamental and Derived Units. For measuring different kinds of quantities, we must obviously have different kinds of units. If these be selected in any arbitrary manner, they will be quite unrelated to each other, and their use will create difficulties and complications in actual practice. They are, therefore, all based on some *fundamental units*, so as to be interdependent and properly related to each other, the guiding principle in their choice being that

(*a*) they are *well-defined* and of a *suitable size*,
(*b*) they are *easily reproducible* at all places,
(*c*) they are *not subject to any secular changes (i.e.*, changes with time),
(*d*) they *do not readily or appreciably vary with varying physical conditions, like temperature, pressure etc.*, and, if they do, their manner of variation is perfectly correctly known.

The *fundamental units chosen*, and internationally employed, are those of **mass, length** and **time** which *C.F. Gauss*, in 1832, termed as *absolute units*. The reason why these alone are chosen as the *'fundamental' units*, and not any others, would seem to be that they represent our elementary scientific notions and cannot be derived from one another ; nor can they be resolved into anything more basic or fundamental. All other units in Mechanics can be derived from them and are, therefore, called *'derived units'*. Thus, the units of area and volume are derived units, for they can both be derived from the unit of length,— the former being the area of a *square*, and the latter, the volume of a *cube*, each of unit length. Similarly, the unit of velocity is a derived unit and is the velocity of a body which covers unit distance or length in unit time, and so on.

1.3. Principal systems of units. There are three principal systems in vogue, *viz.*, *(i) the Metric or the C.G.S. system, (ii) the British or the F.P.S. system* and *(iii) the M.K.S.* or the *S.I. system*.

(i) **The C.G.S. System.** On this system, used in scientific work all over, the unit of length is the *centimetre*, that of mass, the *gramme* and that of time, the *second*.

The Centimetre is *one-hundredth part of a* **metre,** *which is the distance, at a temperature 0°C, between two lines on a platinum-iridium bar, preserved at the International Bureau of Metric Weights and Measures at Sevres, near Paris*.

The International Bureau of Weights and Measures has constructed a line standard *metre*, known as the *Prototype Metre*, copies or replicas of which have been supplied to various governments.

As a safeguard against accidental loss or destruction of the *Metre*, it has been evaluated or 'standardized' in terms of the standard wave-length of a suitable monochromatic light which can be determined with the highest precision. Thus, in terms of the red line of the spectrum of cadmium, in dry air at 15°C temperature and 760 *mm.* pressure, the metre is found to be equal to 1553,164.1 wave-lengths, and can always be accurately reproduced.

The Gramme *is one-thousandth part of a lump of platinum-iridium, called a* **Kilogramme,** made by *Borda*, in accordance with a decree of the French Republic, and also *preserved at Sevres. It is equal to the mass of water, whose volume is one cubic centimetre at 4°C, when it has its maximum density*, (*viz.*, 1 *gm/cc*).

The Second, or the *mean solar second*, as it is called, is $1/24 \times 60 \times 60$*th* or 1/86400*th part of the mean solar day, which is the average value, for one year, of the solar day, or the time which elapses between two consecutive transits of the Sun across the meridian, at any place on the Earth's surface.*

Another unit of time, used in Astronomy, is the *mean sidereal second*, which is 1/86400*th* part of the *sidereal day*, or the true period of revolution of the Earth on its axis, *i.e.*, the interval which elapses between two consecutive passages of a fixed star across the meridian. (1 *mean solar day* = 1.002788 *sidereal days*).

In view of the extreme precision in the measurement of time demanded by the present day requirements of science, a second has, in the year 1964, been defined more accurately in terms of the atomic clock as the *time taken by 9192631770 oscillations of a caesium atom.* Correct to within 1 second in 30,000 years as this is, a still more exact definition is well in the offing in terms of the *HYDROGEN MASER CLOCK* which will, it is hoped be accurate to 1 second in as many as 30 million years.*

(ii) **The F.P.S. System.** On this system, confined more or less to Great Britain, the unit of length is the *foot*, the unit of mass, the *pound* and the unit of time, the *second*.

The foot *is one-third of the distance between two transverse lines, at a temperature of* 62°F, *on two gold plugs in a bronze bar (now replaced by a platinum-iridium one), kept at the Standards Office of the Board of Trade, London.*

---

*Our own National Physical Laboratory in New Delhi maintains a *quartz clock* and broadcasts highly accurate time signals which are then re-broadcast for general information by the All India Radio.

The Pound *(avoirdupois) is the mass of a platinum-iridium cylinder, marked "P.S., 1844, 1 lb." also kept at the Standards Office of the Board of Trade, London.*

The Second, *or the mean solar second,* is the same as defined above.

Other units, derived from those given above, are called the *Board of Trade units* or the **B.O.T. units.**

(iii) **The M.K.S. or the S.I. System.** This is a comparatively new variation of the Metric system in which the *units of length, mass and time are the* **Metre,** *the* **Kilogramme** *and the* **Second** *respectively, the abbreviation S.I. being used for the 'System International d' Unites.'*

The only *new* unit in mechanics, on this system, is that of *force, viz.,* the **newton,** denoted by *N* (or *n*) and defined as *the force required to produce an acceleration of 1 metre per second per second in a mass kilogramme;* so that, 1 *newton* = 10^5 *dynes.*

The advantage of this system is that the units of work (or energy), and power etc. come out to be the same as the familiar practical units on the *C.G.S. system,* namely, the *joule* and the *watt* etc. So that, instead of having to use two sets of units (the absolute and the practical), one set alone suffices for all purposes.

As its very name indicates, it is the system now coming increasingly into vogue internationally, particularly for engineering purposes.

It may as well be mentioned here that we generally choose our units to suit the quantity to be measured. Thus, for example, for the measurement of very small lengths or distances we have successively smaller units of length, *viz.,* the **micron** (μ) = 10^{-3} *mm,* the **millimicron** ($m\mu$) = 10^{-6} *mm* and the **Angstrom unit** (*A.U.* or *A*) = 10^{-7}*mm.* And, for the measurement of very large distances, like those of interstellar space, we have correspondingly larger units like the **light year** or the distance covered by light in vacuo (with a velocity of 2.99×10^{10} *cm/sec*) in one full year. Similar is the case with the units of mass and time.

In general, it has been agreed internationally to adopt the following *multiples* and *submultiples* of the various units.

Multiples		**Submultiples**	
deca (da or D)	– 10^1	*deci (d)*	– 10^{-1}
hecto (h)	– 10^2	*cento (c)*	– 10^{-2}
kilo (k)	– 10^3	*milli (m)*	– 10^{-3}
mega (M)	– 10^6	*micro (b)*	– 10^{-6}
giga (G)	– 10^9	*nano (n)*	– 10^{-9}
tera (T)	– 10^{12}	*pico (p)*	– 10^{-12}

1.4. Dimensions—Dimensional Formulae and Equations

(*a*) **Dimensions.** The units of mass, length and time are denoted by the capital letters, [*M*], [*L*] and [*T*]*, which merely indicate their *nature* and *not their magnitude.* And, since the unit of area is the product of two unit lengths, we have the unit of area represented by [*L*] × [*L*] or [L^2];

*The square brackets merely indicate '*dimension of*'. Once this is understood, they may as well be omitted as we shall quite often do.

and similarly, the unit of volume, being the product of three unit lengths, is represented by $[L]\times[L]\times[L]$ or $[L^3]$. *We express this by saying that the unit of area is of two dimensions in length, and the unit of volume of three dimensions in length.** Since neither the unit of area nor that of volume depends upon mass and time, their dimensions are said to be zero in both mass and time and we may, therefore, represent these units as $M^0L^2T^0$ and $M^0L^3T^0$, respectively.

"The dimensions of a derived unit may thus be defined as the powers to which the fundamental units of mass, length and time must be raised to represent it." Thus, if a derived unit depends upon the *n*th power of a fundamental unit, it is said to be of *n dimensions* in that fundamental unit. For example,

$$velocity = \frac{ditance\ or\ length}{time} = \left[\frac{L}{T}\right] = LT^{-1}$$

and hence the dimensions of the unit of velocity are 1 *in length and* –1 *in time.* Since it is independent of mass, *its dimension in mass is zero*, and we may, therefore, represent it by M^0LT^{-1}

Again, since $acceleration = \frac{velocity}{time}$, *the dimensions of the unit of acceleration are*

$$\frac{M^0LT^{-1}}{T} = M^0LT^{-2}, \text{ and so on.}$$

It will thus be seen that the dimensions of a phyical quantity are obtained by simply defining it in terms of those physical quantities whose dimensions in mass, length and time are known,—the value of a derived unit depending upon the values of the fundamental units from which it is derived. Thus, if we take a *yard* as our unit of length, instead of a *foot*, the units of area and volume will respectively be 3^2 and 3^3 times as big as their units in the ordinary system. So that, *the dimensions of a physical quantity show how its nature and the value of its unit depend upon the fundamental units chosen.*

(*b*) **Dimensional Formulae and Equations.** *A dimensional formula is an expression, showing how and which of the fundamental units enter into the unit of a physical quantity.* Thus, all the expressions in the *Table* opposite, indicating the relation between the derived and fundamental units, are dimensional formulae. For example, the *dimensional formula* for work is ML^2T^{-2}. But when we put it in the form, $W = ML^2T^{-2}$, it is called a *dimensional equation* for work.

This idea of dimensional formulae for physical quantities, as we know it today, was first clearly given by **Fourier**, in the year 1822, although it originated initially with **Newton**, who refers to the *principle of similitude* in his famous and well celebrated *Principia*, (II, Proposition 32).

The student is no doubt aware that in Physics we come across two types of quantities, *viz.*, **variables** and **constants**, which may both be dimensional or non-dimensional (*i.e.*, dimensionless). Thus, we have

(*i*) *Dimensional Variables.* These are quantities like acceleration, velocity, force and most of the others which the Physicist has to deal with, at every step. These are, so to speak, his 'current coin'.

(*ii*) *Dimensional Constants.* Quantities which have a constant value and yet have dimensions are called dimensional constants. As examples of these may be cited *G*, the

* Which is the abbreviated form of '*exponent of dimension*' but is now commonly used and well understood.

Gravitational Constant and *c*, *the Velocity of Light* in vacuo, whose dimensions are $M^{-1} L^3 T^{-2}$* and $M^0 LT^{-1}$ respectively.

(iii) Non-dimensional Variables. These are quantities which are variables and yet have no dimensions; as, for example, *specific gravity, strain* or an *angle*, (see *Table* below)

Here, we also meet with groups of dimensional variables (with or without dimensional constants) such that *their dimension is zero in each of the fundamental quantities, i.e., in length, mass and time.* Thus, for example, the quantity $t\sqrt{g/l}$ has no dimensions; and so also the quantity $v\rho l/\eta$, called *Reynold's number*, can be shown to have zero dimensions in mass, length and time. Such quantities were given the name **'numerics'** by *James Thomson.*

(iv) Non-dimensional Constants. These are mere numbers like 3, 2, π etc.

Thus, *numerics, pure numbers* and quantities like *heat, electricity, temperature and dielectric constant* have no dimensions in *MLT*.

The following *Table* shows at a glance the dimensional formulae of some important physical quantities.

	Physical quantity	**Dimensional formula**
1.	Area = $(\text{length})^2$	$[M^0 L^2 T^0]$ or simply $[L^2]$
2.	Volume = $(\text{length})^3$	$[M^0 L^0 T^0]$ or simply $[L^3]$
3.	Velocity (v) = length/time	$[L/T]$, or $[M^0 L T^{-1}]$ or $[LT^{-1}]$
4.	Acceleration = velocity/time	$\frac{L}{T \times T} = \frac{L}{T^2}$ or $[M^0 LT^{-2}]$ or $[LT^{-2}]$
5.	Momentum = $m \times v$	$M \times L/T$, or $[MLT^{-1}]$
6.	Force = (mass × acceleration) = rate of change of momentum.	$\frac{M \times L}{T \times T} = \frac{M \times L}{T^2}$, or $[MLT^{-2}]$
7.	Impulse = (force × time)	$[MLT^{-2}] \times T = MLT^{-1}$
8.	Work = (force × distance or length)	$\frac{M \times L \times L}{T^2} = [ML^2 T^{-2}]$
9.	Couple (torque) or moment of a force = (force× length)	$M \times (L/T^2) \times L = [ML^2 T^{-2}]$
10.	Kinetic Energy** $= \frac{1}{2} m \times v^2$	$M \times (L^2/T^2) \times L = [ML^2 T^{-2}]$
11.	Potential Energy** = mgh	$M \times \frac{L}{T^2} \times L = ML^2 T^{-2}$
12.	Power (*or rate of doing work*) = work/time	ML^2T^{-2}/T, or $[ML^2 T^0]$
13.	Radius of gyration (K) = distance	L or $[M^0 L T^0]$
14.	Moment of Inertia (I) = MK^2	ML^2 or $[ML^2 T^0]$
15.	*Posisson's* ratio = lateral strain ÷ longitudinal strain = a ratio	No dimensions
16.	Gravitational Constant (G) = force × distence²/ mass²	$\frac{MLT^{-2} \times L^2}{M^2} = M^{-1} L^3 T^{-2}$

* See worked Example 1 (*b*), page 13.

** It will be noted that the dimensions of *couple, kinetic energy* and *potential energy* are the same as those for *work*, because they *are mutually convertibe* and *energy is just work*. Same is the case with *pressure* and *stress*.

	Physical quantity	Dimensional formula
17.	Gravitational potential (V) = work/mass	$M L^2 T^{-2} / M = L^2 T^{-2}$
18.	Density = mass/voume	M / L^3, or $[ML^{-3} T^0]$ or $[ML^{-3}]$
19.	Specific gravity = a *mere ratio*	*No dimensions*
20.	Pressure = force/area	MLT^{-2}/L^2 or $[ML^{-1} T^{-2}]$
21.	Stress = force/area	MLT^{-2}/L^2 or $[ML^{-1} T^{-2}]$
22.	Strain $= \frac{\text{length}}{\text{length}}$ *or* $\frac{\text{volume}}{\text{volume}}$ = *a mere number*	*No dimensions*
23.	Coefficient of Elasticity = stress/strain = stress/a ratio	$[ML^{-1} T^{-2}]$
24.	Coefficient of Viscosity $= \frac{\text{force}}{\text{area}} \div \frac{\text{velocity}}{\text{distance}}$	$\frac{MLT^{-2}}{L^2} \div \frac{LT^{-1}}{L} = [ML^{-1}T^{-1}]$
25.	Surface tension = force/length or = energy/area	MLT^{-2}/L *or* $ML^2 T^{-2}/ L^2 = [ML^0T^{-2}]$ *or* $[MT^{-2}]$
26.	Frequency = 1/time	$1/T = T^{-1}$, or $[M^0 L^0 T^{-1}]$
27.	Angle = length/length = a *number.*	*No dimensions*
28.	Angular velocity = angle/time.	$1/T = T^{-1}$
29.	Angular acceleration $= \frac{\text{angular velocity}}{\text{time}}$	$[1/T] \div T = T^{-2}$
30.	Planck's Constent (h) = energy/frequency.	$ML^2 T^{-2} / T^{-1} = ML^2 T^{-1}$

1.5. Uses of Dimensional Equations. A careful examination of the dimensions of the various physical quantities involved in a relation, *i.e.*, *an analysis of their dimensions*, is of great help to us in more ways than one,—the process being known as **dimensional analysis**. Its three chief uses are the following :

(*a*) *conversion of one system of units into another*,
(*b*) *checking the results arrived at*,
and (*c*) *deriving a correct relationship between different physical quantities*. Let us consider these in some detail.

(*a*) **Conversion of one system of units into another.** It is seen that a physical quantity is expressed in terms of an appropriate unit of the *same nature*, its value being equal to the product of a number and that particular unit. Further, as shown in § 1.1, its value remains the *same* on all systems of units. This affords us an easy method of changing over from one system of units to another.

Thus, suppose there is a physical quantity of dimensions a, b and c in mass, length and time respectively, *i.e.*, whose dimensional formula is $M^a L^b T^c$. Then, if its numerical value be n_1 in one system in which the fundamental units are M_1, L_1 and T_1, it is clearly equal to $n_1[M_1^a L_1^b T_1^c]$

Also, if its numerical value be n_2 in another system of fundamental units M_2, L_2 and T_2, it is equal to $n_2 [M_2^a L_2^b T_2^c]$ in this system. So that,

$$n_2 [M_2^a L_2^b T_2^c] = n_1 [M_1^a L_1^b T_1^c], \text{ whence, } n_2 = n_1 \frac{M_1^a L_1^b T_1^c}{M_2^a L_2^b T_2^c}.$$

Or, $$n_2 = n_1 \left[\frac{M_1}{M_2}\right]^a \times \left[\frac{L_1}{L_2}\right]^b \times \left[\frac{T_1}{T_2}\right]^c \qquad ...(i)$$

So that, knowing the fundamental units in the two systems and the numerical value of the quantity in one of them, its numerical value in the other system can be easily determined. Care must, however, be taken to apply relation (*i*) above, *after expressing the given quantity in absolute units*. Let us consider an example or two.

(1) *To convert a poundal,* (the unit of force in the *F.P.S.* system), *into dynes,* (the unit of force in the *C.G.S.* system).

We know that *force* has dimensions MLT^{-2} and that

$1\ lb = 453.6\ gm$, and $1\ ft = 12 \times 2.54\ cm = 30.48\ cm$. So that,

M units in *F.P.S.* system = $453.6\ M$ units in *C.G.S.* system,

L units in *F.P.S.* system = $30.48\ L$ units in *C.G.S.* system,

and T units in *F.P.S.* system = T units in *C.G.S.* system,

the fundamental unit of time being the same, *viz., the second*, in the two systems.

$$\therefore \quad MLT^{-2}\ poundals = (453.6M)\,(30.48\,L)\,T^{-2}$$

Or, $$1\ poundal = \frac{453.6M \times 30.48L \times T^{-2}}{M \times L \times T^{-2}} = 453.6 \times 30.48$$

$$= 1.382 \times 10^4 \text{ units in the } C.G.S. \text{ system.}$$

Thus, $1\ poundal = 1.392 \times 10^4\ dynes.$

(2) *To convert one Horse Power (F.P.S. system) into Watts (C.G.S. system).* — We know that $1\ H.P. = 550\, ft\text{–}lb\,/\,sec = 550 \times 32.2\, ft\text{-}poundals/sec.$

and $g = 32.2\ ft/sec.^2$

Again, as shown in Ex. (1),

M units in *F.P.S.* system = $453.6\ M$ units in *C.G.S.* system,

L units " " = $32.48\ L$ " "

and T units " " = T " "

Since the dimensional formula for power is ML^2T^{-3}, we have

$$ML^2T^{-3}\ H.P. = 550 \times 32.2\,(453.6M) \times (30.48L)^2 \times T^{-3}$$

$$\therefore \quad 1\ \text{H.P.} = 550 \times 32.2 \times \frac{453.6M \times (30.48L)^2 \times T^{-3}}{M \times L^2 \times T^{-3}}$$

$$= 550 \times 32.2 \times 453.6 \times (30.48)^2 \ ergs/sec.$$

$$= \frac{550 \times 32.2 \times 453.6 \times (30.48)^2}{10^7}\ joules/sec. = 746.4\ watts.$$

[Because, as we know, 1 *joule/sec.* = 1 *watt.*]

Thus, 1 *Horse Power* = 746.4 *watts.*

(*b*) **Checking the results arrived at.** This depends upon what is called the *principle of homogeneity of dimensions,* again due to *Fourier,* according to which *the dimensions of all the terms on the two sides of an equation must be the same.* This follows at once from the fact that it is not possible to compare two physical quantities of different natures, and that only quantities of the

*** This ratio M_1/M_2, if M_1 be the unit of mass in the *C.G.S.* system, M_2, that in the *F.P.S.* system, *i.e.*, the ratio '*gram to the pound*' is called conversion factor.**

same nature can be added up together, their resultant being also of the *same* nature. Thus, for example, we cannot compare *mass* with *length* or *time*, Further, only mass can be added to mass and velocity to velocity, the resultant in the former case being a mass but in the latter a velocity. *If, therefore, in a given relation each term on either side, has the same dimensions, the relation is a correct one*. If not, there is a flaw somewhere, which must be diligently sought out.

Let us again take a couple of examples.

(1) *To check the accuracy of the relation,* $t = 2\pi\sqrt{l/g}$, *for a simple pendulum.*

Here, the term t on the left hand side has only *one dimension in time, i.e.,* (T), its dimensions in both mass and length being zero.

And, on the right hand side, 2π has no dimensions, being just a number; l has one dimension *in length, i.e.*, its dimension is [L], those in mass and time being *zero*; and the dimensions of g, the acceleration due to gravity, are LT^{-2}, that in mass being *zero.* Hence the dimensions of the term $2\pi\sqrt{l/g} = \sqrt{L/LT^{-2}}$ or $\sqrt{T^2}$ = [T] *it has only one dimension in time, the same as the term on the left hand side*. The relation $t = 2\pi\sqrt{l/g}$ is, therefore, a correct one.

(2) *To check the relation* $S = ut + \frac{1}{2}at^2$, *for the distance covered in t seconds by a body, having an initial velocity u and an acceleration a.*

Here, the dimension of the term S on the left hand side is *one in length*, or [L], and taking the terms on the right hand side we have

	(*i*)	dimensions of u (*velocity*)	$= LT^{-1}$
	(*ii*)	dimensions of t (*time*)	$= T$
	(*iii*)	dimensions of $\frac{1}{2}$ (*a number*)	= *Nil*
	(*iv*)	dimensions of a (*acceleration*)	$= LT^{-2}$
and	(*v*)	dimensions of t^2 (*time*^{2})	$= T^2$
∴		dimensions of the term, $ut + \frac{1}{2}at^2$	$= LT^{-1} \times T + LT^{-2} \times T^2$
			$= L + L,$

i.e., the dimension of each term on the right hand side is the same as that of the term on the left hand side; hence the given relation is correct.

A similar dimensional homogeneity will be observed in the case of any other relation, representing a physical phenomenon. The method of dimensions has thus a very definite *mnemonical value** and enables the beginner to resolve his confusion between two alternative possibilities occurring to him regarding a particular half-forgotten formula, as, for example, whether the time-period of a simple pendulum is given by $t = 2\pi\sqrt{g/l}$ or by $t = 2\pi\sqrt{l/g}$, or whether $4\pi r^2$ gives the surface area or the volume of a sphere etc., etc.

(*c*) **Deriving a correct relationship between different physical quantities.** The principle of homogeneity of dimensions also enables us to deduce a relationship between different physical quantities or, at any rate, a preliminary form of such a relationship. For, knowing the factors on which a physical quantity may possibly depend**,— and this requires a little *physical insight* and a

* *i.e.*, value as an aid to memory.

** It is absolutely necessary to take into account all possible *major factors* on which our result may reasonably be expected to depend, though one or more of these factors may get eliminated later. The method, however, ceases to give any worthwhile result if the number of variables included is more than six.

certain amount of '*horse sense*' —an expression for it can be obtained in terms of these factors, such that the dimensions of the terms on the two sides of the expression are the same, *the only acceptable form of the relationship being the one which remains true irrespective of the system of units employed.* A few examples will illustrate the point.

(1) *To deduce an expression for the time-period of a simple pendulum.*

The factors on which the time-period (t) may possibly depend are the following :

(i) the mass of the bob (m), (ii) the length of the pendulum (l), (iii) acceleration due to gravity (g) and (iv) the angle of swing of the pendulum (θ).

Let t be proportional to m^a, l^b, g^c and θ^d. So that,

$t = K.m^a\, l^b\, g^c\, \theta^d$, where K is a *constant of proportionality.*

Taking dimensions of the terms on either side of the sign of equality, we have

Or, $[T] = [M^a]\,[L^b]\,[LT^{-2}]^c = M^a\, L^b\, L^c\, T^{-2c}$.   **K and θ having no dimensions.**

$$T = M^a\, L^{b+c}\, T^{-2c}.$$

Since the dimensions of the terms on the two sides must be the *same*, we have, equating the indices of M, L and T,

$a = 0, b + c = 0$ and $-2c = 1$,

whence, $c = -\frac{1}{2}$ and $\therefore\ b - \frac{1}{2} = 0$, or, $b = \frac{1}{2}$.

Therefore, $t = K.l^{\frac{1}{2}}.\ g^{-\frac{1}{2}}$ Or, $t = K\sqrt{l/g}$.

The value of K can be found out experimentally*, and comes to 2π ; so that, the required relation is $t = 2\pi\sqrt{l/g}$.

It will easily be noted from the above that

(*i*) *the time-period of the pendulum is independent of its mass,*—a fact we know to be true by actual experience;
and (*ii*) *the expression* $t\sqrt{g/l}$, *has no dimensions,* as it is equal to the *dimensionless constant K* and is thus a *numeric.*

(2) *To deduce a relationship for the velocity of sound in a material medium, the temperature of the medium remaining constant.*

The velocity v may depend upon *(i) the elasticity of the medium E* and *(ii) the density of the medium,* ρ ; so that,

$v = K.E^a.\ \rho^b$, where K is a constant.

Again, taking dimensions of the terms on both sides, we have

$$M^0LT^{-1} = \left[\frac{MLT^{-2}}{L^2}\right]^a \left[\frac{M}{L^3}\right]^b$$

$\because$ ***elasticity = stress / strain = $\frac{force/area}{a\ ratio}$ and density = mass/volume.***

$$= [ML^{-1}\,T^{-2}]^a\,[ML^{-3}]^b$$

$$= M^a\,L^{-a}\,T^{-2a}\ M^b\,L^{-3b} = m^{a+b}L^{-a\,-3b}\,T^{-2a}.$$

*** The value of *K* can be determined easily by substituting in the relation obtained, the observed value of *t*, for known values of *l* and *g*.**

Since the dimensions on the two sides must be the *same*, it is clear that $a + b = 0; -a - 3b = 1$, and $-2a = -1$,

whence, $a = \frac{1}{2}$ and $b = -\frac{1}{2}$

Hence, $v = K.E^{\frac{1}{2}}.\ \rho^{-\frac{1}{2}}$ or $v = K\sqrt{E/\rho}$.

The value of K is again determined by experiment, and is found to be 1 in this case; so that, $v = \sqrt{E/\rho}$.

So far only simple cases have been considered. In other cases, the method used above may not always be applicable. Let us consider one such typical example by way of illustration of the method adopted in such cases.

(3) *To obtain a relation between the distance travelled by a body in time t, if its initial velocity be u and acceleration a.*

Let the distance covered by the body in time t be represented by $S = K.\ u^a.\ a^b.\ t^c$. Then, taking dimensions, we have

$$[L] = \left[\frac{L}{T}\right]^a \left[\frac{L}{T^2}\right]^b [T]^c = [LT^{-1}]^a\ [LT^{-2}]^b\ T^c.$$

Or, $[L] = L^a T^{-a} \times L^b T^{-2b} \times T^c = L^{a+b} T^{-a-2b+c}$

Since the dimensions on the two sides must be the *same*, we have $a+b = 1$ .....(*i*)

$-a - 2b + c = 0$; or, $a + 2 - bc = 0$ ....(*ii*)

These two equations alone are *not enough* to give us the values of a, b and c. Hence we proceed as follows:

Suppose the body has no acceleration. Then,

$S = K'\ u^a t^c$, where K is another constant.

Taking dimensions, we have

$L = [LT^{-1}]^a T^c = L^a T^{-a} T^c = L^a T^{c-a}$,

whence, $a = 1$ and $c - a = 0$. Or, $c = a = 1$.

$\therefore\ S = K'.\ ut.$ ...(*A*)

Now, suppose the body has no initial velocity. Then,

$S = K''\ a^b t^c$, where K'' is yet another constant.

Again, taking dimensions, we have

$L = [LT^{-2}]^b T^c = L^b T^{-2b} T^c = L^b T^{c-2b}$

$\therefore\ b = 1;\ c - 2b = 0$. Or, $c = 2b = 2$.

Hence $S = K''.\ at^2$. ...(*B*)

If, thereore, a body has both initial velocity as well as acceleration, its equation of motion contains both expressions (*A*) and (*B*); so that, we have

$$S = K''ut + K''at^2,$$

where the constants K'' *and* K'' can be determined experimentally, and are found to be equal to 1 to and 1/2 respectively. Thus, the required relation comes to be $S = ut + \frac{1}{2}at^2$

In addition to the three chief uses of dimensional analysis, discussed above, mention may also be made here of a couple of others.

(*i*) *It is helpful in selecting experiments likely to give some useful information and avoiding others.* In this connection, *Rayleigh's* remark is worth quoting. Says he, 'I have often been impressed by the scanty attention paid even by original workers in Physics to the great principle of similitude. It happens not infrequently that results in the form of *'laws'* are put forward as novelties on the basis of elaborate experiments which might have been predicted *a priori* after a few minutes' consideration'. How true, indeed!

(*ii*) Then, again, it is a powerful aid to mathematical analysis, when the problem happens to be a complex one and when no experiments to solve it are possible. Even if the number of variables involved in the problem be a large one, dimensional analysis does help obtain at least a partial solution of it.

1.6. Dynamical Similarity An important deduction emerges from point (*ii*) under (1) above (page 9) *viz.*, that if two pendulums having different lengths, (l_1 and l_2) oscillating at two different places, where the values of the acceleration due to gravity are g_1 and g_2 respectively, take time τ_1 and τ_2 to describe *equal arcs*, they may have

$$\tau_1 \sqrt{g_1/l_1} = \tau_2 \sqrt{g_2/l_2},$$

i.e., the value of the *non-dimensional expression or the numeric* $\tau \sqrt{g/l}$ may be the *same* for both.

And, if this be so, it means that the two pendulums pass through *exactly the same phase* for the same value of $\tau \sqrt{g/l}$. This is a case of what is called **dynamical similarity** and all moving systems, such that they pass through an identical phase for the same value of the dimensionless expression (or numeric) consisting of respective corresponding quantities, are said to be *dynamically similar*.

A very interesting and a *classic example of this principle* is the comparison of the speeds of fully grown animals with those of their young.

Very reasonably, *taking the density of the two animals to be the same and their muscular strengths directly proportional to the cross-section of their limbs*, we have the ratio between their *densities* equal to one and similarly that between their *strengths per unit area of cross-section of their limbs*, also equal to *one*; so that, if subscripts 1 and 2 refer to the adult animal and its young respectively, we have

$$\text{ratio of their densities, i.e., } \frac{M_1}{L_1^3} \bigg/ \frac{M_2}{L_2^3} = 1$$

and also *ratio between their muscular strength per unit area, i.e.,*

$$\left(\frac{M_1 L_1}{T_1^2} + L_1^2\right) \bigg/ \left(\frac{M_2 L_2}{T_2^2} + L_2^2\right) = 1$$

From these two relations then, we easily get $\dfrac{L_1}{T_1} = \dfrac{L_2}{T_2}$,

where L_1/T_1 is the speed of the full grown animal and L_2/T_2, that of its young.

The speeds of the two animals are thus the same, — a result which, at first sight, appears to be simply ridiculous. And yet it is an actual fact, *the shorter strides of the young being taken faster than the longer ones of the adult.*

1.7. Limitations of Dimensional Analysis. It will be readily seen from the examples given above, that the method of dimensional analysis is after all not quite so simple or straight in its application, except in obviously easy cases. Very helpful, as far as it goes, it has also its own limitations. Thus, for example:

(*i*) Its one obvious drawback is that *it gives little or no information about pure numerics* (like $t\sqrt{g/l}$ in Ex. 1) *and non-dimensional constants* (like K in Ex. 2), *involved in various physical relations*, and which, therefore, have to be determined by separate calculation or experiment.

(*ii*) Another equally obvious drawback is that its use is by the very nature of things, *restricted solely to physical relationships which are power functions.* Relationships involving exponential and trigonometric functions thus lie quite outside its purview.

(*iii*) Then, again, since at best *only three* equations can be obtained by equating the dimensions of $[M]$, $[L]$ and $[T]$, *the method is of no avail in deducing the exact form of a physical relation which happens to depend upon more than three quantities*, say, n of them. For, clearly, of the n quantities involved, the indices of only three can be expressed in terms of the rest, thus leaving us with a relation between $(n-3)$ non-dimensional groups of terms, giving at best an equation in terms of an undertermined function.

(*iv*) As we have seen in example (3) above, the method cannot directly be applied to obtain the exact form of a physical relationship consisting of more than one term.

(*v*) Finally, quantities having identical dimensions may be of entirely different nature. For example, whereas moment of a force or couple has the same dimensions as work or energy, the two are entirely different in nature, the former being a vector and the latter a scalar quantity.

It will thus be clear that while the method of dimensional analysis remains unrivalled and almost unique, in so far as conversion from one system of units into another and checking the correctness of physical relations are concerned, its use is not quite so safe or certain when it comes to establishing a definite or exact relationship between a given set of physical quantities and, particularly so, in the hands of beginners.

More often than not, the success of the method depends upon the proper choice of *dimensional constants* (like G or c), which have to be introduced as additional variables. And, it needs a trained, subtle and intuitive mind, with the solid background of a mature and comprehensive knowledge of the subject, to decide, on the basis of analysis or experience or perhaps just on that of some sort of inspirtion of the moment, what particular variables to select, and how, when and where to introduce them. A very apt illustration in support of these remarks is perhaps *Rayleigh's explanation, by the method of dimensions, as to why the sky is blue.*

That the colour of the sky is due to the scattering of light by suspended drops of moisture and dust particles etc. (of molecular size) in the atmosphere is fairly well known. From this basic fact, Rayleigh proceeds as follows:

Let A_s be the *amplitude of the scattered wave.* Then, the *possible* factors on which it may depend are

(*i*) A_i, *the amplitude of the incident wave of light,*
(*ii*) l, *the linear dimension of the scattering particle,*
(*iii*) r, *the distance from the particle,*
and (*iv*) λ, *the wavelength of light.*

So that, expressing A_s in terms of all these variables, we have $A_s = K . A_i^{\,a}\, l^b\, r^c\, \lambda^d$,

where K is a constant of proportionality.

Or, taking dimensions, we have $L = L^a . L^b . L^c . L^d = L^{a+b+c+d}$

for, obviously, the dimensions of all these quantities are the same, *viz.*, L, and those of $K = 0$.

We, therefore, have $1 = a+b+c+d$.

Now, we know that the amplitude of the scattered light is (*i*) directly proportional to that of the incident light and (*ii*) inversely proportional to its distance from the scattering particle. This at once gives us $a = 1$ and $c = -1$. And, therefore,

$$1 = 1+b-1+d, \text{ whence, } d = 1-b.$$

So that, $$A_s = K.A_i\, l^b\, r^{-1}\, \lambda^{1-b} = K.\frac{A_i\lambda}{r}\left(\frac{l}{\lambda}\right)^b.$$

Now, as Rayleigh remarks, 'from what we know of the dynamics of the situation (*i.e.*, the ratio of the respective amplitudes of the incident and reflected light), l varies directly as the volume of the scattering particle. And, therefore, $b = 3$.

Hence $$A_s = K.\frac{a_i\, l^3}{r\lambda^2}. \text{ Or, } A_s \propto 1/\lambda^2.$$

And since *intensity* $\propto$ (*amplitude*)^2, we have

intensity of scattered light, $I_s \propto 1/\lambda^4$.

It thus follows, as a natural consequence, that the wave-length of blue light being roughly half that of red light the intensity of scattered blue light is sixteen times that of scattered red light and that the sky, therefore, appears to us to be blue.

The student will appreciate how, in capable hands, the method of dimensional analysis can be made to yield results beyond the pale of elementary analysis.

WORKED EXAMPLES

1. Deduce the dimensions of (*a*) the Coefficient of Viscosity, and (*b*) the Constant of Gravitation (G). Obtain a formula for the time of swing of a simple pendulum from a knowledge of the dimensions of the physical quantities involved. *(Panjab)*

(*a*) We know that the coefficient of viscosity (η) of a liquid is given by the relation, $\eta = \pi\, Pr^4/8vl$, where P is the *pressure difference* between the two ends of the capillary tube; r, its *radius*; l, its *length* and v, the *rate of flow of the liquid* through it, or the *volume of liquid flowing out per second*.

Therefore, taking dimensions of all these quantities, we have

dimensions of $P = ML^{-1}T^{-2}$ [see *Table* one page 5

" " $r^4 = L^4$

" " $v = L^3T^{-1}$ [$\because$ *rate of flow = volume/time*.

" " $l = L$

" " π and 8 are *zero*. [Both being *numbers*.

Hence, dimensions of coefficient of viscosity η are

$$\frac{ML^{-1}T^{-2} \times L^4}{L^3T^{-1} \times L} = ML^{-1}T^{-1}$$

(*b*) We know that, in accordance with *Newton's law of Gravitation*, (*see* Chapter VII) the gravitational force of attraction between two masses m and m', a distance r apart is given by

$$F = \frac{m.m'}{r^2}.G, \text{ were } G \text{ is the gravitational constant.}$$

We, therefore, have $$G = \frac{F.r^2}{m.m'}.$$

Taking the dimensions of the quantities involved, we have

dimensions of $F = MLT^{-2}$ [see *Table* on page 5.

" " $r^2 = L^2$

" " $m, m' = M$, each

Hence, dimensions of G are $\frac{MLT^{-2}.L^2}{M.M} = M^{-1}L^3T^{-2}$.

The student may, as an additional exercise, try to deduce the dimensions of G from the relation $G = C\theta.d^2/M.m.l$, obtained in Cavendish's experiment for the determination of the value of G (page 235).

For answer to the second part of the question, *see* page 9, (Ex. 1).

2. Find the unit of length if one minute be the unit of time; one stone, the unit of mass; and one pound-weight, the unit of force. Given that $g = 32.2$ *ft. per sec. per sec.*

We know that 1 *lb. wt* = *g poundals* = 32.2 *poundals*, and that the dimensions of force are MLT^{-2}.

Now, M units in the new system = $(14M)$ units in the ordinary system, [$\because$ 1 *stone* = 14 *lb*.

T „ „ „ = $(60T)$ units in the ordinary system

and let L „ „ „ = (xL) units in the ordinary system.

Then, $32.2\ MLT^{-2} = 14.xL \times [60\ T]^{-2}$ units in *this* system

$= 14M.xL.60^{-2}T^{-2}$ units in system

Or, $32.2 = \frac{14x}{60^2}$. And $\therefore\ x = \frac{32.2 \times 60^2}{14} = 8283$.

Hence, the *unit of length in the new system would be xL*

$= 8283 \times L$, *i.e.*, equal to 8283 *ft*. [Since $[L]=1$ *ft*

3. If the acceleration due to gravity be represented by unity and one second be the unit of time, what must be the unit of length?

In the ordinary system, in which the unit of length $(L) = 1$ *cm*, and that of time $(T) = 1$ *sec*, we have

unit of acceleration, (dimensions LT^{-2}) = 1 cm/sec^2, and acceleration due to gravity equal to 981 cm/sec^2 = 981 LT^{-2}.

If the unit of length *in the new system* be L_1, we have

acceleration due to gravity, in this system = $1 \times L_1/sec^2 = L_1T^{-2}$, the unit of time being the *same*, *i.e.*, 1 *second*, in this system also.

$\therefore$ $L^1T^{-2} = 981\ LT^{-2}$ [§1

Or, $L_1 = 981\ L$; *that is* $L_1 = 981$ *cm*, since $L = 1$ *cm*

Thus, the unit of length in the new system must be 981 *cm*.

4. Given that the unit of power is one million ergs per minute, the unit of force is 1000 dynes and the unit of time, 1/10 sec., what are the units of mass and length?

Here (*a*), unit of power, [dimensions ML^2T^{-3}] = 1000,000 *ergs/min*.

= 1000,000/60 *ergs/sec*.

(*b*) unit of force, [dimensions MLT^{-2}] = 1000 *dynes*

and (*c*) unit of time [dimension T] = 1/10 *sec*.

$\therefore$ multiplying (*a*) by (*c*), we have

unit of power $\times$ unit of time = unit of work.

$= [ML^2T^{-3}][T] = ML^2T^{-2}$.

$$= \frac{1000,000}{60} \times \frac{1}{10} = \frac{10^4}{6}$$

Dividing *this unit of work by the unit of force*, we have

unit of distance or length $= \frac{10^4}{6} \times \frac{1}{1000} = \frac{10}{6} = \frac{5}{3}$ *cm*

Now, from (*b*) we have

unit of mass $= MLT^{-2} \div \frac{L}{T^2} = MLT^{-2} \times \frac{T^2}{L}$ $\left[\because\ mass = \frac{force}{acc}\right]$

Or, *unit of mass* $= \frac{1000(1/10)^2}{5/3} = \frac{1000 \times 3}{100 \times 5} = 6$ *gm*.

[∵ *unit of force* = 1000 *dynes and unit of acc.* = $\frac{5/3}{(1/10)^2}$]

Therefore, the units of mass and length, in the given system, are 6 gm and 5/3 cm, respectively.

5. If the fundamental units are the velocity of light in air, the acceleration of gravity at Greenwich, and the density of mercury at 0°C, find the units of mass, length and time. (*velocity of light* = 3×10^{10} *cm/sec; acceleration of gravity at Greenwich* = 9.81×10^2 *cm/sec ; density of mercury* = 13.6 *gm per c.c.*)

Here, (*a*) *unit of velocity,* (dimensions LT^{-1}) = 3×10^{10} *cm/sec*

(*b*) ,, ,, *acceleration* (,, LT^{-2}) = 9.81×10^2 *cm/sec*^{2}

(*c*) ,, ,, *density* (,, ML^{-3}) = 13.6 *gm/cm*^{3}

∴ dividing (*a*) by (*b*), we have

$$\text{unit of time } (T) = \frac{3 \times 10^{10}}{9.81 \times 10^2} = \frac{3 \times 10^8}{9.81}$$

$$= 3.058 \times 10^7 \text{ seconds.}$$

Substituting this value in (*a*), we have unit of length, $(L) = LT^{-1}.T.$

$$= 3 \times 10^{10} \times 3.058 \times 10^7 = 9.174 \times 10^{17} \text{ cm}$$

And, from (*c*), we have unit of mass, $(M) = ML^{-3} \times L^3 = 13.6 \times (9.174 \times 10^{17})^3$

$$= 1.051 \times 10^{55} \text{ gm.}$$

Thus, the required units of mass, length and time are 1.051×10^{55} *gm*, 9.174×10^{17} *cm and* 3.058×10^7 *seconds, respectively.*

6. If the units of length and force be each increased four times, show that the unit of energy is increased sixteen times.

We have *unit of energy* = *unit of force* × *unit of distance*

$$= [MLT^{-2}] \times L = ML^2T^{-2}$$

If now, the unit of force and distance be made *four times each*, they would be 4 (MLT^{-2}) and $4L$ respectively, and, therefore, the *new unit of energy* would be $4MLT^{-2} \times 4L = 16\,ML^2T^{-2}$, which is sixteen times ML^2T^{-2}, the ordinary unit.

Thus, we see that by increasing the unit of force and length four times each, the unit of energy is increased sixteen times.

7. Show by the method of dimensions that the relation, $C = n\pi r^4/2l$ for the couple per unit twist of a wire, of length l, radius r and coefficient of rigidity n, is a correct one.

Let us take the dimensions of the terms on the two sides of the sign of equality and see if they are the same. Thus,

dimensions of *C* (couple) = ML^2T^{-2}

,, ,, (*rigidity*) = $ML^{-1}T^{-2}$ [*Same as for elasticity.*

,, ,, r^4 (*radius*)^4 = L^4

,, ,, *l* (*length*) = L

and, π and 2, being *numbers*, have no dimensions.

Therefore, the dimensions of the term $n\pi r^4/2l$, are

$$\frac{ML^{-1}T^{-2} \times L^4}{L} = ML^2T^{-2},$$

the same as for C, on the left hand side

Hence, the relation $C = n\pi r^4/2l$ is a correct one.

8. Test by the method of dimensions the accuracy of the relation $t = 2\pi\sqrt{(k^2 + l^2)/lg}$ for the time period of a compound pendulum.

If the relation be correct, the dimensions of the terms on either side of the sign of equality must be the same.

Let us put the relation as $t = 2\pi\sqrt{\frac{k^2}{lg} + \frac{l}{g}}$

Now the dimensions of $t = [T]$

,, ,, ,, $K^2 = [L^2]$, *K* being the *radius of gyration.*

,, ,, ,, $l = [L]$

,, ,, ,, $g = [LT^{-2}]$.

Therefore, the dimensions of the terms on the right hand side are

$$= \sqrt{\frac{L^2}{LLT^{-2}} + \frac{L}{LT^{-2}}} = \sqrt{T^2 + T^2} = \sqrt{T^2} = T.$$

Thus, we see that *the dimensions of the terms on either side are the same, viz.*, $[T]$. The relation is, therefore, a correct one.

9. Find the dimensions of velocity and acceleration. Assuming that when a body falls from rest under gravity the velocity v is given by $Kg^p h^q$, where h is the distance fallen through, g, the acceleration of gravity and K, p and q are constants. Show by a consideration of the dimensions involved, that $v = K\sqrt{gh}$.

(London Higher School Certificate)

For answer to part one of the question, see *Table* on page 5.

We are given that $v = Kg^p h^q$. Taking dimensions, therefore, we have

dimension of $v = \left[\frac{L}{T}\right] = LT^{-1}$

" " $g^p = \left[\frac{L}{T^2}\right]^p = [LT^{-2}]^p = L^p T^{-2p}$

dimensions of $h^q = [L]^q = L^q$.

" " $K = 0$, for it is a *constant or a mere number*.

Therefore, the dimensions of the term $Kg^p h^q$

$$= L^p T^{-2p} L^q = L^{p+q} T^{-2p}.$$

Since dimensions on both sides of the sign of equality must be the same, we have $LT^{-1} = L^{p+q} T^{-2p}$.

Or, $p+q = 1$ and $-2p = -1$, whence, $p = \frac{1}{2}$ and $q = \frac{1}{2}$

∴ $Kg^p h^q = Kg^{\frac{1}{2}} h^{\frac{1}{2}} K\sqrt{gh}$

$$v = K\sqrt{gh}$$

10. The frequency of vibration (n) of a stretched string is a function of the tension (T), the length (l) and the mass per unit length (ρ) Prove that

$$n \propto \frac{1}{l}\sqrt{\frac{T}{\rho}}$$

Let $n \propto T^a l^b \rho^c$. Then, taking dimensions, we have

dimensions of $n = 1/t = M^0 L^0 T^{-1}$, or $= T^{-1}$

dimensions of T^a, $(force)^a = [MLT^{-2}]^a = M^a L^a T^{-2a}$,

" " " l^b, $(length)^b = L^b$,

" " " ρ^c, $\left(\frac{mass}{length}\right)^c = \left(\frac{M}{L}\right)^c = M^c L^{-c}$

∴ dimensions of the term $T^a L^b \rho^c = M^a L^a T^{-2a} . L^b . M^c L^{-c}$.

Or $T^{-1} = M^{a+c} . L^{a+b-c} . T^{-2a}$.

Since dimensions of n must be the same on both sides, we have $a+c = 0$; $a+b-c = 0$ *and* -1, or $2a = 1$, *i.e.*, $a = \frac{1}{2}$, and hence $\frac{1}{2} + c = 0$, or $c = -\frac{1}{2}$

Also $\frac{1}{2} + b - (-\frac{1}{2}) = 0$, or $\frac{1}{2} + b + \frac{1}{2} = 0$, *i.e.*, $1+b = 0$, $b = -1$.

Therefore, $T^a L^b \rho^c = T^{\frac{1}{2}} . l^{-1} . \rho^{-\frac{1}{2}} = \frac{1}{l}\sqrt{T/\rho}$

Hence $n \propto \frac{1}{l}\sqrt{T/\rho}$

11. The time of oscillation (t) of a small drop of liquid under surface tension depends only on the density (ρ), the radius (a), and the surface tension (T). Show that the period of oscillation is $K\rho^{\frac{1}{2}} . a^{\frac{3}{2}} . T^{-\frac{1}{2}}$ *where K is a numeric.*

Let $t = K\rho^a a^b T^c$. So that, taking dimensions, we have

dimensions of $t = T$, of $\rho^a = [M/L^3]^a = M^a L^{-3a}$, of $a^b = [L]^b = L^b$, and of $T^c = [MT^{-2}]^c = M^c T^{-2c}$, [See page 5.

K having no dimensions being a numeric.

$\therefore$ dimensions of the term $K\rho^a a^b T^c = M^a L^{-3a} L^b M^c T^{-2c}$

Or, $= M^{a+c} L^{-3a+b} T^{-2c}$.

Since the dimensions of the terms on both sides must be the same, we have

$a + c = 0, -3a + b = 0$, and $-2c = 1$.

Or, $c = -\frac{1}{2}$ and $\therefore$ $a = \frac{1}{2}$ and $b = \frac{3}{2}$.

$$\therefore \quad K\rho^a a^b T^c = K\rho^{\frac{1}{2}} . a^{\frac{3}{2}} T^{-\frac{1}{2}}$$

Hence *the time of oscillation of the drop*, $t = K\rho^{\frac{1}{2}} . a^{\frac{3}{2}} . T^{-\frac{1}{2}}$.

12. Explain the principle of homogeneity of dimensions in a physical equation.

Assuming that the mass M of the largest stone that can be moved by a flowing river depends on V, the velocity, ρ, the density of water and on g, show that M varies with the sixth power of the velocity of flow in the river. *(Panjab)*

Let M depend upon V^a, ρ^b and g^c.

So that, $M = KV^a \rho^b g^c$. [K being a constant.

Taking dimensions, we have

dimensions of $M = [M]$, of $V^a = [L/T]^a = L^a T^{-a}$,

Of $\rho^b = [M/L^3]^b = M^{-3b}$, of $g^c = [LT^{-2}]^c = L^c T^{-2c}$

and K has no *dimensions*, being a mere *number*.

$\therefore$ $M = L^a T^{-a} M^b L^{-3b} L^c T^{-2c}$. Or, $M = L^{a-3b+c} M^b T^{-a-2c}$

Since the dimensions on both sides must be the same, we have

$a - 3b + c = 0, b = 1$, and $-a - 2c = 0$, or $a + 2c = 0$;

So that, $a + c = 3b = 3$ and $a + 2c = 0$. And $\therefore$ $c = -3$ and $a = 6$.

Hence $M = KV^6 \rho . g^{-3}$.

Or, $M \propto V^6$

i.e., the mass M varies with the sixth power of the velocity of flow.

EXERCISE I

N.B. Questions marked with an asterisk are of a rather advanced character and may be attempted when confidence has been gained with others.

1. If 10 kg be the unit of mass, 60 *sec*, the unit of time, and the acceleration due to gravity 9.81 *m/sec*^{2}, the unit of acceleration, what would be the unit of energy in *ergs*? **Ans.** 3.46 $\times 10^{12}$ *ergs*.

2. Convert by the method of dimensions, 4.21×10^7 *ergs* into *foot-poundals* (1 *ft* = 30.48 *cm* and 1 *lb* = 453.6 *gm*). **Ans.** 96.6 *ft poundals*.

3. Deduce the dimensions of (*i*) *specific gravity*, (*ii*) *surface density* and (*iii*) *angular velocity*.

Show that the kinetic energy of a body of mass m, moving with a velocity r, is given by Kmv^2, where K is a constant. **Ans.** (*i*) $M^0 L^0 T^0$; (*ii*) $ML^{-2} T^0$; (*iii*) $M^0 L^0 T^{-1}$.

4. Test, by the method of dimensions, the accuracy of the following relations:

(*i*) $v^2 - u^2 = 2aS$, connecting *initial velocity u*, *final velocity v*, *acceleration a* and *distance S* covered by a body.

(*ii*) $S = ut + \frac{1}{2} at^2$, connecting *distance S* with *initial velocity u*, *time t* and *acceleration a* of the body.

(*iii*) $\rho = \frac{3\,g}{4\,\pi rG}$ where ρ is the *mean density of the earth*, r, its *radius* g, the *acceleration due to gravity* and G, *the gravitational constant*.

5. Assuming that the excess pressure (p) inside a soap bubble depends on (*i*) the surface tension (T) of the soap film and (*ii*) its radius (r), show, by your knowledge of dimensions, that it is directly proportional to the former and inversely proportional to the latter.

[**Hint.**—*Simply show that p=k . T/r., whence it follows that p varies directly as T and inversely as r.*]

6. A drop of liquid is suspended in another liquid of the *same* density but with which it is *immiscible*. If the drop is distorted from the spherical shape and released, deduce, by dimensional methods, a formula for its period of oscillation (t), given that the latter depends on *surface tension T, density ρ and drop radius r*.

Ans. $t = k\sqrt{\rho r^2/r}$, where k is a constant.

7. What do you mean by a dimensional equation? What are its uses?

The frequency 'n' of vibration of a stretched string is a function of its tension 'F', its length 'l' and its mass per unit length 'm'. Show that $n \propto \frac{1}{l}\sqrt{\frac{F}{m}}$, making use of the knowledge of dimensions.

8. Show that when bodies of geometrically similar form and of the same material, differing only in dimensions, vibrate in the same manner, the vibrations being due to the elasticity of the material, their periods are proportional to their dimensions.

[**Hint.**— *Proceeding in the usual manner, show that* $t = kl/\sqrt{\rho/E}$, *where l is the linear dimension of the body, ρ, its density and E, the elasticity of the material. Since ρ and E are the same for all bodies, t varies directly as l.*]

9. Calculate, by the method of dimensions, the number of *foot-pounds* in *one calorie*. (Given that 1 *calorie* = 4.2×10^7 *ergs*; $g = 32\,ft/sec^2$; $1\,lb = 453.6\,gm$, and $1\,inch = 2.54\,cm$) **Ans.** 3.115.

10. If in a system of units, the unit of length be 1 *mile* and that of time, 1 *hour*, what will be the value of 'g'?

Ans. 14.88 *miles/sec*^{2}

11. The time of oscillation t of a small drop of a liquid under surface tension depends upon the density d, radius r and surface tension S. Prove dimensionally that $t \propto \sqrt{dr^3/S}$. *(Panjab)*

12. Explain what you mean by the dimensions of a physical quantity; calculate the dimensions of Young's modulus.

Assuming that the period of vibration of a tuning fork depends upon the length of the prongs, and on the density and Young's modulus of the material, find, by the method of dimensions, a formula for the period of vibration. *(Calcutta)*

Ans. $[ML^{-1}T^{-2}]$; $t \propto l\sqrt{d/Y}$,

(where t is the *period of vibration*; l, the *length*; d, *the density and* Y, the value of *Young's Modulus* for the material of the fork).

*13. Using the method of dimensions, obtain an expression for

(*i*) the acceleration of a particle moving with a uniform speed v, in a circle of radius r;

(*ii*) the tension T in a uniform circular wire of radius r and mass m per unit length, rotating in its own plane with an angular velocity ω, about an axis passing through its centre and perpendicular to its plane;

(*iii*) the mass M of a planet round which a satellite completes its orbit of radius r, in a time-interval T.

Ans. (*i*) $K.v^2/r$; (*ii*) $K.mr^2\omega^2$, where K is a constant; (*iii*) $M \propto r^3/GT^2$.

*14. Obtain an expression for the height h to which a liquid, of density ρ and surface tension T, will rise in a capillary tube of radius r, given that $h \propto 1/r$. **Ans.** $h = k.T/r.\rho.g$, (k being a constant).

*15. Assuming that the viscosity η of a gas is proportional to the mean free path λ of its molecules, show that, if the temperature be kept constant, it is independent of the density ρ of the gas.

[**Hint.**—First obtain an expression for η, in terms of ρ, λ, c, (*the root mean square velocity of the molecules*) and D, the diameter of a molecule. Then, since $\eta \propto \lambda$, we shall have $\eta = k.\rho.c.\lambda$ where k is a constant). Again, since ρ is inversely proportional to λ, *i.e.*, $\rho = k/\lambda$ (where k' is another constant), we shall have $n = k.k'.c$, showing that n is independent of ρ).

*16. Show that if the linear dimensions of the whole of Cavendish's or Boys' method for the determination of G be changed, the sensitiveness of the apparatus remains the same. [See Boys' method for G (Chapter VII)].

*17. Show that the volume of a liquid, of coefficient of viscosity η, flowing per second through a tube of circular cross-section is given by $V = \pi pr^4/8\eta l$ where p is the excess pressure between the ends of tube, r, its radius and l, its length.

*18. If the resistance of a liquid to the motion of a body through it with velocity v, be proportional to v^2, show that it is quite independent of the viscosity of the liquid.

*19. A Nicholson's hydrometer of mass m, floating in a liquid of density ρ, is given a slight downward displacement and then released. Obtain an expression for the time-period T of its oscillation. (Assume the area of cross-section of its neck to be a.) **Ans.** $T = 2\pi\sqrt{m/a\rho}$.

*20. A U-tube of uniform cross-section contains mercury up to a height h in either limb. The mercury in one limb is depressed a little and then released. Obtain an expression for its time-priod of oscillaion. **Ans.** $T = 2\pi\sqrt{h/g}$.

[**Hint.** — Just put $T = k.d^a h^b g^c$ and show that $T = k.(d/h)^a\sqrt{h/g}$., where T is the time-period of oscillation of mercury and d, its initial displacement, K being the usual constant of proportionality. For small values of d, $a = 0$ and experiment gives $K = \pi\sqrt{2}$. Substitute their values and obtain the result].

21. Maxwell has said, "the whole system of civilized life may be fitly symbolized by a foot-rule, a set of weights and a clock." Comment.

22. What are dimensional equations ? How are they useful in the investigation of physical problems? Employing the method of dimensions, obtain an expression for the acceleration of a particle in uniform circular motion. *(Panjab).*

CHAPTER II

MOTION ALONG A CURVE—THE PROJECTILE

2.1. Rotation. Suppose we have a rigid body, with a fixed axis, within or without it. Then, if a force be applied to it, it cannot move bodily, as a whole, relatively to the axis, *i.e.*, no *motion of translation* is possible; but it simply moves round or *rotates* about the axis, such that every particle of it undergoes the *same* angular displacement. A body, so rotating about a fixed axis, is said to perform *rotatory motion.*

The force, producing rotatory motion about the fixed axis, called the *axis of rotation*, is said to have a *moment about that axis* which is measured by the *product of the force and the perpendicular distance between its line of action and the axis of rotation.* Obviously, therefore, if either of these be zero, *the moment, or the turning tendency of the force*, will be zero, for the product of the force and perpendicular distance between the axis and the line of action of the force is, then, zero.

If the rotation produced be *anti-clockwise*, the moment of the force is said to be *positive*, and if it be in the *clockwise* direction, the moment is said to be *negative*. And, since the moment of a force is a *vector quantity*, it follows that *if a number of forces act simultaneously on a body, the algebraic sum of their individual moments about the given axis of rotation will be equal to the moment of their resultant about it.*

2.2. Angular Velocity. Let a body rotate about a fixed axis through O, (Fig. 1). Then, the particles composing it, at any distance from O, such as at A, B, C, etc. complete one rotation in the *same* time, *i.e.*, they describe the *same angle* in the *same time*, and, therefore, the angle described by them per unit time is the same. This angle described by a rotating body per unit time is called its *angular velocity* and is usually denoted by the Greek letter ω. Thus, if the rate of rotation of a body be *uniform, i.e.*, if its angular velocity be constant, and it describes an angle θ (*radians*) in time t (*seconds*), we have

angular velocity of the body $\omega = \theta / t$ *radian / sec.*

If the body makes n rotations in time t, the angle described by it is equal to $2\pi n$. Or, $\theta = 2\pi n$.

And, therefore, its *angular velocity* $\omega = 2\pi n/t$ *rad./sec.*

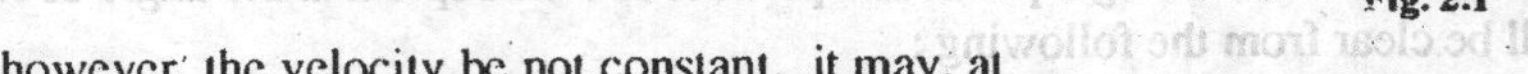

Fig. 2.1

If, however, the velocity be not constant, it may, at a given instant, be expressed in the form $\omega = d\theta/dt$, where $d\theta$ is the small angle described by it in the small interval of time dt.

Now, although the angle θ described by all the particles of the body in a given time t is the *same, the linear distances* travelled by them are *different.* Thus, the particles at A, B and C (Fig. 2.1), cover the linear distances AA', BB' and CC' respectively, (which are *arcs of radii OA, OB, OC*) depending upon their respective distances from the axis of rotation through O.

If $OA = r_1$, $OB = r_2$ and $OC = r_3$,

clearly, *arc* $AA' = r_1\theta$, $BB' = r_2\theta$ [∵ *arc = radius × angle subtended by it.*

and $$CC' = r_3\theta$$

$\therefore$ linear velocity of $A = r_1\theta / t$, that of $B = r_2\theta / t$, and that of $C = r_3 \ \theta / t$.

Or, in general, *linear velocity v of a particle at a distance r from the axis of rotation is rθ/t.*

Or $$v = r\omega \qquad [\because \theta/t = \omega.$$

i.e., linear velocity = distance from the axis of rotation × angular velocity.

2.3. Angular Acceleration. If the angular velocity of a rotating body be not constant, it is said to have *an angular acceleration*, which is defined as the *rate of change of angular velocity.* It is usually denoted by the symbol $d\omega / dt$. Thus, if the angular velocity of a particle about a given axis changes from ω to ω' in time t, its rate of change of angular velocity, or its angular acceleration is, clearly $(\omega' - \omega) / t$, or $d\omega / dt$, if $d\omega$ be the change in angular velocity in time dt.

Now, if the distance of the particle from the axis of rotation be r, its linear velocity changes from $r\omega$ to $r\omega'$ in time t, and, therefore, rate of change of its linear velocity, or its *linear acceleration*, is given by

$$a = \frac{r\omega' - r\omega}{t} = \frac{r(\omega' - \omega)}{t} = r.\frac{d\omega}{dt}.$$

Thus, *linear acceleration = distance from axis of rotation × angular acceleration.*

2.4. Couple. When two *equal, opposite, parallel and non-collinear* forces act on a body, (Fig. 2.2), bringing about rotation, (with no motion of translation), they are said to constitute *a couple*, the turning moment of the couple being measured by the *product of one of the forces and the perpendicular distance between them, or the arm of the couple* as, it is called.

Thus, *moment of a couple, C*
= one of the forces × arm of the couple,
expressed in *dyne-cm* in the *C.G.S* system and in *newton-metre* in the *M.K.S.* or the *S.I.* system.

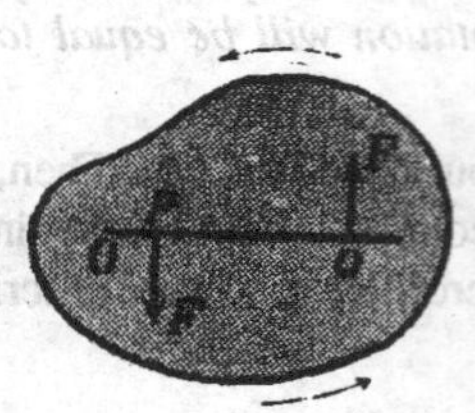

Fig. 2.2

The moment of the couple (also sometimes referred to as the **torque**), *acting upon a body, is quite independent of the position of the axis of rotation.* For, if the two forces F and F, (Fig. 2.2), constituting a couple, act at points P and Q, and if the axis of rotation passes through P, there is no moment of the force acting at P about it and the moment of the force acting at Q is $F \times PQ$, and therefore, the moment of the couple is $F \times PQ$. And, if the axis of rotation passes through any other point O, the moment of the couple about it is equal to the algebraic sum of the moments of the forces P and Q about it, *i.e.*, equal to $(F \times OQ) - (F \times OP) = F \times PQ$, as before.

The same will be true for any other position of the axis.

2.5 Work done by a couple. Work is done by a couple in rotating the body on which it acts, the amount of work done being equal to the product of the couple and the angle of rotation of the body, as will be clear from the following :

Let the axis of rotation of a body, acted upon by a couple, pass through P, the point of application of one of the forces constituting the couple, (Fig. 2.3).

Now, if the body rotates through an angle $d\theta$, the point Q moves through a distance $PQ.d\theta$, where PQ is the perpendicular from P on to the line of action of the force F acting at Q. Therefore, the work done by this force is equal to $F \times PQ.d\theta$. And, since the point P does not move, no work is done by the force at P. Thus, the work done by the two forces, *i.e., by the couple*, in rotating the body through an angle $d\theta$, is equal to $F \times PQ.d\theta + 0 = F \times PQ.d\theta$.

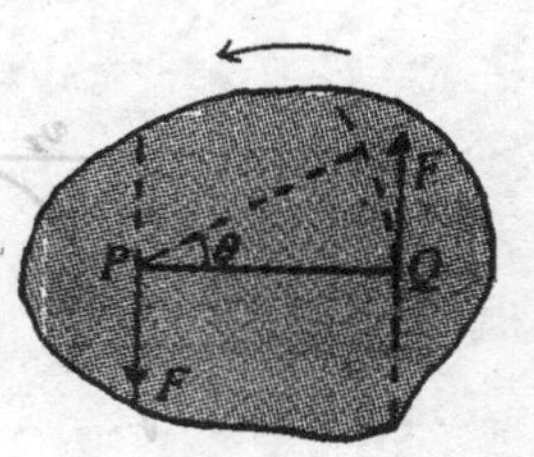

Fig. 2.3

Hence, work done by the couple in rotating the body through the whole angle θ is given by

$$W = \int_0^\theta F . PQ . d\theta = F . PQ \int_0^\theta d\theta = F.PQ.\theta.$$

Now, $F.PQ$ is the moment of the couple C, acting on the body.

∴ *work done by the couple in rotating the body through angle θ,*

i.e., $W = C.\theta =$ *couple × angle of rotation.*

Since, in one complete rotation, the body describes an angle 2π, *work done by the couple in one full rotation of the body* = $2\pi C$. And, ∴ *work done by the couple in n full rotations of the body* = $2\pi nC$.

2.6. Relation between Couple and Angular Acceleration. When the resultant couple acting on a body is *not zero*, it produces an angular acceleration in the body. Let us deduce the relation between the two.

In Fig. 2.3, the couple C, acting on the body, causes it to rotate about the axis of rotation through P.

Breaking up the couple and the body into small elements, let an element δC of the couple cause the rotation of an element of mass δm of the body situated at Q. Then, since *couple = force × distance*, the *force* acting on the mass m at Q is = $\delta C/r$, where r is the arm of the couple. And, since a couple consists of two equal, opposite and parallel forces, it follows that an equal, opposite and parallel force is also acting at P.

Again, since *force = mass × acceleration*, the linear acceleration of the particle δm at $Q = \delta C/r.\delta m$. But, if angular acceleration of the particle be $d\omega/dt$, its linear acceleration is *also* equal to $r.d\omega/dt$, (see §9).

$$\therefore \quad \frac{\delta C}{r.\delta m} = r . \frac{d\omega}{dt} \quad Or, \ \delta C = \frac{d\omega}{dt} .r^2.\delta m.$$

$$\therefore \quad \Sigma\, \delta C = \Sigma\, (d\omega/dt).\, r^2\, \delta m.$$

Or, $$C = (d\omega/dt)\ \Sigma .r^2.\delta m.$$

Now, $\Sigma\, r^2.\ \delta m = I$, the *moment of inertial of the body* about the axis of rotation, (see §3.3)

$$\therefore \quad C = I.d\omega/dt$$

Or, *Couple = moment of inertia × angular acceleration.*

2.7. The Hodograph. When a body describes a *curvilinear path,* so that its motion is accelerated and also changes in direction, its acceleration and its path may easily be determined by means of what is called the *hodograph* of its motion.

The hodograph may be defined as an auxiliary curve, obtained by joining the free ends of a moving vector representing the velocity of a moving particle along any path.

For instance, if a point P moves along a curve ABC, [Fig. 2.4 (*a*)] such that its velocities are v_1, v_2, and v_3.... respectively at A, B and C etc., then, if we take any point O and draw straight lines, *i.e., vectors, Oa, Ob* and *Oc*, [Fig. 2.4(*b*)], representing the velocities of P at A, B and C, in magnitude as well as in direction, the curve passing through a, b and c is the ***hodograph*** of the motion of P.

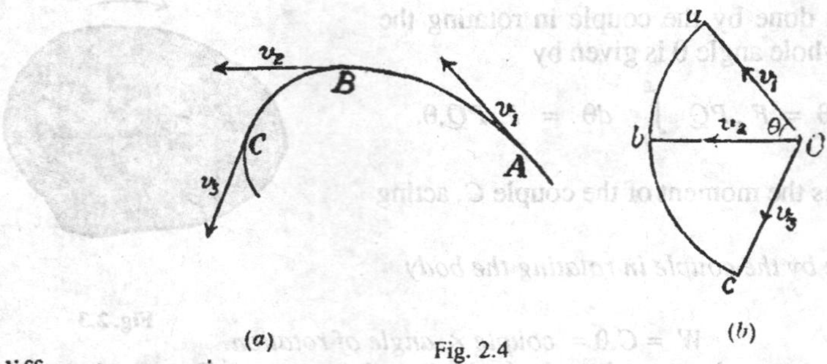

Fig. 2.4

Now, different cases arise :

(*i*) If point P be moving with a *uniform velocity along the same direction*, points a, b, c etc. will all lie in the same place and the hodograph will therefore, be a *single point*.

(*ii*) If point P be moving with a *variable velocity*, but *in the same direction*, the hodograph will be a *straight line*, passing through O. For example, in the case of a body falling freely under the action of gravity, the hodograph will be a *vertical line*, passing through O.

(*iii*) If P be projected with a horizontal velocity, the path described will be a *parabola*, (see §2.14), and both the direction and the magnitude of the velocity will change. The horizontal velocity will throughout remain constant and equal to the initial horizontal velocity, because the acceleration due to gravity acts vertically downwards. The points a, b, c, etc. will, therefore, always be at the same horizontal distance from O, and the hodograph, in this case, will thus be a *vertical line, not passing through O*.

(*iv*) If the path of P be a *closed curve*, the hodograph will also be a *closed curve*. For example, if P moves in a circle with a uniform speed v, the hodograph will also be a circle of radius v, because all the lines, Oa, Ob, Oc, etc. will be of the same length v. If, on the other hand, it moves in a circle with a *variable speed*, the hodograph might be an *oval curve* about the point O.

2.8. Velocity in the Hodograph. An important property of the hodograph is that the acceleration of P at any point on the curve ABC is represented, in magnitude as well as direction, by the velocity of the corresponding point on the hodograph, as can be seen from the following :

Let A and B be two points, close together, [Fig. 2.4(*a*)], and let P move from A to B in time t, such that its velocity v_1 at A is changed to v_2 at B.

Further, let another point p describe the hodogrpah abc, [Fig. 2.4(*b*)], while P describes the curve ABC.

Then, clearly, the point p moves from a to b in time t, and its velocity is, therefore, equal to ab/t.

But, since oa represents the velocity of P at A and ob, that at B, ab represents, in accordance with the law of triangle of velocities, the change in velocity of P in time t, and, therefore, *the rate of change of velocity*, or the *acceleration* of P, is represented by ab/t, *i.e.*, by the *velocity* of p in the hodograph.

We thus see that, *at any instant, the acceleration of P is given by the velocity of p in the hodograph of its motion.*

2.9. Uniform Circular Motion. The above affords us a very simple method of determining the acceleration of a body or a particle, moving with a constant speed along a circle. Such a circular motion is essentially a translatory motion, with only the path of the body or the particle being along the circumference of a circle and the force acting on it directed towards the centre of the circle, as will be clear from the following.

Let a particle P move in a circle, with centre O and radius r, with a uniform speed v, [Fig. 2.5(*a*)]. Then, the hodograph is also a circle, of radius r, [Fig. 2.5(*b*)].

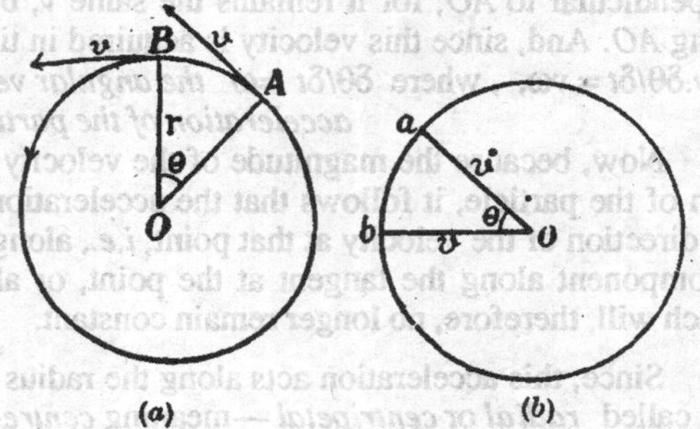

Fig. 2.5

Now, the velocity of P at any instant is *at right angles* to the radius of its circular path, passing through P. Therefore, *oa is perpendicular to OA* and *ob is perpendicular to OB* and $\angle AOB = \angle aob = \theta$ (in circular measure).

If P takes time t to describe the arc AB, its velocity $v = AB/t = r\theta/t$, whence, $\theta = vt/r$.

And, the velocity of the corresponding point p, in the hodograph, is $ab/t = v\,\theta/t$

Since the velocity of p in the hodograph gives the acceleration of P in its actual path, we have

$$acceleration\ of\ P = \frac{v\theta}{t} = \frac{v}{t} \times \frac{vt}{r} = \frac{v^2}{r}$$

And, since ab is small, it is, *in the limit, perpendicular to oa*, or *parallel to AO*.

Thus, the *acceleration* of P is v^2/r and is *directed along the radius or towards the centre of the circular path* in which it is moving.

Further, since $v = r.\,\omega$ (where ω is the *angular velocity* of P), we have

$$acceleration\ of\ P\text{, also} = r^2.\,\omega^2/r = r\omega^2.$$

Alternative Method. The acceleration of a body, executing a uniform circular motion may also be found out directly as follows :

Let a particle move with a uniform linear velocity v, in a circle of radius r, (Fig. 2.6), and let it cover the small distance from A to B in a small interval of time δt, describing an angle δt. Then, clearly, its angular velocity $\omega = \delta\theta/\delta t$.

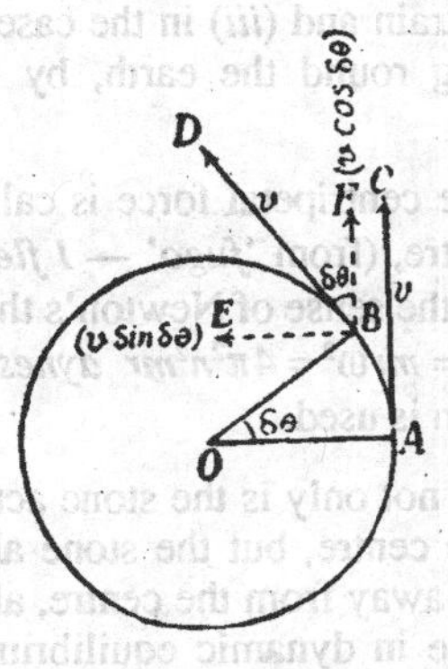

Fig. 2.6

The direction of the *linear velocity* is, at every point, *tangential to the circle at that point* and is, therefore, represented by the tangent AC at the point A, and by the tangent BD at the point B, where $AC = BD$.

Now at A, the entire linear velocity is along AC, there being no component of it along AO, which is at right angles to AC. And, resolving the velocity at B into two rectangular components, one along AO and the other, at right angles to it, we have the component along or parallel to AO, represented by $BE = v\sin\delta\theta$, and the component at right angles to AO, represented by $BF = v\cos\delta\theta$.

If $\delta\theta$ be very small, $\sin\delta\theta = \delta\theta$ (*in radians*) and $\cos\delta\theta = 1$.

So that, component BE, parallel to $AO = v.\,d\theta$

and component BF, perpendicular to $AO = v$.

Thus, if B be very close to A, there is no change in the velocity of the particle along the perpendicular to AO, for it remains the same v, but an additional velocity $v.\delta\theta$ is acquired by it along AO. And, since this velocity is acquired in time δt, the acceleration imparted to the particle is $v.\delta\theta/\delta t = v\omega$, , where $\delta\theta/\delta t = \omega$ *the angular velocity* of the particle. Or,

$$\textit{acceleration of the particle} = v\omega = v.v/r = v^2/r.$$

Now, because the magnitude of the velocity remains the *same* at every point on the circular path of the particle, it follows that the acceleration must be acting in a direction perpendicular to the direction of the velocity at that point, *i.e.*, along the radius of the circle, or else it will also have a component along the tangent at the point, or along the direction of the velocity at that point, which will, therefore, no longer remain constant.

Since, this acceleration acts along the radius of the circle, or towards the centre of the circle, it is called *radial* or *centripetal* —meaning *centre-seeking*—acceleration, (from '*peto*' —*I seek*).

Thus, *centripetal acceleration* = $v^2/r = (r\omega)^2/r = r\omega^2$.

And, if n be the number of revolutions made by the particle per unit time, we have $\omega = 2\pi n$.

$\therefore$ *centripetal acceleration*, also $= r.(2\pi n)^2 = 4\pi^2 n^2 r$

Even if the path be not exactly a circle, but any other curve, the value of the acceleration is v^2/r, where v is the linear velocity, and r, the radius of curvature of the path *at the point considered.*

2.10. Centripetal Force. According to Newton's first law of motion, a body must continue to move with a uniform velocity in a straight line, unless acted upon by a force. It follows, therefore, that when a body moves along a circle, some force is acting upon it, which continually deflects it from its straight or linear path; and since the body has an acceleration towards the centre, it is obvious that the force must also be acting in the direction of this acceleration, *i.e.*, along the radius, or towards the centre of its circular path. It is called the *centripetal force*, and its value is given by the product of the mass of the body and its centripetal acceleration. Thus, if m be the mass of the body, we have

$$\textit{centripetal force} = mv\omega = mv^2/r = mr\omega^2 = 4\pi^2 n^2 mr,$$

in *dynes* if m be in *gm.*, v in *cm/sec* and r in *cm* and in *newtons* if m be in *kg*, v in *m/ sec* and r in *metres.*

Numerous examples of centripetal force are met with in daily life. Thus, (*i*) in the case of a stone, whirled round at the end of a string whose other end is held in the hand, the centripetal force is supplied by the tension of the string ; (*ii*) in the case of a motor car or a railway train, negotiating a curve, it is supplied by the push due to the rails on the wheel of the train and (*iii*) in the case of (*a*) the planets revolving round the sun, or (*b*) the moon revolving round the earth, by the gravitational attraction between them.

2.11. Centrifugal Force. The equal and opposite *reaction* to the centripetal force is called *centrifugal force*, because it tends to take the body away from the centre, (from '*fugo*' — *I flee*). Centripetal force and centrifugal force being just *action* and *reaction* in the sense of Newton's third law of motion, the numerical values of the two are the same, *viz.*, $mv^2/r = mr\omega^2 = 4\pi^2 n^2 mr$ *dynes* or *newtons* according as the *C.G.S.* system or the *M.K.S* (or the *S.I.*) system is used.

Thus, in the case of a stone, whirled round at the end of a string, not only is the stone acted upon by a force, (this centripetal force), along the string towards the centre, but the stone also exerts an equal and opposite force, (the centrifugal force), on the hand, away from the centre, also along the string, the two balancing each other and keeping the stone in dynamic equilibrium, constraining it to move with a constant speed along the tangent to its circular path at every point. In the event of the string getting snapped (or loosened), this pair of constraining forces disappears and the stone flies off tangentially to the circular path at the point where the string snaps.

2.12. Motion of a particle in a vertical circle. As we have seen under §2.11 above, when a particle is whirled round in a *horizontal circle* at the end of a string, the tension in the string supplies the centripetal force and its value remains the same throughout, because the weight of the particle, acting vertically downwards, has no component along the string. The particle, as a result, continues to move with a uniform speed.

If, however, the particle be whirled in a vertical circle, the whole of the weight of the particle will be acting in its top and bottom positions in the circle and some component of it in all other positions, except when it lies in a level with the centre of the circle and the string is horizontal. Obviously, therefore, the tension, in the string does not remain constant in this case and hence also the velocity of the particle along the circle. Let us discuss it a little more fully.

Let *OA* be a string of length *l*, fixed at *O* and suspended freely, with a particle of mass *m* at *A*, (Fig. 2.7). Let the particle be given a horizontal velocity *u* towards the right when at rest at *A*. It will naturally move along an arc of radius *l*, in the anticlockwise direction.

Let the particle be at *A′* at any given instant and let its velocity there *v*. Drop a perpendicular *A′P* from *A* on to the diameter *AB* of the circle and let $A'P = h$ and $\angle AOA' = \theta$. Then, clearly,

$$v^2 = u^2 - 2gh,$$

since the particle has been raised up from its initial position at *A* through a distance $AP = h$.

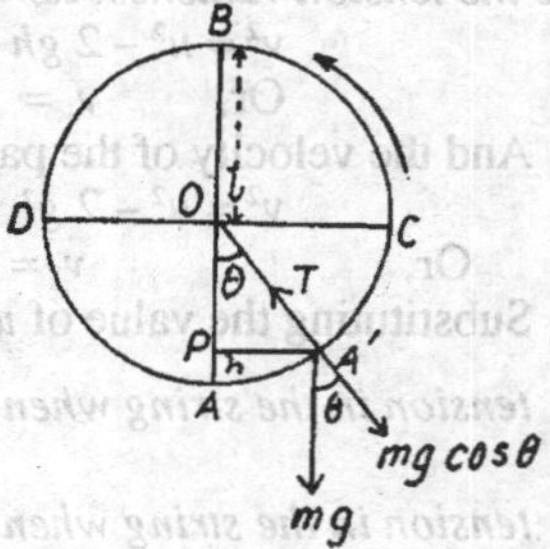

Fig. 2.7

The weight *mg*, of the particle naturally acts vertically downwards at *A′* and the centripetal force mv^2/l along the string in the direction *A′ O*.

Obviously the component $mg \cos\theta$ of the weight of the particle also acts along the string in the direction *OA′*, opposite to that of the centripetal force. If, therefore, *T* be the *tension in the string* in this position, we have

$mv^2/l = T - mg \cos\theta$. or, $T = (mv^2/l) + mg \cos\theta$ ...(*i*)

Now, $\cos\theta = OP/OA' = (l - h)/l$, whence, $h = l(1 - \cos\theta)$

Substituting the values of v^2 and $\cos\theta$ in relation (*i*), therefore, we have

$$T = \frac{m(u^2 - 2gh)}{l} + mg\left(\frac{l-h}{l}\right) = \frac{m}{l}[u^2 - 2gh + g(l-h)]$$

or, $T = \frac{m}{l}[u^2 + gl - 3gh]$

or, substituting $l(1 - \cos\theta)$ for *h*, we have

$$T = \frac{m}{l}[u^2 + gl - 3gl(1 - \cos\theta)] \quad \text{....}(ii)$$

This is the general expression for the tension in the string and the following special cases arise.

(*i*) **When the particle is in the lowermost position A.** Here, obviously, $h = 0$ and also $\theta = 0$. And, therefore,

$$T = \frac{m}{l}[u^2 + gl] = \frac{mu^2}{l} + mg \quad ...(iii)$$

(ii) When the particle is in the topmost position B. In this case, $h = 2l$ and $\theta = 180°$, and we, therefore, have

$$T = \frac{m}{l}[u^2 + gl - 3g.2l] = \frac{mu^2}{l} - 5mg \quad ...(iv)$$

(iii) When the particle is at *C* in a level with centre *O*. Here $h = l$ and $\theta = 90°$, so that

$$T = \frac{m}{l}[u^2 + gl - 3g.l] = \frac{mu^2}{l} - 2mg \quad ...(v)$$

Now, it follows at once from relation (*iv*) above that *T* will vanish if $mu^2/l = 5\,mg$ or $u^2 = 5\,gl$. If this be so, the particle will no longer keep to its circular path. The essential condition for the particle to continue along its circular path, therefore, is that *T should not be zero. i.e.*, $u^2 > 5\,gl$ or $u > \sqrt{5gl}$.

In case $u = \sqrt{5gl}$ or $u^2 = 5\,gl$, *the particle will just be able to reach the topmost position B, where the tension vanishes*. Its velocity will then be given by

$$v^2 = u^2 - 2\,gh = u^2 - 2g.2l = 5\,gl - 4\,gl = gl \qquad [\because u^2 = 5gl, \text{ here}$$

Or, $v = \sqrt{gl}$

And the velocity of the particle at *C* will obviously be given by

$$v^2 = u^2 - 2\,gl = 5\,gl - 2\,gl = 3\,gl. \qquad [\because h = l \text{ here}$$

Or, $v = \sqrt{3gl}$

Substituting the value of u^2 obtained above in relations (*iii*), (*iv*) and (*v*), we have

tension in the string when the particle is at A $= \frac{m}{l}.5gl + mg = 6mg$;

tension in the string when the particle is at C $= \frac{m}{l}.5gl - 2mg = 3mg$

with *the tension in the string at B* = 0

Thus, the tension in the string decreases from 6 *mg* at *A* to 3 *mg* at *C* and finally to *zero* at *B*. (see worked example 5 at the end of the chapter)

2.13 Practical Applications of Centripetal and Centrifugal Forces

(1) **Road Curves.** The centripetal force being directly proportional to the square of the linear velocity of the body and inversely proportional to the radius of its circular path, the radii of curvature of road curves must be large and the speed of the vehicles negotiating them solved down, in order to keep the value of the centripetal force required within reasonable limits.

(2) **Rotating Machinery.** The centrifugal force being proportional to n^2, where *n* is the number of rotations made by the body per second, the spokes of a wheel, joining its outer revolving parts to the axis of rotation, experience an outward force, *away from the centre*, and are, therefore, in a state of tension, and may give way if the value of *n* is very large. So is the case with the parts of other rotating machinery, connecting its outer revolving parts to its axis of rotation. In other words, *there is a limit set to the value of n by the tension these connecting parts can withstand.* This fact is always kept in view while designing highly rotating machinery, like *armature of motors and dynamos* etc.

Let us, as a specific example, discuss the case of a belt or a string rotating at a high speed over a pulley etc.

Let the string rotate in a circle of *radius r*, (Fig. 2.8), and let its *angular velocity* be ω. Consider a small portion *AB* of the string, of length *l* and subtending an angle 2θ at the centre *O* of the circle. This portion is obviously subjected to a tension T, at either end, by the rest of the string as shown. Resolving these tensions T and T at A and B into two rectangular components along and at right angles to PO, (where PO passes through the mid-point of AB), we find that the components $T \cos \theta$ at right angles to PO are *equal and opposite* and thus neutralize each other, but the components $T \sin \theta$ along PO act in the *same* direction. So that, we have

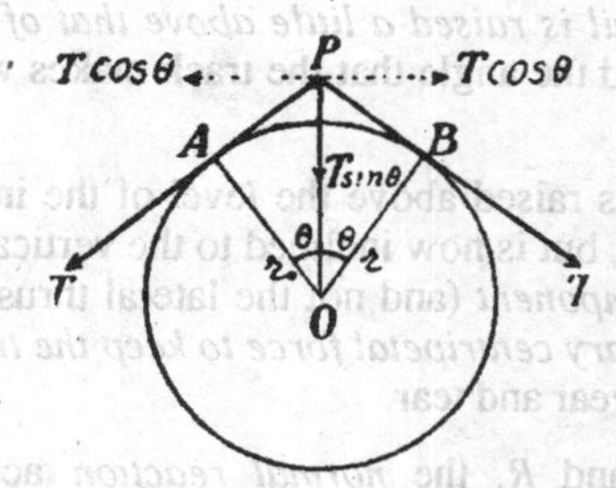

Fig. 2.8

resultant tension on portion AB of the string

$= 2T \sin \theta$ in the direction PO

And, the *centrifugal force acting on portion AB* of the string

$=$ *mass of AB* $\times r\omega^2$, in the direction OP.

If m be the *mass per unit length* of the string, clearly,

mass of $AB = m \times l$.

And $\therefore$ *centrifugal force acting on portion AB* of the string

$= m \times l \times r\,\omega^2$, in the direction OP.

For equilibrium, therefore,

$$2T \sin \theta = mlr\omega^2 = m.2r\theta.r\omega^2$$

[$\because$ clearly, $l = AB = r.2\theta$]

If θ be small, we have $\sin \theta = \theta$. So that,

$$2T\theta = m.2r\theta.r\,\omega^2$$

whence $T = mr^2\,\omega^2$

It will thus be seen that due to centrifugal force, the tension in the string is very high. Indeed, *if the rapidly rotating chain or belt be pushed off the pulley, it will run along like a rigid hoop*.

The same is true about other rotating bodies which are always under a state of *elastic stress*. It is this stress which sets a limit to the speed up to which the flywheels can be rotated safely. Again, it is as a consequence of this stress that the tyres of racing cars get stretched and there is a danger of their being cast off the rims and flung out at very high speeds.

(3) **Revolution of Planets and the Length of the Year.** In the case of a planet revolving round the sun, it is the gravitational force of attraction between the two which supplies the centripetal force, necessary to keep it moving in its nearly circular orbit. Now, the gravitational force between two bodies is directly proportional to the product of their masses and inversely proportional to the square of the distance between them; so that, if m and M be the masses of the planet and the sun respectively and r, the distance between them (or the radius of the planet's orbit round the sun), we have

$$\textit{gravitational pull} = \frac{m.M}{r^2}.G = \frac{k}{r^2}$$

[Putting $m.\ MG = k$, a constat.

Or, $4\pi^2n^2rm = k/r^2$, whence, $n^2 = k/4\pi^2r^3m$.

Or, $$n = \frac{1}{2\pi}\sqrt{\frac{k}{mr^3}}.$$

i.e. $t = 1/n = 2\pi\sqrt{mr^3/k} = 2\pi\sqrt{mr^3/mMG} = 2\pi\sqrt{r^3/MG}$,

where t is the *time taken by one revolution of the planet round the sun, or the length of the year* for that planet.

Thus, t varies as $\sqrt{r^3}$, *i.e.*, the smaller the value of r, or the smaller the distance of the planet from the sun, the smaller the value of t, or the length of the year for it. *A planet will, therefore, have a shorter year if nearer to the sun than when at a distance from it.*

(4) **Banking of Railway Lines and Roads.** When a railway train goes round a level curve on a railway track, the necessary centripetal force is provided only by the force between the *flanges* or the rims of the wheels and the *rails*, the normal reaction of the ground or the track acting vertically upwards and supporting its weight. This results in a grinding action between the wheels

and the rails, resulting in their wear and tear. Not only that, it may also prove dangerous in the sense that it may bring about a displacement of the rails and hence a derailment of the train.

To avoid these eventualities, *the level of the outside rail is raised a little above that of the inside one*. This is known as the **banking** of railway lines, and the angle that the track makes with the horizontal is called the **angle of banking.**

With the track thus *banked, i.e.*, with the outer rail thus raised above the level of the inner one, the reaction R acts perpendicularly to the track, as before, but is now inclined to the vertical at an angle equal to the angle of banking and *its horizontal component* (and not the lateral thrust of the wheel flanges on the outside rail) *now supplies the necessary centripetal force to keep the train moving along the curve*, thereby eliminating all unnecessary wear and tear.

Thus, if θ be the *angle of banking* [Fig. 2.9(*a*)] and R, the *normal reaction* acting perpendicular to it, we have

vertical component of $R = R\cos\theta$

and *horizontal component of* $R = R\sin\theta$

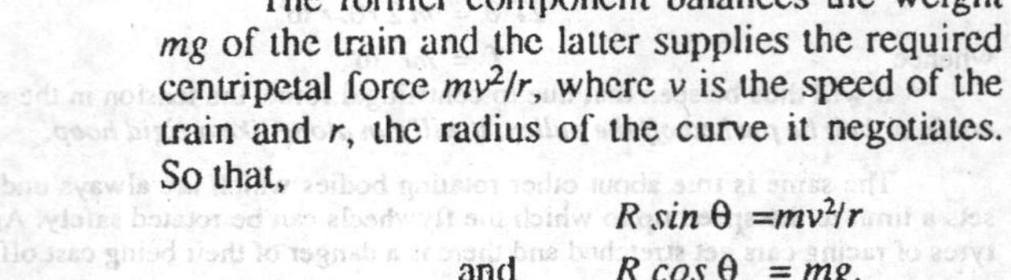

The former component balances the weight mg of the train and the latter supplies the required centripetal force mv^2/r, where v is the speed of the train and r, the radius of the curve it negotiates. So that,

$$R\sin\theta = mv^2/r$$

and $$R\cos\theta = mg.$$

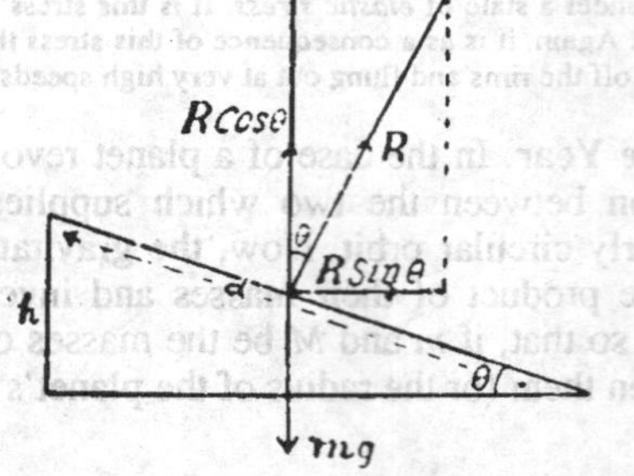

Fig. 2.9(*a*)

$$\therefore \quad \frac{R\sin\theta}{R\cos\theta} = \frac{mv^2/r}{mg}.$$

Or, $$\tan\theta = v^2/rg \qquad \text{...(i)}$$

Or, $$\theta = \tan^{-1}(v^2/rg).$$

The angle of banking thus depends upon the *speed* (v) *of the train* and the *radius* (r) *of the curve* of the track. Obviously, therefore, a track can be *banked* correctly only for a particular speed of the train, – in practice, naturally for its *average* speed, given by $v = \sqrt{rg\tan\theta}$ from relation (*i*) above. At higher or lower speeds than this, there is again a lateral thrust due to the wheel flanges on the outer or the inner rail of the track respectively.

Clearly, the angle that the track makes with the horizontal is equal to θ, *i.e.*, equal to the angle of inclination of the train with the vertical, (Fig. 2.8).

Further, it will be readily seen that if the distance between the rails be d and the height of the outer rail above the inner one be h, we also have $\sin\theta = h/d$. Or, *sine of the angle of banking*

$$= \frac{\textit{height of the outer rail over the inner one}}{\textit{distance between the rails}}$$

In the above discussion, we have neglected the frictional force between the wheels and the rails. If this too be taken into account, the centripetal force will be supplied not only be the horizontal component $R\sin\theta$ of the normal reaction R but also by that of the frictional force F, as shown in Fig. 2.9 (*b*). So that, we now have

$$R\sin\theta + F\cos\theta = mv^2/r$$

and $$R\cos\theta - F\sin\theta = mg$$

Since the maximum value of F is μR, we have

$$R \sin\theta + \mu R \cos\theta = mv^2/r$$

and $$R \cos\theta + \mu R \sin\theta = mg$$

So that,

$$\frac{\sin\theta + \mu\cos\theta}{\cos\theta - \mu\sin\theta} = \frac{\tan\theta + \mu}{1 - \mu\tan\theta} = \frac{v^2}{rg},$$

Whence $$v = \sqrt{rg.\frac{\tan\theta + \mu}{1 - \mu\tan\theta}},$$

which gives the maximum speed of the train while negotiating the curve.

Fig. 2.9 (b)

Similarly, in the case of a car moving round a level corner, the centrifugal force is largely provided by the friction between the road and the tyres of the wheel. That is why, when the road is slippery and the frictional force not enough, the car begins to slide or skid. Here, too, therefore, the roads are '*banked*', the slope being generally steeper outwards, — more or less like a saucer — *the outer parts being meant to be used at higher speeds and the inner ones, at lower speeds*.

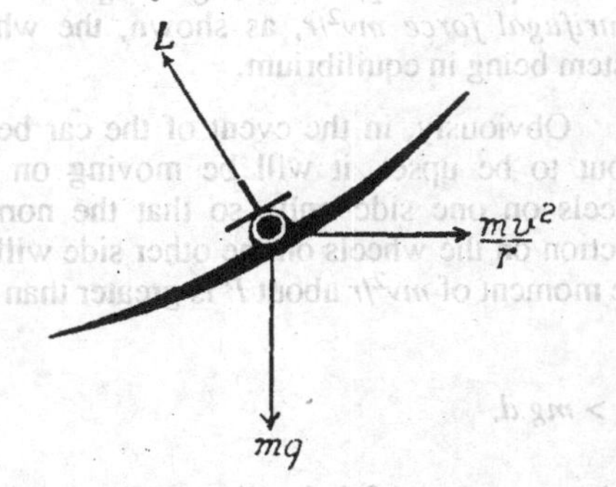

Fig. 2.10

Again, an aeroplane, in order to turn, must also bank, the centripetal force here being supplied by the horizontal component of the lift L, (Fig. 2.10)

The same applies to a cyclist, when negotiating a curve or a corner, and he has to lean *inwards*, (*i.e.*, towards the centre of the curve), by an angle $\theta = \tan^{-1} v^2/rg$; so that the faster his speed and the sharper the curve, the more must he lean over. This will be clear from the following :

Let Fig. 2.11 represent a cyclist turning to the left in a circle of radius r, at a speed v. Then, the *normal reaction* R of the ground acts vertically upwards, with the *force of friction* F between the ground and the tyres and the *centrifugal force* mv^2/r in the direction shown, where

$$R = mg \text{ and } F = mv^2/r.$$

Then, for equilibrium, clearly, we have moment of mg about P *equal and opposite* to moment of mv^2/r about P.

Or, $$mg \times PQ = (mv^2/r) \times GQ.$$

Or, $$mg \times PG.\sin\theta = (mv^2/r).\, PG.\cos\theta$$

whence, $$\sin\theta / \cos\theta = \tan\theta = v^2/rg.$$

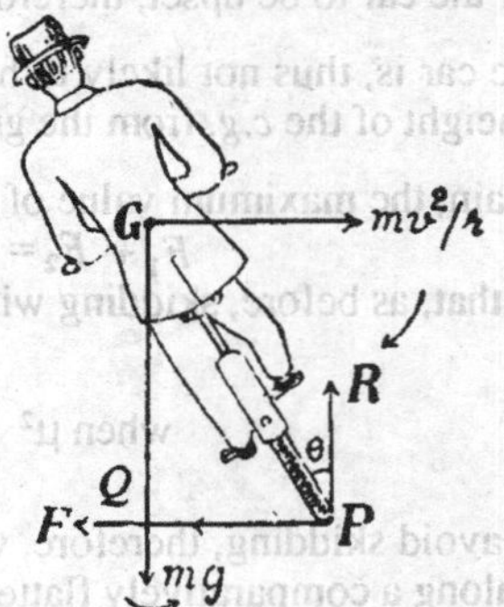

Fig. 2.11

In other words, in order to keep himself in equilibrium, the cyclist must lean inwards from the vertical at an angle $\theta = tan^{-1} (v^2 / rg)$.

If he were to remain vertical, his weight would act through P, having no moment about it, so that the moment of mv^2/r about P would remain unbalanced. In fact it will be readily seen that the system of forces acting on the cyclist form two pairs of couples, one due to F and mv^2/r and the other due to R and mg. So that, in the event of the latter couple vanishing (*i.e.*, if the cyclist were vertical), the former alone will remain operative, resulting in the cyclist toppling over. (*See* also under "*Gyroscope*", Chapter III.)

Further, since the maximum value of $F = \mu mg$, (where μ is the coefficient of friction between the ground and the tyres), the cyclist will skid when $mv^2/r > \mu mg$, or when $v^2 > \mu rg$.

Thus, *skidding will occur (i) if v is large, i.e., if the speed of the cyclist is large; (ii) if μ is small, i.e., if the road is slippery and (iii) if r is small, i.e., if the curve is sharp.*

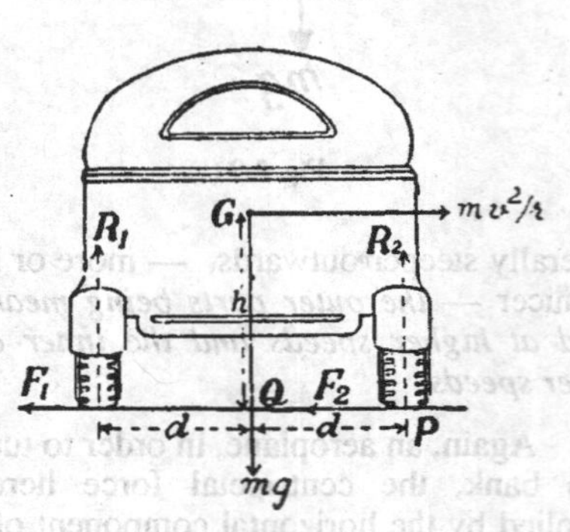

Fig. 2.12

Similar conditions apply in the case of a motor car or any other vehicle. For, here too, if we imagine it to be turning to the left (Fig. 2.12), the various forces acting on it are the *normal reactions* R_1 and R_2, the *frictional forces* F_1 and F_2, its *weight, mg* and the *centrifugal force* mv^2/r, as shown, the whole system being in equilibrium.

Obviously, in the event of the car being about to be upset, it will be moving on the wheels on one side only, so that the normal reaction on the wheels on the other side will be zero; say $R_1 = 0$. So that, it will overturn as soon as the moment of mv^2/r about P is greater than the opposing moment of mg about P, *i.e.*,

$$mv^2/r \,.\, GQ > mg.\, PQ. \text{ Or, } (mv^2/r).h > mg.d,$$

where h is the height of the *c.g.*, G, of the car above the ground and $2d$, the distance between the two wheels.

For the car to be upset, therefore, we have $v^2 > d.rg/h$.

The car is, thus not likely to be upset if $2d$, the distance between the two wheels is large and if h, the height of the *c.g.* from the ground is small.

Again, the maximum value of the total frictional force

$$F_1 + F_2 = F = \mu.mg.$$

So that, as before, skidding will occur when $mv^2/r > \mu mg$.

Or, when $\mu^2 > \mu rg$.

To avoid skidding, therefore, while taking a turn at a fast speed, the corner must be cut so as to move along a comparatively flatter curve than that of the actual turning.

2.14. Other Effects and Applications of Centrifugal Force

1. **Rotation of the Earth — Its Effect.** As we know already, the earth rotates or spins about its axis once during a day. It is this rotation of it which is responsible for its getting flattened at the poles and its bulging out at the equator—a direct consequence of the centrifugal force $m\omega^2 R$ acting on each particle of mass m of it, where ω is its angular velocity about the axis of rotation and R, the distance of the particle from this axis. The value of ω is obviously *same* for each particle, but the distance R increases from *zero* for particles at the poles to a *maximum* for those at the equator. The centrifugal force pulling the earth outwards, as it were, is thus *zero at the poles* and *the maximum at the equator* and it is this force which has made the earth (behaving like a plastic body) to bulge out at the equator and to flatten at the poles, thus bringing about an increase of about 13 miles in its equatorial, as compared with its polar radius.

This effect of the rotation of the earth had been first predicted by *Newton* and was duly verified by a French expedition to Lapland under the leadership of *Maupertius*, whose undue pomposity provoked *Voltaire* into making the caustic remark that he 'behaved as though he had flattened the poles himself'.

2. **The Centrifuges.** These are simple devices used to separate out substances of different densities suspended in a liquid by rapidly rotating the liquid, when particles, whose density is greater than that of the liquid, are driven away from the axis of rotation, whereas those, with a density lower than that of the liquid, are drawn inwards towards it. Thus, for example, in the familiar cream separator, when the vessel containing milk is rotated fast, the cream, being *lighter*, **collects**in a cylindrical layer round about the axis, whence it can be easily drawn off. Similarly, in the now common **centrifugal drying machine**, which is essentially a cylindrical vessel with perforated walls, as the latter is **rotated** fast, the centrifugal force acting on the wet clothes inside forces the water out through the perforations and the clothes get dried up quickly.

Since the centrifugal force ($m\omega^2 r$) increases with r, the pressure on the rotating liquid progressively increases as we move away from the axis, with the result that, on a heavier particle, the centrifugal force outwards is greater than the inward thrust of the liquid, whereas, on a lighter one the reverse is the case. *The problem is more or less akin to the sinking or floating of a body in a liquid at rest, depending* upon the difference in the magnitude of the forces acting on the two sides of the body, *the inward thrust on it corresponding to the upthrust in the case of a stationary liquid.*

In centrifuges having speeds of rotation as high as 40,000 *rev/min* or more, the difference between the outward and inward forces on the heavier and lighter particles exceeds more than a thousand times their difference of weights, so that quick and effective separation results. Sediments, precipitates and bacteria etc. may all be thus separated speedily. And, in the more recent **ultracentrifuges** the speed of rotation reaches the staggering figure of 80,000 or even 100,000 *rev./minute*. They are, therefore, used to separate proteins and hormones etc. from liquids, as also to separate gases of different densities from a given mixture of them.

3. **The Centrifugal Pump**. Also known as the *Turbine Pump*, it consists of three essential parts *viz.*, *(i)* an outer drum-shaped *'casing'*, having an *inlet* near its axis and an *outlet* near its periphery, *(ii)* a paddle wheel, (*i.e.*, a hollow wheel, fitted with vanes), called the *'impeller'*, which can be rotated inside the casing, and *(iii)* the *'spindle'* through which energy is transmitted from the driving motor to the impeller.

If the casing be filled with water and the impeller rapidly rotated, it sets the water into similar rapid rotation which, due to its centrifugal force, exerts high pressure against the outer wall of the casing, forcing the water out through the outlet at the periphery into the rising discharge–tube connected to it. At the same time, there is decreased pressure near the axis, so that the atmospheric pressure forces fresh water in from the reservoir, through the inlet. This is then again flung out through the outlet, in the manner explained, and the process goes on repeating itself over and over again.

The pump starts working only when the casing is full of water, but, once it has started working, it gives a *continuous* supply of water, unlike the ordinary piston pump, where we get only an *intermittent* supply. Further, as there are no valves to operate,the pump can be used safely even if the water contains sledge or any other suspended matter, including sand or even small-sized stones etc.

Among other applications of rotatory motion may be mentioned as *conical pendulum* (§ 6.15) and the *steam engine governor* .

2.15. The Projectile—Motion of a Projectile in a non-resisting medium. Before Galileo's time, it was supposed that a body thrown horizontally, travelled in a straight line until it had exhausted its force and then fell vertically down. It was he who first showed that it must take a *parabolic path**, realising as he did, the physical independence of its horizontal and vertical motions, so that each could be considered separately.

Such a body, subjected simultaneously to a uniform horizontal motion and a uniform vertical acceleration, is called a **projectile**, and *the path it describes is called its* **trajectory.** Let us study its motion in some detail.

Let a body be projected upwards with a velocity u, at an angle θ with the horizontal. Then, resolving u into two rectangular components, along the vertical** and along the horizontal, we have *(i) the vertical component* (along OY) = $u \sin\theta$ and *(ii) the horizontal component* (along OX) = $u \cos\theta$. The latter component, being perpendicular to the direction of gravity, is not accelerated, and hence

$$dx/dt = u \cos\theta.$$

Since at $t = 0$, $x = 0$, we have $x = ut \cos\theta$ ...(*i*)

And, because the vertical component is subjected to a downward acceleration due to gravity, we have $d^2y/dt^2 = -g$,

integrating which, we have $dy/dt = -gt + C_1$,

where C_1 *is another constant of integration.*

Now, at $t = 0$, $dy/dt = u \sin\theta$; so that, $C_1 = u \sin\theta$.

$$\therefore \quad dy/dt = u \sin\theta - gt.$$

Integrating this again, we have $y = ut \sin\theta - \frac{1}{2}gt^2 + C_2$,

where C_2 *is another constant of integration.*

Since $y = 0$ at $t = 0$, we have $C_2 = 0$

So that, $y = ut \sin\theta - \frac{1}{2}gt^2$ ...(*ii*)

Now, from relation (*i*), we have $t = x/u \cos\theta$

Substituting this value of t in relation (*ii*), we have

$$y = u.\frac{x}{u\cos\theta}.\sin\theta - \frac{1}{2}g.\left(\frac{x}{u\cos\theta}\right)^2$$

Or

$$y = x\tan\theta - \frac{1}{2}g.\frac{x^2}{u^2\cos^2\theta}$$

---

* He dropped objects from masts of moving ships, which fell vertically in motion to the ship but along parabolic paths in relation to the sea.

** *i.e.*, along the direction in which the force due to gravity acts.

This is clearly an equation of the second degree in x and the first degree in y and thus represents a parabola, with its axis vertical. The trajectory of the particle is thus a *parabola.*

2.16 . Horizontal Range of a Projectile. Clearly, the time taken by the body to reach the maximum height = $u \sin \theta/g$*, its vertical velocity being $u \sin \theta$.

And, *since time of ascent is equal to time of descent*, the total time taken by the body for the whole flight = $2u \sin \theta/g$.

During this time, the *horizontal distance covered by the body*, with its uniform horizontal velocity $u \cos \theta$, is given by

$$u \cos \theta . \frac{2u \sin \theta}{g} = \frac{2u^2 . \sin \theta . \cos\theta}{g} = \frac{u^2 \sin 2\theta}{g}. \qquad \left[\because 2 \sin \theta \cos \theta = \sin 2\theta.\right.$$

This horizontal distance covered by a projectile is called its **horizontal range**, or, more usually, simply, its **range.** Denoting it by R, therefore, we have $R = u^2 \sin 2\theta/g$

2.17. Maximum Height attained by a Projectile. We have the kinematic relation, $v^2 - u^2 = 2aS$, where the symbols have their usual meanings.

Here, $a = -g$ (the body being projected upwards), and, at the highest point, obviously, $v = 0$ So that, if the *maximum height attained by the projectile be h, (i.e., S = h)*, we have

$$0 - (u \sin \theta)^2 = 2.(-g).h,$$

whence,
$$h = \frac{u^2 . \sin^2\theta}{2g} \qquad \left[\because \text{the initial upward velocity here is } u \sin \theta \text{ and not } u.\right.$$

2.18. Angle of Projection for Maximum Range. It is obvious that for a given initial velocity (u) of the body, its horizontal range (R) will depend upon its angle of projection (θ).

The horizontal range R, as we know, is given by $R = \dfrac{u^2 . \sin 2\theta}{g}$.

Putting x for R, we have $x = \dfrac{u^2 \sin 2\theta}{g}$.

It is thus clear that the value of θ for maximum (horizontal) range of the projectile would be that for which $\sin 2\theta = 1$, *i.e.*, when $2\theta = 90°$, and, therefore, $\theta = 45°$.

Thus, for maximum range, the angle of projection should be 45°.

N.B. The following interesting result follows from the relation $R = u^2 \sin 2\theta/g$.

We know that the sine of an angle is the same as that of its supplement. And, therefore,

$$\sin 2\theta = \sin (180-2\theta)$$

from which it is clear that the *projectile will have the same range* (not the maximum), for the angles of projection θ and $(90-\theta)$, —the two paths taken being, however, different, and called the *high* (*H*) and the *low* (*L*) trajectories respectively, as shown in Fig. 2.13, in view of the different maximum heights attained by the body.

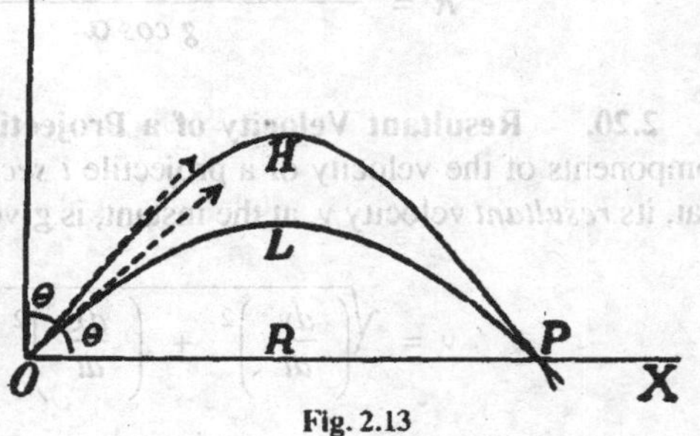

Fig. 2.13

2.19. Range on an Inclined Plane. We have seen above, (§ 2.15) that the equation to the trajectory of a projectile is

---

* We have the kinematic relation, $v = u + at$, where u is the initial velocity (here equal to $u \sin \theta$), $a = -g$ and v, the final velocity, equal to 0 (at the highest point). So that, $0 = u \sin \theta - gt$; or, $gt = u \sin \theta$. Or, $t = u \sin \theta$.

$$y = x \tan\theta - \frac{1}{2} g. \frac{x^2}{u^2 \cos^2\theta}$$

Let us consider a plane inclined to the horizontal at an angle α, (Fig. 2.14); so that,

$$y = x \tan\alpha$$

Now, to obtain the range on the inclined plane, we must determine the point where the trajectory of the projectile will meet the plane; and to do this, we must solve the above two equations. So that, substituting $y = x \tan\alpha$ in the equation for the trajectory, we have

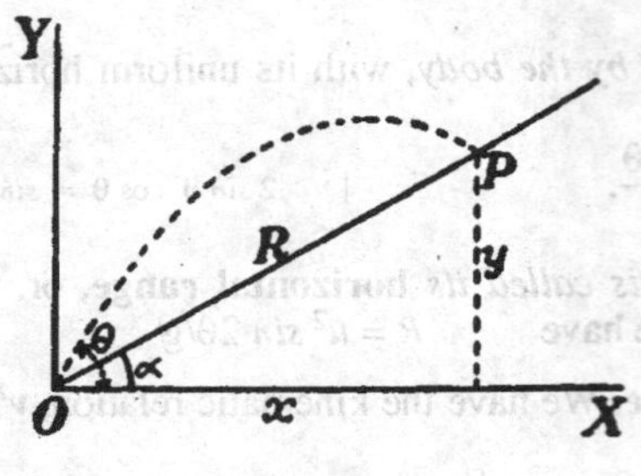

Fig. 2.14

$$x \tan\alpha = x \tan\theta - \frac{1}{2} \cdot \frac{gx^2}{u^2 \cos^2\theta}$$

Or, $$\frac{gx}{2u^2\cos^2\theta} = \tan\theta - \tan\alpha.$$

Or, $$x = \frac{2(\tan\theta - \tan\alpha).u^2\cos^2\theta}{g}$$

And, therefore, substituting *this* value of x in $y = x \tan\alpha$,

we have $$y = \frac{2(\tan\theta - \tan\alpha).\, u^2\cos^2\theta)}{g} . \tan\alpha.$$

So that, the range R is clearly given by the relation, $R^2 = x^2 + y^2$

$$= \left[\frac{2(\tan\theta - \tan\alpha).\, u^2\cos^2\theta}{g}\right]^2 + \left[\frac{2(\tan\theta - \tan\alpha)\tan\alpha.u^2\cos^2\theta}{g}\right]^2$$

$$= \left[\frac{2(\tan\theta - \tan\alpha).u^2\cos^2\theta}{g}\right]^2 [1 + \tan^2\alpha]$$

Now, $$(1 + \tan^2\alpha) = \sec^2\alpha = 1/\cos^2\alpha.$$

$$\therefore \quad R^2 = \left[\frac{2(\tan\theta - \tan\alpha).\, u^2\cos^2\theta}{g}\right]^2 \Big/ \cos^2\alpha.$$

$$R = \frac{2(\tan\theta - \tan\alpha).\, u^2\cos^2\theta}{g\cos\alpha}$$

2.20. Resultant Velocity of a Projectile at a given instant. The vertical and horizontal components of the velocity of a projectile t *sec.* after its projection are clearly dy/dt and dx/dt. So that, its *resultant* velocity v, at the instant, is given by the relation,

$$v = \sqrt{\left(\frac{dy}{dt}\right)^2 + \left(\frac{dx}{dt}\right)^2}$$

Or, putting the values of dy/dt and dx/dt from above, (§ 2.15), we have

$$v = \sqrt{(u\sin\theta - gt)^2 + (u\cos\theta)^2}$$

$$= \sqrt{u^2 sin^2 + g^2 t^2 - 2ugt.sin\,\theta + u^2 cos^2\theta}$$

$$= \sqrt{u^2(sin^2\theta + cos^2\theta) + g^2 t^2 - 2ugt.sin\,\theta}$$

Or, $v = \sqrt{u^2 - 2ugt\,.\,sin\,\theta + g^2t^2}$ $[\because\ sin^2\theta + cos^2\theta = 1.$

And, the angle, β, say, that this resultant velocity makes with the horizontal, is obviously given by

$$tan\,\beta = \frac{dy}{dt}\Big/\frac{dx}{dt} = \frac{u\,sin\,\theta - gt}{u\,cos\,\theta} = tan\,\theta - \frac{gt}{u\,cos\,\theta}.$$

Or, $$\beta = tan^{-1}\left[tan\,\theta - \frac{gt}{u\,cos\,\theta}\right].$$

N.B. For small values of *t*, tan β, and, therefore, β is *positive* but, for large values of *t*, it acquires a *negative* value. *And, obviously, when* β = 0, *the body is moving horizontally, i.e., at the very peaks of the parabola.* In this case, since *tan* β = 0, we have

$$tan\,\theta - \frac{gt}{u\,cos\,\theta} = 0.$$

Let this value of t be denoted by t'. Then, $\frac{gt'}{u\,cos\,\theta} = tan\,\theta = \frac{sin\,\theta}{cos\,\theta}$

Or, $gt' = u\,sin\,\theta$, whence, $t' = u\,sin\,\theta/g$

Clearly, the vertical distance covered by the body at an instant t is given $y = ut + \frac{1}{2}at^2$

∴ $y = h$, the *maximum height* attined by the body, when $t = t' = u\,sin\,\theta/g$.

Substituting these values in the above relation, we have

$$h = u\,sin\,\theta.t' - \frac{1}{2}gt'^2 \quad \text{[initial upward velocity} = u\,sin\,\theta, \text{ and } a = -g\,]$$

$$= u\,sin\,\theta.\frac{u\,sin\,\theta}{g} - \frac{1}{2}g\left(\frac{u\,sin\,\theta}{g}\right)^2$$

$$= \frac{u^2\,sin^2\,\theta}{g} - \frac{u^2\,sin^2\,\theta}{2g}. \quad \text{Or, } h = \frac{u^2\,sin^2\,\theta}{2g},$$

the same result, as obtained in § 2.17 (page 33).

Again, the *horizontal range* may be easily obtained by equating the value of y to 0.

Let $t = t''$, when $y = 0$. Then, we have, from above,

$$0 = u\,sin\,\theta.t'' - \frac{1}{2}gt''^2$$

Or, $u\,sin\,\theta\,t'' = \frac{1}{2}gt''^2$ Or $u\,sin\,\theta = \frac{1}{2}gt''$

Or, $gt'' = 2u\,sin\,\theta$, whence, $t'' = 2u\,sin\,\theta/g$*

Now, as we know, the *initial horizontal velocity* = $u\,cos\,\theta$.

And, therefore, the horizontal range R is given by the relation,

$$R = u\,cos\,\theta.\,t'' = u\,cos\,\theta.\,2\,u\,sin\theta/g$$

$$= \frac{u^2.\,2\,sin\,\theta\,cos\,\theta}{g} = \frac{u^2.\,sin\,2\,\theta}{g}, \quad [\because 2\,sin\,\theta\,cos\,\theta = sin\,2\theta.$$

the same result as obtained before in § 2.16 (page 33).

And, finally, if we substitute this value of t, (*i.e.*, $t'' = 2\,u\,sin\,\theta/g$), in the expression for *tan* β above, we have

$$tan\,\beta = tan\,\theta - \frac{gt''}{u\,cos\,\theta} = tan\,\theta - \frac{g}{u\,cos\,\theta}.\frac{2\,u\,sin\,\theta}{g}.$$

Or, $$tan\,\beta = tan\,\theta - 2\,tan\,\theta = -\,tan\,\theta,$$

---

* Another possible value of t'' is 0. We reject it, however, as it refers to the time when body is just starting on its trajectory.

showing that the projectile comes back to the horizontal surface at the *same* angle at which it was projected upwards. And, a further simple deduction will show that its *tangential velocity at this moment is the same as at the instant of projection.*

It must be emphasized again, however, that the above treatment applies to the motion of a projectile, only in a *non-resisting* medium *i.e., in vacuo*. The presence of a medium, like air, offers a frictional resistance to its motion, which depends, to a great extent, upon the velocity of the body and is, for moderate velocities*, directly proportional to the square of the velocity, in accordance with the law of resistance given by *Newton*, in the year 1687. This alters the very character of the trajectory of a projectile, which no longer remains a parabola but becomes what is called a **ballistic curve****, *with its descending part much steeper than its ascending part and the height and range of the projectile considerably reduced, particularly at high initial velocities.*

At higher altitudes, well above the average level of the earth's surface, however, the air pressure, and hence also the air resistance, becomes much smaller and, therefore, if a projectile be shot up to such great maximum heights, it is quite possible to obtain a high range for it. This is perhaps the most probable explanation of the long range German cannon, used with such conspicuous success in the historic Great war, and which could fire shots to a maximum heights of about 54 kilometres and hence had a range of about 130 kilometres.

WORKED EXAMPLES

1. A particle, moving in a circle of radius 105 cm, has its velocity increased in one minute by 120 rotations per minute. Calculate (*i*) its linear acceleration (*ii*) its angular acceleration.

Change in velocity of the particle $= 120/60 = 2$ rotations/*sec.*

Now, in 1 rotation, distance covered $= 2\pi r$.

∴ „ „ 2 rotations„ „ „ „ $= 2\times 2\pi r = 2\times 2\times \pi\times 105$ *cm.*

∴ change in velocity of the particle *per minute* $= 420\times\pi$ *cm/sec.*

Or, change in velocity *per second* $= 420\times\pi/60 = 7\pi$ *cm/sec.*

Or, *rate of change of velocity* or *linear acceleration*, $a = 7\times 22/7 = 22$ *cm./sec*^{2}.

Now, linear acceleration, *i.e.*, $a = r.d\omega/dt$,

where r is the radius, and $d\omega/dt$, the *angular acceleration.*

So that, *angular acceleration* $d\omega/dt = a/r = 22/105 = .2095$ *radian/sec.*

Hence, the *linear* and *angular accelerations* of the particle are 22.0 *cm./sec*^{2} and .2095 *radian/sec*^{2}, respectively.

2. If the Earth be one-half its present distance from the sun, how many days will there be in one year ?

Let the present distance of the Earth form the Sun, *i.e.*, the radius of its orbit round the Sun, be R.

So that, half its present distance form the sun would be $R/2$.

If a_1 and a_2 be the acceleration (*linear*) of the earth in the two cases respectively, we have

$$\frac{a_1}{a_2} = \frac{(R/2)^2}{R^2} = \frac{R^2}{4R^2} = \frac{1}{4}. \qquad [\because a \propto 1/R^2$$

If v_1 and v_2 be the linear velocities of the Earth, corresponding to these two of the acceleration, we have

$$a_1 = \frac{u_1^{\,2}}{R} \quad \text{and} \quad a_2 = \frac{v_2^{\,2}}{R/2} = \frac{2v_2^{\,2}}{R}.$$

So that, $\dfrac{v_1^{\,2}/R}{2v_2^{\,2}/R} = \dfrac{1}{4}$ Or, $\dfrac{v_1^{\,2}}{R}\times\dfrac{R}{2v_2^{\,2}} = \dfrac{1}{4}$.

Or, $v_1/v_2^{\,2} = 1/2$. Or, $v_2^{\,2} = 2v_1^{\,2}$.

∴ $v_2 = \sqrt{2v_1^{\,2}} = \sqrt{2}.v_1$

---

* For Example, from a velocity of a few metres per second to about the velocity of sound, in the case of air.

** *Ballistics* is the special name given to the science of the motion of projectiles.

Now, $v_1 = \frac{2\pi R}{365}$ per day. $\left[\text{ i.e., } = \frac{\textit{circumference or Earth's orbit}}{\textit{time taken}} \right.$

And, therefore, $v_2 = \sqrt{2v_1} = \sqrt{2} \times 2\pi R / 365$ per day

Again, circumference of the Earth's orbit in the *second case* = $2\pi . R/2 = \pi R$.

And, since one year is the time taken by Earth to go once round the Sun, we have

$$\text{one year, in the second case } = \pi R / v_2$$

$$= \frac{\pi R \times 365}{\sqrt{2} \times 2\pi R} = \frac{365}{2\sqrt{2}} = 129 \textit{ days.}$$

Hence, the number of days in one year, when the Earth is half its present distance from the Sun, is 129.

3. Assuming that the Moon revolves uniformly in a circle round the Earth's centre, calculate the acceleration due to gravity at the Earth's surface from the following data :

Radius of the Earth = 6.4×10^8 cm

Radius of the Moon's orbit = 3.84×10^{10} cm

Period of rotation of the Moon = 27.3 days.

(The force of gravity on a particle is inversely proportional to the square of its distance from the Earth's centre)

(Oxford and Cambridge Higher School Certificate)

Here, *velocity of the Moon*, $v = \frac{\text{circumference of the orbit}}{\text{time taken}}$

i.e., $v = \frac{2\pi \times \text{radius of its orbit}}{\text{time taken}} = \frac{2\pi \times 3.84 \times 10^{10}}{27.3 \times 24 \times 60 \times 60}$ *cm/sec*

$\therefore$ *acceleration of the Moon* towards the centre of the Earth, say $g_m = v^2/r$, where r is the radius of the Moon's orbit.

Or, $$g_m = \left[\frac{2\pi \times 3.84 \times 10^{10}}{27.3 \times 24 \times 3600}\right]^2 \times \frac{1}{3.84 \times 10^{10}} = \frac{4\pi^2 \times 3.84 \times 10^{10}}{(27.3 \times 24 \times 3600)^2} \textit{ cm / sec.}^2$$

If the acceleration due to gravity on the earth's surface be g *cm/sec*^{2}, we have

$$g \propto \frac{1}{(\text{radius of earth})^2} \quad \text{and} \quad g_m \propto \frac{1}{(\text{radius of Moon's orbit})^2}$$

$$\therefore \quad \frac{g}{g_m} = \frac{(\text{radius of Moon's orbit})^2}{(\text{radius of Earth})^2}$$

Or, $$g = \left[\frac{(3.84 \times 10^{10})^2}{(6.4 \times 10^8)^2}\right] \times g_m = \left[\frac{3.84 \times 10^{10}}{10^8}\right]^2 \times \left[\frac{4\pi^2 \times 3.84 \times 10^{10}}{(27.3 \times 24 \times 36000)^2}\right]$$

$$= \frac{4\pi^2 \times (3.84 \times 10^{10})^3}{(6.4 \times 10^8 \times 27.3 \times 24 \times 3600)^2} = 980.9 \textit{ cm. / sec}^2$$

Hence, acceleration due to gravity at the Earth's surface = 980.9 *cm./sec.*

4. The radius of curvature of a railway line at a place when the train is moving with a speed of 20 miles per hour is 800 yd., distance between the rails being 5 ft. Find the elevation of the outer rail above the inner rails so that there may be no side-pressure on the rails. ***(Bombay)***

Here, we have $v = 20 \textit{ m.p.h.} = \frac{20 \times 1760 \times 3}{60 \times 60} = \frac{88}{3}$ *ft / sec.*

and $r = 800 \textit{ yd} = 800 \times 3 = 2400 \textit{ ft.}$

So that, $\tan\theta = \frac{v^2}{rg} = \frac{88 \times 88}{3 \times 3 \times 2400 \times 32} = \frac{121}{10800} = .0112.$

Or, $\theta = \tan^{-1} .0112 = 38.33'$

Now, $d = 5$ ft and $\sin 38.33' = .0112$, [θ being small]

and, $\sin\theta = h/d.$ And $\therefore$ $h = d \sin\theta$

Or, $h = 5 \times .0112 = .0560 \textit{ ft.} = .0560 \times 12 = .6720 \textit{ inch.}$

Therefore, the outer rail should be raised .6720 *inch* above the inner one.

5. The centre of gravity of a fully loaded motor van is 1 metre above the ground and its wheel track is 1.62 metre. Assuming the road to be level, calculate the maximum speed at which it can be driven without overturning while negotiating a curve of 25 metre radius.

Referring to Fig. 2.12 (page 30), the maximum speed v of the van while negotiating a curve of radius r on a level road is given by the relation $(mv^2/r).h = mg.d$, where h is the height of the *c.g.*, above the ground and d, the half wheel track of the van.

Since $h = 1\ m$, $r = 25\ m$, $d = 1.62/2 = 0.81\ m$ and $g = 9.81\ m/sec^2$,

we have, $$v^2 = \frac{25 \times 9.81 \times 0.81}{1}.$$

Or, $$v = \sqrt{25 \times 9.81 \times 0.81} = 14.11\ m/sec \quad \text{or} \quad \frac{14.11 \times 3600}{1000} = 50.74\ km/hr.$$

6. A stone of mass 10 lb, is revolving in a vertical circle at the end of a string, 8 ft., long, the other end of which is fixed. When the stone is at the top of the circle, the velocity is 16 *ft.*, per sec. Assuming *g* to be 32 *ft./sec*², find the stretching force in the string when the stone is (*i*) at the top, (*ii*) at the bottom, (*iii*) at a level with the centre.

Here, mass of the stone, $m = 10\ lb$; radius of the circle, $r = 8\ ft$.

and velocity at the *top* of the circle $= 16\ ft/sec$.

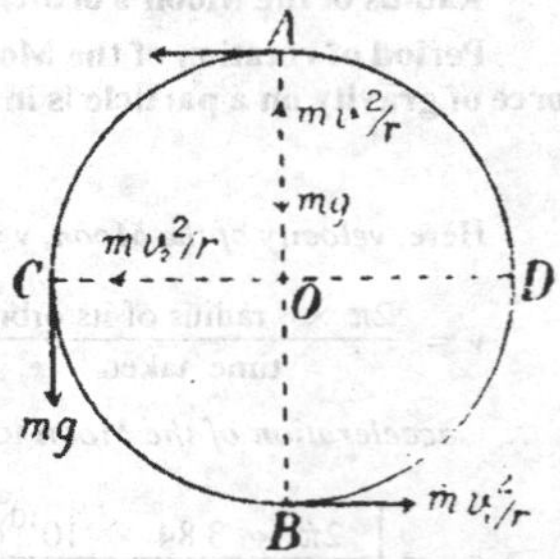

(*i*) Therefore, *force acting outwards, i.e., upwards along the string*, or along OA, (Fig. 2.15)

$$= \frac{mv^2}{r} = \frac{10 \times 16 \times 16}{8} = 320\ poundals.$$

And, downward force due to weight of the stone

$$= mg = 10 \times 32 = 320\ poundals.$$

∴ *resultant stretching force along the string* $= 320 - 320 = 0$.

Fig. 2.15

(*ii*) By the time the stone reaches the *bottom* of the circle, it has acquired additional kinetic energy due to its having fallen a vertical distance $16\ ft$. (the diameter of the circle).

Therefore, *kinetic energy* of the stone at the bottom of the circle

$$= \frac{1}{2}mv^2 + mgh.$$

$$= [\tfrac{1}{2} \times 10 \times 16 \times 16] + [10 \times 32 \times 16] \qquad [\because\ h = 16\ ft.$$

$$= 1280 + 5120 = 6400\ ft.\ poundals.$$

This should be equal to $\frac{1}{2}mv_1^2$, where v_1 is the velocity of the stone at the bottom of the circle. So that,

$$\frac{1}{2}mv_1^2 = 6400, \quad \text{Or,} \quad mv_1^2 = 6400 \times 2 = 12800.$$

Or, $$v_1^2 = 12800/10 = 1280.$$

Now, centrifugal force acting downwards along the string, *i.e.*, along $OB = \frac{mv_1^2}{r}$ and, therefore,

$$= \frac{10 \times 1280}{8} = 1600\ poundals.$$

And the downward force due to the *weight* of the stone is equal to $10 \times 32 = 320$ *poundals* (as before).

∴ total stretching force in the string $= 1600 + 320 = 1920$ *poundals*

$$= 1920/32 = 60\ pounds\ weight.$$

(*iii*) Here, the additional *K.E.* acquired by the stone is due to the fall through a distance of $8\ ft.$, the radius of the circle, and is

$$\therefore \quad = 10 \times 32 \times 8 = 2560\ ft.\ poundals.$$

total *K.E.* of the stone $= \frac{1}{2}mv^2 + 2560 = \frac{1}{2} \times 10 \times 16 \times 16 + 2560$

$$= 1280 + 2560 = 3840\ ft.\ poundals.$$

Obviously, this must be equal to $\frac{1}{2}mv_2^{\ 2}$,where v is the velocity of the stone, in this position.

$$\frac{1}{2}mv_2^2 = 3840, \text{ whence, } v_2^2 = 384 \times 2/10 = 768.$$

centrifugal force acting outwards along the string, *i.e.*, along OC

$$= mv_2^2/r = 10 \times 768/8 = 10 \times 96 = 960 \textit{ poundals.} = 960/32 = 30 \textit{ lb .wt}$$

And, since the weight of the stone is acting vertically downwards, it is acting at *right angles* to this centrifugal force and has no component along it. Therefore, the stretching force in the string, in this case, is equal to 960 *poundals* or 30 *lbwt.*

7. Four masses, each of 10 *lb*, are fastened together with four strings, each 3 *ft.* long, so as to form a square. This square rotates in a horizontal plane on a smooth table at a speed of one revolution per second; find in pounds-weight the tension in the string. *[Cambridge Higher School Certificate]*

Let m_1, m_2, m_3 and m_4 be the four masses of 10 *lb* each, forming a square as shown in Fig. 2.16

The radius of the circle in which they rotate will obviously be equal to $Om_1 = Om_2 = Om_3 = Om_4 = r$, where O is the point of intersection of the diagonals of the square.

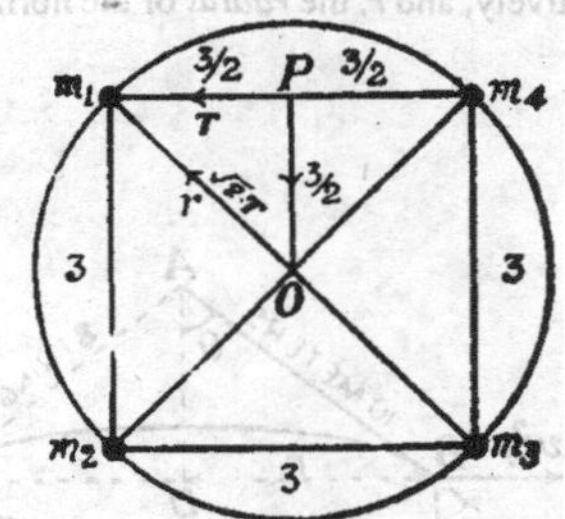

Fig. 2.16

If P be the mid-point of m_1, m_4, we have $OP = Pm_1 = 3/2\,ft.$

So that, $r = \sqrt{(3/2)^2 + (3/2)^2}$

$= \sqrt{18/4} = \sqrt{9/2} = 3/\sqrt{2}\,ft.$

Let T be the tension in each string, when the square is rotating on the smooth table.

Then, representing T, in magnitude and direction, by the straight line Pm_1, we have

force acting along $Om_1 = \sqrt{T^2 + T^2} = \sqrt{2T^2} = \sqrt{2}\,T$

This, clearly, represents the *centrifugal* force acting along Om_1.

Now, the *centrifugal* force acting along Om_1 is also equal to $mr\omega^2$, *i.e.*,

$$= 10 \times \frac{3}{\sqrt{2}} \times (2\pi)^2 = \frac{30 \times 4\pi^2}{\sqrt{2}}$$

where m is the mass of $m_1 = 10\,lb$, $r = 3/\sqrt{2}\,ft$, and $\omega = 2\pi$, because the square *makes one revolution per sec, i.e.*, describes and angle of 2π *radians per sec.*

$\therefore \qquad \sqrt{2}T = (30 \times 4\pi^2)/2 = 120\,\pi^2/2.$

Or, $\qquad 2T = 120 \times \pi^2$, $\qquad$ whence, $T = 60\pi^2$ *poundals*

$$= 60\,\pi^2/32\ lb.\ wt. = 15\,\pi^2/8 = 18.51\ lb.\ wt.$$

The tension in each string is, therefore, equal to 18.51 *lb. wt.*

8. A certain string will break under a load of 50 kg. A mass of 1kg is attached to the end of a piece of the string, 10 metres long, and is rotated in a horizontal circle. Find the greatest number of revolutions per minute which the weight can make without breaking the string.

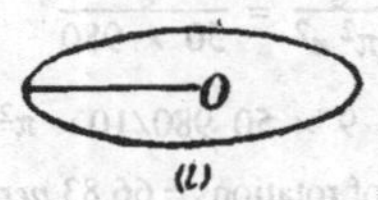

(i)

Here two cases arise, *viz.*

(*i*) when the fixed end of the string is itself the centre of the circle in which the load is rotating, *i.e.*, the *radius of the circle is the length of the string*, [Fig. 2.17(*i*)], and

(*ii*) when the string hangs vertically and a circular motion is given to the load at its end, the circle described being in the horizontal plane [Fig. 2.17 (ii)]

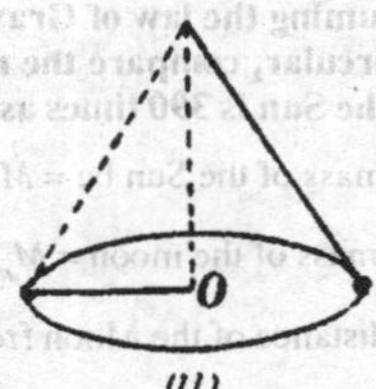

(ii)

Fig. 2.17

First case — The centrifugal force, acting on the string is given by $\qquad F = mv^2/r$

Now, the maximum value of F is 50 *kg. wt.* (given)

$= 50 \times 1000 \times 980$ dynes

So that, $50 \times 1000 \times 980 = 1000 \times v^2/1000 = v^2$

Therefore, $v^2 = \sqrt{50000 \times 980} = \sqrt{49000000}$

$= 7000$ cm / sec.

$= 7000 \times 60$ cm/min

$= 42000$ cm / min

where $m = 1$ kg $= 1000$ gm

$r = 10$ metres. $= 1000$ cm

Now, the distance covered by the load in 1 rotation $= 2\pi r$

$= 2\pi \times 1000 = 2 \times 3.1416 \times 1000$

$= 6283.2$ cm $= 6283$ cm., say

$\therefore$ number of revolutions made by the load *per minute* without breaking the string $= v/2\pi r = 420000/6283 = 66.83$

Or, the number of *complete* rotations made by the load is 66 *per minute*.

Second case. — Here, two forces are acting on the load, *viz.*,

(*i*) mv^2/r, *(centrifugal force) horizontally* $= mr\omega^2$, since $v = r\omega$, where v and ω are the *linear* and *angular velocities* of the load respectively, and r, the *radius* of the horizontal circle in which it rotates.

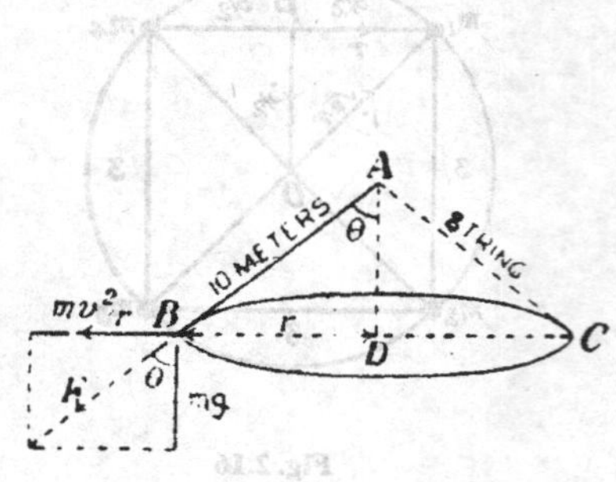

Fig. 2.18

(*ii*) mg, *(weight of the load) vertically downwards.*

Obviously for the load to be in steady motion, the resultant of these two must act along the string, as shown in Fig. 2.18.

Let θ be the angle that the *resultant force F (the tension in the string)* makes with the vertical. *Then,* $\tan\theta = mr\omega^2/mg = r\omega^2/g$

Now. $r = AB \sin\theta = 1000 \sin\theta$, $m = 1$ kg $= 1000$ gm and the maximum value of the tension of the string *i.e.,*

$T = 50 \times 1000 \times 980$ *dynes.*

And, if n be the number of rotations made by the load per minute, the angular velocity $\omega = 2\pi n$ *per min.*

$= 2\pi n/60 = \pi n/30$ *per sec.*

Substituting the values of r and ω in relation (1) above, we have

$$\tan\theta = \frac{\pi^2 n^2 . 1000 \sin\theta}{900\ g}. \quad \text{Or,} \quad \frac{\sin\theta}{\cos\theta} = \frac{\pi^2 n^2 . 1000 \sin\theta}{900\ g}$$

$$= 10. \frac{\pi^2 n^2 \sin\theta}{9g}, \quad \text{whence,} \quad \cos\theta = \frac{9}{10}.\frac{g}{\pi^2 n^2}$$

But, from the figure, we have $\cos\theta = \dfrac{mg}{F} = \dfrac{1000\ g}{50 \times 1000 \times 980} = \dfrac{g}{50 \times 980}$.

$$\therefore \quad \frac{9}{10}.\frac{g}{\pi^2 n^2} = \frac{g}{50 \times 980}. \quad \text{Or,} \quad 9/10.\pi^2 n^2 = 1/50 \times 980,$$

whence, $n^2 = 9 \times 50\ 980/10 \times \pi^2$. Or, $n = \sqrt{9 \times 5 \times 980}/\pi = 66.83$

Hence, number of rotations = 66.83 *per minute*, or number of *complete* rotations made by the load *per minute* is 66.

9. Assuming the law of Gravitation, and taking the orbit of the Earth round the Sun, and of the Moon round the Earth as circular, compare the masses of the Sun and the Earth, given that the Moon makes 13 revolutions per year and that the Sun is 390 times as distant as the Moon.

Let mass of the Sun be $= M_s$, mass of the Earth $= M_e$

and mass of the moon $= M_m$.

And, let distance of the Moon from the Earth $= R$,

so that " " " Sun " " " " $= 390\ R$.

Then, force of attraction between the sun and the Earth $= \dfrac{M_s \times M_e}{(390\ R)^3}.G$, where G is the *gravitational constant.*

$\therefore$ *centripetal acceleration of the Earth* $= \dfrac{force}{mass} = \dfrac{M_s \,.\, M_e}{(390\,R)^2\,M_e}\,.\,G = M_s/(390\,R)^2$

Similarly, force of attraction between the Earth and the Moon $= \dfrac{M_e \,.\, M_m}{R}\,.\,G.$

And, $\therefore$ *centripetal acceleration of the Moon* $= \dfrac{M_e . M_m}{R^2 . M_m}\,.\,G = \dfrac{M_e}{R^2}\,.\,G.$

Let ω_e and ω_m be the angular velocities of the Earth and the Moon respectively. Then, clearly,

centripetal acceleration of the Earth is also $= 390\,R.\,\omega_e^{\,2}$

" " " Moon " $= R\omega_m^{\,2}$.

$\therefore \quad \dfrac{[M_s/(390.R)]^2\,G}{(M_e/R^2)\,G} = \dfrac{390.R.\omega_e}{R.\omega_m^{\,2}}$ Or, $\quad \dfrac{M_s}{(390)^2.\,M_e} = \dfrac{390\omega_e^{\,2}}{\omega_m^{\,2}}$

Now, the Earth goes round the Sun only *once in one year;* and, therefore, angular velocity of the *Earth* = 2π *per year.*

And the moon goes round the Earth 13 *times* in one year; so that angular velocity of the *Moon* $= 13 \times 2\pi$ *per year.*

Therefore, $\quad M_s/(390)^2.\,M_e = 390 \times (2\pi)^2/(13 \times 2\pi)^2$

Or, $\quad M_s/(390)^2\,M_e = 390/13^2 \quad$ Or, $\quad M_e\,M_e = 390^3/13^2.$

i.e., $\quad$ Mass of the Sun : Mass of the Earth :: $(390)^3 : (13)^2$.

10. Show that in the case of liquid, rotating with a uniform angular velocity, (*i*) the pressure varies directly as the distance from the axis of rotation and (*ii*) the free surface of the liquid is a paraboloid of revolution.

(*i*) Imagine a *closed, vertical, cylindrical* vessel, *just* full of a liquid of density ρ, to be rotating about its axis with a *uniform angular velocity* ω.

Now, consider a *ring* of the liquid, of *radius x, width* δx and *vertical height* δh, with its centre at *O on the axis of rotation*, (Fig. 2.19). Then, if the pressures in the liquid at distances x and $x + \delta x$ from the axis of rotation be p and $(p + \delta p)$ *outwards* and *inwards* respectively, we have

resultant inward thrust on the ring $= \delta p.\, 2\pi x\,.\,\delta h,$

and *centrifugal force outward on the ring* $= m\omega^2 r,$

where m is the *mass* of the liquid in the ring.

Clearly, $\quad m$ = *volume of liquid in the ring* $\times\ \rho = 2\pi x.\,\delta x\,.\,\delta h.\,\rho.$

So that, *centrifugal force outward on the ring* $= 2\pi x.\,\delta x\,.\delta h.\,\rho.\omega^2\,x$

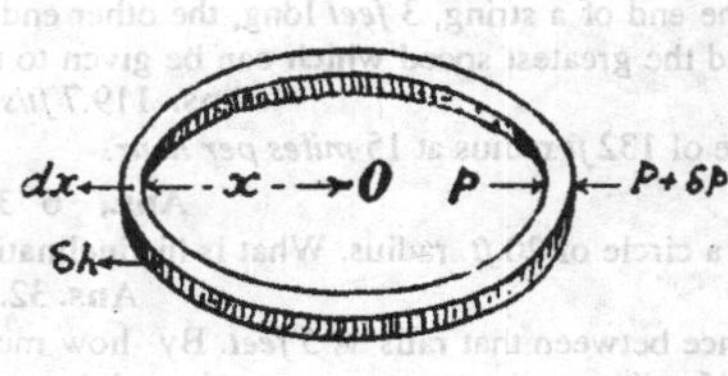

Fig. 2.19

$\because\ r = x$, here,

And, therefore, $\delta p.\,2\pi x.\delta h = 2\pi x\,.\,\delta x.\,\delta h.\,\rho.\,\omega^2 x,\quad$ whence, $\delta p = \rho\omega^2\,.\,x\,.\,\delta x.$

Integrating this expression, we have $\int dp = \int \rho\omega^2 x\,.\,dx = \rho\omega^2 \int x\,.\,dx.$

Or, $\quad p = \dfrac{1}{2}\rho\omega^2 x^2 + C \quad$ | C being a constant of integration.

This is then the expression for the pressure at a point distant x from the axis of rotation in a rotating liquid.

The value of C is obviously equal to that of p at $x = 0$ *i.e.,* **equal to the pressure at O, which as we know, is *zero* at** the surface of the liquid, but increases with depth as in the case of a liquid at rest.

(*ii*) Consider a particle P in the liquid surface, of mass m, whose coordinates are **(x, y) with respect to the axes OX and OY**, (Fig. 2.20), the liquid being supposed to be rotating with a ***uniform angular velocity*** **ω in the direction shown,** about OY.

The forces acting on this particle P are, clearly, *(i) its weight mg, vertically downwards*, and *(ii) the centrifugal force* $m\omega^2 x$, *outwards*. The resultant R of these two forces must act at right angles to the liquid surface, since there is no flow of the liquid taking place, and, obviously, it is counterbalanced by the thrust due to the rest of the liquid on P.

If, therefore, θ be the angle that the centrifugal force makes with the tangent at P, we have

$$\tan\theta = m\omega^2 x/mg = \omega^2 x/g.$$

Now, obviously, $\tan\theta$ = *alope at* $P = dy/dx$.

$dy/dx = \omega^2 x/g$. Or, $dy = (\omega^2 x/g).dx$.

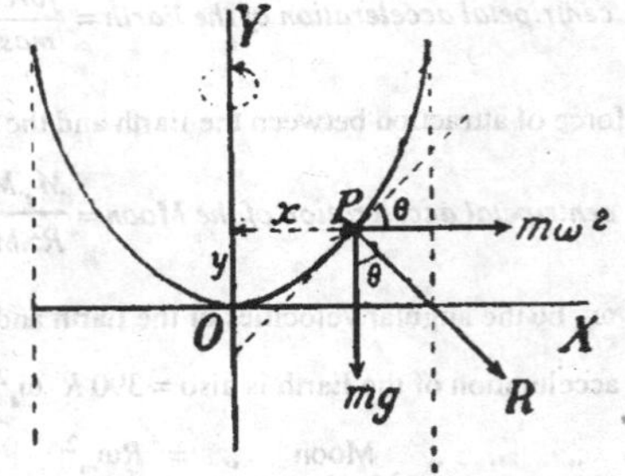

Fig. 2.20

Integrating this, we have

$$\int dy = \int \frac{\omega^2 .x}{g} . dx = \frac{\omega^2}{y} \int x.dx.$$

$$y = \frac{\omega^2 x^2}{2g} + C,$$

where C is again a constant of integration.

Since $y = 0$, when $x = 0$, we have $C = 0$ And $\therefore$ $y = \omega^2 x^2 / 2g$, which is the equation to a *parabola*.

The free surface of a uniformly rotating liquid is thus clearly a paraboloid of revolution.

11. A projectile is fired with an initial velocity of 600 metre/sec at an angle of 30° to the horizontal. Calculate its horizontal range.

As explained under § 2.16 (page 33), the horizontal range of a projectile is given by the relation $R = u^2 \sin 2\theta/g$. Here, $u = 600$ *m/sec*, $\theta = 15°$ and $g = 9.81$ *m/sec*^{2}. So that, $R = 600 \times 600 \times \sin 30°/9.81 = 18350$ *m* or 18.35 *km*.

EXERCISE II

1. A particle of mass 1 *lb*, is whirled uniformly at the end of a string, 2 *ft*. long, and makes 3 revolutions in 1.2 *sec*. Find the tension in the string. **Ans.** 15.43 *lb-wt*.

2. A half-pound weight is being Whirled in a horizontal circle at the end of a string, 3 *feet* long, the other end of the string being fixed. If the breaking tension of the string is 112 *lb wt*, find the greatest speed which can be given to the weight. **Ans.** 119.7 *ft/sec*.

3. At what angle should a cyclist lean over, when negotiating a curve of 132 *ft* radius at 15 *miles per hour*. **Ans.,** 6° 32′.

4. A person skating on ice at the rate of 20 *ft. per second* describes a circle of 20 *ft*. radius. What is his inclination to the vertical ? **Ans.** 32.0°.

5. A train is going round a curve of 1000 *ft* radius and the distance between that rails is 5 *feet*. By how much should the outer rail be raised above the inner one, so that a train running at 45 *miles per hour* may exert no lateral thrust on the outer rail ? **Ans.** 8.094″.

6. A stone is suspended from the roof of a railway carriage by means of a string 5 *ft* long. The angle through which the mass moves from the vertical is 10°, when the train moves along a curve of radius 600 *ft*. Calculate the speed of the train. **Ans.** 58.1 *ft/sec*.

7. Assuming that the moon describes a circular orbit of radius 3.84×10^8 *metres* in 27.3 days and the outer satellite of Mars describes a circular orbit of radius = 2.35×10^7 *metres* in 1.26 *days*, find the ratio of the mass of Mars to the mass of the Earth. *(Cambridge University)* **Ans.** .1076 : 1.

8. A curve on a railway line has a radius 1600 *ft* and the distance between the inner and outer rails is 5 *ft*. If the outer rail be 6″ above the inner one, calculate the maximum speed of a train going along the curve, so that no side-thrust is exerted on the outer rail. **Ans.** 48.89 *miles/hr*.

9. Calculate the increase in length of an elastic string of original length 10 *ft* at the end of which a stone of mass 5 *lb* is whirled at the rate of 4 revolutions *per second*, if a load of 25 *lb* increases the length of the string by 2 % **Ans.** .8576 *ft*.

10. A merry-go-round is revolving in a horizontal circle of radius 3 *ft*. at the rate of 7 revolutions in 11 *seconds*. A child of weight 20 *lb* rides a wooden horse suspended by a vertical string. Find the tension in the string and its inclination to the vertical. [**Ans.** *(i)* 36 *lb wt*. nearly, *(ii)* $\tan^{-1} 3/2$]

[**Hint.** *See worked example* 8 —*(Second case, page* 40.)

11. A sea-plane of total mass 1000 *lb* (including the pilot) rounds a pylon in a circular arc of radius half a mile at a speed of 300 *m.p.h.* Draw a diagram showing the forces acting on the sea-plane, and calculate the resultant force at right angles to its direction of motion exerted upon it by the air. Assuming that the pilot with 12 stones, calculate the force with which he is pressed against his seat during the "turn". *(Cambridge Higher School Certificate)*

Ans. 8×10^4 *poundals*; 30 stone *wt.*

12. Calculate the angle at which a curve of radius 352 *ft.* should be banked so as to avoid side-slip when a motor car is travelling round it at a speed of 30 *m.p.h.* **Ans.** 9° 45′.

13. A road over a bridge has the form of a vertical arc of radius 60 *ft.* What is the greatest speed in *m.p.h.* at which a car can cross the bridge without leaving the ground at the crest of the road? **Ans.** 30 *m.p.h.*

14. A skater is moving on one foot in a circle of radius 20 *ft* at 10 *m.p.h.* At what angle with the vertical will the line passing through his centre of gravity and the edge of the skate be inclined? **Ans.** 18° 35′.

15. In a 'loop-the-loop' railway, descending a steep incline, run round the inside of a vertical circular track, 6.096 *m* in diameter, making a complete turn over. Assuming there is no friction, find the minimum height above the top of the circular track from which the cars must start. **Ans.** 1.524 *m.*

16. A symmetrically loaded lorry weights 5 *tons* and the height of its centre of gravity is 5 *ft* above the ground in a vertical plane midway between the wheels. The breadth of the wheel base may be taken to be 6 *ft* 3 *in.* If there is no sideslip, what is the maximum speed at which the lorry can take a curve of mean radius 6 *yards* without beginning to overturn? **Ans.** 13 *m.p.h.*

[**Hint.** *It will overturn only when the moment of the centrifugal force about the wheels on one side is greater than the moment of the weight about them.* *(see* § 2.13, case 4, page 27]

17. An India rubber band has a mass of 4 *gm* per metre when stretched on the circumference of a wheel of 10 *cm* radius, the stretching force being 20,000 *dynes.* Find how many revolutions per second the wheel must make so that the band may not press upon the wheel. *(Nagpur)*

Ans. 11.3

[**Hint.** See § 2.13, case 2, page 26]

18. Discuss the possibility of a motor cyclist riding round the inside surface of a vertical cylinder.

(Cambridge Scholarship)

19. Explain why a motor-cycle combination (side car on left) is liable to overturn when taking a left hand corner at speed. Assuming that the centre of mass of the combination is 2 *ft from the ground and 1 ft* to the left of the motor cycle, calculate the maximum speed of the combination in a circle of radius 50 *feet.* Assume that the road surface is horizontal and that there is no skidding. *(Oxford Scholarship)*

Ans. 19.3 *m.p.h.*

20. A closed cylindrical can of radius *a* and height *h* is first filled with a fluid of density ρ. It is then rotated with angular velocity ω about its own axis, held vertical. Prove that the total thrust on the top of the can will be $\pi\rho\omega^2 a^4/4$.

21. Prove that the square of the period of revolution of a planet rotating in a circular orbit round the sun is proportional to the cube of its distance from the sun. *(Bombay)*

CHAPTER III

MOMENT OF INERTIA — GYROSCOPIC MOTION

3.1. Moment of Inertia and its Physical Significance—Radius of Gyration. We know that, according to Newton's *first law of motion*, a body must continue in its *state of rest* or of *uniform motion along a straight line*, unless acted upto by an external force. *This inertness or inability of a body to change, by itself, its position of rest, or of uniform motion, is called* **inertia,*** *and is a fundamental property of matter*. Thus, it is by virtue of its *inertia* that a body at rest resists or opposes being put into motion, and a body, in linear or translatory motion, opposes not only being brought to rest but also any changes in the magnitude and direction of its motion. And, we know, by experience, that the greater the mass of a body, the greater its inertia or opposition to the desired change; for, the greater is the force required to be applied for the purpose. *The mass of a body is thus taken to be a measure of its 'inertia for translatory motion'*, as it is *this* that opposes the acceleration, (positive or negative), desired to be produced in it by the applied force.

Exactly in the same manner, in the case of *rotational motion* also, we find that a body, free to rotate about an axis, opposes any change desired to be produced in its state of rest or rotation, showing that it possesses '*inertia*' for this type of motion. And, obviously, the greater the *couple or torque*, (see § 3.3), required to be applied to a body to change its state of rotation, *i.e.*, to produce in it a desired angular acceleration, the greater its opposition to the desired change, or the greater its '*inertia for rotational motion*'. *It is this 'rotational inertia' of the body which is called its* **moment of inertia** ** *about its axis or rotation*, —this name being given to it on the analogy of the moment of the couple which it opposes.

It will thus be seen that the *moment of inertia* of a body, in the case of rotational motion, plays the same part as, or is the analogue of, the *mass* of a body in the case of translatory motion; and we may, therefore, for purposes of analogy, describe the moment of inertia of a body, in rotational motion, as the '*effectiveness of the mass*'. Or, pushing the analogy of a little further, we may define *mass* as the '*coefficient of inertia**** *for translatory motion*', *and the moment of inertia, as the 'coefficient of rotational inertia'*.

Yet, with all this seeming similarity, there is all the difference between the two cases. For, *in the case of translatory motion, the inertia of the body depends wholly upon its mass* and is, therefore, measured in terms of it alone. *In the case of rotational motion, on the other hand, the rotational inertia, or the moment of inertia, of the body depends not only upon the mass (M) of the body but also upon the 'effective distance', (K) of its particles from the axis of rotation and is measured by the expression* MK^2, (see next Article).

This '*effective distance*' (*K*) of the particles of a body from its axis of rotation is called its **radius of gyration** about the axis, and is equal to the *root mean square distance* of the particles from the axis, *i.e.*, equal to the square root of their *mean square distance* (not the square of their mean distance) from it. Or, to give it a clear cut definition, *the radius of gyration of a body, about a given axis of rotation, may be defined as the distance from the axis at which, if the whole mass of*

---

* That is why the comparative slackness or sluggishness of people of Eastern countries, —a consequence of climatic conditions—is dubbed by the Westerners as the '*Inertia of the East*'.

** It is also sometimes referred to as the '**Spin Inertia**' of the body about its axis of rotation.

*** The mass of a body being usually referred to as its **inertia coefficient**.

the body were to be concentrated, the moment of inertia of the body about the given axis of rotation would be the same as with its actual distribution of mass. (See also footnote on page 56).

Now, it is obvious that a change in the position of inclination of the axis of rotation of a body will bring about a corresponding change in the relative distances of its particles, and hence in their '*effective distance*' from the axis, *i.e.*, in the value of the *radius of gyration* of the body about the axis. And, so will the transference of a portion of the matter (or mass) of the body from one part of it to another, or a change in the distribution of the mass about the axis, the total mass of the body remaining the same, in either case.

Thus, whereas the mass of a body remains the *same*, irrespective of the location or inclination of the axis of rotation, *the value of its radius of gyration about the axis depends upon*

(i) the position and direction of the axis of rotation, and

(ii) the distribution of the mass of the body about this axis; so that, its value for the same body is different for different axes of rotation.

Further, it follows, as a converse of the above, that the radius of gyration of a body about a given axis of rotation gives an indication of the distribution of the mass of the body about it and hence, also the effect of this distribution of mass on the moment of inertial of the body about that axis.

3.2. Expression for the Moment of Inertia. Suppose we have a body of mass M, (Fig.3.1), and any axis YY'. Imagine the body to be composed of a large number of particles of masses m_1, m_2, m_3, etc., at distances r_1, r_2, r_3,.... etc. from the axis YY'. Then the moment of inertia of the particle m_1 about YY' is $m_1r_1^2$, that of the particle m_2 is $m_2\,r_2^2$, and so on; and, therefore, the moment of inertia, I, of the whole body, about the axis YY', is equal to the sum of $m_1r_1^2$, $m_2\,r_2^2$, $m_3\,r_3^3$ ...etc.

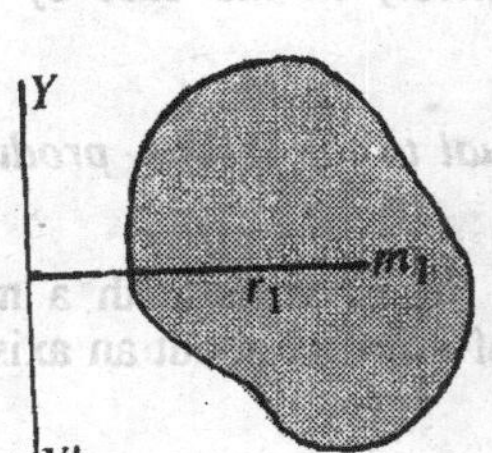

Fig. 3.1

Thus, $$I = m_1r_1^2, m_2r_2^2, m_3r_3^3 + \ldots$$

$$= \Sigma\, m\, r^2.$$

Or, $$I = KM^2,$$

where M is the mass and MK^2 the summation ΣMr^2 for the whole body.

K being the *radius of gyration* of the body about the axis YY'.

3.3. Torque. If we wish to accelerate the rotation of a body, free to rotate about an axis, we have to apply to it a couple. *The moment of the couple, so applied, is called* **torque**, and we say that a torque is applied to the body.

Obviously, the *angular acceleration* of all the particles *irrespective of their distances* from the axis of rotation, is the *same*, but because their distances are different from the axis, their *linear acceleration are different*, (the linear acceleration of a particle being the product of the angular acceleration and the distance of the particle from the axis of rotation).

If, therefore, $d\omega/dt$ be the angular acceleration of the body, or its particles, we have

linear acceleration of the particle distant r_1 from the axis $= r_1.d\omega/dt$,

" " " " " " r_2 " " $= r_2.d\omega/dt$,

and so on.

Hence, if m be the mass of each particle of the body, the forces on the different particles are $mr_1.d\omega/dt$, $mr_2.d\omega/dt$, *etc.*, and the *moments* of these forces about the axis of rotation will, therefore, be

$$(mr_1.d\omega/dt) \times r_1, \quad (mr_2.d\omega/dt) \times r_2 \quad \text{and so on.}$$

Therefore, total moment for the *whole* body

$$= (mr_1.d\omega/dt) \times r_1 + (mr_2.d\omega/dt) \times r_2 + \ldots\ldots\ldots$$

$$= mr_1^2.d\omega/dt + mr_2^2.d\omega/dt + \ldots\ldots\ldots$$

$$= \Sigma mr^2.d\omega/dt,$$

$$= (d\omega/dt).\Sigma mr^2. \qquad [dw/dt \text{ being constant.}$$

But $\Sigma mr^2 = I$, the moment of inertia of the body about the axis of rotation. And, therefore,

moment for the whole body $= I.d\omega/dt$.

This must be equal to the *torque* applied to the body.

So that, *torque* $= I.d\omega/dt$.

It will at once be clear that this relation corresponds to the familiar relation, *force* $= m \times a$, in the case of linear motion, where m is the *mass* and a, the *acceleration* of the body.

Thus, in the case of rotatory motion, torque, moment of inertia and angular acceleration are the analogues of force, mass and linear acceleration respectively in the case of linear or translatory motion.

Now, if $d\omega/dt = 1$, clearly, *torque* $= I$.

Or, *the moment of inertia of a body about an axis is equal to the torque, producing unit angular acceleration in it about that axis.*

Incidentally, the expression for torque, obtained above, furnishes us with a method of deducing an expression for the moment of inertia of a particle of mass m, about an axis distant r from it.

For, if F be the force applied, we have

torque $= F \times r$.

But *torque* is also $= I\ dw/dt$.

And, therefore,

$$I = \frac{torque}{d\omega/dt} = \frac{F \times r}{d\omega/dt} \qquad \ldots(i)$$

Now, $F = m \times a$,

where a is the *linear acceleration* of the particle.

And, since $a = dv/dt$, (where v is the *linear velocity* of the particle), we have

$$F = m.dv/dt.$$

Again, since $v = r\omega$, where ω, is the *angular velocity* of the particle, we have

$$F = m.\frac{d(r\omega)}{dt} = m.\left(\frac{d\omega}{dt}.r + \frac{dr}{dt}.\omega\right).$$

Now, the component, $(dr/dt).\omega$, plays no part in the rotation of the body and may, therefore, be ignored; so that, $F = mr.dw/dt$.

Substituting this value of F in relation (i) above, we have

$$I = \frac{mr.(d\omega/dt).r}{d\omega/dt} = mr^2$$

Thus, the moment of inertia of a particle of mass *m*, about an axis distant *r* from it, is equal to mr^2.

For yet another definition of moment of inertia, see § 3.13 (page 77)

3.4. General Theorems on Moment of Inertia. There are two general theorems of great importance on moment of inertia which, in some cases, enable us to determine the moment of inertia of a body about an axis, if its moment of inertia about some other axis be known. We shall now proceed to discuss these.

(*a*) The Principle or Theorem of Perpendicular Axes

(*i*) **For a Plane Laminar Body.** According to this theorem, the *moment of inertia of a plane lamina about an axis, perpendicular to the plane of the lamina, is equal to the sum of the moments of inertia of the lamina about two axes at right angles to each other, in its own plane, and intersecting each other at the point where the perpendicular axis passes through it.*

Thus, if I_x and I_y be the moments of inertia of a *plane lamina* about the perpendicular axes, *OX* and *OY*, which lie in *the plane of the lamina* and intersect each other at *O*, (Fig. 3.2), the moment of inertia about an axis passing though *O* and *perpendicular to the plane of the lamina*, is given by

$$I = I_x + I_y.$$

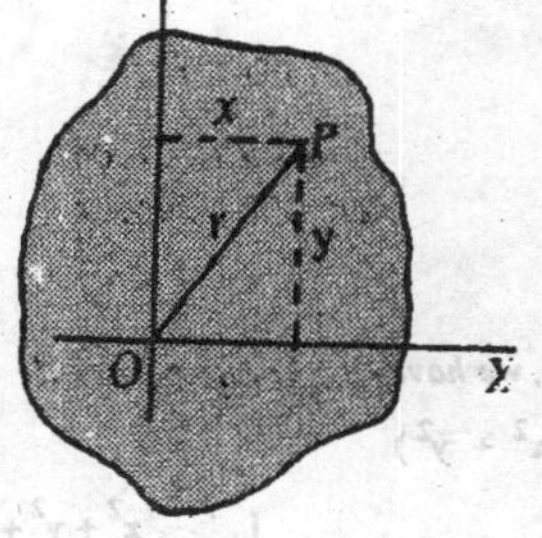

Fig. 3.2

For, considering a particle of mass *m* at *P*, at distances *x* and *y* from *OY* and *OX* respectively, and at distance *r* from *O*, we have

$$I = \Sigma mr^2, \quad I_x = \Sigma my^2 \text{ and } I_y = \Sigma mx^2.$$

So that, $I_x + I_y = \Sigma my^2 + \Sigma mx^2 = \Sigma m(y^2 + x^2)$.

$$= \Sigma mr^2 \qquad [\because y^2 + x^2 = r^2.$$

Or, $$I_x + I_y = I.$$

(*ii*) **For a Three-Dimensional Body.** Suppose we have *a cubical* or a *three-dimensional* body, shown dotted in Fig. 3.3, with *OX*, *OY* and *OZ* as its three mutually perpendicular axes, representing its *length, breadth* and *height* respectively.

Consider a mass *m* of the body, at a point *P*, somewhere inside it. Drop a perpendicular *PM* from *P* on the *xy* plane to meet it in *M*. Join *OM* and *OP*, and from *M* draw *MQ* parallel to the *x*-axis and *MN*, parallel to the *y*-axis; also, from *P* draw *PR*, parallel to *OM*. Then, clearly, the coordinates of the point *P* are

$$x = ON = QM, \; y = OQ = NM, \; z = MP = OR.$$

Since the plane *xy* is perpendicular to the *z*-axis, any straight line drawn on this plane is also perpendicular to it, and, therefore, *OM* and *PR* are both perpendicular to the *z*-axis, ($\because$ *PR* is drawn parallel ot *OM*).

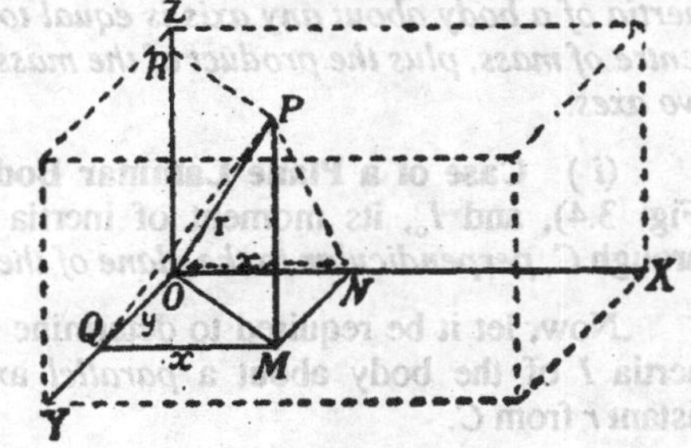

Fig 3.3

Obviously, therefore, $\angle OMP$ is a right angle, because *OM* is parallel to *PR* and *PM* is parallel to *OR*. Hence, we have

$$OM^2 + MP^2 = OP^2. \qquad \text{Or,} \qquad OM^2 + z^2 = r^2 \text{, where, } OP = r. \qquad ...(i)$$

But $$OM^2 = QM^2 = OQ^2 = x^2 + y^2,$$

because $\angle OQM$ is a right angle, being the angle between the axes of *x* and *y*.

Therefore, substituting the value of OM^2 in relation (*i*) above, we have

$$x^2 + y^2 + z^2 = r^2 \qquad ...(ii)$$

Join PN and PQ. Then, PN and PQ are the respective normals to the axes of x and y. For, $\angle PMN$ is a right angle, being the angle between the axes y and z, and, therefore,

$$PN^2 = MN^2 + PM^2 = y^2 + z^2$$

So that $\quad x^2 + PN^2 = x^2 + y^2 + z^2 = r^2$ [From (ii) above.

Or, $\quad ON^2 + PN^2 = r^2$, $\quad x = ON$.

from which it is clear that $\angle PNO$ is a right angle, and, therefore, PN is perpendicular to the x-axis.

Similarly, in the right-angled ΔPMQ, we have

$$PQ^2 = MQ^2 + PM^2 = x^2 + z^2$$

But $\quad x^2 + y^2 + z^2 = r^2$.

$\therefore \quad PQ^2 + y^2 = r^2$. Or, $\quad PQ^2 + OQ^2 = OP^2$,

because $\quad y = OR$ and $r = OP$

Or, $\quad \angle POQ$ is a right angle, *i.e.*, PQ is perpendicular to the y-axis.

Now, moment of inertia of mass m at P about the z-axis

$$= m \times PR^2 = m.OM^2,$$

because $PR = OM$ is the perpendicular distance of the mass from the axis.

$\therefore$ moment of inertia of the whole body about the z-axis, *i.e.*,

$$I_z = \Sigma\, m.OM^2 = \Sigma\, m\,(x^2 + y^2).$$

Similarly, the moment of inertia of the body about the y-axis, *i.e.*,

$$I_y = \Sigma\, m..PQ^2 = \Sigma\, m\,(x^2 + y^2);$$

because PQ is the perpendicular distance between the mass and the axis.

And, the moment of inertia of the body about the x-axis, *i.e.*,

$$I_x = \Sigma m.PN^2 = \text{—}\Sigma m\,(y^2 + z^2)$$

because PN is the perpendicular distance between the mass and the axis.

$\therefore$ adding up the moments of inertia of the body *about the three axes, we have*

$$I_x + I_y + I_z = \Sigma m\,(y^2 + z^2) + \Sigma m\,(x^2 + z^2) + \Sigma m\,(x^2 + y^2)$$

$$= 2\,\Sigma\, m\,(x^2 + y^2 + z^2) = 2\,\Sigma\, mr^2. \qquad [\because x^2 + y^2 + z^2 = r^2.$$

Hence the sum of the moments of inertia of a three-dimensional body about its three mutually perpendicular axes, is equal to twice the summation Σmr^2 *about the origin.*

(*b*) **The Principle or Theorem of Parallel Axes.** This theorem (due to **Steiner**) is true both for a plane laminar body as well as a three-dimensional body and states that *the moment of inertia of a body about any axis is equal to its moment of inertia about a parallel axis through its centre of mass, plus the product of the mass of the body and the square of the distance between the two axes.*

(*i*) **Case of a Plane Laminar Body.** Let C be the *centre of mass* of a body of mass M, (Fig. 3.4), and I_o, its moment of inertia about an axis through C, *perpendicular to the plane of the paper.*

Now, let it be required to determine the moment of inertia I of the body about a *parallel* axis through O, distant r from C.

Consider any particle P of the body of mass m, at a distance x from O.

Then, the moment of inertia of the body about O is given by

$$I = \Sigma\, mx^2 \qquad [\text{Since } OP^2 = x^2$$

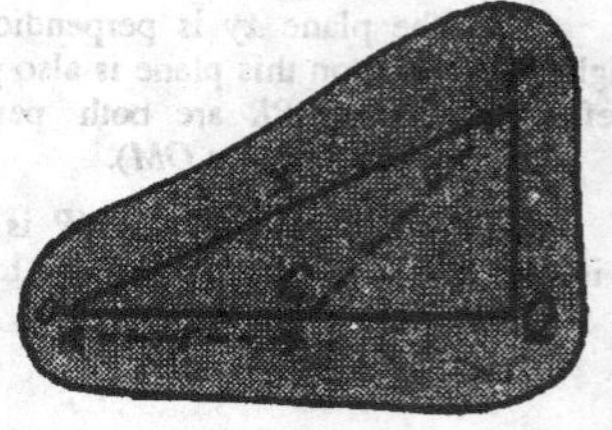

Fig. 3.4

From P drop a perpendicular PQ on to OC produced, and join PC. Then;

$$OP^2 = CP^2 + OC^2 + 2OC.CQ.$$ [By simple geometry.

And $\therefore$ $m.OP^2 = m.CP^2 + m.OC^2 + 2m.OC.CQ.$

$\therefore$ $\Sigma mx^2 = \Sigma m.CP^2 + \Sigma mr^2 + 2r\Sigma m.CQ.$ [$\because OP = x$ & $OC = r$.

Hence $I = I_o + Mr^2 + 2r\,\Sigma m.CQ.$ [$\because \Sigma m.CP^2 = I_o$.

Now, since a body always balances about an axis passing through its centre of mass, it is obvious that the *algebraic sum* of the moments of the weights of its individual particles about the centre of mass must be *zero*. Hence, here, $\Sigma mg.CQ$, (the algebraic sum of such moments about C) and, therefore, the expression $\Sigma m.CQ$ is equal to 0, g being constant at a given place. Consequently, $2r.\,\Sigma m.\,CQ = 0$.

So that, $I = I_o + Mr^2$.

(*ii*) **Case of a Three-Dimensional Body.** Let *AB* be the axis about which the moment of inertia of a body (shown dotted) is to be determined, (Fig 3.5). Draw a parallel axis *CD* through the centre of mass *G* of the body, at a distance *r* from it.

Imagine a particle of mass *m* at any point *P*, outside the plane of the axes *AB* and *CD* and let *PK* and *PL* be perpendiculars drawn from *P* on to *AB* and *CD* respectively and *PT*, the perpendicular dropped from *P* on to *KL* produced.

Put $PL = d$, $LK = r$, $LT = x$ and $\angle PLK = \theta$

Then, if I be the moment of inertia of the body about the axis *AB* and I_o its moment of inertia about the axis *CD* (through *G*), we clearly have

$$I = \Sigma m.PK^3 \quad \text{and} \quad Io = \Sigma m.PL^2 = \Sigma m.d^2$$

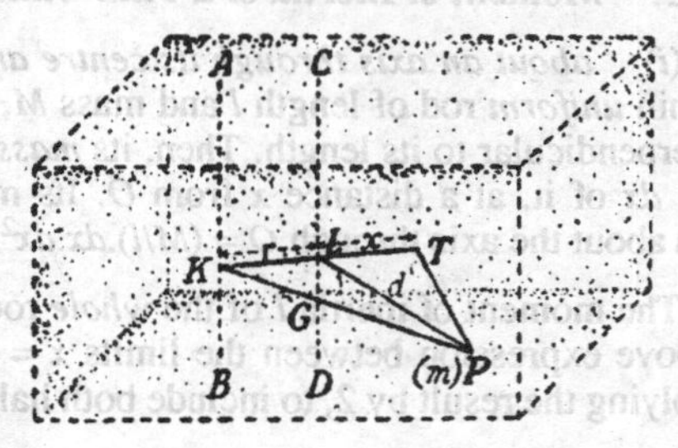

Fig. 3.5

Now, from the geometry of the figure, we have

$$PK^2 = PL^2 + LK^2 - 2PL.LK\ \cos PLK$$
$$= d^2 + r^2 + 2d.r\cos\theta$$

And, in the right-angled ΔPTL, we have $\cos PLT = LT/PL$

where $\angle PLT = (180° - PLK) = (180° - \theta)$. So that

$\cos(180° - \theta) = x/d.$ Or, $-\cos\theta = x/d,$ whence, $d\cos\theta = -x.$

Substituting this value of $d\cos\theta$ in the expression for PK^2 above, we have

$$PK^2 = d^2 + r^2 + 2rx.$$

And, therefore, $I = \Sigma m.PK^2 = \Sigma m(d^2 + r^2 + 2rx) = \Sigma md^2 + \Sigma mr^2 + 2r\Sigma mx.$

Or, $I = I_o + \Sigma Mr^2 + 2r\,\Sigma mx,$

because $\Sigma mr^2 = Mr^2$, where M is the mass of the *whole* body and r, the distance between the two paralled axes and hence a constant. Clearly, $\Sigma mx = 0$, being the total moment about an axis through the centre of mass of the body.

We, therefore, have $I = I_o + Mr^2$,

the same result as obtained above in case (*i*) for a plane laminar body.

3.5. Calculation of the Moment of Inertia of a Body. — Its Units etc. In the case of a *continuous, homogeneous* body of a definite *geometrical shape*, its moment of inertia is calculated by first obtaining an expression for the moment of inertia of an *infinitesimal mass* of it about the given axis – *by multiplying this mass (m) by the square of its distance (r) from the axis, and then integrating this expression over the appropriate range, depending upon the shape of the body concerned, making full use of the theorems of perpendicular and parallel axes, wherever necessary.*

In case, however, the body is not homogeneous or of a definite geometrical shape, the safest thing to do is to determine its moment of inertia by actual experiment, as explained later in § 3.9 and in Chapter VIII.

Now, it will be seen that since the moment of inertia of a body about a given axis is equal to MK^2, where M is its *mass* and K, its *radius of gyration* about that axis, its dimensions are 1 *in mass* and 2 *in length*, its dimensional formula being $[ML^2]$. If the mass of the body and its radius of gyration be measured in the *C.G.S.* units, *i.e.*, its mass in *grams* and radius of gyration in *centimetres*, the moment of inertia of the body is expressed in *gram-centimetre*^{2} (*i.e.*, in *gm* cm^2). If the two quantities be measured in the *F.P.S.* units, *i.e.*, the mass of the body in *pounds* and its radius of gyration in *feet*, the moment of inertia is expressed in *pound-feet*^{2}, (*i.e.*, in $lb.ft^2$). And, in *M.K.S* or *S. I.* units, it will obviously be expressed in *kg-metre*^{2}

And, finally, it must be carefully noted that since the moment of inertia of a body, *about a given axis, remains* unaffected by reversing its direction of rotation about *that axis*, it is just a *scalar quantity**. Thus, the total moment of inertia of a number of bodies, about a given axis, will be equal to the sum of their individual moments of inertia about that axis, in exactly the same manner as the total mass of a number of bodies will be equal to the sum of their individual masses.

3.6. Particular Cases of Moments of Inertia

1. Moment of Inertia of a Thin Uniform Rod

(*i*) *about an axis through its centre and perpendicular to its length.* — Let *AB*, (Fig. 3.6), be a thin *uniform* rod of length l and mass M, free to rotate about an axis *CD* through its centre *O* and perpendicular to its length. Then, its *mass per unit length is M/l*. Consider a small elements of length dx of it, at a distance x from O. Its mass is clearly equal to $(M/l).dx$, and its moment of inertia about the axis through $O = (M/l).dx.dx^2$.

The moment of inertia I of the *whole* rod about the axis is, therefore, obtained by integrating the above expression between the limits $x = -l/2$ and $x = l/2$; or between $x = 0$ and $x = l/2$ and multiplying the result by 2, to include both halves of the rod.

Fig. 3.6

Thus, $$I = 2\int_0^{l/2} \frac{M}{l}.x^2.dx.$$

$$= \frac{2M}{l}\left[\frac{x^3}{3}\right]_0^{l/2}$$

$$= \frac{2M}{l}\left[\frac{(l/2)^3}{3} - 0\right]$$

Or, $$I = \frac{2M}{l}.\frac{l^3}{24} = \frac{Ml^2}{12}.$$

(*ii*) *about an axis passing through one end of the rod and perpendicular to its length.*— The treatment is the same as above, except that, since the axis *CD* here passes through one end *B* of the rod, (Fig. 3.7), the expression for the moment of inertia of the element dx of the rod is now to be integrated between the limits, $x = 0$, at *B* and $x = l$, at *A*.

Thus, if I be the moment of inertia of the rod about *CD*, we have

$$I = \int_0^l \frac{M}{l}.x^2.dx = \frac{M}{l}\left[\frac{x^3}{3}\right]_0^l$$

Or, $$I = \frac{M}{l}.\frac{l^3}{3} = \frac{Ml^2}{3}.$$

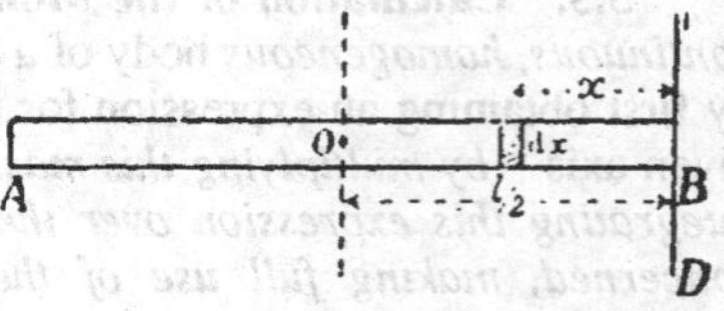

Fig. 3.7.

* *Scalar quantities* are those which *possess only magnitude*, but *no direction*, *e.g., mass, time etc.* On the other hand, *vector quantities* are those which *possess both magnitude as well as direction, e.g., acceleration, velocity, force, etc.*

Or, we could have arrived at the same result by an application of the *principle of parallel axes*, according to which the moment of inertia of the rod about the axis through B is equal to the *sum* of its moment of inertia about a parallel axis through its centre of mass and the product of its mass and the square of the distance between the two axes.

Thus, $$I = \frac{Ml^2}{12} + M\left(\frac{l}{2}\right)^2 = \frac{Ml^2}{12} + \frac{Ml^2}{4} = \frac{Ml^3}{3}.$$

2. Moment of Inertia of a Rectangular Lamina

(*i*) *about an axis through its centre and parallel to one of its sides.*— Let $ABCD$ be a rectangular lamina and let l and b be its length and breadth respectively, (Fig. 3.8). Let the axis of rotation YY' pass through its centre and be parallel to the side AD or BC.

If M be the mass of the rectangle, (supposed *uniform*), its *mass per unit length* will be M/l.

Consider a small *strip*, of width dx of the lamina parallel to the axis. The mass of the strip will obviously be $(m/l).dx$, and, therefore, the moment of inertia of the *strip* about the axis YY' will be $(M/l).dx.x^2$.

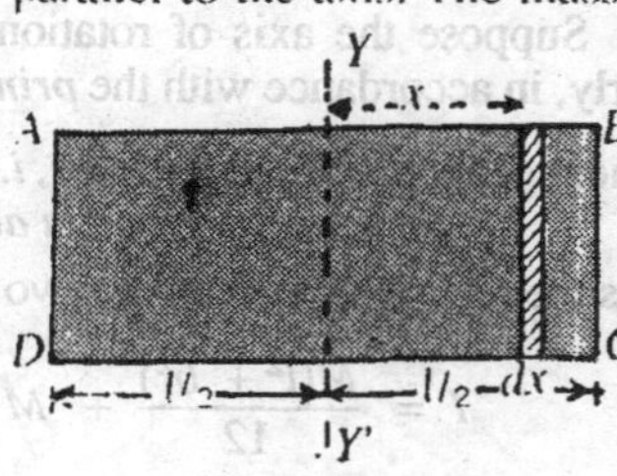

The whole lamina may be supposed to be composed of *such like strips*, parallel to the axis, and, therefore, the moment of inertia I of the *whole* rectangular lamina about the axis YY' is obtained by integrating the expression $(M/l).dx.x^2$, for the limits $x = 0$ and $x = l/2$ and multiplying the result by 2.

Fig. 3.8

i.e., $$I = 2\int_0^{l/2} \frac{M}{l}.x^2.dx = \frac{2M}{l}\int_0^{l/2} x^2.dx = \frac{2M}{l}\left[\frac{x^3}{3}\right]_0^{l/2}$$

Or, $$I = \frac{2M}{l}.\frac{l^3}{24} = \frac{Ml^2}{12}$$

It will be seen at once that *if b be small*, the rectangular lamina becomes a *rod*, of length l, whose moment of inertia, about the axis YY' passing through its centre and perpendicular to its length, would be $Ml^2/12$, [as obtained above, § 3.6 case 1, (*i*)].

(*ii*) *about one side.* — We may proceed as above (in case we want an independent proof) except that the expression $(M/l).dx.x^2$ may here be integrated for $x = 0$ and $x = l$. Thus, the moment of inertia of the rectangular lamina about the side AD or BC is given by

$$I = \int_0^l \frac{M}{l}\,x^2.dx = \frac{M}{l}\int_0^l x^2.dx.$$

Or, $$I = \frac{M}{l}\left[\frac{x^3}{3}\right]_0^l = \frac{M}{l}.\frac{l^3}{3} = \frac{Ml^2}{3}$$

Alternatively, proceeding on the basis of the previous article, we may apply *the principle of parallel axes*, according to which the moment of inertia of the rectangular lamina about side AD or BC is given by

$$I = M.I. \text{ about a } \parallel \text{ axis through its centre} + M\left(\frac{l}{2}\right)^2$$

Or, $$I = \frac{Ml^2}{12} + \frac{Ml^2}{4} = \frac{Ml^2}{3}$$

(*iii*) *about an axis passing through its centre and perpendicular to its plane.*— This may be obtained by an application of the *principle of perpendicular axes* to case (*i*) above, whence the moment of inertia of the rectangular lamina about an axis through its centre O, perpendicular to its plane, is equal to its moment of inertia about an axis through O, parallel to its breadth b, *plus* its moment of inertia about a perpendicular axis through O, parallel to its length l, *i.e.,* $$\mathbf{I} = \frac{Ml^2}{12} + \frac{Mb^2}{12} = \frac{\mathbf{M(l^2 + b^2)}}{\mathbf{12}}$$

The above relation is equally valid in the case of *thin* (*i.e., laminar*) or *thick* rectangular plates or bars, no stipulation with regard to its thickness having been made in deducing it. And, after all, a *thick rectangular plate or bar may be regarded as just a pile of thin (or laminar) plates or bars, placed one above the other.*

The same argument will hold good in all other cases of a similar type, [see cases (*iv*) and (*v*) below].

(*iv*) *about an axis passing through the mid-point of one side and perpendicular to its plane.—*

(*a*) Suppose the axis of rotation passes through the mid-point of *AD* or *BC*, (Fig. 3.8). Then, clearly, in accordance with the *principle of parallel axes*, we have

moment of inertia about *this axis, i.e.,*

I = moment of inertia about a parallel axis through its centre + $M \times (l/2)^2$,

where $l/2$ is the distance between the two parallel axes;

i.e., $$I = \frac{M(l^2 + b^2)}{12} + M.\left(\frac{l}{2}\right)^2$$

$$= \frac{M(l^2 + b^2)}{12} + \frac{Ml^2}{4} = M.\left(\frac{l^2 + b^2}{12} + \frac{l^2}{4}\right)$$

$$= M\left[\frac{(l^2 + b^2 + 3l^2)}{12}\right] = \frac{M.(4l^2 + b^2)}{12}$$

Or. $$\mathbf{I = M\left(\frac{l^2}{3} + \frac{b^2}{12}\right)}$$

(*b*) Similarly, if the axis of rotation passes through the midpoint of *AB* or *DC*, we have

$$I = M\frac{(l^2 + b^2)}{12} + M\left(\frac{b}{2}\right)^2$$ ∵ the distance between the two ∥ axes is now $b/2$.

Or, $$\mathbf{I = M\left(\frac{l^2}{12} + \frac{b^2}{3}\right)}$$

3. Moment of inertia of a solid uniform bar of rectangular cross-section about an axis, perpendicular to its length and passing through its middle point.*

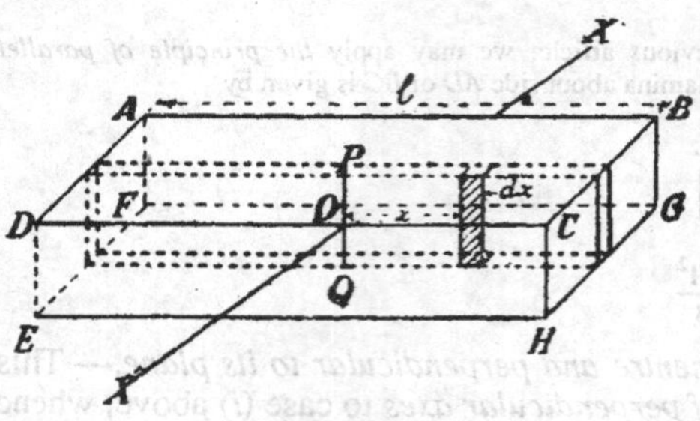

Fig. 3.9.

Let *ABCDEFGH* (Fig. 3.9) be the rectangular uniform bar of length *l*, breadth *b* and thickness *d*, whose moment of inertia about the axis *XX'*, passing through its centre and perpendicular to its length, is desired.

Imagine the whole bar to be made up of a large number of thin rectangular sheets, parallel to the face *CDEH* and perpendicular to the axis *XX'* passing through the centre of mass of each sheet. Consider one such sheet, (shown dotted), of mass *m*, of length and breadth *l* and *d* respectively, and centre of mass *O*, through which the axis *XX'* is passing perpendicular to its plane.

Then, the *M.I. of this* sheet about *XX'* = its *mass* × $(l^2 + d^2)/12$, as can be seen from the following :

* This is really covered by case 2 above, but is given here for a clearer understanding of student.

Let PQ be an axis through O, *in the plane of this sheet, and parallel to its breadth CH* or DE.

Take a thin strip of width dx of *this* strip, *parallel to*, and at *distance x from*, the axis PQ. Then, mass of the strip = $(m/l).dx$ and, therefore, its *moment of inertia about the axis* $PQ = (m/l)\, dx.x^2$

$\therefore$ moment of inertia I of the *whole* sheet about PQ is given by

$$2\int_0^{l/2} \frac{m}{l}.x^2 dx = \frac{2m}{l}\int_0^{l/2} x^2.dx.$$

$$= \frac{2m}{l}\left[\frac{x^3}{3}\right]_0^{l/2} = \frac{2m}{l}\cdot\frac{l^3}{8.3}$$

Or, $$I = ml^2/12.$$

Similarly, the moment of inertia of the sheet about an axis through O, *in its own plane and perpendicular to PQ, i.e., parallel to its length DC or EH will be* $md^2/12$.

Therefore, *by the principle of perpendicular axes*, the moment of inertia of the sheet about the axis XX' through O and perpendicular to its plane

$$= \frac{ml^2}{12} + \frac{md^2}{12} = m\left(\frac{l^2 + d^2}{12}\right).$$

Hence, *moment of inertia of the whole bar about the axis XX'*

$$\textit{mass of the bar} \times \left(\frac{l^2 + d^2}{12}\right).$$

Or, $$I = M\left(\frac{l^2 + d^2}{12}\right)$$ [M being the *mass* of the bar.

4. **Moment of Inertia of a Thin Triangular Plate or Lamina about one side.** — Let ABC, (Fig. 3.10), be a thin, triangular plate or lamina, of *surface density* or *mass per unit area*, ρ, whose moment of inertia is to be determined about the side BC.

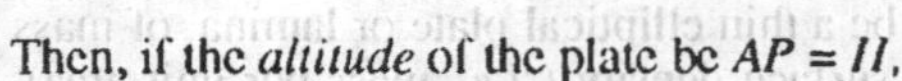

Fig 3.10

Then, if the *altitude* of the plate be $AP = H$,

its *area* $= \frac{1}{2}$ *base* $\times$ *altitude*.

$= \frac{1}{2}.a.H,$ [$\because BC = a$,

and, therefore

its *mass* $M = \frac{1}{2}.a.H.\rho$ ...(i)

Now, let us imagine the triangular plate to be made up of a number of thin *strips*, parallel to BC, and placed side by side; and, let us consider one such strip DE, of width dx, at a distance x from the base BC. Then, clearly, the *area of this* strip, (which may be considered to be almost rectangular, its width being infinitesimally small) $= DE.dx$. And, therefore,

mass of the strip, $= DE.dx.\rho$. ...(ii)

Now, in the similar triangles AQD and APB, we have

$DQ/BP = AQ/AP = h/H$, whence, $DQ = BP.\,h/H$. [$\because AQ = h$.

Similarly, from the similar triangles AQE and APC, we have

$QE/PC = AQ/AP = h/H$, whence, $QE = PC.h/H$.

And, therefore, $DQ + QE = (BP.h/H) + (PC.h/H)$.

Or, $DE = (BP + PC).h/H = BC.h/H = a.h/H$.

$\therefore$ *mass of the strip* $= a.\frac{h}{H}.dx.\rho$. [From (ii) above.

Now, clearly, moment of inertia of DE about the side BC

$= mass\ of\ the\ strip \times x^2 \quad = a \,.\, \frac{h}{H} .dx.\rho x^2.$

$$= a \,.\left(\frac{H - x}{H}\right).dx.\,\rho\,.\,x^2. \qquad [\because \quad h = (h - x)$$

And, therefore, moment of inertia of the *whole* triangular plate about *BC* is equal to the integral of this expression, between the limits $x = 0$ and $x = H$. So that,

M.I. of the plate about BC, i.e., $I = \int_0^h a.\left(\frac{H - x}{H}\right)\rho x^2.dx.$

$$= \frac{a.\rho}{H}\int_0^H (H - x).x^2dx. = \frac{a.\rho}{H}\int_0^H (Hx^2 - x^3).dx.$$

$$= \frac{a.\rho}{H}\left[\frac{Hx^3}{3} - \frac{x^4}{4}\right]_0^H = \frac{a.\rho}{H}\left[\frac{H^4}{3} - \frac{H^4}{4}\right]$$

$$= \frac{a.\rho}{H}\left(\frac{4H^4 - 3H^4}{12}\right) = \frac{a.\rho}{H}.\frac{H^4}{12} = \frac{a.\rho.H^3}{12}$$

$$= \frac{1}{2}a.H.\rho\,.\,\frac{H^2}{6}.$$

But $\frac{1}{2}a.H.\rho = M$, *the mass of the plate*, [See (*i*) above.

M.I. of the triangular plate about side *BC*, *i.e.*, $\mathbf{I = M.H^2/6}$.

5. **Moment of Inertia of an Elliptical Disc or Lamina.** (*i*) *about one of its axes.* —Let *XYX'Y'* be a thin elliptical plate or lamina, of mass *M*, and surface density (*i.e.*, mass per unit area), ρ, and let its *major axis XX'* and *minor axis YY'* be equal to 2*a* and 2*b* respectively. (Fig. 3.11)

Consider a strip *PQSR* of the plate of width, *dx*, parallel to the minor axis *YY'* and at a distance *x* from it. Then, if 2*y* be the length of the strip, its area is clearly equal to 2*y.dx* and, therefore, its *mass* equal to 2*y.dx*.ρ.

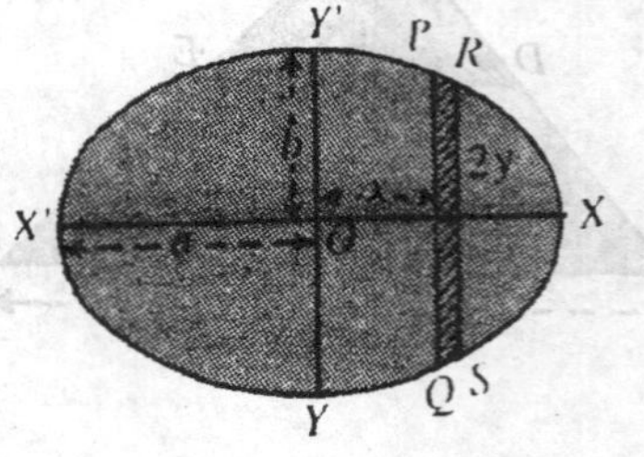

Fig. 3.11

Obviously, then, *M.I.* of the strip about the minor axis $YY' = 2y.dx.\ \rho x^2$; and, therefore, *M.I.* of the *whole* elliptical plate about the axis *YY'* is equal to twice the integral of the above expression, between the limits $x = 0$, and $x = a$. Or, denoting it by I_y, we have

$$I_y = 2\int_0^a 2y.\rho.x^2.dx = 4\rho\int_0^a y.\,x^2dx. \qquad ...(I)$$

Now, with the centre of the ellipse as the origin, and with the co-ordinate axes coinciding with its major and minor axes respectively, we have $\frac{x^2}{a^2} + \frac{y^2}{b^2} = 1$, as the equation to the ellipse; whence

$$\frac{y^2}{b^2} = 1 - \frac{x^2}{a^2}, \text{ or, } y^2 = b^2\left(1 - \frac{x^2}{a^2}\right)$$

So that, $\qquad y = b\sqrt{1 - x^2/a^2}$

Substituting this value of *y* in relation (I) above, we have

$$I_y = 4\rho \int_0^a b\sqrt{(1-x^2/a^2)}\ .\ x^2.dx$$

$$= 4\rho b \int_0^a x^2.\ \sqrt{(1-x^2/a^2)}\ .\ dx \qquad \text{(II)}$$

Now, putting $x = a \sin\theta$, we have $dx/d\theta = a\cos\theta$.

Or, $\qquad dx = a\cos\theta.\ d\theta.$

Substituting these values of x and dx in expression (II) above, we have

$$I_y = 4\rho\ .\ b \int_0^{\pi/2} a^2\ .\sin^2\theta \sqrt{1 - \frac{a^2\ .\sin^2\theta}{a^2}}\ .a\cos\theta\ .\ d\theta\ .$$

$$= 4\rho.b \int_0^{\pi/2} a^2.\sin^2\theta\ \sqrt{1-\sin^2\theta}.a\ \cos\theta.d\theta.$$

$$= 4\rho.b \int_0^{\pi/2} a^2.\sin^2\theta.\cos\theta.a\cos\theta.d\theta. \qquad \left[\because \sqrt{1-\sin^2\theta} = \cos\theta\right]$$

$$= 4\rho b.a^3 \int_0^{\pi/2} \sin^2\theta.\ \cos^2\theta.d\theta.$$

$$= 4\rho b.a^3.\frac{1}{4}\int_0^{\pi/2} (2\sin\theta.\cos\theta)^2.d\theta.$$

$$= \rho b.a^3 \int_0^{\pi/2} \sin^2 2\theta.d\theta.$$

$$= \rho b.a^3. \int_0^{\pi/2} \frac{1-\cos 4\theta}{2}.d\theta \qquad \left[\because \cos 2\theta = 1 - 2\sin^2\theta.\right]$$

$$= \rho b.a^3.\ \frac{1}{2}\left[\theta - \frac{\sin 4\theta}{4}\right]_0^{\pi/2}$$

$$= \rho b.a^3.\frac{1}{2}.\frac{\pi}{2} = \frac{\pi}{4}\rho b.a^3 = \pi.ab.\rho.\frac{a^2}{4}.$$

Now, $\quad \pi.a.b.\rho = M$, the *mass of the elliptical plate.*

$\therefore \qquad$ $\mathbf{I_y = M.a^2 / 4.}$

Similarly, the moment of inertia (I_x) of the elliptical plate about the major axis XX' is given by the expression,

$$\mathbf{I_x = Mb^2 / 4.}$$

(*ii*) *about an axis passing through the centre of the plate or lamina and perpendicular to its own plane.* —The axis in this case will pass through O, (Fig. 3.11), and will be perpendicular to the plane of the paper, (or the plane of its two axes, XX' and YY'). Hence, if I be the moment of inertia of the elliptical plate about *this axis*, we have, by the *principle of perpendicular axes.*

$$I = I_x + I_y = I_y + I_x = Ma^2/4 + Mb^2/4$$

Or, $$\mathbf{I = M.\left(\frac{a^2 + b^2}{4}\right)}$$

6. Moment of Inertia of a Hoop or a Circular Ring.

(*i*) *about an axes through its centre and perpendicular to its plane.* — Let the radius of the hoop or circular ring, (supposed to be *infinitely thin*), be R, and its mass, M, (Fig. 3.12).

Consider a particle of it, of mass m. Then, the moment of inertia of this particle about an axis through the centre O of the hoop, and perpendicular to its plane, will obviously be mR^2.

And, therefore, the moment of inertia I of the *entire* hoop about the axis will be ΣmR^2

Or, $\quad\quad$ $\mathbf{I = MR^2}$.* $\quad\quad$ [$\because \sum m = M$ and R is the same for all particles.

(*ii*) *about its diameter*. Let it be required to determine the moment of inertia of the hoop about the diameter AB, (Fig. 3.12). Obviously, the moment of inertia of the hoop will be the same about all the diameters. Thus, if I be the moment of inertia of the hoop about the diameter AB, it will also be its moment of inertia about the diameter CD, perpendicular to AB

Fig. 3.12

Then, by the *principle of perpendicular axes*, its moment of inertia about the axis passing through centre O, and *perpendicular to it's plane*, is equal to the *sum* of its moments of inertia about the perpendicular axes AB and CD, *in its own plane*, and intersecting at O.

And, $\quad$ therefore, $I + I = MR^2$, $\quad$ Or, $\quad 2I = MR^2$.

Or, $\quad\quad$ $\mathbf{I = MR^2/2}$

7. Moment of inertia of a Circular Lamina or Disc. —

(*i*) *about an axis through its centre and perpendicular to its plane*. Let M be the mass of the disc and R, its radius. Then, since the area of the disc is πR^2, its *mass per unit area* will be $M/\pi R^2$.

Consider a ring of the disc, distant x from O, *i.e.*, of radius x, and of width dx, (Fig. 3.13). Its area is clearly equal to its circumference, multiplied by its width, or equal to $2\pi\, x \times dx$, and its *mass* is thus

$$= \frac{M}{\pi R^2}\, 2\pi x.dx = \frac{2Mx.dx}{R^2}.$$

Hence, moment of inertia of *this* ring about an axis through O and perpendicular to its plane

$$= \frac{2Mx.dx}{R^2} x^2 = \frac{2Mx^3 dx}{R^2}.$$

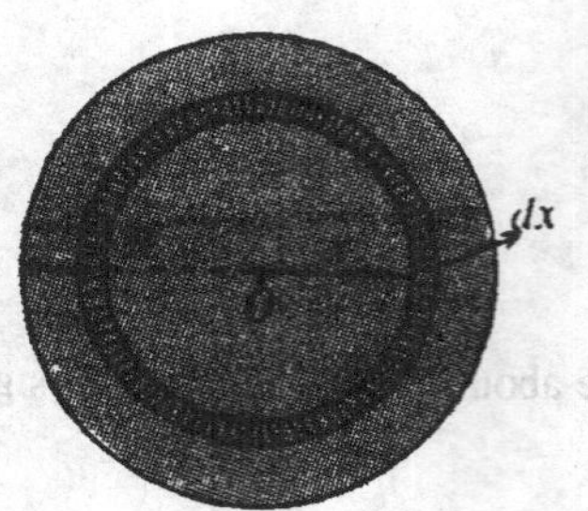

Fig. 3.13

Since the whole disc may be supposed to be made up of such like concentric rings of radii ranging from 0 to R, we can get the moment of inertia I of the disc by integrating the above expression for the moment of inertia of the ring, for the limits $x = 0$, and $x = R$.

$$\therefore \quad M.I. \text{ of the disc} = \int_0^R \frac{2M}{R^2}.x^3.dx = \frac{2M}{R^2}\left[\frac{x^4}{4}\right]_0^R = \frac{2M}{R^2}.\frac{R^4}{4}.$$

Or, $\quad\quad$ $\mathbf{I = MR^2/2.}$

* Since $I = MK^2$, we have here $MK^2 = MR^2$, whence, $K = R$, *i.e.*, the radius of the hoop or the ring is equal to its radius of gyration about a perpendicular axis through its centre. This gives us another defintion of the *radius of gyration of a body about a given axis*, *viz.*, that *it is numerically equal to the radius of a hoop or a ring, infinitely thin, of the same mass as the body* and *having* the same moment of inertia about the axix passing *through it's centre and perpendicualr to it's plane as the body has about the axix*.

(*ii*) *about its diameter.* —Let AB and CD be two perpendicular diameters of a circular disc of radius R and mass M, (Fig. 3.14). Since the moment of inertia of the disc about one diameter is the same as about any other diameter, the moment of inertia about the diameter AB is equal to the moment of inertia about the diameter CD, perpendicular to AB. Let it be I.

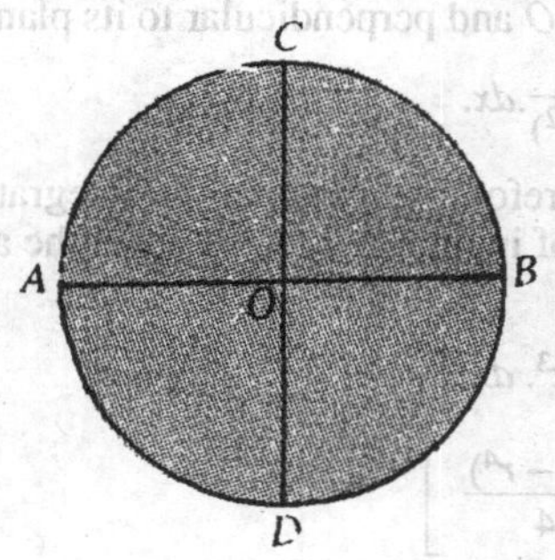

Fig. 3.14

Now, we have, by the *principle of perpendicular axes, M.I.* of the disc about AB + its *M.I.* about CD = its *M.I.* about an axis through O and perpendicular to its plane.

Or, $I + I = MR^2/2$, or $2I = MR^2/2$.

Or, $\mathbf{I = MR^2/4}$.

(*iii*) *about a tangent to the disc in its own plane.*—Let AB be the tangent to the circular disc of radius R and mass M, about which its moment of inertia is to be determined, (Fig. 3.15). Let CD be a diameter of the disc, parallel to the tangent AB. The moment of inertia of the disc about this diameter is, clearly, equal to $MR^2/4$.

So that, by the *principle of parallel axes,* we have *M.I.* of the disc about AB = *M.I.* of the disc about $CD + MR^2$.

Or, $\mathbf{I} = (MR^2/4) + MR^2 = \mathbf{5MR^2/4}$

(*iv*) *about a tangent to the disc and perpendicular to its plane.*—This tangent will obviously be parallel to the axis through the centre of the disc and perpendicular to its plane, the distance between the two being equal to the radius of the disc. Hence, *by the principle of parallel axes,* we have

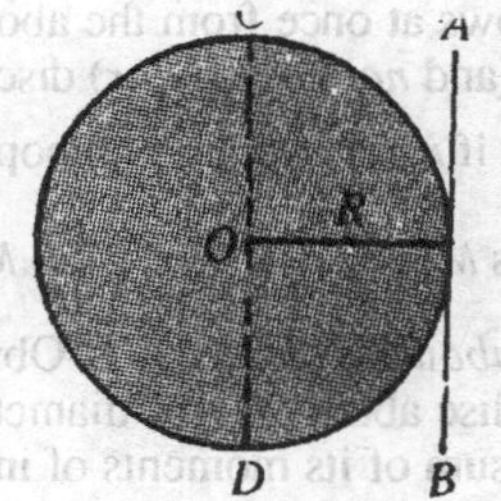

Fig. 3.15

M.I. about the tangent = M.I. about the perpendicular axis + MR^2.

Or. $\mathbf{I} = (MR^2/2) + MR^2 = \mathbf{3\ MR^2/2}$

8. Moment of Inertia of an Annular Ring or Disc. —

(*i*) *about an axis passing through its centre and perpendicular to its plane.*—An annular disc or ring is just an ordinary disc from which a smaller co-axial disc is removed, so that there is a concentric circular hole in it. Let R and r be the outer and inner radii of the disc, (Fig. 3.16), and M, its mass. Then, clearly,

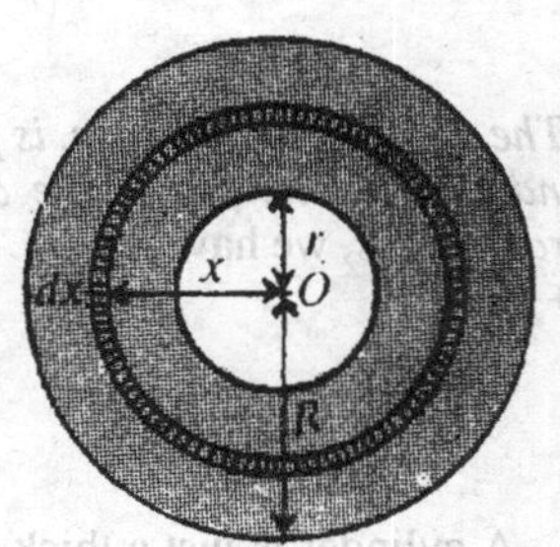

Fig. 3.16

face-area of the annular disc = face-area of disc of radius R – face-area of disc of radius r, *i.e.,*

$$= \pi R^2 - Rr^2 = \pi(\pi^2 - r^2).$$

And ∴ *mass per unit area of the disc* $= M/\pi\ (R^2 - r^2)$

Imagine the disc to be made up of a number of thin circular rings, and consider one such ring of radius x and of width dx.

Then, *face-area of this ring* $= 2\pi x.dx$,

and its $mass = \frac{M}{\pi(R^2 - r^2)}.2\pi x.dx = \frac{2Mx}{(R^2 - r^2)}.dx.$

And, therefore, its moment of inertia about an axis through O and perpendicular to its plane

$$= \frac{2Mx}{(R^2 - r^2)}.dx.x^2 = \frac{2Mx^3}{(R^2 - r^2)}.dx.$$

The moment of inertia of the *whole* annular disc may, therefore, be obtained by integrating the above expression for the limits $x = r$ and $x = R$. Or, moment of inertia of the disc about the axis through O and perpendicular to its plane

$$= \int_r^R \frac{2Mx^3}{(R^2 - r^2)} . dx = \frac{2M}{(R^2 - r^2)} \int_r^R x^3. dx.$$

$$\frac{2M}{(R^2 - r^2)} \left[\frac{x^4}{4} \right]_r^R = \frac{2M}{(R^2 - r^2)} \left[\frac{(R^4 - r^4)}{4} \right]$$

$$= \frac{2M}{(R^2 - r^2)} \left[\frac{(R^2 + r^2).(R^2 - r^2)}{4} \right]$$

Or, $$\mathbf{I = M\,(R^2 + r^2)\,/\,2.}$$

It follows at once from the above that if $r = 0$, *i.e.*, if there is no hole in the disc, or that it is just a *plane* (and *not an annular*) disc, its moment of inertia is $MR^2/2$. [Case 7 (*ii*), above.

Again, if $r = R$, we have a hoop or a circular ring, of radius R

and its $M.I. = M(R^2 + R^2)/2 = M.2R^2/2 = MR^2$. [Case 6(*i*), above.

(*ii*) *about its diameter*. —Obviously, due to its *symmetrical* shape, the moment of inertia of the annular disc about the one diameter will be the same as that about any other diameter. Let it be I. Then, the sum of its moments of inertia about two perpendicular diameters will, by the *principle of perpendicular axes*, be equal to its moment of inertia about an axis through O and perpendicular to its plane, *i.e.*, equal to $M\,(R^2 + r^2)/2$.

Or, $I + I = M\,(R^2 + r^2)/2$, *i.e.*, $2I = M\,(R^2 + r^2)/2$,

whence, $$\mathbf{I = M(R^2 + r^2)/4}$$

Now, if $r = 0$, *i.e.*, if the disc be a *plane* one, we have

M.I. of the disc about a diameter $= MR^2/4$. [Case 7 (*ii*) above.

Or, if $r = R$, we have a *hoop* or a *circular ring* of radius R, and its moment of inertia about its diameter $= M\,(R^2 + R^2)/4 = MR^2/2$ [Case 6 (*ii*) above.

(*iii*) *about a tangent, in its own plane*. —The tangent being parallel to a diameter of the ring or disc, and at a distance R from it, we have, applying the *principle of parallel axes*,

M.I. about the tangent = *M.I.* about the diameter + MR^2.

Or, $\mathbf{I} = M\,[(R^2 + r^2)/4] + MR^2 = \mathbf{M}(5R^2 + r^2)/4.$

(*iv*) *about a tangent, perpendicular to its own plane. —The tangent, in this case, is parallel to the axis through the centre of the ring or disc and perpendicular to its plane, the distance between the two being equal to R.* Hence, by the *principle of parallel axes*, we have

M.I. about the tangent = *M.I.* about the perpendicular axis + MR^2.

$$\mathbf{I} = M\,[(R^2 + r^2)/2] + MR^2 = \mathbf{M\,(3R^2 + r^2)/2.}$$

9. Moment of inertia of a Solid Cylinder.

(*i*) *about its own axis, or its axis of cylindrical symmetry*.—A cylinder is just a thick circular disc, or a *number of thin circular discs, piled one upon the other*, and, therefore, its moment of inertia about its axis is the same as that of a circular disc or lamina about an axis through its centre and perpendicular to its plane, *i.e.*, equal to $\mathbf{MR^2/2}$, where M is its mass and R, its radius.

[Case 7 (*i*), above.

(ii) about an axis passing through its centre and perpendicular to its own axis of cylindrical symmetry. — Let M be the mass of the cylinder, R its radius and l, its Length (Fig. 3.17). Then, obviously, if it be homogeneous in composition, its *mass per unit length* will be M/l. Let YY' be the axis, passing through its centre O and perpendicular to its own axis XX', about which the moment of inertia is to be determined.

Imagine the cylinder to be made up of a number of thin discs and consider one such disc at a distance x from O, and of thickness dx.

Obviously, *the mass of the disc* is $(M/l).dx$ and its *radius*, equal to R; so that, its *moment of inertia about its diameter is equal to mass of the disc* $\times$ *(radius)^2/4, i.e.,*

$$= \frac{M}{l}.dx.\frac{R^2}{4}$$

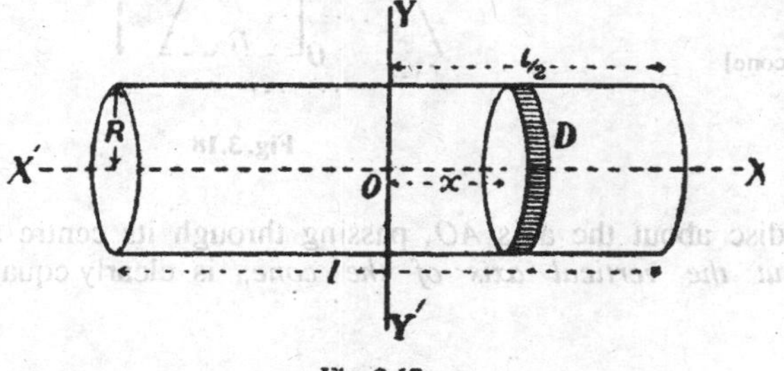

Fig. 3.17

And, its moment of inertia about the axis YY', by the *principle of parallel axes*

$$= \frac{M}{l}.dx.\frac{R^2}{4} + \frac{M}{l}.dx.x^2.$$

Therefore, the moment of inertia of the *whole* cylinder about the axis $Y\,Y'$ may be obtained by integrating this expression for the limits $x = 0$ and $x = l/2$ and multiplying the result by 2. Thus,

M.I. of the cylinder about the axis YY'

$$= 2\int_0^{l/2}\left(\frac{M}{l}.\frac{R^2}{4}.dx + \frac{M}{l}.x^2.dx\right) = \frac{2M}{l}\int_0^{l/2}\left(\frac{R^2}{4}.dx + x^2.dx\right)$$

$$= \frac{2M}{l}\left[\frac{R^2x}{4} + \frac{x^3}{3}\right]_0^{l/2} = \frac{2M}{l}\left[\frac{R^2}{4}.\frac{l}{2} + \frac{l^3}{8\times 3}\right]$$

Or, $$\mathbf{I} = \frac{2M}{l}\left(\frac{R^2 l}{8} + \frac{l^3}{24}\right) = \mathbf{M}\left(\frac{\mathbf{R^2}}{\mathbf{4}} + \frac{\mathbf{l^2}}{\mathbf{12}}\right)$$

(iii) about a diameter of one of its faces.—It is an easy deduction from the above; for, by the *principle of parallel axes*, we have

M.I. of the cylinder about the diameter of one face

$$= M\left(\frac{R^2}{4} + \frac{l^2}{12}\right) + M\left(\frac{l}{2}\right)^2.$$

$$= \frac{MR^2}{4} + \frac{Ml^2}{12} + \frac{Ml^2}{4} = \frac{MR^2}{4} + \frac{Ml^2}{3}$$

Or, $$\mathbf{I} = \mathbf{M}\left(\frac{\mathbf{R^2}}{\mathbf{4}} + \frac{\mathbf{l^2}}{\mathbf{3}}\right)$$

10. Moment of Inertia of a Solid Cone. —

(i) about its vertical axis. — Let mass of the cone be M, its vertical height, h and radius of its base, R, (Fig. 3.18)

Then, its *volume* $= \frac{1}{3}\pi R^2 h$.

And, if ρ be the *density* of its material, its *mass*

$M = \frac{1}{3}\pi R^2 h\rho$, whence, $\rho = \frac{3M}{\pi R^2 h}$.

Imagine the cone to be made up of a number of discs, parallel to the base, and placed one above the other. Consider one such disc at a distance x from the vertex, and of thickness dx.

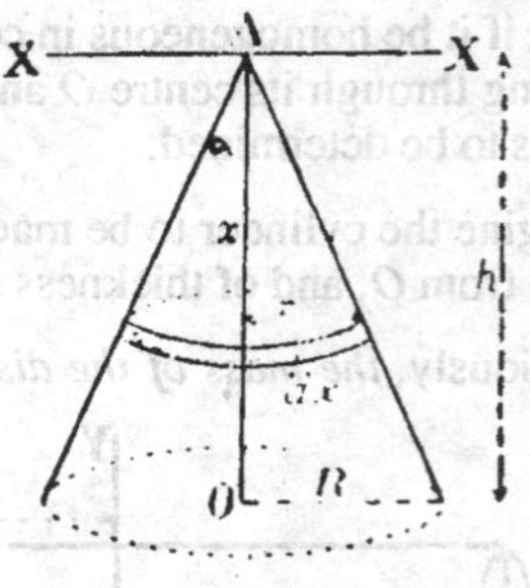

Fig. 3.18

If r be the radius of *this* disc, we have

$$r = x \tan\alpha$$

[where α is the semi-vertical angle of the cone]

And *volume* of the disc $= \pi r^2.dx$.

$\therefore$ its *mass* $= \pi r^2.dx.\rho$.

Now, moment of inertia of the disc about the axis AO, passing through its centre and perpendicular to its plane, *i.e., about the vertical axis of the cone,* is clearly equal to

$$\frac{\textit{its mass} \times (\textit{its radius})^2}{2}$$

$$= \pi r^2.dx.\rho.\frac{r^2}{2} = \frac{\pi.\rho.r^4}{2}.dx = \frac{\pi.\rho.x^4.\tan^4\alpha}{2}.dx.$$

And, therefore, the moment of inertia of the *whole* cone about its vertical axis AO will be the integral of this expression, for the limits $x = 0$ and $x = h$.

i.e., $M.I.$ of the cone about its vertical axis is given by

$$I = \int_0^h \frac{\pi\rho x^4.\tan^4\alpha}{2}.dx = \frac{\pi\rho.\tan^4\alpha}{2}\int_0^h x^4.dx.$$

$$= \frac{\pi\rho.\tan^4\alpha}{2}\left[\frac{x^5}{5}\right]_0^h = \frac{\pi\rho.\tan^4\alpha}{2}.\frac{h^5}{5}.$$

$$= \frac{\pi\rho.R^4}{2.h^4}.\frac{h^5}{5}. \qquad [\because \tan\alpha = R/h.$$

Or, substituting the value of ρ from above, we have

$$\mathbf{I} = \frac{\pi.3M.R^4}{\pi R^2.h.2h^4}.\frac{h^5}{5} = \frac{\mathbf{3MR^2}}{\mathbf{10}}$$

(*ii*) *about an axis through its vertex and parallel to its base.* — Again, considering the disc at a distance x from vertex of the cone, we have $M.I.$ of the disc about its diameter

$$= \pi r^2.dx.\rho.r^2/4.$$

And, therefore, by the *principle of parallel axes*, its moment of inertia about a parallel axis XX', passing through the vertex of the cone

$$= \pi r^2.dx.\rho.\frac{r^2}{4} + \pi r^2.dx.\rho.x^2.$$

Therefore, the moment of inertia of the *whole* cone about the axis passing through the vertex and parallel to the base, *i.e.,* about XX', is obtained by integrating this expression for the limits, $x=0$ and $x = h$. Thus,

$M.I.$ of the cone about XX'

$$= \int_0^h\left[\frac{\pi(x\tan\alpha)^4.\rho}{4}.dx + \pi(x\tan\alpha)^2.x^2.\rho.dx.\right]$$

$$= \frac{\pi\rho \tan^4 \alpha}{4} \int_0^h x^4.dx + \pi\rho \tan^2 \alpha \int_0^h x^4.dx.$$

$$= \frac{\pi\rho}{4} \cdot \frac{R^4}{h^4} \left[\frac{x^5}{5} \right]_0^h + \pi\rho . \frac{R^2}{h^2} \left[\frac{x^5}{5} \right]_0^h$$

$$= \frac{\pi\rho}{4} \cdot \frac{R^4}{h^4} \cdot \frac{h^5}{5} + \pi\rho . \frac{R^2}{5} + \pi\rho . \frac{R^2}{h^2} \cdot \frac{h^5}{5}.$$

Or, substituting the value of ρ, we have

$$M.I. \text{ of the cone about } XX' = \frac{\pi.3M}{\pi R^2 h.4} \cdot \frac{R^4}{h^4} \cdot \frac{h^5}{5} + \frac{\pi.3M}{\pi.R^2 h} \cdot \frac{R^2}{h^2} \cdot \frac{h^5}{5}.$$

Or, $$\mathbf{I} = \frac{\mathbf{3MR^2}}{\mathbf{20}} + \frac{\mathbf{3Mh^2}}{\mathbf{5}}.$$

11. Moment of Inertia of a Hollow Cylinder.

(*i*) *about its own axis.*—A hollow cylinder may be considered to consist of a large number of annular discs or rings of the given internal and external radii, placed one above the other, the axis of the cylinder passing through their centres and being perpendicular to their planes.

The moment of inertia of the hollow cylinder about its own axis is, therefore, the same as that of an annular disc of the given external and internal radii about an axis through its centre and perpendicular to its plane, *i.e.*, equal to $\mathbf{M(R^2 + r^2)/2}$, where M is the mass of the cylinder, R and r, its external and internal radii respectively.

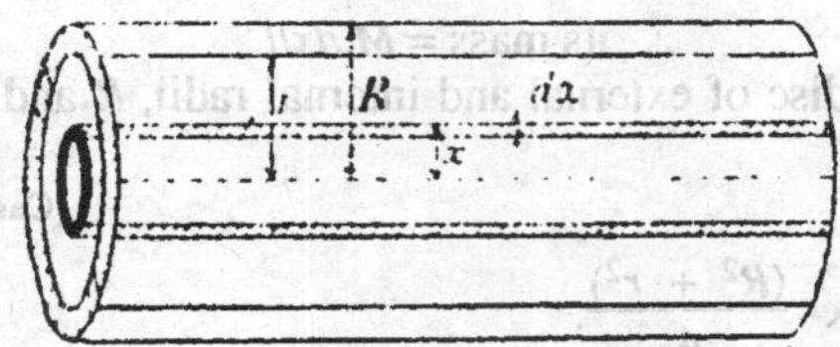

Fig. 3.19

Alternative Proof.— Let M be the mass of the cylinder; R and r, its external and internal radii and l, its length (Fig. 3.19).

Then, *face-area* of the cylinder $= \pi (R^2 - r^2)$.

And, *volume* of the cylinder $= \pi (R^2 - r^2).l$;

so that, *its mass per unit volume* $= \dfrac{M}{\pi(R^2 - r^2)l}$.

Now, imagine the cylinder to be made up of a large number of thin co-axial cylinders, and consider one such cylinder of radius x and thickness dx.

Then, its *face-area* $= 2\pi x.dx$, its *volume* $= 2\pi x. dx. l$,

and its *mass* $= \dfrac{M}{\pi (R^2 - r^2) l} \times 2\pi x.dx.l = \dfrac{2Mx.dx}{(R^2 - r^2)}$.

Since all its particles are equidistant from the axis, its moment of inertia about the axis

$$= \frac{2 Mx.dx}{(R^2 - r^2)} x^2 = \frac{2M. x^3.dx}{(R^2 - r^2)}.$$

And, therefore, the moment of inertia of the *whole* cylinder may be obtained by intergating the above expression for the limits, $x = r$ and $x = R$.

Or, *M.I.* of the cylinder about its axis, *i.e.*,

$$I = \int_r^R \frac{2M.x^3}{(R^2 - r^2)} .dx = \frac{2M}{(R^2 - r^2)} \int_r^R x^3.dx.$$

$$= \frac{2M}{(R^2 - r^2)}\left[\frac{x^4}{4}\right]_r^R = \frac{2M}{(R^2 - r^2)}\left[\frac{(R^4 - r^4)}{4}\right]$$

$$= \frac{2M}{(R^2 - r^2)}\left[\frac{(R^2 + r^2)(R^2 - r^2)}{4}\right].$$

Or,
$$\mathbf{I = M}\left[\frac{\mathbf{R^2 + r^2}}{\mathbf{2}}\right].$$

(*ii*) *about an axis passing through its centre and perpendicular to its own axis.* —As before, let M be the mass of the cylinder, l, its length, and R and r, its external and internal radii respectively; and let YOY' be the axis through its centre O, and perpendicular to its own axis XX', (Fig.3.20). Then, *face-area* of the cylinder $= \pi(R^2 - r^2)$.
and *its volume* $= \pi(R^2 - r^2)l$.
∴ *its mass per unit volume*
$= M/\pi(R^2 - r^2)l$.

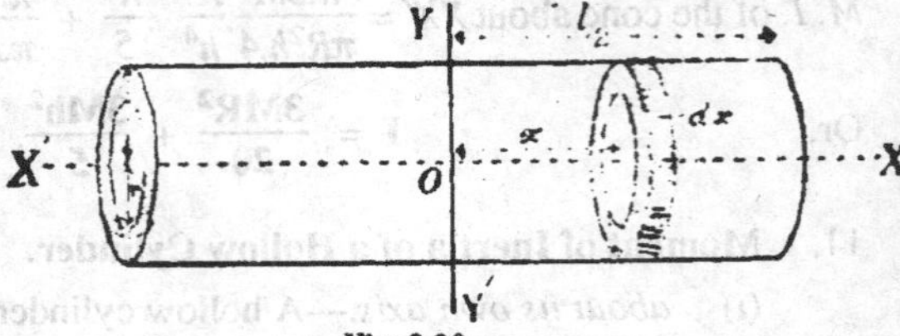

Fig. 3.20

Imagine the cylinder to be made up of a large number of annular discs of external and internal radii R and r, placed one by the side of the other, and consider one such disc at a distance x from the axis YY', and of thickness dx. Then, clearly,

surface area of the disc $= \pi(R^2 - r^2)$,
its volume $= (R^2 - r^2).dx$. and ∴ its mass $= M.dx/l$.

Now, moment of inertia of an annular disc of external and internal radii, R and r, about its diameter

$= its\ mass \times (R^2 + r^2)/4$ [Case 8 (*ii*), above

∴ $M.I.$ of the disc about its diameter $= \frac{M}{l}.dx \times \frac{(R^2 + r^2)}{4}$.

And, therefore, its moment of inertia about the parallel axis YY' is, by the *principle of parallel axes*, given by

$$\frac{M}{l}.dx \times \frac{(R^2 + r^2)}{4} + \frac{M}{l}.dx.x^2.$$

And, clearly, therefore, moment of inertia I, of the *whole* cylinder, about the axis YY', is given by twice the integral of this expression, for the limits, $x = 0$ and $x = l/2$,

i.e.
$$I = 2\int_0^{l/2}\left[\frac{M(R^2 + r^2)}{4l}.dx + \frac{M}{l}.x^2.dx\right]$$

$$= \frac{2M}{l}\int_0^{l/2}\left[\frac{(R^2 + r^2)}{4}.dx + x^2.dx\right] = \frac{2M}{l}\left[\frac{(R^2 + r^2)x}{4} + \frac{x^3}{3}\right]_0^{l/2}$$

Or,
$$\mathbf{I} = \frac{2M}{l}\left[\frac{(R^2 + r^2)l}{4 \times 2} + \frac{l^3}{8 \times 3}\right] = \mathbf{M}\left[\frac{\mathbf{(R^2 + r^2)}}{\mathbf{4}} + \frac{\mathbf{l^2}}{\mathbf{12}}\right].$$

It follows, therefore, that if $r = 0$, *i.e.*, if the cylinder be a *solid one*, we have

M.I. of the cylinder (solid) about an axis through its centre and perpendicular to its own axis $= \mathbf{M(R^2/4 + l^2/12)}$. [Case 9 (*ii*)

12. Moment of Inertia of a Spherical Shell.

(i) *about its diameter.* —

First Method. — Let $ABCD$ be the section of a spherical shell through its centre O and let the mass of the shell be M, and its radius R, (Fig. 3.21)

Then, area of the shell is clearly equal to $4\pi R^2$, and, therefore,

mass per unit area of the shell = $M/4\pi R^2$.

Let it be required to determine its moment of inertia about the diameter AB.

Consider a *thin slice* of the shell, lying between two planes EF and GH, perpendicular to the diameter AB, and at distances x and $(x + dx)$ respectively from its centre O.

This *slice* is obviously a *ring* of radius PE, and width EG, (*not PQ*, which is equal to dx).

$\therefore$ area of *this* ring = its circumference × width = $2\pi.PE \times EG$

and, hence its *mass* = its *area* × $M/4\pi R^2$

$$= 2\,PE \times EG \times M/4\,R^2. \quad ...(i)$$

Fig. 3.21

Join OE and OG, and let $\angle COE = \theta$ and $\angle EOG = d\theta$.

Then, $PE = OE.\cos\ OEP = R\cos\theta$ [∵ $\angle OEP = \angle COE = \theta$.

Similarly, $OP = R\sin\theta$. Or, $x = R\sin\theta$ [∵ $OE = R$ and $OP = x$

Now, differentiating the expression for x with respect to θ, we have

$$dx/d\theta = R\cos\theta$$

Or, $dx = R\cos\theta\,.\,d\theta = PE.d\theta$ [∵ $R\cos\theta = PE$.

And, $EG = R.d\theta$ [∵ arc = radius × angle subtended by the arc.

$\therefore$ *mass of the ring* $= 2\pi\,.PE.R.d\theta\,.\,M/4\pi R^2$. [from (*i*)

$= M.dx/2R$. [∵ $PE.d\theta = dx$.

Hence, moment of inertia of *the ring* about AB, (an axis passing through its centre and perpendicular to its plane), is equal to *its mass* × *(its radius)*^{2}, *i.e.*, $= (M.dx/2R).\,PE^2$, where $PE^2 = (R^2 - x^2)$.

$\therefore$ moment of inertia of *the ring* about $AB = (M.dx/2R).\,(R^2 - x^2)$.

And, therefore, the moment of inertia I of the *whole* spherical shell about AB = twice the integral of $(M.dx/2R)(R^2 - x^2)$, between the limits, $x = 0$ and $x = R$.

Or, $$I = 2\int_0^R \frac{M}{2R}(R^2 - x^2).dx = \frac{2M}{2R}\int_0^R (R^2 - x^2).dx.$$

$$= \frac{M}{R}\left[R^2x - \frac{x^3}{3}\right]_0^R = \frac{M}{R}\left[R^3 - \frac{R^3}{3}\right]$$

i.e., $$\mathbf{I} = \frac{M}{R}\cdot\frac{2}{3}R^3 = \frac{2}{3}\mathbf{MR^2}$$

Second Method. — Let M be the mass of the shell and R, its radius.

Consider a particle, of mass m, anywhere on the shell. Then, since the thickness of the shell is negligible, the distance of the particle from the centre of the shell is the same as the radius of the shell, *i.e.*, R.

Obviously, therefore, the summation I_0, for all the particles of the shell about its centre O, is given by the relation

$$I_0 = \Sigma mR^2. \quad \text{Or,} \quad I_0 = MR^2$$ [All particles being at distance R from O and $\Sigma m = M$

Now, if I be the moment of inertia of the shell about one diameter, it will be the same about any other diameter also, from the sheer symmetry of the shell. Hence, in accordance with the *principle of perpendicular axes for a three-dimensional body*. [§ 3.4(*a*), (*ii*), page 47], the sum of the moments of inertia of the shell about its three mutually perpendicular diameters must be equal to twice the summation I_0 for all its particles about their point of intersection, *i.e.*, the centre O of the shell; so that,

$$I + I + I = 2I_0 \quad \text{Or,} \quad 3I = 2MR^2, \qquad [\because I_0 = MR^2$$

whence, $I = 2MR^2/3$.

(ii) about a tangent. — Obviously, a tangent, drawn to the shell at any point, must be parallel to one of its diameters, and at a distance from it equal to R, the radius of the shell. Hence, applying the *principle of parallel axes*, we have

M.I. of the shell about a tangent

= its *M.I.* about a diameter + MR^2

Or, $$I = \frac{2}{3}MR^2 + MR^2 = 5\,MR^2/3.$$

13. Moment of Inertia of a Solid Sphere.

(i) about its diameter. — Let Fig. 3.22 represent a section of the sphere through its centre O.

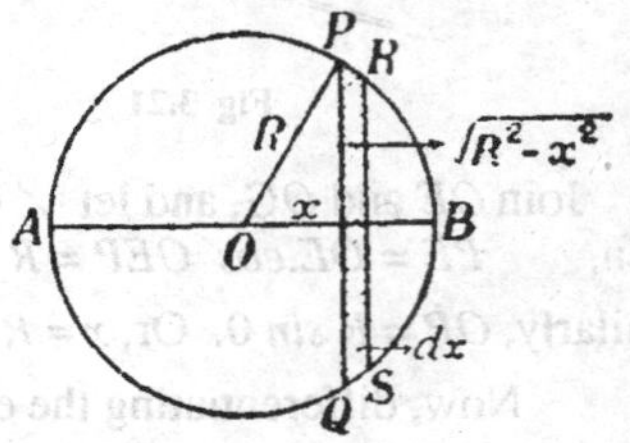

Fig. 3.22

Let *mass* of the sphere be M and its radius, R.

Then, clearly its *volume* = $4\,\pi R^3/3$.

And $\therefore$ its *mass per unit volume*

$$= M / \tfrac{4}{3}\pi R^3 = 3M / 4\pi R^3.$$

Consider a thin circular *slice* of the sphere at a distance x from the centre O and of thickness dx.

This slice is obviously a disc of radius $\sqrt{R^2 - x^2}$, and of thickness dx.

$\therefore$ *surface area of the slice* = $\pi(\sqrt{(R^2 - x^2)})^2 = \pi(R^2 - x^2)$

and its *volume* = *area* × *thickness* = $\pi(R^2 - x^2).dx$.

And, $\therefore$ *its mass* = its *volume* × *mass per unit volume* of the sphere

$$= \pi\,(R^2 - x^2)\,.dx \times \frac{3M}{4\pi R^3} = \frac{3M.(R^2 - x^2)}{4R^3}.dx.$$

Now, the moment of inertia of *this* disc about AB, (an axis passing through its centre and perpendicular to its plane)

= *its mass* × *(radius)*^{2}/2.

$\therefore$ moment of inertia of the disc about AB

$$= \frac{3M.(R^2 - x^2)}{4R^3}.dx.\ \frac{(R^2 - x^2)}{2} = \frac{3M\,(R^2 - x^2)}{8R^3}.dx. \qquad ...(1)$$

$\therefore$ moment of inertia I of the *sphere* about the diameter AB is equal to twice of integral of expression (1) between the limits $x = 0$ and $x = R$.

i.e., $$I = 2\int_0^R \frac{3M\,(R^2 - x^2)^2}{8R^3}.dx = \frac{2 \times 3M}{8R^3}\int_0^R (R^2 - x^2)^2.dx.$$

$$= \frac{3M}{4R^3}\int_0^R (R^4 - 2R^2x^2 + x^4).dx = \frac{3M}{4R^3}\left[R^4x - 2R^2\frac{x^3}{3} + \frac{x^5}{5}\right]_0^R$$

$$= \frac{3M}{4R^3}\left(R^5 - \frac{2}{3}R^5 + \frac{1}{5}.R^5\right) = \frac{3M}{4R^3}\left[\frac{15R^5 - 10R^5 + 3R^5}{15}\right]$$

Or. $$I = \frac{3M}{4R^3} \times \frac{8R^5}{15} = \frac{2}{5}MR^2.$$

Now, the moment of inertia of the sphere about one diameter is the *same* as about another diameter; so that, we have moment of inertia of a solid sphere about any diameter given by

$$I = \frac{2}{5} MR^2.$$

Alternative Method. — Let M be the mass of the sphere and ρ, the density of its material.

Imagining the whole sphere to be made up of a number of thin, concentric spherical shells, one inside the other, and considering one such shell of *radius* x and *thickness* dx, we have

surface area of the shell $= 4\pi x^2$,

and *volume* of the shell $= 4\pi x^2.dx$ and its *mass* $= 4\pi x^2.dx.\rho$.*

$\therefore$ moment of inertia of the *shell* about a diameter

$$= \frac{2}{3} \times (its\ mass) \times (its\ radius)^2 = \frac{2}{3}.4\pi x^2.dx.\rho \times x^2 = \frac{8}{3}\pi.\rho x^4.dx.$$ [Case 11, (*i*)

And, therefore, the moment of inertia I, of the *whole* sphere, about its diameter, is obtained by integrating the above expression between the limits $x = 0$ and $x = R$,

i.e., $$I = \int_0^R \frac{8}{3}.\pi\rho x^4.dx = \frac{8}{3}\pi\rho \int_0^R x^4.dx.$$

$$= \frac{8}{3}\pi\rho\left[\frac{x^5}{5}\right]_0^R = \frac{8}{3}.\pi\rho.\frac{R^5}{5} = \frac{8}{15}\pi\rho.R^5.$$

$$= \frac{4}{3}.\pi R^3.\rho.\frac{2}{5}R^2.$$

But $4\pi R^3/3$ = *volume* of the sphere; and, therefore, $4\pi R^3/3 = M$, its *mass*.

$\therefore$ *M.I.* of the sphere about its diameter, *i.e.*, $I = 2MR^2/5$

(*ii*) *about a tangent.* — A tangent, drawn to the sphere at any point, will obviously be parallel to one of its diameters and at a distance from it equal to R, the radius of the sphere.

Therefore, in accordance with the *principle of parallel axes*, we have *M. I.* of the sphere about a tangent

$$= \text{its } M.I. \text{ about a diameter} + MR^2.$$

Or, $$I = 2MR^2/5 + MR^2 = 7\,MR^2/5.$$

14. Moment of Inertia of a Hollow Sphere or a Thick Shell.

(*i*) *about its diameter.* —A hollow sphere is just a solid sphere from the inside of which a smaller concentric solid sphere has been removed. And so, the moment of inertia of the hollow sphere is equal to the moment of inertia of the bigger solid sphere *minus* the moment of inertia of the smaller solid sphere removed from it, (both about the *same* diameter). If R be the radius of the bigger sphere and r, that of the smaller sphere, *i.e.*, if R and r be the external and internal radii of the hollow sphere, and ρ, the density of its material, we have

volume of the bigger sphere $= \frac{4}{3}\pi R^3$ and $\therefore$ its mass $= \frac{4}{3}\pi R^3 \rho$

and „ „ „ smaller „ $= \frac{4}{3}\pi r^3$, and „ „ „ $= \frac{4}{3}\pi r^3 \rho$

$\therefore$ volume of hollow sphere $= \frac{4}{3}\pi(R^3 - r^3)$ and its *mass* $= \frac{4}{3}\pi(R^3 - r^3)\rho$.

And $\therefore$ *M.I.* of the bigger sphere about it's diameter $= \frac{2}{5}.\left(\frac{4}{3}\pi R^3\rho\right).R^2$

and *M.I.* of the smaller sphere *about the same diameter* $= \frac{2}{5}.\left(\frac{4}{3}\pi r^3\rho\right)r^2$

$\therefore$ *M.I.* of the hollow sphere about that diameter

---

* Or, we could obtain the mass of the shell by multiplying the volume of the shell by the mass per unit volume of the sphere, when it would be equal to $(4\pi x^2.dx) \times (3M/4\pi R^3) = (3M/R).x^2.dx$. The rest of the procedure remains the same.

$$= \frac{2}{5}\cdot\left(\frac{4}{3}\pi R^3.\rho\right).R^2 - \frac{2}{5}\left(\frac{4}{3}\pi r^3.\rho\right).r^2 = \frac{2}{5}.\frac{4}{3}\pi\rho.(R^5 - r^5). \quad ....(1)$$

Now, mass of the hollow sphere, $M = \frac{4}{3}.\pi(R^3 - r^3).\rho.$

Or, $3M = 4\pi \times (R^3 - r^3).\rho.$ and $\therefore \quad \rho = 3M/4\pi(R^3 - r^3)$

Substituting this value of ρ in relation (1) above, we have moment of inertia I of the hollow sphere about its diameter

$$= \frac{2}{5}.\frac{4}{3}\pi.\frac{3M}{4\pi(R^3 - r^3)}.(R^5 - r^5).$$

Or,
$$\mathbf{I = \frac{2}{5}M.\frac{(R^5 - r^5)}{(R^3 - r^3)}}$$

Alternative Method. As in the case of the solid sphere, so also here, we can imagine the sphere to be made up of a number of thin, concentric spherical shells, and considering one such spherical shell of radius x and thickness dx, we have, as before, mass of the shell $= 4\pi x^2.dx.\rho$,

Where ρ is the density of the material of the sphere.

$M.I.$ of the sphere about a diameter $= \frac{2}{3}.4\pi x^2.dx.\rho.x^2$

$= \frac{8}{3}\pi.\rho.x^4.dx.$

Hence, the moment of inertia of the *whole* sphere about its diameter is the integral of the above expression between the limits $x = r$ and $x = R$.

Or, $M.I.$ of the sphere about a diameter, *i.e.*, $I = \int_r^R \frac{8}{3}\pi.\rho.x^4.dx.$

$$= \frac{8}{3}\pi\rho\int_r^R x^4.dx = \frac{8}{3}.\pi\rho\left[\frac{x^5}{5}\right]_r^R$$

$$= \frac{8}{3}\pi.\rho\left[\frac{R^5}{5} - \frac{r^5}{5}\right] = \frac{8}{15}.\pi\rho[R^5 - r^5].$$

$$= \frac{4}{3}\pi(R^3 - r^3).\rho.\frac{2}{5}\frac{(R^5 - r^5)}{(R^3 - r^3)}$$

But $\frac{4}{3}\pi(R^3 - r^3).\rho = M$, the mass of the sphere. [See case (*i*) above.

$$\therefore \quad \mathbf{I = \frac{2}{5}M(R^5 - r^5)\,/\,(R^3 - r^3)}$$

(*ii*) *about a tangent.* —Again, as in the case of a solid sphere, the tangent to the sphere, at any point, will be parallel to one of its diameters, and at a distance equal to its *external radius* R from it. Hence, by the principle of *parallel axes*, we have

$M.I.$ of the hollow sphere about a tangent

$= $ its $M.I.$ about a diameter $+ MR^2$.

Or,
$$\mathbf{I = \left[\frac{2}{5}M(R^5 - r^5)/(R^3 - r^3)\right] + MR^2}$$

15. **Moment of Inertia of a Flywheel and Axle.** A flywheel is just a *large heavy wheel*, with a long, cylindrical *axle*, passing through its centre. *Its centre of gravity lies on its axis of rotation*; so that, when properly mounted over *ball-bearings* (to minimise friction), it may continue to be at rest in any desired position.

Let M be the mass of the flywheel, and m, that of the axle; and let R and r be their respective radii.

Then, for our present purpose, we may regard the flywheel to be a disc, or a small cylinder, from which a smaller, *concentric* disc or cylinder, equal in radius to that of the axle, has been cut off. In other words, we may take it to be an *annular ring*, (or a hollow cylinder) with an *outer radius equal* to R and an *inner radius equal to r*, whose moment of inertia is to be determined about an axis passing through its centre and perpendicular to its plane.

The face area of this wheel or annular disc is clearly equal to the area of the whole disc of radius R *minus* the area of the disc of radius r, *i.e.*, *face-area* of the wheel

$$=\pi R^2 - \pi r^2 = \pi (R^2 - r^2).$$

And, if its mass be M, clearly,

mass per unit area of the wheel $= M / \pi (R^2 - r^2)$.

Now, consider a thin circular ring, at a *distance x* from the centre, and of width dx.

Then, *face-area of the ring = its circumference × its width* $= 2\,\pi x.dx$

And, therefore, its mass $= 2\pi x.dx.M / \pi (R^2 - r^2)$.

Now, since the moment of inertia of a ring about an axis through its centre and perpendicular to its plan is equal to its *mass* × (*radius*)^2, we have

$$M.I. \text{ of the wheel about its axis} = \int_r^R \frac{M}{\pi (R^2 - r^2)} 2\pi x.dx.\, x^2.$$

$$= \frac{2\pi M}{\pi (R^2 - r^2)} \int_r^R x^3.dx = \frac{2M}{(R^2 - r^2)} \left[\frac{x^4}{4}\right]_r^R$$

$$= \frac{2M}{(R^2 - r^2)} \left[\frac{R^4}{4} - \frac{r^4}{4}\right] = \frac{2M}{(R^2 - r^2)} \left[\frac{R^4 - r^4}{4}\right]$$

Or, $$M.I. \text{ of the wheel about its axis} = \frac{M}{2}(R^2 + r^2) = M \cdot \frac{(R^2 + r^2)}{2}$$

The axle, again, is just a disc, (or a solid cylinder), and its moment of inertia about its axis is, therefore, just the same as that of a disc or a cylinder about its axis, *i.e.*, = its *mass* × (*radius*)^2/2.

So that, *M.I. of the axle* $= m.r^2/2$.

Hence, *M.I. of the wheel and axle = M.I. of the wheel + M.I. of the axle.*

Or, $$\mathbf{I = [M (R^2 + r^2)/2] + mr^2/2.}$$

3.7. Table of Moments of Inertia. The values of *moments of inertia* for the cases discussed above, together with some other important ones are given in the *Table* below for ready reference of the student, the mass of the body being taken to be M, in all cases.

BODY	AXIS (Position and Direction)	MOMENT OF INERTIA
1. **Thin uniform rod, of length *l***	(*i*) Through its centre and perpendicular to its length.	$Ml^2/12$
	(*ii*) Through one end and perpendicular to its length.	$Ml^2/3$
2. **Thin and rectangular sheet or lamina, of sides *l* and *b*.**	(*i*) Through its centre and parallel to one side. (*ii*) About one side. (*iii*) Through its centre and perpendicular to its plane. (*iv*) Through the mid-point of one side (*l* or *b*) and perpendicular or to its plane.	$Mb^2/12$ or $Ml^2/12$ $Mb^2/3$ or $Ml^2/3$ $M(l^2+b^2)/12$ $M(b^2/3+l^2/12)$ $M(l^2/3+b^2/12)$
3. **Thick uniform rectangular bar, of length *l* and thickness *d*.**	Through its mid-point and perpendicular to its length.	$M\left(\frac{l^2+d^2}{12}\right)$
4. **Thin triangular plate or lamina, of altitude H.**	About one side.	$MH^2/6$
5. **Elliptical disc or lamina, of major and minor axes 2*a* and 2*b*.**	(*i*) About one of the axes, (major or minor). (*ii*) Through its centre and perpendicular to its plane.	$Mb^2/4$ or $Ma^2/4$ $M(a^2+b^2)/4$
6. **Hoop or circular ring, of radius *R*.**	(*i*) Through its centre and perpendicular to its plane. (*ii*) About a diameter. (*iii*) About a tangent, in its own plane. (*iv*) About a tangent, perpendicular to its plane.	MR^2 $MR^2/2$ $3MR^2/2$ $2MR^2$
7. **Circular lamina or disc, or radius *R*.**	(*i*) Through its centre and perpendicular to its plane. (*ii*) About a diameter. (*iii*) About a tangent in its own plane. (*iv*) About a tangent perpendicular to its plane.	$MR^2/2$ $MR^2/4$ $5MR^2/4$ $3MR^2/2$
8. **Annular ring or disc of outer and inner radii *R* and *r*.**	(*i*) Through its centre and perpendicular to its plane. (*ii*) About a diameter. (*iii*) About a tangent, in its own plane. (*iv*) About a tangent perpendicular to its plane.	$M(R^2+r^2)/2$ $M(R^2+r^2)/4$ $M(5R^2+r^2)/4$ $M(3R^2+r^2)/2$

BODY	AXIS (Position and Direction)	MOMENT OF INERTIA
9. **Solid cylinder of length l and radius R.**	(*i*) About its axis of cylindrical symmetry.	$MR^2/2$
	(*ii*) Through its centre and perpendicular to its axis of cylindrical symmetry	$M\left(\frac{l^2}{12}+\frac{R^2}{4}\right)$
	(*iii*) About a diameter of one face.	$M\left(\frac{l^2}{l^3}+\frac{R^2}{4}\right)$
10. **Solid cone, of altitude h and base-radius R.**	(*i*) About its vertical axis	$3MR^2/10$
	(*ii*) Through its vertex and parallel to its base.	$\frac{3MR^2}{20}+\frac{3Mh^2}{5}$
11. **Hollow cylinder, of length l and external and internal radii R and r.**	(*i*) About its own axis, (*i.e.*, about its axis of cylindrical symmetry.)	$M(R^2+r^2)/2$
	(*ii*) Through its centre and perpendicular to its own axis.	$M\left[\frac{R^2+r^2}{4}+\frac{l^2}{12}\right]$
12. **Spherical shell, of radius R.**	(*i*) About a diameter.	$2MR^2/3$
	(*ii*) About a tangent.	$5MR^2/3$
13. **Solid sphere, of radius R.**	(*i*) About a diameter.	$2MR^2/5$
	(*ii*) About a tangent.	$7MR^2/5$
14. **Thick shell or hollow sphere, of external and internal radii R and r.**	(*i*) About a diameter.	$\frac{2}{5}M\left(\frac{R^5-r^5}{R^3-r^3}\right)$
	(*ii*) About a tangent.	$\frac{2}{5}M\left(\frac{R^5-r^5}{R^2-r^2}\right)$
15. **Flywheel, with radii of wheel and axle, R and r—(mass of axle, m)**	About its own axis.	$\frac{M(R^2+r^2)}{2}+\frac{mr^2}{2}$
16. **Spheroid of revolution, of equatorial radius R.**	About polar axis.	$2MR^2/5$
17. **Ellipsoid of axes $2a$, $2b$ and $2c$.**	About one axis, $(2a)$	$\frac{M(b^2+c^2)}{5}$
18. **Rectangular parallelopiped of edges l, b and d.**	Through its centre and perpendicular to one face, (say, face l—b)	$\frac{M(l^2+b^2)}{12}$
19. **Rectangular prism, of dimensions $2l$, $2b$ and $2d$.**	About axis $2l$.	$\frac{M(b^2+d^2)}{3}$
20. **Very thin hollow cylinder, of length l and mean radius R.**	Through its centre and perpendicular to its own axis.	$M\left(\frac{l^2}{12}+\frac{R^2}{2}\right)$

3.8. Routh's Rule. This rule states that *the moment of inertia of a body about any one of the three perpendicular axes of symmetry passing through its centre of mass is given by*

(*i*) *the product of its mass and one-third of the sum of the squares of the other two semi-axes,* **in the case of a rectangular lamina or parallelopiped;**

(*ii*) *the product of its mass and one-fourth of the sum of the squares of the other two semi-axes,* **in the case of a circular or an elliptical lamina;**

(iii) the product of its mass and one-fifth of the sum of the squares of the other two semi-axes, **in the case of a sphere or a spheroid.**

Quite a few of the cases dealt with in the preceding pages, may be easily deduced by an application of this rule. Thus, for example,

(i) moment of inertia of a uniform rectangular lamina (of mass M, length l and breadth b), about an axis passing through its centre) O and perpendicular to its plane

$$= M\left(\frac{(l^2/4) + (b^2/4)}{3}\right) + M\left(\frac{l^2 + b^2}{12}\right),$$

for, here, the two *semi-axes* of the lamina are clearly, $l/2$ and $b/2$ respectively, (Fig. 3.23)

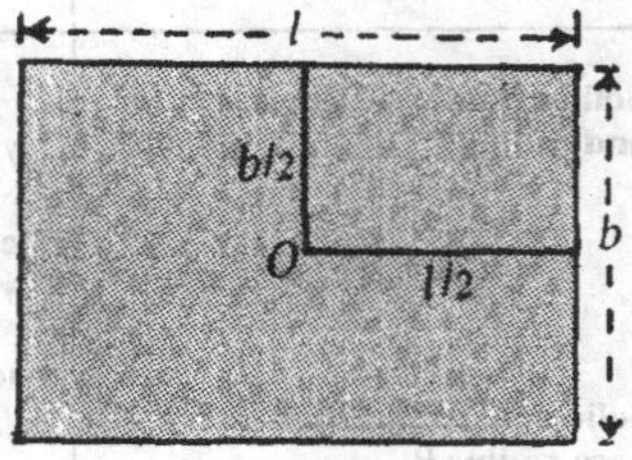

Fig. 3.23

(ii) Moment of inertia of a uniform circular lamina or disc, (of mass M and radius R), about an axis, passing through its centre and perpendicular to its plane is equal to

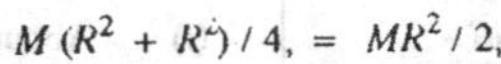

$$M(R^2 + R^2)/4, = MR^2/2,$$

because here the two semi-axis of the lamina or disc are obviously R and R, (Fig. 3.24).

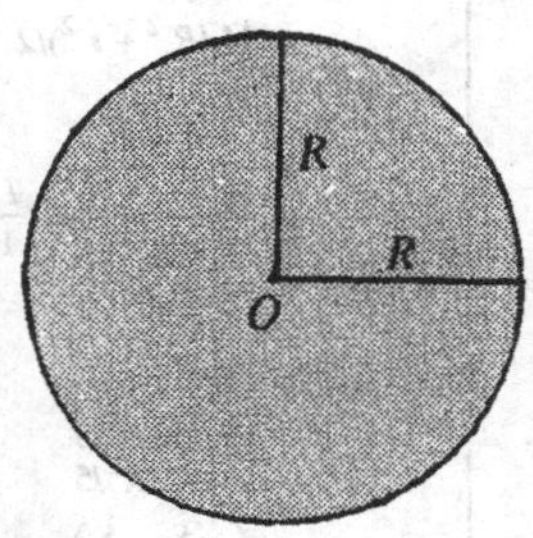

Fig 3.24

And, again, *moment of inertia of a uniform elliptical lamina,* (of mass M and with $2a$ and $2b$, as its major and minor axes respectively), about a perpendicular axis passing through its centre, is equal to $M(a^2 + b^2)/4$,

because a and b are the two semi-axes of the lamina, (Fig. 3.25).

(iii) Moment of inertia of a solid sphere (of mass M and radius R) about its diameter is equal to

$$M(R^2 + R^2)/5 = 2MR^2/5,$$

because here, the two semi-axes of the sphere are R and R.

$M(a^2 + b^2)/4$.

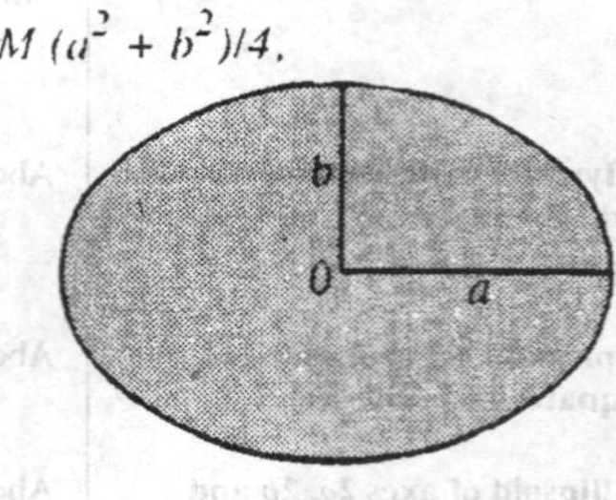

Fig. 3.25

3.9. Practical methods for the Determination of Moments of Inertia. The principle underlying the experimental determination of the moment of inertia I of a body, about a given axis, is to apply a known couple C to it and to measure the angular acceleration $d\omega/dt$ produced in it. Then, from

the relation, $C = I.\ d\omega/dt$ we have $I = \dfrac{C}{d\omega/dt}$,

whence *I may be easily calculated.*

(i) Moment of Inertia of a Flywheel

First Method. The flywheel, whose moment of inertia is to be determined, is mounted on ball-bearings (to minimise friction), and its axle is arranged to lie in the horizontal position at a convenient height from the ground, (Fig. 3.26)

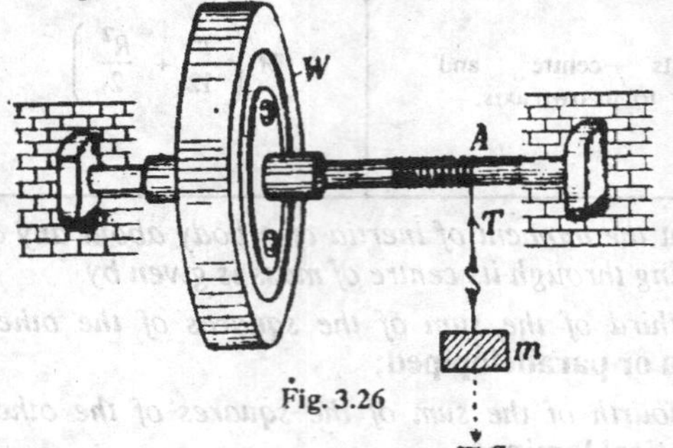

Fig. 3.26

A small loop at the end of a small piece of fine cord is then slipped on to a tiny peg on the axle and *the entire length of the cord wound evenly round the latter,* with a suitable mass m suspended from its free end, and properly held in position.

As the mass is released and allowed to fall under the action of its own weight, the cord starts unwinding itself round the axle, thereby setting the wheel in rotation. The length of the cord is so adjusted that the moment of mass reaches the ground, the entire length of it gets *just* unwound from the axle and slips off the peg on it.

Obviously, the rotation of the wheel, (with the descent of the mass), is due to a *couple T.r*, where T is the tension in the cord and r, the radius of the axle*

If, therefore, I be the *momentof inertia* of the flywheel about *its axis* of notation and $d\omega/dt$, the *angular acceleration* produced in it, we have $I.d\omega/dt = T.r$.

The downward force due to the *weight* of the mass, when it has no acceleration, is mg; but when it has a vertical acceleration a, the force due to it is equal to $m.a.$, and this must clearly be equal to $mg-T$. Or, $m.a = mg-T$, whence, $T = m(g-a)$.

$$\therefore \qquad I.\,d\omega/dt = m.(g-a)\,r.$$

But $\quad d\omega/dt = a/r$, And $\therefore\ I.a/r = m\,(g-a)\,r$. $\qquad [\because\ a = r.\,d\omega/dt$

$$\text{Or,} \qquad I = mr^2.\frac{(g-a)}{a} = mr^2\left(\frac{g}{a} - 1\right) \qquad ...(1)$$

The time-interval between the release of the mass and the slipping of the cord from the axle is carefully noted. Let it be t, and let the distance through which the mass falls down during this interval be S. Then, since the mass starts from rest, we have

$$S = \frac{1}{2}at^2, \text{ whence } a = 2S/t^2.$$

So that, substituting this value of a in relation (1) above, we have

$$I = mr^2.\left(\frac{g}{2S/t^2} - 1\right) = mr^2.\left(\frac{gt^2}{2s} - 1\right),$$

whence, I, the moment of inertia of the flywheel about its axis of rotation, can be easily calculated.

N.B. It will be noted that we have not taken account here of any friction that the wheel encounters at the bearings. In a rough and ready manner, the error due to it may be eliminated by noting before or after the experiment as to what small weight attached to the cord will keep the wheel turning *without acceleration*. Let it be x. Then, we replace m by $(m-x)$ in our expression for I above.

Second Method. — Proceeding, as above, *the loss of potential energy of the falling mass is equated against the gain in kinetic energy of the wheel, the K.E. of the mass itself and the work done against friction*. Thus, when the mass falls through distance S, the potential energy lost by it is equal to $mg.S$. And, if ω be the angular velocity of the wheel at the time, *K.E.* gained by it is $\frac{1}{2}I\omega^2$, the *K.E.* acquired by the mass being $\frac{1}{2}mv^2$, where v is its velocity on descending through distance S.

$$\therefore \qquad mg.S = \frac{1}{2}I\omega^2 + \frac{1}{2}mv^2 + \text{the work done against friction.} \qquad ...(2)$$

To determine the work done against friction, we note the number of turns made by the wheel before coming to rest *after the mass has been detached from the axle*. Then, obviously, the kinetic energy $\frac{1}{2}I\omega^2$, of the wheel, is used up in overcoming the frictional forces at the bearings. If the couple due to friction be C and the number of turns made by the wheel before coming to rest be n, work done by *this* couple is equal to $2\pi n \times C$ ($\because$ *work done = couple × angle*, and the angle described by the wheel in *one* rotation is equal to 2π). So that,

$$\frac{1}{2}I\omega^2 = 2\pi nC. \qquad \text{Or,} \quad C = I\omega^2/4\pi n.$$

The couple due to friction being thus determined, we can easily calculate the work done against friction during the descent of the mass through distance S. For, clearly, the number of turns

* If the cord be appreciably thick, half of its thickness, added to the radius of the axle, gives the effective value of r.

made by the wheel during the fall of the mass through this distance is $S/2\pi r$ and, therefore, the total angle turned through by it is equal to $2\pi S/2\pi r = S/r$.

Hence, work done against friction is equal to $C.S/r = SI\omega^2/4\pi nr$.

$\therefore$ our *energy equation* (2) now becomes

$$mg.S = \frac{1}{2}I\omega^2 + \frac{1}{2}mv^2 + \frac{SI\omega^2}{4\pi nr} = \frac{1}{2}I\omega^2\left(1 + \frac{S}{2\pi nr}\right) + \frac{1}{2}mv^2$$

$$= \frac{1}{2}I.\frac{v^2}{r^2}\left(1 + \frac{S}{2\pi nr}\right)\frac{1}{2}mv^2. \qquad [\because \omega^2 = v^2/r^2$$

Or, $$I.\frac{v^2}{r^2}\left(1 + \frac{S}{2\pi nr}\right) + mv^2 = 2mg.S.$$

Or, $$I.\frac{v^2}{r^2}\left(1 + \frac{S}{2\pi nr}\right) = 2\,mg.S - mv. \qquad ...(3)$$

Now, if t be the time taken by the mass to fall through the distance S, its *average velocity* = S/t; and since average velocity = (*initial velocity* + *final velocity*)/2, we have

$$\textit{final velocity}, v = 2S/t. \qquad \text{Or,} \quad v^2 = 4S^2/t^2.$$

[$\because$ the initial velocity is *zero*, the mass starting from rest.]

Substituting this value of v in expression (3) above, we have

$$\frac{I}{r^2}\cdot\frac{4S^2}{t^2}\left(1 + \frac{S}{2\pi nr}\right) = 2mg.S - m.\frac{4S^2}{t^2}.$$

$$= 2mS\left(g - \frac{2S}{t^2}\right) = 2mS\left(\frac{gt^2 - 2S}{t^2}\right).$$

$$I = 2ms\left(\frac{gt^2 - 2S}{t^2}\right)\frac{r^2.t^2}{4s^2\,(1 + S/2\pi nr)}$$

Or, $$I = \frac{mr^2\,(gt^2 - 2S)}{2S\,(1 + S/2\pi nr)} \qquad ...(4)$$

Alternative Calculation. — Let the number of rotations made by the wheel, before the cord and the mass slip off the axis, (*i.e.*, after the mass has fallen through distance S), be N*. Then, *taking the fractional force to be uniform*, and the work done against it *per rotation* of the wheel to be w, we have

work done against friction during N rotations of the wheel = $N.w$.

Thus, our energy equation (2) becomes

$$mg.S = \frac{1}{2}I\omega^2 = \frac{1}{2}mv^2 + N.\omega. \qquad ...(5)$$

Now, after the detachment of the mass from the axle, the wheel comes to rest after n rotations, and therefore, work done against friction during these n rotations of wheel = $n.w$; and this must obviously be equal to $\frac{1}{2}I\omega^2$, the *K.E.* of the wheel *at the instant that the mass gets detached from it*. Thus,

$$n.w = \frac{1}{2}m\omega^2, \qquad \text{whence,} \qquad w = \frac{1}{2}I\omega^2/n.$$

Substituting this value of w in equation (5) above, we have

$$mg.S = \frac{1}{2}I\omega^2 + \frac{1}{2}mv^2 + \frac{1}{2}NI\omega^2/n = \frac{1}{2}mv^2 + \frac{1}{2}I\omega^2\,(1 + N/n).$$

Or, $$2mg.S = mv^2 + I\omega^2\,(1 + N/n)$$

Or, $$2mg.S - mv^2 = I\omega^2\,(1 + N/n)$$

whence, $$I = \frac{2mg.S - mv^2}{\omega^2\,(1 + N/n)} = \frac{2mg.S - mr^2\omega^2}{\omega^2\,(1 + N/n)} \qquad [\text{Since } v = r\omega.$$

* This is obviously equal to the number of turns of the cord on the axle at the very start.

Or, $$I = \frac{(2mg.S/\omega^2) - mr^2}{1 + N/n} = \frac{m[2g.S/\omega^2) - r^2]}{1 + N/n} \quad ...(6)$$

Now, the angular velocity of the wheel at the instant that the mass gets detached from it is ω, and becomes zero when the wheel comes to rest after time t, say. Hence, if the frictional force *uniformly* retards the rotation of the wheel, its *average* angular velocity, during this interval of time t', may be taken to be equal to $(\omega + 0)/2$, *i.e.*, equal to $\omega/2$. And, since the wheel makes n rotations before coming to rest, it describes an angle equal to $2\pi n$ in time t'.

$\therefore \quad \omega/2 = 2\pi n/t', \quad$ whence, $\quad \omega = 4\pi n/t'$.

So that, substituting this value of ω in relation (6) above, we have

$$I = \frac{m[(2g.S \times t'^2 + 16\pi^2 n^2) - r^2]}{1 + N/n} = \frac{m[(g.S \times t'^2/8\pi^2 n^2) - r^2]}{(n+N)/n}$$

$$= m \times \left(\frac{n}{n+N}\right)\left(\frac{g.S.t'^2}{8\pi^2 n^2} - r^2\right)$$

Or, $$I = \frac{m}{n+N}\left(\frac{g.S.t'^2}{8\pi^2 n} - nr^2\right) \quad ...(7)$$

whence I, the moment of inertia of the flywheel about its axis of rotation, can be easily calculated.

Accurate value of ω — In the above treatment, the angular velocity w of the wheel has been obtained on the supposition that the frictional force remains constant during time t' that the value of ω falls to *zero* after the detachment of the mass from the axle. Obviously, this is by no means a valid assumption because, as we know, the frictional force decreases with increase of velocity; so that, the value of I, the moment of inertia of the wheel, deduced on the basis of the above calculation, cannot possibly be quite accurate.

If we aim at accuracy, therefore, we must adopt a *sensitive* method for determining the value of ω, and the one method, which at once suggests itself, is to make use of a tuning fork, as explained below:

A tuning fork, of a known frequency n, is arranged horizontally, (Fig. 3.27), with a slightly bent metallic style, attached to one of its prongs, such that, when desired, it can be made to lightly press against, or taken off, a strip of smoked paper, wrapped round the rim of the wheel.

Now, with the style kept off the paper-strip, the mass m is allowed to fall, thus setting the wheel in rotation, and just a second or so before the mass is due to get detached from the axle, the tuning fork is set into vibration (by smartly drawing a bow apressed lightly on to the strip, taking care to take it off soon after the detachment of the mass. A long wavy curve is thus traced out by the style on the smoked strip. The *mean* wavelength of this wave is then determined by dividing the total distance occupied by the wavy curve by the total number of waves constituting it.

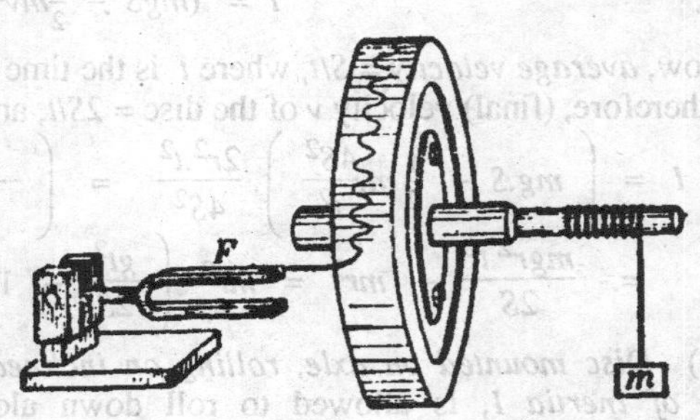

Fig. 3.27

Since one wave is traced out by the style during one vibration of the prong or the fork, we have linear distance covered by the wheel during *one* vibration of the fork $= \lambda$. So that, distance covered by the wheel during n vibrations of the fork $= n\lambda$.

Again, since n vibrations are made by the fork in *one second*, it follows that *distance covered by the wheel in 1 second, i.e., its linear velocity* $v = n\lambda$.

But $v = R\omega$, where R is the radius of the wheel and ω, its angular velocity; so that, we have $R\omega = n\lambda$, whence, $\omega = n\lambda/R$

Thus, knowing n, and R, we can easily calculate the value of ω for the wheel. This value of ω, substituted in relation (6) above, then gives a much more accurate value of I, the moment of inertia of the flywheel about its axis of rotation.

Note. — The student may, as an intersecting exercise, show that expression (4) above can also be reduced to the same form as expression (7). This may be easily done by remembering *(i)* that when the wheel makes one full turn, the mass descends through a distance $2\pi r$, the circumstance of the axle and therefore, when the mass descends through a distance S, the number of rotations made by the wheel is equal to $S/2\pi r$; so that, $S/2\pi r = N$; and further *(ii)* that $t = 2S/v = 2S/r\omega$, where $\omega = 4\, n/t'$, (*see page* 72).

(*ii*) **Moment of inertia or a disc about an axis passing through its centre and perpendicular to its plane :**

(*a*) *Disc suspended by two parallel threads.* — The disc, with a metal axle, is supported on two cords, wound uniformly on the axle on either side, (Fig. 3.28). On releasing the disc, it begins to fall down until the whole cord is unwound from the axle, say through a distance S.

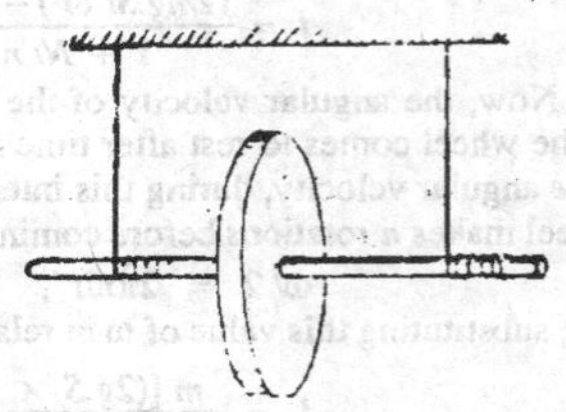

Fig. 3.28

Then, clearly, $P.E.$ lost by the disc $= mg.S$, where m is the mass of the disc and the axle. This energy will obviously be gained by the disc in the form of kinetic energy of rotation and translation.

If ω be the angular velocity acquired by it after falling through this distance S, its $K.E.$ of rotation will clearly be $\frac{1}{2}I\omega^2$, where I is its moment of inertia about an axis passing through its centre and parallel to the axis, (*i.e.*, perpendicular to its plane); and its kinetic energy of translation will be $\frac{1}{2}mv^2$

$$\therefore \quad mg.S = \frac{1}{2}I\omega^2 + \frac{1}{2}mv^2, \quad (\text{ } v \text{ being its final } \textit{linear velocity}),$$

$$= (\tfrac{1}{2}I.v^2/r^2) + \tfrac{1}{2}mv^2. \qquad [\because \omega^2 = v^2/r^2 \quad \text{where } r = \text{radius of the disc.}$$

Or. $$\frac{1}{2}\,I\,v^2/r^2 = (mg.S - \frac{1}{2}mv^2),$$

whence, $$I = (mgS - \tfrac{1}{2}mv^2).2r^2/v^2.$$

Now, *average velocity* $= S/t$, where t is the time taken by the disc in falling through distance S; and, therefore, (final) velocity v of the disc $= 2S/t$, and $\therefore v^2 = 4S^2/t^2$. So that,

$$I = \left(mg.S - \frac{1}{2}m.\frac{4S^2}{t^2}\right).\frac{2r^2.t^2}{4S^2} = \left(\frac{mg.S.t^2 - 2mS^2}{t^2}\right).\frac{r^2t^2}{2S^2}$$

$$= \frac{mgr^2.t^2}{2S} - mr^2 = mr^2\left(\frac{gt^2}{2S} - 1\right).$$

(*b*) *Disc mounted on axle, rolling on inclined rails.* — Here, the disc, of *mass m* and *moment of inertia I*, is allowed to roll down along inclined rails, as shown in Fig. 3.29. Let it acquire a *linear velocity* v and an *angular velocity* ω, when it descends a *vertical distance h*, as it rolls down a distance S *along* the rails.

Then, clearly, *loss of P.E. of the disc*
= K.E. of translation gained by disc
+ K.E. of rotation gained by disc.

Or, $$mgh = \frac{1}{2}mv^2 + \frac{1}{2}I\omega^2.$$

So that, $$mgh = \frac{1}{2}mv^2 + \frac{1}{2}I.\frac{v^2}{r^2},$$

[where r = *radius of axle* and $\therefore \omega = v/r$]

Fig. 329

Or, $$I.\frac{v^2}{2r^2} = m\left(gh - \frac{1}{2}v^2\right), \text{ whence, } I = \frac{2mr^2}{v^2}\left(gh - \frac{1}{2}v^2\right)$$

Or, $$I = \frac{mr^2}{v^2}(2gh - v^2)$$

Or, substituting the value of $v = 2S/t$, (see above), where t is the time taken by the disc to cover the distance S, we have

$$I = \frac{mr^2.t^2}{4S^2} \cdot \left(2gh - \frac{4S^2}{t^2}\right) = \frac{mr^2\,t^2.gh}{2S^2} - mr^2,$$

whence, the value of I, the moment of inertia of the disc, can be easily calculated.

Note : For other methods for the determination of moment of inertia, see under Torsional Pendulum (Chapter VIII).

3.10. Angular Moment and Angular Impulse.— In the case of linear motion, the *momentum* of a body, as we know, is the product of its mass and velocity. On the same analogy, we have, in the case of rotational motion, *the product of the moment of inertia and the angular velocity as the angular momentum of a rotating body.*

Thus, *angular momentum* $= I.\omega$,
where I is the moment of inertia and ω, the angular velocity of the body about the axis of rotation.

For, suppose we have a body, rotating about an axis with a velocity ω. Then, all its particles will have the *same* angular velocity ω, but their linear velocities will depend upon their respective distances from the axis of rotation, being equal to the product of the angular velocity and the distance from the axis. Thus, the linear velocity of a particle, distant r_1 from the axis, will be $r_1\omega$; of that distant r_2 from the axis will be $r_2\omega$ and so on.

And, therefore, if m be the mass of each particle, we have, *linear momentum* of the particle, distant r_1 form the axis, equal to $m.r_1.\omega$ and, therefore, the moment of *its momentum* about the axis would be $m.r_1.\omega \times r_1 = m.v_1^2.\omega$ Similarly, *the moment of momentum* of the particle, distant r_2 from the axis will be $m.r^2\omega$ and so on.

Therefore, *the moment of momentum* of the *whole* rotating body about the axis

$$= mr_1^2\,\omega + mr_2^2 + \ldots\ldots\ldots = \Sigma\, mr^2\omega = I.\omega, \qquad [\because \Sigma mr^2 = I$$

where I is the *moment of inertia* of the body about the axis of rotation.

Thus, the *angular momentum* of a rotating body about its axis of rotation is *the sum of the moments of the linear momenta of its particles about that axis.* For this reason, it is also referred to as its *moment of momentum* about the axis.

Now, we have $I.d\omega/dt = C$ Or, $I.d.\omega = C.dt$,
where C is the *torque* or the *couple* acting on the body.

Integrating this with respect to t, between the limits 0 and t, we have

angular momentum, $$I.\omega = \int_0^t C.dt,$$

an expression which is true, however C may vary with time.

If C be constant, we have $I.\omega = \int_0^t C.dt = C.t$,

which gives the *angular-momentum* acquired by the body *in time t.*

If t be very small and C quite large, the expression $\int_0^t C.dt$ stands for the *angular impulse* given to the body, which again becomes equal to $C.t$, if C be constant.

3.11. Law of Conservation of Angular Momentum. Just as we have the law of conservation of momentum for linear motion, we have, for rotational motion also, the *law of conservation of angular momentum,* which states that *the angular momentum of a rotating body about an axis remains constant, if no external torque be applied to it.*

For, uppose the angular velocity of a body is changed by $d\omega$, by a torque C applied to it for a very smal interval of time dt.

Then, we have $C.dt = I.d\omega$,

where I is the *moment of inertia* of the body about the given axis.

Hence, $C = I.d\omega/dt$, assuming I to remain constant.

If, however, I also changes, we have $C = d(I\omega)/dt$,

i.e., the torque is equal to the rate of change of angular momentum.

Obviously, therefore, if $C = 0$, *i.e.*, if there be no external torque applied to the body, $d\omega/dt$ or $d(I\omega)\ dt$ *is also equal to zero*, or the rate of change of angular momentum remains constant.

It is obvious from the above that, in the case when I is not constant, and no external torque is applied to the body, the angular velocity must change in the inverse ratio to I, in order to keep its angular momentum constant.

This may be clearly seen by whirling round a stone tied to one end of a string, whose other end is held in the hand. On stopping the application of any force to it, *i.e.*, on removing the external torque, the string begins to wind itself on the hand, with continuously increasing velocity, because as the distance of the stone from the hand decreases, its moment of inertia about its axis of rotation also *decreases*, resulting in a proportionate *increase* in its angular velocity.

Another good illustration is provided by an acrobat executing a somersault. For, as we know, he instinctively curls himself up in air, thereby decreasing his moment of inertia and consequently increasing his speed of rotation. But, therefore his feet touch the ground, he slows it down by straightening himself up and increasing his moment of inertia.

Again, the higher speed of the earth when nearest to the Sun is due to its decreased moment of inertia about the Sun in that part of its orbit.

3.12. Laws of Rotation. Corresponding to Newton's three laws in the case of linear motion, we have also three laws of rotational motion, *viz.*,

1. *Unless an external torque is applied to it, the rate of rotation of a rigid body, about a fixed axis in it, remains unaltered.*

An obvious example of this is the *constant rotation of the Earth about its axis.* The force of attraction due to the Sun is certainly there, but it acts at the centre of the earth and hence produces no effect on its rotation.

2. *The rate of change of rotation of a body, about a fixed axis in it, is directly proportional to the external torque applied and takes place in the direction of the torque.*

3. *If a torque be applied by one body upon another, an equal and opposite torque is applied by the latter upon the former, about the same axis of rotation. In other words, a change in the angular momentum of one body brings about an equal and opposite change in the angular momentum of the other body.*

It is useful to remember that *moment of inertia (I) and angular velocity (ω) in rotational motion, correspond respectively to mass (m) and to linear velocity (v), in the case of linear motion.*

The following *Table* gives the linear and rotational analogues at a glance.

Linear Motion	Rotational Motion
1. Mass..........m	Moment of Inertia I
2. Distance covered............S	Angle described ....θ
3. Velocity .... v or dS/dt.	Angular velocity ..... ω, or $d\theta/dt$
4. Acceleration.... a or d^2S/dt^2.	Angular acceleration ....$d\omega/dt$., or $d^2\theta/dt$
5. Force $F = m.a$.	Torque C, or (moment of the couple).
6. Impulse $= F.dt$	Angular impulse $= C.dt$
7. Linear $K.E. = \frac{1}{2}mv^2$	Angular $K.E. = \frac{1}{2}I\omega^2$
8. Linear momentum $= mv$.	Angular momentum or moment of momentum $= I\omega$

9.	Work done by force, *i.e.*, $W = F.S$	Work done by couple, *i.e.*, $W = C.\theta$
10.	$v = u + at$	$\omega_2 = \omega_1 + (d\omega / dt).t.$
11.	$S = ut + \frac{1}{2}at^2$	$\theta = \omega_1 t + \frac{1}{2}(d\omega / dt)t^2.$
12.	$v^2 - u^2 = 2aS.$	$\omega_2^{\ 2} - \omega_1^{\ 2} = 2(d\omega / dt)\ \theta,$ where ω_1 and ω_2 are the initial and final angular velocities and $d\ /dt$, the angular acceleration.

3.13 Kinetic Energy of Rotation.

(*a*) *Kinetic energy of a body about an axis through its centre of mass.* — Suppose we have a body of mass M rotating about an axis AB, passing through its centre of mass O. (Fig. 3.30). It, obviously, possesses kinetic energy due to its motion; this energy of the body is called its *energy of rotation*, because it is due to its motion of rotation.

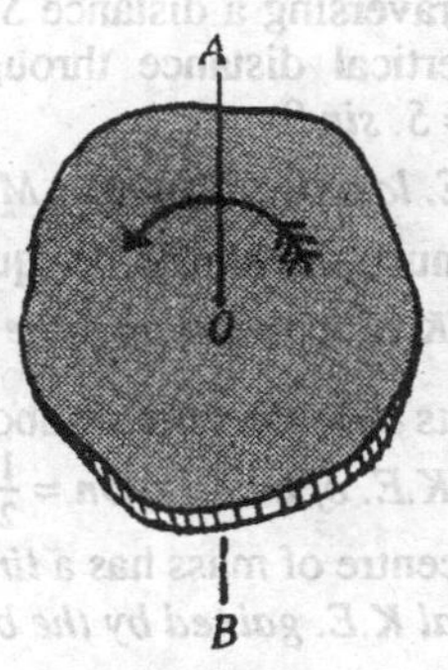

Fig. 3.30

Imagine the body to be divided up into a large number of small particles, of masses m_1, m_2, m_3, etc., at distances r_1, r_2, r_3...... etc., respectively from the axis AB. Then, we have

linear velocity of $m_1 = r_1\omega = v_1$;

of $m_2 = r_2\omega = v_2$; of $m_3 = r_3\ \omega = v_3$ and so on.

Kinetic energy of mass $m_1 = \frac{1}{2}m_1v_1^2$; of mass $m_2 = \frac{1}{2}m_2v_2^2$, of mass $m_3 = \frac{1}{2}m_3v_3^2$ and so on.

Or, total $K.E.$ of the body

$$= \frac{1}{2}m_1r_1^2\omega^2 + \frac{1}{2}m_2r_2^2\omega^2 + \frac{1}{2}m_3r_3^2\omega^2 + \ldots\ldots\ldots$$

$$= \frac{1}{2}\omega^2\ [\ m_1r_1^2 + m_2r_2^2 + m_3r_3^2 + \ldots\ldots\ldots]$$

$$= \frac{1}{2}\omega^2\ \Sigma m r^2 = \frac{1}{2}\omega^2\ MK^2 \qquad [\because\ \Sigma mr^2 = MK^2.$$

Since $MK^2 = I$, the $M.I.$ of the body about the axis AB,

$$\textbf{K.E. of the body} = \frac{1}{2}I\omega^2,$$

in *ergs* in the $C.G.S$ system and in *joules* in the $M.K.S$ (or the $S.I.$ system)

Now, if $\omega = 1$, then, obviously, $K.E.$ of the body $= \frac{1}{2}I$.

Or, $\qquad I = 2\,K.E.$

This gives us another definition of moment of inertia, *viz.* that *the moment of inertia of a body, rotating with unit angular velocity, is equal to twice its kinetic energy of rotation.*

(*b*) *K.E. of body which is not only rotating but whose centre of mass has also a linear velocity v.* —A body which is rotating as well as moving forwards with a velocity v, has both types of kinetic energy, *viz.*, (*i*) *energy of rotation*, because of its rotation about a perpendicular axis through its centre of mass, and (*ii*) *energy of translation*, because of its linear motion. And, clearly, therefore, we have *K.E. of rotation* of the body $= \frac{1}{2}$, and its *K.E. of translation* $= \frac{1}{2}Mv^2$.

$\therefore\qquad$ *total K.E. of the body = K.E. of rotation + K.E. of translation,*

$$= \frac{1}{2}I\omega^2 + \frac{1}{2}Mv^2 = \frac{1}{2}MK^2\omega^2 + \frac{1}{2}Mv^2$$

$$= \frac{1}{2}M(K^2\omega^2 + v^2) = \frac{1}{2}MK^2\ (v^2 / R^2) + \frac{1}{2}Mv^2,$$

because $\omega^2 = v^2/r^2$ where R is the radius of the body.

Or, *total kinetic energy of the body* $= \frac{1}{2}Mv^2 [(K^2/R^2) + 1]$.

3.14. Acceleration of a body rolling down an inclined plane. Let a body like a disc or sphere etc. (*i.e.*, a body having circular symmetry, of mass M roll freely down an inclined plane, of inclination θ to the horizontal, (Fig. 3.31). The plane is supposed to be rough enough, so that there may be *no slipping*, and hence no work done by friction.

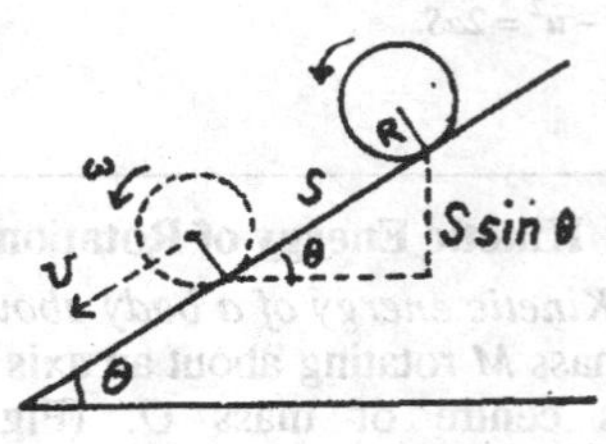

Fig. 3.31

Then, if v be the velocity acquired by the body after traversing a distance S along the plane, we have vertical distance through which it has descended $= S. \sin\theta$

$\therefore$ *P.E. lost by the body* $= Mg. S \sin\theta$

This must, obviously, be equal to the *K.E.* gained by the body.

Now, *K.E. of rotation of the body* $= \frac{1}{2}I\omega^2$,

where ω is its *angular velocity* about a perpendicular axis through its centre of mass.

And its *K.E. of translation* $= \frac{1}{2}Mv^2$,

because its centre of mass has a *linear velocity* v.

$\therefore$ *total K.E. gained by the body* $= \frac{1}{2}M\omega^2 + \frac{1}{2}Mv^2$

$$= \frac{1}{2}Mv^2 [(K^2/R^2) + 1].$$ [See § 3.13

Since *gain in K.E. of the body is equal to the loss in its P.E*, we have

$$\frac{1}{2}Mv^2 [(K^2/R^2) + 1] = Mg. S \sin\theta.$$

Or, $$Mv^2 [(K^2/R^2) + 1] \; 2Mg. \sin\theta.S.$$

Or, $$v^2 (K^2 + R^2)/R^2 = 2g.\sin\theta.S,$$

whence, $$v^2 = 2[R^2/(K^2 + R^2)].g \sin\theta \; S.$$

Comparing this with the *kinematic relation*, $v^2 = 2aS$, for a body starting from rest, we have

acceleration of the body down the plane, i.e.,

$$a = [R^2/(K^2 + R^2)].g \sin\theta.$$

Or, the acceleration is proportional to $R^2/(K^2 + R^2)$ for a given angle of inclination θ.

This shows that

(*i*) *the greater the value of K, as compared with R, the smaller the acceleration of the body coming down the plane* and, therefore, the greater the time it takes in rolling down along it and *vice versa;*

(*ii*) *the acceleration and, therefore, the time of descent, is independent of the mass of the body.*

Let us consider some **particular cases.**

(*i*) **Solid Sphere.** Here $K^2 = \frac{2}{5}R^2$ and, therefore, *acceleration down the plane*,

$$a = (R^2/K^2 + R^2)g \sin\theta = \left[R^2 \Big/ \left(\frac{2}{5}R^2 + R^2\right)\right] g \sin\theta = \frac{5}{7} g \sin\theta.$$

(*ii*) **Spherical Shell.** In this case, $K^2 = \frac{2}{3}R^2$ and therefore,

$$a = [R^2/(\tfrac{2}{3}R^2 + R^2)]g \sin\theta = \tfrac{3}{5}g \sin\theta.$$

(*iii*) **Disc or cylinder.** In this case, K^2 (about its own axis) = $R^2/2$, and, therefore,

$$a = \left[R^2 \Big/ \left(\frac{R^2}{2} + R^2\right)\right] g \sin\theta.$$

(*iv*) **Ring or Hoop.** Here, obviously, $K^2 = R^2$ and, therefore,

$$a = [R^2/(R^2 + R^2)]g \sin\theta = \tfrac{1}{2}g \sin\theta.$$

It follows at once from the above that if a solid sphere, a spherical shell, a disc (or cylinder) and a hoop, *all of the same mass*, be allowed to roll down an inclined plane (without slipping) they will come down the plane in this very order, *viz*, the solid sphere first, then, the spherical shell, followed next by the disc (or the cylinder) and lastly the hoop.

And, since K^2 for a hollow sphere about its diameter is greater than that for a solid sphere of the *same mass and radius*, they can be distinguished from each other by allowing them to roll down the plane. Obviously, the solid sphere will roll down faster than the hollow one, (See worked example 14 at the end of the chapter), The same test may be applied in the case of a hollow and a solid cylinder etc.

3.15 Function of the flywheel in stationary engines. As we know; in the case of a steam engine there are two *'dead centres'*, *viz*., at the end of each stroke, when the crank and the piston rod are in the same line, with the piston momentarily at rest, *producing no turning tendency in the shaft*. And, in two other positions the crank and the piston are at right angles to each other, *when the turning effect on the shaft is the maximum*.

Similarly, in the case of an internal combustion engines, the power is transferred to the shaft only owing the fourth or the working stroke, *i.e., only once during its four strokes and not continuously*.

There are thus, in either case, large variations in the torque applied to the shaft and its rotation is, as a result, neither smooth nor regular. In consequence, the machinery complex to it is driven in jerks. It is precisely to *remove* this and to make the running of the machine smoother and steadier that a flywheel is fitted on to the shaft of stationary engines. For, an account of its large moment of inertia, *it serve as a reservoir of energy*, as it were, absorbing the excess of energy during part of its revolution. When the torque applied to the shaft is large and supplying energy during the rest of its revolution when the torque applied to the shaft is small. It thus continues to rotate with practically a uniform speed, ensuring a smooth running of the machines compiled to the shaft.

Obviously, the larger the moment of inertia (MK^2) of the flywheel, the better its smoothing action. This is the reason why practically the whole mass of the flywheel is concentrated in its rim (to increase K) with the spokes connecting the rim to its centre being just of sufficient thickness to be able to withstand the centrifugal force on due to rotation of the wheel. A slight curvature is given to the spokes in order to destribute the shearing stress more or less evenly on them.

In the case of a railway or a motor car engine, a flywheel is dispersed with because the moving mass of the locomotive or the car itself conserves motion. Further in the case of the car, the smoothness of the motion is also partly obtained by using a multicylinder engine, with the cylinders arranged to deliver power to the shaft one after the other in succession. This, incidentally, also increases the power of the engine.

3.16 Graphical Representation of Plane Vectors. We are already familiar with the two types of physical quantities, *viz.*, (*i*) *scalar* and (*ii*) *vector*,—the former possessing *only magnitude*, but no direction; and the latter, possessing *both magnitude and direction*, (see footnote on page 50). These latter can, as we know, be represented by a straight line, drawn to a chosen scale, whose length and direction respectively represent the magnitude and direction of the quantity.

Any other quantity, either derived from a vector, or obtained by combining a vector with a scalar quantity, is also vectorial in nature. Thus, for example, the acceleration of a body, depending upon the velocity of the body, (a vector quantity), is also a vector quantity.

The vector quantities referred to above are, strictly speaking, *linear vectors*, and must be clearly distinguished from what are called *plane vectors* — a term applied, in rotational dynamics, to such quantities as angular velocity, angular momentum and torque etc., which are all directional in the sense that they are confined to one plane.

Such a *plane* or *two-dimensional* vector is also represented by a straight line, drawn *normal to its plane of rotation, or parallel to its axis of rotation, —its magnitude being represented by the length of the straight line, drawn to a chosen scale*, and its clockwise or anticlockwise rotation being indicated, according to an agreed and established convention, by the straight line being directed towards, or away from, the observer respectively.

This latter convention as to the sense of rotation of the vector is best remembered with the help of what is called the *right-handled screw rule*, stated as follows :

Imagine the stem of a right-handed screw to be along the axis of rotation of the vector and imagine further the screw head to be turned in the direction of rotation of the vector itself. Then, accordingly, as the screw moves forwards or backwards along the axis, the straight line representing the vector (in magnitude) is shown pointing towards or away from the observer.

Thus, if disc *D* in Fig. 3.32 (*a*) rotates in the clockwise direction about the axis *PQ* passing through its centre and perpendicular to its plane, we may imagine a right-handed screw (shown dotted) to lie along *PQ*, with the screw-head also turned clockwise, when, clearly, the screw advances forwards along *PQ*. We may, therefore, represent the angular velocity of the disc by the straight line *OA*, (Fig. 3.32 (*b*)], drawn parallel to the axis, according to a chosen scale, such that *its length represents the magnitude or the numerical value of the angular velocity* (ω) *and its direction indicates the forward motion of the imaginary screw along the axis and hence the clockwise direction of rotation of the disc about it.*

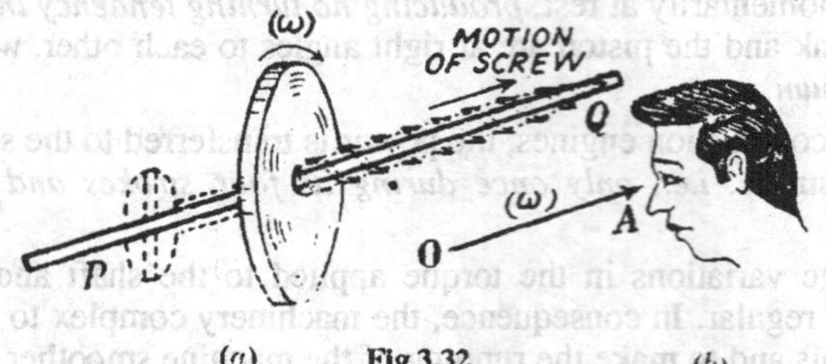

Fig 3.32

Further, corresponding to the parallelogram law for the composition of linear vectors, we have, here, a modified form of it to determine the resultant of two plane vectors, *viz.*, that

"If there be two plane vectors acting simultaneously on a body in two different planes, such that they can be represented in magnitude and direction by the two adjacent sides of a parallelogram, drawn perpendicular to those planes, their resultant is represented completely by the diagonal of the parallelogram, passing through their point of intersection, —this diagonal representing a plane vector in a third plane, perpendicular to itself."*

Thus, if *OA* and *OB*, (Fig 3.33) represent two *couples, in two different planes*, acting simultaneously on a body, where *OA* and *OB* are drawn *perpendicular to those planes*, their resultant is given completely by the diagonal *OC* of the parallelogram, which represents a couple in *a third plane*, perpendicular to itself.

And, obviously, what is true about the composition of couples is equally true for the composition of any other plane or two-dimensional vector quantities.

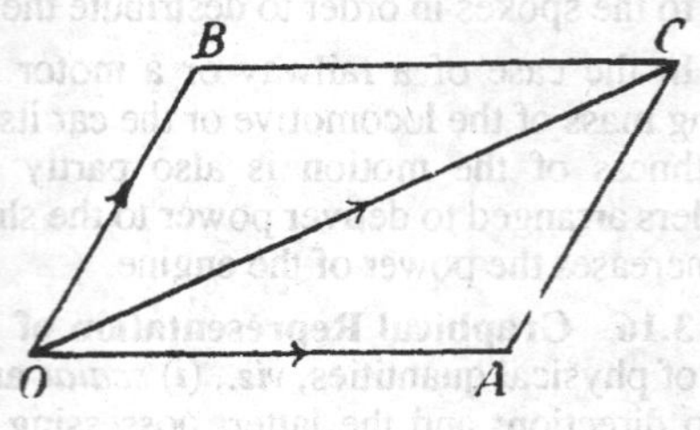

Fig. 3.33

* In accordance with the convention, stated above.

3.11 Precession. Just as in the case of linear motion, we may have a constant acceleration acting on a body, without changing its constant speed, (*e.g.*, the *centripetal* acceleration acting on a body, moving with a uniform speed in its circular orbit), so also, in rotational motion, *we may have a constant angular acceleration acting on a body, having a constant angular speed. This is rendered possible by the plane of rotation changing direction at a given rate*, without, in any way, affecting the rate of rotation of the body about its axis of rotation, or *axis of spin*, as it is also sometimes referred to. *This change in the plane of rotation is called* **'precession'** and is caused by a couple or torque, called the *precessional torque*, acting in a plane, perpendicular to the immediate or instantaneous plane of rotation (or spin) of the body. In other words, *the axis of the torque is, at any given instant, perpendicular to the rotation-axis of the torque is, at any given instant, perpendicular to the rotation-axis of the body*, as will be clear from the following :

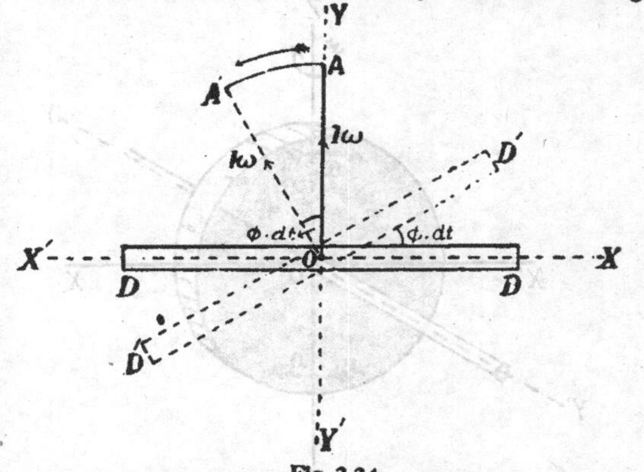

Fig. 3.34

Consider a disc revolving about its geometric axis with a constant angular velocity ω, and let *DD*, (Fig. 3.34), be the edge of the disc. Then, if its moment of inertia about this axis be *I*, its angular momentum will clearly be *I*ω. Let this be represented by the straight line *OA*, drawn perpendicular to the plane of rotation of the disc.

Now, let the axle of the disc also rotate, *i.e*, let there be a *precessional motion*, about an axis, perpendicular to the plane of the paper at a (precessional) rate φ; so that, after a small interval of time *dt*, the disc takes up the position *D'D'* making an angle φ *.dt* with its original position. Its angular momentum, again equal to *I*, is now represented by the straight line $OA' = OA^*$. The angular momentum thus changes only in direction and not in magnitude, just as in the case of circular motion of a body, only the direction of its tangential velocity changes and not its magnitude.

The change in the angular momentum of the disc is thus represented *vectorially* by

$$AA' = I\,\omega\,.\varphi.dt. \qquad [\because \text{ arc } = \text{radius} \times \text{angle}$$

This change has, clearly, been brought about in time *dt*, and, therefore,

rate of change of momentum of the disc $= I.\omega.\phi.dt\,/\,dt = I\omega.\phi$

And, since the rate of change of momentum of a rotating body is equal to the torque applied to it, we have

$$T_1 = I\omega.\phi, \quad \text{whence,} \quad \phi = T_1 / I\omega$$

where T_1 is the *precessional torque* applied to the disc.

Thus, the rate of precession is directly proportional to the applied *torque* T_1 and inversely proportional to the *angular momentum I*ω of the body.

Now, since the change in the angular momentum of the disc is along *AA'*, it is clearly parallel to its plane of rotation, or perpendicular to its axis of rotation, and *AA'* is thus the axis of the torque applied. In other words, the axis of the torque lies along *OX*.

Thus, we see that *if the axis of rotation of a body be along OY and the axis of the applied torque OX, the body* **'precesses'** *about the third mutually perpendicular axis OZ*. This will be readily understood from Fig. 3.35 (*a*), which shows the disc *in perspective*.

* Because the precessional rate is much smaller than the rate of rotation of the disc.

Here, *OY is the axis* of rotation and, therefore, *XOZ* is the plane of rotation; *OX is the axis of the torque or couple applied, and, therefore, YOZ is the plane of the torque and, since the axle of the disc turns towards OX, i.e., about the axis OZ, the plane of precession is XOY.* In other words, the axis of rotation (*OY*) turns in this plane, which is, clearly, perpendicular to the first two planes; —in fact, the three planes are mutually perpendicular. The direction of rotation of *OY* or its precessional motion, (towards *OX* here), obviously depends upon the direction of rotation of the disc and that of the torque or the couple applied and may be easily obtained from the fact that the angular momentum vector (or the spin vector), along the axis of rotation *YY'*, and the torque or the couple vector, along the axis *X'X*, both lie in the plane of precession *XOY* and since the axle of the disc (or the axis of rotation) (*OY*) turns towards *OX*, it follows that *the precessional motion occurs in a direction such that it tends to bring the spin vector (along YY') into the position of the couple vector (along X'X) by the shortest possible route.*

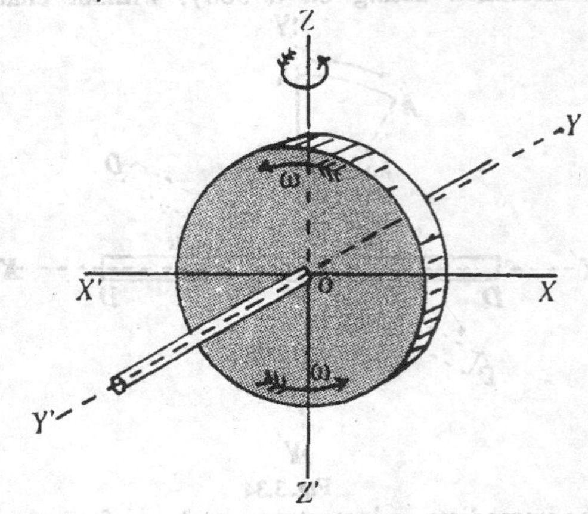

Fig. 3.35 (*a*)

3.18. Principle underlying a Gyrostat. A gyrostat is essentially a disc or a flywheel, having a large moment of inertia, rotating at a high speed about an axle, passing through its centre of mass, and mounted in some such manner as shown in Fig. 3.35 (*a*), so that the wheel and the axle are both free to turn, as a whole, about any axis, perpendicular to the axle itself, *i.e.*, about any of the three mutually perpendicular axes. In other words, it has three degrees of freedom.

As explained above in § 3.17, if a torque or couple be applied to the wheel, with its axis perpendicular to the axis of rotation of the wheel, the wheel '**precesses**' about the third mutually perpendicular axis at a precessional rate given by $\phi = T_1/I\omega$, ,where T_1 is the torque or couple applied, I, the moment of inertia of the wheel about its rotation-axis and ω, its angular velocity about this axis.

Clearly, therefore, for a given torque (T_1) applied to the wheel, *the precessional rate is inversely proportional (i) to the momentum of inertia of the wheel about its axle, and (ii) to the angular velocity of the wheel;* so that, *the larger the momentum of inertia of the wheel about its axle, and the higher its angular velocity, the smaller the rate of precession of the axle, and vice versa.*

The following simple experiment beautifully illustrates the above results :

Take a fairly large and heavy disc, [Fig. 3.35 (*b*)], free to rotate about the axis *YY'* passing through its centre, and fitted inside two sockets at the ends of the horizontal diameter of a bigger ring, suspended by means of a string vertically above its centre of gravity.

(*i*) Now, if with the disc quite stationary, a weight *Mg* be suspended at *Y'*, the torque due to it will tilt the ring, the end *Y'* moving down and the end *Y* moving up, *i.e.*, the ring will turn about *OX*.

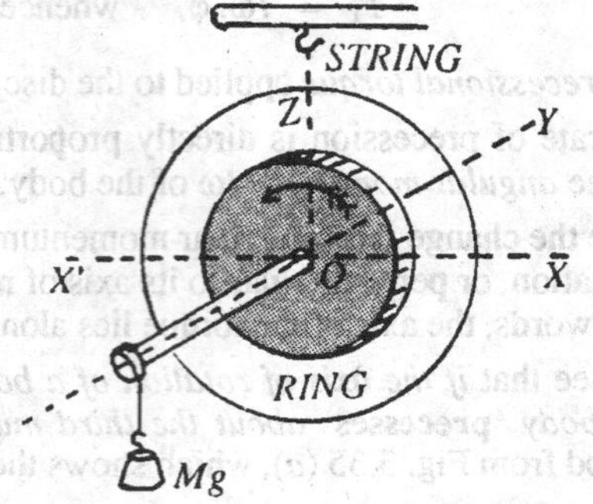

Fig 3.35 (*b*)

But, if instead of suspending the weight, the ring be simply pushed horizontally at Y' from in front or behind, it will turn about OZ.

(*ii*) Let the disc be now set into rotation about its axle, in the direction shown, with the weight Mg kept properly supported, so as to exert no downward pull at Y'. It will be found that the ring remains quite steady and a twist, given to the string either way, hardly produced any tendency in it to rotate about OZ, as it certainly would, if the disc were stationary.

(*iii*) With the disc in motion let the weight Mg be released, so as to exert a downward pull at Y', thus producing a torque about OX. It will be found that the ring at once rotates about OZ, with the end Y' slightly titled downwards. On pushing the ring horizontally at Y' as before, the axle, instead of turning more rapidly about OZ, as might be expected, simply gets titled a little, raising the weight Mg slightly upwards, *clearly showing thereby that the horizontal rotation of the axle* YY' opposes the torque due the Mg, which, therefore, descends comparatively slowly now. So long, however, as the downward descent of the weight continues, just so long does the rate of rotation of the axle about OZ also continue to increase thereby increasingly opposing the torque due to the weight, until a stage is reached where the two exactly balance each other. After this, the downward descent of the weight naturally ceases, and the ring continues to turn about OZ at a constant rate, with the axle YY' slightly tilted.

(*iv*) It will be found that the greater the moment of inertia of the disc about its rotation-axis or the axle, and the higher its angular velocity about it, the smaller the rate of rotation of the axle about OZ, *i.e.*, the smaller the precessional motion about it.

(*v*) Since the torque or a couple is needed to produce this precessional motion of the rotation-axis of the disc, it is clear that a rotationing body offers resistance to a precessional motion of its axis. This resistance to a precessional motion is called *gyrostatic resistance* and *is equal and opposite to the precessional torque, i.e., equal to* $I, \omega\phi$ (see page 81).

A common example of such *gyrostatic resistance* is met with in the case of a cyclist negotiating a curve on a level road. For, obviously, there is a regular precession of the axle of the wheel or the gyrostat) along the curved path taken by the cyclist. The rotating wheel naturally opposes this precessional motion of the axle and offers a *gyrostatic resistance* equal to $I\omega.\phi$ where I is the moment of inertia of the wheel about its axis of rotation, ω, its angular velocity and ϕ, the rate of precession (or the precessional velocity) of the axle. If v be the linear velocity of the wheel (or the cyclist) and R, the radius of the wheel, clearly, $\omega = v/R$. And, if r be the radius of the curved path negotiated by the cyclist, $\phi = v/r$. So that, we have

$$\textit{gyrostatic resistance offered by the wheel} = I\,(\,v\,/\,R\,)\,.\,(\,v\,/\,r\,) \;=\; I.v^2\,/\,R.r.$$

It is in an effort to overcome this gyrostatic resistance of the wheel that the cyclist leans inwards from the vertical, thereby producing an opposing couple due to his weight, as explained in 2.13, (4). So that, he leans inwards partly to counteract the couple due to the centrifugal force mv^2/r and partly to overcome the gyrostatic resistance of the wheels.

Exactly the same thing applies to an aeroplane or a motor boat taking a sudden swift turn, or to a motor car or a railway train negotiating a curve on however, besides being helpful in the manner explained in § 2.13, (4), also helps mitigate the effect of the gyrostatic resistance of the wheels.

3.19. The Gyrostat Proper. In actual practice, the gyrostat usually takes the form of a heavy circular disc, free to rotate at a high speed about an axle passing through its centre of mass and perpendicular to its plane, *i.e.*, coincident with its axis of symmetry, and is so mounted that the disc, together with the axle, can turn freely about any of the three mutually perpendicular axes.

Thus, in Fig. 3.36, disc G is the gyrostat, free to rotate about its axle, coincident with its axis of symmetry XX' and carried by a circular ring H which itself is free to turn about the axis YY' inside another ring V which, in its turn, is free to rotate about the axis ZZ' within a rigid framework F, —the three axes being mutually perpendicular and intersecting in the same point; so that, for any given position of frame F, the disc or the gyrostat can turn freely about any one of them.

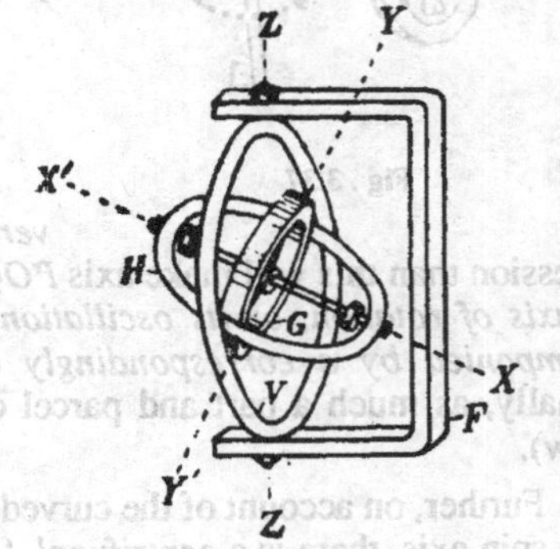

Fig. 3.36

Now from what we have discussed above, it is clear that, with the disc set rotating fast, (*i*) by any rotation of the frame (*F*), —the position and direction of the axis of rotation of the disc remains unaffected with respect to it, and (*ii*) any torque applied to the axis of rotation of the disc results in its precession, but, the rate of precession being inversely proportional to $I\omega$, the greater the moment of inertia of the disc and the higher its angular velocity, the smaller the precession of its axis. It follows, therefore, that *a heavy or a massive disc, rotating at a high speed, will make for greater stability of its axis of rotation.* Hence the use of gyrostats as stabilizers in ships and aeroplanes etc. (see § 3.23)

3.20. Gyroscope. In a majority of cases, a body, subject to precessional motion, is supported at a point, *away from the vertical line through its centre of gravity*, unlike those considered above, where the axis of rotation is free to turn about the *c.g.* of the body.

A gravitational torque or couple thus acts upon the body, which, *in its stationary condition*, simply tends to rotate it into a position of lower potential energy, *i.e.*, simply tends to lower its centre of gravity. But, *if the body be rotating about some axis*, this gravitational torque supplies the necessary processional torque equal in value of its own provided there is no other couple acting on the body. The rate of precession ϕ, maintained by this gravitational torque T_2, is given by the relation,

$$\phi = T_2 / I\omega,$$

where I and ω stand, as usual, for the moment of inertia of the body and its angular velocity about its axis of rotation.

Such a body, with its axis of spin supported at a point away from its c.g., and with the precessional rate of its spin-axis maintained by the gravitational torque (due to its weight) about that point is called a **gyroscope** *or a* **top**—such a precessional motion, maintained by a gravitational torque, being, therefore, appropriately termed *'gyroscopic'*.

Thus, consider a heavy disc *D*, revolving with a high angular velocity ω about its physical axis *POQ*, itself resting on a vertical pivot at *P*, (Fig. 3.37)

Then, clearly, its weight Mg, acting vertically downwards at its *c.g.*, *O*, exerts a *gravitational torque* T_2 on it, given by $T_2 = Mg.\ OP = Mg.l$. [Putting $OP = l$].

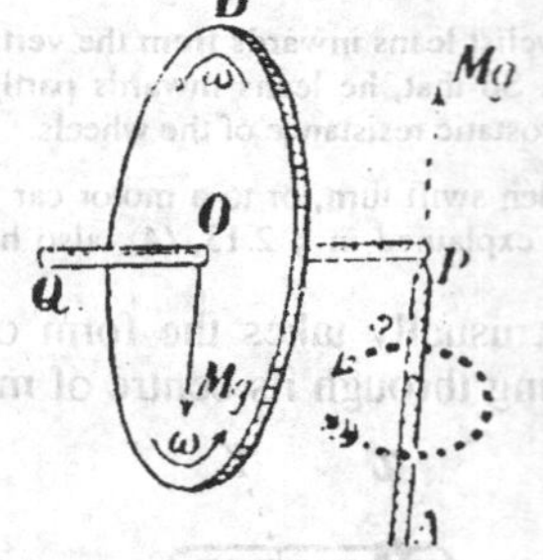

Fig . 3.37

So that, if ϕ be the *rate of precession of the disc* maintained by it, we have

$$\phi = T_2 / I.\omega = Mg.l / MK^2.\omega = gl / K^2.\omega,$$

putting $I = MK^2$, where K is the *radius of gyration* of the disc about the axis *POQ*.

Hence, if t be the time-period of its precessional motion, *i.e.*, if it takes time t to complete its one full cycle of processional motion, we have

$$t = \frac{2\pi}{\phi} = \frac{2\pi}{gl / K^2.\omega} = 2\pi\frac{K^2.\omega}{gl}$$

This precession, once started, can be maintained, *at this very rate*, by the gravitational torque *alone*. A higher rate of precession than this will make axis *POQ* rise and a lower rate will make it fall. *This rise and fall of the axis of rotation, or its oscillation up and down about its position of dynamic equilibrium, accompanied by a correspondingly changing precessional rate, is termed* **nutation** and is, normally, as much a part and parcel of gyroscopic motion as rotation and precession (see note below).

Further, on account of the curved path taken by the disc as a result of the precessional motion of its spin axis, there is a *centrifugal* force acting on the disc along *POQ* and an equal *centripetal* force in the opposite direction *QOP*, their net effect, if they act along the same line, being to increase the frictional resistance at the pivot P. If, however, their lines of action be different, we have yet another couple T_2, formed by them, aptly known as the *centrifugal torque*.

In order to prevent the disc, or the processing body, in general, from moving outwards from the centre of precession, it is necessary that the centrifugal torque on it must be balanced by an equal and opposite centripetal torque, this balancing effect being supplied by part of the gravitational torque, the remaining part of it producing precession. Thus, if T_3 be the centripetal torque and T_1 and T_2, the gyrostatic and gravitational torques, we have

$$\mathbf{T_2 - T_3 = T_1} \quad ...(1)$$

where the different torques are given their proper signs, (*i.e.*, anticlockwise, positive and clockwise, negative,) all acting in the same direction in the case shown.

A general rule to determine the sense of the torque, producing precession in a given direction, which is called **Lanchester's rule**, stated as follows :

If the gyroscope be viewed from a point in its own plane, with the line of sight perpendicular to the axis of the given precession, it is seen to describe an ellipse, the sense of whose path gives the direction of the precessional torque with the line of sight as its axis.

Note. We have said above that nutation also is *normally* a part and parcel of gyroscopic motion, like rotation and precession. For, it is quite possible to have precession, *without nutation*, in particular cases, - such cases being referred to as those of *pure precession*. Thus, for example, if , in Fig. 3.37 the outer end Q of the axle of the rotating disc be given a push in the normal direction of its precession, it would result in an increased pressure of the axle on pivot P and a consequent increase in the reactional force there. This force equals Mg, so that the vertical forces balance each other, with the rotation-axis of the disc attaining its position of dynamic equilibrium. If the end Q of the axle be released at this particular instant, it will continue to process without any rise or fall about this position. In other words, we shall have *precession, without nutation*.

3.21. The Gyrostatic Pendulum.

A gyrostatic pendulum is a small and heavy disc or gyrostat (D), revolving with uniform angular velocity (ω) about a light rigid rod, (SD) as axis, this axis coinciding with its axis of symmetry and *precessing* about the vertical (SO) at a uniform rate (ϕ) as shown in Fig. 3.38.

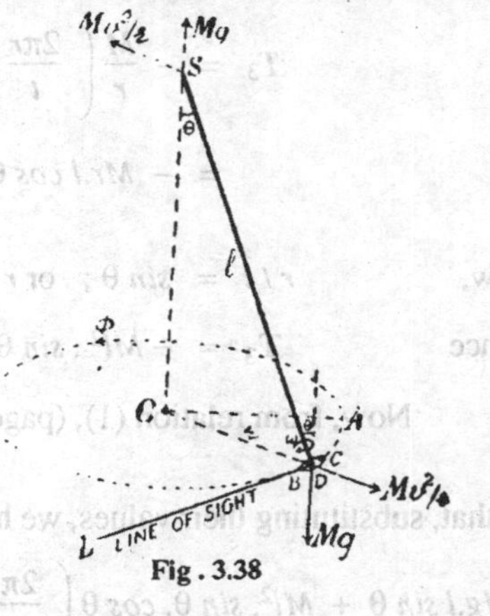

Fig. 3.38

Obviously, there are the following torques acting on the pendulum :

(*i*) **A gyrostatic torque, T_1**, due to the *gyrostat* or disc D possessing two simultaneous rotatory motions.

Since the plane of rotation of D is always perpendicular to the rod SD, *its rate or precession* ϕ is the same as the angular velocity of SD, *i.e.*, equal to v/l. For, in time dt, SD traces out an arc $v.dt$, where v is its *line are velocity* in the horizontal circle of radius r, which is described by it in its precessional motion about SO; or, the angle described by it in time dt is equal to $v.dt/l$, and hence the angle described by it in unit time is clearly equal to $(v.dt/l) \div dt = v/l$.

Thus, $\phi = v/l$.

But, we know that $\phi = T_1/I\omega$. So that, $T_1/I\omega = v/l$, [See page 88

whence, $T_1 = I.\omega.v/l = MK^2.\omega.v/l.$

where K is the radius of gyration of the gyrostat about the axis SD.

To determine the direction of this torque, let us apply *Lanchester's rule, i.e.*, let us look at a point B on the edge of the disc along LB, where LB is perpendicualr to both OD and SD, when B clearly appears to move in the anticlockwise direction, indicating that the direction of the torque T_1 is *anticlockwise*.

If t be the *periodic time of precession* of the disc or the gyrostat, *i.e.*, the *time-period of the gyrostatic pendulum*, or the time taken by the disc to complete one round along the horizontal circle, we have $v.t = 2\pi r$, or $v = 2\pi r / t$. So that, substituting this value of v in the relation for T_1 above, we have $T_1 = MK^2\omega.2\pi r / t.l$.

And, since $r/l = \sin\theta$, we have $\mathbf{T_1} = + MK^2.\omega.\sin\theta.(2\pi / t)$,

the +ve sign indicating that its direction is *anticlockwise*.

(*ii*) **A gravitational torque $\mathbf{T_2}$**, due to the weight Mg of the gyrostat acting vertically downwards at D, (where its whole mass is supposed to be concentrated).

Clearly, the moment of this gravitational torque $= Mg \times BO = Mg.SD \sin\theta = Mg.l \sin\theta$, where BO is the perpendicular distance between Mg and an equal and opposite reaction at S ; l, the length of the rod SD and θ, the angle that it makes with the vertical.

So that, $\mathbf{T_2} = - mg.l \sin\theta$,

the –ve sign indicating that its direction is *clockwise*.

(*iii*) **A centripetal torque $\mathbf{T_3}$**, due to the centrifugal force Mv^2/r, acting on the gyrostat, outwards along OD.

And, the *moment of this torque* T_3 is obviously equal to $\dfrac{Mv^2}{r}.SO$, where SO is the perpendicular distance between Mv^2/r, and an equal and opposite reaction at S. Or, since $SO = l\cos\theta$, we have

$$T_3 = -\frac{Mv^2}{r}.l\cos\theta,$$

the – ve sign again indicating that the direction of the torque is *clockwise*.

Or, substituting the value of $v = 2\pi r/t$ [see under (*i*) above], we have

$$T_3 = -\frac{M}{r}\left(\frac{2\pi r}{t}\right)^2.l\cos\theta = -\frac{M}{r}.r^2\left(\frac{2\pi}{t}\right)^2.l\cos\theta$$

$$= -Mr.l\cos\theta\left(\frac{2\pi}{t}\right)^2$$

Now, $r/l = \sin\theta$; or $r = l\sin\theta$.

Hence $$\mathbf{T_3} = -Ml^2.\sin\theta.\cos\theta\left(\frac{2\pi}{t}\right)^2$$

Now, from relation (1), (page 85). we have $-\mathbf{T_2} - (-\mathbf{T_3}) = +\mathbf{T_1}$,

Or, $-\mathbf{T_2} + \mathbf{T_3} = \mathbf{T_1}$.

So that, substituting their values, we have

$$-Mg.l\sin\theta + Ml^2.\sin\theta.\cos\theta\left(\frac{2\pi}{t}\right)^2 = MK^2\omega.\sin\theta\left(\frac{2\pi}{t}\right)$$

Or, $$-gl + l^2\cos\theta\left(\frac{2\pi}{t}\right)^2 = K^2\omega\left(\frac{2\pi}{t}\right).$$

Now, putting $(2\pi/t) = p$, we have $-gl + p^2l^2\cos\theta = pK^2\omega$.

Or, $p^2l^2\cos\theta - pK^2\omega - gl = 0$, a quadratic equation in p

Therefore $$p = \frac{+K^2\omega \pm \sqrt{K^4\omega^2 + 4gl^3.\cos\theta}}{2l^2.\cos\theta}$$

which, obviously, gives two values of p.

To decide between the two values, we put $\omega = 0$, so that there is no rotation of the disc about *SD*, and *the whole arrangement reduces to* a *conical pendulum*, with

$$p = \frac{\pm\sqrt{4gl^3 \cdot \cos\theta}}{2l^2 \cdot \cos\theta} = \pm\sqrt{\frac{4gl^3 \cdot \cos\theta}{4l^4 \cdot \cos^2\theta}} = \pm\sqrt{\frac{g}{l\cos\theta}}$$

i.e.
$$\frac{2\pi}{t} = \pm\sqrt{\frac{g}{l\cos\theta}}.$$

But, since $2\pi/t$ must necessarily be *positive* for a conical pendulum, the negative value becomes inadmissible and, we, therefore, have $2\pi/t = \sqrt{g/l\cos\theta}$

It follows, therefore, that, in the expression for p above, only the positive value must be taken. So that,

$$p = \frac{2\pi}{t} = \frac{+K^2\omega + K^4\omega^4 + 4gl^3 \cdot \cos\theta}{2l^2 \cdot \cos\theta}$$

whence,
$$t = \frac{4\pi l^2 \cos\theta}{K^2\omega + \sqrt{K^4\omega^2 + 4gl^3 .\cos\theta}}.$$

3.22. Case of a Rolling Disc or Hoop. – A simple and a familiar example of gyroscopic motion is that of a thin circular *disc* or a *hoop*, set rolling over a plane horizontal surface. If its velocity be large enough, it continues to roll along a straight path in a stable vertical position. But, as its velocity decreases, due to friction between *t* and the horizontal surface, its plane inclines progressively to one side and its path becomes curved towards the '*side of lean*', the curvature of the path constantly increasing with decrease in its velocity, so that it follows a spiral path until, finally, it falls *flat* on the surface. Let us study this motion of the disc in some detail.

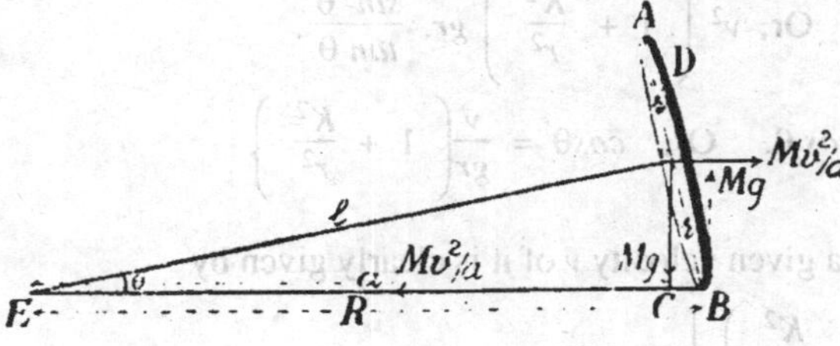

Fig. 3.39

Let D be the circular disc, (Fig. 3.39), of mass M and radius r, rolling along a horizontal surface with v as the *linear velocity* of its centre O, and with its plane AB inclined at an angle θ to the vertical. Then, the three torques acting on it are :

(*i*) *Gyrostatic torque* $\mathbf{T}_1$, due to its simultaneous rotation about its point of contact and about E, such that $T_1 = I.\omega.\theta$, where I is its moment of inertia about the axis OE through its centre O and perpendicular to its plane; ω, its *angular velocity about the same axis* and ϕ, *its rate of precession.*

Now, $I = MK^2$, where K is its *radius of gyration* about OE,

$\omega = \upsilon / r$ and $\phi = \upsilon / l$ [See § 3.21 (*i*)

So that,
$$\mathbf{T}_1 = -MK^2 \cdot \frac{v}{r} \cdot \frac{v}{l} = -MK^2 \cdot \frac{v^2}{r^2} \cdot \frac{r}{l}$$
$$= -MK^2 \cdot \frac{\upsilon^2}{r^2} \cdot \tan\theta,$$

whwere $r/l = \tan\theta$, and the – *ve* sign of $\mathbf{T}_1$ shows that the torque is *clockwise*.

(*ii*) *Gravitational Torque* T_2, due to its weight Mg, acting vertically downwards at O, such that

$$T_2 = -Mg.CB = -Mg.r \sin\theta$$

[∵ in the right-angled ΔOCB, $\sin\theta = CB/OB = CB/r$]

also acting in the clockwise direction.

(*iii*) *Centrifugal torque* T_3, due to its rotation about E, such that

$$T_3 = \frac{-Mv^2}{a}.OC = \frac{-Mv^2}{l\cos\theta}.r\cos\theta,$$

[∵ in the right-angled ΔOCE, $\cos\theta = a/l$, and in the rt. angled ΔOCB $\cos\theta = OC/r$.]

where $EC = a$.

Or, $T_3 = -Mv^2 . r/l = -Mv^2 . \tan\theta$,

the *–ve* sign again indicating the *clockwise* direction of the torque.

Substituting these values of T_1, T_2, and T_3 in relation (*I*), page 85 we have, for equilibrium,

$$-Mg.r\sin\theta - (-M^2 . \tan\theta) = -MK^2 . \frac{v^2}{r^2} . \tan\theta$$

Or, $$-Mg.r\sin\theta + mv^2 . \tan\theta = MK^2 . \frac{v}{r^2} . \tan\theta$$

Or, $$v^2\tan\theta + K^2 . \frac{v^2}{r^2} . \tan\theta = gr.\sin\theta.$$

Or, $v^2 \tan\theta\left(1 + \frac{K^2}{r^2}\right) = gr.\sin\theta$. Or, $v^2\left(1 + \frac{K^2}{r^2}\right) gr. \frac{\sin\theta}{\tan\theta}$.

Or, $v^2\left(1 + \frac{K^2}{r^2}\right) = gr.\cos\theta$. Or, $\cos\theta = \frac{v}{gr}\left(1 + \frac{K^2}{r^2}\right)$

whence, the angle of '*lean*' of the disc for a given velocity v of it is clearly given by

$$\theta = \cos^{-1}\left[\frac{v^2}{gr}\left(1 + \frac{K^2}{r^2}\right)\right] \qquad \ldots (A)$$

and its velocity by $v^2 = \frac{gr\cos\theta}{1 + K^2/r^2}$ for equilibrium in the *leaning* position.

Now, for the **critical velocity** v_c, ***i.e., the minimum velocity at which the disc can move along a straight path, with its plane vertical,*** clearly $\theta = 1$, so that $\cos\theta = 1$.

And, $\therefore\ v_c{}^2 = \frac{gr}{1 + K^2/r^2}$ Or, $v_c = \sqrt{\frac{gr}{1 + K^2/r^2}}$

For a value of v less than v_c, the upright position would obviously be unstable; for, on the slightest displacement, it will be tilted over by the force of gravity until θ attains the value given above, (by expression A), corresponding to the leaning position.

Now, for a (*uniform*) *disc*, $K^2 = r^2/2$. Hence,

for a *disc*, $$v_c = \sqrt{\frac{gr}{1 + r^2/2r^2}}\sqrt{\frac{gr}{3/2}} = \sqrt{\frac{2}{3}}\,gr.$$

And, for a *hoop*, $$v_c = \sqrt{\frac{gr}{1 + r^2/r^2}} = \sqrt{\frac{gr}{2}}$$ [∵ for a *hoop*, $K^2 = r^2$.]

Let us now calculate the radius of curvature R of the *path of the disc on the horizontal surface*. It is clearly equal to $EB = R$.

And, in the right-angled triangle EOB, we have

$$\sin\theta + OB/EB = r/R \; ; \text{ so that, } r = R\sin\theta.$$

$$\therefore \quad R = r/\sin\theta = r/\sqrt{1-\cos^2\theta}.$$

Or, substituting the value of cos θ deduced above, we have

$$R = \frac{r}{\sqrt{1 - \frac{v^4}{g^2r^2}\cdot\left(1+\frac{K^2}{r^2}\right)^2}}$$

So that, for a disc, for which $K^2 = r^2/2$, we have

$$R = r/\sqrt{1 - 9v^4/4g^2r^2},$$

and for a hoop, for which $K^2 = r^2$, we have

$$R = r/\sqrt{1 - 4v^4/g^2r^2}.$$

Note. In the above discussion, while calculating the gravitational and centrifugal torques, we have, for the sake of simplifying the treatment, somewhat tentatively assumed that the mass of the body is concentrated at its centre. This is obviously not generally true. For, only those points that lie on the axis of rotation describe circular paths about the axis of precession, with the radius of the path of each point different. The motion of other points is complicated on account of their simultaneous motion about the axis of rotation as well as that of precession. For a fullar and a more rigorous treatment of the problem, therefore, all this much be taken into proper account. It can be shown that calculated thus, *the true value of the critical velocity for a straight path* is given by

$$v_c = \sqrt{gr^3/2(K^2 + r^2)}$$

So that for a disc $(K^2 = r^2/2)$, $v_c = \sqrt{gr/3}$

and for a harp, $(K^2 = r^2)$, $v_c = \frac{1}{2}\sqrt{gr}$.

3.23. Gyrostatic and Gyroscopic Applications. The tendency of a rapidly rotating disc or wheel (and, in fact, any rigid body), to preserve its axis of rotation endows a gyrostat with a stability of direction, which is made use of in a number of ways for the steadying or *stabilising of motions*. The basic principle underlying this is that the gyroscope must be made to precess in such a manner that the reaction couple due to the rotor (*i.e*, the disc or the wheel) opposes the disturbing couple acting on the frame. Completer stabilisation can be obtained by ensuring the equality of the two couples at every instant. Among the more important and familiar examples of gyroscopic stabilisation may be mentioned the following :

(*i*) The Gyrostatic or the Gyro-Compass. It is a special type of compass used in aeroplanes and ships, and, more particularly, in submarines. In essentials, it consists of a disc or flywheel, of a large moment of inertia, (*i.e.*, a gyrostat), suspended in friction less *gimbals* inside a supporting frame, which is kept rotating at a high speed by means of an electric motor about a horizontal axis, lying in the geographic meridian (*i.e*, in the vertical plane passing through the geographic north and south of the earth). Its directional stability and the conservation of its angular momentum make its axis always lie in the direction of the meridian, *i.e.*, *along the geographic north and south*. And, since the arrangement in such that the disc or the flywheel has three degrees of freedom, irrespective of the position of the supporting frame, a movement of the latter produces no deflecting torque or couple on it, and *this particular direction of its axis continues to be maintained in space all the time, despite any changes in the direction of the ship or the submarine, or any tossings or patchings of it*. It is, therefore, preferred to the ordinary magnetic compass and is more dependable than the latter, in view of the additional advantage of its remaining altogether unaffected by any type of magnetic disturbances.

The Pendulum Gyro-compass. The above arrangement, with a small mass, suitably suspended below the rotating disc or flywheel constitutes what is called the *pendulum Gyro-compass—the small mass supplying the necessary restoring torque* to bring its axis back to its original direction, should it get displaced due to some disturbance. In the absence of this simple but ingenious device, the instrument would lack its restorative action, due to the inherent stability of a gyrostat in any position.

The essential features of the construction of the *Pendulum Gyro-compass* will be clear from Fig. 3.40 where the rotating disc or gyrostat *D* has its axle *PQ* mounted in a horizontal ring *R*, free to rotate about the axis *EF* inside a vertical ring *C* which, in its turn, rotates freely about the axis *AB* within a frame-work *M*, carried on horizontal gimbals, (of which *GG* forms one pair), to ensure the fullest freedom of movement.

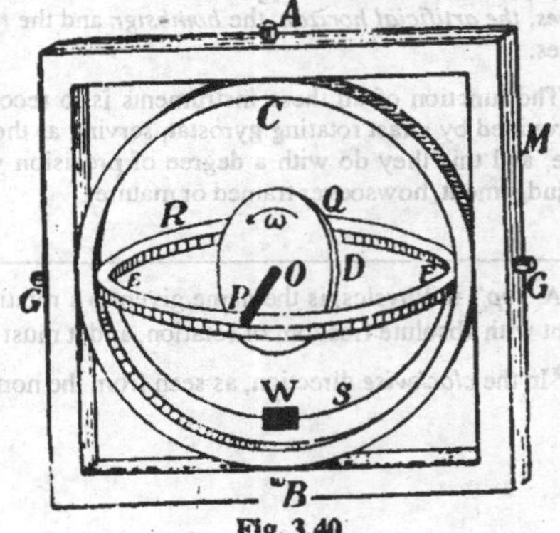

Fig. 3.40

The horizontal ring *R* has a stirrup *S*, fixed rigidly to it, which is loaded with a weight *W*, immediately below *O*, the centre of the disc or the gyrostat.

It can be shown that this arrangement would be stable, at any given place, *only along the true north and south, i.e.*, when the end *P* points truly north,.—any accidental displacement of it calling into play a directive force (due purely to the rotation of the earth), restoring it back to its original direction, and making it a self-righting instrument.

(*ii*) **Rifling of barrels of Guns and Rifles.** This another well-known application of the directional stability of a rapidly revolving body. For, it is found that if a shot or a bullet be given a rapid *'spin'*, about an axis along its direction of motion, its uniformity of flight is greatly improved by making it less responsive to small deflective forces during its passage through air. This is achieved by *'rifling'* the barrel, *i.e.*, by cutting spiral grooves inside it so that the shot or the bullet is first forced to move along these before it emerges out into the air, thus acquiring the necessary *'spin'* to ensure an almost uniform linear motion.

(*iii*) **Riding of Bicycles and Rolling of Hoops or Discs.** These are both cases of what is called *statical instability*; for, neither of the two, at rest, can possibly remain in equilibrium in the position in which it does when it is in motion. Here, again, it is the gyroscopic action that does the trick, by appropriately deflecting their axes of rotation and thereby changing their planes of rotation, to counterbalance of disturbing effect due to gravity.

Thus, when a person rides a bicycle, without holding its handle, he has simply to tilt to one side in order to turn to *that* side; of, by so doing, he produces a couple about the horizontal direction of motion of the front wheel of his bicycle, *which here acts as a rotating gyrostat.* This couple, then, turns the axle of the wheel about the vertical, and hence its plane of rotation, into the desired direction.

(*iv*) **Stabilising of Ships.** Although a ship is intrinsically stable, it is subject to disturbing oscillations in a rough or a heavy sea, both about a transverse and a longitudinal axis. The former type of oscillation is called **pitching** and the latter, **rolling**. Since the amplitude in the case of pitching is much smaller than in the case of rolling, *it is desired to minimise rolling rather than pitching*.

The couple that causes rolling in a ship is a direct consequence of the difference in the buoyancy on the two sides of the centre line of the ship when the ship happens to lie on a wave slope. Periodic in nature, the couple has its maximum value when the ship lies on either side of a wave at the point of maximum slope and has its minimum or zero value when the ship lies on the crest or the trough of a wave. In order that the ship may remain on an even keel, the gyroscope must, at every instant, exert a reaction couple on the ship equal and opposite to that exerted by the wave on it. The exact balancing of the two couples is not, however, possible at every instant, so that the rolling cannot be completely eliminated and the best that can be done is to limit its amplitude. This is brought about by arranging the vertical shaft of the gyroscope to lie in a casing supported in bearings, fixed to the frame of the ship, thus allowing for a precessional motion of the axis of rotation through about 60° on either side of the vertical, the casing itself being driven by an electric motor at an angular velocity which remains constant along most of the path of precessional motion.

(*v*) **Precession of the Equinoxes.** The earth, as we know, is not an exact sphere, but bulges out slightly at the equator, (or has the shape of a *'flattened ellipsoid of revolution'*). Further, the *Sun* and the *Moon* do not usually lie its equatorial plane but rather in the plane of the ecliptic, which is inclined at an angle of 23.5^0 to the former, with the result that the gravitational attraction due to the Sun and the Moon on this equatorial bulge gives rise to a torque, bringing about the precession of the axis of the earth which, acting at a gigantic top*, describes a *cone*** *, relative to the fixed stars, e.g., the pole stars, similar in manner to the cone described by the axis of a precessing top, due to its weight*,—the phenomenon being spoken of as the *'precession of the equinoxes'*. This couple on the earth due to the attractive force of the Sun and the Moon is, however, very small, so that it takes 25,800 years for the earth's axis to describe the complete cone, at which rate of rotation, the star *Vega* will be the pole star in about 12,000 years hence.

It is interesting to observe that atoms too have the mechanical properties of tops, and, at least in one special case their gyrostatic moment has been demonstrated experimentally by *Einstein* and *De Hass*.

(*vi*) **Other Recent Applications.** The modern aircraft appliances, like the *automatic pilot, the steering of torpedoes, the artificial horizon*, the *bombsign* and the *turn and bank indicators ect.*, all depend for their action of *gyrostatic* principles.

The function of all these instruments is to record the effects of a change or orientation between a relatively fixed plane provided by a fast rotating gyrostat, serving as the reference or the *datum plane*, and some other movable plane in the machine, and this they do with a degree of precision which makes their indication far more safe to rely upon than mere human judgement, howsoever trained or mature.

---

* A *'top'*, in Physics, is the name given to a rotating body, either completely free to move, or fixed at the most at just *one* point with absolute freedom of rotation, and it must not, therefore, be confused with the toy that goes by that name.

**In the *clockwise* direction, as seen from the north.

WORKED EXAMPLES

1. A flywheel, of mass 500 kg and 2 metres diameter, makes 500 revolutions per minute. Assuming the mass to be concentrated at the rim, calculate the angular velocity, the energy and the moment of inertia of the flywheel.

(*i*) No. of revolutions made by the flywheel *per minute* = 500

Angle described in one rotation = 2π *radians*.

$\therefore$ Angle described by the flywheel *per minute* = $2\pi \times 500$ *radians*.

Hence, *angle described per second* or *angular velocity* ω of the flywheel

$$= 2\pi \times 500/60 = 50\,\pi/3 \; rad/sec.$$

(*ii*) Mass M of the flywheel = 500 *kg* and its radius of gyration $K = 2/2 = 1$ *metre* because its mass is concentrated at a distance *equal to its radius* from the axis of rotation, which passes through its centre.

$\therefore$ *Moment of inertia of the flywheel*, $I = Mk^2 = 500 \times 1 = 500\ kg\ m^2$.

(*iii*) The energy possessed by the flywheel is due to its *radians i.e., it possesses only rotational energy* which is equal to $\frac{1}{2}I\omega^2$

So that, *energy of the flywheel* $= \frac{1}{2} \times 500 \times (50\,\pi/3)^2 = 6.855 \times 10^4$ *joules*

2. A flywheel weighs 10 tons, and the whole of the weight may be considered as concentrated at a distance 3 ft from the axis. What is the amount of energy stored in the flywheel when rotating at a speed of 100 revolutions per minute ? *(Punjab)*

Here, $M = 10$ *tons* $= 10 \times 2240$ *lb*; $= 100 \times 2\pi$ *radians/min.* $= 100 \times 2\pi/60 = 10\,\pi/3$ *radians/sec.*

and $K = 3\,ft.$ (*Given*)

Since *K.E.* of rotation of a body $= \frac{1}{2} \times I\omega^2 = \frac{1}{2}MK^2\omega^2$, we have

K.E. of the flywheel $= \frac{1}{2} \times 10 \times 2240 \times 3^2 \times (10\,\pi/3)^2$.

$$= 5 \times 2240 \times 9 \times 100 \times \pi^2/9\,ft \text{ poundals}$$

$$= \frac{5 \times 2240 \times 9 \times 100 \times \pi^2}{9 \times 32} \text{ ft. pounds.}$$

Or, kinetic energy stored up in the flywheel $= 34.55 \times 10^4$ *ft lb.*

3. A flywheel of mass 100 kg and radius of gyration 20 cm is mounted on a light horizontal axle of radius 2 cm, and is free to rotate on bearings whose friction may be neglected. A light string wound on the axle carries at its free end a mass of 5 kg. The system is released from rest with the 5 kg mass hanging freely. Prove that the acceleration of this mass is g/2001 cm/sec².

If the string slips off the axle after the weight has descended 2g metres, prove that a couple of moment 31.8 kg wt. θ = cm. (approximately) must be applied in order to bring the flywheel to rest in 5 revolutions. *(Cambridge H.S. Certificate)*

(*i*) The *mass* of the flywheel (M) = 200 *kg* = 100×1000 *gm*.

and its *radius of gyration* (K) = 20 *cm*.

So that, *Moment of inertia* (I) $= MK^2 = 100 \cdot 1000 \times 20^2$

$$= 100 \times 1000 \times 400 = 4 \times 10^7\ gm.\ cm^2.$$

Let angular acceleration of the flywheel be $= d\omega/dt$.

Then, linear acceleration of the mass of 5 *kg* $= r.d\omega/dt$,

where, r is the radius of the axle.

Or, acceleration of the mass, *i.e.*, $a = r.d\omega/dt$,

whence, $d\omega/dt = a/r = a/2$.

$\therefore$ rotational couple acting on the flywheel $= I.d\omega/dt$

$$= 4 \times 10^7 \times a/2\ gm.\ wt.\ \text{-}cm. = 2 \times 10^7 \times a\ gm\ wt.\ cm.$$

This must, obviously, be equal to the couple applied to the wheel by the tension in the string.

If T be the *tension* in the string, the couple due to it $= T.r$.

If the mass of 5 *kg*. had no acceleration, the tension in the string would be equal to its *weight* $= 5 \times 1000 \times g$ *dynes*.

But, since it has an acceleration a, we have

$ma = mg - T$. Or, $T = m(g-a)$.

$\therefore$ *Couple applied to the wheel* $= T.r = m(g-a)\,r$.

$$= 5 \times 1000\,(g-a) \times 2 \text{ dynes cm.}$$

Now, $I.\omega/dt = T.r.$

$\therefore \quad 2\times 10^7 a = 5\times 1000\,(g-a)\times 2 = 10000g - 1000a.$

Or, $2\times 10^7 a + 10000\,a = 10000\,g.$ Or, $a(2\times 10^7 + 10000) = g\times 10^4.$

whence, $a = \dfrac{g\times 10^4}{(2\times 10^7 + 10000)} = \dfrac{g\times 10^4}{2001\times 10^4}$. Or, $a = g/2001\ cm/sec^2.$

Thus, the acceleration of the mass = $g/2001\ cm/sec^2$.

(*ii*) When the weight has descended 2 *metres*, it has lost some *P.E.* This must be equal to the gain in *K.E.* of the wheel and the weight; so that,

K.E. of the wheel + K.E. of the weight = P.E. lost by the weight.

Now *P.E.* lost by the weight = work done by it in falling through 2 *metres* distance = *mgh*, (where *h* = 2 metres = 200 *cm*) = $5\times 1000\times 981\times 200$ *ergs.*

$\therefore$ *K.E.* of the wheel and the weight = $1000000\times 981 = 981\times 10^6$ *ergs.*

This must, therefore, be the work that the couple applied to the wheel must do in order to stop it. If *C* be the couple required for the purpose, we have — work done by the couple = $C\,\theta$, (where θ is the angle of ratation)

Since the wheel comes to rest after 5 revolutions, it describes an angle $\theta = 2\theta\times 5$ *radius.*

So that, *work done by the couple*; $= 2\pi\times 5\times C$. And, $2\pi\times 5\times C = 981\times 10^6$.

Or, $C = \dfrac{981\times 10^6}{2\pi\times 5} = \dfrac{981}{\pi}\times 10^5\ dynes\text{–}cm. = \dfrac{981\times 10^5}{981\times\pi} + \dfrac{10^5}{\pi}\ gm.\ wt\ cm$

$= \dfrac{10^5}{1000\times\pi} = \dfrac{10^2}{\pi}\ kg.wt\text{–}cm. = 31.83\ kg\ wt\text{–}cm.$

Hence, a couple of moment 31.83 *kg wt-cm* will bring the flywheel to rest in 5 revolutions.

4. A flywheel of weight 200 lb, which may be regarded as a uniform disc of radius 1 ft, is set rotating about its axis with an angular velocity of 5 revolution per second. At the end of 40 sec, this velocity, owing to the action of a constant frictional couple, has dropped to 4 revolutions per second. What constant couple must now be applied so that in further 20 sec, the angular velocity will be 8 revolutions per second? Find the total angle turned through during the minute.

(Cambridge Higher Secondary School Certificate)

Here mass of the flywheel, *M* = 2000 *lb*, and its radius *r* = 1 *ft*.

Since it is a uniform circular disc, its moment of inertia about its axis,

i.e., $I = \frac{1}{2}Mr^2 = \frac{1}{2}.200\times l^2 = 100\ lb\ ft^2.$

Its angular velocity, to start with = 5 *revolutions/sec.* = $2\pi\times 5$ *radians/sec.*

and „ „ „ after 40 *sec* = 4 *revolutions/ sec.* = $2\pi\times 4$ *radians/sec.*

$\therefore$ change in angular velocity in 40 *sec.* = $2\pi\times 5 - 2\pi\times 4 = 2\pi$ *radians/sec.*

$\therefore$ rate of change of angular velocity = $2\pi/40 = \pi/20$ *radians/sec*^{2}.

Or, angular *retardation, i.e.,* $d\omega/dt = \pi/20$ *radian/sec*^{2}.

Now, couple = $I.d\omega/dt$.

$\therefore$ frictional couple acting on the wheel = $100\times\pi/20 = 5\pi$ *poundal-ft.*

Again, the velocity of the wheel is now desired to be raised form 4 to 8 revolutions *per sec.*, in 20 *seconds*.

$\therefore$ initial angular velocity = $2\pi\times 4$ *radians/sec.*

And, final angular velocity in 20 *sec* = $2\pi\times 8$ *radians/sec.*

$\therefore$ change in velocity in 20 *sec.* = $2\pi\times 8 - 2\pi\times 4 = 8\pi$ *radians/sec.*

And, $\therefore$ rate of change of velocity = $8\pi/20 = 2\pi/5$ *radians/sec*^{2}.

Or, angular acceleration, $d\omega/dt = 2\pi/5$ *radians/sec*^{2}.

And, the couple required to produce this acceleration = $100\times 2\pi/5 = 40\,\pi$ *poundal ft.* [$\because$ *couple* = $I.d\omega/dt$.

$\therefore$ *total couple required to be applied = this couple of* 40π + *a couple of* 5π (*to overcome frictional couple*).

Or, *total couple required* = $40\pi + 5\pi = 45\pi = 141.4$ *poundal–ft.*

Now, let the angle described by the wheel in the first 40 *sec*, be θ_1

Then, from the relation $\theta_1 = \omega t + \frac{1}{2}.\frac{d\omega}{dt}.t^2$, we have [See page 76.

$\theta_1 = (2\pi\times 5)\times 40 - \frac{1}{2}.\frac{\pi}{20}.(40)^2 = 400\pi - 40\pi = 360\pi$ *radians*

And, if θ_2 be the angle described by the flywheel in the next 20 *sec.*, we have, as above,

$\theta_2 = (2\pi\times 4)\times 20 + \frac{1}{2}.\frac{2\pi}{5}.(20)^2 = 160\pi + 80\pi = 240\pi$ *radians.*

$\therefore$ The *total angle turned through by the flywheel in one full minute*

$$= \theta_1 + \theta_2 = 360\pi + 240\pi = 600\,\pi \text{ radians.}$$

5. A pulley of radius 2 ft has hanging from it, a rope with masses of 60 lb and 52 lb attached to its two ends, the masses being kept at rest initially by holding one of them. If the moment of inertia of the pulleys be 320 lb-ft², what will be the velocity of the masses, when they have moved a distance of 6 ft from their position of rest? It may be assumed that there is no slip between the rope and the pulley and that friction at the axle of the pulley is negligible.

Here, obviously, the *motive force, i.e.,* the force which makes the masses and the pulley move, is the *weight of the excess mass of* (60–52) or 8 *lb wt* at one end of the rope = 8 × 32 = 256 *poundals*, (Fig. 3.41).

When the masses have moved through a distance of 6 *ft the loss of potential energy* suffered by this excess mass is clearly = 256 × 6 = 1536 *ft. poundals.*

This loss of *P.E.* of the excess mass is equal to the gain in the *K.E.* of the system consisting of the two masses and the pulley.

Let v be the velocity of the masses at this instant. Then, *K.E.* of the two masses

$$= \frac{1}{2}(60+52)\,v^2 = \frac{1}{2} \times 112 \times v^2 = 56v^2 \text{ ft. poundals.}$$

And, *K.E.* of the pulley

$$= \frac{1}{2}\,I\omega^2 = \frac{1}{2} \times 320 \times v^2/r^2 = \frac{1}{2} \times 320 \times v^2/4.$$

$$= 40v^2 \text{ ft poundals} \quad [\because\ \omega = v/r \text{ and } v = 2\,ft.$$

$\therefore$ total gain in *K.E.* of the system = $56v^2 + 40v^2 = 96v^2$ *ft poundals.*

Since *gain in K.E. of the system = loss in P.E. of the excess mass,*

we have $96v^2 = 1536$. Or, $v^2 = 1536/96 = 16$,

whence, $v = \sqrt{16} = 4$ *ft/sec.*

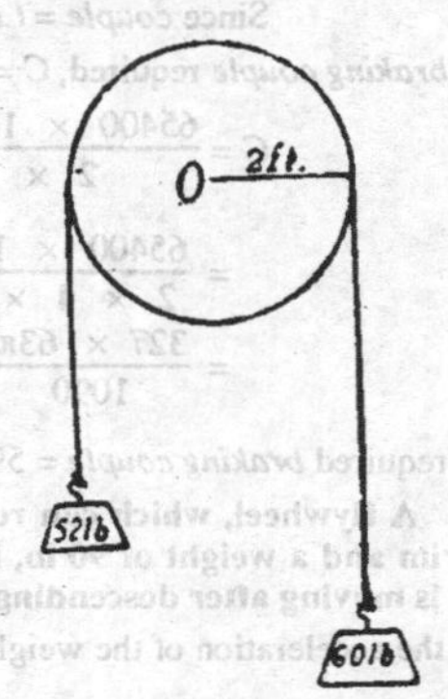

Fig. 3.41

Thus, the velocity of the masses when they have moved through a distance of 6 *ft* will be 4 *ft/sec.*

6. A flywheel of mass 65.4 kg is made in the form of a circular disc of radius 18 cm; it is driven by a belt whose tensions at the points where it runs on and off the rim of the wheel are 2 kg and 5 kg weight respectively. If the wheel is rotating at a certain instant at 60 revolutions per minute, find how long will it be before the speed has reached 210 revolutions per minute. While the flywheel is rotating at this latter speed, the belt is slipped off and a brake applied. Find the constant braking couple required to stop the wheel in 7 revolutions.

(Cambridge Higher School Certificate)

(*i*) Here, obviously,

tension T_1, where the belt runs *on* the rim = 2 *kg wt.*

and " T_2, " " " " off " " = 5 *kg w t*

the resultant tension in the belt $= T_2 - T_1 = (5-2)$ *kg.wt.*

$= 3$ *kg.wt* $= 3 \times 1000 \times 981$ *dynes.*

$\therefore$ moment of the couple due to this tension = 3 × 1000 × 981 × 18 *dynes. cm.*

And, if $d\omega/dt$ be the angular acceleration of the wheel, the *couple acting on wheel*

Now, $I = M.r^2/2 = (65.4 \times 1000 \times 18^2)/2 = 65400 \times 18 \times 9 \times$ *gm-cm²*.

$\therefore$ moment of the couple = 65400 × 18 × 9 × $d\omega/dt$ *dynes cm.*

This must be equal to the moment of the couple due to tension in the belt. Hence,

$$65400 \times 18 \times 9 \times d\omega/dt = 3 \times 1000 \times 981 \times 18$$

Or, $$d\omega/dt = \frac{3 \times 1000 \times 981 \times 18}{65400 \times 18 \times 9} = 5 \text{ radians/sec}^2.$$

Now, we have the relation $\omega_2 = \omega_1 + (d\omega/dt)t$, ...(*i*) [See page 76.

where ω_2 is the *final angular velocity*; ω_1, the *initial angular velocity*, $d\omega/dt$, the *angular acceleration* and t, the *time*.

Here, $\omega_2 = 210$ *rev./min.* $= 210 \times 2\pi/60 = 7\pi$ *radians/sec.*,

$\omega_1 = 60$ " " $= 60 \times 2\pi/60 = 2\pi$ " "

and $d\omega/dt = 5$ *radians/sec²*.

$\therefore$ from relation (*i*) above, we have $7\pi = 2\pi + 5t$. Or, $5\pi = 5t$,

whence, $t = \pi = 3.142$ *sec.*

So that, the flywheel will obtain a speed of 210 revolutions per minute after 3.142 *seconds*.

(*ii*) Let the angular retardation produced in the wheel by the *braking* couple be $d\omega/dt$, the angle turned through by it before coming to rest being equal to $7\times 2\pi = 14\pi$ *radians*. Then, applying the relation $\omega_2^2 - \omega_1^2 = 2\,(d\omega/dt)\,\theta$

[See page 76

we have $\theta^2 - (7\pi)^2 = 2\,(d\omega/dt) \times 14\pi$. Or, $28\pi.d\omega/dt = -49\pi^2$ $\quad [\because \theta = 14\pi$

Or, $d\omega/dt = -49\pi^2/28\pi = -7\pi/4$ *radians / sec*^{2}.

Or, the angular *retardation* required = $7\pi/4$ *radians/sec*^{2}.

Since *couple* = $I.d\omega/dt$, we have

braking couple required, $C = \frac{1}{2} \times 65400 \times 18 \times 18 \times 7\pi/4$ *dynes-cm.*

Or, $C = \dfrac{65400 \times 18 \times 18 \times 7\pi}{2 \times 4 \times 981}$ *gm wt cm.*

$= \dfrac{65400 \times 18 \times 18 \times 7\pi}{2 \times 4 \times 981 \times 1000}$ *kg wt-cm*

$= \dfrac{327 \times 63\pi}{1090} = 59.37$ *kg.wt-cm*

Hence, the required *braking couple* = 59.37 *kg.wt-cm.*

7. A flywheel, which can revolve on a horizontal axis weighs 900 lb and its radius is r ft. A rope is coiled round its rim and a weight of 90 lb, hung from its free end, turns the wheel by its descent. Find the speed at which the weight is moving after descending 20 ft from rest.

Let the acceleration of the weight be *a ft per sec*^{2}, and let *M* be the mass of the weight suspended, (Fig. 3.42)

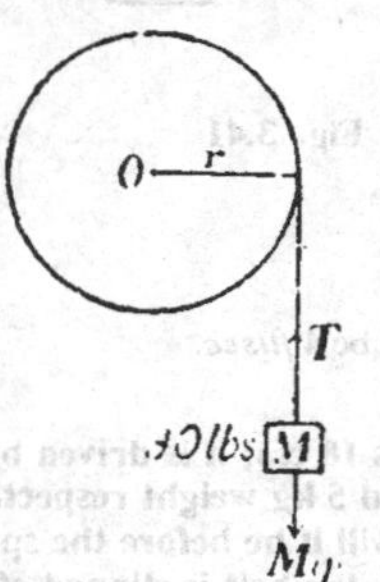

Fig. 3.42

Then, if *T* be the tension in the string, we have

$M.a = (Mg-T)$. Or, $T = M\,(g-a) = 90\,(32-a)$ *poundals.*

Now, moment of the couple acting on the wheel due to tension *T* in the rope = *T. r.*

$\therefore T.r = 90(32-a).r$ *poundal -ft.*

Also, rotational couple on the flywheel = $I.d\omega/dt$.

Here, $I = \frac{1}{2}Mr^2 = \frac{1}{2}900\,.r^2$. Considering the flywheel to be $= 450\,r^2$ *a uniform form circular disc of mass M and radius r.*

And $d\omega/dt = a/r$.

$\therefore$ rotational couple on the wheel = $450\,a.r$. Or, $90\,(32-a) = 450\,a$.

Or, $2880-90a = 450a$. Or, $540a = 2880$.

whence, $a = 2880/540 = 16/3$ *ft/sec*^{2}

It *v* be the velocity of the weight after it has descended 20 *ft*, we have

$v^2-u^2 = 2aS = 2 \times (16/3) \times 20.$ $\quad [\because s = 20\,ft.$

Or, $v^2 = 640/3$, whence, $v = \sqrt{640/3} = 14.61$ *ft sec.*

Therefore, the speed of the weight would be 14.61 *ft/sec.*

8. (*a*) A sphere of mass 50 gm, diameter 2 cm, rolls without slipping with a velocity of 5 cm per sec. Calculate its total kinetic energy in ergs.

(*b*) A solid cylinder (*a*) rolls, (*b*) slides from rest down are inclined plane. Neglect friction and compare the velocities in both cases when the cylinder reaches the bottom of the incline. (*Punjab*, 1961)

(*a*) Here, mass of the sphere, $M = 50$ *gm.*, and radius of the sphere, $r = 1$ *cm.*

Now, moment of inertia of the sphere (solid) is given by

$I = \frac{2}{5}Mr^2 = \frac{2}{5} \times 50 \times 1 = 20 gm-cm^2,$

As the sphere rolls, it rotates about its own diameter as axis as well as its centre of mass moves with a velocity of 5 *cm/sec*. It has, therefore, both kinetic energy of rotation as well as kinetic energy of translation; and, therefore, its total energy is the sum of both.

Now, *K.E. of rotation* $= \frac{1}{2}I.\omega^2 = \frac{1}{2}I\,.v^2/r^2 = \frac{1}{2} \times 20 \times 5^2/1^2 = 250$ *ergs.* $\quad [\because \omega = v/r$

And *K.E. of translation* $= \frac{1}{2}Mv^2 = \frac{1}{2} \times 50 \times 5^2 = 25 \times 25 = 625$ *ergs.*

$\therefore$ *total kinetic energy of the sphere* = 250 + 625 = 875 *ergs.*

(*b*) In case (*a*) when the cylinder *rolls* down the plane, it requires both kinetic energy of translation as well as that of rotation and its acceleration down the plane is $a = [R^2/(K^2+R^2)]\,g \sin\theta$. Or, since $K^2 = R^2/2$

$a = (R^2/\frac{3}{2}R^2)\,g \sin\theta = \frac{2}{3} \sin\theta$. *If, therefore, V be the velocity of the cylinder on coming down the plane, of length l,* say, we have $V = 2al = 2.\frac{2}{3}g\ sin\ \theta.l = \frac{4}{5}g\ sin\ \theta.l$

And in case (*b*) when the cylinder slides down the plane, it requires only kinetic energy of translation and its acceleration is $a = g\ sin\theta$ So that, if V_1 be the velocity of the cylinder on coming down the plane, we have

$V' = 2a.l. = 2.al = 2g.\ sin\ \theta.\ l$

$\therefore \quad V' : V :: 2.g\ sin\ \theta.\ l : \frac{4}{3} g\ sin\ q\ l$ Or as 1.225 : 1

9. Prove from first principles that out of an infinite number of straight lines which may be drawn parallel to a given direction, the movment of inertia of a body is least about the one passing through its centre of gravity. *(Vikram)*

We know that, in accordance with the principle of parallel axes, the moment of inertia I of a body about an axis is equal to the sum of its moment of inertia. I_0 about a parallel axis through its centre of mass or *c.g.* and the product of its *mass M* and the square of the distance r between the two axes, *i.e.*, $I = I_0 + Mr^2$.

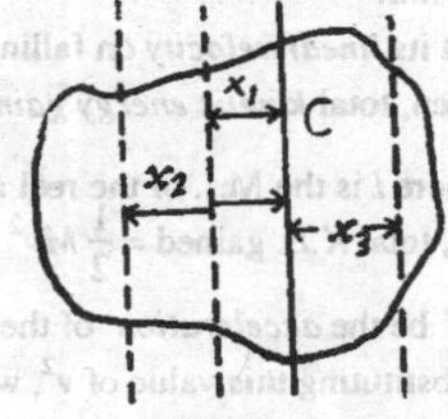

Fig. 3.43

Clearly, therefore, if the different axes all parallel to each other, be distant x_1, x_2, x_3 etc. from the one passing through the *c.g.* (*C*) of the body (Fig. 3.43) *i.e.*, when $x = x_1, x_2, x_3$ *etc.*; their moment of inertia about them will respectively be $I_o + Mx_1^2$, $I_0 + Mx_2^2$, $I_0 + Mx_3^2$, *etc.*

The least value of the moment of inertia will obviously be I_0 *i.e.*, when $x = 0$ or the axis of rotation passes through the *c.g.*, of the body.

10. A flat thin uniform disc of radius *a* has a hole of radius *b* in it at a distance *c* from the centre of the disc, [$c < (a-b)$]. If the disc were free to rotate about a smooth circular rod of radius *b* passing through the hole, calculate the moment of inertia about the axis of rotation. *[Punjab Supplementary]*

Let the disc be of *mass M* and *radius a* (Fig. 3.44), with a hole of radius *b* at distance *c* from its centre *O*.

Obviously, if the hole were concentric with the disc, *i.e..*, if it were at *O*, the centre of the disc as shown dotted in the figure, the *moment of inertia* of the disc about the axis passing through it would be $Io = M(a^2 + b^2)/2$.

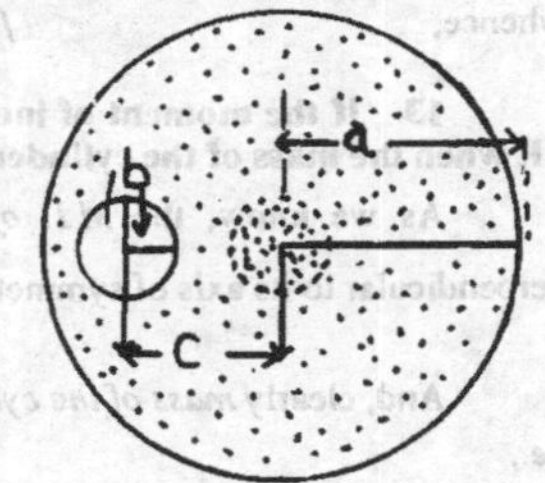

Fig. 3.44

Since the hole is at a distance *c* from O, the centre *e.g.* or the *c.g.* of the disc, the *moment of inertra of the disc about the axis passing through the hole* is, in accordance with the *principle of parallel axis*, given by

$$I = I_0 + Mc^2 = M\frac{(a^2 + b^2)}{2} + Mc^2$$

11. If the pulley in an Atwood's machine be of moment of inertia 1500 c.g.s. units and radius 5 cm., what should be the acceleration of the system in which the weights at the two ends of the string passing over the pulley be 200 and 250 gm respectively? (Given that g = 981 cm/sec²)

Let *a* be the acceleration of the system and *v*, the velocity of the weights, when they have moved a distance *S cm.* from the starting positing, (Fig. 3.44)

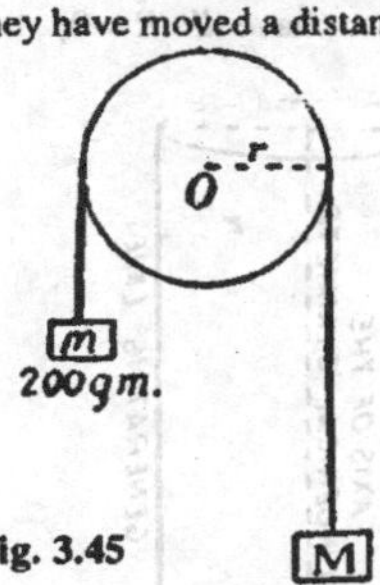

Fig. 3.45

Then, clearly, $v^2 - u^2 = 2aS$. Or, $v^2 = 2aS$

And $\therefore \quad a = v^2/2s$ $\quad [\because u = 0.$

Now, *loss in P.E. of the heavier weight M, falling through distance* $S = Mg.S$ $= 250 \times 981 \times S$ *ergs.*

and, *gain in P.E. of the lighter weight* $m = mg.S = 200 \times 981 \times s$ *ergs.*

$\therefore$ *net loss in P.E. of the system* $= (250\times 981\times S) - (200\times 981 \times S) = 50 \times 981 \times S$ *ergs.*

This must be equal to the gain in *K.E.* of the pulley as well as the weights themselves.

Clearly, gain in *K.E.* of the pulley $= \frac{1}{2}I\omega^2 = \frac{1}{2}I.v^2/r^2$.

$\therefore v = r\,\omega$, where ω is the angular velocity and *r*, the radius of the pulley.

Or, *gain in K.E. of the pulley* $= \frac{1}{2}I.v^2/25 = \frac{1}{2} \times 1500 \times v^2/25 = 30v^2$ *ergs.*

And, *gain in K.E. of the two weights* $= \frac{1}{2}Mv^2 + \frac{1}{2}mv^2 = \frac{1}{2}(M + m)v^2$.

$= \frac{1}{2} \times 450 \times v^2 = 225v^2$ *ergs.*

$\therefore$ total gain in *K.E.* of the pulley and the weights,

$= 30v^2 + 225v^2 = 225v^2$ *ergs*

Since this gain in *K.E.* = loss in *P.E.* of the system and the weights,

we have $225v^2 = 50\times 981 \times S$. Or, $\quad v^2 = \frac{50 \times 981 \times S}{255}$

Now, the acceleration of the system, *i.e.*, $a = v^2/2S$.

$$a = \frac{50 \times 981 \times S}{255 \times 2S} = \frac{327 \times 5}{17} = 96.18 \ cm/sec^2.$$

Or, the acceleration of the system of weights in 96.18 cm/sec^2.

12. The moment of inertia of a reel of thread about its axis is MK^2. If the wose end of the thread is held in the hand, and the reel is allowed to unroll itself, while falling down under the action of gravity, show that it falls down with an acceleration $ga^2/(a^2+k^2)$, where a the radius of the reel. (Nagpur)

As the reel falls under the action of gravity, it also unrolls itself, *i.e.*, it also rotates about its own axis. It thus acquires both *kinetic energy of translation* and *kinetic energy of rotation* at the expense of the potential energy that it loses during the fall.

Let its *linear velocity* on falling through distance h be v and hence its angular velocity $w = v/a$.

Then, total *kinetic energy gained by the reel* $= \frac{1}{2}Mv^2 + \frac{1}{2}I\omega^2$,

where I is the M.I. of the reel about its axis, equal to MK^2.

Or, total $K.E.$ gained $= \frac{1}{2}Mv^2 + \frac{1}{2}MK^2.v^2/a^2 = \frac{1}{2}Mv^2(1 + K^2/a^2)$

It f be the *acceleration* of the reel during its fall, we have $v^2 = 2fh$, since $u = 0$ (the reel starting from rest).

Substituting this value of v^2, we have

total K.E. gained by the reel $= \frac{1}{2}M.2\text{-}gh(1 + K^2/a^2)$,

$$= Mfh\,[(a^2 + K^2)/a^2]$$

And, clearly, *P.E. lost by the reel* $= Mgh$.

Hence, equating the two, we have $Mah[(a^2 + K^2)/a^2] = Mgh$,

whence, $$f = g.\frac{a^2}{a^2 + K^2}$$

13. If the moment of inertia of a cylinder about an equatorial axis is to be a minimum, determine the ratio l/R when the mass of the cylinder is kept constant. *(Punjab, 1966)*

As we know, the *M.I. of a cylinder about its equatorial axis* (*i.e.* the axis passing through its centre and perpendicular to its axis of symmetry) is given by $I = M\left(\frac{l^2}{12} + \frac{R^2}{4}\right)$

And, clearly *mass of the cylinder = its volume × density of its material*,

i.e., $$M = \pi R^2.l.c.$$

differentiating which we have $dM/dR = \pi\rho\,(2Rl + R^2.\,dl/dR) = 0$, because M is a constant.

i.e., $$2Rl + R^2.\,dl/dR = 0, \text{ Or, } dl/dR = -2l/R.$$

Now, clearly, for the moment of inertia of the cylinder to have the minimum value, we showns have $dI/dR = 0$,

Or, $$M\left(\frac{2l}{12}\cdot\frac{dl}{dr} + \frac{2R}{4}\right) = 0, \text{ whence, } \frac{dl}{dr} = -\frac{3R}{l}.$$

So that, equating the two values of dl/dR, we have $-2l/R = -3R/l$ which gives the required ratio l/R to be equal to $\sqrt{3/2}$.

14. Find the moment of inertia of a homogeneous circular cylinder of length, $2l$, radius of cross-section r, about (*i*) the axis of the cylindrical symmetry; (*ii*) a generating line; (*iii*) a diameter of cross-section at a distance x, l, and $2l$ (or θ) from one base.

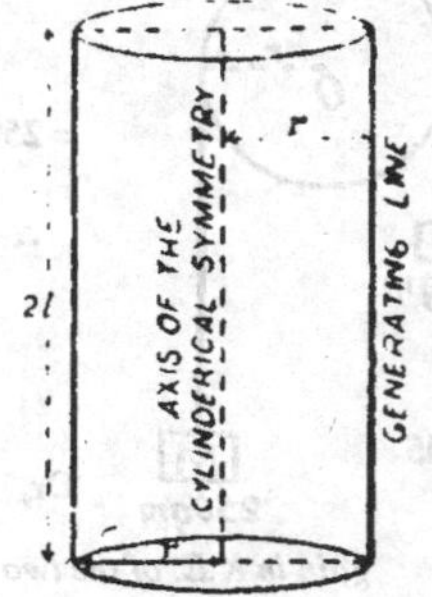

Fig. 3.46

(*i*) The moment of inertia of the cylinder about the axis of cylindrical symmetry is the same as that of a disc about an axis passing through its centre and perpendicular to its plane, (for a cylinder is nothing but a thick disc), and is equal to $MR^2/2$, where M is the *mass* of the disc or cylinder and R, its *radius*.

Therefore, if M be the mass of the cylinder, and r, its radius of cross-section, (Fig. 3.46), we have *moment of inertia of the cylinder about its axix of cylinderical symmetry* equal to $Mr/2$.

(*ii*) The generating line is parallel to the axis of symmetry, passing through the *c.g.* of the cylinder, and is at a distance r from it. Therefore, by the *principle of parallel axes*, moment of inertia of the cylinder *about* the generating line is equal to its moment of inertia about the axis of symmetry *plus* mass of the cylinder × (distance from the axis)^2, *i.e.*,

$$I = \frac{1}{2}Mr^2 + Mr^2 = 3Mr^2/2.$$

(*iii*) (*a*) The moment of inertia of the cylinder about an axis, passing though its centre and perpendicular to its length

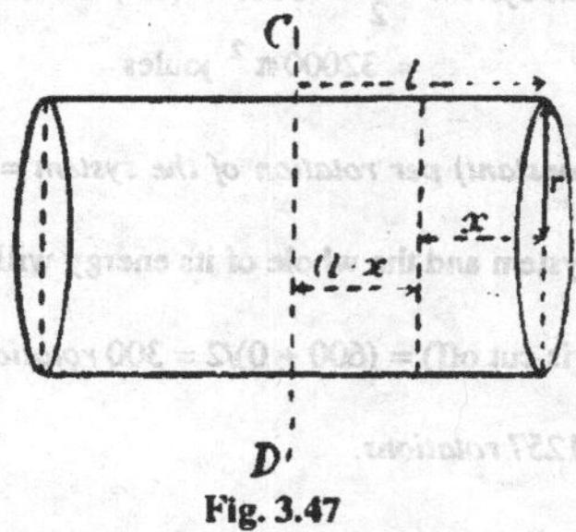

Fig. 3.47

$= M [(length^2/12) + (radius^2/4)]$

$= M [4l^2/12) + (r^2/4)]$ [$\because$ *length* = 2*l*].

$= M [l^2/3 + r^2/4]$

(*b*) Since an axis at a distance *x* from one base is at a distance (*l–x*) from the axis through the centre, we have, by the *principle of parallel axes,*

M.I. about this axis = *M.I.* about a parallel axis through the centre + $M(l–x)^2$.

$= M [(l^2/3) + (r^2/4) + (l - x)^2]$

$= M [(4l^2/3) + (r^2/4) + (x^2 - 2lx)]$.

(*c*) Similarly, by *the principle of parallel axes,*

M.I. about the diameter of cross-section $= M[(l^2/3) + (r^2/4)] + Ml^2 = M[(4l^2/3) + (r^2/4)]$.

15. Find the moment of inertia of a sphere about a diameter.

You are given two spheres of the same mass and size and appearance, but one of them is hollow at the centre and the other is solid throughout. How will you find which is hollow and which is solid ? *(Delhi)*

For answer to first part, see § 3.6 (case 13), page 64.

The moment of inertia of a solid sphere about its diameter is, as we know, $2MR^2/5$, where *M* is its *mass* and *R*, its *radius*, and that of a hollow sphere $\frac{2}{5}[M(R^5-r^5)/(R^3-r^3)]$, where *R* and *r* are its *outer* and *inner radii* respectively, *R* and *M* being the *same* in the two cases.

$\therefore$ radius of gyration for the *solid sphere* = $\sqrt{2R^2/5}$ and radius of gyration for the *hollow sphere*

$$= \sqrt{\frac{2}{5}[(R^5 - r^5)/(R^3 - r^3)]}, \text{ their masses being the same.}$$

Now acceleration of body rolling down an inclined plane is given by

$$a = [R^2/(K^2 + R^2)].g \sin\alpha$$

where α is the angle of inclination of the plane (see page 787

It is thus clear that the greater the value of K^2 as compared with R^2, the less the acceleration of the body.

Now, K^2 for the *solid sphere* $= 2R^2/5$,

and K^2 for the *hollow sphere* $= \frac{2}{5}\left[\frac{R^5 - r^5}{R^3 - r^3}\right]$

$$= \frac{2}{5} \cdot \frac{R^5[1 - r^5/R^5]}{R^3[1 - r^3/R^3]} = \frac{2}{5} \cdot R^2 \left[\frac{1 - r^5/R^5}{1 - r^3/R^3}\right]$$

Obviously, $r^5/R^5 < r^3/R^3$. And $\therefore$ $(1 - r^5/R^5) > (1 - r^3/R^3)$.

So, that the fraction , $\frac{1 - r^5/R^5}{1 - r^3/6R^3} > 1$.

And $\therefore$ $K^2 = \frac{2}{3} \cdot R^2 \times$ *a quantity greater than* 1.

Or, K^2 *for a hollow sphere is greater than* $2R^2/5$.

Thus, the fraction $[R^2/(R^2 + K^2)]$ is less for a hollow sphere than for a solid sphere and, therefore, the acceleration of the hollow sphere is *less* than that of a solid sphere. In other words, the solid sphere will come down the inclined plane faster than the hollow sphere, and the two can thus be easily distinguished from each other.

16. The mass of a flywheel and shaft is 1000 kg and the radius of gyration 40 *cm*. The torque required to turn it is 40 newton-metre. If the driving power of the system be cut off when it is making 600 revolutions per minute, how many revolutions will it make before coming to a stop and how long will it take to do so. Imagine the frictional resistance to remain the same at all speeds.

Let ω be the *angular speed* of the system at the time the driving power is cut off and *I*, its moment of inertia, so that the *K.E. of the system* $= \frac{1}{2}I\omega^2$

Since $I = MK^2 = 1000 \times (40/100)^2 = 1000 \times (0.4)^2$ *kg-m*^{2} and $\omega = 600 \times 2\pi/60 = 20\,\pi$ *radians per second*, we have *K.E. of the system* $= \frac{1}{2} \times 1000 \times (0.4)^2 \times (20\,\pi)^2$

$= 32000\,\pi^2$ joules

Now, *work done = torque × angle described.*

Hence, *work done against frictional resistance, (supposed constant) per rotation of the system* $= 40 \times 2\pi = 80\pi$ *joules.*

This much energy is thus used up during one rotation of the system and the whole of its energy will thus be used up in $32000\,\pi^2/80\pi = 400\,\pi = 1257$ rotations.

And, since average speed of rotation (after the driving power is cut off) = (600 + 0)/2 = 300 *rotations/min, the time* taken to complete 1257 revolutions = 1257/300 = 4.19

The system will thus come to a stop in 4.19 *min. after making* 1257 *rotations.*

17. A cyclist is negotiating a curve of 200 cm radius on a level road at a speed of 5 metres per second. If the M.I. of the front wheel of the cycle be 5×10^5 gm-cm^2 and its radius 50 cm, calculate the gyrostatic resistance encountered.

As we know, the gyrostatic resistance is the resistance to precessional motion and is equal and opposite to the precessional torque. It is also, therefore, equal to $I\omega\phi$ (see page 82, 83), where I, is the *M.I.* of the gyrostat (*i.e.*, disc or wheel), ϕ its *angular velocity* and ϕ, the *rate of precession* (or precessional velocity) of the axle.

Here $I = 5 \times 10^5$ *gm-cm*^{2}, $\omega = v/R = 5 \times 100 = 10$ *radiuan/ sec.*

and $\phi = v/r = 5 \times 100/200 = 5/2$ *radian/sec.*

$\therefore$ *gyrostatic resistance encountered* $= 5 \times 10^5 \times 5/2 = 1.25 \times 10^7$ *dynes-cm.*

18. The angular velocity of a gyrostat is 5 radians/sec. and its M.I. about the axis of rotation, 4×10^5 *gm-cm*^{2}. What will be the precessional velocity of its axle under a precessional torque of 2×10^6 *dynes-cm* ?

Here, *precessional torque* $= I\omega\phi = 2 \times 10^6$ *dynes-cm.*

$I = 4 \times 10^5$ *gm-cm*^{2} and $\omega = 5$ *radians/sec.*, so that angular momentum

$I\,\omega = 4 \times 10^5 \times 5 = 2 \times 10^6$ *gm-cm*^{2}*-sec*$^{-1}$.

So that, *precessional velocity of the axle, i.e.,* $\phi = I\,/\,\omega 4/I\omega = 2 \times 10^6/2 \times 10^6 = 1$ *radian/sec.*

19. Obtain, on the basis of a simplified treatment of the problem, the limiting velocity for a metal hoop of radius 2 ft to be able to roll along a rough horizontal surface, with its plane vertical.

The *limiting* or critical velocity for a hoop to be able to roll along a horizontal surface, with its plane vertical is given by $v_c = \sqrt{gr/2}$, where r is the *radius of the hoop* (see page 88).

Here $r = 2$ *ft* and, therefore, taking $g = 32$ *ft/sec*^{2}, we have *limiting velocity of the hoop* $= \sqrt{32 \times 2/2} = 4$ *ft/sec*^{2}

EXERCISE III

1. Define Moment of Inertia and radius of Gyration. Explain their physical significance. State the laws of (*i*) parallel and (*ii*) perpendicular axes and prove any one of them. (*Bombay ; Poona*)

2. Calculate the moment of inertia of a thin circulate disc of mass M and radius r (*i*) about its diameter (*ii*) about a parallel axis to the diameter and tangential to the disc. **Ans.** $Mr^2/4$; $5Mr^2/4$.

3. (*a*) Define Moment of Inertia of a body about an axis. Show that the moment of inertia of a body about an axis through the centre of gravity is less than that about any other parallel axis. (*Madras*)

(*b*) Calculate the moment of inertia of a right circular cylinder about an axis perpendicular to it geometrical axis and passing through the centre of gravity. (*Agra, Vikram*)

Ans. $M\left(\frac{l}{12} + \frac{R^2}{4}\right)$

4. A flywheel of mass $2\frac{1}{2}$ *tons* and diameter 8 *ft* makes 250 revolutions *per minute*. Find(*i*) its *angular velocity* (*ii*) its *energy* (*iii*) its *moment of inertia*. Assume its mass to be concentrated at the rim.

Ans. (*i*) $25\pi/3$ *radians/sec* (*ii*) 3.07×10^7 *ft poundals* (*iii*) 89600 *lb-ft*^{2}

5. Show that the acceleration of a disc rolling down an inclined plane of angle, θ is $2g \sin\theta/3$, while that of a ball is $5g \sin\theta/7$.

6. A uniform rod 4 *ft* long and weighing 9 *lb*, revolves 60 times a minute about one end. Calculate its *kinetic energy*. **Ans.** 29.61 *ft lb.*

7. A hoop of mass 5 *kg* and radius 50 *cm* rolls along the ground at the rate of 10 *metres per second*. Calculate its kinetic energy in *ergs*. **Ans.** 5×10^9 *ergs.*

8. (a) Explain clearly what you understand by 'Moment of Inertia' and 'Angular momentum'. State the principle of conservation of angular momentum, illustrating your answer by an example. *(Agra)*

(b) Find the moment of inertia of a circular lamina about a tangent in its plane. *(Patna)* **Ans.** $5Mr^2/4$.

9. (a) Define moment of inertia and radius of gyration obtain an expression for the moment of inertia of a solid sphere about (i) its diameter and (ii) a tangent. *(Delhi, 1971)* **Ans.** $2Mr^2/5$; $7Mr^2/5$.

(b) A solid spherical ball rolls on a table. What fraction of its total kinetic energy is rotational? **Ans.** 2/7th.

10. Show that the *K.E.* of a uniform cylinder or disc of mass *M*, rolling so that its centre has a velocity *v*, is 3/4 Mv^2. In the case of a sphere, show that the *K.E.* would be $7Mv^2/10$.

11. A thin hollow cylinder, open at both ends and of mass *M* (a) *slides* with a velocity *v without rotating*, (b) *rolls without slipping*, with the *same* speed. Compare the kinetic energies it possesses in the two cases. **Ans.** 1 : 2.

12. Define (i) radius of gyration, (ii) moment of inertia. Find the moment of inertia of a circular disc about the axis perpendicular to its plane.

A circular disc of mass *m* and radius *r* is set rolling on a table. If ω is its angular velocity, show that its total energy *E* is given by $E = \frac{3}{4} mr^2\omega^2$. *(Panjab)*

13. Derive an expression for the kinetic energy of a body rotating about an axis.

A flywheel is in the form of a uniform circular disc; its radius is 2 *ft* and mass 2 *lb*. Find the work which must be done on the flywheel to increase its speed of rotation from 10 to 20 revolutions per second. *(Madras, B.A.)*

Ans. $\frac{1}{2} I\omega^2$; 2.366×10^4 *ergs.*

14. Five masses, each of 2 *kg* are placed on a horizontal circular disc, (of negligible mass) which can be rotated about a vertical axis passing through its centre. If all the masses be equidistant from the axis and at a distance 10 *cm*. from it, what is the moment of inertia of the whole system? **Ans.** 0.1 $kg\text{-}m^2$.

[**Hint** : *M.I.* of each mass about the axis = Mr^2,
and ∴ total *M.I.* of the system = sum of the *M.I.* of the masses.

15. Define Moment of Inertia' and 'Radius of Gyration'. State the law of parallel axes and prove it.

A wheel of radius 6 *cm* is mounted so as to rotate about an horizontal axis through its centre. A string of negligible mass, wrapped round its circumference carries a mass of 200 *gm* attached to its free end. When let fall, the mass descends through 100 *cm* in the first 8 seconds. Calculate the angular acceleration of the wheel and its moment of inertia. *(Bombay)*

Ans. Angular acceleration = $\frac{4}{8}$ *radians/sec*^{2} ; *M.I.* = 8.748×10^4 $gm\text{-}cm^2$.

16. Describe an experiment for determining the moment of inertia of a flywheel. Derive the necessary formula. *(Gujarat (Subsidiary), 1962)*

17. The free end of a string wrapped round the axle of a flywheel, of moment of inertia : 27.61×10^5 $gm\text{-}cm^2$ carries a weight of 5 *kg* which is allowed to fall. What is the number of revolutions *per second* made by the wheel, when the weight has fallen through 1 *metre?* The kinetic energy of the weight may be neglected. **Ans.** 3.

18. If, in question 17, the wheel be mounted on an axle of half its moment of inertia (*i.e.*, 13.80×10^5 $gm\text{-}cm^2$) and radius 5 *cm*, and the *K.E.* of the weight be taken into account, what will be the number of revolutions per second made by the wheel? **Ans.** 2.413.

19. Masses of 95 *gm* and 105 *gm* hanging freely are connected by a light string which passes over a pulley of mass 20 *gm*. When released, the system moves with an acceleration of 46.7 *cm per sec*^{2}. Calculate a value of *g* if the mass of the pulley is (*a*) neglected, (*b*) taken into account. Regard the pulley as a simple disc of moment of inertia $\frac{1}{2} Mr^2$, and assume that no kinetic energy is lost in friction. *(Northern Universities Higher School Certificate)*

Ans. (*a*) 934 $cm\,sec^{-2}$ (*b*) 980.7 $cm\,sec^{-2}$.

20 Four spheres, each of diameter 2*a* and mass *m*, are placed with their centres *on* the four corners of a square of side *b*. Calculate the moment of inertia of the system about one side of the square. *(Panjab)*

Ans. $\frac{2}{5} m(4a^2+5b^2)$.

21. Describe the experiment to determine the moment of inertia of a flywheel. Derive the formula used in the experiment, without neglecting the friction at the bearings of the flywheel. *(Allahabad ; Gujarat)*

22. A flywheel, which can turn about a horizontal axis, is set in motion by a 500 *gm weight* hanging from a thin string that passes round the axle. After the wheel has made 5 revolutions, the string is detached from the axle and the weight drops off. The wheel then makes 7 revolutions before being brought to rest by friction. The radius of the axle is 2.0 *cm*, and at the instant when the weight drops off the angular velocity of the wheel is 10 radians per sec. Assuming that the work done against friction in each revolution is always the same, calculate the moment of inertia of the flywheel about its axis of rotation. *(Oxford and Cambridge Higher School Certificate)*

Ans. 3.59×10^5 $gm\,cm^2$.

23. A pair of rails is supported in a horizontal position and the axle of a wheel rests on the rails. A thread is wrapped round the axle and a weight hung on the end of the thread. As the weight falls, then wheel moves along the rails. How would you determine the moment of inertia of the wheel with this arrangement?

24. **A circular disc, starting from rest, rolls (without slipping) down an inclined plane of 1 in 8, and covers a distance of 5.32 *ft* in 2 *sec*. Calculate the value of '*g*'.** **Ans.** 31.92 *ft/sec*^{2}

25. **Two gear wheels, of equal thickness, of the same material and having radii in the ratio 2 : 1, are mounted on parallel friction less spindles but are separated so as not to mesh with one another. The larger wheel is set spinning at a speed of 10 *rev per sec*, and the wheels are then brought into mesh. What is the resulting speed of each wheel?** *(Cambridge Scholarship Examination)*

Ans. 8.9 and 17.8 *rev sec*$^{-1}$.

26. (*a*) **Find the moment of inertia of a solid cylinder about a line parallel to the axis and touching its surface.** *(Vikram, B.Sc. (Part II), 1961)*

(*b*) **What must be the relation between *l* and *R* of the moment of inertia of a cylinder about its axis is to be the same as its moment of inertia about the equatorial axis?** **Ans.** (*a*) $3MR^2/2$; (*b*) $R = l\sqrt{3}$.

27. **Calculate the moment of inertia of a sphere about its diameter and also about a tangent.** *(Delhi (Hons.); Vikram Gujrat ;)*

Ans. $2MR^2/5$; $7MR^2/5$

28. Explain the terms : Moment of Inertia and Radius of Gyration. State and prove the theorem of parallel axes.

Two thin discs, each of mass 100 *cm* and radius 5 *cm*, are placed at either end of a rod 20 *cm* long and 1*cm* in diameter. What is the moment of inertia of the system about an axis passing through the centre of the rod and perpendicular to its length ? Density of the material of the rod is 7.8 *gm/c.c.* *(Poona)*

Ans. 25334 *gm cm*^{2}

29. Calculate the moment of inertia of a solid sphere about a diameter.

A solid sphere of mass 100 *gm* and radius 2.5 *cm* rolls without sliding with a uniform velocity of 10 *cm per sec* along a straight line on a horizontal table. Calculate its total energy in *ergs*. *(Vikram, B.Sc. (Part II))*

Ans. 7000 *ergs*.

30. What do you understand by the term "*precession*" ? Show that if the axis of the torque applied to a body be perpendicular to its axis of rotation, the body precesses about an axis perpendicular to either of the first two axes.

31. What is (*i*) a gyrostat and (*ii*) a *gyroscope* ? Describe suitable experiments to illustrate their action. What is meant by the term *rotation* ?

32. Explain the theory underlying a *gyrostatic pendulum* and obtain an expression for its time-period.

33. Discuss in detail the case of a thin disc or hoop set rolling over a plane horizontal surface and obtain expressions for (*i*) its *critical velocity*, and (*ii*) the *radius of curvature of its path on the surface.* *(Bombay)*

34. Write short notes on the following:

(i) Gyro-compass, (ii) Pendulum Gyro-compass, (iii) Rifling of barrels of guns and (iv) Precession of the Equinoxes.

35. Explain what is meant by a precessional motion and obtain a value for the precession torque by using simplified considerations.

State *Lanchester's rule* and apply it to consider fully the motion of a rolling disc. *(Poona)*

36. State and explain the theorems of perpendicular and parallel axes for moment of inertia. Calculate the moment of inertia of a sphere about a diameter. *(Delhi, 1966)*

37. Define moment of inertia. Describe how you would determine experimentally the moment of inertia of a flywheel about its usual axis of rotation. Discuss briefly the part played by a flywheel in a machine. *(Delhi, 1964)*

38. A grinding wheel of mass *kg* has a dimeter 0.20 metre. Calculate its kinetic energy of rotation when it is rotating at 3600 revolutions per minute. Also calculate the height from which it has to fall freely to aquire the same kinetic energy on striking the ground. (g = 9.80 *m/sec*^{2}) *(Delhi)*

39. Define an expression for the moment of inertia of a cylinder about its axis. Hence deduce the moment of inertia of the cylinder about a line perpendicular to the axis and bisecting it (the outer radius of the cylinder is '*a*' whereas the internal radius is '*b*').

$$\textbf{Ans. } M\left[\frac{a^2+b^2}{4}+\frac{l^2}{12}\right]$$

40. A reel of thread unrolls itself falling down under gravity; neglecting mass of the thread, find an expression for the acceleration of the reel. *[Bombay (Subsidiary)]*

Ans. $2g/3$

41. Define moment of inertia. Calculate the moment of inertia of a thin disc about the axis normal to its plane and passing through its centre of gravity. Describe how you would determine experimentally the moment of inertia of a flywheel about its axis. *(Delhi)*

42. An engine is designed to develop a power *a* of 10 kilowatts at the average speed of 1000 revolutions per minute. If the speed variation be 2% of the average speed and the energy fluctuation 0.8 of the work done per revolution calculate the moment of inertia of the flywheel that would be suitable. **Ans.** 1.093 *kg-m*^{2}

[**Hint.** Variation of speed 2%. Hence *max. speed* 1020 *rev/m* in and *min speed* 980 *rev/min*. Or, $\omega = 1020 \times 2\pi/60$ *rev/sec* and $\omega_2 = 980 \times 2\pi/60$ *rd/sec*. So that *fluctuation of energy* $= \left(\frac{1}{2} I\omega_1^2, -\frac{1}{2} I\omega_2^2\right)$ This must be equal to fluctuation of the work done per revolution, *i.e.* 0.8 of $\frac{work\ done\ /\ min}{rev\ /\ min} = \frac{0.8 \times 10 \times 1000 \times 60}{1000}$, whence, I can be easily evaluated.]

43. An annular area is enclosed by two concentric circles irites having radii a and b respectively, $a > b$; find the *M.I.* about any diameter. Find b such that when $a = 15$, the radiuis of gyration is 8.5. *(A.M.I.E.)*

Ans. $\pi(a^4 - b^4)/4$; $b = 8$.

44. Two masses m_1 and m_2 are joined by a weightless root of length r. Calculate the moment of inertia of the system about an axis passing through the centre of mass and perpendicular to the rod:

Ans. $\frac{m_1 m_2}{m_1 + m_2} \cdot r^2$

[N.B. Here, $m_1 m_2/(m_1+m_2)$ is called the reduced, mass of the system]

45. Explain the terms precessional motion and precessional torque. Obtain an expression for the periodic time of the precessional motion of a gyroscope. *(Poona)*

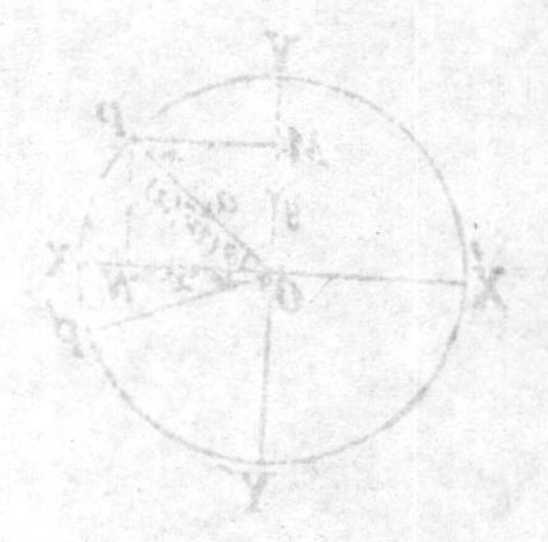

CHAPTER IV

SIMPLE HARMONIC MOTION

4.1. Definitions. A simple harmonic motion is a *particular case of periodic motion, i.e.*, a motion which repeats itself over and over again after regularly recurring intervals, called its *time-period*, and is so called because of its association with musical instruments. Common in nature, it is in fact the most fundamental type of periodic motion, as all other periodic motions, (harmonic as well as non-harmonic) can be obtained by a suitable combination of two or more simple harmonic motions.

If the acceleration of a body be proportional to its displacement from its position of equilibrium, or any other fixed point in its path, and be always directed towards it, the body is said to execute a **simple harmonic motion,** *(written, for short, as S.H.M).*

Now, a simple harmonic motion may be *(i)* **linear**, or *(ii)* **angular,** according as the body moves along a linear path, under the action of a constraining force constantly acting upon it, or rotates about an axis, under the action of a constant torque or couple.

The time-period of a body, executing a *S.H.M.*, is quite independent of the extent of its motion to either side of its mean position, (*i.e.*, of its amplitude), and the motion is, therefore, said to be **isochronous.**

Mathematically, *a linear S.H.M. may be regarded as the projection of a uniform circular motion, or of a rotating vector, on the diameter of the circle, or any other fixed line in the plane of the circle, this circle being referred to as the* **circle of reference,** *and may, in many a case, be purely imaginary.*

Thus, if a particle P (Fig. 4.1) moves with a *uniform speed* v along a circle of radius a, and another particle M, along the diameter YOY', *such that when* P is at X, M is at O, and, as P starts along the circle in the *anticlockwise* direction, M starts along OY, so that when P reaches Y, M also reaches Y. As P continues to travel further along YX', M starts back towards O.

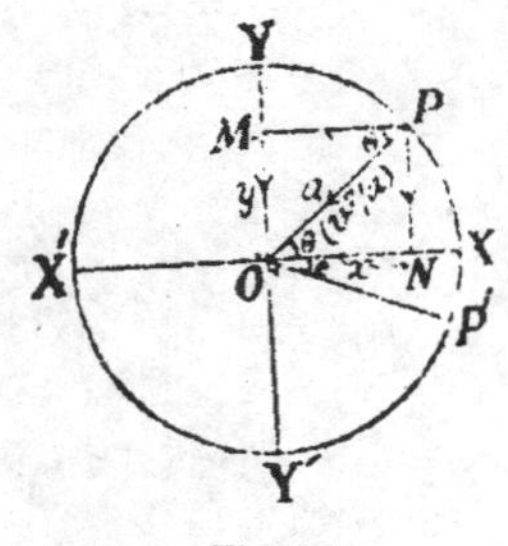

Fig. 4.1

And, when P reaches X', M reaches O. As P now traverses the lower half of the circle along $X'Y'$, M proceeds downwards along OY', so that both P and M reach Y' together and, finally, when P travels further on along $Y'X$, M starts back along $Y'O$, reaching its mean position O when P reaches X.

Thus, the particle M moves along the diameter YOY' from O to Y, from Y to Y' and back to O *(i.e.*, completes one vibration), in the *same time* in which P moves *once* round the circle, such that, *at any given instant, the line joining the positions of P and M is perpendicular to the diameter YOY'* ; *or, the position of M on the diameter YOY' corresponding to the position of P on the circle of reference, at any instant, is given by the foot of the perpendicular from P on to YOY'.*

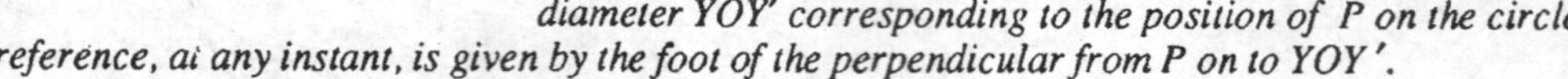

This particle M is said to be performing a linear S.H.M. along YOY' and is, obviously, the *projection* of the particle P, moving uniformly along the *circle of reference XYX'Y'*, or is the projection of the rotating vector OP on the diameter YOY' of the circle.

If the motion of M be due, *not to P*, but to any other force *along its path*, the circle of reference will, as indicated above, be purely an *imaginary* one. Further, the path of M need not necessarily be straight and may as well be curved.

Since a force acting on a body is proportional to the acceleration it produces in it, it is obvious that the force acting on a body executing *S.H.M.* must correspond to the changes in its acceleration. In other words, it must also be *proportional to the displacement of the body from its mean position and must always be directed towards it.*

Some familiar examples of simple harmonic motion

(*a*) Linear :

(*i*) *The up and down oscillations of the piston of a cylinder containing a gas, when suddenly pressed down and released,* (see worked example 1).

(*ii*) *The oscillations of mercury or water contained in a U-tube, when the column in one limb is depressed and released,* (see worked example 7).

(*iii*) *The vertical oscillations of an elastic string* (or *a spiral spring*) *suspended from a rigid support and loaded at its lower end,* (see worked example 10).

(*b*) Angular :

(*i*) *The oscillations of a pendulum, provided the amplitude be small,* (See Chapter VI)
(*ii*) *The oscillations of a magnet suspended in a magnetic field,* (see worked example 8)
(*iii*) *Torsional oscillations, in general* (see Chapter VIII).

4.2. Characteristics of a Linear S.H.M.

1. Amplitude. The maximum distance covered by the body on either side of its mean or equilibrium position is called its *amplitude*. It is obviously, *equal to the radius of the circle of reference.*

Thus, the amplitude of the particle M, in the case above, is $OY = OY' = a$, the radius of the circle $XYX'Y'$.

2. Displacement. The distance of a body from its mean position, at any given instant, *measured along its path*, gives its *displacement* at that instant.

Thus, the displacement of M, in the position shown, (Fig. 4.1) is equal to OM or y, such that

$OM = OP \sin\theta$, Or, $y = a \sin\theta$. [where $\angle XOP = \angle MPO = \theta$]

If ω be the angular velocity of P and t, the time taken by it in traversing the distance OP along the circle, *i.e.*, in describing the angle θ, we have $\theta = \omega t$; so that, $y = a \sin \omega t$

This relation, giving the value of the displacement of a body executing a *S.H.M.*, in terms of its amplitude and the angle velocity of the rotating vector, (or of the particle in the circle of reference) is referred to as its *equation of motion*. Thus, the equation of motion of M along YOY' is $y = a \sin\omega t$.

If we consider the motion of a particle N along the diameter XOX' such that both P and N are together at X and as P goes round the circle in the anticlockwise direction, N starts along XOX', so that when P reaches Y, N reaches O, and when P reaches X', N also reaches X', and when P goes along the lower half of the circle, N starts back along $X'O$, reaching O when P reaches Y' and finally both arrive together at X, then the motion of N along XOX' is also a *S.H.M.* And its displacement $ON = x$ is clearly given by

$ON = OP \cos\theta$ Or, $x = a \cos\theta$ [$\because$ $ON = x$ and $OP = a$

$\therefore$ its *equation of motion is* $x = a \cos \omega t$. [$\because$ $\theta = \omega t$

The position of the particle M executing a *S.H.M.* along the diameter YOY' may, at any time, be found with the help of its **displacement curve,** *which is a graph, showing the relation between the time that elapses since the particle was at its mean position O, and its displacement from O during this time.*

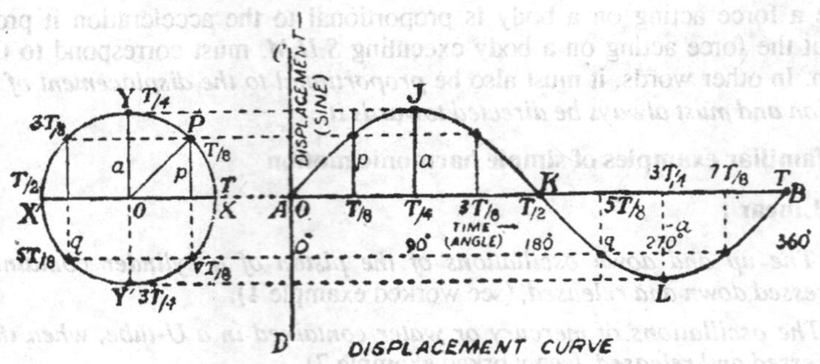

Fig. 4.2

Let time be represented along the horizontal axis *AB* and displacement along the vertical axis *DC*, (Fig. 4.2)

Let the circle *XYX'Y'* be divided up into an equal number of parts, say 8, representing equal intervals of time *T*/8, where *T* is the time taken by the particle *P* to go *once* round the circle. Let these intervals of time be also marked along the axis *AB*, taking *A* as the origin or the starting point. Then, the perpendiculars drawn at the points on the circle on to *XOX'* give the displacements of *M* along *YOY'*, corresponding to the intervals of time represented by them, as tabulated below :

Time	0	*T*/4	*T*/2	3*T*/4	*T*
Displacement	0 (*min.*)	*a* (*max*)	0 (*min.*)	− *a* (*max.*)	0 (*min.*)

Ordinates equal to these perpendiculars are then erected at the corresponding points on *AB*. Thus, the ordinate *p* represents the displacement of *M* after a time *T*/8 of its starting from *O*; the ordinate *a*, after time *T*/4; the ordinate *q* after time 5*T*/8, and so on. The extremities of all these ordinates are then joined and a smooth curve *AJKLB* is obtained, which is a *harmonic* or a *sine curve*, because it is of the same form as would be obtained for the relation between angles from 0° to 360° and their sines, the maximum value, *viz.*, 1, being at 90°, and –1 at 270° and the least, *i.e.*, zero at 0°, 180° and 360°. The displacement curve shows at a glance how the displacement of the particle *M* changes along the diameter *YOY'* and its value can be readily obtained from the curve at any given instant.

3. Velocity. The velocity of the particle *M* is clearly given by the component of the velocity of *P* along the diameter *YOY'*, (Fig. 4.3)

Now, the velocity of *P* is *v* in a direction *tangential to the circle* at *P*.

Representing it in magnitude and direction by the straight line *PK*, we may resolve it into two rectangular components along and perpendicular to the diameter *YOY'*. Then, the component perpendicular to *PM* represents the velocity of *M* along *YOY'*.

Now, clearly, *this* component $= v \sin KPM$

$= v \cos \phi = v \cos \omega t$.. $\quad [\because \phi = \omega t$

Since $\cos \phi = PM/OP$

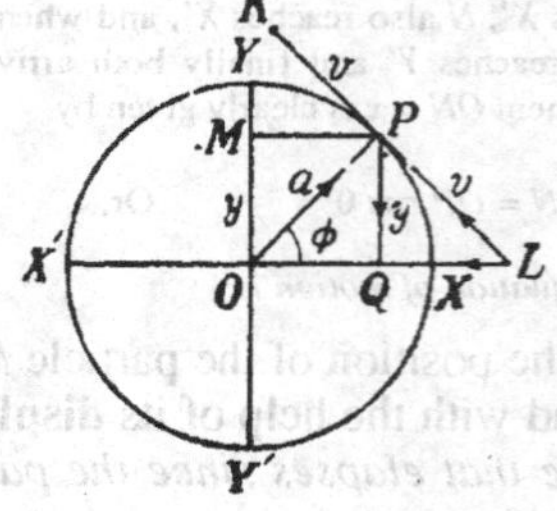

Fig. 4.3

$= \sqrt{OP^2 - OM^2} / OP = \sqrt{a^2 - y^2} / a,$ [∵ $OP = a$ and $OM = y$.

we have *component along* $YOY' = v \cdot \sqrt{a^2 - y^2}/a = a\omega \sqrt{a^2 - y^2}/a,$ [∵ $v = a\omega$

Or, *velocity* of $M = \omega \sqrt{a^2 - y^2}$.

Alternatively, the velocity of M may be obtained by differentiating its displacement y with respect to t (because *velocity is rate of change of displacement*). Thus, since $y = a \sin \omega t$, we have

$dy/dt = a\omega \cdot \cos \omega t.$ Or, $dy/dt = a\omega \cdot \sqrt{a^2 - y^2}/a = \omega\sqrt{a^2 - y^2}$.

Or, *velocity of* $M = \omega\sqrt{a^2 - y^2}$.

Thus, the velocity of M would be different at different points along its path, depending upon its displacement or distance from its mean position O, *being a maximum when y is a minimum*, and a *minimum, when y is a maximum*.

So that (*i*) when $y = 0$, *i.e.*, when M is at O, (or, its displacement is *zero*) its velocity = $\omega \sqrt{a^2} = \omega.a = v$, *the same as that of P*; and (*ii*) when $y = a$, *i.e.*, when M is at Y or Y', (*i.e.*, its displacement is *maximum*), its velocity $= \omega \sqrt{a^2 - a^2} = 0$.

Or, *the velocity of the particle varies inversely as its displacement*.

It will thus be readily seen that the velocity of M varies from a maximum (v) at O to a minimum (*zero*) at Y, then increases from *zero* to v at O, decreasing to *zero* again at Y' and again becoming v when it comes back to O. In other words, at time 0, it is maximum; at time $T/4$, a minimum; at $T/2$, a maximum; at $3T/4$, again, a minimum, and finally at time T, again a maximum, as shown in the table below, where the velocities of M are shown at different times.

Time	0	T/4	T/2	3T/4	T
Velocity	v (*max.*)	0 (*min.*)	$-v$ (*max.*)	0 (*min.*)	v (*max.*)

If, therefore, a graph be plotted between time and velocity of the particle M, we get the **velocity curve** of the particle, shown in Fig. 4.4, where time is shown along the horizontal axis and velocity along the vertical axis. The curve obtained is a *cosine curve*, for it is of the same form as the curve plotted between angles from 0^0 to 360^0 and their cosines.

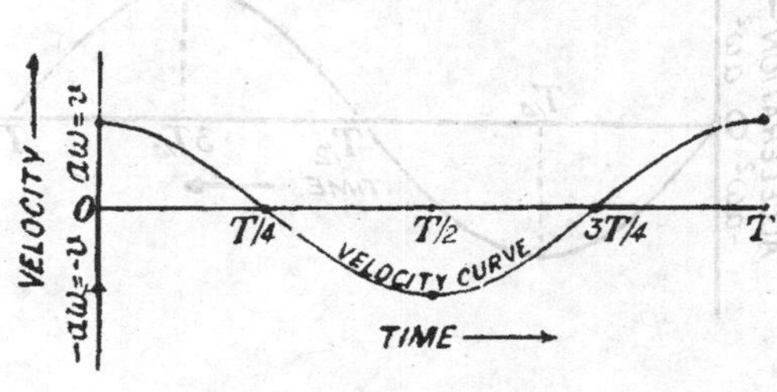

Fig. 4.4

It should be noted that the maximum value of the velocity of the particle is aω or v, and occurs when it passes its mean position to either side, or when its displacement is zero; and its velocity is zero, when it has attained its maximum displacement on either side.

4. Acceleration. As in the case of velocity, so also here, the acceleration of M is the resolved part of the acceleration of P along YOY', (Fig. 4.1). Now, the acceleration of P is v^2/a or $a^2\omega^2/a$, or $a\omega^2$, and is directed towards O. Resolving it into two rectangular components, along PM and MO, we have the component along MO equal to $a\omega^2 . \sin \theta$ or $= a\omega^2 . y/a = \omega^2 y$. And, as is clear from the figure, it is directed towards O, the mean or equilibrium position of M.

Thus, *acceleration of* $M = -\omega^2 . y$

the negative sign being put to indicate that it is directed towards O, in a direction opposite to that of y, its displacement.

Alternatively, we may obtain the acceleration of M by differentiating its velocity with respect to time, for *acceleration is the rate of change of velocity.*

Thus, since $\qquad dy/dt = a\omega . \cos \omega t,$ (See page 105).

we have $\qquad d^2y/dt^2 = -a\omega^2 . \sin \omega t = -\omega^2 . a \sin \omega t = -\omega^2 . y$

i.e., $\qquad$ *acceleration of* $M = -\omega^2 . y.$

Or, we may put it as *acceleration of* $M = -\mu y$*,

where $\omega^2 = \mu$, *a constant of proportionality.*

So that, $\qquad$ *acceleration of* $M \propto y.$

Thus, we see that acceleration of particle M is proportional to y, its displacement from O, and is always directed towards it.

Obviously, *this acceleration has its maximum value,* $a\omega^2$, *the same as that of P, at the extreme positions Y and Y′, where* $y = a$, *and its maximum value, i.e., zero, at O, where* $y = 0$.

It should be noted that the constant of proportionality μ is equal to ω^2, or the *square of the angular velocity* of particle P in the circle of reference.

Further, if $y = 1$, acceleration of $M = \mu$. Thus, μ may be defined as the *acceleration per unit displacement of M.*

Tabulating acceleration of M against time, we have

Time	0	$T/4$	$T/2$	$3T/4$	T
Acceleration	0 (*min.*)	$-a\omega^2$ (*max.*)	0 (*min.*)	$a\omega^2$ (*max.*)	0 (*min.*)

If, therefore, a graph be plotted between time and acceleration, we get the **acceleration curve** of the particle, as shown in Fig. 4.5, which is of a type similar to the *displacement curve*, (Fig. 4.2), but is reciprocal in form, *for acceleration is directed in the opposite sense to displacement.*

5. Time-Period and Frequency. The time taken by particle M in completing one vibration, (or one cycle), *i.e.*, in going from O to Y, Y to Y' and finally back to O, is called its *time-period, period of vibration* or *periodic time*, usually denoted by the letter T. Obviously, it is the *same* as the time-period of particle P, *i.e.*, equal to the time taken by P in making *one* full round of the circle, (from X back to X), or in describing an angle 2π.

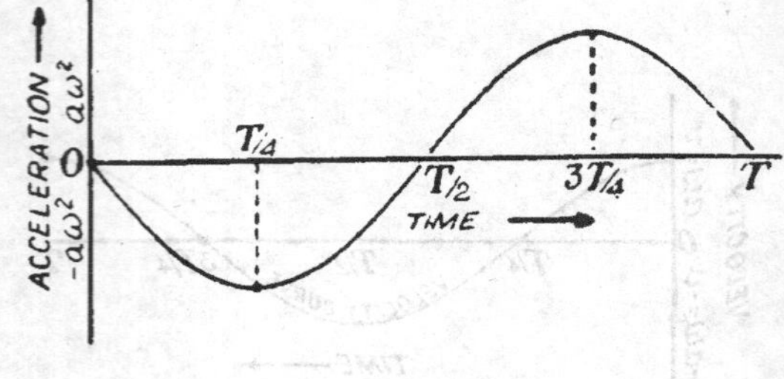

Fig. 4.5

Therefore, $\qquad T = \dfrac{2\pi}{\omega} = \dfrac{2\pi}{\sqrt{\omega^2}} = \dfrac{2\pi}{\sqrt{\mu}},$ $\qquad \left[\because \omega^2 = \mu\right.$, *the acceleration of M per unit displacement.*

whence, $\qquad T = 2\pi\sqrt{1/\mu}.$

*Similarly, in the case of angular *S.H.M.*, we shall have *angular acceleration of the particle* $= \mu . \theta$, where θ is its angular displacement.

So that, here, $\qquad$ *angular acceleration* $\propto \theta$

Or, $\text{time-period of } M = 2\pi\sqrt{\dfrac{1}{\text{acceleration per unit displacement}}}$

The number of vibrations made by the body per second is called the *frequency* of vibration of the body, and is denoted by the letter n. Clearly, therefore, frequency, $n = 1/T = \sqrt{\mu}/2\pi$.

$$= \frac{\sqrt{\text{acceleration per unit displacement}}}{2\pi}$$

6. Phase. The term '*phase*', applied to a vibrating particle, has a meaning similar to the one associated with it when we talk of the '*phases*' of the moon. Just as the phase of the moon—*i.e.* whether it is a *crescent* (or new moon), *half moon* or *full moon*— tells us about its position etc., so also the phase of a particle, executing a *S.H.M.*, enables us to form an idea about its state of vibration.

Thus, the phase of a vibrating particle, at any given instant, may be defined as its state or condition as regards its position and direction of motion at that instant. It tells us in what stage of vibration the particle is at the given instant.

It is indicated either (*i*) in terms of the angle θ, described by the rotating vector, measured as a fraction of the whole angle 2π that it describes in one full rotation, or (*ii*) in terms of time t that has elapsed since the particle last passed its mean position *in the positive direction*, measured as a fraction of its time-period T.

Thus, taking O as the starting position of the particle M (Fig. 4.1), if its phase be zero, it indicates that the particle is at O, tending to move towards Y. And, if the phase be $\pi/2$, or $T/4$, it indicates that it is at Y, the position of the maximum positive displacement; for, the radius of the circle of reference, or the rotating vector, has, up to this instant, described one-fourth of the total angle 2π, *i.e.*, an angle $2\pi/4$ or $\pi/2$, or that one-fourth of the time-period, *i.e.*, $T/4$, has elapsed since the particle last passed its mean position O in the positive or upward direction, towards Y.

Hence, when we talk of a '**phase difference**' between two simple harmonic motions, we mean to indicate *how much the two are out of step with each other*, or by how much angle (measured as a fraction of 2π), or by how much time (measured as a fraction of T) one is ahead of the other.

Now, because the phase of a vibrating particle merely indicates its *actual stage of vibration*, it is clear that two vibrating particles, if they happen to be in *identical stages* of their respective vibrations, at any given instant, will be said to be in the *same phase* (or '*in phase*' with each other) at that particular instant , irrespective of their amplitudes and velocities being the same or different.

Thus, for example, they will be in the same phase, if they both simultaneously attain their maximum displacements, positive or negative; or, when the two pass through their respective mean positions at the same time and in the same direction. Similarly, if one of the particles attains its maximum *positive* displacement simultaneously with the other particle attaining its maximum *negative* displacement or when the two cross each other simultaneously in *opposite* directions at their mean positions, they are said to be in *opposite phases* (or '*out of phase*').

7. Epoch or Initial Phase. We have deduced the relation $y = a \sin \omega t$ for the displacement of the particle M, executing a *S.H.M.*, [§ 4.2,(2)] on the assumption that the starting position of the rotating vector is OX, or that the starting point of the particle in the circle of reference is X, *i.e.*, we start counting time when P crosses the axis of x at X.

Sometimes, however, the starting position of the rotating vector, or the position of particle P in the circle of reference, is fixed *not* in some standard position, as on the axis of x or y, but anywhere *in an arbitrary manner*, such as at P', (Fig. 4.1), *i.e.*, the time is counted from the instant when P is at P', such that the angle $XYP' = e$.

Then, clearly, $\angle POP' = \omega t = \theta + e$. Or, $\theta = (\omega t - e)$.

So that, $y = a \sin \theta = a \sin(\omega t - e)$, where ωt is the *phase angle* of P. This angle e is called the **'epoch'** or the **'initial phase'** of the particle. It may also be measured in terms of the time taken by the particle P in describing this angle, *i.e.*, by the time $(T/2\pi).e$.

It should not be confused with '*phase*'; for, whereas the '*phase*' of the particle *continuously changes with time, its epoch or initial phase remains the same all through.*

4.3 Equation of Simple Harmonic Motion. Let y be the *displacement* from its mean position of a particle executing a *S.H.M*. Then, if v be its velocity at that instant, we have $v = dy/dt$.

So that, *acceleration of the particle* $= \dfrac{d^2y}{dt^2} = \dfrac{dv}{dt} = \dfrac{dv}{dy} \times \dfrac{dy}{dt} = v \cdot \dfrac{dv}{dy}$

Now, acceleration is also given by $-\mu.y$, where μ is the constant of proportionality and is equal to ω^2, (ω being the angular velocity of the rotating vector, or the particle, in the circle of reference).

Thus, $d^2y/dt^2 = -\omega^2.y.$

Or, $v.dv/dy = -\omega^2.y.$

whence, $v.dv = -\omega^2.y.dy.$

[The negative sign indicating that acceleration and displacement (y) are *oppositely directed.*

Integrating this expression, we have

$$\int v.dv = \int -\omega^2.y.dy = -\omega^2 \int y.dy.$$

Or, $\frac{1}{2}v^2 = -\frac{1}{2}\omega^2 y^2 + C,$ ...(*i*)

where C is the *constant of integration*, and has to be determined from the condition of the particle at the instant considered.

Obviously, *the velocity of the particle is zero, when it has its maximum displacement a* or $-a$, *i.e.*, $v = 0$, when $y = a$ or $-a$.

So that, we have, from relation (*i*), above,

$$0 = -\frac{1}{2}\omega^2 a^2 + C. \quad \text{Or,} \quad C = \frac{1}{2}\omega^2 a^2.$$

Hence, $\frac{1}{2}v^2 = -\frac{1}{2}\omega^2 y^2 + \frac{1}{2}\omega^2 a^2 = \frac{1}{2}\omega^2(a^2 - y^2).$ Or, $v^2 = \omega^2(a^2 - y^2).$

And $\therefore$ $v = \omega\sqrt{a^2 - y^2}$ ...(*ii*) [See § 4.2 (3).

Thus, the velocity of the particle can be determined for any value of its displacement y. Clearly, *the value of v is the maximum, i.e., = ωa, when y = 0, i.e., when the displacement is zero, or the particle is in its mean or equilibrium position; and, it is a minimum, i.e., 0, when y=a or, −a, i.e.*, when the *particle has its maximum displacement, positive or negative.*

Now, since $v = dy/dt$, we have, from (*ii*) above,

$$dy/dt = \omega\sqrt{a^2 - y^2}.$$

Or, $\dfrac{dy}{\sqrt{a^2 - y^2}} = \omega.dt.$

Integrating this, we have $\displaystyle\int \frac{1}{\sqrt{a^2 - y^2}}\, dy = \int \omega.dt.$

[Put $y = a \sin\theta$; then, $dy = a\cos\theta.\,d\theta$ and $\sqrt{a^2 - y^2} = a\cos\theta$. So that, $\displaystyle\int \frac{dy}{\sqrt{a^2 - y^2}} = \int d\theta = \theta \sin^{-1}\frac{y}{a}$

Or, $\sin^{-1} y/a = \omega t + C'$

where C' is anohter *constant of integration.*

Or, $y = a \sin(\omega t + C').$ ...(*iii*)

(*a*) Now, *if we start counting time when y = 0, i.e., when the particle is in its mean position, moving in the positive direction, i.e.,* $y = 0$, when $t = 0$, we have, from relation (*iii*) above

$$0 = a \sin (0 + C')$$

Or, $a \sin C' = 0$, whence, $C' = 0$.

∴ Substituting this value of C' in (*iii*) above, we have $y = a \sin \omega t$.

(*b*) *If, on the other hand, we start counting time when the particle has its maximum displacement, i.e.,* $t = 0$, *when* $y = a$, we have

$a = a \sin(0 + C')$, or, $\sin C' = a/a = 1$ or $C' = \pi / 2$;

so that, in this case, $y = a \sin (\omega t + \pi/2)$. Or, $y = a \cos \omega t$.

(*c*) Again, *if we start counting time from an instant t′ before the particle has passed through its mean or equilibrium position,* we have $t=t'$, when $y=0$. Therefore, from (*iii*) above

$$0 = a \sin (\omega t' + C'), \text{ or, } \omega t' + C' = 0,$$

whence, $C' = -\omega t' = -e$,

where e is the **epoch** of the particle in the circle of reference.

Substituting this value of C' in (*iii*), we have $y = a \sin (\omega t - e)$ ...(*iv*)

(*d*) And, *if we start counting time from an instant t′ after the particle has passed through its mean position,* we have

$$y = 0, \text{ when } t = -t', \text{ or, } 0 = a \sin (-\omega t' + C'),$$

whence, $-\omega t + C' = 0$, or, $C' = \omega t' = e$.

And, therefore, $y = a \sin (\omega t + e)$.

A mere glance at the above relations for y indicates that these simple harmonic vibrations of the particle are a case of *periodic motion*.

(*e*) Now, *if the time t be increased by* $2\pi/\omega$, we have, from relation (*iv*) above,

$y = a \sin [\omega(t + 2\pi/\omega) - e] = a \sin (\omega t + 2\pi - e)$,

whence, $y = a \sin (\omega t - e)$, *i.e., the same as before*, in (*iv*),

showing that the position and direction of motion of the particle is the same as $2\pi/\omega$ *seconds* earlier, *i.e., the particle repeats its movements after every* $2\pi/\omega$ *seconds*. In other words, *the time-period of the particle is* $2\pi/\omega$. Further, since this value of the time-period is quite independent of a and e, it is clear that the vibrations or oscillations of the particle are also *isochronous*.

The above results will also be true for *angular S.H.M.*, if we consider *angular displacement, acceleration, velocity*, etc., in place of the linear ones.

Important Note. — We have seen how the acceleration of a particle executing a *S.H.M.* is given by $d^2y/dt^2 = -\omega^2 . y$.

The general solution of this equation is of the form

$$y = a \sin \omega t + b \cos \omega t.$$

Thus, if the displacement of a vibrating particle be given by a relation of the form $y = a \sin \omega t + b \cos \omega t$, it is executing a *Simple Harmonic Motion*.

Clearly, as t takes up the values 0, $2\pi/\omega$, $4\pi/\omega$, $2n\pi/\omega$, etc., ωt assumes the values 0, 2π, 4π, $2n\pi$, etc., with y assuming the same value over and over again. In other words, the *time-period of the motion* = $2\pi/\omega$.

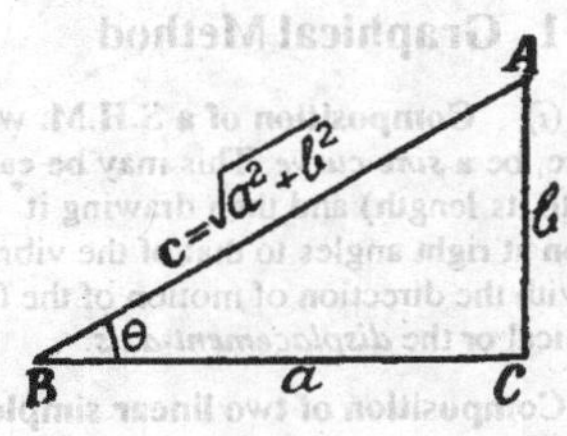

Fig. 4.6

Further, this equation can easily be reduced to the simple *sine form* as follows :

Let $\theta = tan^{-1} (b / a)$, (Fig. 4.6)

Then, clearly, $a = c \cos\theta = \sqrt{(a^2 + b^2)} \cos\theta$.

and $b = c \sin\theta = \sqrt{(a^2 + b^2)} \sin\theta$.

Now, $y = a \sin\omega t + b \cos\omega t$.

So that, substituting the values of a and b, obtained above, we have

$$y = \sqrt{(a^2 + b^2)} \sin\omega t \cos\theta + \sqrt{(a^2 + b^2)} . \cos\omega t \sin\theta .$$

Or, $$y = \sqrt{a^2 + b^2} (\sin\omega t \cos\theta + \cos\omega t \sin\theta) .$$

Or, $$y = \sqrt{a^2 + b^2} \sin(\omega t + \theta) ,$$

which is the usual form of the expression for displacement of a body executing *S.H.M.*

Obviously, the displacement will have the maximum value $\sqrt{a^2 + b^2}$, when $(\omega t + \theta) = 90°$ and, therefore, $\sin(\omega t + \theta) = 1$

And, *since the maximum displacement of the particle is equal to its aptitude*, we have *amplitude of the vibrating particle, here* = $\sqrt{a^2 + b^2}$.

Again, the velocity of the particle is given by

$$v = dy/dt = \sqrt{a^2 + b^2} . \omega \cos(\omega t + \theta) ;$$

So that, the *maximum value of* $v = \sqrt{a^2 + b^2} . \omega$.

And, finally, *acceleration of the particle* = d^2y/dt^2

$$= -\omega^2 \sqrt{a^2 + b^2} . \sin(\omega t + \theta).$$

Hence, *maximum value of acceleration of the particle* = $\omega^2 \sqrt{a^2 + b^2}$.

4.4. Composition of Two Simple Harmonic Motions. Just as a particle may be subjected to two forces or two velocities simultaneously, so also we may have a particle under the action of two simple harmonic motions *at the same time.* Its final motion will then be the resultant of the two simultaneous simple harmonic motions impressed upon it. It does not, of course, mean that it will execute both the motions simultaneously, any more than a particle, having two velocities impressed upon it, will move in both the directions at the same time. All it means is that its resulting motion would be one *as though it were executing the two motions simultaneously.*

It should be clearly understood, however, that the simultaneous execution of two rectilinear simple harmonic motions by a particle is no guarantee that the resultant motion of the particle will necessarily be rectilinear or harmonic. Indeed, if their time-periods be incommensurable, it may not even answer to the definition of a vibration.

We shall now take up first the simpler case of the composition of a *S.H.M.* along one direction with a linear motion in a perpendicular direction and then pass on to the composition of simple harmonic motions along the same straight line and at right angles to each other, both graphically and analytically.

1. Graphical Method

(*i*) **Composition of a S.H.M. with a Uniform Linear Motion perpendicular to it.** The resultant motion will, in this case, be a *sine curve*. This may be easily seen by attaching a small spike or style to the prong of a tuning fork (at right angles to its length) and then drawing it* uniformly over a smoked plate of glass, with the style just touching the plate in a direction at right angles to that of the vibrations of the fork. It will be found that a series of sine curves are traced out on the plate, with the direction of motion of the fork as the horizontal or the *time-axis* and the direction of vibration of the prong as the vertical or the *displacement-axis*.

Composition of two linear simple harmonic motions in the same direction. If two simple harmonic motions take place in the same direction, their resultant is also a simple harmonic motion, defined by the resultant of the vectors which

* Or, holding the fork in position and moving the plate.

define the two motions, this *resultant vector* being obtained by the ordinary law of vector addition. This will be clear from the following :

Let two simple harmonic motions, having the *same time-period* but *different amplitudes and phases*, be represented by the projections of the vectors OP and OQ respectively, on the axis of y (Fig. 4.7), and let their equations of motion be

$$y = a \sin \omega t \quad \text{and} \quad y = b \sin (\omega t + \phi),$$

where a and b are their amplitudes, and ωt and $(\omega t + \phi)$ equal to angles AOP and AOQ respectively.

Then, if $DEFG$ be the *sine-curve* for the first motion, and $DHJK$ for the second, *we obtain the since-curve DLMN* for the *resultant motion, by adding up the ordinates of the two curves at all points*, because the displacements of the two are in the same direction and can be added up algebraically.

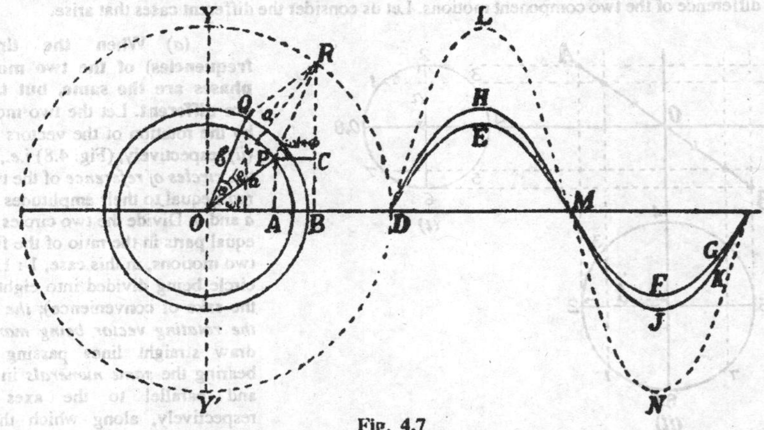

Fig. 4.7

Now, the curve $DLMN$ is the same as would be obtained for the rotation of the resultant vector OR, whose projection on the axis of y, therefore, gives the resultant of the two motions.

For, $\quad RB = CB + RC = PA + RC = OP.\ \sin AOP + PR.\sin CP = a \sin \omega t + b \sin (\omega t + \phi),$

because $OP = a$, $PR = OQ = b$, $\angle AOP = \omega t$ and $\angle CPR = \angle AOQ = \omega t + \phi$.

Thus, the resultant motion is also a S.H.M. and takes place along the same line and, (since the rotating vectors OP, OQ, and OR have the same velocities) *it has the same time-period as the two component motions.*

The amplitude a' of the resultant motion is, clearly, equal to OR.

Now, $\quad OR^2 = OP^2 + OQ^2 + 2\ OP.OQ \cos POQ.$ **By the ‖ gram law of vectors.**

Or, $\quad a'^2 = a^2 + b^2 + 2ab. \cos \phi.$

$$a' = \sqrt{a^2 + b^2 + 2ab. \cos\theta}$$

Now, if $\phi = 0$, *i.e.* if there be no phase difference between the two motions, $\cos \phi = 1$, and therefore,

$$a' = \sqrt{a^2 + b^2 + 2ab} = (a + b),$$

i.e., equal to the algebraic sum of the amplitudes of the two component motions.

The *phase angle* of the resultant motion is, obviously, given by the angle ROB, such that

$$\tan ROB = \frac{RB}{OB} = \frac{CB + RC}{OA + AB} = \frac{PA + RC}{OA + PO}.$$

Or, $$\tan ROB = \frac{a \sin \omega t + b \sin (\omega t + \phi)}{a \cos \omega t + b \cos (\omega t + \phi)}$$

Now, $$\angle ROB = \angle POA + \angle POR = \omega t + e,$$

where $\angle POR = e$, the *phase angle* by which the resultant motion is ahead of the first motion.

$$\therefore \quad \tan ROB = \tan (\omega t + e) = \frac{a \sin \omega t + b \sin (\omega t + \phi)}{a \cos \omega t + b \cos (\omega t + \phi)}.$$

$$= \frac{a \sin \omega t + b \sin \omega t \cos \phi + b \cos \omega t \sin \phi}{a \cos \omega t + b \cos \omega t \cos \phi - b \sin \omega t \sin \phi}$$

Now, *just at the start*, $t = 0$, and $\therefore \ wt = 0$; so that,

$$\sin \omega t = 0, \quad \text{and} \quad \cos \omega t = 1.$$

Hence $$\tan e = \frac{b \sin \phi}{a + b \cos \phi} \quad \text{Or,} \quad e = \tan^{-1} \frac{b \sin \phi}{a + b \cos \phi}.$$

Thus, the resultant motion is ahead of the first motion by a phase angle e,

$$\text{where } e = \tan^{-1}\frac{b \sin \phi}{a + b \cos \phi}$$

(*ii*) **Resolution of a S.H.M. into two components in the same direction.** The converse of the above is also true, *viz., that a simple harmonic motion may be resolved into two, by resolving its rotating vector into two vectors, in accordance with the law of resolution of vectors, each vector defining a component simple harmonic motion.*

(*iii*) **Composition of two linear simple harmonic motions at right angles to each other. Lissajous' figures.** The resultant of two *S.H.M.'s*, impressed simultaneously on a particle, along directions at right angles to each other, is *a curve lying in the plane containing the two motions* and its character depends upon the amplitudes, time-periods (or frequencies) and the phase difference of the two component motions. Let us consider the different cases that arise.

(*a*) **When the time-periods (or frequencies) of the two motions and their phases are the same, but their amplitudes are different.** Let the two motions be defined by the rotation of the vectors in circles (*i*) and (*ii*) respectively, (Fig. 4.8) *i.e.*, let (*i*) and (*ii*) be the *circles of reference* of the two motions, with radii equal to their amplitudes respectively, say *a* and *b*. Divide the two circles into a number of equal parts in the ratio of the frequencies of the two motions, in this case, 1 : 1, as shown, (each circle being divided into eight equal parts, for the sake of convenience); *the starting point of the rotating vector being marked zero.* Then, draw straight lines passing through points bearing the *same numerals* in the two circles, and parallel to the axes *OX* and *OY* respectively, along which the motions take place in the two cases.

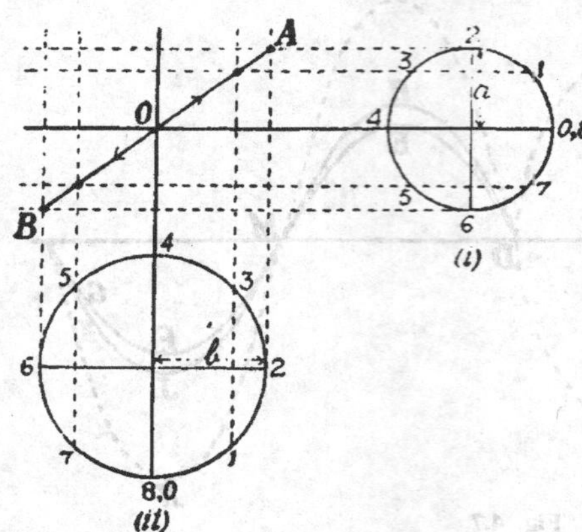

Fig. 4.8

Mark the points where these lines intersect and join all these points of intersection. It will be found that, in this case, the **straight line** *AB* is obtained as the path along which the resultant motion takes place, the arrow heads indicating the direction of motion about *O*. And, as will be readily seen, this straight line is the diagonal of the rectangle with sides 2*a* and 2*b* and the amplitude of the resultant vibration of the particle, *i.e.*, *OA* or *OB* is, therefore, clearly

$$= \sqrt{a^2 + b^2}.$$

(*b*) **When the time-periods or frequencies are the same, amplitudes are different and phase difference is π.** Here, the starting position (Fig. 4.9), of the vector in circle (*ii*) is at the top, as shown, instead of at the bottom, (as in the first case), the second motion being ahead of the first by a distance *equal to half its path*, —the other numerals being shifted accordingly. Again, drawing straight lines through the same numerals and parallel to the corresponding axes *OX* and *OY*, along which the two motions take place, and joining their points of intersection we get the **straight line** *AC*, *inclined in the opposite direction*, showing that th resultant motion is again *a straight line motion*, about *O*, but inclined the other way, (*i.e.*, the other diagonal of the rectangle *ABCD* of sides 2*a* and 2*b*), the direction of motion of the particle being as indicated by the arrow-heads.

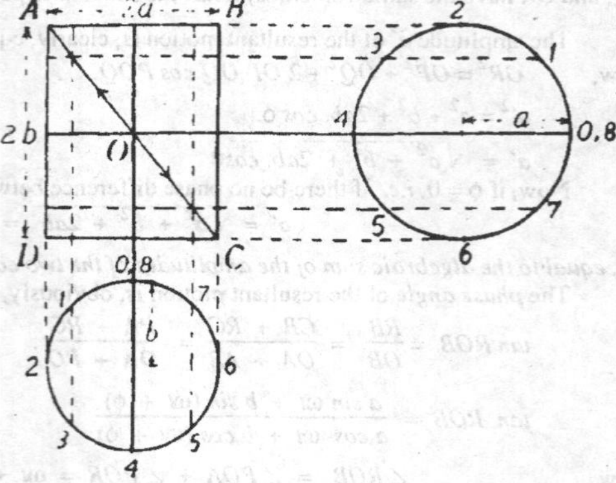

Fig. 4.9

(c) When the time periods are the same, amplitudes different, and the phase difference is $\pi/4$. We again proceed exactly as above, with the only difference that, here, we shift the zero, or the starting position of the radius vector, by *one-eighth of its path in the case of second (lower) circle of reference, (Fig. 4.10), the second motion being ahead of the first by $\pi/4$.*

Joining smoothly the points of intersection of the straight lines through the same numerals parallel to the two axes respectively, we get an **oblique ellipse,** inscribed in a rectangle of sides $2a$ and $2b$, as the resultant path of motion of the particle, the direction of motion along it being indicated by the arrow-heads.

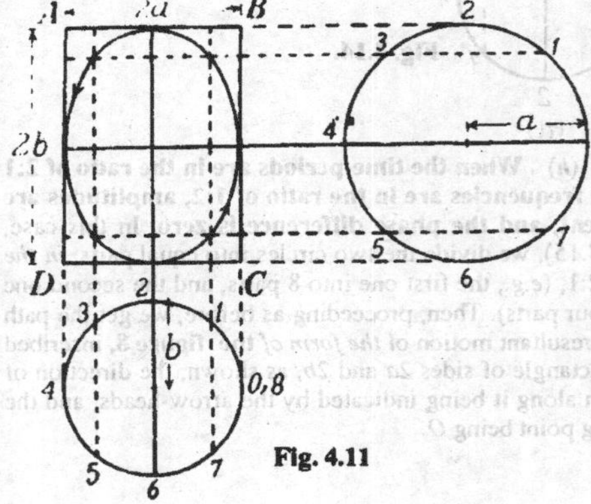

Fig. 4.11

(e) When the time-periods or frequencies are the same, amplitudes different, and the phase difference is $3\pi/2$ Here, (Fig. 4.12), the starting point of the radius vector in the second circle is taken *three-fourths of its path* ahead of the original position and we get an **ellipse,** *similar to that in the last case,* as the resultant path, the direction of motion being now *clockwise,* as shown, and the starting point, O.

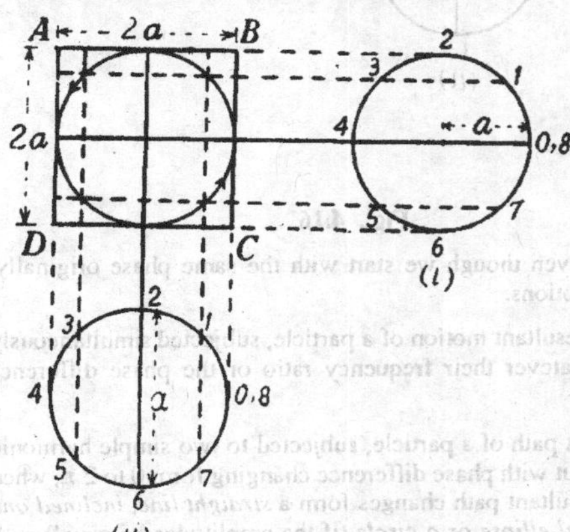

Fig 4.13

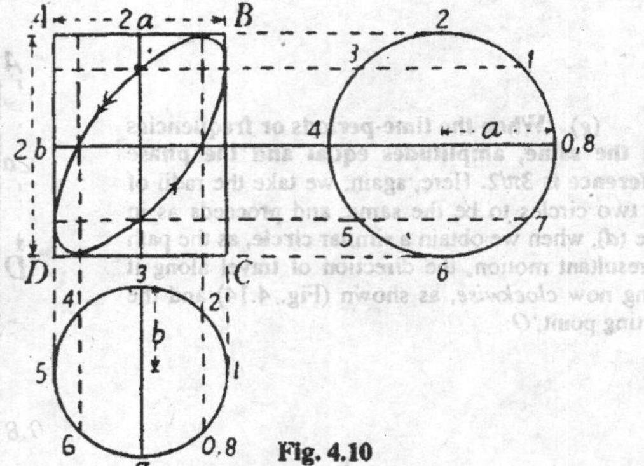

Fig. 4.10

(d) When the time-periods are the same, amplitudes different and the phase difference is $\pi/2$. In this case, the starting point of the radius vector in the second circle of reference is taken a *quarter of its path* ahead of its original position, the phase difference being $\pi/2$.

Then, proceeding as before, we again get an **ellipse** inscribed in a rectangle of sides $2a$ and $2b$, as the path of the resultant motion but *with its axes coincident with the directions of the component motions,* the *starting* point being O and the direction *anticlockwise,* as shown in Fig. 4.11

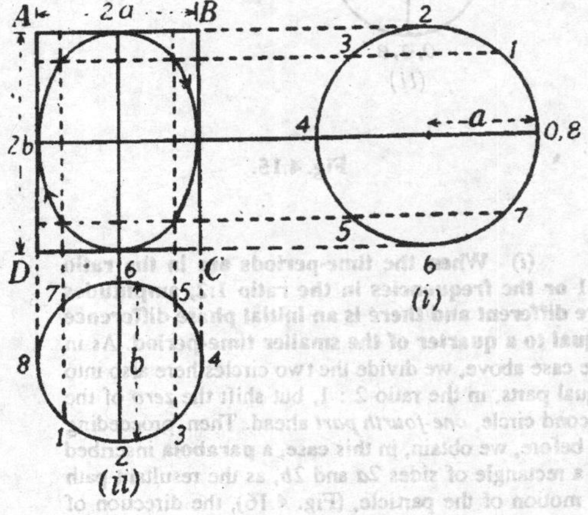

Fig. 4.12

(f) When the time-periods or the frequencies are the same, amplitudes equal and the phase difference is $\pi/2$. In this case, (Fig. 4.13), we take both the circles of reference *of the same radii* and proceed as in case (c), *when a* **circle** (inscribed in a rectangle of sides $2a$ and $2b$) is obtained as the path of the resultant motion, the direction of travel along it being *anticlockwise* and the starting point, O.

(g) **When the time-periods or frequencies are the same, amplitudes equal and the phase difference is 3π/2.** Here, again, we take the radii of the two circles to be the same, and proceeds as in case (*d*), when we obtain a similar circle, as the path of resultant motion, the direction of travel along it being now *clockwise*, as shown (Fig. 4.14) and the starting point, *O*.

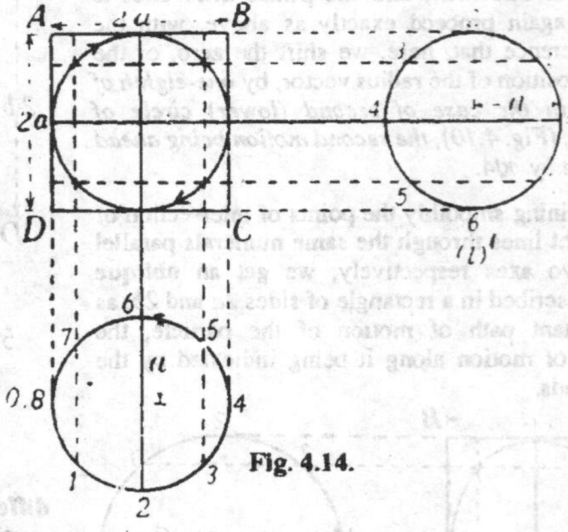

Fig. 4.14.

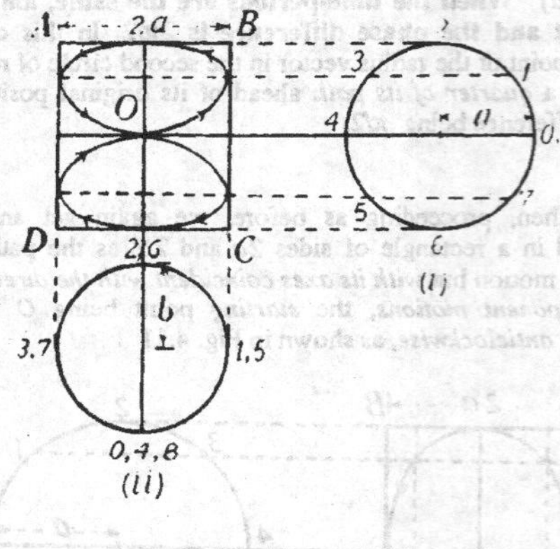

Fig. 4.15.

(*h*) **When the time-periods are in the ratio of 2:1 or the frequencies are in the ratio of 1:2, amplitudes are different, and the phase difference is zero.** In this case, (Fig. 4.15), we divide the two circles into equal parts, *in the ratio* 2:1, (*e.g.*, the first one into 8 parts, and the second one into four parts). Then, proceeding as before, we get the path of the resultant motion of *the form of* **the figure 8**, inscribed in a rectangle of sides 2*a* and 2*b*, as shown, the direction of motion along it being indicated by the arrow-heads, and the starting point being *O*.

(*i*) **When the time-periods are in the ratio 2:1 or the frequencies in the ratio 1:2, amplitudes are different and there is an initial phase difference equal to a quarter of the smaller time-period.** As in the case above, we divide the two circles here also into equal parts, in the ratio 2 : 1, but shift the *zero* of the second circle, *one-fourth part* ahead. Then, proceeding as before, we obtain, in this case, a **parabola** inscribed in a rectangle of sides 2*a* and 2*b*, as the resultant path of motion of the particle, (Fig. 4.16), the direction of motion being as indicated by the arrow heads and the starting point being *D*.

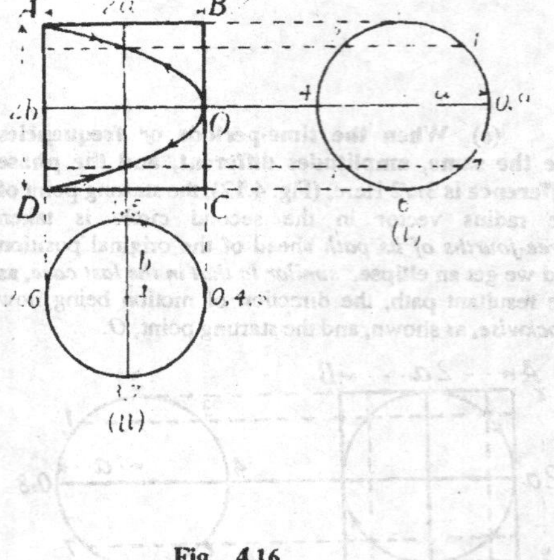

Fig. 4.16

N.B. The epithet *'initial'* has been deliberately used here with *phase difference* to emphasize that the time-periods being different, *the phase does not remain constant*, even though we start with the same phase originally. Inevitably, therefore, a phase difference comes in between the two motions.

Precisely in the same manner, we can obtain the path of the resultant motion of a particle, subjected simultaneously to two simple harmonic motions, perpendicular to each other, whatever their frequency ratio or the phase difference between them.

The student may, as an exercise, try to determine the resultant path of a particle, subjected to two simple harmonic motions, at right angles to each other, with their time periods equal, but with phase difference changing form 0 to 2 π, when he will find that, as the phase difference changes form 0 to π the resultant path changes form a *straight line, inclined one way*, through an *oblique* ellipse inclined the same way, *a symmetrical ellipse* or *a circle* (if the amplitudes be equal) and, again, an *ellipse, inclined the other way*, to finally, a *straight line, inclined at right angles to the first one*, as shown in Fig. 4.17. And, as the phase difference changes from π to 2 π, the same figures are repeated in the reverse order, as indicated.

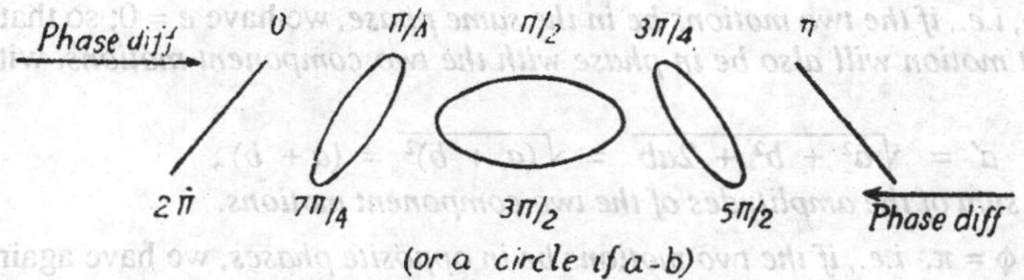

Fig.4.17

The superposition of such rectangular vibrations is of particular importance in the subject of *sound*, since it serves as a test for the equality of the periods of two vibrating bodies like tuning forks etc. The method was first adopted by *Lissajous* and hence the various curves thus obtained, *viz*., those in Fig. 4.8 to 4.16 and others, are usually referred to as **Lissajous figures.**

Analytical method

(1) Composition of two linear simple harmonic motions along the same line. Let two simple harmonic motions, having the *same time-period*, be represented by the equations

$$y_1 = a \sin \omega t \quad \text{and} \quad y_2 = b \sin (\omega t + \phi)$$

where ϕ is the phase angle by which the second motion is ahead of the first.

The phase difference will throughout remain constant, because the time-periods of the two motions are the same.

Now, since the two displacements are along the same line, the resultant displacement y will, at any give instant, be equal to the algebraic sum of the displacements of the two component vibrations.

Thus,
$$y = y_1 + y_2 = a \sin \omega t + b \sin (\omega t + \phi) .$$
$$= a \sin \omega t + b \sin \omega t \cos \phi + b \cos \omega t \sin \phi .$$

Or,
$$y = \sin \omega t (a + b \cos \phi) + \cos \omega t . b \sin \phi .$$

Putting $(a + b \cos \phi) = a' \cos e$ and $b \sin \phi = a' \sin e$, we have

$$y = a' \sin \omega t \cos e + a' \cos \omega t \sin e.$$

Or,
$$y = a' \sin (\omega t + e) ,$$

i.e., the resultant motion is also a S.H.M., along the same line, and has the same time-period, its amplitude being a', *and its phase angle* e, *by which it is ahead of the first motion.*

The value of a' and e may be deduced as follows :

We have $\quad a' \sin e = b \sin \phi \quad$ and $\quad a' \cos e = (a + b \cos \phi)$,

So that, squaring and adding the two, we have

$$a'^2 \sin^2 e + a'^2 \cos^2 e = b^2 \sin^2 \phi + a^2 + b^2 \cos^2 \phi + 2ab \cos \phi .$$

Or,
$$a'^2 (\sin^2 e + \cos^2 e) = a^2 + b^2 (\sin^2 \phi + \cos^2 \phi) + 2ab \cos \phi .$$

Or,
$$a'^2 = a^2 + b^2 + 2ab \cos \phi .$$

Or,
$$a' = \sqrt{a^2 + b^2 + 2ab \cos \phi} .$$

And
$$\tan e = \frac{\sin e}{\cos e} = \frac{a' \sin e}{a' \cos e} = \frac{b \sin \phi}{a + b \cos \phi} .$$

Or, $$e = \tan^{-1} \frac{b \sin \phi}{a + b \cos \phi}$$ (See page 112.

Now, (*i*) if $\phi = 0$, *i.e., if the two motions be in the same phase*, we have $e = 0$; so that, $y = a' \sin \omega t$. Or, *the resultant motion will also be in phase with the two component motions*, with its *amplitude* given by

$$a' = \sqrt{a^2 + b^2 + 2ab} = \sqrt{(a + b)^2} = (a + b), \quad [\because \cos 0 = 1$$

i.e., equal to the sum of the amplitudes of the two component motions. [See page 111.

And (*ii*) if $\phi = \pi$, *i.e., if the two motions be in opposite phases*, we have again $e = 0$. So that, again, $y = a' \sin \omega t$.

Or, the resultant motion will be in phase with the first motion, with its *amplitude* now given by

$$a' = \sqrt{a^2 + b^2 - 2ab} = \sqrt{(a - b)^2} = (a - b), \quad [\because \cos \pi = -1$$

i.e., equal to the difference between the amplitudes of the two motions.

Further, in this case, if $a = b$, *i.e., if the amplitudes of the two component motions be also equal*, we shall have $a' = a - b = 0$.

Or, *the amplitude of the resultant motion will be zero.* In other words *there will be no redultant motion at all.*

Note. — In the above treatment we have, for the sake of simplicity, taken one motion a phase angle ϕ ahead of the other.

The same result may be obtained, however, if we take the phase angles of the two motions to be ϕ_1 and ϕ_2 respectively. For, in this case, we have

$$y_1 = a \sin (\omega t + \phi_1) \text{ and } y_2 = b \sin (\omega t + \phi_2).$$

So that, $$y = y_1 + y_2 = a \sin (\omega t + \phi_1) + b \sin (\omega t + \phi_2).$$

$$= a \sin \omega t \cos \phi_1 + a \cos \omega t \sin \phi_1 + b \sin \omega t \cos \phi_2 + b \cos \omega t \sin \phi_2.$$

$$= \sin \omega t (a \cos \phi_1 + b \cos \phi_2) + \cos \omega t (a \sin \phi_1 + b \sin \phi_2). = a' \sin i$$

Now, putting $(a \cos \phi_1 + b \cos \phi_2) = a' \cos e$ and $(a \sin \phi_1 + b \sin \phi_2) = a' \sin e$.

we have $$y = a' \sin \omega t \cos e + a' \cos \omega t \sin e.$$

Or, $y = a' \sin (\omega t + e)$, as before;

i.e., the resultant motion is also *simple harmonic*, with the same time-period as that of the two component motions.

And, $$\tan e' = \frac{a' \sin e}{a' \cos e} = \frac{a \sin \phi_1 + b \sin \phi_2}{a \cos \phi_1 + b \cos \phi_2}.$$

Again, the value of the amplitude a' of the resulting motion may be obtained in exactly the same manner, as before.

Thus, $$a'^2 \sin^2 e + a'^2 \cos^2 e = (a \sin \phi_1 + b \sin \phi_2)^2 + (a \cos \phi_1 + b \cos \phi_2)^2.$$

Or, $$a'^2 (\sin^2 e + \cos^2 e) = a^2 \sin^2 \phi_1 + b^2 \sin^2 \phi_2 + 2ab \sin \phi_1 \sin \phi_2 + a^2 \cos^2 \phi_1 + b^2 \cos^2 \phi_2 + 2ab \cos \phi_1 \cos \phi_2.$$

Or, $$a'^2 = a^2 (\sin^2 \phi_1 + \cos^2 \phi_1) \; b^2 (\sin^2 \phi_2 + \cos^2 \phi_2) + 2ab (\sin \phi_1 \sin \phi_2 + \cos \phi_1 \cos \phi_2)$$

Or, $a'^2 = a^2 + b^2 + 2ab \cos (\phi_1 - \phi_2)$ whence, $a' = \sqrt{a^2 + b^2 + 2ab \cos (\phi_1 - \phi_2)}$.

which, when $(\phi_1 - \phi_2) = \phi$, gives the same result, as above.

Proceeding again, as before, if (*i*) $\phi_1 = \phi_2$ or, $(\phi_1 - \phi_2) = 0$, *i.e., if the two motions be in the same phase*, we have $a' = (a + b)$;

(*ii*) if $\phi_1 - \phi_2 = \pi$, *i.e., if the two motions be in opposite phases*, $a' = (a - b)$;

and, further, if $a = b$, *i.e., if the amplitudes of the two motions be also equal*, we have $a' = 0$; *i.e.*, the same results as above.

(2) Composition of two linear simple harmonic motions at right angles to each other. Let the two simple harmonic motions be along the axes of co-ordinates XOX' and YOY', and let a and b be their amplitudes respectively and ϕ, *the phase difference* between them. Then, if their displacements at any instant t be x and y, we have

$$x = a \sin \omega t \;...(i) \quad \text{and} \quad y = b \sin (\omega t + \phi), \quad ...(ii)$$

$\therefore \quad sin\ \omega t = x/a$. and $cos\ \omega t = \sqrt{1 - cos^2\ \omega t} = \sqrt{1 - x^2/a^2}$.

Now, $\quad y = b\,(sin\ \omega t\ cos\ \phi + cos\ \omega t\ sin\ \phi)$, [from (*ii*), above

Or, $\quad y/b = sin\ \omega t\ cos\ \phi + cos\ \omega t\ sin\ \phi$.

Or, $\quad \frac{y}{b} = \frac{x}{a}\ cos\ \phi + \left(1 - \frac{x^2}{a^2}\right)^{1/2} sin\ \phi$.

Or, $\quad \left(\frac{y}{b} - \frac{x}{a}\ cos\ \phi\right) = \left(1 - \frac{x^2}{a^2}\right)^{1/2} sin\ \phi$.

So that, squaring both sides, we have

$$\left(\frac{y}{b} - \frac{x}{a}\ cos\ \phi\right)^2 = \left(1 - \frac{x^2}{a^2}\right)^{1/2} sin^2\ \phi .$$

Or, $\quad \frac{y^2}{b^2} + \frac{x^2}{a^2}\ cos^2\ \phi - \frac{2xy}{ab}\ cos\ \phi = sin^2 \phi - \frac{x^2}{a^2}\ sin^2\ \phi$.

Or, $\quad \frac{y^2}{b^2} + \frac{x^2}{a^2}\ cos^2\ \phi\ \ \frac{x^2}{a^2}\ sin^2\ \phi - \frac{2xy}{ab}\ cos\ \phi = sin^2\ \phi$.

Or, $\quad \frac{y^2}{b^2} + \frac{x^2}{a^2}\ (cos^2\ \phi + sin^2\ \phi) - \frac{2xy}{ab}\ cos\ \phi = sin^2\ \phi$.

Or, $\quad \frac{y^2}{b^2} + \frac{x^2}{a^2} - \frac{2xy}{ab}\ cos\ \phi = sin^2\ \phi$ . ...(*iii*)

This is the general equation to an *ellipse,* inclined to the axes of the coordinates, and inscribed in a rectangle of sides 2*a* and 2*b*.

Now, a number of **special cases arise :**

(*a*) **When $\phi = 0$, i.e., when there is no phase difference between the two motions.** In this case, $sin\ \phi = 0$, and $cos\ \phi = 1$. So that, substituting these values in relation (*iii*) above, we have

$$\frac{y^2}{b^2} + \frac{x^2}{a^2} - \frac{2xy}{ab} = 0 .$$

Or, $\quad \left(\frac{y}{b} - \frac{x^2}{a}\right) = 0.\quad$ Or, $\pm\left(\frac{y}{b} - \frac{x}{a}\right) = 0.$

Or, $\quad \pm\, y = \pm\, \frac{b}{a} . x .$

This equation represents a pair of coincident *straight lines, passing through the origin and lying in the first and the third quadrants, such that it meets the axis XX′ at an angle* $tan^{-1}\ b/a$, [see case I, (*iii*), (*a*)].

The resultant motion is, therefore, along the straight line *BD*, (Fig. 4.18), *i.e., the particle describes a S.H.M. along this line, with the same time-period as that of the two component motions* and an amplitude equal to $\sqrt{a^2 + b^2}$, *the line being covered twice in one time-period in the direction shown.*

If amplitudes of the two motions be equal, the straight line *BD* is *inclined at an angle of* 45° to the axes of *x* and *y*.

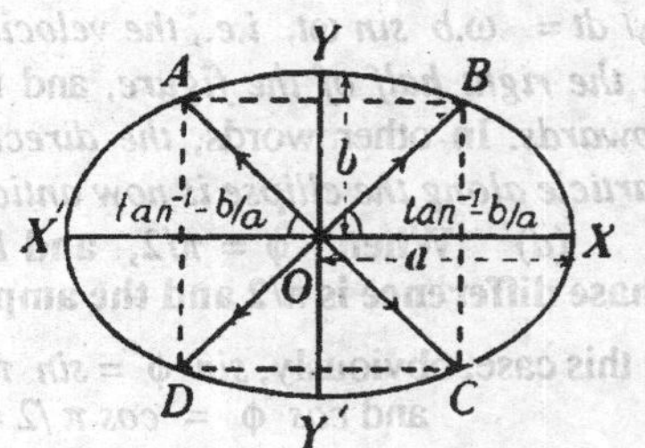

Fig. 4.18

(*b*) **When $\phi = \pi$, *i.e.*, when the phase difference between the two motions is π.** Here, $\sin\phi = 0$ and $\cos\phi = -1$. So that, from (*iii*) above, we have

$$\pm\left(\frac{y}{b} + \frac{x}{a}\right) = 0. \quad \text{Or,} \quad \pm y = \mp \frac{b}{a} x.$$

This equation again represents a pair of two coincident *straight lines passing through the origin and lying in the second and the fourth quadrants, such that it is inclined to the x-axis at an angle* $\tan^{-1} -b/a$; so that, *the resultant motion is again a S.H.M., with the same time-period but along the straight line AC*, (Fig. 4.18), inclined the other way, [see case I, (*iii*), (*b*)], the amplitude being again $\sqrt{a^2 + b^2}$, and the line, again, being covered twice in one time-period in the directions indicated.

(*c*) **When $\phi = \pi/2$, i.e., when the phase difference between the two motions is $\pi/2$.**

Here $\sin\phi = 1$ and $\cos\phi = 0$.

Substituting these values of $\sin\phi$ and $\cos\phi$ in relation (*iii*) above,

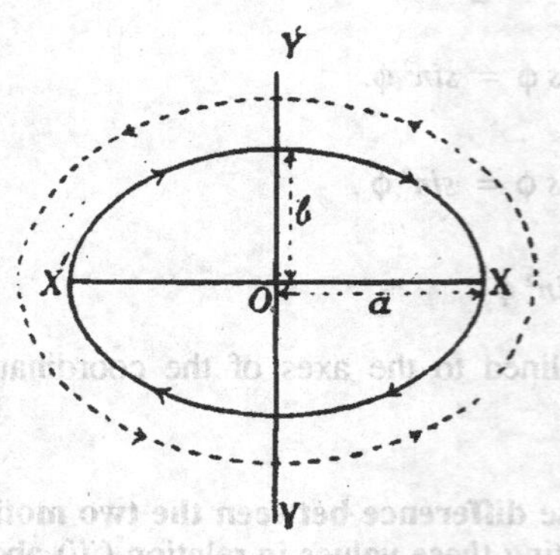

Fig. 4.19.

we have $$\frac{y^2}{b^2} + \frac{x^2}{a^2} = 1.$$

This is the equation to an ellipse, whose major and minor axes coincide with the directions of the two given motions, and whose semi-axes are equal to b and a respectively. The resultant path, is, therefore, an ellipse, (Fig. 4.19), which it describes once in the time-period of each component *S.H.M.*, [see case I, (*iii*), (*d*), above]

The direction of motion of the particle along the ellipse may be determined as follows :

Since $\phi = \pi/2$ and $x = a \sin \omega t$,

we have $y = b \sin(\omega t + \phi) = b \sin(\omega t + \pi/2)$, whence, $y = b \cos \omega t$.

$\therefore$ differentiating the expression for y with respect to t, we have

velocity of the particle $= dy/dt = -\,\omega.b \sin \omega t$.

Now, x and, therefore, $\sin \omega t$, is positive in the right half of the figure. And, therefore, the negative sign of dy/dt means that the velocity of y is *negative, i.e., it is directed downwards in the right half of the figure*. In other words, *the direction of the particle along the ellipse is clockwise*.

If, on the other hand, $\phi = -\pi/2$, we have, again $\dfrac{y^2}{b^2} + \dfrac{x^2}{a^2} = 1$.

i.e., the resultant motion is again an *ellipse*.

But, since $y = b \sin(\omega t - \pi/2) = -b \cos \omega t$, we have $dy/dt = \omega.b \sin \omega t$, *i.e., the velocity of y is now positive in the right half of the figure*, and is, therefore, *directed upwards*. In other words, *the direction of motion of the particle along the ellipse is now anticlockwise*.

(*d*) **When $\phi = \pi/2$, and $b = a$, i.e., when the phase difference is $\pi/2$ and the amplitudes are equal.**

In this case, obviously, $\sin\phi = \sin \pi/2 = 1$
and $\cos\phi = \cos \pi/2 = 0$.

$\therefore$ substituting these values in relation (*iii*) above, we have

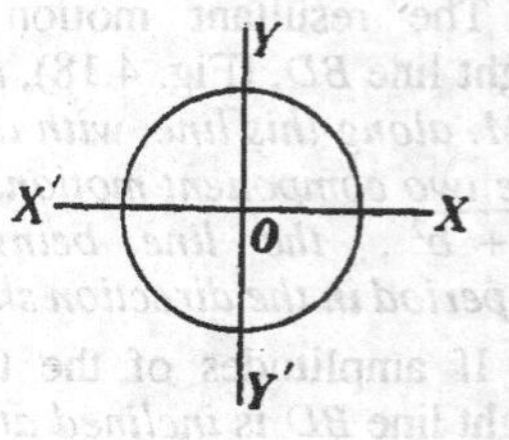

Fig. 4.20

$$\frac{y^2}{a^2} + \frac{x^2}{a^2} = 1. \quad \text{Or,} \quad \frac{y^2 + x^2}{a^2} = 1,$$

whence, $\quad y^2 + x^2 = a^2$

This is the equation to a *circle, whose radius is equal to the amplitude of either simple harmonic motion*; so that, in this case, *the particle describes a circle*, (Fig. 4.20), *once in the same time as that taken by any one of the two component motions*, [see case I, (*iii*), (*g*) above] — the direction of motion along the circle being determined as explained above in the case of the ellipse.

A uniform circular motion may thus be regarded as a combination of two equal or similar simple harmonic motions, at right angles to each other, and differing in phase by $\pi/2$

(*e*) **When $\phi = \pi/4$, i.e., when the phase difference between the two motions is $\pi/4$.**

Here, $\quad \cos\phi = \cos\frac{\pi}{4} = \frac{1}{\sqrt{2}}$ and, so also, $\sin\phi = \frac{1}{\sqrt{2}}$.

Hence, from relation (*iii*) above, we have

$$\frac{y^2}{b^2} + \frac{x^2}{a^2} - \frac{2xy}{ab}\cdot\frac{1}{\sqrt{2}} = \frac{1}{2}. \quad \text{Or,} \quad \frac{y^2}{b^2} + \frac{x^2}{a^2} - \frac{\sqrt{2}xy}{ab} = \frac{1}{2},$$

which is the equation to an *oblique ellipse*. [See case I (*iii*), (*c*) above.] So that, the resultant motion, in this case, is an *oblique ellipse*.

Thus, we see that the two perpendicular linear simple harmonic motions compound into a straight line motion, when they differ in phase by 0 or π, and into an ellipse or a circle, when the phase difference is π/2.

For any other phase difference, the motion is still an *ellipse* with its major and minor axes no longer coinciding with the directions of the two component motions but being inclined to them.

(*f*) **When the amplitudes are different and the time-periods or frequencies are nearly equal.** In the cases, dealt with above, where the time-periods of the two component vibrations are *identical*, the elliptical paths of the resultant vibrations remain fixed. But if the two time-periods differ slightly from each other, there comes about a gradual but progressive change in the relative phase (ϕ) of the two vibrations and the elliptical path consequently undergoes a corresponding cycle of changes, *whose frequency is equal to the difference between the frequencies of the two component vibrations.*

Thus, (*i*) *when $\phi = 0$, the ellipse coincides with one diagonal of the rectangle of sides 2a and 2b, within which the ellipse lies*; for, here,

$$\frac{x}{a} - \frac{y}{b} = 0.$$

(*ii*) *When ϕ increases from 0 to $\pi/2$, the ellipse opens out to the form* $\frac{x^2}{a^2} + \frac{y^2}{b^2} = 1$, *passing through intermediate oblique positions.*

And if, $a = b$, the *ellipse is reduced to the form of a circle.*

(*iii*) *When ϕ increases from $\pi/2$ to π, the ellipse closes up again and finally coincides with the other diagonal of the rectangle;* for now

$$\frac{x}{a} + \frac{y}{b} = 0.$$

And, the same changes take place in the reverse order, when ϕ increases from ϕ to 2π, the ellipse ultimately coinciding with the same first diagonal as in case (i), —all these changes being shown in Fig. 4.17, above.

(*g*) **When the frequencies are in the ratio 1 : 2, or the periods are in the ratio 2 : 1, and the amplitudes are different.** In this case, the angular velocity of the particle in the circle of reference of the second motion will be double of that of the particle in the circle of reference of the first.

$\therefore$ if the two motions be represented by

$$x = a \sin \omega t \quad \text{and} \quad y = b \sin (2\omega t + \phi),$$

where ϕ is the *phase angle* by which the second motion is ahead of the first, we have

$$x/a = \sin \omega t \text{ and } \therefore \cos \omega t = \sqrt{1 - \sin^2 \omega t} = \sqrt{1 - x^2/a^2}.$$

And, $y/b = \sin (2\omega t + \phi) = \sin 2\omega t \cos \phi + \cos 2\omega t \sin \phi$.

Or, $y/b = 2 \sin \omega t \cos \omega t \cos \phi + (1 - 2\sin^2 \omega t) \sin \phi$.

$\therefore$ $\sin 2\omega t = 2 \sin \omega t \cos \omega t$ and $\cos 2\omega t = (1 - 2 \sin^2 \omega t)$

substituting the values of $\sin \omega t$ and $\cos \omega t$ from above, we have

$$\frac{y}{b} = 2\frac{x}{a}\left(1 - \frac{x^2}{a^2}\right)^{\frac{1}{2}} \cos \phi + \left(1 - \frac{2x^2}{a^2}\right) \sin \phi.$$

$$= \frac{2x}{a}\left(1 - \frac{x^2}{a^2}\right)^{\frac{1}{2}} \cos \phi + \sin \phi - \frac{2x^2}{a^2} \sin \phi.$$

Or, $$\left(\frac{y}{b} - \sin \phi\right) + \frac{2x^2}{a^2} \sin \phi = \frac{2x}{a}\left(1 - \frac{x^2}{a^2}\right)^{\frac{1}{2}} \cos \phi.$$

Squaring both sides, we have

$$\left(\frac{y}{b} - \sin \phi\right)^2 + \frac{4x^4}{a^4} \sin^2 \phi + 2\left(\frac{y}{b} - \sin \phi\right)\frac{2x^2}{a^2} \sin \phi$$
$$= \frac{4x^2}{a^2}\left(1 - \frac{x^2}{a^2}\right) \cos^2 \phi.$$

Or, $$\left(\frac{y}{b} - \sin \phi\right)^2 + \frac{4x^4}{a^4} \sin^2 \phi + \frac{4x^2}{a^2}.\frac{y}{b} \sin \phi - \frac{4x^2}{a^2} \sin^2 \phi$$
$$= \frac{4x^2}{a^2} \cos^2 \phi - \frac{4x^4}{a^4} \cos^2 \phi.$$

Or, $$\left(\frac{y}{b} - \sin \phi\right)^2 + \frac{4x^4}{a^4} \sin^2 \phi + \frac{4x^4}{a^4} \cos^2 \phi \, \frac{4x^2}{a^2}.\frac{y}{b} \sin \phi$$
$$- \frac{4x^2}{a^2} \sin^2 \phi - \frac{4x^2}{a^2} \cos^2 \phi = 0.$$

Or, $$\left(\frac{y}{b} - \sin \phi\right)^2 + \frac{4x^4}{a^4} (\sin^2 \phi + \cos^2 \phi)$$
$$+ \frac{4x^2}{a^2}\left[\frac{y}{b} \sin \phi - (\sin^2 \phi + \cos^2 \phi)\right] = 0.$$

Or, $$\left(\frac{y}{b} - \sin \phi\right)^2 + \frac{4x^4}{a^4} + \frac{4x^2}{a^2}\left(\frac{y}{b} \sin \phi - 1\right) = 0.$$

Or, $$\left(\frac{y}{b} - \sin \phi\right)^2 + \frac{4x^2}{a^2}\left(\frac{x^2}{a^2} + \frac{y}{b} \sin \phi - 1\right) = 0 \qquad \text{... (A)}$$

This is the general equation for a curve having two loops, for any values of phase difference and amplitude.

Let us now take some particular cases :

(*i*) **If the phase difference, i.e., ϕ, be equal to 0 or π.**

Here, *sin* $\phi = 0$, and, therefore, $\dfrac{y^2}{b^2} + \dfrac{4x^2}{a^2}\left(\dfrac{x^2}{a^2} - 1\right) = 0,$

which is the equation to a curve of the form as shown in Fig. 4.15 (Case *I (iii) h* above), *inscribed in a rectangle of sides 2a and 2b and symmetrical about the axis of x.*

(*ii*) **If the phase difference $\phi = \pi/2$, *i.e. sin* $\phi = 1$, we have**

$$\left(\frac{y}{b} - 1\right)^2 + \frac{4x^2}{a^2}\left(\frac{x^2}{a^2} + \frac{y}{b} - 1\right) = 0.$$

Or,
$$\left(\frac{y}{b} - 1\right)^2 + \frac{4x^4}{a^4} + 2.\frac{2x^2}{a^2}\left(\frac{y}{b} - 1\right) = 0.$$

Or,
$$\left[\left(\frac{y}{b} - 1\right) + \frac{2x^2}{a^2}\right]^2 = 0.$$

This *represents two coincident parabolas,* each having the equation

$$\left(\frac{y}{b} - 1\right) + \frac{2x^2}{a^2} = 0.$$

Or,
$$\frac{2x^2}{a^2} = -\left(\frac{y}{b} - 1\right) = -\frac{1}{b}(y - b). \quad \text{Or,} \quad x^2 = -\frac{a^2}{2b}(y - b).$$

If, however, the frequencies differ slightly from the ratio 2 : 1 (*i.e.* the time-periods differ from 1 : 2), the variation in the resultant path of the particle may be obtained by substituting the consequent changes in the value of ϕ in the general relation above. The changes occurring when ϕ changes from 0 to π get reversed when ϕ changes from π to 2π.

4.5 Composition of two equal circular motions in opposite directions. Let two particles P_1 and P_2 move with equal velocities along the same circle $XYX'Y'$ of radius $2a$, in opposite directions, as shown, (Fig. 4.21), such that when P_1 passes the point X, P_2 passes X'.

Let the positions of P_1 and P_2 be as shown, at any given instant t, after starting from X and X' respectively; so that, $XOP_1 = X'OP_2 = \phi = \omega t$, where ω is the angular velocity of P_1 and P_2.

Now, *we know that a circular motion is equivalent to two equal linear simple harmonic motions with a phase difference $\pi/2$ and along perpendicular directions to each other,* (see page 119).

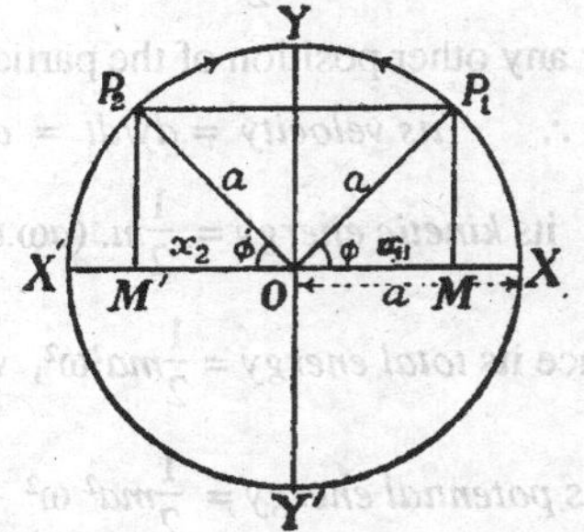

Fig. 4.21

The circular motions of both P_1 and P_2 may, therefore, be resolved along perpendicular directions XX' and YY'. Then, the displacements x_1 and x_2, of P_1 and P_2, will, at the given instant, be equal in magnitude but *opposite in direction* along XOX' and will, therefore, cancel each other out; but their displacements y_1 and y_2 along YOY' will equal and *in the same direction,* so that the resultant displacement is given by $y = y_1 + y_2$ along YOY'

Since $\qquad y_1 = y_2 = a \sin\phi = a \sin\omega t,$ we have

$$y = y_1 + y_2 = a \sin \omega t + a \sin \omega t + 2a \sin \omega t.$$

Or, *the resultant motion is a linear simple harmonic motion along the diameter YOY′ at the extremities of which the particles P_1 and P_2 cross each other as they describe their circular motions.*

And, clearly, the *amplitude of the resultant motion is* 2*a*, and *its time-period the same as that of the two constituent circular motions.*

4.6 Energy of a Particle in Simple Harmonic Motion. The acceleration of a particle, executing a *S.H.M.*, is, as we know, directed *towards its equilibrium position, or in a direction opposite to that in which y, the displacement of the particle, increases. Hence, work is done during its displacement, or the particle has potential energy.* Also, *the particle possesses velocity and, therefore, has kinetic energy.* Thus, it has both potential as well as kinetic energy, or *its energy is partly potential and partly kinetic.* And, if there be no dissipative force at work, *i.e.*, if the energy is not dissipated away in any way, the sum total of the two remains constant, although as the displacement increases and the velocity decreases, the potential energy increases and the kinetic energy decreases.

Now, when the particle has its maximum displacement, positive or negative, its velocity is zero and, therefore, its kinetic energy is then *zero;* so that, in this position, *the whole of its energy is present in the form of potential energy.* And, *when the particle is in the equilibrium position,* its displacement is *zero* and its velocity, maximum; so that, *the whole of its energy is now present in the form of kinetic energy.*

If *m* be the mass of the particle; *a*, its amplitude and $2\pi/\omega$, its time-period *(T), i.e.*, if ω be the angular velocity of the rotating vector, or that of the particle in the circle of reference, we have

velocity of the particle in its equilibrium position = $a\omega$, *a maximum.*

And ∴ its *kinetic energy* $= \frac{1}{2} m. (a\omega)^2 = \frac{1}{2} ma^2\omega^2,$

and, its *potential energy* $= 0,$ [∵ its displacement is zero.

Or, its *total energy* $= \frac{1}{2} ma^2\omega^2 + 0 = \frac{1}{2} ma^2\omega^2.$

In other words, the whole of its energy, here, is present in the *kinetic form.* Similarly, when the particle has its maximum displacement, the whole of its energy is in the *potential form*, which, therefore, is also equal to $\frac{1}{2} ma^2 \omega^2$ in *this* position of the particle.

For any other position of the particle, its displacement is given by $y = a \sin \omega t$

And ∴ *its velocity* $= dy/dt = a\omega \cos \omega t$

Hence, its *kinetic energy* $= \frac{1}{2} m. (a\omega.\cos \omega t)^2 \; \frac{1}{2} m.a^2\omega^2.\cos^2 \omega t.$

And, since its *total energy* $= \frac{1}{2} ma^2\omega^2$, we have

its *potential energy* $= \frac{1}{2} ma^2 \omega^2 - \frac{1}{2} ma^2\omega^2.\cos^2\omega t = \frac{1}{2} ma^2\omega^2.\sin^2\omega t.$

Alternatively, we may proceed as follows :

The acceleration of the particle, executing a *S.H.M*, is given by $d^2y / dt^2 = -\omega^2 y.$

∴ if *m* be the mass of the particle, the force *F* required to maintain this displacement *y* is equal to $m. \omega^2 y.$
And, therefore, work done by the force for a small displacement *dy* is equal to $F. dy = m. \omega^2 y.dy.$
Now, this work is also a measure of the potential energy of the particle at this displacement.

∴ *P.E.* of the particle for a displacement $dy = F.dy = m\omega^2 y \,. dy.$

Hence, total work done for displacement y and, therefore, *total P.E. of the particle for a displacement y*

$$= \int_0^y m\omega^2 . y.dy = m\omega^2 \int_0^y y.dy = \frac{1}{2} m\omega^2.y^2,$$

i.e., Potential Energy $\propto y^2$.

Thus, the *P.E. of a particle, executing a S.H.M. is, at any given instant, directly proportional to the square of its displacement from its mean or equilibrium position at that instant.*

Now, *velocity of the particle at displacement* $y = dy/dt$.

$$= \frac{d}{dt}(a \sin \omega t) = a\omega \cos \omega t.$$

And $\therefore$ *K.E. of the particle* $= \frac{1}{2}m\,(a\omega \cos \omega t)^2 = \frac{1}{2}ma^2\omega^2 . \cos^2\omega t.$

$$= \frac{1}{2}ma^2\omega^2 (1 - \sin^2\omega t) = \frac{1}{2}ma^2\omega^2 (1 - y^2/a^2).$$

$$= \frac{1}{2}ma^2\omega^2 (a^2 - y^2)/a^2 = \frac{1}{2}m\omega^2/(a^2 - y^2).$$

Hence *total energy* of the particle = *its potential energy + its kinetic energy.*

$$= \frac{1}{2}m\omega^2 y^2 + \frac{1}{2}m\omega^2 (a^2 - y^2) = \frac{1}{2}m\omega^2 a^2.$$

And, since $\omega = 2\pi/T = 2\pi n$, where T is the *time-period* of the particle and n, the *frequency* of its vibration, we may also say that

$$\textit{total energy of the particle} = \frac{1}{2}m.\left(\frac{2\pi}{T}\right)^2 a^2 = \frac{2\pi^2 ma^2}{T^2} = 2\pi^2 n^2 ma^2.$$

Now, since in any *conservative system* the sum total of the kinetic and potential energies of the system must be *a constant*, it is clear that the former can only increase at the expense of the latter, and, therefore, attains its maximum value, when the latter is reduced to its minimum value or *zero* and *vice versa*. Thus, the maximum value of any one of the two forms of energy measures the total energy of the system.

4.7 Average Kinetic and Potential Energies of a Particle in S.H.M. We have seen above that, at any given instant, the *P.E.* of a particle is equal to $\frac{1}{2}m\,a\omega^2 \sin^2 \omega t$ and *K.E.* equal to $\frac{1}{2}ma^2\omega^2 \cos^2 \omega t$. Now, the mean or average value of both $\sin^2 \omega t$ and $\cos^2 \omega t$ for a whole cycle, (from 0 to 2π), *i.e.*, for a whole time-period is equal to $\frac{1}{2}$ * and, therefore, the mean or average *K.E.* of the particle over the whole of its period of vibration is equal to its mean of average *P.E.* over the whole period, each being equal to $\frac{1}{2} \times \frac{1}{2}ma^2\omega^2 = \frac{1}{4}ma^2\omega^2$.

Thus, *average K.E. of the particle = its average P.E.* $= \frac{1}{4}ma^2\omega^2$.

And $\therefore$ *total energy of the particle* $= 2 \times \frac{1}{4}ma^2\omega^2 = \frac{1}{2}ma^2\omega^2$.

We may express this by saying that *the energy of a particle, executing a S.H.M., is on an average half kinetic and half potential in form,* – the whole being present in the kinetic form at its mean or equilibrium position, and in the potential form at its extreme position, on either side.

---

* Because the mean value of $\sin^2 \omega t$ for a whole cycle from 0 to 2π is given by

$$\int_0^{2\pi} \sin^2 \omega t.dt \Big/ \int_0^{2\pi} d\omega t$$

$$= \int_0^{2\pi} \frac{1 - \cos^2\omega t}{2}\, d\omega t/2\pi = [\,\omega t/2 - 2\omega t/4\,]_0^{2\pi} / 2\pi$$

$$= \pi/2\pi = 1/2$$

Similar is the case with $\cos^2 \omega t$ for whole time-period.

The results obtained above for linear *S.H.M.* are equally valid for angular *S.H.M.* Only, the linear displacement x or y of the particle or the body, and its mass m, are replaced by their rotational analogues, *viz.*, the angular displacement and its moment of inertia I about the axis of rotation, respectively.

WORKED EXAMPLES

1. A quantity of gas is enclosed in a cylinder, fitted with a smooth heavy piston. The axis of the cylinder is vertical. The piston is thrust downwards to compress the gas, and then let go. Is the ensuring motion of the piston a S.H.M. ? If so, what is its time-period ?

Let original *volume* of the gas be V and *its pressure*, P.

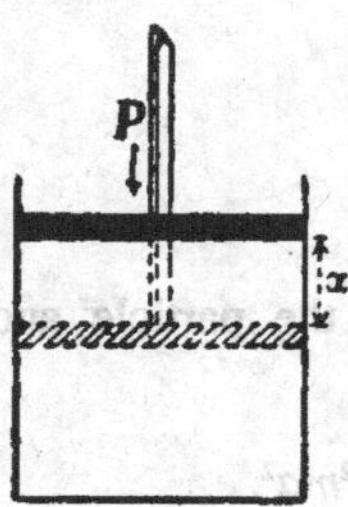

Fig. 4.22

Let a be the *area of cross-section* of the piston, (and cylinder). Then, if the piston be displaced through a distance x, (Fig. 4.22), *the change in volume* produced is given by $x.a$, the corresponding *change in pressure* being p. By *Boyle's law*, therefore,

$$PV = (P + p) \times (V - x.a) = PV - P\,x.a, + pV - p\,x.a.$$

$$\text{Or, } 0 = -P.x.a. + p\,V - p.x.a$$

Neglecting $p.x.a.$ as the product of very small quantities, compared with the other terms in the expression, we have

$$0 = -P.x.a. + pV. \text{ Or, } pV = P.x.a., \text{ whence } p = \left(\frac{P.x}{V}\right)x.$$

Now, the *restoring force on the piston*, which is equal to the disturbing force, is obviously equal to change in pressure *into* area of cross-section of the piston $= p.a.$

$$p.a. = \frac{P.a}{V}.x.a = \frac{P\,a^2}{V}.x.$$

Since, *acceleration = force/mass*, the *acceleration of the piston* $= p.a/m$.

$$= \frac{P.a^2}{V}.\frac{x}{m} = \frac{P.a^2}{Vm}.x. \qquad \text{[Substituting the value of } p.a.\text{, from above.}$$

Or, *acceleration of the piston* $= \mu x$

where $P.a^2/Vm = \mu$ *a constant of proportionality which is equal to the acceleration of the piston per unit displacement* (*i.e.*, when $x = 1$).

Or, *acceleration of the piston is proportional to x, its displacement*.

Hence, the motion of the piston is a *S.H.M.*

And $\therefore$ its *time-period*,

$$T = 2\pi\sqrt{\frac{1}{\text{acceleration per unit displacement}}}$$

$$= 2\pi\sqrt{\frac{1}{\mu}} = 2\pi\sqrt{\frac{1}{P.a^2/V.m}} = 2\pi\sqrt{\frac{V.m}{P.a^2}}$$

or,

$$T = \frac{2\pi}{a}\sqrt{\frac{V.m}{P}}$$

2. A body describing a simple harmonic motion executes 100 complete vibrations per minute, and its speed at its mean position is 15 ft. per second. What is the length of its path ? What is its velocity when it is half-way between its mean position and an extremity of its path ?

Here, *time-period T* of the body $= 1/100$ *min* $= 60/100 = .6$ *sec.*

and *velocity of body at its mean position* $= 15$ *ft/sec*

Since velocity of a body executing a *S.H.M.* $= a\omega$, *at its mean position* where a is its *amplitude and* ω, its *angular velocity, we have*

$$a\,\omega = 15. \quad \text{Or, } a.2\pi/T = 15. \qquad [\because \omega = 2\pi/T.$$

whence, $a = \dfrac{15 \times T}{2\pi} = \dfrac{15 \times .6}{2\pi} = \dfrac{9}{2\pi}$. Or, $a = 1.432$ *ft*.

Now, *length of path of the body = twice its amplitude*, because it goes the same distance on either side of its mean position.

Hence, *length of path* $= 2 \times 1.432 = 2.864$ *ft*

Again, velocity of a body at displacement y is given by $v = \omega\sqrt{a^2 - y^2}$.

Here, displacement of the body *i.e.*, $y = a/2 = 1.432/2 = .716$ *ft*

So that, $v = \frac{2\pi}{T}\sqrt{(1.432)^2 - (.716)^2} = \frac{2\pi}{T}\sqrt{(1.432 + .716)(1.432 - .716)}$

$$= \frac{2\pi}{.6}\sqrt{2.148 \times .716} = 12.99\, ft/sec.$$

Thus, the *length of path* of the body is 2.864 *ft* and its *velocity*, when it is half way between its mean position and an extremity of its path, is 12.99 *ft/sec.*

3. If the earth were a homogeneous sphere, and a straight hole were bored in it thróugh its centre, show that a body dropped into the hole will execute a S.H.M. and calculate the time-period of its vibration. [Radius of the earth = 4000 miles, and value of g on its surface = 32 ft. per sec]

We know that the force with which a body is attracted by the earth towards its centre is equal to the weight of the body, (mg), and also equal to $G.m.\ M/R^2$, where m is the *mass of the body, M,* the *mass of the earth; R the radius of the earth, g,* the *acceleration due to gravity* on the surface of the earth and G, the *gravitational constant.*

$$mg = G.m.M/R^2 \quad \text{Or,} \quad g = G.M/R^2.$$

Since the earth is supposed to be a sphere, its *volume* $= 4\pi R^3/3$, and therefore, if Δ be its *density*, we have its *mass*, $M = 4\pi\ R^3.\Delta/3$. So that, $g = \frac{4}{3}.\frac{\pi R^3\Delta}{R^2}.G = 4\pi R\Delta.G/3.$ ...(*i*)

If the value of acceleration due to gravity at a distance r below the surface of the earth. (Fig. 4.23), be g', we have, as above,

$$g' = \frac{4}{3}.\pi.\frac{(R-r)^3}{(R-r)^2}.\Delta.G = 4\pi.(R-r).\Delta G/3. \qquad ...(ii)$$

Dividing (*ii*) by (*i*), we have

$$g'/g = \frac{4}{3}.(R-r)\ \Delta.\ G/\frac{4}{3}.\pi.R.\Delta G.$$

Or, $g'/g = (R-r)/R.$

Or, $g' = g(R-r)/R = (R-r).g/R.$

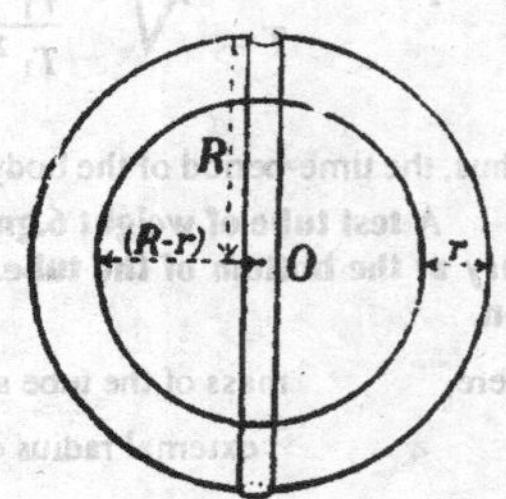

Fig. 4.23.

Thus, the acceleration of a body at a distance ($R-r$) from the centre of the earth inside it is equal to $(R-r).g/R$; and since g/R is constant, *this acceleration is proportional to (R–r), the displacement of the body* from the centre O of the earth. The body, therefore, *executes a S.H.M.*, and its time-period is given by,

$$t = 2\pi\sqrt{\frac{1}{g/R}} = 2\pi\sqrt{\frac{R}{g}}$$

Now $R = 4000$ *miles* $= 4000 \times 1760 \times 3\, ft.$ and $g = 32\, ft./sec^2$.

$\therefore$ Substituting the values of R and g in relation (*iii*) above, we have

$$t = 2\pi\sqrt{\frac{4000 \times 1760 \times 3}{32}} = 2\pi\sqrt{\frac{12 \times 176 \times 10^4}{32}}$$

$$= 5105\ sec = \frac{5105}{60} = 85.07\ minutes.$$

Thus, the *time-period* of vibration of the body would be 85.07 *minutes.*

4. If a body executes a simple harmonic vibration in time T_1, under one constraining force, and in Time T_2, under another, what will be its time-period under both forces together ?

Let mass of the body be m,

and let its acceleration under the *first force* $F_1 = a_1$,

" " " " " *second force* $F_2 = a_2$,

and " " " " " *both the forces* $F_1 + F_2 = a$.

Then, clearly, $a_1 = F_1/m;\ a_2 = F_2/m;\ a = (F_1 + F_2)/m.$ $\left[\because\ acc. = \frac{force}{mass}\right]$

Also, the acceleration of a body executing *S.H.M.* is proportional to its displacement x from the equilibrium position, *i.e.*, acceleration $a \propto x$

$$\therefore \quad a_1 = \mu_1 x;\ a_2 = \mu_2 x \text{ and } a = \mu x,$$

where μ_1, μ_2 and μ are the constants of proportionality in the three cases respectively.

Again, because the time-period of a body executing *S.H.M.* is given by $T = 2\pi\sqrt{1/\mu}$, where μ is a constant of proportionality, we have

$$\sqrt{\frac{1}{\mu}} = \frac{T}{2\pi} \text{ or, } \frac{1}{\mu} = \left(\frac{T}{2\pi}\right)^2. \text{ Or, } \mu = \left(\frac{2\pi}{T}\right)^2.$$

So that, $a_1 = \frac{F_1}{m} = \mu_1 x. = \left(\frac{2\pi}{T_1}\right)^2 . x$; $a_2 = \frac{F_2}{m} = \mu_2 . x = \left(\frac{2\pi}{T_2}\right)^2 . x$

and $$a = \frac{(F_1 + F_2)}{m} = \mu . x = \left(\frac{2\pi}{T}\right)^2 x,$$

where *T is the time-period of the body under both the forces acting together.*

Since acceleration a, under both the forces acting together, will obviously be equal to the sum of the accelerations under the individual forces, we have

$$a = a_1 + a_2$$

And, therefore, $$\left(\frac{2\pi}{T}\right)^2 . x = \left(\frac{2\pi}{T_1}\right)^2 . x + \left(\frac{2\pi}{T_2}\right)^2 . x.$$

Or, dividing by $(2\pi)^2 x$ throughout, we get

$$\left(\frac{1}{T}\right)^2 = \left(\frac{1}{T_1}\right)^2 + \left(\frac{1}{T_2}\right)^2$$

Or, $$\frac{1}{T^2} = \frac{1}{T_1^2} + \frac{1}{T_2^2} = \frac{T_2^2 + T_1^2}{T_1^2 + T_2^2}. \quad \text{Or, } T^2 = \frac{T_1^2 . T_2^2}{T_1^2 + T_2^2}$$

Or, $$T = \sqrt{\frac{T_1^2 . T_2^2}{T_1^2 . T_2^2}} = T_1 . T_2 \sqrt{\frac{1}{T_1^2 + T_2^2}}.$$

Thus, the time-period of the body under both the forces together will be $T_1 T_2 \sqrt{1/(T_1^2 + T_2^2)}$.

5. A test tube of weight 6 gm. and of external diameter 2 cm is floated vertically in water by placing 10 gm of mercury at the bottom of the tube. The tube is depressed by a small amount and then released. Find the time of oscillation. *(Oxford and Cambridge Higher School Certificate)*

Here, mass of the tube and mercury = 10 + 6 = 16 *gm*

and external radius of the tube = 2/2 = 1 *cm*

∴ area of cross-section of the tube $= \pi r^2 = \pi . 1^2 = \pi . sq\ cm$

Let the tube be depressed through a distance *x cm.*

Then *volume of the water displaced* $= \pi \times x = \pi x$ *c.c.*

and, weight „ „ „ $= \pi x 1 = \pi x$ *gm. wt* [taking the density of water to be 1*gm/c.c.*

Therefore, *upward thrust* experienced by the tube is equal to the weight of the water displaced, *i.e.*, $\pi . x . g.$ *dynes*.

Hence *acceleration* of the tube $= \pi . x . g . / 16 = (\pi g / 16) x.$ [∵ acc = force / mass.)

Since $\pi g/16$ is a constant, say, μ, we have

acceleration of the tube $= \mu x$, *i.e., acceleration of the tube is proportional to its displacement*; and therefore, it executes a *S.H.M.*

Hence, its *time-period* is given by $t = 2\pi\sqrt{1/\mu}$.

$$= 2\pi\sqrt{\frac{1}{\pi g/16}} = 2\pi\sqrt{\frac{16}{\pi g}} = 2\sqrt{\frac{\pi^2 . 16}{\pi g}}$$

$$= 2\pi\sqrt{\pi . 16/g} = 2\sqrt{\pi . 16/981} = .4527 \text{ sec.}$$

Or, *the time of oscillation* of the tube ≑ .4527 *sec.*

6. A particle executing a S.H.M. has a maximum displacement of 4 cm and its acceleration at a distance of 1 cm from its mean position is 3 *cm/sec*^{2}. What will its velocity be when it is at a distance of 2 cm from its mean position ?

Here, amplitude of the particle = 4 *cm*

and its acceleration, when its displacement is 1 *cm* is equal to 3 *cm/sec*^{2}.

Now, acceleration $= x . \omega^2$. Or, $3 = 1 . \omega^2$,
where x is the *displacement* and ω, the *angular velocity* of the particle.

Or, $\omega^2 = 3$, where $\omega = \sqrt{3}$ *radians/sec.*

Now, the velocity of a particle executing a *S.H.M.* is given by $v = \omega . \sqrt{a^2 - x^2}$
where a is the amplitude and x, the displacement of the particle,

$\therefore$ when the displacement of the particle, *i.e.*, $x = 2$ *cm*, we have

$$v = \sqrt{3} . \sqrt{4^2 - 2^2} = \sqrt{3} . \sqrt{16 - 4} = \sqrt{3} . \sqrt{12}.$$

$$= \sqrt{36} = 6 \ cm/sec.$$

$\therefore$ the velocity of the particle at a distance of 2 *cm* from its mean position will be 6 *cm / sec.*

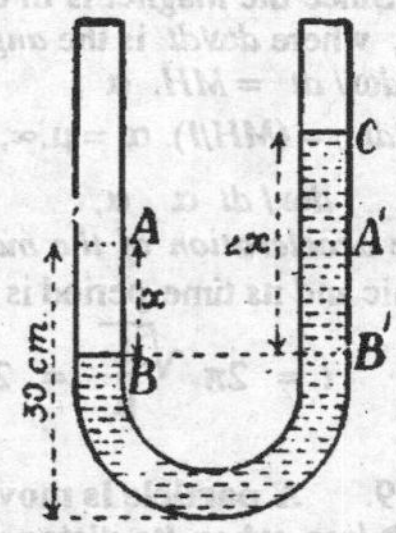

Fig. 4.24.

7. A vertical U-tube of uniform cross-section contains water up to a height of 30 cm. Show that if the water on one side is depressed and then released, its motion up and down the two sides of the tube is simple harmonic, and calculate its time-period. *(Delhi)*

Let AA', (Fig.4.24), represent the level of the water in the two limbs of the U-tube, to start with, and let the column on the left be depressed up to B, through a distance x *cm*. Then, the column on the right will naturally go up, say to the level C, such that the difference of levels in the two limbs is now, $B'C$ where B' is at the same level as B.

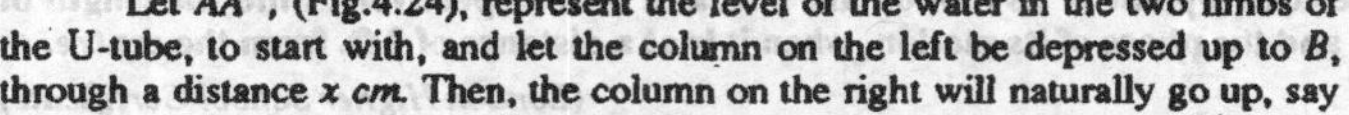

The weight of this column of length $2x$ will now act on the mass of water in the tube, as a result of which it will oscillate up and down.

Now, obviously, the weight of this column of water

$= $ its *mass* $\times g =$ its *volume* $\times$ its *density* $\times g$

$= a \times 2x \times 1 \times g$ *dynes.*

Or, *force* acting on the mass of water in the tube

$= 2.x.a.g$ *dynes.*

And, *mass of water* in the tube (in both its limbs)

$= 2 \times 30 \times a \times 1 = 60\, a$ *gm.*

$\therefore$ *acceleration* produced in the mass of water in the tube

$$= \frac{2x . a . g}{60\, a} = \frac{x . g}{30} = \frac{g}{30} . x = \mu . x, \text{ where } g/30 = \mu, \text{ a constant.}$$

Or, *acceleration is proportional to x, the displacement of the water column.*

Hence, the motion of the water column is *simple harmonic*, and its *time-period* is given by

$$t = 2\pi \sqrt{\frac{1}{\mu}} = 2\pi \sqrt{\frac{1}{g/30}} = 2\pi \sqrt{30/g}, \quad \text{or,} \quad t = 2\pi \sqrt{30/981} = 1.098 \ sec.$$

The water in the U-tube will thus oscillate with a time-period equal to $2\pi \sqrt{30/g}$ or 1.098 *sec.*

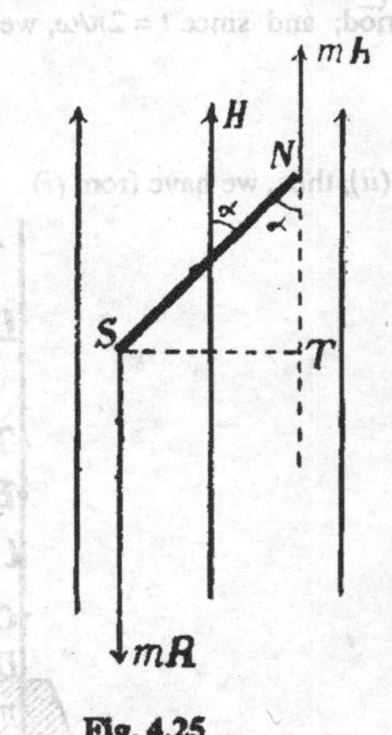

Fig. 4.25

8. Show that the time-period for the swing of a magnet in the earth's field is given by $t = 2\pi \sqrt{I/MH}$, where M is the magnetic moment of the magnet, I, its moment of inertia about the axis of suspension and H, the earths field. *(Sagar)*

Let a magnet NS, of pole strength m, be suspended so as to make an angle α with the earth's field H, (Fig.4.25)

Then, clearly, the forces acting on its two poles are mH and mH, as shown. These two forces being *equal, opposite* and *parallel*, constitute a *couple*, whose moment is equal to the product of one of the forces and the perpendicular distance between them.

So that, *couple C*, acting on the magnet $= mH \times ST$.

---

*It should be noted that the time-period is the same as that of a simple pendulum *of the same length as the height of the water column, i.e.*, of length equal to 30 *cm*.

$= mH \times NS \sin\alpha$ $[\because ST = NS \sin\alpha$

Or, $C = MH \sin\alpha$

$\therefore$ $m \times NS = M$, *the magnetic moment* of the magnet.

If α be small, we have $\sin\alpha = \alpha$ *(in radian measure)*

So that $C = MH.\alpha$.

Since the magnet is in equilibrium, thus must be balanced by the restoring couple set up in the suspension, *i.e.*, by $I.d\omega/dt$, where $d\omega/dt$ is the *angular acceleration of the magnet* and I its *moment of inertia* about the axis of suspension. So that, $I.d\omega/dt = MH.\alpha$,

Or, $d\omega/dt = (MH/I).\alpha = \mu.\alpha$, where $\mu = MH/I$, a constant.

Or, $d\omega/dt \propto \alpha$,

i.e., the *acceleration of the magnet is proportional to its displacement*. The motion of the magnet is, therefore, simple harmonic and its time-period is given by

$$t = 2\pi\sqrt{\frac{1}{\mu}} = 2\pi\sqrt{\frac{1}{mH/I}} = 2\pi\sqrt{\frac{I}{MH}}.$$

9. **A particle is moving in a straight line with simple harmonic motion. Its velocity has the values 5 ft./sec. and 4 ft./sec, when its distance from the centre-point of its motion are 2 ft. and 3 ft. respectively. Find the length of its path, the frequency of its oscillation, and the phase of its motion, when it is at a distance of 2 ft. from the centre.**

(London Higher School Certificate)

Length of path. We know that $v = \omega\sqrt{a^2 - x^2}$.

So that, in the *first case*, when $v = 5$ *ft./sec.* and $x = 2$ *ft.*, we have

$$5 = \omega\sqrt{a^2 - 2^2} \quad \text{or,} \quad 25 = \omega^2\sqrt{a^2 - 4}. \qquad ...(i)$$

And, in the *second case*, when $v = 4$ *ft./sec.*, and $x = 3$ ft., we have

$$4 = \omega\sqrt{a^2 - 3^2} \quad \text{or,} \quad 16 = \omega^2\sqrt{(a^2 - 9)} \qquad ...(ii)$$

$\therefore$ dividing *(i)* by *(ii)*, we have $\dfrac{25}{16} = \dfrac{a^2 - 4}{a^2 - 9}$. Or, $25(a^2 - 9) = 16(a^2 - 4)$.

Or, $25a^2 - 225 = 16a^2 - 64$. Or, $25a^2 - 16a^2 = 225 - 64$.

$9a^2 = 161$ Or, $a^2 = 161/9$ And $\therefore$ $a = \sqrt{161/9} = 4.23$ *ft.*

Or, the amplitude of the particle is 4.23 *ft.*

Since the length of the path traversed by the particle is twice the amplitude, (as it travels equal distance on either side), we have

length of path of the particle $= 2 \times a = 2 \times 4.23 = 8.46$ *ft.*

[where x is the *displacement* of the praticle, a, its *amplitude* and $\theta = \omega t$, the *phase angle*.

Frequency. The frequency of the particle, $n = 1/t$, where t is its time-period; and since $t = 2\pi/\omega$, we have π/ω we have

$$n = \frac{1}{2\pi/\omega} = \frac{\omega}{2\pi} \qquad ...(iii)$$

The value of ω may be obtained by substituting the value of a in either *(i)* or *(ii)*; thus, we have from *(i)*

$$\omega^2 = \frac{25}{a^2 - 4}. \quad \text{Or,} \quad \omega = \frac{5}{\sqrt{a^2 - 4}} = \frac{5}{\sqrt{(4.23)^2 - (2)^2}}$$

$$\text{or,} \quad \omega = \frac{5}{\sqrt{(4.23 + 2)(4.23 - 2)}} = \frac{5}{\sqrt{6.23 \times 2.23}}$$

And, therefore, substituting the value of ω in *(iii)* above, we have

$$n = \frac{5}{2\pi\sqrt{6.23 \times 2.23}} = 0.2135.$$

Thus, the *frequency* of the particle is 0.2135.

Phase angle. We have the relation, $x = a \sin\theta$

Here, $x = 2$ *ft.* and $a = 4.23$ *ft.*

$\therefore$ $2 = 4.23 \sin\theta$.

Or, $\sin\theta = 2/4.23 = .4729$

Or, $\theta = \sin^{-1} .4729 = 28°13'$.

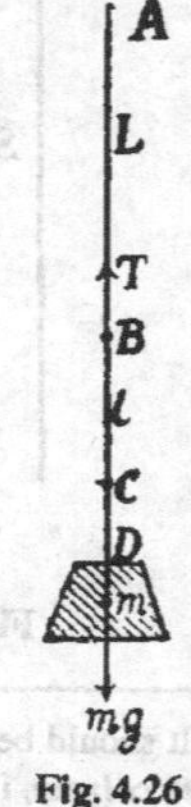

Fig. 4.26

Hence, the phase of the motion of the particle, when its distance is 2 *ft.* from the centre, is 28° 13′.

10. A light elastic string is suspended vertically from a point and carries a heavy mass at its lower free end, which stretches it through distance *l cm*. Show that the vertical oscillations of the system are simple harmonic in nature, and of a time-period equal to that of a simple pendulum of length *l cm*.

Let original length of the string *AB*, (Fig. 4.26), be = *L cm* and let the mass attached to its lower end be *m gm*.

Then, the downward force acting on it = *mg dynes*.

And, if *T* be the tension of the string (upward), we have

$T = mg$, because the string is in equilibrium.

Now, *Young's modulus* for the string, *i.e.*, *Y* = *stres/strain*.

Or, $stress = Y \times strain$.

But, obviously, $stress = T/a$, and $strain = l / L$.

where *a* is the *area of cross-section* of the string.

So that, $\dfrac{T}{a} = Y.\dfrac{l}{L}$. Or, $T = Y.\dfrac{l.a}{L}$.

Since $mg = T$, we have $mg = \dfrac{Y.a}{L}.l$.

Or, $mg/l = Y.a/L$. ...(*i*)

If the string be pulled down a little through a distance *x*, the tension in the string acting upwards will, clearly, be

$$= \frac{Y\ a}{L}.\ (l + x) = \frac{mg}{l}.\ (l + x).$$ [See (*i*) above

And, since the downward force acting on the string = *mg*, the resultant upward force acting on it will now be

$$\frac{mg\ (l + x)}{l} - mg.$$

Or, $resultant\ upward\ force = \dfrac{mg\ (l + x) - mgl}{l} = \dfrac{mg\,x}{l}$.

Now, acceleration $= \dfrac{force}{mass} = \dfrac{mg\,x}{l.m} = \dfrac{g\,x}{l} = \mu.x$, where $g / l = \mu$, a constant.

Or, *acceleration* ∝ *displacement*.

∴ the oscillations are *simple harmonic* in nature; and since the *time-period* of body executing a *S.H.M.* is given by $t = 2\pi\sqrt{1/\mu}$, we have

$$= 2\pi\sqrt{\frac{1}{g/l}} = 2\pi\sqrt{\frac{l}{g}},$$

i.e., the same as that of a simple pendulum of length l cm.

11. A particle moving in a straight line with simple harmonic motion, of period 2π/ω, about a fixed point *O*, has a velocity $\sqrt{3}\ b\omega$, when at a distance *b* from *O*. Show that its amplitude is 2*b* and that it will cover the rest of its distance in time π/ 3ω.

We know that the velocity of a particle executing *S.H.M.*, at a displacement *y*, is given by $\omega\sqrt{a^2 - y^2}$, where *a* is its amplitude.

Here, displacement of the particle is *b*, and its velocity at this displacement is $\sqrt{3}\ .\ b\omega$,

$\sqrt{3}\ \ b\omega, = \omega\sqrt{a^2 - b^2}$. Or, $\sqrt{3}.b = \sqrt{a^2 - b^2}$ [∵ $y = b$.

Or, $3b^2 = a^2 - b^2$, whence, $a^2 = 4b^2$.

Or, $a = 2b$, *i.e.*, the *amplitude* of the particle is 2*b*.

Now, we have $y = a \sin \omega t$

Here, $y = b$, and $a = 2b$. And, therefore, $b = 2b\ \sin \omega t$.

Or, $\sin \omega t = b / 2b = \dfrac{1}{2}$. Or, $\omega t = \sin^{-1}\dfrac{1}{2} = \pi/6$.

Or, $t = \pi / 6\omega$.

Hence the time taken by the particle in covering the distance *b* from *O* is equal to π / 6ω.

Now, since it takes time 2π/ω to complete one vibration, it will take $\dfrac{1}{4}$ of 2π/ω or, time π/2ω to complete $\dfrac{1}{4}$ *th* of its vibration, or its amplitude, 2*b*. And time *already taken by it is* π / 6 ω.

∴ the time it will take to cover the *rest of its distance* is clearly equal to

$$\frac{\pi}{2\omega} - \frac{\pi}{6\omega} = \frac{3\pi - \pi}{6\,\omega} = \frac{2\pi}{6\omega} = \frac{\pi}{3\omega}.$$

12. A body whose mass is 0.005 kg is in simple harmonic motion with a period of 0.04 sec and an amplitude of 0.01 m. (*a*). What is its acceleration when it is 0.005 m from its equilibrium position ? (*b*) What is the force acting on it at that point ? (*c*) What is its maximum acceleration and what is the maximum force acting on it ?

We know that $T = 2\pi / \omega$ and $\therefore\ \omega = 2\pi/T = 2\pi / .04 = 50\,\pi$ *rad / sec.*

Since *acc.* = $-\omega^2 \times$ displacement from mean position, we have (*a*) *acc. at distance .005 m from mean position* $=(50\pi)^2 \times .005 = -123.4$ *m/sec*^{2}

And $\therefore$ (*b*) *force acting on the body at this point = mass × acc.* = .005 × 123.4 = 0.617 *newton.*

(*c*) clearly, *max. acc.* $= -\omega^2 \times$ *amplitude* $= -(50\pi)^2 \times .01 = 246.8$ *m/sec*^{2}.

And, $\therefore$ *max. force acting on the body* = .005 × 246.8 = 1.234 *newton.*

13. The equation of motion of a body executing *S.H.M.* is $y = .01 \cos(\pi t + \pi/4)$, where y is in metres and t in seconds. Calculate (*i*) the time-period and (*ii*) the amplitude of the motion. What is its displacement at t = 0 and t = 2.75 sec ?

(*i*) The time-period of a body executing *S.H.M.* is given by $T = 2\pi/\omega$. Here $\omega = \pi$. so that, $T = 2\pi / \pi = 2$ *sec.*

And (*ii*) from the equation of motion itself, it is clear that the *amplitude* = .01 *metre*.

Now, when $t = 0$, we have $y = .01 \cos \pi/4$ Or, *displacement* $y = .01 \times 1/\sqrt{2} = .00707$ *metre*

And at $t = 2.75$ *sec.*, we have $y = .01 \cos(2.75\,\pi + \pi/4)$

or, $y = .01 \cos 3\pi = .01 \times (-1) = -.01$ *metre.*

14. A hydrogen atom has a mass 1.68×10^{-24} gm. When attached to a certain massive molecule, it oscillates as a classical oscillator with a frequency of 10^{14} cycles/sec and with an amplitude of 10^{-9} cm. Calculate the force acting on the hydrogen atom. (*Delhi, 1966*)

We know that $T = 2\pi/\omega$ and,

therefore, $n = 1/T = \omega / 2\pi$.

Since $n = 10^{14}$, we have $10^{14} = \omega/2\pi$. or $\omega = 2\pi \times 10^{14}$ *rad/sec.*

Now, $acc. = \omega^2.y$ Since here, $y = a = 10^{-9}$ *cm*, we have

acc. of the hydrogen atom $= -\omega^2.a = -(2\pi \times 10^{14})^2 \times 10^{-9}$ *cm/sec*^{2}.

And, since *force = mass × acc.*, we have *force acting on the hydrogen atom towards its mean positio*
$n = 1.68 \times 10^{-24} \times (2\pi \times 10^{14})^2 \times 10^{-9} = 66.34 \times 10^{-5}$ *dynes.*

EXERCISE IV

1. Define simple harmonic motion, explaining the meaning of the terms period, amplitude and phase. (*Calcutta; Delhi*)

2. Deduce expressions for the velocity, acceleration and periodic time of a point mass executing *S.H.M.* (*Ranchi, 1963*)

3. Deduce the equation for the simple harmonic motion of a particle. Two simple harmonic motions, having the same period but differing in phase and amplitude, are acting in the same direction on a particle. Show that the resultant motion is simple harmonic, and deduce the expression for the resulting amplitude and phase. (*Calcutta*)

4. Find the resultant of two mutually perpendicular *S.H.* motions which agree in period but differ in phase. Consider the important cases for phase difference varying from 0 to 2π. (*Punjab*)

5. A particle executes a *S.H.M.* of period 10 *seconds* and amplitude 5 *ft*. Calculate its maximum acceleration and velocity. **Ans.** 1.974 *ft./sec*^{2}; 3.142 *ft./sec.*

6. The path of a body executing a *S.H.M.* along a straight line is 4 *cm*, long and its velocity, when passing through the centre of its path, is 16 *cm/sec*. Calculate its time-period. **Ans.** 0.7854 *sec.*

7. The maximum velocity of a particle undergoing a *S.H.M.* is 8 *ft./sec* and its acceleration at 4 *ft*. from the mean position is 16 *ft./sec*. What is (*i*) its *amplitude* and (*ii*) its *period of vibration*. **Ans.** 4 *ft*; 3.142 *sec.*

8. Explain the characteristics of a simple harmonic motion and show how to find the velocity at any phase of the motion.

A particle executes simple harmonic motion of period 16 *sec*. Two *seconds after it passes the centre of oscillation*, its velocity is found to be 4 *ft per second*. . Find the amplitude. (*Madras*)

Ans. 14.41 *ft*.

9. Define simple harmonic motion and show that if the displacement of a moving point at any time is given by an equation of the form.

$$x = a \cos \omega t + b \sin \omega t,$$

the motion is simple harmonic. If $a = 3$, $b = 4$, and $\omega = 2$, determine the period, amplitude, maximum velocity and maximum acceleration of the motion. *(Madras)*

[**Hint.** — See *Important note*, page 121].

Ans. *(i)* 3.142, *(ii)* 5, *(iii)* 10, *(iv)* 20.

10. Define Simple Harmonic Motion and give its chief characteristics. Give three examples of *S.H.M.* A particle executing *S.H.M.* has an acceleration of 1/2 cm per sec^2, when the displacement is 2 *cm*. Find its period. *(Vikram, 1962)*

Ans. 12.57 *sec*.

11. Define a *S.H.* motion, explaining the meaning of the terms, period, amplitude and phase.

A particle is subjected simultaneously to two *S.H.* vibrations of the same period but of different amplitudes and phases, in perpendicular directions. Find an expression for the resultant motion and show that the path traced by the particle is an ellipse.

For what conditions may the path be a circle and a straight line ? *(Calcutta)*

12. A mass of 15 *lb.* is suspended from a fixed point by a light spring. In the equilibrium position the spring is extended by 8 *inches*. The mass is then pulled down by 4 *inches* and released from rest. Show that it executes a *S.H.M.* and calculate its time-period. Also calculate the energy of the mass. **Ans.** .9069 *sec;* 40 *ft. poundals*.

13. A particle executes *S.H.M.* of period 12 *seconds* and amplitude 8 *cm*. Find the time it takes to travel 3 *cm* from the positive extremity of its oscillation. *[Poona (Lower Physics), 1962]*

Ans. 1.71 *sec*.

14. Calculate the time-period of a circular disc of radius *r*, oscillating about an axis through a point, distant *r*/2 from its centre and perpendicular to its plane. **Ans.** $2\pi \sqrt{3r/2g}$.

15. A particle is moving with simple harmonic motion in a straight line. When the distance of the particle from the equilibrium position has the values x_1 and x_2, the corresponding values of the velocity are u_1 and u_2. Show that the period is $2\pi\,[\,(x_2^{\,2} - x_1^{\,2}) / (u_1^{\,2} - u_2^{\,2})\,]^{\frac{1}{2}}$.

Find also *(i)* the maximum velocity and *(ii)* the amplitude. *(Madras)*

16. A particle moves with uniform speed in a circle. Show that the motion may be resolved into two simple harmonic motions at right angles to each other. How do they differ in phase and amplitude ? Show how the potential and kinetic energies of a particle executing simple harmonic motion vary. *(Calcutta)*

17. Show that the total energy of a particle executing simple harmonic motion is proportional to *(a)* the square of its amplitude, *(b)* the square of its frequency.

Show how, on an average, *its average*, its energy is half *kinetic* and half *potential* in form.

18. In the *HCl* molecule the force required to alter the distance between the atoms from its equilibrium value is 5.4×10^5 *dynes per cm*. What is the fundamental frequency of the vibration of the molecule, assuming the vibration to be simple harmonic, and the mass of the *Cl* atom to be infinite compared to that of the *H* atom which is 1.66×10^{-24} *gm.* ?

(Cambridge Scholarship Certificate)

Ans. 9.078×10^{13}.

19. Find graphically the resultant of two simple harmonic motions at right angles to one another *(a)* when the amplitudes and periods are equal and one vibration differs in phase by $\pi/2$ from the other, *(b)* when the amplitudes are equal, the period of one is twice that of the other and the slower vibration is $\pi/2$ ahead of the other.

20. The total energy of a particle executing a *S.H.M.* of period 2π *sec.* is 10,240 *ergs* $\pi / 4$ sec. after the particle passes the midpoint of the swing, its displacement is $8\sqrt{2}$ *cm*. Calculate the amplitude of the motion and the mass of the particle.

(Oxford and Cambridge Higher School Certificate).

Ans. 16 *cm;* 80 *gm*.

21. A particle oscillates along a straight line with *S.H.M.* Its greatest acceleration is $5\,\pi^2 cm\ sec^{-2}$ and when its distance from its zero position is 4 *cm*, its velocity is $3\,\pi\ cm\ sec^{-1}$. Calculate the period of oscillation of the particle and its amplitude. **Ans.** 2 *sec;* 5 *cm*.

22. What is meant by *S.H.M.* ? Describe its characteristics.

A vertical *U*–tube of uniform cross-section contains water up to a height '*h*' cm. Show that if the water on one side is depressed and then released, its motion up and down the tube is *S.H.M.* and calculate its period. *(Panjab, 1965)*

Ans. $T = 2\pi \sqrt{h/g}$.

23. Find the resultant of two simple harmonic motions at right angles to each other and having their periods in the ratio 1 : 2. Also discuss the cases if their initial phases differ by *(i)* 0, *(ii)* $\pi/2$ *(iii)* π *(Vikram)*.

24. Calculate the displacement of a body, executing *S.H.M.*, in terms of its amplitude (*a*) at which its kinetic energy is equal to thriee its potential energy.

CHAPTER V

MEASUREMENT OF MASS—THE BALANCE

5.1. Mass and Weight. The *mass* of a body is the *quantity* of *matter* contained in it and is an *inherent, invariable* and *fundamental* property of it, quite independent of the presence or absence of any other neighbouring bodies or of the place where the body happens to be situated. Thus, the mass of a given body will be the same at the equator, at the poles of the earth, or, for that matter, anywhere else in the whole of the universe.

The above traditional definition of mass applies more appropriately to what is now referred to as the *'rest mass'* of a body, *i.e.*, at mass which is at rest, usually denoted by m_0. For, in accordance with the *theory of relativity*, the mass *(m)* of a body depends upon the velocity *(v)* with which it is moving (with respect to the observer), being given by the relation $m = m_0/(1-(v^2/c^2)$, where c is the velocity of light in vacuum. On account of the extremely large value of c (*viz.*, 3×10^{10} *cm/sec*), however, the change in the value of the mass of a body from its rest-mass value is hardly very appreciable, even at the highest velocity with which a material body can possibly move. Except, therefore, when we are concerned with intra-atomic particles (*electrons, proton*, etc.), whose velocities may approximate that of light, the mass of a body, for all our ordinary purposes, may well be deemed to be invariable.

For a further discussion of the subject and the distinction between *'inertial'* and *'gravitational'* mass of a body, see § 7.19 (Chapter VII).

The *weight* of a body, on the other hand, is the *force* with which it is attracted by the earth towards its centre, and is equal to the *product of its mass and the acceleration due to gravity*.

Thus, if m be the mass of a body, and g, the acceleration due to gravity, its weight is given by $w = m.g$.

Since the value of g changes from place to place, being inversely proportional to the square of the distance from the centre of the earth (see chapter VI) the weight of the *same* body differs from one place to another, being about half a per cent greater at the poles than at the equator, twenty-eight times its weight on earth on the sun and about one-sixth its weight on earth on the moon.

It will thus be seen that the weight of a body in a *variable* property of it, depending not only upon its own mass but also on its distance from the centre of the earth, *i.e. on its position relative to the earth*.

Then, again, since the *mass* of a body endows it with the *property of inertia* or of *reluctance to change its state of both rest and motion*, we may also define it as the *degree of resistance of matter to changes of motion*. As against this, the weight of a body, being a force, directed towards the centre of the earth, *tends to accelerate its motion in that direction*. Thus, *whereas the one resists, the other tends to produce, motion*.

Nevertheless, at a place, since g is constant* - at any rate, within a small space–the weights of two bodies are directly proportional to their masses. For, if w and w' be their *weights* and m and m', their *masses*, we have

$$w = mg \text{ and } w' = m'g. \quad \text{So that,} \quad w/w' = mg/m'g = m/m'.$$

It follows, therefore, that the common physical balance may be used to compare masses. For, although, strictly speaking, it really compares weights—indicating a measure of their equality or want of equality—but since the value of g for the body as well as the standard weights placed in its two pans respectively is the *same*, the forces exerted at the two ends of the beam, in its equilibrium position, are evidently equal, thus *indirectly* establishing the equality of the two masses, irrespective of the value of g.

---

* This was first shown by *Galileo* in 1590, by his famous experiment of dropping simultaneously two *unequal* masses from the top of the *Leaning Tower of Pisa*, when they reached the ground together. The same fact was confirmed by *Newton* and later by *Bessel*, by using pendulums with hollow *bobs*, filled with materials of different densities and, observing no variations in the value of g beyond those within experimental error. And finally, it has been shown conclusively by *Eotvos* by his experiments with an ingenious modification of the Torsion Balance.

If it be desired, however, to determine the weight of a body, we make use of a *spring balance*, the stretch of the spiral spring of which, if not unduly large, is proportional to the force applied to it by the weight of the body suspended from it, and this, as can be readily seen, will be different for the *same* body at different places, depending on the value of g.

5.2. The Common Balance. It is, in essentials, *an equi-arm lever of the first order* and depends, for its action, on the principle of moments.

The essential feature of its construction is a symmetrical *rigid* beam, usually in the form of a *triangular lattice girder*, as shown in Fig. 5.2 (to ensure lightness with strength), pivoted centrally, so as to be free to rotate in the vertical plane about the horizontal axis provided by a knife-edge of steel or agate, resting on an agate plane carried by a stout vertical pillar. A long and light pointer, fixed at right angles to it moves over a small ivory scale below, whose central division marks its normal position, when the beam is in equilibrium or at rest. A screw, worked upwards and downwards, at the top of the pointer, enables the *c.g.* of the beam (together with that of the pointer) to be raised or lowered, as desired*.

Two other knife-edges, similar to, and equidistant from, the central one, are carried by the beam itself on either side, with two *identical scale pans, of equal mass*, suspended from the agate planes resting on them.

The whole instrument is enclosed in a glass case, with side-windows and a sliding front, to safeguard against disturbance due to air draughts or temperature variations, –all weighings being carried out with the glass case properly closed on all sides.

The body** to be weighed is placed in the left-hand pan and standard weights from the weight box, in the right hand pan***, starting with the seemingly heavier ones, until the pointer swings evenly on either side of the central mark on the ivory scale. If the *arms* (*i.e.*, the two halves of the beam, on either side of the central knife edge) be of the *same length* and the scale pans be of the *same weight*, the beam will come to rest in the horizontal position, but if the weights of the scale pans differ even so slightly, it will be tilted towards the side of the heavier weight, with the pointer moving correspondingly over the scale below.

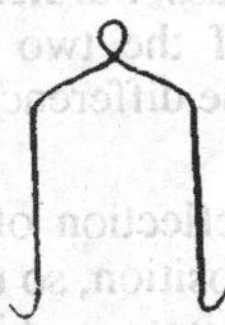
Fig. 5.1

The use of the Rider. Since the weight boxes are not provided with weights smaller than 1 *milligram*, the final adjustment for the equilibrium of the beam is made with the help of what is called a *'rider'*, which is just a piece of wire, weighing 1 *centigramme*, and bent into the form shown, (Fig. 5.1), and can be moved over the right half of the beam by a lever-device, manipulated from outside the case—this arm of the balance being graduated into 100 equal divisions from the central to the end knife-edge. With the rider at the 100th division, the effect is equivalent to placing a centigram weight in the right-hand pan ; so that, when it is, say, at the *nth* division, the effect is equivalent to adding a weight of *n*/100 *centigramme* or 10*n*/100 or *n*/10 *milligramme* to the pan.

5.3. Essentials or Requisites of a Good Balance. There are three essentials of a good physical balance, *viz.*, (*i*) *Truth*, (*ii*) *Sensitiveness* (or *Sensitivity*), and (*iii*) *Stability* or *Quickness*.

1. Truth. *A balance is said to be true when, with its scale pans unloaded, or equally loaded, the beam remains horizontal.*

Let a and b be the lengths of the two arms of the balance, (Fig. 5.2), and S and S', the masses of its two scale pans.

Then, with the scale pans unloaded, the beam will remain horizontal when the moments on either side of the central knife-edge C balance each other, *i.e.*, when

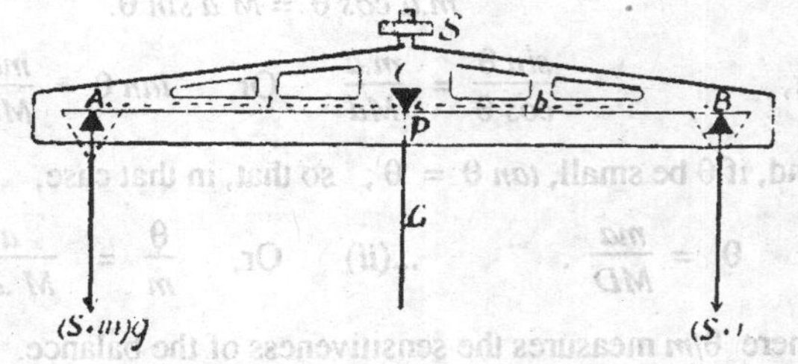

Fig. 5.2

$$Sg \times a = S'g \times b. \quad ....(i)$$

* If the beam were to be pivoted exactly at its *c.g.*, it would be in neutral equilibrium and will remain at rest at any angle with the horizontal. Its *c.g.* is, therefore, arranged to be below the central knife-edge, because as it tilts one way or the other, the *c.g.* rises upward, and the beam is thus in stable equilibrium.

** Too heavy bodies, likely to break or bend the beam, should be avoided.

*** It is purely a matter of convenience, with no principle involved in it.

Now, let the scale pans be loaded with *equal* masses **m** and **m**.

Then, for equilibrium, we have $(S+m)\, g.a = (S' + m)\, g.b.$ ...(*ii*)

Subtracting relation (*i*) from (*ii*), we have $m.a = m.b$,

whence, $a = b.$ ...(*iii*)

Substituting this in relation (*i*) we have $S = S'$ ...(*iv*)

Thus, a *true* balance must have (*i*) *arms of equal lengths* and (*ii*) *pans of equal weights.*

2. Sensitiveness. *A balance is said to be sensitive when, for a small difference of load in the two scale pans, the beam (and, therefore, the pointer) swings through an appreciable angle,* it being assumed that the balance is true.

Thus, the ratio between the deflection of the beam or the pointer and the difference of load, (usually 1 *m gm.*), causing it, measures the sensitiveness or the sensitivity of the balance. So that, the greater the displacement of the pointer for a given difference of load,—or, conversely, the smaller the difference of load required to produce a given displacement of the pointer,—the greater the sensitiveness of the balance. *Usually, a balance is regarded to be quite sensitive when a difference of 1 m gm in load causes the pointer to be displaced through 1 division on the scale.*

(*i*) **Case of a Balance with the three knife-edges in one plane.** Let Fig. 5.3 represent a vertical section of the balance through the centre of the beam, passing through the three knife-edges at *A, B* and *C, all in one plane.*

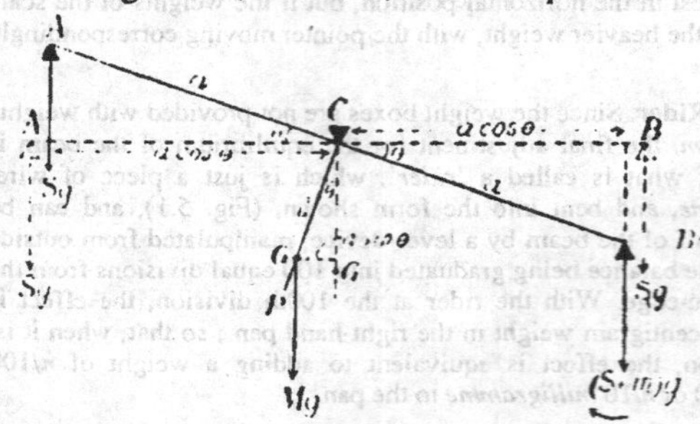

Fig. 5.3.

Let *a* be the length of each arm, *d*, the depth of the *c.g.* (*G*) of the beam from the central knife-edge *C* and *M*, its mass. Further, let *S* and $(S + m)$ be the masses of the two scale pans together with their loads, the difference of load *m* between them being small.

Then, if θ be the deflection of the beam from its initial horizontal position, so that it takes up the position *A'B'*. with its *c.g.* shifted to *G'*, we have, taking moments about *C*,

$$(S + m)\, g.a \cos\theta = Sg.a \cos\theta + Mg.d \sin\theta$$

Or, $$(S + m).\, a \cos\theta = S.a \cos\theta + M.d \sin\theta .$$

$$m.a \cos\theta = M.d \sin\theta.$$

Or, $$\frac{\sin\theta}{\cos\theta} = \frac{m.a}{Md}. \quad \text{Or,} \quad \tan\theta = \frac{ma}{Md}. \qquad ...(i)$$

And, if θ be small, $\tan\theta = \theta$; so that, in that case,

$$\theta = \frac{ma}{MD}. \qquad ...(ii) \qquad \text{Or,} \quad \frac{\theta}{m} = \frac{a}{M.d}, \qquad ...(iii)$$

where θ/*m* measures the sensitiveness of the balance.

It is thus clear that *to increase the sensitiveness* of the balance,

(*i*) *a* must be large, *i.e., the arms* (*or the beam*) *must be long,*

(*ii*) *M* must be small, *i.e., the beam must be light*, and

(*iii*) *d* must be small, *i.e., the c.g. of the beam must be close to the central knife-edge.*

Now, *a* cannot be increased beyond reasonable limits. For, as *Birge* correctly pointed out, the bending of a beam being proportional directly to the cube of its length and inversely to the cube of its thickness or depth, its thickness will also have to be increased in the same proportion with its length if its original stiffness is to be maintained, and this will inevitably increase its mass in a much greater proportion, thereby seriously impairing sensitiveness.

Nor can the beam be made light beyond a limit, or else it will either break or bend permanently. So that, the only workable alternative is to decrease *d*. This may be done with the help of the vertical movement of the screw, provided at the top of the pointer, (Fig. 5.2), though, carried to an excess, this too has its own drawbacks, *viz*., *(i) loss of stability*, and *(ii) a longer period of swing of the beam*. We have, therefore, to content ourselves with a judicious compromise among all these factors.

Further, if l be the length of the pointer, its displacement s on the scale will obviously be $l\theta$. And, therefore, if θ be small, so that $\tan\theta = \theta$, we have from relation *(ii)* above,

$$S = l \cdot \frac{m.a}{M.d}$$

To determine the displacement of the pointer, it is by no means necessary to wait until it actually comes to rest; it can be easily estimated from its swings to the right and the left.

Thus, suppose we have a scale with its zero at one end and the successive turning points of the pointer occur on it at the 10*th*, the 2*nd* and the 8*th* divisions, (Fig.5.4). Then, clearly, if S be its resting point, *i.e.*, the division where it will eventually come to rest, it is clear that the successive displacement of the pointer from *this* division are $s_1 = (10-S)$, $s_2 = (S-2)$ and $s_3 = (8-S)$.

Now, although, theoretically, as we shall soon see, (page 137), the oscillations of the pointer must be simple harmonic, it really seldom happens that the oscillations of any vibrating system remain truly so. The oscillations always die down and their amplitude goes on progressively decreasing due to air-resistance and other causes, but the ratio between the successive swings to the left and the right–very aptly called **'decrement'**—is found to remain constant.

Thus, with s_1, s_2 and s_3 as the successive swings of the pointer, we have

$$s_1/s_2 = s_2/s_3, \quad \text{and so on.}$$

So that, $\dfrac{10-S}{S-2} = \dfrac{S-2}{8-S}$.

Or, $(S-2)^2 = (10-S)(8-S)$

Or, $S^2-4S+4 = S^2-18S+80$

Or, $14S = 76$

whence, $S = 76/14$

$= 5.43.$

Thus, the pointer will ultimately come to rest at the 5.43*rd* division on the scale.

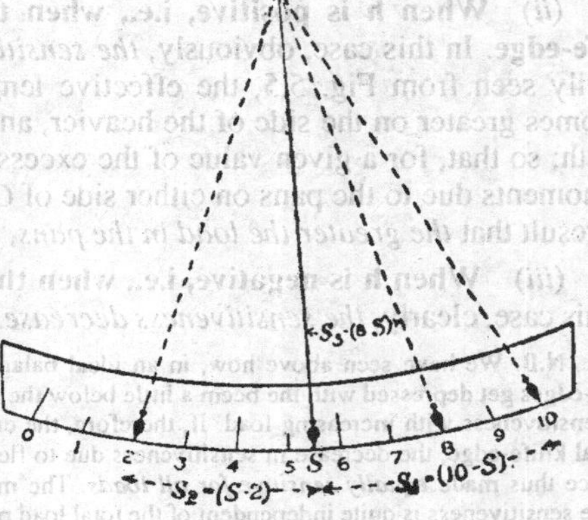

Fig. 5.4

(ii) Case of a Balance with the end knife-edges in a different plane from the central one. The balance, discussed above, with all the three knife-edges lying in the same plane, is really only an *ideal* one,—this condition being hardly ever attainable in ordinary balances. For, the beam does yield, however so little, to the forces acting at its two ends, so that the end knife-edges do get depressed a little below the central one, and no longer remain in the same plane with it.

Let us see how that sensitiveness is affected when the three knife-edges are *not* co-planar

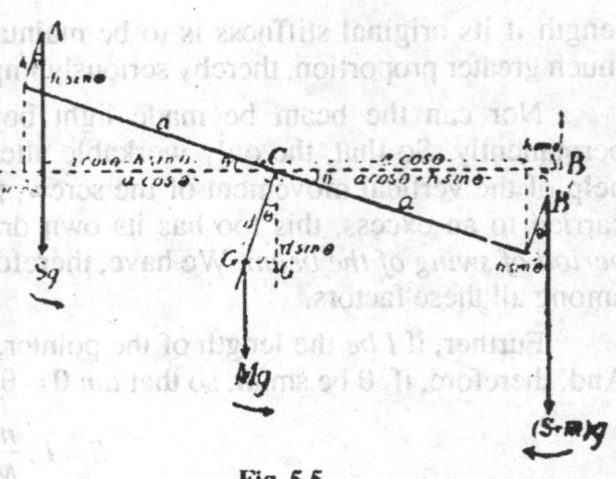

Fig. 5.5

Let h be the height of the end knife-edges A and B, above the central one, C, (Fig. 5.5), and let the beam be deflected through an angle θ, as before, for an extra mass m in the right-hand pan. Then, for equilibrium, we have, here,

$$(S+m).g.(a\cos\theta + h\sin\theta) = S.g.(a\cos\theta - h\sin\theta) + Mg.d\sin\theta.$$

Or,
$$(S+m)(a\cos\theta + h\sin\theta) = S\ (a\cos\theta - h\sin\theta) + M.d\sin\theta.$$

Or,
$$S.a\cos\theta + S.h\sin\theta + m.a\cos\theta + m.h.\sin\theta = S.a\cos\theta - S.h\sin\theta + M.d\sin\theta$$

Or,
$$2S.h\sin\theta + m.a\cos\theta + m.h.\sin\theta = M.d\sin\theta.$$

Now, if θ be small, so that $\sin\theta = \theta$ and $\cos\theta = 1$, we have

$$2S.h.\theta + ma + mh\theta = M.d.\theta.$$

Neglecting $mh\theta$ as the product of very small quantities, we have $2S.h.\theta + ma + M.d.\theta$ Or, $ma = M.d.\theta - 2S.h.\theta = \theta\,(M.d - 2S.h)$

whence, **sensitiveness,** $\dfrac{\theta}{m} = \dfrac{a}{M.d - 2S.h}$ ...(*iv*)

Now, three possible cases arise :

(*i*) **When h=0, i.e., when the three knife-edges are co-planar.** Here, $\theta/m = a/M.d$, [see relation (*iii*) above], and *the sensitiveness is quite independent of the total load* (2S).

(*ii*) **When h is positive, i.e., when the end knife-edges are higher than the central knife-edge.** In this case, obviously, *the sensitiveness increases with the total load.* But, as will be readily seen from Fig. 5.5, the effective length of the arm, in the tilted position of the beam, becomes greater on the side of the heavier, and smaller on the side of the lighter pan than its true length; so that, for a given value of the excess load m, and for a given deflection θ, the difference of moments due to the pans on either side of C is greater for a heavier than for a lighter load, with the result that *the greater the load in the pans, the longer the beam takes to attain equilibrium.*

(*iii*) **When h is negative, i.e., when the end knife-edges are lower than the central one.** In this case, clearly, *the sensitiveness decreases with increase in the total load.*

N.B. We have seen above how, in an ideal balance, with its three knife-edges in the same straight line, the end knife-edges get depressed with the beem a little below the central one when the pans are loaded. This results in a decrease in the sensitiveness with increasing load. If, therefore, the end knife-edges could be arranged at the correct height above the central knife-edge, the decrease in sensitiveness due to flexure could be just offset by its increase due to the latter, and the balance thus made *equally sensitive for all loads*. The method has actually been used with success in building balances whose sensitiveness is quite independent of the total load placed in their pans.

3. Stability or Quickness. *A balance is said to be stable (or quick) if, with the pans unloaded or equally loaded, the beam be disturbed, its time of swing is small and it comes back to rest quickly,* thus making for convenience in weighing.

Now, as we have seen above, with the three knife-edges in the same horizontal plane, the condition for equilibrium is that

$$(S+m)g.a\cos\theta = S.g\,a\cos\theta + Mg.d\sin\theta.$$ [See page 134

Therefore, if the two pans be *equally loaded, i.e.*, if $m=0$, or there be no extra load in the right hand pan, the only *restoring moment* about C, tending to bring the beam back to its original position, is $Mg.d\sin\theta$, or $Mgd.\theta$, if θ be small. This, obviously, tends to accelerate the motion or the swing of the beam; so that, if α be the *angular acceleration* it produces in the beam and I, the moment of inertia of the moving system about the central knife-edge C, we have

$$\alpha = Mg.d\theta/I.$$

And, clearly, I = *moment of inertia of the beam about C + moment of inertia of the two scale pans about C.*

Or, $$I = Mk^2 + 2Sa^2,$$

where K is the *radius of gyration* of the beam about C.

Thus, $$\alpha = \frac{Mg.d}{MK^2 + 2Sa^2}.\theta. \quad \text{Or, } \alpha = \mu\theta,$$

[Putting $\frac{Mg.d}{MK^2 + 2Sa^2} = \mu$, a constant for a given balance.]

where μ is a constant.

Or, $$\alpha \propto \theta,$$

i.e., the angular acceleration of the beam is proportional to its angular displacement. The swing of the beam is thus a *simple harmonic motion*, and its *time-period t* is, therefore, given by the relation,

$$t = 2\pi\sqrt{1/\mu}$$

Or, $$t = 2\pi\sqrt{1\Big/\frac{Mg.d}{MK^2 + 2Sa^2}} = 2\pi\sqrt{\frac{MK^2 + 2Sa^2}{Mg.d}}$$

Or, $$t = 2\pi\sqrt{\frac{K^2}{gd} + \frac{2Sa^2}{Mg.d}}$$

In order, therefore, that t' be small, *i.e.*, the balance be *stable, K,S and a should be small and M and d should be large*. We thus see that *a balance would be stable when*

(*i*) *its arms are short;*

(*ii*) *its beam is heavy, with its c.g. far below the central knife-edge;*

(*iii*) *the radius of gyration of the beam about the central knife-edge is small;*

and that the *stability diminishes with increasing load.* It will be seen at once that *almost all these conditions are opposed to those for sensitiveness.* So that, sensitiveness and stability of a balance are, to a great extent, *mutually exclusive*, and we have, therefore, to strike a working balance between the two.

5.4. Faults in a Balance—Determination of True Weights

1. Arms unequal in lengths and pans unequal in weights. The commonest fault in a balance is that it may appear to be true, *i.e.*, the beam may swing evenly on either side of the central knife-edge, with the moments on either side balancing each other, and yet the arms may have unequal lengths and the pans, unequal weights.

Thus, if S_1 and S_2 be the weights of the two scale pans and ***a* and *b***, the lengths of the two arms respectively, we have

$$S_1.a = S_2.b. \qquad ...(i)$$

The true weight w of a body may be determined with such a balance by the method of *double weighing i.e.*, by weighing the body first in one pan and then in the other.

Let the counterpoising weight in the right hand pan be w_1, when the body is placed in the left hand pan. Then, clearly,

$$(w + S_1).a = (w_1 + S_2).b \qquad ...(ii)$$

And $\therefore$ subtracting relation (*i*) from (*ii*) we have $w.a = w_1.b$. ...(*iii*)

Now, let the body be placed in the right hand pan and let the counterpoising weight required in the left hand pan be w_2. So that,

$$(W + S_2).b = (w_2 + S_1).a \quad ...(iv)$$

Again, subtracting relation (*i*) from (*iv*), we have $w.b = w_2.a$. ...(*v*)

Multiplying relations (*iii*) and (*v*), we, therefore, have

$$w^2.ab = w_1.w_2.ab, \text{ whence, } w^2 = w_1.w_2.$$

Or, $$w = \sqrt{w_1 . w_2},$$

i.e., the true weight of the body is the geometric mean of its apparent weights in the two scale pans.

The same will be true if the scale pans be equal in weight and the arms slightly different in lengths.

Note. If we multiply relations (*iii*) and (*v*) above, crosswise, we have

$$w.w_2.a^2 = w.w_1.b^2. \quad \text{Or,} \quad w_2.a^2 = w_1.b^2$$

Or, $$a^2/b^2 = w_1/w_2 \quad \text{Or,} \quad a/b = \sqrt{w_1/w_2}$$

And, since from relation (*i*) above, $a/b = S_2 / S_1$, we have

$$a/b = S_2/S_1 = \sqrt{w_1/w_2}$$

Thus, we can easily determine the ratio between the lengths of the two arms or that between the weights of the two scale pans.

2. Scale pans unequal in weights. Another common fault in a balance is that whereas the arms may be equal in length, the scale pans may not be truly equal in weights, so that the beam does not remain perfectly horizontal.

To determine the true weight of a body with such a balance, we again resort to *double weighing, i.e.*, to weighing the body first in one pan and then in the other.

Let its apparent weights in the two scale pans respectively be w_1 and w_2. Then, if the length of each arm be a and the weights of the two scale pans, S_1 and S_2, we have

in the first case, $(S_1+w).a=(S_2+w_1).a$, or, $S_1+w = S_2 + w_1$ ...(*i*)

And, *in the second case*, $(S_2+w).a = (S_1+w_2).a$, or, $S_2+w = S_1+ w_2$ ...(*ii*)

Adding relations (*i*) and (*ii*), we have

$$S_1+S_2+2w = S_1 + S_2 + w_1+w_2. \quad \text{Or,} \quad 2w = w_1 + w_2,$$

whence, $$w = w_1 + w_2 / 2$$

i.e., the true weight of the body is now the arithmetic mean of its apparent weights in the two scale pans.

3. Inaccuracy of the Brass Weights. A possible source of error may also be the inaccuracy of the brass or '*standard*' weights, supplied in the weight box, due to their getting worn out by use or getting slightly rusted by disuse. The probability of error due to such causes is presumably the least in the case of the larger weights and the greatest in the case of the smaller ones. So that, assuming the largest among them to be accurate, others, of smaller denominations, are counterpoised against it; these smaller ones are then counterpoised against others smaller than them, the process being continued up to the very smallest ones, and, in this manner, the errors in the smaller weights are easily detected. Thus, for example, a weight of 100 *gm* is first counterpoised against 50+20+20+10; then, the weight of 50 *gm* against 20+10+10+5+2+2+1, and again the weight of 20 *gm* against 10+5+2+2+1 and so on. To make sure, the weight assumed to be correct must also be tested against a weight known accurately in terms of the *International Standard*.

4. Blunting of the Knife-edges. Due to constant use, the knife-edges get blunted or rounded off in course of time; so that, with the tilt of the beam, the point of contact with the plane of support may shift slightly. This tantamount to a slight change in the lengths of the arms and must also be corrected for.

5.5. Correction for Buoyancy. Ordinarily, we make all our weighings in air. But air, in common with all other gases and liquids, exerts an upward thrust on a body immersed in it, in strict obedience to the *Principle of Archimedes*. So that, the body to be weighed, as well as the brass weights, against which it is weighed, are subject to this upthrust or buoyancy due to the air displaced by them, which is equal to the weight of the displaced air, in either case.

If the body weighed happens to have the *same* density as that of the material of which the standard weights are made, their volumes too would obviously be the *same*, when they are counterpoised against each other, and the volume of air displaced by both, and hence the buoyancy or upthrust due to it, would just be counter-balanced and the standard weights used would straightway give the true weight of the body in vacuum. This is, however, rarely the case. But it makes it clear why, in really true and accurate balances, we insist upon the arms and the pans being identical in length, volume and mass.

More often than not, the density of the body is quite different from that of the material of the standard weights and, therefore, even when they counterpoise each other, their volumes, and hence also the weights of the air displaced by them, are altogether different. Let us, therefore, deduce the necessary correction in this commonly occurring case.

Let the *true* mass of the body weighed be M and its density, ρ. And, let the mass of the weights required to counterpoise it be M' and the density of their material, ρ'. And, finally, let the density of air be δ.

Then, clearly, *volume* of the body and hence the volume of the air displaced by it to $= M/\rho$.

And, so the *weight* of this displaced air and, therefore, the upthrust due to it $= M.\delta.g/\rho$. So that, the *apparent* (or *observed*) *weight of the body*

$$= \left(Mg - \frac{M}{\rho}\,\delta.g\right)$$

Similarly, the volume of the standard weights and, therefore, of the air displaced by them $= M'/\rho'$;

and the upward thrust due to this displaced air $= M'.\delta.g/\rho'$.

So that, *their apparent weight* $= \left(M'g - \dfrac{M'}{\rho'}.\delta.g\right).$

Since the body and the standard weights counterpoise each other in air, their apparent weights must be equal. And, therefore,

$$\left(Mg - \frac{M}{\rho}.\delta.g\right) = \left(M'g - \frac{M'}{\rho'}.\delta.g.\right)$$

Or,

$$M(1 - \delta/\rho) = M'(1 - \delta/\rho'),$$

whence,

$$M = M'\left(\frac{1-\delta/\rho'}{1-\delta/\rho}\right) = M'\left(1 - \frac{\delta}{\rho'}\right)\left(1 + \frac{\delta}{\rho}\right)$$

$$= M'\left[1 + \left(\frac{\delta}{\rho} - \frac{\delta}{\rho'}\right)\right].$$

[Neglecting the product of δ/ρ and δ/ρ', compared with the other terms.

Or,

$$M = M\left[1 + \delta\left(\frac{1}{\rho} - \frac{1}{\rho'}\right)\right]$$

From the above it follows at once that

$$M >=< M' \text{ according as } \rho <=> \rho',$$

i.e., the true weight of a body (*i.e., its weight in vacuum*) is greater than, equal to, or less than, its observed or apparent weight in air, according as its density is less than, equal to, or greater than, that of the material of the standard weights used.

WORKED EXAMPLES

1. The arms of a balance are unequal in length but, without the scale pans, the beam and the scale-pan holders are correctly balanced. The scale pans A and B are of weights $2w_1$ and $2w_2$ respectively. A body placed in pan A has an apparent weight W_1 and placed in pan B has an apparent weight W_2. Show that the true weight of the body is

$$[W_1W_2 + 2(w_1W_1+w_2W_2) + (w_1+w_2)^2] - (w_1+w_2).$$

(London Higher School Certificate)

Let the true weight of the body be W and the lengths of the left-hand and right-hand arms be a and b respectively.

Then, since equilibrium is attained with the body in the left-hand pan and a weight W_1 in the right-hand pan, the moments on either side of the central knife-edge must be equal; so that, neglecting moments due to the pan holders, which already balance each other, we have

$$(2w_1+W).a = (2w_2+W_1).b. \quad ...(i)$$

Again, equilibrium is attained with the body in the right-hand pan and a weight W_2 in the left-hand pan; so that,

$$(2w_2+W).a = (2w_2+W_2).a. \quad ...(ii)$$

Multiplying the corresponding sides of relation *(i)* and *(ii)*, we have

$$(2w_1+W)(2w_2+W) = (2w_2+W_1).(2w_1+W_2).$$

Or, $4w_1w_2+2w_1W+2w_2W+W^2 = 4w_1w_2+2w_2W_2+2w_1W_1+W_1W_2$.

Or, $W^2+2W(w_1+w_2) = W_1W_2+2(w_1W_1+w_2W_2)$.

Adding $(w_1+w_2)^2$ to both sides we have

$$W^2+2W(w_1+w_2)+(w_1+w_2)^2 = W_1W_2+2(w_1W_1+w_2W_2)+(w_1+w_2)^2.$$

The left hand expression is clearly the complete square of $W+(w_1+w_2)$:

so that, $[W+(w_1+w_2)]^2 = W_1W_2+2(w_1W_1+w_2W_2)+(w_1+w_2)^2$.

Or, $W+(w_1+w_2) = \sqrt{[W_1W_2+2(w_1W_1+w_2W_2)+(w_1+w_2)^2]}$

And ∴ $W = \sqrt{[W_1W_2+2(w_1W_1+w_2W_2)+(w_1+w_2)^2]} - (w_1+w_2)$.

2. The arms of a balance are similar and of equal length, *a*. The scale pans are similar and of equal weight, P. When the beam of the balance is horizontal, the central knife-edge is a distance *x* vertically above the middle of the line joining the knife-edges of the scale pans, and the c.g. of the balance is a distance *y* vertically below the same point. Assuming that the weight of the moving system of the balance is W, derive an expression for the angle of deflection of the beam when weighs w_1 and w_2 are placed on the scale pans. [$w_1 > w_2$]

Let *AB*, (Fig. 5.6) be the position of the beam, when the pans are yet unloaded, *C*, that of the central knife-edge, *P* and *P*, of the scale pans and *p*, that of the pointer, with *G* as the *c.g.* of the beam.

Let the heavier weight w_1 be now placed in the right-hand pan and the lighter weight w_2, in the left-hand pan and let the beam, and, therefore, also the pointer, deflect through an angle θ, into the positions *A′ B′* and *p′*, with the *c.g.* of the beam at *G′*, where $OG' = y$, (*O* being the mid-point of the line joining the knife-edges of the two scale pans). Let the central knife-edge be now at *C′*, where $OC' = x$ (given).

Then, the different forces, acting vertically downwards on the beam are (*i*) $(P + w_1)$ at *B′* (*ii*) $(P + w_2)$ at *A′* and (*iii*) *w*, the weight of the moving system, at *G′*.

Since the beam is in equilibrium in this position, the moments about the knife-edge *C″*, on either side must be equal.

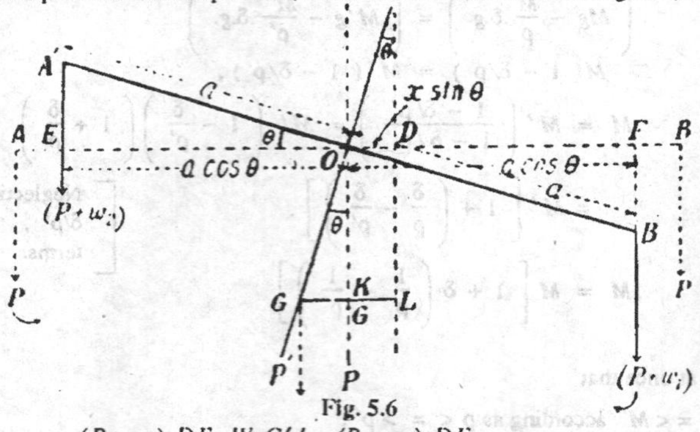

Fig. 5.6

or, $(P+w_2).DE = W.G'L = (P+w_1).DF$.

But $DE = OE + OD = OA' \cos\theta + OC' \sin\theta = a\cos\theta + x\sin\theta$;

$G'L = G'K + KL = G'K + OD = OG' \sin\theta + x \sin\theta$

$= y\sin\theta + x\sin\theta = (x+y)\sin\theta$

and $DF = OF - OD = OB' \cos\theta - OC'' \sin\theta = (a\cos\theta - x\sin\theta)$

So that, $(P + w_2).(a\cos\theta + x\sin\theta) + W.(x + y).\sin\theta = (P + w_1).(a\cos\theta - x\sin\theta)$

Or, $P.a\cos\theta + P.x\sin\theta + w_2a\cos\theta + w_2.x\sin\theta + W.(x + y).\sin\theta$

$= P.a\cos\theta - P.x\sin\theta + w_1.a\cos\theta - w_1.x\sin\theta,$

Or, $2P.x\sin\theta + w_1.x\sin\theta + w_2.x\sin\theta + W.(x+y).\sin\theta$

$= w_1.a\cos\theta - w_2.a\cos\theta$

Or, $\sin\theta\,[2Px + x(w_1+w_2) + W(x+y)] = a\cos\theta\,(w_1 - w_2)$.

Or, $$\frac{\sin\theta}{\cos\theta} = \tan\theta = \frac{a(w_1 - w_2)}{2Px + x(w_1 + w_2) + W(x + y)}$$

whence, $$\theta = \tan^{-1}\frac{a(w_1 - w_2)}{2Px + x(w_1 + w_2) + W(x + y)}$$

This, then, is the angle of deflection of the beam, when weights w_1 and w_2 are placed on the two scale pans.

3. With a balance of which the arms were 10 cm long it was found that 0.010 gm. extra load on one pan deflected the beam of mass 20 gms through 1° and that this deflection was independent of the loads placed on the pans. What can you deduce from these measurements ? *(Oxford Local H.S. Certificate)*

Since the deflection of the beam for the given extra load of .01 *gm* is quite independent of the loads placed on the pans, it is clear that *the three knife-edges are co-planar*.

And since in the question every other factor is given except the depth of the *c.g.*, of the beam below the central knife-edge or the centre of the beam, we are obviously expected to determine its value. Let it be *h*.

Since the beam is in equilibrium at angle θ = 1° from the initial horizontal position, it is evident that the moment about the central knife-edge due to its own weight *W* is just balanced by the opposite moment about *C* due to the extra load, .01 *gm* (Fig. 5.7).

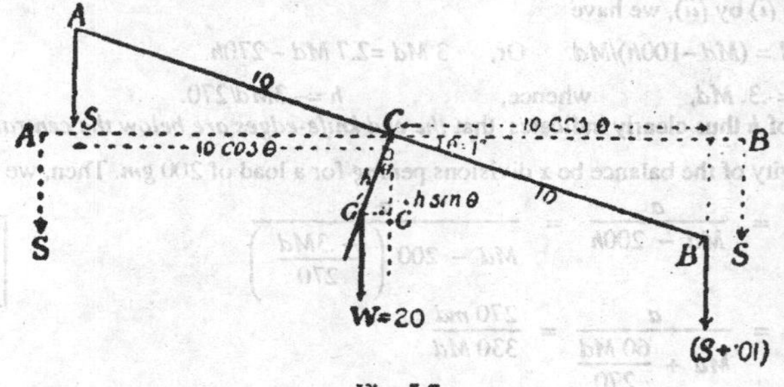

Fig. 5.7

Or, $W \times h \sin\theta = (0.1).\ 10 \cos\theta$. Or, $W \times h \times \sin 1° = (.01) \times 10 \cos 1°$.

Or, $20 \times h \times .0175 = .01 \times 10 \times .9998$. Or, $.35h = .09998$.

whence, $h = \dfrac{.09998}{.35} = .2856\ cm.$

Thus, the *c.g.* of the beam lies .2856 *cm* below its mid-point.

4. A shopkeeper, with a faulty balance, places a weight W in the left-hand pan and weighs out a commodity against it in the right-hand pan. Then, he places the same weight W in the right hand pan and weighs out more of the commodity against it in the left-hand pan. He delivers the total to the customer, charging him for a weight 2W. Does he gain or lose ?

Let the lengths of the two arms of the balance be *a* and *b* and let the weights of the corresponding pans be *S* and *S'*; so that,

with the pans unloaded, we have, for equilibrium, $S.a = S'.b$. ...*(i)*

Now, with the weight *W* in the left-hand pan, if the true weight of the commodity weighed against it in the other pan be W_1, clearly,

$$(S + W).a = (S' + W_1).b \qquad ...(ii)$$

Subtracting relation *(i)* from *(ii)*, we have

$$Wa = W_1 b. \quad \text{Or,} \quad W_1 = W.a/b$$

Similarly, with the weight *W* in the right-hand pan, if the true weight of the commodity weighed against it in the left-hand pan be W_2, we have

$$Wb = W_2 a,. \quad \text{Or,} \quad W_2 = W.b/a.$$

Thus, what the shopkeeper really gives to the customer is a weight

$$(W_1 + W_2) = W.\frac{a}{b} + W.\frac{b}{a} = W\left(\frac{a}{b} + \frac{b}{a}\right) = W.\left(\frac{a^2 + b^2}{ab}\right)$$ of the commodity and *not* 2*W*.

Obviously, therefore, he gains or loses according as $W(a^2 + b^2)/ab$ is less than or greater than 2*W*, for which he actually charges the customer:

Subtracting the latter from the former, we have $W\left(\dfrac{a^2 + b^2}{ab}\right) - 2W$

$$= \frac{W(a^2 + b^2) - 2Wab}{ab} = \frac{W(a^2 + b^2 - 2ab)}{ab} = \frac{W(a - b)^2}{ab},$$

which must always be a *positive quantity*, since the square of even a negative quantity is positive.

And thus, the shopkeeper is, for once, a *loser* in the bargain by a weight $W(a-b)^2/ab$ of the commodity.

5. It is found that the sensitivity of a balance for no load is 3.00 scale divisions per mg and for a load of 100 gm. is 2.70 scale divisions per mg. Find the sensitivity for a load of 200 gm. assuming the beam to be rigid. Indicate the relation between the knife-edge positions.

We know that the sensitivity of a balance is given by the relation,

$$\frac{\theta}{m} = \frac{a}{Md - 2S.h}, \text{ where } 2S \text{ is the total load.}$$ [See p. 136.

Now, *in the first case*, $\theta = 3.0$, $m = 1$ $mg = .001$ *gm. and* $2S = 0$.

And $\therefore$ $3/.001 = a/Md$ ...(i)

In the second case, $\theta = 2.70$, $m = 1$ $mg = .001$ gm and $2S = 100$ gm

So that, $2.7/.001 = a/(Md - 100h)$ ...(ii)

$\therefore$ dividing relation (i) by (ii), we have

$3/2.7 = (Md - 100h)/Md$. Or, $3\,Md = 2.7\,Md - 270h$.

Or, $270 = -.3.\,Md$, whence, $h = -.3Md/270$.

The negative value of h thus clearly indicates that *the end knife-edges are below the central knife-edge*

Now, let the sensitivity of the balance be x divisions per mg for a load of 200 gm. Then, we have

$$\frac{x}{.001} = \frac{a}{Md - 200h} = \frac{a}{Md - 200\left(\frac{-.3Md}{270}\right)}$$ [Substituting the value of h from above.

Or, $$\frac{x}{.001} = \frac{a}{Md + \frac{60\,Md}{270}} = \frac{270\,md}{330\,Md}$$

$\therefore$ dividing relation (iii) by (i), we have

$$\frac{x}{.001} \times \frac{.001}{3} = \frac{270a}{330Md} \times \frac{Md}{a}. \quad \text{Or,} \quad \frac{x}{3} = \frac{270}{330}, \quad ...(iii)$$

whence, $x = 270 \times 3/330 = 27/11 = 2.455$.

Thus, the sensitivity of the balance for a load of 200 gm is 2.455 divisions *per mg*.

6. A piece of metal weighs 300 gm in air. If the densities of the metal and the brass weights used be 19 gm/cc and 8 gm/cc respectively and that of air .00123 g/cc calculate the true mass of the piece in vacuum and the correction due to buoyancy

We know that the true mass of a body in vacuum is given by the relation,

$$M = M'\,[1 + \delta\,(1/\rho - 1/\rho')].$$ [See page 139.

where M is the true mass of the body, ρ, its density, M', its apparent mass, ρ', the density of the weights used and δ, the density of air.

Here, $M' = 300$ gm, $\rho = 19$ gm/cc, $\rho' = 8$ gm/cc and $\delta = .00123$ gm/cc. Substituting these values in the relation for M above, we have

$$M = 300[1 + .00123(1/19 - 1/8)]$$
$$= 300(1 - .00123 \times 11/152) = 300\,(1 - .01352/152)$$
$$= 300 - .026706 = 299.973294\ gm.$$

Thus, the true mass of the body in vacuum will be 299.973294 gm, the buoyancy correction being obviously .026706 gm.

EXERCISE V

1. What are the essentials of a good balance and how are they secured in actual practice ?

Show that the sensitiveness of a balance is independent of the load in the two scale pans, if the three knife-edges be co-planar. How does the position of the centre of gravity of the beam affect the working of the balance ?

2. The arms of a balance are each 7 cm long, the length of the pointer is 12 cm and the mass of the beam is 50 gm. If the knife-edges are in a plane and the centre of gravity of the beam is 0.02 cm below the central knife-edge, how much will the end of the pointer be deflected when the difference in load in the pans is 1 *milligramme* ?

(Cambridge Local Higher School Certificate)

Ans. 0.84 cm.

3. A certain balance has a beam weighing 200 gm, with knife-edges carrying the pans 15 cm from the central knife-edge. What is the depth of the centre of gravity of the beam below this knife-edge, if a weight of 1 mg placed on one of the pans displaces the end of the pointer through a distance of 0.5 cm. the pointer being 15 cm long ? Ans. 0.0225 mm.

4. Two balances, made of the same material, are alike in all respects except that the linear dimensions of one are n times those of the other. Compare the angular deflections of the beams for a given difference in load. **Ans.** $1/n^2 : 1$.

5. A body is weighed first in the left and then in the right hand pan of a balance, the respective weights being 9.842 *gm* and 9.838 *gm*. Find the true weight of the body and the ratio of the lengths of arms of the balance.

Ans. *True weight* 9.837 *gm*, *Ratio of arms*, 1.0005 : 1.

6. Discuss the points to be taken into consideration in the design of an accurate, sensitive and convenient balance. If the arms are of unequal lengths, show how the error on this account can be avoided. How would you expect the sensitiveness to vary with the load ? *(Bombay)*

7. What are the requisites of a balance ? Obtain the general expression used for determining the conditions for these requisites and show that the conditions for the two of these are mutually contradictory. *(Punjab)*

8. Sketch the essential parts of a balance in which the two end knife-edges are *h cm* below the central knife-edge and discuss the conditions of its sensitiveness.

9. Obtain an expression for the true mass of a body in vacuum, when its apparent mass in air is *M′ gm*, its density ρ, the density of the standard weights used, ρ' and the density of air, δ. Would the same treatment be applicable if the body be weighed in a liquid ?

10. Given the apparent weights of a body in two different liquids, of known densities, similar standard weights being used in both the cases, how will you proceed to calculate the density of the body ?

11. A glass stopper (density 2.5 *gm/cc*) is first weighed in water of density .995 *gm/cc* and then in oil. Its apparent weights in the two liquids are found to be 18.6 *gm* and 20.4 *gm* respectively, the brass weights used being similar in either case. Calculate the density of oil. **Ans.** 0.8489 *gm/cc*.

12. Explain what is meant by the terms sensitivity and stability of a balance. Derive an expression for the sensitivity of a beam balance and obtain the condition for which the sensitivity is independent of the load.

Describe a method of finding the true weight of a body with a balance of unequal arms. *(Punjab)*

13. From successive observation of the 'turning points' of a balance pointer, show how its rest position may be deduced. The scale across which the end of a balance pointer swings has its zero at one end. Successive turning points occur at 11, 3 and 8 divisions. Deduce the rest point. **Ans.** 6.08.

CHAPTER VI

ACCELERATION DUE TO GRAVITY

6.1. Acceleration due to Gravity. *Galileo* was the first person to have performed in 1590, the then bold and spectacular experiment of dropping a cannon and a musket ball from the *Leaning Tower of Pisa*, which, contrary to the teachings of *Aristotle*, reached the ground simultaneously. He thus clearly showed that, *at any given place*, all bodies, big or small, when dropped so as to fall freely, do so at the *same uniform rate*, neglecting, of course, the resistance to their motion due to air. That is to say, *all bodies, irrespective of their mass or nature, falling freely in vacuum, will have the same acceleration at a given place*. This acceleration is called the *acceleration due to gravity*, as it is due to the gravitational attraction of the body by the earth towards its centre, (see Chapter VII). It is denoted by the letter g, and is *numerically equal to the force with which a unit mass is attracted by the earth towards its centre, i.e., equal to the weight of unit mass**.

The value of g differs from place to place (See § 6.18), being the *greatest at the poles* and the *least at the equator*. Its value, for all practical purposes, is however taken to be 981 cm/sec^2 in the C.G.S. system, [g.81m/sec^2 in the M.K.S. (on the *S.I.*) system] and 32ft/sec² in the F.P.S. system. Due to this comparatively large value of g, bodies fall much too quickly to the surface of the earth, when dropped freely, and hence it becomes difficult to measure it directly with any great accuracy. It is, therefore, determined indirectly with the help of a simple or a compound pendulum, or by other methods. We shall now proceed to consider some of these in proper detail.

6.2. The Simple Pendulum. A *simple* (or, *a mathematical*) *pendulum* is just a heavy particle, (*ideally, only a point-mass*) suspended from one end of an *inextensible, weightless* string, whose other end is fixed to a rigid support,— the point where the string is fixed to the support being known as *the point of suspension* of the pendulum. In practice, we usually take a small and a heavy metallic spherical bob, tied to a *fine* silk thread.

*The motion of the pendulum is simple harmonic and isochronous, i.e., its time-period is quite independent of its amplitude for small swings***, *and is given by the relation* $t = 2\pi\sqrt{l/g}$, where l is the *length* of the pendulum, (or the *distance between its point of suspension and the centre of gravity of the bob*), and, *g, the acceleration due to gravity at the place*. This will be clear from the following :

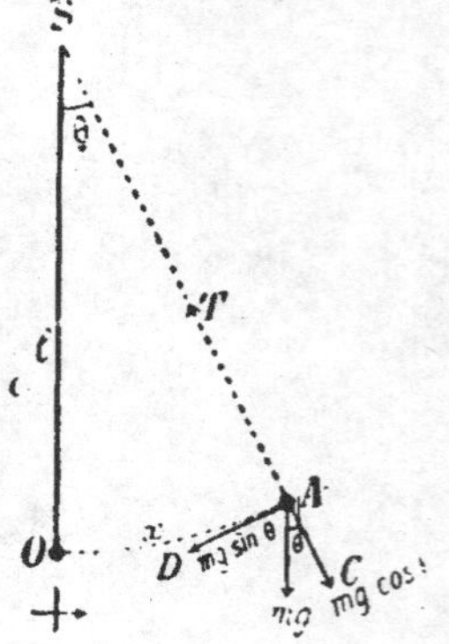

Fig. 6.1

Let S be the point of suspension and O, the mean or equilibrium position of the bob, (Fig. 6.1). Then, if the bob be given a *small* angular displacement θ, *in the vertical plane*, (or the plane of the pendulum itself), so as to occupy the position A, it is clear that it will be under the action of two forces, *viz.*, *(i)* its weight *mg*, *acting vertically downwards*, (*m* being the *mass* of the bob), and *(ii) tension T of the string acting along the string, towards its point of suspension S*.

Resolving the weight (*mg*) of the bob, into two rectangular components, we have

---

* This is so, because the force with which a body of mass m is attracted by the earth towards its centre is equal to its weight mg, and, therefore, if $m = 1$, *i.e.*, if the body be of *unit* mass, *this force of attraction* on it, or its *weight*, is equal to g.

** For the discovery of this property of the pendulum, we are again indebted to *Galileo*, who noticed a swinging lamp on the cathedral at Pisa and timed its oscillations against his own pulse beats. The time taken for each swing was found to be the same and, as far as Galileo could judge, quite independent of the size of the swing.

(*a*) component $mg \cos\theta$, acting along the string, as shown, and

(*b*) component $mg \sin\theta$, at right angles to it, *i.e.*, along the tangent *AD* to the arc *OA*, in the direction shown.

Obviously, the former component ($mg \cos\theta$) is just balanced by the tension (*T*) of the string, there begin no motion along it either way; so that the only force left acting on the bob is $mg \sin\theta$, towards its mean or equilibrium position *O*.

Now, *acceleration=force/mass.*

And, therefore, *acceleration of the bob* $= mg \sin\theta/m = g \sin\theta$, in the direction *AO*, or towards the mean or equilibrium position *O* of the bob. And if θ be small, we have $\sin\theta = \theta$.

Hence *acceleration of the bob* $= g\theta$, directed towards *O*

Again, $\theta = OA/SO = x/l$, [∵ *angle = arc/radius.*

where $OA = x$, the *displacement of the bob*, and *l*, the *length of the pendulum.*

∴ *acceleration of the bob* $= g \cdot \dfrac{x}{l} = \dfrac{g}{l} \cdot x.$

Or, putting $g/l = \mu$, a constant, for a given pendulum at a given place, we have

acceleration of the bob $= \mu x$.

Or, *acceleration of the bob* $\propto x$, *and directed towards O,*

i.e., the acceleration of the bob is proportional to its displacement from its mean or equilibrium position and is always directed towards that position. The bob thus executes a *simple harmonic motion*, and its *time-period t* is given by the relation,

$$t = 2\pi\sqrt{\frac{1}{\mu}} = 2\pi\sqrt{\frac{1}{g/l}} = 2\pi\sqrt{\frac{l}{g}}$$

Alternative Methods

Method (i). Let the bob be given a small angular displacement ∝ in the vertical plane, so as to occupy the position *A*, (Fig. 6.2). In other words, let ∝ be the *angular amplitude* of the bob.

Then, clearly, the *c.g.* of the bob has been raised up through a vertical distance *OC*, where *AC* is the perpendicular dropped from *A* on to *SO*.

∴ *potential energy of the bob at A = mg.OC.*

As the bob is released at *A*, it starts moving towards *O*, thus acquiring *kinetic energy* (due to its motion) at the expense of its potential energy.

Consider the bob to be at *B*, on its way towards *O*, and let its angular displacement here be θ. Then, clearly,

potential energy of the bob at B = mg.OD,

where *BD* is the perpendicular dropped from *B* on to *SO*.

∴ *loss in P.E. of the bob in moving from position A to position B* $= mg.\,OO - mg.OD = mg(OC-OD)$.

$= mg[(SO - SC) - (SO - SD)]$.

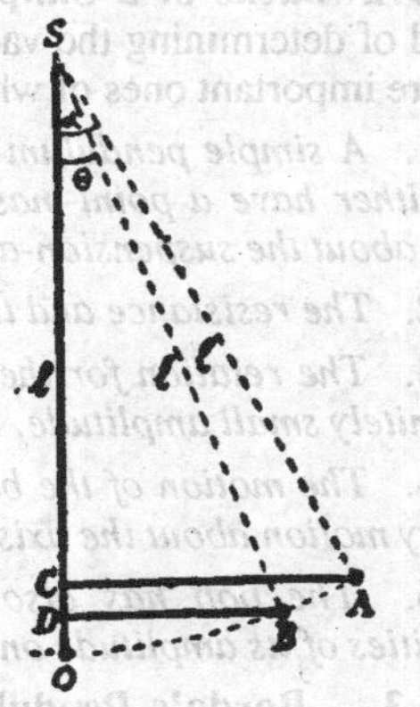

Fig. 6.2

Now, $SO = SA = SB = l$, *the length of the pendulum.*

And ∴ $SC = SA \cos\alpha = l\cos\alpha$ and $SD = SB \cos\alpha = l\cos\theta$

So that, *loss in P.E. of the bob* $= mg\,[(l - l\cos\theta) - (l - l\cos\theta)] = mgl\,(\cos\theta - \cos\alpha)$

This must, therefore, be equal to the gain in *K.E.* of the bob at *B*, or equal to the bob at *B*,(since its *K.E.* at *A* was equal to zero). If *I* be the moment of inertia of the bob about the axis of suspension through *S*, and ω, its angular velocity at *B*, we have

K.E. of the bob at B $= \frac{1}{2} I\omega^2$. And ∴ $\frac{1}{2} I\omega^2 = mgl\,(\cos\theta - \cos\alpha)$. ...(*i*)

Differentiating this expression with respect to time (t), we have

$$I\omega.\ d\omega/dt = -mgl.\sin\theta\ .\ d\theta/dt\ ,$$

the angular amplitude (α) of the bob being a constant quantity.

Now, $d\theta/dt = \omega$, the angular velocity of the bob at B.

And $\therefore$ $d\omega/dt$ is its angular acceleration, here.

Thus, $I\omega\ .\ d\omega/dt = -mgl.\sin\theta\ .\ \omega.$ or, $I.d\omega/dt = -mgl.\sin\theta$

whence, $d\omega/dt = -mgl.\sin\theta/I = -(mgl/I).\theta$ ...(ii) [θ being small

the –ve sign merely indicating that it is directed towards O, opposite to that in which the angular displacement (θ) increases.

Or, neglecting the —*ve* sign., and putting $mgl/I = \mu$, a constant for a given pendulum at a given place, we have *acceleration of the bob* $= \mu.\theta$,

and, therefore, proportional to θ, its *angular displacement* from its mean or equilibrium position.

The bob thus executes a *S.H.M.*, and its time-period is given by

$$t = 2\pi\sqrt{\frac{1}{\mu}} = 2\pi\sqrt{\frac{1}{mgl/I}} = 2\pi\sqrt{\frac{I}{mgl}}$$

Now, if r be the radius of the bob, we have, by the principle of parallel axes, $I = 2mr^2/5 + ml^2$.

And, therefore, $t = 2\pi\sqrt{\dfrac{2mr^2/5 + ml^2}{mgl}} = 2\pi\sqrt{\dfrac{\frac{2}{5}r^2 + l^2}{gl}}$

Since the bob is a small one, its radius r is negligible compared with l, the length of the pendulum; so that, we have

$$t = 2\pi\sqrt{l^2/gl}. \quad \text{Or,} \quad t = 2\pi\sqrt{l/g}.$$

Method (ii). See Borda's pendulum (next article).

Calculation for g. Squaring the expression for the time-period of the pendulum, we have

$$t^2 = 4\pi^2 l/g, \quad \text{whence,} \quad g = 4\pi l^2/t^2.$$

Thus, knowing l, the *length* of the pendulum,and t, its *time-period*, we can easily calculate the value of g at the given place.*

Drawbacks of a Simple Pendulum. Though simple in theory and easy to perform, this method of determining the value of g is not quite an accurate one, due to its numerous drawbacks, the more important ones of which are the following :

1. *A simple pendulum is just an ideal conception, not realizable in actual practice, for we can neither have a point-mass, not a weightless string; so that, the string too has a moment of inertia about the suspension-axis.*

2. *The resistance and the buoyancy of air appreciably affect the motion of the bob.*

3. *The relation for the time-period (t), obtained above, is true only for oscillations having an infinitely small amplitude.*

4. *The motion of the bob is not, strictly speaking, a motion of translation, for it also has a rotatory motion about the axis of suspension passing through the point of suspension.*

5. *The bob has also a relative motion with respect to the suspension-thread at the extremities of its amplitude on either side.*

6.3. Borda's Pendulum. In this pendulum, the bob is a sphere of *large radius* and, assuming, that it is rigidly fixed to the string and oscillates only about the axis of suspension, (there being no relative motion between the bob and the string), its time-period is given by

---

* The earliest determinations of the value of g were all made by means of simple pendulums. Thus, **Picard**, in 1669, used a pendulum in which a copper *bob* of diameter 1 *inch* was suspended by an *aloe fibre* (which remains unaffected by moisture) and in the year 1792. **Borda and Cassini** used one, with a platinum ball of diameter 1.5 *inches*, suspended by an iron wire about 12.75 *ft* long.

$$t = 2\pi\sqrt{\frac{l + \frac{2}{5}r^2/l}{g}}$$

where l is the length of the pendulum and r, the radius of the bob.

This relation for t may be deduced either by alternative method *(i)* for simple pendulum above or as follows :

Let SO be a *Borda's pendulum* (Fig. 6.3), suspended from the point of suspension S and let m be the *mass* of the bob, of *radius* r.

Imagine the bob to oscillate in the plane of the paper, and let it be displaced from its original position A to the position B, through an angle θ, at any given instant.

The only restoring force on the bob in this position is its weight mg, acting vertically downwards, which has, obviously, *a moment about* the axis of suspension through S, (perpendicular to the plane of the paper) $= \times O'D = mg.l \sin\theta$, where l is the *length* of the pendulum, and

$\therefore\ OD = l \sin\theta$.

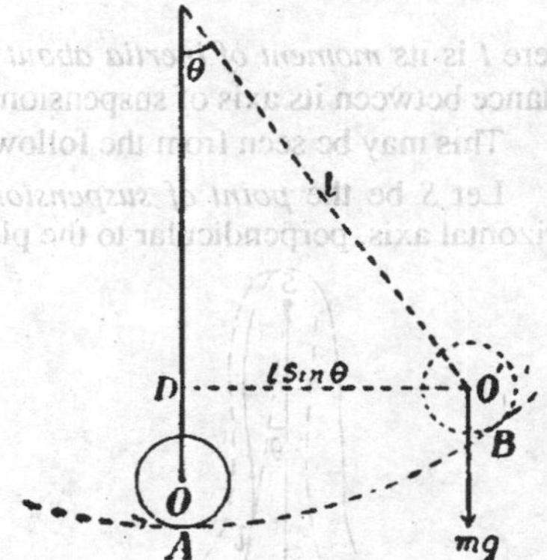

Fig. 6.3

This, then, is the *restoring couple*, acting on the bob in the position B, tending to bring it back to its original position.

If $d\omega/dt$ be the *angular acceleration* produced in the bob and I, its *moment of inertia* about the axis of suspension through S, the *couple* is also equal to $I.d\omega/dt$.

$\therefore\quad I.d\omega/dt = -mg\ l \sin\theta = -mg.l\theta$.

Or, $\quad d\omega/dt = -\theta.mg\ l/I = -\mu\theta$,

where $\quad mgl/I = \mu$, a constant.

Or, $\quad d\omega/dt \propto \theta$,

[∵ if θ be small, $\sin\theta = \theta$.
∵ the couple due to mg is clockwise and ∴ negative.

i.e., the angular acceleration of the bob is proportional to θ, *its angular displacement ; so that, its motion is simple harmonic;* and, therefore, the *time-period* of the pendulum is given by

$$t = 2\pi\sqrt{\frac{1}{\mu}} = 2\pi\sqrt{\frac{1}{mgl/I}} = 2\pi\sqrt{\frac{I}{mgl}}$$

Now, I = the *M.I.* of the bob about an axis through its *c.g.* and parallel to the axis of suspension + ml^2. [Principle of ∥ axes.

$= M.I.$ *of the bob about its diameter* + ml^2.

Or, $\quad I = \frac{2}{5}mr^2 + ml^2$, (∵ *M.I.* of a sphere of mass m, about its diameter is equal to $\frac{2}{5}\ m\omega^2$).

$$t = 2\pi\sqrt{\frac{ml^2 + \frac{2}{5}mr^2}{mgl}} = 2\pi\sqrt{\frac{l^2 + \frac{2}{5}r^2}{gl}} = 2\pi\sqrt{\frac{l + \frac{2}{3}r^2/l}{g}}$$

Thus, as will readily be seen, *the time-period of this pendulum is the same as that of a simple pendulum of length* $(l + \frac{2}{5}r^2/l)$, which, for this reason, is called the *length of an equivalent simple pendulum*, or the *reduced length* of the pendulum.

Further, if r be equal to zero, *i.e., if the bob be just a point*, or the pendulum be a *simple pendulum*, we have, substituting $r = 0$, $t = 2\pi\sqrt{l/g}$, the same expression as deduced above (§ 6.2) for the time period of a simple pendulum.

This pendulum too cannot give an accurate value of g, as, in the first place, *the string has also a moment of inertia about the axis of suspension* and secondly, *there is relative motion between the bob and the string*, the bob oscillating about it at each extreme end.

6.4 Compound Pendulum. *A compound, rigid* (or a *physical*) *pendulum is just a rigid body, capable of oscillating freely about a horizontal axis passing through it.* Its vibrations are also simple harmonic and its time-period is given by the relation

$$t = 2\pi\sqrt{\frac{I}{mg.l}}$$

where I is its *moment of inertia about the axis of suspension; m*, its *mass* and l, its *length* (or, the distance between its axis of suspension and its centre of gravity).

This may be seen from the following :

Let S be the *point of suspension* of the body, (or the pendulum), through which passes a horizontal axis, perpendicular to the plane of the paper, about which the body oscillates. Its *c.g.*, G, will obviously lie vertically below S, in its normal position of rest, (Fig. 6.4).

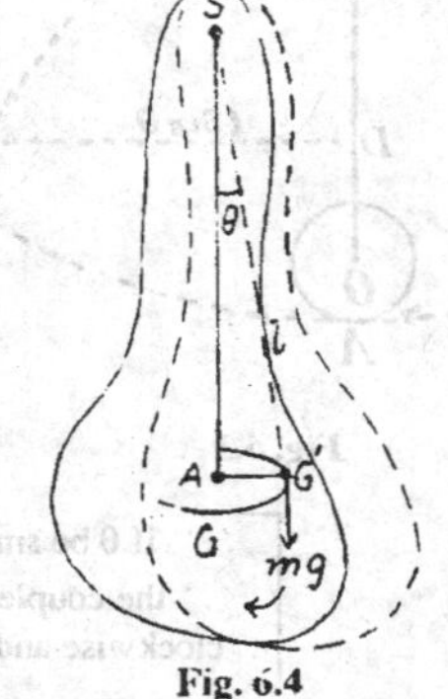

Fig. 6.4

Let the body be displaced through an angle θ, into the dotted position shown; so that, its *c.g.* is now at G'. Then, the *couple* acting on the body due to its weight mg will obviously, be $mg.l \sin\theta$, tending to bring it back into its original position. (SG' being equal to l, the length of the pendulum).

If the angular acceleration produced in the body by this couple be $d\omega/dt$, the couple will also be equal to $I.d\omega/dt$, where I is the *moment of inertia of the body about an axis through the point of suspension S, and perpendicular to the plane of the paper.*

So that, $I.\, d\omega/dt = -mg.l \sin\theta = -mg.l\theta$ ...(A)[θ being small.

Or, $$\frac{d\omega}{dt} = -\frac{mg.l}{i}.\theta = -\mu\theta.$$ [where $mg.l/I = \mu$, a constant.

Thus, the angular acceleration ($d\omega/dt$) of the body is proportional to its angular displacement (θ). The body, therefore, executes a *S.H.M.*, and its time-period is given by

$$t = 2\pi\sqrt{\frac{1}{\mu}} = 2\pi\sqrt{\frac{1}{mgl/I}} = 2\pi\sqrt{\frac{I}{mgl}}. \quad ...(B)$$

If I_0 be the *M.I.* of the body *about an axis through G, its c.g.*, and parallel to the axis of suspension through S, we have, *by the principle of parallel axes*, $I = I_0 + ml^2$.

And, if k be the *radius of gyration* of the body about the axis through G, obviously, $I_0 = mk^2$; so that, $I = mk^2 + ml^2$.

∴ substituting the value of I in relation (B) above, we have

$$= 2\pi\sqrt{\frac{mk^2 + ml^2}{mgl}} = 2\pi\sqrt{\frac{k^2 + l^2}{lg}} \quad ...(C)$$

Or, $$t = 2\pi\sqrt{\frac{k^2/l + l}{g}}$$

i.e., the period of vibration is the same as that of a simple pendulum of length $\frac{k^2}{l} + l$, or $\frac{k^2 + l^2}{l}$ which is, therefore, **the length of an equivalent simple pendulum,** or the **reduced length** of the pendulum. It is sometimes denoted by the letter L.

Since k^2 is always greater than *zero, this length of an equivalent simple pendulum is always greater than l.*

Centre of Oscillation. A point O, on the other side of G, at a distance k^2/l from G, is called the *centre of oscillation*, and a horizontal axis through it, parallel to the axis of suspension, is known as the *axis of oscillation* of the pendulum. Thus,

$GO = k^2/l$, [Fig. 6.5(*a*)]. Putting it equal to l', we have $SO = l + l' = l + k^2/l$.

And,

$$\therefore t = 2\pi\sqrt{\frac{l + k^2/l}{g}} = 2\pi\sqrt{\frac{L}{g}}.$$

Thus, the point of oscillation (O) lies at distance L from the point of suspension (S) or the distance between the two gives the length of the equivalent simple pendulum.

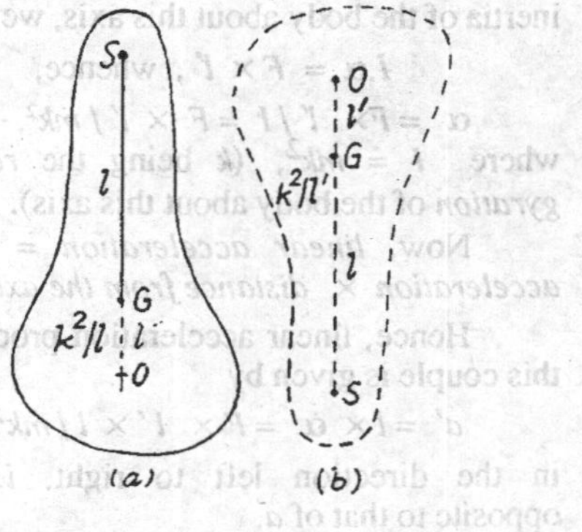

Fig. 6.5

6.5. Interchangeability of Centres of Suspension and Oscillation. If the pendulum be inverted and suspended about the *axis of oscillation* through O, as shown in Fig. 6.5 (*b*), its time-period of vibration will obviously be given by $t = 2\pi\sqrt{\frac{k^2 + l'^2}{l'g}}$.

And, since $k^2/l = l'$, we have $k^2 = l.l'$

So that, the expression for the time-period t becomes

$$t = 2\pi\sqrt{\frac{l.\,l' + l'^2}{l'g}} = 2\pi\sqrt{\frac{l + l'}{g}} = 2\pi\sqrt{\frac{l + k^2/l}{g}}$$

which is the same as about the axix of suspension through S.

Thus, *the centres of suspension and oscillation are interchangeable, i.e., are reciprocal to each other* —a property of the pendulum, first discovered by **Huyghens.**

In other words, we get the same values for the time-period and the length of the *equivalent simple pendulum* whether the pendulum be suspended at S or at O, *i.e, at a distance l* from the *c.g.*, (G), or at a distance *k^2/l from it.*

If, therefore, we can determine, by experiment, these two points, S and G we can easily find out the length L or ($l + k^2/l$) of the *equivalent simple pendulum*, and hence the value of g at the given place, with the help of the relation, $t = 2\pi\sqrt{L/g}$, where t is the time-period of the pendulum.

6.6. Centre of Percussion. Fig. 6.6 shows a section of a rigid body, of mass m, by a vertical plane passing through its *c.g.*, G, with S, as its point of suspension, the axis of suspension through which is perpendicular to the plane of the paper.

Let an impulsive force F be applied at O, in the direction shown, so as to be perpendicular to both the line SGO and the axis of suspension through S. Then, this force is equivalent to (*i*) *an equal and a like parallel force F* at G, *and* (*ii*) *a clockwise couple, formed by the force F* at O *and an equal and opposite force F at G*, the moment of which is clearly equal to $F \times l'$, where l' = *distance GO.*

Now, this force F at G tends to produce linear acceleration a, say, in all the particles of the body, including the centre of suspension S; so that, $a = F/m$, in the direction of the force, *i.e.*, from right to left.

This, then, is the acceleration produced at S by the force F at G. The couple, on the other hand, tends to produce an angular acceleration α, say, in the body, about a parallel axis through G. If I be the moment of inertia of the body about this axis, we have

$$I.\alpha = F \times l', \text{ whence,}$$

$$\alpha = F \times l'/I = F \times l'/mk^2,$$

where $I = mk^2$, (k being the *radius of gyration* of the body about this axis).

Now, *linear acceleration = angular acceleration × distance from the axis.*

Hence, linear acceleration produced by this couple is given by

$$a' = l \times \alpha = F \times l' \times l/mk^2,$$

in the direction left to right, *i.e.*, opposite to that of a.

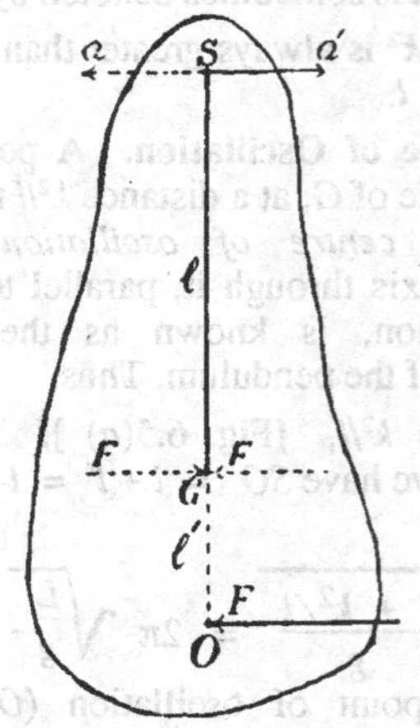

Fig. 6.6

In order, therfore, that the force F applied at O may produce no effect at S, the linear acceleration a due to force F at G must be equal and opposite to the linear acceleration a/due to this couple, *i.e.*, $a=a'$. In other words, $F/m = F \times l/mk^2$,

whence, $l' \times l = k^2$. Or, $l' = k^2/l$.

This, therefore, is the distance of the point O from the *c.g.* of the body, and the point O is thus the centre of oscillation, (see page 149), and is here called the *centre of percussion*, with respect to S.

It is thus clear that *if a body be struck at the centre of percussion, or the centre of oscillation, in a direction perpendicular to its axis of suspension, it does not move bodily, as a whole, at its point of suspension, but simply turns about the axis passing through it.*

This explains why when a ball strikes against a bat such that the point where it strikes the latter is the centre of oscillation, or the centre of percussion, corresponding to thepoint where it is held in the hand as the point of suspension, no *sting* or *shock* of any kind, is felt. The bat being held in the hands of the player has really no-well-defined axis about which it turns, but the player instinctively knows where to receive the ball on it so that no sting or shock is felt. Similarly, a good hammer should be so constructed that its centre of percussion lies in a line with the driving force.

6.7. Other points, collinear with the c.g., about which the time period is the same. Squaring the expression, $t = 2\pi\sqrt{(l^2 + k^2)/lg}$, for the *time-period* of a compound pendulum, where l is its *length* and k, its *radius of gyration* about the *c.g.*, we have

$$t^2 = \frac{4\pi^2(l^2 + k^2)}{l.g}. \quad \text{Or,} \quad 4\pi^2(l^2 + k^2) = l.g.t^2.$$

Or $$4\pi^2 l^2 + 4\pi^2 k^2 = l.g.t^2.$$

Dividing both sides by $4\pi^2$, we have $l^2 + k^2 = \frac{gt^2}{4\pi^2}.l.$

Or, $l^2 - \frac{gt^2}{4\pi^2}.l + k^2 = 0$, which is cleraly a *quadratic equation in l.*

So that, $$l = \frac{\dfrac{gt^2}{4\pi^2} \pm \sqrt{\dfrac{g^2t^4}{16\pi^4} - 4k^2}}{2}$$

Thus, l has two values, *viz.*,

$$\frac{gt^2}{8\pi^2} + \sqrt{\frac{g^2t^4}{64\pi^2} - k^2} = l_1 \quad \text{and} \quad \frac{gt^2}{8\pi^2} - \sqrt{\frac{g^2t^4}{64\pi^2} - k^2} = l_2,$$

such that, $l_1 + l_2 = 2.\, gt^2 / 8\pi^2 = gt^2 / 4\pi^2$ and $l_1 l_2 = k^2$, both positive quantities, showing that l_1 and l_2 are both positive.

There are thus two points, distant l_1 and l_2 (k_2 / l_1) from the *c.g.*, of the pendulum on the same side of it about which its time-period is the same.

Obviously, if we draw two circles with l_1 and k^2 / l_1 (or l_2) as radii and the centre of gravity (G) as centre, the period of vibration of the pendulum will be the same for any point of suspension lying on either of the two circles. Theoretically speaking, therefore, there can be an infinite number of points about which the pendulum will have an identical time-period. Since, however, a vertical line drawn through the *c.g.* (G) will cut the circles in only four points S, S' O' and O as shown Fig. 6.7,. *we have actually only these four points, collinear with the c.g., about which the time-period of the pendulum is the same**

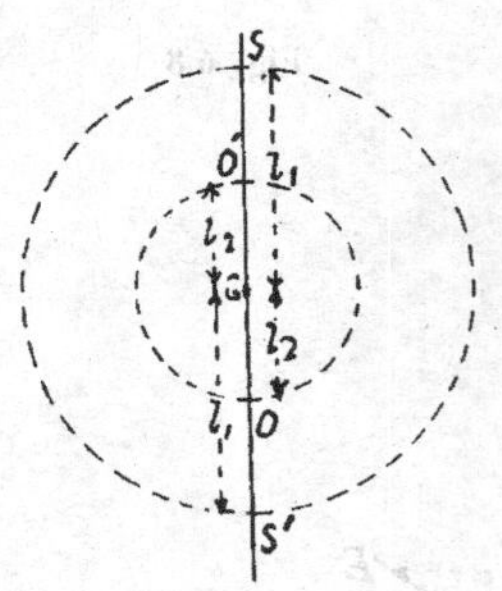

Fig. 6.7

As will be readily seen, for the point of suspension S, the point of oscillation is O and, similarly, for the point of suspensions S' the point of oscillation is O'. So that, we have $SO = SO' = l_1 + l_2 = l_1 + k^2 / l_1 = L$, *the length of the equivalent simple pendulum.* So that, we have

$$(l_1 + l_2) = L = gt^2/4\pi^2,$$

whence the value of g can be easily obtained.

6.8 Conditions for Maximum and Minimum Time-Periods. Re-arranging the expression for t^2 above, we have

$$t^2 = \frac{4\pi^2 (k^2 + l^2)}{lg} = \frac{4\pi^2}{g}\left(\frac{k^2 + l^2}{l}\right) = \frac{4\pi^2}{g}\left(\frac{k^2}{l} + l\right).$$

Differentiating *this* expression with respect to l, we have

$$2t \cdot \frac{dt}{dl} = \frac{4\pi^2}{g}\left(-\frac{k^2}{l^2} + 1\right).$$

Clearly, *t will be a maximum or a minimum when $dt/dl = 0$, i.e.*, when $l^2 = k^2$ or $l = \pm k$ *or when $l = k$*, (the —*ve* value of k being meaningless). Further, since $d t^2/dl^2$ comes out to be *positive*, this means that *t* **is a minimum**, *when $l = k$, i.e., when the distance between the axis through the point of suspension and the c.g., is equal to the radius of gyration of the pendulum about a parallel axis through the latter*. Obviously, the length of the equivalent simple pendulum $(l + k^2/l)$ will then be $(k + k^2/k)$ Or $2k$.

And, from the expression for t, we have $t = \infty$, when $l = 0$ or ∞. Again, ignoring $l = \infty$ as absurd, we have t, **a maximum,** *when $l = 0$, i.e., when the point of suspension passes through*

* All these four points may not all be found to lie on the body of the pendulum itself. Two of them may be well outside it.

the c.g. of the pendulum. This is so because the pendulum will then be in a state of neutral equilibrium, with no restorative action due to gravity on it.

6.9 Bar Pendulum. Perhaps the simplest and the most convenient form of a compound pendulum is what is called the *bar pendulum*, (Fig. 6.8), which is just a uniform metal, bar, (*AB*) having holes drilled along its length, *symmetrically* on either side of its centre, or its *c.g.*, *C*, so that any desired hole may be slipped on to a horizontal knife-edge *K*, and the bar made to oscillate about it in the vertical plane.

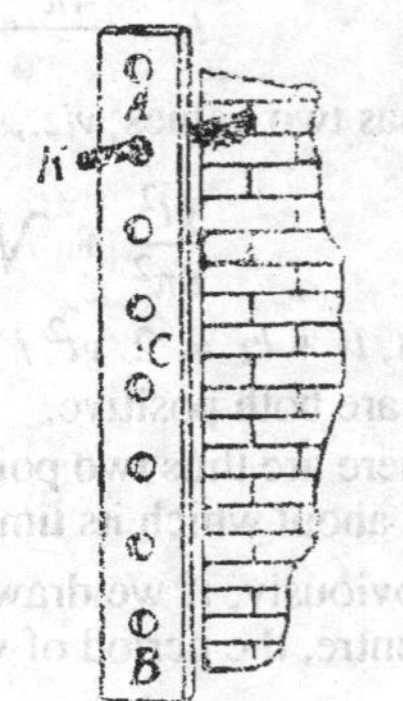

Fig. 6.8

The time-periods of the swings of the bar are determined with the bar supported on the knife-edge successively through the holes from *A* to *C*, the distance of the knife-edge, (*i.e.*, the point of suspension), to the *c.g.*, (*C*) of the bar being noted in each case.

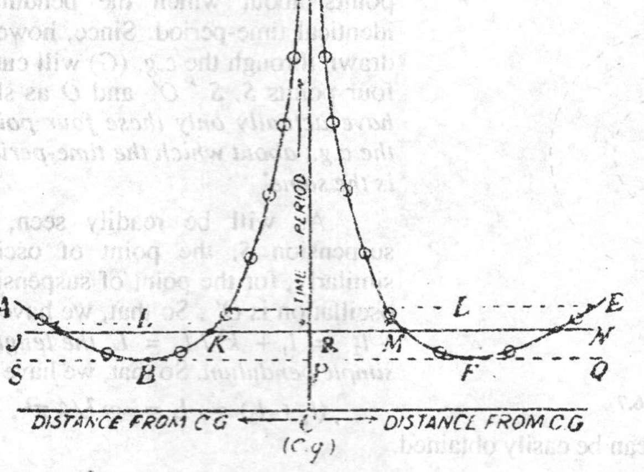

Fig. 6.9

A graph, *ABD*, is then plotted between distance from the *c.g.*, (*C*), along the *x*-axis, and the time-period, along the *y*-axis, (Fig. 6.9). The experiment is repeated, with the bar inverted and swung as before, with the holes from *B* to *C* supported successively on the kinfe-edge, and a similar graph *EFH** plotted alongside the first one—the two lying symmetrically about the line *CG* which corresponds to the *c.g.*, (*C*) of the bar. It will at once be clear from the curves that, in either case, *the time-period decreases at first, acquires a minimum value and then increases as the c.g. of the bar is approached, finally becoming infinite at the c.g.*

Now, if a horizontal line *JN* be drawn, it cuts the two curves at points, *J*, *K*, *M*, and *N*, about which, obviously the time-period of the bar would be the *same*. These, four points correspond to the four points, *S*, *O′*, *O* and *S′* in Fig. 6.7. It follows, therefore, that *JM* = *KN* = *L*, *the length of the equivalent simple pendulum.*

---

* The graph may be drawn straightway, symmetrically with the first one, about the line *CG*, without performing the experiment with the inverted bar, because, due to the symmetrical positions of the holes on either side of *C*, the two graphs are bound to be identical, *one being a laterally reversed image of the other in the axis of time-period (PG) as the mirror, as it were.*

This is so because *points J and M lie on opposite sides of the c.g. and at an equal distance from it. They therefore correspond to the points of suspension and oscillation of the pendulum respectively, the distance between them thus giviing directly the length of the equivalent simple pendulum. L.,* And so does the distance between the points K and N. These distance (LM and KN) are therefore carefully noted and their mean taken as the correct value of L, or the length of the *equivalent simple pendulum.* And, therefore, if t be its time-period, (given by the common ordinate of the points, J, K, M or N), we have

$$t = 2\pi\sqrt{L/g}, \quad \text{whence,} \quad g = 4\pi^2 L/t^2.$$

So that, knowing L and t, we can easily calculate the value of g at the given place.

Determination of k. If we draw the tangential line SQ, touching the two curves at the points B and F respectively, (Fig. 6.9), then, at B and F, the centres of suspension and oscillation coincide with each other, with the time-period of the pendulum, the minimum.

So that, $\quad l = k^2/l, \quad$ or, $\quad k^2 = l^2, \quad$ whence, $\quad k = l.$

Hence, each of the distance BP and FP is equal to the *radius of gyration k*. Thus, $BF = 2k$, or, $k = BF/2$.

So that, if we note this minimum time-period t_m (represented by CP), corresponding to either of the points B and F,

we have $\quad t_m = 2\pi\sqrt{2k/g}, \quad$ whence again, $\quad g = 8\pi^2 R/t_m^2$

It is not, however, very easy to correctly locate the points B and F, where the tangential line just touches the two curves. We, therefore, determine k from the relation $k^2/l = l'$ or $k^2 = \sqrt{ll'}$ whence, $k = \sqrt{ll'}$.

Here, clearly, JR and NR correspond to l and KR and RM to l' so that,

$$k = \sqrt{JR.KR} = \sqrt{NR.MR},$$

from which k for the bar can be easily determined.

Note. In view of the difficulty of determining the *c.g.* of the bar correctly, it is advisable to plot the *time-period*, not against the distance of the point of suspension from the *c.g.*, *but from one end of the bar*, as shown in Fig. 6.10. As will be readily seen, the curve here consists of two parts or branches, symmetrical with each other and similar in form to those in the previous case. The values of g and k are then calculated in exactly the same manner, as explained.

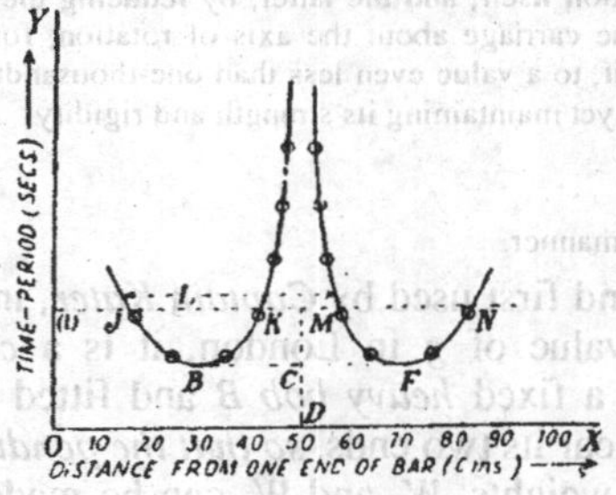

Fig. 6.10

Further, the position of the *c.g.* of the bar can also be easily known by noting the length OD of the bar, corresponding to the mid-point C of distance BF.

Even this method, however, does not lend itself to any desirable degree of accuracy. For, firstly, the two curves, obviously drawn free hand, cannot possibly compete with a straight line curve for deductions to be made from them of the type we are making here. And, secondly, as we have seen, the minima of the curves are not sharp enough for the exact positions of B and F to be pinpointed on them.

Ferguson, therefore, suggested, in the year 1928, that instead of the curves above, a graph may be plotted between lt^2, along the axis of x and l^2, along the axis of y, which, from the relation $l^2 + k^2 = \frac{lt^2}{4\pi^2}.g$, (§ 6.7), must be a straight line, as shown in Fig. 6.11, with its *slope* equal to $g/4\pi^2$ and its intercept on the axis of y equal to $-k^2$. The values of both g and k can thus be obtained much more accurately.

The *c.g.* of the bar will, of course, have to be determined in this case, in order to obtain different values of l for the graph, but, then any slight inaccuracies involved in doing so by the usual method of balancing, get automatically eliminated as the graph is smoothed out into its straight-line form.

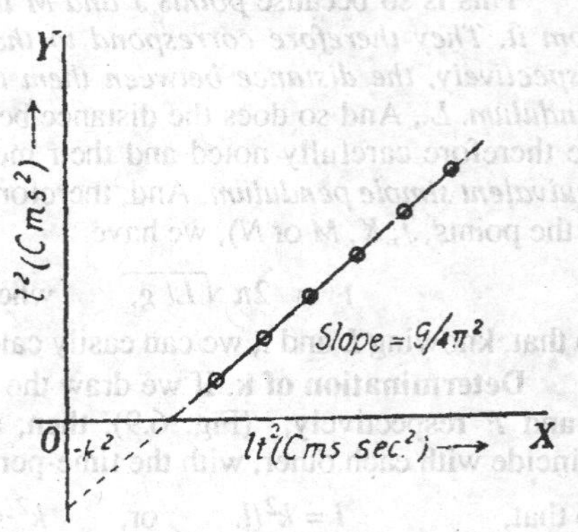

Fig. 6.11

Owen's modification of the bar pendulum. The usual form of the bar pendulum, discussed above, suffers from two main defects *viz.*, (*i*) on account of the short length of the bearing, *it has a tendency to wobble as it swings* about the axis of suspension passing through any one of the holes drilled along its length and (*ii*) *there is no means of ensuring that the axes of the holes are perpendicular to its plane.* As a result, the timing of its oscillations is at best only approximate.

Owen has been able to successfully tide over both these difficulties in his modified form of the bar pendulum, in which he has altogether dispensed with the holes. Instead, he has provided it with a carriage C, (Fig. 6.12), on which is rigidly mounted the knife-edge E, and which can be made to smoothly slide up and down the bar and clamped in any desired position on it.

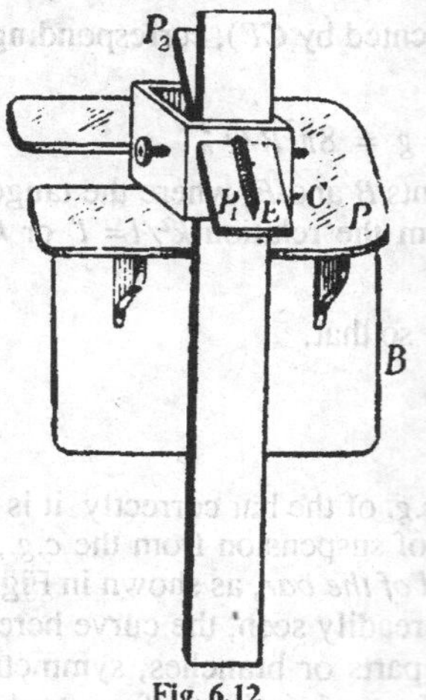

Fig. 6.12

The two projecting portions of the knife-edge, on the two opposite sides of the carriage, rest on two glass plates P_1 and P_2, mounted on either side of a *rigid horizontal* platform P, supported on a carefully set and stout wall bracket B. The platform has a rectangular slot, cut into its middle, to enable the bar to oscillate freely in the vertical plane about the knife-edge.

Now, the presence of the carriage might well be expected (*i*) to increase the gravitational couple on the swinging bar, tending to restore it back to its mean or equilibrium position and (*ii*) to increase the *M.I.* of the whole system. Both these effects are, however, made quite negligible, the former by so arranging the carriage that its *c.g.* lies on the line of the knife-edge or the axis of rotation itself, and the latter, by reducing the moment of inertia of the carriage about the axis of rotation, for any given position of it, to a value even less than one-thousandth of that of the bar, and yet maintaining its strength and rigidity.

The values of g and k are, of course, determined in the usual manner.

6.9. Kater's Reversible Pendulum. Devised and first used by *Captain Kater*, in the year 1817, to make the celebrated determination of the value of g in London, it is a *compound pendulum*, consisting of a brass or steel bar AB, with a fixed *heavy bob B* and fitted with two *adjustable* and mutually facing knife-edges F_1 and F_2* near its two ends, *so that the pendulum may be suspended from either* (Fig. 6.13). Two cylindrical weights, W_1 and W_2 can be made to slide along the length of the bar and clamped in the position desired, the smaller weight, W_2 having a *micrometer screw arrangement* M for finer adjustment of its position. The position of the *c.g.* of the pendulum can be altered by changing the relative positions of the two weights, their positions being, however, so chosen that *the c.g. always lies in between the two knife-edges.***

The pendulum is first suspended from the knife-edge F_1, and its time-period determined. It is then suspended from the knife-edge F_2, and its time-period determined again. If there be a divergence in the two values of the time-period, the heavier weight W_1 is moved up or down and a proper position found for it so that the time-period is very nearly the same, whether the pendulum

* It is gratifying to observe that in Kater's own pendulum, these were made in India, from a special variety of steel, called '*wootz*'.

** In fact, the heavier weight W_1 is there to ensure that this is so.

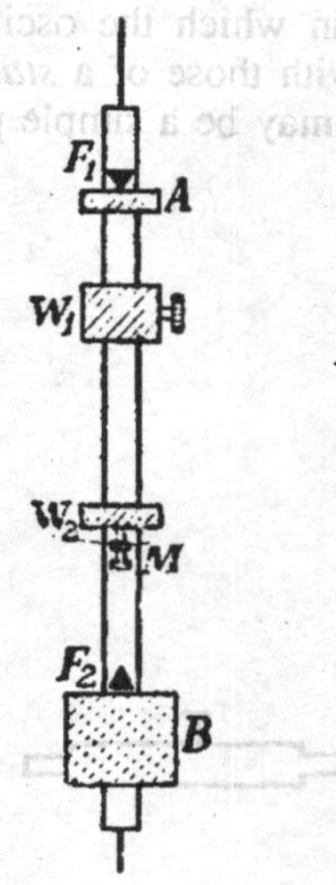

Fig. 6.13

be suspended form F_1 or F_2. The smaller weight W_2 is then adjusted by means of the micrometer screw M, until the time–periods, in the two cases, are as nearly equal as possible (say, differing only by .01 *sec.* or less, *i.e.*, until the number of oscillations made by the pendulum in 24 hours, in the two cases, differs by just a fraction of one full oscillation). *When this is so, we have, obviously, one knife-edge at the centre of oscillation of the other.* The distance between the two knife-edges is measured carefully*. This gives the length L of the *equivalent simple pendulum*** and the value of g is then calculated from the relation $g = 4\pi^2L/t^2$, (see page 153), where t is the mean of the time-periods about the two knife-edges, which *Kater* determined by the method of coincidences. (see § 6.10).

The values of L and t may, however, be determined more easily and accurately as follows :

First, a near equality in the time-periods of the pendulum is obtained about either knife-edge, by adjusting the positions of W_1 and W_2 as explained above, the time-period being slightly greater, say about F_1 than about F_2, *i.e.*, when the bob B is down than when it is up. The weights are now kept fixed in their positions and one of the knife-edges, say F_1, moved up or down a bit to further narrow down the discrepancy in the two time-periods until a position is attained when a little more displacement of F_1 makes the time-period greater about it (with the bob up) than about F_2, (with the bob down), *i.e.*, a reversal in the relative magnitudes of the time-periods about the two knife-edges takes place. The time-periods of the pendulum for two slightly different distances between F_1 and F_2 are noted, *just before this happens*, and similarly for two slightly different distances, between them, *after this happens*. These four distances between F_1 and F_2 are measured accurately and the time periods corresponding to them plotted on an exaggerated scale, as shown in Fig. 6.14, where a and b represent the distances and time-periods for oscillations about F_1 in the *first two* cases and c and d in the *second two cases*. The coordinates of the point of intersection O of ab and cd then give the true length L of the *equivalent simple pendulum* and the true time-period t corresponding to it.

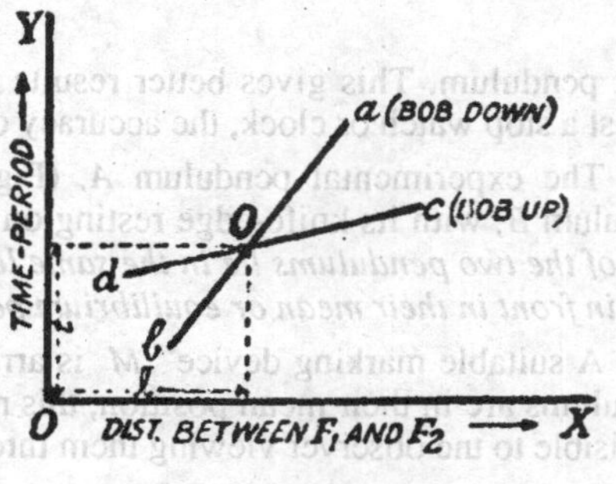

Fig. 6.14

* This is done by means of a travelling microscope. The pendulum is laid horizontally on a table, with its knife-edges lying alongside, and on a level with, a standard steel scale, and the positions of the knife-edges read on the scale with the help of the microscope. Or, a better method is to use a vertical *comparator* (an instrument carrying two microscopes fitted on two massive stone slabs), when the cross-wires of the microscope, are first focussed on the two knife edges of the suspended pendulum, and then on a standard steel scale, fixed vertically in place of the pendulum. The distance between the positions of the two cross-wires then gives the distance between the two knife-edges.

** This deduction from the reciprocal nature of the points of suspension and oscillation was first pointed out by *Bobnenberger*, in the year 1811.

6.10. Kater's Method of Coincidences. Kater determined the time-period of his pendulum by what is known as the *method of coincidences*, in which the oscillations of the experimental pendulum, (Kater's or any other), are compared with those of a *standard seconds pendulum (i.e., a pendulum of time-period two seconds)*, which may be a simple pendulum or a

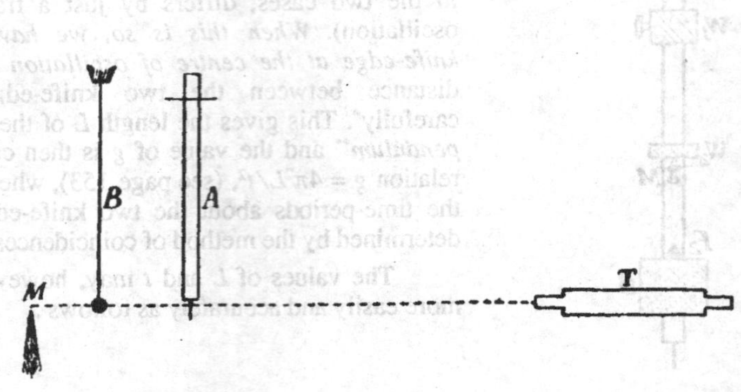

Fig. 6.15

clock pendulum. This gives better results than those obtained by simply timing the oscillations against a stop watch or clock, the accuracy of which can hardly be expected to go beyond .5 *sec.**

The experimental pendulum *A*, (Fig. 6.15) here is suspended in front of the second's pendulum *B*, with its knife-edge resting on a rigid support, *care being taken to see that the lower ends of the two pendulums lie in the same level and exactly coincide with each other, when viewed from in front in their mean or equilibrium position.*

A suitable marking device** *M* is arranged behind the pendulum *B*, such that when the two pendulums are in their mean position, this mark *M* is *just* covered by their lower ends and is thus not visible to the observer viewing them through a telescope *T* some distance away.

The whole idea is to enable the observer, watching the oscillations, to judge as accurately as possible as to when exactly do the two pendulums come into coincidence, *i.e., as to when exactly do they simultaneously pass a particular reference point in the same direction*. And this is perhaps best done by using a cross-wire in the eye-piece of the telescope itself, (in which case the marking device *M* becomes quite unnecessary).

The two pendulums are set oscillating, and, if they start together and have identical time-periods, they continue to oscillate '*in step, i.e.*, they appear to the observer as *just 'one'* pendulum—pendulum *B* being *just* hidden behind pendulum *A*, all the time. But if their time-periods differ ever so slightly, they soon *get 'out of step'*, and their oscillations are watched carefully until they both simultaneously pass the reference point fixed upon, (say, their lowest positions), in the same direction. When this happens, a '*coincidence*' is said to occur, and the mark *M* behind the pendulums is *just not visible* to the observer. After this, the pendulums again get out of step and *the next coincidence occurs when one of them gains or loses a whole swing or oscillation over the other*. The oscillations made by both between two successive coincidences are carefully counted.

* This means that the time taken for several thousand of swings will have to be noted if an accuracy of 1 in 10,000 is to be achieved.

** A white paper pointer or just a black (iron) stand, with a white chalk mark on it, would do.

Let n and $(n+1)$ be the oscillations made by the *seconds* and *the experimental* pendulums respectively.

Then, if t' and t be their respective time-periods, we have

$$t(n+1) = t'.n,$$

whence,
$$t = t'\left(\frac{n}{n+1}\right) = t'.\left(\frac{1}{1+1/n}\right) = t'\left(1+\frac{1}{n}\right)^{-1}$$
$$= t'\left(1-\frac{1}{n}+\frac{1}{n^2}-\frac{1}{n^3}+\ldots\ldots\ldots\right) = t'\left(1-\frac{1}{n}\right), \textit{ very nearly,}$$

neglecting the terms involving the second and the higher powers of n; for, with t' and t nearly the same, n is sufficiently large, (about 500 or more), and these terms become negligibly small.

Now, $t' = 2$ *sec.*; so that, $t = 2\ (1 - 1/n)$,

pendulum B being a *second pendulum.*

Thus t, the time-period of the experimental pendulum, can be easily calculated. This value of t is then used in the expression for g, above.

It will be easily seen that with all the care taken, it is not really possible to determine with absolute certainty the particular oscillations at which an *exact* coincidence occurs. Luckily, however, even a difference of a couple of oscillations, this way or that, hardly matters if *n be fairly large*, as it usually is. For, supposing there is an error of *two* whole oscillations made in judging the point of coincidence. Then clearly,

$$t = t'\left(1-\frac{1}{n\pm 2}\right) = t'\left[1-\frac{1}{n\,(1\pm 2/n)}\right]$$

Or,
$$t = t'\left[1-\frac{1}{n}\left(1\pm\frac{2}{n}\right)^{-1}\right] = t'\left(1-\frac{1}{n}\pm\frac{2}{n^2}\right).$$

The error introduced is thus $\pm 2/n^2$, which, if $n = 500$, works out to +.0008% and is, therefore, not of much consequence.

Note. In Kater's own determination, made at Portland Place, the pendulum was set in front of a standard clock and the swings of the two pendulums observed by means of a telescope from a distance of 9 feet. The standard clock was checked every 24 hours by stellar measurements, so that the rate of swing of the pendulum was, in effect, compared with the rate of rotation of the earth itself.

Successive coincidences occurred every 530 *sec.*, during which time the reversible pendulum completed 528 swings or half oscillations. The error thus worked out to 1 part in 100,000 and the length of seconds pendulum at sea level, in the latitude of London, came to 39.13829 inches.

6.12. Computed Time—Bessel's Contribution. *Kater*, in his reversible pendulum, made the time-period about the two axes *exactly* equal, which, as we have seen, is an extremely tedious process. But **Bessel** showed that it was by no means necessary to make them *exactly equal* and that it was enough to make them only *nearly equal.*

Thus, suppose the time-periods about the two axes are t_1 and t_2 respectively, (very nearly equal), and that l and l' *are their respective distances from the c.g. of the pendulum. Then, we have*

$$t_1 = 2\pi\sqrt{\frac{k^2+l^2}{lg}} \quad \ldots(i) \qquad \text{and} \qquad t_2 = 2\pi\sqrt{\frac{k^2+l'^2}{l'g}} \quad \ldots(ii)$$

where k is the radius of gyration of the pendulum about the *c.g.*

Squaring and re-arranging (*i*) and (*ii*), we have

$$t_1^2 lg = 4\pi^2(k^2+l^2), \quad \ldots(iii) \text{ and } \quad t_2^{\,2} l' g = 4\pi^2(k^2+l'^2). \quad \ldots(iv)$$

So that, subtracting (*iv*) from (*iii*), we have

$$g.(t_1{}^2.l - t_2{}^2.l') = 4\pi^2\,(k^2+l^2) - 4\pi^2\,(k^2+l'^2)$$

$$= 4\pi^2\,(l^2 - l'^2) = 4\pi^2\,(l+l')\,(l-l').$$

$$\therefore \quad \frac{4\pi^2}{g}.(l+l') = \frac{t_1{}^2.l - t_2{}^2.l}{(l-l')} = \frac{2(t_1{}^2.l - t_2{}^2.l')}{2(l-l')}.$$

$$= \frac{(t_1{}^2 + t_2{}^2)\,(l-l') + (t_1{}^2 - t_2{}^2)\,(l+l')}{2(l-l')}.$$

Or,
$$\frac{4\pi^2}{g}.(l+l') = \frac{t_1{}^2 + t_2{}^2}{2} + \frac{t_1{}^2 - t_2{}^2}{2}\cdot\frac{l+l'}{l-l'}.$$

This quantity, $\dfrac{t_1{}^2 + t_2{}^2}{2} + \dfrac{t_1{}^2 - t_2{}^2}{2}\cdot\dfrac{l+l'}{l-l'} = \tau^2$, $\left[cf.\ \dfrac{4\pi^2}{g} = t^2 \right.$ *for a simple pendulum.*

where τ is called the **computed time** of the pendulum.

The distance $(l + l')$ between the two knife-edges can be determined accurately, but $(l - l')$, the difference between the distances of the two axes from the *c.g.* cannot be determined to any high degree of accuracy, because of the uncertainty in locating the position of the *c.g.* of the pendulum. Since, however, $(t_1{}^2 - t_2{}^2)$ is much too small a quantity compared with $(t_2{}^2 - t_2{}^2)$ and $(l_1 - l_2)$ fairly large, the quantity $\dfrac{t_1{}^2 - t_2{}^2}{2}\cdot\dfrac{l+l}{l-l'}$ becomes negliging and we have $\dfrac{4\pi^2}{2}\,(l+l')$ $= \dfrac{t_1{}^2 + t_2{}^2}{2}$, whence $g = \dfrac{8\pi^2\,(l+l')}{t_1{}^2 + t_1{}^2}$.

The value of g can thus be easily obtained.

6.13. (a) Errors in the Compound Pendulum and their Remedies. Besides the difficulty of adjusting the time-periods to be *exactly the same* about either knife-edge, and of correctly measuring the distance between them, there are a number of other sources of error in a Kater's pendulum (or the compound pendulum, in general) for which proper corrections must be applied to obtain an accurate result, the chief among them being the following :

(*i*) The Finite Amplitude of the Pendulum. The expression for the time-period has been deduced on the assumption that the amplitude of swing of the pendulum is *vanishingly small*; for, then alone, will its motion be truly simply harmonic in nature. In actual practice, however, it has always a finite value. This reduces its acceleration and thus increases its time-period, as was first pointed out by *Bernoulli*, in the year 1749, in respect of a simple pendulum. The observed time-period may be corrected for this error by multiplying it with $(1 - \alpha_1.\alpha_2/16)$, where α_1 and α_2 are *half-swings* (in radians) at the beginning and at the end of the experiment respectively as can be seen from the following :

Proceeding in a manner similar to that discussed in connection with a simple pendulum [Alternative method (*i*), page 145], we have

$\dfrac{1}{2}I\omega^2 = mgl\,(\cos\theta - \cos\alpha)$, Or, $I\omega^2 = 2mgl\,(\cos\theta - \cos\alpha)$, where α is the angular amplitude of the pendulum and θ, its *angular displcement at time t*.

Now, $\quad I = mk^2 + ml^2 = m(k^2 + l^2)$ and $\omega = d\theta/dt$.

So that, $\quad m(k^2 + l^2)\,(d\theta/dt)^2 = 2mgl\,(\cos\theta - \cos\alpha)$.

Or, $$\left(\frac{d\theta}{dt}\right) = \sqrt{\frac{2gl}{k^2+l^2}(\cos\theta - \cos\alpha)}$$

Or, $$dt = \frac{d\theta}{\sqrt{\frac{2gl}{k^2+l^2}(\cos\theta - \cos\alpha)}} = \sqrt{\frac{k^2+l^2}{2gl}} \cdot \frac{d\theta}{(\cos\theta - \cos\alpha)^{1/2}}$$

Integrating this expression for the limits 0 to $t/4$ and 0 to α, where t is the observed time-period of the pendulum, we have

$$\int_0^{t/4} dt = \sqrt{\frac{k^2+l^2}{2gl}} \int_0^{\alpha} \frac{d\theta}{(\cos\theta - \cos\alpha)^{1/2}}.$$

$$= \sqrt{\frac{k^2+l^2}{2gl}} \int_0^{\alpha} \frac{d\theta}{\left(1 - 2\sin^2\frac{\theta}{2} - 1 + 2\sin^2\frac{\alpha}{2}\right)^{\frac{1}{2}}}$$

$$= \sqrt{\frac{k^2+l^2}{2gl}} \cdot \frac{1}{\sqrt{2}} \cdot \int_0^{\alpha} \frac{d\theta}{\left[2\left(\sin^2\frac{\alpha}{2} - \sin^2\frac{\theta}{2}\right)\right]^{\frac{1}{2}}}$$

$$\frac{t}{4} = \sqrt{\frac{k^2+l^2}{gl}} \cdot \frac{1}{\sqrt{2}} \cdot \frac{1}{\sqrt{2}} \cdot \int_0^{\alpha} \frac{d\theta}{\left(\sin^2\frac{\alpha}{2} - \sin^2\frac{\theta}{2}\right)^{\frac{1}{2}}}$$

Putting $\sin\frac{\theta}{2} = \sin\frac{\alpha}{2}\sin\phi$, we have $\frac{1}{2}\,d\theta\cos\frac{\theta}{2} = d\phi\sin\frac{\alpha}{2}\cos\phi$.

Or, $$d\theta = 2d\phi\sin\frac{\alpha}{2}\cos\phi/\cos\frac{\theta}{2}$$

Substituting the value of $d\theta$ in the above expression, we have

$$\frac{t}{2} = \sqrt{\frac{k^2\ l^2}{gl}} \int_0^{\pi/2} \frac{2\sin\frac{\alpha}{2}\cos\phi \,.\, d\phi}{\cos\frac{\theta}{2}\left(\sin^2\frac{\alpha}{2} - \sin^2\frac{\theta}{2}\right)^{\frac{1}{2}}}$$

Now, $$\left(\sin^2\frac{\alpha}{2} - \sin^2\frac{\theta}{2}\right)^{\frac{1}{2}} = \left(\sin^2\frac{\alpha}{2} - \sin^2\frac{\alpha}{2}\sin^2\phi\right)^{\frac{1}{2}}$$

$$= \left[\sin^2\frac{\alpha}{2}\left(1 - \sin^2\phi\right)\right]^{\frac{1}{2}} = \left(\sin^2\frac{\alpha}{2}\cos^2\phi\right)^{\frac{1}{2}} = \sin\frac{\alpha}{2}.\cos\phi.$$

So that, $$\frac{t}{2} = \sqrt{\frac{k^2+l^2}{gl}} \int_0^{\pi/2} \frac{2\sin\frac{\alpha}{2}\cos\phi \,.\, d\phi}{\cos\frac{\theta}{2}\sin\frac{\alpha}{2}\cos\phi}.$$

$$= \sqrt{\frac{k^2+l^2}{gl}} \int_0^{\pi/2} \frac{2d\phi}{\cos(\theta/2)}. \quad \text{Or,} \quad t = 4\sqrt{\frac{k^2+l^2}{gl}} \int_0^{\pi/2} \frac{d\phi}{\cos(\theta/2)}$$

Since $\cos\theta/2 = \sqrt{\cos^2\theta/2} = \sqrt{1-\sin^2\theta/2} = \left(1 - \sin^2\frac{\alpha}{2}\sin^2\phi\right)^{\frac{1}{2}}$

We have $$t = 4\sqrt{\frac{k^2+l^2}{gl}} \int^{\pi/2} \frac{d\phi}{\left(1 - \sin^2\frac{\alpha}{2}\sin^2\phi\right)^{\frac{1}{2}}}$$

$$= 4\sqrt{\frac{k^2+l^2}{gl}} \int_0^{\pi/2} \left(1 - \sin^2\frac{\alpha}{2}.\sin^2\phi\right)^{-\frac{1}{2}} d\phi$$

$$= 4\sqrt{\frac{k^2 + l^2}{gl}} \int_0^{\pi/2} \left(1 + \frac{1}{2} sin^2 \alpha . sin^2 \phi + \frac{1 \times 3}{2 \times 4} sin^4 \frac{\alpha}{2} sin^4 \phi + ...\right) d\phi.$$

Or, $$t = 4\sqrt{\frac{k^2 + l^2}{gl}} . \frac{\pi}{2}\left[1 + \left(\frac{1}{2}\right)^2 sin^2 \frac{\alpha}{2} + \left(\frac{1 \times 3}{2 \times 4}\right)^2 sin^4 \frac{\alpha}{2} + ...\right]$$

Or, $$t = 2\pi\sqrt{\frac{k^2 + l^2}{gl}}\left[1 + \frac{1}{4} sin^2 \frac{\alpha}{2} + ...\right]$$

Now, $2\pi\sqrt{\frac{k^2 + l^2}{gl}}$ would be the time period of the pendulum, for an infinitely small amplitude. Denoting it by t_0, we have

$$t = t_0\left(1 + \frac{1}{4} sin^2 \frac{\alpha}{2}\right)$$

If α be small, $sin \ \alpha/2 = \alpha/2$, and we have

$$t = t_0\left(1 + \frac{1}{4} . \frac{\alpha^2}{4}\right) = t_0\left(1 + \frac{\alpha^2}{16}\right).$$

Since the amplitude (or half swing) does not remain constant but decreases from α_1 in the beginning to α_2 at the end, both being small, we replace α^2 by $\alpha_1\alpha_2$. So that,

$$t = t_0\left(1 + \frac{\alpha_1\alpha_2}{16}\right), \quad \text{whence,} \quad t_0 = t \Big/ \left(1 + \frac{\alpha_1\alpha_2}{16}\right).$$

Or, $$t_0 = t\left(1 - \frac{\alpha_1\alpha_2}{16}\right)$$

(*ii*) **Air-effects.** There are *three* distinct ways in which the presence of air affects the time-period of a pendulum.

(*a*) **Buoyancy of the air.** This tends to reduce the restoring couple acting on the pendulum, due to slight decrease in its weight, similar to the one produced in a body immersed in a liquid. For, if m be the mass of the pendulum and m', that of the air displaced by it*, the restoring couple is reduced from the value $mg.l \ sin \ \theta$ to $(ml - m'h) \ g \ sin \ \theta$, where h** is the distance between the *c.g.* of the displaced air and the axis of rotation of the pendulum, and can be obtained sufficiently accurately from its physical dimensions.

The equation of motion of the pendulum thus becomes

$$m(k^2 + l^2) . d^2\theta/dt^2 = - g \ sin \ \theta \ (ml - m'h).$$

Or, $$d^2\theta/dt^2 = -\frac{(ml - m'h)g}{m(k^2 + l^2)} \theta. \qquad | \text{Putting } sin \ \theta = \theta.$$

This is obviously the equation of a simple harmonic motion, of time-period t, given by

$$t = 2\pi\sqrt{\frac{m(k^2 + l^2)}{(ml - m'h)g}},$$

which is clearly greater than $t = 2\pi \sqrt{(k^2 + l^2)/lg}$, the expression for the time-period in vacuo.

The time-period of the pendulum is thus slightly increased due to buoyancy of the air.

This was the only correction taken into account by *Newton*, followed by *Kater*, and it was left to *Bessel* to show that other corrections due to air-effects were also called for.

(*b*) **Some air being dragged along with the pendulum, during its motion (Du Buat's Correction).** The pendulum, during its 'to and fro' motion "*carries air with it*" and this increases

* This can easily be obtained from the volume of the pendulum and the density of the air at the time.

** The value of h may not be the same as that of l, unless the pendulum has a uniform density.

its *effective mass*, and hence its moment of inertia, making the observed time-period greater than the true one, as will be clear from the following :

That the pendulum does 'carry' some air with it, as it oscillates, can be shown by a simple experiment, *viz.*, by attaching a feather to its bob, in a direction at right angles to its direction of motion. It will be found that the feather tilts in a direction opposite to that of the motion of the bob, showing that the air surrounding it is at rest. *If, however, the feather be sufficiently close to the bob, it is not found to tilt at all, clearly indicating that the air in immediate contact with the bob moves along with it, or that the bob 'carries air with it'.*

Let the mass of this air 'carried' by the pendulum by m'' and let the distance of the centre of mass from the point of suspension of the pendulum be d. Then, clearly, the *effective* moment of inertia of the pendulum, *i.e.*, the moment of inertia of the pendulum and the adherent mass of air with it, is equal to $m(k^2 + l^2) + m''d^2$. And, therefore, the equation of motion of the pendulum now becomes

$$\left[m(k^2 + m''d^2 \right] d^2\theta/dt^2 = -(ml - m'h)\,g\,.\theta.$$

Or,
$$\frac{d^2\theta}{dt^2} = \frac{(ml - m'h)g}{m(k^2 + l^2) + m''d^2}.\theta.$$

The time-period of the pendulum is thus given by $t = 2\pi\sqrt{\dfrac{m(k^2 + l^2) + m''d^2}{(ml - m'h)g}}$

It follows at once from the above that

$$\frac{gt^2}{4\pi^2} = \left[m(k^2 + l^2) + m''d^2 \right] \frac{1}{ml} \left[1 - \frac{m'h}{ml} \right]^{-1}$$

$$= \frac{1}{ml}\left[m(k^2 + l^2) + m''d^2 \right] \left[1 + \frac{m'h}{ml} \right] \qquad \text{[Neglecting higher terms.}$$

Or,
$$\frac{gt^2}{4\pi^2} = \frac{k^2 + l^2}{l} + \frac{m''d^2}{ml} + \frac{k^2 + l^2}{ml}.\frac{m'h}{ml}, \qquad \text{[Neglecting second order terms.}$$

assuming the time-period here to be already corrected for the finite arc of its swing.

Now, if l_1 and l_2 be the distances of the two points of suspension from the centre of gravity, on either side, such that the time-periods in the two cases are *nearly the same*, then, if h_1, h_2 and d_1, d_2 be the respective distances from the points of suspension of the centres of buoyancy and the centres of mass of the air adhering to the bob in the two positions, we have

$$\frac{gt_1^{\,2}}{4\pi^2} l_1 = l_1^{\,2} + k^2 + \frac{m''d_1^{\,2}}{m} + \frac{k^2 + l_1^{\,2}}{ml_1}.m'h_1 \qquad ...(i)$$

and
$$\frac{gt_2^{\,2}}{4\pi^2} l_2 = l_2^{\,2} + k^2 + \frac{m''d_2^{\,2}}{m} + \frac{k^2 + l_2^{\,2}}{ml_2}.m'h_2. \qquad ...(ii)$$

So that, subtracting relation (*ii*) from (*i*), we have

$$\frac{g}{4\pi^2}(t_1^{\,2}l_1 - t_2^{\,2}l_2) = (l_1^{\,2} - l_2^{\,2}) + \frac{m''}{m}(d_1^{\,2} - d_2^{\,2}) + \frac{(k^2 + l_1^{\,2})}{l_1}.\frac{m'h_1}{m} - \frac{(k_2 + l_2^{\,2})}{l_2}.\frac{m'h_2}{m}.$$

Since t_1 is very nearly equal to t_2, we have $k^2 = l_1l_2$. And, therefore,

$$\frac{g}{4\pi^2}\frac{(t_1^{\,2}l_1 - t_2^{\,2}l_2)}{l_1 - l_2} = (l_1 + l_2) + \frac{m''}{m}\left(\frac{d_1^{\,2} - d_2^{\,2}}{l_1 - l_2}\right) + \left(\frac{l_1 + l_2}{l_1 - l_2}\right)\frac{m'}{m}(h_1 - h_2).$$

Here, $\dfrac{(t_1^{\,2}l_1 - t_2^{\,2}l_2)}{l_1 - l_2}$ is the square of the *computed time*, (see § 6.12, p. 157).

Denoting it by τ, we have

$$\frac{g\tau^2}{4\pi^2} = (l_1 + l_2) + \frac{m''}{m}\left(\frac{d_1^{\,2} - d_2^{\,2}}{l_1 - l_2}\right) + \left(\frac{l_1 + l_2}{l_1 - l_2}\right)\frac{m'}{m}(h_1 - h_2). \qquad ...(iii)$$

Thus, the obvious method to eliminate this correction is to make $d_1 = d_2$ and $h_1 = h_2$ (*i.e.*, to make the pendulum symmetrical in shape), which will reduce the two expressions on the right-hand side of relation (iii) to zero. This is precisely what has been done in *Repsoid's reversible pendulum*, (see § 6.14, page 167)

Both effects (*a*) and (*b*) due to air can, however, be made almost negligible by arranging to swing the pendulum in reduced pressure — a procedure now being increasingly adopted — for the residual effect in low pressures is found to be a linear function of the pressure. *The required correction can thus be directly obtained by plotting a graph between pressure and time-period and obtaining the value of the latter by extrapolating the graph to zero pressure.*

(*c*) **Viscosity of the air.** The viscous drag due to air produces a damping effect on the pendulum and tends to reduce its amplitude, thereby increasing its time-period.

For, taking the viscosity-drag for small velocities to be proportional to velocity, the equation of motion would be of the type

$$\frac{d^2\theta}{dt^2} + r \cdot \frac{d\theta}{dt} + \mu\theta = 0.$$

Let the solution of the equation be $\theta = Ae^{\omega t}$. Then, clearly,

$$A\omega^2 + Ar\omega + \mu A = 0. \text{ Or, } \omega^2 + \omega r + \mu = 0,$$

which is a quadratic equation in ω.

So that, $$\omega = \frac{-r \pm \sqrt{r^2 - 4\mu}}{2} = \frac{-r}{2} \pm j\sqrt{\mu - \frac{r^2}{4}} \qquad \left[\text{where } j = \sqrt{-1} \text{ and } \mu > r.\right.$$

Hence the general solution is

$$\theta = Ae^{\left[-\frac{r}{2} + j\sqrt{\mu - r^2/4}\right]t} + Be^{\left[-\frac{r}{2} + j\sqrt{\mu - r^2/4}\right]t}$$

$$= e^{-\frac{r}{2}t}\left[(A+B)\cos\sqrt{\left(\mu - \frac{r^2}{4}\right)}.t + j(A-B)\sin\sqrt{\left(\mu - \frac{r^2}{4}\right)}t.\right.$$

$$= A_0 e^{\frac{r}{2}t}\cos\left(\sqrt{\mu - \frac{r^2}{4}}t + \phi\right).$$

which is a simple harmonic motion of decaying amplitude, of a time-period $t = 2\pi/\sqrt{(\mu - t^2/4)}$.

Thus, $$t = \frac{2\pi}{\sqrt{\mu}}\left[1 - \frac{r^2}{4\mu}\right]^{-\frac{1}{2}} = \frac{2\pi}{\sqrt{\mu}}\left(1 + \frac{r^2}{8\mu}\right)$$

Now, $2\pi/\sqrt{\mu} = t_0$, the time-period of the pendulum in the absence of any viscous drag. So that,

$$t = t_0\left[1 + \frac{r^2}{8\mu}\right].$$

And, if we make use of the approximate relation $t' = 2\pi/\sqrt{\mu}$, for the time-period of the pendulum, we have $\mu = 4\pi^2/t'^2$. So that, substituting this value of μ in the expression for t above, we have

$$t = t_0\left[1 + \frac{r^2}{8 \times 4\pi^2/t'^2}\right] = t_0\left[1 + \frac{r^2 t'^2}{32\pi^2}\right].$$

So that, $$t_0 = t\left[1 - \frac{r^2 t'^2}{32\pi^2}\right], \text{ very nearly.}$$

This correction due to viscosity is, however, much too small, being of the order of 10^{-9}, and is therefore, usually neglected.

(*iii*) **Non-rigidity of the Sunport.** Due to the yielding of the support, the time-period tends to be greater than the correct value, as explained below :

It might, at first sight, appear improbable that the support should yield by the mere swinging of a pendulum suspended from it. This is, however, not so, for the simple reason that *no support is perfectly rigid*. In fact, any ordinary support we consider to be rigid, would yield under a weight of 100 *kgm* or so. True, a pendulum is seldom as heavy as that, but in view of the fact that we can measure lengths and time-periods to an accuracy of one in several thousand, it is only in the fitness of things that we must take into account even this slight yielding of the support, if we really aim at a high degree of precision in our work.

Again, it is also true that we can adopt ways and means of eliminating this error altogether (as well as that due to the presence of air) in so far as the pendulum is concerned, as explained in § 6.14 below, we should, nevertheless, acquaint ourselves with the method of deducing a proper correction for it, should it becomes necessary in other similar cases, where its outright elimination is not feasible or possible.

Now, then let us pass on to a brief consideration of it.

We know that a vibrating body tends to set into vibration any other body in contact with it, the degree of response of the latter depending upon how nearly its natural time-period agrees with that of the vibrating body, — the closer this agreement between the two, the greater the response and *vice versa*. In the case of the pendulum, therefore, the support carrying it also yields a little to its vibrations and is forced to oscillate co-peridically with it.

This oscillation of the support may be resolved into two rectangular components.(*i*) *along the vertical* and (*ii*) *along the horizontal*, the latter having a more pronounced effect on the time-period of the peiodulum than the former.

Thus, if OG be the mean or equilibrium position of the pendulum, (Fig. 6.16), of length l, with the axis of suspension passing through O, then, as it swings through an angle into the position OG', there are two forces acting upon it, (*i*) one along the arc of its swing, to which its motion is due, and (*ii*) the other at right angles to it, *i.e.*, along $G'O$ (the centripetal force). So that,

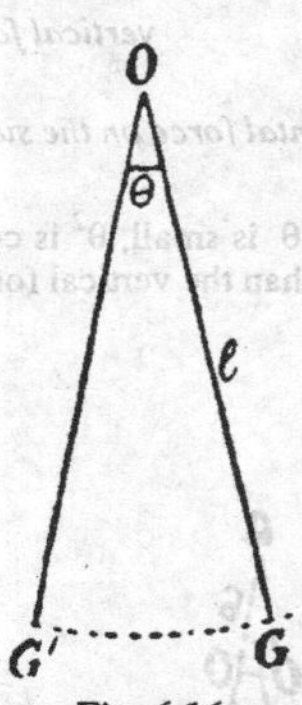

Fig. 6.16

acceleration of the pendulum along the arc

$$= \frac{d^2s}{dt^2} = \frac{d}{dt}\cdot\frac{ds}{dt} = l.\frac{d^2\theta}{dt}, \qquad [\text{ where } s = l\theta$$

and, *acceleration along the length of the pendulum*

$$= \frac{v^2}{l} = \frac{1}{l}\cdot\left(\frac{ds}{dt}\right)^2 = l\left(\frac{d\theta}{dt}\right)^2. \qquad [\because ds/dt = d\theta/dt$$

These acceleration, obviously, act at the support O, with the

$$\text{component in the horizontal plane } = l\,\frac{d^2\theta}{dt^2}.\cos\theta + l\left(\frac{d\theta}{dt}\right)^2.\sin\theta,$$

$$\text{and, \textit{the component in the vertical plane}} \quad = l\,\frac{d^2\theta}{dt^2}.\sin\theta - l\left(\frac{d\theta}{dt}\right)^2.\cos\theta$$

So that, if θ be small, we have

$$\textit{horizontal component of the acceleration} \quad = l.\frac{d^2\theta}{dt^2} + l\left(\frac{d\theta}{dt}\right)^2.\theta \qquad ...(i)$$

$$\text{and \textit{vertical component of the acceleration}} \quad = l.\frac{d^2\theta}{dt^2}.\theta - l.\left(\frac{d\theta}{dt}\right)^2 \qquad ...(ii)$$

because $\cos\theta = 1$ and $\sin\theta = \theta$

Now, from the equation of the pendulum, $I.\frac{d^2\theta}{dt^2} = mg.l.\theta$, we have

$$\frac{d^2\theta}{dt^2} = \frac{mgl}{I}.\theta = \frac{mgl}{m(k^2+l^2)}.\theta = \left(\frac{gl}{k^2+l^2}\right).\theta.$$

And, from the energy equation of the pendulum, *viz.*,

$$mgl(\cos\theta - \cos\alpha) = \frac{1}{2}m(k^2+l^2).(d\theta/dt)^2, \text{ we have}$$

$$\left(\frac{d\theta}{dt}\right)^2 = \frac{2gl(\cos\theta - \cos\alpha)}{k^2+l^2},$$

whence, expanding $\cos\theta$ and $\cos\alpha$ and retaining only the first two terms, in view of the small values of θ and α, we have

$$\left(\frac{d\theta}{dt}\right)^2 = \frac{gl(\alpha^2 - \theta^2)}{k^2+l^2} \qquad \left[\because \cos\theta = 1 - \frac{\theta^2}{2!} + \frac{\theta^4}{4!} - \frac{\theta^6}{6!}\right.$$

Substituting these values of $d^2\theta/dt^2$ and $(d\theta/dt)^2$ in relations (*i*) and (*ii*) above, we therefore have

$$\textit{horizontal acceleration} = l.\left(\frac{gl}{k^2+l^2}\right).\theta + l\theta.\frac{gl(\alpha^2-\theta^2)}{k^2+l^2}$$

and $$\textit{vertical acceleration} = l.\left(\frac{gl}{k^2+l^2}\right).\theta^2 - l.\frac{gl(\alpha^2-\theta^2)}{k^2+l^2}$$

Or, neglecting the second term as being extremely small, we have

$$\textit{horizontal acceleration} = \left(\frac{gl^2}{k^2+l^2}\right).\theta \text{ and vertical acceleration } = \left(\frac{gl^2}{k^2+l^2}\right).\theta^2$$

And, therefore, $$\textit{horizontal force on the support} = \left(\frac{mgl^2}{k^2+l^2}\right).\theta$$

and $$\textit{vertical force on the support} = \left(\frac{mgl^2}{k^2+l^2}\right).\theta^2,$$

i.e., *horizontal force on the support* $\alpha\,\theta$ and the *vertical force* on it $\alpha\theta^2$.

Since, θ is small, θ^2 is comparatively very much smaller. In other words, the horizontal force on the suppert is very much greater than the vertical force on it and the latter may, therefore, be easily ignored.

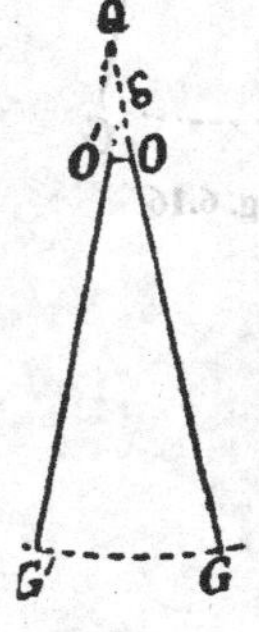

Fig. 6.17

Now, as the pendulum moves from G to G', the point of suspension moves from O to O', say, so that its displacement is equivalent to shifting its axis of suspension to Q, (Fig. 6.17).

Then, if β be the displacement of the support *per unit force, in the horizontal direction*, when displaced through an angle θ, *its displacement due to a horizontal force* $[mgl^2/(k^2+l^2)].\theta$ will clearly be equal to $[mgl^2\beta/(k^2+l^2)]\theta$.

Since the *effective* axis of suspension passes through Q, $\theta = \angle GQG'$ and hence the displacement of the point of suspension is OO'

$$= OQ.\theta = \delta.\theta \qquad [\text{Putting } OQ = \delta.$$

We, therefore, have $\delta.\theta = \dfrac{mgl^2\beta}{k^2+l^2}.\theta$, whence, $\delta = \dfrac{mgl^2\beta}{k^2+l^2}$

It follows, therefore, that the *effective length* of the pendulum, due to this yielding of the support, is $(l+\delta) = l + \left[mgl^2\beta/(k^2+l^2)\right]$

And, therefore, the time-period of the pendulum is now given by

$$t = 2\pi\sqrt{1\Big/\frac{g(l+\delta)}{k^2+(l+\delta)^2}} = 2\pi\sqrt{\frac{k^2+(l+\delta)^2}{g(l+\delta)}} \qquad \left[\text{Writing } (l+\delta) \text{ for } l.\right.$$

From this it follows at once that

$$\frac{gt^2}{4\pi^2} = \frac{k^2+(l+\delta)^2}{(l+\delta)} = (l+\delta) + k^2.\frac{1}{l}\left(1+\frac{\delta}{l}\right)^{-1} = (l+\delta) + \frac{k^2}{l}\left(1-\frac{\delta}{l}\right).$$

So that, if l_1 and l_2 be the lengths of the pendulum for which the time-periods t_1 and t_2 are very nearly the same, we have

$$\frac{gt_1^2}{4\pi^2} = (l_1+\delta_1) + \frac{k^2}{l_1}.\left(1-\frac{\delta_1}{l_1}\right)$$

[where δ_1 and δ_2 are the corresponding additions to the two lenghts of the pendulum.

and $$\frac{gt_2^2}{4\pi^2} = (l_2+\delta_2) + \frac{k^2}{l_2}.\left(1-\frac{\delta_2}{l_2}\right)$$

whence, $$\frac{g}{4\pi^2}\left(\frac{t_1^2l_1 - t_2^2l_2}{l_1-l_2}\right) = (l_1+l_2) + \frac{\delta_1l_1-\delta_2l_2}{l_1-l_2} + k^2\left(\frac{\delta_2}{l_2}-\frac{\delta_1}{l_1}\right)\left(\frac{1}{l_1-l_2}\right).$$

Since t_1 is very merely equal to t_2, we have $k^2 = l_1 l_2$. And therefore,

$$\frac{g}{4\pi^2}\left(\frac{t_1^{\,2} l_1 - t_2^{\,2} l_2}{l_1 - l_2}\right) = (l_1 + l_2) + \frac{\delta_1 l_1 - \delta_2 l_2}{l_1 - l_2} + \frac{\delta_2 l_1 - \delta_1 l_2}{l_1 - l_2}$$

$$= (l_1 + l_2) + \frac{(\delta_1 + \delta_2)\,(l_1 - l_2)}{l_1 - l_2} = (l_1 + l_2) + (\delta_1 + \delta_2).$$

Now,
$$\delta_1 + \delta_2 = \frac{mgl_1^{\,2}\beta}{k^2 + l_1^{\,2}} + \frac{mgl_2^{\,2}\beta}{k^2 + l_2^{\,2}} = mg\beta\left(\frac{l_1^{\,2}}{l_1 l_2 + l_1^{\,2}} + \frac{l_2^{\,2}}{l_1 l_2 + l_2^{\,2}}\right) \quad \left[\because\ k^2 = l_1 l_2.\right.$$

$$= mg\beta\left(\frac{l_1}{l_1 + l_2} + \frac{l_2}{l_1 + l_2}\right)$$

Or,
$$\delta_1 + \delta_2 = mg\beta. \quad \left[\because\ \frac{l_1}{l_1 + l_2} + \frac{l_2}{l_1 + l_2} = \frac{(l_1 + l_2)}{l_1 + l_2} = 1.\right.$$

So that, putting $(t_1^{\,2} l_1 - t_2^{\,2} l_2)/(l_1 - l_2) = \tau^2 =$ (where τ is the *computed time*),

we have
$$g\tau^2/4\pi^2 = l_1 + l_2 + mg\beta = g\tau^2/4\pi^2 = L + mg\beta,$$

for $(l_1 + l_2) = L$, the length of the *equivalent simple pendulum.*

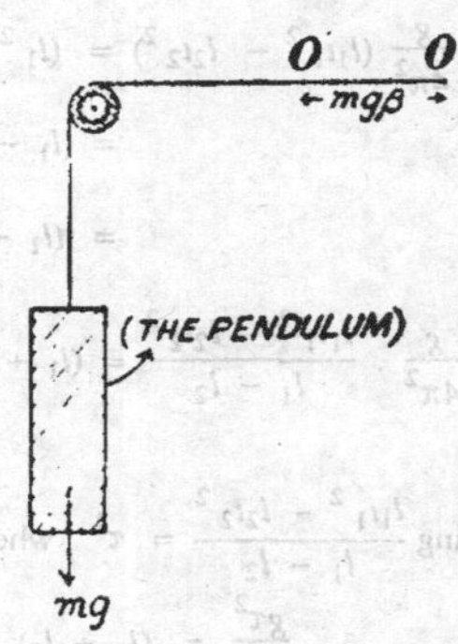

Fig. 6.18

Clearly, here, mg is the weight of the pendulum; so that, *the correction factor $mg\beta$ is the displacement of the support due to a horizontal force equal to the weight of the pendulum* and can be determined directly by suspending the pendulum from a string passing over a pulley and attached horizontally to the support at O, as shown, (Fig. 6.18), when the displacement $OO' = mg\beta$ can be read accurately be means of a microscope.

Vening Meinesz suggested a method by which this correction could be considerably reduced, *viz.*, that of using two pendulums, swinging from the same support but in opposite phases with each other. This involves, however, the difficulty of having to adjust their time-periods to very near equality. The correction is thus best eliminated as explained in § 6.14 (page 167).

(*iv*) **The knife-edges not being perfectly sharp,** (*but more or less rounded*). Due to this also, the effective length of the pendulum is altered.

For, the axis of suspension is *not a mathematical line* we have so far tacitly assumed it to be. In actual practice, it has a definite shape, —generally symmetrical—with a finite radius of curvature.

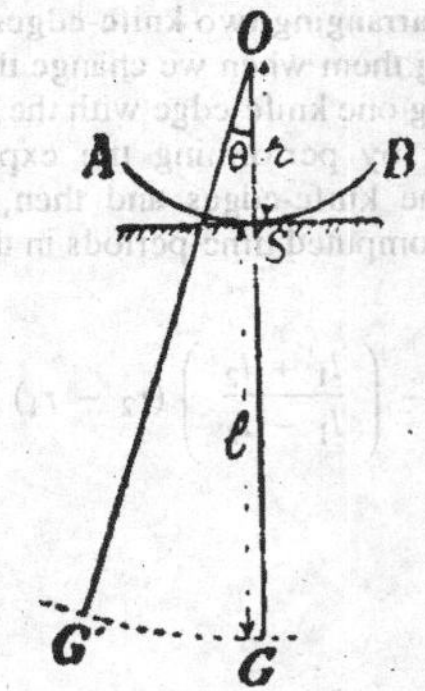

Fig. 6.19

Let us, as a first approximation, assume the edge to be the part of a cylinder,-a *cylindrical cone*—as shown in Fig.6.19. Then, if O be the *centre of curvature* of the edge, a line perpendicular to the plane of the paper and passing through O represents the axis of the cylindrical edge.

As the pendulum (of length l) swings, the edge also moves along with it about *this* axis through O, so that the axis of suspension is, in effect, shifted from S to O, *i.e.*, through the distance $SO = r$, the *radius of curvature of the degree*, and *the effective length of the pendulum thus becomes* $(l + r)$.

Since, however, the instantaneous axis of rotation of the pendulum still passes through the bottom of the knife-edge (S), the moment of inertia (I) is still to be taken about this axis. In other words, we still have $I = m.(k^2 + l^2)$, where m is the *mass* of the pendulum.

The equation of motion of the pendulum thus becomes

$$m\,(k^2 + l^2)\,d^2\theta/dt^2 = -mg\,(l + r)\sin\theta = -mg\,(l + r)\theta. \quad [\theta \text{ being small.}$$

And, therefore, its time-period is given by $t = 2\pi\sqrt{\dfrac{k^2 + l^2}{g(l + r)}}$

whence, $\frac{gt^2}{4\pi^2} = \frac{k^2 + l^2}{l + r} = \frac{k^2 + l^2}{l(1 + r/l)}$. Or, $\frac{gt^2 l}{4\pi^2} = \frac{k^2 + l^2}{1 + r/l} = (k^2 + l^2)\left(1 - \frac{r}{l}\right)$.

If we find two lengths of the pendulum, say l_1 and l_2 on the two sides of its *c.g.*, such that the time-periods t_1 and t_2 for them are nearly the same, (with, of course, l_1 not equal to l_2), then, if r_1 and r_2 be the radii of the two knife-edges respectively, we have

$$\frac{gl_1t_1^2}{4\pi^2} = (k^2 + l_1^2)\left(1 - \frac{r_1}{l_1}\right) \text{ and } \frac{gl_2t_2^2}{4\pi^2} = (k^2 + l_2^2)\left(1 - \frac{r_2}{l_2}\right).$$

So that, subtracting the second expression from the first, we have

$$\frac{g}{4\pi^2}(l_1t_1^2 - l_2t_2^2) = (l_1^2 - l_2^2) - (l_1^2 + k^2)\frac{r_1}{l_1} + (l_2^2 + k^2)\frac{r_2}{l_2}.$$

Since t_1 is very nearly equal to t_2, we have $k^2 = l_1l_2$ And, therefore,

$$\frac{g}{4\pi^2}(l_1t_1^2 - l_2t_2^2) = (l_1^2 - l_2^2) - (l_1^2 + l_1l_2)\frac{r_1}{l_1} + (l_2^2 + l_1l_2)\frac{r_2}{l_2}.$$

Or, $$\frac{g}{4\pi^2}(l_1t_1^2 - l_2t_2^2) = (l_1^2 - l_2^2) - (l_1 + l_2)r_1 + (l_2 + l_1).r_2.$$

$$= (l_1 - l_2)(l_1 + l_2) - (l_1 + l_2)(r_1 - r_2).$$

$$= (l_1 - l_2)\left[(l_1 + l_2) - \left(\frac{l_1 + l_2}{l_1 - l_2}\right)(r_1 - r_2)\right].$$

Or, $$\frac{g}{4\pi^2}\cdot\frac{l_1t_1^2 - l_2t_2^2}{l_1 - l_2} = (l_1 + l_2) - \left(\frac{l_1 + l_2}{l_1 - l_2}\right)(r_1 - r_2).$$

Again, putting $\frac{l_1t_1^2 - l_2t_2^2}{l_1 - l_2} = \tau^2$, where τ is the *computed time of the pendulum*, we have

$$\frac{g\tau^2}{4\pi^2} = (l_1 + l_2) - \left(\frac{l_1 + l_2}{l_1 - l_2}\right)(r_1 - r_2).$$

Here, clearly, the correction term $\left(\frac{l_1 + l_2}{l_1 - l_2}\right).(r_1 - r_2)$ becomes zero, only if $r_1 = r_2$, *i.e.*, only if the two knife-edges have the same radius of curvature.

Since it is difficult to the two knife-edges of exactly the same radius of curvature, the suggestion at once comes to the mind that the same knife-edge may be used at both the points of suspension. But this may affect the position of the *c.g.*, which might be different for the two positions of the knife-edge. And, then, it would disturb the symmetry of the pendulum, necessitating the troublesome air-corrections. This difficulty may be tided over by arranging two knife-edges of the same shape and mass, and by *using only one of them for suspension, i.e.*, by interchanging them when we change the side of the pendulum. Here too, however, an error may creep in if we do not succeed in replacing one knife-edge with the other exactly in its true or original position. This difficulty too may be got over, however, by performing the experiment *four times*,—first, taking the two observations for l_1 and l_2 with one position of the knife-edges and then, two similar observations with the knife-edges interchanged. Thus, if τ_1 and τ_2 be the respective computed time-periods in the two cases, we have

$$\frac{g\tau_1^2}{4\pi^2} = (l_1 + l_2) - \left(\frac{l_1 + l_2}{l_1 - l_2}\right)(r_1 - r_2) \text{ and } \frac{g\tau_2^2}{4\pi^2} = (l_1 + l_2) - \left(\frac{l_1 + l_2}{l_1 - l_2}\right)(r_2 - r_1)$$

So that, adding the two, we have $\frac{g}{4\pi^2}(\tau_1^2 + \tau_2^2) = 2(l_1 + l_2)$,

because the sum of the correction terms (involving r_1, and r_2) = 0.

Or, $$\frac{g}{4\pi^2}\frac{(\tau_1^2 + \tau_2^2)}{2} = (l_1 - l_2).$$

But, even this correction does not help much. For, as the pendulum swings to and fro about its mean position, the edges invariably get chipped off, resulting in the loss of weight of the pendulum.

The one and only way of eliminating this correction, now being increasingly used, is to replace the two knife-edges in the pendulum by just plane bearings, *i.e.*, by flat plates, and to provide a fixed knife-edge on the support—the latter being carefully ground to a sharp edge and the former being accurately plane or flat and always placed in the same position on the knife-edge.

N.B. In a bid for an extremely high degree of accuracy, the effect of the elasticity of the pendulum was also investigated at *Potsdam* under the supervision of **Helmert**, *viz.*, its periodic extension under the varying longitudinal strain and its flexure under the changing bending moment to which it is subjected as it swings—the latter begin the more important of the two and resulting in a reduction in the effective length of the pendulum.

Thanks to the work of *Clark, Heyl* and *Cook*, an accuracy of one in a million is now more or less easily attainable.

(*v*) **Change of Temperature during the Experiment.** This results in a corresponding change in the length and hence the time-period of the pendulum.

A correction for it can, however, be readily applied if we know the coefficient of expansion of the material of the pendulum. Or, the error may be eliminated altogether by using what are called *invariable pendulums*, [see § 6.20 (*i*), 1].

(*vi*) **Other Errors.** We have considered above the errors and corrections in so far as they relate to the pendulum itself. To obtain an accurate value of g at a place, however, certain other corrections must also be applied,*viz.*, the corrections (*a*) for rotation of the earth, (*b*) for *latitude*, (*c*) for *altitude* (*d*) for *elevated masses* and (*e*) for *the terrain* or the *topography of the place*, all of which are discussed under § 6.18.

(*b*) **Superiority of a compound pendulum over a simple pendulum.** Despite the various corrections involved, a compound pendulum is definitely superior to a simple pendulum as a method for determining the value of g at a place, For,

(*i*) whereas a simple pendulum is just an *ideal* conception, not realizable in actual practice, *the length of an equivalent simple pendulum, and hence the value of g, can be easily and accurately determined with its help;*

(*ii*) *it vibrates as a whole, there being no lag between the bob and the string*, as in the case of a simple or a Borda's pendulum ;

(*iii*) *the length to be measured here is clearly defined, viz., the distance between the two knife-edges, and can thus be easily and accurately measured,* whereas, in the case of a simple pendulum, the point of suspension and the *c.g.* of the bob are both more or less *indefinite* points and hence its true length can hardly be expected to be determined correctly;

(*iv*) *due to its large mass, the compound pendulum keeps on oscillating for a fairly long time, thus enabling its time-period to be determined with accuracy.* In a simple pendulum, on the other hand, the oscillations die down much too soon due to the comparatively small mass of the bob, and it becomes difficult to determine the time-period to an equivalent degree of accuracy.

The one obvious disadvantage in the case of a compound pendulum, however, is that during its vibrations to and fro, about its mean position, some air is dragged along with it, as mentioned above, thus increasing its *effective* mass and hence its moment of inertia. But it has been clearly shown by Bessel that if it be of a form *symmetrical about the centre of its geometrical shape (which is not the same thing as its centre of gravity)*, this error is automatically eliminated. This explains the symmetrical shapes of various types of compound pendulums we use, though, theoretically, a rigid body of any shape whatever would do.

6.14. Other Improvements due to Bessel. Not only has Bessel done away with the trouble and the tedium of having to make the time-periods about the two axes identical, but he has also succeeded in removing quite a few other important errors. Thus, for example.:

(*i*) The error due to some air being dragged along with the pendulum is removed by the *symmetrical physical form* of the instrument, as suggested and shown by him.

(*ii*) The error due to the knife-edges not being perfectly sharp, for which a correction, proportional to their radii of curvature, would be necessary, (unless they be of the same radii of curvature, [see § 6.13 (*iv*), page 165], has also been eliminated by him. For, he has shown that this error would automatically vanish if the two knife-edges, at the two ends of the pendulum, could be made interchangeable. Then, if t_1 and t_2 be the *computed times*, before and after the interchange of knife-edges, the true time-period t is given by the relation $t = \sqrt{(t_1^{\,2} + t_2^{\,2})/2}$.

Bessel, unfortunately, died before he could put his theory into actual practice but, later, **Repsold** did actually construct in the year 1860, a *reversible pendulum of this* type and used it with success.

Repsold's Pendulum is more or less a Kater-type pendulum but is *symmetrical in geometrical form about its mid-point*, (Fig. 6.20). Here, we have a rod R fixed on to two rings R_1 and R_2 at its two ends, which, in their turn, have two short rods screwed into them, terminating in knife-edges E_1 and E_2 inside the rings and carrying two bobs B_1 and B_2,— one *solid* and the other *hollow*.*

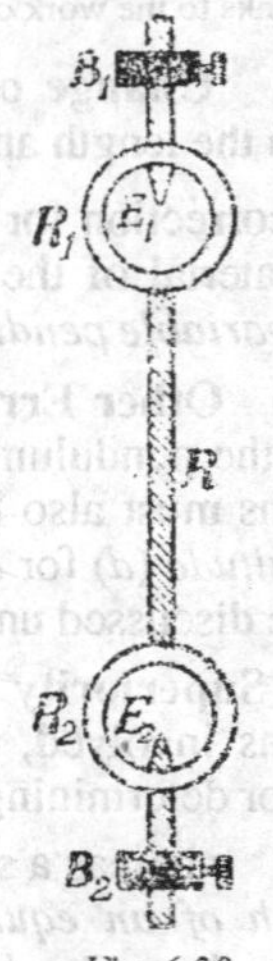

Fig. 6.20

The time-period of the pendulum can be made nearly equal about either knife-edge by moving the bobs up and down and screwing them into the desired position.

With the symmetrical form of the pendulum, the error due to air-effects is automatically eliminated, as explained in § 6.13 *(ii)* above.

And, *(iii)* finally, the error due to the yielding of the support has been eliminated by **Defforges**, by using two reversible *Repsold type* pendulums, of the *same mass* but *different lengths, the same ratio of l to l′* and having a *common pair of knife-edges* (to be used with either of them). He has shown that if L_1 and L_2 be the *reduced lengths* of the two reversible pendulums (*i.e.*, the lengths between the knife-edges) *and* τ_1 and τ_2, their computed times, then $\sqrt{(\tau_1^{\,2} - \tau_2^{\,2})}$ gives the correct time-period of a pendulum of length $(L_1 - L_2)$, as can be seen from the following :

We have $$\frac{g\tau_1^{\,2}}{4\pi^2} = L_1 + mg.\beta - \left(\frac{l_1 + l_2}{l_1 - l_2}\right).(r_1 - r_2)$$ See pages 165 and 166

and $$\frac{g\tau_2^{\,2}}{4\pi^2} = L_2 + mg.\beta - \left(\frac{l_1' + l_2'}{l_1' - l_2'}\right).(r_1 - r_2).$$

where l_1 and l_2 are the two lengths on the two sides of the *c.g.* in the case of one pendulum and l_1' and l_2' in that of the other.

So that, subtracting the second expression from the first, we have

$$\frac{g}{4\pi^2}(\tau_1^{\,2} - \tau_2^{\,2}) = L_1 - L_2 + \left[\frac{l_1 + l_2}{l_1 - l_2} - \frac{l_1' + l_2'}{l_1' - l_2'}\right](r_2 - r_1).$$

Clearly, the correcting term, (*i.e.*, the second term) in this expression can be made zero if $l_1 : l_2 :: l_1' : l_2'$ and this is easily done by adjusting the positions of the bobs of the two pendulums. With this adjustment made, we have $\frac{g}{4\pi^2}(\tau_1^{\,2} - \tau_2^{\,2}) = L_1 - L_2$.

Or, if we use only one knife-edge, (fixed on the support, as explained in § 6.13 *(iii)*, page 162), we have $\frac{g\tau_1^{\,2}}{4\pi^2} = L_1 + mg.\beta$ and $\frac{g\tau_2^{\,2}}{4\pi^2} = L_2 + mg.\beta$ [See page 165

where L_1 and L_2 are the reduced lengths and τ_1 and τ_2 the computed time-periods of the two pendulums respectively.

* This is to ensure that the lengths l_1 and l_2 of the pendulum on the two sides of the *c.g.* are not very nearly equal, or else the correcting terms for the error due to air-effects, (page 161), will not be small.

Thus, subtracting the second expression from the first, we have, straightaway,

$$\frac{g}{4\pi^2}(\tau_1^{\,2} - \tau_2^{\,2}) = L_1 - L_2.$$

This removes at one stroke the errors due to yielding of the support and curvature of the knife-edges, as also those due to air-effects, and we have

$$g = \frac{4\pi^2\,(L_1 - L_2)}{\tau_1^{\,2} - \tau_2^{\,2}}.$$

This is about the most accurate method of determining the value of g at a given place.

6.15. Conical Pendulum. A simple conical pendulum is just *a simple pendulum, (i.e., a small heavy bob attached to a light, inextensible string)*, which is given such a motion that *the bob describes a horizontal circle and the string traces out a cone.* The length of the pendulum is the distance between the point of suspension and the *c.g.* of the bob.

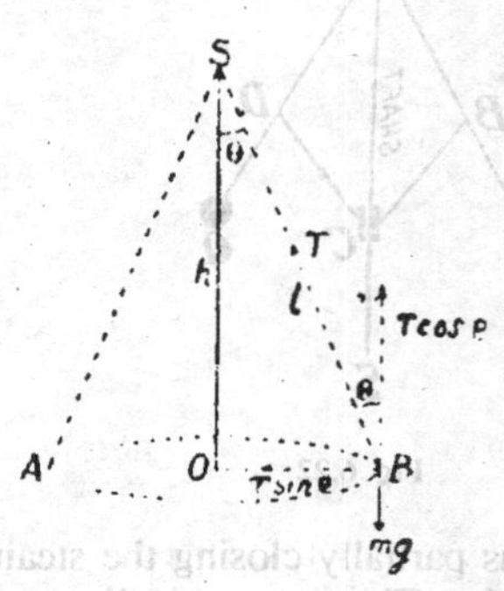

Fig. 6. 21

Let m be the mass of the bob B, v, its velocity and r, the radius of the circle it describes, (Fig. 6.21). Then, its centripetal acceleration towards O, the centre of the circle, is equal to v^2/r, and the centripetal force on it is, therefore, mv^2/r in that direction.

Let SO be equal to h.

Clearly, the forces acting on the bob are *(i) its weight, mg*, vertically downwards, and *(ii) the tension of the string T*, in the direction BS.

The weight mg is balanced by the vertical component $T\cos\theta$, of the tension T of the string, and its horizontal component $T\sin\theta$ provides the centripetal force mv^2/r towards O, where θ is the *semivertical angle* of the cone. Thus,

$$T\sin\theta = mv^2/r \text{ and } T\cos\theta = mg$$

Or, $$\frac{T\sin\theta}{T\cos\theta} = \tan\theta = \frac{mv^2/r}{mg} = \frac{v^2}{rg}.$$

Or, $v^2/rg = \tan\theta$. Or, $v^2 = r.g.\tan\theta$.

Since $v^2 = r^2.\omega^2$, where ω is the angular velocity of the bob,

we have $r^2.\omega^2 = r.g.\tan\theta = r.g.r/h = r^2 g/h$, $[\because \tan\theta = r/h.$

whence, $\omega^2 = g/h$, and $\therefore\ \omega = \sqrt{g/h}$. ...(i)

Now, the time-period of the pendulum is given by

$$t = \frac{2\pi}{\omega} = \frac{2\pi}{\sqrt{g/h}} = 2\pi\sqrt{\frac{h}{g}}.$$

$$t = 2\pi\sqrt{\frac{l\cos\theta}{g}}, \quad ...(ii) \quad [\because h = l\cos\theta.$$

where l is the length of the pendulum.

*It will thus be noted that the time-period is the same as that **of a simple pendulum of length h**, the **axial height** of the cone.*

If θ be very small, $\cos\theta$ is nearly equal to 1; so that, **$h = l$, *i.e.*, the time-period is almost** idependent of θ. *In other words, the time period remains **the same whether the bob moves along a** circular or a linear path.*

6.16. Steam Engine Governor. It will be seen from relation (*i*) above, (§ 6.15), that the angular velocity (ω) of the bob of the conical pendulum varies inversely as the square root of the depth (h) of its *c.g.* from the point of suspension; or, conversely, *that the depth of the c.g. of the bob, below its point of suspension, varies inversely as square of its angular velocity.*

This is made use of in the construction of what is called the 'governor' in a steam engine which is just a device to maintain the speed of the engine constant by regulating, or *'governing'*, the supply of steam from the boiler to the steam-chest.

In essentials, it is just a combination of two similar conical pendulums, mounted on either side of the vertical shaft (with a common point of suspension), rotated by the engine, and consists of two rods OP and OQ hinged together at their upper end O to the shaft OS, and carrying two spherical metallic bobs P and Q at their lower ends (Fig 6.22). Two other smaller rods connect OP and OQ to a metallic collar C, which slides freely along the shaft, thus operating a lever which controls the *throttle valve*, or the *steam valve*, opening it partially or fully, according as the collar moves up or down the shaft.

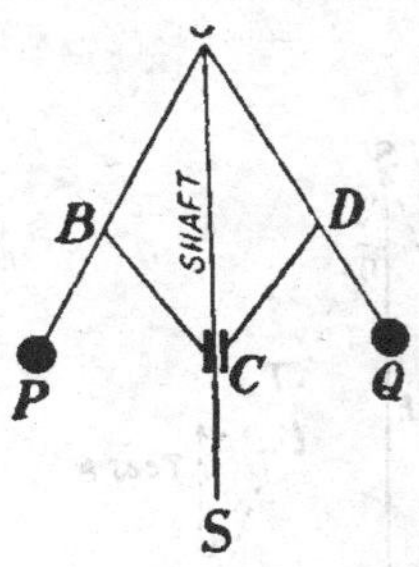

Fig. 6.22

Now, when due to a greater supply of steam to the cylinder, the shaft, and, therefore, the bobs rotate faster, *i.e.*, ω increases h proportionately decreases, or the bobs rise up, thus partially closing the steam valve, thereby partially cutting off the supply of steam to the cylinder. This automatically results in a falling off of the speed of rotation (ω) of the shaft or the bobs, and when this happens, h increases, *i.e.*, the collar slides down with the bobs, thus opening the steam valve more fully, allowing more steam into the cylinder, which then increases the rate of rotation of the shaft. So that, by proper adjustment, the rate of supply to the steam chest or cylinder, and, consequently, the rate of rotation of the shaft, can be maintained at any constant value desired.

The *sensitiveness* of the device, however, decreases with the increasing speed of the engine. For, we have the relation $\omega^2 = g/h$, ...(*iii*)

differentiating which, we have $2\omega \,.\, d\omega = -\dfrac{g}{h^2} \,.\, dh$. ...(*iv*)

And, therefore, dividing relation (*iv*) by (*iii*), we have

$$\frac{2\omega\, d\omega}{\omega^2} = -\frac{g.dh}{h^2} \times \frac{h}{g}. \text{ Or, } \frac{2d\omega}{\omega} = \frac{-dh}{h},$$

whence, $$dh = -\frac{2h}{\omega} \,.\, d\omega.$$

Or, substituting the value of h, from relation (*iii*) above, we have

$$dh = -\frac{2g}{\omega^3} \,.\, d\omega.$$

Thus, it is clear that $dh \propto 1/\omega^3$, *i.e.*, dh decreases as ω increases. In other words, the change in the position of the *c.g.* decreases with increasing angular velocity of the bobs or the shaft, thus slowing down the up and down motion of the collar along the shaft or decreasing the sensitiveness of the device.

6.17. Other methods for the determination of 'g'. The following are a few other methods that may be used to determine the value of g at a place. Although they do not compare favourably with the pendulum methods in point of accuracy or ease of performance, they are, nevertheless, valuable laboratory exercises, affording good illustrations of the various principles employed for the purpose.

(1) **The Inclined Plane.** We have seen before, in 3.14, (Page 78), how the acceleration of a body, rolling down an inclined plane, (without slipping), is given by the expression,

$$a = [r^2/(k^2 + r^2)]g \sin \alpha,$$

where r is the radius of the body; k, its *radius of gyration* about its axis of rotation; α, the angle of inclination of the plane, and g, the *acceleration due to gravity* at the place.

It follows, therefore, that $$g = \frac{a(k^2 + r^2)}{r^2 \sin \alpha}$$

So that, knowing r, *(by means of a vernier callipers)*, k *(from the geometrical shape of the body)*, $\sin \alpha$, *(from the height and length of the plane)* and a, *(by direct experiment, as explained below)*, we can easily calculate the value of g at the place.

The value of a can be easily and accurately obtained by noting the distances covered by the body, down along the plane, in successive equal intervals of time*, and plotting the distance-time curve for it. The equation of the curve being $S = \frac{1}{2}at^2$, (u being zero, because the body starts from rest), it will, obviously, be *parabolic* in form, with its axis coinciding with the distance-axis. So that, substituting the coordinates of any suitable point on the curve in the relation $S = \frac{1}{2}at^2$, the value of a can be easily determined.

(2) **The Dynamical Spherometer**. Let S be the centre of curvature of a spherical surface, arranged horizontally *with its concavity upwards*, and R, its radius, (Fig. 6.23); and let a steel ball, of mass m and radius r, be allowed to roll to and fro on it, *without slipping*. Then, clearly, the ball oscillates on the inside of the surface *as though it were a compound pendulum, with its centre of suspension at S and its centre of oscillation of O, the centre of the ball; so that,* $SO = (R-r)$.

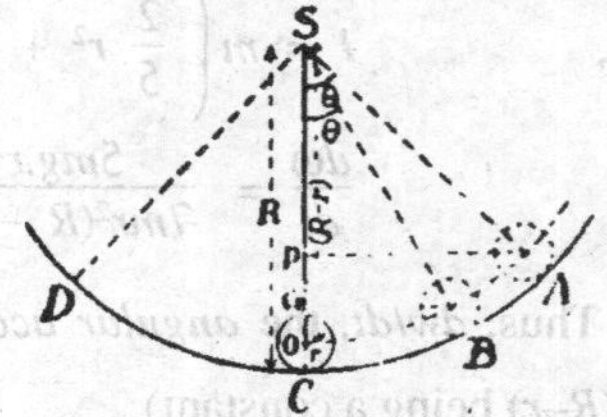

Fig. 6.23

Let the angular amplitude of the ball be θ', (*i.e.*, the angle that it makes at S, when in its extreme positions). Let it be in the position B, at any given instant, such that the angle it now makes at S is θ. Then, clearly, *work done on the ball by the force of gravity in bringing it down from A* to *B* is equal to *weight of the ball* × *the vertical distance through which the ball has fallen.*

This must, therefore, be clearly equal to the loss in its potential energy, *i.e.*, equal to $mg \times PQ$, where mg is the weight of the ball.

Or, *loss in P.E.* $= mg\,(SQ - SP)$.

Now, $SQ = (R-r).\cos\theta$ and $SP = (R-r).\cos\theta'$

And ∴ $SQ - SP = (R-r).(\cos\theta - \cos\theta')$.

So that, *work done by the force of gravity in moving the ball from A* to *B*

$= mg.\ (R-r).(\cos\theta - \cos\theta') =$ loss in potential energy of the ball.

---

* This may be easily done if the angle of inclination (a) of the plane be small; for, then the acceleration of the body will also be small and the time taken by it in rolling down the plane will be fairly accurately measured by means of a stop watch.

This must, clearly, be equal to the *gain in the kinetic energy* of the ball, *i.e.*, equal to $\frac{1}{2} I . \omega'^2$, where I is the moment of inertia of the ball and ω', its angular velocity about the line of contact.

So that, $$mg.(R - r).(\cos\theta - \cos\theta') = \frac{1}{2} I . \omega'^2 .$$

If ω be the angular velocity of the ball about a horizontal axis through S, we have

$$(R - r).\omega = r\omega' . \text{ Or, } \omega' = \left(\frac{R - r}{r}\right).\omega.$$

$$\therefore \quad mg.(R - r).(\cos\theta - \cos\theta') = \frac{1}{2} I \left[\left(\frac{R - r}{r}\right).\omega\right]^2,$$

whence, *differentiating with respect to time*, we have

$$mg.(R - r).\sin\theta.\frac{d\theta}{dt} = 2 \times \frac{1}{2}.I\left(\frac{R - r}{r}\right)^2 \omega.\frac{d\omega}{dt} \quad \left[\theta' \text{ is a constant.}\right.$$

And, since $d\theta/dt = \omega$, we have

$$mg.(R - r).\sin\theta.\omega = I.\frac{(R - r)^2}{r^2}.\omega.\frac{d\omega}{dt}$$

Or, $$mg.(R - r).\sin\theta = I.\frac{(R - r)^2}{r^2}.\frac{d\omega}{dt}.$$

Or, $$\frac{d\omega}{dt} = \frac{mg.r^2 (R - r)\sin\theta}{I(R - r)^2} = \frac{mg.r^2.\sin\theta}{I(R - r)} = \frac{mg.r^2}{I(R - r)}.\theta.$$

Since, $$I = m\left(\frac{2}{5}r^2 + r^2\right) = \frac{7mr^2}{5}, \text{ we have}$$

$$\frac{d\omega}{dt} = \frac{5mg.r^2}{7mr^2(R - r)}.\theta = \frac{5g}{7(R - r)}.\theta.$$

Thus, dw/dt, the *angular acceleration of the ball* $= \dfrac{5g}{7(R - r)}.\theta$, *i.e.*, proportional to θ, ($5g/7(R-r)$ being a constant).

The ball thus executes a *simple harmonic motion*, and its time-period is given by

$$t = 2\pi\sqrt{\frac{1}{5g/7(R - r)}} = 2\pi\sqrt{\frac{7(R - r)}{5g}}$$

whence, $$g = \frac{28\pi^2(R - r)}{5t^2}.$$

Thus, knowing the radii of the concave surface and the ball (with the help of a spherometer and a vernier callipers respectively), and noting the time-period of oscillation of the ball, we can easily calculate the value of g at the given place.

N.B. Re-arranging the expression for t^2, obtained above, we have $5gt^2 = 28\pi^2R - 28\pi^2r$. Or, $28\pi^2R = 5gt^2 + 28\pi^2 r$, whence

$$R = \frac{5gt^2 + 28\pi^2r}{28\pi^2} = \frac{5gt^2}{28\pi^2} = +r.$$

So that, knowing r, the *radius of the ball*, t, *its period of oscillation*, and the value of g, we can easily calculate the *radius of curvature* (R) of the given spherical surface. — hence the name *dynamical spherometer*.

(3) **The Atwood's Machine.** In the ribbon-type of machine, [Fig. 6.24 (*a*)], a *paper strip* or *ribbon R* is passed round the *flat* rim of a light and *frictionless pulley* (running on ball-bearings), with two equal masses *M* and *M* at its two ends, one of which is initially kept resting on a platform *P*.

An identical ribbon is attached to the lower ends of the two masses, as shown, so that, when the system is set into motion, an additional length of the first ribbon, passing on to the right side of the pulley, is exactly balanced by an *equal* length of the second, passing on to its left side, *thus ensuring that no extra mass is transferred from one side of the pulley to the other.*

A steel strip or *vibration V*, of a known time-period *T*, is clamped horizontally at one end, and carries a light *style*, or an *inked brush B*, which *just* touches the paper ribbon going round the pulley.

A small *rider r*, of mass *m*, is placed on mass *M*, resting on platform *P* to make it slightly heavier than the other.

Then, with the masses not yet in motion, the brush is moved across the paper ribbon to mark a horizontal line on it, *indicating the starting point.*

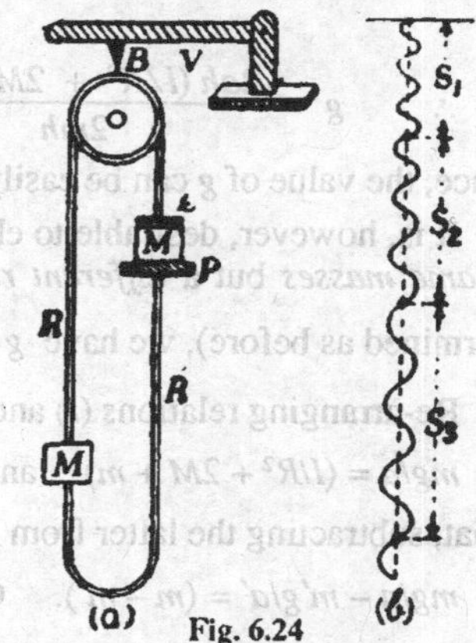

Fig. 6.24

The platform *P* is now suddenly made to fall, (by means of (*'trigger releases'*) and the vibrator simultaneously set vibrating. Naturally, the mass, loaded with the rider, moves *down* and the other, *up*, with a *common acceleration*, say, *a*. And, as the ribbon runs past the brush, a wavy curve, due to the transverse vibrations of *V*, gets traced on it, and goes on gradually lengthening out, Fig. 6.24 (*b*), on account of the accelerated motion of the masses, and hence that of the ribbon.

Since *one wave* is traced out on the ribbon *during one vibration of V*, the distances occupied by successive waves represent the distances covered by the masses during successive time-periods of it.

Thus, if S_1, S_2, S_3 etc., be the distances covered by the masses during the first, second and third etc., vibrations of *V*, we have

$$S_1 = \frac{1}{2}aT^2 \qquad [\because u = 0, \text{ the masses starting from rest.}$$

and $$S_2 = \frac{1}{2}aT.T + \frac{1}{2}aT^2 = aT^2 + \frac{1}{2}aT^2, \qquad [\because \text{ here, } u = aT, \text{ the velocity after time } T.$$

i.e., $$S_2 = 3aT^2/2.$$

And, similarly, $$S_3 = 2aT.T + \frac{1}{2}aT^2 \qquad [\because \text{ now } u = 2\,aT, \text{ the velocity after time } 2T.$$

Or, $$S_3 = 2aT^2 + \frac{1}{2}aT^2 = 5\,aT^2/2.$$

So that, $$S_2 - S_1 = S_3 - S_2 = aT^2 = x, \text{ say} \quad \text{Or, } a = x/T^2.$$

Thus, *T* being known, the value of *a* can be easily calculated.

Now, if *v* be the velocity acquired by the masses, when they have covered a distance *h*, we have

gain in K.E. of the pulley and the masses (together with the rider)

= loss in P.E. of the masses and the rider,

i.e., $$\frac{1}{2}I\omega^2 + \frac{1}{2}(2M + m)v^2 = (M + m)gh - Mgh = mgh,$$

where *I* is the *moment of inertia of the pulley* about its axis of rotation and ω, its *angular velocity* at the time.

Now, if R be the *radius* of the pulley, $v = R\omega$, or $\omega = v/R$.

Hence, $\frac{1}{2} I . v^2/R^2 + \frac{1}{2}(2M + m) v^2 = mgh$ Or, $Iv^2/R^2 + (2M + m)v^2 = 2mgh$

Or, $v^2 (I/R^2 + 2M + m) = 2mgh$, whence, $g = \frac{v^2(I/R^2 + 2M + m)}{2mh}$.

But $v^2 = 2ah$. [∵ $u = 0$ and $S = h$.

$$\therefore \quad g = \frac{2ah(I/R^2 + 2M + m)}{2mh} = \frac{a(I/R\,v^2 + 2M + m)}{m},$$

whence, the value of g can be easily obtained.

It is, however, desirable to eliminate I from this expression by repeating the experiment with the *same masses* but a *different rider*, of mass m'. If a' be now the acceleration of the masses, (determined as before), we have $g = \frac{a'(I/R^2 + 2M + m')}{m'}$

Re-arranging relations (*i*) and (*ii*), we have

$mg/a = (I/R^2 + 2M + m)$ and $m'g/a' = (I/R^2 + 2M + m')$.

So that, subtracting the latter from the former, we have

$mg/a - m'g/a' = (m - m')$. Or, $g(m/a - m'/a') = (m - m')$,

whence, $$g = (m - m') \Big/ \left(\frac{m}{a} - \frac{m'}{a'}\right)^*$$

Thus, knowing m, m', a and a', g can be easily evaluated.

A possible source of error, here, is the *frictional force* encountered by the pulley as it rotates about its axle which, obviously, tends to lower its angular velocity. This may be easily remedied by placing another *auxiliary* rider on the loaded mass, such that, with the main rider (r) removed from it, if an initial velocity be given to it to set the system in motion, it continues to move with the *same uniform velocity, i.e.*, (its motion is neither accelerated nor retarded). Obviously, then, the *weight* of this auxiliary rider exactly counteracts the retarding force due to friction. If, therefore, kept on the mass throughout the experiment, it completely eliminates the error due to friction, and, clearly, neither its weight nor the frictional force need enter into our calculations.

(4) **The Dropping Plate.** A plate of glass, P, smoked by holding it over burning camphor, is suspended with its plane vertical, by means of a thread, as shown in Fig. 6.25 (*a*), and a tuning fork F, of a known frequency N is mounted close to it, so that a light aluminum style, (or better still, a *hog's bristle*, such as may be obtained from a discarded hair brush), attached to one of its prongs, *just* touches the surface of the plate.

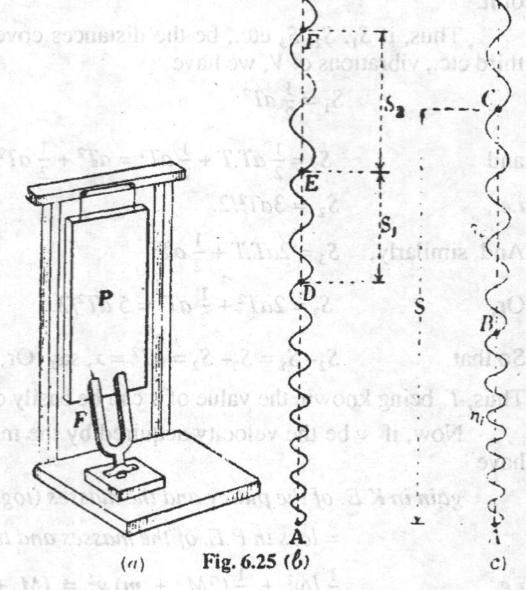

(*a*) Fig. 6.25 (*b*) (*c*)

* Or, we could use *different masses M* and *M'* but the *same rider*, in which case we shall have $g(2M - M')/m\ (1/a - 1/a')$, where a' *is the acceleration of the masses in the second case*.

The fork is set vibrating by lightly drawing a bow across it, or by simply pinching it strongly, and the thread, supporting the plate, burnt or cut *simultaneously*. Thus released, the plate starts falling, with an acceleration, equal to the value of g at the place, and the style traces out a wavy line on it, of the form shown in Figs, 6.25 (*b*) and (*c*) —the waves being smaller and closer together at first but gradually lengthening out and getting further apart, due to the accelerated motion of the plate, though *the time taken to trace each wave remains the same, viz., equal to* $1/N$, *the time-period of the fork.*

Three points D, E and F, [Fig.6.25(*b*)], are then marked on this wavy line, such that DE and EF contain the *same number of waves*, say, n, each. Let distances DE and EF be S_1 and S_2 respectively, as measured by means of a travelling microscope, both being covered by the plate in the *same interval of time* $t = n/N$, taken by the fork to complete n vibrations. Then, clearly,

$$S_1 = ut + \frac{1}{2}gt^2. \quad \text{Or,} \quad 2S_1 = 2ut + gt^2. \qquad ...(i)$$

where u is the velocity of the plate at D.

And, $$(S_1 + S_2) = 2ut + \frac{1}{2}g.(2t)^2,$$

because, here, distance = (S_1+S_2), and time = $2t$.

Or, $$(S_1+S_2) = 2ut + 2gt^2 \qquad ...(ii)$$

So that, subtracting equation (*i*) from (*ii*), we have

$$(S_1 + S_2) - 2S_1 = 2ut + 2gt^2 - (2ut + gt^2).$$

Or, $$(S_2 - S_1) = gt^2, \quad \text{whence,} \quad g = \frac{(S_2 - S_1)}{t^2} \qquad ...(iii)$$

Or substituting the value n/N for t, in relation (*iii*), we have

$$g = \frac{S_2 - S_1}{(n/N)^2} = \frac{N^2(S_2 - S_1)}{n^2}. \qquad ...(iv)$$

Thus, knowing N, n and $(S_2 - S_1)$, we can easily calculate the value of g at the place.

N.B. It will be readily seen that the mass of the style (or the *hog's bristle*), attached to the prong of the fork, together with the friction it encounters at the plate, will slightly lower its frequency, so that it will actually be somewhat less than N. For greater accuracy, therefore, the frequency of the fork (with the style attached to it) must be determined by the method of '*beats*', by sounding it with another fork of an accurately known frequency. The frequency, thus determined, should then replace N in relation (*iv*) for g, above.

Alternative Calculation. The following is a comparatively *more accurate method* of calculating the value of g, because, here, the possible error in correctly counting n is eliminated.

Three points A, B and C are marked on a portion of the wavy line, [Fig. 6.25(*c*)], where the waves are clearly visible and can be distinctly counted. Let there be n_1 waves (and, therefore, n_1 vibrations) made by the fork in between A and B, and n_2 waves (or n_2 vibrations) in between B and C, and let the total distance AC be S.

Then, clearly, (n_1+n_2) waves or vibrations are made by the fork in time $(n_1+n_2)/N$. So that,

$$S = \frac{1}{2}g\left(\frac{n_1 + n_2}{N}\right)^2$$

Or, $$\sqrt{S} = \sqrt{\frac{g}{2}} \cdot \frac{n_1 + n_2}{N},$$

because $u = 0$, $a = g$ and $t = (n_1 + n_2)/N$.

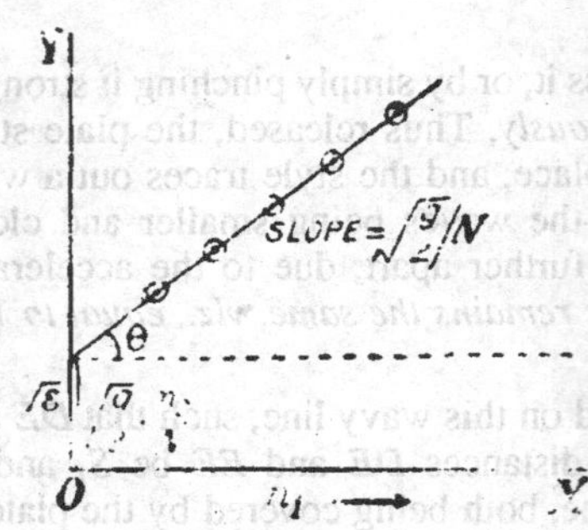

Fig. 6.26

Plotting $\sqrt{S}$ against n_1, therefore, we obtain a straight line, (Fig. 6.26), of slope $\sqrt{\frac{g}{2}}/N$, from which the value of g can at once be calculated, *without knowing n.*

Incidentally, the dropping plate method also shows that a freely falling body is subjected to a constant acceleration due to gravity, — a fact, not easy to demonstrate otherwise.

(5) Vertical Oscillations of a Flat Spiral Spring. A spiral spring is just a uniform wire or ribbon designed to have, in its normal, unstrained condition, the form of a regular *helix,* such as may be obtained by winding the wire closely and uniformly round a cylinder of a diameter much greater than its own.

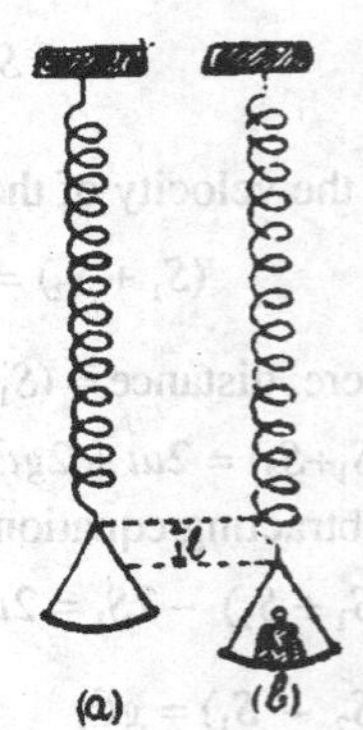

Fig. 6.27

If the plane of each coil of the spiral is perpendicular to the axis of the cylinder, it is called a *flat spiral*, but if it be inclined at a small angle to this axis, it is spoken of as an *inclined spiral.* We shall concern ourselves here only with the *flat spiral of a wire of circular cross-section.*

If a small force be applied to a flat spiral *along its axis, (i.e., along the straight line passing through the centre of each coil of it)*, and perpendicular to its plane, it increases in length a little but still preserves its helical form, as will be clear from Figs. 6.27 (*a*) and (*b*).

Let us consider a flat spiral of *length L* and *radius R*, (where *R* is much greater than its pitch), suspended from a rigid support, with its upper and lower ends (*A* and *B*), bent as shown in Fig. 6.28 (*a*), so as to lie along its axis.

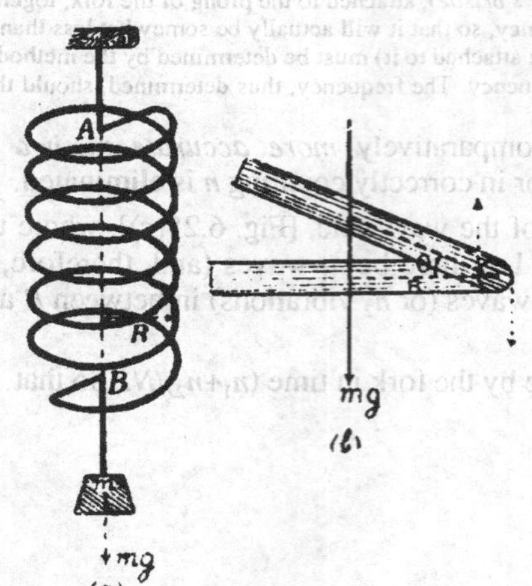

Fig. 6.28

Then, if a mass *m* be suspended from its lower free end, a force equal to *mg* (the weight of the mass) acts vertically downwards along its axis, producing a *statical extension l* in its length (*i.e.*, an extension with the mass *m* at rest).

The effect of this force *mg*, acting along the axis of the spiral is to produce a *turning moment* equal to *mg.R*, at every section of it, [Fig. 6.28(*b*)]. And this, in its turn, subjects the wire to a uniform twist θ, say, *per unit length* of it.

Now, the *twisting or torsional couple* per unit length of the wire* is equal to $n\pi r^4\theta/2$, where *n* is the *coefficient of rigidity* of the material of the wire; *r*, its *radius* and θ, its *angle of twist.*

* See § 8.22 where it is shown that the twisting couple on a cylinder (or wire) is equal to $n\pi r^4\theta/2l$, where θ is the *angle of twist*, *l*, its *length* and *r* its *radius*.

This, then, is the *torsional resistance*, opposing the turning moment $mg.R$ due to the weight mg. For equilibrium, therefore, we have $mg.R = n\pi r^4\theta/2$, ...(*i*)

Now, a twist θ per unit length of the wire corresponds to a displacement or extension $R.\theta$ per unit length of it [Fig. 6.28(*b*)]. Clearly, therefore, the extension produced in the whole length of the wire is equal to $L.R.\theta$.

Or, $l = L.R.\theta$, whence, $\theta = l/LR$ [$\because$ l is the total extension produced.

Substituting this value of θ in relation (*i*) above, we have

$$mg.R = \frac{\pi nr^4}{2}\cdot\frac{l}{LR} = \frac{\pi nr^4 l}{2LR}, \quad \text{whence,} \quad mg = \frac{\pi nr^4.l}{2LR^2}. \qquad ...(ii)$$

The expression $\pi nr^4.l/2LR^2$ thus represents the force of *elastic reaction* for an increase l in the length of the spiral, and, therefore, the *elastic reaction per unit increase* in the length of the spiral $= \pi nr^4/2LR^2$ (because $l = 1$). Denoting this by K, we have

$$mg = K.l, \text{ whence, } K = mg/l.$$

If the suspended mass (m) be now displaced or pulled vertically downwards through a distance x and then released, so as to produce *vertical oscillations* in the spiral, the restoring force F, acting on the mass, may, *for small oscillations*, be taken to be directly proportional to its displacement. So that, $F = K.x$.

And, if d^2x/dt^2 be the acceleration of the mass at the given instant, when its displacement is x, the *inertial reaction* of the mass is clearly equal to $m.d^2x/dt^2$.

Hence, by Newton's third law of motion, we have

$$m.d^2x/dt^2 = -Kx, \text{ whence, } d^2x/dt^2 = -x.K/m.$$

Now, since K and m are both constant quantities for the given spiral, we have K/m = a constant, μ, say.

So that, $d^2x/dt^2 = -\mu.x$. Or, $d^2x/dt^2 \propto x$,

i.e., the acceleration of the mass is directly proportional to its displacement x; and it, therefore, executes a simple harmonic motion, its time-period being given by the expression.

$$t = 2\pi\sqrt{\frac{1}{\mu}} = 2\pi\sqrt{\frac{1}{K/m}} = 2\pi\sqrt{\frac{m}{K}} = 2\pi\sqrt{\frac{m}{mg/l}}.$$

Or, $t = 2\pi\sqrt{l/g}$; ...(*iii*)

i.e., the time-period is the same as that of a simple pendulum of length l, the extension produced in the spiral.

Squaring and re-arranging this expression for t, we have $g = 4\pi^2.l/t^2$, whence the value of g at the given place can be easily calculated.

In the above treatment, we have not taken into consideration the mass of the spring, assuming it to be negligible, compared with the suspended mass m. For greater accuracy, however, it must also be taken into account. It can be shown that if the *mass* of the spring be m, the total mass acting downwards along the axis of the spring becomes $m + m_s/3$, and the expression (*iii*) above, for the time-period of the spiral, becomes $t = 2\pi\sqrt{(m + m_s/3)/K}$

It is, however, best to eliminate m_s altogether. This is done by performing the experiment with two different suspended masses, m_1 and m_2. Then, if t_1 and t_2 be the respective time-periods of oscillation in the two cases, we have $t_1 = 2\pi\sqrt{(m_1 + m_s/3)/K}$ and

$$t_2 = 2\pi\sqrt{(m_2 + m_s/3)/K}$$

---

* See worked example 11 at the end of the Chapter VIII (page 306).

So that, squaring and subtracting the second from the first, we have

$$(t_1^2 - t_2^2) = \frac{4\pi^2}{K}(m_1 - m_2) = 4\pi^2 . \frac{(m_1 - m_2)}{K} . \qquad ...(iv)$$

Now, if l_1 and l_2 be the *statical* extensions corresponding to the two masses, we have

$$m_1.g = Kl_1 \quad \text{and} \quad m_2.g = Kl_2,$$

whence, $m_1.g - m_2.g = Kl_1 - Kl_2$. Or, $(m_1 - m_2).g = K(l_1 - l_2)$.

Or, $$\frac{(m_1 - m_2)}{K} = \frac{(l_1 - l_2)}{g}.$$

Substituting this value of $(m_1 - m_2)/K$ in expression (*iv*) above, we have

$$(t_1^2 - t_2^2) = 4\pi^2 . \frac{(l_1 - l_2)}{g}, \quad \text{whence,} \quad g = 4\pi^2 . \left(\frac{l_1 - l_2}{t_1^2 - t_2^2}\right).$$

Thus, observing t_1 and t_2 directly, and measuring l_1 and l_2, by noting the position of a light pointer, attached to the spiral, on a vertical centimetre scale fixed alongside it, the value of g can be easily calculated.

(6) **The Bifilar suspension**. If a heavy and uniform bar or cylinder, (or, in fact, any rigid body), be suspended horizontally by means of *two equal, vertical, flexible* and *inelastic* threads, *equidistant from its centre of gravity*, the arrangement consititutes what is called a *bifilar suspension*. On being displaced a little *in its own plane*, (*i.e.*, in the horizontal plane), and then released, the bar or cylinder executes a simple harmonic motion about the vertical axis through its centre of gravity.

Now, two cases arise, (*i*) *when the two suspension threads are parallel*, and (*ii*) *when they are not*. Let us consider both.

(*i*) **Bifilar suspension, with parallel threads.** Let AB, [Fig. 6.28(a)], represent the original or equilibrium position of a cylinder, of mass m, and with its *c.g.* at O, where its weight mg acts vertically downwards. Let the two suspension threads PA and QB be *parallel to each other*, and distance $2d$ apart; and let the length of each be l.

Now, if the cylinder be displaced a little into the position $A'B'$ through a small angle θ, about the vertical axis through O, the suspension threads take up the positions PA' and QB' at an angle with their original positions, where ϕ is small.

Let T be the tension in *each* thread, acting upwards along it. Then, resolving it into its two rectangular components, we have

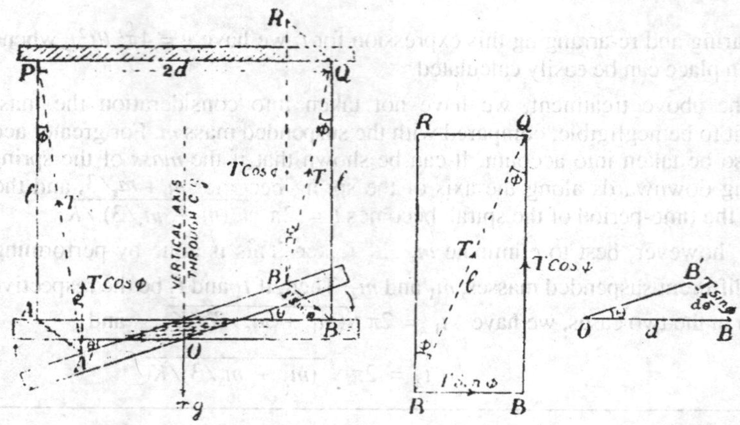

Fig. 6.28

(*i*) *component T cos* ϕ, *acting vertically upwards* ;

and (*ii*) *component T sin* ϕ, *acting horizontally along B'B and A'A*, [Fig. 6.28(*b*)].

Obviously, the vertical components support the weight of the cylinder. Hence,

$$2T \cos\phi = mg. \quad \text{Or,} \quad T \cos\phi \; mg/2.$$

And, since ϕ is small, $\cos\phi = 1$, very nearly.

So that, $T = mg/2$.

The components, $T \sin\phi$, (acting at A' and B') on the other hand, being *equal, opposite* and *parallel*, constitute a *couple*, tending to bring the cylinder back into its original position. And, since $A'A$ and $B'B$ are practically at right angles to $A'B'$, we have

moment of this restoring couple $= T \sin\phi . 2d = T.\phi . 2d.$ [being small

Now, $\theta = BB'/OB = BB'/d$; so that, $BB' = \theta.d.$ [Fig. 6.28 (c)

And, therefore, $\phi = BB'/l = \theta.d/l.$ [Fig. 6.28 (b).

Hence, *restoring couple* $= T . \dfrac{\theta.d}{l} . 2d = \dfrac{mg}{2} . \dfrac{\theta.2d^2}{l} = \dfrac{mg.d^2}{l} . \theta.$

But, the restoring couple is also $= I. d^2\theta / dt^2$,

where I is the *moment of inertia* of the cylinder about the vertical axis through O, (its *c.g.*)., and $d^2\theta / dt^2$, its *angular acceleration*.

$$\therefore \quad I . \frac{d^2\theta}{dt^2} = \frac{-mg.d^2}{l} . \theta, \quad \text{whence,} \quad \frac{d^2\theta}{dt^2} = \frac{-mg.d^2}{I.l} . \theta .$$

Now, $mg.d^2/Il$ is a constant quantity in a given case, and, therefore, putting it equal to μ, we have

$$d^2\theta/dt^2 = \mu.\theta. \quad \text{Or,} \quad d^2\theta/dt^2 \propto \theta,$$

i.e., the angular acceleration of the cylinder is proportional to its angular displacement, and is clearly directed towards its mean position. The cylinder, therefore, executes a simple harmonic motion and its time-period is given by

$$T = 2\pi \sqrt{\frac{1}{\mu}} = 2\pi \sqrt{\frac{1}{mgd^2/Il}} .$$

Or,

$$T = 2\pi \sqrt{\frac{I.l}{mgd^2}} = 2\pi . \frac{1}{d} \sqrt{\frac{I.l}{mg}} . \quad ...(i)$$

Or, if we put $I = mk^2$, where k is the *radius of gyration* of the cylinder about the vertical axis through O, we have

$$T = 2\pi . \frac{1}{d} \sqrt{\frac{mk^2.l}{mg}} = 2\pi . \frac{1}{d} \sqrt{\frac{k^2 l}{g}},$$

whence,

$$T = 2\pi . \frac{k}{d} \sqrt{\frac{l}{g}} . \quad ...(ii)$$

Now, squaring relation ,(*i*) or (*ii*), and re-arranging, we have,

from relation (*i*) $\quad g = 4\pi^2 \; I.l/m.d^2 . T^2$ ...(*iii*)

and, from relation (*ii*), $\quad g = 4\pi^2 . k^2 . l/d^2 . T^2$ ..(*iv*)

And, thus the value of g at the given place may be easily obtained.

(ii) **Bifilar suspension with non-parallel threads.** Let the rod or cylinder AB, [Fig. 6.29 (*a*)], be suspended symmetrically by two *equal* but *non-parallel* threads*, each of length l, and let the distance between the threads at the top and at the bottom be $2d_1$, and $2d_2$ respectively, where $(d_2 - d_1) = x$.

If the cylinder be displaced through an angle θ *in its own plane*, into the position $A' B'$, the suspension threads take up positions PA' and QB'. *Then, as before, tension* T acts upwards along each thread, and may be resolved into two rectangular components, *viz.*,

(i) $T \cos \theta$, *acting vertically upwards, as shown*, and *(ii)* $T \sin \phi$ *acting horizontally, along* $A'K'$ *and* $B'L$ [Fig. 6.29 (*a*)], *where* ϕ is the angle that *each thread makes with the vertical or the perpendiculars PK and QL, from P and Q on to AB.*

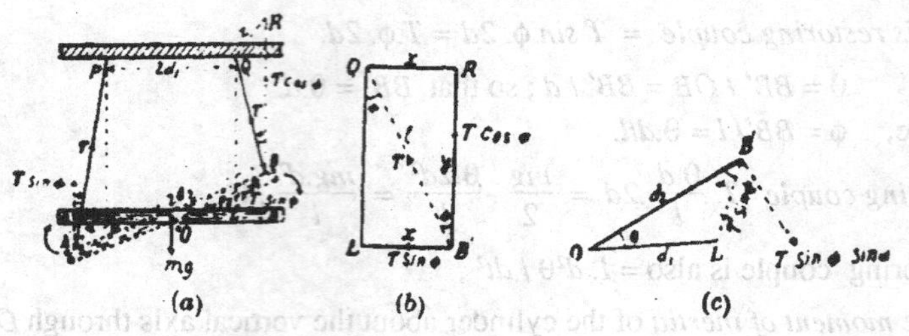

Fig. 6.29

The vertical components $T \cos \phi$ support the weight mg of the cylinder, acting vertically downwards at its *c.g.*, O; and, therefore,

$$2T \cos \phi = mg. \quad ...(v)$$

Now clearly, $LB' = \sqrt{d_2^{\,2} + d_1^{\,2} - 2d_2 d_1 \cos \theta}$. [Fig. 629 (*c*)

$= \sqrt{d_2^{\,2} + d_1^{\,2} - 2d_2 \cdot d_1}$ approximately. [θ being small, and $\therefore \cos \theta = 1$, nearly.

$= \sqrt{(d_2 + d_1)^2}$.

Or, $LB' = (d_2 - d_1) = x$, approximately.

And* $\cos \phi = \dfrac{B'R}{B'Q} = \dfrac{B'R}{l} = \dfrac{\sqrt{B'Q^2 - RQ^2}}{l} = \dfrac{\sqrt{l^2 - x^2}}{l}$ [Fig. 6.29 (*b*).

Now, $T = \dfrac{mg}{2 \cos \phi} = \dfrac{mg}{2\sqrt{l^2 - x^2/l}} = \dfrac{mg.l}{2\sqrt{l^2 - x^2}}$ ...(*vi*) [From relation (*v*) above.

And, resolving forces $T \sin \phi$, *acting at* A' and B' along $A'K'$ and $B'L$ respectively into their rectangular components along and at right angles to $A'B'$, we have

components at right angles to $A'B' = T \sin \phi \cdot \sin \alpha$ [Fig. 6.29(*c*)

Since these two components act in opposite direction at A' and B', they constitute a *couple*, tending to rotate the cylinder back into its original position AB; and, clearly,

moment of this restoring couple $= T \sin \phi \cdot \sin \alpha \cdot A'B'$

Or, *restoring torque on the cylinder* $= T \sin \phi \cdot \sin \alpha \cdot 2d_2$

$$= 2d_2 \cdot \frac{mg.l}{2\sqrt{l^2 - x^2}} \cdot \frac{x}{l} \sin \alpha = \frac{d_2 \cdot mgl}{\sqrt{l^2 - x^2}} \cdot \frac{x}{l} \cdot \sin \alpha .$$ [$\sin \phi = x / l$ from Fig. 6.29 (*b*)

* The threads, in this position of the rod, are not shown to avodid complicating the figure.

Now, the sides in a triangle, being proportional to the sines of the angles opposite to them, we have, from Fig. 6.29(*c*),

$$d_1/\sin\alpha = LB'/\sin\theta.$$

So that, $\dfrac{d_1}{\sin\alpha} = \dfrac{x}{\theta}$. And $\therefore \quad \sin\alpha = \dfrac{d_1}{x}.\theta$. $\left[\because LB' = x \text{ and } \theta \text{ is small}\right.$

Hence, *restoring torque on the cylinder*

$$\frac{d_2 .mg . l}{\sqrt{l^2 - x^2}} . \frac{x}{l} . \frac{d_1}{x} . \theta = \frac{d_1 d_2 . mg}{\sqrt{l^2 - x^2}} . \theta .$$

But *restoring torque is also* $= I . \dfrac{d^2\theta}{dt^2}$,

where I is the *moment of inertia of the cylinder* about the vertical axis through its *e.g.*, and $d^2\theta/dt^2$, its *angular acceleration.*

$$\therefore \quad I. \frac{d^2\theta}{dt^2} = -\frac{d_1 .d_2 .mg}{\sqrt{l^2 - x^2}}. \theta. \quad \text{Or,} \quad \frac{d^2\theta}{dt^2} = -\frac{d_1 . d_2 . mg}{I\sqrt{l^2 - x^2}} . \theta = -\mu\theta,$$

where, $\dfrac{d_1 \; d_2 . mg}{I . \sqrt{l^2 - x^2}}$ = a constant μ.

Thus, $$\frac{d^2\theta}{dt^2} \propto \theta.$$

Or, *the angular acceleration of the cylinder is proportional to its angular displacement.* It, therefore, executes *a simple harmonic motion* and its *time-period* is given by

$$T = 2\pi\sqrt{\frac{1}{\mu}} = 2\pi\sqrt{\frac{1}{d_1 .d_2 .mg/I. \sqrt{l^2 - x^2}}}$$

$$T = 2\pi\sqrt{\frac{I. \sqrt{l^2 - x^2}}{d_1 . d_2 . mg}} . \quad ...(vii)$$

And, if we put $I = mk^2$, where k is the *radius of gyration* of the cylinder about the vertical axis through its *c.g.*, we have

$$T = 2\pi\sqrt{\frac{mk^2\sqrt{l^2 - x^2}}{d_1 . d_2 . mg}} = 2\pi\frac{k}{\sqrt{d_1 d_2}} . \sqrt{\frac{\sqrt{l^2 - x^2}}{g}} .$$

Now, if y be the *vertical* distance between the two ends of each suspension thread, we have

$$y = B'R = LQ = \sqrt{l^2 - x^2} \quad \text{[Fig. 6.29 (b).}$$

So that, $$T = 2\pi\frac{k}{\sqrt{d_1 . d_2}}\sqrt{\frac{y}{g}} . \quad ...(viii)$$

Again, squaring and re-arranging relations (*vii*) and (*viii*), we have

from relation (*vii*), $$g = \frac{4\pi^2 . I . \sqrt{l^2 - x^2}}{d_1 . d_2 . m . T^2} . \quad ...(ix)$$

and, from relation (*viii*), $$g = \frac{4\pi^2 . k^2 . y}{d_1 . d_2 . T^2} . \quad ...(x)$$

The value of g at the place can thus be calculated from either of these relations.

It will be readily seen that if $d_1 = d_2 = d$, and $y = l$, so that $x = 0$, *i.e., when the two threads are parallel and vertical* in the original equilibrium position of the cylinder, we have

(*a*) relation (*vii*) reduced to $T = 2\pi \cdot \frac{1}{d} \sqrt{\frac{I.l}{mg}}$, the same as relation (*i*) for parallel threads, and

(b) relation (*viii*) reduced to $T = 2\pi \cdot \frac{k}{d} \sqrt{\frac{l}{g}}$, the same as relation (*ii*) for parallel threads, (see page 179).

Note. It will be clear from the above that the bifilar suspension may also be use to determine the value of I for the suspended cylinder etc. For,

relation (*iii*) above, re-arranged, gives $I = \frac{m.d^2.T^2.g}{4\pi^2 l}$,

and relation (*ix*), when re-arranged, gives $I = \frac{m.d_1.d_2T^2.g}{4\pi^2.\sqrt{l^2 - x^2}}$.

In fact, this method is more suitable for determining the moment of inertia of a body than for determining the value of g.

6.18. The value of g at a given place is affected by a number of factors, *viz.*, (*i*) *latitude* of the place; (*ii*) *altitude* and (*iii*) *depth*. We shall now proceed to study the effect due to each of these factors a little in detail.

(*i*) **Effect due to latitude.** The effect of *latitude* on the value of g may be considered under two headings, *viz.*, (*a*) the *effect of the rotation of the earth*, and (*b*) *the effect due to the bulge at the equator*.

Let us consider each separately.

(*a*) **Effect of rotation.** We know that the earth is rotating about its axis *from west to east*. If it were at rest, and were a homogeneous sphere, the acceleration due to gravity would be the same for a body at all points on its surface and would be directed towards its centre. Due to its rotation, however, part of the force of gravity on the body is used up in overcoming the centripetal force acting on it, and thus the resultant acceleration on it is different at different places both in magnitude and direction, *i.e., the apparent value of 'g' is different in different latitudes.* This will be clear from the following :

In Fig. 6.30, let *NWSE* represent the earth (supposed to be a perfect sphere) and let its radius be R. Then, if ω be its *angular velocity* about the axis of rotation *NS*, all points on its surface rotate about this axis with angular velocity ω. The linear velocity of each point will depend, however, on its distance from the axis. Thus, the linear velocity of a particle at points E, N, W and S will be $R\omega$, and that at a point P, distant $PM = x$ from the axis, will be $x\omega$, where x is the radius of the circle that the point P describes as it rotates with the earth.

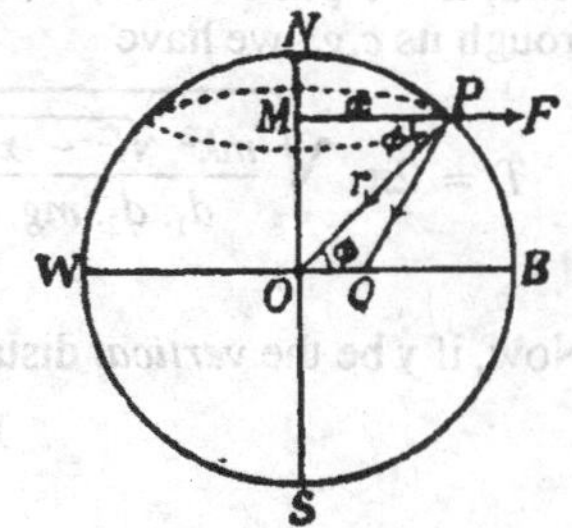

Fig. 6.30

Let ϕ be the latitude in which the point P is situated. Then, since the radius PM of the circle described by P is $x = R \cos \phi$, the linear velocity of P $= x\,\omega = R \cos \phi\, \omega$; so that, the centrifugal force acting on P, *away from the centre (M) of the circle it describes, and acting along MP, is clearly given by* $m R \cos \phi\, \omega^2$. Let it be represented in magnitude as well as direction by the straight line PF.

The force of gravity mg, which would act on the body if the earth were at rest, (g being the acceleration due to gravity, with the earth at rest), would obviously act towards the centre of the earth, O. Let it be represented in magnitude as well as direction by the straight line PO.

Thus, there are two forces acting simultaneously at the point *P*, *viz.*, (*a*) the *centrifugal force* $m R \cos \phi\, \omega^2$, *along PF*, and (*b*) the *force mg due to gravity along PO*. Completing the triangle of force *POQ*, where *PO* represents the gravitational force *mg*, and *OQ*, the centrifugal force $m R \cos \phi\, \omega^2$, we have the resultant force at *P* represented by the third side *PQ* of the triangle, both in magnitude and direction, where

$$PQ = \sqrt{PO^2 + OQ^2 - 2PO.OQ \cos POQ}$$
$$= \sqrt{m^2g^2 + m^2R^2 \cos^2\phi\omega^4 - 2mg\, mR \cos^2\phi\, \omega^2 \cos \phi}$$
$$= \sqrt{m^2g^2 + m^2R^2 \cos^2\phi\omega^4 - 2m^2g\, R \cos^2\phi\, \omega^2}.$$

Now, the value of $R\omega^2$ comes to be about 3.39 cm/sec^2, or about 1/288 of the value of *g*; for $R = 6.378 \times 10^8$ *cm* and ω, equal to 2π/86164 (where 86164 is the number of *mean solar seconds* in one day).

Thus, the expression $m^2R^2.\cos^2\phi\,.\,\omega^4$ is negligible, compared with the other terms involving *g*, and, therefore,

$$PQ = \sqrt{m^2g^2 - 2m^2gR \cos^2\phi\, \omega^2} = \sqrt{m^2(g^2 - 2gR \cos^2\phi\, \omega^2)}$$
$$= \sqrt{m^2g^2\left(1 - \frac{2R \cos^2\phi\omega^2}{g}\right)} = mg\left(1 - \frac{2R \cos^2\phi\, \omega^2}{g}\right)^{\frac{1}{2}}$$

Or, $$PQ = mg\left(1 - \frac{1}{2} \times \frac{2R \cos^2\phi\, \omega^2}{g} + \textit{some other negligible terms}\right)$$

Or, $$\textit{the resultant force on } P = mg\left(1 - \frac{R\,\omega^2 \cos^2\phi}{g}\right)$$

∴ If acceleration at the point *P*, in latitude φ, be *g*, we have

$$g = g\left(1 - \frac{R\omega^2 \cos^2\phi}{g}\right) = g - R\omega^2 \cos^2\phi \qquad \text{...(I). [}\because acc. = force/mass.$$

This value is obviously smaller than *g*, and is *directed towards Q*, and *not towards O*, the centre of the earth; the angle *OPQ* or the change in direction of the gravitational force is, however, very small.

For points on the equator, since $\phi = 0°$, and, therefore, $\cos \phi = 1$, the value of the centrifugal acceleration = $R\omega^2$, *i.e.* a *maximum*. And, *for points on the poles*, because $\phi = 90°$, and, therefore, $\cos \phi = 0$ the value of the centrifugal acceleration is *zero, i.e., a minimum.*

It follows therefore that *the apparent acceleration of a body is the least at the equator and the greatest at the poles with intermediate values in between.*

Substituting the value of $R\omega^2 / g$ = 3.372/981 = 1/291 in the expression for *g* above, we have

$$g_\phi = g\left(1 - \frac{\cos^2\phi}{291}\right).$$

This is a result, not quite in agreement with the experimental value. The discrepancy may, however, be ascribed *(i) to the ellipticity of the earth*, its radius increasing as we proceed from the poles towards the equator, so that points in the higher altitudes are nearer to its centre than those near the equator; *(ii) to the non-homogeneity of its composition*, the density of its different layers being different, *with the inner layers comparatively much denser than* (about more than twice as dense as) *the outer ones.*

(b) **Effect of bulge at the equator.** It was *Richer*, whose experiments in 1672 first showed a variation in the value of '*g*' at two different places.

Determining the length of a seconds pendulum at *Cayenne* (in French Guiana) and at *Paris* he found its length at *Paris* to be just over one-tenth of an inch greater than at *Cayenne*, clearly showing the value of g to be greater at *Paris*.

Newton soon explained this variation on the assumption that the earth behaved as though it were a '*uniformly gravitating fluid globe*' so that, by virtue of its very rotation, it was bound to have a spheroidal shape, with a bulge or a protuberance at the equator, and comparative flattening of the poles under the influence of the centrifugal force acting on it, the value of which varies from *zero* at the poles to a maximum at the equator. In fact, even if the earth were perfectly rigid, it should have assumed this shape before it actually cooled down. As a consequence, *the equatorial radius is about 13 miles greater than its polor radius.* Hence, all bodies in the equatorial regions are farther from the centre than those in the polar regions, and the force of attraction on the latter is, therefore, greater than that on the former.

It can be shown that the true value of g at a place in latitude λ is given by the relation

$$g = (980.61 - 025\cos 2\lambda)\ cm/sec^2.$$

These changes in the value of g due to latitude are of great help in determining the figure or the shape of the earth.

(ii) **Effect of altitude.** The correction for altitude we really owe to *Laplace and Stokes* — particularly to the latter.

Let g be the value of acceleration due to gravity on the surface of the earth and g', its value at a height h above the surface. Then, if the earth be considered to be a sphere of *homogeneous composition*, the acceleration due to gravity at any point above its surface will vary inversely as the square of the distance of that point from its centre; so that

$$\frac{g}{g'} = \frac{(R+h)^2}{r^2} = \frac{R^2+2rh+h^2}{R^2} = 1 + \frac{2h}{R} + \frac{h^2}{R^2}.$$ [where R is the *radius of the eath*.

If h be small, compared with r, the quantity h^2/r^2 will be negligible small, and we shall, therefore, have

$g/g' = 1 + 2h/R$. Or, $g'/g = 1/(1 + 2h/R) = 1 - 2h/R$ approx.

Or, $$g' = g\,(1 - 2h/R),$$

i.e., the greater the value of h, the smaller the value of g'.

Or, *the value of g decreases with altitude.*

The general expression for the acceleration due to gravity at altitude h and in latitude λ, thus becomes

$$g' = (1 - 2h/R)\,(980.61 - .025\cos 2\lambda)\ cm/sec^2.$$

(iii) **Effect of elevated masses.** The correction term $(1-2h/r)$ for altitude would only be valid when there is nothing but space between the surface of the earth and the point h above, *e.g.*, for an observer in an aeroplane at height h. But if we consider the point to lie on the top of a mountain of height h, a complication comes in due to the effect of the attraction by the mountain.

Bouguer suggested the correction $\left(1 - \dfrac{2h}{R} + \dfrac{3.h.\rho}{2R\Delta}\right)$, known as *Bouguer's Rule*, where Δ is the mean density of the earth and ρ, that of the mountain.

It is now found, however, that *Bouguer* somewhat over-estimated the effect of the mountain and his correction*, therefore, gives the upper limit, as it were, of its effect, the lower limit being that in which its attraction is neglected altogether. The *Board of Trade* have, therefore, adopted the following relation for the combined effect of latitude and altitude :

$$g' = (980.61 - .025\cos 2\lambda)\,(1 - 5h/4R)\ cm/sec^2.$$

* This correction by *Bouguer* was prompted by the same idea which inspired his Mountain experiment for the determination of the Gravitational Constant G. (See page 206), *viz.*, that the attraction on a mass due to the mountain could simply be added up to that due to the earth, (taken to be a sphere).

(iv) Effect of depth. Again, *imagining the earth to be a homogeneous sphere*, let us take a body of mass *m, inside the earth*, at a depth *h* below the surface, so that its distance from the centre of the earth is $(R-h)$, where *R* is the *radius of the earth*. Imagine a sphere of this radius $(R-h)$ to be drawn concentric with the earth. (Fig. 6.31)]

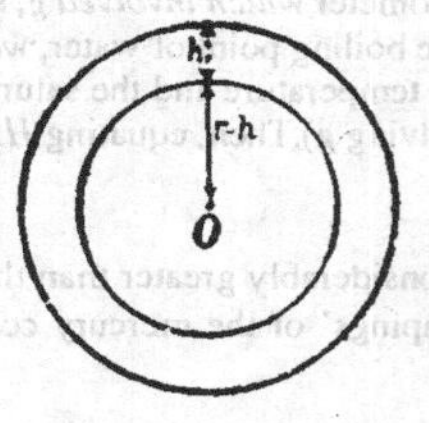

Fig. 6.31.

Then, clearly, the body lies on the surface of this inner sphere, and inside the outer hollow spherical shell, of thickness *h*. Let *g* and *g'* be the accelerations due to gravity at the surface of the earth and at a depth *h* below it, respectively. And since the force of attraction on a body inside a hollow shell is *zero*, the only force of attraction on the body is that due to the inner solid sphere, of radius $(R-h)$, and is directed towards its centre, the magnitude of the force being clearly given by

$$mg' = \frac{\textit{mass of the sphere} \times \textit{mass of the body}}{(R-h)^2} \times G,$$

where *G* is the *gravitational constant.*

Now, *mass of the sphere = its volume* $\times$ *its density* $= \frac{4}{3}\pi(R-h)^3\,\Delta$,

where Δ *is the density* of the earth, *supposed uniform*.

$\therefore$ force of attraction on the body at a depth *h* inside the surface of the earth is given by

$$mg' = \frac{4}{3}\,\frac{\pi(R-h)^3\,\Delta\, m}{(R-h)^2} \times G = \frac{4}{3}\pi\,(R-h)\,\Delta\, m\, G,$$

whence, $$g' = \frac{4}{3}\,\pi\,\Delta\, G\,(R-h) \qquad \ldots II$$

And, if the body were kept on the surface of the earth, the force of attraction towards the centre of the earth would be given by

$$mg' = \frac{4}{3}\,\frac{\pi R^3\,\Delta\, m}{R^2}\, G, \text{ whence, } g = \frac{4}{3}\pi\,\Delta,GR \qquad \ldots III$$

Dividing relation II by relation III, we have

$$\frac{g'}{g} = \frac{\frac{4}{3}\pi\,\Delta\, G\,(R-h)}{\frac{4}{3}\pi\,\Delta\, GR} = \frac{R-h}{R} = \left(1 - \frac{h}{R}\right).$$

Or, $$g' = g\left(1 - \frac{h}{R}\right), \qquad \ldots IV$$

i.e., the value of g also decreases with depth from the surface of the earth, the rate of decrease being half that with altitude. Further, it follows at once, from relation IV above, that at the centre of the earth, where $h = R$, the value of g will be zero; *i.e., the acceleration due to gravity and, therefore, the weight of a body, at the centre of the earth will be zero.*

(v) Effect of terrain — (Topographical correction). This correction consists in reducing the result at any given station to that we would obtain if the land or the terrain in which it is situated were just a horizontal plane, instead of its actual form.

Obviously, some parts of this terrain would be above and others below the horizontal plane, so that the former would exert an upward attraction, thus decreasing the value of *g* and the latter, a downward attractive force, thereby increasing the value of *g*.

It so turns out, however, that *this correction is always a positive one.*

6.19 Determination of the value of g at Sea. Until comparatively recently, the value of g at sea was determined indirectly, because it was not considered possible to use a pendulum on board a ship. The method, suggested by **Hecker and Duffield**, and usually adopted, was to determine the atmospheric pressure in two different ways, one of which involved g and the other did not; so that, by equating the two, the value of g could be easily calculated.

Thus, for example, the atmospheric pressure P could be obtained (*i*) from a barometer *which involved* g, because $P = H.\rho.g$. where H is the height of the mercury column and ρ its density, and (*ii*) from the boiling point of water, *which did not involve* g, because it could be calculated from the *Tables,* giving the relation between temperature and the saturated vapour pressure of water vapour, or directly from an *aneroid barometer,* (again, without involving g). Then, equating $H.\rho.g.$ against P, as obtained from method (*ii*), we have

$$g = P / H.\rho.$$

The result obtained by this method gives an error of about $\pm$ .01 cm/sec^2 which is considerably greater than that given by pendulum methods on land, the chief source of error being the oscillations or 'bumpings' of the mercury column in the barometer, caused by the movement of the vessel — the ship or the submarine.

Vening Meinesz has shown, however, that pendulums can be used for the purpose with far greater accuracy, particularly in a submerged submarine. His argument runs as follows :

A pendulum is subject to four types of disturbances on board a ship, *viz., (i) the point of suspension having a horizontal acceleration, (ii) vertical acceleration of the support, (iii) angular movement of the support or 'rocking' of the plane of oscillation, and (iv) slipping or sliding of the knife-edges on their agate planes.*

Of these, the first disturbance is the most marked, but it can be completely eliminated by simultaneously oscillating two identical *half-second pendulums**, suspended from the *same support*, oscillating in the *same vertical plane*, but with *different phases*, and noting their angular displacements θ_1 and θ_2 . Then, it can be easily shown that ($\theta_1 - \theta_2$) gives the angular displacement of a pendulum, altogether unaffected by this disturbance.

The vertical acceleration of the support can, however, not be eliminated without eliminating g itself, but the disturbance due to this can be greatly minimised by taking the mean of a large number of observations. For, the value of g seems to be affected only by the mean value of the vertical acceleration during the whole period of observation. And, since the vertical motion is alternately up and down the zero positon, the mean value of this acceleration becomes almost inappreciable.

The error due to *'rocking'* can be easily corrected for by noting different values of the rocking angle and computing the necessary correction, which is usually quite small.

And, with all these errors eliminated, or minimised to an extent that the total angular deviation due to them does not exceed 1°, the fourth error, *viz.*, the slipping of the knife-edges automatically eliminated.

What is done, in actual practice, is that *three* half-second pendulums are suspended from the same support and set oscillating, and continuous photographic records of (*i*) the difference ($\theta_1 - \theta_2$) , between the angular displacements of the first and the second, and (ii) the difference ($\theta_1 - \theta_2$) between those of the second and the third pendulum, are obtained on a semsotozed paper, by means of a suitable optical arrangement. The value of g is then calculated from each of these two set of observations and their mean taken.

The whole system is suspended in *gimbals*, to avoid external disturbances due to small angular movements of the ship or the submarine; and, further, to avoid any possible errors due to any slight change in temperature, the whole apparatus is kept properly thermally insulated and any small correction, still necessary, applied. And, finally, to make sure that no magnetic disturbances affect the result, the pendulums are made, not of *invar-steel***, (which would be so helpful in minimising any temperature corrections), but of brass.

The probable error in the value of g thus obtained is claimed to lie within $\pm$ .0018 cm/sec^2 —obviously, a marked improvement over *Duffield's* earlier indirect method.

A recent and comparatively much more accurate method consists in measuring the change in the frequency of transverse vibrations of a wire under tension, due to a weight suspended from it. For, whereas, any variations in the value of g produce next to no effect on the density of the wire (density being the ratio of mass to volume), they do naturally affect the pull of the earth on the suspended weight and hence the tension in the wire, resulting in corresponding change in its frequency of vibration. These can be easily detected to just a fraction of a vibration in a frequency of several thousand, by comparison with the vibrations of a quartz crystal oscillator by the method of beats. This explains the high accuracy of the method, which is obviously equally applicable to the measurement of the value of g on land, particularly at places where it is difficult or impracticable to use the usual method, as, for example, at the bottoms of boreholes etc.

6.20. Local and temporal changes in the value of g. The value of g at a point is also affected, to some slight extent, by local causes, like *small geological deposits near about the*

---

* *i.e.*, pendulums, whose full time-period is *one* second.

** It is an alloy of nickel and steel, whose coefficient of thermal expansion is *exceptionally low*, and which is, therefore, used in the construction of what are called *invariable pendulums*, *i.e.*, pendulums whose length remain practically unaffected by temperature variation,—the name *'invar'* for the alloy being suggested by the word *'invariable'*.

topography of the region or even by *masses like buildings etc. in the neighbourhood.* It is also affected by time, because deformations in the earth's surface take place periodically, thus bringing about a change in the equipotential surface, and hence in the direction of the force of gravity, which is always perpendicular to this surface. These changes are, however, much too small to be meassured by ordinary pendulum methods, whose accuracy is limited to within 10^{-4} *cm/sec*^{2}, or 10^{-1} *milligal*, (where 1 *cm/sec*^{2} = 1 *gal* = 1000*milligals**). More sensitive methods have, therefore, to be used for the purpose. A detailed study of these is beyond the scope of this volume, and we shall, therefore, deal with them only briefly here.

(i) **Local Changes.** Small changes in the value of g due to local causes are measured with the help of (*i*) what are called *invariable pendulums* and (*ii*) *gravity-meters* or *balances*. The former are suitable only for the measurement of place-to-place variations in the value of g in regions free of all marked local abnormalities, and the latter, for changes due to abnormal conditions like irregularities in the density of surface constituents and such other causes. For the most accurate determination of small variations in the value of g, however, a still more sensitive instrument, *viz.*, the *Eotvos gravity balance* must be used.

(1) **The invariable pendulums.** These pendulums are so called, because of their being standardised to such an extent that *their time-periods (t) vary solely due to variations in the value of g and to no other factor*.

They are usually rigid pendulums of *invar steel*, suspended from a massive tripod in a partially evacuated chamber, with a *specified air pressure* inside it, to make all air corrections constant. And the variations due to temperature, already small on account of the use of invar-steel (with its negligible coefficient of expansion), are further corrected for by a direct determination of the change in time-period with temperature.

The time-period of such a pendulum is first determined at a chosen *base station, i.e.*, at a place where the value of g is known and then at the field station, *i.e.*, at the place where it is to be determined. Then, clearly, the *gravity ratio*, or the ratio between the values of g at the two stations, will be given by the inverse ratio of the squares of its time periods there, since

$$t \propto 1/\sqrt{g} \quad \text{and} \quad \therefore \; g \propto 1/t^2$$

The only error possible after all this standardisation, is that in noting the time-periods of the pendulum at the two stations, or in the *'timing operation'*, as it may be called, and the utmost accuracy is attempted to be secured here by arranging to have precise time-signals broadcast at frequent intervals. In the ultimate analysis, however, the results obtained will be restricted to the same order of accuracy to which the time-period of the pendulum and the other constants involved have been determined at the base station.

The use of the time-signals at the field station may be obviated by the technique used by **Bullard** in his determination of the value of g in East Africa (in 1933), *viz.*, that of using *two* pendulums, one at the base station —in his case, Cambridge — and the other at the field station and recording an agreed *Morse signal*, alongside the oscillations of the pendulum, at each station, on a photographic film, repeating the same an hour or so later. The time-periods of the pendulums can then be compared with the *equal* time-intervals given by the Morse signals, and a high degree of accuracy thus attained in their measurement.

(2) **Gravity Meters.** Next in sensitivity come the *gravity meters*, various forms of which are now in commercial use as *prospecting instruments* and otherwise. We shall consider here only a few of them.

The Threlfall and Pollock Gravity Meter. Used first by *Threlfall* and *Pollock*, in 1899, it is perhaps the earliest *gravity meter* and consists of a fine *quartz thread AB*, (Fig. 6.32), stretched *horizontally*, with the end A fixed or *'anchored'* and the end B attached to an axle which can be rotated, in line with the thread (the latter being thus twisted) by means of a pointer (or a vernier) which moves over a circular scale S. A small metal rod R is fused athwart the thread near about its mid-point and is so *weighted* (by a bob or a weight w) that its *c.g.* lies on one side of the thread.

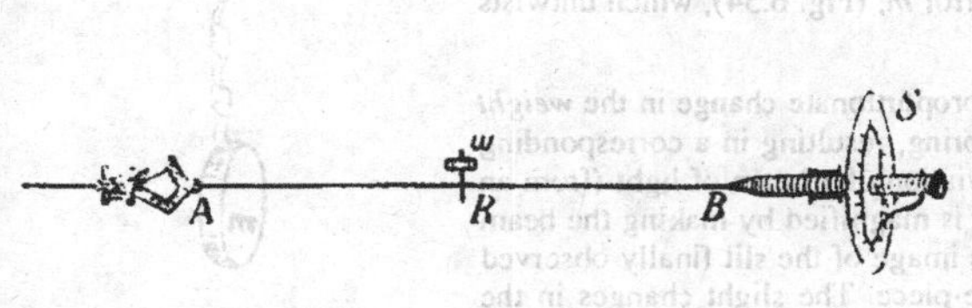

Fig. 6.32

* The *milligal* is a new unit, now increasingly being used to express small changes in the value of g, particularly in geophysics

The end B is twisted by means of the pointer, *until the rod becomes horizontal*—about three full turns of the thread being necessary for the purpose–in which position it is *just* stable under the balancing forces due to the tension of the thread and the gravitational pull on itself, *i.e., when the torsional couple due to the thread just balances that due to the pull of the earth.* The position of the pointer is now read off on the scale, — the slightest further movement of it making the rod lose its precarious balance and turn right over. This is safeguarded against by a suitable stop or arrester, but the very fact of this tending to occur enables its position of approaching instability to be readily determined*. Thus, with a change in the value of g, the rod will no longer remain horizontal and the end B of the thread will have to be twisted to restore it to that position. The angular twist thus given to the thread can be read on the scale from the position of the pointer and is a measure of the variation in the value of g, the pressure being kept constant and proper correction for temperature effects (*i.e.*, for expansion and change in the rigidity of the thread etc.) being made.

The instrument is made direct-reading by first noting the positions of the pointer at two stations where the value g is accurately known, its variation with temperature being determined at one of them. So that, if now the instrument be carried from place to place, the various positions of the pointer indicate the values of g on the scale straightaway.

With proper precautions taken, this simple appliance can yield results of fairly high degree of accuracy.

(*ii*) **The Boliden Gravity Meter**. A later form of gravity-meter, shown in Fig. 6.33, is due to *Boliden* (1938), in which two pieces of spring S, S, support a mass M which ends in two flat plates D and E *above and below, each forming one plate of the parallel plate condensers AD* and BE, whose other plates A and B are properly insulted from the framework of the instrument by means of insulating slabs F and G. The condenser AD above forms part of an oscilatory (or LC) circuit, whose frequency (N) is compared with a standard oscillator.

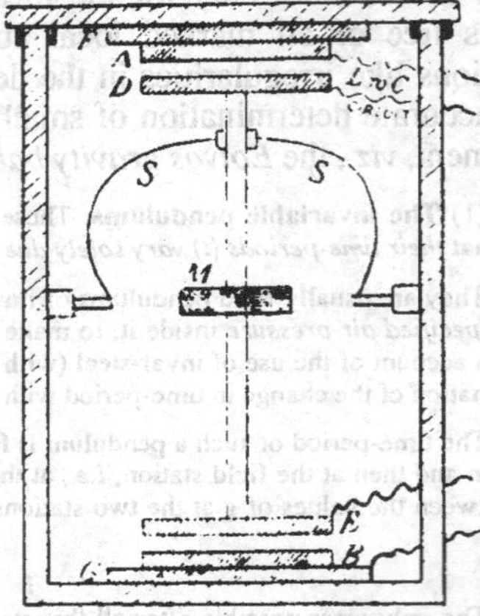

Fig. 6.33

A change δg in the value of g brings about a change in the flexure of the springs and hence a proportionate change δx in the air gap (d) between the condenser plates A and D. This, in its turn, results in a change δC in the capacitance (C) of the condenser, such that $\delta C/C = \delta x/d$. And, finally, the change in the capacity of the condenser is then responsible for a corresponding change δN in the frequency (N) of the oscilatory circuit**.

So that, $\delta g \propto \delta x \propto \delta N$

whence, δg can be easily calculated.

The instrument is calibrated by applying known potential difference to the plates B and E of the lower condenser, calculating the force of attraction between them —(and hence on D) —and the attendant frequency change of the oscillatory circuit, and plotting a graph between the latter two.

The sensitivity of the instrument is rather low, being only about 1×10^{-3} *cm/sec*^{2}, or just 1 *milligal*.

(*iii*) **The Gulf Gravity Meter**. This is a more recent (1941) and sensitive type of gravity meter and depends upon the same principle as a spring balance, *viz.*, that the weight of a body is proportional to the acceleration due to gravity, so that a mass suspended from a spring will exert a different pull on it for different values of g, *the stretch of the spring thus indicating the varations in g*.

The method fails in the case of the ordinary spring balance purely for want of requisite sensitiveness. In the case of the present instrument, however, this sensitiveness is well assured, as much by the choice of a suitable type of spring as by the accuracy of the means of observation.

We use here a *flat, metallic ribbon-spiral spring*, fastened to a torsion head at the top and carrying a load at its free end below, including a mirror m, (Fig. 6.34), which untwists the spring by about 8 full revolutions.

Any change in the value of g will bring about a proportionate change in the *weight* of the suspended mass and the consequent pull on the spring, resulting in a corresponding rotation of the mirror, which can be measured by the deviation of a beam of light (from an illuminated slit) reflected from it. The angle of deviation is magnified by making the beam travel four times between m and a fixed reflector and the image of the slit finally observed by means of a microscope, fitted with a micrometer eye-piece. The slight changes in the value of g corresponding to these deviations can thus be easily determined.

Fig. 6.34

* For, with the approach of the position of instability, the net couple acting on the thread varies only slowly with the change in its inclination and hence the time-period of the torsional vibrations of the thread about its equilibrium position goes on increasing.

** Because the frequency of an oscillatory circuit depends upon the *capacitance (C)* and the *inductance (L)* included in it.

The sensitivity of the instrument is found to be about 5×10^{-5} *cm/sec*^{2} or 5×10^{-2} *milligals*.

(iv) **Eotvos balance.** None of the above appliances possess the necessary sensitivity to be able to measure the small changes in the value of g due to neighbouring buildings or small geological deposits etc. Instruments far more responsive to small variations in the value of g must be employed for these delicate measurements and the gravity balance, devised by **Baron Eotvos**, admirably anwers the requirement. It is not only used for a comparative or an absolute determination of g, but also for the measurement of other important quantities connected with the earth's gravitational field and for purposes of gravitational survey, the accuracy claimed for the instrument being 10^{-9} *cm/sec*^{2}, or 10^{-6} *milligals*.

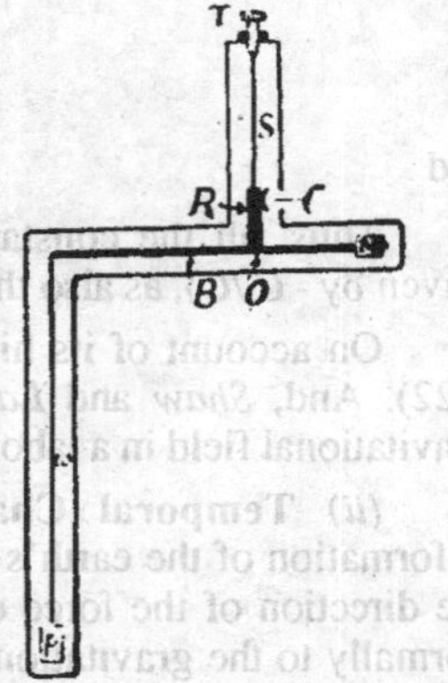

Fig. 6.35

In essentials, the **Eotvos balance** consists of a *rectangular torsion beam B*, (Fig. 6.35), of aluminium, about 40 *cm* in length and between 3×10^{-4} to 4×10^{-4} *mm* in diameter, suspended from a torsion head T, by means of a fine suspension S, about 60 *cm* long, of the alloy *platinum-iridium*, through an aluminium rod R, fixed on to the beam at its *c.g.*, O. The rod carries a small concave mirror C, to enable the deflections of the beam to be read by the lamp and the scale method, with the help of a telescope. A small cylindrical weight P, of platinum, gold or silver, of mass about 30 *gm*, is suspended from one end of the beam, by means of a fine wire (w) of platinum, and a counterpoise weight M, of mass about 25 *gm*, is slid on to, or suspended from, the other end of the beam, as shown.

If the instrument be taken to a place where the value of g varies from point to point, its suspended system experiences a *couple*, producing a twist in the wire and deflecting the beam from the position (not known) that it would occupy if the value of g were constant.

Let the beam, *in its equilibrium position*, make an angle θ with the x-axis, (lying along the north-south direction) —*i.e.*, let θ be the '*azimuth angle*', as it is called —and let S_1 be the reading on the scale in this position. Then, if S_0 be the scale reading in the (unknown) position in which there would be no gravitational torque on the beam, it can be shown that

$$S_1 - S_0 = A\left[\frac{\partial^2 U}{\partial y^2} - \frac{\partial^2 U}{\partial x^2}\right]\sin 2\theta + 2A\left[\frac{\partial^2 U}{\partial x . \partial y}\right] . \cos 2\theta$$

$$-C\left[\frac{\partial^2 U}{\partial z . \partial x}\right]\sin\theta + C\left[\frac{\partial^2 U}{\partial y . \partial z}\right]\cos\theta, \qquad ...(1)$$

where A and C are the *constants of the instrument*, U, *the gravitational potential*, and $U/\partial x$, $U/\partial y$ and $U/\partial z$ the values of the gravitational attraction along the *North*, the *East* and the *vertical* directions respectively, (*this last one being the value of g*). The origin of the three axes, *along these three directions*, is taken to be the mid-point O of the beam.

Putting $$A\left[\frac{\partial^2 U}{\partial y^2} - \frac{\partial^2 U}{\partial x^2}\right] = A', \quad 2A\left[\frac{\partial^2 U}{\partial x . \partial y}\right] = B',$$

$$-C\left[\frac{\partial^2 U}{\partial z . \partial x}\right] = C' \text{ and } C\left[\frac{\partial^2 U}{\partial y . \partial z}\right] = D',$$

we have, from relation (1) above,

$$S_1 - S_0 = A'\sin 2\theta + B'\cos 2\theta + C'\sin\theta + D'\cos\theta . \qquad ...(2)$$

Now, taking $\theta = 0°, 60°, 120°, 180°, 240°$ and $300°$, in equation (2) above, and taking the corresponding values of S_1 to be $S_1, S_2, S_3, S_4, S_5, S_6$, we have

$$S_1 + S_3 + S_5 = S_2 + S_4 + S_6 = 3S_0,$$

$$2\sqrt{3}\,A' = (2S_2 - S_3) - (S_1 - S_4),$$
$$2\,B' = (S_1 + S_4 - 2S_0),$$
$$2\sqrt{3}\,C' = 2(S_2 - S_5) - (S_1 - S_4).$$
and
$$2\,D' = S_1 - S_4.$$

Thus, all the constants of relation (1) being known, the rate of change of g northwards, (given by $-C'/C$), as also that in the eastern direction, (given by D'/C), can be easily determined.

On account of its high sensitivity, the balance is used for geophysical prospecting (see § 6.22). And, *Shaw* and *Lancaster Jones* have successfully mapped out with its help the local gravitational field in a laboratory.

(ii) **Temporal Changes—The Horizontal pendulum.** As we have seen above, a deformation of the earth's surface and its gravitational equipotential surface results in a change in the direction of the force of gravity and hence in that of g. Since a plumb line always sets itself normally to the gravitational equipotential surface of the earth, it is clear that measuring a change in the direction of g at a point is tantamount to measuring a change in the direction of the plumb line at that point. These changes, however, are much too small —seldom exceeding 1" — to permit of their accurate measurement by means of a plumb line. The most commonly used device to measure these is what is called the **horizontal pendulum**, devised by **Hengler**, in the year 1832.

This *horizontal pendulum* essentially consists of a rod AB, (Fig. 6.36), carrying a knob or bob at B, with G as its *c.g.* It is supported in an inclined position, by means of two pieces of a light string, AP and CQ, attached to a *rigid* support at P and C respectively, such that the straight line CP, joining the two, meets AB in O, and makes an angle ϕ with the direction of the force of gravity. *The pendulum thus takes up a position in a plane parallel to the force of gravity.* On giving the bob a slight lateral displacement (towards or away from the observer), it begins to oscillate slowly with a small amplitude, along an arc with O as its centre and OG as radius. Its period of vibration is deduced as follows :

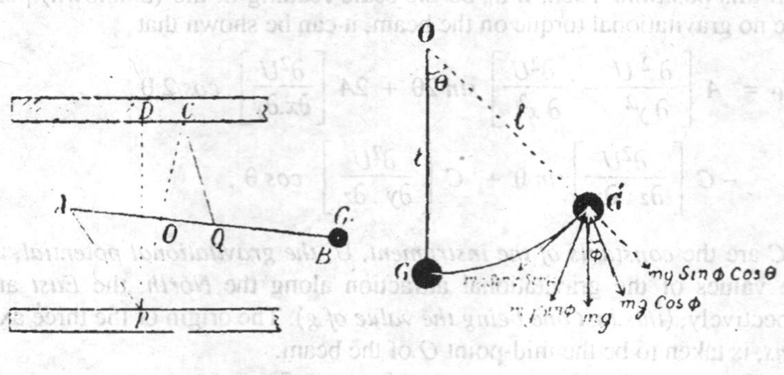

Fig.6.36 Fig. 6.37

Let the bob oscillate along an arc GG', (Fig. 6.37), which lies in a plane, making an angle ϕ with the normal to the plane of the force of gravity.

If the bob be displaced through an angle θ *in this inclined plane* into the position shown, its weight mg acts at *its c.g.*, G' ($\because$ G is now at G') *in the direction of the force of gravity.* Resolving it into its two rectangular components, *(i) in the inclined plane of its rotation* and *(ii) perpendicular to it*, we have the former component = $mg \sin\phi$, and the latter = $mg\cos\phi$.

Further, resolving the component $mg \sin \phi$ into two rectangular components, along and perpendicular to OG', we have

the component along $OG' = mg \sin \phi \cos \theta$

and *the component perpendicular to* $OG' = mg \sin \phi \sin \theta$.

This latter component ($mg \sin \phi \sin \theta$) has, clearly, a restoring, moment about the inclined axis $= mg \sin \phi \sin \theta . l$, (where $OG' = OG = l$), tending to bring the pendulum back to its original position.

If θ be small, $\sin \theta = \theta$, very nearly. And, therefore,

the restoring moment $= mg \sin \phi . \theta . l$

And, if the *angular acceleration* of the pendulum be $d\omega/dt$, and its *moment of intertia about* O be I, we have

restoring moment (or torque) also $= I . d\omega/dt$.

So that neglecting frictional and viscous forces, we have, for equilibrium,

$$I . d\omega/dt = mg.l. \sin \phi . \theta, \quad \text{whence,} \; d\omega/dt = \frac{mg.l . \sin \phi}{I} . \theta .$$

or, putting $mg.l. \sin \phi / I = \mu$, a constant, we have

$$d\omega/dt = \mu . \theta \qquad \text{Or,} \qquad d\omega/dt \propto \theta,$$

i.e, the *angular acceleration of the pendulum is proportional to its angular displacement.*

It is, therefore, a case of *simple harmonic motion*, and its time-period T is given by the relation, $T = 2\pi \sqrt{1/\mu}$.

$$\text{Or,} \qquad T = 2\pi \sqrt{1 \Big/ \frac{mg.l.\sin \phi}{I}} = 2\pi \sqrt{\frac{I}{mg.l. \sin \phi}}$$

But $I = mk^2$, where k is the *radius of gyration* of the pendulum about O.

$$\text{Hence} \qquad T = 2\pi \sqrt{\frac{mk^2}{mgl \sin \phi}} = 2\pi \sqrt{\frac{k^2}{g.l. \sin \phi}}$$

In practice, to make T large, ϕ is made as small as possible.

Now, if $\phi = 90°$, $\sin \phi = 1$, and, therefore, the time-period T', in this case, is given by the relation $T' = 2\pi \sqrt{k^2/g.l.}$

Hence $\qquad T/T' = \sqrt{1/\sin \phi} = (1/\sin \phi)^{1/2}$ And $\therefore \; T^2/T'^2 = 1/\sin \phi$,

whence, $\qquad \sin \phi = T'^2/T^2$.

Thus, knowing T' and T, we can easily calculate $\sin \phi$ and hence ϕ *which represents the change in the direction of the force of gravity, and, therefore, that of g, in the equilibrium plane of the pendulum.*

And, to determine the change in the direction of g in the perpendicular plane, two such horizontal pendulums are mounted in two perpendicular azimuths and to the values ϕ, say ϕ_1 and ϕ_2 obtained in them.

6.21 Gravity Survey. The purpose of a gravity survey is two-fold, *viz.*, (*i*) the main one being *to determine the value of the force of gravity and its direction at various points of the sea-level surface of the earth, or the 'geoid'*, as it is called, and (*ii*) a secondary one being to deduce form it *the possible distribution of matter in the earth*, and thus to form an idea about its structure and internal condition.

Now, in any gravity survey, it is found necessary to introduce two new quantities connected, with the earth's gravitational field. These are (1) the *gravity gradient*, denoted by the letter G and (2) the *horizontal directive tendency* (written as *H.D.T.*, for brevity), denoted by the letter R. Let us try to understand their meaning.

(1) The Gravity Gradient (G). It is a convenient abbreviation for *"maximum gradient of g in a horizontal direction, near a point"* where g is the *vertical gravitational intensity* at the point, *i.e.*, $g = \partial U/\partial z$. The gravity gradient may, therefore, be denoted by $\delta g/\delta s$, *the rate of variation of g per unit distance in the direction of the maximum rate of change in its value*, and it is thus obviously *a vector quantity*.

Now, if the gravity gradient, G, (*i.e.*, $\partial g/\partial z$) makes an angle with the axis of x (or the north-south direction), and if its components along the axes of x and y be $\partial g/\partial x$ and $\partial/\partial y$ respectively, then, clearly,

$$\partial g/\partial x = G \cos\phi \quad \text{and} \quad \partial g/\partial y = G \sin\phi$$

And, since $g = \partial U/\partial z$, we have $\partial g/\partial x = \partial^2 U/\partial x\,\partial z = U_{xz}$

and $\partial g/\partial y = \partial^2 U/\partial y\,\partial z = U_{yz}$

So that, $U_{xz} = G\cos\phi$ and $U_{yz} = G\sin\phi$

$\therefore$ squaring and adding the two, we have

$$G^2\cos^2\phi + G^2\sin^2\phi = U^2_{xz} + U^2_{yz} \quad \text{Or,} \quad G^2(\cos^2\phi + \sin^2\phi) = U^2_{xz} + U^2_{yz}.$$

Or, $$G^2 = U^2 xz + U^2 yz, \quad \text{whence,} \quad G = (U^2 xz + U^2 yz)^{1/2}.$$

Now, U_{xz} and U_{yz} can be esily determined by means of a Eotvos *Balance*, (see page 189), and thus the value of G can be calculated from the above expression.

(2) The Horizontal Directive Tendency (R). It is what is called a *'curvature vector'*, *i.e.*, a directed quantity, though not a true vector. Its value at a point is given by the relation,

$$R = g(1/r_1 - 1/r_2),$$

where r_1 and r_2 stand for the maximum and the minimum radii of curvature of the level surface, or the gravitational equipotential surface at the point. Its direction, according to an agreed convention, is taken to be the directon *in which the level surface has the least downward curvature and, therefore, the maximum radius of curvature.*

If the direction of *H.D.T.* makes an angle θ with the axis of x, or the north-south direction, it can be shown that

$$R\sin 2\theta = 2U_{xy}$$
$$R\cos 2\theta = U^2_{xx} - U^2 yy.$$

where, $U_{xy} = \partial^2 U/\partial x\partial y$, $U_{xx} = \partial^2 U/\partial x^2$ and $U_{yy} = \partial^2 U/\partial y^2$.

The dimensions of both G and R are $[T]^{-2}$ and they are generally expressed in what are called *Eotvos units* — where one *Eotvos unit* is equal to $10^{-9}/sec^2$.

In survey maps, the *gravity gradient* at a point is represented, in magnitude and direction, by an arrow head drawn from the point, whereas the *horizontal directive tendency* is just represented, in magnitude and direction, by a straight line passing through that point, without any arrowhead or feathered tail, as shown in Fig. 6.38, where O is the point in question.

Further, points where the value of g is the same are joined by curves, which are called **isogams** —G being *always directed along the normals to these.*

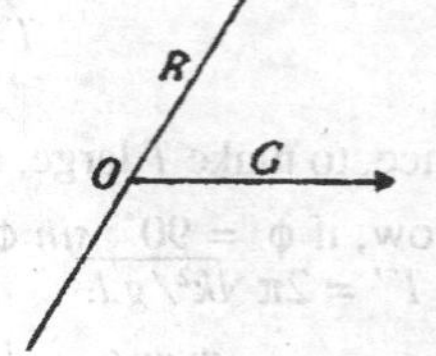

Fig. 6.38

6.22. Geophysical Prospecting. We have seen above how, due to the presence of local geological deposits (*i.e.*, minerals etc.) inside the earth, small variations are produced in the value of g. Similar changes are produced, by their presence, in the normal values of the other quantities, magnetic, electrical, seismic* etc., associated with it; so that, by measuring these variations with the help of specially designed instruments, we can detect their presence. This is technically called '*geophysical prospecting*'. We are concerned here only with the gravitational methods adopted for the purpose, the principle underlying which is to measure the *gravity gradient* G, and the *horizontal directive tendency* R at various points in the region under survey with the help of a sensitive instrument like an *Eotvos gravity balance*, as explained above in § 6.20 *(iv)* —the instrument, when so used, being called a '**gradiometer**'.

---

* See aritcle on *'Earthquakes'* in the next chapter.

Thus, if θ be the angle that R makes with the axis of x, or the *north-south direction* at a point, we have

$R \sin 2\phi = 2U_{xy}$, and $R \cos 2\theta = U^2_{xx} - U^2_{yy}$, | § 6.21, (2), above,

And, therefore, $\tan 2\theta = 2U_{xy} / (U^2_{xx} - U^2_{yy})$,

the two solutions of which, θ_1 and θ_2, differ by $2/\pi$ and give the directions of the two principal axes of curvature of the equipotential surface (*i.e.*, the level surface) at the point. The value of G and R at various points are then plotted, *the direction of* θ_2 *being such that secant* 2θ *and* $(U^2_{xx} - U^2_{yy})$ *are of opposite signs.*

A graphical representation of the variations of g over the region, under examination, is thus obtained, and closed curves or *isogams*, [see § 6.21, (2), above], are then clearly marked out on it, (which, as we know, are at every place perpendicular to G), so that we have an **isogam chart** of the region in question.

Interestingly enough, the physical form of these *isogams* almost faithfully represents the physical form of the subterranean deposits. Thus, for example a *uniformly monoclinic* type of region would give *isogams,* which are all *parallel and equally spaced*, whereas if the subterranean deposits form a dome-like structure, the isogams obtained also resemble the outline or the contour of a surface dome, as it were.

This method can however succeed only in the hands of those well-trained in the use of the delicate instruments employed and in the proper interpretation of the results obtained from them.

WORKED EXAMPLES

1. A metal disc oscillates in its own plane about an axis passing through a point on its edge. What is the length of the equivalent simple pendulum ?

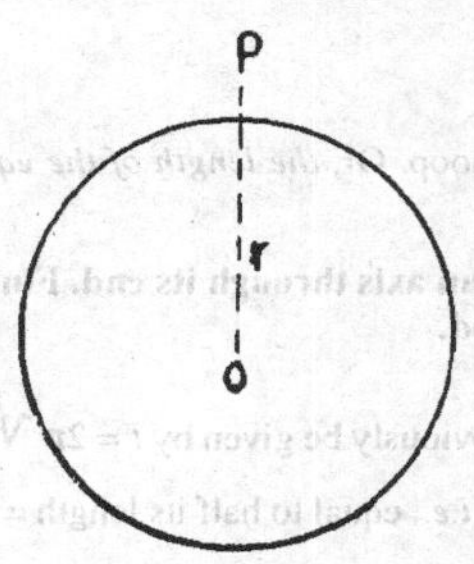

Fig. 6.39

Let the disc of radius r oscillate about an axis through the point P on its edge, (Fig. 6.39)

Then, clearly, the time-period of the disc is given by the relation, $t = 2\pi\sqrt{I/Mg.l}$,

where I is its $M.I.$ about the axis through P; M, its *mass* and l, its *length.*

Or, $t = 2\pi\sqrt{I/Mg.r}$, for $l = r$ the distance between the point of suspension (P) and the $c.g.$ (O) of the disc.

Now, $I = I_0 + Mr^2$

where I_0 is the $M.I.$ of the disc about a parallel axis through O, *i.e.*, $I_0 = Mr^2/2$.

So that, $I = (Mr^2/2) + Mr^2 = 3\,Mr^2/2$.

And, therefore, $$t = 2\pi\sqrt{\frac{3Mr^2/2}{mg.r}} = 2\pi\sqrt{\frac{3r/2}{g}}$$

i.e., the same as that of a simple pendulum of length $l = 3r/2$.

Or, the length of the equivalent simple pendulum is 3/2 times the radius of the disc.

2. (*a*) A circular disc of radius 20 cm oscillates as a pendulum about a point on its circumference. Calculate the period of oscillation. ***(Bihar)***

(*b*) A disc of metal of radius R with its plane vertical can be made to swing about a horizontal axis passing through any one of a series of holes bored along a diameter. Show that the minimum period of oscillation is given by $T = 2\pi\sqrt{1.414\,R/g}$. ***(Sagar)***

(*a*) As we have seen above under worked example 1, the time-period of the disc about an axis passing through a point on its edge is given by $T = 2\pi\sqrt{\frac{3r/2}{g}}$.

Here, r being 20 *cm*, we have $T = 2\pi\sqrt{\frac{3\times 20/2}{g}} = 2\pi\sqrt{30/g}$

Or, taking $g = 980\ cm/sec^2$, $T = 2\pi\sqrt{30/980} = 1.099\ sec.$

(*b*) We know that a compound pendulum has the *minimum time*-period when its lenght (l) is equal to its radius of gyration (*k*) about a perpendicular axis through its c.g., when we have $T = \sqrt{2K/g}$ (see page 153). And since K^2 *for a disc* = $R^2/2$, we have $K = R/\sqrt{2}$

$\therefore$ *Minimum time-period of the disc*, $T = 2\pi\sqrt{\dfrac{2R/\sqrt{2}}{g}} = 2\pi\sqrt{\dfrac{\sqrt{2}\,R}{g}}\ 2\pi\sqrt{\dfrac{1.414\,R}{g}}$

3. Find the period of small oscillations of a rigid body, free to turn about a fixed horizontal axis, and also find a formula for the length of the equivalent simple pendulum.

Three particles of the same mass *m* are fixed to a uniform circular hoop of mass *M* and radius *a* at the corners of an equilateral triangle. The hoop is free to move in a vertical plane about the point on the circumference opposite to one of the masses *m*. Prove that the equivalent simple penduum is equal in length to the diameter of the circle. *(London Higher School Certificate; Patna)*

For first part, see § 6.4 page 148)

Let the three equal masses, *m* and *m*, be fixed to the hoop, of *radius* , as shown (Fig. 6.40), so as to lie at the corners of an equilateral triangle. Since they are all *equidistant from the centre, the c.g. of the triangle is at O, the centre of the circle.* The whole arrangement is thus equivalent to a hoop of mass $(M + 3m)$, with its centre of gravity at its centre *O*.

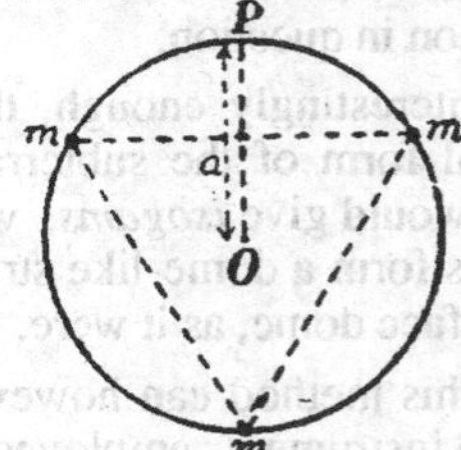

Fig. 6.40

Clearly, then, the moment of inertia of this loaded hoop about *O*, *i.e.*, about an axis through *O* and perpendicular to its plane) = $(M + 3m)a^2$. And, therefore, its moment of inertia about a parallel axis through the point of suspension *P* is *(by the principle of parallel axes)* given by

$I = (M + 3m)a^2 + (M + 3m)a^2 = 2(M + 3m)a^2.$

Now, the *time-period* of the hoop about *P* is given by

And, therefore, $t = 2\pi\sqrt{\dfrac{I}{(M + 3m)\,ga}} = 2\pi\sqrt{\dfrac{2(M + 3m)a^2}{(M + 3m)ga}}$ $\quad \because\ l = a$

$t + 2\pi\sqrt{\dfrac{2a}{g}}$,

..e., the same as that of a simple pendulum of length $2a$, the diameter of the hoop. Or, *the length of the equivalent sinple pendulum is equal to the diameter of the circle.*

4. A thin uniform bar of length 120 cm is made to oscillate about an axis through its end. Find the period of oscillation and other points about which it can oscillate with the same period. *(Raipur)*

The time-period of the bar, oscillating as a compound pendulum, will obviously be given by $t = 2\pi\sqrt{\dfrac{K^2/l + l}{g}}$

where *l* here is *distance between the point of suspension and the c.g. of the bar, i.e.*, equal to half its length = 120/2 = 60 cm. And, K^2 = length^2/12 (see & 3.6, page 50) = 120 × 120 / 12 = 1200

$\therefore\ t = 2\pi\sqrt{\dfrac{1200/60 + 60}{g}} = 2\pi\sqrt{\dfrac{80}{980}} = \dfrac{2\pi \times 2}{7} = \dfrac{4\pi}{7} = 1.796\ sec.$

Obviously, *length of equivalent simple pendulum* = $(k^2/l) + l = (1200/60) + 60 = 80\ cm$.

$\therefore$ the bar will also oscillate with the same time-period about a point *O*, 80 *cm* from the upper end or its point of suspension *S*, (Fig. 6.41) *i.e., O is the point of oscillation corresponding to S.*

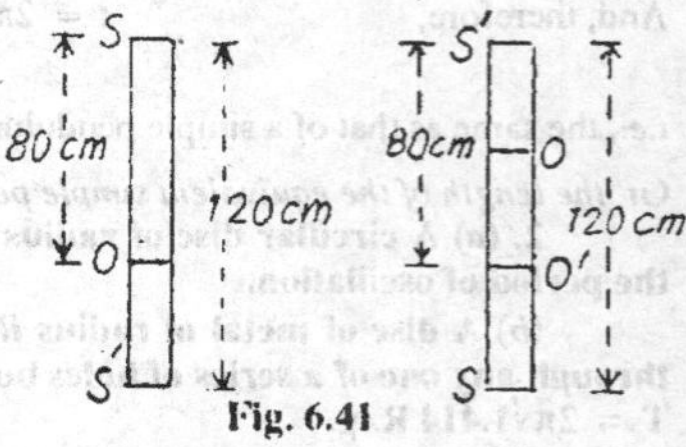

Fig. 6.41

Similarly, on account of being uniform and symmetrical, it will oscillate with the same time-period when suspended from the other end *S'*, *i.e.*, on reversing it, as also about a point *O' 80 cm from S', i.e. O' is the point of oscillation corresponding to S'*. Thus there are two points *S, O', O* and S', distant 0 *cm*, 20 *cm*, 80 *cm* and 120 *cm* from the end *S* about which the bar will oscillate with the same time-period 1.796 sec.

5. How much faster than its present rate should the earth revolve about its axis in order that the weight of a body on the equator may be zero, and how long would it take to make one revolution then ? What would happen if (*i*) the rotation became faster still, (*ii*) the rotations were stopped altogether ? ($g = 978\ cm/sec^2$).

We have seen, (page 182), that the value of *g* is different in different latitudes due to the rotation of the earth, and that, assuming the earth to be a perfect sphere,

$$g_\phi = g\left(1 - \frac{R\cos^2\phi.\omega^2}{g}\right)$$ where g_ϕ is the value of 'g' in latitude ϕ, (see page 183)

Now, at the equator, $\phi = 0$ and $\therefore$ $\cos^2\phi = 1$; so that,

$$g_0 = g\left(1 - \frac{R.\omega^2}{g}\right),$$

where g_0 is the value of 'g' at the equator.

The value of $r.\omega^2$ comes out to be 3.39 (see worked problem 6), and, therefore, we have

$$r\omega^2/g = \frac{3.372}{981} = \frac{1}{291}.$$

Thus, in order that the weight of a body may be *zero*, the value of g_0 should be *zero, i.e.,* $R.\omega^2/g$ should be equal to 1, or the value of $R\omega^2$ should be 291 times greater than its present value, R being a constant. It follows, therefore, that ω should be $\sqrt{291}$ times, *i.e.*, 17.06 times greater than its present value. When this is so, *the outward centrifugal force on the body will, obviosly, be just balanced by the inward force due to gravity.*

If, therefore, the earth rotates 17.06 times, or 17 times, faster than its normal rate, the weight of the body at the equator will be zero.

Now, the earth makes one complete revolution in 24 hours, *i.e.*, describes an angle of 2π in 24 hours. But, in the case considered, *viz.*, when the weight of the body at the equator is *zero*, it rotates 17 times faster, and will, therefore describe an angle $17 \times 2\pi$ in 24 hours, or an angle 2π in 24/17 hours or 1.412 hours.

$\therefore$ *the earth will then make one rotation in* 1.42 *hours.*

If the rotation became *faster still, i.e.*, faster than 16.97 or 17 times its normal rate, obviously, all objects kept loose on the equator will start leaving the surface of the earth; for, the increased centrifugal acceleration on them will be greater than that due to gravity and, therefore, *a resultant force will be acting on them outwards, away from the centre of the earth.*

If, on the other hand, the rotation of the earth about its axis were stopped altogether, we shall have $\omega = 0$; so that, substituting this value of ω in the relation, $g_0 = g\,(1 - R\omega^2/g)$ for the value of 'g' at the equator, we have

$$g_0 = g\,(1-0) = g,$$

i.e., the value of g increases by $(g - g_0) = r^2.\,\omega^2/g$ times $g = (3.372/981)$ times $g = 1/288$ times $g = g/291$.

Thus, *if the motion of the earth were stopped altogther, the value of g would increase by 1/291 of its normal value.*

6. Assuming that the whole variation of the weight of a body with its position on the earth's surface is due to the rotation of the earth, find the difference in the weight of a gram as measured at the equator and at the poles. (Radius of the earth = 6.378×10^8 *cm*).

We have the relation, $g_\phi = g\left(1 - \dfrac{R\omega^2.\cos^2\phi}{g}\right)$, for the value of '$g$' in latitude ϕ.

Since $$R\omega^2 = 6.378 \times 10^8 \times \left(\frac{2\pi}{86164}\right)^2 = 3.39 \quad \text{and} \quad \therefore \quad R\omega^2/g = \frac{3.372}{981}\ \ 1/291.$$

we have $$g_\phi = g\,(1 - \cos^2\phi/291)$$

Now, at the equator, $\phi = 0$. so that, $\cos o$ and $\therefore$ $\cos^2 o = 1$

Hence $$g_0 = g\,(1 - 1/291)$$

where g_0 is the value of 'g' at the equator.

And, at the poles, $\phi = 90°$, so that, $\cos\phi$ and $\therefore$ $\cos^2\phi = 0$.

Hence $$g_p = g\,(1 - 0) = g.$$

where g_p is the value of 'g' at the poles.

Since the weight of the body is mg, where m is its mass, we have

weight of 1 *gm at the equator* $= w = 1 \times g_0$

$$= 1 \times g(1 - 1/291) = g(1 - 1/291). \quad \ldots(i)$$

And *its weight at the poles* $= w' = 1 \times g_p = 1 \times g = g$. $\quad \ldots(ii)$

Hence, *the differece in the weights of the mass at the poles and the equator*

$$= w' - w = g - g\,(1 - 1/291) = g/291 = 981/291 = 3.49 \textit{ dynes}.$$

Or, the difference in *weights* of a *gram* at the poles and at the equator is 3.49 *dynes*.

7. The mass of a railway train is 100 tons. What will be its weight when *(a)* stationary, *(b)* travelling due east, *(c)* travelling due west, along the equator at 60 miles per hour ? Radius of the earth is 4000 miles. *(Punjab)*

(a) **When the train is stationary.** When the train is at rest, its *apparent weight* is 100 *tons wt* = 100 × 2240 = 224 × 10^3 *lb wt*, (because 1 *ton* = 2240 *lb*).

(b) **When the train is moving east.** When the train travels due east, its angular velocity about the axis of rotation of the earth *increases*, because the earth itself is rotating about the axis from west to east. The centrifugal force on the train, therefore, increases, (being proportional to $r.\omega^2$), and hence the apparent force of gravity on it and, therefore, the apparent acceleration towards the centre of the earth, *i.e., apparent acceleration due to gravity decreases*. And, since *weight* = *mass* × *acceleration due to gravity*, the apparent weight of the train decreases. Let us see by how much.

The radius of the earth, R = 4000 *miles* = 4000 × 1760 × 3 *ft*, and therefore, linear velocity (v) of a point on the earth = $2\pi R/24 \times 60 \times 60$ *ft/sec*, since the earth makes one complete rotation in 24 *hours*.

A point on the equator will also thus describe a distance $2\pi R$ in 24 *hours*; its velocity is, therefore, given by

$$v = \frac{2\pi \times 4000 \times 1760 \times 3}{24 \times 60 \times 60} = 1536\ ft/sec.$$

∴ *centrifugal acceleration* of the point, when at rest relative to the earth, is given by

$$\frac{v^2}{R} = \frac{1536 \times 1536}{4000 \times 1760 \times 3} = 0.1116\ ft/sec^2$$

$$\text{Since velocity of the train} = 60\ mi/hr = \frac{60 \times 1760 \times 3}{60 \times 60} = 88\ ft/sec.$$

the resultant velocity of the train, say, v' = 1536 + 88 = 1624 *ft/sec*.

And ∴ centrifugal acceleration on the train moving at 60 *mi/hr* = v'^2/R

$$= \frac{1624 \times 1624}{4000 \times 1760 \times 3} = 0.1248\ ft/sec^2.$$

∴ increase *in centrifugal acceleration, or decrease in acceleration towards the centre of the earth, i.e., in the acceleration due to gravity* = 0.1248 − 0.1116 = 0.00132 *ft/sec*^{2}.

∴ decrease in the *weight* of the train = mass × decrease in acceleration due to gravity.

= 100 × 2240 × 0.0132 *poundals* = 100 × 2240 × 0.0132/32 *lb wt*.

= 100 × 2240 × 0.0132/32 × 2240 = 100 × 0.0132/32 *ton wt* = 0.0412 *ton wt*.

∴ *apparent weight of the train* = 100 − 0.0412 = 99.9588 *ton wt*.

(c) **When the train is moving west.** In this case, since the train is moving from east to west, *opposite to the direction of rotation of the earth*, its anguar velocity about the axis of rotation of the earth *decreases* and, therefore, the centrifugal acceleration on it also decreases, with the result that the acceleration towards the centre of the earth, *i.e., the acceleration due to gravity, increases*. The apparent weight of the train on the equator, therefore, increases. Let us calculate this apparent increase.

As before, velocity of a point on the equator, *i.e*,

$$v = \frac{2\pi R}{24 \times 60 \times 60} = \frac{2\pi \times 4000 \times 1760 \times 3}{24 \times 60 \times 60} = 1536\ ft/sec.$$ | See above

∴ centrifugal acceleration of the point, when at rest relative to the earth

$$= \frac{5v^2}{R} = \frac{1536 \times 1536}{4000 \times 1769 \times 3} = 0.1116\ ft/sec^2.$$

Hence, the resultant velocity of the train, say v'' = 1536 — 88 = 1448 *ft/sec*.

And ∴ centrifugal acceleration of the train moving at 60 *mi/hr* is clearly given by

$$\frac{v''^2}{R} = \frac{1448 \times 1448}{2000 \times 1760 \times 3}\ 0.0993\ ft/sec^2$$

∴ *decrease in centrifugal acceleration, or increase in acceleration towards the centre of the earth, i.e., increase in acceleration due to gravity* = 0.1116 — 0.0993 = 0.0193 *ft/sec*^{2}.

And ∴ increase in apparent weight of the train

= 100 × 2240 × 0.0123 *poundals* = 100 × 2240 × 0.0123/32 *lb wt*

= 100 × 2240 × 0.0123/32 × 2240 = 0.03884 *ton wt*.

Thus, the apparent weight of the train − 100 + 0.03884 = 100.03884 = 100.04 *tons weight*.

8. A body of mass *m* is raised to a height *h* from a point on the surface of the earth, where the acceleration due to gravity is *g*. Prove that the loss of weight due to variation of *g* is approximately 2 *mgh/R*, where *R* is the radius of the earth. *(Agra)*

As we know, the value of the acceleration due to gravity at a height *h* from the surface of the earth is given by $g' = g(1-2h/r)$ approximately, where *g* is its value on the surface of the earth [see §6.18 (*b*) (*ii*)]. So that, $g-g' = 2gh/R$. And, therefore, *decrease in the weight of the body* $= mg - mg' = m(g-g') = 2\,mgh/R$.

9. A helical spring, fixed at its upper end, stretches by 1 mm for every 0.981 newtons of force applied axially to it. If a mass of 10 kg be suspended from its lower free end, pulled down a little and released, calculate the time-period of the ensuring oscillation of the mass.

Clearly, *force applied to the lower end of the spring* $= 10 \times 9.81 = 98.1$ *newtons.*

$\therefore$ *extension produced in the spring* $= \dfrac{98.1}{0.981} = 100\ mm = 0.1 m.$

Now, when the mass is pulled down and released, it executes a *S.H.M.*, whose time-period is given by

$$T = 2\pi\sqrt{\frac{extension\ produces}{g}}$$ [see § 6.17 (5) (page...

So that *time-period of the oscillating mass* $= 2\pi\sqrt{\dfrac{0.1}{9.81}} = 0.6345\ sec.$

EXERCISE VI

1. Derive from first principles the equation for the period of oscillation of a simple pendulum. In what way is a compound pendulum better than a simple pendulum for determining *g* at a place.

Derive an equation relating the two constants *G* and *g*. *(Delhi, B.E.I.)*

2. What is a compound pendulum ? Find an expression for its periodic time and show that the centre of suspension and the centre of oscillation can be interchanged. How do you measure the acceleration due to gravity with a compound pendulum ? *(Punjab)*

3. Deduce the formula for the time of vibration of a compound pendulum and show that this is a minimum when the length of the compound pendulum equals its radius of gyration about a horizontal axis through the centre of gravity of the compound pendulum. *(Punjab)*

4. Distinguish between a simple and a compound pendulum.

For a given compound pendulum, show that the centres of oscillation and suspension are interchangeable.

How is the value of '*g*' determined with the help of a compound pendulum ? *(Agra)*

5. Give the theory of Kater's pendulum and find an expression for the acceleration due to gravity in terms of two nearly equal periods of oscillation about the two parallel knife-edges.

Indicate the sources of error in an experimental determination of *g*. *(Bombay; Punjab)*

6. A Borda's pendulum has a bob of radius 12 *cm* which is suspended by a fine wire, 94 *cm* long. Calculate the length of the equivalent simple pendulum. **Ans.** 100.144 *cm*.

7. State the various corrections used to get the correct periodic time of a compound pendulum. Discuss in detail the correction for finite arc. *(Gujarat)*

8. Explain the Dropping Plate method for the determination of the value of *g*. If there be an error of 1% in measuring the distance covered by the plate as also in measuring the frequency of the fork, how would it affect the result ?

Ans. *The result will be wrong by* 3%.

9. The length between the knife-edges of a Kater's pendulum is 0.8928 *m*, while the times of oscillation about the two edges are 1.920 *sec* and 1.933 *sec*, respectively. The *c.g.* of the pendulum is about 54.4 *cm* from one edge. What is the value of *g* ? **Ans.** 9.79 m/sec^2

10. An Atwood's machine has a pulley of radius *a* and moment of inertia *I*; the masses attached to the ends of the string are each *M* and the rider is of mass *m*. Prove that the acceleration *f* of the masses is given by

$$\frac{g}{f} = 1 + \frac{2M}{m} + \frac{1}{ma^2}$$

assuming that the string does not slip on the pulley, and neglecting axle friction. *(Madras)*

11. What are the points of superiority of a compound pendulum over a simple pendulum ? Describe the method for determining *g* accurately, using a compound pendulum. Prove the necessary formula. *(Delhi B.Sc. (Hons))*

12. A solid cylinder, of radius 4 *cm* and mass 250 *gm*, rolls down an inclined plane, with a slope of 1 in 10. Find the acceleration and the total energy of the cylinder after 5 *sec*. *(Bombay)*

Ans. 65.4 cm/sec^2; 4.799 *joules*.

13. A cylinder, of mass 100 *lb* and diameter 12 *inches*, rolls from rest down a smooth inclined plane of 1 in 8 and 20 *feet* long. Calculate the total kinetic energy and its energy due to rotation, when it reaches the bottom. *(Madras)*

Ans. (*i*) 8.0×10^2 *ft poundals*. (*ii*) 2.6×10^2 *ft poundals*.

14. Define 'centre of suspension' and 'centre of oscillation'. Show that in a compound pendulum they are interchangeable.

What is the distance between the centre of suspension and the centre of oscillation on a uniform cylindrical metal bar used as seconds pendulum?

(Diameter of the bar = 1 *cm*, its density, 8 *gm/cc* and g = 978 *cm/sec*^{2}) *(Allahabad)*

Ans. 99.19 *cm*.

15. Obtain an expression for the time-period of a compound pendulum, and show that

(*i*) there are four points, collinear with its *c.g.*, about which its time-period is the same;

(*ii*) its time-period remains unaffected by the fixing of a small additional mass to it at its centre of suspension.

16. (*a*) Discuss the theory of Kater's pendulum. What is the effect of finite arc of swing on the period of the pendulum ? *[Poona, (Physics–Principal)]*

(*b*) The length between the knife-edges of a Kater's pendulum is 89.28 *cm*, while the periods of oscillation about the two knife-edges are 1.920 *sec* and 1.933 *sec* respectively. The centre of gravity of the pendulum is 54.4 *cm* from one knife-edge. What is the value of g ? *(Vikram, 1965)*

Ans. 979 *cm/sec*^{2}.

17. Derive an expression for the period of oscillation of a circular disc, supported on a horizontal rail passing through a narrow hole, which is bored through the disc half-way between the centre and the periphery. *(Bombay)*

Ans. $T = 2\pi\sqrt{3r/2g}$.

18. A uniform rectangular sheet of metal is supported by frictionless hinges, attached to one edge which is horizontal. Determine the period of oscillation of the sheet if l denotes the length of the side of the rectangle which hangs downwards *(Patna)*

Ans. $T = 2\pi\sqrt{13l/12g}$

19. A metal bar is suspended in turn from two parallel axes on the same side of its *c.g.*, and its time-periods are found to be 1.42 *sec* in each case. If the distance of the two axes be 10.8 *cm* and 39.2 *cm* respectively from the *c.g.*, calculate the value of g and the radius of gyration of the bar about a parallel axis through the *c.g.*

Ans. g = 979.2 *cm/sec*^{2}, and K = 20.58 *cm*

20. A narrow uniform bar of mass 1000 *gm* oscillates about an axis, 40 cm from the centre, with a period of 7.48 *sec* and about a parallel axis, 10 *cm* from the centre, with a period of 1.67 *sec*. Find the value of g, the moment of inertia of the bar about its *c.g.* and the length of the bar. **Ans.** 990 *cm/sec*^{2}; 6.02×10^5 *gm*, *cm*^{2}; 85 *cm*.

21. What is meant by a simple equivalent pendulum ? If the periods of a Kater's pendulum in the erect and inverted positions are equal, prove that the distance between the knife-edges is equal to the length of the simple equivalent pendulum.

A uniform circular rod, with a radius of 2 *cm* oscillates when suspended from a point on its axis at a distance of 4 *cm* from one end. If the length of rod is one *metre*, find the point or points from which, if suspended, the periodic time would remain unaltered. **Ans.** At 31.87 *cm* and 68.13 *cm* also at 96.0 *cm from the same end*.

22. Obtain an expression for the time-period of a conical pendulum. If a mass m is suspended from the end of a string of legth l and a conical pendulum is formed, show that the angular velocity ω of the mass is related to the breaking strength T of the string by the relation $\omega = T/ml$. *(Punjab)*

23. Describe a conical pendulum and derive an expression for its frequency. Explain how it is used to regulate the speed of steam engines. Show that the sensitiveness of the pendulum used as a governor increases with diminishing speed. *(Bombay)*

Ans. $n = 1/2\pi\sqrt{h/g}$

where h is axial height of the cone described by it, and equal to $l\cos\theta$, where l is the length of the pendulum and θ, its angular displacement, (see § 6.15)

24. (*a*) What is a steam engine governor ? Explain clearly the principal underlying its action, and discuss its limitations.

(*b*) Would it be correct to say that even if the earth were a perfect sphere, the experimental value of g would be the least at the equator. If so, why ?

25. (*a*) If the earth were to cease rotating about its axis, what will be the change in the value of g at a place of latitude 45°, assuming the earth to be a sphere of radius 6.38×10^8 *cm* ?

(*b*) How far away from the earth does acceleration due to gravity become 1 per cent of its value at the earth's surface" Assume that the earth is a sphere of radius 6.38×10^8 *cm*. *(Punjab)*

Ans. (*a*) 1.8895 *cm/sec*^{2}; 57.42×10^8 *cm*.

26. Explain how a suitable compound pendulum is a more suitable apparatus for determining of g than a simple pendulum. Discuss the correction due to viscous drag and buoyancy of air in a reversible pendulum. *(Delhi B.Sc. (Hons)*

27. Give the theory of the compound pendulum and show that the centres of suspension and oscillation are reversible.

In a reversible pendulum, the periods about the two knife-edges are t and $(t + T)$, where T is a small quantity. The knife-edges are distant l and l' from the centre of gravity of the pendulum. Prove that

$$l + l' = \frac{gt}{4\pi^2}\left(t + \frac{2l'}{l'-l}\,T\right)$$

(Madras)

28. A heavy uniform rod, 0.3 *m* long, oscillates in a vertical plane, about a horizontal axis passing through one end. When a concentrated mass is fixed on to it at a distance *x* from its point of suspension, its time-period remains unaffected. Calculate the value of *x*. **Ans. 0.2 *m*.**

29. Explain how the length of the simple pendulum which has the same period as a given compound pendulum may be found experimentally.

A unifrom cube is free to turn about one edge which is horizontal. Find in terms of a seconds pendulum, the length of the edge, so that it may execute a complete oscillation in 2 *sec*. *(Central Welsh Board Higher School Certificate)*

Ans. $3\sqrt{2}\,l$.

30. A body of mass 200 *gm* oscillates about a horizontal axis at a distance of 20 *cm* from its centre of gravity. If the length of the equivalent simple pendulum be 35 *cm*, find its moment of inertia about the axis of suspension. *(Patna)*

Ans. 14×10^6 *gm–cm*^{2}.

31. A pendulum, whose period slightly exceeds 2 *sec*, is compared with a standard seconds pendulum by the metnhod of coincidence. Successive coincidences occurred at times 0 *min*, 58 *sec*, 5 *min* 48 *sec*, 8 *min* 48 *sec*. Find the exact period of the pendulum. *(Patna)*

32. A thin rod is suspended by means of two threads parallel to each other and tied to its two ends. Compare the time-period of the rod when it oscillates thus in its own plane with that when it oscillates as a compound pendulum about a horizontal axis, passing through one of its ends. **Ans. 1 : 1.414.**

33. Give the theory of the compound pendulum and show that the centres of suspension and oscillation are interchangeable.

A uniform thin rod *AB*, of mass 0.1kg and length 1.2*m*, can swing in a vertical plane about *A*, as a pendulum, A particle of mass 0.2 kg, is a attached to the rod at a distance x from A. Find x such that the period of vibration is a minimum. **Ans. 0.02748 *m*.**

34. Find the effect of *yielding* of support and currature of kinfe-edges on the period of oscillation of a compound pendulum and explain how these are eliminated in the reversible pendulum *(Delhi, B. Com.)*

35. Discuss the influence of rotation of earth, the depth and the altitude on the acceleration due to gravity.

Obtain a general relation for it. Does it agree with observed values ? If not, explain why ?

36. Obtain an expression for the time-period of oscillation of a compound pendulum. *(Punjab)*

A heavy uniform rod of length l swings in a vertical plane about a horizontal knife-edge passing through one end. Find at what point a concentrated mass can be placed, so that the time of swing may remain unaltered. *(Delhi)* **Ans. $2l/3$**

37. Give the theory with experimental details for the determination of acceleration due to gravity by means of katev's pendulum. Discuss the sources of error. *(Delhi)*

38. Derive expression for (*a*) the period of a compound pendulum and (*b*) the length of the equivalent simple pendulum.

A uniform circular disc of radius 25 *cm* oscillates in a vertical plane about a horizontal axis. Find the distance of the axis of rotation from the centre for which the period is minimum. What is the value of this period ?

(Poona (Subsidiary))|

Ans. $25/\sqrt{2}$ *cm*; $t = 2\pi\sqrt{1.414 \times 25/g}$ *sec*.

39. Give the theory of the Eotvos gravity balance. How has it been used for determining the variation of gravity in any horizontal direction over a small region of the surface of the earth ? [*(Delhi, B.Sc. (Hon.)*]

40. Derive an expression for the period of oscillation of a compound pendulum. Show that the points of oscillation and suspension are interchangeable. Discuss the variation of g with altitude. *(Delhi, 1967)*

41. Describe, giving theoretical details how you would determine the variation of gravity over a small region of the surface of the earth. What is the practical importance of such studies ? *(Delhi*, B.Sc. (Hons.)

42. Write short notes on :

(*a*) Gravity metres and balances, (*b*) Gravity survey and (*c*) Geophysical prospecting.

CHAPTER VII

GRAVITATION

7.1. Historical. The celestial bodies have been an object of interest to scientists all through the ages, and the first astronomical observations, of which we have any definite knowledge, were perhaps made by the Chinese, as far back as 2,000 *B.C.*, though the Babylonian astronomers are credited with having mapped out the constellations even earlier, near about 2700 *B.C.* The first authoritative treatise on the subject, however, was due to **Ptolemy**, working in Alexandria about 100 *A.D.*, who formulated his theory on the basis of the catalogue showing the nightly positions of planets and some 1000 stars, prepared earlier by the Greek astronomer *Hipparchus*. Ptolemy's book, the *Almagest*, enjoyed the authority of the Bible and reigned supreme for 1400 years. According to him, the whole of the heavens, carrying the stars, revolved round the earth, supposed stationary. The forward and retrograde motion of the planets* among the stars was explained by postulating that the planets revolved in circles, with their centres revolving in larger circles round the earth, —the former circles being termed *epicycles* and the latter ones, *deferents*. And, it stands to his credit that, with a suitable choice of radii and velocities, he could explain quite accurately the observed facts of the day.

The Ptolemaic theory was first challenged in 1543, by the famous Polish monk, **Nicolaus Copernicus,** in his book *'Concerning the Revolutions of the Heavenly bodies'***, his geometrical solution being much neatrer than that of Ptolemy, involving only *thiry-four* epicycles as against the *eighty* of the latter. In it he propounded his **heliocentric theory**+, according to which the planets moved in perfect circles round the Sun, which was supposed to be fixed. The theory was, however, received with reserve and scepticism, being objected to on the ground that (*i*) the rotation of the earth should result in bodies being hurled from its surface, and (*ii*) with greater justification, that no parallax (or relative motion) could be noticed between stars as was always observed between objects at different distances from the moving ship. This parallax has since been shown to *actually* exist, and was first measured by **Bessel,** in 1938. It is, however, extremely small, on account of the enormous distances of the stars from the earth.

As it so well known, **Galileo** was compelled to recant his belief in it, a century later, and legend has it that blind and helpless, he was thrown back into prison for murmuring *'And still it moves'*, (*E pur si muove*), until he died nine years later, and that **Giordano Bruno** was actually mercilessly burnt at the stake for refusing to do so.

Then, appeared on the scene, twenty five years later, in 1569, **Tycho Brahe**, an imperious nobleman and a brilliant astronomer++, who rejected the *Copernician theory* and made careful observations of the motions of heavenly bodies, on every clear night for thirty long years, particularly of the motion of the planet *Mars*, from his observatory in Denmark, with his celebrated wooden quadrant, (about 10 *ft* in radius), carrying a brass scale. In view of the fact that the

* From the Greek word, meaning *'wanderer'*, — because a planet moves forwards and backwards or 'wanders' about among the stars.

**He hesitated and deferred publishing book until he was dying, dedicating it to the Pope, who, not taking it seriously, expressed himself pleased with it. And Martin Luther was positively contemptuous towards it. 'Dud not Joshua, (in the Bible) command the Sun to stand still and not the Earth ? he asked.

+ This was really a revival of the theory, first propunded by the Greek Astronomer *Aristarchus*, that the earth was not the centre of the universe but revolved round the Sun, as also did the other planets.

++ He was reputed to be 'an unsurpassed practical astronomer' and made his own instruments for his well-equipped laboratory at *Uraniborg*, built for him by Frederick II, King of Denmark.He had, however, a violent temper and lost part of his nose in a duel, while still young, going about for the rest of his life with this lost part replaced by an artificial one of an alloy of silver. On the death of Frederick he had to flee and seek asylum at Prague, under the patronage of *Rudolph II*, Emperor of Bohemia. It was here that Kepler joined him as his assistant.

telescope was yet to come some forty years later, we cannot but marvel at the unprecedented accuracy of his observations. No wonder, they were used by navigators for centuries together, much in the manner of the *Nautical Almanak* today. With all his great methematical and experimental skill and his 'infinite capacity for taking pains', however, Tycho Brahe could not somehow piece his results together into a proper theory. But later **Kepler***, his assistant at the Royal Observatory at Prague—an impecunious but a gifted mathematician—into whose hands passed all his data on the subject, carried on the work and, accepting the Copernician theory**, which his chief had rejected, worked on the latter's figures and finally succeeded, after twenty-two years of ceaseless work, in evolving the famous three laws, known after him—the first two in the year 1609 and the third, ten years later, in 1619+.

7.2. **Kepler's Laws.** The following are the three laws, formulated by *Kepler*.

1. *The path of a planet is an elliptical orbit, with the Sun at one of its foci.* This is fittingly known as the **Law of elliptical orbits.**

2. *The radius vector, drawn from the Sun to a planet, sweeps out equal areas in equal time, i.e., its areal velocity (or, the area swept out by it per unit time) is a constant. This is referred to as the* **Law of areas.**

3. *The square of a planet's year (i.e., its time-period, or its time of revolution round the Sun) is proportional to the cube of the major semi-axis of its orbit.* This is known as the **Harmonic Law.**

Unfortunately, Kepler was not aware of the property of inertia, and so could not proceed any further. For him, it was necessary to suppose a power acting continuously on a body, in order to make it move. Most of the fellows of the Royal Society++, which included among others, men like *Robert Boyle, Edmund Hailey* and *Samuel Papys*, were convinced, by the year 1685, that a planet could move in an elliptical orbit, only if it were attracted by the Sun with a force varying inversely as the square of its distance from the Sun, but they could not prove it mathematically. **Newton**, who was also a member, was at this time *Lucesian Professor of Mathematics* at Cambridge and seldom attended the meetings of the Society, mostly held at London and Oxford. *Edmund Hailey*, therefore, went all the way to Cambridge to ask him if he could furnish the required proof and was simply astonished to learn that he had already done so years earlier, but had somehow lost his papers.

Realising that no other member of the Royal Society could hope to provide the required proof and also that Newton had really already achieved something much more than this, Hailey pleaded with him to reproduce his papers in book form and, though not a rich man himself, offered to bear all the cost of publication of the same, which ultimately resulted in the appearance of the celebrated *Principia* in the year 1687.

Newton knew that both rest and uniform motion along a straight line were equally natural and after a careful study of Kepler's laws, he showed (*i*) that it follows from his second law that only a *central force* acts on the planet and is directed towards the Sun, it alone being responsible for keeping the planet in its orbital path+++, (*ii*) that it can be deduced from his first and third laws

---

* He actually succeeded *Tycho Brahe*, who died after a little over one year of his migration to Prague, under the impressive designation of 'Imperial Mathematician' at a high salary which was, however, seldom paid. 'He had, therefore, to supplement his income by practising astrology, 'the foolish and disreputable daughter of astronomy, without which the wise old mother would starve'. He was also the founder of Geometrical Optics.

** And for this he had to migrate to a Protestant country to save himself from persecution..

\+ He was so filled with ecstasy at his success in enunciatiang his third law that he declared 'I will indulge in my sacred fury; I will triumph over mankind by the honest confession that I have stolen the golden vases of the Egyptians to build up a tabernnacle for my God'.

++ The Royal Society, the fellowship of which today is considered to be a very high honour, really grew out of informal group meetings of men interested in natural philosophy, about the year 1645, and received its Royal Charter, in 1662, from Charles II, who, according to Samuel Papys, 'mightily laughed at them for spending time only in weighing air'.

+++ See '*Newton's deductions from Kepler's laws*' § 7.4 page 203.

that this force between the planet and the Sun is inversely proportional to the square of the distance between them*, and (*iii*) that it is an easy further deduction from the above that this force of attraction between the two is also directly proportional to the product of their masses.

He further proceeded to verify these deductions from Kepler's laws of comparing the value of acceleration of the Moon towards the Earth, calculated on their basis, with its value obtained experimentally, the two values showing a close agreement with each other, as will be seen from the following :

If g_m be the acceleration of the Moon towards the Earth ; v, its *linear* velocity in its orbit about the Earth and R, the distance between the centres of the Earth and the Moon, we have

$$g_m = v^2/R = (\omega R)^2/R = \omega^2 R,$$

where ω = is the *angular velocity* of the Moon.

Since $\omega = 2\pi/T$, where T is the time taken by the Moon in going *once* round the Sun, we have $g_m = (2\pi/T)^2 \times R$.

Now, $T = 27.3$ *days* $= 27.3 \times 24 \times 60 \times 60$ *sec.*, and $R = 60$ *times the radius of the Earth.*

$= 60 \times 4000$ *miles* $= 60 \times 4000 \times 1760 \times 3.$ *ft.* [radius of the Earth is equal to 4000 *miles*]

Hence, $$g_m = \frac{4\pi^2 \times 60 \times 4000 \times 1760 \times 3}{(27.3 \times 24 \times 3600)^2} = .00899\ ft/sec^2$$

Again, if the acceleration due to gravity be g on the surface of the Earth, its value at the distance of the Moon from it would, in accordance with deduction (*ii*) above, be equal to $g/60^2$,

i.e., $$g_m = g/60^2.$$

So that, taking the value of g to be $32.2\ ft/sec^2$., on the surface of the Earth, we have

$$g_m = 32.2/60^2 = .00894\ ft/sec^2,$$

which is practically the same as the one deduced above, thus fully vindicating the deductions made by Newton, and convincing him of the existence of a *universal* and *mutual* force of attraction between any two masses.

Not only this, but Newton also put to test his assumption that, in so far as the attraction at external points is concerned, both the Earth and the Moon behave *as though their masses were concentrated at their respective centres.* He actually showed that the force of attraction exerted at an external point, by a uniform sphere, or by a sphere consisting of a number of concentric uniform shells, one inside the other, is the same as that exerted by an *equal point-mass*, occupying the same position as its centre. In other words, *the sphere behaves as though the whole of its mass were concentrated at its centre.*

Thus fortified with a clear and complete confirmation of his deductions and assumptions, Newton announced to the world, in the year 1687, his celebrated *Law of Gravitation* in his monumental work, the *Principia***, which the entire scientific world later hailed, in the words of *Lagrange*, as the *'greatest production of the human mind'*.

7.3. Newton's Law of Gravitation. This law, triumphantly announced by *Newton* in the year 1687, as mentioned earlier, states that *every particle of matter in the universe attracts every other particle with a force which is directly proportional to the product of their masses, and inversely proportional to the square of the distance between them.*

---

* 'Newton's deductions from Kepler's laws' (§ 7.4 Page 203)

** Having lost his papers, as mentioned already, Newton had to recreate the whole, step by step, all over again and accomplished the almost superhuman feat in only 18 months. He used geometrical methods, partly due to his admiration for ancient geometers and partly to avoid being baited by 'little smatterers in mathematics'. Perhaps he had *Robert Hooke* in mind, who had claimed priority in the dicovery of the Inverse Square Law; the Royal Society had, however, sided with Newton.

Thus, if m and m' be the masses of two particles, distance r apart, and F, the force of attraction between them, we have

$$F \propto m.m'/r^2. \quad \text{Or,} \quad F = G.m.m'/r^2,$$

where G is a *universal constant*, called the **Universal Gravitational Constant**, or, simply, the *Gravitational Constant*.

Obviously, if $m = m' = 1$ *gm*, and $r = 1$ *cm, then,* $F = G$.

Thus, *the* **Gravitational Constant** *is equal to the force of attraction between two unit masses of matter, unit distance apart*. Its dimensions are $M^{-1}L^3T^{-2}$ and its latest, accurate value, (as determined by *Heyl* in 1930), is taken to be 6.669×10^{-8} *dyne. cm²/gm²* in *C.G.S. units* or 6.669×10^{-11} *N.m²/kg²* in *M.K.S.*(or *S.I*) *units*.

The gravitational constant is also sometimes referred to as the **astronomical unit of force.**

The law is *universal* in the sense that it holds good, right from huge interplanetary distances to the smallest terrestrial ones. The minimum distance up to which it is valid is probably not yet known with absolute certainty, but it seems to break down at molecular distances, which are as small as 10^{-7} *cm*. We shall discuss later, in this chapter, some of the overwhelming evidence in favour of this law as well as the *small deviations from it* and the proper explanation for them, on the basis of the new ideas put forth by **Einstein.**

7.4. Newton's deductions from Kepler's laws.

1. Let A be the position of a planet, (Fig. 7.1) at a given instant t in its elliptical path round the sun S *supposed to be fixed in its position** and situated at one of its foci. Then, if the planet moves on to B in a small interval of time dt, the area swept out by the radius vector SA, in this interval of time, is equal to the area of the triangle SAB,

$$\text{i.e., equal to } \tfrac{1}{2}\, SA\,.\,AB = \tfrac{1}{2}\, R\,R\,.d\,\theta = \tfrac{1}{2}\, R^2.d\,\theta$$

because $SA = R$ and $AB = R.d\theta$

$\therefore$ *areal velocity of the planet* $= \frac{1}{2} R^2.d\,\theta/dt$,

But this, according to Kepler's second law, must be a constant. Putting it equal to $h/2$, therefore, we have $R^2.d\theta/dt = h$

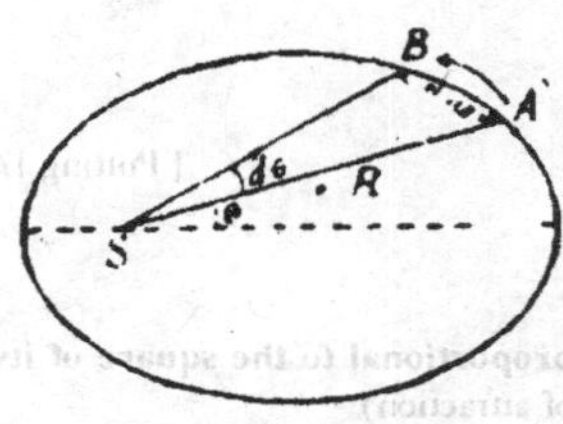

Fig. 7.1

Now, the fact that the planet moves in a curved path and thus continually changes its direction means, in accordance with Newton's first law of motion, that it must be under the action of a force and must consequently be possessing an acceleration in the direction of the force. Resolving this acceleration into its two rectangular components, along and at right angles to the radius vector, we have

(*i*) component a_1, *along the radius vector, i.e.,* the *radial acceleration of the planet*, given by

$$a_1 = \frac{d^2R}{dt^2} - R\left(\frac{d\theta}{dt}\right)^2$$

[a result obtained from simple dynamical consideration.

and (*ii*) component a_2. *at right angles to the radius vector, i.e.,* the *transverse acceleration of the planet*, given by

$$a_2 = \frac{1}{R}\cdot\frac{d}{dt}\left(R^2.\frac{d\theta}{dt}\right).$$

But, we have seen above that $R^2.\theta/dt$ is a constant (h), and hence its differential coefficient must be equal to zero. It follows, therefore, that $a_2 = 0$. In other words, the planet has no transverse acceleration; so that, the only acceleration it has is the *radial one*. Obviously, therefore, *the only force acting on* **the planet** *is towards the Sun.*

2. Now, since $R^2.d\theta/dt = h$ it is clear that $d\theta/dt = h/R^2$ Or, putting $1/R = u$, (or $R = 1/u$), we have $d\theta/dt = hu^2$

* In point of fact, the sun and the planet both describe their own ellipses about their common centre of mass as one of the foci. On account, however, of the sun having an enormously greater mass than the planet, the ellipse described by it almost negligibly small in size compared with that described by the planet. So that we can almost imagine the sun to remain stationary in its position with respect to the planet.

It follows, therefore, that

$$\frac{dR}{dt} = \frac{d}{dt}\left(\frac{1}{u}\right) = -\frac{1}{u^2}\cdot\frac{du}{dt} = -\frac{1}{u^2}\cdot\frac{du}{d\theta}\cdot\frac{d\theta}{dt} = -h\cdot\frac{du}{d\theta},$$

because $\frac{1}{u^2}\cdot\frac{d\theta}{dt} = R^2\frac{d\theta}{dt} = h.$

Differentiating this again with respect to t, we have

$$\frac{d^2R}{dt^2} = -h\frac{d^2u}{d\theta^2}\cdot\frac{d\theta}{dt} = -h^2u^2\cdot\frac{d^2u}{d\theta^2}.$$

$\left[\because \frac{d\theta}{dt} = \frac{h}{R^2} = hu^2.\right.$

Substituting the value of $d\theta/dt$ and d^2R/dt^2 in the expression for a_1 above, we have

$$a_1 = -h^2u^2\cdot\frac{d^2u}{d\theta^2} - R\,(hu^2)^2 = -h^2u^2\left(R\cdot u^2 + \frac{d^2u}{d\theta^2}\right).$$

$$a_1 = -h^2u^2\left(\frac{1}{u}\cdot u^2 + \frac{d^2u}{d\theta^2}\right) = -h^2u^2\left(u + \frac{d^2u}{d\theta^2}\right) \qquad ...(i)$$

Now, let the equation of the *elliptical orbit* of the planet be

$$l/R = 1 + e\cos\theta. \quad \text{Or,} \quad lu = 1 + e\cos\theta, \qquad ...(ii)$$

where l is its *semi latus rectum* and e, its *eccentricity*.

Differentiating this expression *twice* with respect to θ, we have

$$l.d^2u/d\theta^2 = -e\cos\theta. \qquad ...(iii)$$

Add, adding relations (ii) and (iii), we have

$$lu + l.\frac{d^2u}{d\theta^2} = 1 + e\cos\theta = 1.$$

Or, $l\left(u + \frac{d^2u}{d\theta^2}\right) = 1$, whence, $\left(u + \frac{d^2u}{d\theta^2}\right) = \frac{1}{l}$.

Substituting this value of $\left(u + \frac{d^2u}{d\theta^2}\right)$ in relation (i), above, we have

$$a_1 = -h^2u^2/l = -\frac{h^2}{l}\cdot\frac{1}{R^2}.$$

| Putting $1/R^2$ for u^2

Or, denoting the constant h^2/l by K, we have $a_1 = -K/R^2$.

Or, $a_1 \propto -1/R^2$. ...(iv)

i.e., the acceleration, and hence **the force acting on the planet is inversely proportional to the square of its distance from the Sun**, (the $-ve$ sign merely indicating that the force in question is one of attraction).

Now, the *time-period (T) of the planet* (*i.e.*, the time taken by it to complete its one full revolution round the Sun) is given by

$$T = \frac{\textit{area of the ellipse}}{\textit{areal velocity of the radius vector}} = \frac{\pi.ab}{\frac{1}{2}.R^2.\frac{d\theta}{dt}},$$

where a and b are the *semi-major* and *semi-minor* axes of the elliptical orbit of the planet respectively.

Or $T = \frac{\pi.ab}{h/2}$. And $\therefore \; T^2 = 4\pi^2.a^2b^2/h^2$. $\left[\because \frac{1}{2}R^2\cdot\frac{d\theta}{dt} = \frac{h}{2}\right.$

Now, clearly, $b^2/a = l$, the *latus rectum* of the ellipse, and, therefore, $b^2 = al$; so that,

$$T^2 = \frac{4\pi^2 l}{h^2}\cdot a^3 = \frac{4\pi^2}{K}\cdot a^3.$$

$\left[\; h^2/l = K.\right.$

But since, in accordance with Kepler's third law, $T^2 \propto a^3$ for every planet, it follows that $4\pi^2/K$ is a constant, or that K is a constant for every planet. In other words, *K is quite independent of the nature of a planet*.

3. Finally if m and M be the respective masses of the planet and the Sun, and F and F', the force of attraction exerted by the Sun on the planet, and the reaction of the planet on the Sun respectively, we have, from relation (iv), above,

$$F = km/R^2 \text{ and } F' = KM/R^2$$

where k and K are constants.

And, since by Newton's third law of motion, action and reaction are equal and opposite, we have $F = F'$; so that,

$k.m = K.M.$ Or, $k/M = K/m$ = a constant, say, G.

So that, $k = M.G.$

Substituting this value of k in the expression of F above, we have

$$F = \frac{m\,M}{R^2}.G,$$

showing that the force of attraction between the planet and the Sun is directly proportional to the product of their masses.

7.5 Deduction of Kepler's laws from Newton's law of gravitation. Let us now see now *Kepler's laws* may be deduced from Newton's law of gravitation.

As we have seen above (§ 7.4) in the case of a body moving along a curve path, *radial acceleration* is given by

$a_1 = \frac{d^2R}{dt^2} - R\left(\frac{d\theta}{dt}\right)^2$ and its *transverse acceleration* by $a_2 = \frac{1}{R}.\frac{d}{dt}\left(R^2.\frac{d\theta}{dt}\right)$

Now, in accordance with Newton's law of gravitation, a planet experiences a *centripetal acceleration* towards the centre of the sun on account of the force of gravitational attraction between the two. Obviously, then, it will have no transverse acceleration at all and we shall, therefore, have

$$\frac{1}{R}.\frac{d}{dt}\left(R^2.\frac{d\theta}{dt}\right) = 0, \text{ i.e., } R^2.\frac{d\theta}{dt} = Constant = h$$

and hence also $\frac{1}{2}R^2.\frac{d\theta}{dt}$ = a constant $\frac{1}{2}h$

Clearly $\frac{1}{2}R^2.d\theta$ is the area described by the radius vector SA (Fig. 7.1) drawn from the Sun to the planet in time dt. So that, *area described/time taken, i.e.*, the areal velocity of the planet around the Sun is constant. This is, obviously, the **law of areas** or the second law of Kepler.

Now, the force of attraction on a planet of mass m due to the Sun of mass M, distant R from it is, in accordance with Newtons law of Gravitation, given by $F = -M.m\,G/R^2$, so that the *centripetal acceleration produced* in the planet towards the centre of the sun $= F/m = -M.m.G/R^{2m} = -G.M/R^2$. This must obviously be equal to the radial acceleration a_2 of the planet, *i.e.*,

$$\frac{d^2R}{dt}2 - R\left(\frac{d\theta}{dt}\right)^2 = \frac{-GM}{R^2}$$

Multiplying throughout by R^3, we have $R^3.\frac{d^2R}{dt}2 - R^4\left(\frac{d\theta}{dt}\right)^2 = -GMR.$ ...I

Since $R^2(d\theta/dt) =$ Constant h, we have $R^3\frac{d^2R}{dt}2 - h^2 = -GMR.$

Or, putting, $R = 1/u$, as before, (§ 7.4) and differentiating with respect to t, we have

$$\frac{dR}{dt} = -\frac{1}{u^2}\frac{du}{dt} = -\frac{1}{u^2}\frac{du}{d\theta}\frac{d\theta}{dt} = -R^2\frac{d\theta}{dt}\frac{du}{d\theta} = -h.\frac{du}{d\theta}$$

Differentiating once again with respect to t, therefore, we have

$$\frac{d^2R}{dt^2} = -h\frac{d^2u}{d\theta^2}.\frac{d\theta}{dt} = -h.\frac{d\theta}{dt}.\frac{d^2u}{d\theta^2} = -h\frac{h}{R^2}.\frac{d^2u}{d\theta^2} = -h^2u^2.\frac{d^2u}{d\theta^2}, \quad ...II$$

because $R^2.d\theta/dt = h$ and $R = 1/u$.

Substituting these values of $R^2.d\theta/dt$, R and d^2R/dt^2 in relation I above, we have

$$\frac{1}{u^3}\left(-h^2u^2.\frac{d^2u}{d\theta^2}\right) - h^2 = \frac{-GM}{u}$$

Or, multiplying both sides by u/h^2, we have $\frac{d^2u}{d\theta^2} + u = \frac{GM}{h^2}$

And, since GM/h^2 is a constant quantity, we have

$$\frac{d^2}{d\theta^2}\left(u - \frac{GM}{h^2}\right) + \left(u - \frac{GM}{h^2}\right) = 0. \text{ Or, } u - \frac{GM}{h^2} = A\cos\theta. \text{ Or, } u + \frac{GM}{h^2} + A\cos\theta.$$

Or, substituting for u, we have $\frac{1}{R} = \frac{GM}{h^2} + A\cos\theta$

So that, $$\frac{1/R}{GM/h^2} = 1 + \frac{A}{GM/h^2} \cos\theta \;, \qquad \text{...III}$$

an expression similar in form to the polar equation of an ellipse, *viz*, $l/R = 1 + e\cos\theta$ (see 7.4 above), where the ***semilatus rectum of the ellipse***, $l = h^2/GM$ and its *eccentricity* $e = Ah^2/GM$, indicating that the ***path of the planet is an elliptical orbit with the sun at one of its foci*** which is clearly the **law of elliptical orbits or the first law of Kepler.**

Finally, if T be the *time-period of the planet around the sun*, we have

$$T = \text{area of elliptical path / areal velocity} = \pi ab \,/\, \frac{1}{2} R^2 \frac{d\theta}{dt},$$

where a and b are the *semi-major* and *semi-minor* axes of the ellipse respectively.

Or $$T = 2\pi ab / R^2 . \frac{d\theta}{dt} = 2\pi ab/h, \qquad \left[\because R^2 \frac{d\theta}{dt} = h \right.$$

whence, $$T^2 = \frac{4\pi^2 a^2 b^2}{h^2} = \frac{4\pi a^2}{h^2} . b^2$$

Now as the *semi-radius rectaum* $l = b^2/a = h^2/GM$, we have $b^2 = ah^2/GM$

Substituting this value of b^2 in the expression for T^2, above, we have

$$T^2 = \frac{4\pi^2 a^2}{h^2} \times \frac{ah^2}{GM} = \frac{4\pi^2}{GM} . a^3$$

Or, since $4\pi^2/GM$ is obviously a constant quantity K, say, whatever the planet, we have

$$T^2 = K.a^3 \text{ Or } t^2 \propto a^3$$

i.e., the square of the time-period of the planet $\propto$ *cube of semi-major axis of the ellipse*. This is clearly the **Harmonic law or the third law of Kepler.**

We thus see how the three laws of Kepler can be easily deduced from Newton's law of gravitation.

7.6. Determination of Gravitational Constant.

7.6. Determination of Gravitational Constant. The methods for the determination of the *gravitational constant*, (and, therefore, also those for the determination of the density and the mass of the earth), may be divided into two categories, *viz.*,

(*i*) *Mountain and Mine Methods* —which involve the measurement of the force of attraction exerted by a large natural mass, like a mountain or the earth's *crust**, on a plumb line suspended on one side of it, which is then compared with the force of attraction on it due to the earth, as a whole.

(*ii*) *Laboratory Methods*—which involve the more delicate measurement of attraction between small masses.

We shall deal here only briefly with the former—more for their historical interest than otherwise; for, the results obtained were not—indeed, they could not be—very accurate. The latter, *i.e.*, the laboratory methods, we shall however study in proper detail.

(I) Mountain methods.

1. Bouguer was the first to have attempted a determination of the value of G. While engaged in geographical measurements in the *Andes* (*Peru*) in the year 1740, he suspected a deflection of his *plumb line* due to large mountain-masses. Deciding to verify this, he selected a mountain, *Chimborazo*, 20,000 *ft* high, (in the *Andes*) for the purpose. Shorn of details, his experiment (performed in 1749) was as follows :

He chose two stations A and B, (Fig. 7.2, (*i*) and (*ii*)], the former *due south* of the summit of the mountain and *close to it*, and the latter, *in the same latitude* and at about the same altitude some distance to its west, *away from its influence*. At station B, he observed a star passing the meridian directly overhead, so that the plumb line

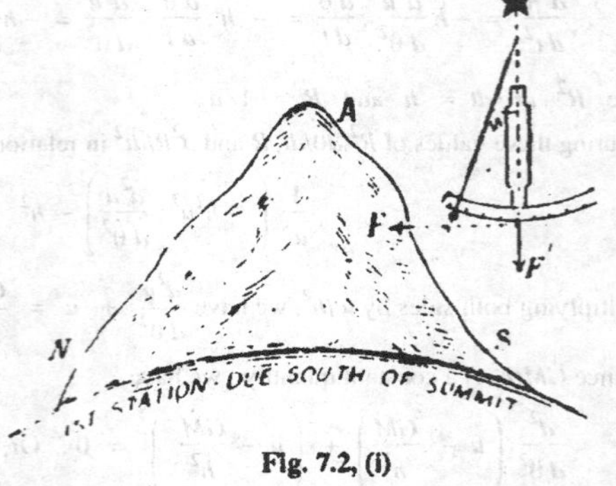

Fig. 7.2, (i)

* The word is probably a relic of the times when the earth was supposed to be a globe of water, bounded by a solid shell or crust. It is now used, however, to signify *the rigid surface layer of the earth*, which is *heterogeneous* and *more or less, in a state of permanent stress and strain.*

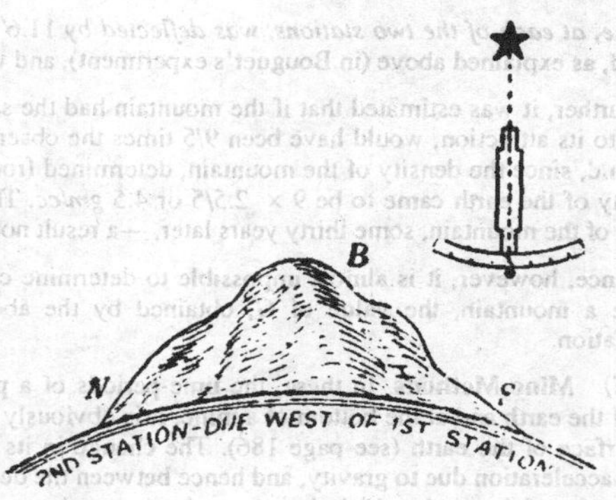

Fig. 7.2 (*ii*)

hung *exactly* vertically parallel to the telescope. But at station *A* he observed that it was attracted by the huge mountain mass, (due to its nearness). He measured this *deflection of the plumb line* at *A* and thus compared the horizontal pull of the mountain with the vertical pull of the earth. For, if *F* and *F'* be the forces of gravitational attraction acting on the plumb line due to the mountain and the earth respectively, and ϕ, its deflection from the vertical, [Fig. 7.2, (*ii*)], we have $\tan \phi = F/F'$.

Now, clearly, $F' = mg$, where *m* is the *mass of the plumb line* and *g*, the *acceleration due to gravity*. So that, $F = mg \tan \phi$.

And, if *V* be the volume and ρ, the density of the mountain, and *r*, the distance of its *c.g.* from the plumb line, we have

$$F = \frac{m \cdot V \cdot \rho}{r^2} \cdot G. \qquad [\because \text{ mass of the mountain } = V.\rho.$$

Hence, $m.V.\rho.G/r^2 = mg \tan \phi$. Or, $G = g.r^2 \tan \phi / V.\rho$.

Thus, a knowledge of the volume, density and shape of the mountain, and hence its *centre of gravity*, was needed to determine the value of *G*. *Bouguer* made an attempt at it but did not quite succeed; for, due to the most adverse conditions of snow the storm under which he had to work, he could not properly survey the mountain, and so the results obtained were very much wide of the mark. Thus, for example, he found that his plumb line was drawn aside by about 8", and his calculation showed that if the mountain were as dense as the earth, itself, the deflection of the plumb line would have been twelve times as great, indicating that the earth was about twelve times as dense as the mountain. And this, as we know, is very much beyond the truth. Nevertheless, he had the satisfaction of showing that the attraction due to the mountain masses did actually exist and that the method was, therefore, possible. Not only that, but he also deserves the fullest credit for proving conclusively that the earth was not just a globe of water or a hollow shell, as was fairly widely supposed at the time.

2. Maskeylne, later, in the year 1774* repeated, at the request of the Royal Society, Bouguer's experiment on the mountain *Schiehallion* (meaning *'constant storm'*), in Perthshire (Scotland), 3547 feet high, an elaborate survey of which was first made to determine as accurately as possible, its volume and density (and hence its mass) and centre of gravity.

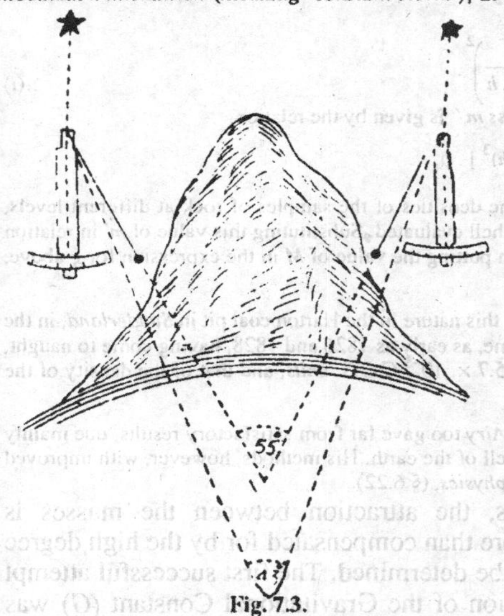

Fig. 7.3.

Two stations were then chosen *at equal distances* from the *c.g.* of the mountain, along the north-south line on its 'steeply sloping' north and south sides respectively, about a quarter of the way up on either side, so as to have the maximum effect on the plumb line, (Fig. 7.3). The *same* star was observed, (as in Bouguer's experiment), by means of a special type of telescope, called the *Zenith Sector***, first at the South Station and, a month later, at the North Station. At the former Station, the star which in the absence of the mountain, would lie directly overhead, appeared to shift slightly to the north, because the plumb line was pulled by the mountain towards the north, (and the *zenith* thus shifted to the south). At the other station, on the other hand, the exact opposite was the case, (the plumb line being pulled towards the south, the *zenith* thus shifting to the north) and the star, therefore, appearing to shift *equally* to the south. Thus, the total shift of the star was double of the deflection of the plumb line at either station, due to its attraction by the mountain. This was carefully measured and was found to be 54.6". Out of this, a shift of 42.94" was calculated to be due to the curvature of the earth's surface; so that, the *net shift or deflection of the plumb line, due to the gravitational pull of the mountain, was* (54.6" − 42.94") = *11.6"*. In other words, *the*

* He was the *Astronomer Royal* at the time and admitted candidly that the hint for such an experiment was given by Newton himself when he wrote "that a mountain of a hemispherical figure, 3 miles high and 6 miles broad, will not, by its attraction, draw the plumb line 2' out of the perpendicular".

** The instrument could rotate about a horizontal (East and West) axis at its object glass end, ponting upwards, and was provided with a plumb line, suspended from this axis, over a scale graduated in degrees, so that the angular distance of the telescope from the vertical could be directly read on it.

plumb line, at each of the two stations, was deflected by 11.6"/25.8" due to the mountain-mass. The value of G was then calculated, as explained above (in Bouguer's experiment), and was found to be 7.4×10^{-8} *C.G.s.* units.

Further, it was estimated that if the mountain had the same density as that of the earth, the deflection of the plumb line, due to its attraction, would have been 9/5 times the observed deflection, showing that the earth was 9/5 times denser than it. And, since the density of the mountain, determined from pieces of rocks composing it, was found to be 2.5 *gm/cc*,. the density of the earth came to be 9 × 2.5/5 or 4.5 *gm/cc*. This was corrected and increased to 5.0 *gm/cc* after a careful re-survey of the mountain, some thirty years later, —a result not very much wide of the mark.

Since, however, it is almost impossible to determine correctly the mass and position of the *c.g.* of a huge natural mass like a mountain, the value of G, obtained by the above methods, is far from reliable and is at best only an approximation.

(ii) **Mine Methods.** In these, the time-periods of a pendulum, (say, ***a seconds pendulum***), are compared on the surface of the earth and at the bottom of a mine. It is obviously smaller in the latter case, the value of g being less there than on the surface of the earth (see page 186). The change in its time-period enables a comparison to be made between the values of acceleration due to gravity, and hence between the density of the layers immediately above the pendulum-bob and that of the rest of the earth, which, in its turn, leads to a determination of M, the mass of the earth, and thus to a calculation of the value of G, as will be clear from the following :

We know that the weight of a body at a place is the force with which it is attracted by the earth towards its centre and is numerically equal to the product of its mass (m) and the acceleration due to gravity at the place. Thus, if g be the acceleration due to gravity on the surface of the earth, we have

$$\textit{wt of the body on the surface of the earth} = mg = \frac{m\,.\,M}{R^2}\,G\,.$$

where M is the mass of the earth and R, its radius.

At the bottom of the mine of depth h, the outer shell of the earth of thickness h will have no effect on the body because, as shall see under § 7.15 *(e)*, there is no gravitational field inside a spherical shell. The only gravitational force on it will, therefore, be due to the inner sphere of the earth of radius $(R-h)$, on the surface of which the body, in effect, lies.

If m' be the mass of the outer spherical shell of the earth of thickness h, clearly, the ***mass of the inner sphere*** $= (M-m')$. If, therefore, g' be acceleration due to gravity on the surface of the sphere (or at the bottom of the mine), we have

$$\textit{weight of the body at the bottom of the mine} = mg' = \frac{m(M-m')}{(R-h)^2}\,G\,.$$

whence,

$$\frac{g'}{g} = \frac{M-m'}{M}\cdot\left(\frac{R}{R-h}\right)^2. \qquad ...(i)$$

And, clearly, if ρ be the density of the outer shell of the earth, its mass m' is given by the relation,

$$m' = \frac{4}{3}\pi\,[\,R^3 - (R-h)^3\,]\,.\,\rho.$$

Now, the average value of ρ was obtained by determining the densities of the samples of rock at different levels, down to the bottom of the mine, and thus the mass (m') of the outer shell evaluated. Substituting this value of m' in relation *(i)* above, the mass (M) of the earth was easily determined. And, then putting the value of M in the expression for g above, the value of G could be calculated straight away.

Airy was the first person to have made a successful attempt of this nature in the Harton coal pit in *Sunderland*, in the year 1854, two earlier attempts made by him, in a Cornish copper mine, as early as 1826 and 1828, having come to naught, due to unfortunate accidents in the mine. *Airy's* value of G came to 5.7×10^{-8} *C.G.S. units*, and that of the density of the earth's surface to 6.5 *gm/cc*.

Like the earlier mountain experiments, these experiments by *Airy* too gave far from satisfactory results, due mainly to the difficulty of determining accurately the density of the outer shell of the earth. His methods, however, with improved modifications, now find a wide and useful scope in the branch of *Geophysics*, (§ 6.22).

(iii) **Laboratory Methods.** In these methods, the attraction between the masses is inevitably feebler due to their smallness. But this is more than compensated for by the high degree of accuracy with which the masses and their sizes can be determined. The first successful attempt at an accurate method of this type for the determination of the Gravitational Constant *(G)* was made by *Cavendish*, in the year 1798, in which he made use of the *Torsion Balance*.

It will be of some interest to recall that *Cavendish* was probably also associated with *Maskelyne* in his Mountain experiment, performed some twenty-five years earlier. He, however, took his cue from *Rev. John Michell*, who had devised an apparatus almost similar to *Cavendish's* own, but was not destined to use it due to his sudden death. His apparatus fell into the hands of *Prof. Wollaston* who passed it on to *Cavendish*.

(*a*) **Cavendish's Method.** The apparatus used by **Cavendish**, and installed in an outhouse in his garden on Clapham Common, was as shown in Fig. 7.4. A long *cross bar PQ*, about 6 *ft* (or about 180 *cm*) long, was suspended from the ceiling of a room and was free to turn about a vertical axis by means of an arrangement manipulated from outside. It carried two large and *equal lead spheres C* and *D*, about 8 to 10 *inches* (or 20 to 25 *cm*) in diameter and weighing about 350 *lb* each, at the ends of two metal rods attached to its two ends.

Immediately below the mid-point of the cross-bar was a *torsion head M* which could also be worked from outside, from which was suspended a *deal-rod RS*, (slightly longer than *PQ*), by means of a *fine torsion wire W*, of silvered copper. Two wires (*w*, *w*) fastened the ends of the deal-rod to a vertical rod *r* in the middle, which was attached to the suspension wire. This increased the strength of the rod, *without increasing its moment of inertia*. Two *small lead balls*, 2 *inches* (or 5 *cm*) in diameter and weighing about a pound and a half (or 680 *gm*) each, were suspended from the two ends of the deal-rod *RS* such that *the centres of the four balls lay in the same horizontal plane*, roughly in a horizontal circle of about 3 *ft* (or 90 *cm*) radius. The arrangement was such that, *when the line joining the centres of the large lead spheres was at right angles to the torsion rod, there was no twist or torsion in the suspension wire W*. Each end of the torsion rod carried a vernier (of five divisions), which moved over a fine ivory scale, fixed to vertical stands, and with each division equal to .05″.

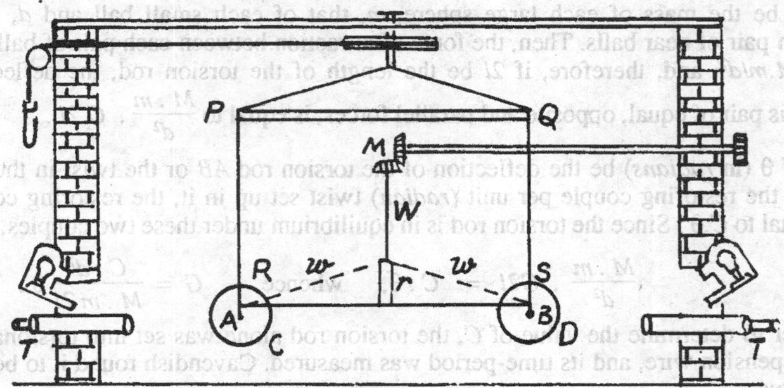

Fig. 7.4

To guard against any changes of temperature, and consequent air-draughts, which would otherwise mask the gravitational effect, the room was closed and observations were taken with the help of telescopes *T* and *T*, fixed into the walls of the room, as shown. And, further, to avoid the effect of any outside electric charges, the whole apparatus was enclosed in a gilded glass case, supported on four levelling screws.

The method of procedure was the following :

The rod *PQ* was rotated until the line joining the centres of the large spheres was at right angles to the torsion [rod Fig. 7.5 (*a*)] *i.e.*, *in the position in which there was no twist in the suspension wire*, and the reading on the verniers, attached to the torsion rod at either end taken. The large spheres were then rotated until they lay on *opposite sides of the* torsion rod and near to the small balls at either end, *i.e.*, in the position *C* and *D*, as shown in Fig. 7.5(b) such that one to the attraction of balls *A* and *B* by the spheres *C* and *D* respectively, the torsion rod *AB* got deflected into the position *A′B′* with the lines joining the centres of *C* and *A′* and of *D* and *B′* were *equal in length* and *perpendicular* to the torsion rod. Obviously, then, the force exerted by the big spheres on the corresponding near small balls were equal and *opposite*, thus constituting a *couple*, tending to rotate the torsion rod. This was resisted by the *torsional couple* set up in the suspension wire, and equilibrium *was attained when the deflecting couple, due to the forces between the two pairs of spheres and balls, was just balanced by the restoring torsional couple, set up in the suspension wire*. The positions of the verniers were again noted on the scales. Since, due to

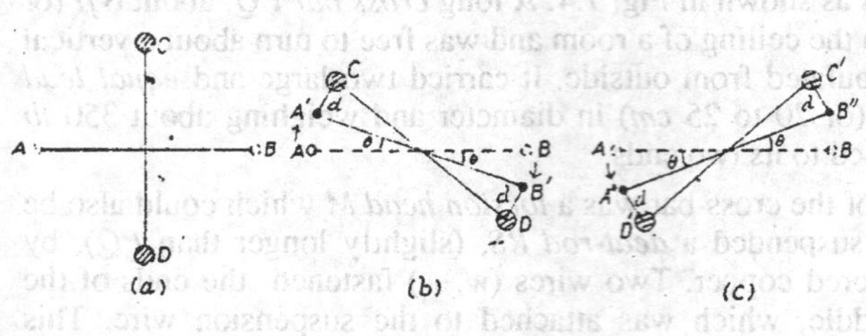

Fig. 7.5

inevitable air currents inside the chamber, the torsion rod could never attain state equilibrium, the position of equilibrium was estimated by the method of oscillation as in the case of an ordinary physical balance. In Cavendish's own experiment, this distance between each pair of balls was 8″ (or 20 *cm*), and the small balls were displaced through .7661″ or (1.915 *cm*). The rod *PQ* was then rotated about its vertical axis the other way, until the large spheres now occupied the positions *C'* and *D'* and the small balls, the positions *A*" and *B*" respectively, [Fig. 7.5 (*c*)] and the same adjustment was made as before, *viz*., that the lines joining the centres of *C'* and *B*″ and pf *D'* and *A*″ were of the *same equal lengths* and *perpendicular* to the torsion rod. The positions of the verniers were again read on the scales and their mean taken as the deflection of the torsion rod. The value of *G* was then calculated as follows :

Let *M* be the mass of each large sphere, *m*, that of each small ball and *d*, the distance between each pair of near balls. Then, the force of attraction between each pair of balls is, clearly, equal to $G.M.m/d^2$; and, therefore, if $2l$ be the length of the torsion rod, the deflecting couple formed by this pair of equal, opposite and parallel forces, is equal to $\frac{M \cdot m}{d^2} \cdot G.2l$.

And, if θ (in *radians*) be the deflection of the torsion rod *AB* or the twist in the suspension wire, and *C*, the restoring couple per unit (*radian*) twist set up in it, the restoring couple due to torsion is equal to $C.\theta$. Since the torsion rod is in equilibrium under these two couples, we have

$$\frac{M \cdot m}{d^2} \cdot G2l = C \cdot \theta, \quad \text{whence,} \quad G = \frac{C \cdot d^2}{M \cdot m\, 2l}\theta .$$

In order to determine the value of *C*, the torsion rod alone was set into torsional vibrations about the suspension wire, and its time-period was measured. Cavendish found it to be 28 *minutes* in his apparatus.

Then, if *I* be the *moment of inertia of the torsion rod* (together with the small balls) about the wire as axis and *t*, the time-period of the rod, we have

$$t = 2\pi \sqrt{I/C}, \quad \text{whence,} \quad C = 4\pi^2 I/t^2 .$$

Substituting, the value of *C*, we have = $G = \frac{2\pi^2 I d^2}{M\, m\, l t^2}$

And if δ be the displacement of each end of the torsion rod, $\theta = \delta / l$

So that, $G = \frac{2\,\pi^2\, I d^2}{M m l^2 t^2}\,\delta.$

Or, *if we ignore the mass of the torsion rod*, $I = 2m(2l/2)^2\, 2ml^2$

And, therefore, $G = \frac{4\pi^2 l d^2}{M t^2}\,\theta = \frac{4\pi^2 \cdot d^2}{M t^2}\,\delta .$

Corrections and Sources of Error in the Experiment. Corrections were applied for the following :

(1) *force of attraction between each large sphere and the distant small ball;*

(2) *force of attraction between the two large spheres and the torsion rod; and*

(3) *force of attraction between the rods carrying the large spheres and the smaller balls.*

The following are the *source of error* in the experiment :

(1) The gravitational forces between a pair of a sphere and a ball being small, the torsion rod had to be made long to increase the deflecting couple. This also minimised the force of attraction between a large sphere and the distant small ball, but required a large chamber and thus *convection currents could not be avoided.* In addition to this, the torsion rod was also disturbed by heavy traffic on a nearby road.

(2) For a given deflecting couple, *the deflection of the torsion rod was small*, because the rather thick suspension wire required a large *torque* per unit deflection.

(3) The torsion wire, being *not perfectly elastic*, did not return to its normal position when the applied forces were removed, and thus *the torque was not strictly proportional to the angle of deflection.*

(4) *The distant large sphere decreased the angle of deflection due to the force of attraction they exerted on the distant small balls, while the rods carrying them increased it due to the force of attraction they exerted on the small balls.*

(5) *The method of measuring the angle of deflection was not sensitive enough.*

(6) *The time-period of the torsion rod, with the attached small balls, was much too big, i.e., its swings were much too sluggish, and impaired, rather than improved, the accuracy of measurement.*

The value of G obtained by Cavendish, as the mean of twenty-nine observations, was 6.754×10^{-8} **dyne cm^2/gm^2* in the *C.G.S. units* or 6.754×10^{-11} *newton–metre^2/kg^2* in *M.K.S.* (or *SI*) units.

Many other attempts were made since Cavendish's time, by *Jolly* and *Polynting*, among others, to obtain the value of G more accurately but the method adopted by *Boys*, with his newly invented quartz fibre used as suspension for the torsion rod, was by far the best of these. We shall, therefore, discuss that first.

(*b*) **Boys' method. Sir Charles Vernon Boys** removed, almost a century later, in the year 1895, all the defects of Cavendish's experiment by

(*i*) reducing greatly the size of the chamber, thus considerably *minimising convection currents* and making it easier to control its temperature ;

(*ii*) arranging the pairs of balls at different levels, thus *making the attraction between the distant large and small balls almost negligible;*

(*iii*) using, for the supension wire, a *quartz-fibre*, which required a compartively very small torque per unit deflection;

(*iv*) *making the angular deflection appreciably large and also proportional to the torque because of the quartz fibre being almost perfectly elastic, besides being fine and strong***; and

(*v*) *measuring the angular deflection by the telescope and scale method,* which greatly enhanced the accuracy of measurement.

Thus, *Boys greatly reduced the size of the apparatus and yet increased its sensitiveness.* This may, at first sight, appear to be an contradiction in terms ; for it is commonly believed that the larger a piece of apparatus, the greater the degree of accuracy obtained with it. Boys clearly showed, however, that this was not so, provided the time of swing of the suspended system remained the same, that the sensitivity of Cavendish's apparatus was thus quite independent of its dimensions and that there was no point, therefore, in attempting a larger version of it. He argued as follows :

* Amusingly enough, Cavendish made a slip in calculating this mean and it was later pointed out by Baily.

** A quartz-fibre is found to be much stronger than a steel wire of the same dimension. Boys was thus able to use fibres having a diameter as small as .0125 *mm*.

The sensitivity of the apparatus $= \dfrac{\textit{deflecting couple}}{\textit{restoring couple}}$.

Clearly, *deflecting couple* $\propto \dfrac{M.m.2l}{d^2}$, *i.e.* $\propto \dfrac{(mass)^2 \times length}{(distance)^2}$

Now, *mass* $\propto$ *volume* $\propto 4\pi\ (radius)^3/3 \propto r^3$

So that, *deflecting couple* $\propto \dfrac{(L^3)^2 \times L}{L^2}$, *i.e.*, $\propto \dfrac{L^6 \times L}{L^2} \propto L^5$

and, *restoring couple* $\propto \dfrac{I}{t^2}$ or $\dfrac{\textit{moment of inertia}}{(\textit{time of swing})^2}$.

Now, there is a practical limit to the time of swing which should not exceed 5 *minutes*, whatever the size of the apparatus, or else the swings become very sluggish, thus impairing the accuracy of measurement. This being so, we have *restoring couple* $\propto MK^2 \propto L^3 \times L^2$ [*t being the same for any apparatus.*

Or, *restoring couple* $\propto L^5$ and $\therefore$ *sensitivity* $\propto L^5/L^5$.

In other words, the sensitivity is independent of the size (L) of the apparatus, (the time of swing remaining the same).

Thus, if for example, we double the dimensions of Cavendish's apparatus, *keeping the time of swing (t) the same*, we find that M is increased by $(2)^3$, because it is proportional to $(radius)^3$, d is increased by (2)3 and l by 2.

The net result is that the value of $\theta = Mt^2\ G/4\pi^2\ ld^2$ remains the same *i.e.*, no advantage is derived by doubling the size of the apparatus, *i.e.*, by increasing the dimensions of all parts in the same ratio.

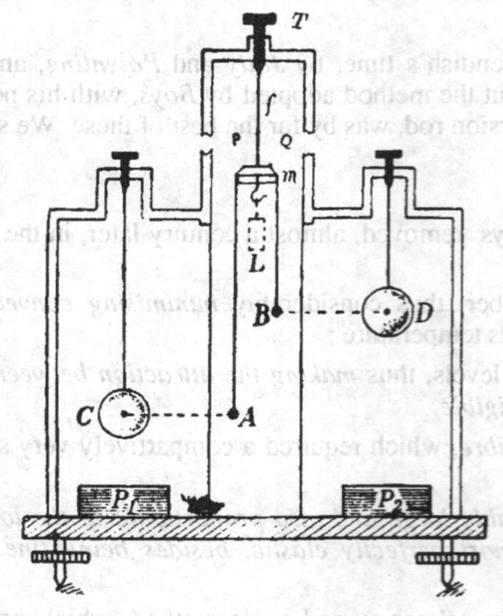

Fig. 7.6

What Boys, therefore, did was to reduce the dimensions of the different parts of the apparatus in *different ratios*, as will be clear from Fig. 7.6.

The moving system here consists of two *gold balls**, A and B, about 0.5 *cm* in diameter and 2.65 *gm* in mass each, suspended by fine gold wires from the two ends of a small beam PQ which, in its turn, is suspended by a fine *quartz* fibre f from a torsion head T inside a glass tube** of circular section, about 4 *cm* in diameter. This serves to protect the system from any outside disturbances. Also, if suspended *centrally* inside he tube, the system is free from effects due to attraction by the tube itself.

Immediately below the beam PQ is a mirror strip m, 2.5 *cm* long and rigidly attached to the beam PQ, with grooves in its two vertical edges, inside which slide the wires carrying the two gold balls. This ensures (*i*) that the *two wires are kept a constant horizontal distance apart and (ii) the effective points of suspension of the gold balls lie on the two edges of the mirror strip which thus acts as the torsion bar*.

In an outer coaxial tube, which can be rotated about the common axis, two large *lead spheres* C and D, each about 11.0 *cm* in diameter, (and about 74 *kgm* in mass), are suspended, as shown,

* On account of the higher density of gold (19.3 *gm/c. c.*) compared with that of lead (11.3 *gm/cc*), the spheres of gold for the same mass are smaller than spheres of lead and thus enable the distance d between the centres of the large and small balls to be reduced.

** In fact, *Boys* had actually used a brass tube.

such that *the centre of C is in level with that of A, and that of D, in level with that of B*, (to ensure greater pecision in the measurement of the distance between each pair), *the distances between the centres of the pairs A and C, and B and D, being exactly equal.* And, to minimise the effect of C on B and that of D on A, one pair (BD) is arranged about 15 *cm* above the level of the other (AC). The whole apparatus rests on a suitable platform, provided with levelling screws, and rubber pads P_1 and P_2 are placed below the two lead spheres as a safeguard against damage, in case they should fall accidentally. The deflection of the torsion bar *m* is measured accurately by the telescope and scale method, a *half millimetre scale* being placed at a distance of about seven metres from it.

The experiments is performed by rotating the outer cylindrical tube until the large (lead) spheres lie on the *opposite sides* of the two gold balls *but not in a line with the mirror strip**, so as to exert the maximum moment on the suspended system, *i.e.*, when the angle of deflection is the largest**. The tube is then rotated, so that the lead spheres now lie on the *other sides* of the gold balls in a *similar* position, again exerting the maximum moment, or producing the greatest deflection. The mean of the two is then taken. Let it be θ.

The calculations for the value of *G* are then made as follows :

Let *A, B, C* and *D*, (Fig. 7.7), be the four balls, when they are *in equilibrium*, in the position of *maximum deflection*. To visualise the balls in these positions, we must remember that, to start with, the centres of all the four balls lie *in the same vertical plane*, there being no twist in the suspension fibre and hence no couple acting on the suspended system. If we now rotate the larger spheres *C* and *D* through a certain angle, the plane containing their centres will also rotate through the *same* angle, with the result that a gravitational couple now comes into play on the suspended system, tending to rotate it into a position of equilibrium in which, once again, the centres of the small balls come to lie in the same vertical plane with the centres of the large spheres, —this being equally true when the suspended system suffers its maximum deflection.

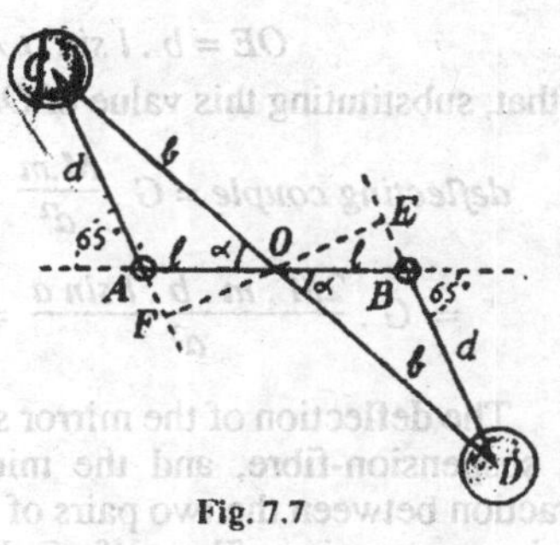

Fig. 7.7

Here, the small balls *A* and *B* are shown in their initial positions, corresponding to θ = 0° and the large balls in their final positions when they have been rotated into a position so as to exert the maximum couple on the suspended system, tending to make the latter suffer its maximum deflection θ. Obviously, equilibrium will be attained again only when the centres of the small balls come to lie in the vertical plane of the centres of the large spheres. To bring the small balls back to their original positions, (shown in the figure), therefore, the torsion head will have to be rotated in a direction opposite to that in which they have been deflected by this couple. In other words, the deflecting gravitational couple exerted by the large spheres on the suspended system, in the position shown, is just balanced by the restoring torsional couple set up in the quartz suspension.

Now, let *O* be the mid-point of the mirror strip, and *l*, its *half-length* (*i.e.*, let $OA = OB = l$). Further let $OC = OD = b$, $AC = BD = d$, $\angle AOC = \angle BOD = \alpha$ and *OE*, the perpendicular drawn from *O* on *DB* produced.

---

* For, in this position, the gravitational forces due to large spheres on the small balls near to them will act in opposite directions along *the same straight line* and thus neutralise each other.

** This position of the lead balls is chosen because when the couple on the suspended system is the maximum, the rate of variation of the couple is small and the relative positions of the balls need not be known with any great accuracy. Boys showed that the couple was the maximum when the line joining the centres of *A* and *C*, as also those of *B* and *D*, were inclined to the line through *AB* (Fig. 7.7) at an angle of 65°.

Obviously, the attraction between the two balls of each of the pairs, A, C and B, D is equal to $M.m.G/BD^2 = M.m.G/d^2$, where M and m are the masses of *each* large sphere and small ball respectively.

These two forces, being *equal, opposite* and *parallel*, constitute a *couple*, tending to rotate the mirror strip; and, quite clearly, the moment of the couple $= G\,.\,\dfrac{M.m}{d^2}\,.\,EF = G.\dfrac{M.m}{d^2}.2.\,OE$, where $EF = 2.OE$ is the *perpendicular distance between the lines of action of the two forces.*

Now, clearly, in the triangle OBD, we have

$$BD = \sqrt{OD^2 + OB^2 - 2OD.OB\ \cos a}\,.$$

Or, $\quad d = \sqrt{b^2 + l^2 - 2bl\ \cos\ a} = (b^2 + l^2 - 2bl\ \cos a\,)^{\frac{1}{2}}$

Also, $\quad \dfrac{\sin\ a}{\sin\ BDO} = \dfrac{BD}{OB} = \dfrac{d}{l};$ $\quad$ [$\because$ in a triangle, the sides are proportinal to the sines of the angles oposite to them.

and, therefore, $\quad \sin BDO = l\ \sin a/d\,.$ ...(*i*)

Now, in the right-angled triangle OED, we have

$$OE = OD.\sin EDO = b.\sin BDO$$

Or, $\quad OE = b\,.\,l\ \sin\alpha\,/d\,.$ [From (*i*) above.

So that, substituting this value of OE, we have

$$deflecting\ couple = G\,.\,\frac{M.m}{d^2}\,.\,2\,.\,\frac{b\,.\,l\ \sin a}{d}$$

$$= G\,.\,\frac{2M\,.\,m\,.\,b\,.\,l\ \sin a}{d^3} = G\,.\,\frac{2M\,.\,m\,.\,b\,.\,l\ \sin a}{(b^2 + l^2 - 2bl\ \cos a)^{3/2}}$$

The deflection of the mirror strip under this couple is resisted by the torsion or twist set up in the suspension-fibre, and the mirror strip comes to rest, when the deflecting couple due to attraction between the two pairs of the balls is just balanced by the restoring torsional couple set up in the suspension fibre. If C be the *torsional couple per unit deflection,* set up in the suspension-fibre*, the total restoring torsional couple is equal to $C.\theta$, and, therefore,

$$\frac{2G.M.m.b.l\ \sin a}{(b^2 + l^2 - 2bl\ \cos\ a)^{3/2}} = C\,.\,\theta\,.$$

Or, $$G = \frac{C\,.\,\theta\,.\,(b^2 + l^2 - 2bl\cos\alpha)^{3/2}}{2M.m.b.l,\ \sin\alpha},\ ^{**}$$ ...(I)

whence the value of G may be easily calculated.

The value of G obtained by *Boys* was 6.6576×10^{-8} *dynes.* cm^2/gm^2 in *C.G.S.* units or 6.6576×10^{-11} m^2/kg^2 in *M.K.S.*(Or *S.I.*) *units*.

Alternative Calculation. The value of G may alternatively be calculated as follows, using the same symbols as above.

---

* This is determined, as in Cavendish's experiment, by oscillating the moving system (consisting of beam PQ, the mirror strip m and gold balls A and B), in the absence of the lead spheres C and D and noting its time-period.

Boys determined it by removing the gold balls, together with their suspension wires from the beam and suspending a small cylinder L (shown dotted in Fig. 7.6) from a hook in it, this cylinder having the same mass as the gold balls and the suspendsion wires and a known moment of inertia about its own axis.

** The value of α is clearly the angle through which the torsin head T must be rotated to bring the small balls back into their original positions and can be easily read on a circular scale attached to it. Since a quartz-fibre is nearly perfectly elastic, the value of α may be taken to be the same as that of θ.

Gravitational pull between each pair of near large and small balls $= \frac{M.m}{d^2}.G$, in the direction A to C and B to D respectively, (Fig. 7.8). Resolving these into their two rectangular components each, along and perpendicular to AB, we have *the latter components of each equal to*

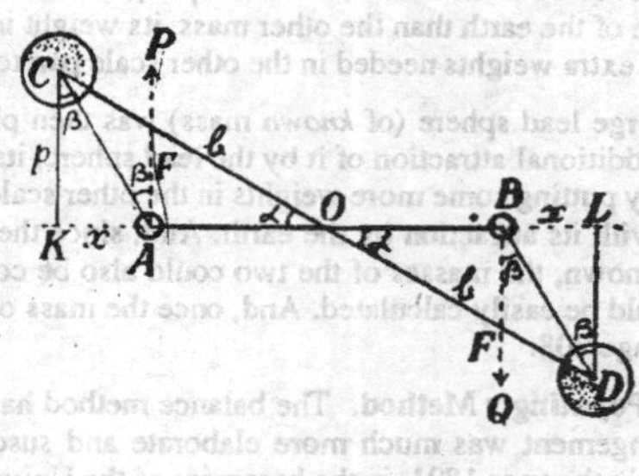

Fig. 7.8

$$F = G.\frac{M.m}{d^2}\cos\beta = G.\frac{M.m}{d^2}.\frac{p}{d}$$

$$= G.\frac{M.mp}{d^3},$$

where $CK = DL = p$ are the perpendiculars drawn from C and D respectively on AB produced.

Now, $\quad AC^2 = BD^2 = d^2 = p^2 + x^2,$

where $\quad p = b\sin\alpha$ and $x = (OK - OA) = (b\cos\alpha - l)$.

So that, $\quad d^2 = p^2 + x^2 = (b\sin\alpha)^2 + (b\cos\alpha - l)^2$

$$= b^2\sin^2\alpha + b^2\cos^2\alpha + l^2 - 2bl\cos\alpha.$$

Or, $\quad d^2 = b^2 + l^2 - 2bl\cos\alpha, \qquad [\because \sin^2\alpha + \cos^2\alpha = 1.$

And, therefore, $\quad d^3 = (d^2)^{3/2} = (b^2 + l^2 - 2bl\cos\alpha)^{3/2}$

Thus, $\quad F = G.\frac{M.m.b\sin\alpha}{(b^2 + l^2 - 2bl\cos\alpha)^{3/2}}$

Hence, the *deflecting couple on AB due to these forces F and F*,

$$= F.AB = F.2l = G.\frac{M.m.b\sin\alpha.2l}{(b^2 + l^2 - 2bl\cos\alpha)^{3/2}}$$

Now, if C be the torsional couple set up in the suspension fibre per unit twist of it, (*i.e.*, for unit deflection of the suspended system), the total restoring torsional couple for a twist θ in it $= C\theta$.

Since, for equilibrium, the two couples must balance each other, we have

$$G.\frac{M.m.b\sin\alpha.2l}{(b^2 + l^2 - 2bl\cos\alpha)^{3/2}} = C\theta, \text{ whence, } G = \frac{C\theta(b^2 + l^2 - 2bl\cos\alpha)^{3/2}}{M.m.b.2l\sin\alpha}$$

the same as expression I in the case above.

N.B. In case the centres of the near large and small balls do not lie in the same horizontal plane, but a vertical distance h apart, as shown in Fig. 7.9, then, if d' be the actual distance between them, we have

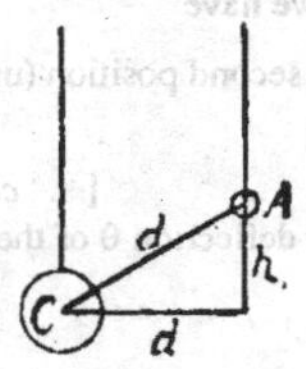

Fig. 7.9

$$\textit{gravitational couple on the suspended system} = \frac{MmGp}{d'^2}$$

But, clearly, $d'^2 = (d^2 + h^2)$, where d is the horizontal (or the perpendicular) distance between the balls, or rather between their centres. So that,

$$\textit{gravitational couple on the suspended system} = \frac{MmGp}{(d^2 + h^2)^{3/2}}$$

And, therefore, proceeding as above, we have

$$G = \frac{C\theta(b^2 + l^2 - 2bl\cos\alpha + h^2)^{3/2}}{M.m.b.2l\sin\alpha} \qquad \text{...(II)}$$

an expression which, when $h = 0$, i.e., when the centres of the near large and small balls lie in the same horizontal plane, is reduced to relation (I), above.

(iii) **Balance Methods.** These methods do not compare in accuracy with *Cavendish's*, *Boys'* or *Heyl's* methods, but are given here only for their historical importance.

(*a*) **Jolly's Method.** As early as 1881, **Von Jolly** had suggested that a common balance could be used to measure directly the gravitational force of attraction exerted on a mass placed in one pan by a large lead mass placed immediately below it.

He actually performed this experiment in Munich, where he had a common balance fixed at the top of a tower, 21 *metres* high, and suspended two long wires from its scale pans, carrying two other pans at their lower ends.

Two equal masses were then placed in the two upper scale pans and balanced against each other. One of the masses was then moved down into the lower scale pan, on the *same* side, so that, being now comparatively (about 20 *metres*) nearer to the centre of the earth than the other mass, its weight increased a little,- -this increase (due to the earth's attraction) being equal to the extra weights needed in the other scale pan to balance the beam.

A large lead sphere (of *known* mass) was then placed immediately below the lower pan carrying the mass, so that due to the additional attraction of it by the lead sphere, its weight again increased a little. This increase was also determined, as before, by putting some more weights in the other scale pan. The attraction of the masses by the lead sphere could thus be compared with its attraction by the earth. And, since the distance between he centres of gravity of the lead sphere and the earth was known, the masses of the two could also be compared. Then, the mass of the lead sphere being known, the mass of earth could be easily calculated. And, once the mass of the earth was obtained, the value of G could be deduced as in § 7.6 (*ii*), page 208.

(*b*) **Poynting's Method.** The balance method has perhaps been used to the best advantage by **Prof. J.H. Poynting**, whose arrangement was much more elaborate and susceptible of a much higher degree of accuracy. He performed his experiment in the year 1891, in the basement of the University of Birmingham.

The apparatus used by him, (shown diagrammatically in Fig. 7.9), consisted of a strong and sensitive bullion type balance, with a gun-metal beam, provided with steel knife edges and planes. The whole apparatus was fully enclosed in the room and all necessary manipulations were made from outside.

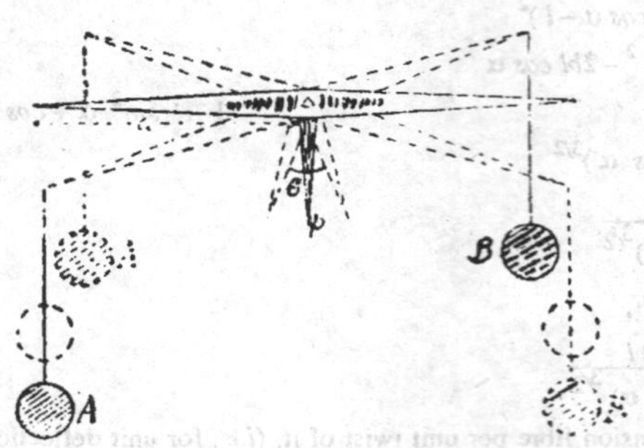

Fig. 7.9

Two equal spherical balls, A and B, of an alloy of lead and antimony, weighing about 50 *lb*, each, were suspended from the two ends of the beam. A large sphere S, of the *same* alloy and weighing about 350 *lb* was arranged below, on a turn-table, which could be turned about its vertical pivot P, so that the sphere could be brought to lie immediately under the ball A or B, as desired. To guard against the tilting of the turn-table due to the weight of sphere S, a smaller sphere S', *of half the mass of* S, wsa placed on the other side of the pivot, a double the distance of S from it, so that in accordance with the principle of moments, the turn-table was kept in equilibrium.

To start with, the sphere S was brought to lie under ball A, so that A was attracted downwards with a force equal to $G.\,M\ \ m/r^2$, where M and m are the respective masses of S and A, and r, the distance between their centres, (which was about one foot).

The turn-table was next rotated about its pivot until sphere S came under A to under B, and the balancing sphere S' moved on to the other side, (into the dotted position shown), as that S now exerted a pull on B instead of on A, resulting in the beam being tilted in the opposite direction to that in the first case, the angle of tilt being now obviously twice* that due to S on a or B. Let this deflection or tilt of the beam be θ.

Then, if a be the length of each arm of the balance, (*i.e.*, if $2a$ be the length of the beam, we have

change in torque or couple due to the shifting of S from the first position (under A) to the second position (under B).

$$= \frac{G.M.m}{r^2}.2a. \qquad [\because \cos\alpha = 1$$

And, if C be the torque or couple required per unit deflection of the beam, the torque for deflection θ of the beam is also equal to $C.\theta$.

Hence $$G.\frac{M.m}{r^2}.2a = C.\theta. \quad \text{So that,} \quad G = \frac{r^2.C.\theta}{M.m.2a}. \qquad ...(i)$$

Thus, knowing M, m, a, r, C and θ, the value of G could be easily calculated.

To determine the value of C, a *centigramme rider* was moved along the arm of the balance and the deflection* of the beam, for a shift l *cm* of it along the arm was noted then clearly

$$.01\,g.\,l = C.\alpha, \quad \text{whence,} \quad C = .01\,g.\,l/\alpha.$$

Substituting this value of C in relation (*i*), we have

$$G = \frac{r^2\theta}{M.m.2a}.\frac{.01\,g.\,l}{\alpha} = \frac{.01\,g.r^2.l}{2M.m.a}.\frac{\theta}{\alpha}.$$

* This is so, because the beam first comes to its original position from its first tilt and then gets tilted *equally* in the opposite direction, due to the attraction of S on B.

whence, G can be easily evaluated.

The effect of S and S' on the beam was eliminated by repeating the experiment with A and B, a *foot* higher up in the dotted positions shown, and proper corrections were also applied for the cross effect of S and S' on A *and* B.

Both the angles, θ and α, being very small, (θ being only about *one second*) were measured by Kelvin's double suspension mirror method, as illustrated in Fig. 7.10, where M is a small mirror, suspended by means of a bifilar suspension (w and w') from two horizontal brackets B and B' in level with each other, with a small adjustable gap between them, - the former being *movable* and attached to the pointer (p) of the balance, and the latter, fixed. Thus, when the beam turned, the wire w also turned with it, making mirror M turn about the stationary wire w'. This, with the gap BB' suitably adjusted, magnified the deflection of the beam about 150 times. The scale, graduated in *half-millimetres*, was arranged about 5 *metres* in front of the mirror, and its image in the latter viewed from the room above by means of a vertical telescope, fixed up in the ceiling.

Fig. 7.10

The effect of the air draughts or currents was eliminated by using what are called damping vanes V, suspended from the mirror and kept immersed in oil.

Poynting obtained the value of G to be 6.6984×10^{-8} *C.G.S. units*.

(c) **Heyl's Method.** The value of G obtained by **P.R. Heyl**, in the year 1930, is taken to be the most accurate one so far. His method is a modification of *Braun's Torsion Balance* experiment, with, in its turn, was a revised version by *Boy's* earlier experiment, referred to above, (page 211).

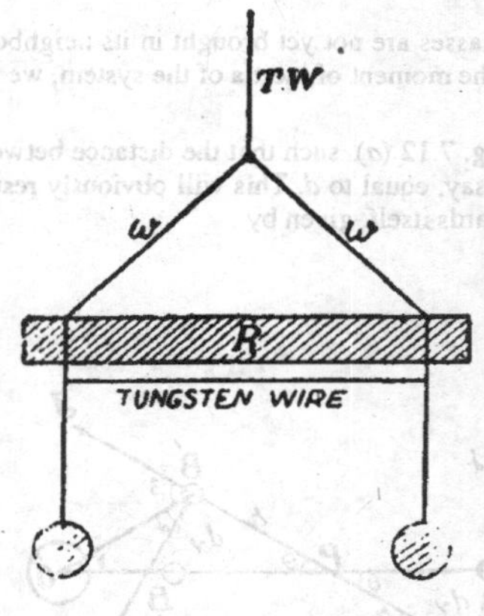

Fig. 7.11

Heyl performed his experiment in a constant temperature enclosure*, with the pressure inside reduced to about 2 *mm*, of mercury column, in order to minimise convection currents. The attracting large masses, used by him, were massive cylinders** of steel (with 0.9% carbon), each of mass about 66.3 *k gm* suspended from a system free to rotate about a vertical axis midway between the two. The smaller masses, each weighing 2.44 *gm* were balls of gold+ platinum and optical glass, in three different sets of experiments respectively, and were suspended from the two ends of a light aluminium torsion rod R, 28.6 *cm* long, (Fig. 7.11), supported by a tungsten thread, *T.W.*, (1 *metre* long and 0.25 *mm* in diameter), and two inclined copper wires, (w, w), so that *almost the whole of the moment of inertia remains in the balls themselves.* He chose as suspension a tungsten thread in preference to a quartz-fibre, because the latter is sometimes found to break quite unexpectedly and for no apparent reason.

This suspension system (*i.e.*, the torsion rod together with the two small masses) was made to oscillate in the gravitational field of the two large masses, which were arranged once with the centres of all the four masses lying *along the same horizontal line*, and then with the horizontal line

---

* It was, in fact, the constant temperature room of the American Bureau of Standards, 35 *ft*. below ground level.

** The reason why the large masses were used in the form of cylinders, rather than spheres (which make the calculations easier), was the difficulty of machining a sphere of such a huge mass. As some one has well remarked, *the burden was thus 'shifted to the broad shoulders of the mathematician'*.

\+ Gold balls were discareded by Heyl when he found that, due to the near vacuum in his enclosure, they absorbed mercury gauges, used to measure the pressure there. He, therefore, tried platinum balls, coated with lacquer and finally, balls of optical glass, in which any internal cavities could be easily detected visually.

joining their centres, *along the right bisector of the torsion rod*, the two positions being referred to as the *'near'* and the *'distant'* positions respectively, [Fig. 7.12 (*a*) and (*b*)] —the gravitational attraction acceleration the oscillations in the first case and retarding them in the second. Or, as a variating of this, the time-period of the suspension system was first determined, *with no other masses in its neighbourhood*, and then with the large masses brought in the near position, shown in Fig. 7.12 (*a*).

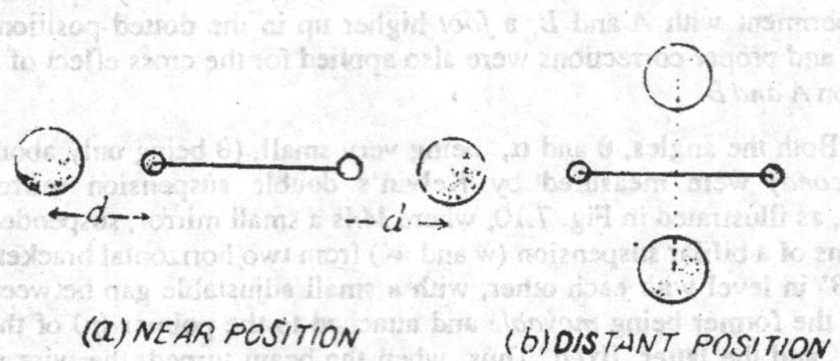

Fig. 7.12

To set the system oscillating, bottles of mercury were brought near the smaller masses for a short while and then removed, when, for an angular displacement of 4°, the system continued oscillating for about 20 hours.

The usual telescope, lamp and scale method was employed to observe the oscillations and the passeges of the lines on the image of a scale across a vertical cross-wire of the telescope were recorded automatically by a pen on a chronograph, another pen marking down on it the *'second'* signals from a standard clock.

From the time-periods of the suspension system, in the two cases, the value of G was then calculated as indicated in brief outline below :

Let T_1 be the time-period of the suspension system, when the large masses are not yet brought in its neighborhood. Then, if C be the torsional couple per unit twist of the suspension wire and I, the moment of inertia of the system, we have

$$T_1 = \sqrt{I/C}.$$

The large masses are then brought into the near position, shown in Fig. 7.12 (*a*), such that the distance between the centres of the neighbouring large and small balls is the same on either side, say, equal to d. This will obviously result in a gravitational pull F by each large mass over the corresponding small one, towards itself, given by

$$F = \frac{M.\,m}{d^2}.G,$$

where M and m are the values of each large and small mass respectively.

Considering the gravitational pull between the neighbouring large and small masses to remain unaltered by any small displacements of the small spheres from their initial or equilibrium positions, the gravitational pull of each large mass over the small one, when the small masses A and B are defle ted a little through an angle θ, into the positions A' and B', (shown in Fig. 7.13), will also be equal to

$$F = \frac{Mm}{d^2}G,$$

towards itself, along the line joining, the centres of the two masses. And, clearly, resolving this gravitational pull on both sides into two rectangular components, *along* and *perpendicular to* $A'B'$, we have the components perpendicular to $A'B'$ (the line joining the centres of the two small masses) equal to $F\cos\alpha$, represented by $A'C$ and $B'E$ respectively, where $\angle OA'C = \angle O'B'E = \alpha$

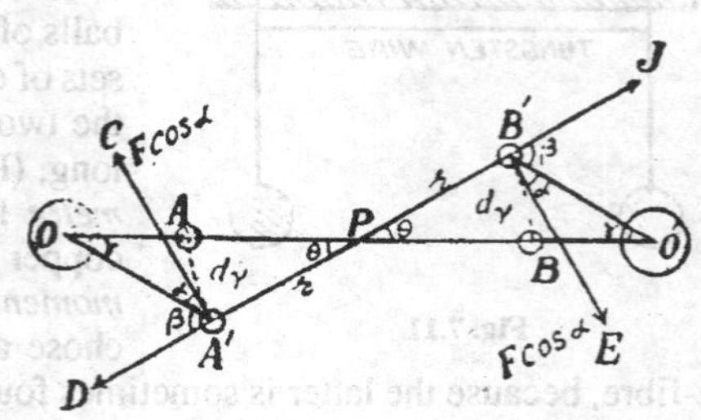

Fig. 7.13

So that, we have $$F\cos\alpha = \frac{MmG}{d^2}.\cos\alpha = \frac{MmG}{d^2}.\sin\beta$$

where $\beta = \angle OA'D = \angle O'B'J = (90-\alpha)$.

Now, since $\angle A'OP = \angle B'O'P = \gamma$ we have $\beta = (\theta + \gamma)$.

And, therefore, $$F\cos\alpha = \frac{MmG}{d^2}\sin(\theta+\gamma). = \frac{MmG}{d^2}.(\theta+\gamma), \quad ...(I)$$

if θ and γ be small, as they are in actual practice.

Now, since $\angle A'OP = \angle B'O'P = \gamma$ we have $\beta = (\theta + \gamma)$.

And, therefore, $F \cos\alpha = \frac{MmG}{d^2} \sin(\theta + \gamma) . = \frac{MmG}{d^2} . (\theta + \gamma)$, ...(I)

if θ and γ be small, as they *are* in actual practice.

Now, clearly, $A'A = B'B = d\gamma = r\theta$, whence, $\gamma = r\theta/d$.

Substituting this value of γ in relation I above, we have

$$F \cos\alpha = \frac{MmG}{d^2} . \left(\theta + \frac{r.\theta}{d}\right) = \frac{MmG}{d^2}\left(1 + \frac{r}{d}\right)\theta.$$

These two forces acting at A′ and B′ obviously form a couple, tending to bring the small balls back into their original position A and B ; and, clearly,

$$moment\ of\ this\ couple = \frac{MmG}{d^2} . \left(1 + \frac{r}{d}\right) . \theta \times 2r. \qquad [\because A'B' = 2r$$

Since C is the torsional couple per unit twist of the suspension wire, the torsional couple set up in the suspension wire, also tending to bring the small balls back into their original positions A and B, is equal to C.θ.

Thus, the *total couple acting on the suspended system of the small balls*

$$= C\theta + \frac{MmG}{d^2}\left(1 + \frac{r}{d}\right)\theta . 2r = \left[C + \frac{2MmG}{d^2}\left(\frac{r.d + r^2}{d}\right)\right].\theta.$$

If, therefore, T_2 be the time-period of the suspension of the system *now*, we have

$$T_2 = 2\pi\sqrt{I/C + \frac{2MmG}{d^2}\left(\frac{r.d + r^2}{d}\right)}.$$

And, thus, $$\frac{T_1^2}{T_2^2} = \left[C - \frac{2MmG(r.d + r^2)}{d^3}\right] \Big/ C = 1 + \frac{2MmG(r.d + r^2)}{C.d^3}$$

Or, $$\frac{2MmG(r.d + r^2)}{C.d^3} = \frac{T_1^2}{T_2^2} - 1 = \frac{T_1^2 - T_2^2}{T_2^2}$$

Or, $$G = \frac{T_1^2 - T_2^2}{T_2^2} . \frac{C.d^3}{2Mm(r.d + r^2)},$$

whence the value of G can be easily calculated.

The mean of Heyl's results (for the different small masses mentioned above) gave the value of G to be 6.70×10^{-8} *C.G.S.* units.

Birge estimated the probable error in the result to be .005. So that, the best value of G obtained so far is $(6.67 \pm 0.005) \times 10^8$ *dynes* $cm^2\ gm^{-2}$.

7.7. (*a*) Density of the Earth. We know that the weight of (or, the force acting on) a body of mass *m*, on the surface of the earth, is equal to *mg*, where *g* is the acceleration due to gravity at the place.

Also, if *M* be the mass of the earth and *R*, its radius, the force acting on the mass is, by Newton's Law of Gravitation, equal to $M.m.G/R^2$, so that, $mg = M.m.G/R^2$. Or, $g = M.G/R^2$.

Taking the Earth to be a *homogeneous sphere*, its volume $V = 4\pi R^3/3$.

Therefore, if Δ be its density, its mass $M = 4\pi R^3 . \Delta/3$

Hence $g = 4\pi R^3\, \Delta . G/3R^2 . = 4.\pi.R.\Delta.G/3$. Or, $\Delta = 3g/4\pi RG$,
whence, the value of Δ may be easily obtained.

N.B. The radius of the earth (*R*) may be estimated by choosing two points P_1 and P_2 on the earth's surface, a known distance *x* apart and in the same *latitude* (Fig.7.14) and measuring the altitude of the Sun at both, simultaneously at about 12.0 noon. If this be ϕ_1 and ϕ_2 at the two points respectively, the angle subtended by them at the centre of the earth is clearly $(\phi_1 \sim \phi_2)$. So that, $\tan(\phi_1 \sim \phi_2) = x/R$. Or, since $(\phi_1 \sim \phi_2)$ is quite small, we have $(\phi_1 \sim \phi_2) = x/R$, whence $R + x/(\phi_1 \sim \phi_2)$

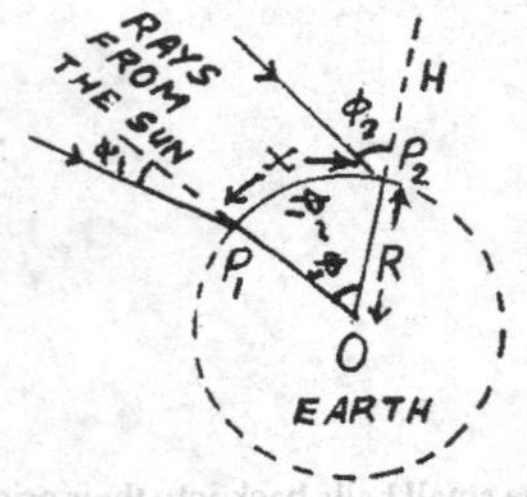

Fig. 7.14

It is interesting to observe how Newton intuitively made a lucky guess at the probable density of the earth, placing it so aptly between 5 and 6, —a truly inspired '*Newtonian*' guess, which stands so amply justified today.

The attempts made by different workers to determine the values of G and Δ are tabulated below in chronological order, for the convenience of the student.

Year	Name of Experimenter	Type of Experiment	Value of G	Value of Δ
1775	Maskelyne	Mountain method	7.4×10^{-8} *dynes* cm^2 gm^{-2}	5.0 *gm/cc*
1898	Cavendish	Torsion Balance	6.754×10^{-8} ,,	5.448 ,,
1854	Airy	Mine method	5.7×10^{-8} ,,	6.5 ,,
1881	Von Jolly	Sensitive Common Balance	6.465×10^{-8} ,,	5.692 ,,
1891	Poynting	,, ,,	6.6984×10^{-8} ,,	5.493 ,,
1895	Boys	Torsion Balance	6.6576×10^{-8} ,,	5.5270 ,,
1896	Eotvos	,, ,,	6.66×10^{-8} ,,	5.53 ,,
1901	Burgess	,, ,,	6.64×10^{-8} ,,	5.55 ,,
1930	Heyl	Oscillation method	6.670×10^{-8} ,,	5.517 ,,

7.7. (*b*) Mass of the Earth and the Sun. Knowing the mean density Δ of the earth, it is but an easy step to determine its mass, which is equal to its *volume* × *density*.

The earth, however, is not actually a sphere, as we know, but a spheroid of revolution about its polar axis; so that, if its equatorial and polar radii be R_1 and R_2 respectively, we have its *volume* $= \frac{4}{3}\pi R_1^2 R_2$ and hence its *mass* $= \frac{4}{3}\pi R_1^2 R_2 \Delta$.

Now, taking the orbit of rotation of the earth around the sun to be circular (though, strictly speaking, it is elliptical), of radius r, it is clear that the *centripetal force* acting on the earth towards the centre of its orbit (*i.e.*, the centre of the sun) is $M_e r\omega^2$*, Or Where M_e is the mass of the earth and ω its *angular velocity* ($2\pi/365 \times 24 \times 60 \times 60$ *radians/sec*).

This centripetal force is obviously supplied by the *force of gravitational attraction* exerted by the Sun on the earth $= M_e M_s G/r^2$, where M_s is the means of the sun. We therefore, have $M_e M_s G/r^2 = M_e r\omega^2$, whence, $M_s = r^3\omega^2/G$ and may easily be evaluated. This, incidentally, enables to obtain an expression for the period of the earth's revolution *(T)* around the sun, in terms of the mass of the sun and the radius of the earth's orbit. For,

$$\omega^2 = M_s G/r^3 \quad \text{and} \therefore \quad \omega = \sqrt{M_s . G/r^3} \text{. Hence } T = 2\pi/\omega \; = \; 2\pi\sqrt{r^3/M_s . G}.$$

7.8. Qualities of Gravitation. We shall now proceed to see whether the gravitational attraction between two bodies is in any way affected by the nature of the intervening medium between them, by the nature of the masses themselves or by physical conditions, like temperature etc.

1. **Permeability.** From the similarity of the formula ($F = G\, m.m'/r^2$) for gravitational attraction between two bodies, with those for magnetic and electrostatic attraction, it might appear that, like the constants μ and k there, the value of G might also depend upon the nature of the intervening medium between them.

* Or, we could take it to be $M_e v^2/r$, where v is the speed of the earth in its orbit round the sun. (See worked example 4).

That *this is not so* has been clearly shown by **Austin** and **Thwing**, who performed a direct experiment with a modified form of *Boys* 'apparatus, in which they placed slabs of different materials in between the two attracting masses and could detect no change whatever in the value of G, within the limits of their experiment.

This stands further confirmed by (*i*) the fact that whereas in the experiments for the determination of G, discussed above, *air* is the intervening medium, in the case of plants, the intervening medium is just *free space*, and yet the astronomical predictions, deduced on the basis of the same law, come out so surprisingly true, showing clearly that the value of G cannot possibly be very different in the two cases; and (*ii*) the very close agreement between the values of G, obtained by different pendulum experiments with their bases of different materials, so that different materials lie between the pendulum and the earth. We, therefore, conclude that *little or no effect is produced in the gravitational attraction between the masses by the nature of the medium interposed in between them.*

2. **Selectivity.** The law simply states that the force of attraction between two masses depends *only upon their magnitude*, having nothing to do with their nature, or their chemical combination etc. This is amply borne out by the large volume of experimental evidence in its favour. For, it has been shown by *Eotvos* and others, by their experiments with *Boys'* apparatus, using a large variety of materials as attracting masses, that the values of G obtained in the different cases all agree admirably, *even in the case of radioactive substances, thus showing clearly that gravitational attraction is by no means a selective phenomenon*.

3. **Directivity.** We know that in he case of *anisotrophic* substances, their physical properties, like refractive index, conductivity for heat and electricity etc., depend upon their orientation, *i.e.*, upon the direction of their crystallographic axes. **Dr. Mackenzie** and a host of other workers, therefore tried to investigate as to whether it had any effect upon the gravitational attraction also, and they all obtained *negative* results. An additional obvious proof of the independence of the value of G of the orientation of crystals is the fact that their *weight* (which is just another name of the force of gravitational attraction between them and the earth) is exactly the same whatever the orientation, *showing that the phenomenon of gravitational attraction is far from directive*.

Poyniting and **Gray** confirmed this fact in an ingenious experiment, in which two quartz crystals or spheres were suspended close to each other, —one enclosed in a case, with the other, outside it, being free to rotate. If there were even a trace of a directive influence in gravitational attraction, the rotation of one crystal would set up forced vibrations in the other; so that, if their time-period agreed, the enclosed crystal or sphere would also be set into appreçiably large resonant or sympathetic oscillations with the outer one. Nothing of the kind, however, was found to occur.

4. **Temperature. Poynting and Phillips**, together with a whole lot of other workers, tried to investigate the effect of temperature on the value of G, and, once again, the result obtained were absolutely negative. Only **Shaw**, experimenting with a *Boys-Cavendish* type of Torsion Balance, observed that the value of G increased slightly with the temperature of the attracting bodies**, the value of the coefficient of increase (α) being negligibly small, —being only about 1.6×10^{-6} between $0°C$ and $250°C$.

All the above mass of evidence thus goes to suggest that *gravitational attraction is purely a function of the masses of the attracting bodies and of the distance between them, being quite independent of all other factors*.

No wonder, then, that Newton's theory of gravitation held such an unquestioned sway over the minds of scientists all over the world. And, for non-scientific people, in general, it had equally spectacular predictions which, when found true, could not but impress them deeply as to its unerring truth.

Thus, for example, **Adams**, in 1845, predicted from his calculations, based on the disturbance of the orbit of *Uranus*, the presence of a hitherto unknown planet. Unfortunately, **Airy**, the then *Astronomer Royal*, did not as much as care to look for this planet, perhaps from sheer scepticism; and **Challis**, the Director of the Cambridge Observatory, though he actually saw a new star, looking like a disc, did not care to verify whether it was the one predicted by *Adams*. A year late, **Leverrier**, a French mathematician, made similar calculations to those of Adams and communicated his results to the Berlin Observatory and, lo and behold, the planet, we call *Neptune*, was there for all to see at the very spot predicted !

And, once again, in 1930, the disturbance of the orbit of Neptune itself led to the discovery, by American astronomers, of the planet *Pluto*.

These two profound discoveries put Newton's theory beyond the pale of any doubt or scepticism and it came to be looked upon as *infallible*. Indeed, it continued to enjoy its 'infallible' status until the arrival on the scene of that genius of modern times, **Albert Einstein**, who showed it

---

* *Anisotropic substances are those whose properties are different in different directions*, —crystals, in general.

** According to **Shaw**, $G = G_0 (1 + \alpha t)$, where G and G_0 stand for the values of the Gravitational Constant at t^0(and $0)°C$ respectively.

to be no better than a close approximation to the actual law of gravitational propounded by him, as we shall see in the next article.

7.9. Law of Gravitation and the theory of Relativity. Although, as we have seen above, the Newtonian Law of Gravitation is found to be valid over a wide range and is supported by a large mass of experimental evidence, there are certain small divergences, not quite in conformity with it. But such, indeed, has been the general faith in the infallibility of the law that any divergences from it were ascribed to some hitherto undiscovered disturbing influences rather than to any possible discrepancy or flaw in the law itself. It was only after **Einstein** put forward his *"Theory of Relativity"** that it came to be realised that Newtons's Law was only an approximation —although an extremely close one—to the true or the fundamental Law of Gravitation.

A detailed discussion of this theory is beyond our present scope, and we shall, therefore, consider here only one or two salient points of it, to show how Newton's formula lays itself upon to criticism in the light of these :

(*i*) One consequence of the theory, fully confirmed experimentally, is what may be called the *'inertia of energy'*, *viz*., that wherever a change in the energy of a body is brought about, a corresponding change takes place in its mass. In other words, *energy and mass are mutually convertible, one into the other*, the relation between the two being the following :

Change in mass (in grammes) = Change of energy (in ergs)/c^2,

where c is the *velocity* (in *cm/sec*) *of light in vacuo*, (equal to 3×10^{10} *cm/sec*). Further, according to this theory, we have

$$m = m_0 \, / \, \sqrt{1 - u^2/c^2},$$

where m is the mass of a body, when moving with a velocity u, (called its ***moving mass***), and m_0, its mass when at rest, (called its **rest mass**).

Thus, the mass of a body is different when in motion from that when at rest, *i.e.*, it changes with its velocity, and *Newton* has not specified which one of these is to be used in his formula for gravitational attraction, —clearly a noticeable omission.

(*ii*) Another consequence of the theory is that *the numerical value of the distance between two points varies according to the system of space-time coordinates chosen*, so that the distance between them changes with the circumstances of the observer, making the measurement.

The effect of these two discrepancies is of course, only slight, but it is there, nevertheless. *Einstein* took both these factors into account and was thus led to the formulation of his famous theory, which explains satisfactorily the deviations from the Newtonian law, among which may be mentioned the following :

(*a*) **Precessional motion of the perpendicular of the orbit of planet Mercury.** On Newtonian theory of gravitation the planetary orbits around the sun must continue to remain fixed in their respective positions, in actual fact, however, they continuously undergo a slight shift, too small to be noticeable except in the case of *Mercury*—due, no doubt, to its comparatively larger orbital velocity. Part of this effect can, of course, be ascribed to the attraction due to other planets, but the rest could be explained only on the basis of the influence of some other hitherto unknown planet, for which even a name— *'vulcan'* had thoughtfully been chosen in advance, but which never could be located.

On *Einstein theory*, on the other hand, the so called *'rostte-shaped'* path of Mercury about the sun, due to a continuous shift in its orbital position, follows as a natural consequence of the *precessional motion of the major axis of the orbit*, with a period of 3×10^6 years, so that it turns through a mere 43 *seconds of arc* in a hundred years. This now stands fully confirmed by actual observations.

(*b*) **Deflection of a ray of light in a gravitational field.** –A ray of light, due to its great velocity, behaves like a *material body*, possessing *mass* and *momentum*, etc., [see (*i*) above], and it must, therefore, suffer deflection in a gravitational field. Thus, a ray of light from a star would get deflected near the edge of the Sun, due to its high gravitational field, resulting in an apparent shift in the position of the star. Calculated on the basis of both these laws, the value of this

---

* See Chapter XVI.

orbit, with a period of 3×10^6 years, so that it turns through a mere 43 *seconds of arc* in a hundred years. This now stands fully confirmed by actual observations.

(*b*) **Deflection of a ray of light in a gravitational field.** –A ray of light, due to its great velocity, behaves like a *material body*, possessing *mass* and *momentum*, etc., [see (*i*) above], and it must, therefore, suffer deflection in a gravitational field. Thus, a ray of light from a star would get deflected near the edge of the Sun, due to its high gravitational field, resulting in an apparent shift in the position of the star. Calculated on the basis of both these laws, the value of this shift by *Einstein's theory*, comes out to be twice that by Newton's law, and actual experimental observation* fully supports the former result.

(c) **Shift of the spectral lines in stellar spectra.** A quantum of light, or a *photon*, from an excited atom on the surface of the stellar body (*i.e.*, the *sun* or a *star*) is naturally attracted by the gravitational field of the latter in travelling through which it loses part of its energy. Its frequency, therefore, decreases in obedience to *Planck's law, resulting in a corresponding increase in its wavelength* and a consequent displacement towards the red end of the spectrum—hence the name '*red shift*' given to it.

In plain and simple language, it means that the wavelengths of spectral lines of an element in a stellar source are somewhat higher and hence they appear to be shifted a little towards the red as compared to the spectral lines of the *same element* in a terrestrial source— this shift being given by the relation $1 - v_s/v_o = GM/c^2R$,

where v_0 and v_s are the *frequencies* of a spectral line of the *same* element on a terrestrial and in stellar body respectively, c, the velocity of light in vacuo (3×10^{10} *cm/sec*), *M, the mass of the stellar body* and , *R, its* radius.

It will be seen at once that the relation is a general one and it follows, therefore, that the percentage shift will be the *same for all frequencies or wavelengths.*

Now, due as much to the extreme smallness of the effect as to the difficulty of determining *M* and *R*, it is not quite so easy to check the result. It has, nevertheless been verified for the *Sun*, in the year 1924, when a shift as small as 0.008 *Angstrom unit*, *i.e.*, 0.008×10^{-8} *cm*, was observed for a spectral line of wavelength 4000 *Angstrom units* Later, in 1925, it was further confirmed by **Adams**, who measured a shift of as much as 0.32 *A.U.*. in the spectrum of the star, known as a '*white dwarf*', companion of *Sirius*, this higher value of the shift being due, obviously, to a stronger gravitational field on the surface of the star than that on the surface of the Sun. From this and the relation given above, the radius of the star was calculated to be 18000 *km*. This tallies well with the value for it obtained by other independent means.

In conclusion, it must be stated, therefore, that strictly speaking, the correct law of gravitation is the one due to *Einstein*, which is true for both strong and weak gravitational fields. But, although, Newton's law is true only for weak gravitational fields, it is a most satisfactorily close approximation to the correct law for our ordinary purposes, engineering and others, any slight discrepancies becoming noticeable at extremely small distances or in some phenomena of the type already referred to.

7.9. Gravitational Field—Intensity of the Field. The area round about a body, within which its gravitational force of attraction is perceptible, (no other body being near about it), is called its *gravitational field*.

The *intensity* or strength of a gravitational field at a point is defined as the *force experienced by a unit mass, placed at that point in the field.* It may also be defined as the *rate of change of gravitational potential, or the potential gradient*, at the point, (see § 7.11). Thus, if *I* be the intensity of a gravitational field at a point, we have

$$E = -dV/dx,$$

where dV is small change of potential for a small distance dx.

N.B. The strength or intensity of a field at a point is often spoken of merely as the *field* at that point.

7.10. Gravitational Potential—Potential Energy. Consider a body *A*, with its gravitational field around it. It will naturally attract any other body *B*, placed at any point in its field, in accordance with the *Law of Gravitation* and this force of attraction will *decrease* with *increase* in the distance of *B* from *A*, so that at an infinite distance from *A* it will be *zero*. But, as *B*

* This was made on the *Island of Principe* (on the African coast) and at Sobral, in Brazil, on May 29th, 1919, during a solar eclipse, (the stars being not visible otherwise). Two well-equipped expeditions at these two places obtained photographs of the portion of the sky, near the Sun, just before the eclipse and again after the eclipse, when the Sun had shifted away from its earlier position. The stars were found to have been displaced with respect to the Sun, the respective values of this displacement at the edge of the Sun obtained by the two expeditions being 1.61" and 1.98". Their mean (1.795"), agreed admirably with the value predicted by *Einstein*.

A similar observation was made again at another solar eclipse, three years later, in 1922,—this time at Wallal in West Australia, when the shifts of as many as eighty stars were observed. The mean shift was found to be 1.74" which was only about .01" short of the calculated value.

the potential decreases, the energy increases, with increase in the distance from *A*). If we, therefore, replace the unit mass by the body *B*, the potential energy of *B* will, clearly, be equal to be product of its mass and the gravitational potential (with its sign reversed) at that point. Thus, *the potential energy of a body at a point in a gravitational field is equal to the product of the mass of the body and the gravitational potential (with the sign reversed) at that point.*

7.11. Gravitational Potential at a Point distant r from a body of mass m. Let a mass *m* be situated at *O* (Fig. 7.15), and let a *unit mass* be situated at *P*. Then, the force of attraction on the unit mass due to *m* is clearly

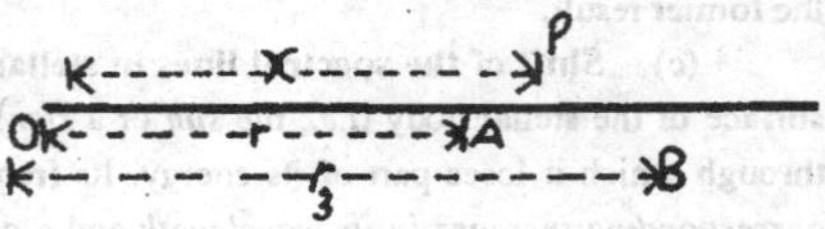

$$\frac{m \times 1}{x^2} \cdot G = \frac{m}{x^2} G,$$

where *x* is the distance of *P* from *O*, the force being directed towards *O*. Therefore, work done when the unit mass moves through a small distance *dx*, towards *O*, is equal to $m.G.dx/x^2$.

Fig. 7.15

And, therefore, work done when it moves from *B* to *A*

$$= \int_B^A \frac{m}{x^2} \cdot G \cdot dx = G.m \int_B^A \frac{1}{x^2} \cdot dx.$$

$$= -G.m \left[\frac{1}{x}\right]_B^A = -G.m \left[\frac{1}{r} - \frac{1}{r_1}\right] = G.m \left[\frac{1}{r_1} - \frac{1}{r}\right],$$

where r and r_1 are the distances of *A* and *B* from *O*. This, obviously, is the **potential difference** between the points *A* and *B*.

If *B* be at infinity, *i.e.*, if $r_1 = \infty$, we have

potential difference between A and $\infty = G.m\left(\frac{1}{\infty} - \frac{1}{r}\right) = -\frac{m}{r} \cdot G.$

But potential difference between *A* and ∞ is equal to the potential *at A*, because the gravitational force at ∞, due to *m*, is *zero* and, therefore, the work done in moving a mass about, there is also *zero*. In other words, the potential at infinity is *zero*. Therefore, the gravitational potential at *A* due to the mass $m = -G.m/r$.

Or, *denoting the gravitational potential at a point distant r from a body of mass m by V*, we have

$$V = -\frac{m}{r} \cdot G.$$

It will be noticed that whereas the value of gravitational potential at an infinite distance from a mass is *zero*, it goes on decreasing as we approach that attracting mass, — *i.e., it is an essentially negative quantity*, its maximum value being zero at infinity, where, at all points, therefore, the potential will be the same.*

7.12. Velocity of Escape. We know that, ordinarily, a body, say, a rifle bullet, projected upwards comes down to the earth due to the gravitational pull of the earth on it. Let us see if it is possible to project it with a velocity such that it will never come back. Obviously, it will be so if it can be given a velocity that will take it beyond the gravitational field of the earth. This velocity of the bullet or the body is called the *velocity of escape*.

Thus, if *m* be the mass of the body and *M*, that of the earth, the force acting on the body at a distance *x* from the centre of the earth is clearly mMG/x^2.

$\therefore$ work done by the body *against the gravitational field*, when it moves a distance *dx* upwards $= \frac{mMG}{x^2}.dx$.

$\therefore$ total work done by the body in escaping away from the surface of the earth

* The *negative sign* is often omitted in writing, but it must always be understood to be there.

$$\int_R^{\infty} \frac{m M G}{x^2} dx = \frac{m.M}{R} G.$$ [where R = radius of the earth

his, as will readily be seen, is m times (*i.e., the mass of the body times*) the gravitational potential on the surface of the irth.

If V_e be the *initial velocity of the body*, (*i.e., its velocity of escape*), its initial kinetic energy (*i.e.*, at the time it starts) ill be $\frac{1}{2} m V_e{}^2$. And this must, therefore, be equal to the work done by it during escape. So that,

$$\frac{1}{2} m V_e{}^2 = m MG/R. \quad \text{Or,} \quad V_e{}^2 = 2MG/R. \quad \text{Or,} \quad V_e = \sqrt{2MG/R} \qquad ...(i)$$

r, *since* $MG/R^2 = g$, we also have $V_e = \sqrt{\frac{2MG}{R^2} R} = \sqrt{2gR}$ ...(*ii*)

On substituting the value of M, G and R in (*i*) or of g and R in (*ii*) we have $V_e = 11.19 \times 10^5$ *cm/sec* $= 11.2 \times 10^5$ *cm ec* or 11.2 *km /sec*, very nearly, which is approximately 25000 *mi/hr*. The smallest value of V_e is for the planet Mercury, eing only 4.2 *km /sec*, and the highest, *viz.*, 61 *km /sec* of the planet Jupiter, (not to speak of the *Sun* for which V_e = 618 *m/sec*, but from which *only atoms*, and no other bodies or rockets etc. can escape.

Let us now see with what velocity to project a body from the surface of the earth to enable it to revolve close around , *i.e.*, to become a satellite* of the earth. Let it be V_o. Then, clearly, the centrifugal force on the body, taking it away from he centre of the earth will be mV_{o2}/R, when m is its mass and R, the radius of its orbit, which may be taken to be practically he same as the radius of the earth itself, the satellite lying close to its surface. This must obviously be balanced by the gravitational pull $Mm\,G/R^2$ on it due to the earth. So that, we have $mV_{o2}/R = m\,MG/R^2$, whence, $V_{o2} = M\,G/R$. or $V_o = \sqrt{mG/R}$.

Or, since $MG/R^2 = g$, we also have or $V_o = \sqrt{(m\,G/R^2).R} = \sqrt{gR}$.

Clearly $\therefore$ $V_0/V_e = \sqrt{gR/2gR} = 1/\sqrt{2}$. Or, $V_0 = V_e/\sqrt{2} = 0.7073\,V_e$.

The following shows at a glance how the path of a satellite depends upon its velocity of projection v in relation to V_o and V_e.

Velocity of satellite	Nature of path
(*i*) $v = V_o$	*Circular path around the earth.*
(*ii*) $v < V_o$	*Elliptical path — return to earth.*
(*iii*) $v > V_o$ but $< V_e$	*Elliptical path around the earth.*
(*iv*) $v = V_e$	*Parabolic path — escape from the earth.*
(*v*) $v > V_e$	*Hyperbolic path — escape from the earth.*

When the orbit of the satellite is elliptical the closest point to the earth is referred to as the perigee and the farthest as the apogee, the speed being the highest at the former $= \sqrt{(GM/a)\left(\frac{1+e}{1-e}\right)}$ and the lowest at the latter point, $= \sqrt{(GM/a)\left(\frac{1-e}{1+e}\right)}$ where a and e respectively stand for the *semi-major axis* and the *eccentricity* of the ellipse.

For further information about a satellite as to how it is launched and what are its uses etc., please see chapter X.

An intresting consequences of escape velocity. An obvious consequence of *escape velocity* is that it enables us to form an idea as to the probable nature of the atmosphere on other planets as also, of course, on our own. For, clearly, the nearer the average velocity of the molecules of a gas to the velocity of escape, the greater the chances of their escaping away from the upper regions of the atmosphere, which may thus get completely denuded of them in course of time.

Thus, considering the case of the *moon*, a landing on which has now been effected, we find that the escape velocity for it is a mere 2.4×10^5 *cm/sec*, as against 11.2 *cm/sec*, for the earth. Is it any wonder, then, that practically all the constituents of the atmosphere, as we know it here on earth, like *oxygen, nitrogen, carbon dioxide* and *water vapour*, the root mean square velocities for which (at 0°*C*) range between $.4 \times 10^5$ to $.8 \times 10^5$ *cm/sec* - not to speak of *hydrogen* and *helium*, whose *r.m.s.* velocities are very much greater, near about 2×10^5 *cm/sec*. – should all be absent from its surface and that the latter two lighter gases (hydrogen and helium) should be so rare even in our own atmosphere on the earth ? Again, the presence of these two lighter gases in the atmosphere of the sun should not surprise us in view of the much stronger gravitational field of the sun and consequently a much higher value of the escape velocity (618 *km/sec*) there.

7.13. Equipotential Surface. A surface, at all points of which the gravitational potential is the same, is called an *equipotential surface*.

---

* See § 10.12, Chapter X.

Thus, if we imagine a hollow sphere, of radius r, with a particle of mass m at its centre, the potential at each point on it will be the same, *viz.*, $-G.m/r$. The surface of the sphere is thus an equipotential surface.

Now, since the difference of potential between any two points on an equipotential surface is *zero*, no work is done against the gravitational force in moving a *unit* (or any other) mass along it. In other words, there is no component of the gravitational field along an equipotential surface or the direction of the field is at every point perpendicular to it.

Thus, consider two points P and Q, a small distance δr apart on an *equipotential surface AB* and let the intensity of the gravitational field at P be I directed along PR at an angle θ with PQ. Then, clearly, component of the field along $PQ = I \cos\theta$ and, therefore, *work done in moving unit mass from P to Q* $= I\cos\theta$. δr Since P and Q. lie on an equipotential surface, $I\cos\theta\,\delta r = 0$. And, since neither I or δr is zero, we have $\cos\theta = 0$ or $\theta = 90^0$ *i.e.*, the field is directed along the perpendicular to the surfaces at P. What is true of point P is also true of all other points on the surface AB. So that, *the direction of the field at every point on an equipotential surface is perpendicular to the surface at that point.*

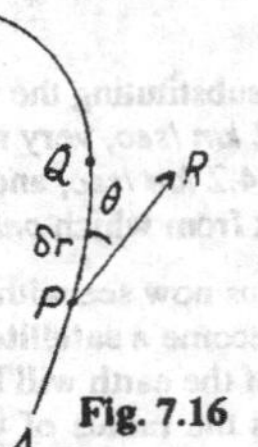

Fig. 7.16

7.14. Gravitational Potential due to a Spherical Shell

(*a*) **At a point outside the shell.** Let P be a point, distant r from the centre O of a spherical shell of *radius R*, (Fig. 7.17), and *surface density*, *(i.e., mass per unit area of surface)*, σ.

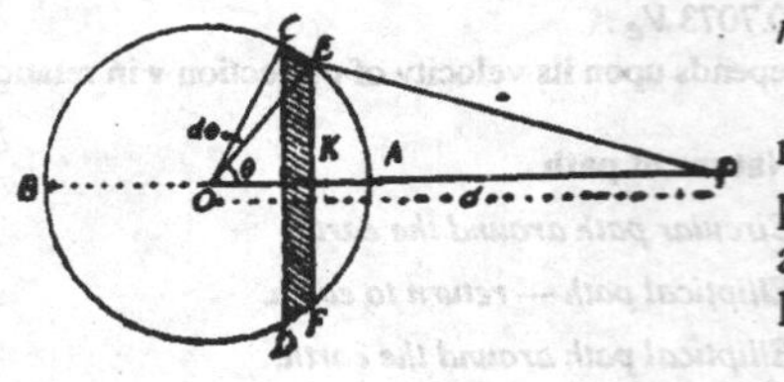

Fig. 7.17

Join OP and cut out a slice $CEFD$, in the form of a ring, by two planes close to each other and perpendicular to the radius OA, meeting the shell in C and D, and in E and F respectively. Let $\angle EOP = \theta$, and let the small $\angle COE = d\theta$.

Clearly, the radius of the ring is $EK = OE \sin\theta = R\sin\theta$; so that, its circumference $= 2\pi R\sin\theta$ and its width $= CE = R\,d\theta$.

$\therefore$ *area* of the ring or slice = its circumference × its width.

$$= 2\pi R.\sin\theta \times R\,d\theta.$$

And $\therefore$ its *mass* $= 2\pi R\sin\theta \times R d\theta \times \rho = 2\pi R^2 \sin\theta\, d\theta\, \rho$

If $EP = x$, every point of the slice is at a distance r from P, and, therefore, the potential at P due to this small slice is given by

$$dV = -\frac{mass\ of\ slice}{x}.G = -\frac{2\pi R^2 \sin\theta\, d\theta\, \rho}{r} G. \quad \text{...(i) [See § 7.11]}$$

Now, in $\Delta\, OEP$, $EP^2 = OE^2 + OP^2 - 2OE.OP\,.\cos\theta$.

Or, $$x^2 = R^2 + r^2 - 2Rr\cos\theta$$

because $OE = R$, the *radius of the shell.*

Differentiating the above expression, we have

$$2x.dx = 0 + 0 + 2Rr\sin\theta\, d\theta = 2R\,r\sin\theta d\theta. \quad \text{[R and r being constants}$$

Hence $$x = \frac{2R\,r\sin\theta d\theta}{2\,dx} = \frac{Rr\sin\theta d\theta}{dx}$$

Substituting this value of x in expression (*i*) above, we have

$$dV = -\frac{2\pi R^2 \sin\theta\, d\theta\, \rho}{Rr\sin\theta\, d\theta} Gdr = -\frac{2\pi R\rho G\,dx}{r}$$

Integrating this between the limits, $x = AP = (r - R)$, and $x = BP = (r + R)$, we get V, the potential due to the whole shell at the point P.

$$\text{Thus, } V = \int_{(r-R)}^{(r+R)} -\frac{2\pi R\rho G dr}{r} = \frac{2\pi R\rho G}{r}\int_{(r-R)}^{(r+R)} dx.$$

$$= -\frac{2\pi R\sigma G}{r}\left[x\right]_{(r-R)}^{(r+R)} = -\frac{2\pi a\rho G}{d}.2a = \frac{4\pi a^2 \rho.G}{d}$$

Now, $4\pi R^2$ is the surface area of the whole shell, and, therefore, $4\pi R^o \rho$ is equal to its mass M. We thus have

$$V = -\frac{M}{r}G.$$

Or, *the potential at the point P due to the whole shell is equal to – M.G/d, i.e., the same as it would be due to a mass M at O.*

The mass of the whole shell thus behaves as though it were concentrated at its centre.

(b) At a point on the surface of the shell. In the above case, if we imagine the point P to lie at A, *i.e., on the surface of the shell* itself, we obtain the potential there by integrating the expression for dV between the limits $x = 0$ and $x = 2R$. So that in this case,

$$V = \int_0^{2a} \frac{2\pi R\rho G}{r}.dr = -\frac{2\pi R\rho G}{r}\left[x\right]_0^{2R} = -\frac{4\pi R^2\rho G}{r}$$

$$= -MG/r = -MG/R \qquad [\because \text{ here, } r = R$$

thus, $$V = -\frac{M}{r}G.$$

Again, therefore, *the whole mass of the shell behaves as though it were concentrated at its centre.*

(c) **At a point inside the shell.** Imagine now the point P to lie inside the spherical shell (Fig. 7.18). Then, proceeding as above, we have

potential at P due to the slice or ring $CEFD$, *i.e.*,

$$dV = \frac{2\pi R\rho G}{r}dx$$

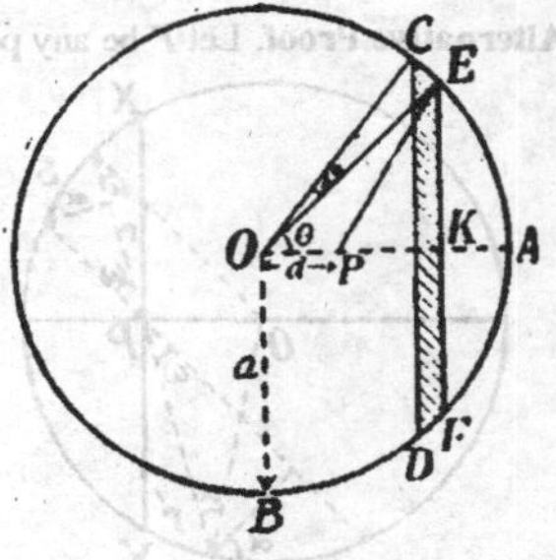

Fig. 7.18.

In this case, the limits of x are $(R-r)$ and $(R + r)$. So that, we have

$$V = \int_{(R-r)}^{(R+r)} -\frac{2\pi R\rho G}{r}dx.$$

$$= -\frac{2\pi R\rho G}{r}\left[x\right]_{(R-r)}^{(R+r)}$$

$$= -\frac{2\pi R\rho G}{d}2r = -4\pi R\rho G.$$

Or, multiplying and dividing by R, we have $V = -\dfrac{4\pi r^2\rho G}{R}$.

Now, $4\pi R^2\rho = M$, the mass of the shell.

Hence $$V = -\frac{M}{R}G,$$

i.e., the *same* as at a point on the shell. [See case *(a)*, above.

Since the above value of V has been obtained for a point P *anywhere* inside the shell, it follows that *the potential at all points inside a spherical shell is the same, and is numerically equal to the value of the potential on the surface of the shell itself, which is also its maximum (negative) value.*

7.15. Gravitational Field due to a Spherical Shell

(*a*) **At a point outside the shell.** We know that the gravitational field at a point is given by the *potential gradient* at that point, or that $E = -dV/dx$.

Now, potential at a point distant r from the centre of the shell, (*i.e.*, $r > R$) is given by $V = -MG/r$. We, therefore have *intensity of the gravitational field at the point, i.e.*,

$$E = -dV/dr = -\frac{d}{dr}\left(-\frac{M}{r}G\right). \text{ Or, } E = \frac{M}{r^2}G,$$

the negative sign indicating that the force is one of attraction *the same as though the whole mass of the shell were concentrated at its centre.*

(*b*) **At a point on the outer surface of the shell.** As we know, the gravitational potential at a point on the outer surface of the shell is given by $V = -MG/R$. We, therefore, have **intensity of the gravitational field at the point,** *i.e.*, $E = -\dfrac{d}{dr}\left(-\dfrac{MG}{R}\right) = -\dfrac{M}{R^2}G.$

(*c*) **At a point inside the shell.** As we have seen in § 7.14 (*c*) above, the gravitational potential at all points inside a spherical shell is the *same*.

Now, the field at a point is given by the potential gradient at that point, *i.e.*, by $E = dV/dx$. Since V is constant for all points inside the shell, $dV/dx = 0$, *i.e.*, the field in the interior of the shell is zero. In other words, **there is no gravitational field inside a spherical shell.**

It follows at once, therefore, that the gravitational field *at a point on the inner surface of the shell will be zero,* whereas on its outer surface, as we have seen, it is $-MG/R^2$. Thus, there is thus a sudden change in its value from $-MG/R^2$ on the outer surface to zero on it inner surface, *i.e., there is a definite, sharp discontinuity in the gravitational intensity at the surface of a spherical shell.*

Alternative Proof. Let P be any point inside a spherical shell or a hollow sphere, of *surface density* (*i.e., mass per unit area*) ρ. Through P draw straight lines so as to form two small cones, with their apices at P and intercepting small areas S and S' on the shell, as shown (Fig. 7.19) on opposite sides of the plane XY, drawn perpendicular to the diameter passing through P. Let S and S' be at distances x and X' from P and let the *solid angles* subtended by the two areas (S and S') at the point P be ω each (see note on solid angles on page 233).

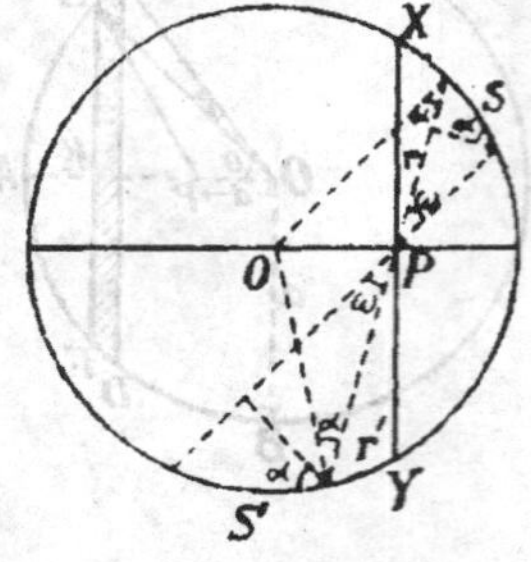

Fig. 7.19

Then clearly, the area of the right section of the cone S is equal to $x^2\omega$ and of that at S' *equal to* $x'^2\omega$

If, therefore α be the angle that the right section of the cones S and S' make with S and S' respectively, we have

$$S \cos\alpha = x^2\omega \quad \text{and} \quad S' \cos\alpha = x'^2\omega$$

And $\therefore$ $\qquad S = x^2 . \omega/\cos\alpha$ and $S' = x'^2\omega/\cos\alpha$.

So that, mass of area $S = x^2\omega\rho/\cos\alpha$. and of $s' = x'^2\omega\rho\,\sigma/\cos\alpha$.

$\therefore$ intensity at P due to $S = \dfrac{x^2\omega\rho}{\cos\alpha\, x^2} = \dfrac{\omega\rho G}{\cos\alpha}$ in the direction PS.

And intensity at P due to $S' = \dfrac{x^2\omega\rho}{\cos\alpha\, x^2} = \dfrac{\omega\rho G}{\cos\alpha}$ in the direction PS'.

These two intensities at P being *equal and opposite*, their resultant is *zero*. Similar is the case for all other pairs of cones on opposite sides of XY into which the shell may be divided, so that, the resultant intensity or field at P due to the whole shell is zero. The same is also true for any other point inside the shell. In other words, there is no gravitational field inside a spherical shell.

It will be noted that this is a result similar to the one we have in *Electrostatics, viz.*, that there is no field of force inside a charged spherical shell.

7.16. Gravitational potential due to a solid sphere

(*a*) **At a point outside the sphere.** Let the point *P* lie *outside* a sphere of radius *R* mass *M*, at a distance *r* from its centre *O*, (Fig. 7.20). Imagine the sphere to consist of a large number of thin shells, concentric with the sphere, one inside the other, as shown and let their respective masses be m_1, m_2, m_3 ... etc.

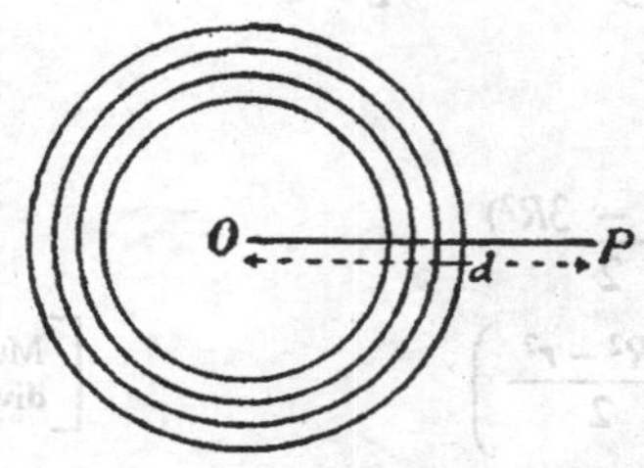

Fig. 7.20.

Then, as we have seen in § 7.14 (*a*), the *potential at P due to each spherical shell will be equal to its mass X G/r, i.e.*, the same as though the mass of each shell is concentrated at its centre. So that, the potential at *P* due to different shells, will be $-m_1.G/r$, $-m_2.G/r$, $-m_3G/r$ and so on.

$\therefore$ *Potential at P due to all the shells, i.e., due to the whole solid sphere, i.e.,*

$$V = -\left[\frac{m_1}{r}.G + \frac{m_2}{r}.G + \frac{m_3}{r}.G + ...\right] = -(m_1 + m_2 + m_3 + ...)\frac{G}{r} = -\frac{M}{r}.G,$$

because $m_1 + m_2 + m_3 + ... = M$, *the mass of the solid sphere.*

Thus, *potential at P due to the solid sphere* $= -MG/r$.

(*b*) **At a point on the surface of the sphere.** If $r = R$, we have

$$\textit{potential at the surface of the sphere} = -\frac{MG}{R}$$

(*c*) **At a point inside the sphere.** Let the point *P* now lie inside the solid sphere of *radius R, mass M* and *volume density* ρ at a distance *d* from its centre *O*, (Fig. 7.21)

The solid sphere may be imagined to be made up of an *inner solid sphere of radius r, surrounded by a number of hollow spheres or spherical shells, concentric with it and with their radii ranging from r to R.* The potential at *P* due to the solid sphere is, therefore, equal to the sum of the potentials at *P* due to the inner solid sphere and all such spherical shells outside it (because potential is a scalar quantity).

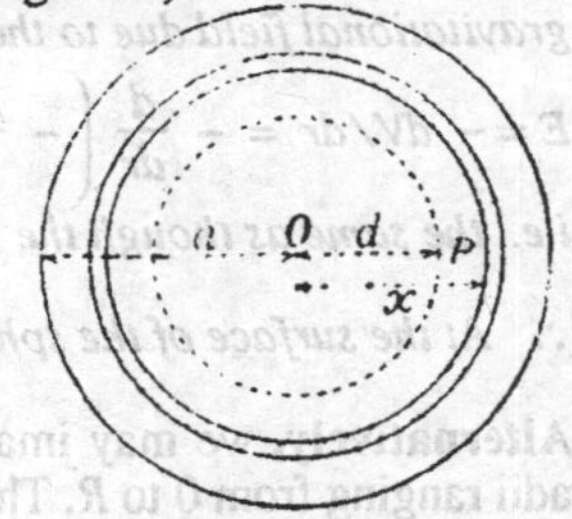

Fig. 7.21

Clearly, *P* lies on the surface of the solid sphere of radius *r* and inside all the shells of radii greater than *r*.

$\therefore$ *Potential at P due to the inner solid sphere of radius*

$$R = -\frac{\textit{mass of sphere}}{r}G = -\frac{4}{3}\pi r^3 \rho G/r = -\frac{4}{3}\pi r^2 \rho G, \qquad ...(i)$$

because mass of the sphere $= \frac{4}{3}\pi r^3 \rho$.

To determine the potential at *P* due to the outer shells, consider a shell of radius *x* and thickness *dx*, so that its *volume = area × thickness* $= 4\pi x^2 dx$ and its mass $= 4\pi x^2 dx\, \rho$.

Now, the potential at any point within a shell is the same as at any point on its surface [§ 7.14 (*b*)] and, therefore, *potential at P due to this shell* $= \dfrac{-4\pi x^2.dx\,\rho G}{x} = -4\pi x\,dx\,\rho G$.

Integrating this for the limits $x = r$ and $x = R$, we get tne potential at *P due to all the shells.* Thus,

$$\textit{potential at P due all the shells} = \int_r^R -4\pi\rho Gxdx = -4\pi\rho G\int_r^R xdx$$

$$= -4\pi\rho G\left[\frac{x^2}{2}\right]_r^R = -4\pi\rho G\frac{(R^2-r^2)}{2}^{*} = -\frac{4}{3}\pi\rho G\,\frac{3(R^2-r^2)}{2}$$

$$= -\frac{4}{3}\pi\rho G\frac{(3R^2-3r^2)}{2} \qquad ...(ii)$$

$$\therefore\quad \textit{total polutial at } P = -\frac{4}{3}\pi r^2\rho G - \frac{4}{3}\pi\rho G.\frac{(3R^2-3R^2)}{2}$$

$$= -\frac{4}{3}\pi\rho G\left(r^2+\frac{3R^2}{2}-\frac{3r^2}{2}\right) = -\frac{4}{3}\pi\rho G\left(\frac{3R^2-r^2}{2}\right)$$

$$= -\frac{4}{3}\pi a^3\sigma G\frac{(3R^2-r^2)}{2a^3} \qquad \left[\text{Multiplying and dividing by } R^3\right.$$

Clearly, $\frac{4}{3}\pi R^3\rho = M$, the *mass of the whole solid sphere.*

So that, potential at *P* due to the solid sphere $= -MG\dfrac{(3R^2-r^2)}{2R^3}$

If $r = 0$, point *P* would lie at the centre of the sphere and we shall have **potential at the centre of the sphere = – 3MG/2R.** It will thus be seen that *potential at the centre of the sphere : potential on the surface of the sphere is as* $-3MG/2R : -MG/R$ or as 3 : 2.

Gravitational field due to a solid sphere

(*a*) **At a point outside the sphere.** We know that the potential at a point outside a sphere, distant *r* from its centre is $-MG/r$, as explained above in § 7.16 (*a*), (page 228). And since gravitational field at a point is equal to the potential gradient there, we have

gravitational field due to the solid sphere at a point distant r from its centre, i.e.

$$E = -\,dV/dr = -\frac{d}{dr}\left(-\frac{MG}{r}\right).\ \text{Or,}\quad E = -\frac{MG}{r^2}$$

i.e., the same as though the whole mass of the sphere were concentrated at its centre.

$\therefore$ *At the surface of the sphere itself* $E = -\dfrac{M}{R^2}G$.

Alternatively, we may imagine the sphere to consist of a number of spherical shells, with their radii ranging from 0 to *R*. Then, considering one such spherical shell of radius *x* and thickness *dx* and, therefore, of mass $4\pi x^2 dx\rho$, we have

Field at a point *P* distant *r* from the centre of the shell $= 4\pi x^2 dx\,\rho G/r^2$

---

* This expression is equal to $-2\pi\rho G\,(R^2-r^2)$; so that, the potential at *P* due to all the outer shells $= -2\pi\rho G\,(R^2-r^2)$; and, the potential at *P* due to the whole solid sphere is, therefore, also

$$= -\frac{4}{3}\pi r^2\rho G - 2\pi\rho G(R^2-r^2) = -2\pi\rho G\left(\frac{2}{3}r^2+R^2-r^2\right)$$

$$= -2\pi\rho G\left(R^2+\frac{2r^2-3r^2}{3}\right) = -2\pi\rho G\left(R^2-\frac{r^2}{3}\right).$$

And $\therefore$ *field at P due to the whole sphere* $= -\dfrac{4\pi\rho G}{3r^2}\displaystyle\int_0^R x^2\,dx = \dfrac{4}{3}\pi R^3 \rho\, G/r^2$.

Clearly, $\frac{4}{3}\pi R^3\rho = M$. the *mass of the sphere*. So that, we have

field at P due to the sphere, i.e. $E = -MG/r^2$.

N.B. This result is of great historical importance in that it enabled *Newton* to apply his law of gravitation to the motion of the moon. For, the radius of the earth not being negligible compared with that of the moon's orbit around it, there would have been no means, in the absence of the above result, of determining what correction terms, if any, would be necessary in the equation of the moon's motion in view of the distribution of matter inside the earth and the finite value of its radius.

(*b*) **At a point inside the sphere.** We have seen in § 7.16 (*c*) above, that the potential at a point inside a solid sphere, a distance *x* from its centre is given by $-MG(3R^2 - r^2)/2R^3$. Since the gravitational field at a point is given by the potential gradient there, we have

gravitational field at distance x from the centre of sphere, i.e.,

$$E = -dV/dr = -\frac{d}{dr}\left[-MG\frac{(3R^2 - r^2)}{2R^3}\right] = -MG^2\frac{r}{2R^3}$$

Or, $\mathbf{E = -M.G.r/R^3}$,* *i.e.*, $E \propto r$,

i.e. the *intensity of the field is directly proportional to the distance from the centre of the sphere*. Obviously, therefore, *at the centre of the sphere, r being zero, the intensity of the gravitational field is zero.*

In other words, *the intensity of the gravitational field inside a solid sphere has its minimum (negative) value at its centre.*

It will also be readily seen that in the expressions for the intensity of the field (*E*) outside and inside a *solid sphere*, deduced above under (*a*) and (*b*), if we put $r = R$, the *radius of the sphere*, we have $E = MG/R^2$ as the field on the surface of the sphere, *in either case, i.e., whether the point P lies on the outer or the inner side of the surface of the sphere*. This shows clearly that *there is a continuity in the intensity of the field at the surface of a solid sphere. (ref. case of a spherical shell* (§ 7.15 (*c*)].

7.11 Intensity and Potential of the Gravitational Field at a Point due to a Circular disc. Let *MN*, (Fig. 7.22), represent a circular disc of radius *R*, with its plane perpendicular to the plane of the paper and let *O* be a point on its axis, distant *r* from its centre *C*, where the intensity and potential due to its gravitational field are to be determined.

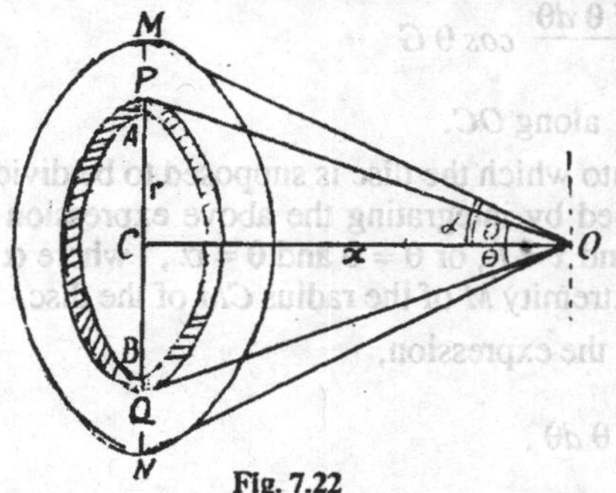

Fig. 7.22

Imagine the disc to consist of an infinite number of concentric rings with *C* as their common centre.

Consider one such ring *PQ* of *radius x* and *thickness dx*.

Join *O* to the extremity *A* of the radius *CA* of the ring. Then, if angle *AOC* be equal to θ we have $x = CA = r\tan\theta$.

And, therefore, differentiating with respect to θ, we have

$dx/d\theta = r\sec^2\theta$. Or, $dx = r\sec^2\theta \, .d\theta$. And $AO = r\sec\theta$.

* Taking the value of *V* at distance *r* from *O* to be $-2\pi\rho G\left(R^2 - x^{2/3}\right)$, (see foot-note page 230), we have *intensity or field at distance from O, i.e.,*

$$E = -\frac{dV}{dr} = -\frac{d}{dr}\left[-2\pi\rho G\left(R^2 - \frac{r^2}{3}\right)\right] = -2\pi\rho G\frac{2r}{3} = -\frac{4}{3}\pi\rho G r.$$

And, therefore, $E \propto r$

Now, *area of the ring = circumference* × *thickness* = $2\pi x dx$.

And, therefore, *mass of the ring* = $2\pi x\, dx\, \rho$

where ρ is *mass per unit area* (or *surface density*) of the disc.

Considering a very small element A of the ring, we have intensity at O due to *this* element = *mass of the element* × G/AO^2, along the direction OA.

This can be resolved into two rectangular components (*i*) along and (*ii*) at right angles to OC. So that,

$$\text{the former component} = \frac{\textit{mass of the element}}{AO^2} G \cos\theta, \text{ along } OC,$$

$$\text{and the latter component} = \frac{\textit{mass of the element}}{AO^2} G \sin\theta, \text{ vertically upwards.}$$

Similarly, for an equal element at B, *diametrically opposite to A*, we have intensity at O, resolved into the same two components, *viz.*,

(*i*) $\dfrac{\textit{mass of element}}{AO^2} G\cos\theta$, *along OC*; and

(*ii*) $\dfrac{\textit{mass of the element}}{AO^2} G\sin\theta$, *vertically downwards.*

Thus, the two vertical components, *being equal and opposite*, cancel out, and the components along OC alone being *effective* are added up, *both acting in the same direction OC*; the same will be the case with other elements into which the ring may be supposed to be broken up; and, therefore, intensity at O due to the whole ring is equal to the sum of the components along OC due to the different elements, *i.e.*,

$$= \frac{\textit{mass of the ring}}{AO^2} G\cos\theta.$$

Or, $\quad$ intensity at O due to the ring $= -\dfrac{2\pi x\, dx\, \rho}{AO^2} G\cos\theta$.

Or, substituting the values of x, dx, and AO, we have

$$\text{intensity at } O \text{ due to the ring} = -\frac{2\pi\rho\, r\tan\theta\, r\sec^2\theta\, d\theta}{x^2\sec^2\theta}\cos\theta\, G$$

$$= -2\pi\rho\, G\sin\theta\, .\, d\theta, \text{ along } OC.$$

Therefore, intensity at O due to all the rings into which the disc is supposed to be divided up, *i.e.*, intensity at O due to the whole disc, is obtained by integrating the above expression for he intensity due to the ring, between the limits $x = 0$ and $x = R$, or $\theta = 0$ and $\theta = \alpha$, where α is the angle between OC and the line joining O with the extremity M of the radius CM of the disc.

Thus, intensity at O due to the disc is given by the expression,

$$\int_0^\alpha -2\pi\rho\, G\sin\theta\, d\theta = -2\pi\rho\, G\int_0^\alpha \sin\theta\, d\theta.$$

$$= -2\pi\rho\, G\Big[-\cos\theta\Big]_0^\alpha = 2\pi\rho\, G\Big[-\cos\alpha - (-\cos 0)\Big]$$

$$= -2\pi\rho\, G(-\cos\alpha + 1)$$

Or, $\quad$ *intensity at O, due to the disc* $= -2\pi\rho\,(1-\cos\alpha)$ $\quad$ ...(*i*)

And, since $\cos\alpha = \dfrac{CO}{MO} = \dfrac{r}{\sqrt{r^2 + R^2}}$, we have

intensity at O, due to the disc $= -2\pi\rho G\left(1 - \dfrac{r}{\sqrt{r^2+R^2}}\right)$ ...(*ii*)

Or, again, because $2\pi(1 - \cos\alpha)$ is the *solid angle*, ω, say, (see *Note on solid angle* below), subtended by the disc at the point *O*, we have, from relation (*i*) above,

intensity at O due to the disc $= \rho G \omega$. ...(*iii*)

Now, potential at *P* due to the ring of radius *x* is equal to the intensity at *P* due to the ring $\times\, r = -2\pi\rho G \sin\theta\, d\theta \times r$, because field at *P* is equal to the *potential gradient* at *P*.

Hence, potential at *P* due to the whole disc, *i.e.*,

$$V = -\int_0^{\alpha} 2\pi\rho G\, r \sin\theta\, d\theta = -2\pi\rho G r \int_0^{\alpha} \sin\theta\, d\theta.$$

$$= -2\pi\rho G x\,(1 - \cos\alpha) \qquad ...(iv)$$

$$= -2\pi\rho G x\left(1 - \frac{r}{\sqrt{r^2+R^2}}\right) \qquad ...(v)$$

$$= -\rho G\,\omega . r \qquad ...(vi)$$

Note on Sodid Angle. Suppose, we have an area *PS*, (Fig. 7.23), as the base of a cone, with its apex at point *O*. Then, if we draw a sphere, with centre *O* and any radius *r*, so that a surface of area *pq* of it is cut off by the cone, then *pq* is proportional to r^2, where *r* is the radius of the spherical surface *pq*; and, therefore, *area* pq/r^2 is constant for any given cone. This quantity, area pq/r^2, is called the *solid angle* of the cone, or the solid angle subtanded by the *area pq* at *O*, and is usually denoted by the letter ω. Obviously it is also equal to *area* PS/OP^2 or the *solid angle* subtended by *area PS at O*, and its numerical value is equal to *area pq*, if $r = 1$ *cm*.

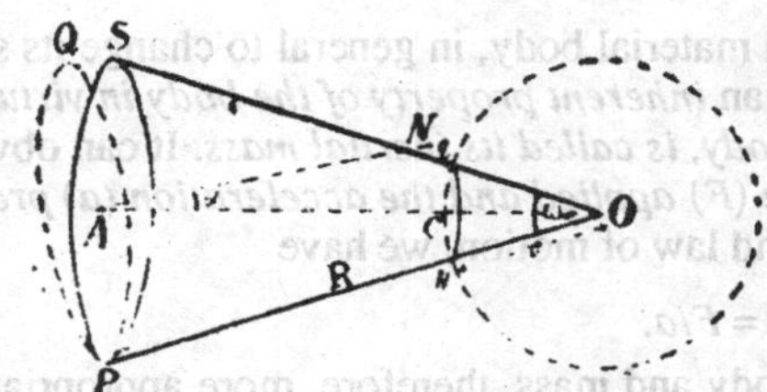

Fig. 7.23.

Now, suppose the given area be *PQ* and not *PS*. Then, if *AN* be the normal to it at its centre *A*, we have *area PS = area PQ cos α* where α is the angle between *AO* and the normal *AN* to the surface *PQ* at *A*.

Thus, $\omega = \dfrac{area\ PQ \cdot \cos\alpha}{OP^2}$.

Now, to determine the solid angle ω, subtended by a circular shell or disc *MN, of radius *R*, at *a* point *O*, distant *x* from its centre *C*, (Fig. 7.24), we draw a sphere, with *O* as centre and $(R^2+x^2)^{1/2}$ as radius, such that its circular face lies on this shell or disc. Then, we have, from above,**

$$\omega = \frac{surface\ area\ of\ slice\ MFN}{(R^2 + x^2)}.$$

Since the area of a sphere lying between two parallel planes is equal to the area of the circumscribing cylinder in between these planes, with its axis perpendicular to them, we have

$$area\ MFN = 2\pi(R^2 + x^2)^{1/2} \times FC$$

$$= 2\pi(R^2 + x^2)^{1/2} \times (FO - CO)$$

$$= 2\pi(R^2 + x^2)^{1/2} \times (MO - x) = 2\pi(R^2 + x^2)^{1/2} \times ([(R^2 + x^2)^{1/2} - x]$$

Fig. 7.24

because $FO = MO$ **and** $CO = x$.

Hence $\omega = \dfrac{2\pi\,[(R^2 + x^2)^{1/2} - [(R^2 + x^2)^{1/2} - x]}{R^2 + x^2} = 2\pi\left(1 - \dfrac{x}{(R^2 + x^2)^{1/2}}\right)$

Or, $\omega = 2\pi(1 - \cos\alpha)$

* ***MN*, here, represents the side-view of the shell or disc.**

where α is the *semi-vertical* angle, subtended by the shell or disc at O.

7.18. Intensity and Potential of the Gravitational Field at a Point due to an Infinite Plane. In the case of the disc, above (§ 7.17), if its radius R becomes infinite, the disc becomes an *infinite plane*. In this case, obviously, α becomes $\pi/2$; so that $\cos\alpha = 0$, and $\omega = 2\pi$.

Thus, if we put $R = \infty$ in expression (*ii*), or $\cos\alpha = 0$ in expression (*i*), or $\cos\omega = 2\pi$ in expression (*iii*) above, we have

$$\textit{intensity at O due to an infinite planes} = -2\pi\rho G.$$

which is, clearly, quite independent of the distance r from it.

Similarly, putting these values of R, $\cos\alpha$ and ω in relations (*iv*), (*v*), (*vi*) above, we have

$$\textit{potential at O due to an infinite plane} = 2\pi\rho G r$$

Inertial and Gravitational mass. (*i*) **Inertial mass.** We are aware of the property of inertia possessed by a material body, *i.e.*, its property of resisting any change in its position of rest or uniform motion along a straight line. Thus, we know that some effort (or force) has to be applied to put into motion a block, of whatever material, lying at rest on a smooth, frictionless horizontal surface, or to stop or even to slow it down or decrease its acceleration, if it be in motion on the surface, and also that the bulkier the block, *i.e.*, the greater its mass, the greater the effort required for the purpose. Obviously, *gravity* does not come into the picture in this case (the motion being perpendicular to the direction in which the gravity acts). In fact, the effort or the force required to set the block in motion or to stop it would still be the same if the experiment were performed in gravity-free space.

This inertness or reluctance of the block, or a material body, in general to change its state of rest or uniform motion along a straight line, is thus an *inherent property of the body in virtue of its mass, which, since it represents the inertia of the body, is called its inertial mass.* It can obviously be measured in terms of the *ratio between the force (F) applied and the acceleration (a) produced in the body.* For, in accordance with Newton's second law of motion, we have

$$F = ma \quad \text{and} \quad \therefore \; m = F/a.$$

Here, m represents the *inertial mass* of the body and mass, therefore, more appropriately be denoted by the symbol m_i.

If we apply the same force F to two bodies of inertial masses m_i and m_i' and if the acceleration produced in them be a and a' respectively, we have

$$F = m_i a = m_i a', \text{ whence } \; m_i/m'_i = a'/a.$$

Thus, *the inertial masses of two bodies are inversely proportional to the accelerations produced in them by a given force.* This relation holds good, irrespective of the magnitude of the force applied.

We can, therefore, measure the *inertial mass* (m_i) of a body by comparing the acceleration produced in it with that produced in a standard mass (M), such as the *standard kilogramme* (preserved at *Sevres*) by the application of the same force F to both If the acceleration produced in the two masses be a and A respectively, we have

$$F = m_i a = MA, \text{ whence } m'/M = A/a \text{ or, } m_i = \frac{A}{a} M.$$

so that, m_i can be easily evaluated.

(*ii*) **Gravitational mass.** We can also measure the mass of a body in terms of the gravitational force of attraction exerted on it by another body, such as the earth. For, as we know, we have to exert an upward force to prevent a body from falling under the gravitational force of the earth and, again the bulkier the body, the greater the force required to hold it. Here, clearly, the property of inertia plays no part but only the action of gravity. *This mass of a body, measured in terms of the gravitational force of attraction on it due to another body, usually, the earth, is called its gravitational mass.* Thus, if F and F' be the forces of attraction exerted by a given body A, of *its gravitational mass.* Thus, if F and F' be the forces of attraction exerted by a given body A, of mass M, on two bodies B and C, of masses m and m' respectively at the *same* distance R from it, we have

$$F = mMG/R^2 \text{ and } F' = m'\,MG/R^2, \text{ whence, } F/F' = m/m'.$$

Here m and m' are the *gravetational masses* of the two bodies and may be denoted by mg and $m'g$ respectively, to distinguish them from their inertial masses m_i and m'_i. So that $F/F' = mg\,/\,mg'$.

Thus, we see that *the gravitational masses of these two bodies B and C are directly proportional to the forcs of attraction exerted on them by the body A.*

If the body A happens to be the *earth*, clearly, $F = w$ and $F' = w'$, where w and w' are the weights of bodies B and C respectively at the given place. We, therefore, have

$$w = \frac{m_g M}{R^2} G \text{ and } w'_{\bullet} = \frac{m_g' M}{R^2} G, \text{ whence, } \frac{w}{w'} = \frac{mg}{m'g}.$$

Or, *the ratio between the gravitational masses of two bodies is equal to the ratio between their weight at a given place.*

It follows, therefore, that the mass of a body, as measure by a spring balance or by balancing it in one pan against a standard mass in the other, of a physical balance, is its *gravitational mass.*

Equality of inertial and gravitational mass. The question naturally arises whether the inertial mass of a body is any different from, or the same as, its gravitational mass. As a first step to enquire into the problem, let us try to see whether the inertial masses of two bodies too are proportional to their weights at a given place. For this, we first determine the inertial masses m_i and m'_i of the two bodies in the manner discussed under (*i*) above. Then, since in free space (or vacuum) all bodies fall, with the same acceleration (g) at a given place, under the action of gravity, we have forces on the two bodies due to attraction by the earth, *i.e.*, their weights w and w' equal to m_ig and m'_{ig} respectively. And, if mg and $m'g$ be the gravitational masses of the two bodies, their weights w and w' are also equal to m_gMG/R^2 and m'_gMG/R^2 respectively, as we have just seen above. So that,

$m_ig = m_gMG/R^2$ and $m'_ig = m_g'MG/R^2$, whence $m_i/m'_i = m_g/m_{g'}$, showing that *the ratio of the inertial masses of two bodies is the same as that of their weight at a given place and hence also the same as that of their gravitational masses.*

The inertial and gravitational masses of a body are thus clearly proportional to each other. To show that they are actually equal to each other, a number of experiments have been performed by *Newton, Eotvos* and others, as briefly explained below.

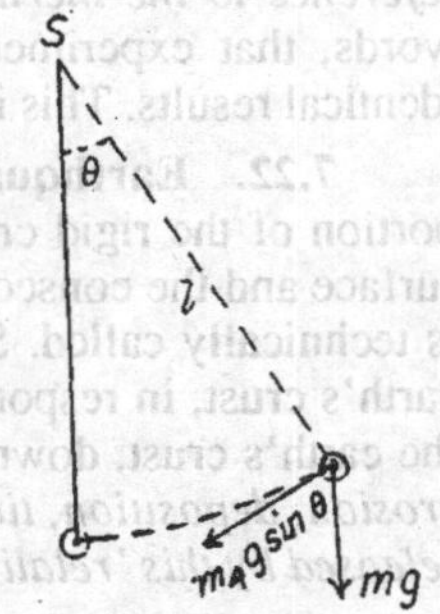

(*a*) Newton's was perhaps the first attempt to test the equality of inertial and gravitational mass by his experiment with the simple pendulum. For, as we know, the acceleration of the pendulum bob for a small displacement x is d^2x/dt^2, (Fig. 7.25). If, therefore, m_i be the *inertial mass* of the bob, the force tending to bring it back to its orgiginal positional is $m_i.d^2x/dt^2$. But, clearly, if m_g be its *gravitational mass*, the restoring force on its is $m_g\,g \sin\theta$. It may be noted that *the gravitational force acting on the gravitational mass m_g of the bob is here producing the acceleration d^2x/dt^2 in its inertial mass m_i.*

We, therefore, have $m_g\,g \sin\theta = m_i\,d^2x\,/dt^2$. or, $m_i\,d^2x\,/dt^2 = m_g g\;\; x\,/l$,

whence, $d^2x\,/dt^2 = \dfrac{m_g}{m_i}.\dfrac{g}{l}x = \mu x$, where, $\dfrac{m_g}{m_i}.\dfrac{g}{l}\,\mu$.

So that, *time-period of the pendulum,* $T = 2\pi\sqrt{\dfrac{1}{\mu}} = 2\pi\sqrt{\dfrac{m_i}{m_g}.\dfrac{l}{g}}$.

Obviously, unless $m_i = m_g$, as instinctively assumed by Newton, the relation for T does not reduce to the simple form $T = 2\pi\sqrt{l/g}$,he derived for it.

In order to confirm this equality of inertial and gravitational mass, Newton used a thin spherical shell as his pendulum bob and determined the time-period of the pendulum, after filling the bob with *exactly the same weight* of

Now, with the force acting on the bob exactly the same in each case, as also the air-resistance to its motion (the external shape of the bob remaining unaltered) any difference in the time-periods of the pendulum in the different cases, could arise only from a change in the inertial mass of the bob due to replacement of one substance by another (of equal weight) inside it. No difference in the time-period of the pendulum could be observed within the limits of accuracy of his experiment, and it was always obtained by the relation $T = 2\pi\sqrt{l/g}$.

This established the equality of the inertial and the gravitational mass of a body to an accuracy of 1 in 1000. The same result was obtained by *Bessel* in his improved version of the pendulum experiment, with an accuracy some 60 times higher.

(*b*) Later, *Eotvos*, in the *year* 1909, devised an ingenious experiment with the torsion balance, in which he suspended two bodies of *different materials* but of the *same weight* from the two arms of the balance. The two bodies were thus acted upon by the gravitational forces of the earth, *viz*., their own weights, and centrifugal force due to the earth rotation. The former forces, as will be readily seen, were proportional to the gravitational masses of the two bodies and the latter, to their inertial masses.

In the event of the inertial and gravitational masses of the bodies being different, the centrifugal forces on the two would have been different and a *torque* would have acted on the beam, tending to twist the suspension fibre, and to turn the beam form its equilibrium position along the east-west direction, and this torque would have been reversed on turning the beam through 180°. No such change in the equilibrium position of the beam of the balance could be observed, thereby establishing the equality of the inertial and gravitational mass of a body to withing 1 part in 5×10^9.

(*c*) Finally, as recently as the year 1961, *R.H. Dicke* performed an even more refined version of the Eotvos torsion balance experiment, at *Princeton* and established the equality of the inertial and gravitational mass of a body to 1 part is 10^{10}. We can thus safely assert that *the inertial mass of a body, is indeed, equal to its gravitational mass.*

This equality of inertial and gravitational mass is of great consequence in astronomy. For example, due to the proportionality of gravitational force to inertial mass, the orbit of a *satellite* around a planet is quite independent of its mass and we can thus '*weigh*' the planet from a mere observation of the orbit of its satellite.

Again, we can determine the mass of one component of a doublestar by observing the orbit of the other around their centre of mass, the required value of G being obtained by any one of the terrestrial methods for it, dealt with under § 7.6.

Principle of equivalence. We have just seen above, how the mass of a body when it is subjected to a gravitational attraction but no acceleration, (*i.e.*, its *gravitational mass*) comes out to be the same as when it is subjected to an acceleration but no gravitational attraction (*i.e.*, its *inertial mass*). *This gave* **Einstein** *the idea that a gravitational field can be initiated by a field of acceleration* and this ultimately, led to the formulation of his *general theory of relativity*, in which he showed that *a non-accelerating* or *inertial frame of reference, in which there is a uniform gravitational field, is physically equivalent to a reference frame accelerating uniformly with reference to the inertial frame but in which there is no gravitational field.* This means, in other words, that experiments carried out in the two frames, under the same conditions will yield identical results. This is called the *principle of equivalence*. (See chapter XVI for a fuller account).

7.22. Earthquakes — Seismic Waves and Seismographs. An earthquake is caused by a portion of the rigid crust* of the earth giving way, or getting fractured, some distance below its surface and the consequent sudden slipping of the resulting portion, or due to '*fault slipping*', as it is technically called. So to speak, it is just a landslide on a large scale, or a re-adjustment of the earth's crust, in response to a change of forces, or more precisely, to changes of pressure deep in the earth's crust, down to a distance of 100 *miles* or so, brought about by a variety of causes, like *erosion, deposition, tidal forces, centrifugal forces*, etc. *An earthquake thus represents the energy released by this 'relative motion of portions of the earth's crust'*.

The place where the actual fracture occurs is called the **focus** of the earthquake, and is not a geometrical point, but an extended region. The point nearest to the focus, on the surface of the earth, is called the **epicentre.**

* See foot note on page 206.

From the focus, (which we may, for our purpose here, regard as just a point), originate a number of different types of waves, collectively called **seismic waves**, which spread on to different points on the surface of the earth and which we *feel* as **'earthquake tremors'**. The general pattern of these seismic waves is as shown in Fig. 7.29, and they consist of the following different types of waves :

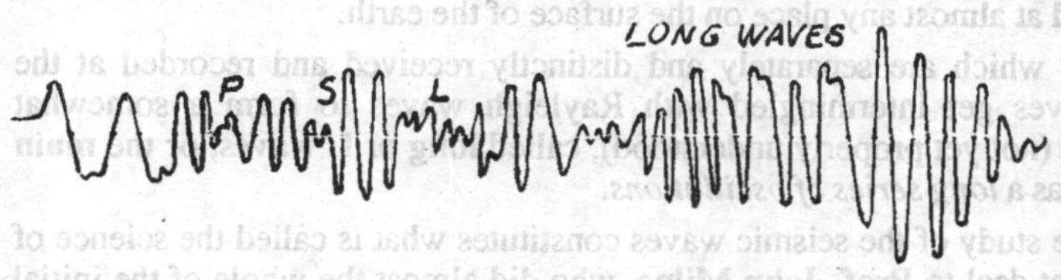

Fig. 7.29

(*i*) *Preliminary waves* and (*ii*) *Surface waves.*

(*i*) **Preliminary waves.** These consist of the *primary* or *P waves* and *secondary* or *S waves.*

(*a*) **The Primary or P Waves.** The first to arrive at the Observing Station, these are *longitudinal waves*, in which the particles of the earth vibrate about their mean position, along the direction of the waves themselves.

If the earth be regarded to be a homogeneous sphere, these waves, starting from the focus, travel along the chord of huge circle of the earth, with a velocity equal to $\sqrt{j/\Delta}$, where j is what is called the '*elongational elesticity*'* of the earth and Δ, its density. These waves are also variously called as '*condensational*', '*irrotational*' and '*push*' waves and their velocity is found to be about 5 *miles per second.*

(*b*) **The Secondary or S Waves.** These are *transverse waves*, in which the particles of the earth vibrate at right angles to the direction of propagation of the waves, thus having no component along this direction. Starting from the focus, these waves also travel along a chord of a huge circle of the earth and are the next to arrive at the Observing Station, with a velocity equal to $\sqrt{n/\Delta}$ where n and Δ represent the modulus of rigidity and the density of the earth respectively. The other names given to these waves are '*distortional*', '*equivoluminal*' and '*shake*' waves, their velocity being about 3 *miles per second.*

(*ii*) **Surface waves** . They also consists of two sets of waves, *viz.*, *Rayleigh waves* and *Love waves.*

(*c*) **Rayleigh Waves.** Discovered by *Lord Rayleigh*, these waves are found to remain confined to a comparatively thin layer in the close vicinity of the earth's surface. Unlike the *P* and *S* waves, *they start from the epicentre* and arrive at the Observing Station along a huge circle of the earth, —the displacement of the particles at any point on the earth's surface, due to them, being in the vertical plane containing their direction of propagation. Resolving this displacement we have (*i*) *a vertical component* and (*ii*) *a horizontal component, along the direction of propagation, there being no horizontal component, at right angles to it.* These waves thus persist over long distances along the surface of the earth, and are almost unique in this respect. If the earth were a homogeneous sphere, these waves also would travel with a constant velocity, but, due to its heterogeneous character, each single wave, starting from the epicentre, gets split up into a number of different sets of waves, each set having a different wavelength, velocity etc.; so that, what we receive at the Observing Station is a series of oscillations instead of one single '*kick* or *throw*' *as would be the case if there were no such splitting up of the original wave, i.e.*, if the earth were really homogeneous in composition.

(*d*) **Love Waves.** The heterogeneity of the layers of the earth is also responsible for another type of *surface waves*, known as *Love waves*, in which the displacement of the earth is horizontal, but transverse to the direction of their propagation. The velocity of these waves is less in the

* The *elongational elasticity* $j = \gamma\ (1 - \sigma)\ /\ (1 + \sigma)\ (1 - 2\sigma)$, where γ and σ stand for *Young's modulus* and *Poisson's ratio* respectively.

earth's crust than in the matter below. Immediately after an earthquake, oscillations, corresponding to these waves, can be detected at almost any place on the surface of the earth.

Unlike *P* and *S* waves, which are separately and distinctly received and recorded at the Observing Station, these waves get intermingled with Rayleigh waves to form a somewhat complicated system of waves, (not yet properly understood), called **long or L waves,** or the **main shock,** registering themselves as a *long series of oscillations*.

7.23. Seismology. The study of the seismic waves constitutes what is called the science of *Seismology*, and it owes a great deal to **Prof. John Milne,** who did almost the whole of the initial pioneering work on the subject. As early as the year 1883, when he was residing in Japan, he predicted that "*every large earthquake might be, with proper appliances, recorded at any point on the land surface of the globe*". And, then, in the year 1889, a curious incident confirmed his prophetic words. For, a delicate horizontal pendulum, set up for the measurement of the gravitational action of the moon, gave recordings, which turned out to be due to an earthquake, with its origin somewhere in Japan. This started a new era of intensive researches on the subject, with Prof. Milne in the very forefront; and, in 1895, he set up his own observatory at Shide in the Isle of Wight, which became the centre of a world-wide seismic survey.

By the year 1901, the main fact as to how the tremors travelled through and round the earth were fully established, again, due in main, to the labours of the eminent Professor himself.

His reports to the British Association on Earthquake Phenomena in Japan from 1881 to 1895, together with those on Seismological Investigation from 1895 to 1913, (the year of his death), form a fascinating and a detailed study of the growth and development of the present-day science of Seismology.

7.24. Seismographs. A *seismograph* (or a *seismometer*), is an instrument used to record the *earth tremors* or the *seismic waves,* to some dynamical function of which, (like displacement, velocity, acceleration, etc.), they respond or react. The record of the vibrations so obtained is called the *seismogram.* The instruments, responding to displacement, are of the mechanical type and we are, therefore, concerned here only with these. The following is, in brief, the theory underlying the mechanical type of seismographs.

All vibrations of the earth may ultimately be resolved into (*i*) *vertical* and (*ii*) *horizontal components**. The problem thus reduces itself to merely recording these vertical and horizontal vibrations. We shall confine our attention here only to the measurement of the *horizontal displacements*, accompanying these latter vibrations. There are two types of instruments in use for the purpose, *viz.*, (*a*) *the vertical pendulum type* and (*b*) *the horizontal pendulum type*.

(*a*) **The Vertical Pendulum Seismographs.** A vertical pendulum is just a rigid body, suspended from a stand resting firmly on the ground; so that, with the horizontal displacement of the ground and the stand with it, the point of support of the pendulum also gets displaced horizontally.

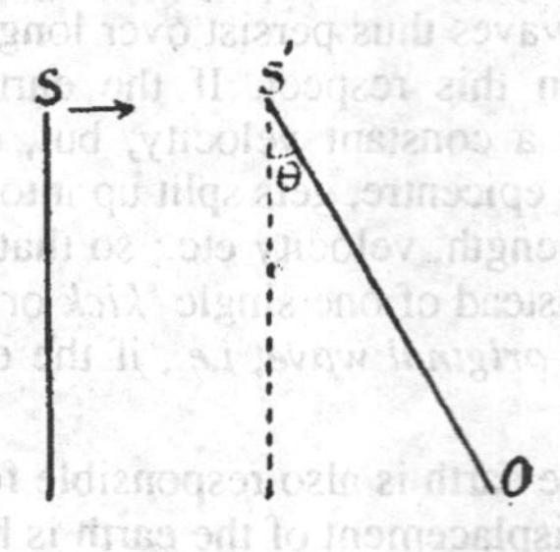

Fig. 7.30

Thus, if the point of support *S* of the vertical pendulum, (Fig. 7.30), is displaced horizontally to *S'*, due to the horizontal displacement of the ground, it can be shown that a style or pen attached to its lower end, reproduces faithfully the movements of the support, with precisely the same frequency, (though on a different scale), it being assumed that the support moves with a definite frequency and amplitude.

These vertical pendulum seismographs, however, suffer from two defects *viz.*, (*i*) they have to be very

* These components may be along East and West or along North and South but will be *horizontal,* nevertheless. these can also be used to measure the horizontal velocity and acceleration of the earth, or rather of the earth's crust.

heavy — as much as 22 *tons* or more — if a good magnification of the vibrations be desired, and (*ii*) their period of vibrations is rather small.

(*b*) **The Horizontal Pendulum Seismographs.** We are already familiar with the horizontal pendulum, [see § 6.20 (*ii*)]*. Only some slight additions to it convert it into a sensitive and a reliable seismograph. With the horizontal movement of the earth, the supports of the pendulum, which are firmly fixed on to it, also share its movement, thus setting its stem or '*boom*' into motion, which can then be magnified mechanically or electrically by various devices.

The best known seismograph of its class is that due to **Prince Boris Galitzin**, in which the greatest care has been taken to see that its indications correspond exactly to the actual movements or vibrations of the earth. We shall, therefore, discuss in some detail only this one instrument here.

7.25. Galitzin's Seismograph. This seismograph measures the horizontal velocity of the earth's crust, and consists of a horizontal pendulum having a *boom* or stem, 28 *cm* long, carrying a cylindrical brass bob, weighing 7 *kgm.* and having its centre at a distance of 14 *cm* from the inner end of the *boom*. The suspension of the pendulum is of the *Zollner type*, (as shown in Fig. 6.33), with a very small inclination of the axes, so that the period of oscillation of the pendulum is about 24 *seconds*. The whole pendulum is built up on a rigid framework, firmly secured to the ground, and consisting of four metal pillars, braced and arranged rectangularly on four points on a metal base or plate, provided with levelling screws.

The recording of the vibrations or *tremors* at the Observing Station is done electromagnetically, and, for this purpose, a flat copper coil is wrapped round a portion of the stem or *boom* of the pendulum, extending beyond its cylindrical bob, and connected to a sensitive moving-coil mirror galvanometer.

With the motion of the stem, (caused by the motion of the ground), the coil moves in the strong magnetic field of a pair of permanent horse-shoe magnets of tungsten-steel. *A current, which is proportional to the angular velocity of the stem*, is thus induced in the coil and produces a deflection in the galvanometer.

A beam of light, reflected from the mirror fixed on the suspension of the galvanometer coil, is passed through a semi-cylindrical lens and allowed to fall on a sensitized (*i.e.*, photographic) paper, wrapped round a rotating drum, worked by a clock-work arrangement and moving uniformly along its axis, with a peripheral speed of 3 *cm per minute*. Time-signals are also similarly recorded on the paper by cutting off, by means of an accurately-timed shutter, the beam of light for two seconds at the beginning of each successive minute. A permanent record of a series of curves, (*i.e.*, the *seisomogram*), is thus obtained on the sensitized paper, from which the time of occurrence of any seismic phenomenon can be determined accurately to within *one* second.

In order that the horizontal displacement of the earth may be correctly calculated from the seismogram thus obtained, it must faithfully correspond to the movements of the earth. To achieve this, damping of both the pendulum and the galvanometer is necessary. Or, else, if the period of oscillation of the pendulum agrees, or nearly agrees, with that of the seismic wave; resonance will occur, producing large deflections, which would give an utterly deceptive picture of the actual movement of the ground. And, if the damping be made *critical*, (*i.e.*, *dead-beat*), the calculations become greatly simplified.

This *damping* is produced by attaching to the outer end of the boom, a horizontal brass plate, which moves in another strong magnetic field, due to a separate pair of horse-shoe magnets, arranged above and below it. The eddy currents, thus induced in the plate, then produce, with proper adjustments, the desired damping effect on the pendulum.

This seismograph has additional advantage of great magnification**, as also of enabling the recording apparatus to be arranged in a separate compartment, away from the pendulum.

---

* The student would do well to refresh his memory by going over this article once again before proceeding further.

** Out of a set of 8 seismographs at the Fordham University in New York city, there are three which magnify the motion of the ground about 2,000 times.

N.B. It will be readily understood that for a large or severe earthquake, less sensitive seismographs are more suitable, while, for smaller, local or nearby earthquakes, the more sensitive ones or the short-period ones, are the more desirable.

7.26. Determine of the Epicentre and the Focus

The Epicentre. To determine the *epicentre* of an earthquake, we determine what are called the *epicentral distances* of it from a network of Observing Stations or Observatories, —*the epicentral distance of an earthquake from a given station being the shortest distance of its epicentre from the station, measured along the surface of the earth, in terms of the angle it subtends at the centre of the earth.* This is done with the help of the *Tables*, compiled by *Zoppritz*, *Turner* and others, which give the relation between the epicentral distances of past earthquakes and the interval between the first arrivals of the Primary (*P*) and the Secondary (*S*) waves at a station, *i.e.*, which express the epicentral distances as functions of the corresponding time-intervals *S–P*.

Thus, from the seismogram of an earthquake, obtained at an Observing Station, we can determine the time-interval *S—P* for it at that station, and the *Tables* then give the epicentral distance of the earthquake from it. This is done at as many stations as possible. Circles are then drawn on a globe, with these different stations as their respective centres and their epicentral distances as the radii. The point of intersection of these circles then gives the most probable position of the epicentre of the earthquake in question. Or, the same may be obtained from the method of least squares.

Usually, three seismographs are used for the purpose, one responding to motions of the ground *along the East-West direction*, the second, along the '*North-South direction*, and the third, having a pendulum suspended by a coiled spring, to respond to the *vertical displacements* of the ground. The information supplied by the three, when pieced together, enables not only the epicentre, but also the character, of the earthquake to be determined fairly accurately.

(*ii*) **The focus.** To determine the position of the focus, immediately below the epicentre, we use what is known as **Seebach's method**, explained belows:

Let *F*, (Fig. 7.31), be the focus of an earthquake, (taken to be a point here), a distance *h* vertically below the epicentre *E*, and let *O* be the position of the Observing Station, a horizontal distance *d* from the epicentre.

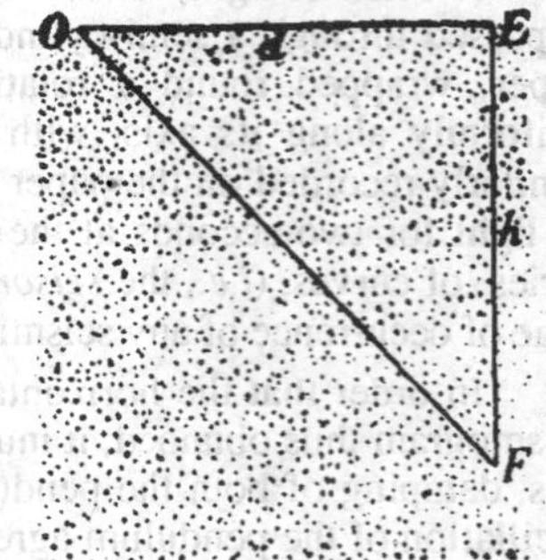

Fig. 7.31

Then, assuming the homogeneity of the medium in between the earth's surface and the surface of a sphere, concentric with it and passing through *F*, the time *t* taken by the *P* waves to travel from *F* to *O* is clearly given by

$$t = FO/v = \sqrt{d^2 + h^2}/v,$$

whence, $(d^2 + h^2) = v^2.t^2$, ...(*i*)

where *v* is the velocity of *P* waves in the medium and can be determined independently by other methods.*

Now, if the time at which the earthquake occurs be τ_0*, and the time at which the first *P* waves arrive at *O* be τ, we have $t = (\tau - \tau_0)$ where τ naturally varies with distance *d*.

Thus, relation (*i*) above may be put in the form,

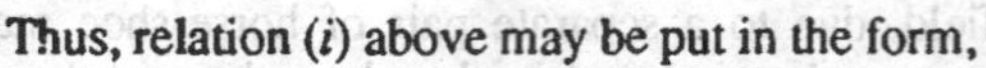

$$(d^2 + h^2) = v^2(\tau - \tau_0)^2.$$

Obtaining the corresponding values of *d* and *t* from a number of different Observing Stations, we plot a graph between *d* and *vt*, which gives a hyperbola, from which *h* can be easily calculated and hence the position of the focus determined.

* This, though not known to us, is certainly constant.

A better method, however, is to calculate, by the method of least squares, the most probable values of h and τ_0.

N.B. It may be of interest to know that the severest earthquakes have their foci about a hundred kilometres below the earth's surface.

7.27. Modern Applications of Seismology. The development of the modern science of seismology has led to its application in four important fields, *viz.*, *(i) investigation of the nature of the interior of the earth, (ii) prospecting for oils and minerals, (iii) construction of quake-proof buildings, and (iv) forecasting of the occurrence of earthquakes.*

(i) **Nature of the interior of the earth.** It is now almost fully established (according to *Jeffreys*) that the earth consists of a *dense core* of a molten mass, mostly of iron, together with some nickel, of a density of about 12.0 *gm/cc* (at the pressure existing there), surrounded by a *solid outer shell* or *crust*, about 3,000 kilometres thick, the density of which decreases from about 5.0 *gm/cc* at its innermost layers to about 2.7 *gm/cc* at the outermost layers, or at the surface of the earth.

The existence of the dense core is decided from the observed refraction of the seismic waves, as they pass through the earth, and is further confirmed by the production and propagation of the secondary or shake waves (*S*) through the core. These waves, as we know, are *transverse* in nature and, as such, can only be produced and propagated in media possessing elasticity of shape or rigidity, *viz.*, in solids.

(ii) **Prospecting for oils and minerals.** Prospecting for oil, coal and other minerals is now being increasingly done with the help of seismic waves*, the process being technically known as *'seismic prospecting'*.

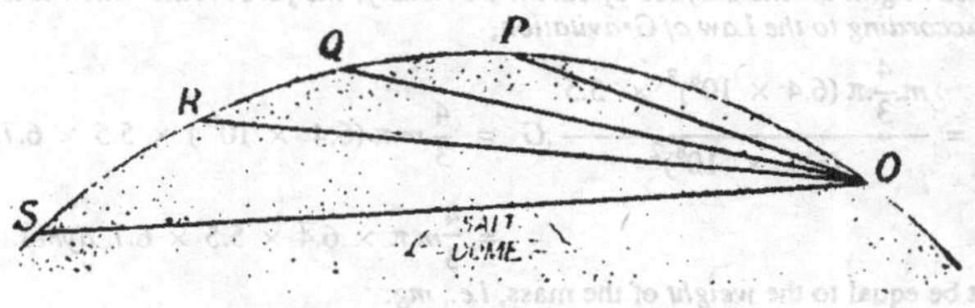

Fig. 7.32

Artificial earthquakes are set up in the ground-region to be surveyed for the purpose by detonating an explosive, like gun-cotton or gelignite, at a point O on the earth'ssurface, (Fig. 7.32), and the time of explosion noted. The time of arrival of the first low frequency longitudinal waves, or the primary waves, thus produced, is noted, with he help of seismographs, at different stations *P*, *Q*, *R* and *S*, etc., all lying in the same plane. The distance from *O*, covered along the chords *OP*, *OQ*, *OR*, *OS* etc. of the earth are carefully measured and the mean velocities of the waves calculated along these different paths or chords.

If one of the paths or chords, say, *OS*, happens to pass through a mineral deposit, like a salt dome, the value of the mean velocity along this particular chord will be different from that along the other chords. The experiment is then repeated along a direction, perpendicular to the first, by exploding a fresh charge of explosives. And, if this confirms the results of the first experiment, a more elaborate survey determine the positions of the top and the sides** of the salt dome.

(iii) **Construction of quake-proof buildings.** It has now been found possible to erect *quake-proof* buildings in California, Japan and other places frequently visited by earthquakes, at

* We have already studied the gravitational methods of prospecting, by means of the Eotvos balance etc. (See § 6.22)

** The locating of the sides of the dome is equally important, because some mineral oil is almost always found to be there.

a surprisingly low additional cost of just 15%. For, it has been shown by *Prof. Suyehiro* that the severest earthquakes of Japan can do but little damage to buildings *designed to resist a horizontal force equal to one-tenth of their total weight.* The day is thus not far off when damage to buildings due to earthquakes will just become a memory of a dreadful past.

(iv) **Forecasting of the occurrence of earthquakes.** The prediction of the occurrence of an earthquake, a good time in advance, is now fast coming into the realm of practical possibility. For, it has been established that the region, where an earthquake occurs, exhibits for quite a few years, before, a 'tilt' or a *gradual rise*, very much like the rubber tube of a pneumatic tyre or a football bladder swelling up before it actually bursts.

There seems to be but little doubt that much sooner than we can imagine at the moment, an earthquake forecast will become as general and universal an affair as the weather forecast is today.

But even as it is, the loss in buildings etc. resulting from the severest earthquakes seldom exceeds 5%, due to their being confined to a very small area, and, quite often, an uninhabited one. The disastrous effects of earthquakes have thus been unduly magnified; and, for all we know, they may be for our own good, designed by a benign Providence by way of safety devices to save us from being blown up all in a heap.

WORKED EXAMPLES

1. Given $g = 6.7 \times 10^{-8}$ *c.g.s.* units the radius of the earth = 6.4×10^8 cm and its mean density, 5.5 gm/cc, calculate the acceleration due to gravity at the earth's surface.

Imagine the earth to be a perfect sphere, we have

$$\text{volume of the earth} = \frac{4}{3}\cdot\pi\,(6.4\times10^8)^3\ cc$$

And $\therefore$

$$\text{mass}\ \ ,,\ \ ,, = \frac{4}{3}\cdot\pi.\,(6.4\times10^8)^3\times 5.5\ gm$$

Consider a mass *m gm. on the surface of earth. Obviously, the force with which it is being attracted by the earth towards its centre is, according to the Law of Gravitation,*

$$= \frac{m.\frac{4}{3}.\pi\,(6.4\times10^8)^3\times5.5}{(6.4\times10^8)^2}.G = \frac{4}{3}m.\pi\,(6.4\times10^8)\times5.5\times6.7\times10^{-8}$$

$$= \frac{4}{3}m.\pi\times6.4\times5.5\times6.7\ dynes.$$

$\therefore$ this must be equal to the *weight* of the mass, *i.e., mg*.

Thus,
$$mg = \frac{4}{3}.\pi.m\times6.4\times5.5\times6.7$$

whence,
$$g = \frac{4}{3}\pi\times6.4\times5.5\times6.7 = 988.3\ cm/sec^2$$

Or, the acceleration due to gravity at the earth's surface is $988.3\ cm/sec^2$

2. Two lead spheres of 20 cm and 2 cm diameter respectively are placed with their centres 100 cm apart. Calculate the force of attraction between the spheres given the radius of the earth as 6.67×10^8 cm and its mean density as 5.33 gm/cc (Sp. gr. of lead = 11.5).

If the lead spheres be replaced by brass spheres of the same radii, would the force of attraction be the same ?

Clearly, force of attraction between the masses = $\dfrac{product\ of\ masses}{(distance)^2}.G$.

$$= \frac{4}{3}.\pi.(10)^3\times\frac{4}{3}\pi.(1)^3\times(11.5)^2.G/100^2 = 16\pi^2\times10^3\times(11.5)^2.G/9\times10^4.$$

$$= 16\pi^2\times(11.5)^2\times G/90 = 8\pi^2\times(11.5)^2\times G/45.$$

Now, force on a mass of *one gram* on the earth's surface = $G\,M/R^2 = g$.

Or, $G.M/R^2 = 980$, taking $g = 980\ cm/sec^2$

Or, $G.\dfrac{4}{3}.\dfrac{\pi R^3}{R^2}\times5.33 = 980.$ Or, $G.4.\pi.R\times5.33 = 980\times3.$

Or, $G = 980\times3/4.\pi.R\times5.33.$

$\therefore$ force of attraction between the load spheres is given by

$$F = \frac{8\pi^2 \times (11.5)^2}{45} \times \frac{980 \times 3}{4\pi R \times 5.33} = 15.28 \times 10^{-6} \text{ dynes}$$

Since the force of attraction between the spheres depends upon their masses, it will naturally be different in the case of brass spheres whose mass will be much less than that of the lead spheres, (the density of brass being much less than that of lead).

3. Calculate the mass of the earth from the following data :

Radius of the earth = 6×10^8 cm; Acceleration due to gravity = 980 cm/sec^2 and Gravitational Constant = 6.6×10^{-8} cm^2 gm^{-1} sec^2. *(Punjab)*

We know that the force with which the earth attracts a *unit mass* towards itself; (*i.e.*, towards its centre)

$$= 1 \times g = 980 \text{ dynes.}$$

Also, the force of attraction between the mass and the earth is given by $M \times G/R^2$, where M is the *mass of the earth*, R, its *radius* and G, the *Gravitational Constant*. Clearly, therefore, $M.G/R^2 = g$.

Or, $$M = g.R^2 / G. \quad ...(i)$$

$\therefore$ substituting the values of g, R and G in relation (*i*), we have

$$M = 980 \times (6 \times 10^8)^2/6.6 \times 10^{-8} = 53.47 \times 10^{26} \text{ gm.}$$

4. Estimate the mass of the Sun, assuming the orbit of the earth round the Sun to be a circle. The distance between the Sun and the Earth is 1.49×10^{13} cm., and G = 6.66×10^{-8} c.g.s. units.

(Bombay, Agra (B.Sc. Part II), Aligarh)

Let the mass of the Sun be M *gm.* and that of the Earth, m *gm.* Then, if R be the radius of the Earth's orbit round the Sun (or the distance between them), we have

$$\textit{force of attraction between the Sun and the Earth} = G.M.m/R^2.$$

This clearly supplies the centripetal force to the earth as it goes round the Sun in its orbit, (here assumed to be circular)

If v be the velocity of the earth in its circular orbit, this *centripetal force* = mv^2/R. And, we, therefore, have

$$G M.m/R^2 = mv^2/r, \text{ whence } M = v^2R^2/RG = v^2R/G$$

Now, $v = 2\pi R/T$, where T is the time–period of the Earth's motion round the Sun, *i.e.*, the time taken by it is describe its orbit once round the Sun. So that, we have

$$M = \frac{(2\pi R)^2.R}{T^2.G} = \frac{4\pi^2.R^3}{T^2.G}$$

Here $R = 1.49 \times 10^{13}$ *cm*, $G = 6.66 \times 10^{-8}$ *C.G.S. units*

and $T = 365$ *days* $= 365 \times 24 \times 3600$ *sec.*

Substituting these values in the expression above, we therefore have

$$M = \frac{4\pi^2 \times (1.49 \times 10^{13})^3}{(365 \times 24 \times 3600)^2 \times 6.66 \times 10^{-8}} = 19.72 \times 10^{34} \text{ gm.}$$

Or, the mass of the Sun is 19.72×10^{34} *gm.*

5. Assuming that a sphere of mass 40 kilograms is attracted by a second sphere of mass 80 kilograms, when their centres are 30 cm apart, with a force equal to the weight of $\frac{1}{4}$ mg.; calculate the Constant of Gravitation.

Here, force of attraction between the two masses

$$= \frac{40 \times 1000 \times 80 \times 1000}{30^2}.G = \frac{32 \times 10^8}{9 \times 10^2}.G = \frac{32 \times 10^6}{9}.G.$$

But this is equal to $\frac{1}{4}$ *mg.wt.*, *i.e.*,

$$= \frac{1}{4} \times \frac{1}{1000} \text{ gm wt} = \frac{1}{4} \times \frac{1}{1000} \times 981 \text{ dynes}, \quad \therefore \frac{32 \times 10^6}{9}.G = \frac{981}{4000}$$

Or, $$G = \frac{981 \times 9}{4000 \times 32 \times 10^6} = \frac{981 \times 9}{4 \times 32 \times 10^9} = 6.898 \times 10^{-8} \text{ C.G.S. units.}$$

6. Two small balls of mass *m* each, are suspended side by side by two equal threads of length *L*. If the distance between the upper ends of the threads be *a*, find through what angle the threads are pulled out of the vertical by the attraction of the balls.

Let the upper ends of the threads be at *A* and *B*, (Fig. 7.33), such that the distance $AB = a$. Then, due to mutual attraction, the balls are drawn towards each other, say, through a distance *x* each, from their original positions.

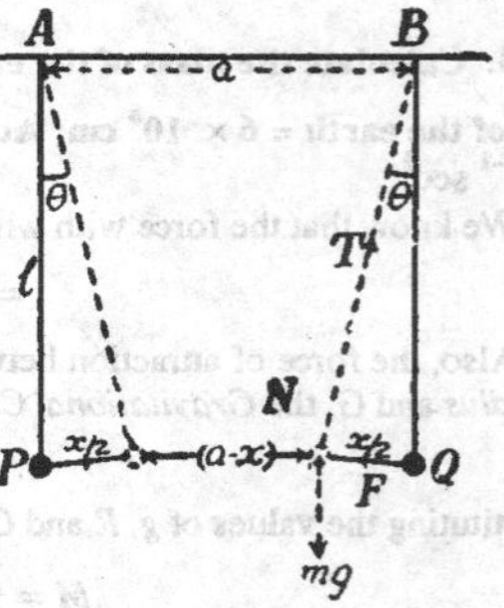

Considering the forces acting on the ball *Q*, which keep it in equilibrium, we have

(i) *weight* of the ball = *mg*, acting vertically downwards; (ii) *force of attraction*, *F* =

$$\frac{m \times m}{(a-x)^2}.G = \frac{m^2.G}{(a-x)^2}$$

and (iii) *tension of the thread T*, along it.

Since the ball is in equilibrium, the three forces can be represented by the three sides of a triangle, taken in order.

If the dotted lines show the positions of the threads when the balls are in equilibrium, the three forces can be represented, in magnitude as well as direction, by the sides of the triangle *BQN*, —the side *BQ*, representing the *weight* mg, the side *QN*, representing *F* and the side *NB*, representing the *tension* T of the string, *in cyclic order*.

Fig. 7.33

So that clearly, $\frac{F}{mg} = \frac{QN}{BQ} = \tan\theta$. Or, $\frac{m^2G}{(a-x)^2.\,mg} = \tan\theta$.

whence, $\tan\theta = \frac{m.G}{(a-x^2).g}$. Or, $\theta = \tan^{-1}\frac{m.G}{(a-x)^2.g}$

Thus, the threads will be pulled out of the vertical through an angle $\theta = \tan^{-1} mG/(a-x)^2.g$.

7. The radius of the Moon's orbit '*r*' is 240,000 miles, and the period of revolution is 27 days, the diameter of the Earth is 8,000 miles and the value of gravity on its surface is 32 ft./sec². Verify the statement that the gravitational force varies inversely as the square of the distance.

Here, distance covered by the Moon in 27 *days* = 2π × 240000 *miles*.

∴ velocity of the Moon, $v = \frac{2\pi \times 240000 \times 1760 \times 3}{27 \times 24 \times 60 \times 60}$ *ft/sec*.

Now, centripetal acceleration of a body moving in a circle = v^2/r. Hence the centripetal acceleration of the Moon towards the centre of the Earth is given by

$$g_m = \left[\frac{2\pi \times 24 \times 176 \times 3 \times 10^5}{27 \times 24 \times 60 \times 60}\right]^2 \times \frac{1}{240000 \times 1760 \times 3}$$

$$= \frac{4\pi^2 \times 176 \times 3 \times 10^5}{(27 \times 24 \times 36)^2 \times 10^4} = \frac{4\pi^2 \times 176 \times 3 \times 10}{(27 \times 36)^2 \times 24} = 0.009189\ ft/sec^2.$$

Then, denoting the acceleration due to gravity on the surface of the Earth by g_e (= 32 *ft per sec*²), and supposing it to be inversely proportional to the *n* th power of the distance, we have

$$\frac{g_m}{g_e} = \left(\frac{R_e}{R_m}\right)^n,$$

where R_e is the radius of the earth, and R_m the distance of the Moon from the Earth.

Or, $$\frac{0.009189}{32} = \left(\frac{4000 \times 1760 \times 3}{240000 \times 1760 \times 3}\right)^n = \left(\frac{1}{60}\right)^n.$$

Or, taking logarithms, we have

$\bar{4}.4582 = n\,(\bar{2}.2218)$. Or, $n = \bar{4}.4582/\bar{2}.2218 = 2$, *nearly*.

Thus, *g* varies inversely as the *second power* of the distance and hence *the gravitational force varies inversely as the square of the distance*.

8. The radius of the earth is 6.37×10^{-8} cm, its mean density, 5.5 *gm./cc.* and the gravitational constant, 6.66×10^{-8} c.g.s. units. Calculate the earth's surface potential. *[Agra (B.Sc., Part I); Bombay.]*

We know that potential $V = GM/x$. (Taking the earth to be a perfect sphere.

Now, mass of the earth, M = volume × density $= \frac{4}{3}.\pi\,(6.37 \times 10^8)^3 \times 5.5$

distance, $x = r$, in this case, $= 6.37 \times 10^8$ *cm.*

and $G = 6.66 \times 10^{-8}$ *C.G.S. units* (given).

$$V = \frac{6.66 \times 10^{-8} \times 4\pi\,(6.37 \times 10^8)^3 \times 5.5}{3 \times 6.37 \times 10^8} = \frac{6.66 \times 10^8 \times 4\pi(6.37)^2 \times 5.5}{3}$$

$$= 2.22 \times 10^8 \times 4\pi\,(6.37)^2 \times 5.5 = 6.27 \times 10^{10}\ ergs/gm.$$

9. Define intensity of gravitational field and gravitational potential; hence show that the intensity and potential at any point of the surface of the earth are 'g' and 'gR' respectively, assuming the earth to be a uniform sphere of radius 'R'. *(Poona).*

For definitions of gravitational field and gravitational potential see articles 7.9 and 7.10 respectively.

Assuming the earth to be a *uniform sphere* of *mass M* and *radius R*, we have *intensity of gravitational field at any point on its surface* $= MG/R^2$.

Now, since the weight of a body of *mass m* on the surface of the earth $= mg = m.M.G/R^2$, we have $g = MG/R^2$. So that, intensity of gravitational field at any point on he earth's surface $= g$.

Now, *potential at any point on the earth's surface, i.e.,*

$V = MG/R$, [see § 7.16 (*b*) [ignoring u

Or, $V = (MG/R^2).R = gR$.

10. Show how the mass of the earth may be compared with that of the sun from a knowledge of the time-periods of the earth round the sun and of the moon round the earth, together with the radii of their respective orbits, (taken to be circular).

Let M_s and M_e be the masses of the sun and the earth respectively and R_e, the radius of the earth's orbit round the sun. Then if ω_e be the angular velocity of the earth, we have

$$\frac{M_s . M_e}{R_e^2} . G = M_e . \omega_e^2 . R_e, \quad \text{whence, } M_s . G = \omega_e^2 . R_e^3. \qquad ...(i)$$

Similarly, if M_m be the mass of the moon, R_m, the radius of its orbit round the earth and ω_m, its angular velocity, we have

$$\frac{M_e . M_m}{R_m^2} . G = M_m . \omega_m^2 . R_m, \quad \text{whence, } M_e . G = \omega_m^2 . R_m^3. \qquad ...(ii)$$

From relations (*i*) and (*ii*), we have $$\frac{M_e}{M_s} = \left(\frac{\omega_m}{\omega_e}\right)^2 . \left(\frac{R_m}{R_e}\right)^3. \qquad ...(iii)$$

Now, if T_e and T_m be the *time-periods* of the earth and the moon respectively, we clearly have $(\omega_m/\omega_e)^2$ in relation (*iii*) above, we have

$$\frac{M_e}{M_s} = \left(\frac{T_e}{T_m}\right)^2 . \left(\frac{R_m}{R_e}\right)^3.$$

Thus, knowing the time-periods of the earth and the moon and the radii of their respective orbits, the mass of the earth can be compared with that of the sun.

N.B. It will be easily seen that (*i*) the method is only an approximate one and (*ii*) it is applicable when one of the planets in question has a satellite revolving about it.

11. Calculate the intensity at a point due to an infinitely long straight wire of line density ρ.

Let *AC* be a portion of the wire, (Fig. 7.34), of line density ρ, and consider an element *AB* of the wire, of length *dl*. Let *AO* be a length *l* of the wire; and let *P* be a point at a distance *x* from *O*.

Join *PA*, and let $\angle APO = \theta$.

Then, clearly, $\tan\theta = l/x$.

Or, $l = x . \tan\theta$

Differentiating it with respect to θ, we have $dl = x \sec^2\theta . d\theta$.

Therefore, mass of this element $AB = \rho \cdot dl = x \cdot sec^2 \theta \cdot d\theta . \rho$.

And $\therefore$ intensity at P due to the element $= \dfrac{x.sec^2 \theta.d\theta.\rho}{AP^2}$. G, in the direction PA.

Or, *intensity* at P due to the element =

$$= \frac{x.sec^2\theta.d\theta.\rho}{x^2.sec^2\theta} . G = \frac{\rho.G.d\theta}{x}, \text{ along } PA.$$

because $AP = x \sec \theta$.

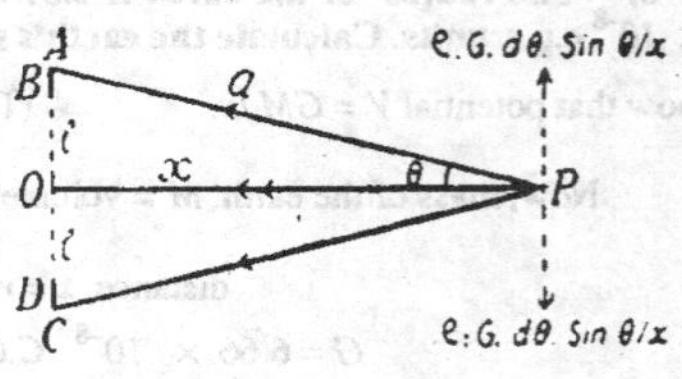

Fig. 7.34

Resolving it into two rectangular components, along PO and perpendicular to it, we have *(i)* the *component along* PO, equal to $\rho \cdot G \cdot d\theta \cdot cos\ \theta/x$ and *(ii)* the component at right angles to it, equal to $\rho.G \cdot d\theta \cdot sin\ \theta/x$.

Similarly, the intensity at P due to an equal element dl at the other end C, at a distance l from O will be $\rho \cdot G \cdot d\,\theta/x$, in the direction PC. Resolving it into two rectangular components, the component $\rho \cdot G \cdot d\theta \cdot sin\ \theta/x$, being equal and opposite to that due to the element AB will cancel out and the componen $\rho.G.d\ \theta.cos\ \theta\ /\ x$ will act along PO, as before, and the two will, therefore, be added up. The same is true for any two similarly situated equal elements of the wire. Therefore, considering the whole wire, we have

$$\text{intensity at } P \text{ due to the whole wire} = 2\int_0^{\pi/2} \frac{\rho.G.cos\ \theta}{x} . d\theta\ 2 = \left[\frac{\rho.G}{x} . sin\ \theta\right]_0^{\pi/2}$$

$$= 2 . \frac{\rho . G}{x} . sin \frac{\pi}{2} = 2 . \frac{\rho.G}{x} .1 = 2 . \frac{\rho . G}{x} .$$

Or, intensity at a distance x due to an infinitely long straight wire is $2\ \rho\ .\ G/x$.

EXERCISE VII

1. Describe an accurate method of determining the value of the gravitational constant. Calculate the minimum velocity required by a body to escape the gravitational field of the earth. *(Delhi*, 1969)

2. What is meant by 'gravitation' constant ? What are its dimensions ? Give an account of the experiments of Cavendish and Boys to determine this constant. *(Banaras)*

3. If $G = 6.66 \times 10^{-11}\ n\ .m^2/kg^2$, what is the force between two small spheres weighing 2 *kg*, each, placed 30 *cm* apart. **Ans.** 2.961×10^{-9} newtons.

4. State and explain Newton's law of gravitation and describe an accurate method of measuring the gravitation constant. What celestial evidence led to the formulation of the law ? Is this law universally correct ? Explain your statement. *(Calcutta)*

5. If the earth were a solid sphere of iron, of radius 6.37 million *metres*, and of density 7.86 *gm./cm*^{3}, what would be the value of gravity at its surface, taking the *gravitational constant* to be 6.658×10^{-8} *c.g.s. units* ? **Ans.** 1396 *cm/sec*^{2}.

6. Give the theory of Cavendish experiment, explaining how the density of the earth is determined. Explain why and how Boys modified the Cavendish method. *(Madras)*

7. Explain how Cavendish determined the value of gravitation constant. Indicate how, from the knowledge of the value of the gravitation constant, it is possible to calculate the mass of the earth. *(Sagar)*

8. If $G = 6.66 \times 10^{-8}$ *c.g.s. units*, and the radius of the earth equal to 6.37×10^8 *cm*, what is the density of the earth ? **Ans.** 5.52 *gm/cm*^{3}.

9. The earth moves round the Sun in a circle of radius 9.288×10^7 *miles*, and completes a revolution in 365 *days*. A satellite of Jupiter moves about the Jupiter in a circle of radius 1.161×10^6 *miles*, completing one revolution in 16.6 *days*. Calculate the mass of Jupiter in terms of the mass of the Sun. **Ans.** 945×10^{-6}.

10. Assuming the law of gravitation, find an expression for the period of revolution of a planet.

The moon describes a circular orbit of radius 3.8×10^5 *km* about the earth in 27 days and the earth describes a circular orbit of radius 1.5×10^8 *km* round the Sun in 365 days. Determine the mass of the Sun in terms of that of the earth. *(Bombay)*

Ans. 33.66×10^5

11. *(a)* State the law of Universal Gravitation and describe a method of determining the Gravitation constant. *(Punjab)*.

(b) A small satellite revolves round a planet of mean density 10 *gm/cc*, the radius of its orbit being slightly greater than the radius of the planet. Calculate the time of revolution of the satellite. ($G = 6.66 \times 10^{-8}$ *c.g.s. units*). *(Bombay)*

12. Describe Poynting's method of determining the gravitational constant. *(Punjab, 1966, Raipur, 1965)*

13. *(a)* Define 'Potential' and 'Potential Energy' of a gravitational field. Derive an expression for the potential due to a sphere of uniform density at an external point. *(Agra)*

(b) If r_1 and r_2 be the radii of the earth and the moon and d_1 and d_2, their mean densities respectively, obtain in the ratio of the acceleration due to gravity on the earth and the moon. [*Vikram (B.Sc. Part I)*, 1960)]

Ans. $r_1 d_1 : r_2 d_2$

14. Explain what do you mean by gravitational potential at a point. How does it differ from the other kinds of potential with which you are familiar?

Find an expression for gravitation potential due to a thin hollow sphere of uniform density at a point outside it. *(Calcutta)*

15. Two balls, each weighing 10 *gm*, are hung side by side by threads, *10 metres* long. If the threads are 1 *cm* apart at the upper ends, by how much is the distance between the centres of the balls less than 1 *cm*. **Ans.** 1.5×10^{-6} *cm*.

16. *(a)* Describe an accurate method for the determination of the Universal Constant of Gravitation. *(Punjab, 1964)*

(b) The star Sirius has a mass of 6.9×10^{33} *gm* and its distance e is 8×10^{13} *km*. The mass of the earth is 6×10^{27} *gm*. The tensile strength* of steel is about 20,000 *kg/cm*^{2}. Calculate the cross-section of a steel bar which could just withstand the gravitational pull between Sirius and the earth. ($G = 6.67 \times 10^{-8}$ *dyne-cm/gm*$^{-2}$). **Ans.** 2.169×10^6 *sq cm*.

17. Prove that the least velocity with which a particle must be projected from the surface of a planet of radius R and density ρ in order that it may escape completely is $R\sqrt{8\pi G\rho/3}$, where G is the gravitational constant. Calculated the velocity in the case of the moon from the following data :

mean density of earth = 5.52 *gm/cc*, *mean density of moon* = 3.36 *gm/cc*,

mean radius of earth = 6380 *km* ; *mean radius of moon* = 1740 *km* ;

acceleration of gravity at earth's surface = 980 *cm per sec per sec*. *(Oxford Scholarship)*

Ans. 2.38×10^5 *cm sec*$^{-1}$.

18. Describe an accurate *'balance-method'* for the determination of the value of G and write a short note on the *qualities of gravitation*.

19. What is Gravitation constant ? Describe in detail an accurate method for determining it.

Calculate the mean density of the Earth, given : $G = 6.6 \times 10^{-8}$ *c.g.s.* units, the radius of the earth = 6.4×10^8 *cm* and $g = 90$ *cm/sec*^{2}. *(Delhi)*

Ans. 5.538 *gm/cc*

20. Explain the terms 'gravitational potential' and 'gravitational field'. Obtain expressions for the gravitational potential and the field at a point *(i)* inside, *(ii)* outside, a hollow spherical shell. *(Vikram (B.Sc. Part II); Bombay; Banaras)*

21. Describe an accurate method of measuring the density of the earth and *'G'*, the Newtonian Gravitational Constant. Derive the necessary formula. *(Gujarat, 1962)*

22. The mass of planet *Venus* is 0.814 times that of the earth (5.98×10^{24} *kg*) and its radius equal to 0.967 of that of the earth (6.38×10^6 *m*). Calculate the velocity of escape from it, taking $G = 6.67 \times 10^{-11}$ *M.K.S (or S.I)* units.

Ans. 10.26 *km/sec*.

Hint. See § 7.12 (page 224)

23. Give the construction and working of a seismograph and indicate the nature of earth movements during an earthquake. [*Punjab (Sept.)*, 1963]

24. What is an earthquake ? How is it caused ? Describe in brief the principle underlying *seismographs*. Why are they so called ?

25. Describe in detail Galitzin's seismograph and explain how the *epicentre* and the *focus* of an earthquake may be determined with its help.

26. What is *geophysical prospecting* ? Write a short descriptive note on *(i)* the *gravitational* and *(ii)* the *seismic* methods used for the purpose.

27. Calculate the limiting velocity required by an artificial satellite for orbiting round the earth, given the radius of the earth as 6.4×10^8 *cm* and $g = 980$ *cm/sec*^{2}. *(Delhi)*

Ans. 7.92×10^5 *cm*.

28. Define and explain gravitational potential. Calculate the gravitational potential due to a sphere at any point *(i)* outside the sphere and *(ii)* inside the sphere. *(Delhi, Bombay)*

29. State and explain Newton's law of gravitation. Describe an accurate method for the determination of the gravitational constant. Discuss how the mean density of the earth can be determined. [*Delhi, (Sept.)*]

---

* See next chapter, § 8.3

30. Define potential and potential energy of a gravitational field. Derive an expression for the gravitational potential due to a uniform sphere at a point *(i)* inside the sphere and *(ii)* outside the sphere. ***(Delhi)***

31. Give briefly the outline of various methods to find the universal gravitation constant and describe in detail the most accurate method you know of. ***(Vikram)***

32. In terms of the acceleration due to gravity on the earth's surface and its radius, obtain an expression for the minimum vertical velocity with which a body may be projected in order to escape completely from the earths influence.

[*B.Sc. (Hons.), Delhi, 1967*]

33. If the earth were to cease rotating about its axis, what will be the change in the value of *g* at a place of latitude 45°, assuming the earth to be a sphere of radius 6.38×10^8 *cm*. ***(Vikram)***

Ans. Increase of 1.686 *cm/sec.*

Hint. See worked example 5.

34. Define gravitational field and gravitational potential.

Obtain expressions for gravitational potential due to a spherical shell at a point *(i)* outside the shell and *(ii)* inside the shell. ***(Delhi)***

CHAPTER VIII

ELASTICITY

8.1. Introductory. All bodies can, more or less, be *deformed* by suitable applied forces. The simplest cases of deformation are those (*i*) in which a wire, fixed at its upper end, is pulled down by a weight at its lower end, *bringing about a change in its length* and (*ii*) in which an equal compression is applied in all directions, so that *there is a change of volume but no change in shape* or (*iii*) in which a system of forces may be applied to a body such that, although there is no motion of the body as a whole, there is relative displacement of its contiguous layers, causing a *change in the shape or 'form' of the body with no change in its volume*. In all these cases the body is said to **be strained or deformed.**

When the deforming forces are removed, the body tends to recover its original condition. For example, the wire in the case above, tends to come back to its original length when the force due to the suspended weight is removed from it, or, a compressed volume of air or gas throws back the piston when it is released, in an attempt to recover its original volume. This property of a material body to regain its original condition, on the removal of the deforming forces, is called **elasticity.** Bodies, which can recover completely their original condition, on the removal of the deforming forces, are said to be **perfectly elastic.** On the other hand, bodies, which do not show any tendency to recover their original condition, are said to be **plastic.** There are, however, no *perfectly* elastic or plastic bodies. The nearest approach to a perfectly elastic body is a *quartz fibre* and, to a perfectly plastic body, is *putty*. But even the former yields to large deforming forces and, similarly, the latter recovers from small deformations. Thus, there are *only differences of degree*, and a body is more elastic or plastic when compared to another.

We shall consider here only bodies or substances, which are (*i*) **homogeneous** and (*ii*) **isotropic,** *i.e.*, which have the *same properties at all points* and *in all directions*. For, these alone have similar elastic properties in every direction, (together with other physical properties like linear expansion, conductivity for heat and electricity, refractive index etc.). Fluids (*i.e.*, liquids and gases), as a rule, belong to this class, but not necessarily all solids, some of which may exhibit different properties at different points and in different directions, *i.e.*, may be *heterogeneous* (or *non-homogeneous*) and *anisortropic* (or *non-isotopic*). Examples of this class of solids are *wood*, and *crystals, in general*, including those **metals** which are crystalline in structure. As a class, however, *metals*—particularly in the form of *rods* and *wires* —may be regarded to be more or less wholly isotropic in so far as their elastic behaviour is concerned.

8.2. Load, Stress and Strain. The term *load*, in the present context, implies *the combination of external forces* acting on a body and its effect is to change the form or the dimensions of the body*. It is thus essentially a *deforming force*.

Now, as a result of the load or the deforming forces applied to a body, *forces of reaction* come into play *internally* in it, due to the relative displacement of its molecules, tending to balance the load and restore the body to its original condition. *The restoring or recovering force per unit area set up inside the body is called* **stress.** Being a *distributed force*, it is measured in the same manner as fluid pressure, *i.e.*, in terms of the load or the deforming force applied per unit area of the body, being *equal in magnitude* but *opposite in direction* to it, until a permanent change has been brought about in the body, *i.e.*, until its elastic limit has been reached, (see § 8.3). If the force

* As, for example, the weight of the body itself, together with those connected with it; centrifugal forces in the case of rotating wheels and pulleys ; forces due to friction or forces due to uequal expansion and contraction on changes of temperature etc. etc.

be inclined to the surface, its component, perpendicular to the surface, measured per unit area, is called **normal stress*** and the component acting along the surface, per unit area, is called **tangential or shearing stress.** Further, the former may be compressive or expansive (*i.e., tensile*) according as a decrease or an increase in volume is involved. Similarly, the stress is compressive or tensile according as a body is subjected to axial thrusts or pulls across any section perpendicular to its length. Obviously, being force per unit area, the dimensions of stress are the same as those of pressure, *viz.*, $ML^{-1}T^{-2}$, (see page 5) and its units, $dynes/cm^2$, $poundal/ft^2$ and $newton/metre^2$ (*i.e.*, N/m^2) in the *C.G.S.*, *F.P.S.* and *M.K.S.* (or *S.I.*) units respectively.

The change produced in the dimensions of a body under a system of forces or couples in equilibrium, is called **strain,** and is measured by the *change per unit length (linear strain), per unit volume, (volume strain), or the angular deformation, (shear strain,* or simply shear)** according as the change takes place in *length, volume* or *shape* of the body. Thus, being just a ratio, (or *an angle*) it is a *dimensionless quantity*, having no units.

It will be readily seen that *for a perfectly elastic body (i) the strain is always the same for a given stress; (ii) the strain vanishes completely when the deforming force is removed and (iii) for maintaining the strain, the stress is constant.*

8.3. Hooke's Law. This fundamental law of elasticity was enunciated by *Robert Hooke* in the year 1679 in the concise phrase '*Ut tensio sic vis*'+ and may be formally stated as follows:

"Provided the strain is small, the stress is proportional to the strain"; so that, in such a case, the ratio stress/strain is a constant, called the **modulus of elasticity,** (a term first introduced by *Thomas Young*), *or the* **coefficient of elasticity.**

Since stress is just pressure, (or tension per unit area) and strain is just a ratio, *the units and dimensions of the modulus of elasticity are the same as those of stress or pressure.*

When the stress is continually increased in the case of a solid, a point is reached at which the strain increases more rapidly than is warranted by Hooke's law. *The stress at which the linear relationship between stress and strain ceases to hold good is referred to as the* **elastic limit** *of the material in question for the type of stress applied.* Thus, if the material happens to be in the form of a bar or a wire under stretch, it will recover its original length on the removal of the stress so long as the stress is below the elastic limit, but if this limit be exceeded, it will fail to do so and will have acquired what is called a **'permanent set'.**

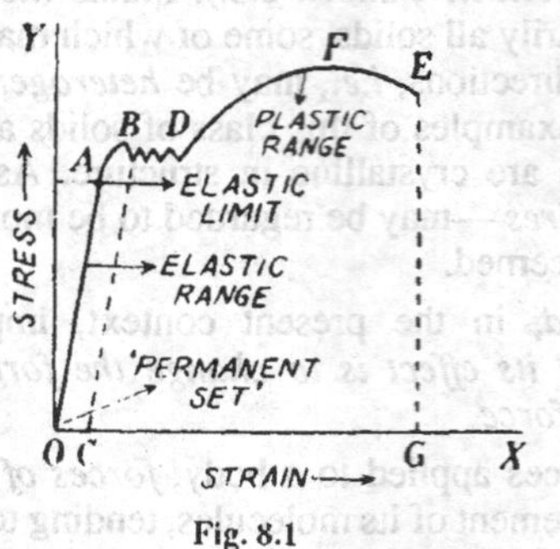

Fig. 8.1

The relationship between *stress* and *strain* may best be studied by plotting a graph between the two for various values of stress and the accompanying strain. This graph in the simple case of a bar or a wire subjected to an increasing tension is, in general, of the form shown in Fig. 8.1 and is known as the **stress-strain diagram** for the bar or the wire.

The straight and sloping part *OA* of the curve clearly shows that the strain produced is directly proportional to the stress applied or that Hooke's law is obeyed perfectly up to *A* and that, therefore, on the removal of the stress (*i.e.*, on the wire being

* The stress is always normal in the case of a change in the length of the wire, or in the case of a change in the volume of a body, but is tangential in the case of a change in the shape of a body.

** This will be dealt with more fully later in § 8.6, (3).

+ Meaning '*As the tension, so the strain*'.

unloaded), it will recover its original condition of zero strain, represented by O. In other words, the wire is perfectly elastic up to the stress represented by A which therefore measures the **elastic limit***of the material of the wire of tensile stress and is often spoken of as the **elastic strength** of the wire under test. The strain up to this limit is really very small, the total increase in length (which is due to simple tension and is thus purely elastic) being of the order of 10^{-3} of the original length of the wire.

As soon as the elastic limit is crossed, the strain increases more rapidly than the stress, and the graph curves along AB, the extension of the wire now being partly elastic and partly plastic. Hence, on being unloaded here, say, at the point B, it does not come back to its original condition along AO, but takes the dotted path BC, so that there remains a **residual strain** OC in it, which is what we really mean by the *'permanent set'* acquired by the wire.

Beyond the point B, for practically little or no increase in stress (or the load applied), there is a large but erratic increase in strain (*i.e.*, in the extension produced) up to D, so that the portion BD of the graph is an irregular wavy line, the stress corresponding to D being less than that corresponding to B. This point B**, where the large increase in strain (*i.e.*, in extension) commences is called the **yield point** (also sometimes referred to as the *commercial elastic limit*), the stress corresponding to it being known as the **yielding stress.**

The yielding ceases at D** and further extension, which now becomes plastic, (being mainly due to shear stress rather than simple tension), can only be produced by gradually increasing the load so that the portion DF of the graph is obtained, the cross-section of the wire decreasing uniformly with extension (or strain) up to F and hence its volume remaining constant. *The maximum load (or force) to which the wire is subjected divided by its original cross-sectional area is called the* **ultimate strength** *or the* **tensile strength** *of the wire* and is also sometimes termed the **'breaking stress'**.

The extension of the wire goes on increasing beyond F without any addition to the load, indeed, even if the load be reduced a little, and the wire behaves as though it were literally 'flowing down'. This is so because of a faster rate of decrease of its cross-sectional area at some section of its length where a local constriction, called a *'neck'* or a *'waist'* begins to develop, with the result that, even if the load be not increased, the load per unit *area* or the *stress* becomes considerably greater there, bringing about a corresponding increase in strain or extension in the wire. The load is, therefore, decreased, *i.e.*, the stress is reduced at this stage and the wire finally snaps or breaks (or, technically speaking, is *'destroyed'*) at E, which thus represents the **breaking point** for it.

Alternatively, we could not plot a **load-extension diagram** for the given specimen of wire, which would really be the same as the *stress strain diagram*, shown in Fig. 8.1, plotted for a different set of scales along the two axes, with the deductions from it being identical with those obtained from the latter.

Now, it is obvious enough that work has to be done in stretching the wire to 'destruction' under an increasing stress or tension. Up to the elastic limit, it remains stored up in the wire *in the form of potential energy*, in the same manner as in a compressed spring, and can be obtained back from it on unloading it. Beyond the elastic limit, however, the crystalline structure of the metal gets affected, so that the work done in now converted into heat energy.+ *The total work done is represented by the area under the load-extension diagram,* whereas *the area, under the stress-strain diagram* (*i.e., area OABDEFG* in Fig. 8.1) represents, by the very nature of its

---

* In quite a few cases, Hooke's law is obeyed only up to a point a little below the elastic limit, represented by A. The portion of the curve from O to *this point* (below A), is then said to indicate the **limit of proportionality** distinguish it from the *eleastic limit.* The two are thus not always identical, though they are generally regarded to be so, in view of the ver small difference between them.

** These points B and D are also quite often referred to as the *upper and the lower yeild - points*, respectively.

+ This can be seen from the fact that, if touched immediately on snapping, it is found to be sensibly hot.

co-ordinate axes, the work done per unit volume (usually *per cu in*) during the process. (See also § 8.9).

Another definition of Elastic limit. The *elastic limit* of a material is also sometimes defined as *the force producing the maximum reversible or recoverable deformation in it*, and may, for a given specimen, be determined by loading and unloading it with a number of different loads and measuring its length after *each unloading*, until it acquires a *permanent set*. The latter is then plotted against the load, and from the curve thus obtained, the particular load at which the permanent set *just* starts can be easily estimated.

8.4 Ductility, Brittleness and Plasticity. If a material can acquire *a large permanent deformation, without fracture or rupture*, it is said to be *ductile* and this property of it is termed ductility, measured by *the percentage elongation at rupture*. Obviously enough the breaking point for such a substance will be very much beyond its yield-point *i.e.*, the portion *DE* will be large on the stress-strain-diagram for it. Examples of ductile substances are *aluminum, brass, annealed copper* and *low-carbon steel*.

The very opposite of the property of ductility is what is called brittleness; so that, a brittle substance is one *which ruptures without any appreciable amount of permanent deformation being produced in it*—the elongation at rupture hardly ever exceeding 2 to 5% and being, in most cases, even less than 1%. There is thus hardly any yield-point on the stress-strain diagram of such a substance or, we may say that it almost coincides with the breaking-point itself. Among examples of brittle substances may be mentioned *glass, stone, brick, cast iron* and *high carbon steel*.

Finally, the property of *plasticity* of a material *is its ability to get deformed continuously and permanently, without any rupture, i.e. its ability to flow during the application of a stress, exceeding the yield-point value for it.* Thus, in the stress-strain diagram of a *plastic* substance, the graph will run parallel to the strain axis immediately, beyond the yield-point. Before the yield-point is reached, however, such substances behave like ordinary elastic substances. A familiar example is that of *pitch*.

Bingham, one of the earliest workers on the plastic state ordinary, has defined an *ideal plastic solid* – referred to as a *Bingham solid*— as that for which the rate of flow is *zero* at stresses below the yield-point but proportional to the excess stress beyond it. The existence of such a solid is highly doubtful, however, in actual practice since the yield point of a substance is never a sharply defined point and there is almost always an intermediate region of *imperfect elasticity* between those of *perfect elasticity* and *plastic flow*.

8.5. Elastic behaviours of Solids in general—Some peculiar traits.

Almost all solids exhibit, in varying degrees the following peculiar traits in their elastic behaviour.

(*a*) **Elastic after-effect.** It is found that, even within the elastic limit, few solids come back to their original condition directly the deforming force is removed. Almost all of them only '*creep*' back to it, (*i.e.*, take some time to do so), though they all do so, ultimately. *This delay in recovering back the original condition, on the cessation of the deforming force, is called elastic after-effect.* Glass exhibits this effect to a marked degree, the few *exceptions* to this almost general rule being *quartz, phosphor-bronze, silver* and *gold*, which regain their original condition as soon as the deforming force ceases to operate. Hence their use in Cavendish's and Boys' experiments for the determination of *G*, in quadrant electrometers and in moving-coil galvanometers etc. etc.

(*b*) **Elastic hysteresis.** As a natural consequence of the elastic after-effect, *the strain in a material, (in glass, for example), tends to persist or lag behind the stress to which it is subjected,* with the result that during a rapidly changing stress, the strain is greater for the same value of stress, when it is decreasing than when it is increasing, as is clear from the curve in Fig. 8.2. *This lag between stress and strain is called elastic hysteresis,* (the term '*hysteresis*' meaning '*lagging behind*'). The phenomenon is similar in its implications to the familiar *magnetic hysteresis*, where the magnetic effects tend to persist, or lag behind, even after the magnetising influence is removed, - the curve referred to above may thus be called the *elastic hysteresis loop*. And, exactly in the same manner, the energy dissipated as *heat* during a cycle of loading and unloading, is given by the *area enclosed by the loop*. There is, however, very little hysteresis in the case of metals or of quartz.

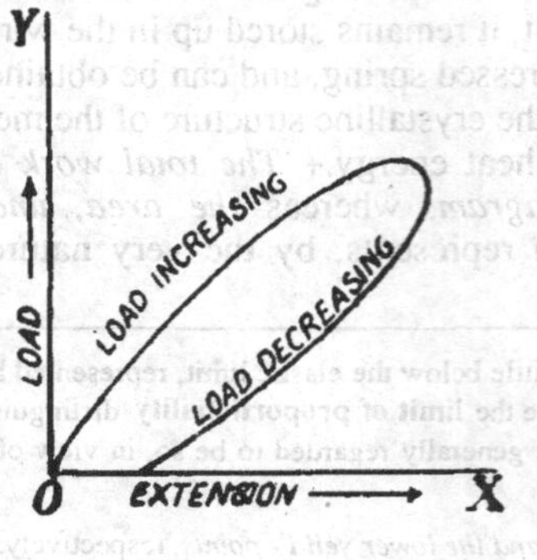

Fig. 8.2

(*c*) **Elastic fatigue.** It was shown by *Lord Kelvin*, during his investigation of the rate of decay of torsional vibrations of wires, that the vibrations died away much faster in the case of wire kept vibrating continuously for some time than in that of a fresh wire. The same happens to any elastic body, subjected to an alternating strain. The continuously vibrating wire gets *'tired'* or *'fatigued'*, as it were, and finds it difficult to continue vibrating. Lord Kelvin fittingly expressed this by the term *'elastic fatigue'*.

A body, thus subjected to repeated strains beyond its elastic limit, has its elastic properties greatly impaired, and may break under a stress less than its normal breaking stress even within its elastic limit. This phenomenon is, obviously, of great importance in cases like those of the pistom and the connecting rods in a locomotive, which as we know, are subjected to repeated tensions and compressions during each revolution of the crank shaft. Great care has, therefore, to be taken in éstimating the stresses that the various parts of a machinery or a structure may be allowed to put up to, consistent with safety as well as economy. This has led to the conception of a *breaking stress* and the *factor of safety* (see § 8.4).

8.6. Working Stress and Factor of Safety. The region within which Hooke's Law is valid for metals is found to be rather small. It is, however, of the utmost importance in that, for most engineering purposes, it is imperative to ensure that the piece of metal, which is not subjected to stresses extending to the plastic range.* Or else, there would be set up a permanent deformation application of the maximum load.

As a necessary safeguard, therefore, the stress on a member-piece is, as a rule, kept far below the ultimate or the breaking stress and is, in fact, never allowed to cross the elastic limit. This maximum stress, within the elastic limit, to which the member-piece is subjected in actual practice is called its **working stress** or **working strength** (or **tenacity**) and the load corresponding to it is spoken of as the **working load.**

The ratio between the ultimate stress (or the breaking stress) and the working stress is aptly known as the *factor of safety*. Thus,

$$\textbf{factor of safety} = \frac{\textbf{tensile strength or breaking stress}}{\textbf{working stress}}$$

Now, although the actual value of the factor of safety is primarily a consequence of judgment and experience, there are certain considerations that must inevitably influence the choice of a particular value for it. Here are a few, by way of illustration :

(i) *Dependability of the material of the piece in question.*

(ii) *Extent of the expected wear and tear of the material, when in actual use.*

(iii) *Magnitude of the maximum load that is ever likely to be put on the piece.*

(iv) *The character of the load, as to whether it is a fixed or a variable one.*

(v) *Possible consequences of a break down, should it unfortunately occur eventually.*

(vi) *Allowance for uncertainties of the problem, e.g., sudden changes of cross-section (resulting in stress concentration), sudden application of the load, the departure of the material from isotropic conditions and its method of manufacture (as to whether it has been forged, cast or machined) etc. etc.*

And, finally it must depend, naturally, on the *standard of workmanship* available, the *boldness* of the designer and his *preparedness to take risks* in the interests of a higher output. This perhaps explains why it is usually 10 in the United Kingdom and only 5, quite often, in the USA.

---

* This does not mean that the engineer has no use for this plastic region. He makes, on the other hand, good, though indirect, use of it in more ways than one. To quote an instance or two, (*i*) he quite often inserts his rivets hot and uses their contraction on cooling for tightening the joints. This is obviously rendered possible because of the existence of a region of plastic flow permitting the rivets to stretch; (*ii*) again, he clean ignores a hole here and a hole there in a metal piece which is to be subjected to a tensile or a bending stress. Why ? Because he knows that despite the regions of high stress around the holes, the plastic flow that occurs in these regions on the application of a load results, more often than not, in a more equitable distribution of stress all through the piece.

It does not, therefore, seem to be quite so unreasonable if the factor of safety is sometimes also referred to as the **'factor of ignorance'**.

8.7. Factors affecting Elasticity. The elastic properties of a material are linked up with the fine mass of its structure. It is now finally established, by careful microscopic examination, that metals are just an aggregation of a large number of fine crystals, in most cases arranged in a random or a chaotic fashion, *i.e.*, their *cleavage planes* (or the planes along which their constituent atoms can easily slide over each other), being distributed haphazardly, in all possible directions. Single crystals, when subjected to deformation, show a remarkable increase in their hardness. Thus, for example, a single crystal of silver, on being stretched to a little more than twice its length, is known to increase to as much as ninety-two times its original strength or stiffness. So that, the following consequences flow :

(*i*) **Effect of hammering, rolling and annealing.** Operations, like hammering and rolling etc., which help break up the crystal grains into smaller units, result in an increase or extension of their elastic properties; whereas operations, like annealing (*i.e.* heating and then cooling gradually) etc., which tend to produce a uniform pattern of orientation of the constituent crystals, by orienting them all in one particular direction and thus forming larger crystal grains, result in a decrease in their elastic properties or an increase in the softness or plasticity of the material.

This is because, in the latter case, slipping (or sliding between cleavage planes), starting at a weak spot proceeds all through the crystal and, in the former, the slipping is confined to one crystal grain and stops at its boundary with the adjoining crystal. Indeed, the former may be compared to a small cut, developing into a regular tear all along a fabric and the latter to the tear stopping as it reaches a seam in the fabric. Thus, *'paradoxically'*, as *Sir Lawrence Bragg* puts it, *'in order to be strong, a metal must be weak'*, meaning thereby that metals with smaller grains are stronger than those with larger ones.

(*ii*) **Effect of impurities.** It is well known that sometimes suitable impurities are deliberately added to metals to help bind their crystal grains better, without affecting their orientation. Thus, for example, carbon and potassium are added in minute quantities to molten iron and gold respectively for this purpose. Such impurities naturally affect the elastic properties of the metal to which they are added, enhancing or impairing them according as they are themselves more elastic or plastic than the metal concerned. In either of the two cases mentioned here, the elastic properties are considerably strengthened.

(*iii*) **Effect of change of temperature.** A change in the temperature also affects the elastic properties of a material, —a rise in temperature usually decreasing its elasticity and *vice versa,* except in certain rare cases, like that of *invar steel,* whose elasticity remains practically unaffected by any changes in temperature. Thus, for example, lead becomes quite elastic and rings like steel when struck by a wooden mallet, if it be cooled in liquid air. And, again, a carbon filament, which is highly elastic at the ordinary temperature, becomes plastic when heated by the current through it, so much so that it can be easily distorted by a magnet brought near to it.

8.8. Three Types of Elasticity. Corresponding to the three types of strain, we have three types of elasticity, *viz.*;

(*i*) *linear elasticity,* or *elasticity of length, called* **Young's Modulus,** corresponding to linear (or tensile) strain :

(*ii*) *elasticity of volume* or **Bulk Modulus,** corresponding to volume strain; and

(*iii*) *elasticity of shape, shear modulus or* **Modulus of Rigidity,** corresponding to shear strain.

(1) **Young's Modulus.** When the deforming force is applied to the body *only along a particular direction**, the change per unit length in that direction is called *longitudinal, linear or elongation strain,* and the force applied per unit area of cross-section is called *longitudinal or linear stress. The ratio of longitudinal stress to linear strain, within the elastic limit, is called the coefficient of direct elasticity or* **Young's Modulus,** *and may be denoted by the letter Y, though now it is increasingly denoted by the letter E.*

Thus, if F be the force applied *normally* to a cross-sectional area a, the *stress* is F/a. And, if there be change l produced in the original length L, the strain is given by l/L. So that,

$$\textit{Young's Modulus}, Y \text{ or } E = \frac{F/a}{l/L} = \frac{F.L}{a.l}$$

Now, if $L = 1$, $a = 1$ and $l = 1$, we have Y (or E) $= F$. In other words, if a material of unit length and unit area of cross-section could be pulled so as to increase in length by unity, *i.e.,* to double its length, the force applied would measure the value of Young's Modulus for it.

Since, however, the elastic limit is exceeded when the extension produce is 10^{-3} *cm/cm,* the material will snap before *this much* extension is produced.

In cases, where elongation produced is not proportional to the force applied, we can still determine Young's Modulus from the ratio $L.d\,F/a.dL$, where dF/a is the infinitesimal increase in the longitudinal stress and dL/L, the corresponding increase in strain.

Or, $$Y \text{ (or } E) = \frac{L}{a} \cdot \frac{dF}{dL}.$$

N.B. The particular case of rubber may, with advantage, be mentioned here, which the beginner finds so confusing, when, in ordinary conversational language, we refer to it as being *'elastic'*. For, he knows well enough that it requires a much smaller force than steel to stretch, it, (and that, therefore, its elasticity is much less than that of steel). In fact, the value of Young's Modulus for rubber is about *one-fiftieth* of that of steel. What we mean when we term it as elastic, therefore, *is just that it has a very large range of elasticity;* for, whereas a crystalline body can be stretched to less than even one per cent of its original length before reaching its elastic limit, rubber can be stretched to about eight times (or 80%) of its original length.

This high extensibility of rubber is due to its molecule containing, on an average, some 4,000 molecules of *isoprene* (C_5H_8), whose 20,000 carbon atoms, spreading out in a chain, make it very *long and thin,* — about 1/4000 *mm* in length.

Rubber, in bulk, has thus been rightly compared to an intertwined mass of long, wriggling snakes—its molecules, like the snakes, tending to uncoil when stretched and getting coiled up again when the stretching force is removed.

(2) **Bulk Modulus.** Here, the force is applied *normally and uniformly* to the whole surface of the body; so that, *while there is a change of volume, there is no change of shape.* Geometrically speaking, therefore, we have here *a change in the scale of the coordinates* of the system or the body. *The force applied per unit area, (or pressure), gives the* **stress** *and the change per unit volume, the* **strain** (or *volume strain), their ratio giving the coefficient of cubical elasticity or the* **Bulk Modulus** *for the body. It is usually denoted by the letter K.*

Thus, if F be the force applied uniformly and normally on a surface area a, the *stress,* or *pressure,* is F/a or P; and, if v be the change in volume produced in an original volume V, the strain is v/V and, therefore,

$$\textit{Bulk Modulus}, K = \frac{F/a}{v/V} = \frac{F.V}{a.v} = \frac{P.V}{v}. \qquad [\because F/a = P.$$

If, however, the change in volume be not proportional to the stress or the pressure applied, we consider the infinitesimal change in volume dV for the corresponding change in pressure dP; so that, we have, $K = dP.V/dV$.

Bulk Modulus is sometimes referred to as **incompressibility** and hence its reciprocal is called **compressibility**; so that, *compressibility* of a body is equal to $1/K$, where K is its *Bulk*

* *i.e., when it is subjected to a simple tension or a simple compression,* (no external forces being applied to prevent the accompanying small lateral contraction or expansion)

Modulus. It must thus be quite clear that *whereas Bulk Modulus is stress per unit strain, compressibility represents strain per unit stress.*

Since fluids (*i.e.*, liquids and gases) can permanently withstand or sustain only a hydrostatic pressure, the only elasticity they possess is Bulk Modulus (K), which is, therefore, all that is meant when we refer to their *elasticity.* This, however, is of two types : *isothermal* and *adiabatic.*

For, when a fluid is compressed, there is always some heat produced. If this heat be removed as fast as it is produced, the temperature of the fluid remains constant and the change is said to be *isothermal;* but if the heat be allowed to remain in the fluid, its temperature naturally rises and the change is then said to be *adiabatic.*

It can be easily shown that the *isothermal elasticity of a gas* (*i.e.*, when its temperature remains constant) *is equal to its pressure P, and its adiabatic elasticity equal to γP, where γ is the ratio between C_p and C_v* for the gas in question,* —its value being 1.41 for air, (see **worked Example** 1 (*b*) at the end of the Chapter).

It will thus be readily seen that the *Bulk Modulus of a gas* (whether isothermal or adiabatic) *is not a constant quantity,* unlike that of a solid or a liquid.

(3) **Modulus of Rigidity (or Torsion modulus).** In this case, while there is a change in the shape of the body, there is no change in its volume. As indicated already, it takes place by the movement of contiguous layers of the body, one over the other, very much in the manner that the playing cards would do when a pack of them, placed on the table, is pressed with the hand and pushed horizontally. Again, speaking geometrically, *we have, in this case, a change in the inclinations of the coordinate axes of the system or the body.*

Consider a rectangular solid cube, whose lower face *aDCc* (Fig. 8.3) is fixed, and to whose upper face a tangential force F is applied in the direction shown. The couple so produced by this force and an equal and opposite force coming into play on the lower fixed face, makes the layers, parallel to the two faces, move over one another, such that the point A shifts to A', B to B', d to d' and b to b' , *i.e., the lines joining the two faces turn through an angle* θ**.

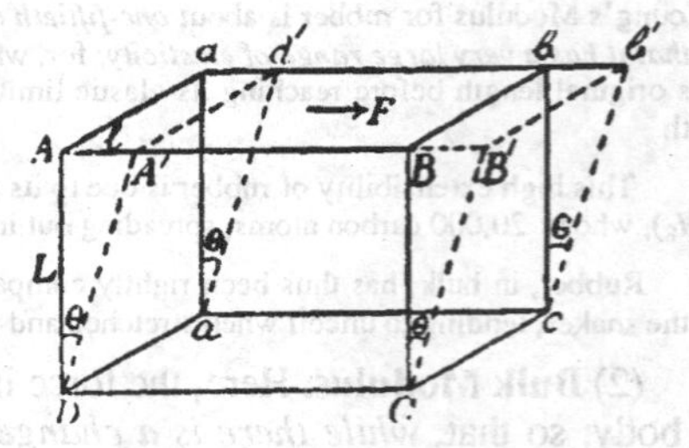

Fig. 8.3

The face ABCD is then said to be **sheared** through *an angle θ. This angle θ (in radians), through which a line originally perpendicular to the fixed face is turned, gives the strain or* the **shear strain,** *or the* **angle of shear,** *as it is often* called. As will be readily seen, $\theta = AA' / DA = l/L$, where l is the *displacement AA'* and L, the length of the side AD or the *height* of the cube; or θ = *relative displacement of plane ABbd/distance from the fixed plane aDCc.* So that, if the distance from the fixed plane, *i.e.*, $L = 1$, we have $\theta = l$ = *relative displacement* of plane *ABbd.*

Thus, *shear strain (or shear) may also be defined as the relative displacement between two planes unit distance apart.*

And, *stress or* **tangential stress** *is clearly equal to the force F divided by the area of the face ABbd, i.e., equal to F/a. The ratio of the tangential stress to the shear strain gives the* **coefficient**

* The symbols C_p and C_v stand for the specific heats of a gas at constant pressure and at constant volume respectively,— their ratio $\gamma = C_p / C_v$ being the highest (1.67) for a monatomic gas, like helium, goes on decreasing with increasing atomicity of the gas but is always greater than 1.

** As a matter of fact, if this were the only couple acting on the body, it would result in the rotation of the body. This is prevented by another equal and opposite couple, formed by the weight of the body (plus any vertical force applied) and the reaction of the surface on which the body rests.

of rigidity *of the material of the body, denoted by n.* (It is also sometimes referred to as the *coefficient of transverse elasticity).*

Thus, *tangential stress* = F/a and *shear strain* = $\theta = l/L$.

And, therefore, *Coefficient of Rigidity*, or *Modulus of Rigidity* of the material of the cube is given by

$$n = \frac{F/a}{\theta} = \frac{F/a}{l/L} = \frac{F.L}{a.l} \quad ...(i)$$

This is a relation exactly similar to the one for Young's Modulus, with the only difference that, here, F/a is the tangential stress, not a linear one, and l, a displacement at right angles to L, instead of along it.

Again, if the shearing strain, or shear, be not proportional to the shearing stress applied, we have

$$n = \frac{dF/a}{d\theta},$$

where $d\theta$ is the increase in the angles of shear for an infinitesimal increase dF/a in the shearing stress.

Further, it is clear from relation (*i*) above, that if $a = 1$, and $\theta = 1$ *radian* (or 57° 18′), we have $n = F$.

*We may thus define modulus of rigidity of a material as the shearing stress per unit shear, i.e., a shear of 1 radian, taking Hooke's law to be valid even for such a large, strain.**

Axial modulus. Another elastic constant, comparatively recently introduced,is what is called the *axial modulus*. It is defined as *the longitudinal stress required to produce unit linear strain, unaccompanied by any lateral strain and is denoted by the symbol* χ.

It is thus similar to *Young's modulus*, with the important difference that here the lateral strain produced (in the form of lateral contraction) is offset by applying two suitable stresses in a direction perpendicular to that of the linear stress. So that, the total stress is Young's modulus plus these two perpendicular stresses.

8.9. Equivalence of a shear to a compression and an extension at right angles to each other. Consider a cube *ABCD*, (Fig. 8.4) with the face *DC* fixed, and let the face *ABCD* be *sheared* by a force, applied in the direction shown, through an angle θ, into the position *A′B′CD*. Then, clearly, the diagonal *DB* is *increased in length* to *DB′* and the diagonal *AC* is *shortened* to *A′C*.

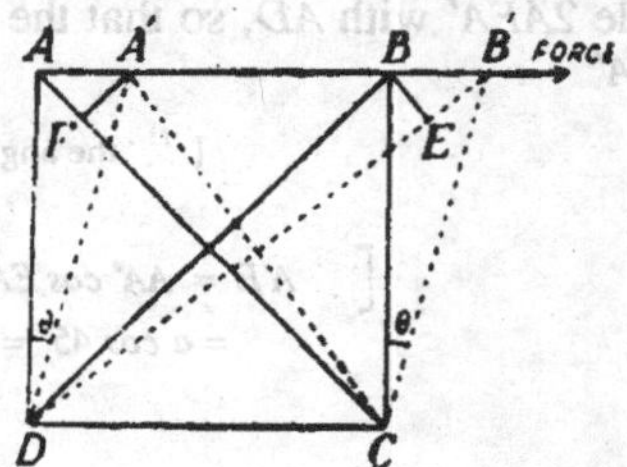

Fig. 8.4

The shear is really very small in actual practice, and, therefore, triangles *AFA′* and *BEB′* are *isosceles right-angled triangles*, (*i.e., right-angled* 45° *triangles*). And, therefore, $EB' = BB' \cos BB'E = BB'.1/\sqrt{2} = BB'/\sqrt{2}$,

$\because \ \ = 45°$ and $\cos 45° = 1/\sqrt{2}$.

If $AB = l$, then, clearly, $DB = DE = l\sqrt{2}$

∴ *extension strain* along diagonal *DB*

$$= \frac{EB'}{DB} = \frac{BB'}{\sqrt{2}} \cdot \frac{1}{l\sqrt{2}} = \frac{BB'}{2l} = \frac{\theta}{2}. \qquad [\because BB'/l = \theta.$$

Similarly, the *compression strain* along the diagonal *AC* is given by

* In the case of metals, however, *Hooke's law* no longer holds even if the shear exceeds 1/200 *radian*, or .33°.

$$\frac{AF}{AC} = \frac{AA' \cos A'AF}{AC} = \frac{AA' . \cos 45°}{l\sqrt{2}}$$

$$= \frac{AA'}{l\sqrt{2} . \sqrt{2}} = \frac{AA'}{2l} = \frac{\theta}{2}. \qquad [\because AA'/l = \theta.$$

Thus, we see that *a simple shear θ is equivalent to two equal strains, an extension and a compression, at right angles to each other.*

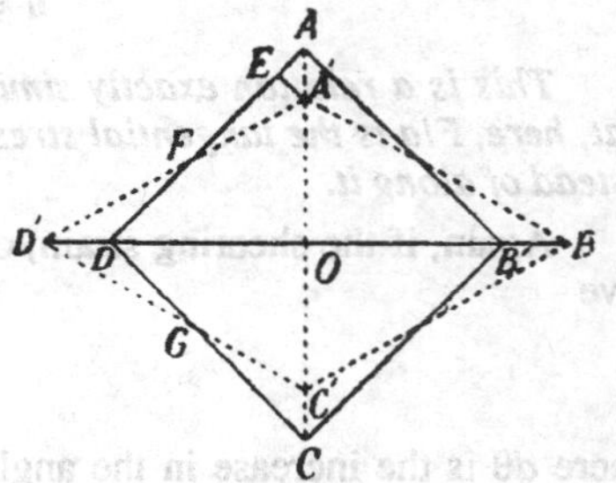

Fig. 8.5

Corollary. *The converse of the above follows as a corollary, viz., that simultaneous equal compression and extension at right angles to each other are equivalent to a shear*, as will be seen from the following;

Let the cube *ABCD*, of side *l*, be compressed along the diagonal *AC* so that the new diagonals become *A'C'* and *B'D'*, (Fig. 8.5)

Let $AA' = BB' = a$.

And since $OA = AB \cos BAO$

$= AB.\cos 45^0 = AB / \sqrt{2} = l / \sqrt{2}$ we have,

$$OA' = OA - AA' = \left(\frac{1}{\sqrt{2}} - a\right) \quad \text{and} \quad OB' = OB + BB' = \left(\frac{l}{\sqrt{2}} + a\right)$$

Clearly ∴ $(A'B')^2 = (OA')^2 + (OB')^2 = \left(\frac{l}{\sqrt{2}} - a\right)^2 + \left(\frac{l}{\sqrt{2}} + a\right)^2$

$$= \frac{l^2}{2} - \frac{2al}{\sqrt{2}} + a^2 + \frac{l^2}{2} + \frac{2al}{\sqrt{2}} + a^2 = l^2 + 2a^2.$$

In practice, $2a^2$ is very small as compared with l^2, and may, therefore, be neglected. So that,

$$(A'B')^2 = l^2 \qquad \text{Or,} \qquad A'B' = l = AB.$$

Thus, *A'B'C'D'* may be rotated through the angle *DGD'* = angle *AFA'*, so that *D'C'* coincides with *DC*. Then, it is obvious that *A'D'* would make an angle 2*AFA'* with *AD*, so that the *angle of shear* is equal to *twice* the angle *AFA'*, *i.e.*, is equal to 2 *AFA'*

Or, *angle of shear* $=2A'E/EF$, [∵ the angle is small.

where *A'E* is the perpendicular from *A'* to *AF*.

Now, $A'E = a/\sqrt{2}$ and $EF = l/2$. [∵ $A'E = AA' \cos EA'A = a \cos 45° = a/\sqrt{2}$

$$\therefore \ 2\angle AFA' = \frac{2a}{\sqrt{2}} + \frac{l}{2} = \frac{2a}{\sqrt{2}} \times \frac{2}{l} = \frac{2a\sqrt{2}}{l}.$$

Denoting this *angle of shear* by θ, we have $\theta = 2a\sqrt{2}/l$

Now, *compression strain* along the diagonal *AC*

$$= \frac{AA'}{AO} = \frac{a}{l/\sqrt{2}} = \frac{a\sqrt{2}}{l} = \frac{\theta}{2}.$$

Or, the *compression strain is half the angle of shear*.

Similarly, it can be snown that the extension strain is also half the angle of shear.

Thus, we see that *simultaneous and equal compression and extension at right angles to each other are equivalent to a shear*, the direction of each strain being at an angle of 45^0 to the direction of shear.

8.10. Shearing stress equivalent to an equal linear tensile stress and an equal linear compressive stress at right angles to each other.

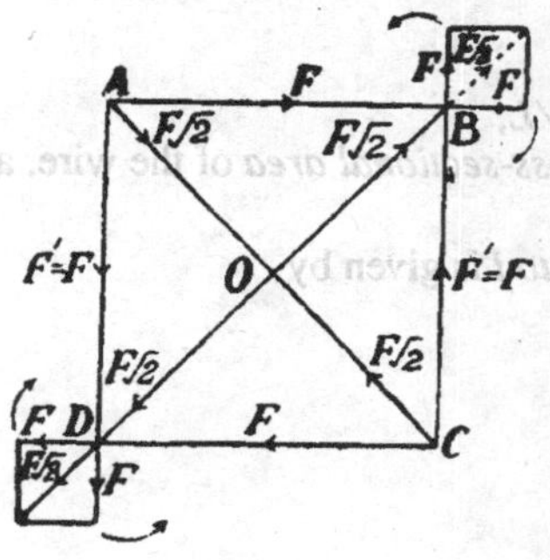

Fig. 8.6

In the case of the cube above, if F were the only force acting on its upper face, it would move bodily in the direction of this force. Since, however, the cube is fixed at its lower face DC, an equal and opposite force comes into play in the plane of this face, giving rise to a couple $F.l$ *, tending to rotate the cube in the clockwise direction, (Fig. 8.6).

Again, since the cube does not rotate, it is obvious that the plane of DC applies an equal and opposite couple $F'.l$, say, by exerting forces F' and F' along the faces AD and CB, tending to rotate it in the anticlockwise direction, as shown. Thus, because the cube is in equilibrium under the two couples, we have

$$F.l = F'.l. \quad \text{Or,} \qquad F = F',$$

i.e., a tangential force F applied to the face AB results in an equal tangential forced acting along all the other faces of the cube in the directions shown.

Clearly, the resultant of the two forces F and F' or F and F along AB and CB respectively is $F\sqrt{2}$ along OB, and of those acting along AD and CD is also $F\sqrt{2}$ along OD. And, thus, an *outward pull* acts on the diagonal DB of the cube at B and D, resulting in its extension, as we have just seen above (§ 8.9). Precisely similarly, an *inward pull* acts on the diagonal AC at A and C, thereby bringing about its compression.

Thus, a tangential force F applied to one face of a cube gives rise to a force $F\sqrt{2}$ *outward along one diagonal* (BD, in the case shown) and an equal force $F\sqrt{2}$ *inward along the other diagonal* (AC) of the cube, resulting in an extension of the former and a compression of the latter.

Now, if the cube be cut up into two halves, by a plane passing through AC and perpendicular to the plane of the paper, each face, parallel to the plane, will have an area $l \times l\sqrt{2} = l^2\sqrt{2}$, and, clearly, the outward force $F\sqrt{2}$ along BD will be acting perpendicularly to it. So that, we have

$$\textit{linear tensile stress along } BD = F\sqrt{2}/l^2\sqrt{2} = F/l^2.$$

Similarly, if we cut the cube into its two halves by a plane passing through BD and perpendicular to the plane of the paper, we shall have an *inward* force $F\sqrt{2}$ along AC acting perpendicularly to a face of an area $l^2\sqrt{2}$ So that, we have

$$\textit{linear compressive stress along } AC = F\sqrt{2}/l^2\sqrt{2} = F/l^2$$

Obviously, F/l^2 is the shearing stress over the face AB of the cube, which produces the shear θ in it, (see page 256).

Thus, it is clear that a *shearing stress is equivalent to an equal linear tensile stress and an equal linear compressive stress at right angles to each other.*

8.11. Work done per unit volume in a strain. In order to *deform* a body, work must be done by the applied force. The energy so spent is stored up in the body and is called the *energy of strain.* When the applied forces are removed, the stress disappears and *the energy of strain appears as heat.*

Let us consider the work done during the three cases of strain.

(*i*) **Elongation Strain— (stretch of a wire).** Let F be the force applied to a wire, fixed at the upper end. Then, clearly, for a small increase in length dl of the wire, the work done will be equal to $F.dl$. And, therefore, during the whole stretch of the wire from 0 to l,

* l being the length of each edge of the cube and hence the perpendicular distance between the two forces F and F'

$$Work\ done = \int_0^l F.dl.$$

Now, Young's modulus for the material of the wire, *i.e.*,

$$Y = F.L/a.l. \quad \text{And} \ \therefore\ F = Y.a.l/L,$$

where L is the *original length*, l, the *increase in length*, a, the *cross-sectional area* of the wire, and F, the *force applied*.

Therefore, *work done during the stretch of the wire from* 0 *to* l is given by

$$W = \int_0^l \frac{Y.a}{L}.l.dl = \frac{Y.a}{L}\int_0^l l.dl.$$

$$= \frac{Ya}{L}.\frac{l^2}{2} = \frac{1}{2}\frac{Y.a.l}{L}.l.$$

$$Y.a.l/L = F, \text{ the } force \text{ applied.}$$

Hence $\quad W = \frac{1}{2}F.l = \frac{1}{2}$ *stretching force* $\times$ *stretch.*

work done per *unit* volume $= \frac{1}{2}\ Fl \times \frac{1}{La}$. $\quad \left[\because \text{ volume of the wire } = L \times a.\right.$

$= \frac{1}{2}.\frac{F}{a}.\frac{l}{L} = \frac{1}{2}$ *stress* $\times$ *strain.* $\quad \left[\because F/a = \text{stress. and } l/L = \text{strain.}\right.$

Alternatively, the same result may also be obtained graphically as follows:

Let a graph OP be plotted between the *stretching force* applied to the wire and the *extension produced* in it, *within the elastic limit*, as shown in Fig. 8.7.

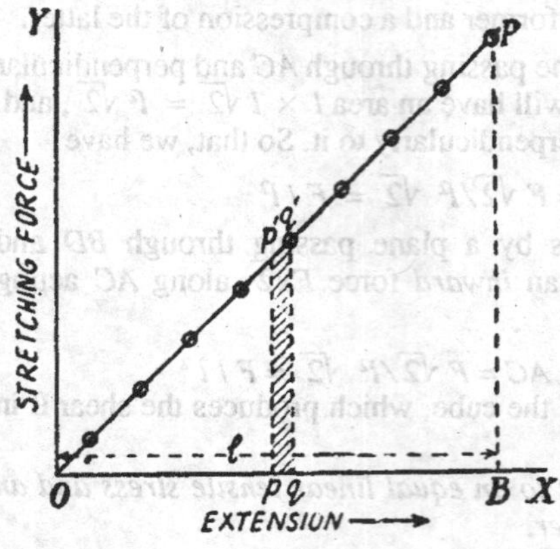

Fig. 8.7

Consider a small extension pq of the wire and erect ordinates at p and q to meet the graph in p' *and* q' respectively, where pp' is very nearly equal to qq' (the extension pq being really small).

Then, clearly, *work done upon the wire or energy stored up in it = stretching force* $pp' \times$ *extension* pq $= pp' \times pq =$ *area of strip* $pp'q'q$.

So that, imagining the whole extension $OB = l$, of the wire, to be broken up into small bits like pq and erecting ordinates at their extremities, we have *total work done upon the wire or total energy stored up in it = sum of the areas of all such strips formed.*

$=$ *area of the triangle* $OBP = \frac{1}{2}\ OB \times BP = \frac{1}{2} l \times F,$

where the total extension $OB = l$ and the stretching force corresponding to it is $BP = F$.

Now, if L be the original length of the wire and a, its area of cross-section, clearly,

volume of the wire $= L \times a$.

$\therefore$ *work done*, or *strain energy, per unit volume of the wire*

$$= \frac{1}{2}lF/La = \frac{1}{2}.\frac{F}{a}.\frac{l}{L} = \frac{1}{2}\ stress \times strain.$$

(*ii*) **Volume Strain.** Let p be the stress applied. Then, over an area a, the force applied is $p.a$, and therefore, work done for a small movement dx, in the direction of p, is equal to $p.a.dx$.

Now, $a.dx$ is equal to dv, the small change produced in volume. Thus, work done for a change dv is equal to $p.dv$.

And, therefore, total work done for the *whole change in volume*, from 0 to v, is given by

$$W = \int_0^v p.\,dv.$$

Now, $K = p.V/v$; so that, $p = K.v/V$,

where V is the original volume and K, the *Bulk Modulus*.

And $\therefore\ W = \int_0^v \frac{K.v}{V}.dv = \frac{K}{V}\int_0^v v.dv\ \frac{K}{V}.\frac{1}{2}v^2.$

$$= \frac{1}{2}.\frac{Kv}{V}.v = \frac{1}{2}p.v = \frac{1}{2}\textit{stress} \times \textit{change in volume}.$$

Or, work done per unit volume $= \frac{1}{2}p.v/V\ \frac{1}{2}\ \textit{stress} \times \textit{strain}.$

(*iii*) **Shearing Strain.** Consider a cube of edge L, (Fig. 8.8) with its lower face DC fixed, and let F be the *tangential force* applied to its upper face in the plane of AB, so that the face $ABCD$ is distorted into the position $A'B'CD$, or *sheared* through an angle θ. Let the displacement AA' be equal to $BB' = l$. Then, work done during a small displacement dl is equal to $F.dl$. And, therefore, work done for the *whole of the displacement*, from 0 to l, is given by

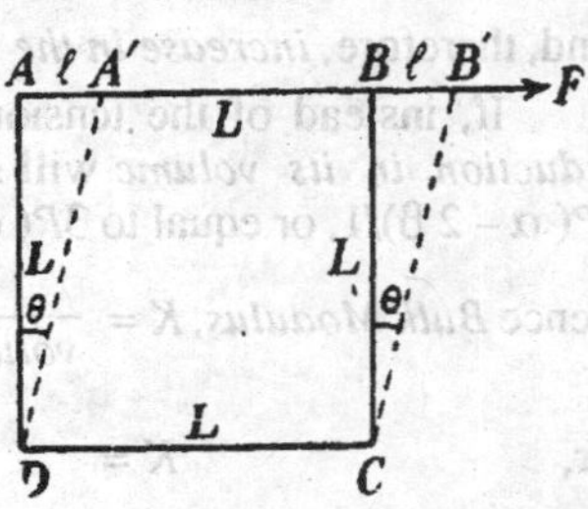

Fig. 8.8

$$W = \int_0^l F.dl.$$

Now, $n = F/a.\theta$, or $F = n.a.\theta$ and $a = L^2$; also $\theta = l/L$,

where L is the length of each edge of the cube.

So that, $F = n.L^2.l/L = n.L.l.$

$\therefore$ work done during the *whole stretch* from 0 to l, *i.e.*,

$$W = \int_0^l n.L.l.dl = \frac{1}{2}n.L.l^2 = \frac{1}{2}.\frac{F}{Ll}.Ll^2. \qquad \left[\because n = F/a\theta = \frac{F}{L^2}.\frac{L}{l} = \frac{F}{L.l}.\right.$$

$$= \frac{1}{2}F.l = \frac{1}{2}\ \textit{tangential force} \times \textit{displacement}.$$

$\therefore$ work done per *unit* volume $= \frac{1}{2}F.l/L^3$. $\qquad \left[\because \textit{volume} \text{ of the cube} = L^3\right.$

$$= \frac{1}{2}.\frac{F}{L^2}.\frac{l}{L} = \frac{1}{2}.\frac{F}{a}.\theta = \frac{1}{2}\ \textit{stress} \times \textit{strain}.$$

Thus, we see that, in any kind of strain, *work done per unit volume is equal to* $\frac{1}{2}$ *stress* $\times$ *strain*.

8.12. Deformation of a Cube—Bulk Modulus. Let $ABDCGHEFA$ be a unit cube and let forces T_x, T_y and T_z act perpendicularly to the faces $BEHD$ and $AFGC$, $ABDC$ and $EFGH$, and $ABEF$ and $DHGC$ respectively, as shown, (Fig. 8.9). Then, if α be the increase per *unit length* per *unit* tension *along the direction of the force* and β, the contraction produced per *unit length* per *unit*

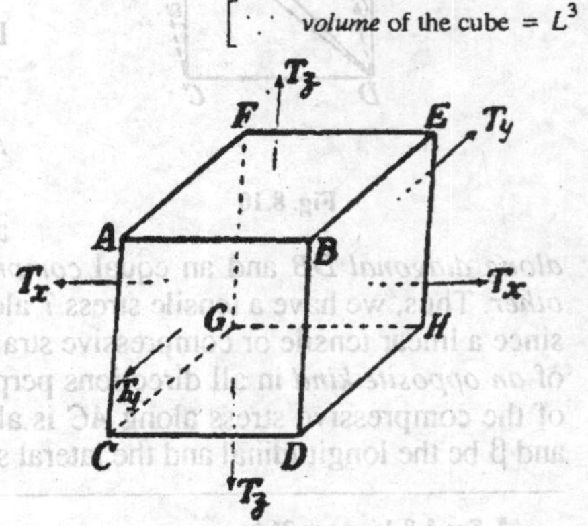

Fig. 8.9

tension, *in a direction perpendicular to the force*, the *elongations* produced in the edges *AB*, *BE* and *BD*, *will obviously be* $T_x.\alpha$, $T_y.\alpha$ and $T_z.\alpha$ respectively, and the *contraction* produced *perpendicular to them will be* $T_x.\beta, T_y.\beta$, and $T_z.\beta$. The lengths of the edges thus become as follows :

$$AB = 1 + T_x.\alpha - T_y.\beta - T_z.\beta.$$

$$BE = 1 + T_y.\alpha - T_x.\beta - T_z.\beta.$$

$$BD = 1 + T_z.\alpha - T_x.\beta - T_y.\beta.$$

Hence the volume of the cube now becomes

$$(1 + T_x.\alpha - T_y.\beta - T_z.\beta)(1 + T_y.\alpha - T_x.\beta - T_z.\beta)(1 + T_z.\alpha - T_x.\beta - T_y.\beta)$$

$$= 1 + (\alpha - 2\beta)(T_x + T_y + T_z),$$

neglecting squares and products of α and β, which are very small compared with the other quantities involved.

If $T_x = T_y = T_z = T$,

the volume of the cube becomes $1 + (\alpha - 2\beta).3T$.

And, therefore, *increase in the volume of the cube* $= 1 + 3T(\alpha - 2\beta) - 1 = 3T(\alpha - 2\beta)$.

If, instead of the tension T outwards, we apply a pressure P, compressing the cube, *the reduction in its volume* will similarly be $3P(\alpha - 2\beta)$, and, therefore, *volume strain, equal to* $3P(\alpha - 2\beta)/1$, or equal to $3P(\alpha - 2\beta)$. [$\because$ ***original volume*** **of the cube = 1.**

Hence *Bulk Modulus*, $K = \dfrac{stress}{volume\ strain} = \dfrac{P}{3P(\alpha - 2\beta)}$.

Or, $$K = \frac{1}{3(\alpha - 2\beta)}. \quad ...(I)$$

And, *Compressibility*, which is the *reciprocal of Bulk Modulus*, is, therefore, equal to $3(\alpha - 2\beta)$.

8.13. Modulus of Rigidity. Let the top face *ABHG* of a cube (Fig. 8.10) be '*sheared*' by a shearing force F, *relative to the bottom face*, such that A takes up the position A' and B, the position B' the angle ADA' being equal to the angle $BCB' = \theta$. Then,

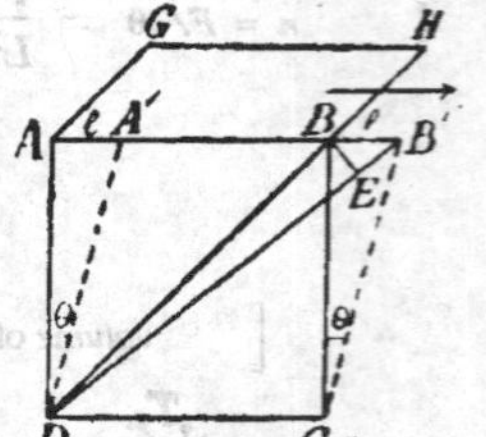

Fig. 8.10

tensile stress $= F/$*area of face ABHG*,

$= F/L^2 = T$, say,

where L is the length of each edge of the cube.

Let the displacement $AA' = BB' = l$.

Then, *shear strain* $= l/L = \theta$.

And *coefficient of rigidity*, $n = T/\theta$.

Now, as we have seen in § 8.10, a shearing stress along AB is equivalent to an *equal linear tensile stress along diagonal DB* and an equal *compressive stress* along *diagonal AC*, *at right angles to each other*. Thus, we have a tensile stress T along DB and *an equal compressive stress T along AC*. And, since a linear tensile or compressive strain is always accompanied by a transverse or a lateral strain of *an opposite kind* in all directions perpendicular to the direction of the applied force*, the effect of the compressive stress along AC is also to increase the length of diagonal DB. If, therefore, α and β be the longitudinal and the lateral strains per unit stress respectively, we have

* See § 8.16, page 264.

extension of diagonal DB due to tensile stress along it $= DB.T.\alpha$ and
its extension due to compressive stress along AC $= DB.T.\beta$.

Therefore, *total extension in length of the diagonal DB*

$$= DB \times (T.\alpha + T.\beta) = DB \times T(\alpha + \beta).$$

$$= L\sqrt{2}.T(\alpha + \beta). \qquad [\because DB = \sqrt{L^2 + L^2} = L\sqrt{2}.$$

Drop a perpendicular BE from B on to DB'.

Then, *increase in length of DB* is practically equal to EB'.

And, clearly, $EB' = BB'.\cos BB'E$.

$$= l\cos = 45^0 = l/\sqrt{2} \qquad [\because \angle BB'E = 45^0, \text{ very nearly, and } \cos 45^\circ = 1\sqrt{2}.$$

$$\therefore \quad L\sqrt{2}.T(\alpha + \beta) = l/\sqrt{2}.$$

Or, $$\frac{L.T}{l} = \frac{1}{2(\alpha + \beta)}. \text{ Or, } \frac{T}{l/L} = \frac{1}{2(\alpha + \beta)}$$

Or, $$\frac{T}{\theta} = \frac{1}{2(\alpha + \beta)} \qquad [\because l/L = \theta.$$

And, since $T/\theta = n$ the *coefficient of rigidity* of the material of the cube, we have

$$n = \frac{1}{2(\alpha + \beta)}. \qquad ...(II)$$

8.14. Young's Modulus. If we now imagine a *cube of unit edge*, acted upon by *unit tension* along one edge, the extension produced is α. Then, clearly,

stress = 1 and *linear strain* = $\alpha / 1 = \alpha$.

Therefore, *Young's Modulus*, $Y = 1/\alpha$ ...(III)

8.15. Relation connecting the Elastic Constants. We have from relation (I), above,

$$\alpha - 2\beta = 1/3K. \qquad ...(i)$$

And, from relation (II), $\alpha + \beta = 1/2n$ ...(ii)

$\therefore$ subtracting *(i)* from *(ii)*, we have

$$3\beta = \frac{1}{2n} - \frac{1}{3K} = \frac{3K - 2n}{6nK},$$

whence, $$\beta = \frac{3K - 2n}{18nK}. \qquad ...(iii)$$

Again, multiplying *(ii)* by 2 and adding to *(i)*, we have

$$3\alpha = \frac{1}{n} + \frac{1}{3K} = \frac{3K + n}{3Kn}. \text{ Or, } \alpha = \frac{3K + n}{9Kn}.$$

Or, $$\frac{1}{Y} = \frac{3K + n}{9Kn}. \text{ Or, } Y = \frac{9Kn}{3K + n}. \qquad ...(a)$$

[$\because \alpha = 1/Y$ from (III), above.

i.e. $$\frac{9}{Y} = \frac{3K + n}{Kn} = \frac{3K}{Kn} + \frac{n}{Kn},$$

Or, $$\frac{9}{Y} = \frac{3}{n} + \frac{1}{K}. \qquad ...(b)$$

This, then is the relation connecting the three elastic constants.

8.16. Poisson's Ratio. It is a commonly observed fact that when we stretch a string or a wire, *it becomes longer but thinner, i.e.*, the increase in its length is always accompanied by a decrease in its cross-section (though not sufficient enough to prevent a slight increase in its volume). In other words, *a linear or a tangential strain produced in the wire is accompanied by a transverse or a lateral strain of an opposite kind in a directions at right angles to the direction of the applied force.** And, of course, what is true of a wire, is true of all other bodies under strain. Thus, for example, when a cube is subjected to an outward force perpendicular to one pair of its faces, there is *elongation produced along this face*, but *a contraction in a direction perpendicular to it*, as we have seen already in § 8.12.

Within the elastic limit, the lateral strain (β) *is proportional to the linear or tangential strain* (α) *for the material of a given body and the ratio between the two is a constant, called* **Poisson's ratio,** *for that material. It is denoted by the letter* σ and, in engineering practice, usually by **1/*m***. Thus, *Poisson's ratio = lateral strain/tangential* (or *linear*) *strain*; or, $\sigma = \beta/\alpha = 1/m$.

Or, *secondary strain/primary strain** $= \sigma = 1/m$.

It follows, therefore, that

(*i*) If a body under tension suffers no lateral contraction, the value of Poisson's ratio (σ or $1/m$) for it is zero, and, because its volume decreases, its density must increase.

(*ii*) Knowing the linear strain in a bar and the Poisson's ratio for its material, it is possible to calculate the changes in all other dimensions of it. For, *if the linear strain in the bar be* α, the lateral strain in it will be $\sigma \times \alpha$ or $\alpha \times 1/m$. So that, we have

(*a*) *for a rectangular bar, of length l, width b and thickness t,*
lateral strain $= \delta b/b = \delta t/t = \sigma\alpha = \alpha/m$, whence δt and δb, the changes in its width and thickness respectively can be easily calculated.

And, (*b*) *for a cylindrical bar, for length l and diameter d,*

$$\text{lateral strain} = \delta d/d = \sigma\alpha = \alpha/m.$$

whence, δd, the change in its diameter may be easily obtained.

Again, to calculate the *volume strain* in the bar, we have

(*a*) *for the rectangular bar*, original volume $V = l \times b \times t$,

change in volume $\delta V = b.t.\delta l + t\delta.l.\ \delta b + l.b.\delta t$,

where δl is the change in its length corresponding to linear strain α.

So that, we have
$$\frac{\delta V}{V} = \frac{\delta l}{l} + \frac{\delta b}{b} + \frac{\delta t}{t}.$$

Or, *volume strain = algebraic sum of the strains along the three dimensions of the bar.*

Now, as we know, the lateral strain is *always* of the *opposite sign* to that of the linear strain. Hence, we have

$$\frac{\delta V}{V} = \alpha - \sigma\alpha - \sigma\alpha = \alpha - \frac{\alpha}{m} - \frac{\alpha}{m} \qquad [\because \delta l/l = \alpha.$$

Or,
$$\delta V/V = \alpha(1 - 2\sigma) = \alpha(1 - 2/m).$$

And (*b*) *for the cylindrical bar*, original volume $V = \pi.\dfrac{d^2}{4}.l$,

$$\text{change in volume } \delta V = \frac{\pi d^2.\delta l}{4} + \frac{\pi}{4} \times 2d.\delta d.l,$$

where the first term gives the change in volume due to change of length and the second, due to change of diameter.

So that, we have
$$\frac{\delta V}{V} = \frac{\delta l}{l} + \frac{2\delta d}{d}.$$

---

* Strains along the direction of the applied force are sometimes referred to primary strains and those at right angles to it as lateral or **secondary strains**, the two being always of opposite kinds (*i.e.*, if the former are *tensile*, the latter will be *compressive* and *vice versa*.

Or, substituting for $\delta l/l$ and $\delta d/d$ and remembering that they bear opposite sign, we have

$$\delta V/V = \alpha - 2\sigma\alpha = \alpha - 2\alpha/m \quad \text{or} \quad \delta V/V = \alpha(1 - 2\sigma) = \alpha(1 - 2/m),$$

i.e., exactly the same as for a rectangular bar.

Thus, in either case, knowing the original volume (V), linear strain (α) and *Poisson's ratio* ($\sigma = 1/m$) we can easily calculate the change in volume of the bar irrespective of its cross-section being rectangular or circular.

8.17. Relations for K and n in terms of Poission's ratio. We have, from relation I (page 262) above,

$$K = \frac{1}{3\alpha\left(1 - \frac{2\beta}{\alpha}\right)} = \frac{1}{3\alpha(1 - 2\sigma)} = \frac{Y}{3(1 - 2\sigma)}, \qquad \left[\because 1/\alpha = Y, \text{[See (III)] above, (page 263)}\right.$$

whence, $Y = 3K(1 - 2\sigma)$. Or, $Y = 3K(1 - 2/m)$* ...(*iv*)

Similarly, from relation (II) (page 263) above, we have

$$n = \frac{1}{2\alpha\left(1 + \frac{\beta}{\alpha}\right)} = \frac{1}{2\alpha(1 + \sigma)} = \frac{Y}{2(1 + \sigma)},$$

whence, $Y = 2n(1 + \sigma)$. Or, $Y = 2n(1 + 1/m)$* ... (*v*)

Now, from relations (*iv*) and (*v*), we have

$$3K(1 - 2\sigma) = 2n(1 + \sigma),$$

whence, $\sigma = 1/m = \dfrac{3K - 2n}{6K + 2n}$,

which gives the value of *Poisson's ratio* in terms of *K* and *n*.

Similarly, if we eliminate σ from (*iv*) and (*v*), we have

$$Y = \frac{9Kn}{3K + n}, \qquad \text{[Same as relation (a), above, (page 263).}$$

whence, $\dfrac{9}{Y} = \dfrac{3}{n} + \dfrac{1}{K}$. [Same as relation (*b*), above, (page 263).

8.18. Limiting values of σ. We have seen above how

$$3K(1 - 2\sigma) = 2n(1 + \sigma),$$

where *K* and *n* are essentially *positive* quantities. Therefore,

(*i*) *if σ be a positive quantity*, the right hand expression, and hence also the left hand expression, must be positive, and for this to be so, $2\sigma < 1$, or $\sigma < 1/2$ or .5, And,

(*ii*) *if σ be a negative quantity*, the left hand expression, and hence also the right hand expression, must be positive, and this is possible only when σ be not less than -1.

Thus, the limiting values of σ are -1 and .5. Or, else, as will be readily seen from relations (*iv*) and (*v*) above, either the bulk modulus or the modulus of rigidity would become *infinite*. Further, a negative value of σ would mean that, on being extended, a body should also expand laterally, and one can hardly expect this to happen, ordinarily. At least, we know of no such substance so far. Similarly, a value of $\sigma = 0.5$ would mean that the substance *is perfectly incompressible*, and, frankly, we do not know of any such substance either.

---

* These relations would not be found to apply in the case of wire specimens of materials for the simple reason that the process of wire-drawing brings about at least a partial alignment of the minute crystals of the substance, which thus no longer remain oriented at random, with the result that the substance loses its isotropic character.

In actual practice, the value of σ is found to lie between .2 and .4. Although Poisson had a theory that the value of σ for all elastic bodies should be .25, this is not borne out by any experimental facts.

8.19. Determination of Young's Modulus (Y). Young's modulus, as we know, is the ratio between *tensile stress* (or tangential force applied oper unit area) and *elongation strain* (or extension per unit length). The extension produced is rather small and it is difficult to measure it with any great degree of accuracy. The different methods used are thus merely attempts at measuring this extension accurately. We shall consider here only two simple methods *for a wire* and *one for a thick bar*.

I. For a Wire. (*i*) Searle's Method. Two wires, *A* and *B*, of the *same* material, length and area of cross-section, are suspended from a rigid support and carry, at their lower ends, two metal frames, *C* and *D*, as shown in Fig. 8.11, one carrying a constant weight *W* to keep the wire stretched or *taut* and the other, a hanger *H*, to which slotted weights can be slipped on, as and when desired.

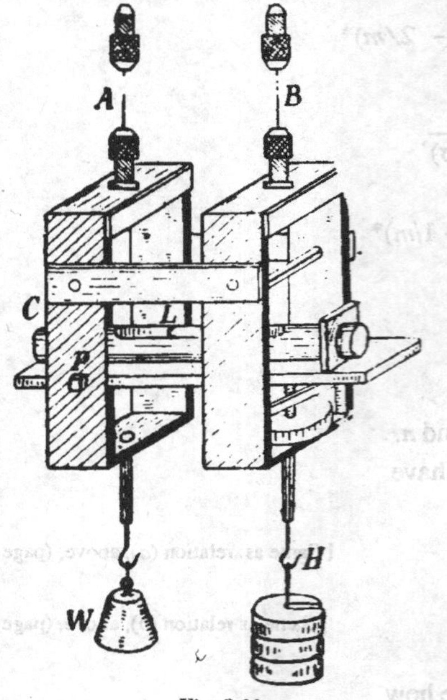

Fig. 8.11

A spirit-level *L* rests *horizontally* at a point *P* in frame *C*, and on the tip of a *micrometer screw* (or spherometer) *S*, working through a nut in frame *D*.

The screw is worked up or down, until the air bubble in the spirit-level is just in the centre, Weights are now slipped on to the hanger, so that the frame *D* moves down a little due to the extension of wire *B*, and the air bubble shifts towards *P*. The screw is now worked up to restore the bubble back to its central position. The distance through which the screw is moved up is read on the vertical scale, graduated in half-millimetres, and fixed alongside the disc of the screw. This gives the increase in length of wire *B*. A number of observations are taken by increasing the weight in the hanger by the *same equal steps* and making the adjustment for the spirit-level for each additional weight. The mean of all these readings of the screw gives the mean increase in the length of the wire for the stretching force due to the given weight. Thus, if *l cm* be the increase in the length of wire *B*, and *L cm*, its original length, we have

$$\textit{elongation strain} = l/L.$$

And, if *W kgm* be the weight added each time to the hanger, the stretching force is equal to $W \times 1000$ *gm. wt* $= W \times 1000 \times 981$ *dynes*, or equal to *F dynes*, say

So that, if *a sq cm* be the area of cross-section (πr^2) of the wire, we have

$$\textit{tensile stress} = F/a.$$

And, $\therefore$ *Young's modulus* for the material of the wire, *i.e.*,

$$Y = \frac{F}{a} \div \frac{l}{L} = \frac{F \times L}{a \times l}.$$

The other wire *A* merely acts as a *reference wire*, its length remaining constant throughout, due to the constant weight suspended from it (which need not be known). Any yielding of the support or change in temperature during the experiment affects both the wires equally, and the *relative increase* in the length of *B* (with respect to *A*) thus remains unaffected by either change.

If a graph be now plotted between the load suspended and the extension produced, it would be found to be a *straight line* (just like *OA* in Fig. 8.1), passing through the origin, showing that the extension produced is directly proportional to the load. *Hooke's law* also can thus be easily verified.

(ii) **Gravesande's method.** The wire *AB*, under test, about 1 metre in length and .5 *mm* in diameter, is arranged horizontally between two heavy and rigid clamps C_1C_2, so as to remain *taut*, as indicated by the dotted horizontal line in Fig. 8.12, with its mid-point at *C*. Let its half length $AC = BC = l_0$ and the tension in it T_0. It is now loaded at *C* with some weights in a scale pan so that the point gets depressed to *D*, through a vertical distance $CD = y$, the half-length *AD* of the wire now increasing to l and its tension to *T*. Let angle $CAD = \theta$.

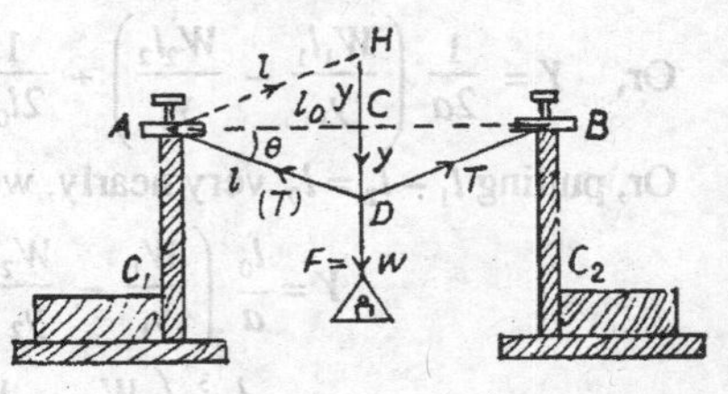

Fig. 8.12

Obviously point *D* is in equilibrium under the action of the three forces *T*, *T* and *W*, acting along *DA*, *DB* and vertically downwards respectively. Drawing *AH* parallel to *DB* to meet *DC* produced in *H*, we, therefore, have the *triangle of forces DAH*, representing the three forces in equilibrium. So that,

$$\frac{T}{l} = \frac{W}{2y} \cdot \text{ Or, } T = \frac{1}{2} W \frac{l}{y}. \quad ...(i)$$

$\left[\because HD = HC + CD = y + y = 2y \right.$

Now, $\quad l = (l_0^2 + y^2)^{1/2} = l_0 (1 + y^2/l_0^2)^{1/2}$

Or $\quad l - l_0 = \frac{1}{2} y^2/l_0 = \textit{increase in the length of the wire}$

Clearly, therefore, *linear strain* $= \dfrac{l - l_0}{l_0}$

And, since tension in the wire has increased from T_0 to *T*, the *tensile stress* $= (T - T_0)/a$, where *a* is the area of cross section of the wire.

$\therefore$ *Young's modulus for the material of the wire* is given by $Y = \dfrac{T - T_0}{a} \div \dfrac{l - l_0}{l_0}$ ...I

Here, unfortunately, T_0 is not known and must, therefore be eliminated or calculated. This may be done in two ways *(a) graphically* or *(b)* by performing the experiment with two different loads at *C*, as explained below.

(a) **Graphically.** Substituting the values of *T* and $(l - l_o)$ in relation I above, we have

$$Y = \frac{1}{a}\left(\frac{l}{2y} \cdot W - T_0\right) \div \frac{y^2}{2l_0^2}.$$

Or, $\quad Yay^2 = l_0^2 \left(\frac{l}{y} . W - 2T_0\right),$

which gives $\quad \dfrac{W}{y} = \dfrac{y_a}{l_0^3} . y^2 + 2\dfrac{T_0}{l_0}$ $\quad$ [putting $l = l_0$ very nearly

Hence, plotting W/y, against y^2, we obtain a straight line graph, the intercept of which gives the value of T_0 or, T_0 may be eliminated altogether by choosing two points on the graph, fairly distant apart.

(b) **By performing the experiment with two different loads.** Let W_1 and W_2 be the loads respectively and let the depressions corresponding to them be y_1 and y_2 and the tensions in the wire T_1 and T_2. Then, we have, from relations *(i)* and *(ii)* above

$$T_1 = W_1 l_1/2y_1,\ T_2 = W_2 l_2/2y_2,\ (l_1 - l_0) = y_1^2/l_0 \text{ and } (l_2 - l_0) = y_2^2/2l_0.)$$

whence, $(l_1 - l_2) = \dfrac{1}{2l_0}(y_1^2 - y_2^2)$

Now, as we know, $Y = \dfrac{T_1 - T_2}{a} \div \dfrac{l_1 - l_2}{l_0}$,

where $(T_1 - T_2)/a$ is the *tensile stress* and $(l_1 - l_2)/l_0$, the *linear strain*.

Or, $\quad Y = \dfrac{1}{2a}\left(\dfrac{W_1 l_1}{y_1} - \dfrac{W_2 l_2}{y_2}\right) \div \dfrac{1}{2l_0^2}\left(y_1^2 - y_2^2\right)$

Or, putting $l_1 - l_2 = l_0$, very nearly, we have

$$Y = \frac{l_0}{a}\left(\frac{W_1}{y_1} - \frac{W_2}{y_2}\right) \times \frac{l_0^2}{y_1^2 - y_2^2}$$

Or, $$Y = \frac{l_0^3}{a}\left(\frac{W_1}{y_1} - \frac{W_2}{y_2}\right) \Big/ \left(y_1^2 - y_2^2\right)$$

whence Y can be easily evaluated.

II. For a thick Bar — Ewing's Extensometer Method. *Ewing's Extensometer* is merely a device to magnify the small extension of the bar under test, and essentially consists of two *clamp-pieces* C_1 and C_2, parallel to each other and fixed on the *test bar B* by means of a pair of set *screws* S_1 and S_2, with the distance between the two screw-points accurately known (Fig. 8.13). This gives the *initial length L* of the bar before it is subjected to any tension.

A *micrometer screw M.S.*, attached to clamp C_1 near one end, has a conical hollow in its tip which engages with a *ball-point B.P.* at the top of an upright clamp V carried by C_2, the ball-point serving as a *fulcrum* about which clamp C_1 can turn when the bar is extended.

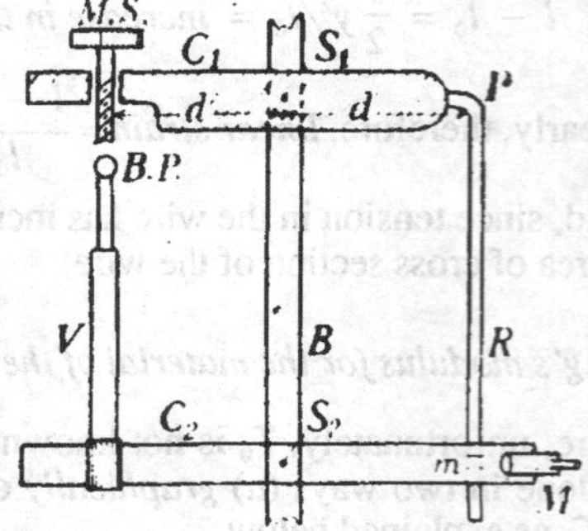

Fig. 8.13

At the other end P of C_1, lying in a level with, and at the same distance d from, the axis of the test bar as the ball-point is pivoted a *vertical rod R* which passes through a slot in clamp C_2. In its portion inside C_2, the rod has a *glass plate* set into a rectangular notch in it, with a horizontal line or mark m engraved on the plate and kept illuminated by reflected light from a small mirror. The movement of m can thus be easily and accurately read with the help of a micrometer eye-piece of a *microscope M*, arranged inside, and co-axially with clamp C_2, as shown.

The microscope is first focussed on mark m on rod R and the micrometer screw *M.S.* adjusted, if necessary, to make its image coincide with a convenient division in the micrometer scale of the eye-piece. The test bar is then stretched by means of a vertical testing machine and as it extends downwards, clamp C_1 turns about its fulcrum *B.P.*, resulting in a downward movement of R and a consequent displacement of mark m. This displacement, read on the eye-piece scale, gives twice the extension l of the bar for the simple reason that *the axis of rod R lies twice as far away from the fulcrum as the axis of the bar*. Half the displacement read thus gives the value of l.

Then, from the length L of the bar between S_1 and S_2, its area of cross-section and the stretching force applied to it, we can calculate the value of *Young's Modulus* for its material in the usual manner.

The apparatus is sensitive enough to be able to measure extensions (l) of the order of 2.5×10^{-5} *cm* with reliable accuracy. Further, it may be used with both vertical and horizontal testing machines and the variation in length of the given specimen of the bar watched continuously, if so desired.

N.B. A modification of Ewing's Extensometer, as shown in Fig. 8.14, is called the *Cambridge Extensometer*, in which there is a vibrating reed R arranged, as shown, the arrangement being such that as the bar B is stretched by the testing machine, *that part of the reed which touches the micrometer screw M, moves downwards through a distance five times the extension of the rod.* Thus, by noting the micrometer screw readings, when the vibrating reed just touches the micrometer screw-point both before and after the rod B has been stretched, we can *directly* obtain the increase l in the length of the rod.

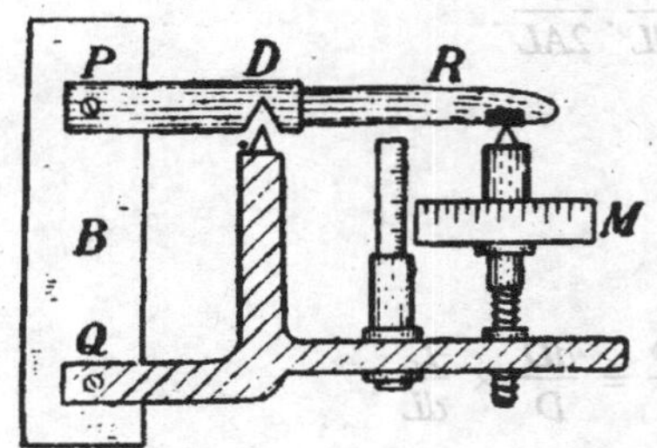

Fig. 8.14

8.20. Determination of Poisson's ratio for rubber. To determine the value of σ for rubber, we take about a metre-long tube AB of it, (Fig. 8.15), such, for example, as the tube of an ordinary *cycle tyre*, and suspended it vertically, as shown, with its two ends properly stoppered with rubber bungs and seccotine (a liquid glue). A glass tube C, open at both ends, about half a metre long and about 1 cm in diameter, is fitted vertically into it through a suitable hole in the stopper at the upper end A, so that a major part of it projects out.

The rubber tube is completely filled with air-free water until the water rises up in the glass tube to a height of about 30 *cm* from A. A suitable weight W is now suspended from the lower end B of the tube. This naturally increases the length as well as the internal volume of the tube, resulting in a fall of the level of the meniscus in C. Both, the increase in length (dL) and the fall (dh) in the level of the meniscus in C, are measured with the help of a travelling microscope, by noting the new positions of point P and the level of the meniscus C, though, for ordinary purposes, the former may be read conveniently on a vertical metre scale, with the help of a pointer, attached to the suspension of W.

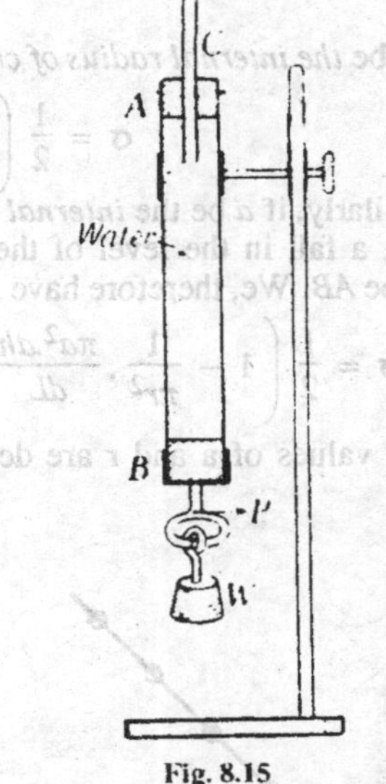

Fig. 8.15

Let the original *length, diameter* and *volume* of the rubber tube be L, D and V respectively.

Then, its *area of cross-section*

$$A = \pi (D/2)^2 = \pi D^2/4, \quad \text{....}(i)$$

differentiating which, we have

$$dA = \frac{\pi D}{2} \cdot dD, \quad \text{whence,} \quad dA = 2A.dD/D, \quad \text{....}(ii)$$

from (i) above, by eliminating π.

Now, if corresponding to a small increase dV in the volume of the rubber tube, the increase in its length be dL, and the decrease in its area of cross-section be dA, we have

$$V + dV = (A - dA)(L + dL) \quad [\because \text{Volume} = \text{area of cross section} \times \text{length.}$$

$$= AL + A.dL - dA.L - dA.dL.$$

Or, $$V + dV = V + A.dL - dA.L,$$

where $AL = V$, the *original volume* of the tube and $dA.dL$ is a very small quantity, compared with the other terms.

So that, $dV = A.dL - dA.L = A.dL - \frac{2AL}{D} dD$. [Substituting the value of dA from *(i)* above

Or, dividing both sides by dL, we have

$$\frac{dV}{dL} = A - \frac{2AL}{D} \cdot \frac{dD}{dL}. \text{ Or, } \frac{2AL}{D} \cdot \frac{dD}{dL} = A - \frac{dV}{dL},$$

whence,
$$\frac{dD}{dL} = \left(A - \frac{dV}{dL}\right) \bigg/ \frac{2AL}{D} = \frac{AD}{2AL} - \frac{dV}{dL} \cdot \frac{D}{2AL}.$$

$$= \frac{D}{2L} - \frac{dV}{dL} \cdot \frac{D}{2AL}$$

Or,
$$\frac{dD}{dL} = \frac{D}{2L}\left(1 - \frac{1}{A} \cdot \frac{dV}{dL}\right).$$

Now, *Poisson's ratio*, $\sigma = \dfrac{\textit{lateral strain}}{\textit{longitudinal strain}} = \dfrac{dD/D}{dL/L} = \dfrac{dD}{D} \times \dfrac{L}{dL}$

Or,
$$\sigma = \frac{L}{D} \cdot \frac{dD}{dL}.$$

Or, substituting the value of dD/dL from relation *(iii)* above, we have

$$\sigma = \frac{L}{D} \cdot \frac{D}{2L}\left(1 - \frac{1}{A} \cdot \frac{dV}{dL}\right) = \frac{1}{2}\left(1 - \frac{1}{A} \cdot \frac{dV}{dL}\right) \quad ...(I)$$

If r be *the internal radius of cross-section of the tube*, clearly, $A = \pi r^2$ So that,

$$\sigma = \frac{1}{2}\left(1 - \frac{1}{\pi r^2} \cdot \frac{dV}{dL}\right).$$

Similarly, if a be the *internal radius of the capillary tube*, we have $dV = \pi a^2.dh$, because, as we know, a fall in the level of the meniscus in C corresponds to an increase in volume of the rubber tube AB. We, therefore have

$$\sigma = \frac{1}{2}\left(1 - \frac{1}{\pi r^2} \cdot \frac{\pi a^2.dh}{dL}\right) = \frac{1}{2}\left(1 - \frac{a^2}{r^2} \cdot \frac{dh}{dL}\right) \quad ...(II)$$

The values of a and r are determined carefully at several points by means of a travelling microscope and a vernier calliper respectively and their mean values taken. And, the *average value of dh/dL* is obtained from the slope of the straight-line graph by plotting a number of corresponding values of dh against those of dL, (Fig. 8.16), on suspending different loads from the tube. The value of σ for the given specimen of rubber can then be easily calculated from relation (II).

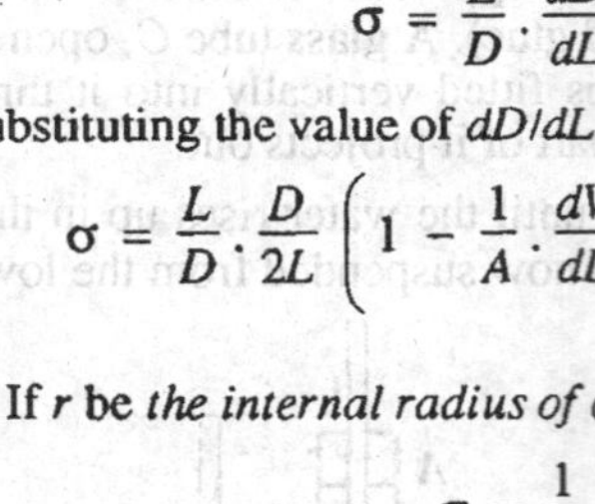

Fig. 8.16

For ordinary purposes, we may use a capillary tube graduated in *cubic centimetres* and read dV directly on it. Then, substituting the values of A (*i.e.*, πr^2) and of dV/dL in relation *I*, σ may be evaluated. Here too, however, it is advisable to plot dV against dL for different loads suspended from the tube and to use the average value of dV/dL given by the slope of the straight-line graph, thus obtained.

Poisson's ratio for glass. An identical method with that used for rubber above, may be used for the determination of the value of σ for glass, but since the change in its volume is comparatively much too small, we have to use a capillary tube, instead of an ordinary glass tube, to measure it to an adequate degree of accuracy.

8.19. Resilience. By the *resilience* of an elastic body we understand its *capacity of resisting a blow or a mechanical shock, without acquiring a permanent set** and we measure it by the *amount of work done in straining the body up to the elastic limit.* Let us consider it for a uniform bar of length L and area of cross-section a.

We know that when the bar is subjected to a stretching force W, so that it increases in length by l, we have

Young's modulus for the material of the bar, Y = stress/strain,

$$i.e., Y = \frac{W/a}{l/L} = \frac{W}{a}.\frac{L}{l} = \frac{F}{strain}. \text{ Or, } strain = F/Y,$$

where F denotes the stress W/a.

Now, *work done per unit volume in elongation strain* = $\frac{1}{2}$ *stress × strain.*

∴ *work done in producing extension* $l = \frac{1}{2}$ *(stress × strain) × volume V.*

$$\frac{1}{2} F.\frac{F}{Y} \times volume\ (V) = \frac{1}{2}.\frac{F}{Y}. = \frac{VF^2}{2Y}.$$

Thus, *work done* or *resilience of the bar*

$$= \frac{VF^2}{2Y} = \frac{volume \times (stress)^2}{2 \times Young's\ modulus}.$$

And ∴ *resilience per unit volume of the bar*

$$= \frac{F^2}{2Y}, \ i.e., = \frac{(stress)^2}{2 \times Young's\ modulus}.$$

Height from which the bar can be dropped without acquiring a permanent set. —Since resilience is a measure of the power to resist a blow or shock without acquiring a permanent set, let us see from what height the bar can be dropped without taking on a permanent set.

This height must obviously be one in falling through which the bar acquires energy equal to its own resilience. Let it be h. Then, if w be the *weight per unit volume* of the bar, clearly,

energy acquired by the war in falling through height $h = Vw.h$.

Equating this against the resilience of the bar, therefore, we have $Vw.h = VF^2/2Y$, whence, $h = F^2/2wY$.

Thus, the bar can absorb a blow or a shock due to a fall from this much height.

Proof Resilience. The maximum amount of energy per unit volume that can be stored in a body or a piece of material, without its acquiring a permanent set, *i.e.*, without it's undergoing a permanent strain, is called its *proof resilience*. Thus, if F_m be the *maximum stress* to which a material, in the form of a wire, can be subjected, *i.e.*, if F be its *elastic limit*, we have

proof resilience of the material $= F_m^2/2Y$.

8.20. Effect of a suddenly applied load. Suppose we have a uniform bar of *length L* and *area of cross-section a*, suspended vertically from one end, with a *collar C* provided at the other, and with a weight W, in the form of a ring, threaded on to it at a height h from the collar, as shown in Fig. 8.17.

* The meaning attached to the word *'resilience'* in our common everyday parlance is different, *viz.*, that the body comes back to its normal condition when the applied forces are removed.

If we now allow the weight to fall freely so as to hit the collar, so that the length of the bar is increased by a small amount l, with the collar taking up the position C', clearly, the total height through which the weight has fallen is $(h + l)$.

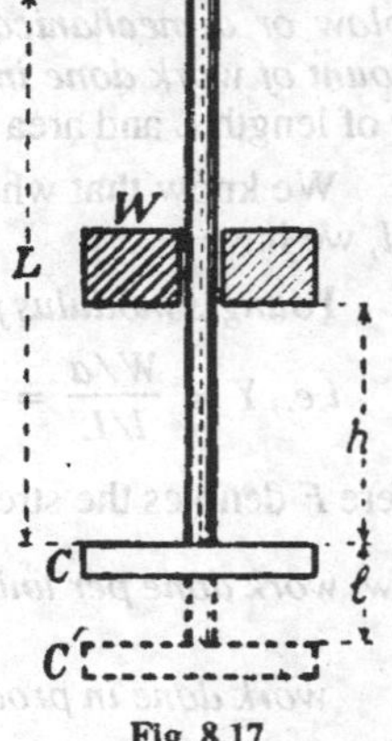

Fig. 8.17

$\therefore$ *potential energy lost by the weight* $= W(h + l)$. This has obviously been utilised in stretching the bar through l and must, therefore, be equal to the work done in so stretching it.

If F_m be the *maximum stress* in the bar, the *resistance offered by the bar, or the restoring force set up in it* $= F_m \cdot a$.

And, since *work done during stretch* $= \frac{1}{2}$ *stretching force* $\times$ stretch,

we have [See page 260

work done in stretching the bar $= \frac{1}{2} \cdot F_m a.l$

And, therefore,

$$W(h+l) = \frac{1}{2} \cdot F_m .a.l \qquad ...(i)$$

Now, as we know, $Y = \dfrac{tensile\ stress}{tensile\ strain} = \dfrac{F_m}{l/L}$, and $\therefore = F_m . L/Y$.

Substituting this value of l in expression (i) above, we have

$$W\left(h + \frac{F_m . L}{Y}\right) = \frac{1}{2} . F_m . a . \frac{F_m . L}{Y} = \frac{1}{2} . \frac{F_m^{\ 2} . a . L}{Y}$$

Or,

$$2W\left(h + \frac{F_m . L}{Y}\right) = \frac{F_m^{\ 2} . a . L}{Y}$$

Or,

$$F_m^{\ 2} . \frac{aL}{Y} - 2\frac{F_m . L}{Y} . W - 2Wh = 0.$$

Solving this quadratic equation, we have

$$F_m = \frac{2W\dfrac{L}{Y} \pm \sqrt{4W^2 . \dfrac{L^2}{Y^2} + 8\dfrac{aL}{Y} . Wh}}{2aL/Y}$$

$$= \frac{W}{a} \pm \frac{\sqrt{W^2L^2 + 2aL.W.h.Y}}{aL}$$

So that, if $h = 0$, we have $F_m = 2W/a$.

This clearly shows that when the full load W is applied to a bar *all at once*, the maximum stress is $2W/a$, *which is clearly twice the value of the maximum stress* W/a set up in the bar when the load (W) is applied gradually to it, as for example, when the bar is stretched in a testing machine.

In other words, *the effect of a suddenly applied load is to produce a stress double that produced by a gradually applied one.*

8.21. Determination of Bulk Modulus (K)

I. For solids. The value of K for a solid is often obtained indirectly by determining the values of Y and n for it and substituting these in the relation $9/Y = 3/n + 1/K$ (page 265)

Among the simpler direct method for determination of K may be mentioned the ones due to *Amagat* and to *Mallock* respectively.

Amagat's method *consisted in measuring the change in the internal volume of a cylindrical tube* (with its radius large compared with its thickness) *when subjected to an internal pressure.* The cylinder was filled with water, with the lower end of a vertical and calibrated capillary tube dipping into it, and a load applied to the lower end of the cylinder. The change dV in the internal volume of the cylinder is thus obtained from the positions of the water meniscus in the capillary tube before and after the cylinder is loaded. Then, if V be the *original (internal) volume* of the cylinder, P, the applied force per unit area, it was shown by *Amagat* that $\delta v/v = P(1 - 2\sigma)/Y = P/3K$, whence, K for the material of the cylinder can be easily evaluated.

Mallock's method consists in subjecting a long, thin-walled tube of the material to internal pressure and the increase in its length magnified and measured by an optical device. Then, if the radius of the tube is large compared with its thickness, the change in its length is solely a function of the bulk modulus of its material, and is quite independent of the other elastic constants. So that, *for an isotropic material*, we have

$$K = \frac{PR_1 \, . \, l}{6(R_2 - R_1) \, . \, \delta l},$$

where l is the original length of the tube, δl, the increase in its length under pressure P and R_1 and R_2 its internal and external radii respectively.

The value of K for the material of the tube can thus be easily obtained and of course, its *compressibility* ($1/K$) straightaway.

Mallock applied pressures within the range of 1 to 30 atmospheres and, within this range, he formed K to be constant.

Bridgman later used much more refined methods and in one series of his experiments extended the pressure range to the staggering figure of 97000 atmospheres. We need not, however, go into his method here.

II. For Liquids. In the case of a *liquid* or a *gas* (*i.e.*, a *fluid*), the only relevant elasticity is *bulk modulus* since, unlike a solid, it cannot *permanently* withstand a shearing stress* and has thus no rigidity.

Now, a perfect liquid is said to be one which is totally incompressible. No such liquid is found in actual practice though, in quite a few cases, any change in volume produced, except under high pressure, is almost negligible.

Since, a liquid must necessarily be contained in some vessel, a few difficulties crop up in designing a suitable experiment for determining its bulk modulus. Thus, *(i) the change in the size of the containing vessel must be accurately know and corrected for*, *(ii) proper arrangements must be made for the application and measurement of high pressures* and *(iii) the measurement of the small changes of volume produced to a high degree of accuracy must be ensured.*

Regnault was the first to have devised a satisfactory method for the purpose, the apparatus used being referred to as a *piezometer* (from *piezo*, meaning *pressure*).

Before proceeding any further, it may as well be pointed out here that the earlier attempts at determining K for liquids failed because of the erroneous assumption, that if the same pressure be applied to the inside and outside of a vessel, say, a cylinder, (supposed to be sufficently thick), there is little or no change in its internal volume. The actual fact is that there is a decrease in volume equal to that which would be caused in a solid block of the material that would just fit the inside of the vessel (i.e., will have the same volume with it) when subjected to the same pressure externally. It follows, therefore, that *the observed or apparent decrease in volume of the liquid = real decrease in its volume – decrease in internal volume of the vessel.*

This will be clear from Fig. 8.18, where the containing vessel has been taken to be cylindrical in form, with plane ends, carrying a calibrated capillary tube at its upper end. When no pressure has yet been applied to the inside or the outside of the vessel, the liquid stands up to level A, say (Fig. 8.18 (*a*)), under atmospheric pressure P, with its volume, as also the internal volume of the vessel up to A, equal to V. On the pressure being increased to $(P + dP)$ on the inside and the outside of the vessel, the internal volume of the vessel will, as we have seen, decrease and will now be less than V; so that if the liquid be, for a moment, supposed to remain unaffected, *i.e.*, be imagined to be incompressible,

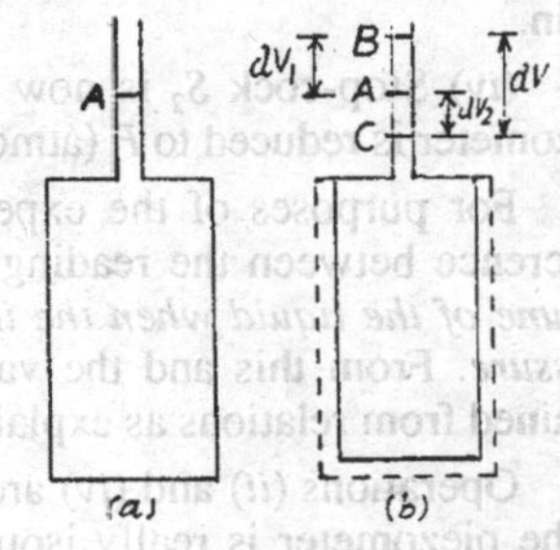

Fig. 8.18

* That is does so temporarily is due to its property of *viscosity* (see § 12.8).

its level will rise up in the vessel, say up to B (Fig. 8.18 (b)], where AB = *decrease in internal volume of the vessel* $\doteq dV_1 = dP.V/K_1$, where K_1 *is the bulk modulus for the material of the vessel.*

Actually, however, the liquid (not being incompressible will) undergo a decrease in volume under the pressure (dP) applied and its level will, therefore, fall to, say, C. So that, what we observe is the *apparent decrease in the volume of the liquid* $AC = dV_2$, *its real decrease in volume being* $BC = dV = dP.V/K$, where K is the *bulk modulus for the liquid.*

Clearly, $BC = AB + AC$, *i.e.*, $dV = dV_1 + dV_2$, or, $dP.V/K = dPV/K_1 + dV_2$,

whence, $dV_2 = dPV/K_2 - dP.V/K_1$, Or, $dV_2/V = dP\ (1/K - 1/K_1)$.

Or, $$\frac{1}{K} = \frac{dV_2}{dP.V} + \frac{1}{K_1}$$

So that, the value of K for the liquid can be easily obtained.

The actual arrangement of the apparatus is shown diagrammatically in Fig. 8.18, where AD is the *piezometer* proper, with a cylindrical bulb A, ending a long and calibrated *capillary tube D*, fitted with a stop-cock S_4 at the top. Filled with a suitable amount of the experimental liquid, it is placed inside a stout *copper tube C*, about 2 *mm* thick, containing water, so that the desired external pressure may be applied to the outside of A through the water. The whole assembley is kept immersed in a large water bath W to maintain its temperature constant during the experiment. A lead tube E, bent twice at right angles, is connected to D and three other stop cocks S_1, , S_2 , S_3 arranged as indicated in the figure, with a compressor put in communication with D through a side tube G.

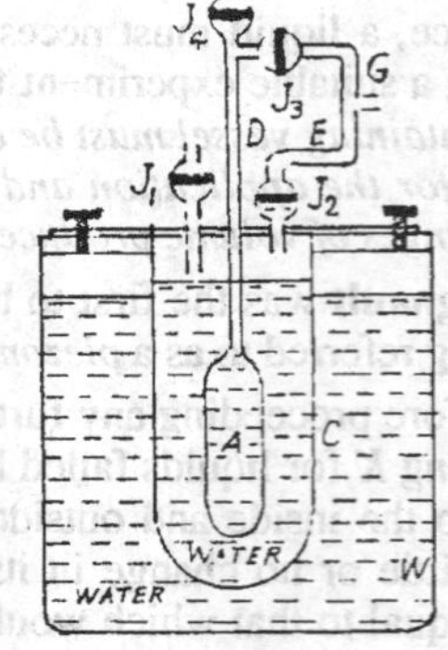

Fig. 8.19

The procedure consists of the following four operations.

(*i*) stop-cocks S_1 and S_4 are opened, so that the pressure inside and outside the piezometer becomes the same, *viz.* atmospheric, say, P, and the position of the liquid level in D is carefully read.

(*ii*) Stop cock S_1, is then closed, S_2 opened to put the copper vessel C in communication with the compressor and the pressure there, *i.e.*, outside the piezometer, is increased to $P + dP$, with the pressure inside it remaining P.

(*iii*) Stop cock S_4 is closed and S_3 thrown open, so that the inside of the piezometer is also put in communication with the compressor and the pressure inside it too increased to $P + dP$, the same as on its outside. The liquid level in capillary D is read again.

(*iv*) Stop-cock S_2 is now closed and S_1 opened, so that the pressure on the outside of the piezometer is reduced to P (atmospheric), whereas that inside it remains $P + dP$.

For purposes of the experiment, actually only operations, (*i*) and (*iii*) are necessary, the difference between the readings obtained in them directly giving dV_2, *the apparent decrease in volume of the liquid when the inside and the outside of the piezometer are subjected to the same pressure*. From this and the values of V, dP and K_1, the bulb modulus K of the liquid can be obtained from relations as explained above*.

Operations (*ii*) and (*iv*) are mean t as a sort of check to find out whether or not the material of the piezometer is really isotropic and of a uniform thickness, as has been tacitly assumed all along. For, in case this is so, the difference between the readings in operations (*ii*) and (*iv*) must

* The relation *I* for K has been deduced on the assumption that the cylindrical bulb A has *plane ends* in Regnant's apparatus, the ends were *hemispherical* and a correction for this had to be applied.

also give the same value (dV_2) for the apparent decrease in the volume of the liquid when the inside and outside of the piezometer are subjected to the same pressure. Needless to say, this hardly ever happens, since glass is not really elastically isotropic. Metallic vessels are certainly more likely to be so but then, they may exit some porosity.

Regnault limited himself to pressures up to 10 atmospheres only. The range was later extended to 2000 atmospheres by *Parsons* and *Cook*, and finally to 12,000 atmospheres by *Bridgman*, whose experiments are now accepted as classical and his results, as standards.

Shorn of experimental detail **Bridgman's method** consisted in using a string chrome - vanadium steel cylinder as the piezometer and subjecting the liquidg in it to pressure by the motion of an accurately fitting and specially designed leak proof piston of a known cross section, the apparent change in the volume of the liquid being obtained from the distance moved through othe piston. The cylinder was kept immersed in a high pressure chamber, the pressure in the vessel being obtained from the resistance of a coil of manganin wire placed in it, an accurate relation between resistance and pressure for the coil having been previously established.

The entire apparatus was maintained at a constant temperature by thermostat control and readings for pressure and volume in the case of water obtained up to 80.0°*C*, the correction for expansion of the containing vessel being determined by partially replacing the liquid by a steel plug and obtaining their combined compressibility.

Repeating the experiment with different liquids it was found that *compressibility decreases rather fast with increasing pressure*. This is probably just as it should be or else the volume will ultimately be reduced to *zero*. This may reasonably be explained on the view that at comparatively low pressures, there is quite a bit of free space in between molecules. Compressibility at such pressure can, therefore, be considerably large on account of the relative ease with which the molecules can be brought nearer together. Once, however, this free space is no longer there, any compressibility can only come at the expense of a reduction in the volume of the molecules themselves.

8.22. Twisting Couple on a Cylinder (or Wire). If we have a cylinder or a wire, clamped at one end, and *twisted at the other through an angle* θ, about *its* axis, it is said to be *under tension*. Due to the elasticity of the material of the cylinder or the wire, *a restoring couple is set up in it, equal and opposite to the twisting couple*.

Consider a cylindrical rod of *length l* and *radius r*, of a material of *coefficient of rigidity n*. Let its upper end be fixed and let a *couple* be applied to the lower end in a plane *perpendicular to its length* (with its axis coinciding with that of the cylinder) twisting it through an angle θ (*radians*).

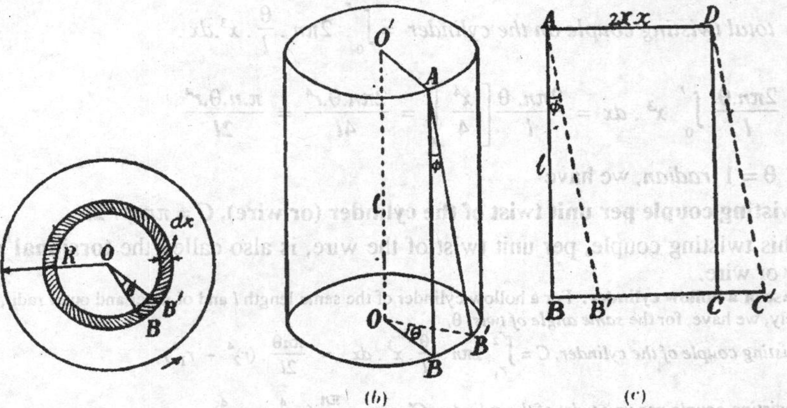

Fig. 8.20

This, incidentally, is an example of what is called a '*pure*' shear; for, the twist produces a change neither in the length nor the radius of the cylinder,—the value of the twist for any cross-section of the cylinder being quite independent of the direction of the couple, a reversal of which also reverses the direction of twist.

Now, *in the position of equilibrium, the twisting couple is equal and opposite to the restoring couple*. Let us calculate the value of this couple.

Imagine the cylinder to consist of a large number of *co-axial, hollow* cylinders, and consider one such hollow cylinder of *radius x*, and *radial thickness dx*, [Fig. 8.20 *(a)*]. Each radius of the lower end is turned through the *same angle* θ, but the *displacement is the greatest at the rim, decreasing as the centre is approached, where it is reduced to zero.*

Let *AB* [Fig. 8.20(*b*)] be a line, parallel to the axis, before the cylinder is twisted. On twisting, since the point *B* shifts to *B'*, the line *AB* takes up the position *AB'*, such that, before twisting, if this hollow cylinder were to be cut along *AB* and flattened out, it will form the rectangular plate. *ABCD*, but, after twisting, it takes the shape of a parallelogram, *AB'C'D*, [Fig. 8.20(*c*)]. The angle through which this hollow cylinder is *sheared* is, therefore, *BAB'* = ϕ, say. Then, clearly,

$$BB' = l\,.\,\phi. \quad \text{Also, } BB' = x\,.\,\theta. \quad \therefore\ \phi = x\,.\,\theta/l. \qquad \text{[See Fig. 8.20 (a) or (b).]}$$

Obviously, ϕ will have the maximum value where x is the greatest, *i.e., the maximum strain is on the outermost part of the cylinder, and the least, on the innermost.* In other words, *the shearing stress is not uniform all through.*

Thus, although the angle of shear is the *same* for any *one* hollow cylinder, it is different for different cylinders,— being the *greatest for the outermost* and the *least for the innermost one.*

Since $$n = \frac{shearing\ stress}{strain\ or\ angle\ of\ shear} = \frac{F}{\phi},$$

we have $$F = n\,.\,\phi = n x.\theta/l.$$

Now, *face area of this hollow cylinder* = $2\pi x.dx$.

And $\therefore$ *total shearing force this on area* = $2\pi x.dx \times \dfrac{n.x.\theta}{l} = 2\pi n\,.\,\dfrac{\theta}{l}\,.\,x^2\,.\,dx.$

Therefore, *moment of this force about the axis OO'* [Fig. 8.20 (*b*)] of the cylinder is equal to $2\pi n.\theta\,.\,x^2.\,dx\ x/l = 2\pi n.\ \theta.x^3.\ dx/l$.

Integrating this expression between the limits, $x = 0$ and $x = r$, we have

$$total\ twisting\ couple\ on\ the\ cylinder = \int_0^r 2\pi n\,.\,\frac{\theta}{l}\,.\,x^3.dx\,.$$

$$= \frac{2\pi n.\theta}{l}\int_0^r x^3\,.\,dx = \frac{2\pi n.\ \theta}{l}\left[\frac{x^4}{4}\right]_0^r = \frac{2\pi n.\theta.r^4}{4l} = \frac{\pi.n.\theta.r^4}{2l}$$

If $\theta = 1$ *radian*, we have

twisting couple per unit twist of the cylinder (or wire), $C = \pi n r^4 / 2l$

This twisting couple, per unit twist of the wire, is also called the **torsional rigidity** of the cylinder or wire.

Case of a hollow cylinder. For a hollow cylinder of the same length l and of inner and outer radii, equal to r_1 and r_2 respectively, we have, for the *same angle of twist* θ,

$$twisting\ couple\ of\ the\ cylinder, C = \int_{r_1}^{r_2} 2\pi n\,.\,\frac{\theta}{l}\,.\,x^3\,.\,dx = \frac{\pi n\theta}{2l}\,.\,(r_2^4 - r_1^4).$$

And $\therefore$ *twisting couple per unit twist of the cylinder, C', say* = $\dfrac{\pi n}{2l}(r_2^4 - r_1^4)$

Comparing the twisting couples in the two cases, therefore, we have

$$\frac{C'}{C} = \frac{(r_2^4 - r_1^4)}{r^4} = \frac{(r_2^2 - r_1^2)\,(r_2^2 + r_1^2)}{r^4} \qquad ...(I)$$

If the two cylinders be also of the *same mass m*, we have

$$m = \pi\,(r_2^2 - r_1^2)\,.\,l.\rho = \pi.r^2.l.\rho,$$

where ρ *is the density of the material of the two cylinders.*

So that, $$(r_2^2 - r_1^2) = r^2.$$

Substituting this value of $(r_2^2 - r_1^2)$ in expression (*I*) above,

we have
$$\frac{C'}{C} = \frac{r^2.(r_2^2 + r_1^2)}{r^4} = \frac{r_2^2 + r_1^2}{r^2}. \quad ...(II)$$

Now, as we have just seen, $(r_2^2 - r_1^2) = r^2$, whence, clearly, $r_2^2 = r^2 + r_1^2$ and, therefore, adding r_{12} to both sides, we have $r_2^2 + r_1^2 = (r^2 + r_1^2) + r_1^2$; from which it is clear that $(r_2^2 + r_1^2)$ is greater than $(r^2 + r_1^2)$ and hence still greater than r^2, or that $(r_2^2 + r_1^2)/r^2$ is greater than 1. It follows from relation (*II*) above, therefore, that $C'/C > 1$. Or, that $C' > C$, *i.e., the twisting couple per unit twist for a hollow cylinder is greater than that for a solid one of the same material, mass and length.*

This means, in other words, that a *hollow shaft, of the same material, mass and length as a solid shaft, is comparatively very much stronger than the latter.*

8.23. Variation of stress in a twisted cylinder (or wire). Let us again imagine a cylinder or wire, of length l and radius r, to consist of a large number of co-axial hollow cylinders and consider both, a cylinder of radius x and the outermost cylinder of radius r, (Fig. 8.21), in which the lines AB and CD respectively are parallel to the axis OO' before the cylinder is twisted, and shift into the positions AB' and CD' after it has been twisted, as explained above. Then, clearly, $\angle BOB' = \angle DOD' = \theta$, and the angles through which the two cylinders are sheared are $BAB' = \phi$ and $DCD' = \phi_m$ respectively, where $\phi = BB'/l$ and $\phi_m = DD'/l$, this latter being the strain on the surface of the cylinder and, therefore, the maximum on it, (because $DD' = r\theta$).

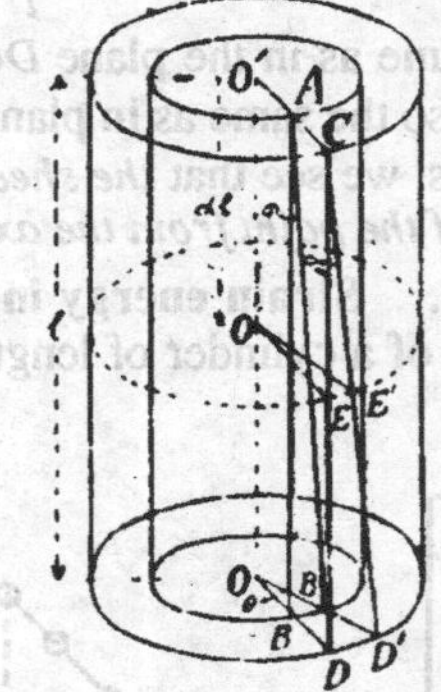

Fig. 8.21

Since $OB = OB' = x$ and $OD = OD' = r$,

we have $\theta = BB'/x = DD'/r$.

So that, $\phi_m = \dfrac{DD'}{l} = \dfrac{r\theta}{l} = \dfrac{r}{l}.\dfrac{BB'}{x} = \dfrac{r}{x}.\dfrac{BB'}{l} = \dfrac{r}{x}.\phi$,

whence, $\phi = \dfrac{x}{r}.\phi_m$.

Or, *strain in the cylinder or wire at distance x from the axis*

$$= \frac{x}{\textit{radius of cylinder or wire}} \times \textit{maximum strain.}$$

Now, $n = F/\phi$ and, therefore, $F = n\phi = n.\dfrac{x}{r}.\phi_m$,

where F is the *shearing stress at distance x* from the axis.

And, since ϕ_m is the *maximum strain*, we have $n.\phi_m = F_m$, *the maximum shearing stress on the wire, i.e., on its surface.*

We, therefore, have $F = \dfrac{x}{r}.F_m$

Or, *shearing stress at distance x from the axis* $= \dfrac{x}{\textit{radius}} \times$ *maximum stress.*

In other words, *both shearing strain and stress go on increasing as we proceed away from the axis of the cylinder or the wire, and acquire their maximum values on its surface.*

Let us now see whether there is also any variation in shearing strain and stress along the length of the cylinder or wire.

Let us, therefore, consider the stress in the plane $EO''E'$ of the cylinder at a distance αl from its upper fixed end, where $\alpha > 0$ and < 1. Here, clearly, $\angle EO''E'$ between the radii $O''E$ and $O''E'$ $= \alpha\,\theta$.

$= EE'/r$ and $\therefore\ EE' = r.\alpha.\theta.$ $[\because O''E = O''E' = r$

$\therefore$ shear strain on the surface of the wire *in this plane*

$$= EE'/CE = r.\alpha\,\theta/\alpha l = r\theta/l.$$

Now, $\theta = \dfrac{l}{r}.\phi_m$, as we have seen above.

So that, *shear strain on the surface of the wire in plane* $E\,\underline{O}''E'$

$$= \frac{r}{l}.\frac{l}{r}\phi_m = \phi_m,$$

i.e., the same as in the plane DOD', as discussed above. Clearly, therefore, shearing stress in this plane is also the same as in plane DOD', namely, $n.\phi_m = F_m$.

Thus, we see that *the shearing stress at a point in a cylinder, or a wire, depends only on the distance of the point from the axis and not its vertical distance from either end of it.*

8.24. Strain energy in a twisted cylinder (or wire). Let C be the couple applied to the lower end of a cylinder of length l and radius r, with its upper end fixed and let θ be the angle of twist produced at the former (*i.e.*, lower) end. Then, if the limit of elasticity is not exceeded, the relation between C and θ is a linear one and we obtain a *straight line graph OP* between the two, as shown in Fig. 8.22.

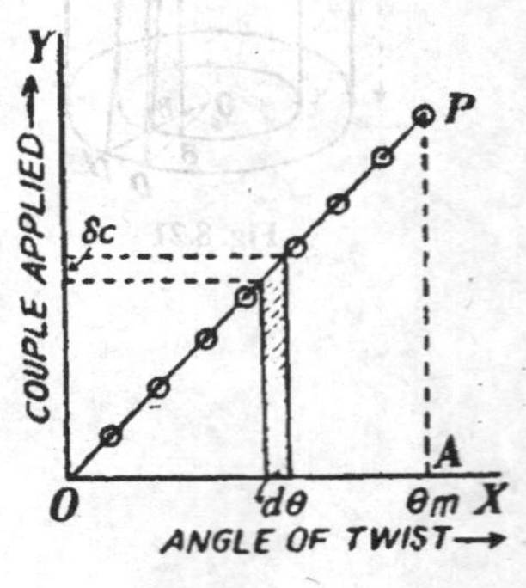

Fig. 8.22

So that, for a small increase δC in the value of the couple, the increase in the angle of twist is $d\,\theta$, and the *work done* on the cylinder, or the *energy stored up* in it, is, therefore, $C.d\theta$, where C is the *average* value of the couple. This is represented by the area of the shaded strip in the figure. And, therefore, the *total work done* on the wire, or *the total energy stored up in it* for the maximum twist θ_m (represented by OA), to which its lower end is subjected, is represented by the whole area OAP.

This strain energy is obviously equal to $E = \int_0^{\theta_m} C\,.\,d\theta$.

Now, C for a twist θ *in* the wire is, as we know, equal to $\pi n\theta r^4/2l$.

$$\text{And therefore, } E = \int_0^{\theta_m}\frac{\pi n\theta r^4}{2l}.d\theta = \frac{\pi n r^4}{2l}.\frac{\theta_m^{\,2}}{2} = \frac{\pi n r^4\theta_m^{\,2}}{4l}$$

$$= \frac{1}{2}.\frac{\pi n r^4\theta_m}{2l}.\theta_m = \frac{1}{2}C_m.\theta_m,$$

where $\pi n r^4\theta_m/2l$ is the value of the maximum couple Cm, corresponding to the maximum twist θ_m; *i.e.*, $C_m = \pi n r^4\theta_m/2l$.

Or, substituting from this value of θ_m in the expression for E above, we have

strain energy in the twisted cylinder, $E = \dfrac{1}{2}C_m.\theta_m = C_m^{\,2}.l/\pi n r^4$.

As will be readily seen, this is half the energy ($C_m - \theta_m$) that would be stored up in the cylinder if the stress in it were to have the same value throughout, equal to its maximum value on the surface of the wire, which, as we know, is not actually the case, — the stress increasing from *zero at the axis* to a *maximum on its surface*, (see § 8.23 above).

8.25. Alternative expression for strain energy in terms of stress.

We know that *strain energy per unit volume*

$$= \frac{1}{2}\ stress \times strain = \frac{1}{2}F.\phi = \frac{1}{2}F.\frac{F}{n} = \frac{1}{2}\frac{F^2}{2n} \quad \text{... (i)}$$

Now, if we consider an element of the cylinder or wire, defined by radii x and $x+dx$, the stress will, as we have seen in § 8.23, be constant at all points in it and its value will be $x.F_m/r_1$, where F_m is the maximum value of the stress in the wire on its surface *i.e.*, $F = x.F_m/r$.

Since the volume of the cylindrical element we are considering is $2\pi x.l.dx$, we have, from relation (*i*) above, *energy of the cylindrical element, i.e.*,

$$dE = \frac{1}{2}.\frac{x^2}{r^2}.\frac{F_m^{\ 2}}{n}\ 2\pi x.l.dx = \frac{\pi l F_m^{\ 2}}{nr^2}x^3.dx.$$

And $\therefore\ E = \dfrac{\pi.lF_m^{\ 2}}{nr^2}\displaystyle\int_0^r x^3.dx = \dfrac{\pi lF_m^{\ 2}}{nr^2}.\dfrac{r^4}{4} = \dfrac{\pi r^2 l}{4n}.F_m^{\ 2}.$

If ϕ_m be the maximum shear strain corresponding to the maximum stress F_m, we have $F_m = n\,\phi_m = n.r\ \theta_m/l$, (see § 8.23), where θ_m is the angle of twist for the maximum value of the couple.

So that, $E = \dfrac{\pi r^2 l}{4n}.\dfrac{n^2 r^2\theta_m^{\ 2}}{l^2} = \dfrac{1}{2}.\dfrac{\pi r^4 n\theta_m}{2l}.\theta_m = \dfrac{1}{2}C_m.\theta_m,$ $[\because \pi nr^4\theta_m/2l = C_m$

i.e., the strain energy is again half the value it would have if all the elements of the cylinder (or wire) were subjected to the same maximum stress F_m.

8.26. Torsional Pendulum. A heavy cylindrical rod or disc, suspended from one end of a fine wire (attached to its centre) whose upper end is fixed, constitutes what is called a *torsional pendulum*, (Fig. 8.23). The rod or disc is turned in its own (*i.e.*, in the horizontal) plane to twist the wire, so that, on being released, it executes *torsional vibrations* about the wire as axis.

Fig. 8. 23

Let θ be the angle through which the wire is twisted. Then, the restoring couple set up in it is equal to $\pi.nr^4.\theta/2l = C.\theta$, where $\pi nr^4/2l = C$ is the *twisting couple per unit (radian) twist* of the wire.

This produces an angular acceleration $d\omega/dt$ in the rod or the disc.

$\therefore$ If I be the moment or inertia of the rod or the disc about the wire, we have

$$I.d\omega/dt = -C.\theta. \quad \text{Or,} \quad d\omega/dt = -(C/I).\theta,$$

i.e., the angular acceleration ($d\omega/dt$) *of the disc or the rod is proportional to its angular displacement* (θ), and, therefore, *its motion is simple harmonic*. Hence, its *time-period* is given by

$$t = 2\pi\sqrt{\frac{1}{C/I}} = 2\pi\sqrt{\frac{I}{C}}.$$

Or, $$t = \sqrt{\frac{\textit{moment of inertia of the disc or rod about the wire}}{\textit{restoring couple per unit twist of the wire.}}}$$

8.27. Determination of the Coefficient of Rigidity (n) for a Wire.

(1) **Statistical Method.** This method is based on the direct application of the expression for the twisting couple on a wire, deduced in § 8.22. There are two different types of apparatus used for the purpose, according as the specimen under test is a rod or a wire. We shall now consider these in detail.

(a) **Horizontal Twisting apparatus for a rod.** Here, a couple, which can be measured directly, is applied to a horizontal rod and equated against the expression for the torsional or twisting couple $\pi n r^4\theta/2l$ whence the value of n for the rod can be easily calculated.

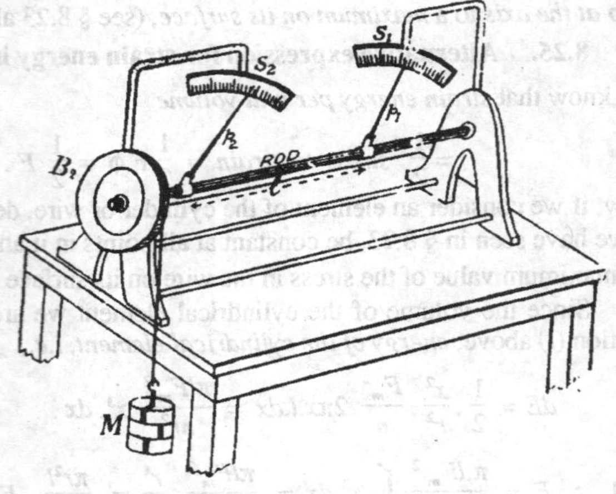

Fig. 8.24

The arrangement of the apparatus is as shown in Fig. 8.24, where one end of the rod under test, about 50 *cm*, in length and of radius about .25 *cm* is secured firmly to a block B_1, with its other end attached to a steel axle of a large pulley B_2.

A cord is wound round the pulley and has a mass M suspended from its lower free end. Thus, a couple acts on the rod, tending to twist it about its own axis.

Two pointers, p_1 and p_2, are clamped on to the rod at two points, a known distance l apart, so as to move freely over the circular scales S_1 and S_2, graduated in degrees, on which the twist produced in the rod at these two points can be read directly.

Now, if R be the radius of the pulley, the couple acting on the rod, due to the suspended mass (M) is, clearly, equal to $Mg.R$.

This couple is balanced by the couple due to the torsional reaction of the rod, equal to $\pi n r^4.(\theta_2 - \theta_1)/2l$, where r is the *radius* of the rod and θ_1 and θ_2, the *angles of twist (in radians**) produced at the two chosen points, as indicated by the two pointers.

So that, $\pi n r^4\ (\theta_2 - \theta_1)/2l = Mg.R.$ Or, $n = \dfrac{2\,Mg.R.l}{\pi r^4\,(\theta_2 - \theta_1)}$

whence, the value of n for the material of the rod can be easily determined.

Usually, the experiment is repeated with different masses and a graph plotted between M and the corresponding twist $(\theta_2 - \theta_1)$. The slope of the straight line graph, then gives the mean value $(\theta_2 - \theta_1)/M$ which is used in the expression for x above.

The apparatus, though quite simple in manipulation, suffers from **two serious drawbacks**, *viz.*,

(i) there being one single pointer moving over the circular scale, an error is caused due to eccentricity of the axis of the rod with respect to it ;

(ii) there being just one pulley, only one single force is applied to the end of the rod attached to it, thus exerting a side-pull on it. This results in friction between the rod and the bearings, thus appreciably hindering the rod from twisting freely.

* To convert *degrees* into *radians*, multiply by $\pi/180$, because $180° = \pi$ *radians*.

(b) Vertical Twisting apparatus for a Wire. This was designed by **Barton**, and here also a couple, which is measured directly, is applied to the lower end of the vertically suspended wire, and the twist produced in it is noted. Then, equating this couple against the expression $n\pi r^4\theta/2l$ for it, the value of n for the wire can be easily calculated.

The wire W, whose *coefficient of rigidity* is to be determined, is clamped at its upper end H, (Fig. 8.25), and has a *heavy* cylinder C attached to it, at its lower end.

Two pieced of cords are wound round the cylinder and, leaving it *tangentially* at either end, pass over two frictionless pulleys, as shown, with equal masses M and M, suspended from their free ends.

The *couple*, formed by the two masses, rotated the cylinder about the wire as axis, and thus twists the wire through an angle θ (*radians*), say which is read directly on the horizontal circular scale S by the movement over it of the pointer attached to the wire, or by the more sensitive lamp and scale method, with the help of a telescope.

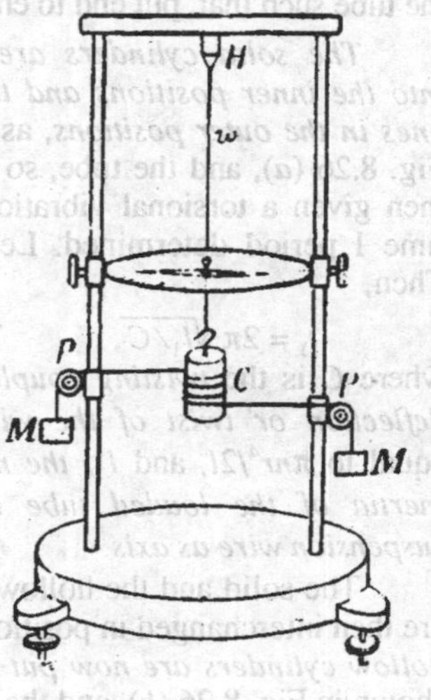

Fig. 8. 25

Then, if d be the *diameter of the cylinder*, we have

$$\textit{twisting couple applied to the wire} = Mg.d.$$

Also, twisting couple for a twist of θ radians $= \pi nr^4\,\theta/2l$, where r is the radius of the wire, l, its length and n, the coefficient of rigidity for its material.

Clearly, therefore, $\pi nr^4\theta/2l = Mg.d$ Or, $n = 2l.Mg.d/\pi r^4\theta$,

whence, n for the material of the wire can be easily obtained.

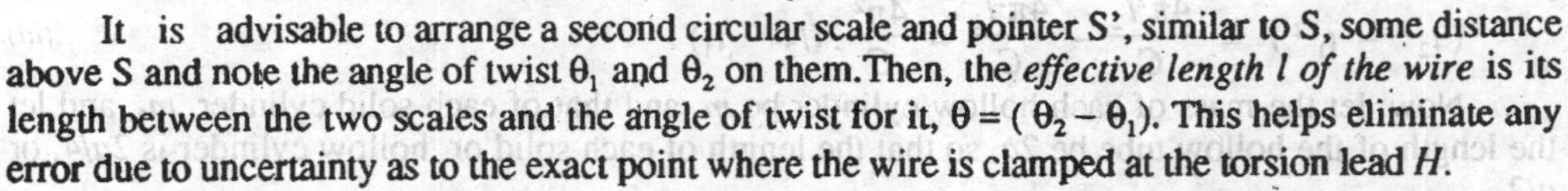
It is advisable to arrange a second circular scale and pointer S', similar to S, some distance above S and note the angle of twist θ_1 and θ_2 on them.Then, the *effective length l of the wire* is its length between the two scales and the angle of twist for it, $\theta = (\theta_2 - \theta_1)$. This helps eliminate any error due to uncertainty as to the exact point where the wire is clamped at the torsion lead *H*.

The two sources of error present in the first method are eliminated here. For, (*i*) *due to the very nature of the arrangement of the pointer and the scale enabling both ends of the pointer to be read, the error due to eccentricity of the axis of the wire does not arise and (ii) due to the use of two pulleys, the side pull on the wire is also avoided.*

N.B. It will be noted that the weak point in the above two methods is the radius r of the rod or of wire, the *fourth power* of which occurs in the expression for n. It must, therefore, be measured most carefully.

(2) Dynamical Method—Maxwell's Vibrating Needle. The dynamical method of determining n for the material of a wire consists in determining, by direct observation, the time-period t of a body, like a disc or a rod, suspended from the wire and executing torsional vibrations about the wire as axis, *i.e.*, of a *torsional pendulum.*

Then, since $t = 2\pi\sqrt{I/C}$, where I is the *moment of inertia* of the body about the wire and C, the *couple per unit (radian) twist* of *the wire*, we can easily obtain from it the value of C. Equating this against the expression $\pi nr^4/2l$ for it, the *coefficient of rigidity (n)* for the wire can be easily calculated.

It is not, however, easy to determine the moment of inertia (I) of the body accurately. *Maxwell*, therefore, devised a method in which the necessity of determining it was altogether obviated. Let us study his method in detail.

Maxwell's Vibrating Needle Method. A hollow tube or cylinder, open at both ends, is *rigidly* fastened in the middle to the wire, the coefficient of rigidity of the material of which is to be determined, and which is suspended vertically from a support and has a small piece of plane mirror attached to it, as shown, (Fig. 8.26), to enable the vibrations of the tube to be observed by the telescope and scale method.

Two *hollow* and *two solid* metal cylinders, *of equal lengths and diameters*, can be fitted into the tube such that, put end to end, they *just fill it completely*.

The solid cylinders are first put into the inner positions and the hollow ones in the outer positions, as shown in Fig. 8.26 (*a*), and the tube, so loaded, is then given a torsional vibration, and its time I period determined. Let it be t_1. Then,

$$t_1 = 2\pi\sqrt{I_1/C}, \qquad ...(i)$$

where C is the *twisting couple per unit deflection or twist of the wire* and is equal to $\pi nr^4/2l$, and I_1, *the moment of inertia of the loaded tube about the suspension wire as axis*.

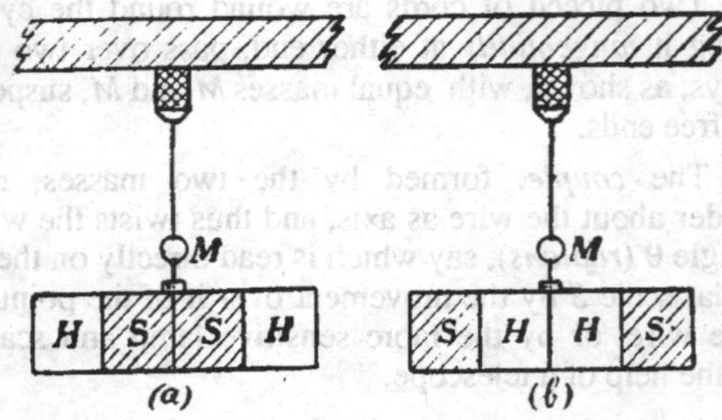

Fig. 8. 26

The solid and the hollow cylinders are then interchanged in position, *i.e.*, the *hollow cylinders are now put in the inner positions and the solid ones in the outer positions*, as shown in Fig. 8.26 (*b*), and the time-period of the torsional vibration of the loaded tube determined again. Let it be t_2. Then, $\quad t_2 = 2\pi\sqrt{I_2/C}, \qquad ...(ii)$

where I_2 is now the *M.I.* of the loaded tube about the suspension wire

Squaring and subtracting (*i*) from (*ii*), we have

$$(t_2{}^2 - t_1{}^2) = \frac{4\pi^2 I_2}{C} - \frac{4\pi^2 I_1}{C} = \frac{4\pi^2}{C}.(I_2 - I_1). \qquad ...(iii)$$

Now, let the mass of each hollow cylinder be m_1 and that of each solid cylinder, m_2, and let the length of the hollow tube be $2a$, so that the length of each solid or hollow cylinder is $2a/4$, or $a/2$.

Clearly, then, the *centres of mass* of the inner and outer cylinders lie at distances $a/4$ and $3a/4$ respectively from the axis of oscillation.

Therefore, *the change from the first adjustment, when the solid cylinders occupy the inner positions to the second adjustment, when they occupy the outer positions, consists in transferring an extra or excess mass* $(m_2 - m_1)$ *from a distance* $a/4$ *to a distance* $3a/4$ *from the axis of oscillation, on either side of it.* The moment of inertia of the loaded tube, therefore, *increases*, and, by the application of the *principle of parallel axes*, we have

$$I_2 = I_1 + 2(m_2 - m_1)\,[(3a/4)^2 - (a/4)^2]$$

Here, we multiply the mass by 2, because the change takes place on both the sides.

Or, $$I_2 = I_1 + 2(m_2 - m_1) \times \left(\frac{9a^2}{16} - \frac{a^2}{16}\right) = I_1 + 2(m_2 - m_1).\frac{a^2}{2},$$

i.e., $\quad I_2 = I_1 + (m_2 - m_1).a^2.\quad$ And $\quad \therefore\ (I_2 - I_1) = (m_2 - m_1).a^2.$*

* Or, this may easily be deduced as follows :

If I be the moment of inertia of the *hollow tube* about the suspension wire, and I_s and I_h, those of the solid and the hollow cylinders about the vertical axix through their respective centres of mass, we have, *by the principle of parallel axes*,

$$I_1 = I + 2\,[I_s + m_2.(a/4)^2 + 2\,[I_h + m_1.(3a/4)^2].$$

and $$I_2 = I + 2\,[I_h + m_1.(a/4)^2] + 2\,[I_s + m_2.(3a/4)^2].$$

So that, $$(I_2 - I_1) = (m_2 - m_1)\ a^2.$$

Substituting this value of $(I_2 - I_1)$ in relations (*iii*) above, we have

$$(t_2^{\,2} - t_1^{\,2}) = \frac{4\pi^2}{C}\,.\,(m_2 - m_1)\,.\,a^2.$$

Further, substituting the value of C, we have

$$(t_2^{\,2} - t_1^{\,2}) = \frac{4\pi^2}{\pi n r^4/2l}\,.\,(m_2 - m_1)\,.a^2 = \frac{4\pi^2 \times 2l}{\pi n r^4}\,.\,(m_2 - m_1)\,.a^2.$$

Or, $$(t_2^{\,2} - t_1^{\,2}) = \frac{8\pi.l.a^2}{n\,.\,r^4}\,.\,(m_2 - m_1)\,.\text{ whence, } n = \frac{8\pi.l.a^2.(m_2 - m_1)\,.}{(t_2^{\,2} - t_1^{\,4})\,.\,r^4}$$

Thus, knowing l, a, m_2, m_1, t_1, t_2 and r, the value of n for the material of the given wire can be easily determined.

The **two clear advantages of the method** are (*i*) *that the total mass suspended from the wire throughout remains the same and hence the value of C remains unaffected (see note on page* 285*) and (ii) that there being no necessity to use the moment of inertia of the needle in our calculations, the question as to the uncertainty of its value does not arise.*

N.B. *The value of n obtained by the dynamical method is slightly higher than that obtained by the statical method,* because, in most cases, the twist produced by a torsional couple depends, to some slight extent, upon the time for which the couple is applied and so, in the dynamical method, where the time of vibration is rater short, the twist (θ) is smaller for the same value of the couple than in the statical method.

Further, since wires are made by squeezing the molten metal through holes, (as in a sieve), their outer layers are invariably tougher than the inner ones, ad hence the value of *n* for a thinner wire needs must be higher than for a thicker wire of the *same material*.

8.28. (a) Determination of Moment of Inertia with the help of a Torsional Pendulum-The Inertia Table. The moment of inertia of a body of a *regular geometrical shape* can be easily calculated from its *mass and dimensions*. But, if it be of an *irregular shape*, it is not possible to do so. In either case, however, it may be determined by making use of a torsional pendulum. The apparatus used for the purpose is called the **Inertia Table,** a common form of which is shown in Fig. 8.27, and consists of a horizontal *aluminium disc D*, about 15 *cm* in diameter, fitted with a pair of small *vertical pillars P and P* at the extremities of one of its diameters, with a *cross bar B* connecting the two at the top. The whole of this assembly is suspended by means of a *long* and thin wire *w* from a torsion head *T* insidea larger *framework F*, mounted on a heavy iron base, provided with levelling screws. A piece of *mirror m*, fixed on to the cross-bar *B* right at its mid-point, enables the torsional vibrations of the inertia table (*i.e.*, disc *D*, together with pillars *P, P*) to be observed by the usual scale and telescope arrangement.

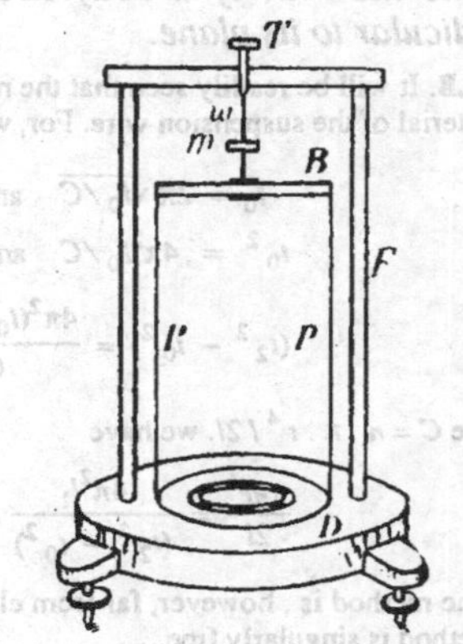

Fig. 8.27

To facilitate levelling of disc *D*, a concentric circular groove is cut along its edge, carrying three *balancing weights*, the respective positions of which inside the groove are suitably adjusted until the disc becomes quite horizontal, as judged from a pair of spirit levels fixed permanently on to it along two perpendicular diameters. And, to help place the given body *centrally* on disc *D*, *such that its c.g. lies on the suspension wire*, a number of concentric circles are drawn on the inner side of the circular groove. The wire thus becomes the axis of rotation of the body through its *c.g.* and perpendicular to its plane. The entire apparatus is enclosed in a glass envelope to avoid any possible disturbance due to air draughts.

The heavy iron base is first levelled with the help of the levelling screws and disc D of the *inertia table* arranged to be horizontal in the manner indicated. The latter is then set into torsional vibration in its own (*i.e.*, the horizontal) plane about wire w as axis and its *time-period* t_0 carefully noted. Then, if I_0 be the *M.I.* of the table (*i.e.*, of disc D, together with the pillars and the cross bar etc.) about the perpendicular axis through its *c.g.* and C, the restoring *torsional couple per unit (radian) twist of the wire*, we have

$$t_0 = 2\pi\sqrt{I_0/C}. \qquad ...(i)$$

The body, whose moment of inertia I is desired to be determined, is now placed *centrally* on the table and its *time-period* t_1, when so loaded, noted again, where

$$t_1 = 2\pi\sqrt{(I_0 + I)/C}. \qquad ...(ii)$$

Finally, the given body is now replaced on the table by one, of a known moment of inertia I_1 and the *time-period* t_2 of the table noted, as before, where

$$t_2 = 2\pi\sqrt{(I_0 + I_1)/C}. \qquad ...(iii)$$

Now, squaring and dividing relation (*ii*) by (*i*), we have

$$\frac{t_1^2}{t_0^2} = \frac{I_0 + I}{I_0}. \quad \text{Or,} \quad \frac{I}{I_0} = \frac{t_1^2 - t_0^2}{t_0^2}. \qquad ...(iv)$$

Similarly, squaring and dividing relation (*iii*) by (*i*), we have

$$\frac{t_2^2}{t_0^2} = \frac{I_0 + I_1}{I_0}. \quad \text{Or,} \quad \frac{I_1}{I_0} = \frac{t_2^2 - t_0^2}{t^2} \qquad ...(v)$$

So that, dividing relation (*iv*) by (*v*), we have

$$\frac{I}{I_1} = \frac{t_1^2 - t_0^2}{t_2^2 - t_0^2}, \text{ whence, } I = \frac{t_1^2 - t_0^2}{t_2^2 - t_0^2} \times I_1,$$

and can be easily evaluated.

Though quite simple to perform and particularly suited for the determination of moments of inertia of bodies of irregular shapes, its one obvious disadvantage is that *it can be used to determine the M.I. of a body about only one axis, viz., the one passing through its c.g., and perpendicular to its plane.*

N.B. It will be readily seen that the method could also be used to determine the value of n, the coefficient of rigidity of the material of the suspension wire. For, we have

$$t_0 = 2\pi\sqrt{I_0/C} \quad \text{and} \quad t_2 = 2\pi\sqrt{(I_0 + I)/C}.$$

whence
$$t_0^2 = 4\pi^2 I_0/C \quad \text{and} \quad t_2^2 = 4\pi^2(I_0 + I_1)/C.$$

Or
$$(t_2^2 - t_0^2) = \frac{4\pi^2(I_0 + I_1)}{C} - \frac{4\pi^2 I_0}{C} = \frac{4\pi^2 I_1}{C}, \text{ whence, } C = \frac{4\pi^2 I_1}{t_2^2 - t_0^2}.$$

And, since $C = n\,.\,\pi\,.\,r^4/2l$, we have

$$\frac{\pi n r^4}{2l} = \frac{4\pi^2 I_1}{(t_2^2 - t_0^2)}. \quad \text{Or,} \quad n = \left(\frac{4\pi^2 I_1}{t_2^2 - t_0^2}\right)\cdot\frac{2l}{\pi r^4} = \frac{8\pi I_1 l}{(t_2^2 - t_0^2)\,.\,r^4}.$$

The method is, however, far from elegant and suffers from both the disadvantages from which Maxwell's vibrating needle method is singularly free.

(*b*) **Comparison of Moments of Inertia.** If, however, it is simply desired to *compare* the moments of inertia of two bodies, we first use one and then the other *as the disc or rod of the torsional pendulum*, and determine the time-periods t_1 and t_2 respectively for their torsional vibration about the wire as axis. Then, if I_1 and I_2 be their respective moments of inertia about this axis and C, the torsional couple per unit twist of the wire, we have

$$t_1 = 2\pi\sqrt{I_1/C} \quad \text{and} \quad t_2 = 2\pi\sqrt{I_1/C}.$$

So that, squaring and dividing one by the other, we have

$$I_1/I_2 = t_1^{\,2} / t_2^{\,2},$$

and thus, knowing t_1 and t_2, the moments of inertia of the two bodies may be easily compared.

Note. In the above cases, the amplitude of vibration need not be small, because it is found that the restoring couple continues to be proportional to the twist θ in the wire, up to fairly large values of θ. The assumption made, however, that even with different bodies suspended from the wire, resulting in a change in its longitudinal tension, the value of C (or the twisting couple per unit twist of the wire) remains the same is found to be only approximately true.

8.29. Bending of Beams. A **beam** is a rod or a bar of *uniform cross-section* (circular or rectangular) of a *homogeneous, isotropic elastic* material, whose length is very great compared with its thickness, so that the shearing stresses over any section are negligibly small and the bending is pure or simple.

When such a beam is fixed at one end and loaded at the other *within the limit of perfect elasticity*, the loaded end sinks a little. As a result, the beam gets bent under the action of the couple due to the load applied, as will be seen from Fig. 8.28 which shows the beam in longitudinal section. The *upper* surface of the beam thus gets stretched and assumes a *convex form* and its *lower* surface gets compressed and assumes a *concave form, the plane of bending being the same as the plane of the couple*.

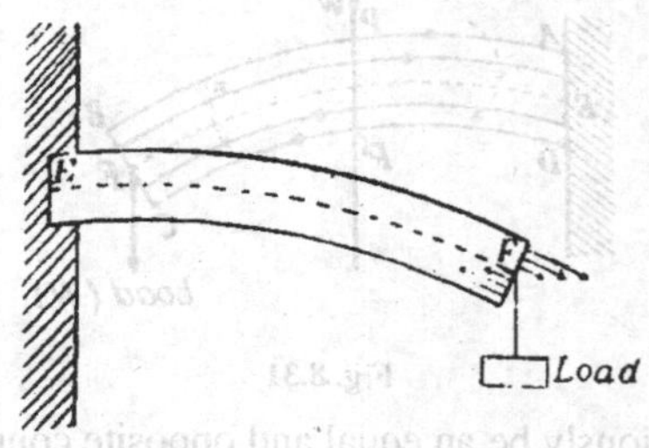

Fig. 8.28

The extensions and contractions are, however, not confined only to the upper and the lower surfaces of the beam respectively but also take place inside the body of the beam, all *longitudinal filaments* in its upper half getting extended and, in its lower half, getting compressed. This is shown in the figure by means of arrows, whose lengths indicate, in rough proportion, the extent of these extensions and contractions of the filaments in the upper and the lower halves of the beam respectively.

As will be readily seen, the extension is the maximum in the uppermost filaments and the compression, the maximum in the lowermost ones, both decreasing progressively as we proceed towards the axis of the beam from either side. So that, in the plane perpendicular to the section of the beam containing the axis EF, the filaments neither get extended or lengthened, nor compressed or shortened. This is only to be expected, for it is in this plane or surface that the extensions in the upper half change sign to become compressions in the lower half of the beam. This surface is, therefore, aptly called the **neutral surface,** shown shaded in a rather exaggerated portion of the beam in Fig. 8.29.

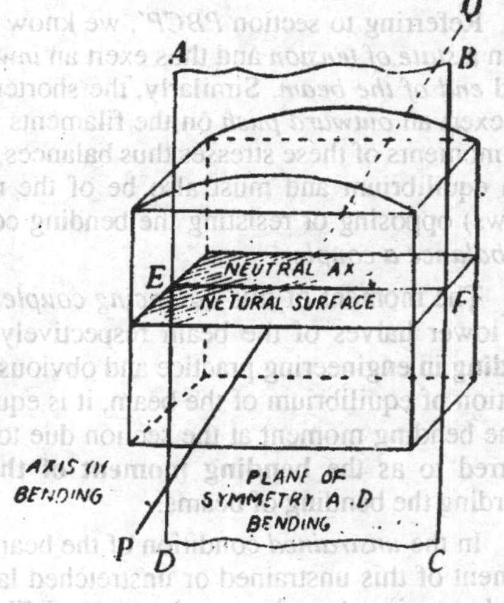

Fig. 8.29

If the bending be uniform, the longitudinal filaments all get bent into circular arcs in planes parallel to the plane of symmetry ($ABCD$) of the beam (Fig. 8.29), which thus itself becomes the **plane of bending.** The straight line PQ, perpendicular to this plane, on which lie the centres of curvature of all bent filaments, is called the **axis of bending,** and the line of intersection (EF) of the plane of bending with the neutral surface (perpendicular to it) is called the **neutral axis.**

Alternatively, we may look at the problem slightly differently and closely examine a portion of the beam before and after bending. We find that whereas before bending sections PP' and QQ' are parallel to each other ,(Fig. 8.30) the latter takes up, after bending, the position Q_1Q_1' as though it has, in consequence of the lengthening and shortening of the filaments above and below the neutral surface respectively, rotated about an axis passing through the neutral axis EF.

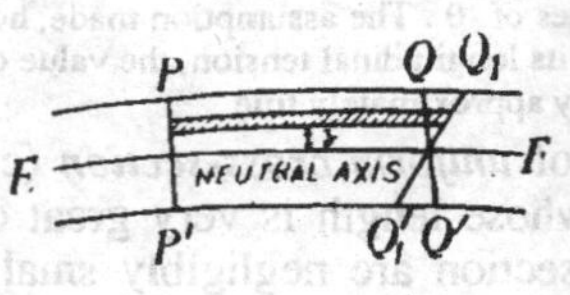

Fig. 8.30

It is clear at once from the figure that *the lengthening or shortening of a filament varies directly as its distance from the neutral surface, being the maximum at the top and the bottom of the beam respectively.*

Bending moment. Let AB be a beam fixed at the A end and loaded at the end B, shown in longitudinal section in Fig. 8.31, with EF as its *neutral axis.* Let us consider the equilibrium of a section $PBCP'$ of it, cut by a plane PP' at right angles to its length and the plane of bending.

An equal and opposite *reactional* force W^* must be acting vertically upwards along $P'P$, due ultimately to the beam being fixed at the end AD. The couple due to these two *equal* and *opposite* forces tends to rotate or bend the beam in the clockwise direction. *This couple, due to the load applied to the free end of the beam, is thus the* **bending couple** and the moment of this couple is called the **bending moment,** M.

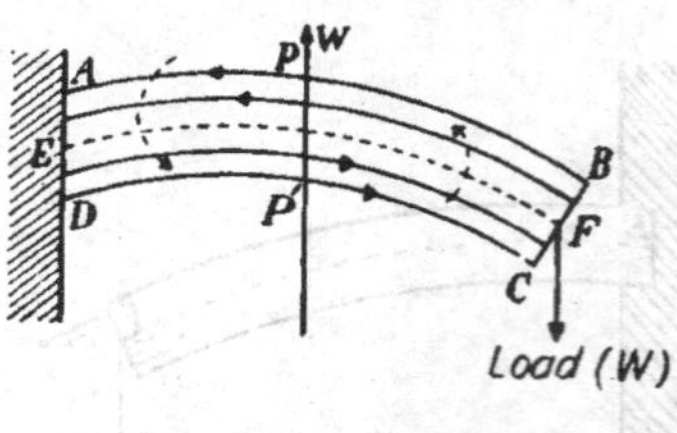

Fig. 8.31

Since the beam is in equilibrium, there must obviously be an equal and opposite couple also acting on the beam. Let us see how this is brought about.

Referring to section $PBCP'$, we know that the extended filaments above the neutral axis EF are in a *state of tension* and thus exert an *inward pull* on the filaments adjacent to them towards the fixed *end of the beam.* Similarly, the shortened filaments below EF are in a *state of compression* and exert an *outward push* on the filaments adjacent to them *towards the loaded end of the beam.* The moments of these stresses thus balances the bending moment M due to the load when the beam is in equilibrium and must also be of the nature of a couple (indicated in the figure by dotted arrows) opposing or resisting the bending couple due to the load; for, as we know, *only a couple can balance a couple.*

The moment of this *balancing couple,* due to tensile and compressive stresses in the upper and lower halves of the beam respectively, is referred to as the **moment of the resistance to bending** in engineering practice and obviously acts in the plane of bending. Since, however, in the position of equilibrium of the beam, it is equal in magnitude to the moment of the bending couple, or the bending moment at the section due to the load, it is usually, though obviously erroneously, referred to as the **bending moment of the beam.** This forms the very basis of the *theory*** regarding the bending of beams.

In the *unstrained* condition of the beam, the neutral surface becomes a plane surface and the filament of this unstrained or unstretched layer or surface, lying in the plane of symmetry of the bent beam, is referred to as the **neutral filament.** It passes through the *c.g.* (or the *centroid*) of every transverse section of the beam.

---

* Neglecting, for the moment, the weight of the beam in comparison with the load.

** The theory is subject to the limitations mentioned in § 8.31 (page 290), which the student would do well to keep in mind.

In the strained condition of the beam, as we have seen, *the change in length of any filament is proportional to its distance from the neutral surface.*

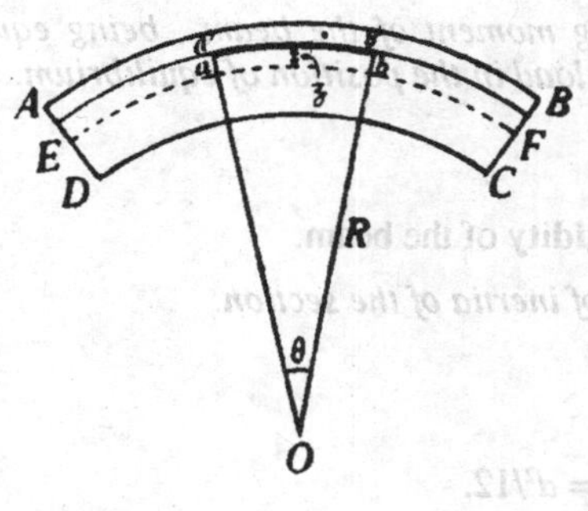

Fig. 8.32

Let a small part of the beam be bent, as shown in Fig. 8.32, in the form of a circular arc, subtending an angle θ at the centre of curvature *O*. Let *R* be the radius of curvature of this part of the *neutral axis*, and let *a'b'* be an element at a distance *z* from the *neutral axis*. Then, $a'b' = (R + z).\theta$.

Since its original length $ab = R.\theta$, we have

increase in length of the filament $= a'b' - ab$.

$$= (R + z).\theta - R.\theta = z.\theta,$$

And $\textit{strain} = \dfrac{\textit{change in length}}{\textit{original length}} = z.\theta/R.\theta = z/R,$

i.e., the strain is proportional to the distance from the neutral axis.

Since there are no shearing stresses, nor any change of volume, *the contractions and extensions of the filaments are purely due to forces acting along the length of the filaments.*

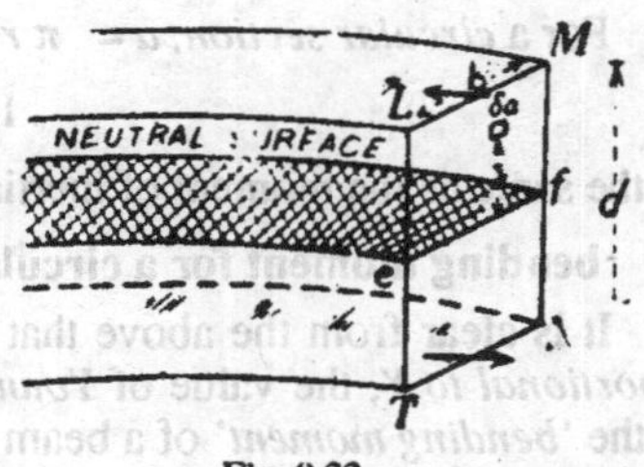

Fig. 8.33

Imagining the beam to be of a rectangular cross-section, for the sake of convenience, and considering a section *LMNT* of it, perpendicular to its length and the plane of bending, (Fig. 8.33), it is clear that the line *ef* lies on the neutral surface and that the internal forces opposing extensions in the filaments act inwards, perpendicularly to the upper half *LMfe* and those opposing contractions act *outwards* perpendicularly to the lower half *efNT* of the section, as shown. Although the two sets of forces thus act in opposite directions, their moments about *ef* are directed the same way, (namely, in the opposite direction to that in which the beam has been bent), tending to bring the beam back to its original condition.

Consider a small area δa distant *z* from the *neutral surface*. The strain produced in a filament passing through this area will be z / R, (see above).

Now, $\quad Y = \textit{stress/strain}$ and $\therefore$ $\textit{stress} = Y \times \textit{strain}.$

Therefore, *stress* at this area $= Y \times z/R$, where *Y* is the value of *Young's modulus* for the material of the beam.

And, therefore, *force on area* $\delta a = \delta a.Y.z / R$,
and, *moment of this force about the line ef* $= Y.z \times \delta a \times z/R$.
$$= Y.\,\delta a.z^2 / R.$$

Since, as mentioned above, the moments of the forces acting on both the upper and the lower halves of the section are in the same direction, the *total moment of the forces acting on the filaments in the section LMNT* is given by

$$\Sigma \frac{Y.\,\delta a z^2}{R} = \frac{Y}{R}.\,\Sigma \delta a.z^2.$$

Now, $\Sigma\, \delta a.z^2$ is the *geometrical moment of inertia* (I_g)* of the section about *ef*, and, therefore, equal to ak^2, where *a* is the *whole area* of the surface *LMNT* and *k*, its *radius of gyration*** about *ef*.

---

* It is so called because it is proportional to the mechanical moment of inertia of a plane lamina of the same shape as the cross-section. It is denoted, here, by the symbol I_g, so that the student may not confuse it with the ordinary mechanical moment of inertia denoted by *I*.

Hence, the *moment of the forces about* $ef = \frac{Y}{R} \cdot a\,k^2 = \frac{YI_g}{R}$.

This, then is the restoring couple or the *bending moment of the beam* **,** ***being equal and opposite to the moment of the bending couple due to the load in the position of equilibrium.***

Thus, bending moment of the beam $= \frac{Y}{R} \cdot I_g$

• The quantity $Y.I_g = Y.a\,k^2$ is called the **flexural rigidity** of the beam.

∴ ***bending moment*** = ***(Y/R)*** × ***geometrical moment of inertia of the section.***

= ***flexural rigidity/R,***

whatever the shape of the cross-section of the beam.

For a *rectangular cross-section*, $a = b \times d$, and $k^2 = d^2/12$.

∴ $I_g = a\,k^2 = b.d^3/12.$+

Hence, bending moment for a rectangular cross-section

= Y. b.d³/12R.

For a *circular section*, $a = \pi r^2$ and $k^2 = r^2/4$.

Ig = a k2 = p r4 / 4,

i.e., the same as the moment of inertia of a disc about a diameter.

∴ **bending moment for a circular cross-section** $= Y \cdot \pi r^4/4R$.

It is clear from the above that the expression for the bending moment of a beam is ***directly proportional to Y***, **the value of** ***Young's modulus*** **for its material. And, as we have seen, what we call the '*bending moment*' of a beam is, in fact, the** ***moment of the resistance to bending*** **offered by the beam. It is thus this resistance to bending which really depends upon Y. That this should be so is only natural, for the extensions and compressions of the filaments, produced by the bending couple applied, are purely linear in nature (there being no shearing stresses in view of the much longer length of the beam compared with its thickness) and, as such, are inversely proportional to Y.**

Obviously enough, the smaller these extensions and compressions of its filaments, or the higher the value of Young's modulus for the material of the beam, the smaller the extent of the bending. It follows, therefore, that ***it is the value of Young's modulus for its material which ultimately determines the resistance to the bending of a beam or even a thin rod.***

Note. We have seen above how strain in a beam is proportional to the distance z from its neutral axis, and is equal to z/R, where R is the radius of curvature of the portion of the neutral axis under consideration. So that, if F be the stress corresponding to the strain z/R, we have

$$Y = \frac{F}{z/R}. \text{ Or, } \frac{F}{z} = \frac{Y}{R}.$$

** Which is the same as in the case of the mechananical moment of inertia, *viz.*, the root mean square distance of the various elements from *ef*.

+ Imagine the section as a rectangular plate of *unit mass per unit area*, (Fig. 8.34). Then, area of the strip *AB*, of length *A* breadth *dz*, is equal to *b .dz*. And, thereore, its mass = *b.d.z.* 1 = *b.dz*.

Hence, geometrical moment of inertia of the strip about $ef = b.dz.z^2$;

and, therefore, moment of inertia of the whole plate or section about *ef*

$$= 2\int_0^{d/2} b.z^2.dz = 2b\left[\frac{z^3}{3}\right]_0^{d/2} = 2b\left[\frac{d^3}{24}\right] = \frac{b.d^3}{12}.$$

Fig. 8.34

If, therefore, F_1, F_2 etc. be the value of stress at distance z_1, z_2... from the neutral axis, we have $F_1/z_1 = F_2/z_2 = Y/R$.

And $\therefore$ *bending moment* $M = Y.I_g/R = \frac{F_1}{z_1}.I_g = \frac{F_2}{z_2}.I_g$ *etc.* $= \frac{I_g}{z_1}F_1 = \frac{I_g}{z_2}F_2$ *etc.* $= Z_1F_1 = Z_2F_2$ *etc.*

where, $Z_1 = I_g/z_1$ and $Z_2 = I_g/z_2$ are called the moduli of the section under consideration.

Thus, modulus of a section $= \frac{\text{geometrical moment of inertia}}{\text{distance from the neutral axis}}$.

Anticlastic Surface. In the case of a flat bar or beam, of *rectangular cross-section*, if the bending be small, there is no change in volume. There does come about, however, a change in the shape of the section, such that all lines in it, originally perpendicular to the plane of bending, get bent into arcs, which are all *concentric* and *convex to the axis of bending*. In other words, the layer of the beam, which was originally plane and perpendicular to the plane of bending, and which contained the neutral filament, *now gets changed into what is called an anticlastic surface* (Fig. 8.35), of radius R in the plane of bending (which, here, coincides with the plane of the paper), and, of radius R' in the plane perpendicular to it, the two centres of curvature lying on either side of the beam, like that in the case of a *horse saddle*. This is what is to be expected, because a transverse bending must, of necessity, be associated with a longitudinal bending of the beam, with the curvature of the former opposite to that of the latter. For, the filaments above the neutral axis, which get extended, must obviously suffer a lateral contraction σ times as great and, similarly, the filaments below the neutral axis, which get compressed, must suffer a lateral extension.

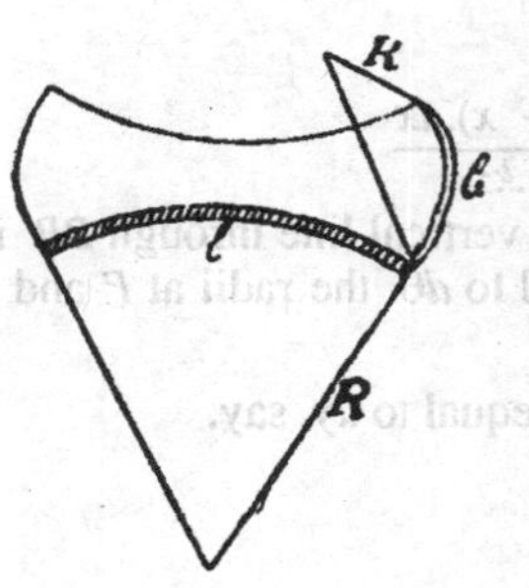

Fig. 8.35

Thus, by way of illustration, if a rectangular piece of Indian rubber be bent longitudinally in the form of an arc, it takes up the form shown in Fig. 8.36, with its longitudinal fibres bent so as to be concave with respect to a point *below*, and the transverse fibres, so as to be concave with respect to a point *above*, the rubber piece, in the case shown. *It is this bending, which occurs in a plane normal to the longitudinal plane, that gives the rubber piece (or the beam) an antielastic curvature.*

And, therefore, as we have seen before, (page 286), the longitudinal and lateral strains in a filament, distant z from the neutral axis, will be given by z/R and z/R' respectively. So that, *Poisson's ratio* σ, for the material of the beam, is given by the expression

$$\sigma = \frac{\textit{lateral strain}}{\textit{longitudinal strain}} = \frac{z/R'}{z/R} = \frac{R}{R'}$$

This, then, gives us a method for the determination of σ for the material of a given beam or bar, the two radii being determined directly by attaching suitable pointers to the rod and noting the distances and angles traversed by them, when a known couple is applied to the beam.

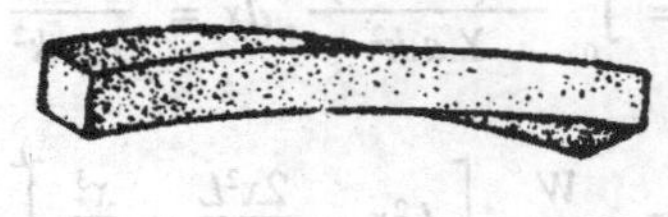

Fig. 8.36

8.30. The Cantilever. A *cantilever* is a beam fixed horizontally at one end and loaded at the other.

(*i*) **Cantilever loaded at the free end.** Here, two cases arise, *viz.*, (*a*) when the weight of the beam itself produces no bending, and (*b*) when it does so. Let us consider both the cases.

(*a*) **When the weight of the beam is ineffective.** Let AB, (Fig. 8.37) represent the neutral axis of a cantilever, of length L, fixed at the end A, and loaded at B with a weight W, such that the

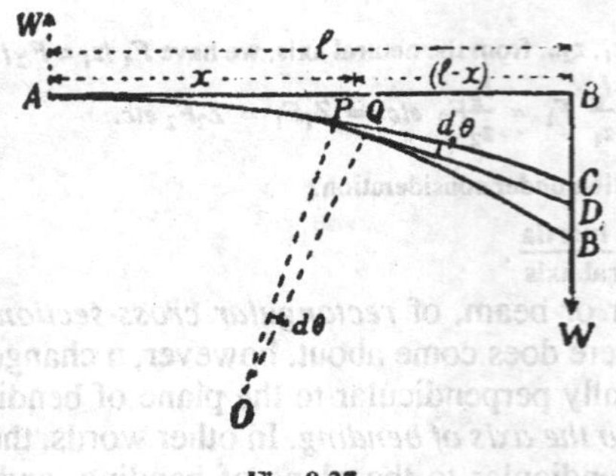

Fig. 8.37

end B is deflected or depressed into the position B' and the neutral axis takes up the position AB' *it being assumed that the weight of the beam itself produces no bending.*

Consider a section P of the beam at a distance x from the fixed end A.

The moment of the external couple at *this* section, due to the load W, or the *bending moment* acting on it $= W \times PB' = W(L - x)$.

Since the beam is in equilibrium, this must be equal to $YI_g/R = Y.a\,k^2/R$, where R is the radius of curvature of the *neutral axis* at P.

Therefore, $W.(L-x) = Y.I_g/R = Y.a.k^2/R.$ ...(*i*)

Since the moment of the load increases as we proceed towards the fixed end A, the radius of curvature is different at different points and decreases as we approach the point A. For a point Q, however, at a small distance dx from P, it is practically the same as at P. So that

$PQ = r.d\theta$. Or, $dx = R.d\theta$, [where $d\theta$ is the $\angle POQ$

whence, $R = dx/d\theta$.*

Substituting the value of R in (*i*) above, we have

$$W.(L-x) = \frac{Y.a.k^2.d\theta}{dx} \quad \text{Or,} \quad d\theta = \frac{W.(L-x).dx}{Y.a.k^2} \qquad ...(ii)$$

Draw tangents to the neutral axis at P and Q, meeting the vertical line through BB' in C and D respectively. Then, the angle subtended by them is also equal to $d\theta$, the radii at P and Q being perpendicular to the tangents there.

Now, clearly, the depression of Q below P is equal to CD, equal to dy, say.

Then, $dy = (L-x).d\theta$. Or, substituting for $d\theta$,

$$dy = \frac{(L-x).W.(L-x).dx}{Y.a.k^2} = \frac{W(L-x)^2.dx}{Y.a.k^2} \qquad ...(iii)$$

Therefore, the depression $y = BB'$ of the loaded end B below the fixed end A, is obtained by integrating the expression for dy between the limits $x = 0$ and $x = L$.

$$\text{Or,} \quad y = \int_0^L \frac{W.(L-x)^2}{Y.a.k^2}.dx = \frac{W}{Y.a.k^2}\int_0^L (L^2 - 2xL + x^2).dx$$

$$= \frac{W}{Y.a.k^2}\left[L^2x - \frac{2x^2L}{2} + \frac{x^3}{3}\right]_0^L = \frac{W}{Y.a.k^2}\left[L^3 - L^3 + \frac{L^3}{3}\right] = \frac{WL^3}{3Y.a.k^2} = \frac{W.L^3}{3Y.I_g}.$$

[Putting back I_g for $a.k^2$

Thus, the free end of the cantilever is depressed by $\dfrac{W.L^3}{3Y.a.k^2} = \dfrac{W.L^3}{3Y.I_g}$.

---

* See worked example 4, page 315, where it is shown that $1/R = d^2y/dx^2$ the rate of change of slope. A mathematical minded student will find the solution given there, with this value R, much neater and also perhaps a trifle easier.

(*b*) **When the weight of the beam is effective.** In this case in addition to the weight W at B, the weight of the portion $(L-x)$ of the beam is also acting at the mid-point, or the *c.g.* of this portion, so that, if w be weight per unit length of the beam, a weight $w(L-x)$ is acting at a distance $(L-x)/2$ from the section PQ. And, therefore, the *total bending moment on the beam*

$$= W(L-x) + w(L-x).(L-x)/2 = W(L-x) + \frac{w}{2}(L-x)^2.$$

Since the beam is in equilibrium, this must be equal to $Y I_g/R$ or $Y.a.k^2/R$.

Thus, $$W(L-x) + \frac{w}{2}(L-x)^2 = \frac{Y.a.k^2}{R} = \frac{Y.a.k^2.d\theta}{dx}.$$

And $\therefore$ $$d\theta = \frac{W(L-x).dx + \frac{w}{2}.(L-x)^2.dx}{Y.a.k^2}$$

Then, $$dy = (L-x).d\theta = \frac{W(L-x)^2.dx + \frac{w}{2}(L-x)^3.dx}{Y.a.k^2}$$

And $\therefore$ $$y = \frac{W}{Y.a.k^2}\int_0^L (L-x)^2\,dx + \frac{w}{2Y.a.k^2}\int_0^L (L-x)^3.dx.$$

$$= \frac{WL^3}{3Ya.k^2} + \frac{wL^4}{8Ya.k^2}.$$

Now $w.L = W_1$, say, the weight of the beam.

So that, $$y = \frac{W.L^3}{3Y.I_g} + \frac{W_1 L^3}{8\,Y.I_g} = \left(W + \frac{3}{8}W_1\right)L^3/\,3YI_g$$

i.e., the beam now behaves as though it is loaded at the end B with a weight W plus 3/8th of the weight of the beam.

(*ii*) **Cantilever loaded uniformly.** Let the uniform load on the cantilever be w per unit length. Then, the weight of the portion of the beam $(L-x)$, *i.e.*, $w(L-x)$ alone produces a bending moment about the section PQ, there being no weight suspended from the end B. And, since the weight $w(L-x)$ acts at a distance $(L-x)/2$ from the section PQ, we have

bending moment due to it $= w(L-x).(L-x)/2 = \frac{w}{2}.(L-x)^2.$

$\therefore$ for equilibrium of the beam $$\frac{w}{2}(L-x)^2 = \frac{Y.I_g}{R} = \frac{Y.I_g.d\theta}{dx},$$

whence, $$d\theta = \frac{w(L-x)^2}{2YI_g}.dx.$$

Substituting this value of $d\theta$ in the relation $dy = (L-x)d\theta$, we have

$$dy = \frac{w}{2\,Y.I_g}.(L-x)^3\,dx.$$

And $\therefore$ $$y = \frac{w}{2\,Y.I_g}\int_0^L (L-x)^3.dx = \frac{w.L^4}{8Y.I_g}.$$

Since $w.L = W$, the total load on the beam, $y = \frac{W.L^3}{8\,Y.I_g}$

It will be seen that this expression could be obtained directly from the result in case (*i*) *b*, above, by putting *W*, the load at the end *B*, equal to zero.

8.31. Limitations of the simple theory of bending. In discussing the above simple cases of bending, we have tacitly made the following assumptions :

(*i*) *That the cross-section of the beam remains unaltered during bending.* This, as we have seen, is not strictly true; for, the extension of the filaments above the neutral surface brings about their lateral contraction, σ times as great, and the contraction of the filaments below the neutral surface brings about their lateral extension. So that, the cross-section of a rectangular beam, bent so as to be *concave downwards along its length, is convex downwards across its length.* Similarly, a circular cross-section may change into an oval form. This change in the shape of cross-section of the beam, due to bending, results in a change in the value of the cross-section and hence in that of I_g for it. Usually, however, it is much too small to be of any practical consequence and may safely be ignored.

(*ii*) *That the radius of curvature of the bent beam, or rather that of its neutral surface, is large compared with its thickness.* This is almost always true for all cases of elastic bending.

(*iii*) *That the minimum deflection of the beam is small compared with its length.* This, while more or less true for ordinary engineering problems, is not strictly so in quite a number of cases. Thus, for example, in the case of a clock spring, the deflection produced is very large even within the elastic limit. We shall, therefore, do well to discuss this particular case here, as representing the more general case of *strongly bent beams.*

8.32. Strongly bent beams. When a beam bends very strongly, its inclination to its original, unbent or unstrained position, and hence the tangent of this inclination, is no longer small. Consequently, its curvature ($1/R$) can no longer be taken to be equal to the rate of change of slope, d^2y/dx^2, as is done in the ordinary cases of small curvatures (see worked example 4, page 315) but is now given by

$$\frac{d^2y}{dx^2} \Big/ \left[1 + \left(\frac{dy}{dx}\right)^2\right]^{3/2}.$$

The problem thus becomes quite complicated in many a case, with the differential equations obtained not being amenable to easy solution.

However, there are also some cases which can be investigated in a much simpler manner and we shall here consider only one of these, *viz.*, that of **flexible cantilever**, like a *clock-spring*, clamped at one point and loaded at its free end. As we pass or 'pay out' more and more of the spring through the clamp, *keeping the load constant*, its free end drops further and further down, as a result of the large amount of bending, *until finally it becomes quite vertical.* The horizontal distance between the clamp and the loaded end of the spring is now the *maximum* and any more of the spring 'paid out' through the clamp merely hangs vertically.

Thus, let *OAB*, (Fig. 8.38), be the bent position of the spring, clamped at *O* and loaded with a weight *W* at the end *B*, such that the tangent at *A*, (coordinate $x = x$) makes an angle θ and that at *B* (coordinate $x = a$), an angle φ, with the axis of *x*. Then, the *bending moment at A* is clearly equal to $W(a-x)$, So that $W(a-x) = Y.I_g/R$, where *R* is the radius of curvature of portion *OA* of the spring.

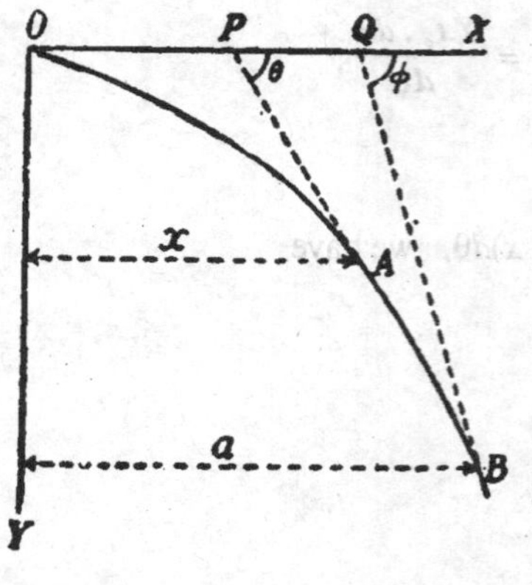

Fig. 8.38

Now, $\frac{1}{R} = \frac{d\theta}{ds}$. [See worked Example 4, page 315]

$$\therefore \quad W(a-x) = Y.I_g\frac{d\theta}{ds} = Y.I_g.\frac{d\theta}{dx}.\frac{dx}{ds}.$$

$$= Y.I_g.\cos\theta.\frac{d\theta}{dx}. \qquad [\because dx/ds = \cos\theta.$$

Or $\quad W(a-x).dx = Y.I_g.\cos\theta.d\theta$

∴ integrating this expression between the limits $x = 0$ and $x = a$, the horizontal distance of the loaded end from *O*, we have

$$\int_0^a W(a-x).dx = Y.I_g\int_0^\phi \cos\theta.d\theta$$

Or, $\quad W\left[ax - \frac{x^2}{2}\right]_0^a = Y.I_g[\sin\theta]_0^\phi$

Or, $\quad \frac{Wa^2}{2} = Y.I_g.\sin\phi \qquad ...(I)$

Now, when the loaded end becomes vertical, $\phi = 90^0$ and the horizontal distance a becomes the maximum, say, a_m. So that, substituting $a = a_m$ and $\phi = 90^0$ in expression (*I*) above, we have

$$Wa_m^2/2 = Y.I_g, \text{ whence, } a_m^2 = 2YI_g/W.$$

Or, $$a_m = \sqrt{2YI_g/W}. \quad ...(II)$$

It will be seen that the value of *y* for the material of the flexible beam can easily be determined from either of the relations (*I*) or (*II*) above, if we know the angle of inclination ϕ of the loaded end of the beam with the horizontal, or the maximum horizontal distance (*i.e.*, *a* or a_m) of the loaded end from the clamp.

8.32. Transverse Vibrations of a Loaded Cantilever. (*i*) Dynamical method of determining Y for the material of the Cantilever. If the loaded free end of a cantilever be depressed a little and then released, it starts moving up and down its original position, *i.e.*, executes *transverse vibrations*. Let us calculate the time-period of these vibrations.

We have seen above [§ 8.30 (*a*), page 289], how

$$y = \frac{WL^3}{3YI_g}, \text{ whence, } W = \frac{3YyI_g}{L^3},$$

which gives the load or the force *W*, required to maintain the cantilever in equilibrium with its free end depressed or *displaced* through *y*.

This, thus, also measures the *elastic reaction* of the cantilever, which is oppositely directed to it.

Now, if $a = d^2y/dt^2$ be the *acceleration* of the mass *M* suspended from the free end of the cantilever, (*i.e.*, if *M* be the mass of the load *W*), the force of *inertial reaction* on it is equal to *M.a*.

Hence, since the cantilever is in equilibrium and there is no other external force acting on it, we have

$$M.a = -\frac{3.\,YyI_g}{L^3}, \text{ whence, } a = \frac{3.YI_g}{ML^3}.y.$$

Or, $$a = -\mu . y,$$

where $3YI_g/ML^3 = \mu$, a constant for the given cantilever, with the given load.

Thus, $$a \propto y,$$

i.e, the acceleration of the mass (or of the free end of the cantilever) is directly proportional to its displacement. It is thus a case of *simple harmonic motion*, and its *time-period* is, therefore, given by

$$t = 2\pi\sqrt{\frac{1}{\mu}} = 2\pi\sqrt{\frac{1}{3YI_g/ML^3}} = 2\pi\sqrt{\frac{ML^3}{3Y.I_g}}.$$

As can be seen at once, this relation for *t* gives a good *dynamical method for the determination of the value of Young's modulus (Y) for the material of a given beam or rod.* It is particularly suitable for beams like a metre stick etc. All that need be done is to plot *M* against t^2 to obtain a straight line graph, the slope of which is given by $4\pi^2L^3/3YI_g$. So that, knowing *L* and I_g, the value of *Y* is easily obtained.

(*ii*) **Statical method.** Here, different loads (*W*) may be applied to the free end of the cantilever and its depressions (*y*) noted. A graph may then be plotted between *W* and *y* which will obviously be a straight line, passing through the origin, as shown in Fig. 8.39, with a slope *y*/*W*

$$= WL^3/3YI_g.W = L^3/3YI_g,$$

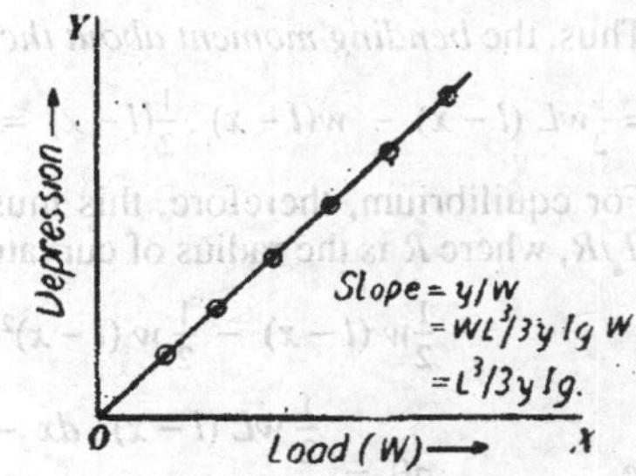

Fig. 8.39

whence, l and l_g being known, Y for the material of the cantilever can be easily evaluated.

8.33. Depression of a Beam supported at the ends.

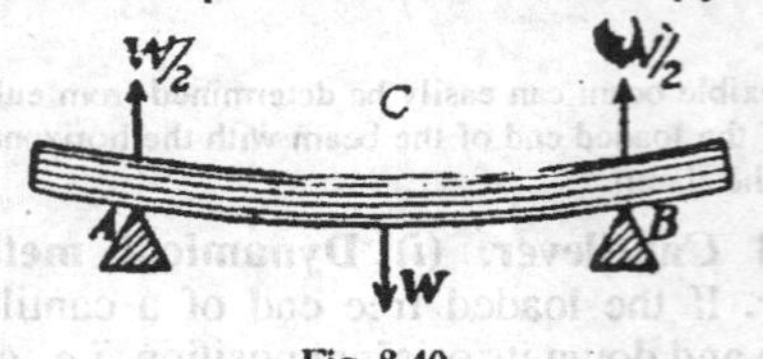

Fig. 8.40

(*i*) **When the beam is loaded at the centre.** Let a beam be supported on two knife-edges at its two ends A and B, as shown in Fig. 8.40, and let it be loaded in the middle at C with a weight W.

The reaction at each knife-edge will clearly be $W/2$, in the *upward* direction.

Since the middle part of the beam is horizontal, *the beam may be considered as equivalent to two inverted cantilevers, fixed at C*, the bending being produced by the loads $W/2$, acting upwards at A and B.

If, therefore, L be the length of the beam AB, the length of each cantilever (AC and BC) is $L/2$, and the elevation of A or B above C or, what is the same thing, the depression of C below A and B is, in accordance with § 8.30 above, given by

$$y = \frac{W/2\,.\,(L/2)^3}{3Y\,.\,ak^2} = \frac{W\,.\,L^3}{48\,Y\,.\,ak^2} = \frac{WL^3}{48\,Y\,.\,I_g}. \qquad [\because\ ak^2 = I_g$$

If the beam be a **circular cross-section,** we have $ak^2 = \pi r^4/4$, where r is the radius of its cross-section; so that, for such a beam,

$$y = \frac{WL^3}{48.Y} \times \frac{4}{\pi r^4} = \frac{WL^3}{12\,.\,Y\,.\,\pi r^4}. \qquad ...(i)$$

And, if the beam be of a **rectangular cross-section**, of breadth b and depth d, we have $ak^2 = bd^3/12$, and therefore, for such a beam,

$$y = \frac{WL^3}{48.\,Y} \times \frac{12}{bd^3} = \frac{WL^3}{4\,Y\,bd^3}. \qquad ...(ii)$$

(*ii*) **When the beam is loaded uniformly.** Here, let w be the load per unit length of the beam, so that the total weight acting downwards at the *c.g.* is $wL = W$, where L is the length of the beam. The reaction at each knife-edge is thus obviously $\frac{1}{2}WL$, acting *upwards*, (Fig. 8.41), with the beam behaving as a system of two cantilevers, fixed at C.

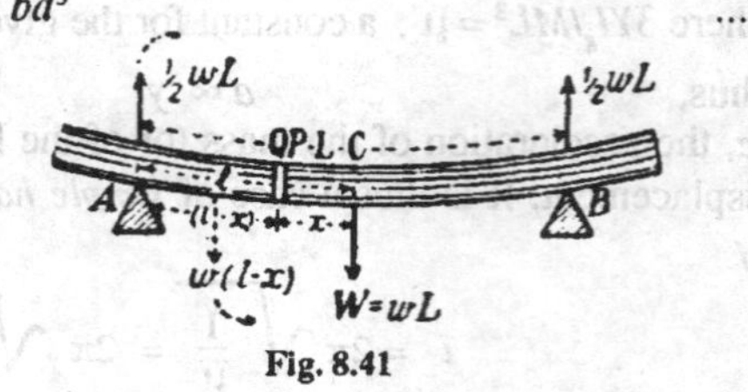

Fig. 8.41

Considering again, a section PQ, distant x from the mid-point C of the beam, and taking half-length of the beam equal to l, the weight of the portion $(l-x)$ of the beam, *i.e.*, a weight $w(l-x)$ acts *downwards* at a distance $(l-x)/2$ from the section PQ.

Thus, the *bending moment about the section*

$$= \frac{1}{2}wL\,(l-x) - w(l-x)\,.\,\frac{1}{2}(l-x) = \frac{1}{2}wL\,(l-x) - \frac{1}{2}w\,(l-x)^2.$$

For equilibrium, therefore, this must be equal to the *moment of the resistance to bending, viz.*, $Y.I_g/R$, where R is the radius of curvature of the neutral axis at PQ.

i.e., $$\frac{1}{2}w\,(l-x) - \frac{1}{2}w\,(l-x)^2 = Y.\,I_g/R = Y.I_g\,d\theta/dx \qquad [\S\ 8.30$$

whence, $$d\theta = \frac{\frac{1}{2}wL\,(l-x).\,dx - \frac{1}{2}w(l-x)^2.\,dx}{Y.\,I_g}$$

$$\therefore \qquad dy = (l-x)\,.d\theta = \frac{\frac{1}{2}wL\,(l-x)^2\,.\,dx - \frac{1}{2}\,w(l-x)^3\,.\,dx}{Y\,.\,I_g}$$

And hence
$$y = \int_0^l \frac{\frac{1}{2}\,wL\,(l-x)^2\,.\,dx - \frac{1}{2}w\,(l-x)^3\,.\,dx}{Y\,.\,I_g}.$$

$$= \frac{1}{2}\cdot\frac{wL}{YI_g}\int_0^l (l^2 - 2xl + x^2)\,.\,dx - \frac{1}{2}\cdot\frac{w}{YI_g}\int_0^l (l^3 - 3l^2x + 3lx^2 - x^3)\,dx\,.$$

$$= \frac{1}{2}\cdot\frac{wL}{YI_g}\left[\,l^2x - 2l\,.\,\frac{x^2}{2} + \frac{x^3}{3}\right]_0^l - \frac{1}{2}\cdot\frac{w}{YI_g}\left[\,l^3x - 3l^2\,\frac{x^2}{2} + 3\,.\,\frac{lx^3}{3} - \frac{x^4}{4}\right]_0^l$$

$$= \frac{1}{2}\cdot\frac{wL}{YI_g}\cdot\frac{l^3}{3} - \frac{1}{2}\cdot\frac{w}{YI_g}\cdot\frac{3l^4}{12} = \frac{wL}{YI_g}\cdot\frac{l^3}{6} - \frac{wl^4}{8YI_g}\,.$$

Now, substituting $L/2$ for l, we have

$$y = \frac{w}{YI_g}\left(\frac{L^4}{48} - \frac{L^4}{128}\right) = \frac{w}{YI_g}\cdot\frac{5L^4}{384} = \frac{1}{YI_g}\cdot\frac{wL.5L^3}{384}\,.$$

But $wL = W$, the total weight on the beam.

Hence
$$y = \frac{5WL^3}{384\,Y\,I_g}.$$

8.34. Determination of Y by bending of a beam. It will be easily seen that if we measure the depression (y) of a beam of known dimensions, supported at the ends and loaded at the centre, as in case (i) § 8.33 above, we can easily determine the value of Y for its material, by applying relation (i) or (ii), as the case may be. In practice, it is convenient to use a beam of rectangular cross-section; so that, knowing W, L, b, d and y, Y can be easily calculated from relation (ii) § 8.33 above.

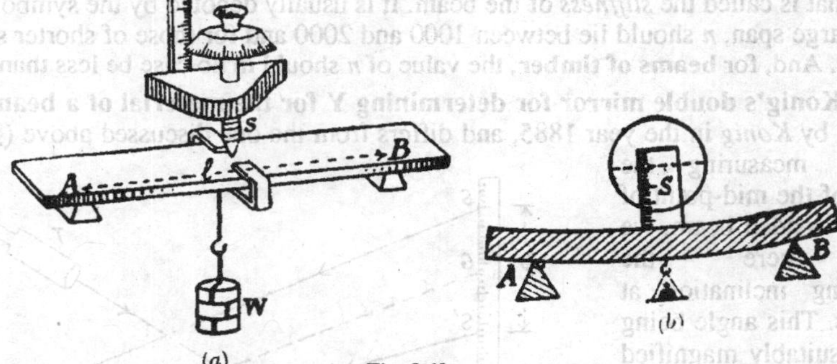

Fig. 8.42

The arrangement is as shown in Figs. 8.42 (a) and (b). The beam is supported *horizontally* and *symmetrically* on two parallel knife-edges, a known distance L apart and the load is applied by placing weights in a scale pan, also supported on a knife-edge, midway between them as shown. The depression y of the mid-point, thus produced, is noted directly with the help of a micrometer screw, [Fig. 8.42 (a)], or more accurately, with the help of a microscope, the eyepiece of which is fitted with cross-wires [Fig. 8.42 (b)].

Readings are taken, first with the load increasing in *equal steps*, and then with the load decreasing in the *same equal steps*, and their mean taken. This gives y. Then, if the load were increased (or decreased) in regular steps of W each, we have, as explained above, $y = WL^3/4Ybd^3$, where b and d are the *breadth* and the *depth* of the beam, and Y, the *Young's modulus* for its material.

Hence $$Y = WL^3/4ybd^3.$$

There are some *weak points* of the experiment, *viz.*, *(i)* while increasing the load on the beam for different observations, the *elastic limit* of its material may inadvertently be exceeded, *(ii)* under a heavy load, the knife-edges may bite into the beam which may, therefore sink a little, giving a wrong value of the depression y and *(iii)* the method of measuring y is susceptible only of a limited accuracy. All these have been removed in *Konigs* method discussed below in § 8.35.

Important Deduction . From the relation $y = WL^3/4Yb.d^3$ above, it is clear that, *for a given load, the depression of the beam is*

(i) *directly proportional to the cube of its length,*

(ii) *inversely proportional to its breadth,*

(iii) *inversely proportional to the cube of its depth, and*

(iv) *inversely proportional to the Young's modulus for its material.*

It follows, therefore, that *in order that the depression of a beam may be small for a given load*, its length should be small, *i.e.*, *its span should be small*, its *breadth and depth should be large* and the *Young's modulus for its material should also be large*.

When a girder is supported at its two ends, its middle part is depressed, and the surfaces above and below the *neutral surface* are respectively compressed and extended, the compression being the utmost at the upper face, and the extension, the maximum at the lower face, the stresses being the maximum there and decreasing as we proceed towards the neutral surface from either side. It follows, therefore, that the upper and the lower faces of the beam should be much stronger than its middle portions. In other words, *the middle portions of a girder may be made of a much smaller breadth than the upper and the lower faces, thus affecting a good deal of saving in the material, with little or no loss in its strength. It is for this reason that the girders are usually manufactured with their cross-section in the form of the letter I.*

Stiffness of a beam. The *ratio between the maximum deflection of a beam and its span* measures what is called the *stiffness* of the beam. It is usually denoted by the symbol $1/n$. For **steel girders** of large span, n should lie between 1000 and 2000 and for those of shorter spans, between 500 and 700. And, for **beams of timber**, the value of n should in no case be less than 360.

8.35. Konig's double mirror for determining Y for the material of a beam. The method was devised by *Konig* in the year 1885, and differs from the one discussed above (§ 8.34) in that, instead of measuring the depression of the mid-point of the centrally loaded beam, we measure here the accompanying inclination at its two ends. This angle being small, it is suitably magnified and measured accurately by the scale and telescope arrangement and, for this purpose, a plane mirror is mounted at either end of the beam.

The theory and working of the experiment will be clear from the following.

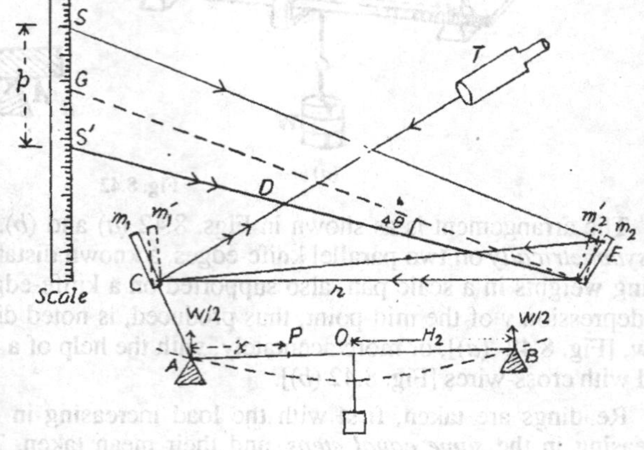

Fig. 8.43

Let AB be the beam of length L resting horizontally on two knife-edges near its two ends and let a load W be suspended from its mid-point O such that it gets bent into the dotted form shown, (Fig. 8.43).

Consider a section of the beam at P, distant x from A. For equilibrium of the section, obviously, the sum of the moments $\frac{W}{2}.x + W\left(\frac{L}{2} - x\right)$ must be balanced by the *bending moment* $YI_g.d^2y/dx^2$, where the symbols have their usual meanings. We, therefore, have

$$\frac{W}{2}.x + W\left(\frac{L}{2} - x\right) = -Y.I_g.\frac{d^2Y}{dx^2}.$$

Or, $$Y.I_g.\frac{d^2y}{dx^2} = -\frac{W}{2}x + \frac{WL}{2} + Wx = -\frac{WL}{2} + \frac{Wx}{2},$$

which, on integration, gives

$$Y.I_g.\frac{dy}{dx} = -\frac{WLx}{2} + \frac{Wx^2}{4} + C,$$

where C is a constant of integration. To determine the value of C, we note that dy/dx is zero at the mid-point O of the beam, *i.e.*, at $x = L/2$. so that,

$$0 = -\frac{WL}{2}.\frac{L}{2} + \frac{W}{4}.\left(\frac{L}{2}\right)^2 + C. \text{ Or, } C = \frac{WL^2}{4} - \frac{WL^2}{16} = \frac{3WL^2}{16}.$$

Substituting this value of C, therefore, we have

$$YI_g.\frac{dy}{dx} = -\frac{WLx}{2} + \frac{Wx^2}{4} + \frac{3\,WL^2}{16}, \text{ whence } \frac{dy}{dx} = \left(-\frac{WLx}{2} + \frac{Wx^2}{4} + \frac{3WL^2}{16}\right)\Big/ YI_g.$$

And ∴ $$\left(\frac{dy}{dx}\right)_{x=l} = \left(-\frac{WL^2}{2} + \frac{WL^2}{4} + \frac{3\,WL^2}{16}\right)\Big/ YI_g = -\frac{WL^2}{16\,YI_g}.$$

But $\left(\frac{dy}{dx}\right)_{x=l}$ is the slope of the beam at the knife-edge B, *i.e.*, $(\pi - \theta)$, where θ is the required angle of inclination of end B of the beam, so that,

$$\left(\frac{dy}{dx}\right)_{x=l} = -\frac{WL^2}{16\,YI_g} = \tan(\pi - \theta) = -\tan\theta. \text{ Or, } \frac{WL^2}{16\,YI_g} = \tan\theta = \theta,$$

because θ is small.

$$\therefore \quad Y = WL^2 / 16\,I_g.\theta.$$

Determination of θ. In the figure, let m_1 and m_2 be the positions of the two mirrors at the ends A and B of the beam respectively before the beam is loded, so that a ray SE from a division S on the vertical scale gets reflected at mirror m_2 at E and then at mirror m_1 at C, along CT, so that the image of S is formed on the cross wires of the eye-piece of the telescope T, which is fixed in the position.

The beam is now loaded with a weight W at its mid-point O, so that due to the slight inclination of the ends of the beam, the mirrors take up the positions m_1^1 and $m_2^{\ 1}$ at an angle θ with their initial positions m_1 and m_2. Let the image of division S' of the scale be formed on the cross-wire of the telescope, after reflection first at m_2 and then at m_1, as shown by the rays with double arrowheads.

Now, if we imagine the course of rays to be reversed, the ray TC will be reflected, in position m_1 of the first mirror, along CE which, on reflection at the second mirror in position m_2, will proceed along ES. The same ray TC will, in position m'_1, of the first mirror, be reflected along CF, which, in position of the second mirror, will be reflected along FS'.

Since, for a given incident ray, the reflected ray turns twice as rapidly as the reflecting surface, we have angle $ECF = 2\theta$, where θ is the angle between position m and m'_1 of the first mirror. If follows, therefore, that $EF = r2\theta$ where r is the distance between the two mirrors. Further, due to rotation of the second mirror also through θ, the angle between the two reflected rays ES and FS' from it corresponding to the incident rays CE and CF, will naturally be $2 \times 2\,\theta = 4\,\theta$. If FG be drawn parallel to ES, therefore we have $\angle\, GFS' = 4\theta$ and hence $GS' = D.4\,\theta$, where $S'F = D$.

Clearly $SS' = SG + GS' = EF + GS' = r\,2\,\theta + D.\,4\,\theta$.

So that, if $SS' = p$, we have $p. = 2\,\theta\,(r + 2D)$, whence, $\theta + p/2\,(r + 2D)$.

Substituting the value of θ in the expression for Y above, therefore, we have

$$Y = \frac{WL^3}{16\,I_g} \times \frac{2\,(r + 2D)}{p} = \frac{WL^3\,(r + 2D)}{8\,I_g \cdot p}$$

Now, for a beam of rectangular cross-section $I_g = bd^3 / 12$. And, therefore,

$$Y = \frac{WL^3\,(r + 2D)}{8p} \cdot \frac{12}{bd^3} = \frac{3WL^3\,(r + 2D)}{2p\,bd^3}$$

Whence the value of Y may be easily obtained.

Advantages of the method. The method has following three obvious advantages over the earlier one (§ 8.34).

(i) Since smaller loads suffice for the purpose, there is no possibility of the elastic limit of the beam being crossed.

(ii) Even if the beam sinks a little into the two knife-edges at its two ends (due to a heavy load having been used), it introduces no error here.

(iii) The angle of inclination of the ends of the beam (θ) can certainly be much more accurately measured (in the manner explained) than the depression (y) of the mid-point of the beam.

8.36. Determination of elastic constants by Searle's method. This is a simple yet elegant method for determining all the elastic constants Y, n, σ and K for a material given in the form of a short length of wire.

The given piece of wire is fastened to the mid-points of two identical metal bars AB and CD (Fig. 8.44), of circular or rectangular cross section, so that their moment of inertia, can be obtained from a knowledge of their dimensions. Then, for the determination of the different constants we proceed as explained below.

(*i*) **Determination of Y.** The two bars are suspended from a rigid support by means of two small vertical lengths of threads, so that, when the wire is straight, the bars are parallel to each other, as shown.

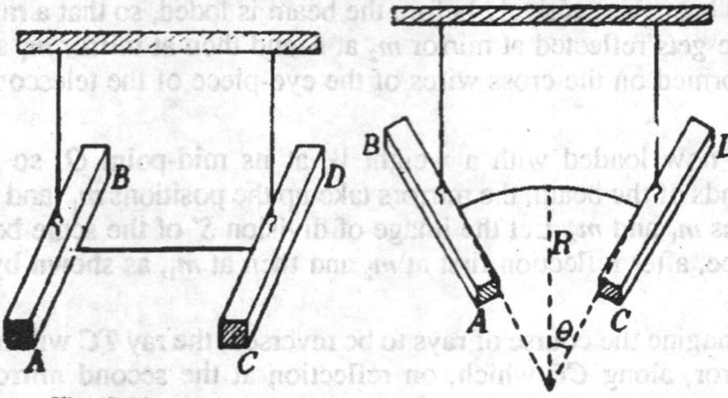

Fig. 8.44 Fig. 8.45

On slightly pulling together the ends A and C of the two bars *symmetrically and through equal distances*, the wire is bent into a circular arc (Fig. 8.45). On releasing the bars, they begin to vibrate in a horizontal plane from a circular arc on one side to a similar arc on the other, due to the *torque* exerted on them by the wire, the midpoint of the bars remaining almost at rest.

If l be the length of the wire and θ, the angle of deflection of each bar from its normal position, the angle subtended by the wire at the centre of curvature of the circular arc into which it is bent, is clearly 2θ, such that $l = R \,.\, 2\,\theta$, where R is the *radius of the arc or*, $R = l/2\,\theta$.

Now, the bending moment of the wire and the couple exerted by it on each bar is, as we know,

$$\frac{YI_g}{R} = \frac{Y \,.\, \pi r^4}{4R} = \frac{Y\pi r^4 \,.\, 2\theta}{4l} = \frac{Y \,.\, \pi r^4 \theta}{2l}. \qquad [\because\ I_g = \pi r^4/4 \text{ for the wire.}$$

This couple produces an angular acceleration $d\omega/dt$ in each bar, and, therefore,

$$I \,.\, d\omega/dt = -\ Y\pi r^4 \,.\, \theta/2l,$$

where I is the *moment of inertia* of each bar about an axis through its mid-point and perpendicular to its length, *i.e.*, about the thread from which it is suspended.

And $\therefore$ $\quad d\omega/dt = \dfrac{Y \,.\, \pi r^4}{2l.I} \,.\, \theta \,.\quad$ Or, $\ d\omega/dt \propto \theta.$

i.e., the acceleration is proportional to the (angular) displacement.

Therefore, the motion is a *simple harmonic* one, and hence the time-period of each bar is given by

$$t_1 = 2\pi\sqrt{\frac{2Il}{Y \,.\pi r^4}} \quad \text{whence, } Y = \frac{8\pi Il}{r^4 \,.\, t_1^{\,2}} \qquad \ldots(I)$$

whence the value of Y for the meterial of the wire is easily obtained.

(ii) **Determination of n.** The suspension threads are now removed, and one of the bars is clamped horizontally, so that the other one hangs vertically below it at the other end of the wire, as shown in Fig. 8.46. The suspended bar is then turned about the wire in the horizontal plane, so as to twist the wire when, on being released, it begins to vibrate *torsionally*. Its time-period t_2 is noted.

Fig. 8. 46

Now, $t_2 = \sqrt{I/C}$ where C is the twisting couple set up in the wire per unit deflection or twist, and is equal to $n\pi r^4/2l$. (§ 8.22)

$$\therefore \quad t_2 = 2\pi\sqrt{\frac{I \,.\, 2l}{n\pi r^4}}, \text{ whence } n = \frac{8\pi\, l \,.\, I}{r^4\, t_2^{\,2}}, \qquad \ldots(II)$$

whence, n, the modulus of rigidity of the material of the wire can be easily obtained.

(iii) *(a)* **Composition of Y and n and** *(b)* **Determination of σ.** *(a)* Dividing relation *(I)* by relation *(II)* obtained above, we have

$$\frac{Y}{n} = \frac{8\pi \,.\, Il}{r^4 \,.\, t_1^{\,2}} \cdot \frac{r^4 \,.\, t_2^{\,2}}{8\pi \,.\, Il} = \frac{t_2^{\,2}}{t_1^{\,2}}. \qquad \ldots(III)$$

This gives the *ratio of Young's modulus and the coefficient of rigidity* of the material of the wire in terms of t_1 and t_2.

(b) Now, *Poisson's ratio,* $\sigma = \dfrac{Y}{2n} - 1$.

So that, substituting the value of $Y/2n$, we have

$$\sigma = \frac{l_2^{\,2}}{2\,l_1^{\,2}} - 1 = \frac{l_2^{\,2} - 2l_1^{\,2}}{2l_1^{\,2}}. \qquad ...(IV)$$

Thus, *Poisson's ratio* for the material can also be easily determined.

(*iv*) **Determination of K.** we know that $K = \dfrac{nY}{9n - 3Y}$, (see § 8.15).

so that, $\quad K = \dfrac{Y}{9 - 3Y/n}$.

Substituting the values of Y and Y/n from relations (*I*) and (*III*) above, therefore, we have

$$K = \frac{8\pi l I}{r^4 l_1^{\,2}} \Big/ 9 - \frac{3\,l_2^{\,2}}{l_1^{\,2}} = \frac{8\pi l I}{3r^4 (3l_1^{\,2} - l_2^{\,2})}.$$

Thus, K for the material of the wire can also be easily obtained.

N.B. As will be readily seen the radius (r) of the wire, the measurement of which is the chief source of error (see page 280), has been eliminated altogether.

8.37. Cornu's method for elastic constants of glass. *Cornu* suggested, in the year 1869, how the usual method of bending for determination of Young's modulus for a substance could, in the case of good reflectors like metals or glass, be made much more accurate by obtaining the necessary data on the deformation of the surface of the beam from the interference fringes formed in between the surface and an optically plane cover glass placed on it. The method is particularly well suited for glass in view of the fact that only small stresses can be applied to it.

Unfortunately, some of Cornu's own results were vitiated due to the two supporting knife-edges being much close together in comparison with the width of the beam. The bending moment applied to the beam was not, therefore, uniform.

Jessop, in 1921, repeated Cornu's experiment, taking good care to ensure uniformity of the bending moment over a large part of the beam. His method, in essential details, will be clear from the following.

The rectangular glass plate or beam B resting on the knife-edges E_1 and E_2, fairly wide apart, is loaded *equally* near either end a little outside the knife-edges, as shown in Fig. 8.47 (*a*). With this method of loading, *there is no shear anywhere in between the two knife-edges and the bending moment over this part of the beam is the same all over,* being equal to the sum of the couples due to the two loads.

An optically plane cover glass G is placed centrally on the beam and light from a sodium flame S, rendered parallel by a lens L and reflected downwards by a glass plate P, inclined at 45°, is allowed to fall *normally* on it. The interference pattern located in the thin layer of air enclosed between the lower surface of G and the upper surface of the glass beam is observed either through a microscope placed vertically

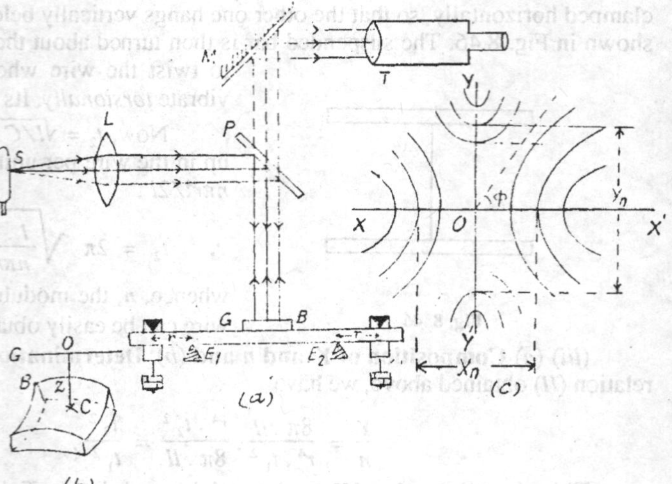

Fig. 8.47

above the glass plate P (not shown) or more conveniently, with the help of a long focus travelling microscope or a short distance telescope T arranged horizontally, as shown, to receive the rays after reflection at 45° at a plane mirror M.

Now, under the constant bending moment applied, the glass beam B undergoes a uniform bending both along and perpendicular to its axis, so as to acquire an *anticlastic curvature* in these two perpendicular directions as indicated in Fig . 8.47(*b*). There are thus two sets of interference fringes formed, consisting of two conjugate systems of ***rectangular hyperbolae***, as indicated diagrammatically in Fig 8.47(*c*).

The distances X_n between the nth pair of fringes and $X_{(n+p)}$ between the $(n + p)$th pair along the direction XOX' *i.e.*, along the axis of the beam, are carefully measured with the help of the travelling microscope (or the telescope) T. So also are measured distances Y_n and $Y_{(n+p)}$ between the nth and the $(n + p)$th pair of fringes respectively along YOY', i.e., in the perpendicular direction to the axis of the beam.

Then, if in Fig. 8.48(*a*), P and P' be the positions of the nth pair of fringes and R_1, the radius of curvature of the bent beam E_1E_2 or, rather, that of its *neutral surface* in case one (*i.e.*, bending along the axis), we have

$$CD \cdot DO_1 = PD.DP = CD.DO_1 = PD^2 = (PP^1/2)^2.$$

$$= CD(2R_1 - CD) = (PP^1/2)^2$$

Or, $CD.2R_1 - CD^2 = (PP^1/2)^2$

Or, neglecting CD^2, compared with the other terms, we have $CD.2R_1 = (X_n/2)^2$ ...(*i*)

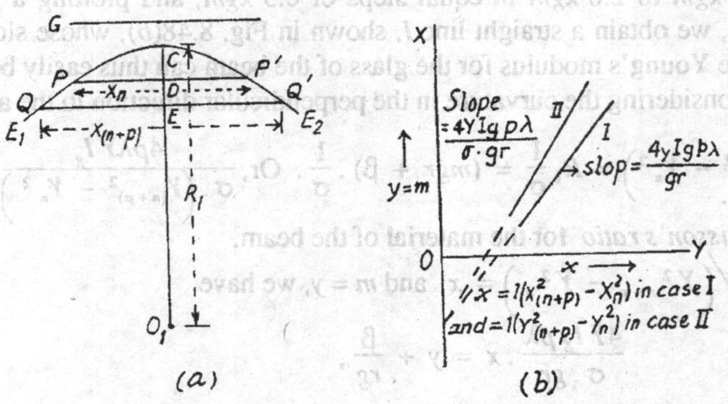

Fig. 8.48

Similarly, if Q and Q' be the positions of the $(n + p)$th pair of fringes and $X_{(n+p)}$, the distance between them, we have, as above, $CE.2R_1 = (X_{(n+p)} / 2)^2$ ...(*ii*)

So that, subtracting (*i*) from (*ii*), we have

$$2R_1 (CE - CD) = (X_{(n+p)}^2 - X_n^2)/4,$$

From the theory of fringes,

$$CE - ED = P\lambda/2$$

So that, , $2R_1 . p\lambda/2 = X_{(n+p}^2 - X_x^2)/4$

whence, $R_1 = \frac{1}{4p\lambda}\left(X_{(n+p)}^2 - X_n^2\right)$ ...I

Similarly, R_2 be the radius of curvature of the bent beam (*i.e.*, of its neutral surface) in the perpendicular direction to its axis, and Y_n and $Y_{(n+p)}$, the distances between *n*th and $(n+p)$th pairs of fringes respectively, we have, proceeding as above.

$R_2 = \frac{1}{4p\lambda}\left(Y_{(n+p)}^2 - Y_n^2\right)$ ...II

It may be noted that due to the beam acquiring an anticlastic curvature, the cover glass *G* is no longer in contact with the glass beam *B* at *C*, the centre of the interferences pattern formed. This does not, however, effect in any way the shape or the position of the fringes, as the above treatment shows.

Now, if the curvature of the beam were purely due to the load applied, say, *mg*, we shall have in case one (bending along the axis), *bending moment* $YI_g/R_1 = mgr$, where *r* is the distance between the point of application of the load and the knife-edge near to it, at either end and $I_g = bd^3/12$, where *b* and *d* are the *breadth* and the *depth* of the beam. Since, however, the beam also attains some curvature due to its own weight, the relationship is of the type $YI_g/R_1 = mgr+\beta$, where β is a constant. Or, substituting the value of R_1 from relation *I* above, we have

$$\frac{4YI_g p\lambda}{(X_{(n+p)}^2 - X_n^2)} = mgr + \beta$$

Putting $m = y$ and $\frac{1}{(X_{(n+p)}^2 - X_2^2)}$ we, therefore, have $\frac{4YI_g p\lambda}{gr}.x = y + \frac{\beta}{gr}$,

which is the equation to a straight line. So that, repeating the experiment with different values of *m*, ranging from 0.5 *kgm* to 2.0 *kgm* in equal steps of 0.5 *kgm*, and plotting a graph, between the values of *x* and *y*, we obtain a straight line *I*, shown in Fig. 8.48(*b*), whose slope is $4YI_g p\lambda/gr$. The value of *Y*, the Young's modulus for the glass of the beam can thus easily be obtained.

Similarly, considering the curvature in the perpendicular direction to the axis, we have

$$R_2 = \frac{1}{4p\lambda}\left(Y_{(n+p)}^2 - Y_n^2\right) = R_1\frac{1}{\sigma} = (mgr + \beta).\frac{1}{\sigma}. \text{ Or, } \frac{4p\lambda YI_g}{\sigma.\left(Y_{(n+p)}^2 - Y_n^2\right)} = mgr + \beta,$$

where σ is the *Poisson's ratio* for the material of the beam.

Again, putting $1\Big/\left(Y^2_{n+p} - Y^2_n\right) = x$ and $m = y$, we have

$$\frac{4YI_g p\lambda}{\sigma.gr}.x = y + \frac{\beta}{rg},$$

which, again, gives straight line *II*, shown in Fig. 8.48(*b*), whose slope is

$4YI_g p\lambda/\sigma gr$, *i.e.*, $\frac{1}{\sigma}$ of the slope of *I*. Thus,

$$\textit{Poissson's ratio for the glass of the beam, } \sigma = \frac{R_1}{R_2} = \frac{\textit{slope of straight line I}}{\textit{slope of straight line II}}$$

Knowing the values of *Y* and σ for the material of the beam, it is just an easy step to obtain those of *n* and *K* for it from the familiar relations.

$$n + Y/2(1 + \sigma) \text{ and } K = Y/3\ (1 - 2\sigma).]$$

N.B. 1. Poisson's ration (σ) for the glass of the beam may also be obtained from relations I and II above. For,

$$\sigma = \frac{R_1}{R_2} = \frac{X_{(n+p)}^2 - X_n^2}{Y_{(n+p)}^2 - Y_n^2}.$$

2. Shape of the interference fringes formed. For purposes of our experiment, it is hardly necessary at all to consider the shape of the interference fringes formed between the cover glass and the glass beam. Just by way of additional information, it may, however, be discussed briefly as follows :

As we know, due to the anticlastic curvature of beam B, its upper surface does not remain in contact with the lower surface of the cover glass G [Fig. 8.47(*b*)]. Let the horizontal tangent plane to the surface of the beam touch it at the point C (the centre of the interference pattern) being, of course, parallel to the lower surface of G, and let OC be the perpendicular distance z between them. Then, the distance of any point on the surface of the beam from a point in the x–y plane of G will be

$z + \frac{x^2}{2R_1} - \frac{y^2}{2R_2}$, as will be clear from Fig. 8.49 (*a*) and (*b*), – *this distance being constant for any given pair of fringes. And, since z is a fixed distance, we have*

$$\frac{x^2}{2R_1} + \frac{y^2}{2R_2} = K, \text{ where } K \text{ is a constant.}$$

Or, $$\frac{x^2}{a^2} + \frac{y^2}{b^2} = 1 \quad \left[\begin{array}{l}\text{where } 2KR_1 = a^2 \\ \text{and } 2KR_2 = b^2\end{array}\right.$$

Fig. 8.49

This is the equation to a *hyperbola*; so that *the fringes are, in general, hyperbolic in shape*. If α be the angle that the asymptotes with the x-axis, we have

$$\sigma = a^2/b^2 = \cot^2 \alpha$$

whence *Poisson's ratio* (σ) for the material of the beam may be easily obtained.

If instead of the plane cover glass, we use a convex lens of a large focal length, we shall obtain *Newton's rings* as the interference pattern before the bending of the beam. These will increasingly assume an elliptical shape as the bending proceeds, so long as $R_2 > R$ radius of curvature of the face of the lens neurest to the glass beam. In case, $R_2 < R$ the fringes assume the hyperbolic form, and if either R_1 or R_2 be equal to R, the system of fringes is a straight one.

8.38. Determination of *n* and *Y* for the material of a spiral spring

(*i*) **Determination of *n*.** We have seen in § 6.17(5), how the vertical oscillations of a spiral spring under a load suspended from its lower free end enable us to determine the value of g at a place. The extension of the spring and its oscillations also provide what may be called a *statical* and a *dynamical* method respectively for the value of n for the material of the spring.

(*a*) **Statical method.** Referring back to § 6.17(5), we have the relation $mg = n\pi r^4 l/2LR^2$ where L, the *total length* of the spiral R, its *radius*, r, the *radius of the wire* and l, the *increase in its length* under a load mg.

So that, we have $n = 2LmgR^2/\pi r^4 . l$.

Or, better still, we obtain the increase in length (l) *for different loads* (mg) and plot a graph between mg and l, which comes out to be a straight line passing through the origin, the slope of which is given by $2LR^2/\pi nr^4$, whence n can be easily evaluated.

(*b*) **Dynamical method.** We have seen how relation (*iii*), page 290 gives

$$t_1^2 - t_2^2 = \frac{4\pi^2}{K}(m_1 - m_2), \quad \text{where } K = n\pi r^4/2LR^2,$$

So that, $$t_1^2 - t_2^2 = \frac{4\pi^2 (m_1 - m_2) \times 2LR^2}{n\pi r^4} = \frac{8\pi(m_1 - m_2)LR^2}{nr},$$

whence, $$n = \frac{8\pi (m_1 - m_2) LR^2}{(t_1^2 - t_2^2) . r^4}.$$

And the value of n can thus be casily obtained.

(*ii*) **Determination of Y.** If we apply a horizontal couple to the lower free end of a spiral and then release it, it executes an *angular oscillation about its own axis.*

Let τ be horizontal couple applied such that the curvature of the spiral changes from $1/R$ to $1/(R-\delta R)$. Then, we naturally have

$$\tau = Y I_g \left(\frac{1}{R-\delta R} - \frac{1}{R}\right) = Y . I_g . \delta R/R^2 \quad \text{very nearly.}$$

And since I_g for the wire $= \pi r^2/4$,we have $\tau = \dfrac{Y . \pi r^4}{4} . \dfrac{\delta R}{R^2}$.

Now, the radius of each coil of the spiral being reducer by δR, if its length is to remain constant, its two ends must turn relative to one another through an angle θ, where $\theta = 2\pi N . \delta R/R$ where N is number of coils in the spiral.

$\therefore$ substituting for $\delta R/R$ in the expression for τ above, we have

$$\tau = \frac{Y\pi r^4}{4} . \frac{\theta}{2\pi NR} = \frac{Yr^4}{8NR} . \theta$$

Since the total length of the spiral *i.e.*, $L = 2\pi RN$, we have $N = L/2\pi R$,

$$\text{whence } \tau = \frac{Y r^4 . \theta}{8R} . \frac{2\pi R}{L} = \frac{\pi Y r^4}{4L} . \theta ,$$

whence the value of Y for the material of the spiral may be obtained.

Or, we have from the above, *couple per unit per unit length of the spiral* $= C = \pi y r^4/4$.

Suppose we attach a small body to the lower end of the spring, the *time-period of the angular oscillation of the spiral* will be given by $T = 2\pi\sqrt{I/C}$, where $I = Mk^2$, the moment of inertia of the body about the suspension axis, *i.e.*, the axis of the spiral, **plus** the moment of inertia of the spiral about the same axis.

Since the *effective mass of the spiral* $= m_s/3$ (see worker example 11)

we have $I = Mk^2 + m_s R^2/3$, and , therefore ,

$$T = 2\pi\sqrt{\frac{Mk^2 + m_s R^2/3}{\pi Y r^4/4}} = 2\pi\sqrt{\frac{4(Mk^2 + m_s R^2/3)}{\pi r^2 . Y}}$$

whence Y may be easily evaluated.

8.39. Strain energy in a bent beam. We have seen before, (page 287), how when a small portion AB of a beam is bent into the form of a circular arc (Fig. 8.50), subtending an angle $\delta\theta$ at its centre of curvature, the strain produced in an element of it at a distance z from the neutral axis is given by z/R, where R is the radius of curvature of the neutral axis.

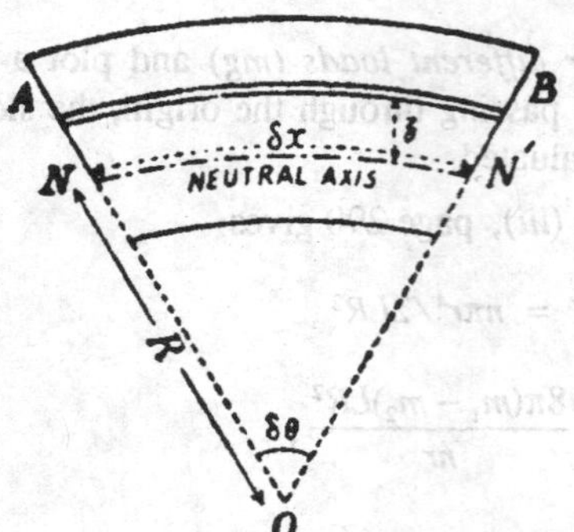

Fig. 8.50

Now, *energy per unit volume in any type of strain* $= \dfrac{1}{2}$ *stress* × *strain.*

So that, *energy associated with the eleme in question* $= \dfrac{1}{2}$ *stress* × strain × *volume of the element.*

Then, if the stress be F, the strain, e, the area of cross-section of the element normal to the plane of the diagram, δA, and the length of the neutral axis, δx, clearly,

energy in the element $= \dfrac{1}{2} F.e.\delta A.\delta x$.

Now, $Y = stress/strain = F/e$ and

$\therefore \quad F = Ye = Y.z/r.$

So that, *energy in the element* $= \frac{1}{2} . Y . \frac{z^2}{R^2} . \delta A . \delta x$.

Integrating this expression over the whole cross-section, we have

strain energy in the entire portion AB of the beam

$$= \frac{1}{2} . \delta x . \frac{Y}{R^2} \int z^2 . \delta A = \frac{1}{2} \delta x . \frac{Y I_g}{R^2},$$

since, as we know, $z^2 . dA = Ig$, the geometrical moment of inertia of the section considered.

Or, *strain energy in portion AB of the beam* $= \frac{1}{2} \left[\frac{Y I_g}{R} \right]^2 . \frac{1}{Y I_g} . \delta x$.

But $Y I_g / R = M$, *the bending moment of the section.*

Hence *strain energy in portion AB of the beam* $= \frac{1}{2} . \frac{M^2}{Y I_g} . \delta x$.

And $\therefore$ *strain energy E in the whole beam of length L* $= \frac{1}{2} \int_0^L \frac{M^2}{Y I_g} . \delta x$.

Or, $$E = \frac{1}{2 Y I_g} \int_0^L M^2 . dx$$

So that, substituting the relevant values of the bending moment, we can easily determine the *strain energy* of the beam in different cases.

Thus, for example, in the *case of a light cantilever, with a load W suspended from its free end*, we have $M = W(L - x)$, (see page 290), and, therefore,

strain energy, in this case $= \frac{1}{2 Y . I_g} \int_0^L W^2 (L - x)^2 . dx = \frac{1}{6 Y I_g} . W^2 L^3$.

8.40. Resilience of bent beams. The work done in deflecting a horizontal beam of whatever type, loaded with a weight W is equal to $\frac{1}{2} W \times$ (*deflection at the loaded point*). Since this work done measures the resilience of the beam, as explained in § 8.19, we have

$$\text{resilience of the beam} = \frac{1}{2} W . y$$

where y is the *deflection* of the loaded point.

Thus, in the case of a light beam, of *length L, breadth b,* and *depth d,* supported at the two ends and loaded with a weight W at the centre, we have $y = WL^3/48 Y I_g$, (page 294) and I_g in this case, is given by $bd^3/12$, (page 288). So that, we have

$$\text{resilience of such a beam} = \frac{1}{2} W . \frac{WL^3}{48 Y I_g} = \frac{W^2 L^3}{96 Y} . \frac{12}{bd^3} = \frac{W^2 L^3}{8 Y b d^3}$$

8.41. Columns, Pillars and Struts. A long beam of an isotropic material used for supporting loads is called a *column*, a *pillar*, or a *strut*. Now, whereas a column or a pillar must always be *vertical* and generally fixed rigidly at its ends, a strut may be *vertical*, *horizontal* or *inclined* and may either have both its ends fixed rigidly or both connected to the surrounding structure through flexible joints, or it may have one end rigidly fixed and the other connected to a joint. The theory underlying the two is, however, the same, the commonest case being that in which the load applied is a compressive one, *i.e.*, acts at one end of the column or strut, along its axis, tending to compress or shorten it lengthwise, though, in some cases, there may as well be a lateral load, in addition. Let us take the case of a column or a pillar first.

8.42. Critical load for a long column. Consider a *long* and *straight* strip AB, Fig.8.51(a), of wood or metal, arranged in a vertical position, representing a column, with both its ends rounded and fitted into metal sockets, as shown magnified in Fig. 8.51(b), so as to allow it freedom to bend all along its length. Let a load be applied to it at the top in the form of a metal cylinder, containing lead shot or mercury, so that its magnitude may be varied at will, with the cylinder moving between two parallel guides GG to ensure its vertical descent.

Now, first, with the load insufficient to bend the strip or the column, we apply a lateral force f_1 at its midpoint O, as shown, to make it bend a little and we find that, on removing the lateral

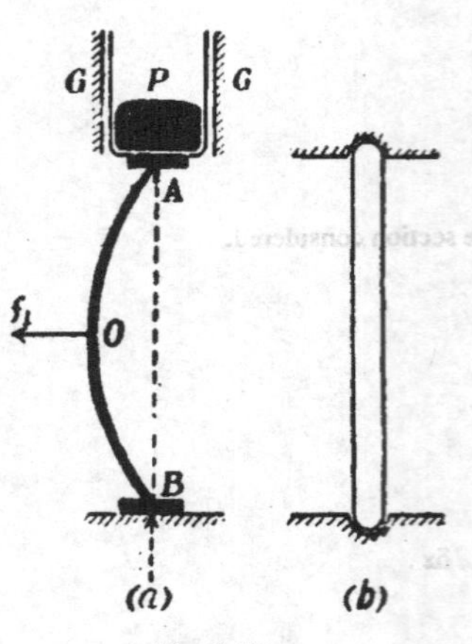

Fig. 8.51

force, the column straightens itself out. We increase the load at the top, apply the lateral force, as before, and then removed it. Perhaps, the column again bends and then straightens itself. In this manner, we go on repeating the experiment with successively increasing loads at the top until we find that when the lateral force is applied and then removed, the column remains bent. At this stage, we find that whatever the deflection we produce in the column by the lateral force, the column continues to retain the *same* on the removal of this force, provided, of course, that the elastic limit has not been exceeded. *This load which just keeps the column bent, but does not bend it further, is called the* **critical** (or quite often, also the **buckling** or **cripping) load** *for it.*

If we increase the load beyond the critical value and give a slight bend to the column, as before, we find that the load now increases the bending further and the column either acquires a permanent set or collapses due to buckling *i.e.*, by bending or bulging out.

Let us see how we may account for this critical load. Let us, therefore, consider the equilibrium of the column AB, under a vertical load P_1 and a lateral force f_l at its mid-point O, with the deflection of the column equal to y_1, as indicated in Fig. 8.52(a).

Since, for equilibrium, the lateral force f_1 must be balanced by two horizontal forces, each equal to $f_1/2$, acting at A and B, in the opposite direction to that of f_l, we have *total bending moment about O*

$$= P_1y_1 + (f_1/2)\,L/2 = P_1y_1 + f_1\,L/4.,$$

so that, for equilibrium, we have

$$P_1y_1 + f_1L/4 = YI_g/R_1,$$

where I_g is the geometrical moment of inertia of the cross-section of the column and R_1, its radius of curvature at the mid-point O, *i.e.*, at $L/2$.

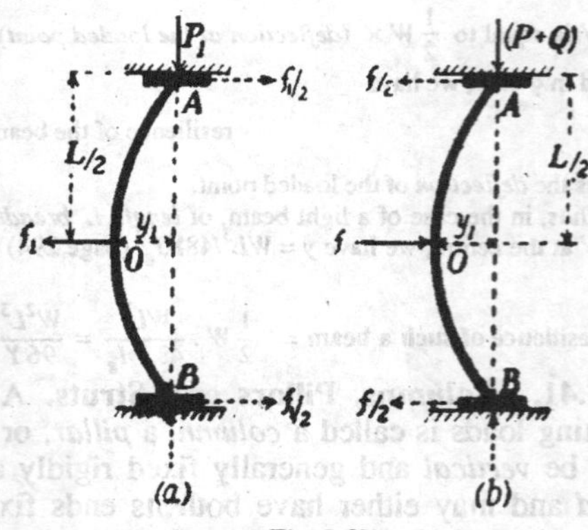

Fig. 8.52

If now we decrease the lateral force f_1 to *zero* and increase load P_1 to P (the *critical load*) so that the deflection of the column remains the same y_1, clearly, the condition for equilibrium demands that

$$P.y_1 = YI_g/R_1, \text{ whence, } P = YI_g/R_1y_1. \qquad [\because f_1 \text{ is now } Zero$$

Again, if the column be in equilibrium when subjected to a vertical load P_2 and a lateral force f_2, with its deflection at O now y_2 and we reduce f_2 to *zero* and increase P_2, to P^1, with theii deflection remaining unaltered at y_2, we have, proceeding as before, $P_1 = YI_g/R^2$ where R_2 is now the radius of curvature at O.

Since the deflection is proportional to curvature, we have $1/R_1y_1 = 1/R_2\,y_2$. And, therefore, $P' = P$, the *critical load*, thus clearly showing that the column will remain in equilibrium under the same critical load P for any value of the deflection we choose to give it within the elastic limit as we have seen already in the experiment discussed above.

If, however, we increase the vertical load to a value beyond its critical value P, say, to $(P + Q)$, as shown in Fig. 8.52 (*b*), the bending moment will clearly increase to $(P + Q).y_1$ or to $Py_1 + Qy_1$. And, since the moment of resistance to bending, *viz.*, YI_g / R_1 balances only the portion Py_1 of it (as we have just seen above), the portion Qy_1 remains unbalanced, resulting in an increase in the bending or deflection of the beam beyond y_1. In order to keep the deflection at the same value y_1, therefore, we shall have to apply a force f, say, at the mid-point O of the column, *in the opposite direction this time*, so as to balance the portion Qy_1 of the bending moment. Since f is supposed to be balanced by two equal forces $f/2$ and $f/2$ acting at A and B, as shown, its moment about $O = (F/2).(L/2) = f.L/4$. So that, to prevent the column from bending further (beyond y_1) we must have $f . L/4 = Q.y_1$.

Now, within the elastic limit, the moment of resistance to bending is proportional to the deflection (y) of the column. But, once the elastic limit is exceeded, the column acquires a permanent set, though it is also possible that, due to the moment of resistance due to bending now increasing more rapidly (as it always does beyond the elastic limit), *the column may acquire a new position of equilibrium under the additional load Q*. But, if this does not happen, the column will continue to bend further and further and finally collapse.

8.43. Euler's Theory of Lond Columns

(*i*) **When the two ends of the column are rounded or hinged.** Let AB represent a long and initially straight column of an isotropic material, of lenght L and of a *uniform cross-section* and *uniform elasticity*, with rounded or hinged ends *so as to be free to bend throughout its length.* Further, let the critical load P *act axially* upon it, *i.e.*, in a line with its axis in its straight unloaded position, and let it be given a slight bend by the application of a lateral force for an instant, (Fig. 8.53).

Now, consider a point C in the column, at a distance x from its mid-point O. If the deflection here be y, clearly, the *bending moment here due to* $P = P.y$. But, if the radius of curvature of the bend at C be R, the *moment of the resistance to bending* there is $Y.I_g/R$. And clearly, therefore,

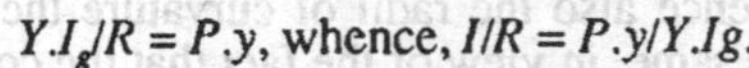

$$Y.I_g/R = P.y, \text{ whence, } I/R = P.y/Y.Ig.$$

But, as we know, $I/R = -d^2y/dx^2$, (the —*ve* sign being given to make R *positive*, for dy/dx decreases as y increases). We thus have the differential equation $P.y/YI_g = -d^2y/dx^2$.

The solution of this equation gives $y = A \sin \lambda x + B \cos \lambda x$. ...I

Fig. 8.53

where $\lambda = \sqrt{P/Y.I_g}$ and A and B are constants to be determined.

Differentiating this, we have $dy/dx = A\lambda \cos \lambda x - B\lambda \sin \lambda x$.

(*a*) Now, when $x = 0$, $dy/dx = 0$ and obviously, $\sin \lambda x = 0$; so that, $A = 0$. From equation *I*, therefore, we have $y = B \cos \lambda x$.

But when $x = 0$, we have $y = y_1$, $x = 0$, and $\therefore B \cos \lambda x = 1$ and $\sin \lambda x = 0$. So that, from equation *I*, we have $y_1 = B$.

(*b*) When $x = L/2$, (*i.e.*, at A), $y = 0$. And $\therefore B \cos \lambda x/2 = 0$ [From (*a*) above

This means that either $B = 0$ or $\cos \lambda L / 2 = 0$. But, as we have seen [in (*a*) above], $B = y_1$. It follows, therefore, that

$$\cos \lambda L/2 = 0, \text{ or, that } \lambda L/2 = \pi/2, 3\pi/2, 5\pi/2 \text{ etc.}$$

Ruling out the other values except the first, since they indicate points of inflection in the column, we have

$$\sqrt{P/Y.I_g}.L/2 = \pi/2, \qquad [\text{for, } \lambda = \sqrt{P/Y.I_g}.$$

Or, $\sqrt{P/YI_g}.L = \pi$. Or, $P/Y.I_g = \pi^2/L^2$.

whence, $$P = \frac{\pi^2.YI_g}{L^2} = \frac{9.87 \times YI_g}{L^2}.$$

This is, therefore, the value of the *critical load*, or the load which can just keep the column bent at the initial curvature given to it. And addition to this makes the column collapse.

It will be clear from this expression for the critical load that, for the same values of Y and I_g, *the smaller the length of the column, the greater the critical load for it.*

(*ii*) **When the two ends of the column are fixed.** Let the ends A and B of the column be now fixed, as shown in Fig. 8.54 (*a*), so that when it gets bent or deflected, the tangents to it at points A, O and B are all vertical, with the line of action of resultant load now no longer passing through the centres of its end-points. It passes, instead, between the initial unbent position AB of the column, and its mid-point O in the bent position, cutting the bent column in points C and D. At these points, therefore, there is no bending moment, so that they are points of opposite flexure.

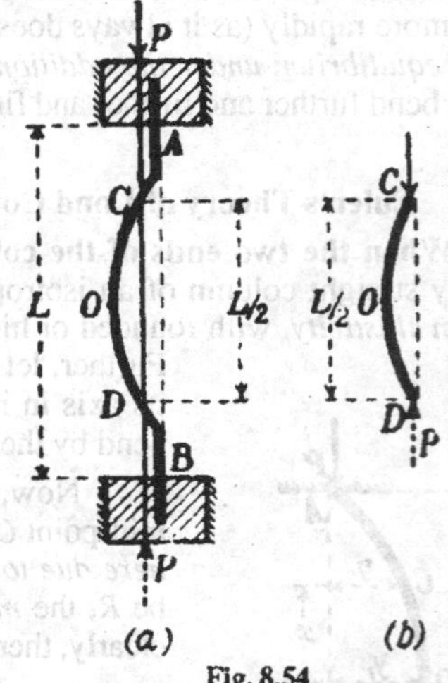

Fig. 8.54

Now, considering the portions CA and CO of the bent column, we observe that the deflections at certain points in the two curves (as measured from the vertical line through C) are equal. At all such points in the two curves, therefore, the bending moments must be equal, and hence also the radii of curvature there must be the same, in view of the fact that the column is of a uniform cross-section. Further, the two curves (CA and CO) have clearly the same slope at C and also at A and O, the tangents at all these points being vertical. Obviously, therefore, the two curves are *equal and similar*. The same is also true of the curves DB and DO. The points C, O and D thus divide the whole column into four equal parts and the length of the portion COD of the column is, therefore, equal to half its total length, *i.e.*, $COD = L/2$.

Clearly, this portion COD of the column, the whole of which is bent, behaves exactly as the cases considered above, i.e., like a column of length $L/2$, with its ends rounded or hinged and carrying an axial load P at C, as shwon in Fig. 8.54(*b*). So that, proceeding as in case(*i*) above, we have

$$P = \frac{\pi^2 Y I_g}{(L/2)^2} = \frac{4\pi^2 Y I_g}{L^2},$$

i.e., the critical load (P) for the column, in this case, is four times that in case (*i*).

thus, a column with its ends fixed, has four times the strength (to resist thrust) that it will have with its ends rounded or hinged. Or, putting it differently, a column, with its ends fixed, can support, without bending, the same load as one half of its length, with its ends rounded or hinged, would do.

(iii) When one end of the column is fixed and the other loaded. This is an easy deduction from case (*i*) above. For, suppose we have a column *AB*, with rounded or hinged ends, and of length L', as shown in Fig. 8.55(*a*), with *P* as the critical load on it. Then, the tangent to it at its mid-point *O* is vertical. If, therefore, we clamp it tightly at *O*, without disturbing the direction of the tangent at that point, the lower half *OB* of the column might as well be removed, without in any way affecting the upper half *OA*. So that, the upper half then behaves as an independent column, of length $L = L'/2$, fixed at its lower end and loaded at the top, as indicated in Fig. 8.55(*b*).

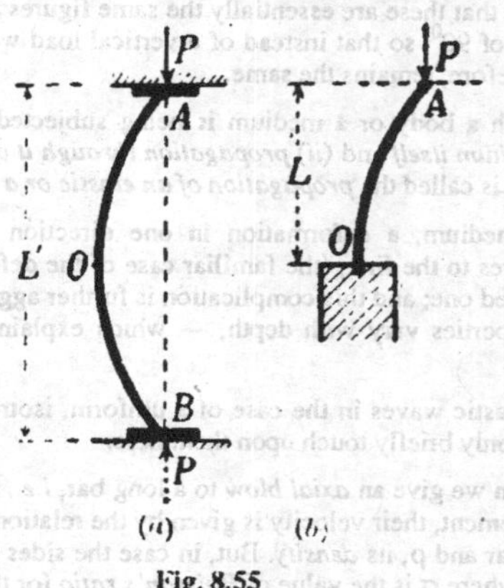

Fig. 8.55

All that we have to do to calculate the critical load, in this case, is to consider the column, of length *L* and fixed at one end, as equivalent to a column of length $L' = 2L$, with both its ends rounded or hinged. Therefore, proceeding in the same manner as in case (*i*), we have the critical load given by

$$P = \frac{\pi^2 Y I_g}{(L')^2} = \frac{\pi^2 Y I_g}{(2L)^2} = \frac{\pi^2 Y I_g}{4L^2},$$

which is clearly one-fourth of the critical load for a column of length *L*, with its ends rounded or hinged.

Thus, we find that *a column, hinged at one end and loaded at the other, has only one-fourth the strength of the same column when hinged at both ends.*

(iv) Maximum height of a column supported under its own weight — Height of a tree. We have just seen under (*iii*) above that the maximum load supported by a column or a pillar, fixed at one end and loaded at the other is given by $P = \pi^2 Y I_g / 4L^2$.

Considering the case of a long column and imagining its cross-section to be uniform all through, we can imagine its weight *W* to be acting at its *c.g., i.e.*, at its mid-point. Thus, for all practical purposes the problem is reduced to this that we have a column of length $L/2$, fixed at the lower end and loaded at the other with a load $P = W$. So that, we have

$$W = \pi^2 Y I_g \Big/ 4\left(\frac{L}{2}\right)^2 = \pi^2 Y I_g \Big/ L^2,$$

whence the length *L* of the column or the pillar may be calculated.

Now, although a tree usually tapers at its upper end, we may, for the sake of simplicity, imagine it too to have a uniform cross section. Then, its maximum height (that can support its own weight) is also given by the above expression for *W*.

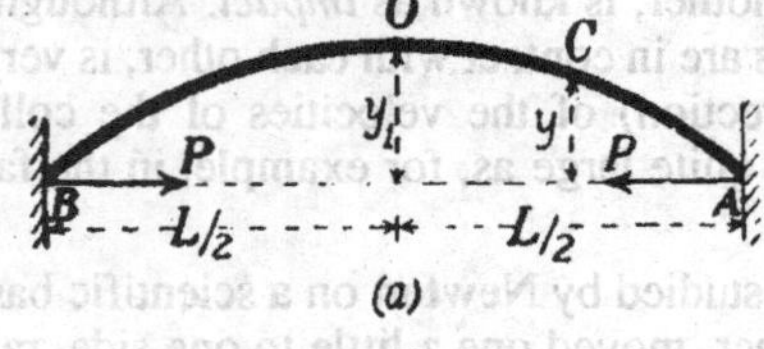

(a)

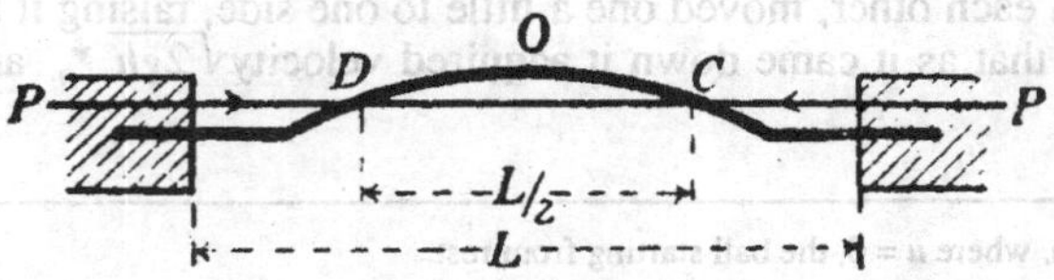

Fig. 8.56

N.B. Exactly a similar treatment to that in cases (*i*) and (*ii*) applies to the corresponding cases of a **strut,** arranged horizontally, so long as the strut is loaded axially, or along its axis, like the columns in cases (*i*) and (*ii*). Thus, if the two ends of the strut be rounded or hinged, so that the whole of it can bend, we can represent its behaviour as in Fig. 8.56 (*a*) and, when its two ends are fixed, as in Fig. 8.56(*b*). It will be noted that these are essentially the same figures as in cases (i) and (*ii*), respectively, but are rotated, as it were, through an angle of 90^0, so that instead of a vertical load we now have a horizontal load. The method of calculation for the critical load, therefore, remains the same.

8.44. Elastic Waves. When a system of stresses, to which a body or a medium is being subjected, is suddenly altered, we have (*i*) *a corresponding motion of the body or the medium itself* and (*ii*) *propagation through it of the changes in stress,* —the two occurring simultaneously and constituting what is called the *propagation of an elastic or a stress wave.*

Now, as we know, even in the case of an isotropic medium, a deformation in one direction is invariably accompanied by deformations in two other directions, at right angles to the first, (the familiar case of the deformation of a cube), so that the theory of elastic waves is really quite a complicated one; and this complication is further aggravated in the case of bodies like the earth, for example, where the elastic properties vary with depth, — which explains at once the complicated pattern of the seismic waves (§ 7.22).

In general, however, we have three types of elementary elastic waves in the case of a uniform, isotropic medium, *viz.*, (*i*) *compressional* (*ii*) *shear* and (*iii*) *flexural waves.* We shall only briefly touch upon them here.

(*i*) **Compressional Waves.** These waves are produced when we give an *axial blow* to a long bar, *i.e.*, strike it along its axis, and, assuming the sides of the bar to have freedom of movement, their velocity is given by the relation $v = \sqrt{Y/\rho}$, *where Y* is the value of *Young's modulus* for the material of the bar and ρ, its *density*. But, in case the sides of the bar too are fixed, *Y* is to be replaced by $Y(1 - \sigma)/(1 + \sigma)(1 - 2\sigma)$,, where σ is the value of *Poisson's ratio* for the material of *the bar*. This expression takes many forms, the simplest among them being $K+4n/3$, where *K* and *n* are the coefficients of bulk and rigidity modulii for the material of the bar. Thus, we have different types of compressional waves, all of which, however, have the common features that (*i*) *the vibrations occur along the direction of propagation of the wave, i.e., the wave is of the longitudinal type, and* (*ii*) *the velocity of the wave is given by*

$$v = \frac{\textit{appropriate modulus of elasticity}}{\textit{density}}$$

Thus, in the case of a liquid or a gas, $n = 0$ and we, therefore, have the *wave velocity* $v = \sqrt{K/\rho}$, which is the well known expression for the velocity of sound in a liquid or a gas-filled tube.

(*ii*) **Shear Waves.** These waves are obviously due to a sudden change of shear produced in a body,—by far the simplest case being that of a torsion applied to a bar of circular cross-section. The velocity of this type of wave is given by $v = \sqrt{n/\rho}$, where *n* is the *modulus of rigidity* of the material of the bar. It will be readily seen that it is an expression similar to that for a compressional wave, with only the modulus of elasticity now appropriately being the modulus of rigidity. They differ, however, from compressional waves in the very important respect of their nature, with the vibrations now occurring at right angles to the direction of propagation of the waves instead of along it. In other words, these are *transverse type* of waves and not longitudinal.

(*iii*) **Flexural Waves.** These are also transverse waves and occur when there is a sudden alteration in the loading of a beam. The expression of their velocity is a complicated one, but *it is found to depend upon the frequency of vibrations, being greater for a higher frequency, and vice versa.* They would seem to resemble light waves travelling through a material medium, the value of the *refractive index* for which, as we know, changes with the frequency of the light wave, and which are also of the transverse type. It can easily be shown that in the case of a beam supported, with its two ends free, *the frequencies with which it can vibrate are proportional to the square of the natural numbers,* so that the various frequencies are integral multiples of the lowest or the fundamental frequency. Use is, therefore, made of such vibrating beams in musical instruments. No such simple relation holds, however, if the beam be fixed at one or both of its ends, and such beams, therefore, find little or no use in musical instruments.

All these elastic waves play their own part in the theory of earthquakes, as we have already seen in § 7.22. But perhaps their particular importance lies in the fact that they have a direct bearing on the proper designing of structures.

8.45. Impact—Coeffient of Restitution. The colliding of two bodies against each other, or the impinging of one body on another, is known as *impact.* Although the duration of impact, *i.e.*, the time for which the two bodies are in contact with each other, is very small, it results in a change in the magnitude (and even direction) of the velocities of the colliding bodies, the change in momentum thus produced being quite large as, for example, in the familiar case of a cricket ball, hit by a bat.

The phenomenon was first studied by **Newton** on a scientific basis, who suspended two balls at a small distance from each other, moved one a little to one side, raising it through a distance *h*, and then released it, so that as it came down it acquired velocity $\sqrt{2gh}$ *, and struck against the

---

* Beacause $v^2 - u^2 = 2gh$, where $u = 0$, the ball starting from rest.

other ball. He again determined the velocities of the balls from the heights to which they rose after the impact.

From his experiments, he came to the conclusion, now usually referred to as **Newton's law of impact,** that *when the collision is 'direct' or 'head on', i.e., when the relative velocities of two colliding bodies, at the instant of collision, lie along the common normal at the point of impact, the relative velocity after the impact bears a constant ratio to that before the impact, but is reversed in direction**. This constant is called the **coefficient of restitution or resilience,** and is usually denoted by the letter e.

The value of e is different for different materials, *depending only upon their nature* and not upon their masses,—*its maximum and minimum values being respectively* 1, *in the case of perfectly elastic bodies**, and zero, in the case of perfectly plastic ones, with intermediate values between 0 and 1 for the rest.*

Thus, if u_1 and u_2 be the velocities of two bodies before the impact and v_1 and v_2, after the impact, we have

$$\frac{(v_1 - v_2)}{(u_1 - u_2)} = e$$

[neglecting the –ve sign of the relative velocity after the impact.]

Or, $$(v_1 - v_2) = e(u_1 - u_2), \quad ...(i)$$

where $(u_1 - u_2)$ and $(v_1 - v_2)$ are their relative velocities, before and after the impact.

It will thus be seen that if $e = 1$, *i.e.*, **if the bodies be perfectly elastic,** $(v_1 - v_2) = (u_1 - u_2)$, *i.e., the relative velocities of the bodies are the same before and after the impact,* (suffering only a reversal of direction, in the latter case).

But if $e = 0$, *i.e.*, **if the bodies be perfectly plastic,** $(v_1 - v_2) = 0$, or $v_1 = v_2$, *i.e., the two bodies move with the same common velocity, after the impact.*

In practice, we come across bodies which are neither perfectly elastic nor perfectly plastic but lie somewhere in between the two. So that, what actually happens is that when one body, of mass m_1, moving with a velocity u_1, collides against another body, of mass m_2, moving with velocity u_2, (where $u_1 > u_2$), the surface between them gets compressed and when this compression or pressure reaches its maximum value, their relative velocity becomes zero. Therefore, the elastic stress between them makes them recede from each other, the compression is released, and the two bodies move away with different velocities, say v_1 and v_2.

Clearly, then, in accordance with the *law of conservation of momentum,* we have

sum total of momenta after the impact = sum total of momenta before the impact.

Now, sum total of momenta after the impact $= m_1v_1 + m_2v_2$,

and sum total of momenta before the impact $= m_1u_1 + m_2u_2$.

So that, $$m_1v_1 + m_2v_2 = m_1u_1 + m_2u_2. \quad ...(ii)$$

From relations (*i*) and (*ii*), we can easily calculate the values of v_1 and v_2. For, multiplying relation (*i*) by m_2, we have

$$m_2v_1 - m_2v_2 = m_2e(u_1 - u_2). \quad ...(iii)$$

And, adding relations (*ii*) and (*iii*), we have

$$m_1v_1 + m_2v_1 + m_2v_2 - m_2v_2 = m_1u_1 + m_2u_2 + m_2e.(u_1 - u_2).$$

---

* In case the collision is not direct, but an oblique one, the same will applty to the components of the relative velocities along the common normal at the point of contact.

** It will thus be readily seen that the nearer the value of e to 1, the more nearly do the two colliding bodies form a perfectly elastic pair. In other words, it is a ready measure of the nearness or its approach to the state of perfect elasticity.

Or, $v_1(m_1 + m_2) = m_1u_1 + m_2u_2 + m_2 . e(u_1 - u_2)$,

Whence, $$v_1 = \frac{m_1u_1 + m_2u_2 + m_2 .e(u_1 - u_2)}{(m_1 + m_2)} \quad ...(iv)$$

Similarly, multiplying relation (*i*) by m_1, and subtracting from relation (*ii*), we have

$$v_2 = \frac{m_1u_1 + m_2u_2 - m_1 . e(u_1 - u_2)}{(m_1 + m_2)}. \quad ...(v)$$

Further, it can also be shown that

$$e = \frac{impulse\ during\ restitution}{impulse\ during\ compression}$$

The value of *e*, however, is found to diminish with increase in the velocities of the approaching bodies, and *vice versa*, and it has been shown by **Sir C.V. Raman**, that its value is very nearly equal to 1, if the collision be very weak.

8.46. Loss of Kinetic Energy on Impact. It can be shown from the above relations that

$$\frac{1}{2}m_1v_1^{\ 2} + \frac{1}{2}m_2v_2^{\ 2} = \frac{1}{2}m_1u_1^{\ 2} + \frac{1}{2}m_2u_2^{\ 2} - \frac{1}{2}.\frac{(1 - e^2) . m_1m_2 .(u_1 - u_2)^2}{m_1 + m_2}$$

where $\frac{1}{2}m_1v_1^{\ 2} + \frac{1}{2}m_2v_2^{\ 2}$ is the total kinetic energy E_2 of the colliding bodies, *after* the impact, and $\frac{1}{2}m_1u_1^{\ 2} + \frac{1}{2}m_2u_2^{\ 2}$, their total kinetic energy E_1, *before* the impact.

Thus, $$E_2 = E_1 - \frac{1}{2}.\frac{(1 - e^2) . m_1 . m_2 . (u_1 - u_2)^2}{m_1 + m_2}.$$

Or, $$E_1 - E_2 = \frac{1}{2}.\frac{(1 - e^2) . (m_1 . m_2) . (u_1 - u_2)^2}{m_1 + m_2}.$$

i.e., loss of energy on impact $= \frac{1}{2}.\frac{(1 - e^2) . (m_1 . m_2) . (u_1 - u_2)^2}{m_1 + m_2}$

an expression, with always a *positive* value, showing that *there is always a loss of energy on impact between two bodies.*

Now, the following special cases arise :

(*i*) *When e* = 1, *i.e., when the colliding bodies are perfectly elastic.* In this case, $(E_1 - E_2) = 0$, *i.e., there is no loss of energy on impact of elastic bodies.*

(*ii*) When *e* = 0, *i.e., when the colliding bodies are perfectly plastic.* Here, $(E_1 - E_2)$ has the maximum value, *viz.,*

$$\frac{1}{2}.\frac{(m_1m_2) .(u_1 - u_2)^2}{(m_1 + m_2)},$$

i.e., there is maximum loss of energy on impact of plastic bodies.

(*iii*) When $u_1 = u_2$, *i.e., when the bodies have the same velocity,* (*in magnitude as well as direction*). In this case, the relative velocity of one body, with respect to the other, is zero so that *no impact takes place at all* between the two bodies, and, therefore, $(E_1 - E_2) = 0$, or again *there is no loss of energy.*

The question now arises as to what happens to this loss of energy on impact ? Until very recently, it was supposed that the energy lost during impact was converted into (*i*) sound, (*ii*) heat, or (*iii*) vibration or rotation of the colliding bodies.

Sir *C.V. Raman's* experiments have shown, however, that the production of sound is in no way related to the energy of impact, being solely due to the impulse set up in the air during the reversal of the motion of the colliding bodies, after impact.

The change in temperature too is almost always very small and hence it appears that an appreciable portion of this energy lost during impact is used up in bringing about a redistribution of the molecules in the surface layers of the colliding bodies, Indeed, it has been shown by **Hertz** that impact produces a definite flattening of the point of contact of the colliding bodies, with a finite common area between them, each body being compressed in its neighbourhood,—the compression increasing at first to a maximum, (which is proportional to the *two-fifth* power of the velocity), then diminishing and finally vanishing altogether, when the bodies get separated from each other.

Prof. Deodhar has also verified this molecular displacement in the surface layers of the impinging bodies, by making different bodies (*'similar and dissimilar'*) impinge upon each other, with *'extremely low velocities'* and measuring their velocities before and after the impact. From his experiments, he has come to the following conclusions :

(*i*) With great velocities of the colliding or impinging bodies, the changes in their surface are *'vivid'*, and a greater portion of their energy of impact is used up in producing these deformations.

(*ii*) With very small velocities of the colliding bodies, on the other hand, the value of e increases, in the limit, to unity–the increase of e with the *'minimal velocities'* being independent of their material.

(*iii*) The rate of change of e is quite independent of the medium in which the impact takes place.

(*iv*) The duration of impact is observed to be greater in water than in air, *i.e.*, it depends upon the density of the medium.

(*v*) A distinct change in the structure of the impinging bodies is noticeable under the microscope, though no trace of it is visible to the naked eye.

He estimates from this that energy, of the order of 1000 *kgm. per sq cm* is used up *'in displacing the molecular aggregates'*. Further, bodies, when strained, take time to recover their original condition and a rapid rise and fall in the stress may result in the dissipation of some energy, provided the elastic limit of the bodies, for gradually varying forces, is not exceeded*.

8.47. Relative mass of colliding bodies. If, in the above example of two colliding bodies or balls, the second one be at rest, so as to have no kinetic energy, we have

total kinetic energy before impact $= \frac{1}{2} m_1 . u_1^{\ 2}$.

And, *loss of energy during impact* $= \frac{1}{2} \cdot \frac{(1 - e^2) . m_1 . m_2 . u_1^{\ 2}}{(m_1 + m_2)}$.

$$\text{So that, } \frac{\textit{loss of energy}}{\textit{total energy}} = \frac{1}{2} \cdot \frac{(1 - e^2).m_1.m_2.u_1^{\ 2}}{(m_1 + m_2)} \bigg/ \frac{1}{2} m_1 u_1^{\ 2}.$$

$$= \frac{(1 - e^2).m_2}{m_1 + m_2} = \frac{(1 - e^2)}{(m_1 + m_2)/m_2} = \frac{(1 - e^2)}{\frac{m_1}{m_2} + 1}$$

$$\text{Or, } \qquad \textit{loss of energy} = \left[\frac{(1 - e)^2}{\frac{m_1}{m_2} + 1} \right] \times \textit{total energy}.$$

Clearly, therefore, the loss of energy will be small if m_1/m_2 be large and *vice versa*.

Thus, *in order to minimize loss of energy, the ratio m_1/m_2 must be made large i.e., the mass of the striking body must be much greater than that of the body struck.* Hence it is that a

* This should not, however, be understood to mean that an exceeding or *'overlapping'* of the elastic limit is necessary for a loss of energy to occur.

slow-moving heavy hammer is more suitable for imparting momentum to a body than a quick moving lighter one, even though the two may possess the same momentum.

On the other hand, *if the loss of energy is to be converted into useful work, the ratio m_1/m_2 must be small, i.e., m_2 must be much greater than m_1*. That is why, while forging instruments etc., we must have heavy anvils underneath them.

8.48. Experimental determination of Coefficient of Restitution. We know that, as a direct consequence of *Newton's law of impact*, the velocity of the first rebound of a body, impinging upon a plane surface of the same material as its own, is given by the relation, *velocity of rebound(v) = e × velocity of impact (u)*,

i.e., $v = e.u$, whence, $e = v/u$,

where *e* is the *coefficient of restitution* of that material.

A simple laboratory method to determine the coefficient of restitution of a material would thus be to drop a small sphere* of the material from a known height on to a plane surface of the same material and to note the height of its (first) rebound, whence the values of *u*, *v* and, therefore, that of *e* may be easily calculated.

Thus, if h_1 be the height from which the sphere is initially dropped and h_2, its height of first rebound after impact with the plane surface, we have

$$u = \sqrt{2gh_1} \quad \text{and} \quad v = \sqrt{2gh_2}\,.$$

And, therefor, $e = v/u = \sqrt{2gh_2} \,/\, \sqrt{2gh_1} = \sqrt{h_2/h_1}\,.$

Alternatively, the value of *e* could also be determined by noting the *total time* taken by the sphere from the moment of its release at height h_1 until it finally comes to rest after successive rebounds from the plane surface, as will be clear from the following :

The sphere being released at height h_1 from the plane surface, its velocity of impact with it is $u = \sqrt{2g\,h_1}$ and hence the *velocity of its first rebound = eu*. This is then also the *velocity with which it falls back for its next impact* with the surface and hence the *velocity of its second rebound* is $e.eu = e^2.u = e^2\sqrt{2g\,h_1}$. Similarly, the velocities of the other successive rebounds are $e^3 . u\sqrt{2g\,h_1}$, $e^4 . \sqrt{2g\,h_1}$ and so on, *i.e.*, they go on decreasing according to the terms of a geometrical progression, (*e* being less than 1).

The time taken for the rebounds of the sphere to cease, *i.e.*, for it to come to rest, after the first impact will, obviously, be equal to twice the sum of the times taken by the successive velocities of rebound (eu, e^2u, e^3u etc.) to be reduced to *zero* each.**

Now, from the kinematic relation $v = u + at$, we have, for the first rebound, *v* (final velocity) = 0, initial velocity = $e.u.$, $a = -g$ and t = say, t_1. So that,

$$0 = eu - gt_1. \quad \text{Or,} \quad t_1 = eu/g = e\sqrt{2gh_1}/g = e\sqrt{2h_1/g}\,.$$

Similarly, the times for the second, third etc., reboun-velocities to be reduced to zero are respectively $t_2 = e^2u/g = e^2\sqrt{2h_1/g}$, $t_3 = e^3u/g = e^3.\sqrt{2h_1/g}$ etc.

The rebounds, following the first impact, thus all cease in time

$$t_1 + t_2 + t_3 + \ldots\ldots\ldots\ldots = e.\sqrt{2h_1/g} + e^2.\sqrt{2h_1/g} + e^3.\sqrt{2h_1/g} + \ldots\ldots\ldots$$

If, therefore, the time taken by the sphere to fall from height h_1 to make the first impact be, say, $t = \sqrt{2h_1/g}$, we have *total time taken by the sphere from the moment of its release until the ceasing of its rebounds given by*

---

* Besides other reasons of convenience, the chief advantage ofa sphere is that its area of contact with the plane surface is the least and hence its rebound much quicker, almost instantaneous.

** Because between two successive rebounds the sphere covers the same vertical distance twice, —first upward until its velocity of the previous rebound is reduced to zero and then the same distance downwards to make the next impact with the same velocity.

$$T = t + 2(t_1 + t_2 + t_3 + ...)$$

$$= (\sqrt{2h_1/g}) + 2(e.\sqrt{2h/g} + e^2.\sqrt{2h_1/g} + e^3.\sqrt{2h_1/g} + ...)$$

$$= \sqrt{\frac{2h_1}{g}}\ [1 + 2(e + e^2 + e^3 + ...)].$$

Since $e < 1$, the infinite series $(e, e^2, e^3,..)$ is a convergent one, with its sum equal to $e/(1-e)$

So that, $$T = \sqrt{\frac{2h_1}{g}}\cdot\left(1 + \frac{2e}{1-e}\right) = \sqrt{\frac{2h_1}{g}}\cdot\left(\frac{1+e}{1-e}\right)$$

Or, $\dfrac{1+e}{1-e} = \dfrac{T}{\sqrt{2h_1/g}} = \sqrt{\dfrac{gT^2}{2h_1}}$, whence, $e = \dfrac{\sqrt{gT^2/2h_1} - 1}{\sqrt{gT^2/2h_1} + 1}$.

Thus, knowing the *total time T* and the *initial height h* of the sphere from the plane surface, the coefficient of restitution *e* for the given material can be easily obtained.

Worked Examples

1. Show that (a) a small and uniform strain v is equivalent to three linear strains $v/3$, in any three perpendicular directions :

(b) the bulk modulus for a gas (i) at constant temperature (i.e., under isothermal conditions) is equal to its pressure and (ii) when the temperature is not constant, (i.e., when the conditions are adiabatic) it is equal to γ times its pressure, where γ is the ratio C_p / C_v for it.

(a) Imagine a *unit cube* to be compressed equally and uniformly on all sides, so that the length of each edge decreases by a length l and its volume by a small amounv. Then, clearly, *volume strain in the cube* $= v/1 = v$, and *linear strain along each edge of the cube* $= l/1 = l$.

Since length of each edge of the cube now becomes $(1 - l)$, the *new volume of the cube* becomes $(1-l)^3$. Therefore, *decrease in volume of the cube, i.e.,*

$$v = 1 - (1-i)^3 = 1 - 1 + 3l - 3l^2 + l^3 = 3l,\ i.e.,\ l = v/3,$$

neglecting l^2 and l^3, the value of l being small.

Thus, a small uniform *volume strain* v is equal to three *linear strains*, each equal to $v/3$, in three perpendicular directions (along the three edges of the cube).

(b) Let P be the pressure and V, the volume of a gas, and let it be compressed by increasing the pressure to $(P + dP)$, so that the volume is reduced by dV and becomes $(V - dV)$.

Then, clearly, *stress = force per unit area = pressure applied* $= dP$, and *volume strain = change in volume/original volume* $= -dV/V$. [dV is a *decrease* in volume.

$\therefore$ *bulk modulus* for the gas, *i.e.,* $K = -\dfrac{dP}{dV/V} = -\dfrac{dP}{dV}.V$ ...(*i*)

(*i*) If the gas is compressed *isothermally**, its temperature remains constant and *Boyle's law* holds good. We, therefore, have

$PV = k$, where k is a constant. Or, $P = k/V$.

Differentiating this, we have $-dP/dV = k/V^2$.

Or, $$-V.dP/dV = V.k/V^2 = k/V = P.$$ ...(*ii*)

---

* This may be done by using a cylinder and a piston of a perfectly conducting material, so that the heat is conducted out into the surrounding air as soon as it is generated and the temperature of the gas remains the same as before.

Comparing relations (*i*) and (*ii*), we therefore have $K = P$, *i.e.*, **the isothermal elasticity of a gas is equal to its pressure.**

(*ii*) If, on the other hand, the gas is compressed *adiabatically**, we have

$PV^{\gamma} = k$, where $\gamma = C_p/C_v$, the ratio of the specific heats of the gas at constant pressure and at constant volume respectively. So that, $\gamma P = K/V^{\gamma} = kV^{-\gamma}$.

Again, on differentiation, we have

$$dP/dV = -\gamma k V^{-\gamma-1} = -\gamma k V^{-\gamma}/V = -\gamma P/V.$$

And, therefore, $K = -V.\, dP/dV = \gamma P$, *i.e.*, **the adiabatic elasticity of the gas is equal to γ times its pressure, or γ times its isothermal elasticity.**

2. (*a*) Show that the Bulk modulus K, Young's modulus Y and the Poisson's ratio σ are connected together by the relation, $K = Y/3(1-2\sigma)$. (*Punjab; Delhi*)

(*b*) A teel wire 1 m long is suspended from a rigid support and carries a 6 kg mass at its lower free end. If the wire be 1 mm square in cross-section and if Y for the wire be 2×10^{11} N/m², calculate the increase in its length.

We have $K = \dfrac{1}{3(\alpha - 2\beta)} = \dfrac{1}{3\alpha\,(1 - 2\beta/\alpha)}$ [See page 262

Now $1/\alpha = Y$ and $\beta/\alpha = \sigma$ [See pages 263 & 264

therefore, $K = Y/3\,(1-2\sigma)$.

(*b*) We know that $Y = F \times L / a \times l$, whence, $l = F \times L / a \times Y$.

Here, $F = 6 \times 9.81$ *newtons*, $L = 1\ M$, $a = 10^{-6}$ *sq.m* and $Y = 2 \times 10^{11}$ newton/m². So that, we have

$$l = \frac{6 \times 9.81 \times 1}{10^{-6} \times 2 \times 10^{11}} = 3 \times 9.81 \times 10^{-5} m = .2943\ mm$$

Thus, increase in the length of the wire = .2943 *mm*

(3.) (*a*) A bronze bar, 1.7 m long and 50 mm in diameter, is subjected to a tensile stress of 70 meganewtons/m². Calculate the extension produced in the bar and the work done during the process. The value of Young's modulus for the material of the bar may be taken to be 85×10^{9} N/m².

(*b*) Calculate the factor of safety in the case of a piston rod of diameter 20 cm, the maximum tensile pull on which, under working conditions, is 1 meganewton and the tensile strength for the material of which is 600 MN/m².

(*a*) Since $Y = \dfrac{stress}{strain}$, we have *strain* = *stress*/*Y*

And ∴ *extension produced* = *strain* × *length*

$$= \frac{stress}{Y} \times length = \frac{70 \times 10^6}{85 \times 10^9} \times 1.7 = 14 \times 10^{-4} m = 1.4\ mm$$

Now, *work done during stretch* $= \frac{1}{2} \times$ *stretching force* × *stretch* [see 8.11

Here *stretching force* = *tensile stress* × *area of cross-section of the rod* = $(70 \times 10^{6}) \times \pi \times 50^2 \times 10^{-6}/4$

$$\therefore \quad work\ done = \frac{1}{2} \times \frac{70 \times 10^6 \times \pi \times 2500 \times 10^{-6}}{4} \times 14 \times 10^{-4}$$

$$= \frac{7 \times \pi \times 35}{8} = 96.23\ joules.$$

(*b*) clearly, *maximum stress under working conditions, i.e., working stress* = *max. load/area*

$$= \frac{1 \times 10^6}{\pi \times 20^2 \times 10^{-4}/4} = \frac{10^8}{\pi}\ N/m^2.$$

so that, *factor of safety* = $\dfrac{tensile\ strength}{working\ stress}$ [§ 8.6

$$\frac{600 \times 10^6}{10^8/\pi} = 6\pi = 18.86$$

* In this case, the cylinder and piston are of a perfectly non-conducting material or the cylinder is placed on a perfect insulator, so that the heat generated on compression of the gas cannot escape out but remains inside the gas itself, thus raising its temperature a little.

4. Obtain an expression for the radius of curvature of a flat curve in terms of the slope of the curve, and use the result to find the value of deflection in the case of a bar fixed horizontally at one end and loaded at the other. *(Bombay)*

Let APQ be a flat curve (Fig. 8.57), and let P and Q be two points on it, a small distance δx apart. Draw tangents to the curve at P and Q, and let o be the centre and R, the radius of curvature of the portion PQ of the curve. Then, if $\angle POQ = \theta$, we have $PQ = R.\theta$.

Or, $\delta x = R.\,\theta.$ ...(*i*)

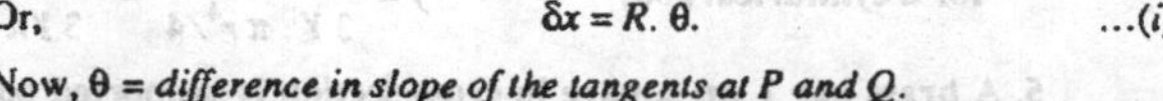

Now, θ = *difference in slope of the tangents at P and Q*.

And since slope of the tangents at a point is measured by dy/dx at the point, we have

$$\theta = \left[\frac{dy}{dx} \text{ at } Q - \frac{dy}{dx} \text{ at } P\right]$$

Now, *the rate of change of slope* is given by the *second differential coefficient*, d^2y/dx^2.

$\therefore$ change in slope from P to Q = $\delta x\,.\,d^2y/dx^2$.

i.e., $\theta + \delta x.d^2/dx^2$. And $\therefore\ \delta x = (R.\delta x.d^2/dx^2)$. [from (*i*) above

i.e., $R.\,d^2y/dx^2 = 1.$

Or, $1/R = d^2y/dx^2$ = *rate of change of slope at P.*

Fig. 8.57

Since in the case of bent rods, or beams, the curve of the neutral axis is very slight, the relation $1/R = d^2y/dx^2$ gives the *radius of curvature* of the axis at any given point.

Now, for a bar fixed horizontally at one end and loaded at the other, (*i.e.*, in the case of a *cantilever*), we have

$W.(L-x) = Y.\,I_g/R$, [see page 290

the axis of x being taken along the horizontal and the axis of Y, vertically downwards.

Here, L is the length of the bar from the fixed to the free end, x, the distance of section of the section PQ from the fixed end, and W, the weight applied at the free end.

Therefore, substituting the value of $1/R$, from above, we have

$$W.(L-x) = Y.I_g.\frac{d^2y}{dx^2} \quad \text{Or,} \quad \frac{Y.I_g}{W}.\frac{d^2y}{dx^2}. = (L-x).$$

Integrating this, we have $$\int \frac{Y.I_g}{W}.\frac{d^2y}{dx^2}.dx = \int (L-x).dx.$$

Or, $$\frac{Y.I_g}{W}.\frac{dy}{dx} = (Lx - x^2/2) + C_1 \qquad ...(ii)$$

where C_1 is a *constant of integration*, to be determined from the conditions of the experiment.

Clearly, dy/dx *is zero at A, i.e.*, $dy/dx = 0$, when $x = 0$.

Substituting these values of dy/dx and x in (*ii*), we have $C_1 = 0$.

$$\therefore \quad \frac{YI_g}{W}.\frac{dy}{dx} = Lx - \frac{x^2}{2}. \qquad ...(iii)$$

Again, integrating this expression (*ii*), we have

$$\int \frac{Y.I_g}{W}.\frac{dy}{dx}.dx = \int\left(Lx - \frac{x^2}{2}\right).dx. \text{ Or, } \int \frac{YI_g}{W}.dy = \int\left(Lx - \frac{x^2}{2}\right).dx$$

Or, $$\frac{YI_g}{W}.Y = \frac{Lx^2}{2} - \frac{x^3}{6} + C_2. \qquad ...(iv)$$

whence C_2 is another *constant of integration*.

To determine this constant, we observe that the depression y of the rod is *zero* at the end A; so that, $y = 0$, when $x = 0$.

Putting these values of y and x in (*iv*), we have $C_2 = 0$.

Hence $$\frac{YI_g}{W}.y = \frac{Lx^2}{2} - \frac{x^3}{6}.$$

Now, to obtain the deflection of the loaded end, let us put $x = L$.

Then, clearly, $$\frac{Y.I_g}{W}.y = \frac{L^3}{2} - \frac{L^3}{6} = \frac{L^3}{3}. \text{ Or, } y = WL^3/3YI_g.$$

If the rod be *rectangular* cross-section, $I_g = b\,d^3/12$, where b and d are its breadth and depth respectively; so that,

for a rectangular rod, $y = \dfrac{W.L^3}{3\,Yb.d^3/12} = \dfrac{4\,W\,L^3}{Ybd^3}$.

And, if the rod be of a *circular* cross-section, (*i.e.*, if it be *cylindrical*), $I_g = \pi r^4/4$, where r is its *radius*; so that,

for a cylindrical rod, $y = \dfrac{W.L^3}{3\,Y.\pi r^4/4} = \dfrac{4W.>L^3}{3\,Y\pi r^4}$.

5. A brass bar 1 cm square in cross-section is supported on two knife-edges 100 cm apart. A load of 1 kg at the centre of the bar depresses that point by 2.51 mm. What is Young's modulus for brass ?

We know that the depression of the mid-point of the bar is given by

$y = WL^3/\,48\,YI_g$. [See page 294.

Now, for a bar of rectangular cross-section, $I_g = b.d^3/12$

Here, $b = d = 1\ cm$, because the bar is $1\ cm$ *square* in cross-section.

$\therefore$ $b.\,d^3 = 1 \times 1 = 1;\ W = 1\ k\ gm\ wt = 1000 \times 981\ dynes.$

$L = 100\ cm$ and $y = 2.51\ mm = .251\ cm.$

Therefore, $y = \dfrac{WL^3}{48\,Y.\,bd^3/12} = \dfrac{12\,WL^3}{48\,Y\,b.d^3}$. Or, $Y = \dfrac{WL^2}{4ybd^3}$

Or, $Y = \dfrac{1000 \times 981 \times 100^3}{4 \times .251 \times 1} = \dfrac{981 \times 10^9}{4 \times .251} = 9.77 \times 10^{11}\ dynes/cm^2.$

6. (*a*) Find the work done in stretching a wire of cross-section 1 sq mm and length 2 metres through 0.1 mm, if Young's modulus for the material of the wire is $2 \times 10^{11}\ N/m^2$.

(*b*) Calculate the twisting couple on a solid shaft, of length 1.5 metre and diameter 120 mm when it is twisted through an angle 0.6^0. The coefficient of rigidity for the material of the shaft may be taken to be $93 \times 10^9\ N/m^2$.

(*a*) As we know, (§ 8.11), *work done in stretihing a wire* $= \dfrac{1}{2}$ *stretihing force* $\times$ *stretch* $= \dfrac{1}{2} F.l = \dfrac{1}{2}\dfrac{Y.a.l}{l}\ l.$

Here, $Y = 2 \times 10^{11}\ N/m^2$, $a = 1\ sq.\ mm = 10^{-6}\ sq.\ metre$, $l = 0.1\ mm = 10^{-4}\ m$ and $L = 2\ m$. So that,

$$work\ done = \frac{1}{2}.\frac{2 \times 10^{11} \times 10^{-6} \times 10^{-4} \times 10^{-4}}{2} = \frac{10^{-3}}{2} = 5 \times 10^{-4}\ joules.$$

(*b*) We know that the twisting couple $C = \pi nr^4.\ \theta/2l.$

Here, $\theta = 0.6° = \pi \times 0.6/180$ *radian*. So that, we have

$$C = \frac{\pi \times 93 \times 10^9 \times (0.120)^4 \times 0.6\pi}{2 \times 16 \times 1.5 \times 180} = 1.322 \times 10^4\ N\text{-}m.$$

Thus, the *twisting couple on the shaft* $= 1.322 \times 10^4$ *newton-metre.*

7. Show that for a homogeneous isotropic substance, $Y/N = 2(\sigma + 1)$, where Y is the Young's modulus, N, the simple rigidity, and σ, the Poisson's ratio.

A gold wire 0.32 mm in diameter, elongates by 1 mm, when stretched by a force of 330 gm-wt and twists through 1 radian, when equal and opposite torques of 145 dyne-cm. are applied at its ends. Find the value of Poisson's ratio for gold.

An isotropic substance is such that two equal, similar portions cut from it, with any orientation, are exactly alike and indistinguishable from one another.

For proof of the relation, $Y/N = 2(\sigma + 1)$, see worked example 3 above (page 316), where it is shown that $N = Y/2(\sigma + 1)$, whence, $Y/N = 2(\sigma + 1)$.

Now, $Y = F \times L/a \times l$

Here, $F = 330 \times 981\ dynes$; $l = 1\ mm = .1\ cm$; and $a = \pi \times (.016)^2\ sq.m$

[because radius $= .032/2 = .016\ cm$ and $a = \pi r^2$

$$Y = \frac{330 \times 981 \times L}{\pi \times .016^2 \times .1}\ dynes/cm^2.$$

Since *couple* acting on the wire = 145 *dynes-cm* and *angle of twist* = 1 *radian*, we have couple per unit twist = 145/1 = 145 *dynes-cm.*

This must be equal to $N.\pi\ r^4/2\ L$, where, r is the radius of the wire.

Thus, $N\pi r^4/2L = 145$, whence, $N = 145 \times 2L/\pi \times (.016)^4$

Or, $$\frac{Y}{N} = \frac{330 \times 981 \times L}{\pi \times (.016)^2 \times .1} \times \frac{\pi \times (.016)^4}{145 \times 2L}.$$

$$= \frac{330 \times 981\ (.016)^2}{2 \times .1 \times 145} = \frac{330 \times 981 \times (.016)^2}{29} = 2.858.$$

Since $Y/N = 2(\sigma + 1)$, we have $2(\sigma + 1) = 2.858$

Or, $(\sigma + 1) = 2.858/2 = 1.429$. And $\therefore\ \sigma = 1.429 - 1 = .429$.

8. A square metal bar of 2.51 cm side, 37.95 cm long, and weighing 826 gm is suspended by a wire 37.85 cm long and 0.0501 cm radius. It is observed to make 50 complete swings in 335.7 sec. What is the rigidity coefficient of the wire ?

Here, time-period of the bar, *i.e.*, $t = 335.7/50 = 6.714$ *sec.*

Now, time-period of a body executing a torsional vibration is given by $t = 2\pi\sqrt{I/C}$, where I is its *moment of inertia about the suspension wire as axis and C, the twisting couple per unit deflection or twist of the wire.*

Here, $$I = mass\left(\frac{length^2 + breath^2}{12}\right) = 826 \times \frac{(37.95)^2 + (2.51)^2}{12}$$

$$= 826 \times \left(\frac{1440 + 6.301}{12}\right) = 826 \times \frac{1446.3}{12} = 99540\ gm.cm^2.$$

$\therefore$ $6.714 = 2\pi\sqrt{99540/C}$, squaring which, we have

$$(6.714)^2 = 4\pi^2 \times 99540/C. \text{ Or, } \quad C = 4\pi^2 \times 99540/(6.714)^2$$

Now, C is also $= n.\pi r^4/2l$ where the *symbols* have their usual meanings.

$$\therefore \quad \frac{n \times \pi\ (.0501)^4}{2 \times 37.85} = \frac{4\pi^2 \times 99550}{(6.714)^2},$$

whence, *coefficient of rigidy,* $$n = \frac{8\pi \times 99540 \times 37.85}{(.0501)^4 \times (6.714)^2} = 3.357 \times 10^{11}\ dynes/cm^2$$

9. A disc of 10 cm radius and mass 1 kg is suspended in a horizontal plane by a vertical wire attached to its centre. If the diameter of the wire is 1 mm and its length is 1.5 metres and the period of torsional vibration of the disc is 5 sec, find the rigidity of the material of the wire. Prove the equation you employ. *(Bombay)*

For proof of the formula, see § 8.26 (page 279), which gives the relation,

$$t = 2\pi\sqrt{I/C}.$$

Here, in this case $t = 5$ *sec* and I, for the disc, about the axis of suspension, is given by $Mr^2/2$, where M is the mass of the disc, and r, its radius. Since $M = 1\ kg = 1000\ gm$ and $r = 10\ cm$, we have

$$I = 1000 \times 10^2/2 = 500 \times 100 = 5 \times 10^4\ gm\text{-}cm^2.$$

Therefore, $t = 5 = 2\pi\sqrt{5 \times 10^4/C}$. Or, $25 = 4\pi^2 \times 5 \times 10^4/C$,

Hence, $C = 4\pi^2 \times 5 \times 10^4/25 = 4\pi^3 \times 10^4/5$.

Hence C is also equal to $n\pi r^4/2l$, we have $n\ \pi r^4/2l = 4\pi^2 \times 10^4/5$

$\therefore$ substituting the values of r and l, we have

$$\frac{n \times \pi\ (.05)^4}{2 \times 150} = \frac{4\pi^2 \times 10^4}{5}$$

$r = .5\ mm = .05\ cm$ and $l = 1.5$ *metres* $= 150\ cm$.)

whence, $$n = \frac{4\pi^2 \times 10^4 \times 2 \times 150}{6 \times \pi \times (.05)^4} = \frac{4\pi \times 10^4 \times 60}{(.05)^4} = 1.206 \times 10^{12}\ dynes/cm^2.$$

10. An elastic string has a mass M suspended at its lower end, the upper being fixed to a support. The mass is pulled down over a short distance and let go. Explain the motion that ensues and find an expression for the time of oscillation.

If a mass m is added to the mass M, the time is altered in the ratio of 5 : 4. Compare the masses m and M. *(Bombay)*

For first part, see worked example 10, Chapter IV, where it is shown that if l be the extension produced in the string in the equilibrium position due to the mass, the string executes a *S.H.M.*, its time-period being given by

$$t = 2\pi\sqrt{l/g}.$$

Let the time-period in the first case, when the mass M is suspended from the end of the string be t_1, and, in the second case, when the mass $(m + M)$ is suspended, be t_2. Then, if l_1 and l_2 be the respective extensions produced by the two masses in the equilibrium position, we have

$t_1 = 2\pi\sqrt{l_1/g}$ and $t_2 = 2\pi\sqrt{l_2/g}$. So that $t_1^{\,2}/t_2^{\,2} = l_1/l_2$.

$\therefore$ if $t_1 = 4$ *sec* and $t_2 = 5$ *sec*, we have $16/25 = l_1/l_2$.

Now, l_1 and l_2 are directly proportional to the stretching force applied *viz.*, Mg and $(M+m)g$ respectively; for, in the 1*st case*, $Y = \dfrac{Mg\,.\,L}{a\,.\,l_1}$, and in the 2*nd case*, $Y = \dfrac{(M+m)\,.\,gL}{a\,.\,l_2}$.

whence, $l_1 = Mg\,.\,L/Ya$ and $l_2 = (M+m)g\,.\,L/Ya$,

where L is the *original length of the string*, a, its *area of cross-section* and Y, the *Young's modulus* for its material.

$\therefore$ $l_1/l_2 = mg/(M+m)\,g = M/(M+m)$.

Hence, $\dfrac{M+m}{M} = \dfrac{25}{16}$, *whence*, $\dfrac{M+m-M}{M} = \dfrac{25-16}{16}$, *i.e.* $\dfrac{m}{M} = \dfrac{9}{16}$

11. Obtain an expression for the time-period of the vertical oscillations of a flat spiral spring under a load suspended from its free end, both when the weight of the spring be neglected and when it is taken into account.

For the first part, see § 6.17 (5)

Second part. Let μ be the *mass per unit length* of the spring and L, its *total length*, and let the displacement of the suspended mass m (Fig. 6.24) from its position of static equilibrium be l at any given instant t. Then, its *velocity* $= dl/dt$ and hence its $K.E. = \dfrac{1}{2}m\,(dl/dt)^2$.

The spring too has kinetic energy, to calculate which, consider a small element dx of it, of mass $\mu\,dx$, at a distance x (measured along the wire) from the fixed end. Since the vertical depression at any point of the spring is proportional to its distance from its fixed end, clearly the depression of this element dx will be $l\,x/L$ and its velocity $= \dfrac{x}{L}\,.\,\dfrac{dl}{dt}$. So that, its $K.E. = \dfrac{1}{2}\mu dx\left(\dfrac{x}{L}\,.\,\dfrac{dl}{dt}\right)^2$.

Taking the spring to be of a uniform composition and of a uniform diameter, we have *K.E. of the whole spring*

$$= \frac{1}{2}\frac{\mu}{L^2}\left(\frac{dl}{dt}\right)^2\int_0^L x^2.dx = \frac{1}{2}\frac{\mu}{L^2}\left(\frac{dl}{dt}\right)^2 - \frac{L^3}{3}$$

$$= \frac{1}{2}\,.\,\frac{\mu L}{3}\,.\left(\frac{dl}{dt}\right)^2 = \frac{1}{2}\,.\,\frac{m_s}{3}\,.\left(\frac{dl}{dt}\right)^2$$

where $\mu L = m_s$, the mass of the srpring

$\therefore$ *total K.E. of the system of the mass and the spring* $= \dfrac{1}{2}\,m\left(\dfrac{dl}{dt}\right)^2 + \dfrac{1}{2}\,.\,\dfrac{m_s}{3}\,.\left(\dfrac{dl}{dt}\right)^2 = \dfrac{1}{2}\left(m + \dfrac{m_s}{3}\right)\left(\dfrac{dl}{dt}\right)^2$.

Now, the weight mg acting axially on the spiral spring exerts a turning moment mgR on it, producing in it a twist θ, say, such that $C\theta = mgR$, where C is the torsional couple per unit twist of the wire and R. the radius of the spring. Therefore, work done in an additional twist $d\theta$ of the wire $= C\theta\,.\,d\theta = n\pi r^4\theta.d\theta/2L$, because, as we know, $C = n\pi r^4/2L$, where r is the *radius of the wire*, L, its *total length* and n, the *coefficient of rigidity* for its material (see page 276)

$\therefore$ *work done during the whole twist* $\theta = \displaystyle\int_0^\theta c\theta\,.\,d\theta = \int_0^\theta \frac{n\pi r^4}{2L}.\,\theta\,.\,d\theta = \frac{n\pi r^4\,\theta^2}{4L}$.

Since $L = R.\theta$ or $\theta = L/R$, we have

$$\textit{Work done during twist } \theta \textit{ of the wire} = \frac{n\pi r^4\,.\,l^2}{4LR^2}.$$

This work is an obvious addition to the potential energy of the spring.

Since, however, the extension l of the spring at its lower free end, lowers the weight mg, suspended from it through a distance l, and the *c.g.* of the spring through distance $l/2$*, the loss of *P.E.* on this account $= mgl + m_s\,g.l/2$.

Thus, the *net change in P.E. of the system of the mass (m) and the spring* $= \dfrac{n\pi r^4 l^2}{4LR^2} - mg.l - m_s g\,.\,\dfrac{l}{2}$.

* Because the extension of spring under its own weight may be taken to be half that due to load mg.

Since in accordance with the *principle of conservation of energy K.E + P.E. = constant*, we have

$$\frac{1}{2}\left(m + \frac{m_s}{3}\right)\left(\frac{dl}{dt}\right)^2 + \frac{n\pi r^4 l^2}{4LR^2} - mgl - m_s g.\frac{l}{2} = \text{constant}.$$

Differentiating with respect to t, therefore, we have

$$\frac{1}{2}\left(m + \frac{m_s}{3}\right).2\left(\frac{dl}{dt}\right)\left(\frac{d^2l}{dt^2}\right) + \frac{n\pi r^4}{4LR^2}.2l.\frac{dl}{dt} - mg.\frac{dl}{dt} - \frac{m_s g}{2}.\frac{dl}{dt} = 0$$

Dividing throughout by $(m + m_s/3)\, dl/dt$, we have $\dfrac{d^2l}{dt^2} + \dfrac{n\pi r^4}{2LR^2(m + m_s/3)}.l - \dfrac{m + m_s/2}{m + m_s/3}.g = 0,$

which represents a simple harmonic motion of **period time T**, given by

$$T = 2\pi\sqrt{\frac{2LR^2(m + m_s/3)}{n\pi r^4}} = 2\pi\sqrt{\frac{m + m_s/3}{K}}$$

putting $n\pi r^4/2LR^2 = K$, the elastic reaction per unit increase in the length of the spiral (see page 200).

Thus, the only effect of taking the mass of the spring m_s also into consideration is as though the suspended mass m has been increased effectively to $(m + m_s/3)$.

Here, too, we have neglected the depression due to vertical shear in comparison with that due to torsional shear. This is however, quite justified in view of the fact that the ratio of the two works out of $r^2/2R^2$, an obviously negligible quantity.

12. (*a*) A cylindrical rod of diameter 14 mm rests on two knife-edges 0.8 m apart and a load of 1 kg is suspended from its mid-point. Neglecting the weight of the rod, calculate the depression of the mid-point if Y for its material be 2.04×10^{11} N/m².

(*b*) Compare the loads required to produce equal depression for two beams, made of the same material and having the same length and weight, with the only difference that while one has a circular cross-section, the cross-section of the other is square. ***(Nagpur)***

(*a*) We know that the depression y of the mid-point of a beam of circular cross-section, supported at the ends and loaded in the middle is given by $y = WL^3/12\pi r^4.Y$, as shown in § 8.33 (i), (page 294)

Here, $L = 0.8\ m$, $r = .014/2 = .007\ m$, $W = 1 \times 9.81\ N$ and $Y = 2.04 \times 10^{11}\ N/m^2$

So that *the depression of the mid-point of the beam* is given by

$$y = \frac{9.81\,(0.8)^3}{12\,\pi\,(.007)^4 \times 2.04 \times 10^{11}} = .000272\ m = 0.272\ mm.$$

(*b*) Let the two beams, supported at their ends, be loaded at their midpoints with loads W and W' respectively, so as to suffer equal depressions. Then, we have

$$y = \frac{WL^3}{12\,Y\pi r^4} = \frac{W'L^3}{4\,Ybd^3} \qquad [\ §\ 8.33\ (i),\ page\ 294.$$

since the second beam has a square cross-section, $b = d$ and, therefore, $\dfrac{W}{3\pi r^4} = \dfrac{W'}{bb^3} = \dfrac{W'}{b^4}$

Now, the weights of the two beams being equal, naturally, their masses and hence their volumes are also equal (their material being the same). So that, $\pi r^2 L = b^2 L$, whence, $b^2 = \pi r^2$.

We thus have $\dfrac{W}{3\pi r^4} = \dfrac{W'}{\pi^2 r^4}$, whence , $W : W' :: 3 : \pi$

The loads required are thus in the ratio $3 : \pi$

13. A copper wire 3 metres long for which Young's modulus ks 12.5×10^{11} dynes per sq cm has a diameter of 1 mm. If a weight of 10 kg is attached to one end, what extension is produced ? If Poisson's ratio is 0.26, what lateral compression is produced ?

Here, original length of the wire (L) = 3 *metres* = 300 *cm.*

Young's modulus of the wire (Y) = 12.5×10^{11} *dynes-cm*$^{-2}$.

radius of the wire (r) $= \frac{1}{2}\ mm = .5\ mm = .05\ cm$,

and ∴ its area of cross-section $= \pi r^2 = \pi \times (.05)^2$ *sqcm*

and force applied (F) = 10 *kg wt* = 10×1000 *gm.wt*

$= 10 \times 1000 \times 981$ *dynes* $= 981 \times 10^4$ *dynes.*

Now, we have the relation, $Y = F \times L / a \times l$, whence, $l = F.L/a.Y$.

Or, $$l = \frac{981 \times 10^4 \times 300}{\pi \times (.05)^2 \times 12.5 \times 10^{11}} = \frac{981 \times 3}{\pi \times (.05)^2 \times 12.5 \times 10^5} = .2997\ cm.$$

Again, *Poisson's ratio*, σ = *lateral strain / longitudinal strain.*

$$\therefore \quad .26 = \frac{\text{lateral strain}}{l/L}$$

Or, lateral strain = $.26 \times l/L = .26 \times .2997/300 = 2.598 \times 10^{-4}$.

This, therefore, gives the value of lateral strain, *i.e.*, d/D, where d is the decrease in diameter and D, the original diameter of the wire.

Hence $$d/D = 2.598 \times 10^{-4}.$$

Since $D = 1\ mm = .1\ cm$ we have

$$d/.1 = 2.958 \times 10^{-4}. \text{ Or, } \quad d = 2.598 \times 10^{-4} \times .1 = 2.598 \times 10^{-5}.$$

Hence, lateral compression produced = 2.598×10^{-5} *cm.*

14. A uniform glass tube is hung from a support and stretched by a weight. It is found that one metre of the tube stretches by .08 cm, but that a column of water 1 metre long contained within the tube lengthens by only 0.4 cm. Find Poisson's ratio for glass.

We know that *Poisson's ratio* $\sigma = \beta/\alpha$

Now, let the stress be = P *dynes/cm*^{2} and internal radius of the tube = r *cm.*

Then, increase per unit length of the tube $= P.\alpha\ cm.$

And, decrease per unit radius of the tube $= P.\beta\ cm.$

$\therefore$ increase in 1 *metre* of 100 *cm* length of the tube = $100.P\alpha\ cm.$

and decrease in the radius of the tube $= P.\beta.r\ cm.$

so that, the radius of the tube is now $= (r - P.\beta.r)\ cm = r(1 - P.\beta)\ cm.$

Now, increase in length of the tube $= .06\ cm.$

$\therefore \quad 100.P.\alpha = .06$, whence $\alpha = .06/100\,P.$

And clearly, initial cross-section of the tube $= \pi r^2\ sq.\ cm.$

and, final cross-section of the tube

$$= \pi[r(1 - P.\beta)^2]\ sq.\ cm.$$
$$= \pi r^2 \times [1 - 2P.\beta + (P.\beta)^2]\ sqcm$$
$$= \pi r^2 \times [1 - 2P.\beta]\ sq\ cm$$

neglecting $(P.\beta)^2$ as a very small quantity.

Therefore, volume of water column *initially* $= 100.\ r^2\ cc$

And volume of water column *finally* $= 100.4 \times r^2(1 - 2P.\beta)\ cc$

($\because$ length of the water column is now $100 + .04\ cm$)

Since *volume of the water column remains the same*, we have

$$100.04 \times \pi r^2(1 - 2P\beta) = 100\pi r^2. \text{ Or, } \ 100.04(1 - 2P\beta) = 100.$$

Or, $$100.4 - 100.04 \times 2P\beta = 100,$$

Or, $$100.04 \times 2P\beta = 100.04 - 100.00 = .04, \text{ whence } \beta = \frac{.04}{2P \times 1000.04}$$

$$\therefore \sigma = \frac{\beta}{\alpha} = \frac{.04}{2P \times 100.04} \times \frac{100\,P}{.06} = \frac{.04 \times 100}{2 \times 100.04 \times .06} = 0.3332.$$

15. A lathe of width 2 cm and thickness 3 mm, supported horizontally on knife-edges 80 cm apart, is loaded with weight of 10 gm hung from its ends which project 15 cm beyond the knife-edges. If the centre of the lathe is thereby elevated 2 mm, calculate Young's modulus for its material.

Let ABC (Fig. 8.58) be the position of the lathe after it has been bent into an arc of radius R in between the two knife-edges, under the bending moment due to a weights of 10 *gm* at each end, equal to $10 \times 981 \times 15$ *dynes-cm.* This must obviously be equal to YI_g/R, where I_g is the *geometrical moment of inertia* of the lathe, equal to

$bd^3/12 = 2 \times (.3)/12 = .0045\ gm\text{-}cm^2$. so that, we have

$$\frac{Y \times .0045}{R} = 10 \times 981 \times 15,$$

whence, $Y = 10 \times 981 \times 15 \times R / .0045$

Now, from the geometry of the figure, we have

$$(2R - .2) \times .2 = 40 \times 40,$$

whence, $R = 40 \times 40 / .2 \times 2 = 4000$ *cm*

(neglecting $CC'^2 = .2 \times .2$ compared with $2R$)

So that, $Y = 10 \times 981 \times 15 \times 4000 / .0045$

$= 1.308 \times 10^{11}$ *dynes/cm*^{2}.

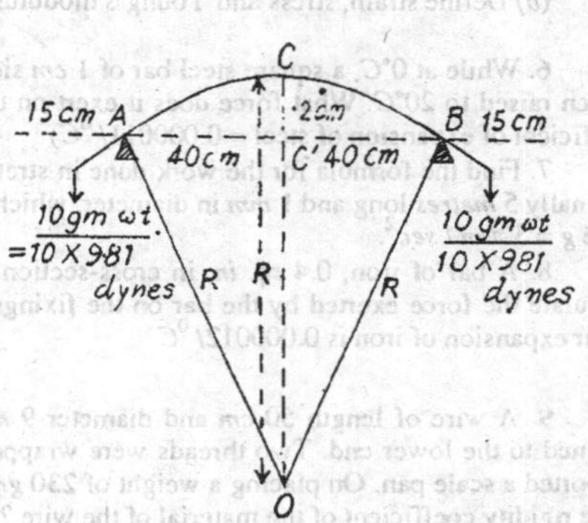

Fig. 8.58

16. A steel wire 2 mm in diameter is just stretched between two fixed points at a temperature of 20°C. Determine its tension when the temperature falls to 10°C. [*Coefficient of linear expansion of steel is .000011 and Young's modulus for steel is* 2.1 × 10¹² dynes per sq. cm]

Let the length of the wire be *L cm*.

Then, on a fall in its temperature, from $20^\circ C$ to $10^\circ C$, its length will decrease by an amount $= L \times .000011 \times 10$ *cm*.

And ∴ *strain* produced in it $= L \times .000011 \times 10 / L = .00001 \times 10 = 11 \times 10^{-5}$

Let *T dynes* be the *tension* in the wire. Then,

$$stress = T/\pi r^2 = T/\pi\,(.1)^2 = T/\pi \times .01 \; sq\; cm,$$

[∵ radius of the wire, $r = 1$ *mm* $= .1$ *cm* and ∴ its area of cross-section

$$= \pi r^2 = \pi \times (.1)^2 \; sq\; cm.]$$

Now, Young's modulus, (Y) = stress/strain. So that,

$$Y = \frac{T}{\pi \times .01} \times \frac{1}{11 \times 10^{-5}} = \frac{T}{\pi \times .11 \times 10^{-5}} = 2.1 \times 10^{12}$$

whence, $T = 2.1 \times 10^{12} \times \pi \times .11 \times 10^{-5} = 2.1 \times 10^{7} \times \pi \times .11 = 7.257 \times 10^{6}$ *dynes*.

17. If one body impinges on another which is at rest, find the relation between (*a*) momenta, (*b*) the kinetic energies of the system before and after impact.

A steel ball is let fall through a height of 64 cm on a plate of steel. The height through which it rebounds is 36 cm. Calculate the coefficient of restitution.

We know that $e = \dfrac{relative\;\; velocity\;\; after\;\; impact}{relative\;\; velocity\;\; before\;\; impact}$

Here, relative velocity *after* impact, say, $v = \sqrt{2.g.\,36}$, [∵ $h = 36$ *cm*.

and " *before* impact, say, $u = \sqrt{2.g.\,64}$ [∵ here, $h = 64$ *cm*

Therefore, $e = \sqrt{2.g.36/2\,g\,.\,64} = \sqrt{36/64} = \sqrt{9/16} = 3/4 = .75$

Thus, the coefficient of restitution = .75.

EXCERCISE VIII

1. (*a*) Define Young's modulus. Show that a shear is equivalent to a compression and an extension. Find an expression for the work done in stretching a wire and hence deduce an expression for the energy per unit volume of the wire. *(Madras, B.A.)*

(*b*) When the barometer reads 76 *cm* show that the isothermal elasticity of air is very nearly 10⁶ *dynes/cm*².

2. A wire 300 *cm* long and 0.625 *sq. cm* in cross-section is found to stretch 0.3 *cm* under a tension of 1200 kilograms. What is the Young' modulus of the material of the wire? *(A.M.I.E., 1961)*

Ans. 2.3 × 10¹² *dynes/sq cm.*

3. Explain the terms : *stress, strain, Young's modulus, Poisson's ratio, bulk and rigidity moduli*. Show that the value of Poisson's ratio must lie between −1 and +1/2.

4. Define Young's modulus, Bulk modulus and modulus of Rigidity. If *E, K* and *n* represent these moduli respectively, prove the relation, $E = 9nK/3K+n$. *(Allahabad : Poona)*

5. (*a*) A solid ball 330 *cm* in diameter is submerged in a lake at such a depth that the pressure exerted by water is 1.00 *kg wt/cm*^{2}. Find the change in volume of the ball. (*K* for the material = 1.00×10^7 *dynes/cm*^{2}) (*Bombay*)

Ans. 1.386 *cc*

(*b*) Define strain, stress and Young's modulus. Find the ratio of the adiabatic and isothermal elasticities of a gas. (*Patna*)

6. While at 0°*C*, a square steel bar of 1 *cm* side is rigidly fixed at both ends so that it cannot expand. Its temperature is then raised to 20°*C*. What force does it exert on the clamps ? (Young's modulus for steel = 2×10^{12} *dynes/sq. cm* and coefficient of expansion of steel = 0.000011/°C) **Ans.** 448 *kg. wt.*

7. Find the formula for the work done in stretching a wire, and apply it to find the elastic energy stored up in a wire, originally 5 *metres* long and 1 *mm* in diameter, which has been stretched by 3/10 *mm* due to a load of 10 *kg*. Take $g = 3\pi\, m/sec^2$. **Ans.** 1.414×10^{-2} *joules*.

8. A bar of iron, 0.4 *sq. in.* in cross-section is heated to 100°*C*. It is then fixed at both ends and cooled to 15°*C*. Calculate the force exerted by the bar on the fixings. Young's modulus for iron is 30,00,000 *lb./sq. in.* The coefficient of linear expansion of iron is 0.000012/°*C* (*Institute of Civil and Electrical Engineering*)

Ans. 5.464 *tons*.

9. A wire of length 50 *cm* and diameter 9 *mm* was fixed at the upper end while a wheel of 10 *cm* diameter was fastened to the lower end. Two threads were wrapped round the wheel and passed horizontally over pulleys ; each thread supported a scale pan. On placing a weight of 230 *gm* on each pan the lower end of the wire was twisted through 45°. What is the rigidity coefficient of the material of the wire ? **Ans.** 7.96×10^{11} *dynes cm*$^{-2}$

[**Hint.** *Convert degrees into radians:* π *radians* = 180°.]

10. Explain what you understand by 'shearing strain'. What are its dimensions? Deduce an expression for the moment of the couple required to twist the lower end of a rod of circular cross-section by 90°, the upper end being clamped. (*Agra*)

Ans. Couple = $\pi^2 . nr^4/4l$.

[**Hint.** θ = 90° = $\pi/2$ *radians*.]

11. What couple must be applied to a wire, 1 *metre* long, 1 *mm* diameter, in order to twist one end of it through 90°, the other end remaining fixed? The rigidity modulus is 2.8×10^{11} *dynes cm*$^{-2}$.

[**Hint.** 90° = $\pi/2$ *radians*]. **Ans.** 4.3×10^6 *dynes. cm*

12. Explain what is meant by 'modulus of rigidity' and find out its dimensions. Describe one method of finding experimentally the modulus of rigidity of a wire and give the theory of the method. Find the force necessary to stretch by 1*mm* a rod of iron 1 *metre* long and 2 *mm* in diameter. Also calculate the energy stored in the stretched rod. [Young's modulus for iron = 2×10^{11} *M.K.S.* (or *S.I*) *units*. (*Patna*)

Ans. (*i*) 64.12 *kg. wt* (*ii*) 3.142×10^{-1} *joules*.

13. (a) Find an expression for the bending moment of a horizontal beam clamped at one end and loaded at the other. (*Baroda*)

(*b*) A vertical rod of circular section of radius 1 *cm* is rigidly fixed in the earth and its upper end is 3 *metres* from the ground level. A think string which can stand a maximum tension 2 *kg* is tied at the upper end of the rod and pulled horizontally. Find how much will the top be deflected before the string snaps. (*Y* for steel = 2×10^{12} *C.G.S.* units, *g* = 1000 *C.G.S. units*). (*Saugar*)

Ans. 11.47 *cm*.

14.(*a*) Show that the depression of the free end of a light cantilever, when the end is loaded, is proportional to the cube of its length. Outline the method of measuring the Young's modulus of the material of the cantilever. [*Punjab (Sept.)*]

(*b*) Explain why steel girders and rails are made in the form of I section. (*Delhi*)

15. A uniform beam is clamped horizontally at one end and loaded at the other. Obtain the relation between the load and the depression at the loaded end.

Compare loads required to produce equal depression for two beams, made of the same material and having the same length and weight, with the only difference that while one has a circular cross-section, the cross-section of the other is a square. (*Saugar*)

Ans. 3 : π

16. A weight is suspended from the free end of a uniform cantilever. Find the equation of the curve into which the cantilever is bent. The weight of the cantilever may be neglected.

A uniform rigid rod 120 *cm* long is clamped horizontally at one end. A weight of 100 *gm* is attached to the free end. Calculate the depression of point 90 *cm* distant from the clamped end. The diameter of the rod is 2 *cm*. Young's modulus of the material of the rod is 1.013×10^{11} *dynes per sq cm* and *g* = 980 *cm/sec*^{2}. (*Bombay*)

Ans. 2.834 *mm*.

17. A light beam of circular cross-section is clamped horizontally at one end and a heavy mass is attached at the other end. Find the depression at the loaded end.

If the mass is pressed, down a little and then released and, show that it will perform simple harmonic motion. Explain how from a knowledge of the period of oscillation, the mass ad the dimensions of the bar, the value of Young's modulus for the material of the bar may be determined. *(Madras)*

18. A vertical wire is loaded (within the limits of Hooke's Law) by weights, which produce a total extension of 3 *mm* and 5 *mm* respectively. Compare the amounts of work necessary to produce these extensions. **Ans.** 9 : 25.

19. Derive a relation between Poisson's ratio, Young's modulus and rigidity modulus for a material. Describe Searles' methods for determining Poisson's ratio. *(Delhi)*

20. A bar, one metre long, 5 *mm* square in section, supported horizontally at its ends and loaded at the middle, is depressed 1.96 *mm* by a load of 100 *gm*. Calculate Young's modulus for the material of the bar. (Take $g = 980$ *cm/sec*^{2})

Ans. 19.99×10^{11} *dynes cm*$^{-2}$.

21. Calculate the time of vertical oscillation of a mass of 1 *kg* hanging by a steel wire 3 *metres* long and 5 *mm* in diameter. (Y for steel $= 2 \times 10^{12}$ *C.G.S.* units). **Ans.** .05 *sec.*

[**Hint.** *Find extension l produced. Then,* $t = 2\pi\sqrt{l/g}$. *See worked exam* *le* 10. Chapter IV]

22. Prove that *Young's modulus Y, the bulk modulus K, the modulus of rigidity n* and *Poisson's ratio* σ satisfy the relations :

(i) $\frac{Y}{2n} = 1 + \sigma$; (ii) $\frac{Y}{3K} = 1 - 2\sigma$; and (iii) $\frac{3}{n} + \frac{1}{K} = \frac{9}{Y}$.

23. Define Poisson's ratio, and show mathematically, from first principles, that it must be less than 0.5 and cannot be less than −1. Calculate Poisson's ratio for, and the rigidity of silver from the following data :

Young's modulus for silver wire $= 7.25 \times 10^{11}$ *dynes cm*$^{-2}$.

But modulus for silver wire $= 11 \times 10^{11}$ *dynes cm*$^{-2}$.

Ans. $n = 2.607 \times 10^{11}$ *dynes cm*$^{-2}$, and = 0.39.

[**Hint.** (1) *From (i) and (ii) above (Ex. 22), we have* $3K(1 - 2\sigma) = 2n(1 + \sigma)$

Since K and n are both +ive, σ *cannot be more than ..5 and less than* −1.

(2) See § 8.15 page 263 , whence, it can be shown that

$$n = \frac{3YK}{(9K - Y)},\ K = \frac{nY}{(9n - 3Y)} \text{ and } Y = \frac{9Kn}{n + 3K}.$$

Also, $$3\beta = \frac{1}{2n} - \frac{1}{3K}, \text{ whence, } \beta = \frac{3K - 2n}{18nK}.$$

and $$\alpha = \frac{1}{Y}, \text{ Or, } Y = \frac{1}{\alpha} = \frac{9Kn}{n + 3K}.$$

$$\therefore \quad \sigma = \frac{\beta}{\alpha} = \frac{3K - 2n}{18nK} \times \frac{9Kn}{n + 3K} = \frac{3K - 2n}{2(n + 3K)}$$

Or, directly form (1) (Ex. 22), $$\sigma = \frac{Y}{2n} - 1.$$

24. A wire, 1 *metre* in length and 1 *mm* in diameter, is stretched by 0.6 *mm* by a load of 10 *kg .wt* and is twisted through 70° by a force of 5 *gm wt* applied to each and of a 20 *cm* rod soldered at its mid-point to the end of the wire. Calculate (1) Young's modulus, (2) Shear modulus, and (3) Bulk modulus of the wire.

Ans. $Y = 20.81 \times 10^{11}$ *dynes cm*$^{-2}$, $n = 8.268 \times 10^{11}$ *dynes cm*$^{-2}$ and $K = 14.35 \times 10^{11}$ *dynes cm*$^{-2}$.

[**Hint.** (*i*) *Twisting couple* $= \pi n\theta r^4/2l = 5 \times 981 \times 20$ *dynes. cm*

and $\theta = 70 \times \pi/180$ *radians* .

(*ii*) $K = nY/(9n-3Y)$. [*See* E_x ×. 23, **Hint.** (*ii*)]

25. A block of soft rubber, 5" square, has one face fixed, while the opposite face is sheared through a distance .5" parallel to the fixed face by a tangential force of 39 *lb wt*. How much work is done per unit volume of the cube to do this ?

Ans. 8.64 *ft lb*.

26. Calculate the depression at the free end of a thin light beam, clamped horizontally at one end and loaded at the other.

For the same mass per unit length, show that a beam of square section is stiffer than one of circular section, the deflections being in the ratio $3/\pi$. *(Bombay)*

27. (*a*) A rectangular bar of iron is supported at its two ends on knife-edges and a load is applied at the middle point. Calculate the depression of the middle point. *(Allahabad)*

(*b*) Explain the terms 'flexural rigidity' and 'torsional rigidity'. Give expressions for these for a rod of uniform circular section. *(Bombay)*

28. Find the value of Young's modulus for copper. In an experiment, the diameter of the rod was 1.26 *cm* and the distance between the knife-edges 70 *cm*. On putting a load of 900 *gm* at the middle point, the depression was 0.025 *cm*. Calculate the Young's modulus of the substance. **(Agra)**

Ans. 20.42×10^{11} *dynes/cm*$^{-2}$.

29. Define Poisson's ratio and describe a method for its determination. Derive the formula used. **(Agra)**

30 .Derive the expression for the bending of a tube supported at the two ends and loaded in the middle. **(Banaras)**

31. If Y, K, n and σ represent the Young's modulus, bulk modulus, coefficient of rigidity and Poisson's ratio, then derive various relations connecting each other. **(Punjab)**

32. Define Young's modulus, rigidity modulus and Poisson's' ratio and obtain a relation connecting the three. Describe a method of determining the modulus of rigidity of a wire. **(Delhi)**

33. Describe how the dimension of a beam are altered when the beam is bent. What is meant by neutral surface ? How is the existence of such a surface experimentally demonstrated ?

A rectangular bar 20 *mm* in breadth and 10 *mm* in depth and 1*m* in length is supported at its ends and a load of 2 *kg* is applied at its mid point. Calculate the depression if the Young's modulus of the material of the bar is 2.0×10^{11} *newton/metre*^{2}. **Ans.** 1.225 *mm*.

34. How do you differentiate between a *column* and a *strut* ? Derive an expression for the critical load for a long column with its ends rounded or hinged.

35. Discuss *Euler's theory of long columns* for case *(i) when both ends of a column are rounded or hinged*, and *(ii) when both ends of the column are fixed.*

36. Show that *(i) a column, with its ends fixed, has four times the strength to resist a thrust than a similar column, with its ends rounded or hinged* and *(ii) a column hinged at one end and loaded at the other has only one-fourth the strength of the same column when hinged at both ends.*

37. Two steel balls of masses 1 and 10 *kg* respectively are moving each towards on another with a relative velocity of 4 *metres* per second. Find the loss of energy after impact and state the reason thereof. **(Bombay)**

Ans. 50290 ergs.

38. A sphere of mass 3 *lb* moving with a velocity of 7 *ft./sec* impinges directly on another sphere, of mass 5 *lb* at rest; after the impact, the velocities of the spheres are in the ratio of 2 : 3. Find the velocities after impact and the loss of kinetic energy. **(London University)**

Ans. *(i)* 2 *ft./sec* and 3 *ft./sec* *(ii)* 45 *ft poundals*.

39. Explain briefly the terms; resilience and stiffness of a beam. What is proof resilience ?

40. Write a brief note on elastic waves.

41. Define the terms : *strain, stress and Young's modulus of elasticity*. Describe the method of measuring Young's modulus by the method of bending of a bar. Deduce the formula you use. **(Delhi)**

42. Explain the terms : modulus of rigidity and Poisson's ratio.

Derive an expression for the moment of the couple required to twist one end of a cylinder, when the other end is fixed. Hence find the periodic time of a torsion pendulum. **(Punjab)**

43. Define modulus of rigidity and find its dimensions. Describe an experiment to determine it for a steel wire. Derive the formula used. **(Delhi)**

44. Define Young's modulus, Bulk modulus and Poisson's' ratio and derive a relation between them.

What will be the density of lead under a pressure of 20,000 *newtons' cm*^{2} ? Density of lead 11.4 *gm/cm*^{3} and Bulk modulus of lead 0.80×10^{10} *N/M*^{2} **(Delhi)**

Ans. 11.685gm/cm

45. What couple must be applied to a wire 1 *metre* long and 2 *mm* in diameter in order to twist one of its ends through 45° when the other remains fixed. Given $n = 5 \times 10^{11}$ *dynes/cm*^{2}. **(Vikram)**

Ans. 61.6 *dynes-cm*

46. Describe an interference method for determining the Young's modulus of a rectangular glass plate. Prove the necessary formula. **(I.A.S, 1962, B.Sc (Hons), Delhi)**

CHAPTER IX

HYDROSTATICS

9.1. Fluids—Liquids and Gases. Hydrostatics deals with the *mechanics of fluids in equilibrium* and our first step, therefore, is to understand clearly as to what exactly do we mean by a 'fluid'.

Unlike a solid, in which, as we have seen, the strain set up under a shearing stress lasts throughout the period of application of the stress, *a fluid may be defined as that state of matter which cannot indefinitely or permanently oppose or resist a shearing stress.* In fact, it constantly and continuously yields to it, though the yield may be rapid in some cases and slow in others. In the former case, the liquid is said to be *mobile*, (like *water, alcohol* etc.) and in the latter, *viscous*, (like *honey, treacle* etc.). In either case, however, a *fluid has no definite shape of its own* and assumes ultimately the shape of the containing vessel.

And yet, with all this seemingly clear-cut distinction between a solid and a fluid, it is not quite so easy to distinguish between the two in many a border-line case. Thus, for example, pitch, which looks so much like a solid that it has to be hammered in order to be broken, is *essentially* a fluid; for, when subjected to the shearing stress of its own weight, by putting a piece of it in a funnel, or by putting a barrel of it on its side, it does begin to yield or flow, although infinitely slowly. On the other hand, metal wires, which are obviously solids, when subjected to an excessive tension, begin to flow in the manner of fluids and , indeed, may be considered to be so, for the duration of the yield. Once the yield is over, however, they behave like solids they in fact are.

Then, again, we have, on the one hand, highly elastic solids, like *quartz*, in which no change of shape is discernible even in millions of years, as is evidenced by the sharpness of its crystals which look as though they had *just* been formed, and, on the other, fluids, like water, the rapid flow of which almost instantaneously does away with any sharpness of its edges and which, in small quantities, assumes a spherical form, with no sharp edges or corners whatever.

The fundamental distinction between the two nevertheless remains, and we declare a substance to be a fluid or a solid according as it does or does not yield to a shearing stress applied to it over a long enough period.

Now, fluids too are further divided into two classes, *viz.*, (*i*) *liquids* and (*ii*) *gases*.

A **liquid** *is a fluid which, although it has no shape of its own, occupies a definite volume which cannot be altered, however great the force applied to it** . *In other words, a liquid is a fluid which is quite incompressible* and *has a free surface of its own*, as, for example, water, alcohol, ether, honey, treacle etc.

A **gas**, *on the other hand, is a fluid, which cannot only be easily compressed when subjected to pressure but, with a progressive reduction of the pressure on it, can also be made to expand indefinitely, occupying all the space made available to it.* Thus, the whole of the gas will escape out from a vessel, if there be the tiniest aperture in it somewhere.

Summarizing then, *a gas is a fluid which has neither a shape nor a free surface of its own* ; as, for example, oxygen, hydrogen, carbon dioxide, air (a mixture of gases) etc. We shall consider first the case of liquids.

---

* Strictly speaking, all liquids do get compressed a little, when subjected to very high pressures. The compression is, however, almost negligible. Thus, water, when subjected to a pressure of about 200 atmospheres, undergoes a reduction of only a hundredth part of its original volume.

9.2. Hydrostatic Pressure. Since a liquid possesses *weight*, it exerts force on all bodies in contact with it, *e.g.*, on the bottom and the walls of the vessel containing it,—the force due to it being always spread over an area*. And, if this force be uniformly distributed over the whole area, *i.e.*, be the *same* on each small equal element of the surface, its value per unit area is called **pressure** or **hydrostatic pressure** of the liquid—meaning *pressure due to the liquid at rest.*** And, if the force is not uniform, the ratio between the small force δF and the area δA on which it acts gives the pressure. Thus,

$$\textbf{pressure} = \delta F / \delta A.$$

So that, when δA is progressively diminishing, we have

$$\textbf{pressure at the point} = \lim_{\delta A \to 0} \frac{\textbf{force}}{\textbf{area}}.$$

Or, denoting pressure at the point by p, we have, in mathematical notation, $p = dF/dA$.

The *total force* exerted by a liquid column on the whole of the area in contact with it is called *thrust*. Thus,

$$\textbf{thrust} = \textbf{pressure} \times \textbf{area}.$$

That a liquid at rest, always exerts a thrust *normally* to the surface in contact with it is obvious. For, if it were not so, there would be a component of the thrust along, or parallel to, the bounding surface, and an equal and opposite thrust on it due to the reaction of the surface would cause it to flow, since it must, by its very nature, yield to a tangential force. It follows, therefore, that since the liquid is at rest, the thrust due to it must be perpendicular to the bounding surface at every point.

In other words, the free surface of a liquid at rest must be at right angles to the forces acting on it. Thus, when the only force acting on it is due to gravity, its surface remains *horizontal*+, being *perpendicular to the force of gravity*, but, with a steady wind blowing, it is slightly inclined, again, however, at right angles to the resultant of the forces due to gravity and the wind.

Further, since every layer of a liquid at rest is in equilibrium, it follows that the *downward* thrust on it, due to the liquid column above it, is just balanced by an equal *upward* thrust due to the liquid column below it. *In other words, at any given level, in a liquid at rest, the downward thrust due to liquid column is equal to the upward thrust on it.*

9.3. Hydrostatic Pressure due to a Liquid Column. Let us now calculate the *hydrostatic pressure* due to a liquid column h. Imagine a narrow metal cylinder, of area of cross-section a and fitted with a frictionless piston, of negligible weight, to be supported in a liquid of density ρ, (Fig. 9.1).

Then, if the *upthrust* on the piston due to the liquid below it be F, obviously, an equal and opposite force F has to be exerted on the piston to keep it in position.

Hence, if the piston be moved down through a distance x, work done on it is clearly equal to $F.x$.

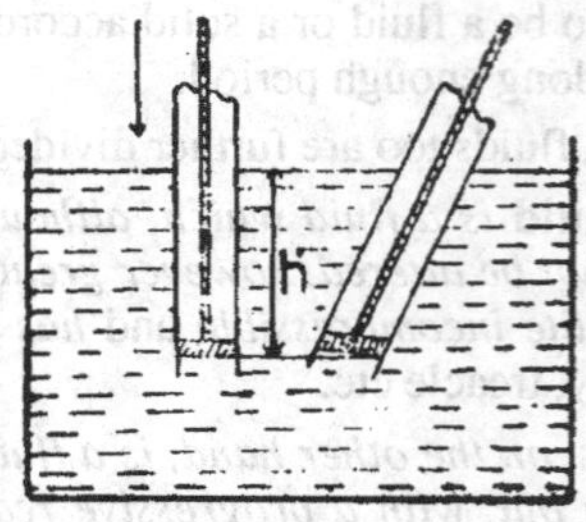

Fig. 9.1

---

* The same is true of the force exerted on a liquid, as, for example, when we press the piston down in a cylinder contining the liquid.

** It is also sometimes called *the pressure in a liquid due to gravity.*

+ In the case of a large expanse of water, the surface is spherical and is thus again perpendicular to the direction of gravity at every point.

This downward motion of the piston will obviously expel a *volume x.a.* of the liquid out of the tube, its *mass* being $x.a.\rho$ and its weight, equal to $x.a.\ \rho\ g$.

Since the level of the liquid in the containing vessel is thus slightly raised, *it is tantamount to this weight of the liquid x.a. ρ g. rising up through a vertical distance h up to the liquid surface*.

In other words, *increase in potential energy of the liquid* = $x.a.\rho.g.h$. And, *this gain in potential energy will obviously be equal to the work done on the piston*. So that, we have

$$F.x = x.a.\ \rho\ .g.h.\ \text{Or,}\ F/a = \ h.\ \rho.\ g.$$

i.e., the hydrostatic pressure due to a liquid column at depth h from the surface is equal to $h.\rho.g$. ($\because$ $F/a = force/area = pressure$).

N.B. The argument remains the same even if the metal tube is inclined and not vertical, so long as the *vertical depth* of the piston remains the same. It will thus be seen that the *pressure due to a liquid column depends only upon its depth and density*, and not only any other factor like the surface area of the containing vessel etc.

9.4. The Hydrostatic Paradox. A remarkable fact follows from the above, *viz.*, that so long as the vertical height of the column of a liquid remains the same, the *pressure* exerted by it remains the same, irrespective of its actual mass or weight.

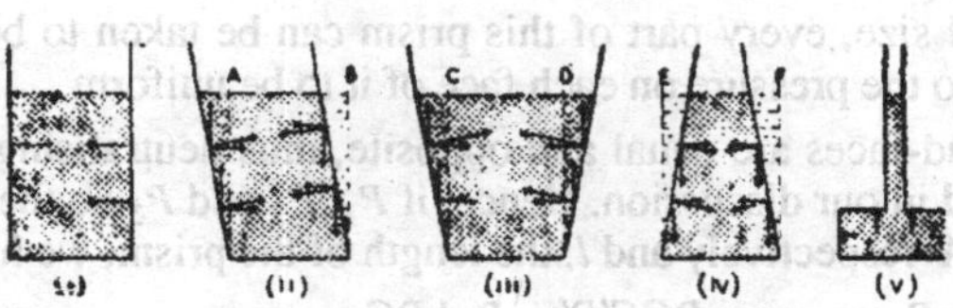

Fig. 9.2

Pascal discussed this result in his *'Treatise on the equilibrium of liquids'* published in the year 1663, with reference to vessels of different shapes, known as *Pascal's vases*, (Fig. 9.2), all having *equal bases* and *containing water upto the same vertical height h*, so that the *pressure on the base of each vessel was equal to h gm. wt/cm²* and, therefore, the *thrust* on it was *h.a gm wt* where *a* is its *area*.

He was perhaps the first person to have pointed out the paradoxical truth that even if vessel *(i)* contains 100 *lb* of water and vessel *(v)* only 1 *oz* of it, the thrust on the bases of both is the *same*. Aptly, therefore, it is called the *hydrostatic paradox*.

Strange as it may seem, but if the water in vessel *(v)* be frozen into ice and detached from its sides, the thrust exerted by this ice, on its base will be only 1 oz wt but once this ice is melted back into water, the thrust again increases to 100 *lb wt*. Th explanation of this seeming paradox is, however, simple. The ice does not exert any upward thrust on the part of the vessel opposite to the base and the latter, therefore, exerts no equal and opposite thrust on it. But the water does exert an upward thrust on it and hence receives back an equal and opposite downward thrust from it.

In case of vessel *(i)*, the thrust on the base is equal to the entire weight of the water on it.

In vessel *(ii)*, the upward component of the thrust due to the left side of the vessel supports the weight of water in it between the left side and the dotted line *A*, while the downward component of the thrust due to the right side of the vessel exerts a downward thrust on it equal to the weight of the water in between the right side of the vessel and the dotted line *B*; so that the thrust on the bases is the *same as due to a vertical column h of water*.

In vessel *(iii)*, the upward components of the thrusts due to both the left and the right sides of the vessel support the extra weight of the water in between the two sides and the dotted lines *C* and *D* and, again, therefore, *the thrust on the base is equal to that due to the cylindrical column h of water in between the dotted lines C and D*.

And, similarly, in vessel *(iv)*, the downward components of the thrust due to the two sides of the vessel exert an *extra* thrust on the base, equal to the weights of the water contained between either side and the dotted lines *E* and *F*; so that, once again, *the total thrust on the base is the same as that due to a cylindrical water column of height h*.

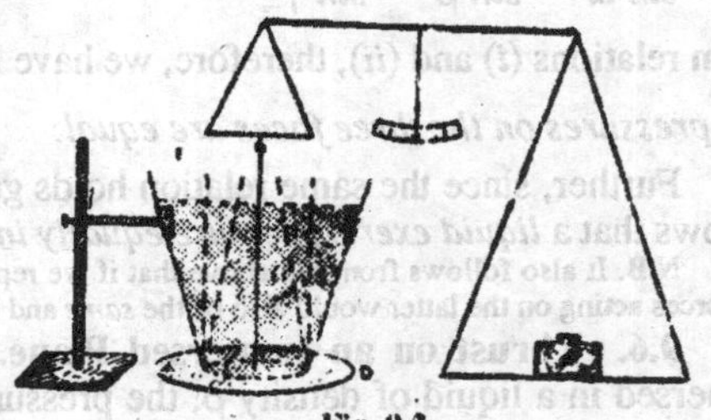

Fig. 9.3

Pascal. experimentally verified the above fact by supporting, by means of a separate stand, bottomless vases of the above shapes, one by one, on a large disc, *D*, (Fig. 9.3), suspended from the shorter pan of a hydrostatic balance and kept pressed against their bases by placing a heavy weight in the longer pan, and pouring water into the vessel. *The disc just got detached from its base as the water reached the same level in each case*, thus clearly demonstrating the equivalence of the thrust on the disc in *each* case and fully vindicating his deductions.

9.5. A liquid transmits pressure equally in all directions —Pascal's Law. Since we do not have any boundary demarcated in the interior of a liquid, *we may define pressure there as the force exerted per unit area across any plane in it*, and it can be easily shown that this pressure is exerted *equally in all directions* in the liquid.

Thus, let us consider a portion of the liquid in the form of a triangular prism, [Fig. 9.4 (*i*)] with its faces *ABC* and *A'B'C'* vertical and its edges *AA'*, *BB'* and *CC'* horizontal.

This triangular liquid prism is obviously in equilibrium under the action of the forces acting on its different faces. Let us study the inter-relation of all these forces.

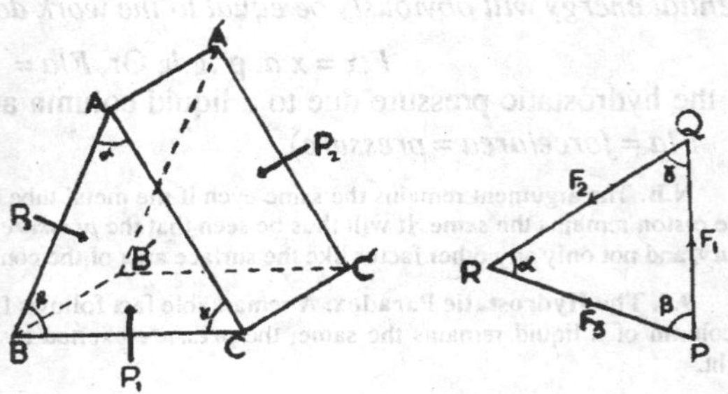

Fig. 9.4

It is clear that due to its small size, every part of this prism can be taken to be at the *same depth* from the liquid surface and also the pressure on each face of it to be uniform.

Now, the forces on the two end-faces are equal and opposite, thus neutralising each other's effect and may, therefore, be ignored in our discussion. Hence, if P_1, P_2 and P_3 be the pressures on the faces *BCC'B'*, *CAA'C'* and *ABB'A'* respectively and *l*, the length of the prism, we have

force F_1 on face $BCC'B' = P_1 \times \text{ area } BCC'B' = P_1.l.BC,$

force F_2 " " $CAA'C' = P_2 \times \text{ area } CAA'C' = P_2..l.CA,$

and force F_3 " " $ABB'A' = P_3 \times \text{area } ABB'A' = P_3.l.AB.$

Since these three forces keep the liquid-prism in equilibrium, they can be represented by the three sides of a triangle, taken in order. Let *PQR* [Fig. 9.4 (*ii*)] be this triangle of forces, with its sides *PQ*, *QR* and *RP* representing F_1, F_2 and F_3 respectively.

Then, clearly, $F_1 / \sin \alpha = F_2 / \sin \beta = F_3 / \sin \gamma$ (*Lame's theorem*)

Or, $$\frac{P_1.l.BC}{\sin \alpha} = \frac{P_2.l.CA}{\sin \beta} = \frac{P_3.l.AB}{\sin \gamma} \quad \text{or,} \quad \frac{P_1 BC}{\sin \alpha} = \frac{P_2.CA}{\sin \beta} = \frac{p_3.AB}{\sin \gamma} \quad ...(i)$$

Now, since angles *A*, *B* and *C*, of the triangle *ABC*, are respectively equal to angles α, β and γ, (the sides *PQ*, *QR* and *RP* being perpendicular to *BC*, *CA* and *AB* respectively), we have

$$\frac{BC}{\sin \alpha} = \frac{CA}{\sin \beta} = \frac{BA}{\sin \gamma} \quad ... (ii)$$ [The sides of a triangle being proportional to the sines of the angles opposite to them.

From relations (*i*) and (*ii*), therefore, we have $P_1 = P_2 = P_3$,

i.e., pressures on the three faces are equal.

Further, since the same relation holds good in whatever position the prism may be rotated, it follows that a *liquid exerts pressure equally in all directions within itself.*

N.B. It also follows from the above that if we replace the liquid-prism by a solid one, of the *same size and weight*, the forces acting on the latter would also be the *same* and hence it would also be in equilibrium.

9.6. Thrust on an Immersed Plane. If we have a plane horizontal surface of area *A*, immersed in a liquid of density ρ, the pressure *P* on it is uniform, since all points on it are at the same depth *h*, say, from the liquid surface which, as we know, is also horizontal and, therefore, parallel to the immersed surface.

So that, $$P = h.\rho.g.$$

Now, *force* or *thrust, i.e.,* F = *pressure* × *area* = $P.A. = h.\rho.g.A.$

If, on the other hand, the plane of the immersed surface be inclined at an angle to the liquid surface, (Fig. 9.5), we must first determine the thrust on a small area dA of the surface and integrate its value over the entire surface.

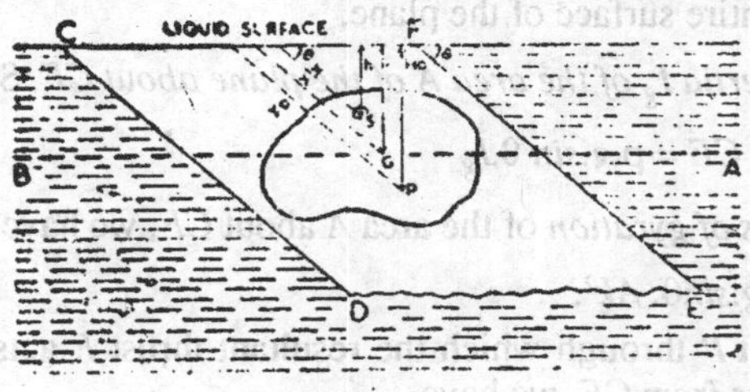

Fig. 9.5

Let h be the *depth* of this element of area dA. Then, the *thrust* dF on this area is clearly equal to $h.\rho.g.dA$.

Now, if x be the distance of this element from the line CF, in which the plane of the immersed surface meets the liquid surface, we have

$$\sin\theta = h/x. \text{ Or, } h = x\sin\theta.$$

So that, $\quad dF = \rho.g.\sin\theta.\ x.dA.$

And, therefore, the thrust on entire area A of the surface is given by

$$F = \int \rho.g.\sin\theta\, x.dA = \rho.g.\sin\theta \int x.d.A \qquad \text{...}(i)$$

Now, $x.dA$ is clearly the moment of the element dA of the surface about the line CF, and , therefore, $\int x.dA$ is the moment of the whole area A about CF, *i.e.,* $\int x.dA = A.X$, where X is the distance of the *centroid** G of the area from CF.

So that, substituting the value of $\int x.\ dA$ in relation (*i*) for F above,

we have, $\quad F = \rho.g.\sin\theta.A.X.$

Again, if H be the depth of the centroid G of the immersed surface from the liquid surface, we have $\sin\theta = H/X$.

And, therefore,
$$F = \rho.g.\frac{H}{X}.A\,X = H.\rho.g.A \qquad \text{...}(ii)$$

Clearly, $H.\rho.g$ is equal to the pressure at the centroid or the centre of area. So that, we have *resultant thrust on the immersed plane = pressure at centroid or centre of area × area of the plane.*

It should be carefully noted that the thrust on the immersed plane is quite independent of its angle of inclination (θ) *with the liquid surface.*

9.7. Centre of Pressure. Having obtained the value of the resultant thrust on the immersed plane, our next step obviously is to determine the point of the plane through which this resultant acts, --this point being known as the *centre of pressure.*

We know that the liquid pressure acts *normally* at every point of the immersed plane. So that, the force $h.\rho.g.dA$, acting on elementary areas of the plane, (like dA), are so many *like parallel forces.* We may, therefore, determine the centre of these parallel forces (*i.e.,* the point through with their resultant acts) by an application of the *principle of moments, viz., that the algebraic sum of the moments of a system of parallel forces about a given axis is equal to the moment of the resultant about the same axis.*

Now, clearly, *moment of the thrust (or force)* $h.\rho.g.dA$ *acting on area* dA about CF, (Fig. 9.5) $= h.e.g.dA.x.$

---

* This term *'Centroid'* or *'Centre of mass'* is ordinarily used synonymously with *'Centre of gravity'* and, in a uniform gravitational field, the two are one and the same point. But, in a non-uniform field, the weights of the particles are not proportional to their masses. In such a case, therefore, the weights may not form a system of parallel forces, reducing to a single resultant force, but may form a couple instead, varying with the different orientations of the body, whereas the centre of mass is quite independent of the gravitational field.

$= \rho \cdot g x \sin\theta . x^2 .dA. x = \rho . g. \sin\theta . x^2.dA.$ [$\because h = x \sin\theta$.

Therefore, *total moment of the forces on such like elementary areas dA of the plane aboutCF*

$$= \int \rho.g.\sin\theta.x^2.dA = \rho.g.\sin\theta \int x^2.dA, \quad ...(iii)$$

where the integration extends over the entire surface of the plane.

Now, $\int x^2.dA$ *is the geometrical moment of inertia* I_g *of the area A of the plane about CF*. So that,

total moment about $CF = \rho.g.\sin\theta.I_g$

And, since $I_g = Ak^2$, where k is the *radius of gyration* of the area A about CF, we have

total moment about $CF = \rho.g.\sin\theta. Ak^2$.

Again, if X_0 be the distance of the point P through which the resultant thrust F acts on the plane, *i.e., the distance of the centre of pressure from CF*, we have

moment of the resultant thrust about $CF = F.X_0$.

And, therefore, by the principle of moments, we have $F.X_0 = \rho.g.\sin\theta .Ak^2$

whence, $$X_0 = \frac{\rho.g.\sin\theta. Ak^2}{F} = \frac{\rho.g.\sin\theta. Ak^2}{H.\rho.g. A} = \frac{\sin\theta}{H}. k^2.$$

Or, $$X_0 = k^2/X. \quad ...(iv) \quad [\because H = X \sin\theta.$$

If k_0 be the *radius of gyration of the area about a parallel axis through its centroid G*, we have, by the principle of parallel axes,

$$I_g = I_0 + AX^2,$$

where I_0 is its $M.I.$ about the axis through G.

Or, $Ak^2 = Ak_0^2 + AX^2$, whence, $k^2 = k_0^2 + X^2$.

Substituting this value of k^2 in relation (*iv*) above, therefore, we have

$$X_0 = (K_0^2 + X^2)/X, \quad ...(v)$$

whence, X_0 may be easily determined.

Alternatively, equating $F.X_0$ against expression (*iii*) above, for total moment about CF due to the thrusts on elementary areas dA, we have $F. X_0 = \rho .g. \sin\theta \int x^2 . dA$. (*vi*)

And, clearly, if H_0 be the *distance of the centre of pressure from the liquid surface*, we have $H_0/X_0 = \sin\theta$; and $\therefore X_0 = H_0/ \sin\theta$

Putting this value of X_0 in relation (*vi*) above, we have

$$F.H_0 / \sin\theta = \rho . g . \sin\theta \int x^2 \theta . dA . \qquad \left[h = x \sin\theta. \text{ and } \therefore x = \frac{h}{\sin\theta} \right.$$

Or, $$\frac{F.H_0}{\sin\theta} = \rho. g. \sin\theta \int \frac{h^2}{\sin^2\theta} d.A .$$

Or, $$H_0 = \frac{\rho . g. \sin^2\theta \int \frac{h^2}{\sin^2\theta} . dA}{F} = \frac{\rho.g. \int h^2.dA}{F}.$$

$$= \frac{\rho. g \int h^2.dA}{H .\rho.g. A} = \frac{\int h^2.dA}{H.A}, \quad [\because F = H.\rho.g.A.$$

the value of the integral $\int h^2.dA$ like the expression $\int x^2.dA$, depending upon the *shape* of the immersed plane.

We thus see that whereas the distance of the centre of pressure from the liquid surface is quite independent of the density of the liquid, it depends upon the shape of the plane.

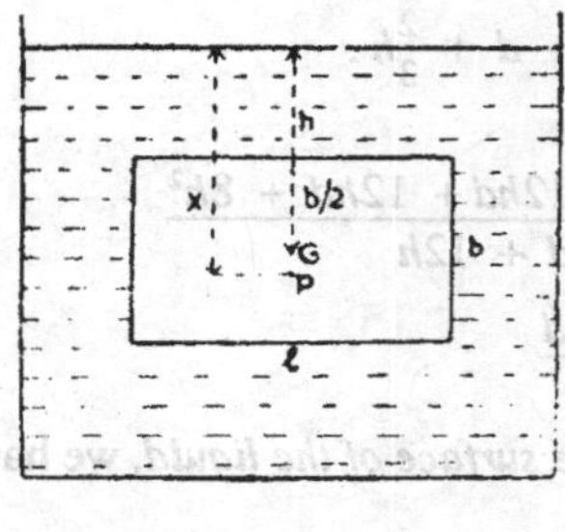

Fig. 9.6

9.8. Particular Cases of Centre of Pressure. Let us now consider some simple cases of centre of pressure on surfaces of definite geometrical shapes.

(*i*) **Centre of Pressure on a Rectangular Lamina.** Suppose we have a rectangular lamina of *length l* and *breadth b*, immersed vertically in a liquid to a depth h below its free surface, (Fig. 9.6). The, since for a rectangular lamina $I_0 = A.k_0^2 = l.b^3/12$, (see page 288), where I is its *geometrical moment of inertia* about an axis through its centre of area (or centroid) and parallel to its length, and k_0, its *radius of gyration* about this axis, we have

$A.k_0^2 = l.b.b^2/12 = Ab^2/12$, whence, $k_0^2 = b^2/12$ [$\because$ $l \times b = A$, the area of the lamina.

And clearly, $X = h + (b/2)$. So that, we have

$$X_0 = \frac{k_0^2 + X^2}{X} = \frac{\frac{b^2}{12}\left(h + \frac{b}{2}\right)^2}{h + (b/2)} = \frac{\frac{2}{3}(b^2 + 3hb + 3h^2)}{2h + b}$$

whence, the position of the centre of pressure for the rectangular lamina may be easily calculated for any depth h.

Clearly, therefore, *if the lamina be just submerged in the liquid, i.e., with its upper edge just lying in the liquid surface*, $h = 0$, and we, therefore, have $X_0 = \frac{2}{3}b^2/b = \frac{2}{3}b$, *i.e., the centre of pressure, in this particular case, lies two-thirds below the top of the lamina.*

(*ii*) **Centre of Pressure on Circular Lamina.** Let the centroid of a circular lamina, of radius r, lie at a depth X from the free surface of the liquid. Then, since k_{02} for a circular lamina *is* $r^2/4$, we have

$$X_0 = \frac{k_0^2 + X^2}{X} = \frac{\frac{r^2}{4} + X^2}{X} = \frac{r^2}{4X} + X$$

Or, the centre of pressure lies at $\left(\frac{r^2}{4X} + X\right)$ below the liquid surface.

If, *however, the lamina be submerged in the liquid, with its edge just touching the liquid surface, we have* $X = r$, and, therefore,

$$X_0 = \frac{r^2}{4r} + r = \frac{r}{4} + r = \frac{5}{4}r.$$

Or, the centre of pressure, in this case, lies at $5r/4$ from the liquid surface.

(*iii*) **Centre of Pressure on a Triangular Lamina.** Here, two cases arise, *viz.*, *(a) when the lamina is immersed upright into the liquid, i.e., with its vertex up and base down and (b) when it is immersed upside down, with its base up and vertex down.*

(*a*) **When the Lamina is immersed upright.** Let h be the *height* of the lamina and let its apex be at a depth d below the free surface of the liquid. Then, clearly,

$$k_0^2 = h^2/18 \quad \text{and } X = d + \frac{2}{3}h$$

its centroid lying $2h/3$ below the apex.

And, therefore, $$X_0 = \frac{\frac{h^2}{18} + \left(d + \frac{2}{3}h\right)^2}{d + \frac{2}{3}h} = \frac{h^2}{18\left(d + \frac{2}{3}h\right)} + d + \frac{2}{3}h.$$

$$= \frac{h^2}{18d + 12h} + d + \frac{2}{3}h = \frac{h^2 + 18d^2 + 12hd + 12hd + 8h^2}{18d + 12h}$$

$$= \frac{9h^2 + 18d^2 + 24hd}{18d + 12h} = \frac{3h^2 + 6d^2 + 8hd}{6d + 4h}$$

If, *however, the vertex of the lamina just touches the free surface of the liquid,* we have $d = 0$. In *this case,* therefore,

$$X_0 = 3h^2/4h = \frac{3}{4}h,$$

i.e., the centre of pressure *now* lies 3/4*ths* down the height of the lamina.

(*b*) **When the lamina is immersed upside down.** Again, let the depth of the base of the lamina be d from the free surface of the liquid. Then, we have $k_0^2 = h^2/18$ and $X = (d + \frac{1}{4}h)$.

And, therefore, $$X = \frac{\frac{h^2}{18} + \left(d + \frac{1}{3}h\right)^2}{\left(d + \frac{1}{3}h\right)} = \frac{\frac{h^2}{6} + 3\left(d + \frac{1}{3}h\right)^2}{3d + h}$$

$$= \frac{3h^2 + 18d^2 + 12hd}{6(3d + h)} = \frac{\frac{1}{2}(h^2 + 6d^2 + 4hd)}{3d + h}.$$

And, *if the lamina be just submerged in the liquid, with its base in the surface of the liquid, we have* $d = 0$. In that case, therefore,

$$X_0 = h^2/2h = \frac{1}{2}h,$$

i.e., the centre of pressure lies a distance $\frac{1}{2}h$ below the liquid surface.

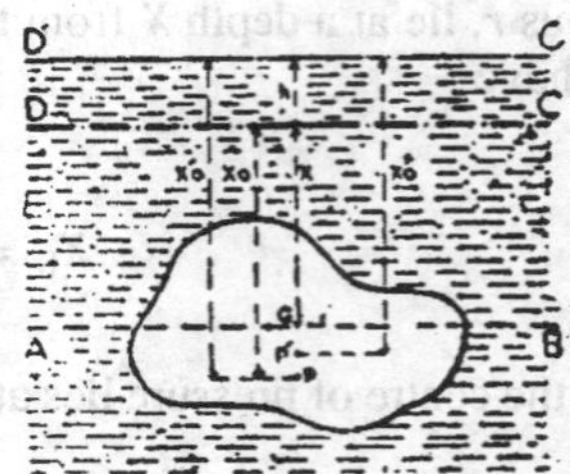

Fig. 9.7

9.9. Change of depth of centre of pressure. Let a plane lamina of area A be immersed in a liquid such that its centroid G is at a depth from the liquid surface CD, (Fig. 9.7).

The, if k_0 be the radius of gyration of the lamina about the axis AB passing through G, in the plane of the lamina and parallel to the liquid surface CD, its radius of gyration about a parallel axis, lying in the liquid surface, will clearly be k_1, such that $k_1^2 = k_0^2 + X^2$.

Now, let the level of the liquid surface be raised through a distance h by adding some more liquid to that already present. Then, clearly, radius of gyration k_2 of the lamina about a parallel axis to AB, and lying in this elevated liquid surface $C'D'$, *is given by*

$$k_2^2 = k_2^2 + (X + h)^2.$$

Subtracting one from the other, we have

$$k_2^2 - k_1^2 = [k_0^2 + (X + h)^2]\ (k_0^2 + X^2).$$

$$= k_0^2 + X^2 + 2Xh + h^2 - k_0^2 - X^2 = 2Xh + h^2.$$

Or, $$k_2^2 - k_1^2 = h(2X + h). \qquad \text{....}(i)$$

If X_0 be the depth of the centre of pressure of the original amount of the liquid from its free surface, we have

$$X_0 = k_1^2/X.$$ [See relation (*iv*), page 332.

So that, after the addition of a liquid layer of thickness *h*, its depth becomes

$$X_0' = X_0 + h = \frac{k_1^2}{X} + h = \frac{k_1^2 + Xh}{X},$$

whence, $k_1^2 + Xh = X \, . \, X_0'$ Or, $k_1^2 = X.X_0' - Xh = X(X_0' - h).$

And, if X_0'' be the depth of the *new* centre of pressure from the raised surface of the liquid, after the addition of the liquid layer, we have $X_0'' = k_2^2/(X + h)$. Or, $k_2^2 = X_0''(X + h)$

So that, $$k_2^2 - k_1^2 = X_0''(X + h) - X(X_0' - h)$$

From relations (*i*) and (*ii*), therefore, we have

$$X_0''(X + h) - X(X_0' - h) + h(2X + h).$$

Or, $$X_0''(X + h) - X X_0' + Xh = 2Xh + h^2.$$

Or, $$X_0''(X + h) = 2Xh + h^2 - Xh + X \, . \, X_0' = h^2 + Xh + X X_0'$$

$$= h(h + X) + X \, . \, X_0'.$$

Or, $$X_0''(X + h) = h(h + X) + X(X_0 + h).$$ $[\because X_0' = (X_0 + h)$.

Dividing throughout by $(X + h)$, we have $X_0'' = h + \dfrac{X.(X_0 + h)}{(X + h)}$.

And, therefore, *the shift in the position of the centre of pressure is given by*

$$X_0' - X_0'' = (X_0 + h) - \left[h + \frac{X(X_0 + h)}{(X + h)} \right]$$

$$= X_0 - \frac{X(X_0 + h)}{(X + h)} = \frac{X_0 h - Xh}{X + h}$$

Or, $$X_0' - X_0'' = \frac{h(X_0 \, X)}{(X + h)}$$

It will also be easily seen that the distance between the depths of the new centre of pressure and the centroid of the lamina is given by

$$X_0'' - (X + h) = h + \frac{X(X_0 + h)}{X + h} - (X + h) = \frac{X \, . \, X_0 - X^2}{X + h}.$$

Or, $$X_0'' - (X + h) = X(X_0 - X) \,/\, (X + 2),$$

which approaches *zero* as *h* approaches infinity.

Thus, *the greater the depth of the liquid, the nearer does the centre of pressure come to the centroid of the lamina; so that, at an infinite distance, the two must just coincide with each other.*

9.10. Principle of Archimedes. Imagine a body *ABCD* to be immersed in a liquid, (Fig. 9.8), and a vertical line *GA* to travel round it, touching

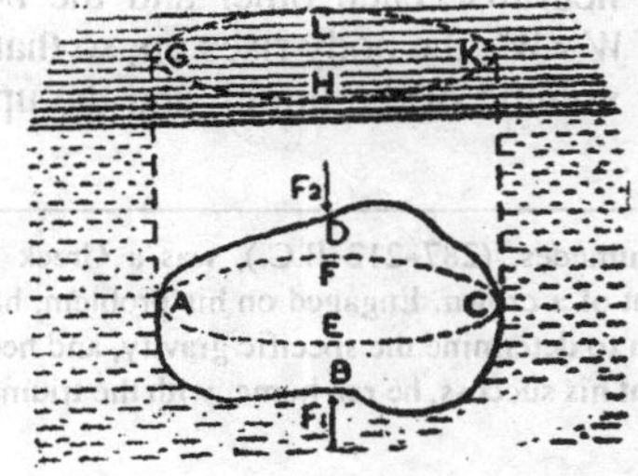

Fig. 9.8

it along the line $AECF$ and meeting the liquid surface in the curve $GIIKL$.

Then, clearly, the resultant upward thrust F_1 on the surface $ABCEA$ and the resultant downward thrust F_2 on the surface $ADCEA$ are given by the weights of the liquid that would occupy the spaces $ABCKG$ and $ADCKG$ respectively, acting through their respective centres of gravity. So that, the resultant thrust on the whole body is given by

$F_1 - F_2$ = *weight of liquid occupying space ABCKG minus weight of liquid occupying space ADCKG*

= *weight of liquid occupying space ABCD*

= *weight of liquid equal in volume to that of the immersed body*

= *weight of liquid displacement by the body.*

In other words, when a body is immersed in a liquid, it experiences an upward thrust equal to the weight of the liquid displaced by it. It can easily be shown that the same is also true for a body which is only partially immersed in the liquid, the upward thrust on it being equal to the weight of the liquid displaced by its *immersed part*. We may, therefore, generalise and state that

when a body is wholly or partly immersed in a liquid, it experiences an upthrust equal to the weight of the liquid displaced by it, (i.e., by its immersed part).

This is known as the *Principle of Archimedes**, for it was he who first enunciated it.

The point where this upthrust acts is obviously the *c.g.* of the displaced liquid, which is called the **centre of buoyancy**, the upthrust being referred to as the **force of buoyancy.**

N.B. The applications of *Archmkedes Principle* are many and various. It gives us the method of determining specific gravities or densities of liquids as well as the instruments, known as *Hydrometers*, with which the Degree students are no doubt already familiar.

9.11 Equilibrium of Floating bodies. A body, immersed wholly or partly in a liquid, is subject to two forces, *viz., (i) its own weight W, acting vertically downwards at its c.g., G,* and *(ii) the upthrust W′, acting vertically upwards at its centre of buoyancy B*, (Fig. 9.9).

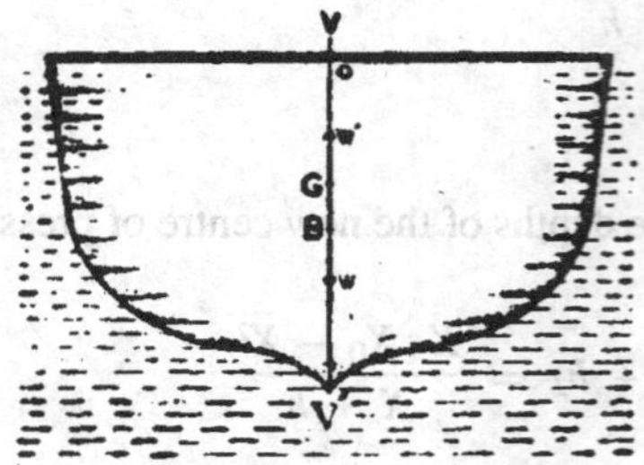

Fig. 9.9

If these two points of application, (G and B), of the two forces respectively, coincide or lie in the same vertical line, called the **centre line**, *the body sinks, just remains suspended (or floating), or rises up, according as W is greater than, equal to, or less than W′*

For, obviously, if $W > W'$, the body sinks further down, displacing more and more of the liquid and thereby increasing the upthrust until the two balance each other, and the body just stays there, *i.e.*, is in equilibrium.

If $W = W'$, then obviously, the two forces on it are equal and opposite and their line of action being the same, they just neuralise each other and the body remains suspended or floating in the liquid. And, finally, if $W < W'$, the body rises up, so that a lesser volume of it is under the liquid, *i.e.*, it displaces a smaller volume of the liquid, and the upthrust on it is now less. This rise of the body continues

* Archimedes, (287–212 B.C.), was a Greek philospoher. He was asked by *King Heiro*, at *Syracuse* to test the gold-content of a crown. Engaged on his problem, he suddenly discovered the law of upthrust, while taking a bath, which enabled him to determine the specific gravity, and hence the quantity of gold in the crown, without in any way damaging it. Overjoyed at his success, he ran home, with the triumphant cry 'Eureka', 'Eureka' — I have found it, I have found it'

until the upthrust is just equal to the weight of the body, and the body then continues to float in that very position.

Thus, two conditions are necessary for the equilibrium of a floating body, *viz.*, (*i*) *its weight must be equal to the weight of the displacement liquid** *and (ii) the c.g. of the body and the centre of buoyancy of the displaced liquid must either coincide with each other*** *or lie in the same vertical line.*

9.12. Stability of Equilibrium. If the floating body be tilted a little to one side or the other from its original equilibrium position, through a small angle θ, (Fig. 9.10), *so that the weight of the displaced liquid or the upthrust on it, remains the same,* then, since *the shape of the displaced liquid changes,* its centre of buoyancy shifts a little, say, into the position *B'*; so that, the vertical line drawn through the *new* centre of buoyancy *B'*, meets the old centre line in *M*. This point *M* is called the **metacentre** of the body and the distance *MG'*, where *G'* is its centre of gravity, is called its **metacentric height.**

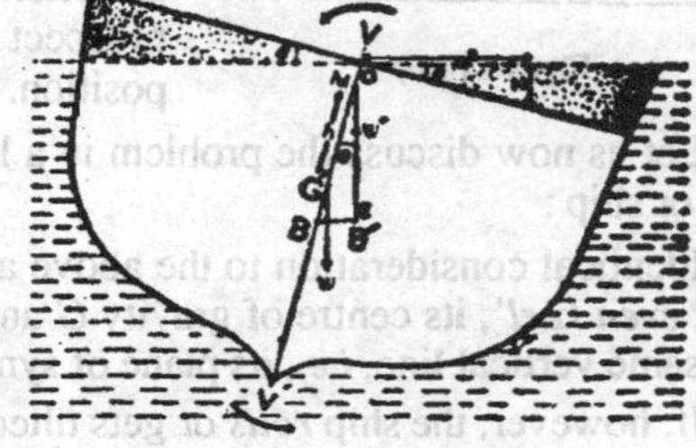

Fig. 9.10

Now, whether the *metacentre* (*M*) coincides with, lies above, or below, the *shifted position of the centre of gravity* (*G'*) of the body, depends upon the shape of the body and determines whether the body will be in *neutral, stable* or *unstable* equilibrium.

(*i*) Thus, in the case of a sphere, [Fig. 9.11 (*i*)], a tilt this way or that brings about no change in the shape of the displaced liquid, the metacentre coinciding with the *c.g.* of the body all the time.

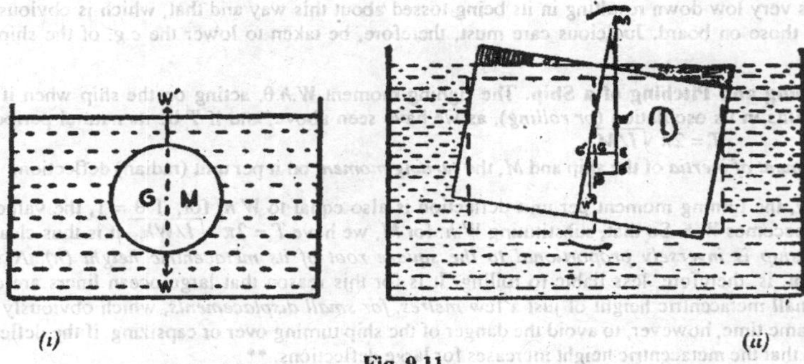

Fig. 9.11

It is, therefore, in *neutral equilibrium* and continues to float in all positions.

(*ii*) In the case of a *rectangular body,* floating in a liquid, as shown in Fig. 9.11 (*ii*). *M* lies above *G'*, the position of the *c.g.* of the body in its tilted position, and a couple is formed by *W* and *W'*, the two *equal, opposite, parallel* and *non-collinear* forces acting at *G'* and *B'*, which represent the shifted positions of the *c.g.* of the body and the centre of buoyancy of the displaced liquid respectively. This couple tends to rotate the body back into its original position, thus making its equilibrium a stable one.

---

* It is for this reason that the weight of a ship or boat is often referred to as its *displacement,* —the weight of the water displaced by it being equal to its own weight.

** As happens in the case of a spherical body.

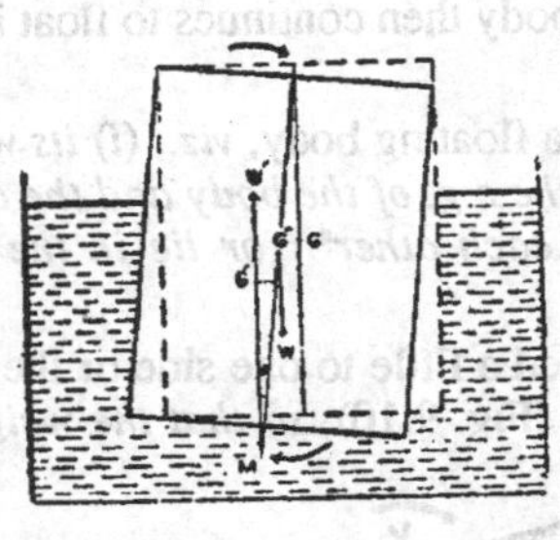

Fig. 9.11 *(iii)*

Obviously, the moment of this *restoring couple* = $W.G'M \sin\theta$. It has been appropriately called the **'righting moment'**, (particularly in the case of ships and other floating vessels), because it tends to bring the body, or to *'right it'* into its original position.

(iii) In the case of a rectangular body, floating in the manner shown in Fig. 9.11 *(iii)*, M lies below G', and the body is, therefore, in unstable equilibrium. For, the couple formed by W and W' tends here to rotate the body further in the *same* direction in which it has been tilted already. There is thus no prospect of its coming back (or being righted) into its original position.

Let us now discuss the problem in a little more detail, with particular reference to a floating vessel or ship :

Identical consideration to the above applied also to a floating ship; so that, when the ship is *'on an even keel'*, its centre of gravity G and the centre of buoyancy B, of the displaced water, lie in the same vertical line, *i.e.*, its plane of symmetry (VV') is vertical (Fig. 9.9).

If, however, the ship *rolls* or gets tilted through an angle θ, (Fig. 9.10), its plane of symmetry (VV') is no longer vertical, and , although this rolling or tilting does not alter its *c.g.* with respect to the ship, the centre of buoyancy shifts to B', giving rise to the *righting moment* $W.G'M \sin\theta$ or $W.h \sin\theta$, where h denotes the metacentric height $G'M$. If θ be small, so that $\sin\theta = \theta$, this righting moment is equal to $W.h.\theta$.

It will thus be clear that *the greater the values of h, the metacentric heigh of a ship (or a floating body, in general), the greater the stability of its equilibrium.* It is for this reason that heavy cargo is stowed as low as possible in a ship or that it is provided with a *leaden keel*, to lower its *c.g.* or *to increase h.**

N.B. The lowering of the *c.g.* is, however, not quite so desirable beyond a certain point. For, due to the waves in the sea, the ship is subject to lateral forces in different direction and the moment of these forces can be quite considerable if the *c.g.* of the ship is very low down resulting in its being tossed about this way and that, which is obviously most unpleasant and annoying to those on board. Judicious care must, therefore, be taken to lower the *c.g.* of the ship within reasonable limits.

9.13. Rolling and Pitching of a Ship. The righing moment $W.h.\theta$, acting on the ship when it is tilted through a small angle θ results in its oscillation *(or rolling)*, as we have seen above, and if T be its natural period of oscillation, we have

$$T = 2\pi\sqrt{I/M}$$

where I is the *moment of inertia* of the ship and M, the *turning moment* on it per unit (radian) deflection.

Obviously, the turning moment per unit deflection is also equal to $W.h$; for, if $\theta = 1$, the value of the righting or turning moment becomes $W.h$. So that, substituting $W.h.$ for M, we have $T = 2\pi\sqrt{I/Wh}$. It is thus clear that *the period of rolling (T) of a ship is inversely proportional to the square root of its metacentric height (h).* A ship, with a small metacentric height, is, therefore, less liable to rolling. It is for this reason that large ocean liners are designed to have a comparatively small metacentric height of just a few *metres, for small displacements*, which obviously makes them much steadier. At the same time, however, to avoid the danger of the ship turning over or capsizing, if the deflections be large, the designing is such that the metacentric height increases for large deflections. **

Similarly, to avoid *'pitching'*, or tilting of a ship in the direction of its length, its metacentric height in this direction also is suitably adjusted.

9.14. Determination of Metacentric Height. The displacement of the ship through an angle θ causes a wedge-shaped portion of the ship, (shown shaded in Fig. 9.10), to be immersed on the right hand side and an equal wedge-shaped portion of it to rise out of the water, on the left hand side. Let these wedge-shaped portions be divided into a number of elementary vertical prisms by

* The stability may also be increased by making the ship quite broad at the water line, *i.e.*, the line along which it is just touched by athe water -surface.

** The rolling motion of a ship can be greatly minimised by the use of a Gyrocompass.

planes perpendicular to the water surface, on either side, and consider one such prism, of height H, at a distance x from O, where the plane of symmetry meets the water-surface.

Then, clearly, $H = x \tan\theta = x.\theta$. [θ being small.

If the base area of the prism be dA, we have

volume of the prism $= x.\theta.dA$, and its mass $= x.\theta.dA.\rho$,

where ρ is the *density* of water.

Clearly, therefore, *weight of the prism*, or *the weight of the water displaced, i.e., the upthrust on the prism* $= x.\theta.dA.\ \rho.g.$ *and its moment about* $O = x.\theta.dA.\rho.g.\ x = \rho.g.\theta\ .\ x^2.dA$.

Similarly, considering the equal wedge-shaped portion on the left hand side, we find that there is a loss of upthrust due to its rising out of the water, whose moment about O is, obviously, also equal to $\rho.g.\ x^2.dA$, the direction being *anticlockwise* in either case. Hence, the moment of the couple acting on the ship (or a floating body, in general), due to its displacement θ, is given by

$$\int \rho.g.\ \theta\ .x^2.\ dA = \rho.g.\theta \int x^2.\ dA = \rho.g.\theta A\ k^2 \qquad ...(i)$$

where $\int x^2.dA = I_g = Ak^2$, the geometrical moment of inertia of the surface-plane of the liquid about the axis through O, k being its *radius of gyration* about this axis.

This displacement (θ) of the ship being small, the volume V of water displaced remains unaffected by it, and the upthrust $\rho.g.V$, due to this displaced water, acts through its new centre of buoyancy (after the displacement θ). The floating body or ship is thus acted upon by a couple equal to

$$\rho.g.V.BM \sin\theta = \rho.g.V.BM.\ \theta. \qquad ...(ii)\ [\ \theta \text{ being small.}$$

Equating the two values of the couple, we have

$$\rho.g.V.BM.\theta = \rho.g.\ \theta.Ak^2.$$

Or $\quad V.BM = Ak^2.\quad$ Or, $\quad BM = Ak^2/V.$

So that, the *metacentric height* h of the ship or the vessel is given by

$G'M = BM - BG' = \dfrac{Ak^2}{V} - BG'$ and may thus be easily determined.

Alternatively, in the case of a ship, its metacentric height may be easily determined by moving a known weight w from one point A to another point B across the deck, say through a distance d, *i.e.*, $AB = d$.

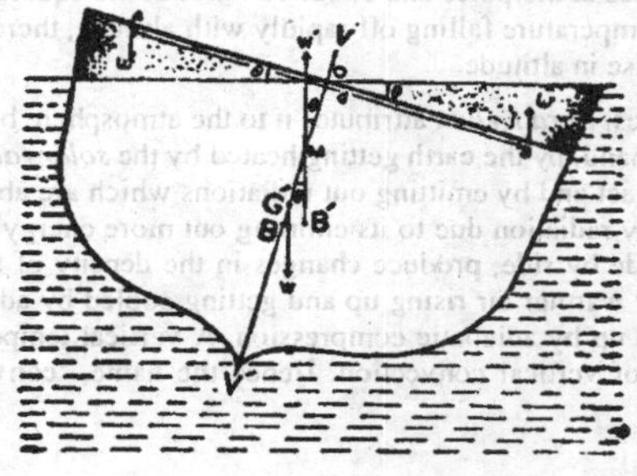

Fig. 9.12

Now, this shift of weight w from A to B is equivalent to an *upward* force w at B and a *downward* force w at A, (Fig. 9.12), thus constituting a *couple*, of moment $w.d \cos\theta$. For equilibrium, therefore, this must be equal to the couple due to the weight W of the ship and an equivalent upward thrust at the new centre of buoyancy B', *i.e.*, equal to the couple of moment $W.G'M \sin\theta$. [See above]

Or, $\quad W.G'M \sin\theta = w.d \cos\theta.$

Or, $\quad G'M = \dfrac{w.d}{W} \cdot \dfrac{\cos\theta}{\sin\theta} = \dfrac{w.d}{W} .\cot\theta = \dfrac{w.d}{W} \cdot \dfrac{1}{\tan\theta}.$

Or, $\quad G'M = \dfrac{w.d}{W.\theta}.\quad$ [θ being small, $\tan = \theta$.

Thus, knowing w, W, d and θ, we can easily calculate the metacentric height of the ship.

9.15. Pressure due to a Compressible Fluid or a Gas. A gas differs from a liquid in that, unlike the latter, it is highly compressible* and therefore, also highly expansible, tending to expand perpetually and indefinitely.

The *pressure exerted by a gas is thus fundamentally different in the nature of its cause from that of a liquid* and cannot be taken to be proportional to the height of the gaseous column, for the simple reason that, except in the case of a small volume of a gas, the density goes on progressively increasing as we go further down the column, due to the layers above pressing down upon the layers below, thus giving rise to a well defined *pressure gradient* all along the column.

So that, whereas a liquid exerts pressure only under the action of gravity, *i.e.*, due to its weight, or due to an external force applied to it, as, for example, when it is pressed down by a piston, *the pressure due to a gas is entirely a consequence of the incessant mobility and the kinetic energy of its molecules, or due to what* **Boyle** *called the 'spring' of the gas.*

The mad and random motion of the gaseous molecules results in their colliding not only against each other but also against the walls of the containing vessel, and it is this bombardment of the walls by the fast and haphazardly moving molecules that causes the pressure. (See Chapter XV).

We are here concerned mainly with the pressure exerted by the gaseous mantle or envelope, surrounding us over land and sea alike and in all latitudes, which we call '*air*' or '*atmosphere*' and which, as we know, is a mixture of a number of gases and vapours. In pursuance of its inherent property of indefinite expansion, this air or atmosphere should expand to an infinite distance above the earth, but the earth's gravitational attraction on a huge mass like it sets a limit to its expansion. Even so, it has been known to exist up to a height of 300 *miles* from the surface of the earth, although even at 25 *miles* or so, its density and pressure start falling off so rapidly that at altitudes above 300 *miles*, it may be said to be as good as non-existent, with just a void or a vacuum beyond.

Now, the atmosphere can be divided up into two very distinct regions, *viz.*, (*i*) a lower region, called the *troposphere* or the *convective zone*, and (*ii*) an upper region, called the *stratosphere* or the *advective zone*, the surface of separation of the two being known as the **tropopause**, which varies with the latitude and falls from a height of about 14 *kilometres* at the equator to about 8 or 10 *kilometres* at the poles, and is found to be higher in summer than in winter.

(*i*) **The Troposphere**. This extends to a height of about 6 miles at the poles and about 10 *miles* at the equator, *with a vertical distribution of temperature as its chief characteristic*, the temperature falling off rapidly with altitude, there being a vertical temperature gradient or a *lapse rate*** of 1°C per 500 *feet* rise in altitude.

This *temperature gradient* is probably due to a variety of causes. *Lord Kelvin* attributes it to the atmosphere being in a state of convective equilibrium, which is brought about, on the one hand, by the earth getting heated by the *solar radiation* falling on it+ , and then warming up the layers above it by direct contact and by emitting out radiations which are absorbed by them, and, on the other, by the lower atmosphere getting cooled by radiation due to its emitting out more energy than it absorbs at the ordinary temperature. The two processes, going on side by side, produce changes in the density of the air, conducive to the setting up of vertical convection currents, the lower warmer air rising up and getting cooled by adiabatic expansion and the upper, colder air coming down and getting heated up by adiabatic compression. A vertical temperature gradient is thus established and maintained throughout this region of vertical convection. Hence the name, **'convective zone,'** also given to it.

---

* An idea of the high compressibility of a gas, compared with that of a liquid, can be had from the fact that whereas the density of sea-water at a depth of 5 miles is about the same as that of the surface layer, the density of the atmosphere at the same height above sea-level is reduced to one quarter of that at the latter.

** Usually, a vertical *temperature gradient* is taken to be the fall in temperature per 100 *metres* rise in altitude and the *lapse rate*, as the fall in temperature per *one kilometre* rise in altitude.

+ Little or no heat is absorbed by the air during the passage of the solar radiation to the earth through it, and whatever little is, is distributed over too large a mass to be able to produce any appreciable rise in its temperature, this absorption being the *same* at all altitudes..

This seems to be amply borne out by the fact that the lapse rate for dry air, calculated on this assumption, comes out to be 3°C per 1000 *ft*, *which, though appreciably higher than the observed value, is quite understandable, considering that the air is really never 'dry'* and the moisture present in it inevitably tends to lower the lapse rate.

(*ii*) **The Stratosphere.** Also known as the **adjective zone**, it is the region above the troposphere, where the vertical convection, referred to above, becomes much too feeble, with the temperature falling to such an extent that the heat radiated out is equal to the heat absorbed from radiations from the earth and the solar radiation passing through it, there being set up a *radiative equilibrium* in the region, the temperature remaining constant at about 55°C,—hence the name **'isothermal layer'** also given to it.

It will thus be readily seen that the stratosphere is a direct consequence of, and is characterised by, the cessation of vertical convection and the setting up of a radiative equilibrium, with the temperature constant at 55°C up to a height of 300 *miles* or so, after which it probably shoots up to 700°C or thereabouts.

9.16. Measurement of Atmospheric Pressure. The instruments used to measure the atmospheric pressure are known as **barometers**, one of the best forms of which is the cistern-type **Fortin's barometer.** Another handy and portable type of barometer is the **Aneroid barometer,** —(from '*a*'—without, and '*neros*'—liquid)—as no mercury or any other liquid is used in it. We have already studied these in good detail in the junior classes and shall not, therefore, repeat them here. Instead, we shall pass on to a consideration of the corrections that must be applied to the readings obtained from them.

9.17. Correction of Barometric Reading. Although the *Fortin's barometer* is quite an efficient instrument, a few corrections have to be applied to its readings for greater accuracy. We shall consider here only two important ones of them, *viz.*,

(*i*) *correction for the expansion of the brass scale*, on which the reading is taken, and is usually calibrated at 0°C.

(*ii*) *correction for expansion of mercury*, and consequent lowering of its density.

(*i*) Let the temperature, at which the reading is taken, be t °C, and let H_t be the observed reading at this temperature. Then, H_t is in fact just the value of the divisions of the scale, *correct only at* 0°C. If, therefore, α be the coefficient of linear expansion of brass, the correct length at t°C, is given by $H = H_t(1+\alpha t)$ *cm*.

(*ii*) Again if v_0 and ρ_0 be the volume and density of a certain mass m of mercury at 0°C, and v_t and ρ_t, its volume and density respectively, at t °C, we have $m = v_0 . \rho_0 = v_t . \rho_t$.

Or, $\dfrac{v_t}{v_0} = \dfrac{\rho_0}{\rho_t}$. Or, $\dfrac{v_0(1+\gamma t)}{v_0} = \dfrac{\rho_0}{\rho_t}$ [where γ is the coefficient of cubical expansion of mercury.

Or, $\rho_0 / \rho_t = 1 + \gamma t$, whence $\rho_0 = \rho_t(1+\gamma t)$

Now, clearly, $H_0 . \rho_0 . g = H . \rho_t . g$,

where H_0 *cm* is the true barometric height at 0°C.

Or, $H_0 . \rho_0 = H.\rho_t$. Or, $H_0 . \rho_t(1+\gamma t) = H . \rho_t + H_t(1+\alpha t) . \rho_t$,

whence, $H_0 = \dfrac{H_t(1+\alpha t) . \rho_t}{\rho_t(1+\gamma t)} = H_t(1+\alpha t)/(1+\gamma t)\ H_t[1 + (\alpha - \gamma)_t]$.

neglecting squares and higher powers of α and γ.

Or, $H_0 = H_t[1 - (\gamma - \alpha)t]$.

For *mercury*, $\gamma = .00018$ and for *brass*, $\alpha = .000019$.

$\therefore$ $H_0 = H_t[1 - (.00018 - .000019)t]$. Or $H_0 = H_t(1 - .000161t)$ *cm*

whence, the barometric height at 0°C can be easily calculated.

Other errors, due to pressure of mercury vapour and capillarity etc., are much too small for tubes of reasonably wide bores, and are, therefore, usually neglected.

9.18. Change of Pressure with Altitude. Consider two points A and B, distance dx apart, vertically below each other, in air, (Fig. 9.13). If A be at a height x above the ground, the height of

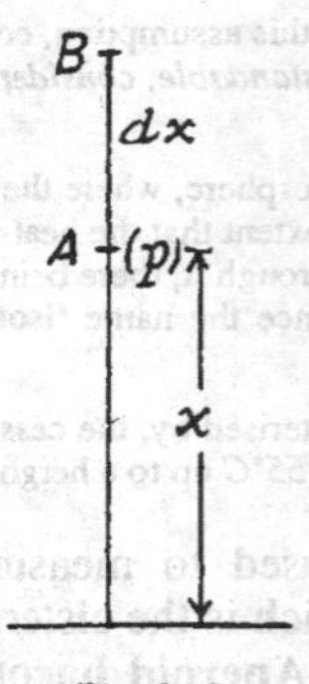

Fig. 9.13

B from the ground is obviously $(x + dx)$. Since the density of air and, therefore, its pressure, decreases with altitude, for a pressure p at A, that at B will be $p-dp$, say. If, therefore, ρ be the density of air between A and B, and g, the acceleration due to gravity, we have $-dp = \rho g.dx$, ...(*i*)

the –ve sign being used, because the pressure decreases with height.

If the temperature of the air be constant, $\rho \propto p$.

Or, $\rho = K.p$, [∵ $1/V \propto p$, (Boyle's law) and $1/V \propto \rho$.]

where K is a constant, equal to ρ/p

Substituting this value of ρ in relation (*i*) above, we have

$$-dp = K.p.g.dx. \quad \text{Or,} -dp/p = K.g.dx$$

Or, $$dp/p + K.g.dx = 0$$

Integrating this, we have $\log_e p + K.g.x = \text{a constant } C.$ ... (*ii*)

Now, if the pressure at heights h and H be p and P respectively, we have, from (*ii*) above,

$\log_e p + K.g.h = C$, ...(*iii*) and $\log_e P + K.g.H = C$ ...(*iv*)

∴ subtracting (*iv*) from (*iii*), we have

$$\log_e p - \log_e P = K.g.H - K.g.h = K.g.(H-h).$$

Or $$\log_e (p/P) = K.g.(H-h).$$ ...(*v*)

Thus, $$(H-h) = \frac{\log_e (p/P)}{K.g}$$

Or, substituting the value of K, *i.e.*, ρ/p, *in* (*vi*), we have

$$(H-h) = \frac{\log_e(p/P).p}{\rho.g} = \frac{(\log_e p - \log_e P).p}{\rho.g}.$$

Thus, knowing p, P, ρ and g, the altitude $(H-h)$ can be easily determined.

In the above treatment, it has been assumed that the temperature of the air, or the atmosphere, remains the *same* throughout. This, as we know, is far from being the case. Nevertheless, the result is accurate enough for the determination of small heights.

If now we have a number of heights, h_1, h_2, h_3 etc., in *arithmetical progression*, then

$$(h_2 - h_1) = (h_3 - h_2) \text{ and so on.}$$

And, if p_1, p_2, p_3 etc. be the pressures at these heights, we have from (*v*) above,

$\log_e (p_1/p_2) = K.g.(h_2 - h_1)$, and $\log_e (p_2/p_3) = K.g.(h_3 - h_2)$.

Since $(h_2-h_1) = (h_3-h_2)$, we have $\log_e (p_1/p_2) = \log_e (p_2/p_3)$.

Or, $$p_1/p_2 = p_2/p_3.$$

i.e., p_1, p_2, p_3 etc. are in *geometrical progression.*

Thus, we see that *as the height or altitude increases in arithmetical progression, the pressure decreases in geometrical progression.*

Note. To convert logarithms to the base e into common logarithms, (*i.e.*, to the base 10), multiply the former by 2.302.

9.19. Units of pressure. Pressure, in general, is measures in *dynes* $/cm^2$ in the *C.G.S system*, in *lb* $/ft.^2$* in the *F.P.S. system* and in *newton* $/m^2$ in the *M.K.S.* (or the *S.I*) system, this last being also called the *pascal.*

High pressures are commonly expressed in ***bars***, where *one bar (b)* = $10^5 N/m^2$, and low pressures in ***millibars***, *one millibar (mb) being equal to* 10^2 *N/m²*.

In industry, they use a still bigger unit, namely, the ***hectobar (hb)*** = $10^7 N/m^2$.

WORKED EXAMPLES

1. A rectangular cistern 2 ***ft*** **long and 5** ***ft*** **wide is filled with water to a depth of 3** ***ft.*** **Taking one** ***cuft*** **of water to weigh 62.5** ***lb,*** **find the magnitude and position of the resultant fluid thrust on each side.**

Fig. 9.14

(*a*) Here, clearly (Fig. 9.14).

depth of water = 3 *ft.*

∴ centre of area for each side of the cistern = 3/2 = 1.5 *ft.*

∴ *pressure at centre of area* = *h*.ρ *g*.

= 1.5 × 62.5 × 32 *poundals*

= 1.5 × 62.5 *lb wt.*

Now, area of each longer side in contact with water = 3 × 3 = 9 *sq ft.*

and area of each smaller side in contact with water = 2 × 3 = 6 *sq ft.*

Since *thrust* = *pressure at centroid or centre of area* × *area,*

we have *pressure on each longer side* = 1.5 × 62.5 × 9 = 843.7 *lb wt.*

and, *pressure on each smaller side* = 1.5 × 62.5 × 6 = 562.6 *lb wt.*

And, *centre of pressure* $= \frac{2}{3} \times depth = \frac{2}{3} \times 3 = 2\,ft.$

2. Find the position of the centre of pressure of a triangular plate immersed in a liquid with its plane vertical and one side in the surface.

ABC is a vertical triangular door in the side of a ship, AB is horizontal, C below AB, and the triangle equilateral of side 5 ft. The door is hinged along AB, and kept shut against the pressure of the water by a fastening at C. If the water rises to the level of AB, find the force on the fastening. (One cu ft of water weighs 62.5 lb).

(Liv. Inter.)

The *centre of pressure* of the triangular plate, with one of its sides in the plane of the liquid surface will be at a depth *h* /2, from the liquid surface, where *h* is the height of the plate, (see page 332)

Let *ABC* be the triangular door hinged along *AB* and having a fastening at *C*, where *AB* lies in the plane of the water surface, (Fig. 9.15).

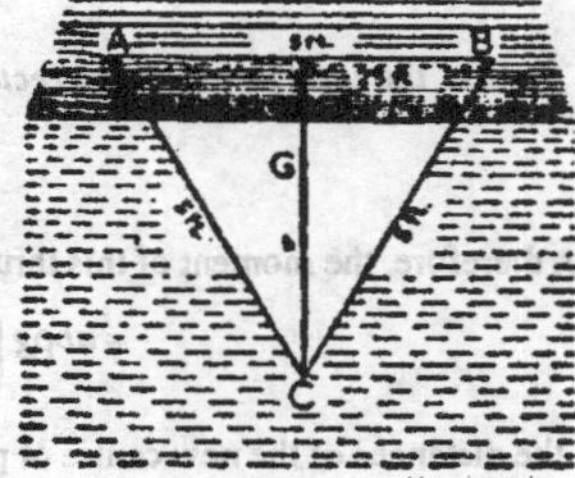

Fig. 9.15

Obviously, height *h* of the triangular door

$= \sqrt{5^2 - 2.5^2} = \sqrt{25 - 6.25} = \sqrt{18.75} = 4.329\,ft.$

Since the centre of a triangular lamina is $\frac{1}{3}$ *rd* of its height below the vertex, its depth below *AB*, or the water surface, is $\frac{1}{3}$ *rd* of its height, *i.e.*, equal to $\frac{1}{3} \times 4.329 = 1.443\,ft.$

Therefore, pressure at the centroid

= *h*.ρ *lb wt* = 1.443 × 62.5 *lb wt.*

And ∴ *thrust on the door* = 1.443 × 62.5 *lb* × *area of the door*

$= 1.443 \times 62.5 \times \frac{1}{2}$ *base* × *altitude* $= 1.443 \times 62.5 \times \frac{1}{2} 5 \times 4.329$ *lb wt*

= 1.443 × 62.5 × 2.5 × 4.329 *lb wt.*

Now, centre of pressure of the triangular door lies at $\frac{1}{2}$ *h*, *i.e.*, at

* It may also be expressed quite often in *lbs / sq.. in.*

$\frac{1}{2} \times 4.329 = 2.164\ ft.$ [See above.

$\therefore$ moment of the thrust about AB

$= $ thrust $\times$ depth of centre of pressure

$= 1.443 \times 62.5 \times 2.5 \times 4.329 \times 2.164$ *lb wt.*

And, if F be the magnitude of the force on the hinge, its moment about $AB = F \times h = F \times 4.329$ *lb wt.*

Clearly, therefore, $F \times 4.329 = 1.443 \times 62.5 \times 2.5 \times 4.329 \times 2.164.$

Or, $F = 1.443 \times 62.5 \times 2.5 \times 2.164 = 487.5$ *lb wt.*

Thus, the force on the fastening at $C = 487.5$ *lb wt.*

3. Find the centre of pressure of a rectangular sheet 'a' in long and 'b' in. wide, of uniform thickness, immersed in a liquid of uniform density, with one side of length 'b' in. in the surface, the plane of the rectangle being inclined at an angle θ to the vertical.

If the rectangular sheet remains in the same position with respect to the vessel containing the liquid, and the depth of the liquid be increased by h in., find the new position of the centre of pressure.

(London Higher School Certificate)

Let $ABCD$ be the rectangle, immersed in the liquid of density ρ, with its side $AB = b$ in the liquid surface EF, and its plane inclined to the vertical at an angle θ (Fig. 9.16).

Then, since the *vertical depth* of the rectangle is clearly $BK = BC \cos\theta = a \cos\theta$ its centroid lies at a vertical depth $\frac{a}{2}\cos\theta$, from the liquid surface.

Fig. 9.16

And, hence, proceeding as in § 9.8 [case (*i*), page 333], we have depth X_0 of the centre of pressure P from the surface of the liquid, clearly given by

$$X_0 = \frac{2}{3} \times \textit{vertical depth} = \frac{2}{3}\, a \cos\theta.$$

Now, let a column of liquid $EE'F'F$, k *in*, thick, be added on the top of the liquid surface to increase its depth by h *in.* and lot 'P'' be the new centre of pressure of the rectangle, (whose position is otherwise unchanged with respect to the vessel) at a distance X_0' from the new surface $E'F'$

Then, clearly, *thrust on the rectangle due to the original liquid column*

$= $ *area of the rectangle* $\times$ *depth of centroid G* $\times$ *density of the liquid* $\times\ g$

$$= ab.\frac{a}{2}.\cos\theta \times \rho \times g = \frac{1}{2}a^2 b \cos\theta \times \rho \times g;$$

and *increase in thrust* on the rectangle due to the new layer h of the liquid added

$= $ *area of rectangle* $\times$ *depth of new liquid column added* $\times$ *density of liquid* $\times g$

$$= ab \times h \times \rho \times g.$$

So that, the *total thrust on the rectangle* $= \frac{1}{2}a^2 b \cos\theta \times \rho \times g + ab \times h \times \rho \times g$

$$= ab\rho g\left(\frac{1}{2}a\cos\theta + h\right).$$

Clearly, therefore, the moment of this thrust about the new liquid surface $E'F'$

$$= ab\rho g\left(\frac{1}{6}a\cos\theta + h\right).X_0'. \quad ...(i)$$

Again, the distances of the new centre of pressure P' and the centroid of the rectangle from the new liquid surface

$E'F' = (X_0 + h) = \left(\frac{2}{3}a\cos\theta + h\right)$ and $\left(\frac{1}{2}a\cos\theta = h\right)$, respectively.

And, therefore, the moment about the new liquid surface $E'F'$ is also equal to

$$\frac{1}{2}a^2 b.\cos\theta.\rho.g.\left(\frac{2}{3}a\cos\theta + h\right) + ab.h.\rho.g\left(\frac{1}{2}a\cos\theta + h\right)$$

$$= ab\rho g\left[\frac{1}{2}a\cos\theta\left(\frac{2}{3}a\cos\theta + h\right) + h\left(\frac{1}{2}a\cos\theta + h\right)\right] \quad ...(ii)$$

Equating (*i*) and (*ii*), therefore, we have

$$a.b.\rho.g\left(\frac{1}{2}\cos\theta + h\right).X_0' = a.b.\rho.g\left[\frac{1}{2}a\cos\theta\left(\frac{2}{3}a\cos\theta + h\right) + h\left(\frac{1}{2}a\cos\theta + h\right)\right]$$

Or, $\left(\dfrac{a\cos\theta + 2h}{2}\right).X_0' = \dfrac{1}{3}a^2\cos^2\theta + \dfrac{1}{2}ah\cos\theta + \dfrac{1}{2}ah\cos\theta + h^2.$

Or, X_0', *i.e.*, the depth of the new centre of pressure of the rectangle from the new liquid surface

$$= \frac{h^2 + a\,h\cos\theta + \frac{1}{3}a^2\cos^2\theta}{\left(\frac{a\cos\theta + 2h}{2}\right)} = \frac{2(h^2\ ah\cos\theta + \frac{1}{3}a^2\cos^2\theta)}{2h + a\cos\theta}\ \textit{inches}.$$

4. Neglecting atmospheric pressure, find the depth of the centre of pressure of a circular lamina just completely immersed with its plane vertical in an incompressible liquid.

A circular door in the vertical side of a tank is 'hinged' at the top and opens inwards, and the tank contains water to a height just sufficient to cover the door. If the diameter of the door is 2 ft., find the magnitude of the force that must be applied normally to the centre of the door in order just to open the door. Find also the reaction at the hinge when this force is being applied. ***(Cambridge Higher School Certificate)***

The *centre of pressure* of the vertical circular lamina, *just* immersed in an incompressible liquid, is equal to $5\,r/4$, as explained in § 9.8 (*ii*), (page 333).

Here, obviously, the centroid of the circular lamina is at its centre, at a depth equal to its radius from the water surface, *i.e.*, at a depth $2\,ft/2$ *or* $1\,ft$ below it, (the lamina being *just immersed* in water).

$\therefore$ *pressure at the lamina* $= h.\rho.g = 1 \times 62.5 \times g$ *poundals* $= 62.5$ *lb wt.*

and *area of the circular lamina* $= \pi r^2 = \pi \times 1 = \pi$ *sq. ft.*

So that, *thrust on the lamina = pressure on the centroid* $\times$ *area of the lamina* $= 62.5 \times \pi lb$ *wt.* $= 196.3$ *lb wt.*

This thrust acts at the centre of pressure of the lamina, whose depth from the water surface, as we know,

$= 5r/4 = 5 \times 1/4 = 5/4\,ft.$

$\therefore$ moment of this thrust about the liquid surface, or about the hing $= 62.5 \times \pi \times 5/4 = 245.5$ *lb wt.*

If *F lb wt* be the force applied to the centre of the door, *just to open it*, we have

moment of *F* above the hinge $= F \times r = F \times 1\,lb\,ft.$

And $\therefore$ $F \times 1 = 245.5$ Or, $F = 245.5\,lb\,ft.$

The force required to be applied to the centre of the door to *just* open it is thus equal to 245.5 *lb wt.*

Thus, the force acting on the two sides of the door are respectively 245.5 *lb wt* and 196.3 *lb wt*; and, therefore, reaction at the hinge $= 245.5 - 196.3 = 42.2$ *lb wt.*

5. If a load of 20 tons, moved 50 ft. across the desk of a ship of 2,000 tons displacement, causes the ship to tilt through $\frac{1}{2}$°, what is its metacentric height ?

Let a load $w = 20$ *tones* be moved across the deck of a ship from *A* to *B* through a distance of 50 *ft.* (Fig 9.17), and let the ship be tilted through an angle $\frac{1}{6}$° into the position shown.

This, as we know, is equivalent to the application of a *couple*, of moment 20× 50 *ton-feet*, tending to turn the ship clockwise, [see § 9.14 (alternative treatment), page 339]

And, the couple, formed by the weight of the ship acting vertically at its *c.g. G′* and the equal weight of displaced water at its *centre of buoyancy B′* tends to restore the ship back into its original position, the moment of the couple being $W \times MP$, where *MP* is the perpendicular from the *metacentre M* on to the vertical line through *G′*.

Since the ship is *in equilibrium* under the action of these two opposite couples, they must obviously be balancing each other. So that,

$W \times MP = 20 \times 50.$ Or, $20{,}000 \times MP = 20 \times 50.$

Or, $MP = 20 \times 50/20{,}000 = 1/20\,ft. = .05\,ft.$

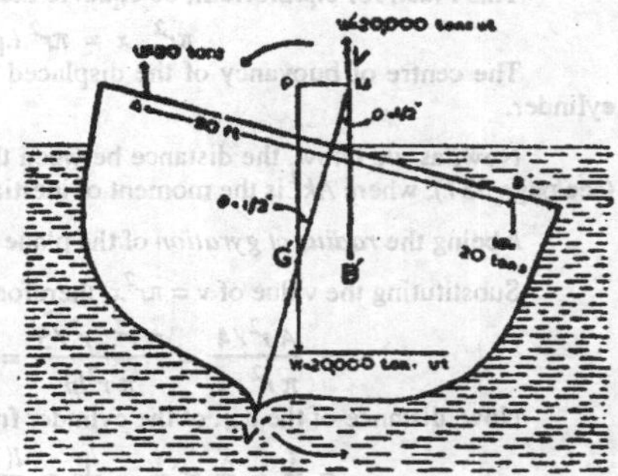

Fig. 9.17

Now, $MP = MG' \sin\frac{1}{2}° = MG'\ .0087,$

whence, $MG' = \dfrac{MP}{.0087} = \dfrac{.05}{.0887} = 5.748\,ft.$

Thus, the *metacentric height* of the ship $= 5.748\,ft.$

6. State the theorem of Archimedes, and explain what you understand by the terms "force of buoyancy", and "centre of buoyancy". A cylinder of radius 1 cm and length 4 cm, made of material of specific gravity 0.75 is floated in water with its axis vertical. It is then pushed vertically downwards so as to be just immersed. Find (*a*) the work done, (*b*) the reduction in the force on the bottom of the containing vessel when the cylinder is subsequently taken out of the water. *(Oxford and Cambridge Higher School Certificate)*

Here, the volume of the cylinder is clearly equal to $\pi r^2 l = \pi.1.4$ *c.c*

And $\therefore$ its *weight* = $\pi \times 4 \times 0.75 = 3\pi$ *gm wt*.

Since it floats vertically in the water, we have

weight of displaced water also = $3\ \pi$ *gm wt*

Let length of the cylinder inside water be = *x cm*.

Then, the volume of the immersed part of the cylinder = $\pi \,.\, 1 \,.\, x = p\,x\,c.\,c.$

i.e. volume of water displaced by the cylinder $= \pi\, x$ *c.c.*

And $\therefore$ weight of this displaced volume of water = $\pi \,.\, x \,.\, 1$ *gm wt.*

Hence, $\pi.x = 3\ \pi$. Or, $x = 3$ *cm.*

i.e., length of the cylinder inside water = 3 *cm*.

And, therefore, length of the cylinder outside water = 4–3 = 1 *cm*.

Thus, to immerse the whole of the cylinder just inside the water, we have to simply push it down through 1 *cm*. Obviously, the volume of the displaced water or upthrust on the cylinder will be $\pi \times 1 \times 4 \times 1 = 4\pi$ *gm wt*.

Hence, increase in the upthrust on the cylinder will be = $4\pi - 3\pi = \pi$ *gm wt*.

Since this increase in the upthrust takes place gradually from 0 to π, as the cylinder is pushed steadily down into the water from its initial position, we may take the average value of the upthrust against which we work in pushing the cylinder down through 1 *cm* to be $(0 + \pi)/2$ or $\pi/2$ *gm wt*.

And, therefore, work done in pushing the cylinder down through 1 *cm* against this average increase in the upward thrust, will clearly be equal to average thrust × distance through which the cylinder is pushed down.

i.e., work done = $(\pi/2) \times 1 = \pi/2 = 1.571$ *gm wt cm*

Now, with the whole of the cylinder immersed in the water, the weight of water displaced

$= 4\pi$ *gm wt* $= 12.57$ *gm wt*

This must also, therefore, be the downward thrust on the bottom of the containing vessel. So that, when the cylinder is removed out of the water, the reduction in the thrust on the bottom of the containing vessel will also be the same, *viz*., 12.57 *gm wt*.

7. Calculate the metacentric height and determine the necessary condition for the stable equilibrium of a cylinder of length *l*, radius *r*, and density ρ, floating vertically in water.

Let a portion *x* of the cylinder be *inside* water.

Then, volume of water displaced by the cylinder, *i.e.*, $v = \pi r^2 . x . c$

And $\therefore$ weight of water displaced or upthrust on the cylinder $\pi r^2 . x . 1 = \pi r^2 . x$

This must, for equilibrium, be equal to the weight of the cylinder, i.e., equal to $\pi.r^2.l.\,\rho$ *gm wt or*,

$\pi r^2 . x = \pi r^2 . l.\rho$ *gm wt.* therefore, $x = l.\rho$.

The centre of buoyancy of the displaced liquid must, therefore, be at a height $x/2 = l\rho/2$ from the bottom of the cylinder.

Now, as we know, the distance between the centre of buoyancy of the displaced liquid and the metacentre is Ak^2/v, (see page 337), where Ak^2 is the moment of inertia of the surface-plane of the cylinder about its diameter. So that, $k^2 = r^2/4$, *k* being the *radius of gyration* of the plane about the surface-line or the diameter of the cylinder.

Substituting the value of $v = \pi r^2 x$, therefore, we have distance between the centres of buoyancy and metacentre

$$= \frac{A.r^2/4}{\pi r^2 . x} = \frac{\pi r^2 . r^2/5}{\pi r^2 l\rho} = \frac{r^2}{4\, lw}. \qquad \left[\begin{array}{l} A = \pi r^2 \\ \text{and } x = l.\,\rho. \end{array}\right.$$

Now, distance of the *c.g.* of the cylinder from the centre of buoyancy

$$= \frac{l}{2} - \frac{x}{2} = \frac{l}{2} - \frac{l\rho}{2} = \frac{l(1-\rho)}{2}$$

And, therefore, distance between the *c.g.* of the body and the metacentre, or the *metacentric height h* of the cylinder = distance between centre of buoyancy and the metacentre *minus* distance between the *c.g.* of the body and the centre of buoyancy.

Or, $$h = \frac{r^2}{4\,l.\rho} - \frac{l\,(1-\rho)}{2} = \frac{r^2 - 2\,l\rho.l(1-\rho)}{4\,l.\rho} = \frac{r^2 - 2\,l^2.\rho(1-\rho)}{4\,l.\,\rho}$$

Now, for the stable equilibrium of the cylinder, the metacentre should be above the *c.g.* of the body, *i.e.*, h should have a *positive* value.

And, obviously, this is possible only when $r^2 > 2l^2 \rho(1-\rho)$. This is, therefore, the necessary condition required.

EXERCISE IX

1. Define pressure at a point in a fluid. Find the total thrust on the sides and vertical ends of a V-shaped trough, 1 *ft.* deep, 2 *ft* wide at the top and 4 *ft long, when filled with water, density 62.5 lb/ c ft.*
(Oxford and Cambridge Higher School Certificate)

Ans. 177 *lb wt.*; 20.8 *lb wt.*

2. Define clearly the term *'Centre of pressure'*. Determine its position for *(i) a circular lamina of radius r*, just immersed vertically, and *(ii) a triangular lamina of height h*, immersed vertically with *(a)* its apex, and *(b)* its base in the surface. **Ans.** *(i)* $5r/4$; *(ii) (a)* $3h/4$; *(b)* $h/2$.

3. A square lamina with its sides 3 *ft* long is just immersed vertically in water with an edge in the surface and is then lowered 10 *ft*. Find the distance of the centre of pressure in the new position from the centre of the square (Neglect the pressure of the atmosphere in each case). *(Joint Matriculation Board and H.S. Certificate)*

Ans. 0.0303 *ft*.

4. A circular area of radius a *ft* is immersed in water, with its plane vertical. The surface of water rises from $2a$ *ft* above the centre of the circle to $4a$ *ft* above it. Neglecting atmospheric pressure, prove that the centre of pressure rises through a distance $a/16$ *ft*. *(London Higher School Certificate)*

5. State the *Principle of Archimedes* and define clearly the terms *(i) Centre of buoyancy*, *(ii) Metacentre*, and *(iii) Metacentric Height*. Discuss in detail the conditions for the stable equilibrium of a floating body, with particular reference to a floating ship.

6. Show that if a floating body be given a small rotational displacement in its plane of symmetry, the distance between the centre of buoyancy of the displaced liquid and the metacentre is Ak^2/V, where A is the area of the surface plane of the body, k, the radius of gyration about the surface-line and V, the volume of the displacement liquid.

Discuss the conditions necessary for a hollow cylinder of height h, and density ρ, open at both ends, with r_1 and r_2 as its internal and external radii, to float vertically in a liquid in stable equilibrium. **Ans.** $(r_1^2 + r_2^2) > 2h^2 \cdot \rho(1-\rho)$.

7. Discuss how the atmospheric pressure changes with altitude above the surface of the earth, the temperature remaining constant, and show how if the altitude increases in arithmetical progression, the pressure decreases in geometrical progression.

8. A mercury barometer is known to be defective and to contain a small quantity of air in the space above the mercury. When an accurate barometer reads 770 *mm*, the defective one reads 760 *mm*. and when the accurate one reads 750 *mm*, the defective one read 742 *mm*. What is the true atmospheric pressure when the defective barometer reads 750 *mm*. ?
(Cambridge Scholarship). **Ans.** 758.8 *mm*.

9. A simple barometer has the glass tube attached to a spring balance. What weight does the balance record when the open end of the tube is just dipping under the surface of the mercury in the reservoir, and what changes occur when the tube is lowered so that more of it dips under the mercury ? *(Oxford Higher School Certificate)*.

Ans. *(i) The balance records the weight of the tube and the mercury column (ii) A progressive decrease in the weight, due to buoyancy of the tube, until the latter is full, when finally it decreases to zero.*

10. A sealed spherical cellophane balloon has a diameter of 5 *metres* and weighs, with the apparatus it carries, 1 *kg*. It contains one-tenth of the volume of hydrogen required to fill it at atmospheric pressure. The balloon is allowed to ascend; if the cellophane does not expand and if the temperature of the atmosphere is assumed to be constant at $0^\circ C$ at all heights, calculate at what height the envelope becomes full and the height to which the balloon rises. The pressure p at height h *(km)* is related to that at the ground (p_0) by the relation $h = 20 \log_{10}(p_0/p)$. (Densities of air and hydrogen at $0^\circ C$ and atmospheric pressure are 1.29 and 0.09 *gm* per litre respectively). *(Oxford and Cambridge Higher School Certificate)*

Ans. *(i)* 20 *km* and *(ii)* 34.5 *km*.

11. *(a)* Prove the formula $x = h \times (k^2/h)$, in the usual notation, for the depth x below the free surface, of the *centre of pressure* of a plane vertical lamina immersed in a homogeneous liquid. Show also that if the lamina is further lowered vertically, the *centre of pressure* approaches the centre of gravity asymptotically.

(b) A cone whose vertical angle is 2α, has its lower generator horizontal and is *filled with* liquid. Prove that the resultant pressure on the curved surface is $\sqrt{1 + 15 \sin^2 \alpha}$ times the weight of the liquid. *[Poona, (Subsidiary)]*

CHAPTER X

JET PLANES, ROCKETS AND SATELLITES

10.1. Jet Propulsion. We are all familiar with the meaning of the word **'jet'** which is just the term applied to a *high-velocity stream of fluid* (liquid or gas) *issuing out of a nozzle* as, for example, a *'jet of water'* or a *'jet of steam'* etc. And, therefore, *jet propulsion* is obviously the method of driving or propelling a body or a machine forwards through the agency of a jet,—the body or the machine thus driven being said to be *jet-propelled.*

That a jet possesses such a motive or tractive force can be easily seen from a number of facts of everyday life, if only we care to stop a while and analyse them. Thus, for example, when a bullet is forced out of the barrel of a rifle by the exploding mixture of gases inside it, the rifle suddenly moves or *'kicks'* back in a direction opposite to that of the bullet and the exploded gases.

So that, if we continuously fire a rifle fastened to the rear of a boat, with its barrel facing outwards, we shall find that the boat continues to move forwards with a jerky motion so long as the firing continues, — each bullet fired producing a push forwards. We, therefore, have here a jet-propelled boat ! In fact, even when we ply the oars, the action is similar. For, what we do is simply to push some volume of water backwards and the boat, as a consequence, moves forwards.

Indeed, if we did nothing else but simply sit quietly in the boat and throw stones into the water, with our face towards the stern of the boat, the boat will still move forwards (*i.e.*, opposite to the direction in which the stones are thrown).

All these examples are, as the student is no doubt already aware, a consequence of the well known *Newton's third law of motion*, according to which *action and reaction are equal and opposite*, or, what follows from it, *viz., the law of conservation of momentum*, which demands the equality of the momentum lost with the momentum gained. So that, the boat, in the cases above, moves oppositely as a result of the reaction to the motion of the bullet (and the gases) or the stones, or because the momentum lost by the bullets or the stones is equal to the momentum gained by the boat. A force such as the one experienced by the boat is called the *reactive force;* and, in the case of a jet, sometimes also the *jet-force.*

Now, does it surprise the student when he is told that even the usual type of airplane, in which we use the ordinary reciprocating (*i.e.*, the piston-type) engine, makes use of a jet for its propulsion forwards*? For, the propeller, as it whirls round at a high speed, throws a jet of air (or in the case of a ship, a jet of water) backwards, as a reaction to which the plane (or the ship) is pushed forwards, against the viscous resistance of the air (or water). The question, therefore, naturally arises as to why then do we not call them jet-propelled planes. The answer is that, technically speaking, the narrower the cross-section of the high-velocity fluid stream, the more nearly does it come up to the definition of a jet, and the term *jet-propelled planes* is, therefore, reserved for 'planes in which the jet is a narrow one, about one foot in diameter, as compared with ten feet or more in the case of the ordinary airplane.

Again, it must not be inferred from what has been said above that a jet must necessarily consist of hot gases. No, it may just as well be of cold air, as in the case of what are called the *ducted-fan type* of 'planes** or as was the case with perhaps the earliest jet-propelled plane, constructed by *Campini* in Italy,—the jet in this latter case being produced by a compressor, driven

---

* The same being the case with a ship.

** In these 'planes, air is sucked in through two-holes or ducts, by two fans, rotating in opposite directions, —the latter thrusting the air away with considerable force, propelling the 'plane forward. This sucking action also helps to buoy the 'plane up. Further, the gyroscopic effect, produced by the oppositely rotating fans, greatly helps in enhancing and ensuring the stability of the 'plane.

by the ordinary reciprocating type of engine.

10.2. Thrust supplied by the jet. Let us now calculate the thrust supplied to an aircraft by the jet produced by the power-unit inside it.

Suppose we have an aircraft travelling with a speed V and fitted with a power-unit which produces a jet of fluid, of velocity u, relative to the aircraft, where u is higher than V, — the velocity u of the jet being measured at a point in it a little away from the nozzle, where the static pressure is the same as that in the surrounding air. Then, if we impose a velocity V on the aircraft in the opposite direction to its own, the aircraft comes to rest, with the air streaming past it with velocity V. So that, if a be the area of cross-section of the jet at the point where its velocity is u, the *volume of the fluid flowing per second in the jet is clearly a.u.* If, therefore, ρ be the density of the fluid, we have

$$\textit{mass-flow of the fluid per second in the jet} = a.u.\rho = m, \text{ say.}$$

And, therefore, *momentum of this fluid in the jet = m.u.*

If this mass (m) of the fluid finally emerges out from the aircraft with a velocity V, its momentum is clearly reduced to mV. It thus suffers a loss of momentum, equal to $(mu-mV)$ or $m(u-V)$ *per second. i.e., its rate of change of momentum* $= m(u-V)$.

And this, therefore, in accordance with Newton's second law of motion, must be the *force* or' the *thrust F* supplied to the aircraft by the jet in the direction opposite. So that,

$$F = m(u-V) = a u \rho.(u-V).$$

10.3. Efficiency of the jet. If we consider the exact state of affairs in the case above, *viz.*, that the aircraft is really not at rest, as we had imagined, but is moving with velocity V, then, the velocity of the final jet, moving in the opposite direction, clearly becomes $(u-V)$ and, therefore,

$$K.E. \text{ of the final jet} = \frac{1}{2} m (u-V)^2 = \frac{1}{2} F(u-V) \qquad [\because m(u-V) = F.$$

Also, the aircraft does FV amount of work per second against the air resistance as it moves forwards. So that,

total energy that must be supplied by the jet propulsion unit,

$$(\textit{i.e., by the power-unit}) = FV + \frac{1}{2}F(u-V).$$

Of this, obviously, *the portion usefully employed is only FV*, the rest being simply a waste, creating a disturbance behind the aircraft. So that, *efficiency of the jet*, or the *Froude efficiency*, as it is commonly known,

$$= \frac{\text{energy converted into useful work}}{\text{total energy supplied}} = \frac{FV}{FV + \frac{1}{2}F(u-V)}$$

$$= \frac{2FV}{Fu + FV} = \frac{2V}{u+V}.$$

Note. Clearly, the efficiency will have the maximum value 1, when $u = V$, *i.e.*, when the initial jet-velocity is equal to the flight velocity of the aircraft; for, then, the energy wasted in the form of *K.E.* of the final jet, *viz.* $[\frac{1}{2}F(u-V)]$, will also become zero. This condition of maximum efficiency is, therefore, not a practicable proposition, just as it is not in any other type of machine either.

10.4. Effect of smaller cross-section of the jet. As indicated earlier, the cross-section of the jet in a jet-propelled plane, should be narrow. Let us see what advantage is to be gained by it.

Apparently, from the relation $F = m(u-V)$ for the thrust supplied to the aircraft by the jet, we find that a reduction in its cross-section will mean a diminution in the value of the mass-flow of the fluid, m, so that, to obtain the *same* thrust F, as before, $(u-V)$ will have to be correspondingly greater. This will naturally mean a higher value of $\frac{1}{2}(u-V)^2$, the *K.E.* of the final jet, which, as we have seen, is a mere waste of energy. Not only that, but, as a natural consequence, the efficiency of the jet, $FV/[FV + \frac{1}{2} F(u-V)]$, will also fall below its previous value. It would thus appear that a decrease in the cross-section of the jet, far from improving matters, does just the reverse, *viz.*, increases the loss of energy and decreases the efficiency for propulsion. In what manner, then, is jet-propulsion a better mode of preopulsion ?

The answer is manifold :

(*i*) Initially, when jet propulsion was just introduced round about the year 1940, it was intended to render auxiliary support to the then prevalent *gas turbine* engine. The materials of the gas turbine could not function satisfactorily at the temperatures obtaining in the earlier reciprocating type of engines and the products of combustion required to be diluted with a large excess of air. This seeming difficulty was actually turned into an advantage by the engine-designers, who used this necessary excess of air as a narrow jet to supply the entire thrust required to be given to the aircraft, thus eliminating the necessity of the propeller and quite a few other accessories. The jet was made to escape through a small turbine which then supplied the necessary power to the generator, the fuel pumps and the compressor etc. Thus, although the introduction of the jet inevitably entailed a loss in efficiency, with the fuel-consumption rate rising higher, it gave the *distinct advantage of reducing the weight of the whole unit for the same value of power*. In view of this smaller weight but higher rate of fuel-consumption, the *turbo-jet* engines, as these engines were aptly christened, came to be considered more suitable for flights of shorter durations, say, of less than 2 hours in those early days when the highest speed was only 400 miles per hour.

(*ii*) It was found that although the efficiency of a narrow jet is rather low at moderate flight-speeds, it increases rapidly with the flight-speed. In fact, if we take into consideration also the other advantages that go with high speed, (*e.g.*, assistance given to the compression-process in the engine, etc.), the overall result is that the power output (FV) increases directly with flight-speed with only a comparatively very small increase in fuel consumption, *i.e.*, $FV \propto V$. Clearly, therefore, F remains practically constant for varying flight-speeds.*

This linear increase in power (FV) with speed (V), with practically a constant fuel-consumption rate, necessarily implies that if the flight-speed be high, the turbo-jet unit will also be about as economical as the ordinary propeller-engine and will, in addition, possess the advantages of (*a*) having *less weight* and (*b*) *capacity of packing large power in a smaller space*.

In fact, both the turbo-jet and the propeller engine will have the *same* efficiency, *i.e.*, their power output for the same fuel-consumption will be the same, at a speed of 700 *m.p.h.*, *provided the propeller engine had a constant power-output upto this speed.* And this is the point where the jet-unit scores over the propeller-unit. For, the power output of the propeller engine does not really remain constant with speed but falls steeply as the flight-speed approaches the speed of sound, *viz.*, 762 *m.p.h.* at ground level and 660 *m.p.h.* at altitudes above 3600 *ft*. This is so, because a propeller may be regarded essentially as a wing, with the difference that whereas the latter provides a lifting force to the aircraft against the force of gravity, the former supplies a similar force in the form of a thrust in the direction of its motion, for which purpose it is rotated in a plane perpendicular to the direction of flight,—the lifting force in the case of the wing, and the forward thrust in the case of the propeller, being always roughly perpendicular to the direction of their respective motions through air, both experiencing an air-resistance or, 'drag' opposing their motion, with some power used up in overcoming the same**. So that, despite all improvements made in the designs of propellers (such as, making their sections near the tips very thin, etc.) their propulsive power falls

---

* That is why, in the case of a turbo-jet unit, *only its thrust (F) is indicated and not its power (FV)*

** The resistance of the compressed air in front of the aircraft is of little consequence at speeds below that of sound, for it simply moves away with the speed of sound. But when the speed of the aircraft is higher than that of sound, the condensed air in front can only move sideways but not forward, with the result that the nose of the aircraft has to carry along a bulk or compressed air, with a consequent large increase in the drag on it.

greatly at high flight-speeds, where as that of a jet-unit rises equally greatly. This comes about because the actual velocity of the propeller blade is the resultant of its velocity of rotation and translation and , as such, is higher than the flight—speed of the aircraft itself,-even a non-rotating wing experiencing a large increase in the drag on it much before the speed of sound is attained.

Thus, from the point of view of overall efficiency, a jet-unit is certainly superior to a propeller-unit at speeds of 600 *m.p.h.* and above.

Then, again, another advantage that a jet-unit possesses over the propeller-unit is that there being no accessories and protubrances like radiators, oil-cookers etc., the drag is comparatively less. And the absence of the propeller which makes for a smoother flow over the entire surface of the plane, cuts down the drag over these surfaces by as much as 20 to 30 per cent.

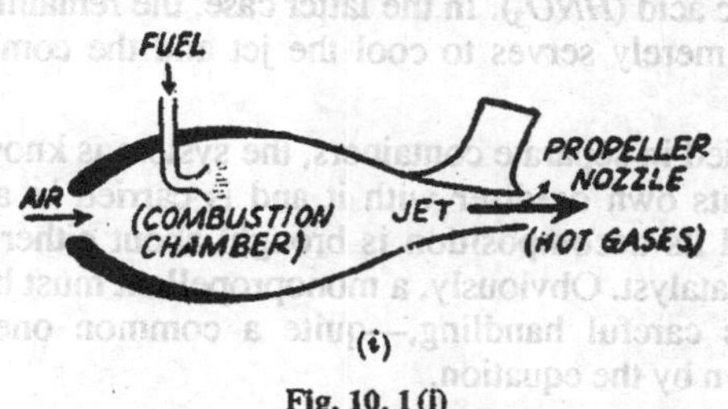

Fig. 10. 1 (I)

10.5. Rocket Planes. The small fire-work rockets, rising pretty high up in the air, to the amusement of onlookers, are a common enough sight everywhere and they are obviously jet-propelled on their own small scale. A rocket plane is merely a large scale version of the same phenomenon. It possesses a higher speed and can rise to a much greater height than even a turbo-jet plane. In fact it is a turbo-jet plane, in which the technique of jet-propulsion is carried a stage further, with the jet still narrower in cross-section and its velocity higher.

Is this difference in degree, then, the only factor that distinguishes it from a jet-plane ? Of course not ; for, the essential difference between the two lies in the method of production of the jet. *In a jet-propulsion unit, the fuel alone is carried on the aircraft, with the oxygen necessary for its combustion being drawn from the surrounding air,* [Fig. 10.1(*i*)]—only a fraction of which is usually consumed, the rest, together with the considerable larger quantity of nitrogen 'swallowed', merely serving to keep the temperature down throughout the jet-propulsion unit. *In the case of a rocket propulsion unit, on the other hand, the fuel as well as the oxidising agent, required for its combustion, are together carried on the aircraft and no air has to be drawn in from the surroundings. It is thus a self-contained unit in itself,* [Fig. 10.1(*ii*)]. This is a point of great importance in that it makes the working of the motor quite independent of the presence or absence of any surrounding air. Thus, whereas a jet plane can only attain a height of 50 to 80 thousand feet up to which it can have its supply of air from the surroundings, there is no such limit to the height of a rocket-plane, which alone is capable of rising up to higher altitudes beyond the earth's atmosphere, where there is obviously no air to be drawn in. So that, its being self-contained, with its own supply of the oxidising agent, while it may be a comparative disadvantage at lower altitudes, is clearly a tremendous advantage at higher ones.

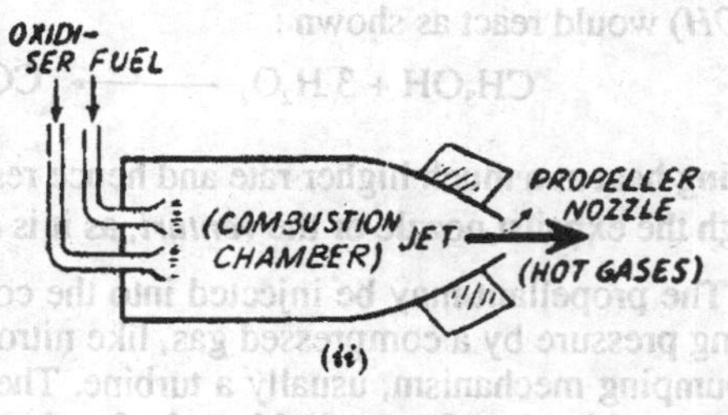

Fig. 10.1 (ii)

Calculating the thrust (F) given to the aircraft by the issuing jet in the same manner as in the case of the jet-propulsion unit, (§ 10.2, page 349), we have $F = m.u$, where m is the *mass-flow* of the fluid per second in the jet and u, its final velocity, its initial velocity here being *zero*, since all the constituents are carried on the aircraft itself. Its *efficiency* thus works out to $2Vu/(u^2+V^2)$, which again, as in the case of jet-propulsion, will have the maximum value 1, when $u = V$.

Further, the power-unit, in the case of the rocket plane, is also much lighter. Thus, while, for supplying a thrust of 1 *lb* at ground level, or of 0.2 *lb* at an altitude of 50,000 *ft* a turbo-jet unit weighs about 0.3 *lb*, a rocket-unit, weighing only 0.1 *lb* can supply the same thrust of 1 *lb at all altitudes.* It follows, therefore, that a rocket plane is more suitable for use only for flights of *short duration* or at *very high speeds.*

10.6. Rocket Fuel. The rocket-unit, in which both the fuel and the required amount of oxygen for its consumption are carried on the aircraft itself, is much simpler than that other one which requires the compression of large quantities of air. For in this case, the only problems with which we are concerned are those of the combustion chamber and the propelling jet.

The oxygen may be carried either in the liquid form, or in the form of oxidisers rich in oxygen, like hydrogen peroxide, (H_2O_2) or nitric acid (HNO_3). In the latter case, the remaining part of the oxidiser, going into the propelling jet, merely serves to cool the jet and the combustion chamber.

Now, if the fuel and the oxidiser are carried in separate containers, the system is known as a **bipropellant rocket,** but if the fuel contains its own oxidiser with it and is carried in a single container, it is known as **monopropellant** and its decomposition is brought about either by the application of heat or through the agency of a catalyst. Obviously, a monopropellant must be some sort of an explosive and, therefore, requires careful handling,—quite a common one being hydrogen peroxide, which decomposes, as shown by the equation.

$$2H_2O_2 \longrightarrow 2H_2O + O_2 + 690\ C.H.U.\ lb^*.$$

The products of decomposition of a monopropellant substance are sometimes themselves rich in oxygen, as we can see in this case of H_2O_2, and the substance can, therefore, also be used as one of the components of a bipropellant. Thus, for example, H_2O_2, used with methyl alcohol (CH_3OH) would react as shown :

$$CH_3OH + 3\,H_2O_2 \longrightarrow CO_2 + 5\,H_2O\ 1702\ C.H.U./lb$$

releasing heat at a much higher rate and hence resulting in a much higher exit, velocity of the gases through the exhaust nozzle or the *venturi*, as it is called, and consequently a *much higher thrust.*

The propellant may be injected into the combustion chamber in one of the two ways: (*i*) by exerting pressure by a compressed gas, like nitrogen or air on the propellant tanks or (*ii*) by means of a pumping mechanism, usually a turbine. The former method admits of no variation or control of thrust and is, therefore, suitable only for short-duration flights or remotely controlled missiles, and the latter is the one commonly used for rocket-propelled aircrafts.

10.7. Specific Impulse. The performance of a rocket motor is measured in terms of what is called the *specific impulse* or the *specific pull*, *I*, which is *the thrust generated by unit rate of fuel-consumption, i.e.,*

$$I = \frac{thrust\ (lb)}{rate\ of\ fuel\ consumption\ (lb/sec)} = \frac{F}{mg}.$$

So that, the dimensions of *I* are the same as those of time. Physically, therefore, it is *the time for which a unit thrust can be generated by a unit weight of fuel.*

Now, as we have seen, the thrust in the case of a rocket is equal to *mu*, where *m* is the *mass-flow* through the nozzle and *u*, the *exhaust velocity* of the gases. So that,

$$I = mu/mg = u/g.$$

And, therefore, *the higher the jet-velocity, the higher the specific impulse and the smaller the fuel-consumption for a given thrust.*

Besides fuel-consumption and thrust, there are quite a few other factors which determine the suitability of various fuels, *e.g.*, the weight of the engine, the temperature in the combustion chamber etc., etc. In modern rocket motors, the total weight of the pump, control and installation etc. must be about one-tenth of the maximum thrust developed. In short, the performance of a

* 1 *C.H.U.* (Centigrade heat unit) is the heat required to raise the temperature of 1 *lb* of water through 1°C.

rocket depends chiefly upon three factors (*i*) *jet-velocity*, (*ii*) *density of the propellant* and (*iii*) *weight of power plant*, which includes that of the propellant tanks and the fuel-supply system etc., into details of which we need not enter in an elementary discussion of the type we are concerned with here.

10.8. Shape of the Rocket. During an upward flight, particularly, through the denser layers of the atmosphere, the components of the rocket are subjected to intense air-pressure, and also a lot of heat is produced due to viscous friction of the air. Both these factors are taken into account while designing a rocket. Its frame is accordingly made of a heat-resisting material and its velocity during the first part of its flight, through the denser layers of the air, kept sufficiently low. Further, it is so designed as to reduce the air-pressure on its each individual part to the very minimum, its overall shape being more or less like that of a cigar.

10.9. The Multistage Rocket. If a rocket is hurled into space beyond the earth's gravitational field, then, supposing that its acceleration takes place in the latter region, where the value of g is 32 *ft/sec*^{2}, the velocity V_0 that it must acquire to escape from the earth's gravitational field or the '*escape velocity*', as it is called, is given by the relation $V_0{}^2 = 2mG/R$, (see § 7.12, page 255) from which the value of V_0 works out to about 11.19×10^5 *cm/sec* or about 36000 *ft/sec*.

Now, at the present stage of rocket development, no single rocket can achieve this velocity. To tide over the difficulty, therefore, we make use of what is called a *multistage rocket*, which is just a combination of rockets, either (*i*) *joined consecutively or in series*, as it were, or (*ii*) *one inside the other* or (*iii*) *with the rear part of one inside the nozzle of the other*, as indicated diagrammatically in Fig. 10.2. In all these three types, the first stage rocket is the largest in both dimensions and weight, and the last stage one, the smallest.

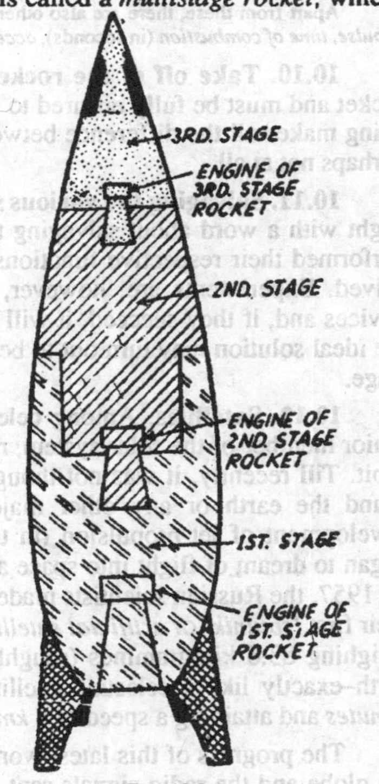

Fig. 10.2

Naturally, the first stage rocket is used first and when it has done its job, it gets detached and is discarded, with the second-stage rocket taking over the task of producing further acceleration. Then, this too is discarded and the third stage rocket takes over and so on. The velocity thus goes on increasing at each stage by the *same amount* as it does in a single stage rocket and each stage has its own propulsion and control system. Obviously enough, the fuel-consumption and the thrust for the first stage rocket are the highest of all, say, about a hundred times the corresponding values for the third stage rocket; and the fuel-stock too in the first stage is about sixty times that in the third stage, the same being the ratio of the total weights carried by the former to that by the latter.

Considerations of both weight and cost demand that the number of stages should not be large and that, therefore, the pay-load of each stage (which includes, in addition to the useful load of the final stage, the weights of the intervening stage rockets to be discarded later) be limited to about 20% of its own weight. Clearly, the useful pay-load of the final stage thus works out to be a very small fraction of the initial overall weight. Thus, for example, if there be n stages in all, this fraction is just $1/5^n$ of the initial total weight. Or, to give a more concrete example, a spaceship of the size of the well-known V2, designed by Dr. Verner Von Braun, would be about just sufficient to land a match box or a packet of cigarettes, by means of a parachute, on the planet Mars.

Each individual rocket of the multi-stage rocket, has its own independent design and basic characteristics, with its function co-related with those of the others. These characteristics include the following :

(*i*) **Net weight.** The net weight for a single stage rocket includes also the weight of the instruments and appliances or the weight of ammunition, if any, etc. And, in the case of a multi-stage rocket, obviously, the total weight of the second stage is the net weight of the first stage and the total weight of the third stage, the net weight of the second and so on, the ratio between the two being usually 3 : 1 for each stage.

(*ii*) **Steering Equipment.** This is necessary to steer the course of the rocket during its flight during the other stages except the first *which only serves as a sort of runway for the rocket*, as it were.

(*iii*) **Design.** This includes the frame of the rocket or of the individual rockets, in the case of a multi-stage rocket, with its fortifications and fastenings etc.

(*iv*) **Rocket-length.** This obviously means the height of the rocket or that of the individual rockets of the multi-stage one. This is an important factor in as much as *the very stability of the rocket in its trajectory depends upon the ratio between its length and its mean diameter* (*i.e.*, the mean diameter of the whole rocket or of each one of the stage-rockets).

(*v*) **Number of Motors.** Each stage rocket has its own separate motors. The first stage rocket, naturally, in view of the highest total weight its has to carry and the greatest resistance of the lower denser layers of air it has to overcome, has more than one motor and the last stage rocket, because of its lightest load and the least resistance to be overcome, is provide with only one motor.

Apart from these, there are also other characteristics of a rocket, like its *fuel consumption, thrust, specific pull or impulse, time of combustion* (in seconds), *acceleration, lift* or *range* etc.

10.10. Take off of the rocket. This is perhaps the most important part in the flight of a rocket and must be fully ensured to be correct. The slightest error in the timing or the accuracy of firing makes all the difference between the rocket returning back in this generation or the next or perhaps not at all.

10.11. Salvaging the various stage rockets. Let us wind up our elementary study of a rocket flight with a word about salvaging the various stage-rockets which are discarded after they have performed their respective functions. This problem cannot yet be said to have been satisfactorily solved. Experiments are, however, being made with various systems of parachutes and other devices and, if they succeed, it will mean a tremendous economy in cost. And for all one knows, the ideal solution may turn out to be the utilisation of the material of one stage as fuel for the next stage.

10.12. Satellites. Among celestial bodies, a satellite is what may be called a minor or a junior member of the solar system, revolving round one of the major planets in its own prescribed orbit. Till recently, it was not thought possible that anything man-made could also be so placed round the earth or any other major planet to revolve in a given orbit. But, then with the development of jet-propulsion (in the year 1940), followed by that of high-speed rockets, man began to dream of flight into space and of inter-planetary travel, when all of a sudden, on October 4, 1957, the Russian scientists made the whole world gasp with wonder and surprise by launching their first *'sputnik'* or *artificial satellite*. This *Sputnik I*, of the form of a ball, 58 *cm* in *diameter* and weighing 83.6 kilogrammes (roughly 185 *lb*) was placed into an elliptical trajectory round the earth–exactly like a celestial satellite—making, in its initial phase, one full revolution in 96.2 *minutes* and attaining a speed of 8 *km* or nearly 5 *miles/sec* at a distance of 950 *km* from the earth.

The progress of this latest wonder was watched with dumb admiration by scientists all over the globe and the radio signals sent out by it listened to attentively as long as its source of power lasted. It existed *as a satellite* for full 58 days, during which it made 1400 revolutions of the earth, thus covering a distance of 39 million kilometres. Its existence, however, continued for 92 days and the entire distance covered by it totalled up to the enormous figure of 60 million kilometres, when, finally, on January 4, 1958, it entered the denser layers of the atmosphere and got burnt out due to the intense heat produced by friction.

This artificial satellite was obviously an *automatic rocket*, hurled into its pre-determined and well-calculated orbit by a multi-stage rocket. Indeed, the rocket carrier too continued to revolve round the earth at about the same height as the sputnik but at a distance of about a thousand kilometres from it; and , then, while descending through the denser layers of the atmosphere, it also began to burn out, with fragments from it falling somewhere in *Alaska* and *North America*.

After almost exactly a month, on November 3, 1957, the Russians put their second artificial satellite *'Sputnik II'* into orbit round the earth, containing scientific equipment for exploratory purposes, as well as the first space traveller, the dog *'Laika'*, in a sealed cabin, which they successfully retrieved back, safe and sound. The total weight of the Sputnik was this time much greater, being 508.3 *kgm* or 1126 *lb* (including the dog). Its distance from the earth was also greater, 1700 *km*, its period of revolution, 102 *sec*, with the angle of tilt of its orbit roughly 65° from the equatorial plane.

This was followed by the first American artificial satellite, *'The Explorer'* on January 31, 1958, though of a comparatively much smaller weight and size.

These sensational events brought still more sensational and breath-taking ones in their wake, with the Russians putting the first cosmonaut of the immortalised name, *Major Yuri Gagarin*, into space on April 12, 1961, in a much larger space-vehicle or spaceship and retrieving him back, with the Americans later repeating the performance.

As if this was not progress enough, the Russians, on March 18th, 1965, made space history by making their astronaut, *Alexei Leonov*, emerge from his spaceship *Voshkod II* out into space, where he kept floating for some ten minutes, at the end of an umbilical cord, before getting back into his cabin and later landing safely on earth with the ship's commander, *Pavel Balyayev*.

The stupendous feat was repeated with better ease and fineness by the American astronaut, *Edward White*, only a couple of months later, on June 3, 1965, when he similarly stepped out of his spacecraft *Gemini* 4 and covered over 100 miles in space over the Pacific Ocean, overstaying his time there by about 8 minutes, despite requests from the Ground Control and the ship's commander, *James McDiwitt*.

These two astronauts were thus virtually human satellites, hurtling along, during their brief sojourn in the weightless void of space, at a tremendous speed, close on 20,000 miles per hour. Verily, the conquest of man over nature is now near-complete with men having landed and walked on the Moon on July 20, 1969, the pride of place among whom goes to **Neil Armstrong**, the American astronaut who rightly said when he stepped on the Moon, 'one small step for man, a giant step for mankind'. The possibility of space travel is obviously now just around the corner.

Let us try to understand the basic principles underlying this great phenomenon.

10.13. Conditions for a satellite to be placed in orbit. It is obvious that an artificial satellite goes round the earth exactly as a celestial satellite goes round a planet; as the moon, which, for all practical purposes, is a satellite of the earth, goes round it, or as the earth and the other planets go round the sun, *i.e.*, in accordance with the laws, first enunciated by *Kepler*, leading to Newton's celebrated Law of Gravitation, which forms the basis of the entire celestial mechanics.

The student is quite familiar with the whirling motion of a stone, tied to one end of a string, the other end of which is held in the hand. Precisely similar is the case with a planet going round the sun, or an artificial satellite going round the earth, with the force of gravitational attraction replacing the tension in the string. There is, however, one fundamental difference between the two, *viz.*, that whereas the tension in the string is, within limit, a variable quantity, permitting a lower or a higher velocity of the stone, *the attractive force of the earth on the satellite is a specific quantity and thus permits only a specific velocity for the satellite, if it is to remain in orbit,* —this velocity, for a satellite close to the earth being, as mentioned already, about 8 *km* or 5 *miles* per second. For in order to be able to revolve around the earth, quite close to it, the satellite must be in dynamic equilibrium under *(i) the centrifugal force, drawing it away from the centre of the earth* and *(ii) the gravitational pull on it due to the earth, drawing it towards the centre of the earth*.

If m be the mass of the satellite, M, the mass of the earth, R, its radius and V, the velocity of the satellite, we have *centrifugal force on the satellite* = mV^2/R, pulling it *outwards*, away from the centre of the earth, taking the radius of the orbit of the satellite to be the same as that of the earth, the satellite being close to it). And, *gravitational pull on the satellite* = $m.M.G/R^2$. pulling it *inwards* towards the centre of the earth.

For dynamic equilibrium of the satellite, therefore,

$$mv^2/R = m.M.G/R^2,$$

whence, $V^2 = MG/R$, or $V = \sqrt{MG/R}$ *

Since $mg = m.GM/R^2$, we have $MG/R = gR$. So that, $V = \sqrt{gR}$.

Since, however, the gravitational force decreases with increase of distance from the centre of the earth, a satellite further away from the earth will need a smaller velocity to remain in its orbit than the one nearer to it, though up to about 1000 *km* above the earth's surface, this reduction in velocity is only nominal. This is clear from the fact that the moon, which is roughly 380000 *km* away from the earth and, therefore, moves in a much larger orbit, has only a velocity of about 1 *km/sec*, which is about one-eighth of a satellite close to the earth; so that, whereas the moon makes only one revolution of the earth in one month, the satellite makes as many as 15 revolutions in one day.

Now, the question is how to hurl the satellite with such a high velocity away from the earth, to enable it to go into orbit around it. As can be seen, not only has the opposing gravitational force to be overcome but also the very considerable air resistance, particularly, in the lower denser part of it. As we have seen above, the least velocity for the purpose is about 8 *km* or 5 *miles/sec.*, called the **first cosmic velocity**. But, if the velocity rises to about 11.2 *km/sec.*, called the **second cosmic velocity** or the **velocity of escape**, the satellite passes right out of the earth's gravitational field and flies away into the cosmos, within the range of the solar system.

This formidable problem can, as mentioned earlier, be easily solved by carrying the satellite on a multi-stage rocket, for no single rocket can possibly (at any rate, not yet) achieve the requisite velocity all by itself alone. We have already discussed the essential features of such a rocket in §10.5. Let us now see how exactly to launch the rocket, carrying the satellite, into the required orbit.

10.14. Launching of the Satellite. Apparently, the shortest route for the satellite to take from the launching base to its assigned orbit would be the vertical one. This, however, is not feasible in actual practice for the simple reason that the gravitational pull of the earth will then be in the directly opposite direction to its motion and counteract the pull of the engines. So that, before it can gather the necessary speed, its limited fuel-stock may get exhausted, resulting in its first coming to a stop and then starting falling down. Vertical launching of the satellite is, therefore, not a practicable proposition.

To ensure that the satellite does not fall back to the earth, it is essential to give it a sufficient horizontal velocity. Its upward flight is, therefore, so arranged that it is brought into its orbit in the shortest possible time, acquiring meanwhile the requisite horizontal velocity. It is thus clear that the particular trajectory that will take the satellite to its assigned orbit has first to be most carefully calculated.

It is usual to arrange the first portion of the flight of the rocket to be vertical, so that it may pass through the first 20 *km* of the denser portion of the atmosphere the earliest. Thereafter, as it enters the rarefied portion of the atmosphere, it is given a gradual tilt by means of a mechanical pilot, so that it emerges into its orbit with a horizontal velocity large enough for the centrifugal force coming into play on it, (on account of its circular motion) to just balance the force due to the gravitational pull. And, the trajectory of its path is so chosen that the loss of velocity entailed, due to air resistance and the earth's pull, is a small percentage of its required or characteristic velocity. In fact to make up for this loss, the actual velocity given to it is a little higher than the computed

* Substituting the values of g and R, we have V = 8 *kms* or 5 *mile* per *sec.*, very nearly.

value of its characteristic velocity. When launched laterally to the earth's rotation, however, an increase in its velocity is automatically obtained at the expense of the velocity of the earth's rotation, depending upon the latitude of the launching site. Thus, for example, this increase is the maximum at the equator, being as much as 400 *metres/sec*, which is higher than that of the fastest fighter planes of the day.

If it be desired to give the satellite an elliptical orbit, instead of a circular one, the rocket carrying it must either be given a higher velocity than the peripheral one or its velocity, immediately after completion of the motor's performance, must not be directed along the tangent to the circular orbit. In the elliptical orbit, the point nearest to the earth is called *'perihe'* and the farthest from it, the *'acme'*. And it is quite possible that the satellite at the former point may be nearer to, and at the latter, farther from, the earth than at any point in its circular path.

In any case, the accuracy demanded in the firing of the rocket into its correct orbital path is really exacting. For, even an error of 1% in the direction of velocity may produce a height variation of te *perihe* and the *acme* which may be as much as 120 *km* or more. This firing accuracy is secured by means of proper steering devices, directing the course of the rocket at every stage of its flight. And, clearly, rudders of the type used in the ordinary jet aircraft, are hardly suitable for the purpose, since they cannot possibly function equally effectively both in the denser and the rarefied regions of the atmosphere.

The necessary steering control can, however, be effected in a number of ways but the one usually resorted to is to so design the rocket as to enable it to change the direction of the escaping jet by a mere tilt of the longitudinal axis of its motor with respect to its own. This is actually the device adopted in most of the present-day long-range rockets.

The manner in which the angle of inclination of the longitudinal axis of the rocket with the horizon is varied, will be clear from Fig. 10.3. As will be readily seen, the trajectory of the rocket from its very start until its longitudinal axis takes up the horizontal position, (*i.e.*, until its outward motion towards its orbit) is split up into a number of stages, indicated by h_1, h_2, h_3 etc., depending upon the height of the orbit. The angles that its horizontal axis makes with the horizon at each stage is carefully calculated beforehand and the control instruments set accordingly, to ensure that the rocket takes its assigned trajectory. And, this very setting of the instruments also regulates the fuel-supply in keeping with the pre-determined requirement at the time.

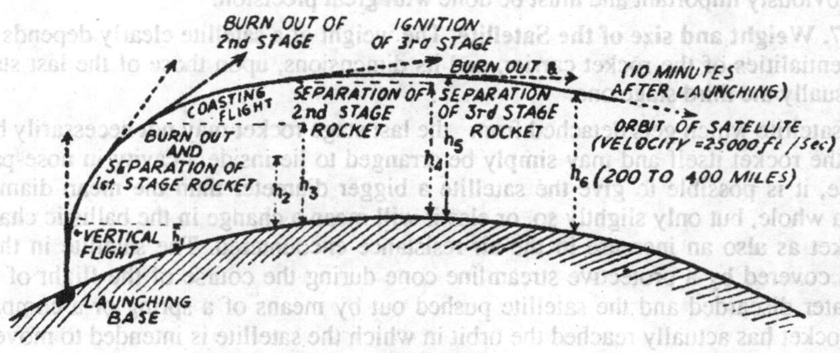

Fig. 10.3

Now, it will be easily understood that, while going round in its allotted orbit, the satellite passes over different parts of the globe in its successive rounds. For, by the time it has completed one round, the earth has also rotated about its axis and hence, in its next round, it naturally passes over other parts that now fall below its orbit. This will always be so except when the satellite goes round an orbit coinciding with the equatorial plane, in which case, obviously, it will always pass over the same parts or countries situated at the equator. It does not mean that we can launch the satellite in any orbit we choose. For, *the orbit must be one such that is plane passes through the centre of the earth and it will, therefore, clearly depend upon the site of launching*.

Not only that, but *even the time of the day and the season at the time of launching matter a great deal.* For, a satellite receives energy direct from the sun through special type of solar batteries fitted into it, a particular side of which must all along be illuminated by the sun. The satellite must, therefore, be launched in an orbit, the plane of which is perpendicular to the rays of the sun, and this is possible only at a particular hour of the day, *viz.*, when the radius of the earth connecting the starting point of the satellite with its centre is perpendicular to the sun's rays. And, the season is important because, with the satellite launched in its orbit, as explained, the earth which also moves round the sun, comes in between it and the sun at a particular time, thus preventing the rays of the sun from reaching it. Account has, therefore, to be taken of this occurrence and the season of launching chosen such that the satellite can get the maximum time to store up enough energy from the sun to suffice for the period when the sun will remain hidden from it later during its flight.

And, finally, it must also be clearly understood that in view of the uneven distribution of the mass of the earth and, therefore, with its *c.g.*, some 500 *km* away from its geometric centre, the satellite in its orbit is subjected to varying forces of attraction at different intervals, with the result that its real course is neither circular nor elliptical. It does not even lie in either of the two planes and is, in fact, a curve of a complicated pattern. For the same reason, there are variations in the velocity of the satellite at different points along its path.

10.15. Stability of the rocket during flight. It is imperative that throughout its flight, along its allotted trajectory, the rocket should not get tilted. This is achieved by means of an auto-pilot [see § 3.23 (*vi*)] and a suitable gyroscopic arrangement.

10.16. Form of the Satellite. In designing a satellite, attention is naturally paid to the geometrical shape it should be given to ensure its smooth motion in its orbit. The present view is that this form should be spherical; for, then, it will always have the same area of resistance and thus help calculation of the air-resistance to its motion at higher altitudes and hence in the assessment of the density of air at those altitudes. Further, with a spherical shape, there will be less chances of its getting overturned than if it were cylindrical or of any other shape. At the same time, a spherical shape is also a drawback, since it does not make for an easy setting of the various instruments and other equipment inside it. For, as will be easily relised, the instruments must be set, not haphazardly but in a definite order so as to ensure both an equitable distribution of the total weight inside the satellite and a specific position of its *c.g.* This '*balancing*' of the satellite, as it is called, is obviously important and must be done with great precision.

10.17. Weight and size of the Satellite. The weight of a satellite clearly depends essentially on the potentialities of the rocket carrier, and its dimensions, upon those of the last stage rocket, which is usually the third stage one.

The satellite which gets detached from the last stage rocket may not necessarily be included as part of the rocket itself and may simply be arranged to lie inside a cavity in nose-part of it. In such a case, it is possible to give the satellite a bigger diameter than the mean diameter of the rocket, as a whole, but only slightly so, or else it will mean a change in the ballistic characteristics of the rocket as also an increase in the air-resistance encountered. The satellite in the cavity is sometimes covered by a protective streamline cone during the course of the flight of the rocket, which is later discarded and the satellite pushed out by means of a spring or a compressed gas, when the rocket has actually reached the orbit in which the satellite is intended to move. This was exactly the case with Russian *Sputnik I*, whereas *Sputnik II* formed part of the third stage rocket itself and did not get detached from it.

10.18. Material of the frame of the satellite. Obviously, the material of the satellite frame must be both *light* and *strong*,—the former from considerations of its weight and the latter, to make sure that the instruments etc. inside it are securely attached and that it can withstand the onslaught of micrometeorites to which it is subjected during its orbital motion in space. Then, the material must also be less sensitive to changes of temperature and must be able to properly reflect radio-waves. It must, therefore, be either *aluminium, magnesium* or one of their alloys, with, in some cases, a suitable outer covering.

If, however, it is desired to study the electric currents in the ionosphere, the frame of the satellite should neither be a conductor of electricity nor should it possess any magnetic properties. So that, in this case, a metallic frame is clearly ruled out in favour of one of a plastic material, some of the modern varieties of which are just as tough and durable as steel.

10.19. Duration of satellite's existence. It is only natural to enquire as to how long can a satellite be expected to stay in its orbit. Well, if the space in which it moves along its orbital path were completely devoid of air, there would be nothing to stop it and it could go on perpetually, like the moon, for instance. But there being air even at a height of 1000 *km* and above, it has to encounter resistance due to it, however small, —this resistance being greater for orbital paths nearer the earth than further away from it. So that, when its velocity is thus sufficiently retarded, it cannot possibly remain in its orbit and starts falling down along a spiral path. In doing so, it either gets burnt up due to the heat produced by friction in the denser atmosphere or drops down to the earth with the help of parachutes.

The actual calculation for its 'life' is rather a complicated one, but it basically depends upon the density of the upper regions of the atmosphere, *i.e.*, on the height of its orbit from the earth.

10.20. Other Essentials. In case a man is to be placed in the artificial satellite or the sputnik, there are quite a few other problems to be tackled, as, for example, provision of an hermetically sealed cabin, with requisite conditions for the sustenance of life, and with windows fitted with the type of glass that absorbs ultraviolet and X-rays, a prolonged exposure to which is harmful in its effects. It is, however, almost impossible to afford any protection to the cosmonaut inside the cabin against cosmic rays which, as we know, can penetrate even through a block of lead, one metre thick. Luckily, although their effect on human or animal life is yet not quite clear, they do not appear to produce any baneful effects. Then, there are other problems, like those of weightlessness etc. All these have now been more or less overcome, as it evidenced by two astronauts, — one Russian and one American —having actually 'walked' in space and come safely back to the earth.

Another very essential item is the special type of dress that an astronaut must wear during his voyage in the cosmos. This is fittingly called the **Astrosuit** and must at once be *air-tight* and *loose-fit* to allow free respiration. In fact, the astronaut needs one type of dress during the take-off of the rocket from the earth, which must be so designed as to free him from the feeling of overload, experienced during a vertical ascent. Then, he should have another lighter dress for free locomotion inside the cabin. A small cylinder, provided in the girdle of this dress, creates an artificial pressure on him to increase his blood pressure (which falls appreciably at greater heights) as also to counteract weightlessness. The dress should also have provision to ensure normal respiration and the requisite body temperature and must not restrict movement.

10.21. Return of Artificial Satellite. For the return of the satellite back to the earth, the main problem is of sufficiently slowing down its motion or braking it. There are two devices for it, *viz.*, (*i*) *utilising air as the resisting medium* and (*ii*) *using rockets*.

If air-resistance is to be used for slowing down the motion of the satellite on its journey back, it must be given the form of a rocket. For, then, as it enters the denser layers of the air, its velocity falls but it rebounds back into the cosmos; it then re-enters the air with a reduced velocity and goes a little deeper than before and there is further reduction in its velocity. This process is repeated a few times and the velocity of the satellite is thus sufficiently reduced to enable it to continue falling on specially provided slide-wings and sliding-planes.

On the other hand, if a rocket is to be used for its downward journey, an automatically-controlled rocket-motor is necessary, the reaction of which is in the opposite direction to that of the motion of the satellite, *i.e.*, it produces an effect opposite to that of the rocket carrier during upward flight. It is, therefore, called a *retro-rocket*, (*i.e.* a rocket, taking the satellite back). The velocity of the satellite is thus reduced and can be controlled by regulating the fuel-supply to the rocket motor. The distance it thus has to cover up to the landing strip being carefully estimated with the help of a radar or some other similar appliances. And, an automatic guidance system is provided to control and manoeuvre the downward descent of the satellite.

Now, the first method is certainly the simpler of the two, from the technical point of view, but its great handicap is that it is extremely difficult to design a landing strip to receive the landing satellite. The second method, although more complicated technically, ensures a smooth and an accurate landing on a properly constructed landing strip.

10.22. Uses of an Artificial Satellite. Ignoring the military uses to which a satellite may be put, we shall concern ourselves here only with its uses for strictly scientific purposes, among the more important ones of which may be mentioned the following :

(i) Proper study of the upper regions of the atmosphere. Despite the fact that the atmosphere is being studied for a long enough time, our present knowledge of it is still much too meagre and superficial, particularly about the region, called the *Ionosphere*, as also about cosmic rays. The artificial satellites will, it is hoped, help to improve this.

(ii) Weather forecasting. This can be made much more accurate and dependable with a number of satellites around the earth in various orbits. Meteorological observations over various countries could then be made simultaneosuly, thereby greatly improving the reliability of weather forecasts.

(iii) Determination of the exact shape and dimensions of the earth. This is the task that scientists all over the earth had set for themselves during the third International Geophysical Year.

(iv) Detailed study of the solar radiation.

(v) Study of meteorites.

(vi) Experimental verification of the theory of relativity.

(vii) Study of propagation-characteristics of radio waves in the upper regions of the atmosphere.

(ix) Astronomical observations, without atmospheric and other disturbances, etc., etc.

WORKED EXAMPLES

1. Calculate the limiting velocity required by an artificial satellite for orbiting round the earth, given the radius of the earth as 6.4×10^8 *cm* and $g = 980$ *cm/sec*^{2}. *(Delhi)*

We know that the maximum velocity of a satellite revolving round and close to the earth is given by

$$V = \sqrt{MG/R} = \sqrt{gR}.$$

So that, substituting the given values of g and R, we have

$$V = \sqrt{980 \times 6.4 \times 10^8} = 10^4\sqrt{98 \times 64} = 8 \times 10^4\sqrt{98}$$

$$= 8 \times 10^4 \times 10 = 8 \times 10^5 \text{ cm./sec.} = 8 \text{ km/sec. very nearly.}$$

Thus, the limiting velocity the satellite must possess is 8 *km./sec.*

2. It is said that the first artificial satellite was revolving round the earth at a distance of 560 miles from it. Estimate its velocity and period of revolution, taking the radius of the earth to be 4000 *miles* and the value of g to be 32 *ft / sec*^{2}.

Here, the distance of the satellite from the centre of the earth = 4000 + 560 miles = R', say.

If, therefore, V be the velocity of the satellite, we have

$$\textit{centrifugal force on the satellite outwards} = mV^2/R'$$

And, $\textit{gravitational pull on the satellite inwards} = m \,.\, M \,.G/R'^2,$

where M is the mass of the earth.

Since the satellite is in dynamical equilibrium under these two forces, we have

$$mV^2/R' = m \,.\, M \,.\, G/R'^2, \text{ whence, } V^2 = MG/R'.$$

Again, because $mg = m.M.G/R^2$, or $MG = gR^2$, we have

$$V^2 = gR^2/R', \text{ whence, } V = \sqrt{gR^2/R'} = R\sqrt{g/R'}.$$

Substituting the given values of g and R′, we therefore, have

$$V = 4000 \times 1760 \times 3\sqrt{32/4560 \times 1760 \times 3} = 24{,}340 \text{ ft/sec.}$$

And, the period of revolution of the satellite, i.e. , the time it takes to complete one orbit round the earth, is clearly eqaual to

$$\frac{\text{circumference of the orbit}}{\text{velocity of the satellite}} = \frac{2\pi \times 4560 \times 1760 \times 3}{24340} = 6207 \text{ sec.}$$

$$= 1 \text{ hr. } 43 \text{ min. } 27 \text{ sec.}$$

Thus, *the veloicty of the satellite is* 24,340 *ft. sec.* and its *period of revolution*, 1 *hr.* 43 *min.* 27 *sec.*

EXERCISE X

1. What do you understand by the term *'jet-propulsion'* ? Give, in brief, an account of jet-propelled planes.

2. What is a *rocket* ? How do *rocket-planes* differ from *jet-planes* ? Explain the principle underlying a multi-stage rocket.

3. What is an *artificial satellite* ? Explain as clearly as you can how a satellite may be placed in its orbit around the earth.

4. Mention the essential pre-requisites and conditions for a satellite to be placed in its orbit and its return back to the earth.

5. What are the scientific uses of an artificial satellite ?

6. Write a brief account of the development of jet and rocket propulsion.

CHAPTER XI

FRICTION AND LUBRICATION—PRINCIPLE OF VIRTUAL WORK AND ITS SIMPLE APPLICATIONS

11.1. Static Friction—Laws of Friction. In all cases of motion of material bodies,counter forces come into play in between their surfaces which tend to nullify or neutralise the effect of the driving forces applied. These counter forces are called *resistances,* the most important among which is *friction.* As we have already learnt in our junior classes, when one solid body is sought to be moved over the surface of another on which it rests, an opposing force, called the *'force of friction'*, comes into play in between the two surfaces, tending to destroy the relative motion between them,* and which is usually measured by the force required to produce *uniform* relative motion between the two surfaces. It is this force which always acts in a direction opposite to that in which the motion is desired and which is called the *force of friction* or rather *static friction***. Experiment shows that friction roughly obeys the following laws, called the *'laws of friction'*, discovered by **Amontons** (1699) and **Coulomb** (1779), and hence sometimes referred to as **Coulomb's laws :**

(*i*) *The frictional force is a self-adjusting force and increases with the applied force, so as to be equal and opposite to it, until motion is just about to ensue; —this maximum frictional force is called the* **'limiting friction'** *and its value is different for different pairs of surfaces.*

Before this limiting value of friction is reached, its magnitude is just enough to preserve equilibrium.+

(*ii*) *The limiting friction between the surfaces of two bodies is directly proportional to the normal reaction of the supporting surface.*

Thus, if R be the *limiting firction* of the supporting surface and F, the *limiting friction* set up between the two surfaces, we have

$$F \propto R, \quad \text{Or,} \quad F/R = \mu$$

where μ is a constant, called the **'Static Coefficient of Friction'** or simply the *coefficient of friction* for the given pair of surfaces,

$$\textit{i.e.,} \qquad \textit{static coefficient of friction} = \frac{\textit{limiting friction}}{\textit{normal reaction}}.$$

Thus, obvioulsy, $F = \mu R$; so that, if $R = 1$, we have $F = \mu$, *i.e.*, the *coefficient of friction for a given pair of surfaces may be defined as the limiting friction coming into play in between them for unit normal force applied to them, or, as the fraction of the normal force that is required to keep the two surfaces in uniform relative motion.*

(*iii*) *The frictional force is independent of the surface areas in contact with each other*++ *and of their relative velocities.*

---

* Strictly speaking, this is not the only force that opposes the relative motion between the two surfaces. There is also another force, called the *'force of adhesion'*, (from the Latin word *'adhaerere'— to stick* or *to cling*, which is molecular in origin and which tends to make the bodies cling together.

** *'Static'*, because the two surfaces are initially at rest with respect to each other.

+ It will be readily seen that this really follows from *Newton's third law of motion*.

++ This is no longer so, if what are called *lubricants*, like grease, graphite, talc etc., are introduced in between the two srufaces. For, the normal force applied is more likely to squeeze out the lubricant from in between the two surfaces, when applied to a small area than when applied to a larger area.

(iv) *The frictional force is independent of the relative velocities of the two surfaces.*

It may be pointed out here that these laws apply only in the case of clean and smooth or well-polished surfaces.

11.2. Sliding Friction. We have already seen how the force of friction between the surfaces of two bodies, one resting over the other, continues to increase with the applied force and is always equal and opposite to it until its maximum or limiting value is reached for that particular pair of surfaces

If, at this state, *viz.*, when the friction is about to attain its limiting value, we apply a force in the form of a gentle push to the body resting over the other, such that it is maintained in *unifrom motion* over the latter, *then, this force measures what is called the sliding friction* between their surfaces, *i.e., the frictional force in between them when motion ensues.* it is also spoken of as *kinetic* or *dynamic* friction, to distinguish it from *static* friction (that comes into play before motion actually takes place) and is found to be somewhat less than the *limiting friction* for the same pair of surfaces. That is why we find it easier to maintain a body in uniform motion over the surface of another than to start it moving.

The ratio between *this* sliding friction and the normal reaction then gives the **coefficient of sliding friction** for the given pair of surfaces and is also obviously less than the coefficient of static friction for them. The difference between the two is however quite small and we usually assume them to be the same for all practical purposes.

11.3 Angle of Friction—Cone of Friction. If we place a body on an inclined plane, [Fig. 11.1 (*a*)], it is clear that it is under the simultaneous action of three forces, *viz.*,

(i) *its weight W, acting vertically downwards at its c.g., G;*

(ii) *the normal reaction R of the plane, acting perpendicularly to the plane; and*

(iii) *the limiting friction F, acting upwards along the plane.*

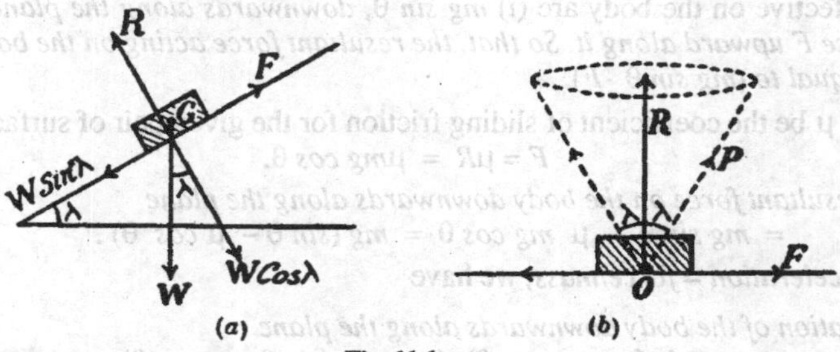

Fig. 11.1

If the angle of inclination of the plane be so adjusted that the body is *just* on the point of sliding down, it is clear that the three forces are in equilibrium and, therefore, concurrent, so as to be represented by the three sides of a triangle, taken in order. This angle of inclination of the plane, λ, at which the body is just on the verge of sliding down, is called the *angle of friction.*

Resolving W into its two rectangular components, we have (*i*) component $W \sin \lambda$, along the plane, tending to move the body down the plane and component $W \cos \lambda$, at right angles to the plane.

Since the body is in equilibrium, we have

$$F = W \sin \lambda \text{ and } R = \cos \lambda .$$

So that,
$$\mu = \frac{F}{R} = \frac{W \sin \lambda}{W \cos \lambda} = \tan \lambda$$

Or, *the coefficient of friction for a given pair of surfaces is equal to the tangent of the angle of friction for them.*

From the above relation for μ, we have $F = R \tan \lambda$, from which it follows that the resultant of F and R must lie along the surface of a *cone*, with λ as its semi-vertical angle, and the direction of the normal reaction, as its axis.

The same is true in the case of two horizontal surfaces, where the friction force F acts along the supporting surface and the normal reaction R, perpendicular to it, [Fig. 11.1 (*b*)]. Their resultant P then makes with the latter the angle of friction λ and, again, therefore, the resultant (P) lies along a cone of semi-vertical angle λ, such that $\tan \lambda = F/R$.

This *cone* is called the **cone of friction,** and it is obvious that no force, whatever its magnitude, with its line of action lying *within* the cone, can possibly produce motion in the body, its component along the surface of contact being less than the limiting friction (F) between them.

11.4. Acceleration down an Inclined Plane. We have just seen in § 11.3, above, that a body placed on an inclined plane will not start sliding down along it until the angle of inclination θ of the plane is equal to the angle of friction λ for the surfaces of the body and the plane; for, at a smaller angle of inclination than this, $\tan \theta < F/R$ or μ, where F is the limiting friction and μ, the *coefficient of friction* for the two surfaces in question. And, when $\theta = \lambda$ clearly $\tan\theta = \tan \lambda = F/R$ and sliding *just* commences.

But when $\theta > \lambda$, $\tan \theta > \tan \lambda$ and hence greater than F/R, the body slides down the plane with accelearted motion. Let us calculate this acceleration of the body.

As before, resolving the weight of the body $W = mg$ (where m is the mass of the body) into two rectangular components, along and perpendicular to the plane, we have

$$\textit{component along the plane} = mg \sin \theta$$

$$\text{and } \textit{component perpendicular to the plane} = mg \cos \theta.$$

Since there is no motion perpendicular to the plane, the normal reaction R of the plane is equal to $mg \cos \theta$ and the two, *being equal and opposite*, neutralise each other. And, thus, the only two forces effective on the body are (*i*) $mg \sin \theta$, *downwards along the plane*, and *(ii) the sliding frictional force F upward along it. So that, the resultant force acting on the body downwards along the plane is equal to* $(mg \sin \theta - F)$.

Now, if μ be the coefficient of sliding friction for the given pair of surfaces, we clearly have

$$F = \mu R = \mu mg \cos \theta.$$

So that, *the resultant force on the body downwards along the plane*

$$= mg \sin \theta - \mu\, mg \cos \theta = mg (\sin \theta - \mu \cos \theta).$$

And, since *acceleration = force/mass*, we have

acceleration of the body downwards along the plane

$$= mg(\sin \theta - \mu \cos \theta)/m = g(\sin \theta - \mu \cos \theta).$$

If the plane be perfectly smooth, the acceleration of the body sliding down the plane would be $g \sin \theta$. Clearly, therefore, the acceleration of the body down the plane is reduced by $\mu g \cos \theta$ due to the frictional force between them.

11.5. Rolling Friction. The frictional force between two surfaces when one *rolls* over the other is called *rolling friction* and is found to be much less than when sliding occurs between the same two surfaces. That is why vehicles are provided with wheels and their axles, with ball-bearings*, the latter converting the chief frictional loss of the wheel that occurs at the axle or the *journal* in the form of sliding friction—here, called *journal* friction—into rolling friction.

It was shown by **Osborne Reynolds** that in rolling, an appreciable amount of slipping or sliding of one surface over the other occurs and that *the frictional resistance to this slipping, or*

---

* Here, there is a ring of small balls between the wheel (pulley or disc etc.) and the axle, so that when the former rotates, the balls all rol also.

sliding, really constitutes rolling friction. As extreme cases of this slipping between two rolling surfaces may be mentioned (*i*) an iron cylinder rolling over a plane rubber surface or (*ii*) a rubber cylinder rolling over a plane iron surface. In the former case, the cylinder covers a distance equal to only *nine-tenths* of its circumstance in its one full turn and, in the latter case, a distance equal to eleven-tenths of its circumference; so that, in either case, there is a slip of one-tenth of its circumference. In ordinary cases too, some slip always occurs between two rolling surfaces, even when the two surfaces are of the same material.

It follows as a natural consequence that rolling friction between two *surfaces* would be *zero* (*a*) when either the sliding friction between them is *zero, i.e., when* μ *for them is zero* (*b*) or when no slipping occurs between them during rolling, *i.e., when* μ *for them is infinite*. For all other values of μ (between 0 and ∞) there must be friction and, for a particular value of μ, it must have its maximum value.

It may as well be mentioned here that *while lubrication of the surfaces always reduces the value of the coefficient of sliding friction* (μ) *for them, it may or may not reduce the rolling friction between them.* Thus, as is so well and so generally known, lubricating ball-bearings only results in increasing friction.*

11.6. Friction and Stability. When a body, say, a block of wood, rests on a plane horizontal surface, it does so because the weight of the block W, acting vertically downwards at its *c.g.*, is just neutralised by the *equal and opposite* normal reaction R acting there. And, when a horizontal force F is applied to the block to move it forwards on the plane, it does not move or slide along it so long as $F < \mu R$, where μ is the ciefficient of friction between the surfaces of the

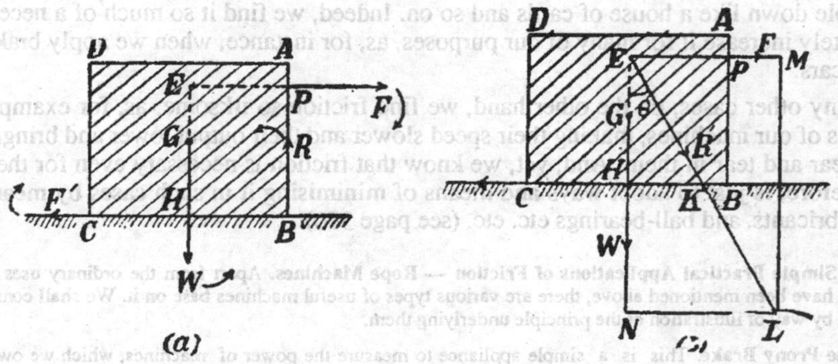

Fig. 11.2

block and the plane. The possibility is, however, there that the block may topple over; for, the moment the horizontal force F is applied at a point P, say, a frictional force F', *equal and opposite* to F, comes into play in between the surfaces of the block and the plane, [Fig. 11.2 (*a*)], thus constituting a *couple* = $F \times PB$, tending to rotate the block (in the clockwise direction, in the case shown), and thus making it topple over.

Now, as F is gradually increased, this couple formed by F and F' makes the centre of reaction of the plane shift from H towards B,** with the force of C progressively decreasing and that at B increasing, until, in the limiting case, the whole reaction R acts at B (that at C being zero). We thus have another couple, formed by W and R, equal to $W \times HB$, tending to rotate the block in the opposite direction to that due to the first couple (here, anticlockwise) which thus tends to

* This might raise a question in the mind of the student as to why then are they lubricated at all ? The simple answer is that it is done only with a view to reducing wear.

** For, with no force F acting on the block, its weight is uniformly distributed over its base B.

restore the block back to its original position. So long, therefore, as this restoring couple $W \times HB$, is greater than the couple $F \times PB$, due to F and F', the block remains at rest and upright, *i.e., in stable equilibrium*, but if $(W \times HB) < (F \times PB)$, it topples over.

Let us now investigate the conditions under which the block will remain upright but slide along the plane. For this, let us first determine the resultant of F and W by the ordinary application of the parallelogram law of forces and then take the moments of this resultant R' and the frictional force F' about B. Let R' be represented by the diagonal EL of the parallelogram $EMLN$, with its adjacent sides EM and EN representing F and W respectively, and let it cut the plane in K, [Fig. 11.2 (*b*)]

Now, clearly, moment of F' about B is *zero*, since its line of action passes through B. So that, so long as $HK < HB$, *i.e.*, so long as the resultant R' passes through the contour of the base of the block, it will have a restoring moment about B and the block will remain upright. But, if the angle θ made by R' with the vertical, or the direction of the normal reaction R, be greater than λ, the *angle of friction* for the surfaces of the block and the plane, (see § 11.3), it will *slide* along the surface. If, however, $HK > HB$, *i.e.*, if the resultant R', of F and W, passes outside the contour of the base of the block, the block will topple over.

Another important case of stability due to friction is that of a fast moving vehicle on a curved track, discussed already in § 2.13 (4).

11.7. Friction, a necessity. Taking most of our daily activities in life as a matter of course, we seldom care to pause and think as to how much they are dependent on the existence of friction. Thus, for example, in the absence of friction, we would find it impossible to walk or to drive on a road, and if we just start moving, we shall not be able to stop; again, it would be impossible to climb a tree, tie a knot or even fix a nail in the wall. Brick would not stand on brick and buildings would tumble down like a house of cards and so on. Indeed, we find it so much of a necessity that we deliberately increase it for many of our purposes, as, for instance, when we apply brakes to our bicycles or cars.

In many other cases, on the other hand, we find friction so irksome; as, for example, in the various parts of our machines, making their speed slower and their output lower and bringing about a greater wear and tear in them. And, yet, we know that friction is necessary even for them. What we do, therefore, is just to adopt ways and means of minimising it in such cases by means of oils and other lubricants, and ball-bearings etc. etc. (see page 370).

11.8. Simple Practical Applications of Friction — Rope Machines. Apart from the ordinary uses of friction, some of which have been mentioned above, there are various types of useful machines base on it. We shall consider here a couple of them by way of illustration of the principle underlying them.

(1) **The Prony Brake.** This is a simple appliance to measure the power of machines, which we owe to *Baron G.C. F. Prony*, a French Mathematician, (1750–1839). It is in fact a **break dynamoinc** and consists of two wooden arms, or , *'cheeks'* A and B (Fig. 11.3) in between which can be clamped the **shaft** of the machine whose power (*i.e.*, rate of working) it is desired to measure. The frictional force between the shaft and the cheeks is regulated by tightening or loosening the screws S and S, provided on the upper cheek A, to which is also attached a small rod R, about .5 to 1.0 metre in length, carrying a scale pan at its other end.

Suppose the shaft of the machine turns in the anti-clockwise direction, as shown. Then, if there be no weights placed in the sacel pan, *i.e.*, if the break be unloaded, it will tend to be carried around with the shaft due to friction between itself and the shaft. But, if the brake be sufficiently loaded, before the shaft starts rotating, *ie.*, before the machine is started, the moment due to the load may be enough to overcome the moment due to friction between the brake and the shaft and the

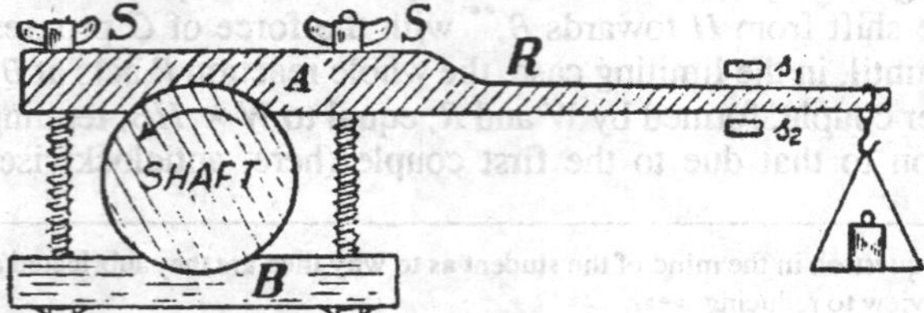

Fig. 11.3

brake may turn in the clockwise direction; so that, it is quite possible to so adjust the weights W in the scale pan, *i.e.*, to so load the brake, that the rod R remains quite horizontal in between, and equidistant from, the stops s_1 and s_2, placed on its two sides, a little distance away from it. When this is so, obviously, the frictional resistance between the brake and the shaft is equal to the force F exerted by the machine on the periphery of the shaft and is exactly balanced by the weights or the *load* W placed in the scale pan. So that, we have

moment of force F (due to machine) on the shaft = moment of weight W on it.

If r be the radius of cross-section of the shaft and l, the length of rod R, we clearly have

$$\text{moment of } F \text{ on the shaft} = F \times r,$$

and, $\quad$ moment of W ,, ,, $= W \times l$.

Thus, $\quad F \times r = W \times l, \quad$ whence, $\quad F = W \,.\, l/r$.

If the shaft makes one full rotation in time T, we have

work done by the shaft, *i.e.*, by the machine in time T

$$= F \times \textit{circumference of the shaft} = F \times 2\pi r$$

And $\therefore$ work done by the machine per unit time, *i.e.*,

$$\textit{power of the machine} = F \times 2\pi r/T.$$

Or, substituting the value of F, obtained above, we have

$$\textit{power of the machine} = \frac{W\,.\,l}{r} \times \frac{2\pi r}{T} = \frac{2\pi}{T}\,.\,W\,.\,l = \frac{2\pi}{T} \times \textit{moment of the load.}$$

Thus, knowing W, l and T, we can easily determine the power of the machine which, as we can see, is clearly proportional to the moment of the load.

N.B. If W be taken in *dynes*, *l in cm* and T in *sec*, we get the power of the machine in *ergs per second*.

(2) **The Rope Brake**. Before discussing any rope-machines, we must be clear in our minds as to what exactly is meant by a rope and what, if any, are the peculiar properties possessed by it that make for its usefulness.

A rope, then, is any flexible body or combination of flexible bodies capable of transmitting tension. Thus, *the string of a violin, a piece of spring, a strap, a band* and *a chain* all come under the definition of a rope.

When one end of a rope is connected to a body and its other end pulled, stresses are caused at every cross-section of it in exactly the same manner as in a rigid rod. The stresses in neighbouring cross-sections neutralise each other and the pull is thus directly exerted on the body at the other end. *A rope can thus transmit a tension, undiminished in magnitude, from one end to the other.* Due to its flexibility, however, *it cannot transmit any compressional forces along it*; but, at the same time, its flexibility is of great advantage in many ways and makes it particularly suitable as a means of transmitting tensional force, unaffected even by a change of direction*, as in the case of a pulley etc.

On the other hand, when coiled round a cylindrical body, a rope can exert a very large couple on the body, due to friction between itself and the body, so much so indeed, that a man pulling at one end may even hold a ship, fastened to its other end ! Let us see how this comes about.

Let $ABCD$ be the cross-section of the cylindrical surface, with its centre at O and let a rope $PABCQ$, coiled round it, leave its surface at points A and C, (Fig. 11.4).

Consider an infinitesimal portion EF of the rope at B and let the mean tension over this portion be T, with the angle subtended by it at O equal to $d\theta$.

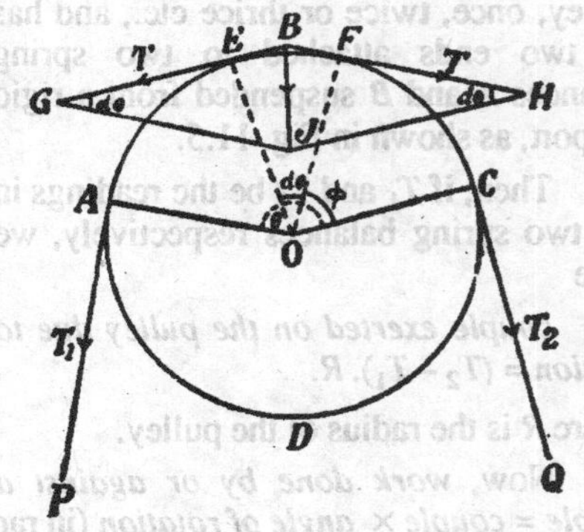

Fig. 11.4

* This is possible only so long as the body, (*e.g.*, the pulley or the ring etc.) over which the rope is passed does not interfere with its freedom of motion, *i.e.*, is perfectly smooth and round.

Then, representing the tensions T, at E and F, by the tangents EG and FH respectively, we have their resultant force represented by BJ (by the simple application of the parallelogram law of forces), where BJ is clearly equal to $BG.\,d\,\theta = T.\,d\,\theta$ [$\because$ BG represents T and $\angle BGJ = d\theta$], in the direction BJ, normal to the section EF, at B.

If, therefore μ be the coefficient of friction for the rope and the cylindrical surface, we have *frictional force between the rope and the surface* $= \mu\,.T\,.\,d\theta^{*}$, in the direction of the rope at B.

Due to this frictional force there comes about a change in the tension at the two ends of the rope, which, in the absence of any friction, would have been the same, (see above). In fact, the difference in the tensions at the two ends is just equal to this frictional force. So that, if dT be the difference in tensions at E and F, we have $dT = \mu.T\,d\,\theta$. Or, $(1/T).\,dT = \mu\,.\,d\theta$, integrating which, we have

$$\int \frac{1}{T}\,.\,dT = \int \mu\,.\,d\,\theta.\ \text{Or,}\ \log_e T = \mu\,\theta + K, \qquad ...(i)$$

where θ is the angle EOA and K, a constant of integration.

Clearly, if $\theta = 0$, $T = T_1$, the tension at A; so that, we have

$$\log_e T_1 = \mu \times 0 + K = K$$

Substituting this value of K in expression (i) above, we have

Or $\log_e T = \mu\theta + \log_e T_1$. Or, $\log_e T - \log_e T_1 = \mu\theta$.

$\log_e T/T_1 = \mu\,\theta$. Or, $T/T_1 = e^{\mu\,\theta}$,

whence it is clear that the *tension T increases as θ increases.*

Obviously, therefore, if ϕ be the total angle AOC, and T_2, the tension at C, we have

$$T_2/T_1 = e^{\mu\phi}.$$

It is thus clear that if ϕ be large, *i.e.*, if the rope be coiled many times round the cylinder, T_2/T_1 is also very large and a small tension applied at P may be made to exert a large pull at Q.

(3) **The Band Brake.** The principle discussed in § 11.8 (2) above, is made use of in many forms of *brakes*, for the measurement of the power (or the rate or doing work) of machines, one of the simplest of which is the *Band brake*.

It is a simple device, consisting of a pulley, fixed on the rotating shaft of the machine whose power is to be determined. A band (or a cord) is passed round the pulley, once, twice or thrice etc., and has its two ends attached to two spring balances A and B suspended from a rigid support, as shown in Fig. 11.5.

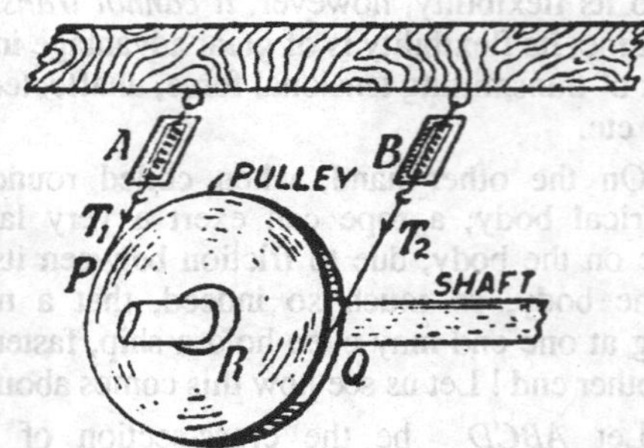

Fig. 11.5

Then, if T_1 and T_2 be the readings in the two spring balances respectively, we have

couple exerted on the pulley due to friction $= (T_2 - T_1).\,R$.

where R is the radius of the pulley.

Now, *work done by or against a couple = couple × angle of rotation* (in radians).

And, therefore, *work done against a couple per second = couple × angle turned through in one second.*

---

*Because frictional force = μ × *normal reaction R*, and here $R = Td.\theta$.

Or, *rate of doing work, i.e., power = couple × angular velocity.*

Thus, *work done per second by the machine against the couple due to friction* $= (T_2 - T_1).R \times 2\pi n$, where n is the number of rotations made by the *shaft* or the axle of the pulley per second and hence $2\pi\, n$, *the angular velocity of the shaft.*

This, then, gives the worm done per second by the machine or the power of the machine, *i.e.*, we have

$$\text{power of the machine} = (T_2 - T_1).\ R \times 2\pi n,$$

which can be easily evaluated.

Further, we have the relation $T_2/T_1 = e^{\mu\theta}$ where μ is the coefficient of friction for the cord or band in question and the surface of the pulley and θ, equal to π, 3π, or 5π, etc. according to the number of times the cord or the band passes round the pulley. So that, knowing T_1, T_2 and θ, we can also easily calculate the value of μ.

11.9. Mechanism of Friction. Since the laws of solid friction were enunciated by *Amontons* and *Coulomb*, much has come to be known as to the how and why of friction, thanks to the work of **Hardy, Bowden and others.**

According to *Dr. Bowden's* lucid exposition, the mechanism of friction may be summed up as follows :

The smoothest or even the most polished surfaces are really not smooth enough, having projections and depressions of larger than molecular dimensions on them; so that, the areas of the two surfaces *in actual contact* is much smaller than the *apparent* one we see to be so, —being less than ten-thousnadth part of the apparent area of contact, in the case of plane steel surface.

It is possible now to form a fairly correct estimate of the actual area of contact between two metals by measuring their electrical resistance, and it is found to be *practically independent of the size and roughness of the surfaces and to depend only upon the load.* The obvious inference is that the projections on the two surfaces get shorn off or crushed down under the load, until an area, enough to support that load, is cleared up and comes into actual contact.

Viewed in *this* light, *Amonton's* law, regarding the independence of the force of friction of the surface area, means no more than that, for a given load, a change in the apparent area makes little or no difference to the area in *actual* contact.

A logical consequence of this smallness of the area of true or actual contact is that the pressure at the points of contact must be enormously high. Indeed, it is estimated to be of the order of 100 *tons per square inch* in the case of mild steel. And this inevitably leads to the result that during sliding between the two surfaces, the temperature at these points of actual contact must rise to equally enormous proportions. All this now stands amply verified by actual experiment. For, it has been found possible to measure the temperature of the points of actual contact in the case of two dissimilar metals by using them as a *thermocouple* and by amplifying and applying the thermo-electromotive force, thus generated, to the deflecting plates of a *cathode-ray oscillograph*. Temperatures, as high as 1000^0C*, have been found to obtain at these actual contact points or '*hot spots*', although they last only for just a split second, —the temperatures thus reached depending upon *(i) the magnitude of the load, (ii) the speed at which sliding takes place, (iii) the thermal conductivity and (iv) the melting points of the metals in question.*

In most cases, therefore, there is welding at these points during sliding, due to melting and consequent flow of metal. The surfaces of both metals thus get damaged by this sliding occurring between them. The surface of the softer metal gets torn and 'ploughed' by the projections on the surface of the harder metal and the latter has the softer metal welded on to it, —the maximum damage occurring in the case of two *similar* metals sliding one over the other, the coefficient of friction being the highest in their case.

The frictional force between two metal surfaces may thus be attributed to the shearing of these temporarily-welded points as well as the 'ploughing' resulting from the projections of the harder metal, —the former being by far the greater of the two forces depending upon *(a)* the actual contact area of the metals and the shear-strength of the softer metal**. To minimise friction, therefore, both these factors should be made as small as possible. In almost all cases, however, if one is less, the other is high; as, for example, if a hard metal is made to slide on steel, factor *(a) i.e.*, the actual area of contact, is small, but factor *(b) i.e.*, the 'ploughing' is greater, and if a softer metal is made to slide on steel, the ploughing, [factor *(b)*] is small, but the contact area [factor *(a)* becomes greater. As a consequence, for most pairs of metals, the coefficient of friction comes to lie within the same common limits, .6 to 1.0.

---

* Even higher temperatures are reached in the case of poor conductors of heat. Thus, for example, glass rubbed against glass, or quartz against quartz, gives tiny sparks of light, changing from reddish to white as the vigour and speed of rubbing is increased.

** *Coulomb* had originally suggested that the frictional force between two surfaces was purely a consequence of their roughness and consequent interlocking of their projections and depressions, so that there should be practically no frictional force between perfectly polished surfaces. This is, however, not found to be so, as we shall presently see.

11.10. Lubricants. A thin film of a soft metal, like lead or iridium*, smeared on a harder surface, like steel, often reduces friction, because of *its reducing the shear-strength without unduly increasing the contact area*. This explains why we frequently use copper or silver bearings, smeared with lead, or iridium in our modern aero-engines. A substance so used to minimise friction is called a **lubricant.**

Even apparently clean metal surfaces in air have a thin smearing of some sort of a lubricant between them, *e.g.*, a thin film of air or a thin film of oxide or some such other contaminant, like moisture etc., which greatly reduces their shear-strength and hence the friction between them. Consequently, the friction between two metal surfaces, is found to be greater in vacuo than in air, —the cause of this enhanced friction in the former case being the *force of adhesion* between the two surfaces which grip or seize each other under the smallest load. Thus, as *Dr. Bowden* so aptly puts it, *it is really fortunate that all metal surfaces are more or less contaminated in air, or else the world would be a very sticky place indeed.*

All cases of friction fall into one or the other of the following categories, *viz.*, (*i*) **dry friction between solid surfaces,** *i.e.*, the frictional force encountered when no lubricant is present between them; (*ii*) **boundary friction,** when only a thin film of a lubricating material is present between the two given solid surfaces; and (*iii*) **fluid or floating friction,** when there is plenty of a liquid lubricant present in between the two surfaces; so that, the friction, in the case, depends almost wholly upon the properties of the lubricant, in particular, upon its viscosity, (see Chapter XII).

We have already studied briefly the essential details of the first two categories, *viz.*, friction between solid surfaces in the absence of any lubricating substance, and 'boundary friction', with a thin film of lubricant present in between them. The ordinary cases of friction between solid surfaces, which, as we have already seen, are almost always contaminated to some slight extent, with oxide or moisture etc., strictly belong to this second category of boundary friction. **Hardy** thoroughly studied this type of friction and, according to him it is the chemical natures of the metal and the lubricant which determine the strength of the boundary film, a metal attracting more strongly the polar group (*COOH*) at one end of the molecules of an oil than the non-polar ones at the other. This results in the layer of molecules nearest the mettal surface being orientated so as to stand upright (*i.e.*, normal to the surface), *with their polar ends inwards*, with possibly further double layers, having their polar ends adjacent, being formed on them, so that, sliding actually occurs between pairs of these molecular layers over their non-polar ends. Obviously, therefore, *the more strongly does the lubricant adhere to the surface, the more effective will it be in minimising friction*.

In the case of *fluid or floatation friction*, the thickness of the film makes most of the molecules lie outside the range of the adhesive force of the solid surface, so that the only force to be overcome now is that due to the viscosity of the lubricant.

Beuchamp Tower, in 1883, experimented on a revolving shaft, resting in a well lubricated bearing and showed that not only was the friction between the two greatly reduced but that the laws obeyed by frictional force were entirely different from those in the case of dry friction. For (*i*) *while in the case of dry surfaces, the frictional force is proportional to the normal reaction between them it is practically quite independent of the normal reaction in the case of lubricated surfaces*, (*ii*) *while, in the case of dry surfaces, the magnitude of the frictional force depends upon the nature of the two surfaces, it is quite independent of the nature of the lubricated surfaces, depending only upon the nature of the lubricant*, and (*iii*) *whereas in the case of dry surfaces, the frictional force is quite independent of their relative velocity, it varies directly with the relative velocity in the case of lubricated surfaces*. Three years later, in 1886, **Osborne Reynolds** put forward his fundamental theory of lubrication. *Tower* had shown that a film of the lubricant was formed in between the shaft and the bearing and *Reynolds* showed that the film, which he thought must be several molecules thick, could be maintained in spite of the enormous pressure between the two surfaces, provided that the diameter of the bearing were a wee bit (*about one-thousandth part*) greater than the diameter of the shaft. This makes the two surfaces slightly eccentric, with the *wedge-shaped* clearance between them, filled in with the lubricating oil, its pressure increasing with the narrowing down of the wedge and being equal to that developed between the sliding surfaces.

The oil is carried round *by the adhesive force on layers, nearer the shaft, and by viscous forces on those away from it*, and the two surfaces are thus kept reasonably apart for the quanity of the lubricant in this wedge to be sufficient to have its normal bulk-properties and thus to be able to resist the shear purely by virtue of its viscosity. *The shearing of the temporarily welded points between the two surfaces is thus replaced by this inter-liquid shear.*

Obviously, therefore, *liquids possessing the two essential properties of viscosity and the adherence to the solid surfaces will be the best lubricants. But, from the practical standpoint, perhaps by far the most important property required of a lubricant is that of chemical stability*, and mineral oils undoubtedly claim a definite superiority over vegetable oils in this respect as well as in supplying the necessary minimum viscosity** required to maintain a multi-molecular film between the two surfaces.

On the other hand vegetable oils excel over mineral oils in their property of forming strongly adhering films on the solid surface, (*i.e.*, the boundary films), in view of the polar groups in their molecule, due to the presence of free fatty acids in them.

---

* Iridum is even softer than lead.

** A high viscosity of the lubricant being also not quite desirable, in view of the resistance it would offer to the sliding motion between two surfaces. Further, since the resistance to motion due to viscosity increases with speed, a higher viscosity at lower speeds and lower viscosity at higher speeds, being deemed to be more deisrable.

To take full advantage of both these properties, therefore, the modern lubricants used are a *mixture of mineral oils with vegatable oils* (like *castor oil*) in proper proportion, —*the former supplying film lubrication and the latter, boundary lubrication*. In addition, it is also quite usual to add to the above mixture of lubricating oils, a measure of *colloidal graphite*, a thin film of which is formed on the solid surfaces, so that even if contact occurs between the two metal surfaces, seizure take place between the graphite layers, offering little or no resistance to any relative motion between them.

In addition to all this, other factors, such as change of viscosity with increase of temperature and pressure between the two surfaces, have also to be taken into consideration.

11.11. Principle of the Virtual Work. The principle of *virtual work* was first enunciated by **John Bernoulli**, in the year 1717, and later, in 1788, quite independently by **Lagrange**, and *applies to all cases of equilibrium*, thus leading to the deduction of all theorems relating to mechanical transmission of force, (of course, neglecting friction). Let us, before coming to the *Principle* itself or its applications, try to understand clearly as to what is meant by the term *virtual work*.

If we imagine a system of forces, acting on a particle, to suffer a slight displacement, consistent with the geometrical conditions, *i.e., compatible with the constraints to which the particle is subjected*, the displacement so imagined is called a *virtual displacement* for it may never really occur in actual practice, and is thus purely arbitrary, existing only in our imagination. *The product of each force of the system and the virtual displacement along its line of action is called* **virtual work***.

The **principle of virtual work** may now be stated as follows :

A system of force acting at points, connected by any mechanism will be in equilibrium, if the total work, accompanying a virtual displacement, is zero.

Let us consider a few simple cases, illustrative of this principle :

(*i*) Case of a body in equilibrium on a smooth inclined plane under the action of a force. Let P be a body of mass m, on a smooth incline of angle θ (Fig 11.6) such that a force F, acting on it at an angle ϕ with the plane, keeps it in equilibrium.

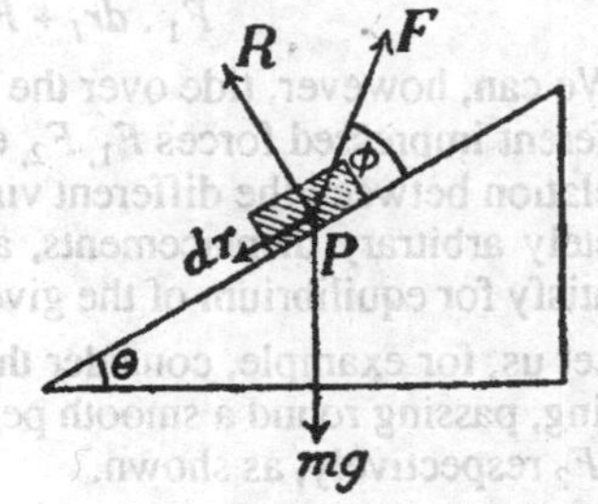

Fig. 11.6

The normal reaction R of the plane, obviously, acts perpendicularly to the plane, as shown. In order to eliminate any reference to it and thus to simplify our calculations, let us imagine a *virtual displacement dr* of the body, along the plane, so as to be perpendicular to R, so that $R.dr$ is *zero*. Then, clearly,

virtual work done *by* the weight mg of the body $= mg \sin\theta.\, dr,$

and *vitual work* done *against* force $F = -F\cos\phi.dr$, where $mg \sin\theta$ and $F\cos\phi$ are the components of mg and F along the plane, the *negative sign*, in the latter case, merely indicating that the work is done *against* the force.

Hence, according to the *principle of virtual work*, the condition that must be satisfied for the equilibrium of the body is that

$$mg.\sin\theta.dr - F\cos\phi.dr = 0.$$

Or, that $\qquad (mg \sin\theta - F\cos\phi)\,.dr = 0.$

Since, obviously, dr is not *zero*, we have

$$mg\sin\theta - F\cos\phi = 0. \text{ Or, } F\cos\phi = mg\sin\theta,$$

which gives the condition for the equilibrium of the body and which, we know, is the one we obtain by the ordinary *conventional method*, in which we also take into consideration the normal reaction of the plane.

* Very aptly so, since the displacement is only virtual or imagined.

(ii) **Case of equilibrium of a body on a rough inclined plane.** Let a body, of mass, *m*, be *just in equilibrium, i.e., just on the point* of sliding down a rough incline of angle θ, when a horizontal force *F* acts upon it, (Fig. 11.7)

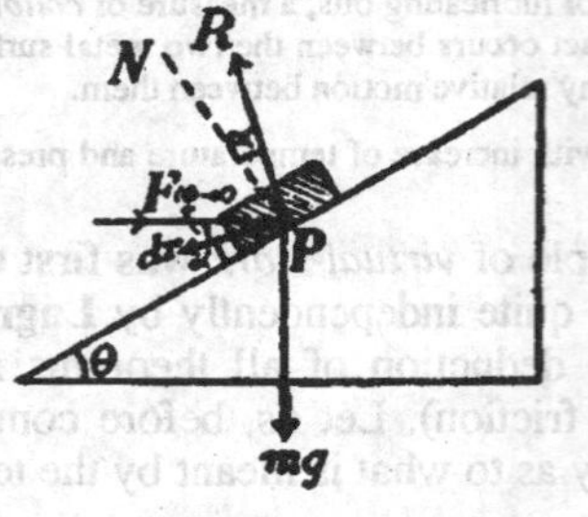

Fig. 11.7

Here, obviously, the reaction *R* of the plane will make an angle, α, say, (the *angle of repose*) with the normal *PN* to the plane.

Again, considering the body to have a *virtual displacement dr*, perpendicular to the direction of *R*, we have $R.dr = 0$. And, therefore, virtual work done by the weight *mg* of the body $= mg \sin(\theta - a) \cdot dr$,

and virtual work done *against* force F

$$= - F \cos(\theta - \alpha)\, dr.$$

So that, by the *principle of virtual work*, we have, for equilibrium,

$$mg \sin(\theta - \alpha) \cdot dr - F \cos(\theta - \alpha) \cdot dr = 0.$$

Again, since *dr* is *not* zero, we have $F \cos(\theta - \alpha) = mg \sin(\theta - \alpha)$.

Or, $$F = mg . \tan(\theta - \alpha),$$

which thus gives the condition for the equilibrium of the body.

(iii) **Case of equilibrium of a system of two or more connected bodies.** Here, a slight complication arises in that, if we imagine the bodies to undergo *virtual displacement* dr_1, dr_2, etc., under the action of forces F_1, F_2, etc., acting on them respectively, these displacements, *since they must be compatible with the constraints of the system*, cannot possibly be completely arbitrary, but will rather be inter-related to each other, although they will satisfy the relation

$$F_1 \cdot dr_1 + F_2 \cdot dr_2 \cdot + \ldots\ldots\ F_n . dr_n = 0. \quad ...(I)$$

We can, however, tide over the difficulty by first writing down the above general relation for the different impressed forces F_1, F_2, etc., and then deducing, from the geometry of the system, the inter-relation between the different virtual displacements. This will then enable us to obtain a set of completely arbitrary displacements, and hence the necessary conditions that the different forces must satisfy for equilibrium of the given system of bodies.

Let us, for example, consider the equilibrium of two bodies, m_1 and m_2, tied to the two ends of a string, passing round a smooth peg or pulley (Fig. 11.8), and subjected to two impressed forces F_1 and F_2 respectively, as shown.

Taking, for convenience, the virtual displacements dr_1 and dr_2 of the two bodies in the downward direction (which is quite compatible with the restraints imposed upon the system), we have, for equilibrium, on the *principle of virtual work*,

$$(F_1 + m_1 g).dr_1 + (F_2 + m_2 g)dr_2 = 0.$$

Now, an examination of the situation at once reveals the relationship between dr_1 and dr_2, *viz.*, that *the string being inextensible*,

$$dr_1 + dr_2 = 0, \text{ Or, that, } dr_2 = - dr_1.$$

So that, the above relation reduces to

$$(F_1 + m_1 g).dr_1 - (F_2 + m_2 g).dr_2 = 0,$$

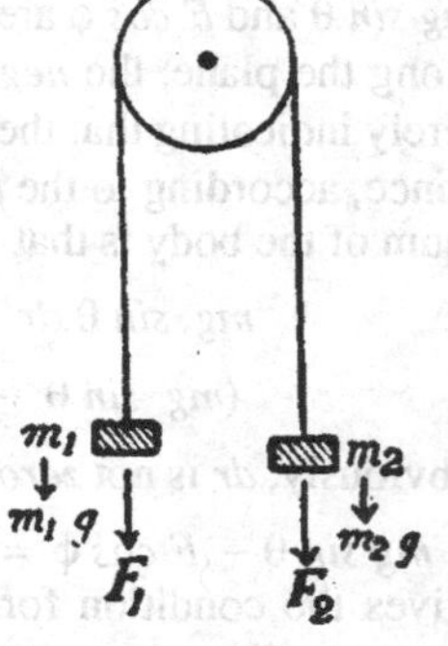

Fig. 11.8

whence, $$F_1 + m_1 g = F_2 + m_2 g,$$

which thus gives the necessary condition for the equilibrium of the two bodies.

It must be noted here that, if we were to use the old conventional method to determine the equilibrium condition of the bodies, we should have had to take into account the tension in the string. Since it is *not an impressed force*, we can clean ignore it in our calcuations on the principle of virtual work.

(iv) **Relation between Equilibrium and Potential Energy.** We may express relation *I*, of case *(iii)* above, in *Cc tesian form*, and write for the *p*th particle in a system of particles,

$$\mathbf{dr}_p = \mathbf{i}dx_p + \mathbf{j}.dy_p + \mathbf{k}.dz_p,$$

and $$\mathbf{F}_p = \mathbf{i}Fx_p + \mathbf{j}Fy_p + \mathbf{k}.Fz_p{}^*,$$

where dx_p, dy_p and dz_p denote the virtual displacements of the particle in the *x, y* and *z* directions respectively, and Fx_p, Fy_p and Fz_p, the components of the applied force $\mathbf{F}_p$ in these directions.

The principle of virtual work may now be expressed in the form

$$\sum_{p=1}^{n} (Fx_p.dx_p + Fy_p.dy_p + Fz_p.dz_p) = 0.$$

If the forces be *conservative* (*i.e.*, for which the total mechanical energy remains constant or *conserved*') so that there exists a potential energy function $(V_p(x_p.y_p:z_p)$, such that

$$Fx_p = -\frac{\partial V_p}{\partial x_p}, \quad Fy_p = -\frac{\partial V_p}{\partial y_p} \text{ and } Fz_p = -\frac{\partial V_p}{\partial z_p}$$

then, the total potential energy of the whole system of particles is given by $\sum_{p=1}^{n} V_p = V$.

The principle of virtual work then takes the form $\sum_{p=1}^{n} dV_p = 0$,

i.e., for the virtual displacement considered, the change in the total potential energy is zero.

In other words, *for the equilibrium position, the potential energy must either be a maximum or a minimum*; for only then, can the variation due to a small displacement be zero.** In the latter case, the equilibrium is said to be stable, and in the former, *unstable.*

Thus, taking the simplest case of a single particle, acted upon by the force of gravity and resting on a smooth surface, we have potential energy given by $V = mg.z$, taking the *z*-axis vertical, (*z* being *positive* upward).

Now, if the particle be in equilibrium on a *concave* surface, the total force acts on it normally to the surface, so taht *z* and, therefore, *V* is a *minimum,* and hence the equilibrium, *stable*; for, any finite displacement of the particle on the surface tends to increase its potential energy. But if it be in equilibrium on a convex surface, we have *z* and, therefore, *V*, a *maximum,* and hence the equilibrium, *unstable*, for, a finite displacement of the particle on the surface *now* tends to decrease its potential energy.

(v) **Tension in a flywheel.** Let us now consider the application of the *principle of virtual work* in a problem on the mechanical transmission of force and calculate the tension in a rotating flywheel.

---

* In vector notation, **i, j** and **k** denote *vectors of unit magnitude* in *x, y* and *z* directions respectively.

** Strictly, mathematically speaking, it is not quite correct to say so, and all we can reasonably assert, therefore, is that the potential energy will have a stationary value. In most mechanical problems, however, it is quite enough to confine our attention to *maxima* and *minima*.

If the radius of the ring-shaped flywheel be r and its angular velocity, ω, it is clear that an element δm of it will experience a *centrifugal force* $\delta m.r\omega^2$ *outwards*, at right angles to its circumference, and *the ring will thus tend to stretch itself, i.e., will be in a state of stress.*

If we imagine each element of the ring to suffer a *radial virtual displacement* dr, the virtual work done by it would be $\delta m.r\omega^2.dr$. And, therefore,

$$\text{total virtual work done } \textit{by the ring}, \text{ as a whole} = \Sigma\,(\delta m.r\omega^2.dr)\,.$$

Obviously, if the ring be cut at any point on its circumference, its two ends will, under *this* centrifugal force, fly apart through a distance equal to $2\pi(r + dr) - 2\pi r = 2\pi.dr$.

In order to hold the two ends of the ring together, obviously, an external force F, *equal to the tension in the ring*, will have to be applied to it *inwards*. And, therefore,

$$\text{virtual work done } \textit{against} \text{ the force } F = -F.2\pi.dr.$$

For equilibrium, therefore, we have, in accordance with the principle of virtual work,

$$\Sigma\,(\delta m.r\omega^2.dr) + F.2\pi.dr = 0\,.\ \text{Or,}\quad \Sigma\,(\delta m.r\omega^2.dr) = F.2\pi.dr.$$

Or, $$r\omega^2.dr\,\Sigma\,\delta m = F.2\pi.dr.$$

Now, $\Sigma\delta m = M$, the mass of the ring or the flywheel.

So that, $$r\omega^2.dr.M = F.2\pi dr.\ \text{Or,}\ F = \frac{r\omega^2\,.\,dr\,.\,M}{2\pi.dr} = \frac{r\omega^2.M}{2\pi},$$

which enables the tangential tension in the flywheel to be easily calculated.

The above examples will, it is hoped, suffice to bring home to the student the great utility and importance of this rightly celebrated principle of virtual work.

WORKED EXAMPLES

1. A gramophone disc is set revolving in a horizontal plane and reaches a steady state of motion of two revolutions per second. It is found that a small coin placed on the disc will remain there if its centre is not more than 5 cm from the axis of rotation. Explain this and calculate the coefficient of friction between the coin and the disc.

(Oxford and Cambridge Higher School Certificate)

When the small coins is placed on the rotating gramophone disc, it is subjected to a centrifugal force tending to pull it outwards, away from the axis of rotation of the disc. Its motion is, however, opposed by the force of friction coming into play in between the surfaces of the disc and its own. But, as the distance of the coin from the axis of rotation increases, the centrifugal force pulling it outwards also increases, (being equal to $mr\omega^2$, where m is its mass, ω, its angular velocity and r, its distance from the axis of rotation) and is just balanced by the limiting friction between itself and the disc, *when it is at a distance of* 5 *cm from the axis of rotation*. Beyond this distance, the centrifugal force on it is greater than the limiting friction between it and the disc and it, therefore, moves outwards.

Here, $r = 5\ cm$. and $\omega = 4\pi$ [$\because$ in 1 rotation, the disc describes an angle 2π.]

So that, if m be the *mass* of the coin, we have

$$\textit{centrifugal force on the coin} = mr\omega^2 = m \times 5 \times (4\pi)^2 = 80\,\pi^2\,.\,m\ dynes\,.$$

And, clearly, *the normal reaction (R) of the disc on it* = its own weight mg.

And, therefore, the *limiting friction between the coin and the disc* = $\mu R = \mu\,.\,mg$.

where μ is the *coefficient of friction* for them.

Since the coin is *just in equilibrium here*, we have $\mu\,.\,mg = 80\,\pi^2 m$. [Taking $g = 980\ cm/sec^2$]

Or, $$\mu g + 80\pi^2,$$

whence, $$\mu = 80\pi^2/g = 80\pi^2/980 = 4\pi^2/49,$$

Or, $$\mu = .8054.$$

Thus, the coefficient of friction for the surface of the coin and the disc is equal to .8054.

2. Define the coefficient of sliding friction.

A uniform ladder of length $2l$ and weight W rests against a vertical wall with its foot on the ground at a distance l from the wall. If the coefficient of friction between the wall and the ladder and between the ground and the ladder is 0.4, find how far up the ladder a man of weight $2W$ can ascend without disturbing equilibrium.

(Joint Matriculation Board)

For definition of *coefficient of sliding friction*, see § 11.2 , page 363.

Here, let PQ be the ladder resting against the wall AB (Fig. 11.9), such that it makes an angle θ with the ground, the ladder being in the vertical plane, perpendicular to the wall, with its weight W acting vertically downwards at its mid-point O.

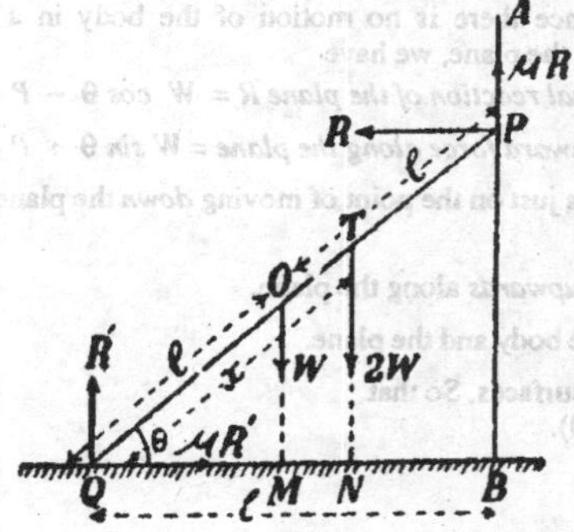

Fig. 11.9

Let the man climb up a distance x on the ladder, up to T, before th ladder starts slipping. Then, clearly, wthe weight $2W$ of the man acts vertically down at T. And since, when the ladder is just on the point of slipping, its upper end P moves downward towards B and its foot Q *away from*, B, the frictional forces at P and Q act *away from and towards B*, respectively. If R and R' be the normal reactions of the wall and the ground at P and Q, and μ, the *coefficient of friction* in either case, we have *limiting frictional force at $P = \mu R$, along the wall, away from B* and limiting frictional force at *$Q = \mu R'$, along the ground, towards B.*

In the equlibrium position of the ladder, clearly,

$$R = \mu R' \;\;.....(i) \quad \text{and} \quad R' + \mu R = W + 2W = 3W \quad .....(ii)$$

So that, substituting the value of R in relation (*ii*), we have

$$R' + \mu^2 R' = 3W \;\; ...(iii) \quad \text{Or,} \quad R'(1 + \mu^2) = 3W, \quad ...(iv)$$

whence, $R' = 3W/(1 + \mu^2)$ ...(*v*) And ∴ $R + 3\mu W/(1 + \mu^2)$. (*vi*)

Now, taking moments about Q, we have

$$W \times QM + 2W \times QN = R \times PB + \mu R \times QB.$$

Or, $W \times l \cos\theta + 2W \times x \cos\theta = R \times 2l \sin\theta + \mu R \times 2l \cos\theta$,

where, $QM = l\cos\theta$, $QN = x\cos\theta$, $PB = 2l\sin\theta$ and $QB = 2l\cos\theta$.

Or, $Wl\cos\theta + 2Wx\cos\theta - 2\mu Rl\cos\theta = 2Rl\sin\theta$

Or, $\cos\theta(Wl + 2Wx - 2\mu Rl) = 2Rl\sin\theta$.

Or, $$\frac{Wl + 2Wx - 2\mu Rl}{2Rl} = \tan\theta = \frac{\sqrt{3}\,l}{l} = \sqrt{3}$$

$\left[\because \tan\theta = PB/QB \text{ and } PB = \sqrt{(2l)^2 - l^2} = \sqrt{3l^2} = \sqrt{3}l.\right.$

So that, $Wl + 2Wx - 2\mu Rl = 2\sqrt{3}Rl$.

Or, $Wl + 2Wx = 2\mu Rl + 2\sqrt{3}Rl = 2Rl(\mu + \sqrt{3})$.

Or, substituting the value of R from relation (*vi*) above, we have

$$W(l + 2x) = \frac{2 \times 3\mu W}{1 + \mu^2} \times l(\mu + \sqrt{3}).$$

Or, $$l + 2x = \frac{6\mu l}{1 + \mu^2}(\mu + \sqrt{3}) = \frac{6 \times .4l}{1 + .16} \times (.4 + \sqrt{3}) \qquad [\because \mu = .4$$

$$= \frac{2.4\,l \times (.4 + 1.732)}{1.16} = \frac{2.4\,l \times 2.132}{1.16}$$

Or, $$2x = \frac{2.4 \times 2.132l}{1.16} - l = l\left(\frac{2.4 \times 2.132}{1.16} - 1\right)$$

Or, $$2x = (4.411 - 1)l = 3.411\,l, \text{ whence}, \; x = \frac{3.411l}{2} = 1.7055\,l.$$

Thus, the man can ascend up a distance $1.7055\,l$ without disturbing the equilibrium of the ladder.

3. A rough plane is inclined to the horizontal at an angle θ , where θ is less than the angle of friction λ, and a body of weight W is placed on the plane. Calculate the minimum horizontal force required to make the body move (*i*) down the plane and (*ii*) up the plane.

(*i*) Let AB be the rough plane, inclined to the horizontal at an angle θ, and let the body of weight W be placed on it, as shown, (Fig. 11.10)

Then, since θ is less than the angle of friction λ for the body and the plane, it will not by itself slip down, (see page 363).

Let P be the least horizontal force required to make the body just move down the plane.

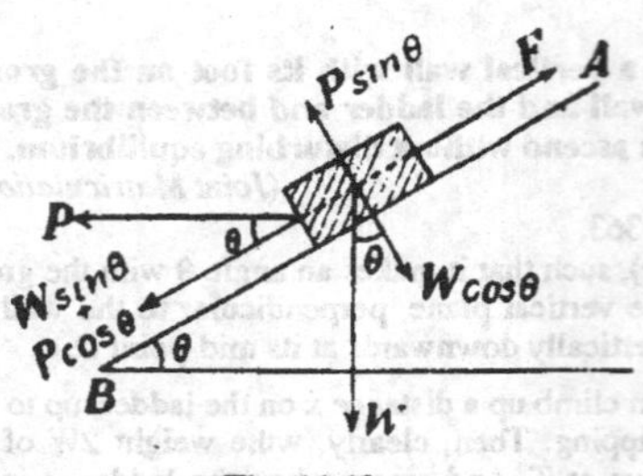

Fig. 11.10

Then, clearly, resolving P into its two rectangular components, we have its componetns $P\ cos\ \theta$, acting downwards, *along the plane*, and its components $P\ sin\ \theta$, acting upwards, *perpendicular to the plane*.

Similarly, resolving W into its two rectangular components, we have its component $W\ sin\ \theta$ acting downwards, *along the plane*, and its component $W\ cos\ \theta$ acting *perpenciculr to the plane*.

Thus, since there is no motion of the body in a direction perpendicular to the plane, we have

normal reaction of the plane $R = W\ cos\ \theta - P\ sin\ \theta$,

and *downward force along the plane* $= W\ sin\ \theta + P\ cos\ \theta$.

Since the body is just on the point of moving *down* the plane, it is clear that $W\ sin\ \theta + P\ cos\ \theta = F$, ...(*i*)

where F is the *limiting friction* between the body and the plane, acting *upwards* along the plane.

Now, $F = \mu R$, where μ is the *coefficient of friction* for the surface of the body and the plane.

Again, $\mu = tan\ \lambda$, where λ is the *angle of friction* for the given pair of surfaces, So that

$$F = \mu R = tan\ \lambda\ (W\ cos\ \theta - P\ sin\ \theta).$$

Substituting this value of F in relation (*i*) above, we have

$$W\ sin\ \theta + P\ cos\ \theta = tan\ \lambda\ (W\ cos\ \theta - P\ sin\ \theta)$$

Or, $P\ cos\ \theta = W\ tan\ \lambda\ cos\ \theta - P\ tan\ \lambda\ sin\ \theta - W\ sin\ \theta$.

Or, $P = W\ tan\ \lambda - P\ tan\ \lambda\ tan\ \theta - W\ tan\ \theta$. [*i.e.*, dividing both sides by $cos\ \theta$.

Or, $P + P\ tan\ \lambda\ tan\ \theta = W\ tan\ \lambda - W\ tan\ \theta$.

$P(1 + tan\ \lambda\ tan\ \theta) = W\ (tan\ \lambda - tan\ \theta)$.

Or, $$P = W \cdot \frac{tan\ \lambda - tan\ \theta}{1 + tan\ \lambda\ tan\ \theta}.$$

Clearly, $(tan\ \lambda - tan\ \theta) / (1 + tan\ \lambda\ tan\ \theta) = tan\ (\lambda - \theta)$.

So that, $$P = W\ tan\ (\lambda - \theta)$$

Thus, the least horizontal force required to make the body move *down* the plane is $W\ tan\ (\lambda - \theta)$.

(*ii*) In this case, again, let P be the least force required to make the body *just* move *up* the plane. Obviously, it will now have to be applied to the opposite direction to that in the first case, as shown in Fig. 11.11, and the frictional force F will now act *downwards* along the plane.

Resolving P, as before, into its two rectangular components, we have

component $P\ cos\ \theta$, *acting upward along the plane,*

and *component* $P\ sin\ \theta$, *acting downwards perpendicular to the plane*.

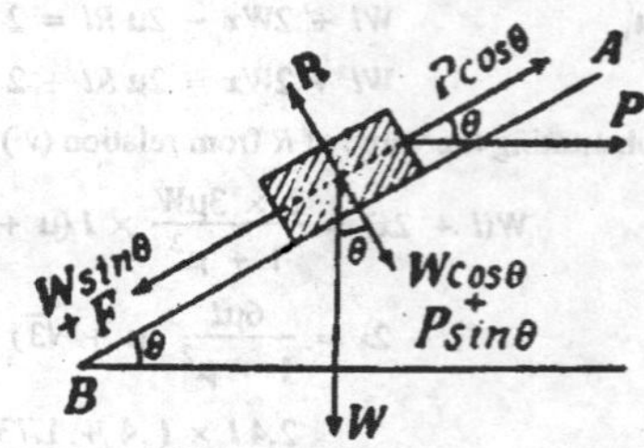

Fig. 11.11

So that, the *normal reaction on the* plane is now

$$R = W\ cos\ \theta + P\ sin\ \theta.$$

∴ for equilibrium of the body, we have

$P\ cos\ \theta = W\ sin\ \theta + F = W\ sin\ \theta + \mu R$.

Or, $P\ cos\ \theta = W\ sin\ \theta + tan\ \lambda\ (W\ cos\ \theta + P\ sin\ \theta)$. [∵ $\mu = tan\ \lambda$ and $R = W\ cos\ \theta + P\ cos\ \theta$.

Or, $P\ cos\ \theta = W\ sin\ \theta + W\ tan\ \lambda\ cos\ \theta + P\ tan\ \lambda\ tan\ \theta$.

Or, $P = W\ tan\ \theta + W\ tan\ \lambda + P\ tan\ \lambda\ tan\theta$.

Or, $P - P\ tan\ \lambda\ tan\ \theta = W\ (tan\ \lambda + tan\ \theta)$.

Or, $P(1 - tan\ \lambda\ tan\ \theta) = W(tan\ \lambda + tan\theta)$,

whence, $$P = W\ \frac{tan\ \lambda + tan\ \theta}{1 - tan\ \lambda\ tan\ \theta}$$

Since $(tan\ \lambda + tan\ \theta)/(1 - tan\ \lambda\ tan\ \theta) = tan\ (\lambda + \theta)$, we have

$$P = W\ tan\ (\lambda + \theta).$$

Or, the minimum horizontal force required to make the body just move *up* the plane is $W\ tan\ (\lambda + \theta)$.

4. The force required to pull a body up a plane inclined to the horizontal at angle of 18° 8′ is 105 N and the force required to pull it down the plane, 10 N. What is the weight of the body and the coefficient of friction between its surface and that of the plane ?

As we know, *the force required to pull a body of weight W up an incline of an angle* $\theta = \mu W \cos\theta + W \sin\theta$,

where μ is the *coefficient of firction* between the surfaces of the body and the plane.

And, the *force required to pull the body down the plane* = $\mu W \cos\theta - W \sin\theta$

We, therefore have

$$105\,N = (\mu W \cos 18°\,18' + W \sin 18°\,18')\,N$$

Or, $105\,N = (.9494\,\mu W + .3140\,W)\,N$ ...(*i*)

And, $10\,N = (\mu W \cos 18°\,18' - W \sin 18°\,18')\,N$

Or, $10\,N = (.9494\,\mu W - .3140\,W)\,N$ ...(*ii*)

Subtracting relation (*ii*) from (*i*), we have

$95\,N = .624.\,W$, whence, $W = 95/.6240 = 152.3\,N$

Or, $W = 152.3/9.81 = 15.52\,kg$

Now, substituting the value of $W = 152.3\,N$ in relation (*i*) above, we have

$105\,N = (.9494\mu \times 152.3)\,N + (.3140 \times 152.3)\,N$

Or, $105 = 144.5\,\mu + 47.79.$

Or, $144.5\,\mu = 57.21$, whence, $\mu = 57.21/144.5 = 0.3959.$

5. A load of 24 kg lies on a plane inclined at an angle of 17° to the horizontal, the coefficient of friction for the surfaces of the two being 0.36. Calculate the least horizontal force required to make the load move (i) up and (ii) down the plane.

If λ be the angle of friction for the surfaces of the load and inclined plane, we have

$\mu = \tan\lambda = .36$, whence, $\lambda = 19°\,48'$. So that, $(\lambda + \theta) = 19°\,48' + 17°\,36°\,48'$ and $(\lambda - \theta) = (19°\,48' - 17°) = 2°\,48'$.

Then, the *least force (applied horizontally) required to move the load move up the plane* is given by

$P = W \tan(\lambda + \theta)$, [See Ex. 3 above.

i.e., $P = 24 \times 9.81 \tan 36°\,48' = 24 \times 9.81 \times .7481 = 172.4\,N.$

And, the *least force (applied horizontally) required to move the load down the plane* is given by $P = W \tan(\lambda - \theta)$

So that, $P = 24 \times 9.81 \tan 2°48' = 24 \times 9.81 \times 0.0489 = 11.27\,N.$

6. (a) Define the angle of friction.

A uniform rod rests in limiting equilibrium in contact with a horizontal floor and a vertical wall, the rod being in a vertical plane which is perpendicular to the wall. If the wall and the floor be equally rough, prove that the angle between the rod and the wall is twice the angle of friction. *(London Higher Secondary Certificate)*

(*b*) What would be the value of the angle in the question above, (i) if the wall and the floor be unequally rough, (ii) if the wall be smooth, (iii) if the floor be smooth, (iv) if both be smooth ?

(*a*) Let *PQ* be the rod, of length 2*l*, resting in limiting equilibrium with its upper end *P* in contact with a vertical wall *AB* and its foot with a horizontal floor, (Fig. 11.12), such that it lies in a vertical plane perpendicular to the wall and makes an angle θ with it.

Then, its weight *W* acts vertically downwards at its midpoint *O*, with the normal reactions *R* and *R′* of the wall and the floor acting at *P* and *Q* respectively, as shown.

Since, the wall and the floor are *equally rough*, the coefficient of friction for either of them and the rod must be the *same*. Let it be μ. Then, the frictional forces at *P* and *Q* are clearly μ*R* and μ*R′*, acting in the directions shown, as explained in answer to Ex. 2 above.

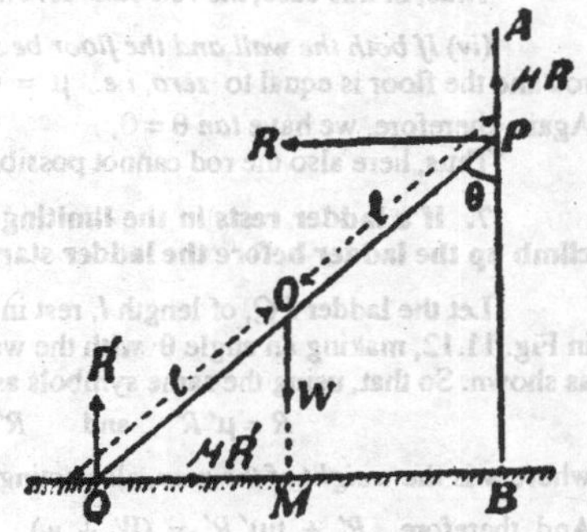

Fig. 11.12

The rod being in limiting equilibrium, we clearly have

$R = \mu R'$ ...(*i*) and $\mu R + R' = W$. ...(*ii*)

So that, substituting the value of *R* from expression (*i*) in expression (*ii*),

we have $R' + \mu^2 R' = W$. Or, $R'(1 + \mu^2) = W$. ...(*iii*)

And $\therefore$ $\mu R'(1+\mu^2) = \mu W = R(1+\mu^2) = \mu W$, ...(iv) [$\because$ $\mu R' = R$.

whence, $R = \mu W/(1+\mu^2)$.

Now, taking moments about Q, we have

$$W \times Q\ M = R \times PB + \mu R \times QB.$$

Or, $W \times l \sin\theta = R \times 2l\cos\theta + \mu R \times 2l\sin\theta$ ...(vi)

Or, $Wl\sin\theta - \mu R . 2l\sin\theta = 2Rl\cos\theta$

Or, $l\sin\theta(W - 2\mu R) = 2Rl\cos\theta$

Or, $\dfrac{\sin\theta}{\cos\theta} = \tan\theta = \dfrac{2Rl}{l(W-2\mu R)} = \dfrac{2R}{W-2\mu R}$

So that, substituting the value of R from relation (v) above, we have

$$\tan\theta = \frac{2\mu W}{(1+\mu^2)\left(W - 2\mu . \dfrac{\mu W}{1+\mu^2}\right)} = \frac{2\mu W}{W(1+\mu^2) - 2\mu^2 W}.$$

Or, $\tan\theta = \dfrac{2\mu}{(1+\mu^2) - 2\mu^2} = \dfrac{2\mu}{1-\mu^2}$ ...(vii)

Now, $\mu = \tan\lambda$, where λ is the *angle of friction.*

And, therefore, $\tan\theta = \dfrac{2\tan\lambda}{1-\tan^2\lambda} = \tan 2\lambda$, whence, $\theta = 2\lambda$.

Thus, *the angle θ that the rod makes with the vertical wall is twice the angle of friction.*

(b) – (i) *If the wall and the floor be unequally rough,* the coefficient of friction for the rod and the wall will be different from that between the rod and the floor. Let these be μ and μ' respectively. Then, clearly,

$$R = \mu' R' \text{ and } R' + \mu R = W.$$

Or, $R' + \mu\mu' R' = W$. Or, $R'(1+\mu\mu') = W$.

So that, $\mu' R'(1+\mu\mu') = \mu' W$. Or, $R(1+\mu\mu') = \mu' W$. [$\because$ $\mu' R' = R$.

Or, $R = \mu' W/(1+\mu\mu')$.

And $\therefore$ $\tan\theta = \dfrac{2\mu' W}{W(1+\mu\mu') - 2\mu\mu' W} = \dfrac{2\mu'}{1-\mu\mu'}$ [See above. ...(viii)

Or, $\theta = \tan^{-1}[2\mu'/(1-\mu\mu')]$.

Thus, *the rod now makes with the wall an angle* $\theta = \tan^{-1}[2\mu'/(1-\mu\mu')]$.

(ii) *If the wall be smooth,* we shall have the coefficient of friction for the rod and the wall equal to *zero, i.e.,* $\mu = 0$.

And, therefore, expression (viii) above becomes

$$\tan\theta = 2\mu' = 2\tan\lambda,$$

i.e., the tangent of the angle that the rod makes with the wall is twice the tangent of the angle of friction.

(iii) *If the floor be smooth,* the coefficient of friction for the rod and the floor is equal to *zero, i.e.,* $\mu' = 0$. So that,

$$\tan\theta = 0. \quad \text{Or,} \quad \theta = 0.$$

Thus, in this case, *the rod can rest only in the vertical position.*

(iv) *If both the wall and the floor be smooth,* the coefficient of friction for the rod and the wall as well as that for the rod and the floor is equal to *zero, i.e.,* $\mu = 0$ and also $\mu' = 0$.

Again, therefore, we have $\tan\theta = 0$, Or, $\theta = 0$.

Thus, here also the rod cannot possibly rest in any inclined position.

7. If a ladder rests in the limiting position against a vertical wall and a horizontal floor, how far can a man climb up the ladder before the ladder starts slipping ?

Let the ladder PQ, of length l, rest in the limiting position against a vertical wall AB and a horizontal floor, as shown in Fig. 11.12, making an angle θ with the wall. Then, the weight W of the ladder acts vertically downwards at its mid-point, as shown. So that, using the same symbols as before, in Ex. 6(b), above, we *now* have

$$R = \mu' R' \quad \text{and} \quad R' + \mu R = (W + w),$$

where w is the *weight of the man,* also acting vertically downwards.

And, therefore, $R' + \mu\mu' R' = (W+w)$. Or, $R'(1+\mu\mu') = (W+w)$

---

* This is obvious otherwise also ; for, in this case, the three forces acting on the rod do not all meet in one point.

Or, $\mu' R' (1 + \mu\mu') = \mu' (W + w)$.

Or, $R (1 + \mu\mu') = \mu' (W + w)$, whence, $R + \frac{\mu' (W + w)}{1 + \mu\mu'}$. [∵ $\mu' R' = R$.

Now, if the man can climb a maximum distance x up the ladder before slipping just occurs, we have, using the same symbols as in Ex. 6(b) above, and taking moments above Q,

$$Wl \sin\theta + wx \sin\theta = R.2l \cos\theta + \mu R.2l \sin\theta.$$

Or, $Wl \sin\theta + wx \sin\theta - \mu R.2l \sin\theta = R.2l \cos\theta$.

$$\sin\theta\,(Wl + wx + 2\mu Rl) = 2Rl \cos\theta.$$

Or, $\frac{\sin\theta}{\cos\theta} = \tan\theta = \frac{2Rl}{Wl + wx - 2\mu Rl}$.

But, $\tan\theta = 2\mu'/(1 - \mu\mu')$, [See Ex. 6 ($b$), above

And, therefore, $\frac{2\mu'}{1 - \mu\mu'} = \frac{2Rl}{Wl + wx - 2\mu Rl}$.

Or, $2\mu' Wl + 2\mu' wx - 4\mu\mu' Rl = 2Rl - 2\mu\mu' Rl$.

Or, $2\mu' Wl + 2\mu' wx = 2Rl - 2\mu\mu' Rl + 4\mu\mu' Rl = 2Rl (1 + \mu\mu')$

Substituting the value of R from above, therefore, we have

$$2\mu' Wl + 2\mu' wx = 2l(1 + \mu\mu') \times \frac{\mu'(W + w)}{1 + \mu\mu'} = 2\mu' l(W + w).$$

Or, $2\mu.' Wl + 2\mu' wx = 2\mu' Wl + 2\mu' wl$ Or, $2\mu'.wx = 2\mu' wl$, whence, $x = l$.

Thus, the man (irrespective of his weight) can climb up the ladder to only a maximum distance equal to half its length before slipping occurs.

8. A framework ABCD, consisting of four uniform, freely jointed rods, each of the same length and weight w, is hung from A, and the corners A, C are connected by a string. Find the tension in the string.

Let AC be the string, connecting the opposite corners of the framework $ABCD$, formed by the four equal and uniform rods, AB, BC, CD and DA, (Fig. 11.13), in which the tension is to be determined.

Clearly, the weight w, of each rod acts vertically downwards at its $c.g.$ $i.e.$, at its midpoint, (the rod being uniform).

Then, if $2l$ be the length of each rod and θ, the angle it makes with AC, we have $\cos\theta = AP/AD = a/2l$, whence, $a = 2l \cos\theta$. (P being the mid-point of AC, and $AP = a$).

So that $AC = 2AP = 2a = 2 \times 2l \cos\theta = 4l \cos\theta$.

Now, let x be the depth below A, of the $c.g.$ of each of the two upper rods. Then, clearly, $x = l \cos\theta$.

And, therefore, $AC = 4l \cos\theta = 4x$.

And, since the $c.g.$ of each of the lower pair of rods lies a distance x above C, its depth below A is $4x - x = 3x$.

If we, therefore, imagine the whole system to be displaced downwards a little, so that the $c.g.$ of the upper rods is shifted through a distance δx, we have

virtual works done by the weights of the two upper rods $= 2w.\delta x$

Similarly, *virtual work done by the weights of the two lower rods*

$$= 2\,w.\,\delta\,(3x) = 6\,w.\,\delta x.$$

And therefore, *total virtual work done by the weights*

$$= 2w.\delta x + 6w.\delta x = 8w.\delta x.$$

And, if T be the tension in the string, we have *virtual work of the string*

$$= -T.\delta(4x) = -4T\delta x,$$

the negative sign indicating that T acts in the opposite direction to the displacement of C.

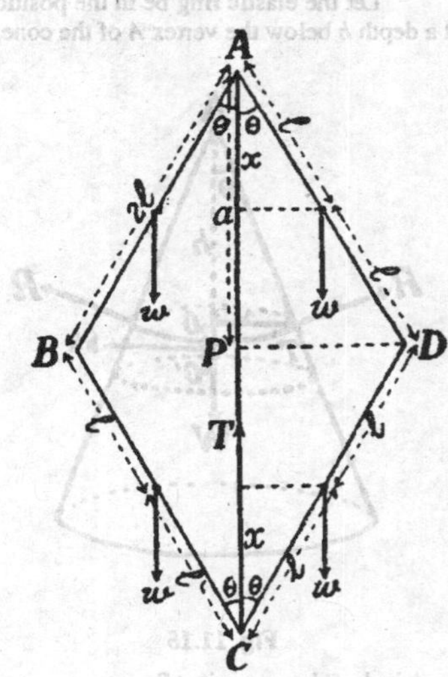

Fig. 11.13

Since the system is in equilibrium, we have *total virtual work* = 0.

Or, $8w.\delta x - 4T.\delta x = 0$. Or, $4T.\delta x + 8w.\delta x$, whence, $T = 2w$.

Thus, the tension in the string is $2w$, $i.e.$, equal to twice the weight of each rod.

9. A uniform rod of length $2l$ lies in equilibrium over a smooth peg, with its lower end against a smooth vertical wall. If the peg be at a distance d from the wall, show that the rod is inclined to the wall at an angle $\sin^{-1}(d/l)^{1/3}$

Let PQ be the rod, resting *in equilibrium* over the smooth peg K and against the smooth vertical wall AB, such that it makes an angle θ with the wall at P, (Fig. 11.14)

Obviously, the weight w of the rod acts vertically downwards at its *c.g. i.e.*, at its mid-point O, at a vertical height $MN = x$, above the peg K, where

$$x = MN = PN - PM \qquad \left[\because PN = OP\cos\theta, \text{ and } PM = KM\cot\theta\right.$$

$$= OP\cos\theta - KM\cot\theta.$$

Or, $$x = l\cos\theta - d\cot\theta. \qquad \left[\because OP = l \text{ and } KM = d.\right.$$

Differentiating this with respect to θ, we have

$$dx = (-l\sin\theta + d\,cosec^2\,\theta).d\theta,$$

Since the rod is in equilibrium, we have $dx = 0$.

And, therefore, $d\,cosec^2\,\theta - l\sin\theta = 0$.

Or, $d/\sin^2\theta = l\sin\theta$, whence, $\sin^2\theta = d/l$.

Or, $$\theta = \sin^{-1}(d/l)^{\frac{1}{3}}$$

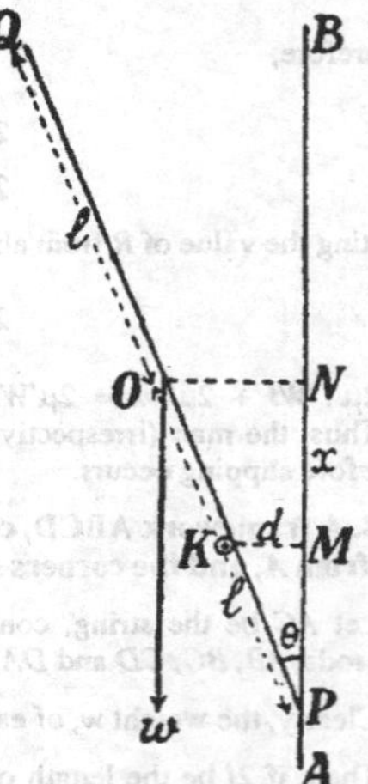

Fig. 11.14

Thus, the rod makes an angle $\sin^{-1}(d/l)^{1/3}$ with the wall, in its position of equilibrium.

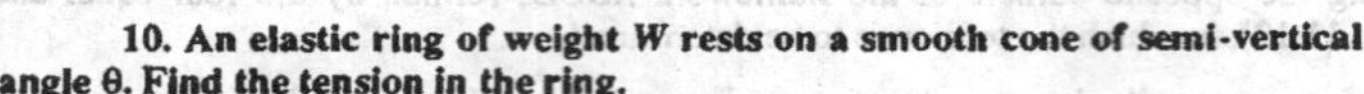

10. An elastic ring of weight W rests on a smooth cone of semi-vertical angle θ. Find the tension in the ring.

Let the elastic ring be in the position shown in Fig. 11.15, with its centre O at a depth h below the vertex A of the cone, so that its radius $r = h\tan\theta$.

Clearly, the forces operating on the ring are *(i) its weight W, acting vertically downwards, at its centre of gravity or centre O., (ii) the tension T of the ring acting along the circumference of the ring and (iii) the normal reactio1 R of the cone, acting at right angles to the surface of the cone*, its surface being smooth (and hence there being no frictional component along its surface).

Now, imagine the ring to be given a small downward displacement δh, so that its distance from the apex A of the cone now becomes $h + \delta h$ and hence its radius become $(h + \delta h)\tan\theta$

Its *radius thus increases by*

$$(h + \delta h)\tan\theta - h\tan\theta = \delta h.\tan\theta,$$

and its circumferences, therefore, increases by $2\pi\delta h.\tan\theta$. Clearly, then, we have

$$\textit{virtual work done by weight } W \textit{ of the ring} = W.\delta h.$$

and vertical work done *against tension T of the ring* $= -T.2\pi\delta h.\tan\theta$, the negative sign indicating that the displacement here take place in the direction *opposite to that of* T.

Fig. 11.15

And, because the ring moves at right angles to the direction of R, *no work is done by or against* R.

Since the ring is in equilibrium, the total virtual work done must be *zero*. We, therefore, have

$$W.\delta h - T.2\pi\delta h.\tan\theta = 0. \quad \text{Or,} \quad T.2\pi\delta h\tan\theta = W.\delta h,$$

whence, $$T = \frac{W.\delta h}{2\pi\delta h.\tan\theta} = \frac{W}{2\pi}.\frac{1}{\tan\theta} = W\cot\theta/2\pi.$$

This, therefore, is the tension in the elastic ring.

EXERCISE XI

1. A man weighing 140 *lb* climbs up a uniform ladder, 20 *ft* long and 70 *lb* in weight, which rests against a rough vertical wall at an angle of 45°. If the coefficient of friction at each end of the ladder is 0.5, how far will the man be able to climb up the ladder before it begins to slip. ***(Northern Universities Higher School Certificate)***

Ans. 13.0 *ft*.

2. A uniform rod is in limiting equilibrium, one end resting on a rough horizontal plane and the other on an equally rough plane inclined at an angle α to the horizon. If λ be the angle of friction and the rod be in a vertical plane, show that the inflination θ of the rod to the horizon is given by

$$\tan\theta = \frac{\sin(\alpha - 2\lambda)}{2\sin\lambda\ \sin(\alpha - \lambda)}.$$ ***(Delhi)***

3. Distinguish between *static* and *sliding* (kinetic) friction and define the *coefficient of sliding friction*.

How would you investigate the laws of sliding friction between wood and iron ?

An iron block, mass 10 *lb* rests on a wooden plane inclined at 30° to the horizontal. It is found that the least force parallel to the plane which causes the block to slide *up* the plane is 10 *lb wt*. Calculate the coefficient of sliding friction between wood and iron. ***(Northern Universities Higher School Certificate)***

Ans. 0.58

4. A uniform ladder, 13 *ft* long and weighing 60 *lb*, rests with its upper end against a smooth vertical wall and with its lower end 5 *ft* from the wall on rough ground. By a graphical method find the magnitude and direction of the reaction at the foot of the ladder. Check your result by calculation. What is the least coefficient of friction between the ground and the ladder necessary to maintain equilibrium ? ***(Cambridge Local Higher School Certificate)***

Ans. 61.3 *lb wt*; at 78° 10′ to the ground ; 0.21.

[Hint. To determine the *magnitude* and *direction* of the reaction at the foot of the ladder, we must remember that the resultant reaction of the ground passes through the point of intersection of the lines of action of *R* and the 50 *lb wt* since, as we know, three forces in equilibrium must all be concurrent. The angle λ (*i.e.*, the angle between the resultant ground reaction and the vertical) is then measured and *tan* λ determined.]

5. Explain the meaning of the term *coefficient of friction* and describe an experimental method of measuring it.

A rectangular block with a square base of side 10 *cm* rests on a horizontal surface. If a horizontal force is applied near the bottom of one vertical face, the block slides. If the force is applied near the top, the block topples over. When the force is applied at 20 *cm* from the bottom, the block sometimes slides and sometimes topples. Find the coefficient of friction between the block and the surface. ***(Cambridge Higher School Certificate)***

Ans. 0.25.

6. A uniform rod of weight *W* rests with its one end against a rough inclined plane *AB*, of inclination α, and the other end against a smooth vertical wall *BD*, *B* being higher than *A*. If θ be the inclination of the rod to the vertical in the limiting position of equilibrium, show that $\tan\theta = 2\tan(\lambda - \alpha)$, where λ is the angle of friction for the plane.

7. Define the *terms : friction, limiting friction, angle of friction* (λ) *and coefficient of friction* (μ), and show that the coefficient of friction (μ) = *tan* λ.

What is meant by the *cone of friction ?*

8. A square framework formed of uniform heavy rods of equal weights *W*, is hung by one corner. A weight *W* is suspended from each of the three lower corners and the shape of the square is preserved by a light rod along the horizontal diagonal. Find its tension ? ***(Allahabad and Delhi)***

Ans. 4*W*.

9. Three equal smooth pencils, each of weight *W*, are tied together by a single loop of fine inextensible cotton so that each touches the other two. Prove that the bundle can rest in stable equilibrium on a smooth table only if the breaking tension of the cotton is not less than $W\sqrt{3}/6$. ***(Oxford Scholarship and Higher School Certificate)***

10. A cylinder of radius 1 *cm* and length 4 *cm* is standing on end on an inclined plane, the angle of which is gradually increased. If the coefficient of friction between the cylinder and the plane is 0.3, find whether the cylinder will slide or topple first. **Ans. *Will slide first*.**

11. A force of 190 newtons is required to pull a load of 25 *kg up* an incline of 1 in 2 with a constant speed, the force being applied parallel to the plane. Calculate the *coefficient of friction* for the surfaces of the load and the inclined plane.

Ans. 0.317

CHAPTER XII

FLOW OF LIQUIDS— VISCOSITY

12.1. Rate of flow of a fluid. Fluids, as we know, include both liquids and gases, their chief characteristic being that they cannot permanently withstand any shearing stresses, howsoever small, (see § 9.1,)

We shall concern ourselves here only with *ideal* liquids and gases, *i.e.* with liquids which are *perfectly mobile* (zero viscosity) and *incompressible* (zero compressibility or infinite bulk modulus) and with gases which perfectly obey *Boyle's* and *Charles' laws.*

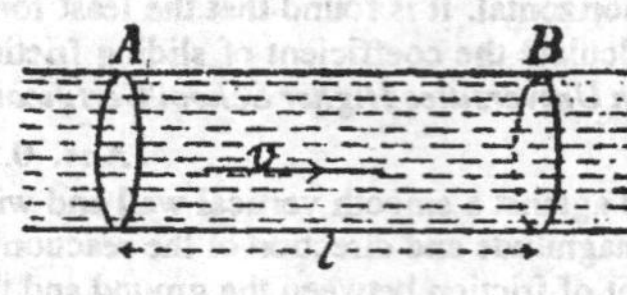

Fig. 12.1

The *rate of flow* of a fluid is defined as *the volume of it that flows across any section of a pipe in unit time.* It is really the *volume rate of flow* of the fluid or its **discharge,** usually represented by the letter Q or V.

Considering the fluid to be incompressible, if its velocity of flow be v, in a direction perpendicular to two sections A and B, (Fig. 12.1) of area a, and distance l apart, and if t be the time taken by the liquid to flow from A to B, we have $vt = l$.

Obviously, the volume of liquid flowing through the section AB, in *this* time, is equal to the cylindrical column $AB = l \times a$, or $= vt \times a$. This, therefore, is the volume of liquid flowing across the section in time t.

$\therefore$ *volume rate of flow of liquid* $\times$ *discharge* Q or $V = vt \times a/t = v \times a$

$= $ *velocity of liquid* $\times$ *area of cross-section of the tube.*

When measured in *cubic feet per sec.* (*F.P.S.* system), the unit is the **cusec** (short for *c.ft/sec*)

Sometimes, the rate of flow of a liquid is also expressed in terms of the *mass of the liquid flowing across any section in unit time* and is referred to as its *mass rate of flow*. Thus,

mass rate of flow of liquid = mass of liquid flowing across any section per unit time

= velocity of liquid $\times$ *area of cross-section* $\times$ *density of liquid* $= v \times a \times \rho$.

12.2. Lines and Tubes of Flow. In a simple flow of fluid *i.e., when it is slow and steady*, the velocity at every point in the fluid remains constant, (in magnitude, as well as direction), the energy needed to drive it being used up in overcoming the *'viscous drag'* between its layers. In other words, each particle follows exactly the *same path* and has the *same velocity* as its predecessor and the fluid is said to have an *orderly* or a **streamline flow.** In such a case, if we consider a line along which the particles of the fluid move, the direction of the line at any point is the direction of the velocity of the fluid at that point. Such a line is called a **streamline.** More correctly, *a streamline may be defined as a curve, the tangent to which at any point gives the direction of flow of the fluid at that point;* for, *it may be straight or curved, according as the lateral pressure on it is the same throughout or different, —in the latter case, the pressure being greater on the convex side than one the concave one. No two streamlines can ever cross one another.*

Fig. 12.2

If we consider two areas A and B at right angles to the direction of flow of the fluid (Fig. 12.2) and draw streamlines through their boundaries, as shown, we obtain *a tubular space AB, bounded by a surface containing streamlines,* called a **stream tube**, or a **tube of flow**. The sides of the tube of flow being everywhere in the direction of the fluid flow, *no fluid can cross the sides as though they were rigid* and must enter and leave through the ends. In other words, there is no intermingling of the fluid in two adjacent tubes of flow across their imaginary walls.

This holds good, however, only so long as the velocity of the liquid does not exceed a particular limiting value, called its **critical velocity** (see § 12.11) beyond which the flow of the liquid loses all its steadiness and orderliness and becomes *zigzag* or *sinnous,* acquiring what is called turbulent motion.

This may be easily seen by introducing a small jet *J* of a colouring matter, like ordinary ink, for example, axially into a tube *AB* through which water may be made to flow with gradually increasing velocities (by raising the waterhead responsible for the flow (Fig 12.3). It will be seen that as long as the velocity of water-flow remains below it critical value, there is only a *thin streak* of the colouring matter along the axis of the tube, as shown in [Fig. (*a*)], signifying a steady, streamline motion. As soon as the velocity attains its critical value, the colouring matter takes a zigzag path. [Fig. (*b*)] indicating haphazard changes of velocity from point to point and consequent distortion of tubes of flow. And, later, when this value is exceeded, the colouring matter spreads out in all directions, filling the entire tube, showing that the tubes of flow have now broken down and the motion has become utterly disorderly or turbulent.

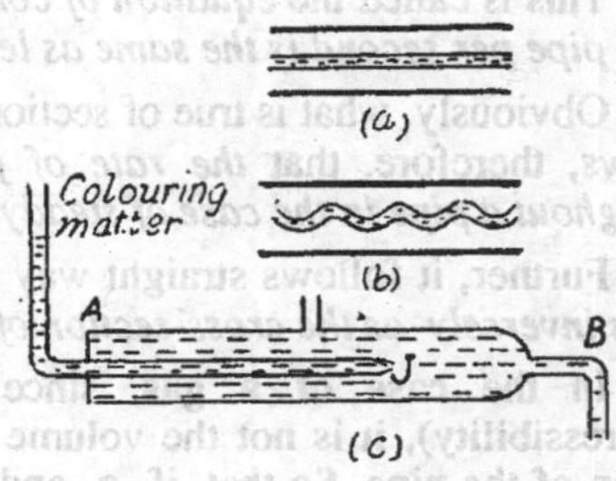

Fig. 12.3

N.B. *Reynold* actually used this very method to determine the critical velocity for water. Obviously, the velocity of the water just before the thin streak of colouring matter becomes zigzag or sinnous [(Fig. (*b*)], directly gives the value of the *critical velocity* v_c for it and is equal to *rate of flow of the liquid, i.e., volume of liquid flowing out of the tube per second, divided by the area of cross-section of the tube.*

12.3. Equation of Continuity of flow. This is a fundamental equation of flow and is a special case of the general physical law of conservation of matter. For an *incompressible fluid i.e., a liquid,* it may be deduced as follows :

Imagine the fluid to be flowing through a pipe *AB*, (Fig. 12.4) with a_1 and a_2 as its ares of cross-section at sections *A* and *B* and consider an infinitesimally small tube of flow (shown dotted) of cross-sectional areas da_1 and da_2 at its two ends and with velocities of the fluid v_1 and v_2 at sections *A* and *B* respectively. Then, if the fluid covers distances ds_1 and ds_2 in time dt, at the two ends and ρ_1 and ρ_2 be the densities of the fluid at *A* and *B*, we have

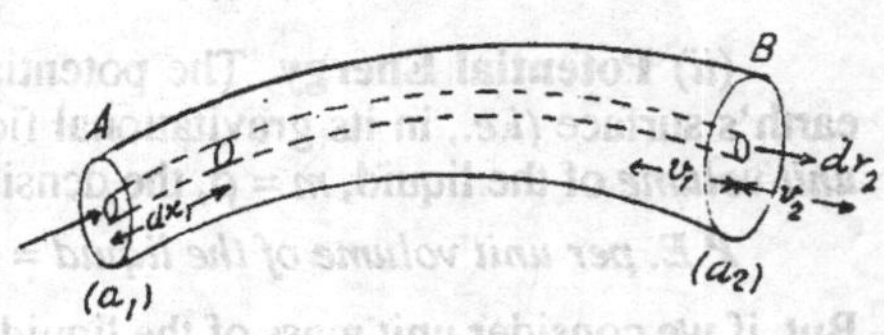

Fig. 12.4

mass of fluid entering flow-tube at end A per unit time $= da_1 \,.\, ds_1 \,.\, \rho_1 / dt$

$$= da_1 \,.\, v_1 \,.\, \rho_1 \qquad [\because ds_1/dt = v_1$$

and *mass of fluid leaving flow tube at end B per unit time* $= da_2 .\, ds_2 .\, \rho_2/dt$

$$= da_2 \,.\, v_2 \,.\, \rho_2 \qquad [\because ds_2/dt = v_2$$

$\therefore$ *mass of fluid entering the whole section A per sec; i.e. mass of flow at A*

$$= \int_0^{a_1} da_1 \,.\, v_1 \,.\, \rho_1 = a_1 \,.\, v_1 \,.\, \rho_1 \,,$$

and *mass of fluid leaving the whole section B per sec, i.e., mass rate of flow at B*

$$= \int_0^{a_2} da_2 \, . \, v_2 \, . \, \rho_2 = a_2 \, . \, v_2 \, . \, \rho_2$$

Since the fluid is incompressible, $\rho_1 = \rho_2$ and since we have no source or sink inbetween sections A and B, we have, from the law of conservation of matter

$$a_1 v_1 = a_2 v_2 = V \text{ or } R, \text{ say,}$$

i.e., the *rate of flow at A = rate of flow at B*.

This is called the *equation of continuity* and states that *the quantity of fluid entering one end of the pipe per second is the same as leaving the pipe at the other end per second.*

Obviously, what is true of sections A and B is also true of all other sections of the pipe too. It follows, therefore, that *the rate of flow of an incompressible and mobile fluid is the same throughout a pipe in the case of steady or streamline flow.*

Further, it follows straight way from the above that $v_1/v_2 = a_2 / a_1$, *the velocity of the fluid varies inversely as the cross-section of the pipe.*

In the case of a gas, since the density changes, with pressure (due to its high compressibility), it is not the volume but the mass of the gas that remains constant through any section of the pipe. So that, if ρ_1 and ρ_2 be the densities of the gas at the two sections A and B respectively in the figure above, we have $a_1 v_1 \rho_1 = a_2 v_2 \rho_2$, or $V_1 \rho_1 = V_2 \rho_2$.

12.4. Energy of a Liquid in flow. Since a liquid has *inertia*, it posesses *kinetic energy*, when in motion. It also exerts *pressure* on the walls of the containing vessel or the pipe and has, therefore, *pressure energy*, and may also have *potential energy*, due to its position. We have thus three types of energy possessed by a liquid in flow, *viz.*, (*i*) *kinetic energy*, (*ii*) *potential energy*, and (*iii*) *pressure energy*.

(*i*) **Kinetic Energy.** Clearly, the kinetic energy of a mass m of a liquid, flowing with a velocity v, is given by $\frac{1}{2} mv^2$. If we consider *unit volume* of the liquid, $m = \rho$, the *density* of the liquid, and, therefore, we have

$$\textit{kinetic energy per unit volume of the liquid} = \frac{1}{2} \rho \, v^2 .$$

And, if we consider *unit mass* of the liquid, $m = 1$, and, therefore,

$$\textit{kinetic energy per unit mass of the liquid} = \frac{1}{2} v^2 .$$

(*ii*) **Potential Energy.** The potential energy of a liquid of mass m at a height h above the earth's surface (*i.e.*, in its gravitational field) is, as we know, equal to mgh. Again, if we consider *unit volume* of the liquid, $m = \rho$, the density of the liquid, and, therefore.

P.E. per unit volume of the liquid $= \rho \, .g.h$

But, if we consider unit mass of the liquid, $m = 1$ and we have

P.E. per unit mass of the liquid $= gh$.

(*iii*) **Pressure Energy.** Consider a tank A, containing a liquid of density ρ, provided with a narrow side tube T, of cross-sectional area a, properly fitted with a piston P that can be smoothly moved in and out, (Fig. 12.5). Let the *hydrostatic pressure* due to the liquid, at the level of the axis of the side tube, be p, so that the force on the piston is $= p.a.$ If, therefore, more liquid is to be introduced into the tank, this much

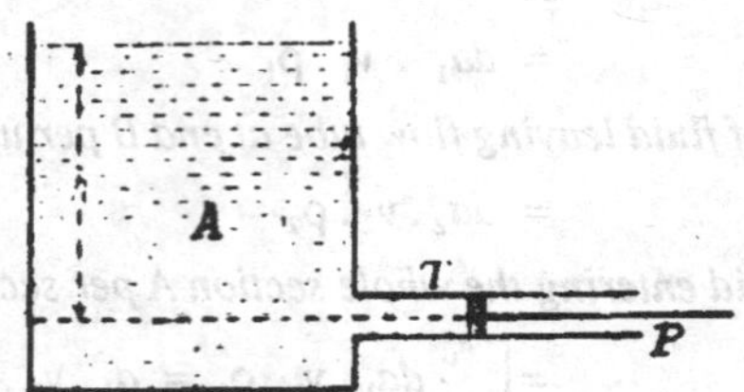

Fig. 12.5

force has to be applied to the piston in moving it inwards. Let the piston be moving slowly inwards through a distance x, so that the velocity of the liquid be very small and *there may be no kinetic energy acquired by it*. Then, clearly, a volume of liquid $a.x.$, or a mass $a.x.\,\rho$ of it, is forced into the tank, and an amount of work $p.a.x.$ is performed to do so. This work, (or energy), $p.a.x$, *required to make the liquid move against pressure p, without imparting any velocity to it*, thus becomes the energy of the mass $a.x.\,\rho$ of the liquid in the tank, for it can do the same amount of work in pushing the piston back, when escaping from the tank. It is referred to as the *pressure energy* of the liquid.

Thus pressure of a mass $a.x.\rho$ of the liquid is equal to $p.a.x$ and, therefore,

$$\textit{pressure energy per unit mass of the liquid} = \frac{p.a.x}{a.x.\rho} = \frac{p}{\rho} = \frac{pressure}{density}$$

Now, if we consider *unit volume* of the liquid, we have

pressure energy of volume $a.x$ of the liquid $= p.a.x$,

and $\therefore$ *pressure energy per unit volume of the liquid* $= \dfrac{p.a.x}{a.x} = p$, *the pressure of the liquid.*

The three types of energy possessed by a liquid under flow are mutually convertible, one into the other. For, consider a liquid of density ρ contained in a vessel, and let its depth be h, (Fig. 12.6). Then, pressure due to the liquid column h at the bottom of the vessel is $p = h.\,\rho.g$. If we take unit mass of the liquid from the bottom B to the surface A, clearly, $h.g.$ units of work has to be done against gravity, and, therefore, the potential energy of the liquid increases by this much amount; or this much work is done by gravity if unit mass of the liquid comes down through a depth h. Hence, potential energy of unit mass of the liquid is equal to $h.g$. And, since pressure at a depth h, is given by $p = h.\,\rho.g$, and pressure energy per unit mass of the liquid = pressure/density, we have *pressure energy per unit mass of the liquid* $= h.\,\rho.g/\rho = h.g =$ *potential energy lost by the liquid in descending through distance h.*

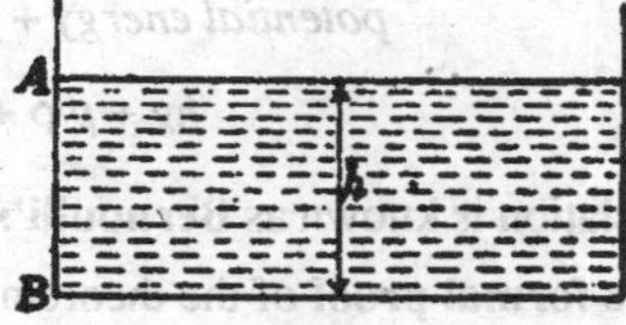

Fig. 12.6

Thus, we see that pressure energy and potential energy are convertible, one into the other, and, therefore, their sum for a liquid at rest is constant.

Again, consider the flow of liquid through a tube, (Fig. 12.7). If the liquid has a constant velocity, there is no resultant force acting upon it. But, if the flow is accelerated, there must be a *pressure gradient* along the tube of flow. Let the change of pressure for a distance dx be dp, *i.e.*, let the pressure gradient be dp/dx, which may be taken to be constant over a short length of the tube.

Fig. 12.7

If the direction of flow be from A to B, the pressure decreases from A to B. If, therefore, p be the pressure at the cross-section B, that at A will be greater by $\delta x\,.\,dp/dx$, , if the small distance AB be dx, *i.e.*, the pressure at A will be $p + \delta x.dp\,/\,dx$. The resultant force on the slice AB of the liquid will, therefore, be , $a\,\delta x\,.\,dp/dx$ where a is the cross-section of the tube, (*force being equal to pressure × area*).

Let the *velocity gradient* along the tube of flow be dv/dx; then, if v be the velocity at A, the velocity at B will be $v + \delta x\,.\,dv\,/\,dx$, because *the velocity increases in the direction A to B*, and, therefore, increase in velocity through the distance δx will be $\delta x\,.\,dv\,/dx$.

If the liquid covers this distance in time δt, we have

$$\delta t = \delta x/v \quad \text{whence} \quad v = \delta x\,/\,\delta t.$$

Or, *in the limit*, $v = dx/dt$.

Now, *acceleration = rate of change of velocity*, and, therefore, acceleration at the section $AB = dv/dt$, and mass of liquid in the section $= a.\delta x.\rho$; so that, force on it $= a.\delta x.\rho.dv\,/dt$, ; (because *force = mass × acceleration*).

But force on this slice of the liquid is also equal to $a.\,.\,\delta x\,.\,dp/dx$.

Therefore, $$-\,a.\frac{dp}{dx}\,\delta x = a.\delta x\,.\,\rho\,.\frac{dv}{dt},$$

The —*ve sign* merely indicating that the pressure and velocity gradients are opposite in sign, i.e., *whereas the pressure decreases*, the *velocity increases along AB*.

Or, $$-\frac{dp}{dx} = \rho.\frac{dv}{dt} = \rho.\frac{dv}{dx}.\frac{dx}{dt} = \rho.v.\frac{dv}{dx}. \qquad [\because dx/dt = v.$$

And $\therefore$ $-\,dp = \rho.v.dv.$ Or, $-\,dp/\rho = v.dv.$

Or, $-\frac{1}{\rho}\int_{p_1}^{p_2} dp = \int_{v_1}^{v_2} v.dv.$ where p_1, p_2 and v_1, v_2 are pressures and velocities at sections 1 and 2, respecively

Or, $-\frac{p_2}{\rho}-\left(-\frac{p_1}{\rho}\right)=\frac{v_2^2}{2}-\frac{v_1^2}{2}.$

Or, $\frac{p_1}{\rho}-\frac{p_2}{\rho}=\frac{1}{2}(v_2^2-v_1^2).$

i.e., pressure energy and kinetic energy are convertible, one into the other.

Since pressure energy is also convertible into potential energy, it follows that the *three types of energy are mutually convertible into each other.*

12.5. Bernoulli's Theorem. This theorem, given by the Swiss engineer *Daniel Bernoulli in the year 1738, states that the total energy per unit mass of a liquid flowing from one point to another, without any friction, remains constant throughout the displacement.*

Essentially *a law of variation of pressure along a streamline*, it is an easy deduction from the *principle of conservation of energy*. For as, we have seen, the pressure and potential energies of a liquid are convertible, one into the other, and so are its pressure and kinetic energies. It follows, therefore, that in any *streamline flow* of liquid, the loss of energy in one form is equal to the gain of energy in another, or that the sum total of its energy, *viz.*,

potential energy + pressure energy + kinetic energy = a constant.

Or, $$hg + p/\rho + \frac{1}{2}v^2 = C, \text{ a constant}$$

This relation is known as **Bernoulli's Equation.**

A more **formal proof** of the theorem may be given as follows :

Consider a pipe of varying cross-section, through which an *incompressible* liquid, of *density* ρ, is in streamline motion, (Fig. 12.8). Let *AB* and *CD* be two sections of the pipe, at right angles to its axis and of cross sectional areas a_1 and a_2 and let the *pressure* and *velocity* of the liquid at these sections be p_1, v_1, and p_2, v_2 respectively.

In accordance with the continuity equation $a_1v_1 = a_2v_2$; we have $v_2 > v_1$, since $a_2 < a_1$, *i.e.*, the flow of the liquid is accelerated towards the narrower end of the pipe. Obviously, therefore, *pressure* p_1 *at section AB > pressure* p_2 *at section CD.*

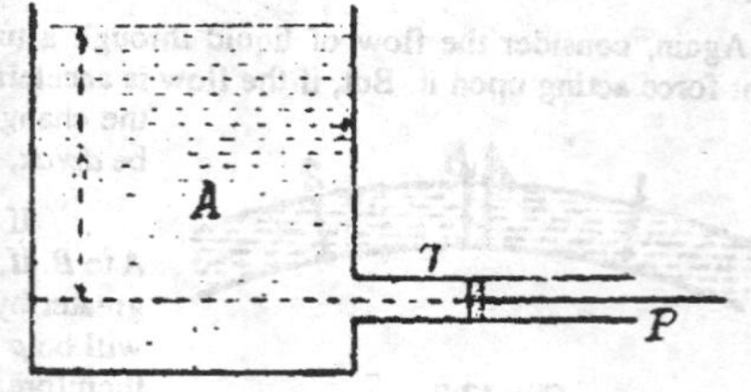

Fig. 12.8

Let the liquid column *ABCD* between the two sections *AB* and *CD*, move, in a small interval of times δt. into the position *A' B'D'C'* so that the face *AB* of it advances through a distance δl_1 into the position *A'B'* and the face *CD* through a distance δl_2 into the position *C'D',* where the volume of the liquid in strip *ABB'A'* $= a_1.\delta l_1$ and of the liquid in strip *CDD'C'* $= a_2.\delta l_2$ Since the liquid is *incompressible*, obviously $a_1\delta l_1 = a_2.\delta l_2 = V$, say.

It will be readily seen that the motion of the liquid column *ABDC* into the position *A'B'D'C'* really amounts to this volume *V* of the liquid being transferred from position *ABB'A'* into position *CDD'C'*. So that, if h_1 and h_2 be the heights of the liquid (*i.e.*, of its *c.g.*) from an arbitrary datum plane (*i.e.*, arbitrary zero-level), we have

$$\textit{loss in P.E. of liquid} = V.\rho g h_1 - V.\rho.g h_2 = V.\rho.g\,(h_1 - h_2)$$

And, since force acting on the liquid column at section *AB* is $p_1 a_1$ clearly, *work done on the liquid* here in moving it through distance δl_1 *forward* $= p_1.a_1 . \delta l_1 = p_1 V$

Similarly, *work done by the liquid* at section *CD* against pressure p_2 is $p_2.a_2.\delta l_2 = p_2 V$. So that, net work done on the liquid $= p_1 V - p_2 V = V(p_1 - p_2)$.

This work is obviously done at *the expense of the pressure energy* of the liquid and thus represents a *loss in the pressure energy of the liquid.* The *total loss in the energy of the liquid* is thus
$$V.\rho.g(h_1 - h_2) + V(p_1 - p_2)$$
Since the velocity $v_2 > v_1$, we have

$$\textit{gain in K.E. of the liquid} = \frac{1}{2}V.\rho v_2^{\,2} - \frac{1}{2}V\rho v_1^{\,2}$$
$$= \frac{1}{2}V\rho(v_2^{\,2} - v_1^{\,2})$$

In accordance with the *law of conservation of energy,*

total loss in energy = total gain in energy.

Or,
$$V.\rho.g(h_1 - h_2) + V(p_1 - p_2) = \frac{1}{2}V\rho(v_2^{\,2} - v_1^{\,2})$$

which gives
$$h_1 g + p_1/\rho + \frac{1}{2}v_1^{\,2} = h_2 g + p_2/\rho + \frac{1}{2}v_2^{\,2},$$

i.e.,
$$\frac{1}{2}v^2 + p/\rho + gh = \textbf{constant.} \quad \text{... (i)}$$

If we divide relation (*i*) by *g*, we have
$$h + \frac{p}{\rho g} + \frac{1}{2}.\frac{v^2}{g} = C', \text{ another constant} \quad \text{...(ii)}$$

Each term of this equation has the same dimensions as length, thus representing a *distance* or a *height.* It is, therefore, called a *head.* Thus, *h* is what is called the **gravitational head,** $p/\rho g$, the **pressure head** and $\frac{1}{2}v^2/g$, the **velocity head.** * Thus,

gravitational head + pressure head + velocity head = a constant.

We may, therefore, also state Bernoulli's theorem in another way *viz.*, that *at all points, in the streamline flow of a liquid, the sum of the gravitational head, the pressure head and the velocity head remains constant throughout.*

It follows at once from relation (*ii*) that if the flow of the liquid be *horizontal*, the gravitational head *h* is a constant; so that, here,
$$\frac{p}{\rho.g} + \frac{1}{2}.\frac{v^2}{g} = \textit{a constant}.$$
Similarly, from relation (*i*), we could have $p/\rho + \frac{1}{2}v^2$ = a *constant*, since the potential or gravitational energy *hg* would be constant.

Or,
$$p + \frac{1}{2}\rho v^2 = \textit{a constant}. \quad \text{...(iii)}$$
Here, *p* is referred to as the **static pressure** of the liquid and $\frac{1}{2}\rho v^2$ as its **dynamic or velocity pressure.** So that, we may express this result by saying that *for a horizontal motion of the liquid, the sum of its static and dynamic pressures remains a constant.*

Thus, if in a liquid, flowing horizontally, the pressure and velocity at one point be p_1 and v_1 and at another, p_2 and v_2 respectively, we have
$$\frac{p_1}{\rho} + \frac{1}{2}v_1^{\,2} = \frac{p_2}{\rho} + \frac{1}{2}v_2^{\,2},$$

* For, the liquid must fall through this much height to attain the velocity *v*.

which shows that pressure and velocity (and, therefore, kinetic energy) can only increase at the expense of one another, *i.e., points of maximum pressure correspond to those of minimum velocity,* and *vice versa.** The principle is made use of in various important practical applications, (see § 12.6)

12.6. Important Application of Bernoulli's Theorem

(*i*) Velocity of efflux of a Liquid—Torricelli's Theorem. Let the surface of the liquid be at a height h above the level of the *circular* and *sharp edged* orifice O in a tank, (Fig. 12.9). If the tank be sufficiently wide, the velocity at the liquid surface may be taken to be *zero*, the pressure there being clearly atmospheric. Since the pressure is also atmospheric at the orifice, where the liquid emerges, it plays no part in the flow of the liquid. If v be the velocity at the level of the orifice, we have, considering a tube of flow beginning at A and ending at O,

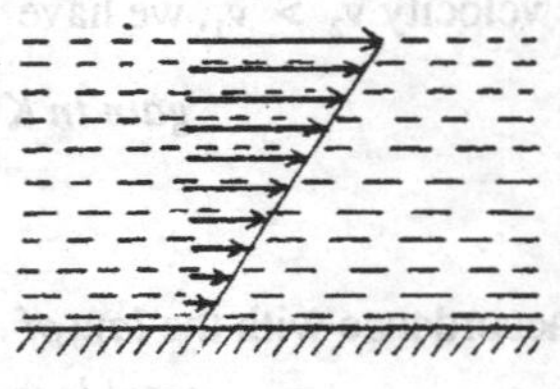

Fig. 12.9

total energy at A = pressure energy + potential energy + kinetic energy $= 0 + hg + 0$,

because *pressure* at $A = 0$, $P.E. = gh$ and $K.E. = 0$. $[\because v = 0.$

And, *total energy at O*, the level of the orifice $= 0 + 0 + \frac{1}{2}v^2$,

because pressure at $O = 0$, $\quad P.E. = 0$, and $K.E. = \frac{1}{2}v^2$.

Since total energy remains the *same*, we have

$$\frac{1}{2}v^2 = hg. \qquad \text{Or,} \qquad v^2 = 2gh, \text{ whence } v = \sqrt{2gh}$$

This, then, is the **velocity of efflux** of the liquid at the orifice O.

This result was first obtained by *Torricelli* (in the year 1644), and hence is known as **Torricelli's Theorem,** or *the Law of Efflux*, and may be stated as follows :

The velocity of efflux of a liquid through an orifice is equal to that which a body attains in falling freely from the surface of the liquid to the orifice.

For clearly, if the liquid had fallen freely through this height h, its velocity would be given by the relation, $v^2 = 2gh$, to be equal to $v = \sqrt{2gh}$, the same as obtained above.

This *ideal* velocity, is, however, seldom attained because of loss of kinetic energy of the liquid due to friction (or viscosity) and the *actual* velocity is $Cv \times$ the *ideal velocity, i.e.*, equal to $C_v\sqrt{2gh}$, where C_v is the *coefficient of velocity*, its value lying in the range 0.95 to 0.99 for water depending upon the shape of the orifice and the head of water, the value usually taken being 0.97.

Now, the liquid jet flows out in the form of a *parabola*, and takes time equal to $\sqrt{2h_1/g}$ to fall through a height h_1 to a plane in level with the bottom of the tank, striking the plane at a distance R, called its *horizontal range* or simply, **range,** given by

$$R = v \times \sqrt{2h_1/g} = \sqrt{(2gh)(2h_1/g)} = 2\sqrt{hh_1}$$

Clearly $2\sqrt{hh_1} = \sqrt{(h+h_1)^2 - (h-h_1)^2}$ and will thus have the maximum value when the term $(h - h_1)^2 = 0$, or $h = h_1$.

Thus, for a given depth $H = (h + h_1)$ of a liquid column, *the range of the liquid jet issuing out of the orifice will be the maximum when* $h = h_1 = H/2$, *i.e., when the orifice is half way down the depth of the liquid column and its value then will obviously be* $2\sqrt{hh} = 2h =$ *twice the depth of the orifice*

---

* This result is also true for compressible fluids and is sometimes referred to as **Hawksbee's law.**

from the liquid surface = 2(H/2) = H, i.e., equal to the total depth of the liquid column in the tank (provided, of course, H be maintained constant).

And, obviously, if the liquid jet issuing out of the orifice be directed upwards, it should *theoretically* rise to the level A of the liquid surface. Again, however, due to air resistance and viscosity, the actual height attained is less than this ideal one.

Vena Contracta. The *whole* of the liquid entering the orifice does not move perpendicularly to it, but comes from all directions, as shown in Fig. 12.10, the streamlines near the edges being curved. The liquid coming from the sides of the vessel, as it enters the orifice, has still a lateral velocity due to inertia, and continues to move inwards towards the centre of cross-section of the jet, until the increasing outward pressure is balanced by the atmospheric pressure at the jet. The liquid jet thus contracts from the mouth of the orifice up to a distance of about half the diameter of the orifice from it, say, up to C, where it forms a *neck*, called the **Vena Contracta** or the **Contracted Vein.** It is here that the streamlines in the jet become parallel to each other and perpendicular to the plane of the orifice, *i.e.*, the jet-becomes *uniform* and the velocity becomes the same throughout and, naturally, has its maximum value because of the least cross sectional area there. It is this velocity which is given by *Torricelli's* equation, (see § 12.6, above)

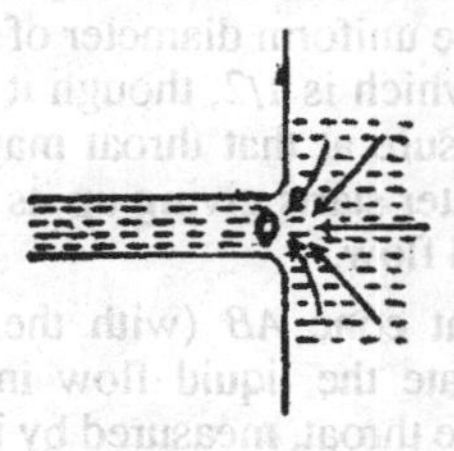

Fig. 12.10

Obviously, the cross-section of the jet at the *Vena Contracta* is smaller than the cross-section of the orifice. The ratio between the two is called the *coefficient of contraction*, denoted usually by C_c. Its theoretical value is $\pi/(\pi + 2) = 0.611$, but it varies from 0.611 to 0.69 in actual practice, depending upon the size and the shape of the orifice and the head of liquid under which the flow occurs. For *circular orifices*, the best experimental value is found to be 0.624 very nearly. So that, the volume of liquid passing out through the orifice in unit time, *i.e.*, the *rate of flow* through it is $C_v.C_c\, a\sqrt{2gh}$, where C_v is the coefficient of *velocity*. For water, the product $C_v.C_c$ called the *coefficient of discharge* $C_D = 0.62$

And $\therefore$ *rate of flow of water through the orifice* $= 0.62\, a\sqrt{2gh}$

With outflow tubes of suitable shapes, it is possible to completely avoid *Vena Contracta*. The diminution in the velocity of efflux, however, does occurs, being quite independent of the *Vena Contracta*.

(ii) **Velocity of efflux of a gas.** Since a gas issues out of an orifice in a reservoir under adiabatic conditions, we have, from *Bernoulli's equation for adiabatic flow of a gas* (§ 12.7, *(ii)*)

$$\frac{1}{2}v^2 + \frac{\nu}{\nu-1}\cdot\frac{p}{\rho} = \frac{1}{2}v_0{}^2 + \frac{\nu}{\nu-1}\cdot\frac{p_0}{\rho_0},$$

where v_0, p_0 and ρ_0 are the velocity, pressure and density of the gas inside the reservoir and v, p and ρ, their corresponding values as the gas issues out of the orifice in the reservoir. So that v is the *velocity of efflux of the gas*.

Since, obviously $v_0 = 0$, we have

$$\frac{1}{2}v^2 = \frac{\nu}{\nu-1}\left(\frac{p_0}{\rho_0} - \frac{p}{\rho}\right). \quad \text{Or,} \quad v^2 = \frac{2\nu}{\nu-1}\left(\frac{p_0}{\rho_0} - \frac{p}{\rho}\right)$$

And, because under adiabatic conditions, $p_0/\rho_0{}' = p/\rho'$, we have

$$v^2 = \frac{2\nu}{\nu-1}, \frac{p_0}{\rho_0}\left[1 - \left(\frac{p}{p_0}\right)^{\frac{\nu-1}{\nu}}\right],$$

whence, the *velocity of efflux of the gas*, v, may be easily obtained.

(iii) **Venturimeter**. It is a simple appliance, based on *Bernoulli's theorem, for measuring the discharge or the rate of flow of a liquid through a pipe.* It is invariably used to determine the rate

of flow of water at any point along a pipe line on the drains of the water supply, in which case it is referred to as the *Venturi water meter*.

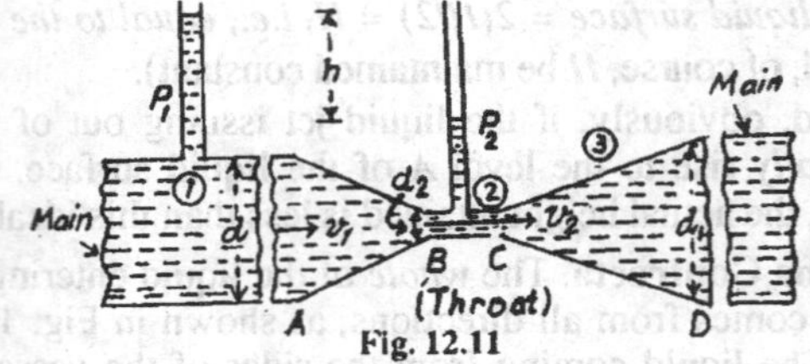

Fig. 12.11

Devised by an American engineer *Clemens Herschel* in the year 1886, it has been named after the Italian engineer and philosopher *G.B. Venturi*, who did most of the initial pioneering work on the subject as early as 1791.

It consists of three distant parts : (*i*) *a convergent cone AB*, with a slope of its sides about 1 in 4 or 1 in 5, (*ii*) a short length *BC*, of uniform diameter, called the *throat* and (*iii*) a *divergent cone*, or *diffuser CD*, with a slope of its sides about i in 10 or 1 in 20.

The whole of the instrument is connected up as shown in Fig. 12.11, to form part of the main pipe through which the rate of flow of the liquid is to be determined. Obviously, the diameter of the *inlet* or the upstream end *A* of the cone *AB* as also that of the *outlet* or the downstream end *D* of cone *CD* is the same as that of the main pipe, say, d_1. Thus, if the uniform diameter of the throat be d_2, the ratio d_2/d_1 is called its *throat ratio*, the usual value of which is 1/2, though it may vary between 1/4 to 3/4. It should not, however, be such that the pressure at that throat may become negative. For, if the pressure here be very low, the liquid or the water starts giving up its dissolved air and thus begins to vaporize, breaking the continuity of the liquid flow.

Further, the comparatively sharper slope of the convergent cone *AB* (with the angle of convergence between 20° to 30°) is designed to rapidly accelerate the liquid flow in order to develop an appreciable pressure difference between the inlet and the throat, measured by the rise of liquid in the two vertical tubes p_1 and p_2 called *piezometer* tubes, fitted in the main, close to the outer end of *AB* and in the throat respectively, as shown.

The smaller slope of the divergent cone *CD* (with the angle of divergence between 5° to 14°) serves to decelerate the liquid flow and to increase the area of cross-section of the liquid stream back to that at the inlet or the main pipe *in a gradual manner*. For, if this latter process be hastened by making the slope sharper, the layer in contact with the wall of the diffuser or cone *CD* may break away from it. (due to what is called the boundary effect), roll itself up in the form of vortices and be left behind. Indeed, it is for this very reason, that the pressure measurements for calculating the rate of flow are made only between the converging cone (*AB*) and the throat (*BC*) and not between the throat and the diverging cone *CD*.

The calculations for the rate of flow are made as follows :

let p_1, v_1 and p_2, v_2 be the pressures and velocities respectively in the inlet and the throat at some such points as (1) and (2). Then, since the pipe is horizontal, we have, in accordance with Bernoulli's theorem,

$$\frac{1}{2}v_1^2 + p_1/\rho = \frac{1}{2}v_2^2 + p_2/\rho,$$

where ρ is the density of the liquid, supposed constant.

So that, $$(p_1 - p_2) = \rho(v_2^2 - v_1^2)/2,$$

Or, multiplying and dividing the right hand expression by v_1^2 we have

$$(p_1 - p_2) = \frac{\rho v_1^2}{2}\left(\frac{v_2^2 - v_1^2}{v_1^2}\right) = \frac{\rho v_1^2}{2}\left(\frac{v_2^2}{v_1^2} - 1\right) \qquad ...(I)$$

Now, in accordance with the equation of continuity of flow, we have $a_1v_1 = a_2v_2$, where a_1 and a_2 are the cross-sections at points (1) and (2). And, therefore, $v_2/v_1 = a_1/a_2$. Substituting in relation *I*, we have $(p_1 - p_2) = \frac{\rho v_1^2}{2}\left(\frac{a_1^2}{a_2^2} - 1\right) = \frac{\rho v_1^2}{2}\left(\frac{a_1^2 - a_2^2}{a_2^2}\right)$

Clearly, $(p_1 - p_2) = h.\rho.g$, , where h is the difference of liquid levels in the two peizometer tubes.

So that, $$h.\rho.g = \frac{\rho v_1^2}{2}\left(\frac{a_1^2 - a_2^2}{a_2^2}\right). \text{ Or, } v_1^2 = \frac{2hg.a_2^2}{a_1^2 - a_1^2}$$

Or, $v_1 = a_2\sqrt{\frac{2hg}{a_1^2 - a_2^2}}$, where h is called the *venturi head*.

And, ∴ *ratio of flow* or *discharge of the liquid.*

i.e., $$Q = a_1v_1 = a_1 a_2\sqrt{\frac{2hg}{a_1^2 - a_2^2}} \quad ...(II)$$

Clearly, the factor $a_1a_2\sqrt{2g}\,(a_1^2 - a_2^2)$ depends upon the dimensions of the venturimeter itself. It is, therefore, called the *constant of the venturimeter*, denoted by the letter C, and is known for a given instrument. So that $Q = C\sqrt{h}$, which gives the *theoretical rate of flow or discharge*, neglecting the slight fall in the pressure head between AB and BC (due partly to friction and partly to shock due to change of section). If this be taken into account, we have

rate of flow or discharge $Q = kC\sqrt{h} = 0.97C\sqrt{h}$,

where k is called the *coefficient of discharge* or the *coefficient of the venturimeter*, its value for water being 0.97, though it is not strictly constant for all velocities—this variation in the value of k being responsible for taking away from the accuracy of the venturimeter at low velocities of the liquid flow.

The above relation for Q holds good, however, even if the venturimeter be used in an inclined or the vertical position only, the proper value of h has to be taken. Thus, the value of h, the *venturi head*, is in general given by $h = \left[\frac{p_1}{\rho} - h_1\right] - \left[\frac{p_2}{\rho} - h_2\right]$

And, in the horizontal position, by $h = (p_1/\rho - p_2/\rho)$, as we have seen above.

(*iv*) **Pitot tube.** This too is a simple device for the measurement of the rate of flow of water or a liquid through a pipe, based on *Bernoulli's* theorem. It takes its name from *Henry Pitot*, a Frenchman, who noticed, in the year 1730, that as he dipped a bent glass tube in the river *Seine*, the water rose up into the tube to a height proportional to the square of the velocity of flow.

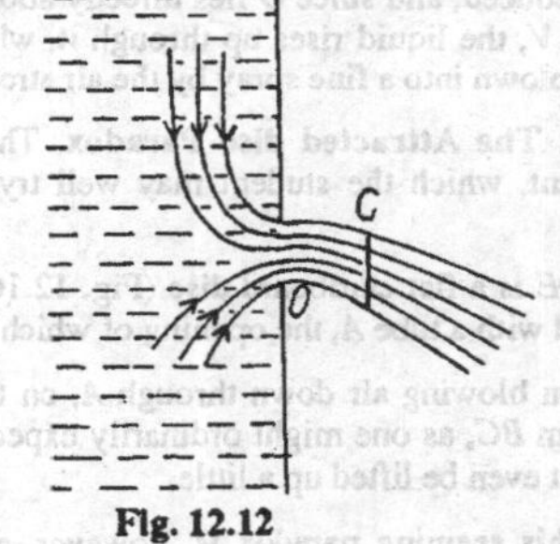

Fig. 12.12

A pitot tube may either be a combination of two vertical tubes PQ and RS with small apertures at their ends, [(Fig. 12.12 (*a*)]either separate or joined together [Fig. 12.12 (*b*)]. In any case, the plane of the aperture of one tube PQ is parallel to the direction of flow and that of the other (RS) faces the flow. The rise of liquid column in tube PQ, therefore, gives the pressure at Q. And, since, the water is stopped in the plane of the aperture S of the tube RS, *its velocity there becomes zero* -- point S is therefore called the *stagnation point*. The kinetic energy of the stream is thus reduced from $\frac{1}{2}v^2$ to *zero* at the point; consequently, its pressure here increases by an equal amount $\frac{1}{2}v^2$, with the water rising to a higher level in the tube RS, then in the tube PQ. If h be difference of the liquid levels in the two tubes, we have

$$\frac{1}{2}v^2 = hg. \quad \text{Or,} \quad v^2 = 2gh,$$

whence, $$v = \sqrt{2gh} \text{ And } \therefore \textit{ rate of flow, } Q = av = \sqrt{2gh}$$

where a is the cross-section of the pipe, where the tubes are placed.

(v) Other Common Applications of Bernoulli's Theorem.

1. The Steam Injector. It is a simple device to accelerate the ejection of the exhaust steam from the cylinder of a steam engine, and consists of a tube A, (Fig. 12.13), narrowing down into a nozzle N at its lower end, inside another tube B, having a side-tube C, which is connected to the cylinder of the engine.

A jet of steam is introduced into A, and as it issues out of the nozzle, N, its velocity is considerably increased, resulting in a corresponding fall in pressure there, and the steam from the engine-cylinder thus rushes into this region of reduced pressure, whence, it is ejected out through the lower end of B.

Fig. 12.13

2. The Filter Pump. It is also based on the same principle and is used to reduce the pressure in a vessel. Here, a stream of water from a tap, flowing through a tube A, (Fig. 12.14), issues out in the form of a jet from its narrow orifice O, which results in a great rise in its velocity and proportionate fall into its pressure, which is thus soon reduced to a value below that of the atmosphere. The air from the vessel, connected through a side-tube B to this region of reduced pressure, then rushes into it, and is carried away by the stream of water as it flows down through C.

Fig. 12.14

In this way, the pressure in the vessel is ultimately reduced to just a little above the vapour pressure of water, in a comparatively very short time.

If the inlet water tube be a twisted, instead of a straight one, the exhaustion proceeds more rapidly, due to the rotating water-jet in the tube breaking up more readily and mixing up easily with the incoming air from the vessel.

3. The Atomizer. The *atomizer* or *sprayer*, used for spraying scents etc., is yet another example of *a fall in pressure due to an increase in velocity*.

Here, (Fig. 12.15), air is blown through a tube T, (usually by compressing a rubber bulb, fitted on the tube at one end), which, when it rushes out of the aperture O, where the tube narrows down, acquires a high velocity. The pressure in the vicinity of O is thus greatly reduced, and since O lies directly above the vertical tube, dipping in the liquid in vessel V, the liquid rises up through it, when, on issuing out of the aperture at the top, it is blown into a fine spray by the air stream from T.

4. The Attracted disc Paradox. The following is a simple and interesting experiment, which the student may well try for amusement at a small gathering at home.

DE is a flat cardboard disc (Fig. 12.16), over which is placed another flat disc BC, fitted with a tube A, the opening of which is in *flush* with BC.

On blowing air down through A, on to DE, the latter, instead of being blown away from BC, as one might ordinarily expect, sticks on to it more and more closely, and might even be lifted up a little.

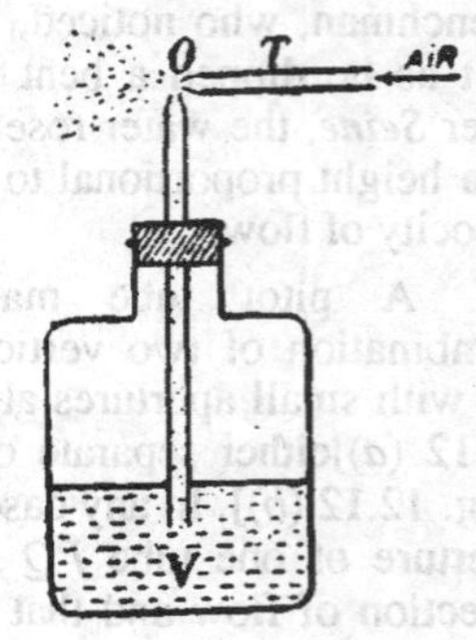

Fig. 12.15

This seeming paradox is, however, easily explained. For, as the air from A rushes through the narrow space in between BC and DE, its velocity increases and consequently the pressure there decreases, so that it soon falls below the atmospheric pressure on DE, which thus pushes it up towards BC.

5. The Bunsen-Burner. This too is familiar example of a fall of pressure due to increased velocity. For, as the gas-issues out with a great velocity from the fine

nozzle, down below, the pressure falls in its immediate neighbourhood, and the air is thus sucked in through the hole *O*, (Fig. 12.17), and gets mixed up with the gas.

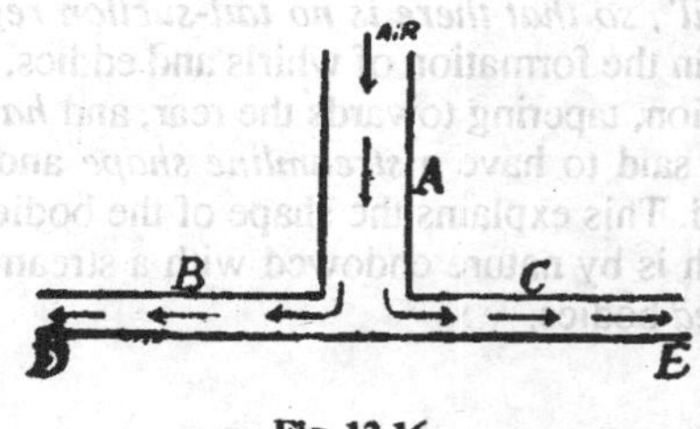

Fig. 12.16

6. The Magnus Effect. If a ball, or a sphere, be spinning about an axis through it, perpendicular to the plane of the paper, the air surrounding it is also set into motion,—the streamlines taking the form of concentric circles in planes, parallel to the plane of the paper, their direction being the same as that of the spin of the ball, shown in Fig. 12.20 (*a*). And, obviously, the rougher the surface of the ball, the thicker the layer of air thus set into motion.

If, however, the ball be given *only a linear forward motion*, it pushes aside the air in front of it; to make room for itself, and, this displaced air then flows along its sides on to its back or the rear end, the form of the streamlines being as shown in Fig. 12.20(*b*) *in the opposite direction to the motion of the fall.*

Fig. 12.17

And, finally, if the ball be given both motions (*a*) and (*b*) simultaneously, it is clear from Fig. 12.18(c) that the streamlines due to the two motions run in opposite directions on the underside of the ball, but in the same direction on its upper side. Thus, there is *a decrease of velocity* or *an increase of pressure* on its under, and *an increase of velocity* or a *decrease of pressure* on its upper side. As a result, the ball experiences a *lift* and takes a *curved path upwards convex to the under or the greater pressure side*. This *effect of the generation of a lift due to spinning* is called *Magnus effect*, since it was first explained by *Magnus* in the year 1852, although it had been observed earlier by *Lord Rayleigh*. The effect is easily observed when a tennis or a golf ball is given a spin.

(*a*) **Fig. 12.18** (*b*)

7. The Cylindrical Shape of a Bullet. We have already seen [§ 3.23, (*ii*), page 90], that for giving directional stability, it is desirable to give the shot or bullet a rapid '*spin*' about an axis, along its direction of motion, and how this object is achieved by '*rifling*' the barrel of a gun or rifle, (*i.e.*, by cutting spiral grooves inside it).

Now, if the bullet were spherical in shape, there will come about, as explained above in case 6, a difference in the lateral pressure on it during its passage through air, on account of its simultaneously possessing a rotatory and a linear motion and the bullet will thus be deflected from its straight path. To avoid this, *the bullet is made cylindrical in shape*, so that the lateral pressure on it remains uniform and it flies undeflected along its path.

8. Streamline Bodies. The student has, no doubt, heard of streamline bodies of automobiles, particularly of racing cars etc. We shall discuss in brief here as to what this *streamlining* or '*fairing*' of a body really connotes in the language of Science.

As a body moves through air, or through a fluid, in general, it carries a part of it along with itself, pushing the rest on to either side. The streamlines of the fluid, directed towards the body, open out on either side to make way for it, as it were, and meet some distance behind it. This fluid at the rear of the body, enclosed by the streamlines meeting there is thus carried by the body as a sort of a '*tail*'. Some extra work has thus to be done by the body in carrying this extra burden, resulting in an appreciable decrease in its kinetic energy and velocity. If fact, the body has to encounter a double opposition to its forward motion, *viz*., (*i*) *an increased pressure in front*, **called the head pressure**, and (*ii*) *a decreased pressure* or the **tail suction behind, (which exerts a backward pull on it).**

Naturally, the surrounding fluid flows into this rear region of decreased pressure or *tail suction*, and is thus thrown up into vortices (*i.e.*, whirls and eddies) there, which results in a further fall in pressure in this region. These vortices are thus responsible for dissipating away a fairly good part of the energy of the body, thus decreasing its velocity or offering resistance to its motion.

If, therefore, the resistance to the forward motion of the body is to be minimised, *it should be given a shape similar to that of the fluid forming its 'tail', so that there is no tail-suction region formed at all at its rear*, and no energy is thus dissipated in the formation of whirls and eddies. The body is thus made with a gradually decreasing cross-section, tapering towards the rear, and *having no sharp corners or edges anywhere*. The body is then said to have a *streamline shape* and the resistance to its forward motion is considerably decreased. This explains the shape of the bodies of big airliners and of most of the modern cars. Since a fish is by nature endowed with a streamline body, it is usual to refer to streamline bodies as *fish-shaped* bodies.

12.7. Viscosity. When a liquid flows slowly and steadily over a fixed horizontal surface, *i.e.*, when its flow is *streamline*, its layer in contact with the fixed surface remains stationary and the velocity of the other layers increases with the distance from the fixed surface, *i.e.*, there is a regular *velocity gradient* set up in the liquid and, therefore, the greater the distance of a layer from the fixed surface, the greater its velocity, with the topmost layer thus moving the fastest. The velocity distribution curve or the profile of the liquid is, therefore a straight line inclined to the fixed horizontal surface, as shown in Fig. 12.19. Such a flow is also referred to as a *laminar flow*, meaning thereby that one layer of the liquid quietly glides over another without any two layers getting mixed up.

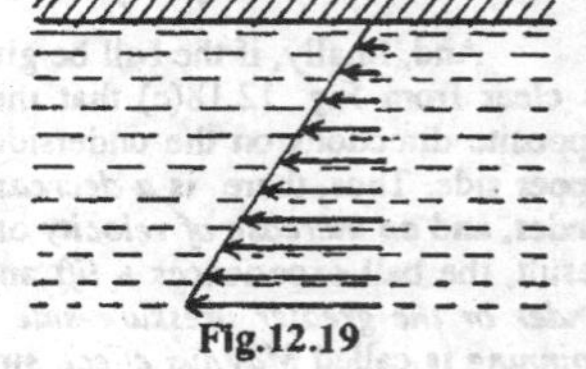
Fig.12.19

If the liquid flows through a tube, its layers are concentric cylindrical shells coaxial with the tube. Now, obviously, the layer in contact with the wall of the tube remains at rest and the velocity increases as we proceed towards the axis, so that the profile of the advancing liquid is a *parabola*, as shown in Fig. 12. 20 (page 398)

Now, considering any particular layer of the liquid (Fig. 12.21) we have the layer immediately *below* it moving *slower* than it, and the one immediately *above* it moving *faster*, than it, so that the former tends to retard its motion and the latter tends to accelerate it. The two layers thus tend to destroy their relative motion, *as though there were a backward dragging force, acting tangentially on the layers.* If, therefore, the relative velocity between the two layers is to be maintained, an external force must be applied to overcome this *backward drag*. In the absence of any such outside force, the relative motion between the layers is destroyed and the flow of the liquid ceases. *This property of a liquid by virtue of which it opposes relative motion between its different layers is known as* **viscosity** *or internal friction of the liquid.*

Coefficient of Viscosity. Newton showed that the backward dragging, or *viscous*, force, acting tangentially on any liquid layer, is directly proportional to its surface area A, and velocity v, and inversely proportional to its distance x from the stationary layer. Denoting this force by F, therefore, we have

$$F \propto A;\ F \propto -\ v;\ F \propto \frac{1}{x},$$

[the —*ve sign* of v merely indicating that the direction of the force is opposite to that of velocity] .

Or $$F \propto -\frac{A.v}{x},\ \ i.e.,\ \ F = -\eta.\frac{A.v}{x},$$

where η is a constant, depending upon the nature of the liquid, and is called its **coefficient of viscosity.**

Putting v/x in the form dv/dx, which gives the *rate of change of velocity with distance* or *the velocity gradient*, we have $F = -\eta \, . \, A . dv/dx$. ...I

This is known as **Newton's law of viscous flow in streamline motion.**

If $A = 1$ and $dv/dx = 1$, we have $F = \eta$.

Thus, the **coefficient of viscosity** *of a liquid may be defined as the tangential force required per unit area to maintain unit relative velocity between two layers unit distance apart.*

In the *C.G.S. system*, therefore, $\eta = 1$ *if a tangential force of 1 dyne is required per sq. cm to maintain a relative velocity of 1 cm/sec. between two layers 1 cm apart.* It is called **Poise**, after **Poiseuille** whose work in viscosity is important.

In the *M.K.S.* (or *S.I.*) *system*, on the other hand, $\eta = 1$ if a tangential *force of 1 newton per square metre is required to maintain a relative velocity of 1 m/sec between two layers one metre apart.*

In the *F.P.S. System*, $\eta = 1$, *if a tangential force of 1 pound (weight) per sq ft is required to maintain a relative velocity of 1 ft/sec between two layers 1 ft apart.*

N.B. This coefficient of viscosity (η) *is referred to as the* **dynamic viscosity** of the liquid, with *Poise* as its *C.G.S. unit*. *Its ratio to the density of the liquid, i.e.,* η/ρ is called the *kinematic viscosity* (*k*) of the liquid and the corresponding *C.G.S. unit* for it is the *Stokes*.

Viscosity in liquids *corresponds to solid friction* in so far as, like the latter, it also opposes relative motion between two layers. *It, however, differs from solid friction in that, unlike the latter, it depends upon (i) surface area of the liquid layer, (ii) its distance from the stationary layer and (iii) its velocity with respect to the stationary layer.*

Dimensions of η. From relation I above, $\eta = \dfrac{F}{A . dv/dx}$.

So that, the dimensions of η are those of $\dfrac{force}{area \times velocity\ gradient}$

$$i.e., \quad \frac{F}{L^2\left[\dfrac{L/T}{L}\right]} = FL^{-2}T \qquad ...(i) \qquad \left[\because \frac{dv}{dx} = \frac{L/T}{L}\right]$$

Or, putting in the dimensions of F, we have

$$dimensions\ of\ \eta = \frac{MLT^{-2}}{L^2\left[\dfrac{L/T}{L}\right]} = \frac{MLT^{-2}}{l^2 T^{-1}} = ML^{-1}T^{-1} \qquad ...(ii)$$

Thus, on the *C.G.S. system*, η may be expressed in *dynes* cm^{-2} *sec* (or Poise) in accordance with (*i*) above, or in *gm.* cm^{-1} sec^{-1} in accordance with (*ii*)

On the *M.K.S.* (or *S.I.*) *system*, similarly, η may be expressed either in *newtons* m^{-2} *sec* or in *kg* m^{-1} sec^{-1}.

And, on the *F.P.S. system*, it may be expressed in *lb (weight)* ft^{-2} sec or in the Engineer's units, *slug* ff^{-1} sec^{-1}. (where *slug is the unit of mass, the weight of a mass of* 1 *slug being* 32 *lb* or the *mass of* 1*lb weight* $= \frac{1}{32}$ *slug i.e., weight in pounds/g (in* ft^2*/sec) = sliugs).*

12.8. Fugitive Elasticity. The expression, for F, above, may be rearranged and put as

$$\eta = \frac{F/A}{dv \, . \, dx}, \qquad ...(a)$$

i.e., coefficient of viscosity = $\dfrac{tangential\ stress}{velocity\ gradient}$*

This is an expression similar to the one for the coefficient of rigidity, *viz.*,

$$n = \frac{F/A}{\theta} = \frac{F/A}{.dy/dx}. \quad ...(b) \qquad \left[\because \theta = \frac{dy}{dx}.\right.$$

$$= \frac{tangential\ stress*}{displacement\ gradient}$$

Maxwell, therefore, considered a liquid to possess a certain amount of rigidity, breaking down continually under a shearing stress. Very fittingly, he imagined viscosity of a liquid to be the limiting case of the rigidity of a solid, when the latter breaks down under the shear applied. *A liquid is thus regarded as capable of exerting and sustaining an amount of shearing stress for a short time, after which it breaks down and the shear is formed over again.* In other words, a liquid offers a *fugitive resistance* to shearing stress, which is continually breaking down, and it may thus be said to possess a fugitive rigidity.

Now, if the rate at which the shear (θ) breaks down be taken to be proportional to shear, we have

rate of the breakdown of shear $\propto \theta$.

Or, " " " " " $= \lambda.\theta$, where λ is a constant.

And, clearly, the rate of formation of the shear $= \dfrac{d\theta}{dt} = \dfrac{d}{dt}\left(\dfrac{dy}{dx}\right) = \dfrac{d}{dx}\left(\dfrac{dy}{dt}\right) = \dfrac{dv}{dx}$.

[v being the velocity in the *same plane*.

Thus, when the motion of the fluid becomes quite steady, the rate of formation of the shear must be the same as that of its breakdown. So that,

$$\lambda.\theta = \frac{dv}{dx}. \text{ Or, } \theta \Big/ \frac{dv}{dx} = \frac{1}{\lambda}. \quad ...(i)$$

Now, dividing relation (*a*) by (*b*), above, we have $\dfrac{\eta}{n} = \dfrac{\theta}{dv/dx}$.

$\therefore$ substituting $\dfrac{\eta}{n}$ for $\theta \Big/ \dfrac{dv}{dx}$ in relation (*c*) above, we have $\dfrac{\eta}{n} = \dfrac{1}{\lambda}$

This quantity $1/\lambda$ is called the **'time of relaxation of the medium'** and gives the time taken by the shear to disappear, provided no fresh shear is applied.

12.9. Critical Velocity. It was *Osborne Reynolds* who first showed by direct experiment that the *critical velocity* v_0 of a liquid is given by the relation, $v_c = k.\eta/\rho\, r$, called *Osborne Reynold's formula,* where η is its *coefficient of viscosity*, ρ, its *density* and r, the *radius of the tube*, the constant k being called **Reynold's number**, its value being about 1000 for narrow tubes.

The expression for v_0 may, however, be easily deduced by the method of dimensions, as explained below :

Since v_0 is found to depend upon (*i*) η, (*ii*) ρ, and (*iii*) r, we have

$$v_o = k.\eta^a \rho^b r^c, \text{ say.} \qquad [k \text{ being a constant.}$$

So that, putting the dimensions of the quantities involved, we have

---

* It will be noted from expressions (*a*) and (*b*) that whereas in a fluid, the viscous drag is proportional to the *velocity gradient, perpendicular to the direction of motion*, the shearing stress, in a solid, is proportional to the *displacement gradient, perpendicular to the direction of shear*.

$$[LT^{-1}] = [ML^{-1}T^{-1}]^a [ML^{-3}]^b [L]^c$$ [k having no dimensions, being a constant.

Or, $$[LT^{-1}] = [M^{a+b} L^{-a-3b+c} T^{-a}]$$

Since the dimensions on the two sides of the equation must be the same, (by the *principle of homogeneity of dimensions*), we have

$a + b = 0$ ...(*i*); $-a - 3b + c = 1$ ... (*ii*); $-a = -1$ ... (*iii*)

So that, adding (*i*) and (*iii*), we have $b = -1$.

Substituting this value of (*b*) in (*i*), we have $a = 1$. [Or, directly from (*iii*).

And, substituting the value of *a* and *b* in (*ii*), we have $c = -1$.

Hence $$v_c = k.\eta/\rho r,$$
where *k* (*Reynold's number*) is, as mentioned above, near about 1000 for narrow tubes. Thus, *for narrow tubes*, $v_c = 1000.\eta/\rho r$

It must be emphasized again that this relation applies only to narrow tubes. For tubes of wide bores, the value of v_c is very much greater, and may be even a thousand times greater than that given by the above relation.

Now, a mere glance at the expression for v_c, deduced above, will show that

$$(i)\ v_c \propto \eta\ ;\ (ii)\ v_c \propto 1/\rho\ ;\ \text{and}\ (iii)\ v_c \propto 1/r,$$

i.e., the critical velocity of a liquid is (i) directly proportional to its viscosity, (ii) inversely proportional to its density and (iii) inversely proportional to the radius of the tube through which it flows.

It follows, therefore, that *narrow tubes, and liquids of high viscosity and low density tend to promote orderly motion*, whereas tubes of wide bores, and liquids of low viscosity and high density lead to turbulence.

Again, if we have a perfectly *mobile* or *inviscid* liquid, *i.e.*, a liquid, for which $\eta = 0$, then, obviously, $v_c = 0$; so that, its flow would be turbulent and not orderly, even for the smallest velocity and in the narrowest of tubes.

Thus, we see that *it is viscosity of a liquid alone due to which its flow may possibly be orderly and thus approximate to that of a perfect fluid.*

12.10. Significance of Reynold's number. From the relation for critical velocity, $v_c = k\,\eta/\rho r$ it is clear that the *Reynold's number* $k = v_c.\rho.r/\eta$, *i.e., a pure numeric*, with no dimensions.

The significant fact about it is that it is what is called a '*critical variable*' and that its value determines the process taking place inside a cylindrical tube during the flow of a liquid, or a fluid, in general, through it. In the case of apparatus, *geometrically similar*, whatever their actual dimensions, turbulence sets in at the same constant value of Reynold's number in all cases of liquid-flow. So that, even though the values of *r*, ρ and η may all be different from each other, *so long as the quantity $v_c\,\rho r/\eta$, i.e., k remains the same, the liquid-flow will be similar in all cases.*

Even in the case of motion of bodies through viscous media, *Reynold's number* leads to the '*law of similarity*'. So that, for geometrically similar bodies, the streamlines are similar if the Reynold's number be the same for them. In other words, the shape of the streamlines and the direction of motion of a body at any point in a viscous medium depends upon the value of the Reynold's number for it. This law has been subjected to rigorous experimental tests and found valid.

The great **technical importance** of the law lies in the fact that it enables the resistance of a medium to the motion of a body, of any shape, through it to be correctly determined from the resistance offered to the motion of a much smaller scale model of it through that medium. Not only that, but *we may also predict the results of the motion of the body through air from those obtained for its motion through water.*

The one great snag is that the necessary conditions for the value of k to be the same in two media are not easily obtainable. Even so, quite a lot of useful information can be had regarding the motion of the body through a viscous medium, *even beyond the velocity range of the scale model.*

12.11. Poiseuille's equation for flow of a liquid through a horizontal capillary tube. Let a liquid be flowing through a horizontal capillary tube of *length l* and *radius r* under a *constant pressure difference P* across its two ends. Then, to obtain an expression for its *rate of flow* through the tube in terms of its coefficient of viscosity, we make the following assumptions.

(*i*) *that the flow of the liquid is steady and streamline, with the stream lines all parallel to the axis of the tube;*

(*ii*) *that there being no radial flow, the pressure is constant over every cross-section of the tube.*

(*iii*) *that the liquid layer in contact with the sides of the tube remains stationary.*

(*iv*) *that the liquid yields to just a small shearing stress, and*

(*v*) *that the tube is horizontal, so that gravity does not affect that flow.*

Now, imagine a cylindrical layer or shell of liquid, coaxial with the tube, and of radius x. The velocity of flow *at all points* on this cylindrical shell will be the same, say, v. Since the velocity of the layer in contact with the wall of the tube is zero and goes on increasing towards the axis, it is clear that the liquid inside the imaginary cylinder is moving faster than that outside it, and the *backward tangential force* due to the outer, slower-moving liquid on the inner faster moving liquid is, in accordance with relation *I* above, given by $\eta.2\pi x.l.dv/dx$, where η is the *coefficient of viscosity* of the liquid, [because, here, *surface area* (*A*) of the cylindrical shell of radius x is equal to $2\pi x.l$, where l is the length of the capillary tube, and dv/dx is the *velocity gradient* there].

Since the difference of pressure at the two ends of the capillary is P, the *forward force* on the cylindrical liquid shell, in the direction of flow, is clearly $P \times \pi x^2$, tending to accelerate the motion of the liquid. If, therefore, the motion of the liquid be steady, we have

$$\eta.2\pi x.l.dv/dx = -P.\pi x^2$$

the *–ve sign* indicating that the two forces act in opposite directions.

And ∴ $$dv = \frac{P.\pi x^2.dx}{\eta.2\pi x.l} = -\frac{P.x.dx}{2\eta l}$$

Integrating this expression for dv, we have $v = -\dfrac{P}{2\eta l}\int x.dx.$

Or, $$v = -\frac{P}{2\eta l}\frac{x^2}{2} + C_1 = -\frac{Px^2}{4\eta l} + C_1,$$

where C_1, is a constant of integration.

Now, $v = 0$, when $x = r$, because the layer in contact with the sides of the tube is stationary.

$$\therefore \quad 0 = -\frac{Pr^2}{4\eta l} + C_1, \quad \text{whence,} \quad C_1 = \frac{Pr^2}{4\eta l}$$

$$\therefore \quad v = -\frac{Px^2}{4\eta l} + \frac{Pr^2}{4\eta l} = \frac{P}{4\eta l}(r^2 - x^2).$$

This, therefore, is the *velocity of flow of the liquid* at a distance x from the axis of the tube, and as a glance at the expression for it will show, the *profile* or the *velocity distribution curve* of the advancing liquid in the tube is a *parabola*, (Fig. 12.20) the velocity increasing from *zero* at the wall of the tube to a maximum along its axis.

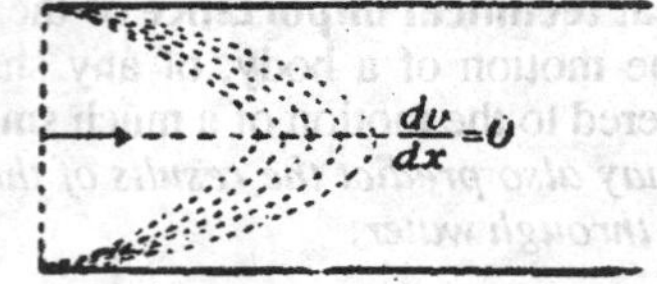

Fig. 12.20

Imagine another coaxial cylindrical shell of radius $(x + dx)$ enveloping the first. Then, the cross-sectional area between the two shells is clearly $2\pi x.dx$ and, since the velocity of the liquid there is v, we have *volume of the liquid flowing through this cross-sectional area* in 1 *sec.*, say, $dV = 2\pi x \,.\, dx \,.\, v$.

Imagine the whole of the capillary tube to consist of such like coaxial cylindrical shells, the *volume of the liquid flowing through the entire tube in one second* is given by

$$V = \int_0^r 2\pi x.dx.v = \int_0^r 2\pi x \,.\, \frac{P}{4\eta l} \,.\, (r^2 - x^2)\, dx \,.$$

$$= \frac{\pi P}{2\eta l}\int_0^r (xr^2 - x^3)\,.\,dx = \frac{\pi P}{2\eta l}\left[\frac{x^2r^2}{2} - \frac{x^4}{4}\right]_0^r$$

$$= \frac{\pi P}{2\eta l}\left(\frac{r^4}{2} - \frac{r^4}{4}\right) = \frac{\pi P.r^4}{8\eta l},$$

whence, $$\eta = \frac{\pi P r^4}{8\,Vl} \qquad ...(II)$$

Thus, if we know P, r, V and l, the coefficient of viscosity of the liquid (η) can be easily determined.

Alternative deductions of Poiseuille's equation.

1. Considering a cylindrical layer or shell of the liquid, coaxial with the tube and of radius x and thickness dx, we have, in accordance with *Newton's law of viscous flow, tangential viscous force acting on the inner side of the layer*, say,

$F = -\eta.2\pi x.l.dv/dx$

and tangetial viscous force acting on the outer side of the layer $= F + \dfrac{dF}{dx}\,.\,dx$

where dF/dx is the rate of change of force with distance x.

Since the layer on the inner side of the cylindrical shell is moving faster, and the layer on the outer side moving shower, than it, the shell experiences a forward push due to the former and a backward pull due to the latter.

$\therefore$ *resultant backward drag experienced by the shell* $= (F + \dfrac{dF}{dx}\,.\,dx) - F = \dfrac{dF}{dx}\,.\,dx$

As the liquid is neither accelerated nor retarded but continues to flow with the same steady velocity, it is clear that this backward dragging force on the shell is just balanced by the forward force $P \times 2\pi x.dx$ due to the pressure difference (P) across the two ends.

We thus have $\dfrac{dF}{dx}\,.\,dx = P \times 2\pi x.dx$. Or, $\dfrac{dF}{dx} = P \times 2\pi x$.

Since $F = -\eta\,.2\pi xl\,.\,dv/dx$, we have $\dfrac{dF}{dx} = \dfrac{d}{dx}\,(-\eta.2\pi xl\,.\,dv/dx) = P \times 2\pi x$.

which, on integration with respect to x gives

$-\eta\,.\,2\pi xl.dv/dx = \pi x^2\,.\,P + C\,.$ [P being independent of x.

where C is a constant of integration.

The velocity being the maximum along the axis of the tube, we have $dv/dx = 0$ when $x = 0$; so that $C = 0$, and we have $-\eta\,.\,2\pi xl.dv/dx = \pi x^2.P$. Or, $dv/dx = -Px/2\eta l$.

Integrating again, therefore, we have

$$v = -\frac{Px^2}{4\eta l} + C_1\,,$$ where C_1, is another constant of integration.

This yields $v = \dfrac{P}{4\eta l}\,.\,(r^2 - x^2)$ as in the first case (page 398)

And, proceeding in exactly the same manner as there, we have *rate of flow of liquid through the tube*, $V = \pi pr^4/8\eta l$.

2. **Method of dimensions.** The *rate of flow* (V) of a liquid in streamline motion through a horizontal capillary tube may reasonably be supposed to depend upon *radius* r of the tube, the *pressure gradient* P/l along it and the *coefficient* of viscosity η of the liquid. We may, therefore, put $V = Kr^a\,(P/l)^b\,\eta^c$, where K is a constat.

So that, taking dimensions, we have

$$L^3T^{-1} = L^a\left(\frac{MLT^{-2}/L^2}{L}\right)^b\left[ML^{-1}T^{-1}\right]^c.$$

Or, $L^3T^{-1} = M^{b+c}L^{a-2b-c}T^{-2b-c}$,

i.e., $b + c = 0, a - 2b - c = 3$ and $-2b - c = -1$,

which gives $a = 4, b = 1$ and $c = -1$.

Hence $$V = K.r^4(P/l)^1\eta^{-1} = K.\frac{r^4P}{\eta l}$$

The value of K, obtained experimentally, comes to be $\pi/8$,

And, therefore, $V = \pi P r^4/8\eta l$.

12.12. Criticism of Poiseuille's equation–Corrections to it. The assumptions made in deducing Poiseuille's equation for the rate of flow through a horizontal capillary tube (§ 12.11) are more or less valid *if the velocity of the flow be small* and *the tube be narrow*, for then alone a streamline motion of the liquid is possible. For tubes of wider bores, the value of critical velocity is small and is soon reached ($v_c \propto 1/r$ § 12.10), so that turbulence sets in and the relation breaks down. This may easily be seen by plotting a graph between the pressure difference P across the two ends of the tube and the *rate of flow V* of the liquid through the tube, as shown in Fig. 12.21.

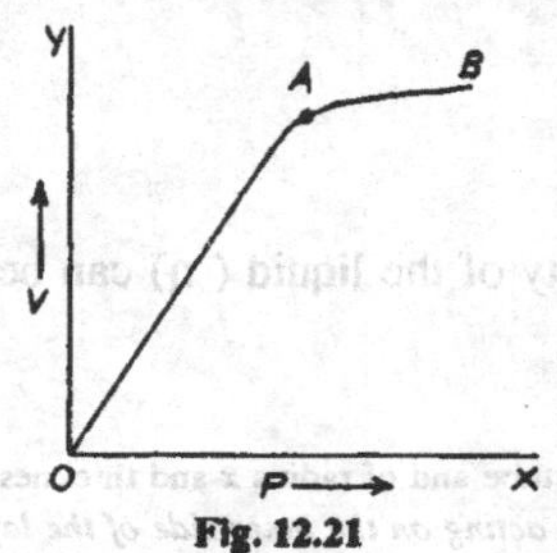

Fig. 12.21

The straight part OA of the graph corresponds to velocities below the critical value and clearly indicates that within the range the rate of flow V is proportional to the pressure difference P and *thus depends chiefly on the viscosity* (η) *of the liquid.* The curved part AB, on the other hand, which corresponds to velocities above the critical value, signifies that the rate of flow is no longer proportional to the pressure difference which is now almost wholly used up in combating the turbulence set up in the liquid and in imparting kinetic energy to it. The rate of flow thus no longer depends upon η, and is, in fact, found to depend mainly on the density (ρ) of the liquid and to be approximetely proportional to $\sqrt{P}$.

Then, again, the assumption regarding the layer of the liquid in contact with the wall of the tube remaining stationary is also found to be valid for really small velocities of flow.

It would thus appear that if the tube be really narrow and the velocity of flow of the liquid so small as to take the form of almost a trickle of drops, *Poiseuille's* equation will be fully applicable.

There are, however, two factors which have not been taken account of in deducing this equation : *(i) part of the thrust due to the pressure difference across the capillary tube imparts kinetic energy to the liquid and only the rest of it is available for overcoming the viscous resistance of the liquid; the effective value of the pressure difference is thus less than P, (ii) the motion of the liquid where it enters the capillary tube is somewhat accelerated with the result that the velocity of flow is not uniform over the first short length of the tube.* Proper corrections have, therefore, to be applied for these.

(*i*) Correction for effective pressure. Usually referred to as the *kinetic energy correction*, it may, following *Wilberforce*, be obtained as follows :

Considering a cylindrical layer or shell of the liquid of radius x and thickness dx, coaxial with the tube, we have *volume of liquid flowing along it in* 1 *sec* $= 2\pi x \,.\, dx \,.\, v$,

where v is the steady velocity of the liquid flow at distance x from the axis of the tube.

And $\therefore$ *mass of the liquid* $= 2\pi x.dx.v.\rho$, where ρ is the *density* of the liquid.

$\therefore$ kinetic energy of this liquid $= \frac{1}{2} \,.\, 2\pi x \,.\, dx.u.\rho.v^2$

$$= \pi x.dx.\rho.v^3 = \pi x.dx.\rho \left[\frac{P}{4\eta l}(r^2 - x^2)\right]^3$$

substituting the value of v. [§ 12.11, page 398.

So that, *K.E. imparted to the whole of the liquid flowing through the tube per second*

$$= \int_0^r \pi x.dx.\rho. \left[\frac{P}{4\eta l}(r^2 - x^2)\right]^3$$

$$= \pi.\rho.\left(\frac{P}{4\eta l}\right)^3 \int_0^r (r^2 - x^2)^3.x.dx = \pi\rho\left(\frac{P}{4\eta l}\right)^3 \cdot \frac{r^8}{8} = \left(\frac{\pi P r^4}{8\eta l}\right)^3 \cdot \frac{\rho}{\pi^2 r^4}$$

Since $\pi P r^4 / 8\eta l = V$, the rate of flow of the liquid, we have

K.E. imparted to the liquid flowing out per sec $= V^3\rho/\pi^2 r^4$

This energy is obviously imparted to the liquid at the expense of part of the pressure difference P across the two ends of the tube. Let the remaining or the effective pressure difference actually used for overcoming the viscous resistance of the liquid be P_1. Then, *work done in overcoming viscous resistance per sec* $= P_1V$.

$\therefore$ *total work done* $= P_1V + V^3\rho/\pi^2 r^4$

This must obviously be equal to PV. So that, we have

$$PV = P_1V + V^3\rho/\pi^2 r^4, \text{ whence, } P_1 = P - V^3\rho/\pi^2 r^4$$

This value (P_1) must, therefore, be substituted for P in relation *II* above for η.

(*ii*) Correction for acceleration imparted to the liquid at the inlet of the capillary tube. This correction consists in imagining the length of the tube to be increased 1.64 r in all cases. In other words, the *effective length of the tube* is taken to be (l + 1.64 r), where r is the radius of the tube.

Taking the above values for *effective pressure* and *effective length*, relation *II* above for η becomes

$$\eta = \frac{\pi(P - V^3\rho/\pi^2 r^4).r^4}{8V(l + 1.64r)} = \frac{\pi P r^4}{8V(l + 1.64r)} - \frac{V.\rho}{4\pi(l + 1.64r)}$$

12.13. Interesting consequences of turbulent liquid flow. From the discussion under § 12.12 *Poiseuille's equation* and *turbulence*, the following interesting facts emerge :

(*i*) Since in turbulent motion, the rate of flow of a liquid is quite independent of its viscosity, it obviously follows that *all liquids, irrespective of their different viscosities, would require the same pressure difference to be driven through a tube at velocities higher than their critical velocities.* Thus, for example, a viscous liquid, like treacle, would require the same pressure difference to be driven through a tube, at a velocity greater than its critical velocity, as would be needed to drive water through it at the same velocity.

(*ii*) Since the critical velocity of a liquid is inversely proportional to the radius of the tube through which it flows, it is clear that *liquids of all viscosities would flow equally readily through tubes of sufficiently wide bores.* Thus, in a wide tube, *treacle will flow just as freely as water*. A typical natural example of this is the *free flow of the highly viscous lava down the sides of an erupting volcano*, –its rate of flow being about the same as we would expect in the case of water.

Verify, in nature, *turbulent motion is the rule and steady or streamline motion, a rare exception.*

12.14. Experimental determination of η for a liquid—Poiseuille's method.* A capillary tube T, of known length l and radius r is fixed *horizontally* near to the bottom of a vessel A, (Fig. 12.22), the liquid level in which can be kept constant at any desired height by means of an overflow arrangement O. A clean and dry beaker, of known weight, is placed below the outer end of tube T to collect the liquid flowing out through it. The liquid is allowed to flow out in a slow trickle and collected in the beaker for a known time, and the beaker is then weighed again. The difference of the two weights gives the *mass* of the liquid flowing out in that time. Then, knowing the density of the liquid its volume can be determined and, dividing it by the time for which the liquid was allowed to flow, its volume V flowing out *per second* is known. Substituting the value of V, so obtained, in relation II, above, the coefficient of viscosity (η) of the liquid can be easily calculated.

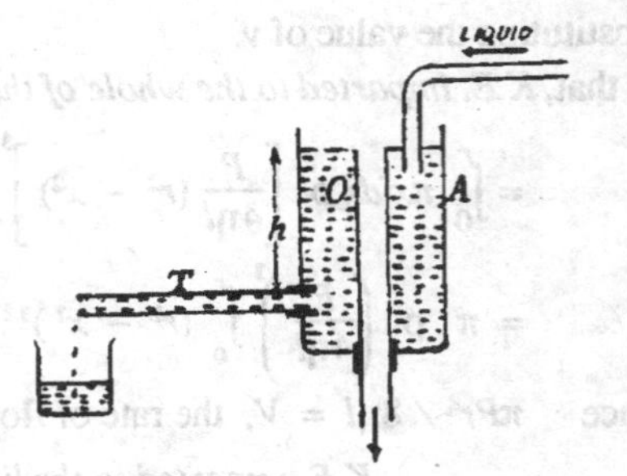

Fig. 12.22

For better accuracy of results, the two corrections mentioned in § 12.12 (for effective pressure across the two ends of the flow-tube and accelerated motion of the liquid in its earlier part) are applied, when we have

$$\eta = \frac{\pi P r^4}{8V(l + 1.64\, r)} - \frac{V.\rho}{8\pi\,(l + 1.64\, r)}.$$

A much better apparatus, however, is the following, in which the flow-tube F is a *long* one, and of a uniform circular cross-section, and the difference of pressure for a length AB of it is given directly by a manometer M, whose limbs are arranged over two fine holes at A and B, as shown, (Fig. 12.23), where A and B lie at a distance of at least 10 *cm* from the two ends of the flowtube respectively, so that the velocity of the out-flowing liquid becomes uniform near about them. This very much minimises the two sources of error referred to above —the second one almost completely. So that, with a slow rate of flow of the liquid and a fairly small size of the holes at A and B, no further corrections are necessary.

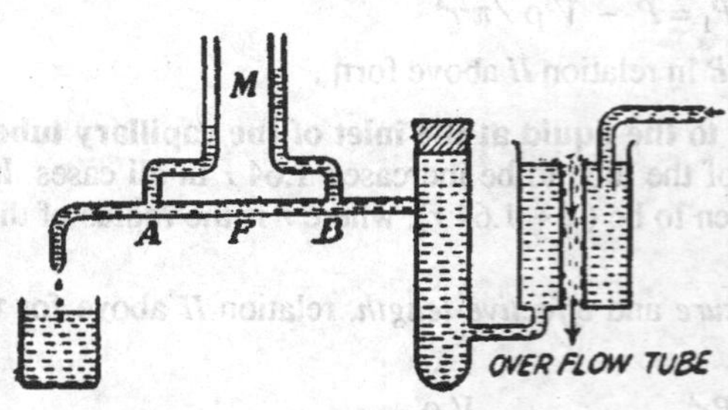

Fig. 12.23

Note. In either of these apparatus, it is essential that the outflow tube should have a *perfectly uniform* bore. The uniformity of the bore may be tested in a manner similar to that employed in construction of a mercury thermometer, *i.e.*, by introducing a small thread of mercury into the tube and measuring its length in the different parts of the tube. In no part should the length vary by more than 5%.

And, since the 4*th* power of the radius occurs in the formula for η, it should be determined most accurately. The tube is, therefore, properly dried and filled with mercury, and the length of the mercury thread measured most carefully by means of a vernier microscope, making the necessary correction for the curvature of the ends of the thread. The mercury is then taken out in a clean, dry and weighted watch glass and its mass determined as accurately as possible. Then, if m be its mass, ρ, its density at the then-temperature, l', the length of its thread in the tube, and r, the radius of the tube, we clearly have

$$\pi r^2 . l' . \rho = m, \quad \text{whence,} \quad r + \sqrt{m/\pi . l'} . \rho.$$

So that, knowing $m.l'$ and ρ, the accurate value of radius of the tube can be calculated.

12.15. Motion in a Viscous Medium—Stoke's law. When a body falls through a viscous medium, its motion is opposed by a force, *frictional in nature*, due to the fact that whereas the

* The method is suitable only for comparatively less viscous liquids like water.

layer of the liquid medium in immediate contact with it is carried along with it, that at an *infinite* distance from it is at rest. Energy is being continually absorbed by the medium and is converted into heat. Possibly also, eddy currents and waves are set up in the medium—particularly when the body moving is a fast one, like high speed cars and air-planes or projectiles—and these absorb still more energy. That is why cars etc., are streamlined these days to minimise the absorption of energy in this way. Even if the body be moving so slowly that no eddy currents or waves are set up, energy is still wasted due to the *viscous drag* it has to overcome.

Now, this opposing force, increases with the velocity of the body, until, in the case of small bodies, it becomes just equal to the *motive* or the *driving force*, and the body then attains a constant velocity, called its **terminal velocity.**

Stokes showed that the *retardation F*, due to the *viscous drag*, for a spherical body of radius r, moving with velocity v, in a medium whose coefficient of viscosity is η, is given by

$$F = 6\pi v r \eta.$$

This relation, known as Stokes' law, may be deduced as follows, by the method of dimensions :

For slow moving bodies, $F \propto$ *velocity v, radius r of the body and coefficient of viscosity* η *of the medium.* So that,

$$F = K v^a r^b \eta^c,$$

where K is a constant and a, b and c, the dimensional coefficients of v, r and η *respectively.*

Now putting the proper dimensions of the different terms, we have

$$[MLT^{-2}] = [LT^{-1}]^a [L]^b [ML^{-1}T^{-1}]^c$$

$$= M^c L^{(a+b-c)} T^{(-a-c)}$$

whence, (*i*) $c = 1$, (*ii*) $a+b-c = 1$, and (*iii*) $9-a-c = -2$,

Since, from relation (*i*), $c = 1$, we have, from relation (*iii*), $a = 1$; so that, from (*ii*), $b = 1$

$\therefore$ $F = k.v.r.\eta$; and the value of K was found by *Stokes* to be 6π. Hence $F = 6\pi v r \eta$, as stated above.

If the density of the spherical body be ρ,

$$\text{its } weight = volume \times \rho \times g = \frac{4}{3}.\pi r^3 \times \rho \times g,$$

And, if σ be the density of the medium, we have

$$upthrust\ on\ the\ body\ due\ to\ the\ displaced\ medium = \frac{4}{3}\pi r^3.\sigma.g.$$

$$\therefore\ resultant\ downward\ force\ on\ the\ body = \frac{4}{3}\pi r^3.\rho.g - \frac{4}{3}\pi r^2.\sigma.g$$

$$= \frac{4}{3}\pi r^3.g.(\rho - \sigma).$$

Equating this against the value of F, we have

$$6\pi v r \eta = \frac{4}{3}\pi r^3.g(\rho - \sigma), \quad \text{whence, } v = \frac{4}{3}\pi r^3.g.(\rho - \sigma)/6\pi r\eta.$$

Or, $$v = \frac{2}{9}.\frac{r^2.g(\rho - \sigma)}{\eta}. \quad ...(i)$$

Thus, the *terminal velocity* of a body, (of course, of a small size), falling through a viscous medium, is (*i*) *directly proportional to the square of its radius* (r^2), (*ii*) *directly proportional to the difference in the densities of the body and the medium,* ($\rho - \sigma$), and (*iii*) *inversely portional to the coefficient of viscosity of the medium* (η).

In arriving at the above result, *Stokes* made the following *assumptions :*

(*a*) That the medium through which the body falls is *infinite in extent.*

(*b*) That the spherical body is *perfectly rigid and smooth.*

(*c*) That there is *no slip between the spherical body and the medium.*

(*d*) That the medium is *homogeneous*, so far as the spherical body is concerned, *i.e.*, the diameter of the spherical body is large compared with the spaces between the molecules of the medium.

(*e*) That there are *no eddy currents or waves set up in the medium due to the motion of the body through it;* —in other words, that the body is moving very slowly through it, or that the motion of the medium is smooth and not turbulent. *Stokes* found that the relation holds good only when v is smaller than $\eta/\sigma r$*, called the critical velocity.

A striking example of a body falling through a viscous medium is that of the tiny drops that form what we call '*clouds*'. These tiny drops of water have a radius as small as .001 *cm.* and their terminal velocity, as they fall through air, for which η = .00018, comes to about 1.2 *cm/sec*, [from relation (*i*) above]. That is why they remain suspended in the air and appear to us to be floating about as clouds.

Bigger rain drops, on the other hand, have a radius about 10 times as great (*i.e.*, = .01 *cm*) and their terminal velocity, therefore comes to about 120 *cm/sec* ; so that, they *fall* through the air, instead of floating in it, (v being proportional to r^2).

Also, if the density of the medium in which the body falls be greater than that of the body itself, *i.e.*, if $\sigma > \rho$, it is clear that the terminal velocity v will have a *negative value*. In such a case, therefore, the body will have an upward terminal velocity. That is why bubbles of air or gas can rise up through water or any other liquid,—the smaller the bubble, the smaller its velocity.

The case finds a useful application in (*i*) determining the coefficient of viscosity of a liquid (see § 12.14) (*ii*) determining the radius of small spherical objects like *rain drops* and *ball bearings* etc., and (*iii*) determining the electronic charge by Millikan's falling drops method.

12.16. Determination of coefficient of viscosity of a liquid —Stokes' falling body *viscometer*. The relation for v obtained above, (page 403), has been used to determine the viscosity of a liquid. The method consists in finding the time of fall of small spheres, such as ball-bearings etc., in the liquid, and then to apply *Stokes*' relation,

$$v = \frac{2}{9}.\frac{r^2.g(\rho-\sigma)}{\eta}, \text{ whence, } \eta = \frac{2}{9}.\frac{r^2 g(\rho-\sigma)}{v}. \qquad ...(ii)$$

Care must, however, be taken to see that the velocity of fall does not exceed. $6\eta/\sigma r$, or else the above relation cannot be used. Thus, for thin machine oils, for which = 1 *c.g.s unit i.e.*, 1 *poise* nearly, spheres of 1 *mm* radius are the maximum size that can be used. For thick oils, or liquids like glycerine, whose coefficient of viscosity is greater than 10 *c.g.s. units*, steel ball-bearings of about $\frac{1}{4}''$ diameter, are the largest that can be employed. For less viscous liquids, much smaller spheres are required, and these may be obtained, in any desired size, from radii of .0002 to 1 *cm* by flowing melted *Wood's metal* into cold water through capillary tubes of different bores. The diameters of these spheres, and hence their radii, are measured accurately by means of a microscope, fitted with a micrometer eye-piece.

The liquid is taken in a tall jar, of a large diameter, and the time taken by a sphere of suitable size, dropped *centrally* into it, to pass three marks at different levels is noted and the velocity calculated for each of the two distances between he three marks. If the velocity over each path is the *same*, it means that it has acquired its constant velocity, or terminal velocity, v. If, however, the velocities be different, being greater over the lower track than over the upper one, a smaller ball-bearing must be tried, until the velocities are the *same*, within experimental error.

---

* It was shown by **Arnold**. however, that in actual practice, v should be less than $6\eta/\sigma r$.

In actual practice, we have two marks A and B, [Fig. 12.24(*a*)] some distance (say, 10 to 12 *cm*) below the top and above the bottom of the jar respectively, and allow small spheres of different known radii to fall through the liquid *centrally*, noting the time taken by each to cross the distance S between the two marks A and B, it being assumed that, due to the small size of the sphere, it has already acquired its terminal velocity before crossing the mark A. Thus, since it moves with a constant velocity over this distance S, its terminal velocity, $v = S/t$.*

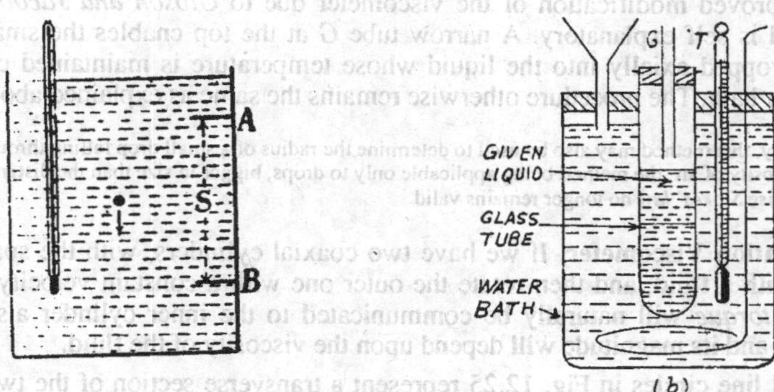

Fig. 12.24

Then, putting $v = S/t$ in the above expression for η, we have

$$\eta = \frac{2}{9} \cdot \frac{r^2 g (\rho - \sigma).t}{S}.$$

Now, since r^2t is constant in the case of a given sphere and a given liquid, provided its temperature remains the same, we plot r^2 for a number of spheres against their corresponding values of $1/t$. The slope of the *straight-line curve*, thus obtained, gives the mean value of $r^2.t$, and this value is then substituted in the above expression for η.

Further, since viscosity of a liquid depends upon its temperature, (see §12.18), and the rate of change is fairly rapid in the case of liquids like oils, a sensitive thermometer must also be put into the jar of the liquid to indicate its temperature, and the experiments with the different spheres performed in *quick succession*.

Corrections. (*i*) *For wall effect*. For greater accuracy, a proper correction must be applied for the finite size of the containing vessel, as the relation for η above was deduced by *Stokes* on the assumption that the medium is infinite.

Ladenburg has shown that if the liquid column in the tall jar be supposed to be divided up into three equal compartments, *lengthwise*, and the sphere be dropped into the central compartment, the velocity v_∞ of the sphere, as it would be in the liquid medium, if it were unbounded by the walls of the containing jar, is given by the relation $v_\infty = v\,(1+2.4r/R)$, where v is the observed velocity of the sphere and R, the radius of the jar. This is called the correction for the *'wall-effect'*.*

(*ii*) *For end effect*. Similarly, he has shown that if the liquid medium were unhindered by the bottom or the end of the containing jar, the velocity of the sphere would be given by

$$v_\infty = v\,(1 + 3.3r/\mu),$$

where μ is the *full depth of the liquid column* in the jar. This is called the correction for the *'end-effect'*.

* A still more accurate correction of this effect is given by *Faxen* in the form

$$v = v_\infty\left[1 - 2{\cdot}104\left(\frac{r}{R}\right) + 2.1\left(\frac{r}{R}\right)^3 - 0.95\left(\frac{r}{R}\right)^5\right],$$ which has been verified by *Bacon*.

Combing the two corrections, therefore, we have the following relation for η, *viz.*,

$$\eta = \frac{2}{9} \cdot \frac{r^2 . g\,(\rho - \sigma)}{v\left(1 + 2.4\dfrac{r}{R}\right)\left(1 + 3.3\dfrac{r}{\mu}\right)} . \qquad ...(iii)$$

A later improved modification of the viscometer due to *Gibson and Jacobs*, is shown in Fig.12.26 (*b*) and is self explanatory. A narrow tube *G* at the top enables the small sphere (ball bearing) to be dropped axially into the liquid whose temperature is maintained constant by the surrounding water bath. The procedure otherwise remains the same as explained above.

N.B. Obviously, this method may also be used to determine the radius of a small drop falling through air, if we know the coefficient of viscosity of air, the method being applicable only to drops, bigger in size than the distance between the air molecules, for otherwise *Stokes' law* no longer remains valid.

12.17. Rotation Viscometer. If we have two coaxial cylinders, with the space in between them filled up with a fluid, and then rotate the outer one with a constant velocity Ω about their common axis, a *torque* will naturally be communicated to the inner cylinder also through this intervening fluid, and its magnitude will depend upon the viscosity of the fluid.

Let the full line circles in Fig. 12.25 represent a transverse section of the two cylinders, of radii *a* and *b*, with their common axis perpendicular to the plane of the paper, and passing through *O*.

Then, as the outer cylinder rotates with its small angular velocity Ω (*within the limits of a streamline flow*) about the inner cylinder, the layers of the fluid in contact with the former also rotate with it with its own velocity Ω, the velocity of the other layers progressively decreasing as we move inwards, until, for the layers in contact with the inner cylinder, it is reduced to *zero*. There is thus brought about a relative motion between the different layers of the fluid.

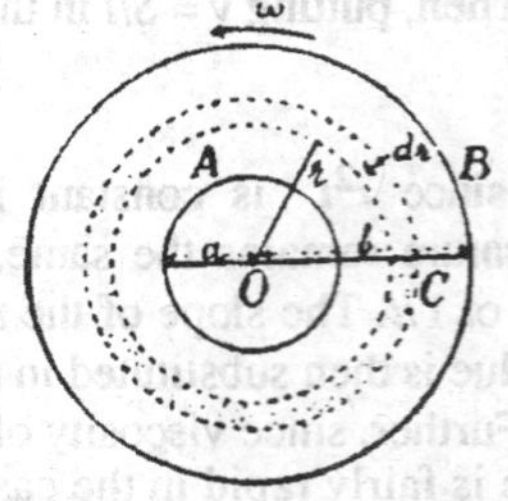

Fig. 12.25

Let us consider a coaxial *cylindrical layer C* of the fluid, at a distance *r* from *O*, and of thickness *dr*. If it be rotating with an angular velocity ω, its *linear velocity* is clearly $r\omega$. And, therefore, if its length or height be *l*, the *viscous force F* acting on it, in accordance with Newton's Law, (see page 394), is given by the relation, $F = its\ area \times \eta \times dv/dr = 2\pi rl.\eta.dv/dr$, where η is the *coefficient of viscosity* of the fluid and dv/dr, the *velocity gradient* at distance *r* from *O*.

Now, $dv/dr = d(r\omega)/dr = \omega + r.d\omega/dr$.

And, since ω is a constant quantity and would represent the velocity of the layer in the absence of any viscous slip, (*i.e.*, if it were to rotate like a rigid body), it does not contribute at all towards the velocity gradient; and we may, therefore, take the velocity gradient, responsible for the viscous drag on the imaginary cylindrical layer *C*, to be simply equal to $r.d\omega/dr$. Hence, this viscous drag or force $F = 2\pi r.l.\eta.r.d\omega/dr$. And, therefore, *moment of this force*, or the *torque T*, acting on the layer *C*, is clearly equal to $r.2\pi r.l.\eta.r.d\omega/dr$. Or, $T = 2\pi r^3.l.\eta.d\omega/dr$, which may be put as $2\pi.l.\eta.d\omega/T = dr/r^3$.

So that, the effect of the whole fluid in between *A* and *B* is obtained by integrating this expression for the limits $\omega = 0$ and $\omega = \Omega$ and $r = a$ and $r = b$. Thus, we have

$$\int_0^{\Omega} \frac{2\pi.l.\eta}{T}\,.\,d\omega = \int_b \frac{dr}{r^3}. \quad \text{Or,} \quad \frac{2\pi.\eta.\Omega}{T} = \frac{1}{2}\left(\frac{1}{a^2} - \frac{1}{b^2}\right),$$

whence, $$T = \frac{4\pi l\eta.\Omega}{(1/a^2 - 1/b^2)} = \frac{4\pi l.\eta\ .\ \Omega\ .a^2.b^2}{b^2 - a^2} \qquad ...(i)$$

The torque T tends to accelerate the motion of the fluid in between C and A. But this fluid being in a state of *steady motion*, the inner cylinder A must also be exerting an equal retarding torque on the fluid in contact with it. And, since action and reaction are equal and opposite, it follows that an equal and opposite torque T is also exerted on the inner cylinder A, tending to rotate it through an angle θ, say, until it is just balanced by a *restoring torsional torque* equal to T, set up in the suspension wire, carrying it.

Now, if C be the torsional couple, set up in the suspension wire per unit twist of it, we have $T = C\theta$. So that,

$$C\theta = \frac{4\pi.l.\eta,\Omega\ .\ a^2\ .b^2}{(b^2 - a^2)}, \text{ whence}, \eta = \frac{C.(b^2 - a^2).\theta}{4\pi.l.\Omega.a^2.b^2}. \qquad ...(ii)$$

In the above discussion, we have not taken into account the torque on the base of the inner cylinder*, so that if this torque be T_b, we have

total torque on the fluid between A and B given by

$$T_1 = C\theta = \frac{4\pi.l.\eta.\Omega.a^2.b^2}{(b^2 - a^2)} + T_b \qquad ...(iii)$$

In actual practice, we eliminate T_b altogether, by repeating the experiment with a different length or height l' of the fluid in between A and B. So that, if now the total torque be T_2 and the angle of rotation of the inner cylinder be θ' we have

$$T_2 = C\theta' = \frac{4\pi.\eta.\Omega.a^2.b^2.l'}{(b^2 - a^2)} + T_b. \qquad ...(iv)$$

∴ subtracting relation (*iv*) from (*iii*), we have

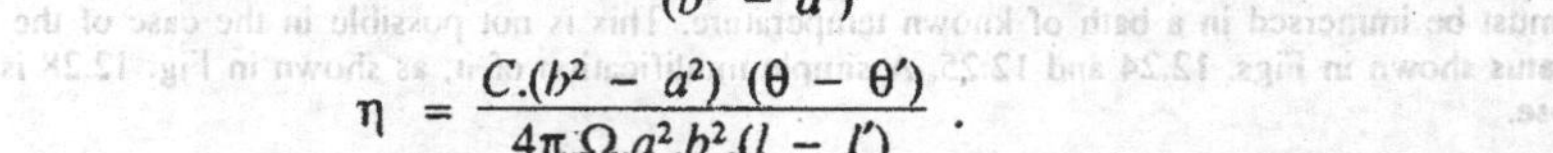

$$(C\theta - C\theta') = C(\theta - \theta') = \frac{4\pi.\eta.\Omega.a^2.b^2}{(b^2 - a^2)}(l - l'),$$

whence, $$\eta = \frac{C.(b^2 - a^2)(\theta - \theta')}{4\pi.\Omega.a^2.b^2.(l - l')}. \qquad ...(v)$$

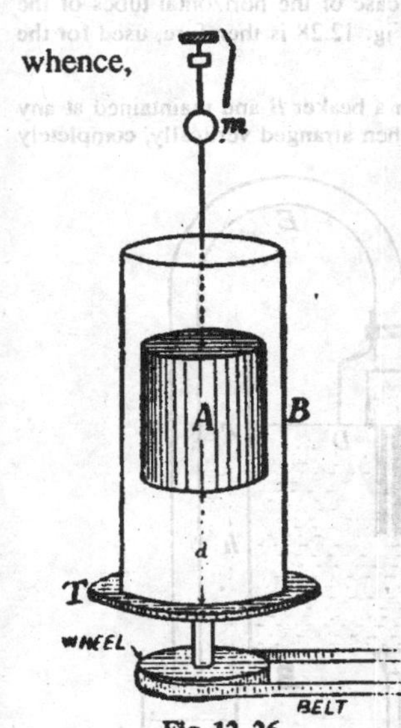

Fig. 12.26

The essential parts of the apparatus used are shown in Fig. 12.26, where the outer cylinder B is clamped on to a table T, so as to be coaxial with its spindle, which can be rotated by a small electric motor at a speed of 20 to 60 rotations *per minute*.

The inner cylinder A is suspended inside B by means of a *long* and *thin* suspension wire, which carries a small mirror m, to enable its angle of rotation, (caused by the constant rotation of B), to be noted accurately by the usual scale and telescope arrangement.

The value of C, the torsional couple per unit twist of the suspension wire, is determined by first setting the inner cylinder alone into torsional vibrations about it, and then with a hollow, metallic disc, of a known moment of inertial I', placed centrally upon it, and noting their time-periods t and t' *respectively*.

Then, as we know,

$$t = 2\pi\sqrt{I/C} \text{ and } t' = 2\pi\sqrt{(I + I')/C},$$

where I is the $M.I.$ of the inner cylinder alone about the suspension wire.

* This torque on the base of the inner cylinder depends not only upon the radii a and b of the two coaxial cylinders, but also upon distance between their bases.

So that, $(t'^2 - t^2) = \dfrac{4\pi^2(I + I' - I)}{C} = \dfrac{4\pi^2 I'}{C}$

whence, $C = 4\pi^2 I'/(t'^2 - t^2)$.

This value of C is then substituted in relation (v) above, when

we have $$\eta = \frac{\pi I'(b^2 - a^2)(\theta - \theta')}{\Omega . a^2 b^2 (t'^2 - t^2)(l - l')}.$$

It must be emphasized again, that the speed of rotation of the outer cylinder must be kept low, or else the fluid-flow becomes turbulent and T is then no longer proportional to Ω. This may be clearly seen from the accompanying curve between Ω and T/Ω (Fig. 12.27). It will be noted that T/Ω remains constant up to Q, beyond which it varies, as shown by the dotted curve, in a somewhat uncertain and irregular manner, and at higher values of Ω, the relation takes the form $(\alpha.\Omega + \beta . \Omega^2)$, where α and β are constants.

Note. The rotation viscometer may be used for the determination of η of both liquids and gases. Only, whereas for liquids, we take two observations with different heights l and l' of the liquid in the outer cylinder, in the case of gas, we use two different inner cylinders, of lengths l and l', the distance between their bases and that of the outer cylinder being kept the same in either case.

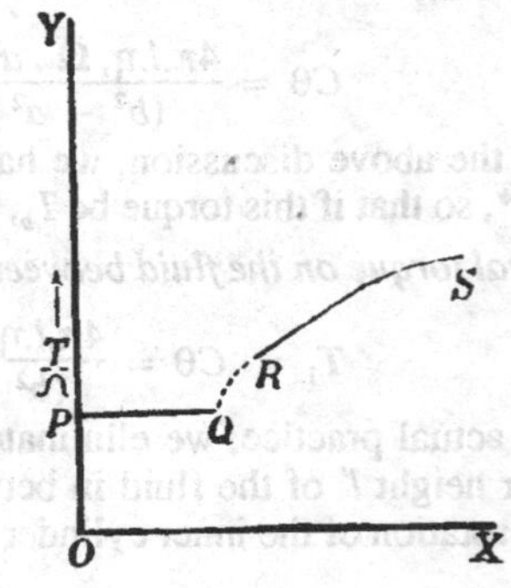

Fig. 12.27

12.18. Variation of viscosity of a liquid with temperature. As we have seen above, to determine the viscosity of a liquid we have to measure its rate of flow through a capillary tube (Figs. 12.22—12.23).

If, therefore, we wish to see how the viscosity of a liquid is affected by a change in its temperature, the capillary tube must be immersed in a bath of known temperature. This is not possible in the case of the horizontal tubes of the apparatus shown in Figs. 12.24 and 12.25. A simple modification of it, as shown in Fig. 12.28 is therefore, used for the purpose.

Here, the liquid, whose coefficient of viscosity is to be determined, is taken in a beaker B and maintained at any desired temperature, which can be noted on the thermometer T. A capillary tube C is then arranged vertically, completely immersed in the liquid, and is connected to a wider tube DEF, bent as shown, to siphon the liquid over *through it* into a weighed vessel J.

The method of procedure is the same as in the case of the horizontal capillary tube, with the difference that here the flow of the liquid through C is caused by the liquid column DF or the head of liquid h; for, clearly, no liquid will flow if the end D of the capillary tube were open to the atmosphere. The pressure due to the column h is, therefore, the effective pressure difference P responsible for the flow of the liquid through C.

Now, as the liquid flows through the tube, the liquid level falls in B, thus altering the head of liquid h. To maintain it constant throughout the experiment, B is gradually raised as the liquid flows through C ad DEF, so that the liquid level in B is always maintained at D. This is easily done by attaching a bent pin to the tube, as shown, and making the liquid level in B always touch its tip at D.

The liquid is thus allowed to flow into vessel J for a known time t and its mass determined; then, dividing it by its density and the time t (in *seconds*), the volume blowing out *per second, i.e.*, its rate of flow V is determined. Substituting the values of V, P, r (radius of the capillary tube C) and l (length of the tube C) in Poiseuille's formula $\eta = P\pi r^4/8Vl$, we can easily calculate the value of η for the liquid at the temperature of the bath.

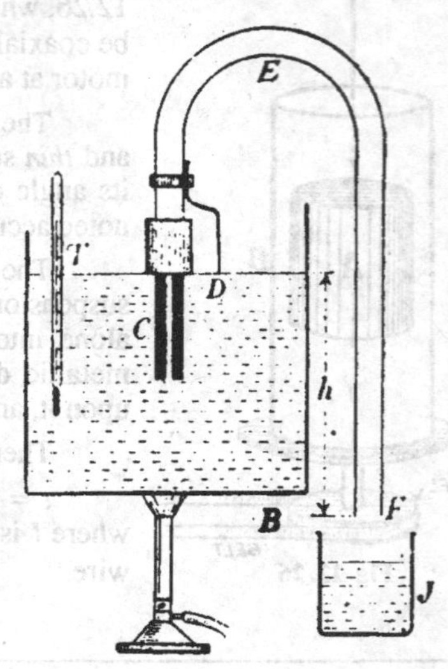

Fig. 12.28

The experiment is repeated with the liquid at different constant temperatures and a graph plotted between t and η. It will be found that η *usually* falls with a rise in temperature of the liquid, though there is no definite or universal relationship between the two, —the variation being more pronounced in some cases than in others.

12.19. Comparison of Viscosities —Ostwald Viscometer. The viscosities of two liquids, or the same liquid at different temperatures, may be conveniently compared with the help of the simple apparatus, shown in Fig. 12.29, and known as the *Ostwald Viscometer*, a well known form of the so called *commercial viscometers*, which avoid the exact measurement of the dimensions of the viscometer, so tedious, yet so essential, for the absolute determination of viscosities.

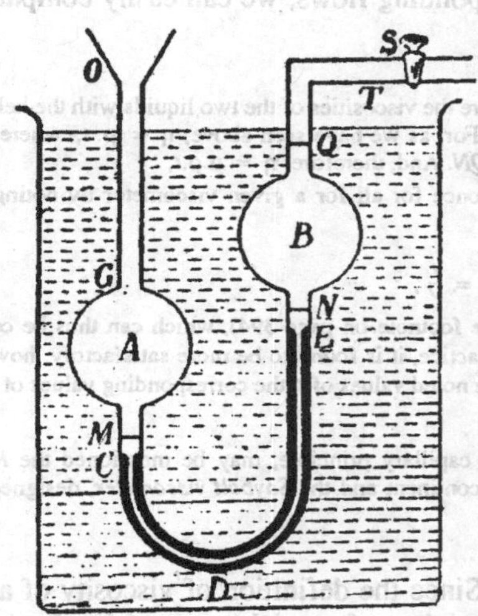

Fig. 12.29

As will be readily seen, this viscometer consists of a U-shaped tube *OGDEQT*, with two bulbs *A* and *B*, a capillary portion *CDE*, a side-tube *T*, fitted with a tap *S*, and with three marks engraved on it at *M*, *N* and *Q* respectively. The whole apparatus can be kept immersed in a suitable bath, to keep its temperature constant.

Each liquid is introduced into bulb *A* through the open end *O* of the tube, until, with the tap *S* closed, it occupies the space *QDM* of its bent portion, thus ensuring that the same volume of the liquid is taken in each case.

Now, first with one liquid thus taken into the tube, some air is abstracted from it by connecting *T* to a suitable pump through the tap *S*, until the liquid column rises above the level of *Q*, and the tap *S* is then duly closed. The tap is now opened again, when, due to the increased pressure on it, the liquid flows from the portion *QE* of the tube into the portion *CG*. The time when the liquid menisicus *just* passes the mark *Q* downwards is carefully noted, and again when it *just* passes the mark *N*. The difference gives the time t, say, taken by the liquid to flow from the position *QM* into the position *NG*, *its rate of flow being determined by the capillary portion EC* of the tube, the wider parts of the tube hardly affecting it, because, as we know, the retarding force due to viscosity varies inversely as the fourth power of the radius of the tube. Thus, if V be its *rate of flow*, we have $V = \pi P r^4/8\eta l$,
where l is the length of the capillary portion EC; P, the pressure difference across it and η, the coefficient of viscosity of the liquid.

The experiment is then repeated with the second liquid, of coefficient of viscosity η', (*at that very temperature*). Then, if V' be its rate of flow, we have $V' = \pi P' r^4/8\eta' l$.

Now, although the pressure difference keeps on changing during the flow of the liquid it is proportional to the density of the liquid for every position of it. And, therefore, if in the corresponding positions of the two liquids, of densities ρ *and* ρ^1 respectively, the pressure differences be P and P', we have

$$P \propto \rho \text{ and } P' \propto \rho'$$

And $\therefore$ *of the first liquid*, $V \propto \pi \rho r^4 / 8\eta l$,

and *for the second liquid*, $V' \propto \pi \rho' r^4 / 8\eta' l$

So that, $$\frac{V}{V'} = \frac{\pi \rho r^4/8\eta l}{\pi \rho' r^4/8\eta' l} = \frac{\rho}{\rho'} \cdot \frac{\eta'}{\eta} \qquad ...(i)$$

Again, the rates of flow, V and V', also vary (because of the variation in the pressure difference P and P) during the flow of the liquid, but the times for the flow of *unit volume* of the two liquids in *corresponding positions*, vary inversely as V and V', and it follows, therefore, that the total times t and t', for the flow of the two liquids respectively must likewise vary inversely as V and V'. In another words, $t/t' = V'/V$

From relations (*i*) and (*ii*), therefore, we have

$$\frac{\rho' . \eta}{\rho . \eta'} = \frac{t}{t'}, \text{ whence } \eta = \frac{\rho}{\rho'} . \frac{t}{t'}$$

Thus, knowing ρ and ρ', the densities of the two liquids, (or of the same liquid at different temperatures), and t and t', the times for their corresponding flows, we can easily compare their coefficients of viscosities, η and η'.

N.B. It will be seen that we may not merely be able to compare the viscosities of the two liquids with the help of this viscometer, *but may also determine the viscosity of a given liquid.* For, as we have seen above, $\eta \propto \rho . t$, where ρ is the density of the liquid and t, the time for its flow through the distance QN. And, therefore, $\eta = \alpha . \rho . t$,

where α is a constant of the viscometer and may be determined once for all for a given viscometer by noting t for a liquid of known density (ρ) and viscosity (η).

Or, from relation (*iii*) above, we have $\frac{\eta/\rho}{\eta'/\rho'} = \frac{t}{t'}$. Or, $\frac{k}{k'} = \frac{t}{t'}$,

where k and k' are the *kinematic viscosities* of the two liquids (see footnote on page 394), which can thus be compared easily. In fact, this was the relation actually used by *Ostwald*. In practice, it is found to be more satisfactory, however, to plot a curve between k and t for a number of liquids, so that, from the noted values of t, the corresponding values of k can be obtained straightway.

Among other commercial viscometers, based on a similar capillary principle, may be mentioned the *Redwood viscometer*, used in England, *Engler viscometer*, used all over the continent and the *Saybolt viscometer*, designed by the Standard Oil Company, and used in America.

12.20. Viscosity of Gases—Meyer's formula. Since the definition of viscosity of a liquid deduced in § 12.9, applies equally well to a gas, it might at first sight appear that Posieuille's formula for the rate of flow of a liquid should also apply in the case of a gas. But the snag is that whereas the density of a liquid is practically independent of the pressure on it, (liquids being almost incompressible), that of a gas varies directly with it; and hence whereas in the case of a liquid, the *volume* (as well as the mass) of it flowing through any section of the tube in a given time can be taken to be constant, in the case of a gas, it is the *mass* of it (and *not the volume*), flowing across a section of the tube in a given time, that alone can be taken to be constant (see § 12.3).

Thus, if V be the volume of a gas flowing across a section *per second*, at a distance x from the inlet-end of the tube, and ρ, its density at the uniform pressure P over that section, we have

$$\rho V = \text{a constant. And, since } \rho \propto P, \quad \text{we have } PV = \text{a constant.}$$

Now, if we consider a section dx of the tube at distance x from the inlet-end, with a pressure difference dP across it, we have, in accordance with Poiseuille's formula, the volume of the gas flowing per second through the section given by

$$V = -\frac{\pi . a . P . r^4}{8 dx . \eta} = -\frac{\pi r^4}{8\eta} . \frac{dP}{dx},$$

the *negative sign* merely indicating that the pressure P decreases as distance x increases.

Or, $$PV = \frac{\pi r^4}{8\,\eta} . \frac{P . dP}{dx} = a\ constant.$$

So that, if V_1 be the volume of the gas entering the tube at pressure P_1, we have $P_1 V_1 = PV$, the cross section of the tube being constant.

Or, $P_1V_1 = -\dfrac{\pi r^4}{8\eta}\cdot\dfrac{P.dP}{dx}$. Or $P_1V_1.dx = -\dfrac{\pi r^4}{8\eta}.P.dP.$

Integrating this expression for the limits $x = 0$ and $x = 1$ and $P = P_1$ and $P = P_2$, where l is the length of the tube and P_1 and P_2, the pressures at its inlet and outlet ends, we have

$$P_1V_1\int_0^l dx = -\frac{\pi r^4}{8\eta}\int_{P_1}^{P_2} P.dP. \quad \text{Or,} \quad P_1V_1.l = \frac{\pi r^4}{8\eta}.\frac{(P_1^{\,2} - P_2^{\,2})}{2}$$

Or, $$P_1V_1 = \frac{\pi r^4(P_1^{\,2} - P_2^{\,2})}{16\eta l}\,* \qquad ...(I)$$

referred to as **Meyer's formula** *for gaseous flow through a capillary tube.*

(*i*) Grindley and Gibson's method. The above method was actually used by *Grindley* and *Gibson* who noted the difference of pressure between the two ends of a flow tube through which the gas was made to flow from one container to another, by forcing water into the former. The volume of the gas passing through per second could thus be easily determined and η for the gas evaluated from relation *I* above.

(*ii*) Rankine's method. The simple apparatus used by *Rankine* for determining the coefficient of viscosity of a gas is shown in Fig. 12.30, where *ABCD* is a glass tube, *one meter* in length, having a *capillary section AB* (of about 2 *mm* bore) and fitted with two stop cocks *I* and *O*, which serve as inlet and outlet for the gas respectively. A small mercury pellet *M* is introduced into the part of the tube opposite to the capillary section *AB*.

Fig. 12.30

When the tube is held vertically in the position shown, the mercury pellet starts falling down under its own weight, forcing some of the gas into the capillary *BA*, and its rate of flow is observed by noting the time taken by the pellet to fall through a measured distance. This is then equated against its calculated rate of flow, whence the value of η for the gas can be easily obtained.

To start with, suppose the tube is laid horizontally on the table, so that the gas acquires a uniform pressure *P* all along the tube. Then, if ρ be the density of the gas *at unit pressure*, and if the total volume of gas enclosed in the tube be *V*, we have

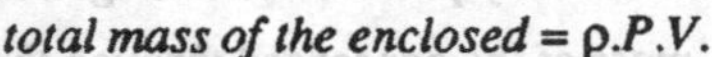

total mass of the enclosed = $\rho.P.V.$

Let the tube be now held vertically, so that the mercury pellet starts falling down. Let it be in the position *M* at any given instant and let the volume of the gas *above it be* v_1 . Then, if p_a be the pressure of the gas *above M*, it is clear that the pressure of the gas *below M* will be $p_a + mg/a$, where *mg* is the weight of the mercury pellet and *a*, the *area of cross-section of tube DC*. Thus,

mass of the gas above M = $\rho.p_a.v_1$

* In deducing this relation, we have assumed that, as in the case of a liquid, so also here, the gaseous layer in contact with the wall of the tube remains stationary. Experimental evidence shows, however, that in the case of a gas there is 'slipping', with the result that the radius of the tube is effectivly increased by λ, called slipping coefficient for the gas and equal to, the *mean free path* for it. We should therefore, substitute $(r+\lambda)$ for *r*, when we obtain the relation

$$P_1V_1 = \frac{\pi(r + \lambda)^4\ (P_1^{\,2} - P_2^{\,2})}{16\,\eta l}$$

Or, since $(r + \lambda)^4 \approx r^4\ (1 + 4\lambda/r)$, we have

$$P_1V_1 = \frac{\pi r^2\,(P_1^{\,2} - P_2^{\,2})}{16\eta l}\cdot\left(1 + 4\frac{\lambda}{r}\right)$$

and $$\textit{mass of the gas below } M = \rho(p_a + mg/a)(V - v_1),$$

where $(V - v_1)$ is the volume of the gas below M.

$\therefore$ $$\textit{total mass of the gas in the tube} = \rho.p_a.v_1 + \rho(p_a + mg/a)(V - v_1).$$

Or, $$\rho.P.V = \rho.p_a v_1 + \rho(p_a + mg/a)(V - v_1).$$

which solves out to $p_a = P - \dfrac{mg}{a} + \dfrac{mg}{a}.\dfrac{v_1}{V}$. ...(i)

And, if p_b be the pressure of the gas below M, we have

$$p_b = p_a + \frac{mg}{a} = \left(P - \frac{mg}{a} + \frac{mg}{a}.\frac{v_1}{V}\right) + \frac{mg}{a}$$

Or, $$P_b = P + \frac{mg}{a}.\frac{v_1}{V}. \quad \text{...(ii)}$$

Let the mercury pellet fall to position M' so that the volume of the gas above M now becomes v_2. Then, if the pressure of the gas now becomes p'_a, here, we have, as before,

$$p'_a = P - \frac{mg}{a} + \frac{mg}{a}.\frac{V_2}{V}. \quad \text{...(iii)}$$

And, if p_b be now the pressure of the gas below the mercury pellet, we have

$$p'_b = P + \frac{mg}{a}.\frac{v_2}{V}. \quad \text{...(iv)}$$

Now, with the mercury pellet at M, the mass of the gas below it up to B is equal to $\rho.p_b(V - v_1)$ and that of the gas below it, when it takes up the position M', is equal to $\rho.p'_b.(V - v_2)$. So that, the difference of these two gives the mass of the gas forced into the capillary tube by the fall of the mercury pellet from M to M'.

Thus, *mass of the gas forced through the capillary tube*

$$= \rho.p_b.(V - v_1) - \rho.p'_b(V - v_2).$$

If the position of the mercury pellet be so arranged that $(v_1 + v_2) = V$, we have

$$\textit{mass of the gas forced through the capillary} = \rho.P.(v_2 - v_1).$$

Hence, if t be the time taken by the mercury pellet to traverse the distance MM', we have

$$\textit{mass of the gas flowing per unit time through the capillary tube} = \rho.P.(v_2 - v_1)/t. \quad \text{...(I)}$$

Now, if v be the volume of the gas flowing per unit time when the pellet is at M, we have

$$\frac{(p_b^2 - p_a^2).\pi r^4}{16\eta l} = p_b.v,$$ [See relation *I*, (page 411).

where r is the *radius* and l the *length* of the capillary section.

And, similarly, if v' be the volume of the gas flowing per unit time when the pellet is at M', we have

$$\frac{(p'^2_b - p'^2_a).\pi r^4}{16\eta l} = p'_b.v'$$

Hence, *the average rate of flow of the gas is proportional to*, $[(p_b^2 - p_a^2) + (p'^2_b - p'^2_a)]/2$ which works out to be equal to $2mg.P/a$. So that, mass of the gas flowing per unit time through the capillary tube.

$$= \rho \times 2mg \cdot \frac{P}{a} \cdot \pi r^4/16\eta l. \qquad ...(II)$$

Thus, from relations (*I*) and (*II*), we have

$$\rho \times \frac{2mg.P}{a} \cdot \frac{\pi r^4}{16\eta l} = \frac{\rho.P(v_2 - v_1)}{t},$$

whence, $$\eta = \frac{mg.\pi r^4.t}{8a.l(v_2 - v_1)},$$

from which the coefficient of viscosity of the given gas may be easily determined.

As can be readily seen, the apparatus may also be used to study the effect of pressure on the viscosity of a gas. Rankine determined the viscosities of many gases with the help of this apparatus and showed that viscosity is quite independent of pressure, as predicted by the kinetic theory of gases.

For *extreme accuracy*, however, Rankine's method is far from suitable, with its inherent defect of a capillary tube experiment. For purposes where a really accurate value of η for the gas is required, as, for example, in Millikan's oil drop method for determining the electronic charge, a rotating cylinder apparatus is found to be more satisfactory and the one form of it almost universally used is that due to *Bearden*, who gave the value $(1.82462 \pm 0.00006) \times 10^{-4}$ *poise* for air at 23°*C*.

(*iii*) **Searle's method.** This method, also used independently by *Edwards*, measures the viscosity of a gas *at constant volume*, unlike the preceding two methods which do so at constant pressure.

The apparatus (Fig. 12.31), which is extremely simple, just consists of a large bulb or vessel *A*, fitted with a mercury manometer *M* on one side and a long vertical capillary tube *C* on the other, as shown.

Tap T_2 is first kept closed and the vessel filled with the given gas (or air) through tap T_1, until the pressure of the gas (as indicated by manometer *M*) is somewhat less than that which would cause a turbulent flow through on opening T_2.

When the mercury columns of the manometer become quite steady, their difference of level is carefully read. Let it be h_1. Then, if *H* be the barometeric height at the time, we have

initial pressure of the gas $= [H + h_1).\rho.g. = P_1$

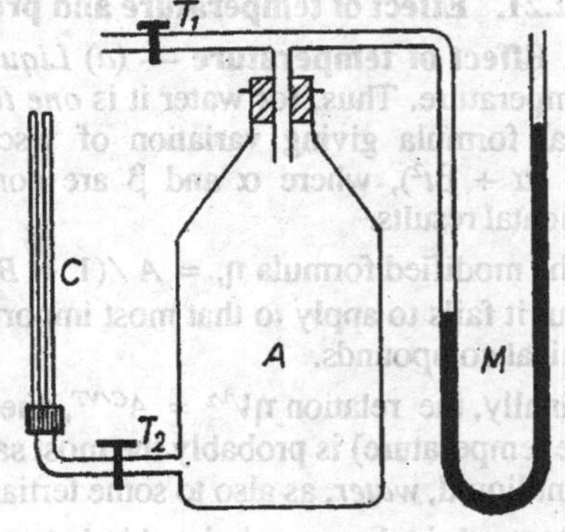

Fig. 12.31

where ρ is the density of mercury.

Tap T_2 is now opened for a known time *t sec* and then closed, making sure that the flow of gas through *C* takes place at a slow rate, *i.e.*, *under isothermal conditions.* If the difference of the mercury levels in the two limbs be now h_2, we have

final pressure of the gas $= (H+h_2).\rho.g = P_2$

Now, if at a given instant after throwing tap T_2 open, the pressure of the gas in the vessel be $P = (H + h).\rho.g$ and the rate of flow of gas into the capillary, V_1, clearly, volume of the gas after a small interval of time *dt*, we shall have *volume of the gas* $= (V + V_1.dt)$, where *V* is the volume of the vessel. If the pressure during this interval changes to $P + dP$, we have, in view of the isothermal conditions of the experiment,

$$PV = (P + dP)(V + V_1 dt), \text{ whence, } PV_1 = -V.dp/dt \text{ nearly}$$

But, as we know, $PV_1 = \pi r^4 (P^2 - P_0{}^2)/16\eta l,$

where P_0 is the atmosphere pressure = $H.'\rho.g$; r, the radius and l, the length of the capillary tube and η, the coefficient of viscosity of the gas.

We, therefore, have $\dfrac{\pi r^4(P^2 - P_0{}^2)}{16\eta l} = -\dfrac{V.dP}{dt}$. Or, $\dfrac{\pi r^4.dt}{16\eta lV} = -\dfrac{dP}{P^2 - P_0{}^2}$.

Or, $\dfrac{\pi r^4}{16\eta lV}\displaystyle\int_0^t dt = -\dfrac{1}{2P_0}\int_{P_1}^{P_2}\dfrac{dP}{P - P_0} - \dfrac{dP}{P + P_0}$. Or $\dfrac{\pi r^4 P_0\, t}{8\eta lV} = -\Big[\log_e(P - P_0) - \log_e(P + P_0)\Big]_{P_1}^{P_2}$

Or, $\dfrac{\pi r^4 P_0\, t}{8\eta lV} = \log_e \dfrac{P_1 - P_0}{P_2 - P_0} - \log_e \dfrac{P_1 + P_0}{P_2 + P_0} = \log_e \dfrac{P_1 - P_0}{P_2 - P_0} \times \dfrac{P_2 + P_0}{P_1 + P_0}$,

whence, $$\eta = \frac{\pi r^4 P_0\, t}{8lV \times 2.3026 \log_{10} \dfrac{P_1 - P_0}{P_2 - P_0}\cdot\dfrac{P_2 + P_0}{P_1 + P_0}},$$

Or, $$= \frac{\pi r^4 .H.\rho.g.t}{8lV \times 2.3026 \log_{10}\left(\dfrac{h_1}{h_2}\cdot\dfrac{2H + h_2}{2H + h_1}\right)}$$

so that η for the gas may be easily evaluated.

N.B. For an expression for η on the basis of the kinetic theory, see § 15.11.

12.21. Effect of temperature and pressure on viscosity of fluids.

1. **Effect of temperature** — (*a*) *Liquids*. The viscosity of liquids, in general, falls rapidly with temperature. Thus, for water it is *one third* at 80°*C* of its value at 10°*C*. Perhaps the earliest empirical formula giving variation of viscosity with temperature is due to *Slotte, viz.*, $\eta_t = \eta_0/(1 + \alpha t + \beta t^2)$, where α and β are constants. This , however, hardly accords with actual experimental results.

The modified formula $\eta_t = A/(1 + Bt)^c$, with A, B and *C* as constants is probably slightly better but it fails to apply to that most important class of liquids, the oils or, in fact to any mixture of chemical compounds.

Finally, the relation $\eta V^{1/2} = A^{C/VT}$, due to *Andrae*, (where *V* is the *specific volume* and *T*, the absolute temperature) is probably the most satisfactory so far but then, it does not apply to the most important liquid, *water*, as also to some tertiary alcohols.

The truth is that no relationship between viscosity and temperature of a general validity for all liquids has so far been established.

(*b*) *Gases*. The viscosity, in the case of gases, *increases with temperature* and the relation given by *Sutherland, viz.*, $\dfrac{\eta_t}{\eta_0} = \dfrac{273 + C}{T + C}\left(\dfrac{T}{273}\right)^{3/2}$, accords well with experimental results. Here *C* is called the *Sutherland's Constant* and its value is different for different gases.

2. **Effect of pressure**. (*a*) *Liquids*. Except for that unique liquid, *water*, the effect of pressure on all liquids is qualitatively similar *viz.*, *a rise in viscosity with pressure*, but quantitatively vastly different, (an almost negligible rise in the case of mobile liquids (*i.e.*, liquids of low viscosity) and a very large rise in the case of highly viscous liquids like mineral oils. Except for water, the effect of pressure increases at higher pressures. For water, at atmospheric pressure, the viscosity decreases, instead of increasing, for the first few hundred atmospheres.

Here, again, there is no satisfactory relationship established between η and pressure which may have a general validity.

(*b*) *Gases*. On the basis of the *kinetic theory*, *Maxwell* has shown that the viscosity of a gas is quite independent of pressure, at ordinary pressures (§ 15.11). This is amply borne out by experiment and is, in fact, a great point in favour of the theory. At low pressures (different for different gases), when the mean free path of the molecules becomes comparable with the dimensions of the containing vessel, viscosity continually falls with decrease of pressure and this too is fully borne out by experiment, the reason probably being the 'slipping' that occurs between the wall of the vessel and the layer adjacent to it. At high pressures, it has been shown by *Bridgeman* that viscosity increases with pressure.

WORKED EXAMPLES

1. A flat plate of area 10 sq cm is separated from a large plate by a layer of glycerine 1 mm thick. If the viscous coefficient of glycerine is 20 gm per cm per sec, what force is required to keep the plate moving with a velocity of 1 cm per second ?

We know that the *viscous force* is given by $F = \eta.A.v/x$.

Here, $\eta = 20\ gm/cm\ sec$, $A = 10\ sq.\ cm$, $v = 1\ cm/sec$ and $x = 1\ mm = .1\ cm$

$$\therefore \quad F = \frac{20 \times 10 \times 1}{.1} = 20 \times 10 \times 10 = 2000\ dynes.$$

Hence, the force required is equal to 2000 *dynes*.

2. (*a*) Water flows along a horizontal pipe of which the cross-section is not constant. The pressure is 1 cm of mercury where the velocity is 35 cm/sec. Find the pressure at a point where the velocity is 65 cm/sec. **(*Poona*, 1963)**

(*b*) Fresh water flows through a short horizontal pipe of a tapered bore. At the broader end of the pipe the velocity of water is 2 m/sec. At the narrower end, the pressure is 16 *kN/m*^{2} less. What is the velocity of water at this end ?

(*a*) Here, $p_1 = 1\ cm = 1 \times 13.6 \times 981\ dynes/cm^2$, $v_1 = 35\ cm/sec$,

$v_2 = 65\ cm/sec$, $\rho = 1\ gm/cc$, $p_2 = ?$

Using *Bernoullis relation* $\frac{1}{2}v_2^{\ 2} - \frac{1}{2}v_1^{\ 2} = p_1 - p_2$,

We have $\frac{1}{2}\left(65^2 - 35^2\right) = (13.6 \times 981) - p_2$

Or, $1500 = 13350 - p_2$, whence, $p_2 = 11850\ dynes/cm^2$

Or, $p^2 = 11850/(13.6 \times 981) = 0.8884\ cm\ of\ mercury$

Thus, the *pressure at the point* is 0.8884 *cm of mercury*

(*b*) In this case, $v_1 = 2m/sec$, $p_1 - p_2 = 16\ kilo\ newton/m^2$

$= 16 \times 10^3\ newtons/m^2$, $\rho = 10^3\ kg/m^2$, $v_2 = ?$

So that, using *Bernoullis relaton*, we have

$$16 \times 10^3 = \frac{1}{2} \times 10^3\ (v_2^{\ 2} - 2^2).$$

Or, $v_{22} = 16 \times 2 + 4 = 36$, whence $v_2 = \sqrt{36} = 6m/sec.$

The velocity of water at the narrower end of the pipe is thus 6 *m/sec*.

3. (*a*) If the diameters of a pipe are 10 cm and 6 cm at the points where a venturimeter is connected and the pressures at the points are shown to differ by 5 cm of water column, find the volume of water flowing through the pipe per second.

(*b*) Water is flowing through a horizontal venturimeter, with a bore of 360 mm at the entrance and of 120 mm at the throat. If the pressure differences across the two be equivalent to 450 mm head of water, calculate the mass flow of water through the meter.

(*a*) We have, for volume flow of a liquid, the relation

$$Q = a_1 a_2 \sqrt{\frac{2hg}{a_1^{\ 2} - a_2^{\ 2}}}$$

Here, $a_1 = \pi(10^2)/4\ sq.cm$, $a_2 = \pi(6^2)/4\ sq\ cm$, $h = 5\ cm$ and $g = 981\ cm/sec^2$. So that,

$$Q = 25\pi \times 9\pi \sqrt{\frac{2 \times 5 \times 981}{(25\pi)^2 - (9\pi)^2}} = 225\pi^2 \sqrt{\frac{9810}{34\pi \times 16\pi}}$$

Or, *volume flow* $= 225\pi \sqrt{\frac{9810}{34 \times 16}} = 3.002cc.$

(*b*) Here, $a_1 = \pi(.360)^2/4\ sq.m$, $a_2 = \pi(0.120)^2/4\ sq.m$, $h = 0.450\ m$ and $g = 9.81\ m/sec^2$.

So that, volume flow $Q = \frac{\pi^2}{16} \times 0.360 \times 0.120 \sqrt{\frac{2 \times 0.450 \times 9.81}{\frac{\pi^2}{16}\left[(0.360)^4 - (0.120)^4\right]}}$

Or, $Q = \frac{\pi}{4}(0.360)^2 \times (0.120)^2 \sqrt{\frac{2 \times 0.450 \times 9.81}{(0.360)^4 - (0.120)^4}} = .03382 m^3/sec$

∴ *mass flow = volume flow × density of water* $= Q \times \rho$

$= .03382 \times 10^3 = 33.82\ kg/sec.$

4. A tube having its two limbs bent at right angles to each other is held with one end dipping in a stream and opposite to the direction of flow. If the speed of the stream be 6 miles/hr, find the height to which water rises in the vertical limb of the tube.

Clearly, the flow of water will be stopped by the tube dipping in the stream and facing the flow, so that the loss of *K.E.* per unit mass of water is v^2. This much must, therefore, be the gain in the pressure energy, *i.e.*, p/ρ.

∴ $p/\rho = \frac{1}{2}v^2$.

Since ρ for water = 1, we have $p = \frac{1}{2}v^2$, or, $hg = \frac{1}{2}v^2$, [∵ $p = h.\rho.g = hg$, as $\rho = 1$.

Now, $v = 6\ miles/hr = \frac{6 \times 1760 \times 3}{60 \times 60} = \frac{88}{10} = 8.8\ ft/sec$

∴ $hg = \frac{1}{2} \times (8.8)^2$. Or, $h = \frac{(8.8)^2}{2 \times g} = \frac{(8.8)^2}{3 \times 32} = \frac{(8.8)^2}{64}\ 1.21\ ft$

Therefore, water rises to a height of 1.21 *ft* in the tube.

5. Calculate the velocity of efflux of kerosene oil from a tank in which the pressure is 50 lbs wt per sq inch above the atmospheric pressure. The density of kerosene is 48 lb per cubic foot. *(Bombay)*

Let *h* be the height of the level of kerosene oil in the tank above the axis of the orifice.

Then, pressure due to it at the level of the axis of the orifice = *h.p.g poundals/ft²*, *i.e.*, = *h.p lb wt/ft²*.

But, pressure = 50 *lb wt per sq inch* = 50 × 144 *lb wt /ft²*.

∴ $h.\rho = 50 \times 144\ lb\ wt/ft^2$ whence, $h = 50 \times 144/p = 50 \times 144/48\ ft.$ [∵ $\rho = 48\ lb/c\ ft$.

velocity of efflux is given by $v = \sqrt{2gh}$

$= \sqrt{2 \times 32 \times 50 \times 144/48} = \sqrt{9600} = 97.97\ ft/sec.$

6. A vertical tube of 4 mm diameter at the bottom has water passing through it. If the pressure be atmospheric at the bottom, where the water emerges at the rate of 800 gms per minute, what is the pressure at a point in the tube 25 cm above the bottom, where the diameter is 3 mm ?

Here, rate of emergence of water = 800 *gm/mt* = 800/60 = 40/3 *cc/sec*.

because, density of water = 1 *gm/cc*.

This will be the *same* across any section of the tube.

Now, radius of the tube at the bottom = 4 / 2 *mm* = .2 *cm*.

∴ area of cross-section of the tube, at the bottom $= \pi \times (.2)^2 = .04\pi\ sq\ .cm.$

Since, volume of water passing through any cross-section = cross-sectional area × velocity, we have *velocity of water passing through the bottom*

$$= \frac{volume\ flowing\ per\ sec.}{cross-sectional\ area} = \frac{40}{3 \times .04\pi}\ cm/sec = v_1,\ say.$$

∴ *K.E.* per unit mass of water at the bottom $= \frac{1}{2}v_1^{\ 2} = \frac{1}{2}\left(\frac{40}{3 \times .04\pi}\right)^2$ *ergs.*

and *pressure energy* per unit mass at the bottom = $p_1/\rho = p_1 = 76 \times 13.6 \times 981$ *ergs* $\quad [\because \rho = 1$ *gm/cc.*

$\therefore$ *total energy* = *K.E.* + *pressure energy*

$$= \frac{1}{2}\cdot\left(\frac{40}{3 \times .04\pi}\right)^2 + (76 \times 13.6 \times 981) \text{ ergs.}$$

Again, radius of the tube, 25 *cm* above the bottom = 3/2 *mm* = .15 *cm.*

And, $\therefore$ area of cross-section of the tube = $\pi \times (.15)^2$ *sq cm.*

So that, velocity of water, here, $= 40/3 \times \pi \times (.15)^2$ *cm / sec.*

$\therefore$ *K.E.* of water per unit mass $= \frac{1}{2} \times \left(\frac{40}{3 \times \pi \times (.15)^2}\right)^2$ *ergs.*

Let the pressure here be p_2.

Then, pressure energy per unit mass, in terms of mercury column

$= p_2 \times 13.6 \times 981$ *ergs.*

Also *P.E.* per unit mass of water, here, = *hg* = 25×981 *ergs.*

And, $\therefore$ *total energy*, here = *K.E.* + *pressure energy* + *P.E.*

$$= \frac{1}{2}\cdot\left(\frac{40}{3 \times \pi \times (1.15)^2}\right)^2 + (p_2 \times 13.6 \times 981) + (25 \times 981).$$

Now, in accordance with *Bernoulli's theorem*, the total energy must be the same at the two places ; so that, we have

$$\frac{1}{2}\left(\frac{40}{3 \times .04\pi}\right)^2 + (76 \times 13.6 \times 981) = \frac{1}{2}\left(\frac{40}{3\pi \times .0225}\right)^2 + (p_2 \times 13.6 \times 981) + (25 \times 981)$$

Or, $$= \frac{800}{(.12\pi)^2} + (76 \times 13.6 \times 981).$$

Or, $$= \frac{80}{(.0675\,\pi)^2} + (p_2 \times 13.6 \times 981) + (25 \times 981)$$

Or, $5627 + 1014000 = 17780 + 13350\,p_2 + 24520$. Or, $13350\,p_2 = 977330$*

Or, $$p_3 = \frac{977300}{13350} = 73.24 \text{ cm.}$$

Thus, the pressure at the point where the diameter of the tube is 3 *mm* is 73.24 *cm* in terms of mercury column.

7. A pitot tube is fixed in a main of diameter 15 cm, and the difference of pressure indicated by the gauge is 4 cm of water column; find the volume of water passing through the main in a minute.

Here, radius of the main = 15/2 = 7.5 *cm.*

$\therefore$ area of cross-section of the main = $\pi \times (7.5)^2$ *sq cm.*

Loss of *K.E.* per unit mass of water due to the stoppage of flow by the tube = $\frac{1}{2}v^2$ *ergs.*

And, gain of pressure energy = $p/\rho = p = h.g = 4 \times 981$ *ergs* $\quad [\because \rho = 1$ *gm/cc.*

$\therefore \quad \frac{1}{2}v^2 = 4 \times 981. \qquad$ Or, $\quad v^2 = 8 \times 981 = 7848$, whence, $v = \sqrt{7848} = 88.57$ *cm/sec.*

$\therefore$ *volume of water flowing per second across the section = area of section × velocity* = $\pi \times (7.5)^2 \times 88.57$ *cm/sec.*

And $\therefore$ volume flowing through the section, or the main per minute = $\pi \times (7.5)^2 \times 88.57 \times 60 = 9.396 \times 10^5$ *cc.*

8. A water main of internal diameter 8" is fed by a pipe of internal diameter 2", which delivers water at the rate of 1 c ft per second. If the pressure of water in the pipe be 50 lb wt per sq inch, calculate the pressure in the main.

Here, radius of the main = 8/2 = 4" = $\frac{1}{3}$ *ft.*

and radius of pipe = 2/2 = 1" = 1/12 *ft.*

---

* The figure is rounded off thus, to be able to use the ordinary four-figure logarithmic table.

$\therefore$ cross-sectional area of the main $= \pi \times \left(\frac{1}{3}\right)^2 = \pi/9$ *sq ft.*

and " " pipe $= \pi \times \left(\frac{1}{12}\right)^2 = \pi/144$ *sq ft.*

Rate of flow of water = 1 *c.ft./sec.*

Now, $\text{velocity of flow} = \dfrac{\text{rate of flow}}{\text{cross-sectional area}}$.

$\therefore$ velocity of flow of water (v_1) in the main $= 1 \Big/ \frac{\pi}{9} = 9/\pi$

and velocity of flow of water (v_2) in the pipe $= \dfrac{1}{\pi/144} = 144/\pi$

$\therefore$ *K.E.* per unit volume (*i.e.*, per 1 *c ft* of water) in the main $= \frac{1}{2}\rho v_1{}^2$, where ρ is the density of water, (equal to 62.5 *lb per c. ft*)

$$= \frac{1}{2} \times 62.5 \times \left(\frac{9}{\pi}\right)^2 = \frac{62.5 \times 81}{2\pi^2} = 256.4 \textit{ ft poundals.}$$

And, *K.E.* per unit volume in the pipe $= \frac{1}{2}\rho v_2{}^2$.

$$= \frac{1}{2} \times 62.5 \times \left(\frac{144}{\pi}\right)^2 = \frac{62.5 \times (144)^2}{2\pi^2} = 65660 \textit{ ft poundals.}$$

Let pressure in the main be $= p_1$ *lb wt/ sq inch* $= p_1 \times 32 \times 144$ *poundals/ft²*.

And, pressure in the pipe, as we know $= p_2 = 50$ *lb wt /sq inch.*

$$= 50 \times 32 \times 144 = 230400 \textit{ poundals/ft}^2$$

$\therefore$ applying *Bernoulli's equation*, We have $p_1 + \frac{1}{2}\rho v_1{}^2 = p_2 + \frac{1}{2}\rho v_2{}^2$.

Or, $p_1 \times 32 \times 144 + 256.4 = 230400 + 65660 = 296060.$

whence, $p_1 = 295803.6/32 \times 144 = 64.19$ *lb wt / inch²*.

Therefore, the pressure in the main = 64.19 *lb wt per sq inch.*

9. Water is escaping from a cistern by way of a horizontal capillary tube, 10 cm long and 0.4 mm in diameter, at a distance of 50 cm below the free surface of water in the cistern. Calculate the rate at which the water is escaping. (*Bombay*)

We know that the rate at which a liquid escapes out from a capillary tubing, *i.e.*, the volume of the liquid flowing out *per second*, is given by $V = \pi.Pr^4/8\eta.l$, where P is the difference of pressure between the two ends of the capillary tube, r its radius, l its length and η, the coefficient of viscosity of the liquid flowing through it.

Here, $P = 50 \times 1 \times 981$ *dynes/cm²*; $l = 10$ *cm*

and, $r = .4/2 = .2$ *mm* $= .02$ *cm.*

Hence $V = \dfrac{\pi \times 50 \times 981 \times (.02)^4}{8\eta \times 10} = \dfrac{3.082 \times 10^{-4}}{\eta}$

where *n* is the coefficient of viscosity of water.

Thus the rate at which the water is escaping from the capillary tube is $3.082 \times 10^{-4}\, \eta/$ *cc per second.*

10. Write down Poiseuille's formula for the rate of flow of a liquid through a capillary. From this show that if two capillaries of radii a_1 and a_2, having lengths l_1 and l_2 respectively are set in series, the rate of flow V is given by

$$V = \frac{\pi P}{8\eta}\left(\frac{l_1}{a_1{}^4} + \frac{l_2}{a_2{}^4}\right)$$

where P is the pressure across the arrangement and η, the coefficient of viscosity of the liquid. (*Rajasthan*, 1963)

Poiseuille's formula for the rate of flow (V) of a liquid is $V = \pi Pa^4/8\eta l$ where a is the radius of the tube and l, its length.

Obviously, the rate of flow of the liquid through each capillary will be the same, in accordance with the equation of continuity.

Let P' be the pressure difference across the first capillary. Then, that across the second will clearly be $P - P'$. So that, equating the rates of flow through them, we have

$$V = \pi P' a_1^4/8\eta l_1 = \pi (P - P') a_2^4/8\eta l_2,$$

whence, $$P' a_1^4/l_1 = P a_2^4/l_2 - P' a_2^4/l_2$$

Or, $$P'\left(\frac{a_1^4}{l_1} + \frac{a_2^4}{l_2}\right) = \frac{P a_2^4}{l_2}. \text{ Or, } P' = \frac{P a_2^4}{l_2} \div \left(\frac{a_1^4}{l_1} + \frac{a_2^4}{l_2}\right)$$

i.e., $$P' = \frac{P l_1}{a_1^4 (l_2/a_2^4 + l_1/a_1^4)} = \frac{P l_1}{a_1^4}\left(\frac{l_2}{a_2^4} + \frac{l_1}{a_1^4}\right)^{-1}$$

Substituting this value of P' in the expression for V, we have $V = \frac{\pi . a_1^4}{8\eta l_1} . \frac{P l_1}{a_1^4}\left(\frac{l_2}{a_2^4} + \frac{l_1}{a_1^4}\right)^{-1}$

Or $$V = \frac{\pi P}{8\eta}\left(\frac{l_1}{a_1^4} + \frac{l_2}{a_2^4}\right)$$

11. A gas bubble of diameter 2 cm, rises steadily through a solution of density 1.75 gm / cc at the rate of .35 cm/sec. Calculate the coefficient of viscosity of the solution. (Neglect the density of the gas).

We have relation, $\eta = \frac{2}{9} . \frac{r^2 . g(\sigma - \sigma)}{v}$, (see page 404), so that, neglecting ρ, (the density of the gas bubble), we have $\eta = 2r^2 . g/9v$

Here, $r = 2/2 = 1\ cm$ $g = 981\ cm/sec^2$, $\sigma = 1.75\ gm/cc$ and $v = .35\ cm/sec$,

Hence, $$\eta = \frac{2 \times 1 \times 981 \times 1.75}{9 \times .35} = \frac{2 \times 981 \times 1.75}{3.15} = 1.09 \times 10^3 \textit{ Poise.}$$

12. A cylindrical vessel of radius 7 cm is filled with water to a height of 50 cm. It has a capillary tube 10 cm long and 0.2 mm radius protruding horizontally at its bottom. If the viscosity of water is 0.01 C.G.S. units and g = 980 cm/sec², find the time in which the level will fall to a height of 25 cm. *(Punjab)*

Let A be the area of cross-section of the vessel, h, the height of water in it at any given instant and let the height fall by a small amount dh in a small interval of time dt. Then, clearly, *rate of flow through the capillary tube* $= -A.\ dh/dt$, the $-ve$ sign indicating that as time increases, the height decreases.

But the rate of flow, as we know, is, in accordance with *Poiseuille's equation*, given by

$V = \pi P r^4/8\eta l$, where $P = h.\rho.g = h.g.$ since ρ for water is 1 cm/cc.

So that, $$-A\frac{dh}{dt} = \frac{\pi . h . g . r^4}{8\eta l}, \text{ whence, } dt = \frac{-8\eta l . A}{\pi g r^4} . \frac{dh}{h}$$

And $\therefore$ $$\int_0^t dt = \int -\frac{8\eta l . A}{\pi g r^4} . \frac{dh}{h}. \text{ Or, } t = \frac{8\eta l . A}{\pi g r^4} . \log_e \frac{h_1}{h_2}$$

$$t = \frac{8 \times .01 \times 10 \times \pi \times 7^2}{\pi \times 980 \times (0.2)^4} \times 2.3026 \log_{10} \frac{50}{25} = 1.734 \times 10^5 \textit{ sec.} = 48.16 \textit{ hr.}$$

Thus, the water level in the vessel will fall in 25 *cm* in 48.16 *hours.*

13. Two horizontal capillary tubes A and B are connected together in series so that a steady stream of fluid flows through them. A is 0.4 mm in internal radius and 256 cm long. B is 0.3 mm in internal radius and 40.5 cm long. The pressure of the fluid at the entrance is 3 inches of mercury above the atmosphere. At the exit end of B, it is atmospheric (30 inches of Hg). What is the pressure at the junction of A and B if the fluid is (i) a liquid, (ii) a gas ? *(I.A.S., 1961)*

(i) **When the fluid is a liquid.** Let the pressure at the junction of the two capillary tubes be *h inches* of mercury. Then, clearly, pressure across tube $A = (33-h)$ *inches* of Hg $= P_1$, say, and that across tube $B = (h-30)$ *inches* $= P_2$, say. And, since rate of flow of the liquid must be the same through both the tubes, say, V, we have

$$V = \frac{\pi P_1 r_1^4}{8\eta l_1} = \frac{\pi P_2 r_2^4}{8\eta l_2}, \text{ whence, } \frac{P_1}{P_2} = \frac{l_1}{l_2}\left(\frac{r_2}{r_1}\right)^4,$$

where the symbols have their usual meanings.

Or, substituting the values of the various quantities, we have

$$\frac{33 - h}{h - 30} = \frac{256}{40.5} . \left(\frac{0.3}{0.4}\right)^4 = \frac{256}{40.5} \times \frac{81}{256} = 2.$$

Or, $33 - h = 2h - 60$. Or, $3h = 93$, whence $h = 31$ *inches.*

Thus, *the pressure at the junction of the two tubes* is 31 *inches* of mercury.

(ii) **When the fluid is a gas.** Here, the mass of the gas passing through each capillary tube per second being the same, we have $\frac{\pi r_1^4 (33^2 - h^2)}{16 \eta l_1} = \frac{\pi r_2^4 (l^2 - 30^2)}{16 \eta l_2}$, where *h*, as before, is the pressure in inches of mercury column at the junction of the two capillary tubes.

$$\text{Or,}\quad \frac{33^2 - h^2}{h^2 - 30^2} = \frac{l_1}{l_2}\cdot\left(\frac{r_2}{r_1}\right)^4 = 2. \quad \text{Or,}\quad 33^2 - h^2 = 2h^2 - 2\times 30^2.$$ [See above

Or, $h^2 = (2 \times 30^2 + 33^2)/3 = 2889/3 = 963,$ whence, $h = \sqrt{963} = 31.03$ *inches.*

Thus, here, the pressure at the junction of the two tubes is 31.03 *inches*.

14. A glass bulb of volume 500 cc has a capillary tube of length 40 cm and radius 0.020 cm leading from it. The bulb is filled with hydrogen at an initial pressure of 86 cm of mercury, density 13.6 gm per cc, and it is found that if the volume of the gas remaining in the vessel is kept constant, the pressure falls to 80 cm of mercury in 25.4 sec. If the height of the barometer is 76 cm and g = 981, find the viscosity of hydrogen.

This is a direct application of Searles method of determining the viscosity of a gas, [§ 12.20, *(iii)*], where the

$$\text{coefficient of viscosity } \eta = \frac{\pi r^4 . H.\rho.gt}{8lV \times 2.3026 \log_{10}\left(\frac{h_1}{h_2}\cdot\frac{2H + h_2}{2H + h_1}\right)}$$

Here, $h_1 = 86-76 = 10$ *cm*, $h_2 = 80-76 = 4$ *cm*; $H = 76$ *cm*, $\rho = 13.6$ *gm/cc*, $g = 981$ *cm/sec*^{2}, $t = 25.4$ *sec*, $l = 40$ *cm*, $r = .02$ *cm* and $V = 500$ *cc*.

Substituting these values in the expression for η, we have

$$\eta = \frac{\pi \times (.02)^4 \times 76 \times 13.6 \times 981 \times 25.4}{8 \times 40 \times 500 \times 2.3026 \log_{10}\left(\frac{10}{4}\cdot\frac{152+4}{152+10}\right)} = \frac{\pi \times (.02)^4 \times 76 \times 13.6 \times 981 \times 25.4}{8 \times 40 \times 500 \times 2.3026 \times 0.3815}$$

$$= 9.204 \times 10^{-5} \textit{ C.G.S. Units}$$

EXERCISE XII

1. Derive the formula for the flow of a liquid through a capillary tube. Why does the formula fail in the case of a wide bore ? *(Agra)*

2. What is meant by the term 'coefficient of viscosity' ? Obtain an expression for the rate of flow of a liquid through a capillary tube of circular cross-section. Note the precautions to be adopted in the experimental determination of this coefficient, using this expression.

3. Define 'coefficient of viscosity' for a liquid. Describe how the coefficient of viscosity is determined for water.

Water is conveyed through a horizontal tube 8 *cm* in diameter and 4 *kilometres* in length, at the rate of 20 *litres per sec*. Assuming only viscous resistance, calculate the pressure required to maintain the flow. (Coefficient of viscosity of water is 0.01 *C.G.S.* units). *(Banaras)*

Ans. 1.274×10^7 *dynes/cm*^{2}.

4. Define coefficient of viscosity of a liquid and find its dimensions.

Describe the way in which the different parts of a viscous liquid move when flowing through a fine tube. What changes take place if the motion is increased ? In an experiment with *Poiseuille's apparatus*, the following figures were obtained : Volume of water issuing *per minute* = 7.08 *cc* ; Head of water = 34.1 *cm*. Length of the tube = 56.45 *cm*. Radius of the tube = .0514 *cm*. Find the coefficient of viscosity. *(Calcutta)*

Ans. 0.01377 *Poise*

5. Calculate the mass of water flowing in 10 *minutes* through a tube 0.1 *cm* in diameter, 40 *cm* long, if there is a constant pressure head of 20 *cm* of water. The coefficient of viscosity of water is 0.0089 *c.g.s. units*. *(A.M.I.E.)*

Ans. 81.19 *gm.*

6. Water, with a coefficient of viscosity η, flows through a horizontal capillary tube of radius *a*, under a constant pressure head *p*. If the length of the tube is *l*, calculate the volume delivered from the tube per second. The correction for the velocity of exist may be neglected. *(Delhi)*

Ans. $V = \pi p a^4/8\eta l$

7. *(a)* Define coefficient of viscosity.

Describe fully what you would do to compare the viscosities of two liquids.

(b) Why do you find cloud particles floating in the sky ? *(Agra)*

8. Explain what is meant by the viscosity of a liquid ? How will you study experimentally the variation of viscosity with temperature ?

What are the 'dimensions' of viscosity in terms of length, mass and time ? *(Allahabad)*

9. Define the coefficient of viscosity of a liquid. What is the effect of temperature upon it ? How would you determine the viscosity of water at different temperature ? Derive the formula you use. *(Punjab)*

10. A square plate of 10 *cm* side moves parallel to another plate with a velocity of 10 *cm per sec.*, both plates being immersed in water. If the viscous force between them is 200 *dynes*, and the viscosity of water is 0.01 *gm/cm sec*, what is their distance apart ? **Ans.** 0.44 *cm.*

11. Enunciate and prove *Bernoulli's theorem*, and mention some of its important applications.

12. Two tubes, with small apertures at their lower ends, are held vertically with their lower ends dipping in a pipe carrying water, such that the aperture of one faces the flow and that of the other has its plane parallel to the direction of flow of water, which rises in the former to a height 10 *cm* above that in the latter. Determine the velocity of flow of water in the pipe. If the pipe has a diameter 20 *cm*, what is the volume of water flowing along the pipe per minute ?

Ans. (*i*) 14.01 *cm/sec* (*ii*) 26.4×10^5 *cc.*

13. Calculate the velocity of efflux of alcohol (*sp. gr.* 80) from a cylinder in which the total pressure is 2 *atmospheres.*

[Hint. Here, pressure due to the alcohol is one atmosphere, equal to 76 *cm* of mercury column.]

Ans. 15.92 *metres/sec.*

14. A venturimeter is connected to two points in a main where its radii are 200 *mm* and 150 *mm* respectively and the levels of the water columns in the two limbs differ by 10 *cm*. How much water in cubic metres flows through the pipe per hour ? **Ans.** 431.2 *cm.*

15. Water flowing in a horizontal main, of a non-uniform bore, has a velocity 100 *cm/sec.. at a point where the pressure is 7/19th* of the atmosphere pressure. What will be the velocity at a point where the pressure is one half of that at the first point ? **Ans.** 251.7 *cm/sec.*

16. Deduce the expression for the rate of steady flow of a liquid through a capillary tube of circular section.

A vessel of cross-section 20 *sq cm* has at the bottom a horizontal capillary tube of length 10 *cm* and internal radius 0.5 *mm*. It is initially filled with water to a height of 20 *cm* above the capillary tube. Find the time taken by the vessel to empty one-half of tis contents, given that the viscosity of water is 0.01 *C.G.S unit*. *(Madras)*

Ans. 9 *min* 36 *sec.*

17. A capillary tube, 1 *mm* in diameter and 20 cm in length, is fitted horizontally to a vessel kept full of alcohol, of density .8 *gm/cc*. The depth of the centre of capillary tube below the surface of alcohol is 30 *cm*. If the viscosity of alcohol is 0.012 *c.g.s.* unit, find the amount that will flow out in 5 *minutes*. Prove the formula you use. *(Bombay)*

Ans. 57.74 *gm*

18. Water at 20°*C* ($\eta = 0.010\ gm\ cm^{-1}\ sec^{-1}$) is escaping from a tank by a horizontal capillary tube, 20 *cm* long and 1.2 *mm* diameter. The water stands 100 *cm* above the tube. At what rate is the water escaping ? **Ans.** 2.5 *cc/sec.*

19. If in question 18, the area of the tank be 10^4 *sq cm* how long will it take the water to fall to 50 *cm* above the tube ? **Ans.** 91 *hours.*

20. A funnel is attached to a vertical tube of length 49.6 *cm*. Water is poured into the funnel and kept at a height of 68.3 *cm* above the lower end of the tube, whilst 78 *cc* are collected in 10 *minutes*. The tube was then removed and entirely filled with mercury; 3.695 *gm* were required. If the density of mercury is 13.56 *gm per cc*, find the viscosity of water at the temperature of the experiment. **Ans.** 0.01247 *gm / cm sec.*

[Hint. *The radius of the tube can be determined from the amount of mercury filling the tube. For,*

$$mass = volume \times density.$$

And $\therefore$ *volume = mass/density. And volume* = $\pi r^2.l$, *, where r and l are the radius and length of the tube respectively, whence r may be evaluated and tis value substituted in the relation* $\eta = \pi r^4/8Vl$

21. Give Poiseuille's method of measuring the velocity of liquids. Derive the formula used. Indicate a method which could be employed to measure the viscosity of liquids at different temperatures. *(Banaras)*

22. What kind or kinds of energy result from the work done by a fluid against viscosity ? How can the viscosity of liquids be determined ? *(Bombay)*

23. Define "coefficient of viscosity".

Derive *Stokes'* formula for the velocity of a small sphere falling through a viscous liquid. Explain how this is utilised to determine the viscosity of liquid like castor oil. Mention one more application of *Stokes'* formula. *(Madras)*

24. Determine the radius of the drop of water falling through air, if the terminal velocity of the drop is 1.2 *cm/sec.* Assume the coefficient of viscosity for air = 1.8×10^{-4} and the density of air = 1.21×10^{-3} *gm/cc* *(A.M.I.E)*

Ans. 0.001 *cm* (approximately)

25. Explain the rotating cylinder method of determining the coefficient of viscosity of a liquid and give its theory.

[Agra (B.Sc. Part II), Delhi (Hons); Poona,

26. Distinguish between streamline and turbulent flow of a liquid and explain the significance of the Reynold's number.

Describe a method of comparing the viscosities of two liquids. ***(Punjab)***

27. Define 'coefficient of viscosity' of a liquid. What is the effect of the pressure on viscosity ?

[Agra (B.Sc. Part II).

28. In what respects is the viscosity of a gas different from that of a liquid ?

Describe an experimental set up for the measurement of coefficient of viscosity of water at different temperatures.

[Punjab (Supp.)].

29. State 'Bernoulli's theorem' and apply it to obtain an expression for the reducation of pressure when water flows through a constriction in a pipe. Describe the device depending on this theorem, for the measurement of liquid flow.

(Punjab)

30. Explain the limitations of *Poiseuille's formula* for the rate of flow of a liquid through a capillary tube. Why does it fail in the case of a gas ?

Explain, with necessary theory, *Rankine's* method for the determination of the viscosity of a gas.

[Poona (Principal paper)]

31. Describe, with necessary theory, Searle's method of determining the viscosity of very viscous fluids. ***(I.A.S.)***

32. Describe the Ostwald viscometer and explain how the coefficient of viscosity of two liquids can be compared by the method of capillary flow. ***(Jodhpur; I.A.S)***

CHAPTER XIII

DIFFUSION AND OSMOSIS

13.1. Diffusion. If we place a solution of, say, copper sulphate, (or even solid crystals of potassium permanganate or potassium dichromate), in a beaker, and cover it with pure water, *disturbing it the least*, so that the line of demarcation between the two is clearly seen, we shall find that on standing for some time, —may be, even days or months,—the characteristic colour of the salt *ascends upwards* and fills the entire space occupied by the liquid, the concentration of the entire solution becoming the *same* all over.

The same is true for gases also. For, if we divide a tall glass jar into two compartments by a movable horizontal glass plate and put a heavier, or a denser gas, like carbon dioxide, in the lower compartment and a lighter or a rarer gas like hydrogen, in the upper one, we find that, on removing the glass plate, the molecules of the two gases intermix with each other, until we get a mixture of the two, of a uniform density throughout.

This process, by virtue of which the molecules of a solute, (in this case, copper sulphate), move upward from the lower portions of greater concentration to the upper ones of lower concentration, or, by virtue of which the molecules of one gas mix with those of another, even against the force of gravity, is called **diffusion.**

The solute is pushed up as though under some pressure, until equilibrium is attained, and the concentration and pressure of the solution become uniform throughout.

Solids too, although they possess a definite crystalline structure, have been known to exhibit the phenomenon of diffusion, if placed in good contact with each other over a long enough time, the diffusion taking place more readily between two different solids than between two portions of the same solid. Thus, for example, the diffusion of gold into lead has been clearly shown by **Robert Austen**, by fusing a small lead cylinder (about 7 *cm* long and 1.4 *cm* in diameter) into a thin gold plate and subjecting it to pressure for well over a month, in a constant temperature enclosure, —the temperature being kept below their lowest melting point. And more recently, it has been clearly established by **Groh** and **Hevesy** that *radioactive* lead can diffuse into ordinary lead, if the two be kept in contact for some reasonably good time, *i.e.*, well over a year. This then, is a case of what is aptly called **self-diffusion,** for we have, here, the case of a substance (lead) diffusing into its own self, (*i.e.*, into lead), for although the atomic weights of the two types of lead are different, they are otherwise identical in their chemical properties.

The phenomenon of diffusion of liquids was first investigated by **Graham**, in the year 1851. It would, therefore, be of interest to give a brief account of his experiments here.

He filled a large-mouthed bottle with the salt solution under investigation, to a level very near the top, floated a cork-disc centrally on the surface of the solution and carefully filled the rest of the bottle with pure water, (the solvent), by lightly squeezing it on the disc from a sponge. Then, covering the mouth of the bottle with a glass disc, lowered it lightly into a larger jar, containing water, so that the mouth of the bottle was some distance below the water surface in the jar. Then, allowing the motion of the two liquids to subside, he carefully uncovered the mouth of the bottle, causing the least possible disturbance to the liquids, and finally covered the jar with a glass plate to minimise evaporation, maintaining the jar at a constant temperature to avoid convection currents.

Due to diffusion of the salt into the pure water above, the concentration of the solution began to change, and this was determined from time to time by withdrawing portions of the solution by means of a pipette. From the change in concentration of the solution, *Graham* estimated the amount of the solute diffusing into the solvent or the water above.

Graham repeated his experiment with different salt solutions, and, although he was unable to formulate any exact law for the rate of diffusion, he arrived at the following conclusions:

(*i*) that solutions of *different salts*, of the *same strength or concentration*, diffuse at *different rates*, and thus, *the rate of diffusion depends upon the nature of the solute* ;

(*ii*) that solutions of the *same salt, of different strengths or concentrations* difuse at rates proportional to their concentrations;

(*iii*) that, in general, diffusion alters the proportion of two salts in a mixture;

(*iv*) that the rate of diffusion increases with temperature;

(*v*) that solutions may be divided into two distinct categories, *viz.*, (*a*) **crystalloids,** consisting of mineral acids, salt solutions, sugar etc., *which diffuse comparatively quickly* and (*b*) **colloids,** consisting of albumen, gum, caramel and gelatine, etc. *which defuse more slowly than the crystalloids.*

Diffusion, in the case of liquids, may easily be explained on the basis of the kinetic theory of liquids. For, according to this theory the molecules of the solvent possess kinetic energy of translation, which is directly proportional to their absolute temperature. This they share with the molecules of the solute present and the latter thus roam about in the solvent, so much so that they even rise up against the force of gravity.

13.2. Fick's Law. Four years later, in 1855, **Fick** formulated a *law of diffusion* based on *Fourrier's law* of conduction of heat due to a close analogy between the process of diffusion of a solute and the flow of heat through a solid. He established the law *that the rate of diffusion in any direction is proportional to the concentration gradient of the solute in that direction.* The law, after him, is known as **Fick's law.**

Thus, if we imagine a rectangular slab of a solution, of thickness *x*, such that the concentrations of the solute all over its two opposite faces are C_1 and C_2 respectively — C_1 being greater than C_2 —then, the amount of solute passing into one face and out at the other, *i.e.*, passing any cross-section of it in a given time, is proportional to the *concentration gradient*, $(C_1 - C_2)/x$.

If we denote the change of concentration for a small distance *dx* by *dC*, the *concentration gradient* is *dc*/*dx*, and the quantity (*Q*) of the solute diffusing *in unit time* through an area *A* is given by the relation,

$$Q = K.A.dC/dx,$$

where *K* is a constant, called **coefficient of diffusion**, or **diffusivity,** *of the solute*, and depends not only on the nature of the solute in a given solvent but also on its concentration, *i.e.*, mass (*in grams*) of the solute *per cc*. And, since the dimensions of *Q* are $[MT^{-1}]$, of *A*, $[L^2]$ and of *dC*/*dx*, $[ML^{-4}]$, we have

$$[MT^{-1}] = [K].[L^2][ML^{-4}], \text{ whence, } [K] = [L^2T^{-1}].$$

Now, if the concentration at the first layer be *C*, the concentration at a layer opposite, distance δx apart, will be $C - \delta x.dC/dx$, because the concentration gradient is dC/dx and, therefore, a change of concentration in a distance δx will be $\delta x.dC/dx$. Thus, the quantity of the solute entering the first layer in *unit time* is $K.A.dC/dx$, and that leaving the second layer in the same time, (*i.e., unit time*), is given by $K.A.\frac{d}{x}\left(C - \frac{dC}{x}.\delta x\right) = K.A.\frac{dC}{dx} - K.A.\frac{d^2C}{dx^2}.\delta x$, *i.e.*, the rate at which the solute is leaving the second layer is lower than that at which it enters the first layer by $K.A.\delta x d^2C/dx^2$. This amount of the solute is, therefore, added in unit time to the volume of the solution between the first and the second layers, *i.e.*, to a volume, $A.\delta x$.

Thus, the *rate of change of concentration* $= K.A.\delta x.d^2C/dx^2/A.\delta x = K.d^2C/dx^2$.. Denoting this rate of change of concentration by dC/dt we, therefore, have $dC/dt = Kd^2C/dx^2$.

This is a relation of fundamental importance in diffusion of liquids and is, in fact *Fick's law*, put in the form of a differential equation.

It is easy to see from the above that when diffusion attains a *steady state*, $dC/dt = 0$, or $d^2C/dx^2 = 0$ which on integration twice, yields $C = ax + b$, where *a* and *b* are constants — a relation experimentally tested by Ficks and found valid.

The analogy between *Fick's law* of diffusion and *Fourrier's law* of conduction of heat on the one hand, and *Ohm's law* of conduction of electricity on the other, will at once be apparent. For, just as heat always flows from a point at a higher temperature to that at a lower temperature, and just as electricity always flows from a point at a higher potential to that at a lower potential, *so also does a solute flow from regions of higher concentration to those of lower concentration.* Not only that, but like conduction of heat and electricity, diffusion too is, to all intents and purposes, an **irreversible process**, because, if the solvent, or a portion of it, gets impregnated with the solute, it can never divest itself of it without external aid.

13.3. Relation between Time of Diffusion and Length of Column. Suppose we have two columns of solution, of lengths l_1 and l_2 respectively, having *identical* concentrations at different points along them at a given instant. Let the two columns again acquire similar concentrations at corresponding points after time-intervals t_1 and t_2 respectively.

Quite obviously, these time-intervals must be proportional (*i*) directly to the distances covered by the solute and (*ii*) inversely to the absolute velocity of the solute at each layer. And, therefore, since the distances covered are in proportion to these lengths of the columns, we have

$$t_1/t_2 = (l_1/l_2).(v_2/v_1), \quad ...(i)$$

where v_1 and v_2 are the solute-velocities at corresponding layers of the two columns.

Now, by *Fick's law* $\quad Q = K.A.dC/dx.$

But $Q = A.vC.$, if v be the velocity of the solute at a given layer, of concentration C

$$\therefore \quad A.v.C. = K.A.\frac{dC}{dx}. \quad \text{Or,} \quad v = \frac{K}{C}.\frac{dC}{dx},$$

i.e., $\quad v \propto dC/dx.$ $\quad$ [K/C being a constant for the given layer.

But dC/dx will be inversely proportional to the length of the column, if the two columns have similar distribution of concentrations. And, therefore,

$$v \propto 1/l, \quad \text{[where } l \text{ is the } \textit{length} \text{ of the column.}$$

Hence, $\quad v_2/v_1 = l_1/l_2.$

So that, substituting the value of v_2/v_1 in relation (*i*) above, we have

$$t_1/t_2 = l_1^2/l_2^2 \; i.e., \; t \propto l^2.$$

Or, the time of diffusion from one distribution of concentration to another, in a given solution, is directly proportional to the square of the length of its column.

13.4. Experimental measurement of Diffusivity. The diffusivity of a solution may be determined by measuring its concentration at a chosen point, from time to time, and applying *Fourrier's theorem* to the relation,

$$dC/dx = K.d^2C/dx^2.$$

Different methods have been used for the purpose. Thus, for example **Kelvin** determined the densities of the different layers of the solution by introducing into it, a series of beads of different but known densities and noting their equilibrium positions. This method, however, is open to two very valid objections, *viz.,* (*i*) some salt from the solution crystallizes on the beads and (*ii*) air bubbles are formed on their surfaces. Both these factors tend to alter the buoyancy of the beads and the results obtained are thus far from reliable.

Other methods used to measure concentration are based on the measurement of (*a*) the refractive index of the solution, (*b*) the contact potential difference, and (*c*) in the case of optically active solutions, like that of sugar, the rotation of the plane of polarization etc., etc. **Littlewood** has succeeded in measuring the concentration of the solution at a given layer within .05 *gm per litre* by an optical method, based on the bending of light rays.

For aqueous solutions of salts like $NaCl$, KCl, KNO_3 etc., however, the method devised by **Clack**, (1942), is perhaps the best. We shall, therefore, study this in some detail.

Clack's method. He took a *'diffusion cell'*, which was just a rectangular tube, about 5 *cm* long, 1 *cm* wide and with a horizontal thickness 4 *cm* made *up of glass plates*, and fitted it vertically into the bottom of a glass box containing an air-free, saturated solution of the salt in question, to maintain which at its saturation point, crystals of the salt, or the solute, were also placed in the solution. Outside the tube, and above the glass box, were arrange compartments carrying pure distilled water, at a level about 5 *cm* higher than the tube. An inlet-tube for the water to enter and an outlet-tube for its exit were also provided, so that *there was a constant slow stream of the distilled water flowing across the upper free end of the cell,* resulting in the incoming water being carried away as a feeble solution of constant concentration, due to the upward diffusing solute from the cell getting mixed up with it. The whole system was allowed to come to a steady state, which took as many as 12 days or more and the flow of water was maintained at about 50 *cc* per day, the temperature being kept constant by means of a suitable *thermostat*. The theory underlying the method will be clear from the following:

Consider a layer P, say, distant l from the top of the top of the cell, and let Q be the *mass* of the solute crossing upwards *here* per second. Then, if C be the concentration of the solution at this layer and A, the area of cross-section of the tube, we have, by *Fick's law*,

$$K = \frac{Q}{A} \cdot \frac{dl}{dC},$$ [See page 424.

where K is the *diffusivity* of the solution.

Since, $\frac{dl}{dC} = \frac{dl}{d\mu} \cdot \frac{d\mu}{dC}$, we have $K = \frac{Q}{A} \cdot \frac{dl}{d\mu} \cdot \frac{d\mu}{dC}$, ...I

where μ is the optical refractive index of the solution at P.

Now, apart from diffusion, the motion of the solute is also affected by the *mass motion* or *bulk motion* of the solvent. (*i.e.*, water) downwards into the tube. If v be the *uniform, downward* velocity of this mass-motion of the solvent, it is obvious that mass $A.v\,C$ of the solute and a mass $A.v.(\rho - C)$ of the solvent flow downwards at P, where ρ is the density of the solution at P and, therefore, $(\rho - C)$, the concentration of the solvent there.

Hence, if M be the mass of the solvent entering the cell *per second*, at the top, and m, that of the solute leaving the cell *per second* there, we have

$$M = A.v.(\rho - C) \qquad ...(i)$$

and $$m = Q - A.v.C. \quad \text{Or,} \quad Q = m + A.v.C \qquad ...(ii)$$

But, since $A.v. = M/(\rho - C)$, [from relation (*i*) above.

we have, from relation (*ii*), $Q = m + [M/(\rho - c)].C$ ...(*iii*)

So that, putting $M/m = \delta$, a constant, we have $Q = m + \left(\frac{\delta m}{(\rho - C)}\right).C = \frac{m(\rho - c) + \delta.mC}{\rho - C} = \frac{m(\rho - C + \delta C)}{\rho - C}$ ...(*iv*)

Now, the net loss in the mass of the system *per second* is clearly equal to *loss in the mass of the solute per second minus gain in mass of the solvent per second*. Denoting this net loss by i, therefore, we have

$$i = m - M = m - \delta.m = m(1 - \delta), \text{ whence, } m = i/(1 - \delta)$$

Substituting this value m, in relation (*iv*) above, we have

$$Q = \frac{i.(\rho - C + \delta C)}{(1 - \delta).(\rho - C)}$$

And, putting this value of Q, in relation I above, we have

$$K = \frac{i(\rho - C + \delta C)}{A(1 - \delta)(\rho - C)} \cdot \frac{dl}{d\mu} \cdot \frac{d\mu}{dC} \qquad ...II$$

The various quantities involved in this relation for K are obtained as indicated below :

(*i*) i and δ are determined either by *chemical analysis* or by drying and weighing the solution.

(*ii*) $dl/d\mu$ is measured by making a narrow horizontal beam of monochromatic light incident on layers of increasing densities, when the beam gets refracted in the vertical plane and emerges out downwards at an angle α with the horizontal, such that $\frac{d\mu}{dl} = \frac{\sin\alpha}{t}$, [$t$ being the horizontal thickness of the cell,

where α is measured by noting the vertical displacement of the central fringe in the interference pattern produced by two narrow horizontal slits, close to each other, and illuminated by the green light from a mercury arc, when this light is allowed to pass through the cell.

(*iii*) $d\mu/dC$ is determined by means of *Rayleigh's interferometer*, for solutions of different concentrations C and

(*iv*) the concentration C at P is computed from the curves $(l, d\mu/dl)$ and $(C, d\mu/dC)$.

Thus, all quantities in relation II above for K being known, we can easily calculate K, the diffusivity for the given solution.

13.5. Graham's Law for Diffusion of Gases. Gases diffuse much more rapidly than liquids, due to their high molecular speed, their diffusion being governed by the following law, known as *Graham's law of diffusion :*

"The rates of diffusion of two gases are inversely proportional to the square roots of their densities."

Thus, if r_1, r_2 and ρ_1, ρ_2 be the rates of diffusion and densities respectively of two gases, we have

$$\frac{r_1}{r_2} = \sqrt{\frac{\rho_2}{\rho_1}}$$

This follows readily from the *kinetic theory of gases,* [see 15.6, *(iii)*]

The difference in the rates of diffusion of gases was used by *Graham* to separate gaseous mixtures, to which process he gave the name **'atmolysis'**.

13.6. Effusion — Transpiration and Transfusion

(*i*) When a gas escapes from a vessel into a vacuum, through a small hole in a thin plate, such that the width of the hole is greater in comparison with its length, the gas is said to *effuse*, and the process is called **effusion**. Graham showed that *the rate of effusion varies directly as the square root of the difference of pressure on the two sides of the hole, and inversely as the square root of its density, and is quite independent of the passage of any other gas at the same time.*

Thus, $\quad \textit{velocity of effusion} \propto \sqrt{\dfrac{\textit{pressure difference}}{\textit{density}}}$

It must be noted that in the process, this gas flows, as a whole, through the plate, there being no separation of a mixture of gases into its constituents.

(*ii*) If, on the other hand, the hole in the plate be not too fine, and the thickness of the plate be greater than the diameter of the hole, the process of escape of the gas through it is called **'transpiration'**. Here, the flow of the gas is controlled by viscosity alone, which is subject to the same laws as are applicable to the flow of a gas through a long tube.

Here, too, the gaseous mixture, as a whole, passes through the plate, there being no separation of it into its various constituents.

(*iii*) If, however, the holes be so fine that their diameters are comparable with those of the molecules of the gas, the passage of the gas is known as **diffusion** or **'inter-diffusion'**.

Here, partial separation of a mixture of gases, *i.e., atmolysis,* take place and its constituents may thus be easily separated.

13.7. Osmosis and Osmotic Pressure. If two liquids, which can diffuse into one another, be initially separated by a membrane, they inter-diffuse, one into the other. *This process of inter-diffusion of two liquids through a membrane is called* **osmosis,** — from '*osmos*', '*a push*'.

To demonstrate the phenomenon of osmosis we use what are called **semipermeable membranes** (a term first used by *Van't Hoff* in 1886) which have the property of **selective transmission** allowing some liquids to pass through them and not others. There are various membranes of animal or vegetable origin, which have this property of selective transmission and are, therefore, semipermeable. Examples of such membranes are : ordinary *animal bladder, cell-walls of plants,* and *certain inorganic precipitates* etc.

For instance, if we fill an animal bladder with alcohol, tie its mouth tightly and place it in a vessel of water, the bladder begins to swell, and may ultimately burst; if, on the other hand, we fill it with water and place it in alcohol, it shrinks— a fact first observed by *Abbe Nollet* as far back as the year 1748. This clearly shows that *the bladder is more permeable to water than to alcohol*. For, water can pass through it easily whereas alcohol cannot. **Graham** made use of the semipermeable properties of pig's bladder to separate crystalloids from colloids,—a process he termed **'dialysis'**. The mixture of crystalloids and colloids is placed in a tray with a bottom of parchment paper and the tray is then floated in water, then the crystalloids pass through the parchment into the water, leaving the colloids behind in the tray. Poisons in the viscera of poisoned animals can be detected in this manner.

It will be readily seen that *osmosis differs from diffusion in that (i) there is only one-way traffic of molecules here, unlike the two-way one in diffusion, (ii) the direction of flow or transmission here is from regions of lower to those of higher concentration—the exact opposite of what happens in diffusion, (iii) it occurs only through the agency of a semipermeable membrane unlike diffusion which is brought about by direct contact between the diffusing substances and,* finally, *(iv) it is a selective phenomenon whereas diffusion is a more general one.*

The systematic study of the laws of osmosis was first made by **Pfeffer,** who used a semipermeable membrane of *precipitated copper ferrocyanide,* (first suggested by *Traube* in the year 1864) supported by the pores of an ordinary battery-pot or porous biscuit-ware, as ordinary parchment membranes are too feeble to withstand high pressures. The pot is first soaked in water to drive out all the air from its pores. It is then filled with a solution of copper sulphate and placed inside a solution of potassium ferrocyanide. Both these salts diffuse into the walls of the pot, where they meet to form a brown semipermeable precipitate of *cupric cyanide.* The pot is then washed with distilled water.

If such a pot be now filled with a salt or a sugar solution closed tightly with a rubber bung, carrying a mercury manometer, as shown, (Fig. 13.1), and then placed in a vessel containing pure water, (the solvent), it is found that water passes into the pot and an increase of pressure is recorded by the manometer, *showing thereby that more water is entering the pot from the outside than is escaping out from the inside.* More and more water thus gets into the pot and the pressure goes on steadily rising until a certain definite pressure is reached. This will happen when equilibrium is attained, *i.e., when water passes in at just the same rate at which it passes out.* This increase of pressure, or the excess pressure inside over the atmospheric pressure on the water outside, now prevents any further passage of water into the pot and measures what is called the **osmotic pressure** of the solution. *Osmotic pressure is thus the pressure which must be applied on the solution-side of the membrane to prevent any flowing in of the solvent through the membrane.* Its value depends upon the *nature, concentration and temperature* of the solution.

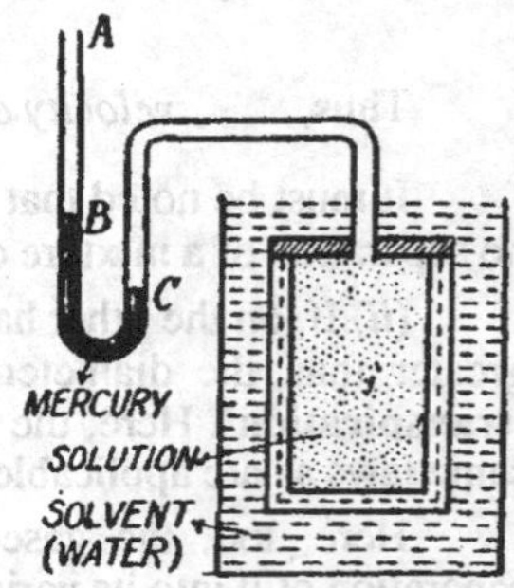

Fig. 13.1

If the manometer be open at the top (*A*), the travel of mercury in it is inconveniently large. To prevent this, the upper end is *closed,* and if the volume of air in *AB* be known, when the pressure is atmospheric, any subsequent reading of the volume of air in it enables the pressure to be calculated.

This method for the measurement of osmotic pressure, however, is not quite a suitable one; for,

(*i*) it can be used only for weak (aqueous) solutions;

(*ii*) due to the solvent entering the pot, *the concentration* is *lowered,* and hence the osmotic pressure measured is for this final *lowered* concentration of the solution, and not for the original one;

(*iii*) *the final concentration of the solution, is not uniform throughout,* and the pressure measured is thus for the concentration of the solution in the immediate neighbourhood of the membrane, which is not quite so easy to determine.

A better method is due to **Berkeley** and **Hartley,** who, in 1906, measured the osmotic pressure of *strong, concentrated solutions* by determining the external pressure necessary to be applied to the solution to *just* prevent any solvent passing into it.

E TO COMPRESSOR AND PRESSURE GAUGE
G
S
SOLVENT
M
N
SOLVENT (WATER)
D
F
GUN METAL CASE

Fig. 13.2

Their apparatus, (Fig. 13.2), consists of a porous tube *MN*, about 15 *cm* long and of external diameter 2 *cm* and internal diameter 1.6 *cm*, (with a membrane of cupric cyanide formed inside its wall close to its outer side), which is filled with the solvent (water) and placed inside a *gun metal vessel G* which is filled with the solution, through the tube *E*. The porous tube *MN* is connected to a *graduated capillary tube D* at one end and a tube *F* connected to a reservoir *R* of water, as shown. Due to water passing through the wall of tube *MN* into the solution in *G*, resulting in the fall of the water meniscus in the capillary tube *D*.

Pressure is, therefore, applied to the solution through the tube *E*, so that no solvent can enter into it from the tube *MN*, *i.e.*, the miniscus of the liquid column in *D* remains stationary, or the condition of equilibrium is attained. *The external pressure applied through E thus gives the osmotic pressure of the solution in B (plus the small and almost negligible hydrostatic pressure of the solvent in the tube MN).*

Osmotic pressures, as high as 130 atmospheres, may thus be easily measured with the help of this apparatus.

13.8. Laws of Osmotic Pressure. Pfeffer's results led to the establishment of the following laws of osmotic pressure **for dilute solutions of non-electrolytes.**

1. *The osmotic pressure (P) of the solution of a given solute is proportional to its concentration (C), provided the temperature remains constant; or,* $P \propto C$

Since concentration is inversely proportional to volume (V), this corresponds to *Boyle's law for gases, viz.*, that $P \propto 1/V$.

2. The osmotic pressure of the solution of a given concentration is proportional to its absolute temperature : or, $P \propto T$. This, obviously, corresponds to *Charles' law for gases.*

Combining the two laws, therefore, we have $P \propto C/T$.

Or, $$P = K.C.T,$$ [where K is a constant.

Or, putting $C = 1/V$, where C is the concentration of the solution, in *gram- molecules* of the solute per *cc.*, and V, its volume containing 1 *gram-molecule* of the solute, we have $P = K.(1/V).T$.

Or, $$PV = KT.$$

This constant K for very dilute solutions, is found to be identical with the *gas constant R*; so that, *for dilute solutions*, we have

$$PV = RT$$ [*i.e.*, the same as the standard equation for a gas.

3. Solutions of non-electrolytes exerting equal osmotic pressures contain the same number of gram-molecules per *cc.*, *i.e.*, are of concentrations proportional to the molecular weights of the solute. This corresponds to *Avogadro's hypothesis for gases. Solutions exerting the same osmotic pressure are called* **isotonic** *or* **osmotic solutions.**

4. Solutions of *electrolytes* exert a greater osmotic pressure than those of non-electrolytes, due to their splitting up into ions. In the case of electrolytes, the relation between P, V, R and T is given by $PV = iRT$, where i is a factor depending upon its *degree of dissociation*, and is equal to the ratio of the observed osmotic pressure to that calculated as above ; so that, knowing i, the *percentage dissociation* of an electrolyte can be calculated.*

13.9. Kinetic Theory of Solutions. Van't Hoff propounded the kinetic theory of solutions, similar to the one for gases, and deduced, for infinitely dilute solutions, the relation $PV = RT$, from purely thermodynamical considerations. This close similarity with the behaviour of gases led him to suggest that the *osmotic pressure of a dilute solution is the same as would be exerted by the solute, if it could exist as a gas, and if it occupied a volume equal to that occupied by the solution, at the same temperature.* This is known as **Van't Hoff law.**

This parallelism between dilute solutions and gases has been amply verified by the accurate results obtained by Berkeley and Hartley.

13.10. Osmosis and Vapour Pressure of a Solution. The vapour pressure of a solution is always less than that of the pure solvent, as will be clear from the following :

---

* See worked example 9.

If we take a solvent enclosed in an evacuated chamber (Fig.13.3), and place vertically in it a long glass tube, containing a solution, with its lower end closed by means of a semipermeable membrane which allows the solvent to pass through it, but not the solute, we find that the liquid rises in the tube up to H, say, until equilibrium is attained. Now, clearly, the vapour at H is in contact with the solution, and at J, with the solvent. Let the maximum pressures at J and H be p and p' respectively. Then, obviously, p is greater than p' by an amount $h.\sigma g$, where σ is the density of the vapour of the solvent and is practically the same as that of the vapour of the solution at H, the column h being small.

Thus, $$p - p' = h.\sigma.g. \qquad ...(i)$$

If P be the osmotic pressure of the solution, we have

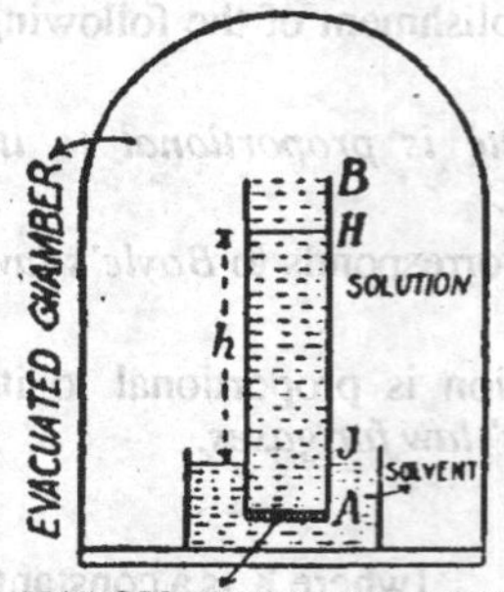

Fig. 13.3

P = *excess pressure inside the tube over that outside at J* on the solvent. $= h.\rho.g. - h.\sigma.g$,

where ρ is the density of the solution.

Or, $$P = h.g.(\rho - \sigma) \qquad ...(ii)$$

whence, $$h.g = \left(\frac{P}{\rho - \sigma}\right)$$

Substituting this value of $h.g.$ in relation (i) above, we have $p - p' = \sigma.[P/(\rho - \sigma)]$

Or, neglecting σ in comparison with ρ, we have $$p - p' = P.\sigma/\rho,$$

i.e., the lowering of the vapour pressure $(p - p')$, *is directly proportional to the osmotic pressure* (P).

Now, $\sigma = 1/V$, where V is the volume of 1 *gm* of vapour at pressure p.

So that, $$p - p' = P/V.\rho. \qquad ...(iii)$$

If P_o be the atmospheric pressure, V_o, the volume of the vapour at this pressure and σ_0, its density, we have

$$p.V = P_0 V_0 = P_0/\sigma_0 \qquad [\because V_0 = 1/\sigma_0.$$

Thus, $$p.\sigma_0 = P_0/V = P_0.\sigma. \quad \text{Or, } \sigma = p.\sigma_0/P_0. \qquad ...(iv)$$

And, clearly, $V = P_o V_o/p$.

So that, substituting this value of V in relation (iii) above, we have

$$p - p' = \frac{P}{\rho} \times \frac{p}{P_0.V_0} = \frac{P.p.\sigma_0}{P_0 \rho}. \qquad [\text{putting } \sigma_0 \text{ for } 1/V0.$$

Or, $$\frac{p - p'}{p} = \frac{P.\sigma_0}{P_0.\rho}.$$

Or, putting $p - p' = dp$, we have $$\frac{dp}{p} = \frac{P}{P_0}.\frac{\sigma_0}{\rho}, \qquad ...(v)$$

where P/P_0 is the osmotic pressure in atmospheres, and σ_0/ρ, the ratio of the densities of the solvent in the gaseous and the liquid states at *N.T.P.**

**N.T.P.* stands for *normal* temperature and pressure, *i.e.*, for 0°C temperature and 76 *cm* pressure, or a pressure of one atmosphere.

This relation thus gives the ratio between lowering of the vapour pressure and the vapour pressure of the solvent. It is true only, however, if h be small, which will be the case, when the concentration, and, therefore, the osmotic pressure of the solution, is small. If these be large, and, therefore, h also fairly large, the relation between p and p' is deduced as follows :

Considering a very small length, dh, of the column, we have

$$dp = -dh.\sigma.g,$$

(the negative sign merely indicating that pressure decreases as h increases)

Now, $\sigma = p.\sigma_0/P_0$. [see (*iv*) above.

$$\therefore \quad dp = -g.\frac{p.\sigma_0}{P_0}.dh. \quad \text{Or,} \quad -dh = \frac{P_0}{g.\sigma_0}.\frac{dp}{p}.$$

Integrating this expression for dh between the limits, p and p', we have

$$\int dh = -\frac{P_0}{g.\sigma_0}\int_p^{p'}\frac{1}{p}.dp.$$

Or, $$h = -\frac{P_0}{g.\sigma_0}.[\log p]_p^{p'} = -\frac{P_0}{g.\sigma_0}.(\log p' - \log p).$$

Or, $$h = \frac{P_0}{g.\sigma_0}.(\log_p - \log p') = \frac{P_0}{g.\sigma_0}.\log\frac{p}{p'}.$$

Or, $$\log\frac{p}{p'} = \frac{h.g.\sigma_0}{P_0}$$

But $P = h.\rho.g$, whence, $h = P/\rho.g$.

$$\therefore \quad \log\frac{p}{p'} = \frac{P}{\rho.g}\times\frac{g\sigma_0}{P_0} = \frac{P}{P_0}.\frac{\sigma_0}{\rho}. \quad ...(vi)$$

This gives the relation between p and p' *for large values of concentration and osmotic pressure of a solution.*

Since a dilute solution behaves as a gas, the volume occupied at *N.T.P.* by one gram-molecule of a substance in solution would be the same as it would occupy under the same conditions, (*i.e.*, at *N.T.P.*) in the gaseous state, *viz.*, 22.4 *litres* ; for, we know that one *gram molecule* of gas at *N.T.P.* occupies 22.4 *litres*.

It follows, therefore, that the osmotic pressures for a solution of one gram-molecule of a substance in a litre would be 22.4 atmospheres.

Now, the volume occupied by 1 *gram-molecule* of water vapour, *i.e.*, 18 *gm* of it, is 22.4 *litres* or 22,400 cc. at *N.T.P.*

$\therefore$ density of the vapour σ_0 at *N.T.P.* = 18/22400 *gm/cc.*

And, density of water, ρ = 1 *gm/cc.* ; so that, *for a molar solution, i.e.*, a solution containing 1 *gram-molecule* of the solute, we have, applying the above relation,

$$\frac{dp}{p} = \frac{P}{P_0}.\frac{\sigma_0}{p}.$$ [See relation (*v*) above.

Here, P/P_0 = 22.4 *atmospheres*, σ_0 = 18/22400 *gm/cc.*, and ρ = 1 *gm/cc.*

$$\therefore \quad \frac{dp}{p} = \frac{22.4\times 18}{22400\times 1}\times 100\% = 1.8\%.$$

Thus, the lowering of the vapour pressure of the molar solution of a non-electrolyte in water is 1.8 %.

As already indicated, the lowering will be greater in the case of an electrolyte, due to its molecules dissociating into ions.

13.11. Osmosis and Boiling Point of a Solution. We know that a liquid begins to boil when its maximum or saturation vapour pressure is equal to the external atmospheric pressure on its surface, *i.e., the boiling point of a liquid is the temperature at which its maximum vapour pressure is equal to the atmospheric pressure.* This value of the maximum vapour pressure for a liquid increases with increase of temperature.

Now, as we have seen, the vapour pressure of a solution, at a given temperature, is lower than that of the pure solvent ; so that, at the temperature at which the solvent begins to boil, its vapour pressure is equal to the atmospheric pressure, but that of the solution will be lower than it, and the solution will, therefore, have to be heated further in order that its vapour pressure may become equal to the atmospheric pressure, *i.e.*, it may begin to boil.

In other words, the boiling point of a solution is higher than that of the pure solvent.

Let us see how this rise or *elevation of the boiling point* of a solution is related to its osmotic pressure.

The elevation of the boiling point of a solution may be obtained with the help of **Claussius-Clapeyron's latent heat equation,** *viz.,*

$$\frac{dT}{T} = \frac{(V_{vap} - V_{liq})}{L} \cdot dp,$$

where V_{vap} and V_{liq} are the volumes of the vapour and the liquid respectively, L, the latent heat of the liquid, T, the boiling point of the solvent, dp, the lowering of the vapour pressure of the solution, and dT, the elevation of the boiling point produced.

The value of dp to be used in the above equations is obtained from relation (*v*), § 13.10, above. *viz.*, $dp/p = P.\sigma_0/P_0 . \rho$.

Or, it may be determined directly as follows:

Let A and B be the solution and the solvent, separated by a semipermeable membrane and enclosed inside a bell jar, (Fig. 13.4), such that both are at their boiling points $(T + dT)^0$ and T^0 Absolute, respectively, so that their vapour pressures are equal, and their vapours are in free communication with each other above the membrane, the system being in equilibrium.

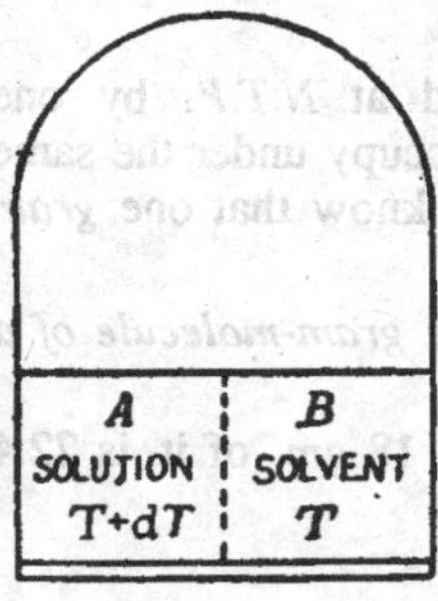

Fig. 13.4

Treating the arrangement as a reversible heat engine, working between the solution and the solvent as source and sink respectively, consider the following cycle of operations:

(*i*) Let a small quantity *Vcc.* of the solvent pass through the membrane into the solution and dilute it. Then, work done in this *expansion* of the solute is equal to the osmotic pressure of the solution × change in volume, *i.e.,*

$$P \times V \text{ ergs},$$

where P is the osmotic pressure for the solution in *dynes per sq.cm.*

(*ii*) Let these V *cc.*, or $V.\rho$ *gm.*, of the solvent be evaporated from the solution in A, where ρ is the density of the solvent.

If L be the latent heat of vaporization of the solvent, in *ergs per gm.*, the total heat absorbed by the mass $V.\rho$. *gm* of the solvent will be equal to $(V.\rho.L + P.V)$ *ergs.*

(*iii*) Let this vapour pass on from the space above A to that above B. Obviously, *no work is done during this passage of the vapour*, as the pressure is the same above A and B.

(*iv*) And, finally, let this vapour condense back into liquid in B, giving up $V.\rho.L$ *ergs* of heat, thus completing the cycle.

Then, we have, from the theory of the reversible heat engine,

$$\frac{heat\ abosorbed\ from\ the\ source\ (A)}{heat\ given\ to\ the\ sink\ (B)} = \frac{temperature\ of\ source}{temperature\ of\ sink}.$$

Or, $$\frac{V.\rho.L + P.V}{V.\rho.L} = \frac{T + dT}{T}$$

Or, $$\frac{V.\rho.L + PV - V.\rho.L}{V.\rho.L} = \frac{T + dT - T}{T}. \text{ Or, } \frac{V.V}{V.\rho.L} = \frac{dT}{T},$$

whence, $$dT = T.\frac{P}{\rho L}.$$

Thus, the elevation of the boiling point of a solution is directly proportional to its osmotic pressure. It follows, therefore, that the boiling points of isotonic solutions, in the same solvent, will be the same.

Let us now calculate the elevation of the boiling point for a solution of one *gram-molecule in one litre of water* at 100°*C*, under the normal atmospheric pressure.

We know that the osmotic pressure of a solution of 1 *gram-molecule in one litre is* 22.4 *atmospheres*, at *N.T.P.*, (see page 431). Therefore, osmotic pressure at 100°*C* and normal atmospheric pressure, (76 *cm* of mercury), will be = 22.4 × 373/ 273 *atmospheres*,

i.e., $$P = \frac{22.4 \times 373}{273} \times 76 \times 13.59 \times 981\ dynes/cm^2.$$

Now, $L = 540\ calories = 540 \times 4.19 \times 10^7\ ergs/gm$, $[\because J = 4.19 \times 10^7\ ergs/cal.$

$\rho = 0.958\ gm/cc$, *at the boiling point* of water,

and $T = 100°C = 100 + 273 = 373°\ Abs.$

Substituting these values in the relation for dT above, we have

$$dT = \frac{373 \times 22.4 \times 373 \times 76 \times 13.59 \times 981}{273 \times 0.958 \times 540 \times 4.19 \times 10^7} = 0.534^0\ C.$$

The elevation of the boiling point of water per 1 gram-molecule in 100 cc of it is known as the **molecular elevation of the boiling point,** *and will clearly be* 0.534 × 10 or 5.34°C, (because $dT \propto C$ and *C*, here, will be 10 times as great).

13.12. Osmosis and Freezing point of a Solution. A liquid will freeze, as we know, when its temperature is reduced until its vapour pressure equals that of its solid. Therefore, since a solution has a lower vapour pressure than the solvent, it will freeze at a lower temperature than the pure solvent. In other words, *the freezing point of a solvent is lowered or depressed, when a solute forms a solution with it.* Let us calculate this depression of the freezing point of a solution.

Consider, again, a solvent *B* and a solution *A*, separated by a semipermeable membrane, and enclosed under a bell jar. Let both be at their freezing points, *T* and *T*– *d T* respectively, the system being in equilibrium. Treating the arrangement, again, *as a reversible heat engine*, we take the solvent here to be the source, and the solution to be the sink, because the former is at a higher temperature than the latter. Now, let the following cycle of operations be performed:

(*i*) Let *V cc*, or *V.ρ gm* of the solvent pass through the membrane into the solution. Then, if *P* be the osmotic pressure for the solution, work done is equal to *P.V ergs*.

(*ii*) Let the solution be frozen, so that *V cc* of pure solvent separates out, giving up heat, equal to (*V.ρ. L* – *P.V*) *ergs*, where ρ is the density of the solvent at temperature *T*° and *L*. its latent heat of liquefaction, or latent heat of fusion of its solid, (*in ergs per gram*).

(*iii*) Let this frozen solid be now transferred to the solvent *B*.

(*iv*) And, finally let this solid be melted back to liquid in *B*, taking up heat equal to $V.\rho.L$ *ergs* during the process.

Then, as before, we have

$$\frac{\textit{heat absorbed from the source (B)}}{\textit{heat rejected to the sink (A)}} = \frac{\textit{temperature of source}}{\textit{temperature of sink}}.$$

Or, $$\frac{V.\rho.L}{V.\rho.L - P.V} = \frac{T}{T - dT}.$$

Or, $$\frac{(T - dT) - T}{T} = \frac{(V.\rho.L - P.V) - V.\rho.L}{V.\rho.L}.$$

Or, $$-\frac{dT}{T} = -\frac{P.V}{V.\rho.L} = -\frac{P}{\rho.L}. \quad \text{Or,} \quad \frac{dT}{T} = \frac{P}{\rho.L}.$$

Or, the *depression of the freezing point*, $dT = T.P/\rho L$,

a result similar to that for the elevation of the boiling point of a solution.

Since the value of *L* is smaller in this case than in the first case, (because the latent heat of solidification of a liquid is always smaller than its latent heat of vaporization), the value of *dT* will be greater here. Let us calculate the depression of the freezing point of water, (*i.e.*, *dT*), when 1 *gram-molecule* of a solute is dissolved in a *litre* of it at 0°*C* and under normal atmospheric pressure, (76 *cm*). We have

$T = 0°C = 273°$ *Abs.*, $P = 22.4 \times 76 \times 13.59 \times 981$ *dynes/cm*²,

$L = 80$ *calories/gm.* $= 80 \times 4.19 \times 10^7$ *ergs/gm.* and

$\rho = 1$ *gm./cc.*

Substituting these values in the relation for dT, above, we have

$$dT = \frac{273 \times 22.4 \times 76 \times 13.59 \times 981}{1 \times 80 \times 4.19 \times 10^7} = 1.85°C.$$

The depression of the freezing point of water produced by 1 *gram-molecule of a solute dissolved in* 100 *cc. of it is called the* **molecular depression of the freezing point**, *and will clearly be equal to* $1.85 \times 10 = 18.5°C$.

13.13. Determination of Percentage of Dissociation of an Electrolyte. It must be emphasized again that *the above relations are true only for non-electrolytes, i.e.*, for substances like sugar, where the solute exists in the form of single molecules. A greater effect, (*i.e.*, elevation of boiling point or depression of freezing point) than that obtained by the above relations is observed in the case of an electrolyte, because of its dissociation into ions. The deviation of the observed effect enables the percentage dissociation of the solute to be determined. For example, if we take a substance, like potassium chloride, sodium chloride, or sodium nitrate, each molecule of which breaks up into two ions, we can calculate its *percentage ionisation* as follows:

Suppose these are α molecules of the solute, and β of them dissociate into two ions each, so that the total number of particles is $(\alpha - \beta + 2\beta) = (\alpha + \beta)$. Then,

$$\frac{\textit{observed effect } dT'}{\textit{calculated effect } dT} = \frac{\alpha + \beta}{\alpha},$$

where dT' and dT are the observed and calculated elevations of boiling point, or depression of freezing point, respectively.

Or, $$dT' = dT\left(\frac{\alpha + \beta}{\alpha}\right) = dT\left(1 + \frac{\beta}{\alpha}\right)$$

Or, $$dT.\beta/\alpha\; dT' - dT \quad \text{whence,} \quad \frac{\beta}{\alpha} = \frac{(dT' - dT)}{dT}.$$

Or, $\quad$ *percentage of ionization* $= \dfrac{dT' - dT}{dT} \times 100.$

Similar treatment will also apply to substances whose molecules dissociate into more than two ions each.

A perfectly dilute solution is supposed to dissociate completely into its ions, and the degree of dissociation decreases with concentration, so that *a concentrated solution of an electrolyte behaves, more or less, like a solution of a non-electrolyte.*

The reverse of the above happens in the case of colloids. The substance in a solution, here, forms clusters or aggregations of molecules, having different sizes in different cases. In such cases, the osmotic pressure, and, therefore, the elevation of the boiling point, or the depression of the freezing point, is very much smaller than in the normal case of a non-electrolyte, where the whole solute exists in the form of single molecules.

13.14. Determination of molecular weight of a substance from elevation of boiling point, or depression of freezing-point of a solution of the substance.

(*a*) **From elevation of the boiling point.** Suppose we have *x gm* of the given substance dissolved in 100 *cc* of water, and the elevation of the boiling point observed is dT. Then, since an elevation of 5.34°*C* is due to 1 *gram-molecule* in 100 *cc* the number of gram-molecules in the solution must be $dT/5.34$. This number of *gram-molecules* is obviously present in *x grams* of the substance taken.

So that, $dT/5.34$ *gram-molecules* of the substance weigh *x gm*. And, therefore, 1 *gram-molecule* of the substance would weigh $5.34 \times x/dT$ *gm.*

Or, $\quad$ the *molecular weight of the substance*, $(M) = 5.34 \times x/dT$ *gm.*

(*b*) **From depression of the freezing point.** Again, if *x* grams of the substance dissolved in 100 *cc* of water lower, or depress, its freezing point by dT, we have

number of *gram-molecules* in *x grams* of the substance $= dT/18.5$, because a lowering of the freezing point by 18.5°*C* indicates the presence of 1 *gram-molecule* in 100 *cc*.

Or, $\quad dT/18.5$ *gram-molecules* of the substance weigh *x gm.*

And $\quad \therefore$ 1 *gram-molecule* of it would weigh $18.5 \times x/dT$ *gm.*

Or, $\quad$ *molecular weight of the substance* $= 18.5\, x/dT$ *gm.*

WORKED EXAMPLES

1. One gram of sugar is dissolved in water, making a solution occupying 1000 *cc*. Find the osmotic pressure for the solution at 0°C and 1000°C. Molecular weight of sugar = 342 gm. ; *R* for 1 gm molecule = 8.4×10^7.

We know that the relation connecting osmotic pressure, volume and temperature of a dilute solution, is $PV = RT$. And, therefore, $P = RT/V$.

(i) **At 0°C.** $\quad$ Here, $R = 8.4 \times 10^7$ *ergs/deg. C*, $\quad T = 0 + 273 = 273°$ *Abs.*,

and $\quad V = 342 \times 1000 = 342 \times 10^3$ *cc.*,

because a mass of 1 *gm* of sugar occupies 1000 *cc.*, a mass of 342 *gm.* would occupy $342 \times 1000 = 342 \times 10^3$ *cc.*

Therefore, $\quad P = \dfrac{8.4 \times 10^7 \times 273}{342 \times 10^3} = \dfrac{8.4 \times 10^4 \times 273}{342} = 6.707 \times 10^4$ *dynes/cm²*

Or, the osmotic pressure for the solution, at 0°*C*, is equal to 6.707×10^4 *dynes/cm²*.

(ii) **At 100°C.** $\quad$ Again, $R = 8.4 \times 10^7$ *ergs/deg. C*, $T = 100 + 273 = 373°$ *Abs.*

and $V = 342 \times 1000 = 342 \times 10^3$ *cc.*

So that, substituting these values in the relation $P = RT/V$, we have

$$P = \frac{8.4 \times 10^7 \times 373}{342 \times 10^3} = \frac{8.4 \times 10^4 \times 373}{342} = 9.162 \times 10^4 \textit{ dynes/cm}^2.$$

2. One gram of salt is dissolved in water to make a solution of 100 cc. Assuming the salt to be entirely dissociated, find, in atmospheres, the osmotic pressure of the solution at 20°C.

Atomic weights.—Na = 23, Cl = 35.5, Density of mercury = 13.6 gm/cc and R = 8.4 × 10⁷ for 1 gm molecule.

Each molecule of the salt (*NaCl*) dissociates into two ions, *viz., Na* and *Cl*. And, since the salt is completely dissociated, *there will be twice as many particles present in 1 cc of the solution as the number of whole molecules*. It follows, therefore, that the osmotic pressure will be double of that obtained by the relation, $P = RT/V$, *i.e.*, the osmotic pressure, in this case, is given by

$$P = 2 \times RT/V.$$

Now, $R = 8.4 \times 10^7$ *ergs/deg. C*; $T = 20 + 273° = 293°$ *Abs.*

and $V =$ *mol wt* of salt $\times$ 100 *cc*.

because 1 *gm* is contained in 100 *cc.*, and, therefore, its *mol. wt.* is contained in *mol.wt.* × 100 *cc.*

Or, $V = (23 + 35.5) \times 100\ c.c = 58.5 \times 10^2\ cc.$

$$\therefore \quad P = \frac{2 \times 8.4 \times 10^7 \times 293}{58.5 \times 10^2} = \frac{2 \times 8.4 \times 10^5 \times 293}{58.5}\ dynes/cm^2.$$

$$i.e., \text{osmotic pressure } P = \frac{2 \times 8.4 \times 10^5 \times 293}{58.5 \times 76 \times 13.6 \times 981} = 8.299 \text{ atmosphere}$$

[$\because$ 1 *atmosphere* $= 76 \times 13.6 \times 981$ *dynes/cm²*.)

3. When 1 gm of iodine is dissolved in ether making 40 cc of solution, the osmotic pressure at 20°C is 2.4 atmospheres. Show that the molecule of dissolved iodine consists of two atoms. (At. wt. of iodine = 127).

Here, observed osmotic pressure = 2.4 *atmospheres*.

Now, let us calculate the osmotic pressure from the given data.

We have $P = RT/V$.

Since $R = 8.4 \times 10^7$ *ergs/deg. C*; $T = 20 + 273 = 293$ *Abs.* and $V = 127 \times 40\ cc.$,

(because iodine being monotomic, its *mol. wt.* is equal to its *atomic weight*),

we have $$P = \frac{8.4 \times 10^7 \times 293}{127 \times 40}\ dynes/cm^2 = 4.778 \text{ or } 4.8\ atm.$$

which is equal to 2×2.4 atmospheres.

Thus, the observed osmotic pressure is half of the calculated osmotic pressure and, therefore, it follows that the two atoms of iodine in solution combine to form one molecule, or that *the molecule of dissolved iodine consists of two atoms*.

4. Calculate the strength of a cane-sugar solution whose osmotic pressure at 270°C is one atmosphere. Mol. wt. of sugar = 342 gm and R = 8.4 × 10⁷ ergs/deg. C.

We have the relation $PV = RT$. Or, $V = RT/P$,

where V is the volume of the solution, containing 1 *gram-molecule* of the substance,—in this case, 342 *gm* of sugar.

Thus, *V cc*, of the solution contain 342 *gm* of sugar,

1000 *c.c.* " " " $\frac{342 \times 1000}{V}$ *gm* of sugar.

Or, strength of the sugar solution $= 342 \times 1000 / V$

$= 342 \times 1000 \times P/RT$ *grams/litre* $\quad V = RT/P$

Here, $P = 1$ *atmosphere* $= 76 \times 13.6 \times 981$ *dynes/cm²*.

$R = 8.4 \times 10^7$ *ergs/deg. C.*, and $T = 27 + 173 = 300°$ *Abs.*

$$\therefore \quad \text{strength of the sugar solution} = \frac{342 \times 1000 \times 76 \times 13.6 \times 981}{8.4 \times 10^7 \times 300} = 13.76\ gm/litre.$$

5. The osmotic pressure of a solution containing 6 gm of cane sugar in 100 cc of water is 307 cm of Hg at 13°C. What is the osmotic pressure of a solution of one-sixth this concentration at 50°C ? *(London Inter-Science)*

Let the *molecular weight* of sugar be = *m gm*.

Then, the volume containing 1 *gm mol* in the first case, is, say, V_1, where

$$V_1 = \frac{m \times 100}{6}\ cc, \qquad [\because 6\ gm \text{ are contained in } 100\ cc \text{ and } \therefore m\ gm \text{ in } \frac{100 \times m}{6}\ cc.$$

and the volume in the second case, say, $V_2 = m \times 100\ cc.$,

(because now only 1 gm is contained in 100 *cc*.)

And temperature T_1 in the *1st case* = 13 + 273 = 286° *Abs*.

T_2 „ „ *2nd case* = 50 + 273 = 323° „

If P_1 and P_2 be the osmotic pressures in the two cases, we have

$$P_1 = RT_1/V_1, \quad \text{and} \quad P_2 = RT_2/V_2,$$

where R is the constant for the solution, — the same as the gas constant.

$$\therefore \quad \frac{P_2}{P_1} = \frac{RT_2}{V_2} \cdot \frac{V_1}{RT_1} = \frac{T_2}{T_1} \cdot \frac{V_1}{V_2}. \quad \text{And} \therefore \; P_2 = \frac{T_2}{T_1} \cdot \frac{V_1}{V_2} \cdot P_1.$$

Substituting the values of T_1, T_2, V_1, V_2 and P_1, we have

$$P_2 = \frac{323 \times m \times 100 \times 307 \times 13.6 \times 981}{286 \times 6 \times m \times 100}$$

because $P_1 = 307$ *cm* of mercury = 307 × 13.6 × 981 *dynes/cm*^{2}

$$\text{Or,} \quad P_2 = \frac{323 \times 307 \times 13.6 \times 981}{286 \times 6} \textit{ dynes / cm}^2.$$

$$= \frac{323 \times 307 \times 13.6 \times 981}{286 \times 6 \times 13.6 \times 981} = \frac{323 \times 307}{286 \times 6} = 57.77 \textit{cm of Hg}.$$

Thus, the osmotic pressure in the second case = 57.77 *cm* of mercury.

6. Find the percentage lowering of the vapour pressure of water produced by dissolving 1 gm of sugar in 100 cc (Mol. wt. of sugar = 342 gm).

We have the relation for lowering of vapour pressure,

$$\frac{dp}{p} = \frac{P}{P_0} \times \frac{\sigma_0}{\rho}.$$

[See page 430 relation (*v*).

$$\therefore \text{ percentage lowering of vapour pressure } = \frac{dp}{p} \times 100 = \frac{P}{P_0} \times \frac{\sigma_0}{\rho} \times 100,$$

where P/P_0 is the osmotic pressure of the solution *in atmospheres*, and σ_0/ρ, the ratio of the densities of the solvent in the *gaseous* and the *liquid* states, at *N.T.P*.

Now, $P = RT/V$,

and, here, $T = 273°$ *Abs*., $V = 342 \times 100\,cc$, [taking the solution at 0°C.

and $R = 8.4 \times 10^7$ *ergs/deg. c*

$$\therefore \quad P = \frac{8.4 \times 10^7 \times 273}{342 \times 10^2} \textit{ dynes/cm}^2.$$

$$\text{And, hence} \quad \frac{P}{P_0} = \frac{8.4 \times 10^5 \times 273}{342 \times 76 \times 13.6 \times 981} \textit{ atmospheres},$$

because $P_o = 76$ *cm* of *Hg* = 76 × 13.6 × 981 *dynes/cm*^{2}.,

$$\sigma_0 = \frac{18}{22400}$$

[∵ 1*gm mol*. of water vapour, or 18 *gm* of water vapour, at *N.T.P*., occupies 22.4 *litres, or 22400 c.c*.

and ρ for water at *N.T.P*. = 1 *gm/cc*.

$$\therefore \quad \frac{dp}{p} \times 100 = \frac{8.4 \times 10^5 \times 273}{342 \times 76 \times 13.6 \times 981} \times \frac{18}{22400} \times 100\% = 0.05316\%$$

∴ lowering of vapour pressures of water is equal to 0.05316%.

7. By how much will the freezing point of the sugar solution in question 6, above, be depressed ?

Latent heat of ice = 80 cal/gm, and atmospheric pressure = 10^6 dynes/cm^2.

We have, here, 1 *gm* of sugar in 100 *cc* of water, or 1/342 *gm molecules* in 100 *cc* of water, and, therefore, 10/342 *gm molecules in* 1000 *cc*, or *one litre* of water.

Now, 1 *gm molecule* in 1 litre of water exerts an *osmotic pressure* = 22.4 *atmospheres at N.T.P*.

∴ 10/342 *gm molecules in* 1 *litre* of water will exert an osmotic pressure

$$P = \frac{22.4 \times 10}{342} = \frac{224}{342} \textit{ atmospheres} = \frac{224}{342} \times 10^6 \textit{ dynes/cm}^2.$$

[1 *atm*. = 10^6 *dynes/cm*^{2}.

Now, $dT = T \cdot \frac{P}{\rho.L.}$ [see page 433

Here, $T = 0°C = 273° \; Abs., P = \frac{224 \times 10^6}{342} \; dynes/cm^2,$

$\rho = 1 \; gm/cc$ and $L = 80 \; cal./gm. = 80 \times 4.2 \times 10^7 \; ergs \; per \; gm.$

∴ depression of the freezing point, *i.e.*, $dT = \frac{273 \times 224 \times 10^6}{342 \times 1 \times 80 \times 4.2 \times 10^7} = \frac{273 \times 22.4}{342 \times 80 \times 4.2} = 0.05322°C.$

8. Determine the elevation of the boiling point of ether produced by solving 10 gm of carbon hexachloride (C_2Cl_6) in 100 *cc* of ether.

The boiling point of ether is 35°C, its density at 35°C = 0.695 gm per cc and its latent heat of vaporization is 81.5 cal/gm. (Take atmospheric pressure = 10^6 dynes/cm^2).

Here, molecular weight of carbon hexachloride ($C_2 Cl_6$) = 24 + 213 = 237 *gm*.

Hence, 100 cc of ether contain 10 *gm* of carbon hexachloride, or (10/237) of 237, or 10/237 *gram-molecules* of carbon hexachloride.

∴ 1000 *cc*. or 1 *litre* of ether will contain $\frac{10}{237} \times \frac{1000}{100}$, or $\frac{100}{237}$ *gram-molecules* of the hexachloride.

Now, osmotic pressure for a solution of 1 *gram-molecule* in 1 *litre* is 22.4 *atmospheres at N.T.P*.

∴ osmotic pressure for a solution of 100/237 *gram-molecules* in 1 *litre at N.T.P*. = 100 × 22.4/237 *atmospheres*.

And ∴ osmotic pressure for a solution of 100/237 *gram-molecules* in 1 *litre*, at 35°C, or (273+35) or 308° *Abs*. is given by

$$P = \frac{100 \times 22.4}{237} \times \frac{308}{273} \; atmospheres = \frac{2240 \times 308 \times 10^6}{237 \times 273} dynes/cm^2.$$

Now, elevation of the boiling point $dT = T.P/\rho.L.$ [See page 433.

Here, $T = 35 + 273 = 308° \; Abs., P = \frac{2240 \times 308 \times 10^6}{237 \times 273} \; dynes/cm^2$

$\rho = 0.695 \; gm/cc$ and $L = 81.5 \; cal/gm = 81.5 \times 4.2 \times 10^7 \; ergs/gm.,$ (taking $J = 4.2 \times 10^7 \; ergs/cal.$).

$$\therefore \quad dT = \frac{308 \times 2240 \times 308 \times 10^6}{237 \times 273 \times 0.695 \times 81.5 \times 4.2 \times 10^7} = 1.38°C.$$

Or, the elevation of the boiling point of ether = 1.38°C.

9. A one per cent solution of potassium chloride (KCl) has an osmotic pressure 6.604 atmospheres at 270°C. Determine its degree of dissociation. (At. wts : K = 39, Cl = 35.5).

Here, molecular weight of potassium chloride is equal to 39 × 35.5 *gm*.

∴ 74.5 *gm* of potassium chloride dissolved in 1000 *cc* of water would exert a pressure of 22.4 atmospheres at *N.T.P.*, if *undissociated*.

The solution given, however, contains 1 *gm* of *KCl* in 100 *cc* or 1/74.5 *gm– molecules* in 100 *cc, or 10/74.5 gm molecules* in 10000 *c.c*.

∴ osmotic pressure of the solution = 22.4/74.5 *atmospheres at N.T.P*.

Hence, osmotic pressure of the solution, at 27°C.

$$= 22.4 \times \frac{10}{74.5} \times \left(\frac{273 + 27}{273}\right) = \frac{224 \times 300}{74.5 \times 273} = 3.303 \; atmospheres.$$

Or, osmotic pressure of the solution, if no dissociation takes place = 3.303 *atmospheres*.

But the observed osmotic pressure = 6.604.

$$\therefore \quad i = \frac{6.604}{3.303} = 1.999, \text{ because } i = \frac{observed \; effect}{calculated \; effect}.$$

This means that instead of there being 1000 particles, 1999 particles are actually exerting pressure. Thus,

1000 particles dissociate into 1999 particles.

And ∴ 100 " " " $\frac{1999}{1000} \times 100 = 199.9$ particles

Or, dissociation = 199.9–100 = 99.9%.

Thus, degree of dissociation of the *KCl* solution = *99.9%*.

10. 31.88 gm of a substance is dissolved in 2 litres of water. The boiling point of the solution is 0.266°C higher than that of pure water, under normal atmospheric pressure. Calculate the molecular weight of the substance.

Here, 2000 *cc* of water contain 31.88 *gm* of the substance.

$\therefore$ 100 *cc* of water contain $\frac{31.88}{2000} \times 100 = \frac{31.88}{20}$ *gm* of the substance.

Now, we know that the molecular weight of a substance, *i.e.*,

$$M = \frac{5.34x}{dT} \, gm.$$ [See page 431.

Here, $x = 31.88/20$ *gm* and $dT = 0.266$°C.

$\therefore$ *molecular weight* of a substance, *i.e.*, $M = \frac{5.34 \times 31.88}{0.266 \times 20} = 32.00$ *gm.*

EXERCISE XIII

1. Enunciate Fick's law of diffusion and find the dimensions of the coefficient of diffusion.

Starting with the same concentration of a given solute at the bottom of columns of the same liquid, show that the times required to set up a given concentration at the top are proportional to the squares of the heights of the columns. *(Madras)*

2. State Fick's law of diffusion in liquids. Compare diffusion in liquids with heat conduction.

Indicate a method of determining experimentally the coefficient of diffusion of a salt in a solution. *(Madras)*

3. What do you understand by osmosis, dialysis and diffusion ? State the laws of osmosis and describe an arrangement for measuring osmotic pressure. *(Calcutta)*

4. State the laws of osmotic pressure.

Give an account of the relation between osmotic pressure and other properties of a solution. *(Madras)*

5. Calculate the osmotic pressure of a sugar solution, (*mol.wt.* = 342 *gm.*), containing 1.5 *gm* of sugar *per litre*, at 37.0°C. **Ans.** 8.478 *cm of Hg.*

6. What is the strength of a sugar solution whose osmotic pressure at 17°*C* is 1.5 *atmospheres.* (1 *atmosphere pressure* $=10^6$ *dynes/cm*^{2}, and $R = 8.4 \times 10^7$ *ergs/gm.mol per degree Centigrade).* **Ans.** 21.06 *gm/litre.*

7. Define osmotic pressure and state its laws. How will you demonstrate this pressure ? Deduce an expression for the elevation of the boiling point of liquid by a non-volatile substance dissolved in it. *(Madras)*

8. Calculate the osmotic pressure of a one per cent solution of sodium chloride (*NaCl*) at 27°*C*, assuming the dissociation of the molecule to be 99.5% (*At. wts : Na* = 23.0, *Cl* = 35.5) **Ans.** 8.013 *atmospheres.*

9. Determine the percentage lowering of the vapour pressure of an aqueous solution of a non-electrolyte, of *mol. wt.* 100 *gm*, when 1 *gm. of it is dissolved in 100 cc* of water at *N.T.P.* ($R = 8.4 \times 10^7$/*gm, mol per degree C*). **Ans.** 0.1818%.

10. Deduce the relationship between osmotic pressure and relative lowering of vapour pressure. How is osmotic pressure measured ? *(Madras)*

11. A solution of 5 *gm.* of an electrolyte in water, total volume 1000 *c.c.*, freezes at- 0.279°*C*. What is the molecular weight of the substance in solution ? (Molecular lowering of the freezing point = 18.6°*C*) **Ans.** 33.3 *gm.*

12. What will be the boiling point of a solution of cane sugar, (*mol.wt.*=342), containing 1 *gram* of sugar per 100 *c.c.* of it, under the normal atmospheric pressure ? Density of water at 100°*C* may be taken to be 0.9580 *g./c.c*

Ans. 100.0156°*C.*

13. 12.5 *gms of sugar (mol. wt.* = 342) dissolved in 1000 *cc.* of water lowers the freezing point by 0.068°*C*. Calculate the molecular lowering of the freezing point for water. **Ans.** 18.6°*C*

14. In terms of the molecular theory of matter, discuss the phenomena associated with (*a*) diffusion, (*b*) osmosis. *(Bombay)*

15. Calculate the freezing point of a salt solution containing 1 *gm.* of salt in 1000 *c.c.* of water, assuming that the salt is completely dissociated into sodium and chlorine ions. (*At. wts. : Na* = 23, *Cl* = 35.5). **Ans.** –0.064oC.

16. The osmotic pressure of a solution of 3 *gm.* of acetone dissolved in water to make 100 *cc* is 7.75 *atmospheres* at 10°*C*. Calculate the molecular weight of acetone. ($R = 8.4 \times 10^7$ *ergs/gm. mol.per deg.C.*). **Ans.** 60.

17. 1.2 *gm.* of a substance dissolved in 24.5 *gm* of water caused a freezing point depression of 1.05°*C*. What is the molecualr weight of the substance ? Molecular lowering of freezing point for water may be taken to be 18.6°C.

Ans. 86.8 *gm.*

18. 12.5 *gm* of sugar (*mol wt* = 360) dissolved in 1 *litre* of water lowers the freezing point by 0.065°*C*. Calculate the molar lowering of the freezing point of water. *(Delhi)*

Ans. 18.72°*C.*

CHAPTER XIV

SURFACE TENSION-CAPILLARITY

14.1. Molecular Forces. Since surface tension is essentially a molecular phenomenon, we had better have first a clear idea as to what forces operate between molecules.

There are two types of molecular forces : (*i*) **forces of adhesion, or adhesive forces, and** (*ii*) **forces of cohesion or cohesive forces.**

(*i*) *Adhesion is the force of attraction between molecules of different substances*, and is defferent for different pairs of substances, *e.g.*, gum has a greater adhesive force than water or alcohol.

(*ii*) *Cohesion, on the other hand, is the force of attraction beween molecules of the same substance*. This force is different from the ordinary gravitational force and does not obey the ordinary inverse square law, the force varying inversely probably as the eighth power of the distance between two molecules and thus decreases rapidly with distance, — in fact it is *'appreciable when the distance between two molecules is inappreciable and becomes inappreciable when the distance is appreciable*.* It is the greatest, in the case of solids, less in the case of liquids and the least in the case of gases, almost negligible at ordinary temperature and pressure, when the molecules lie very much further apart for it to be appreciable. This explains at once why *a solid has a definite shape, a liquid has a definite free surface and a gas has neither*.

14.2. Molecular range – Sphere of Influence. The maximum distance up to which the force of cohesion between two molecules can act is called their *molecular range*, and is generally of the order of 10^{-7} *cm* in the case of solids and liquids, being different for different substances. *A sphere drawn around a molecule as centre, with a radius equal to its molecular range is called the sphere of influence*** *of the molecule*. Obviously, the molecule is affected only by the molecules inside this sphere, *i.e.*, it attracts and is, in turn, attracted by them, remaining unaffected by the molecules outside it, as they lie beyond its range of attraction. **Laplace** (1806) and **Gauss** (1830) were the first to have evolved this theory of cohesive force between molecules in order to satisfactorily explain the various effects of surface tension, like capillarity etc.

14.3. Surface Tension. It is a general experience that a liquid in small quantity, free from any external force, like that due to gravity, will always assume the form of a spherical drop - *e.g.*, rain drops, small quantities of mercury placed on a clean glass plate etc.

Now, for a given volume, a sphere has the least surface area. Thus, a liquid always tends to have the least surface area. The following experiments beautifully illustrate this tendency of a liquid to decrease its surface area.

1. If we take a wire-ring and dip it in a soap solution, we find that, on taking it out, a thin film is formed across the ring. If we place a moistened cotton loop lightly on the film, it will remain in any form or position in which it is placed, as shown in Fig. 14.1 *(a)*; for, the loop-film lies both idside and outside the loop and at every point on the loop, therefore, there are equal and opposite forces, tending to pull it outward (due to the outside film) and inwards (due to the inside film), thus cancelling each other out. But if the film inside the loop be pricked, it disappears and the loop at once gets stretched into the form of a circular ring, as shown in Fig. 14.1*(b)*, because the inward forces having all

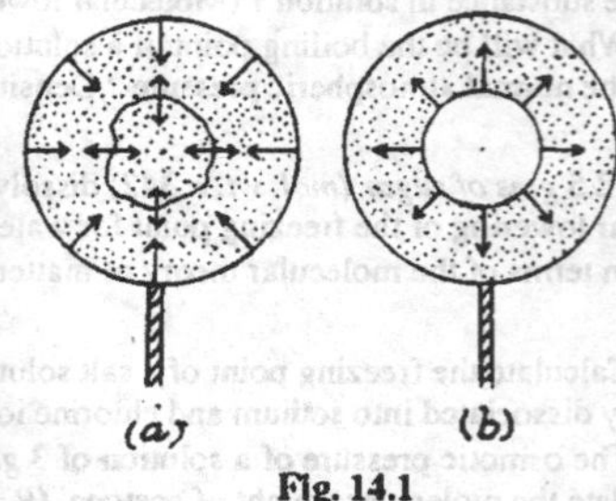

Fig. 14.1

*** As *N.K. Adam* remarked in his *'Physics and Chemistry of Surfaces,* ' if the force were gravitational in character and obeyed the inverse square law, 'the surface tension of the ocean would be far greater than that of a cup full of water, because the distant parts would act with sensible efeect.'**

****It is also sometimes referred to as th *sphere of molecular attarction, or of molecular activity.***

vanished, only the outward forces are left acting normally on it at every point. Now, *for a given perimeter, a circle encloses the greatest area;* so that the loop now encloses the *maximum area.* In other words, the area of the film left between the loop and the wire-ring is now *reduced to minimum*, clearly showing that the film has a tendency to contract or shrink, or that there is *tension* in it.

2. Let a drop of olive oil be placed, with the help of a pipette, inside a mixture of alcohol and water, of the *same density as that of the oil*, but with which it does not mix. It will be found that the drop will keep on floating in the mixture *with a perfectly spherical form*. The reason obviously is that the weight of the drop having bean exactly balanced by the upthrust of the displaced mixture, the action of gravity on it is completely eliminated and *it is thus subject to the force of surface tension alone*, which compels it to acquire a shape having the least surface area. And, since for a given volume, a sphere has the minimum area, the drop acquires this particular shape. This is referred to as *Platean's experiment*.

3. If we place a greased needle (N) on a piece of blotting paper and put the paper lightly on the surface of water, the blotting paper will soon sink to the bottom, but the *needle will remain floating* on the surface, (Fig. 14.2). Careful observation will show that there is a small depression formed below and around the needle and that the free surface of water is slightly extended. *The weight mg of the needle is here supported by the resultant tension in the depression acting upwards, as shown.* If one end of the needle be made to pierce the surface of water, it rapidly goes slantingly down to the bottom.

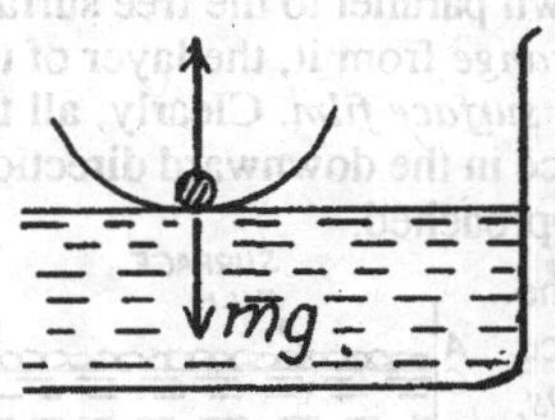

Fig. 14.2

4. If we immerse an ordinary camel-hair paint brush in water, its hair all spread out, presenting a sort of a bushy appearance, but the moment it is withdrawn, they all come closer together in a more or less compact mass, as though bound down by some sort of a contracting membrane.

5. Yet another beautiful experiment, often performed for fun by junior students, is the rapid movement of a *camphor scorpion* on water. What they do is simply to arrange pieces of camphor together, in the shape of a scorpion, and put it on water, when, due to the reduction in the surface tension of water, on account of the camphor gradually dissolving into it, the camphor is drawn or pulled aside by the surrounding uncontaminated water of a higher surface tension. And, since we have camphor dissolving more rapidly at some points than at others, this force due to surface tension is not uniform all round, with the result that the 'scorpion' scampers about haphazardly in different directions.

If, however, the water be already contaminated with some grease etc, its surface tension may be reduced to an extent that the camphor has no further possibility of reducing it. In such a case, therefore, the movement of the camphor may altogether stop.

The above experiments clearly show that the surface of a liquid behaves as though it were covered with an elastic skin or membrane, having a natural tendency to contract, with the important difference, however, that whereas in the case of the membrane or skin, the tension increases as the skin is stretched, or its surface area is increased, in accordence with Hooke's Law, it is quite independent of the area of the surface in the case of a liquid, unless the liquid film is reduced in thickness to less than 10^{-7} cm, when the tension in it decreasesw rapidly. This tension or pull in the surface of a liquid is called its **surface tension**, and may be defined as *the force per unit length of a line drawn in the liquid surface, acting perpendicularly to it at every point and tending to pull the surface apart along the line.*

14.4. Explanation of Surface Tension. An extension of the kinetic theory of gases to the case of liquids easily explains the phenomenon of surface tension and the credit for it must go to Laplace, who first attempted it,

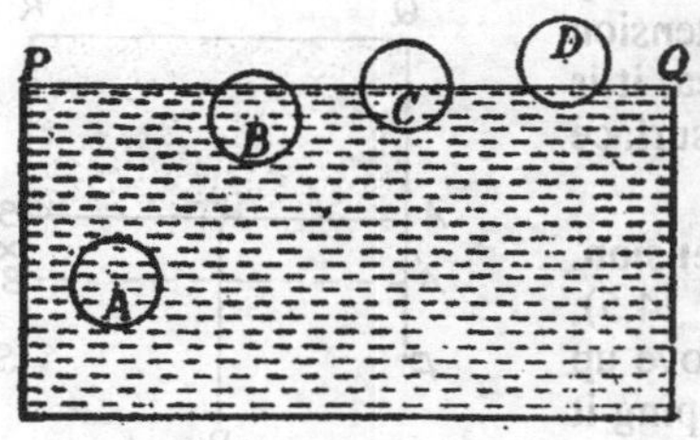

Fig. 14.3

Consider four molecules, A, B, C, and D of a liquid, with their spheres of influence drawn around them, as shown in Fig. 14.3,–sphere A *being well inside* the liquid ; B, *near the free surface of the liquid;* C, *just on the free surface and* D, *above the free surface.*

Since the sphere of influence of A lies wholly inside the liquid, it is attracted equally in all directions by the other molecules lying within its sphere of influence, so that there is no resultant cohesive force on it one way or the other, and it, therefore, merely possesses its *thermal velocity*.

The sphere of influence of molecule B, on the other hand, lies partly outside the liquid, and this part contains only a comparatively few molecules of the gas or vapour above the liquid, so that the upper half of the sphere contains fewer molecules attracting it upwards, than the lower half, attracting it downwards, and so there is a *resultant downward force* acting on B.

The molecule C lies on the surface of the liquid, so that full one-half of its sphere of influence lies above the surface of the liquid, containing only few molecules of the gas or vapour, whereas there are liquid molecules in its entire lower half, and thus, the resultant downward force in this case is the *maximum. This downward or inward force exerted per unit area of a liquid surface is called its* ***internal, intrinsic or cohesion pressure,*** and is the cause of cohesion. It is this presure which is represented by the term a/v^2 in **van der Waal's** modified gas equation.

In the case of the molecule D, which has passed out of the liquid surface, only a part of the sphere of influence lies inside the liquid, so that downward force on it decreases, and when the sphere of influence passes entirely outside the liquid surface, there is no downward force on the molecule at all, and it is free to wander about as a molecule of the vapour or gas. We thus see that all over *the surface of the liquid there is a downward pull due to attraction between the molecules.*

14.5. Surface Film and Surface Energy. *If a plane CD* be drawn parallel to the free surface layer AB of a liquid (Fig.14.4) *at a distance equal to the molecular range* from it, the layer of the liquid, lying in between the free surface and this plane, is called the *surface film*. Clearly, all the molecules in the film are being acted upon by a resultant cohesive force in the downward direction, the magnitude of which increases as the free surface of the liquid is approached.

When a molecule is brought from the interior of the liquid to the surface-film, work has to be done against the downward cohesive force acting upon it, and its potential energy is thus increased. *It follows, therefore, that the potential eneryt of the molecules in the surface film is greater than that of those inside the liquid.*

Fig. 14.4

Since, however, *the potential energy of a system tends towards a minimum,* (*i.e.*, a system tends to have the lowest possible potential energy), *the film tends to have the least surface area, (its thickness, equal to the molecular range, being already fixed) in order that the number of molecules in it may be a minimum. The potential energy per unit area of the surface film is called its* ***surface energy.***

Now, before a molecule can be brought to the surface of the liquid, room has to be made for it, and this is done by separating the adjacent molecules on the surface. *Work is, therefore, done against the cohesive forces between the molecules in the surface, and not against the internal or downward cohesive force,* for the molecules are moved in a direction perpendicular to this latter force. It thus appears that *in increasing the surface area of a liquid, work is done as though the surface under a state of tension were being extended, very much like a rubber sheet being extended.* The analogy, however, does not go far, as already stated, whereas in the case of a rubber sheet, the tension increases with extension or increase in its surface area, it is quite independent of the surface area in the case of a surface film, and is the same at all points in it.

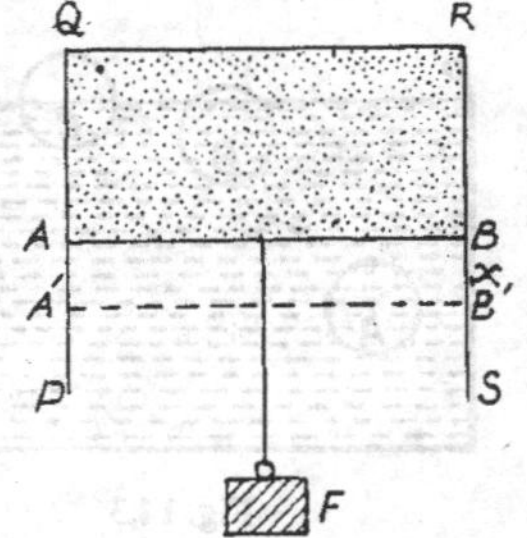

Fig. 14.5

14.6. Free Energy of a Surface and Surface Tension. Take a rectangular framework of wire $PQRS$, (Fig. 14.5), with a horizontal wire AB placed across it, free to move up and down, and form a soap-film across $AQRB$, by dipping it in a soap solution. The wire AB is pulled upwards by the surface tension of the film, acting perpnedicularly to the wire and in the plane of the film. To keep the wire in position, therefore, a force has to be applied downwards, equal and opposite to the upwards force due to surface tension. Let this downward force be equal to F including the weight of the wire AB, which is also acting downwards. Then, if T be the surface tension of the film, *i.e.*, if T be the force *per unit lenght of the film* and l, the length of the wire AB, we have upward force acting on the wire $AB = 2l.T$, because the film has *two surfaces* and *each has a surface tension T.*

Since the film is in equilibrium, it is clear that $2.l.T = F$.

Now, if the wire AB be pulled downwards through a small distance x into the position $A'B'$, *i.e.*, if the film be extended by an area $l.x$. on each side, we have

$$\textit{work done} = F \times x = 2.l.T.x.$$

The film gets cooled on being stretched, because the drawing out of the molecules from the interior against the attractive force results in a retardation of their thermal agitation, with a consequent lowering of temperature,*. It, therefore, takes up heat from the atmosphere to come to its original temperature. *This heat absorbed, together with the mechanical work done, forms the energy of the new surface area 2lx of the film formed.*

If, therefore, E be the *surface energy* of the film and Q *ergs* of heat be absorbed per unit area of the *new* surface formed (See N.B. below), we have $E \times 2l.x = 2.l\,T.x + Q.2.l.x$.

Or, $$E = T + Q.$$ [Dividing throughout by $2lx$.

Or, $$T = (E - Q) = (\textit{surface energy–heat energy per unit area})$$

$$= \textit{potential energy per unit area.}$$

i.e., $$T = \textit{work done in creating unit area of the skin of the film.}$$

*Thus, the surface tension of a liquid may be defined as the amout of work done in increasing the surface area of the liquid-film by unity, or as the mechanical part of the surface energy of the liquid film. This mechanical part of the surface energy of a liquid-film is free energy; so that, the surface tension of a liquid is equal to the free energy of the liquid-film or surface.***

N.B. To detemine the value of Q, we *may* to subject the film to a *Carnot Cycle as* explained hereunder.

(i) First, let the film be stretched *isothermally*, at temperature $\theta_1°$, A, through distance x, so that increase in its surface area $= 2lx$. If, therefore, T_1 be the surface tension of the film at temperature θ_1A, we have

$$\textit{work done on the film} = T_1 .\ 2l\ x.$$

And, if Q be the *quantity of heat energy (in ergs)* absorbed per unit area of the new surface formed, we have

$$\textit{heat absorbed} = Q\,2l\,x.$$

(ii) Let the film be now given an extremely small *adiabatic expansion*, so that its temperature falls to θ_2°A ,with no work done and with a quantity $C\,(\theta_1 - \theta_2)$ of heat evolved, where C is the *thermal capacity* of the film.

(iii) Let the film be now allowed to contract by an area $2lx$ isothermally at temperature $\theta°_2 A$. Then, if T_2 be surface tension of the film at $\theta°_2\,A$, we have

$$\textit{work done by the film} = 2\,T_2\,l\,x$$

(iv) Finally let the film be allowed to undergo extremely small *adiabatic contraction* to bring ti back to its original condition at temperature θ_1°A,. Again, no work is done and a quanity $C\,(\theta_1 - \theta_2)$ of heat is absorbed.

The two adiabatic procesecs being extremely small and their effects counter-balancing each other, we have

net work done by the film $= 2T_2 lx - 2T_1 lx$

And, net quantity of heat supplied $= Q\,2\,l\,x$

$$\therefore\ \textit{efficiency of the carnot cycle}\ \eta = \frac{2\,(T_2 - T_1)l\,x}{2l\,x\,Q} = \frac{(T_2 - T_1)}{Q} = \frac{\theta_1 - \theta_2}{\theta_1}$$

In the limit, when $(\theta_1 - \theta_2)$ is very small, we have

---

*This is clearly an example of *Le Chatelier's principle, viz., that "if one the factors of any system in equilibrium is changed, thus disturbing the equilibrium, the effect produced tends to restore that factor to its original value."* Thus, a lowering of the temperature of the surface results in a rise in its tension, which increases the force opposing enlargement of its surface. ** Surface tension of a liquid is generdly, but erroneously, defind as the surface-energy of the liquid surface. But obviously, the surface energy can be equal to the surface-tasion *only* if the heat absorbed by the film be ignored.

$$Q = \theta \cdot \frac{dT}{d\theta},$$

where $(dT/d\theta)$ is the *rate of change of surface tension with temperature.*

14.7. Units and dimensions of surface tension. As we have seen above, surface tension may be expressed as *force per unit length* or as *energy per unit area.* Its units, therefore, are

(i) *dyne/cm* or *erg/sq. cm in the C.G.S. system.*

(ii) *poundal/ft* or *ft. pounial/sq ft. in the F.P.S. system*

and (iii) *newton/metre (N/m)* or *jou'e/sq.metgre* in the M.K.S. (S.I) *system.*

And, its dimension are those of *force/length* or *work or energy/area,*

i.e.,
$$\frac{MLT^{-2}}{L} \quad \text{or} \quad \frac{ML^2T^{-2}}{L^2},$$

both of which come to MT^{-2}.

Thus, the dimension of surface tension are ML^0T^{-2} or MT^{-2}.

Fig. 14.6

14.8. Pressure Difference Across a Liquid Surface. (*i*) Suppose the free surface of a liquid is *plane*, as shown in Fig. 14.6. (*i*). Then, the resultant force due to surface tension on a molecule on its surface is zero, and the cohesion-pressure, (*i.e*, the downwards or inward force per unit area of the liquid surface) is therefore, just *nominal* or *negligible.*

(*ii*) If the free surface of the liquid be *concave,* [Fig. 14.6 (*ii*)], the resultant force on a molecule on the surface would be upwards, and cohesion pressure is, therefore, *decreased.*

(*iii*) And, finally, if the liquid surface be *convex*, as shown in Fig. 14.6 *(iii)*, the resultant force due to surface tension on a molecule on the surface will be directed downwards, so that the cohesion pressure is, in this case, *increased.*

Case of Drops and Bubbles.

(a) Excess Pressure inside a Liquid Drop. It must be clear from the above that the molecules near the surface of a drop,(which is a *convex* surface), experience a resultant pull *inwards*. The pressure inside it must, therefore, be greater than the pressure outside it. Let this excess pressure inside over the pressure outside the drop be p. Then, if r be the radius of the drop, and T, its surface tension, we have, considering the equilibrium of one-half of the drop–say, the upper half, or the upper hemisphere, –the upward thrust on the plane face $ABCD$, (Fig. 14.7), due to the excess poressure p is equal to $p \pi r^2$.

Fig. 14.7

And, force due to surface tension, acting downwards on it and round its edge, is equal to $T.\ 2\pi r$. Since the hemisphere is in equilibrium, we have
$$p.\pi r^2 = T.2\pi r,$$

whence,
$$p = \frac{T.2\pi r}{\pi r^2} = \frac{2T}{r}.$$

(b) Excess Pressure Inside a Soap bubble. If instead of a drop of liquid, we consider a bubble, there are *two surfaces* to be considered, and *not one*, because it is like a spherical shell or a hollow sphere, having an inner and an outer surface; so that, the force due to surface tension, in this case, is $2 \times 2\,\pi rT$ (*i.e.,* $2\,\pi r.T.$ due to each surface). Therefore, for equilibrium of the hemisphere, we have

$$p.\pi r^2 = 2 \times 2\,\pi r.T = 4\,\pi r.T,$$

whence $p = \dfrac{4\,\pi rT}{\pi r^2} = \dfrac{4T}{r}$

In case the bubble happens to lie inside a liquid as for example, an **air bubble inside water**, it will have *only one free surface* (*viz.* the inner one) and will thus behave take a *drop*. The *excess pressure inside* it therefore will be the same as inside a drop, *i.e.,* $2T/r$

Alternatively, we could obtain the above expressions for excess pressure inside a drop and a bubble as follows :

Suppose, on account of the excess pressure P inside a liquid *drop* of radius r, the liquid surfaces is pushed outwards through a very small distance dr, such that there is little or no increase in volume and the excess pressure continues to remain p.

Then, clearly, *increase in surface area of the drop* $= 4\,\pi\,(r + dr)^2 - 4\,\pi r^2 = 8\,\pi\,r\,dr$

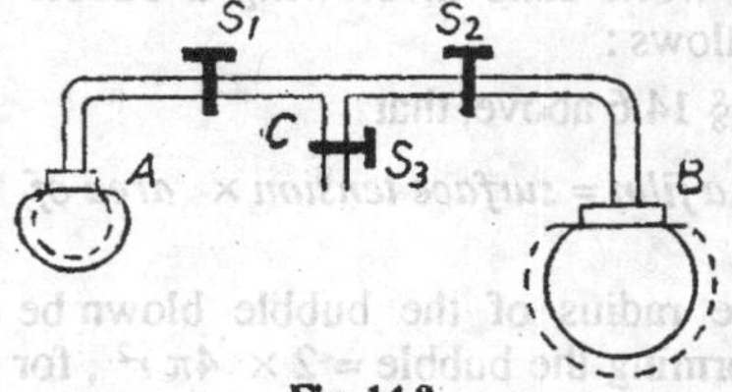

Fig. 14.8

neglecting $(dr)^2$ as a very small quantity.

$\therefore$ *increase in surface energy of the liquid surface* $= T \times 8\,\pi\,rdr$

This must obviously be equal to the work done in pushing the liquid surface outwards through dr, *i.e.,* equal to $4\,\pi r^2 \times p \times dr$.

Equating the two, therefore, we have $4\,\pi r^2 p\,dr = 8\,\pi\,Tr\,dr$

whence $p = \dfrac{2T}{r}$

In the case of a bubble, there being two surfaces in contact with air, the total increase in surface area of the liquid film is $2 \times 8\,\pi\,r.dr$ and, therefore, increase in surface energy of the film $= T \times 16\,\pi\,r.dr$

So that, $4\,\pi r^2 p.dr = 16\,T \times \pi\,r\,.dr,$ whence $p = 4T/r$.

It will thus be seen that the excess pressure inside a drop or a bubble is inversely proportional to its radius (i.e., $p \propto 1/r$); so that, the smaller the drop or the bubble, the greater the excess pressure inside it.

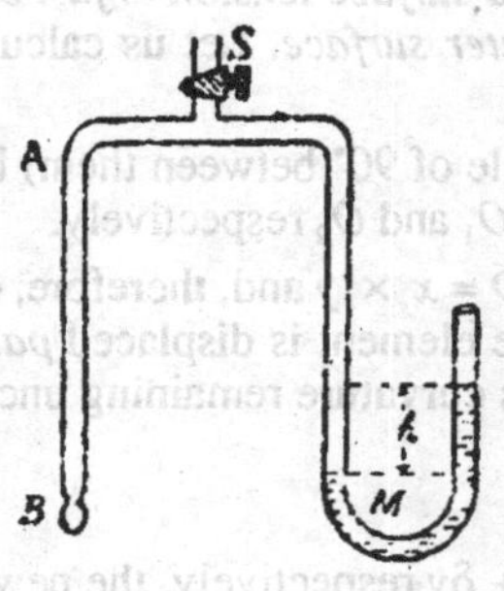

Fig. 14.9

This may be beautifully demonstrated by the following simple experiment.

The two ends of the tube T, (Fig. 14.8) carrying small cups at their ends, are dipped in some soap solution and then taken out. Stop cock S_2 is first closed, S_1 and S_3 opened and air blown through tube C, so that a small bubble is formed at A. S_1 is then closed and S_2 opened and again, blowing air through C, a larger bubble is formed at B. S_3 is then closed and S_1 and S_2 opened, so that the two bubbles are put in direct communication with each other. Air at once starts passing from the smaller into the larger bubble, because the pressure inside the former is greater than that inside the latter. As a result, the smaller bubble goes on shrinking and the larger one swelling until the smaller one is reduced to a *hemisphere* ; for then, any further shrinkage would mean an increase in its radius and, therefore, a decrease in the pressure inside it. Equilibrium between the two bubbles is attained when the curvatures of the two become the same.

14.10. Determination of the Surface Tension of a Bubble. The expression for the excess pressure inside a bubble, deduced in 14.9, above affords us a simple method of determinating the surface tension of a bubble. For, if we have a vertical glass tube *AB*, (Fig. 14.9) with a fine orifice or aperture at its lower end *B*, connected to a manometer *M*, and a stopcock *S* ; then, on dipping *AB* into the experimental liquid, a liquid film is formed at the orifice. This film is then blown into a bubble at *B*, by opening the stop-cock for a while and allowing some air to come into *AB*. The difference *h* in the heights of the liquid columns in *M* is noted. Then, if ρ be the density of the *manometer liquid*, the exess pressure inside the bubble is clearly equat to $h.\ \rho.g$. But, as we know, the excess pressure inside the bubble is also equalt to $4T/r$, where T is its surface tension. Hence, equating the two, we have

$$4T/r = h.\rho.g\,, \qquad \text{whence,} \qquad T = r.h.\rho.g/4\,,$$

where *r* is the radius of curvature of the bubble, which is obtained from its diameter, measured accurately by means of a travelling microscope.

It will easily be seen that for accuracy of the result, the value of *h* should be large, *i.e.*, the excess pressure inside the bubble should be large. This would be so, if the bubble be small or the size of the aperture at *B* be small.

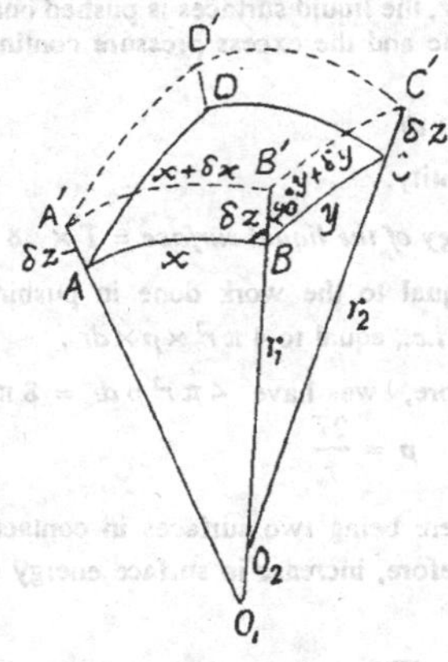

Fig. 14.10

14.11. Work done in Blowing a Bubble. If, for the sake of simplicity, we neglect the cooling produced when a film is stretched, the work done in blowing a bubble is easily calculated as follows :

We know, from § 14.6 above, that

work done in creating a film = surface tension × area of the film formed.

If, therefore, the radius of the bubble blown be *r*, the area of the film forming the bubble = $2 \times 4\pi r^2$, for it has two surfaces, an inner and an outer one, each of surface area $4\pi r^2$

Therefore, *work done in blowing the bubble*

$$= T \times 8\pi r^2 = 8\pi r^2 T\,.$$

14.12. Curvature, Pressure and Surface Tension. Imagine a *small curvilinear* element *ABCD* (Fig. 14.10) of a liquid surface lobe at rest. This can, obviously be possible only *when the inward pressure on it due to surface tension is just balanced by the outward excess pressure on its inner, over that on its outer surface.* Let us calculate this *excess pressure p.*

Let the length of adjacent of sides *AB* and *BC* (with an angle of 90° between them) be *x* and *y*, their radii of curvature, r_1 and r_2 and their centres of curvature O_1 and O_2 respectively.

Then, clearly, *surface area* of the curvilinear element *ABCD* = $x \times y$ and, therefore, outward force on it due to excess pressure *p* is equal to $p.xy$. Suppose the element is displaced *parallel to itself*, through a small distance δz into the positon *A′ B′ C′ D′*, its curvature remaining unchanged. Then,

work done during the displacement = $p.xy.\ \delta z$.

If the lengths of sides *A′B′* and *C′D′* be now $x + \delta x$ and $y + \delta y$ respectively, the new area of the element is obviously $(x + \delta x)(y + \delta y)$

So that, *increase in surface area of the element*

$$= (x + \delta x)(y + \delta y) - xy = x\,.\,\delta y + y\,.\delta x$$

Increase in surface energy of the element = $T(x\,\delta y + y\,\delta x)$

where T is the surface tension of the liquid.

This must obviously be tension of the liquid.

We thus have $\qquad p.x.y.\delta z = T(x\delta y + y\,.xy\,.\delta x)$

whence $$p = T\left(\frac{1}{y}\cdot\frac{\delta y}{\delta z} + \frac{1}{x}\cdot\frac{\delta x}{\delta z}\right), \quad ...(i)$$

Now, the element being small and $A'B'$ being parallel to AB, we may take $A'B'O$ and ABO as two similar trianlges, so that, $A'B'/AB = B'O_1/BO$

Or, $(x+dx)/x = (r_1+dr)/r_1$

or, $\delta x/x = \delta r/r_1$

i.e., $$\frac{1}{x}\cdot\frac{\delta x}{\delta z} = \frac{1}{r_1}.$$

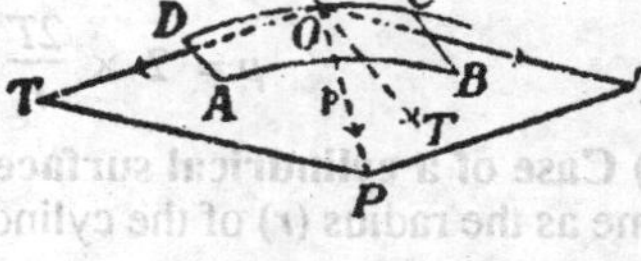

Fig. 14.11

Similarly, considering the two similar triangles $B'C'O_2$ and BCO_2, we have $\frac{1}{y}\cdot\frac{\delta y}{dz} = \frac{1}{r_2}$

Now, substituting these values in relation (*i*) above, we have $p = T\left(\frac{1}{r_1} + \frac{1}{r_2}\right)$

Alternatively, we may proceed as follows :

Consider a protion $ABCD$ of a liquid surface, (Fig. 14.11), *cylindrical in form, i.e.*, curved only in one direction. Then, the force of surface tension T acts at right angles to every unit length of its boundary. The forces over AB and CD are equal and opposite and hence cancel out, but those over AD and BC, though equal in magnitude, are inclined to each other, and have thus a resultant p, normal to the surface, as shown.

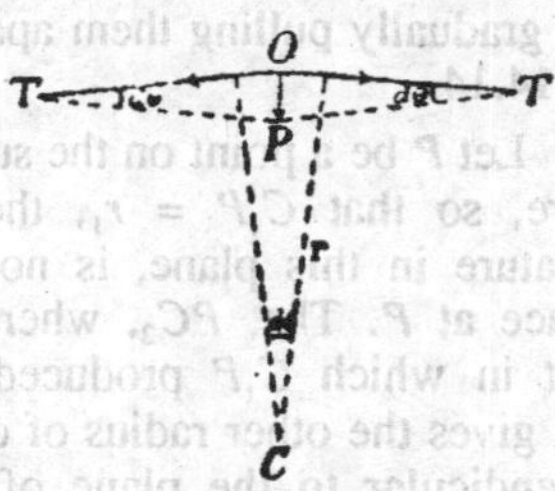

Fig. 14.12

To determine the value of p, consider unit length of the surface at O. (Fig. 14.12), taken along the direction fo DA. Forces T and T act at the point O, as shown, their resultant being $OP = T.d\theta$.

Thus, a force $T.d\theta$, acts inwards on an area equal to $1 \times r.d\theta$, where r is the radius of curvature of the surface, (because we are considering unit length of the surface of breadth $r.d\theta$). If a resultant pressure p acts outward on this area, the total force directed outward on the area is equal to $p.r.d\theta$.

For this element of the liquid surface to be at rest, therefore, we have

$$p.r.d\theta = T.d$$

whence, $p = T/r$ dynes/cm^2.

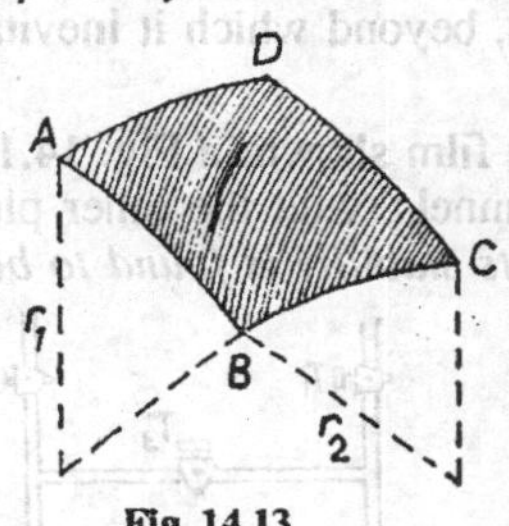

Fig. 14.13

This resultant pressure is, therefore, the difference of pressure on the two sides of the surface, which is required to balance the effect due to surface tension. *It follows, therefore, that the pressure must be greater on the concave side than on the convex side, to balance the effect of surface tension on the surfaces.*

If the surface be curved in two directions, as shown in Fig. 14.13, and the radii of the two curvatures be r_1 and r_2 respectively, the pressure due to the curvature of AB and CD will be T/r_1 and that due to the curvature of AB and BC will be T/r_2; and, therefore, the total difference of pressure on the two sides of the surface will be given by

$$p = \left(\frac{T}{r_1} + \frac{T}{r_2}\right) = T\left(\frac{1}{r_1} + \frac{1}{r_2}\right)$$

Here, O_1 and O_2 lie on the *same* side of the element $ABCD$, so that r_1 and r_2 have the same sign or the surface is *symelastic*. Even if the surface be *antielastic*, however, i.e., even O_1 and O_2, lie on the *opposite* sides of the element so that r_1 and r_2 bear opposite signs, the above relation for p still holds good if we take proper care of the signs of r_1 and r_2.

Now, let us consider a few special cases :

(*i*) **Case of a spherical surface.**

(*a*) A liquid drop, or an air-bubble inside a liquid. Here, we have $r_1 = r_2 = r$, say ; so that excess pressure inside it is given by

$$p = T\left(\frac{1}{r}+\frac{1}{r}\right) = \frac{2T}{r}$$ [see § 14.8

(*b*) **A Soap Bubble.** In this case, there are two surfaces, an *inner* and an *outer* one and we, therefore, have

$$p = 2 \times \frac{2T}{r} = \frac{4T}{r}$$

(*ii*) **Case of a cylindrical surface.** In this case, one of the radii is infinite, whereas the other is the same as the radius (r) of the cylinder so that,

(*a*) **for one single surface,** *i.e., for a cylindrical column of liquid or for a cylindrical bubble in a liquid,* $p = T/r$

and (*b*) **for two surfaces,** *i.e., for cylindrical bubble or film,* $p = 2T/r$
which is clearly half that for a spherical bubble of the *same* radius.

In other words, *the excess pressure inside a cylindrical bubble or film is the same as that inside a spherical bubble of twice its radius.*

(*iii*) **Case of a Catenoid :** Here, the surface being one of revolution, with no difference of pressure across it, we have $p = 0$ and, therefore, $(1/r_1 + 1/r_2) = 0$, or, $r_1 = -r_2$.

An example of such a surface is that of a soap-film supported between two parallel rings, with its two ends burst to make the pressure difference zero, or, better still, between the mouths of two funnels (obtained by first dipping them in the soap solution, bringing them into contact and then gradually pulling them apart), with their narrower ends open to the atmosphere, as shown in Fig. 14.14.

Let P be a point on the surface of the film and C_1, the centre of curvature in the plane of the figure, so that $C_1P = r_1$, the radius of curvature in this plane, is normal to the surface at P. The, PC_2, where C_2 is the point in which C_1P produced meets the axis, gives the other radius of curvature r_2, perpendicular to the plane of the figure, such that $r_1 = r_2$.

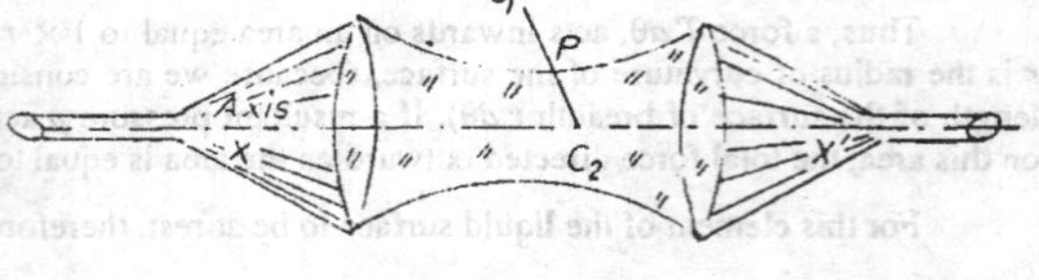

Fig. 14.14

This catenoidal surface of the film about the ; axis XX_1 remains intact, however, only up to a certain distance between the two funnels, beyond which it inevitably gets ruptured.

(*iv*) **Nearly cylindrical films and their stability.** If in the film shown in Fig. 14.14 above, some air be blown in through the narrow end of one of the funnels, with the other plugged, it acquires a more or less cylindrical form, and *the condition of its stability is found to be that its length must be smaller than its circumference.* This may be illustrated beautifully by a simple experiment due to *Rucker* and *Reinold.*

An *H*-shaped tube, fitted with three taps, T_1, T_2 and T_3, arranged in the manner indicated [Fig. 14.15) has two equal-sized cups A and B, of radius R, at tis two lower ends, such that almost cylindrical soap – films of circumference $2\pi R$ can be formed between them and identical cups C and D below, the distance between the pairs of cups A, C and B, D, *i.e., the lengths of the films* being adjustable, as desired.

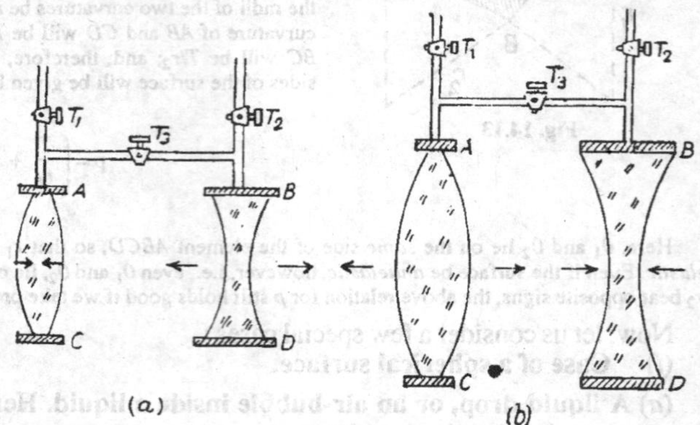

Fig. 14.15

Consider case (a) first, where *the length of the film is adjusted to be less than half its circumference* (πR). Let a little extra air be blown into film AC and a little of air removed from film BD, so hat the former bulges out or swells and the latter bulges in or shrinks, in the middle, so as to be respectively convex and concave out-wards, as shown. On opening tap T_3 to put the two film AC and BD in communication with each other, we find that air flows from the former to the latter, clearly indicating that the *excess pressure inside AC is higher than that inside BD*. As a result, the respective outward and inward bulges of the two films get reduced and they both tend to acquire a uniform cylindrical shape. Obviously, therefore, *the films are both quite stable here*.

Consider now case (b), where *the lengths of the films AC and BD are arranged to be longer than half their circumference* (πR). On making film AC convex and film BD concave outwards, as in case *(a)* and opening tap T_3 to put them in communication with each other, we find that air now flows from film BD to AC, showing that *the excess pressure is now higher inside BD than inside AC*. The air flow thus results in further swelling of AC and shrinking of BD until the former bursts and the walls of the latter touch each other and it collapses. Obviously, then, *the two films are quite unstable in this case*.

Finally, let us imagine the two films of Fig 14.15 *(a)* to be joined end to end, so as to form *a long film* of a length less than its circumference $2\pi R$. As will be readily seen, air will rush from the higher pressure inside the swollen half to the lower pressure inside the shrunken half to the lower pressure inside the shrinken half of the film, until the entire film acquires a near perfect cylindrical form. *This long film, with its length less than its circumference, is thus quite stable*.

If we now imagine the two films of Fig 14.15 (*b*) to be joined end to end so as to form n long film, of a length greater than its circumference, we shall find that air now rushes from the shrunken half of the film into the swollen half, thus further increasing their inward, and outward bulges respectively, leading to their collapse. Clearly, then *this film with its length greater than its circumference is not a stable one*.

Theoretically, the above results may be deduced as follows :

Consider the film AC of Fig. 14.15 (*a*), which is swollen or bulging outwards, to be formed between the mouths of two equal — sized funnels, each of radius R, as shown in Fig. 14.15 (*c*) with its length l less than half the circumference πR of the ideal cylindrical film of radius R, shown dotted.

Fig. 14.15 (c)

Leaving alone the points of inflexion A and C, let us take a point P on the surface of the film. If C_1 be the centre of curvature and, therefore, $C_1P = r_1$, *the radius of curvature* of the film at P, *in the plane of the figure*, C_2 will be the centre of curvature, and, therefore, $C_2\ P = r_2$, *the radius of curvature, in the plane perpendicular to that of the figure*. So that, *excess pressure inside the film i.e.,*

$$p = 2\,T\left(\frac{1}{r_1} - \frac{1}{r_2}\right) \text{ or } , \frac{1}{r_1} - \frac{1}{r_2} = \frac{p}{2\,T},$$

because, as we know, r_1 and r_2 bear opposite signs.

Now, if dy/dx be small compared with unity, we have

$$\frac{1}{r_1} = \frac{d^2y}{dx^2} \text{ and } r_2 = C_2\,P \approx OQ = y,$$

where, $AQ = QC = l/2$.

$$\therefore \quad \frac{d^2y}{dx^2} - \frac{1}{y} = \frac{p}{2T}. \text{ or, } \frac{d^2y}{dx^2} - \frac{1}{R+z} = \frac{p}{2T},$$

because $y = OD + DQ = R + z$, where R is the radius of the ideal cylindrical film or that of the mouth of each funnel and z = distance DQ.

Since $\dfrac{1}{R+z} = \dfrac{1}{R} - \dfrac{z}{R^2}$, neglecting z^2 as a very small quantity, we have

$$\frac{d^2y}{dx^2} - \left(\frac{1}{R} - \frac{z}{R^2}\right) = \frac{p}{2\,T}. \text{ Or, } \frac{d_2y}{dx^2} + \frac{z}{R^2} = \frac{1}{R} + \frac{p}{2\,T} \quad \text{...(I)}$$

This is the equation of a *simple harmonic oscillator*, a general solution of which is

$z = A \cos(x/R) + B \sin(x/R)$, where A and B are *constants*.

Since a particular solution of equation (I) is obviously $z = R + pR^2/2T$, its *complete solution* is

$$z = R + \frac{pR^2}{2T} + A \cos\left(\frac{x}{R}\right) + B \sin\left(\frac{x}{R}\right) \quad ...(II)$$

To determine the constants A and B, we note that $y = R$ or $z = 0$, at $x = 0$ and $x = l$. So that, the two constants work out to be $A = -R - \frac{pR^2}{2T}$ and $B = -\left(\frac{1 - \cos(l/R)}{\sin(l/R)}\right)\left(R + \frac{pR^2}{2T}\right)$

$$...(III)$$

This means that $B/A = \tan(l/2R)$, from which it follows that $dz/dx = 0$ at $x = l/2$, or that in other words, *z has its maximum or minimum value at $x = l/2$.*

Differentiating equation (II) twice with respect to x, we have

$$\frac{d^2z}{dx^2} = -\frac{A}{R^2}\cos\left(\frac{x}{R}\right) - \frac{B}{R^2}\cos\left(\frac{x}{R}\right),$$

whence, putting $x = l/2$ and substituting the valves of A and B, we have

$$\left(\frac{d^2z}{dx^2}\right)_{x=l/2} = -\left(\frac{p}{2T} - \frac{1}{R}\right)\left(\sec\frac{l}{2R}\right) \quad ...(IV)$$

Clearly, the second expression on the right hand side of the equation will be *positive* or *negative*, according as $l < \pi R$ or $l > \pi R$. And, although the solution for $l > \pi R$ implies that there are a number of *maxima* and *minima* of z along the length of the film, we shall simply rule it out on purely physical grounds (in the same manner as we did in the case of a pillar in § 8.42 (*i*).) For, it has little or no physical significance in view of the fact that the film actually gets ruptured before this can ever happen.

Let us now examine both these cases a little more closely.

(*i*) **When $l < \pi R$.** Here, since [$\sec(l/2R)$] is always *positive*, it follows that

(*a*) **If d^2z/dx^2 be negative,** *(i) z has its maximum value at l/2*, *i.e.*, the film is bulging out or is convex outwards and *(ii) the excess pressure inside it is greater than 2T/R.*

Thus we see that *if the length of a film be smaller than πR, the semicircumference of an ideal cylindical film of radius R, and if it be convex outwards, the excess pressure inside it is greater than 2T/R, which is the excess pressure inside the cylindical film.*

(*b*) **If (d^2z/dx^2) be positive,** it clearly means that (*i*) *z has its minimum value* at $l/2$ *i.e.*, the film is bulging inwards or is conceave outwards and (*ii*) the excess pressure p inside it is less than $2T/R$.

In other words, if the length of a film is less than πR, the semi-circumference of an ideal cylindrical film of radiu, R, and if it be concave outwards, the excess pressure (p) inside it is less than $2T/R$, which is the excess pressure inside the ideal cyluidrical film.

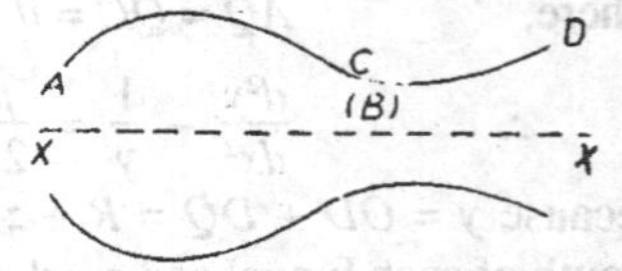

Fig. 14.15 (*d*)

(*ii*) **When $l > \pi R$.** In this case, the sign of *sec* ($l/2R$) is always *negative*, *i.e.*, opposite to that in case (*i*). All the results of case (*i*), therefore, get reversed here. So that, the excess pressure (p) inside a film longer then πR and *convex outwards*, is less than and that inside a film longer than ΠR, and *convex outwards*, more than $2T/R$ (the excess pressure inside an ideal cylidual film of radius R).

Condition for Stability. From the results obtained under (*i*) and (*ii*) above, we can easily deduce the condition for the stability of in nearly cylinctrical film, such as the one shown in Fig. 14.15 (*d*).

Such a film may be easily obtained by first forming a catenoidal film in between the mouths of two equal-sized funnels of radius R, as explained above, and then blowing air into it through the open end of one funnel, keeping that of the other plugged. As will be readily seen, it is of the same form as though the two films AC and BD of Fig. 14.15 were placed end to end, its length AD being less than or greater than, the circumference $2\pi R$ of the ideal cylindrical film of radium R according as the two films AC and BD correspond to those of Fig. 14.15 (*a*) or 14.15 (*b*).

In case the length of the film AD is less than $2\pi R$, each half of it, AC or CD, has a length less than half the circumference πR of the ideal cylindrical film. So that, in accordance with the discussion under (*i*) above, the excess pressure inside the swollen half AC is greater than that inside the shrunken half CD, and air therefore flows from the former into the latter, thereby reducing their outward and inward bulges respectively and tencling to give the whole film a uniform cylindrical shape of radius R. *The film AD is thus a stable one.*

If, on the other hand, the length *(AD)* of the film be greater than $2\pi R$, each half of the film is longer than half the circumference πR of the ideal cylindrical film of radius R. In this case, therefore, the excess pressure inside the swollen half of the film, AC, is, in accordance with the discussion under (*ii*) above, less than that inside shrunken half CD, so that, air flows from the latter into the former, thereby increasing their outward and inward bulges respectively, resulting in the ultimate collapse of the film, with only a large and a small spherical bubble left behind. Obviously, therefore, *the film AD is an unstable one here.*

The condition for the stability of a nearly cylindrical film thus turns out to be that its length must be less than its circumference $2\pi R$, where R may be taken to be its average radius or the radius of the ideal cylindrical film between the two funnels.

The same is true for the stability of a long cylindrical liquid column, the excess pressure inside which is, as we have seen, given by $p = T/R$.

It is in consequence of this inherent instability of a long cylindrical column or film that a jet tends to break up with a number of drops, big and small, and a liquid drop, breaking away from a tube, acquires a critical radius determined by the stage at which its neck becomes unstable.

Again, if we dip a wire in water and take it out in an attempt to obtain a long and thin column of water, the water at once collects in almost equally spaced droplets along the length of the wire, showing the instability of the water column.

Only the property of visciosity of the liquid opposes or resists this 'breaking up' effect due to surface tension. That is why only highly viscous liquids, like molten glass or quartz, can possibly be drawn into long and thin filaments.

14.13. Layer of liquid between two plates. It is common experience that if two plates of glass be placed one over the other there is no difficulty in separating them, but if a drop of liquid, say, water, be placed in between them and squeezed into a thin layer, it may require considerable force to pull them apart. The reason will be clear from the following :

The thin layer or film of water wets the plates over an almost circular area, which is concave outwards, as shown in Fig. 14.16.

If d be the thickness of the water layer in between the plates, the radius of curvature of the two concave edges of the liquid layer is *nearly* $d/2$. Thus, if r be the radius of the circular area of film of water, the excess pressure p *inside* the liquid-film over the *outside* atmospheric pressure is given by

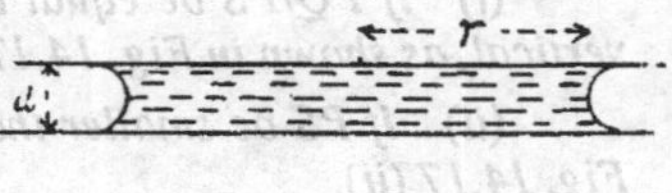

Fig. 14.16

$$p = T\left(\frac{1}{r} - \frac{2}{d}\right),$$

where T is the surface tension of the water-film.

Since r is very large compared with d $1/r$ is almost negligible in comparison with $2/d$; so that, we have

$$p = T\left(0 - \frac{2}{d}\right). \quad \text{Or, } p = -\frac{2T}{d}.$$

Thus, the pressure inside the film is less than the outside atmospheric pressure, (as is clear from the $-ve$ sign of p), by $2T/d$ and, therefore, the excess pressure of the atmosphere on the two plates pushes them closer together, making d still smaller and r larger, thus further increasing p.

Now, if A be the circular area over which the water wets the plates, the *total force* which squeezes them together is equal to

$$2\frac{T}{d} \times A = 2\frac{AT}{d} = \frac{2\pi r^2 . T}{d}$$

This much force, perpendicular to their surface, will, therefore, be needed to pull them apart. Obviously, the smaller the value of d, *i.e., the thinner the layer of water, the greater the force required to pull them apart.*

N.B. The force needed to separate the two plates (viz, $F = \eta A \frac{dv}{dx}$) would be proportional to $1/d$.

14.14. Shape of Liquid Meniscus in a Capillary Tube. We know that a liquid meniscus in a capillary tube is concave for a liquid like water and convex for a liquid like mercury. Let us see what determines its shape.

Let a capillary tube of glass be dipped vertically in a liquid, meeting its surface at P (Fig. 14.17). Then, a liquid molecule at P, in contact with the tube there, will be attracted outward by the solid molecules of the tube near to it, due to the *force of adhesion*, and attracted inwards by the near molecules of the liquid, due to the *force of cohesion* ; the force actually acting on it will, therefore, be the resultant of these two forces of adhesion and cohesion.

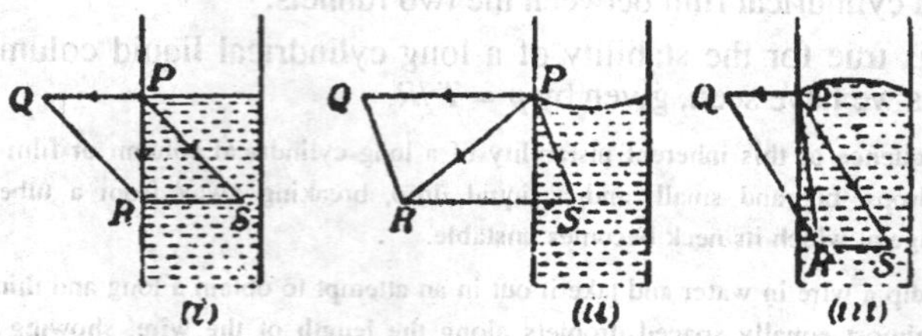

Fig. 14.17

Now, the resultant force of adhesion acts at right angles to the tube at the point P. Let it be represented by the straight line PQ ; and, the resultant force of cohesion acts at an angle of 45° to the vertical, and is represented by the straight line PS ; so that, the two forces acting on the molecules are inclined to each other at an angle of 135°. Their resultant, obtained by the ordinary law of parallelogram of forces, is represented by the diagonal PR of the parallelogram $PQRS$ with PQ and PS as adjacent sides. Its direction will naturally depend upon the relative magnitudes of the two forces PQ and PS, and the following different cases arise :

(i) *If PQ/PS be equal to $1/\sqrt{2}$, i.e., if $PS = \sqrt{2}.PQ$, the resultant, (PR) will lie along the vertical, as shown in Fig. 14.17 (i).*

(ii) *If PS be smaller than $\sqrt{2}.PQ$, the resultant (PR) will lie outside the liquid, as shown in Fig. 14.17 (ii).*

(iii) *If PS be greater than $\sqrt{2}.PQ$ then, PQ, will lie inside the liquid, as shown in Fig. 14.17 (iii).*

What happens to molecule P, happens to all other molecules in contact with the glass of the tube. And since a liquid cannot permanently withstand a shearing stress, *its surface at every point will be at right angles to the resultant force there,* ; *when the liquid attains the position of equilibrium.*

Thus, in the first case, when the resultant force PR acts along the vertical, *i.e.*, when $PS = \sqrt{2}.PQ$, or, *the cohesive force is $\sqrt{2}$ times the adhesive force*, the molecules of the liquid are neither raised nor lowered and *the liquid surface remains flat or plane.*

In the **second case**, when PR lies outside the liquid, *i.e.*, when $PS < \sqrt{2}\,PQ$, or *the cohesive force is less than* $\sqrt{2}$ *times the adhesive force*, the molecules of the liquid near the walls of the tube are raised up against the tube, those in the middle remaining practically unaffected, thus making the liquid surface **concave upwards**, as in th case of water and other liquids which *wet* the walls of the tube.

And, in the **third case** when PR lies inside the liquid, *i.e.*, when $PS > \sqrt{2}\,PQ$, or, *the cohesive force is greater than* $\sqrt{2}$ *times the adhesive force*, the liquid molecules near the walls of the tube are depressed, making the surface **convex upwards**, as in the case of mercury and other liquids which *do not wet* the walls of the tube.

14.15. Angle of Contact

(*i*) **Case of a liquid in contact with a solid.** When a liquid meets a solid, its surface near its plane of contact with the solid is, in general, curved. *The angle between the tangent to the liquid surface at the point of contact and the solid surface, inside the liquid is called the* **angle of contact** *for that pair of solid and liquid.*

This angle may have any value between 0° and 180°. For most liquids and glass, it is less than 90°; for mercury and glass, it is about 140°. *It really depends upon the nature of the liquid and the solid and is not altered by a change in the inclination of the solid.*

In the figures shown, [Fig. 14.18 (*i*) and (*ii*)], $\angle PQR$ is the *angle of contact*. It is *acute* in (*i*) and *obtuse* in (*ii*) ; for, in the former case the liquid rises up a little alongside the glass plate, dipped in the liquid, and the angle between QR, the tangent to the liquid surface, and the part QP of the plate, *inside the liquid*, is acute whereas, in the latter case, the liquid is depressed a little where it comes into contact witth the glass plate, and the angle between the tangent PR to the liquid surface and the part QP of the plate, *inside the liquid*, is obtuse.

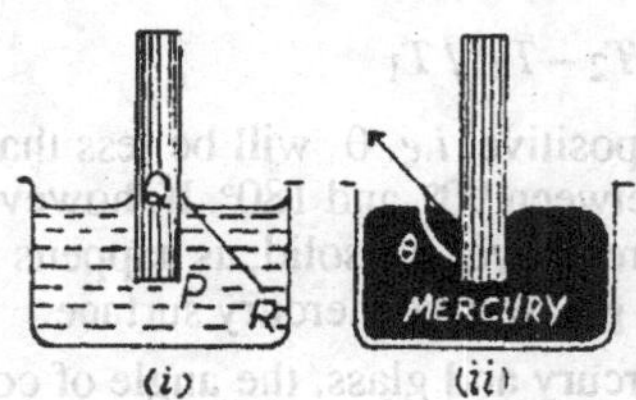

Fig. 14.18

For *pure water* and *clean glass*, the angle of contact is 0°. For ordinary water and glass, it is about 18°; and if the surface of the glass be *contaminated with grease*, its value may be as much as 35°.

(*ii*) **Case of two liquids in contact with each other and with air.** If two liquids, *not miscible with each other*, be brought into contact, as at O, (Fig. 14.19), both being in contact with air, three surface tensions are to be taken into consideration, (*a*) that of the surface between air and liquid I, *viz.*, T_1; (*b*) that of the surface between air and liquid II, *viz.*, T_2; and (*c*) that between liquid I and liquid II, *viz.*, T_3.

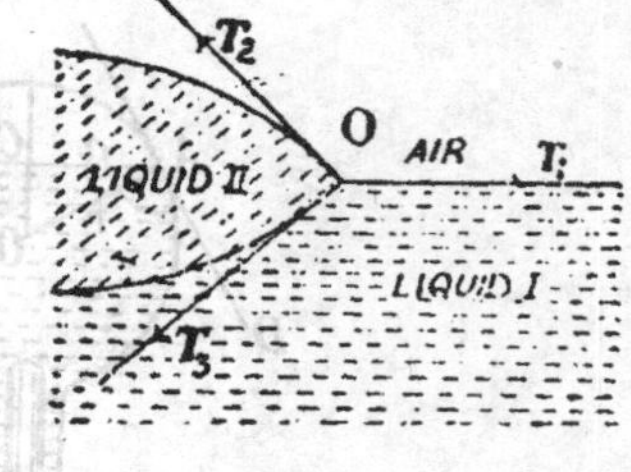

Fig. 14.19

For equilibrium, T_1, T_2 and T_3 should be represented by the three sides of a triangle, taken in order. This triangle of forces is konown as **Neumann's triangle.** In actual practice, we come across no two pure liquids for which the *Neumann's trianlge* may be constructed, – one of the surface tensions being always greater than the other two ; so that, the equilibrium condition shown in the figure is never attained. Thus, for ecample, in the case of water, mercury and air, the water drop, when placed over mercury, spreads all over its surface, provided both water and mercury are pure. This is so, because the surface tension of mercury is about 550 *dynes /cm.*, and that of water, only 75 *dynes /cm.* But, if the mercury surface be contaminated with grease, it surface tension decreases and some water drops may stay on it, so that, in that case, the construction of *Neumann's triangle* can be possible.

(*iii*) **Case of solid, liquid and air in contact.** This is a more important case than the previous one, for we have to consider three surface tensions, *viz.*, T_1 for *air-liquid*, T_2 for *air-solid* and T_3 for *liquid-solid surfaces* respectively, (Fig. 14.20).

Fig. 14.20

Let θ be the angle of contact of the liquid with the solid, – *acute*, in case (*a*) and *obtuse* in case (*b*). For equilibrium, therefore, T_3 *plus* the component $T_1 \cos\theta$ of T_1, in the direction of T_3, must be balanced by T_2, *i.e.*, $T_3 + T_1 \cos\theta = T_2$.

Or, $T_1 \cos\theta = T_2 - T_3$, whence, $\cos\theta = (T_2 - T_3)/T_1$

Clearly, therefore, if T_2 is greater than T_3, $\cos\theta$ will be positive, *i.e.* θ will be less than 90° and if T_2 is less than T_3, $\cos\theta$ will be nagative, and θ will lie between 90° and 180°. If, however, $T_2 > (T_1 + T_3)$, there will be no equilibrium, and the liquid will spread over the solid, as happens when a water drop is placed over a perfectly clean plate of glass, or a grease-free mercury surface.

14.16. Measurement of the Angle of Contact. For mercury and glass, the angle of contact may be determined by the following simple method, due to *Gay Lussac*.

A small round-bottomed glass flask is nearly filled with mercury (more than 3/4 *ths* of it), and its mouth closed tightly by a rubber bung through which passes a *glass rod R* to adjust the level of

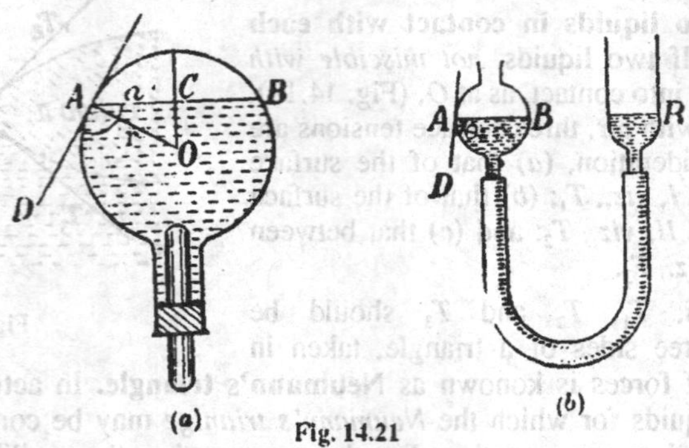

Fig. 14.21

mercury in it. The flask is then clamped in the inverted position, as shown in Fig. 14.21 (*a*), and the rod (*R*) is moved in or out, *until the surface of mercury is plane where it meets the glass, i.e., no meniscus or curved portion* (or *capillary curve*, as it is sometimes called) *appears there*. This can be tested by observing the image of a printed sheet of paper held against the flask, by the light reflected at *grazing incidence* from the mercury surface. An undistorted image indicates that the surface of mercury is *plane* where it meets the glass.

The mercury surface in the flask forms a circular sheet of diameter *AB*, which is measured by means of a pair of calipers, whence the radius *AC* of the sheet is known. Let it be denoted by *a*. Then, if *r* be the radius of the *bulb* of the flask, and $\angle BAO = \phi$, we have

$$\cos\phi = \frac{a}{r}. \quad \text{Or,} \quad = \phi = \cos^{-1}\left(a/r\right) \qquad ...(i)$$

Now, if a tangent *AD* be drawn to the spherical bulb at *A*, the angle of contact θ, for mercury and glass, is clearly equal to the angle *BAD*.

Or $\quad \theta = \angle BAD = \angle BAO + \angle OAD = (\theta + 90^\circ) \qquad ...(ii)$

So that, knowing ϕ, from relation (*i*) above, the value of θ can be easily determined.

A better arrangement for making the surface of mercury plane is the one shown in Fig. 14.21(*b*). Here, the level of mercury in the spherical bulb, which is open at one end, is adjusted by raising or lowering the reservoir of mercury (*R*), connected to it by an India rubber tubing. The procedure otherwise is the same as in the first experiment.

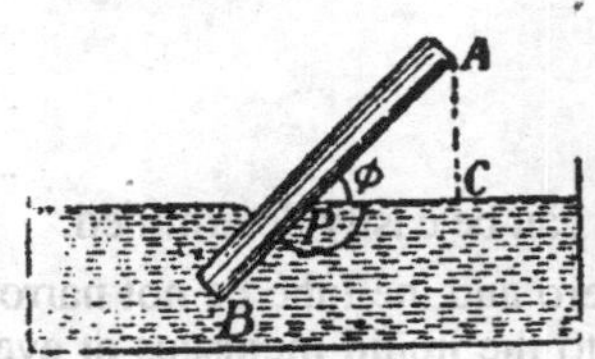

Fig. 14.22

Another simple method to determine the angle of contact for mercury and glass is to insert a small slanting glass plate *AB* into mercury, as shown in Fig. 14.22 and to adjust its inclination until the mercury meets the glass at *P* without curvature, *i.e.*, the surface of mercury is horizontal there. Then, the angle of contact for mercury and glass is equal to $(180 - \phi)$, where $\angle APC = \phi$.

To measure the angle ϕ, a plumb line is dropped from *A*.

Then, clearly, $\quad \tan\phi = AO/PC \quad$ Or, $\quad \phi = \tan^{-1}\left(AC/PC\right)$

For measuring the angle of contact between water and glass, *coated with wax*, a similar method was used by Adam. A glass trough is coated on the inside with wax, so that it may be filled with water above the level of the sides, and is supported by two screws, *S* and *S*, (Fig, 14.23), such that by working them up or down, the top of the trough is made *perfectly horizontal*. It is then filled with water and any impurities on its surface skimmed off by moving lightly across it a wax = coated glass plate.

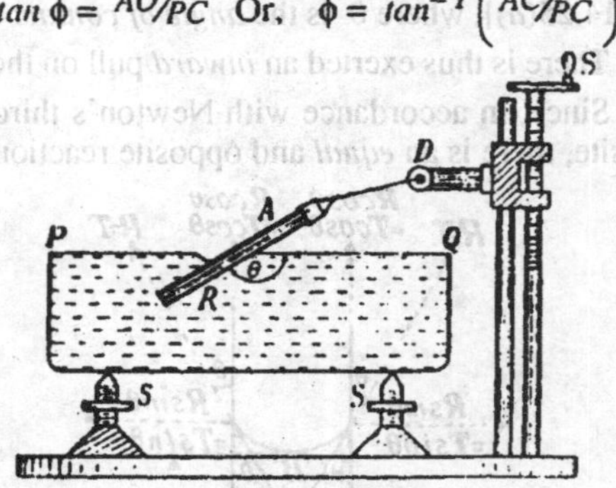

Fig. 14.23

A glass plate *A*, also coated with wax, is clamped,as shown, so that it can be moved up and down in the vertical plane, about a horizontal axis at *D*, by means of a screw *S'*. It is then dipped into the water, and its angle of inclination adjusted until the water surface touching it is perfectly horizontal. Then, the angle θ between the lower part of the plate and the horizontal water surface is the angle of contact for water and the wax-coated glass plate, and is measured as in the above case for mercury and glass.

The angle θ, in this case, is found to depend on whether the plate is lowered into the water or raised up and hence two readings are taken–one, when the plate is lowered and the other, when it is raised, and their mean is taken as the correct value of θ, or the required *angle of contact*.

14.17. Rise of liquid in a Capillary tube. One of the most striking effects of surface tension is to raise a liquid in a capillary tube dipped into it, — a capillary tube being just a tube of a very fine bore (from the Latin word, *capillus*—a hair). It is for this reason that surface tension is also sometimes called capillarity.

When a capillary tube is dipped in a liquid like water, which wets it and for which the angle of contact may be taken to be *zero*, the liquid immediately rises up into it, and if the tube be a *fine* one, the shape of the liquid meniscus is spherical and concave upwards, as shown at *B*, (Fig. 14.24).*

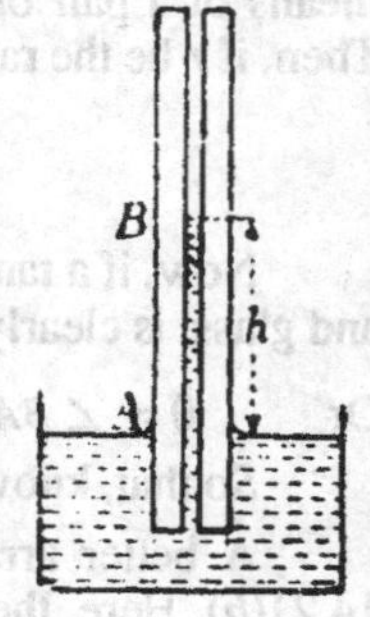

Fig. 14.24.

Let r be the radius of the tube at B, the point up to which the liquid rises into it. Then, it will be practically the same as the radius of the concave meniscus, so that the excess pressure above the meniscus over that immediately below it is $2T/r$, *i.e.*, the pressure in the liquid, just below the meniscus, is less than the atmospheric pressure above it by $2T/r$. And, since the pressure on the liquid surface, outside the tube, is atmospheric, the liquid will be forced up into the tube, until the hydrostatic pressure of the liquid column in the tube equals this excess pressures $2T/r$. If the liquid rises to a height h, the hydrostatic pressure due to the liquid column in the tube on the surface of the liquid will clearly be $h.\rho.g$, where ρ is the density of the liquid.**

$\therefore \quad 2T/r = h.\rho.g. \quad$ Or, $\quad 2T = r.h.\rho.g.$

whence, $\quad T = \dfrac{r.h.\rho.g}{2}$. ...(*i*)

Thus, knowing r, h, ρ and g, the surface tension T of the liquid can be easily determined.

In case, however, the angle of contact for the liquid be not zero and the tube be not narrow, the surface tension T of the liquid acts *inwards* along the tangent to the liquid meniscus at every point of its contact with the inner surface of the tube, making an angle θ with the wall of the tube, [Fig. 14.25(*a*)], where θ is the *angle of contact* for the liquid and the glass of the tube.

There is thus exerted an *inward* pull on the glass in *this* direction at all these points.

Since, in accordance with Newton's third law of motion, action and reaction are equal and opposite, there is an *equal* and opposite reaction R exerted by the glass on the liquid. This reaction

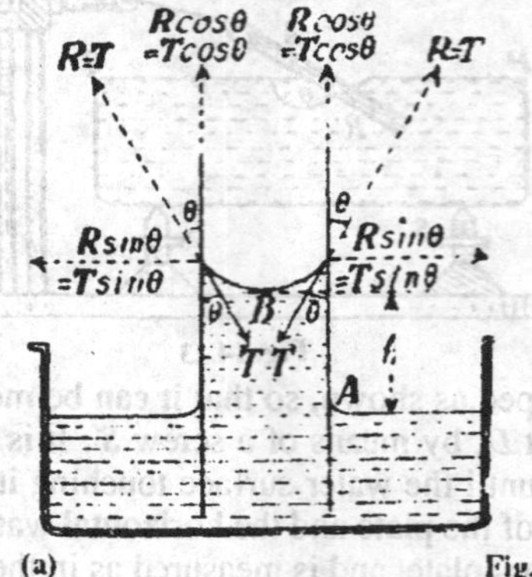

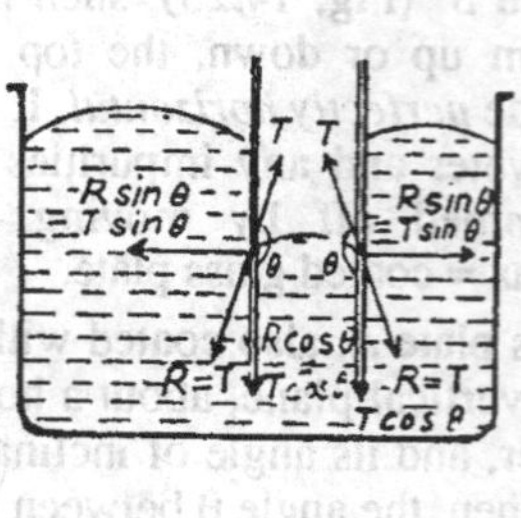

(a) Fig. 14.25 (b)

R (equal to T) may be resolved into two rectangular components-(*i*) $R \cos\theta = T \cos\theta$ *per cm, along the vertical, in the upward direction and (ii)* $R \sin\theta = T \sin\theta$ *per cm, at right angles to it, in the outward direction, as shown.* Taking the whole meniscus into consideration, the horizontal or

* The rise of a liquid in a capillary tube was first explained by **Hawksbee** in a book of his, published in the year 1709, as being due to the attraction of the glass for the liquid. Observing that a liquid does not rise higher in a thick-walled tube than in a thin-walled one, he naturally concluded that only the molecules close to the surface of the glass must be concerned in this attraction. As mentioned already, we now know that the forces between molecules causing surface tension are precisely the same as those operating in the cases of cohesion, solution or chemical reaction.

** It will thus be seen that *the capillary tube acts like a manometer*, giving the difference of pressure above and immediatley below the liquid meniscus.

outward components all cancel each other out, and only the vertical components are effective, which are thus added up.

Now, if r be the *radius of curvature of the tube* at the height of the meniscus, then, obviously, the meniscus touches it along a length $2\pi r$, the *circumference of the circle of radius r*; so that, the *total upward force on the liquid in the tube is* $2\pi r.T\cos\theta$. It is *this force* which supports the weight of the column h of the liquid in the tube, (where h is the length of the column from the horizontal surface A of the liquid, outside the tube, to the bottom of the meniscus at B) *plus* the weight of a volume v of the liquid, *in the meniscus itself,– i.e.*, the weight of a total volume of the liquid, $\pi r^2.h + v$, (where $\pi r^2 h$ is the volume of the liquid column h).

$$\therefore \qquad 2\pi r.T\cos\theta = (\pi r^2.h + v).\rho.g,$$

where ρ is the density of the liquid.

$$\text{Or,} \qquad T = \left(\frac{\pi r^2.h + v}{2\pi r.\cos\theta}\right).\rho.g.$$

If the volume of the liquid in the meniscus be negligible, in comparison with that in the column h, *i.e., if the tube be of a very fine bore*, we have

$$T = \frac{\pi r^2.h.\rho.g}{2\pi r\cos\theta} = \frac{r.h.\rho.g.^*}{2\cos\theta}, \qquad ...(ii)$$

the relation being known as ***Jurin's Equation.***

N.B. In case the capillary tube is not vertical, but inclined at an angle α to it, we take into consideration only the *vertical height* (h) of the liquid column in it, which will obviously be equal to $l\cos\alpha$ where l is the length of the tube occupied by the liquid. So that, here,

$$T = r.\rho.g.l\cos\alpha/2\cos\theta.$$

If, however, v be not negligible, its value may be determined by taking the meniscus to be *hemispherical* in shape,(since the tube is narrow) of radius nearly equal to r, *i.e.*, the *same* as that of the tube at that place, so that *the volume of the liquid in the meniscus is equal to the difference between the volumes of a cylinder of radius r and length r and a hemisphere of radius r.*

$$\text{Or,} \quad v = \pi r^2.r - \frac{2}{3}.\pi r^3 = \pi r^3 - \frac{2}{3}.\pi r^3.$$

[$\because$ volume of a cylinder $= \pi r^2 l$; of a sphere $= 4\pi r^3/3$ and of a hemisphere $= 2\pi r^3/3$.]

$$\text{Or,} \qquad v = \frac{1}{3}\pi r^3.$$

In this case, therefore,

$$2\pi r.T\cos\theta = (\pi r^2.h + \frac{1}{3}\pi r^3)\rho.g = \pi r^2.\left(h + \frac{1}{3}r\right).\rho.g.$$

$$\text{Or,} \qquad T = \frac{\pi r^2.\left(h + \frac{1}{3}r\right).\rho.g.}{2\pi r.\cos\theta}.$$

[*i.e., the effective height of the liquid column is now* $h + \frac{1}{3}r$, *and not* h.]

And, in case $\theta = 0$ $\cos\theta = 1$; so that, we then have

$$T = \frac{r\left(h + \frac{1}{3}r\right).\rho.g}{2} \text{ dynes/cm}.$$

But *if θ be greater than* 90° [Fig. 14.25(*b*)] $\cos\theta$ is *negative* **and the liquid column in the tube is depressed below the liquid level outside it, *i.e.*, h is *negative*. Hence it is that we find it so difficult to introduce mercury, (for which $\theta = 140°$ nearly), into a fine capillary tube.**

* For experimental details of the method, see page 469

Again, if we introduce into water, a capillary tube, *with its inside coated with paraffin wax* to make $\theta > 90°$, the surface of the water column also, inside the tube, lies below that in the outer vessel.

And, further, we may come across two liquids for which the values of T are the same but whose angles of contact (θ) are different. They will naturally rise to different heights when the same capillary tube is introduced into them.

14.18. Rise of liquid in a tube of insufficient length. We have seen above, how a liquid rises up into a capillary tube, dipped into it, until the weight of the liquid in the tube is just balanced by the force due to its surface tension. If θ be the angle of contact between the liquid and the tube, and R, the *radius of the liquid meniscus* in the tube, we have $r = R \cos \theta$ (Fig. 14.26), where r is the *radius of the tube*; so that, relation (*ii*) above now becomes

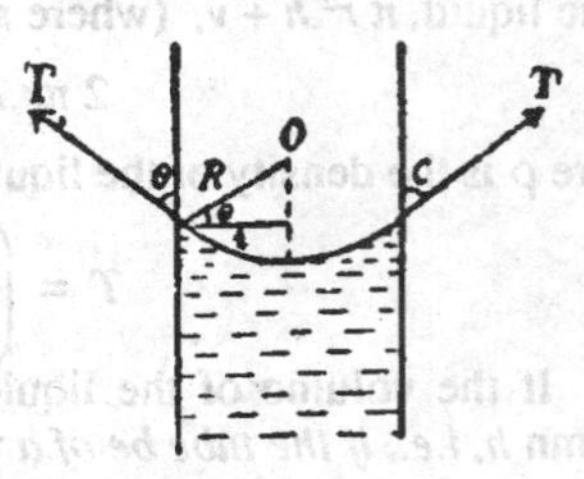

Fig. 14.26

$$T = \frac{R \cos \theta . h . \rho . g}{2 \cos \theta} = \frac{R . h . \rho . g}{2},$$

where h is the height of the liquid column in the tube.

Hence, clearly, $R.h = 2\,T/\rho.g$, a constant.

Now, with the tube sufficiently longer than h, it is the value of h alone that changes to satisfy the above relation for T. But, if the tube be smaller than the calculated value of h, the only variable in the above relation is R, because now $h = l$, the *length of the tube* (a eonstant) and so is θ a constant for the given liquid and the tube. *The liquid thus just spreads over the walls of the tube at the top and its meniscus acquires a new radius of curvature R', such that $R'.l = 2T/\rho g$, or, that $R'.l = R.h = $ a constant.* And, since l is smaller than h, $R' > R$, *i.e., the meniscus becomes less curved.*

14.19 Rise of Liquid in a Conical Capillary Tube. Suppose we take a conical capillary tube, (*i.e.,.a capillary tube with a fine conical bore*) ABC, with a hole at its upper end, (Fig. 14.27), and dip it in a liquid, like water, which wets its walls. We shall find that the liquid rises up into it to a height h, above the liquid surface, in accordance with the relation,

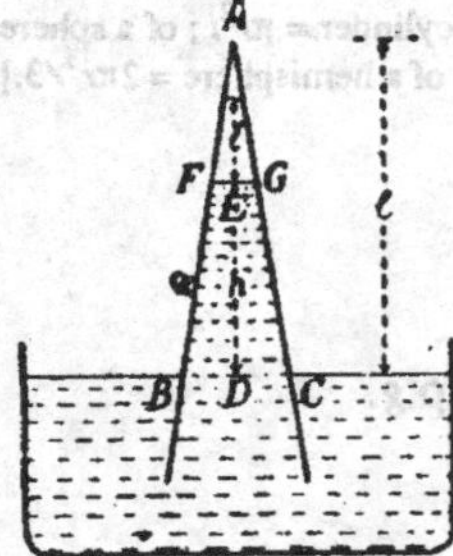

Fig. 14.27

$$h = 2T \cos \alpha / R\, \rho . g. \quad ...(i)$$

where α is the semi-vertical angle of the tube, and R, the radius of the tube at the liquid meniscus FG.

Let the vertical height AD of the apex of the cone from the liquid surface be denoted by l, and the radius of the tube, where it meets the liquid surface, (*i.e.*, of the portion BC of it),. by r. Then, from the similar right-angled triangles AEF and ADB, we have $FE/AD = AE/AD$.

Or, $\frac{R}{r} = \frac{l-h}{l}$, whence, $R = \left(\frac{l-h}{l}\right).r.$

Putting this value of R in relation (*i*), above, we have $h = \frac{2l . T . \cos \alpha}{(l-h) . r . \rho . g}$

So that, $h(l-h) = \frac{2l . T . \cos \alpha}{r . \rho . g}$. Or, $h^2 - hl + \frac{2l . T \cos \alpha}{r . \rho . g} = 0$

which is a quadratic equation in h.

So that,
$$h = \frac{+l \pm \sqrt{l^2 - 4 \times \frac{2l . T \cos \alpha}{r . \rho . g}}}{2}$$

$$= \frac{l}{2} \pm \sqrt{\frac{l^2}{4} - \frac{2l.T\cos\alpha}{r.\rho.g}} \qquad (ii)$$

whence the height h, up to which the liquid rises into the tube, can be easily determined.

And, if the tube be only slightly conical, we have $\cos\alpha = 1$, very nearly. In that case, therefore,

$$h = \frac{l}{2} + \sqrt{\frac{l^2}{4} - \frac{2lT}{r.\rho.g}}$$

N.B. The –ve sign is not usually indicated.

Further, in such a tube, R is practically equal to r, *i.e.*, it is practically a uniform capillary tube, as shown in Fig. 14.24 and therefore, for the liquid to rise up to the top into the tube, the minimum value of T should be greater than $r.l.\rho.g/2$.

14.20. Energy required to raise a liquid in a capillary tube. We have seen above how when a capillary tube is dipped vertically into a liquid which wets the walls of the tube, there is a rise of the liquid inside the tube. *The rise, obviously, takes place against the action of gravity and the liquid, therefore, must gain in potential energy.* The question, therefore, arises as to where does it get this increase in its potential energy from. For, according to the law of conservation of energy, energy can only be converted from one form into another, but cannot be created. The explanation is, however, simple.

We have three surfaces of separation to consider when a capillary tube is immersed in a liquid, *viz., (i) an air-liquid surface (ii) an air-glass surface* and *(iii) a glass-liquid surface*, each having its own surface tension, different from the others, and *equal to its free surface energy per unit area.*

Now, as the *plane* liquid surface in the tube acquires a curvature, (*i.e.*, becomes concave), the air-liquid surface increases and, as the liquid rises in the tube, the glass-liquid surface increases, the air glass surface decreasing by an equal amount. Thus, the surface energy of the air-liquid and the glass-liquid surfaces increases while that of the air-glass surface decreases by the same amount. *In other words, the energy required to raise the liquid in the capillary tube is obtained from the surface energy of the air-glass surface.*

On the other hand, a liquid, which does not wet the walls of the tube, gets depressed inside it, below its level outside the tube. In this case, obviously, the glass-liquid surface decreases, whereas the air-glass surface increases by an equal amount, resulting in a *net increase* in the surface energy of the whole system. *This energy is derived form the depression of the liquid inside the tube, whose gravitational potential energy is thus decreased by an equal amount.*

14.21. Rise of a liquid between Two Parallel Plates. An almost similar case to the above is that of two vertical plates kept parallel and close to each other in a liquid, when the liquid rises in between them, (if it wets the plates). Let us calculate the height to which it rises.

If d be the distance between the two plates and, r, the radius of the meniscus, (which is *cylindrical*, in this case), we have $\frac{d/2}{r} = \cos\theta$, whence, $2\,r\cos\theta = d$, or $r = d/2\cos\theta$, where θ is the angle of contact for the liquid and the plates.

Now, we know that the pressure just below the meniscus is less than the pressure just above it by an amount equal to $T\left(\frac{1}{r} + \frac{1}{r'}\right)$, where r and r' are the radii of the two curvatures, at right angles to each other. Since the meniscus is cylindrical, one of the curvatures has a radius equal to that of the cylinder and the other, a radius equal to ∞, (the surface being plane), so that $r = \infty$, and, therefore, $1/r' = 0$, *and hence the excess pressure just above the meniscus over that just below it is equal to T/r.* Or, substituting the value of r, we have *excess pressure above the meniscus* $= \frac{T}{d/2\cos\theta} = \frac{2T.\cos\theta}{d}$. Therefore, the liquid column will rise up in between the two plates, until the hydrostatic pressure due to it becomes equal to $2T.\cos\theta/d$. Thus, if h be the height up to which the liquid rises, and ρ, its density, we have $2T\cos\theta/d = h.\rho.g$.

Or, $$h = \frac{2T.\cos\theta}{d.\rho.g}.$$

Or, h is inversely proportional to d, *i.e., the rise* of the liquid is inversely proportional to the distance between the two plates.*

In case, however, the two plates be not parallel but inclined at a small angle to each other, *meeting along a common vertical edge,* the liquid does not rise, uniformly in between them. But, at any point on the liquid column, the distance (d) between the plates is proportional to the distance x of the point from their common edge. And, since, as we have seen, $h \propto 1/d$, it follows at once that $h \propto 1/x$.

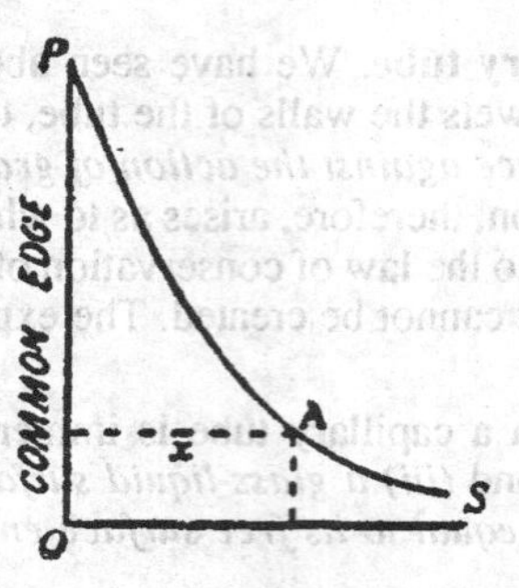

Fig. 14.28

Now, from the relation for h, deduced above, we have $h.d = 2T\cos\theta/g$ = a constant.

It is, therefore, clear that

$h.x$ = a constant, also. $[\because d \propto x$

In other words, *the liquid surface in between the two plates will, in this case, be a part of a rectangular hyperbola,* (Fig.14.28).

14.22. Force between bodies partly immersed in a liquid. It is a common observation that pieces of cork and such other light bodies, when floating in water, cling and collect together into clusters. This is due to capillary action, *i.e.*, the rise of water into the small spaces in between the pieces.

There are three cases to be considered, *viz.*, *(i) when the liquid wets the two bodies, (ii) when it does not, and (iii) when it wets one and not the other.*

In the *first case*, say, for example, two glass-plates, partly immersed in a liquid, like water, [Fig. 14.29 (*i*)] the liquid rises in between them to a level higher than that outside them. And, since the pressure in the liquid just below the meniscus in between them is smaller than the pressure due to the atmosphere at the same level outside them, they get pushed towards each other, *i.e.*, **they seem to attract each other.**

In the *second case*, as for example, when the two plates are partly immersed in a liquid, like mercury, (*which does not wet them*), the liquid is depressed between them below the level outside them, [Fig.14.29 (*ii*)], and the atmospheric pressure above the liquid meniscus, in between them, is

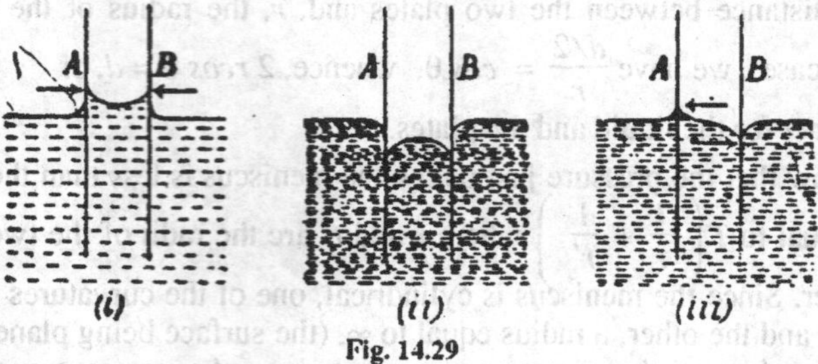

Fig. 14.29

smaller than the pressure in the liquid at the same level outside them, and so this excess pressure

* In case θ > 90° cos θ will be *negative* and the liquid will get depressed in between the two plates instead of rising up.

on their outside again pushes them towards each other, and again, therefore, **they seem to be attracting each other.**

In the *third case*, as for example, when one of the glass plates, say, *A*, be clean so that it can get wet with water, and the other *B*, be coated with wax, (so as not to get wet with water), and both be partly immersed in water, the liquid meniscus is concave near to the plate *A*, *wetted* by the liquid and concave, near to the other plate *B*, *not wetted* by it. Due to the pull of one meniscus on the other, therefore, the concave meniscus on the inside of *A* is a little lower than the concave meniscus on the outside, and the concave meniscus on the inside of *B* is al little above the convex meniscus on its outside, [Fig. 14.29 (*iii*)]. Since the pressure just below the concave meniscus outside of *A* is lower than the atmospheric pressure at the same level and above the concave meniscus on its inside, it is pushed outward, away from *B*, as shown by the arrow head. Again, since the atmospheric pressure above the convex meniscus outside *B* is lower than the pressure in the liquid below the convex meniscus at the same level on its inside, it is also pushed outwards, away from *A*, as indicated by the arrow head, and thus both *A* and *B* move away from each other, *i.e.*, **they seem to repel each other.**

This explains why a small piece of wood, (wetted by water) and a needle (not so wetted), when floated close together in water, steadily get apart, – the former drifting towards the edge and the latter moving towards the centre.

It should be noted that the above is true only when the two floating bodies are near together. For, if the distance between them be large, the portion of the meniscus where it changes from convex to concave, or *vice versa*, will become quite straight or horizontal, *i.e.*, in a level with the rest of the liquid outside them, and there will, therefore, be no resultant force acting on them, one way or the other. On the other hand, if they be very close to each other, the point of inflextion of the meniscus disappears altogether, resulting in the rise of liquid in between them, and they get pushed towards each other.

14.23. Shape of liquid Drop on a Horizontal Plate. There are two forces acting on the drop of a liquid, placed on a horizontal plate, – *viz.*, *(i) the force due to gravity* and *(ii) the force due to surface tension*.

Whereas the former tends to flatten it and to spread it out until its surface becomes horizontal, *i.e.*, tends to '*squash*' the drop, as it were, the latter, by trying to *decrease* its surface area, tends to gather it up into the form of a sphere. *It is, therefore, a case f gravity versus surface tension.* Now, the gravitational force depends upon the mass of the drop, which is proportional to its volume, *i.e.*, to the *cube* of its linear dimensions, and the surface tension depends upon the surface area, *i.e.*, upon the *square* of its dimensions. *Therefore, the effect of gravitation is more pronounced on a large drop and it gets flattened, whereas the surface tension has the upper hand, in the case of a small drop and gives its a spherical shape.* That is why the *small* dew and rain drops, or those of mercury or oil etc., are all spherical in shape*. This can also be readily seen by placing small and large drops on a plate of glass or water drops on paraffin wax, when it will be faound that small drops assume a spherical shape, but the large ones get flattened until their upper surface becomes horizontal. In the case of mercury on a glass plate, a large drop will assume a long elliptical sort of shape, its upper surface, in the middle, being plane, with the edges protruding on either side, the angle of contact being about 140 °, (Fig. 14.33).

It is possible to calculate the surface tension of mercury or the angle of contact, by considering the various forces that keep it in equilibrium. The method was developed by **Quincke**, and later improved by **Edser**, [see § 14.24 (4)].

14.24. Experimental Determination of Surface Tension. We shall now consider some of the usual methods employed to determine the surface tension of liquids.

*** A striking proof of the rain drops being perfect spheres is the natural phenomena of the *rainbow* and the *halos*, whose arrangement of colours and shape can only be explained on this basis. The slightest deviation from the spherical shape of the drops would materially affect both these.**

1. Searl's Torsion Balance Method. This is perhaps the *simplest, most direct* and the *quickest* method for the determination, or comparison, of surface tension of liquids, *particularly of those for which the angle of contact is zero. Its accuracy is, however, only limited.*

The essentials of the apparatus used are as indicated in Fig. 14.30, where R is a rigid rod, fixed to a fine torsion wire w, (about 1.25 *mm* in radius), stretched horizontally across a framework, as shown. The rod terminates in a pointer, moving over a vertical scale S at one end, and carrying a sliding weight W at the other, with help of which it is counterpoised, so as to remain horizontal.

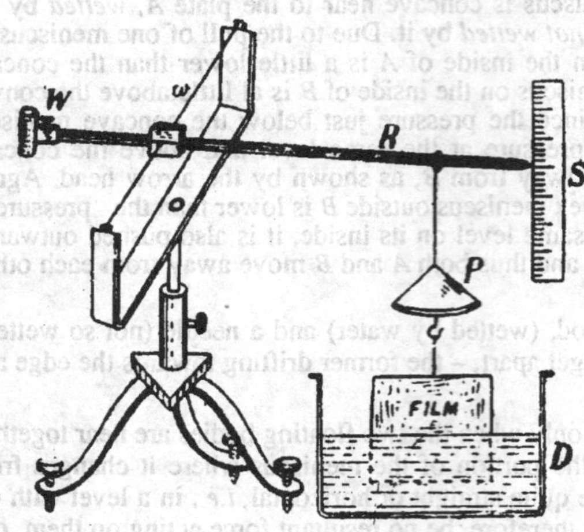

Fig. 14.30

From a point, near the front end of the rod, is suspended a scale pan, which, in its turn, carries a rectangular wire-frame or a glass plate, (about 10 *cm* long and .02 to .05 *cm* in thickness), fitted in a metal socket or clip, with its lower edge *perfectly horizontal.*

A vessel D, containing the experimental liquid, is placed on an adjustable table, which can be raised up or lowered down, as desired.

To start with, the vessel D is raised up until the wire-frame dips into it or until the lower edge of the glass plate just dips a little into the liquid. It is then lowered down until this lower edge of the wire-frame, or the plate, lies exactly in a level with the liquid surface. A film of the liquid is thus formed in the wire-frame, or in between the plate and the liquid surface, exerting a downward pull on the frame, or the plate, due to the surface tension. This results in the pointer-end of the rod also deflecting downwards and its position is read on the scale. The vessel is removed from under the frame or the glass plate. The film is now punctured, or the plate allowed to dry up. The liquid film thus disappears and with it also the downward pull on the frame or the plate, and hence on the rod, which, therefore, returns back to its initial position.

Weights are now placed in the scale pan until the pointer-end of the rod is again deflected downward to the same extent as before. Clearly, then, *the downward force on the rod due to this weight, mg, say, is the same as the downward pull on it due to surface tension, in the first case.*

Now, if l be the *length of the wire-frame or the plate*, and t, its thickness, it is in contact with the liquid along a total length equal to $2(l + t).\ cos\ \theta$, where θ is the angle of contact for the liquid. Hence, if T be the surface tension of the liquid, the downward pull on the frame, or the plate, due to surface tension is equal to $2(l + t).\ cos\ \theta.T$.

Hence $\quad 2(l + t).\ cos\ \theta.\ T = mg.\quad$ Or, $\quad T = mg / 2\ (l + t).\ cos\ \theta.$

In the case of the wire-frame, t is negligible, and, therefore,

$$T = mg/\ 2\ l\ cos\ \theta.$$

And, if θ be *zero* for the given liquid, we have $cos\ \theta = 1$; so that, $T = mg\ /\ 2\ (l + t)$, *in the case of the plate and* $T = mg/2l$, *in the case of the wire frame.*

And thus, knowing mg, l and t, the value of surface tensions (T) for the liquid can be easily calculated.

As will be readily seen, the method may also be used to compare the surface tensions of two given liquids, θ for which is zero.

N.B. It happens sometimes that, despite all care, the lower edge of the plate, instead of being just in a level with the liquid surface, dips a little, say to a depth h inside it. If this be so, it is obvious that downward pull on the plate, due to the surface tension, is reduced by an amount equal to the upward force on it due to the buoyancy of the displaced liquid which, as we know, is equal to the weight of the liquid displaced by the plate, *i.e.*, equal to $(l \times t \times h \times \rho \times g)$. where ρ is the density of the liquid and g, the acceleration due to gravity at the place.

Thus, we have $2(l + t).\cos\theta.T - (l \times t \times h \times \rho \times g) = mg.$

Or, $2(l + t).\cos\theta.T = mg + l.t.h.r.g$, whence, $T = \dfrac{mg + l.t.h.\rho.g}{2(l + t)\cos\theta}$

2. Jaeger's method (Or the maximum bubble - pressure method). We know that the *excess pressure inside an air bubble in a liquid* is equal to $2T/r$, [§ 14.9], where T is the surface tension of the liquid and r, the radius of the bubble. Jaeger has, by a simple and ingenious method, measured this excess pressure p, necessary to produce such a bubble. So that, knowing p and r, the surface tension (T) of the liquid can be easily determined.

The apparatus simply consists of a long thin glass tube AB, (Fig. 14.31), with its lower portion ending in a fine jet of about .2 to .5 *mm* in diameter, and with its tip cut quite smooth and square*, so as to be perpendicular to its axis. This dips in the experimental liquid, contained in a vessel, with about 4 to 5 *cm* of its length inside the liquid,– the diameter of the vessel being at least 8 *cm to ensure the flatness of the liquid surface in it.* It is then connected to a manometer M and a Woulff's bottle, fitted with a dropping funnel F containing water (or mercury) as shown. The liquid used in the manometer M is *Xylol* (a liquid hydrocarbon) in preference to water, – because of its *lower density*, – so that the difference of level in the two limbs may be large.

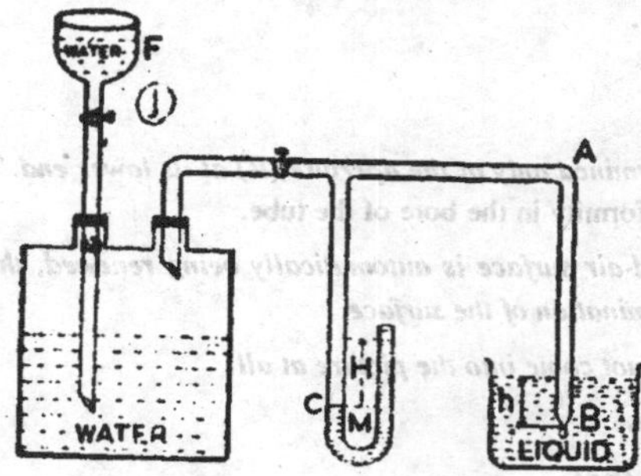

Fig. 14.31

Due to capillary action, some liquid rises up into the tube AB, the shape of its meniscus being nearly *hemispherical*. Some air is now forced into the tube by dropping water into the Woulff's bottle, which displaces its own volume of air from it. The liquid column in AB thus slowly moves down until it reaches B, when a bubble is formed there. The radius of curvature of the bubble gradually decreases with increasing pressure inside it, until it reaches the *minimum* value, and the bubble acquires a more or less hemispherical shape, with radius r, equal to that of the aperture at B, *the pressure inside being now the maximum*, as indicated by the difference of levels (H) in the two limbs of the manometer.

The bubble now becomes unstable ; for, any further growth of it tends to increase its radius, which results in a corresponding decrease in the pressure inside it due to surface tension, thus destroying the equilibrium between its internal and the constant external pressure. It, therefore, gets detached from the tube, and the whole process starts all over again.

It is imperative that the bubbles form *(i) only one at a time and (ii) so slowly that it takes at least 10 sec to form one bubble. This can be easily arranged (a) by suitably reducing the volume of the air space above water in the Woulff's bottle; so that as soon as a bubble gets detached at B, the pressure in the apparatus falls to a value lower than that required to blow another*, and *(b)* by *properly regulating the flow of water from the dropping funnel into the Woulff's bottle.*

Now, clearly, just before the bubble breaks away from B, the *pressure inside it* is equal to that at C, *i.e.*, equal to $P + H.\rho.g$, where P is the *atmospheric pressure* and $H.\rho.g$, the pressure due to the liquid column H in the manometer, (ρ being its density).

And, when the bubble just breaks away from B, *the pressure on it* is equal to that at the level of B in the beaker, *i.e.*, equal to $P + h.dg$, where h is the depth of the tip B from the surface of the liquid in the beaker and d, its density.

* This must be so, not only to the naked eye but even under a microscope, with no trace of any roughness or raggedness at its inner or outer edges.

And, therefore, the excess pressure inside the bubble

$$= (P + H.\rho.g) - (P + h.d.g) = g\,(H.\rho - h.d).$$

But the excess pressure inside the bubble, we know, must be equal to $2T/r$, where T is the surface tension of the liquid in the beaker. Hence, $2T/r = g\,(H.\rho - h.d)$. or, $T = r.g\,(H.\rho - h.d)/2$.

Thus, noting H, h, ρ and d, determining r with the help of a microscope, fitted with a micrometer eye-piece, we can easily calculate the value of T for the given liquid.

An obvious **drawback** of the method is that, despite all care, there is no absolute certainty as to the radius of the bubble when it gets detached from the tube, and *it may not be hemispherical and of quite the same radius as the aperture at B*. In fact, quite the contrary. For, it can be hemispherical only in the case of *extremely narrow tubes*, and its radius is found to be always a function of the radius of the aperture. For a greater accuracy in the result, therefore, the following relation is often used: $T = \frac{rg}{2}.\,[H.\rho - d(h + 2r/3)\,]$.

This method has a number of **points in its favour :**

(i) *The internal radius of the capillary tube is to be determined only at the aperture (B) at its lower end.* There can thus be no question as to errors due to want of unfomity in the bore of the tube.

(ii) *Since a fresh bubble is formed every time, the liquid-air surface is automatically being renewed, thereby greatly reducing the possibility of errors due to contamination of the surface.*

(iii) *The angle of contact for the liquid-glass surface does not come into the picture at all.*

(iv) The quantity of the liquid required is *not unduly large.*

Apart from this, the method has its own **importance** in that

(a) *it can be used to study the variation of surface tension with temperature,* since the temperature of liquid in the containing vessel can be easily controlled and the bubble is formed inside the liquid itself;

(b) by using capillary tubes of platinum etc., *the method can also be used to determine surface tension of molten metals;*

(c) *the variation of surface tension of a solution with different concentrations of the solute may also be studied with its help;*

(d) *the surface tensions of different liquids can be easily compared by this method;* and

(e) *the molecular aggregation of the liquid, i.e., the number of atoms in its molecule, may also be easily studied from the slope of the curve between temperature and surface tension.*

Sugden's improvement over Jaeger's method. An inherent officially in maximum bubble-pressure methods is that the exact mode of variation of the density of the liquid with temperature must be known for a correct determination of its surface tension. This is unfortunately not always available. Sugden, therefore, suggested that the necessity for it could be obviated by forming bubbles at the ends of two different capillary tube of different radii r_1, and r_2. Then, if the same lengths of the two capillaries be inside the liquid, we have $2T/r_1 = g(\rho H_1 - h\,d.)$ and $2T/r_2 = g(\rho H_2 - h\,d.)$ whence $2T\left(\frac{1}{r_1} - \frac{1}{r_2}\right) = g\rho(H_1 - H_2)$. So that, it can be easily evaluated. Another advantage is that the containing vessel need not necessarily be wide for the liquid surface to be flat in it, the term involving h, the height of the liquid surface from the aperture (B) being no longer there in the final relation for T.

3. The Drop-Weight Method. This is a simple, though perhaps not quite so accurate a method for determining the surface tension of a liquid by considering the vertical forces that keep a small drop of liquid in equilibrium, *just before it gets detached* from the end of a *vertical* glass tube of *circular aperture*.

At the instant the drop gets detached, it assumes a *cylindrical* shape at the orifice of the tube, (Fig. 14.32), so that, if T be the surface tension of the liquid and r, the radius of the orifice, we have *excess pressure* (p) *inside the drop over the outside atmospheric pressure*

$= \quad T/r.$

(See § 14.12.)

Hence the *downward force* on the drop due to it $= \pi r^2 . T/r$.

And, since the *weight mg* of the drop also acts vertically downwards on it, we have

total downward force on the drop $= \pi r^2 . T/r + mg$.

Now the liquid touches the tube all along its circumference $2\pi r$, and hence the *upward force* acting on the drop *due to surface tension* is equal to $2\pi r . T$.

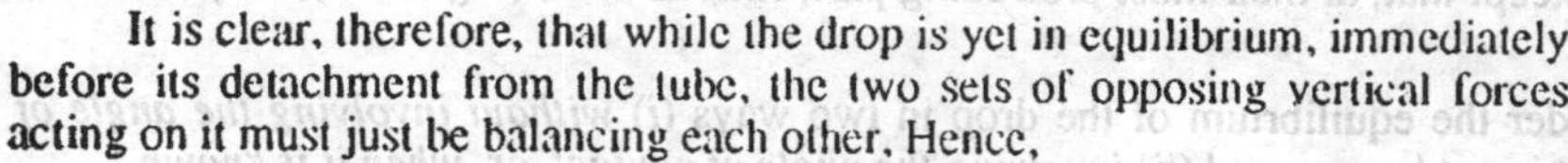

Fig. 14.32

It is clear, therefore, that while the drop is yet in equilibrium, immediately before its detachment from the tube, the two sets of opposing vertical forces acting on it must just be balancing each other. Hence,

$$2\pi . r T + \pi r^2 . T/r + mg = \pi r . T + mg.$$

Or, $2\pi r . T - \pi r . T = mg$. Or, $\pi r . T = mg$, whence, $T = mg/\pi r$.

Thus, knowing the mass of the falling drop and the radius of the lower end of the tube, we can easily calculate the value of T, the surface tension of the liquid in question.

The actual experiment is carried out as follows :

A burette is fitted vertically in a suitable clamp and a thin, clean and dry tube of glass (of about 4 *mm* bore) is attached to its nozzle by a piece of India-rubber tubing, carrying a pinch-cock on it. The burette is filled with the experimental liquid and its flow through the glass tube regulated by the pinch-cock, *so that small drops* form slowly at its lower end*, — their rate of detachment from it being about *one per minute*, when an accuracy of about 2% may possibly be attained.

The drops, as they fall, are received in a clean, dry and an accurately weighed beaker and the *average mass* (m) of a drop determined by weighing the beaker again with its contents. The diameter of the orifice of the tube is also determined carefully by means of a travelling microscope.

As indicated above, the method is far from accurate, for the simple reason that the liquid drop seldom gets detached from the end of the tube, under the ideal statical conditions assumed,- the radius of the 'neck' of the drop as well as the *amount* of it that actually gets detached being more or less uncertain quantities, even if the end of the glass tube be smeared with wax, thus making the whole problem a complicated dynamical one.

N.B. *Lord' Rayleigh* has suggested the relation $T = mg/3.8r$, which yields better results, and *Harkins* and *Brown* have experimentally shown by using liquids of known values of T, that the relation $T = mg.F/r$ holds true, where F is a factor related to v/r^3 ** —the symbols v and r standing for the volume of the drop and the radius of the tube respectively.

4. Quincke's Method. This method is applicable in the case of liquids *which do not wet the surface in contact with them*, as for example, *mercury*. Not only can it be used to determine the surface tension of such a liquid but also its angle of contact with the solid surface on which it rests. We shall consider here the case of mercury, for which the method is so frequently used.

It has been indicated already that in the matter of the shape of a liquid drop resting on a solid surface, which it does not wet, it is a tussle between surface tension and gravity, the former having the upper hand in the case of small drops, which thus assume a spherical shape and the latter holding sway in the case of the larger ones, tending

Fig. 14.33

* The drops formed at the narrow orifice of the tube need must be small, for, as they grow in size, the *liquid skin* there is unable to support their weight.

** This relation too is found to be true only up to the limit where the maximum value of v/r^3 does not exceed 10.

to spread them out until their surface becomes horizontal. This explains the gradual flattening up of a drop as it grows in size, until its top becomes quite horizontal, beyond which no more flattening occurs due to the limit imposed by the angle of contact of the liquid and the solid in question. So that, the shape ultimately acquired by a large drop of the liquid, (in our case, mercury), is as shown, (Fig. 14.33); with the central part of the top surface horizontal and the two ends protruding outwards, such that the tangents to them make an angle α each with the horizontal solid surface, where α is the supplement of the angle of contact (θ) for the liquid and the solid, — its value being about 40° in the case of mercury and glass, (the angle of contact for them being nearly 140°). The actual shape of these protruding edges of the drop is of little or no concern to us for our present purpose, except that, at their most protruding part, such as at *D*, (Fig.14.39), they are more or less *vertical*.

Let us consider the equilibrium of the drop in two ways (*i*) *without involving the angle of contact, or, when it is not known, and (ii) involving the angle of contact, or, when it is known.*

(*i*) **Without involving the angle of contact.** Imagine a large drop with its central part horizontal, to be cut into two halves by a vertical plane *AGFJ*, perpendicular to its length and passing *centrally* through it, and one of the halves to be further cut by two parallel vertical planes *ABNJ* and *GCEF*, perpendicular to the first, so as to obtain a thin slice of it, shown in Fig. 14.34, with a horizontal width $GA = FJ = b$.

As is evident from the very symmetry of the drop, (*a*) *the forces acting on the two sides, GCEF and the one opposite to it at the back, must be equal and opposite, and (b) so also must be the pressure immediately above and below the horizontal part of it at the top.*

Let us consider the portion or the slice, lying above the horizontal plane *DLKM* a distance *h* below the top and passing through *D*, *where the drop protrudes most* and which, therefore, represents *the maximum horizontal surface, i.e., the surface having the maximum area in the section of the drop.*

The forces acting on this slice are the following :

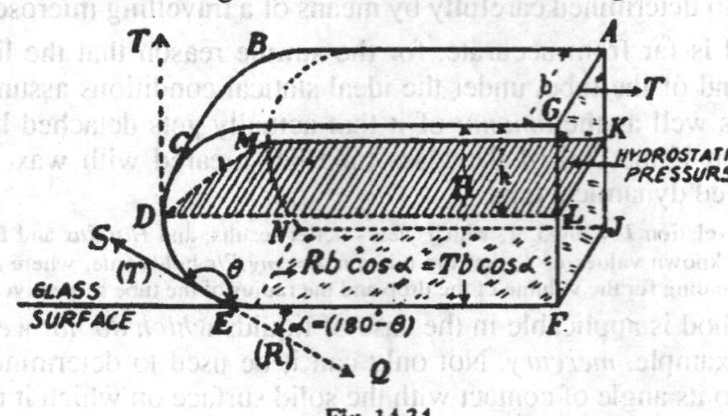

Fig. 14.34

(*i*) **the surface tension at *D*, *which acts* vertically *upwards* and has, *therefor*, no component along the horizontal;**

(*ii*) **the laterals hydrostatic pressure (P) on the plane face GLKA** ***due to the neighbouring part of the liquid on its right,*** **acting towards the left;**

(*iii*) **the pull due to surface tension over GA,** ***acting perpendicularly to it,*** **towards the right.**

Thus, the only horizontal forces acting on the slice are *the latter two, in oppose directions,* and since the slice is in equilibrium, they must be *equal in magnitude.*

Now, *lateral hydrostatic pressure due to a liquid being equal to the pressure halfway down,* we have

$$\text{hydrostatic } pressure\ P \text{ over } GLKA = \frac{1}{2}GL.\rho.g = \frac{1}{2}h.\rho.g,$$

where $GL = h$; ρ, the *density* of the liquid and g, the *acceleration due to gravity.*

$\therefore$ *hydrostatic thrust* over $GLKA = P \times$ area $GLKA$.

$$= \frac{1}{2} h.\rho.g \times h \times b = h^2 . \rho.g.b/2.$$

And, the pull on *GA due to surface tension* = *T.b.*, where *T* is the surface tension of the liquid ; so that,

$$T.b = h^2 . \rho.g.b/2. \quad \text{Or,} \quad T = h^2 . \rho.g/2, \quad ...(A)$$

whence, *T*, the *surface tension* of the liquid may be easily determined, if *h* be measured (see below), *without involving the angle of contact for the liquid and the solid in question.*

(*ii*) **Involving also the angle of contact.** In this case, we consider the equilibrium of the *whole* slice. So that, the surface-tension pull *T.b.*, due to mercury on glass; at *E, i.e.*, on the line of contact *EF* of glass and mercury, acts tangentially to the slice in the direction *ES*, with an equal reaction of the glass on the slice, *in the opposite direction EQ*. The horizontal component of this *reactional force due to glass*, along *EF*, is clearly equal to $T.b. \cos\alpha$, where α is the supplement of the angle of contact for the given liquid and the solid surfaces, and acts *towards the right.* And, so does the pull due to surface tension over *GA*, *viz.*, *T.b*, acting perpendicularly to *GA* ; so that, the total horizontal force towards the right $= T.b\cos\alpha + T.b.$

This is opposed by the horizontal hydrostatic thrust over the face *GFJA*, acting *towards the left*, — this thrust being now equal to

$$\frac{1}{2} H.\rho.g \times H \times b = \frac{1}{2} H^2 \rho.b.g.,$$

where *GF* = *H*, the *total height* of the slice.

So that, there being no other horizontal forces to be considered, we have, for the equilibrium of the slice,

$$T.b\cos\alpha + T.b = \frac{1}{2} H^2 . \rho . b.g. \quad \text{Or,} \quad T\cos\alpha + T = \frac{1}{2} H^2 . \rho.g.$$

$$\text{Or,} \quad 2T + 2T\cos\alpha = H^2 . \rho.g. \quad \text{Or,} \quad 2T(1 + \cos\alpha) = H^2 \rho.g.$$

$$\text{Or,} \quad T = \frac{H^2.\rho.g}{2(1 + \cos\alpha)}, \quad ...(B)$$

whence, *T* may be calculated, if *H* and α be known.

The quantities *h*, *H* and α, involved in relations (*A*) and (*B*) above may be measured as follows:

(*i*) *Measurement of h.* For the measurement of *h*, *the distance between the top of the drop and the horizontal surface of maximum area*, we must determine the point *D*, where the free surface of the drop is *perfectly* vertical. This has been beautifully done by *Edser*, by means of the arrangement, shown in Fig. 14.35.

A fairly large drop of mercury on a clean glass plate is placed on a table, provided with levelling screws, and its top arranged to be *perfectly horizontal*. A special vernier microscope, having a plane glass plate *P*, cemented on to its objective at an angle of 45°, is then arranged horizontally (i.e., with its axis horizontal), right in front of the most protruding part of the drop. Light from an incandescent lamp *S* is focused by a lens *L* and the plate glass *P*, acting as a mirror, on to the edge of the drop, when a *bright, thin, horizontal line* is seen at *D*, where the drop protrudes out most and the surface is vertical. The aperture of the microscope is adjusted until this bright horizontal line

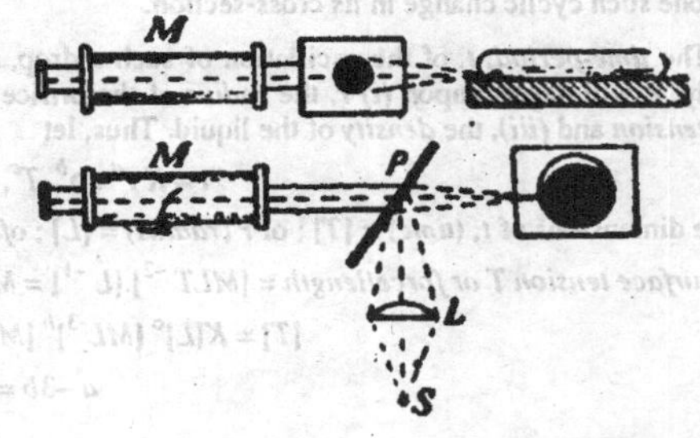

Fig. 14.35

coincides with the horizontal cross-wire of the eye-piece, and its reading is noted. The microscope, which is capable of both a vertical and a horizontal movement, is then raised up and moved towards the drop until the image of the top *flat* surface of the drop coincides with the horizontal cross-wire of the instrument. This latter adjustment is greatly facilitated by sprinkling a little lycopodium powder over the top of the drop and making the image of a speck or two of the powder coincide with the horizontal cross-wire of the eye-piece. The distance through which the instrument has had to be raised up directly gives the required distance $GL = h$.

(*ii*) **Measurement of H.** This may be easily done by means of a spherometer, or a microscope, the latter being focussed on the top, (*i.e.*, on a speck of the lycopodium powder on its surface, as before), and then on the point of the edge at the bottom which is in contact with the surface of the glass plate.

(iii) **Measurement of α, and hence also the angle of contact θ.** We have, from relations *A* and *B* above,

$$\frac{h^2.\rho.g}{2} = \frac{H^2.\rho.g}{2(1+\cos\alpha)} \quad \text{Or,} \quad h^2 = \frac{H^2}{(1+\cos\alpha)}.$$

Or, $\quad 1+\cos\alpha = H^2/h^2$, whence, $\cos\alpha = (H^2/h^2) - 1$.

So that, knowing *H* and *h*, we can easily calculate the value of $\cos\alpha$ and hence that of α.

Clearly, *this also gives us a method of measuring* θ, *the* **angle of contact for the given liquid and solid** ; *for*, $\theta = (180° - \alpha)$ *the two angle* θ *and* α *being supplementary angles.*

Important corollaries. (i) Relation (*A*) above tells us that *h*, the distance of the top flat surface of the drop from the horizontal surface of maximum area, is quite independent of the size of the drop, but depends only upon the nature of the liquid and the solid in contact, (*i.e.*, upon the angle of contact for them).

(*ii*) We have, from relation (*B*) above,

$$H^2 = \frac{2T(1+\cos\alpha)}{\rho.g} = \frac{4T\cos^2(\alpha/2)}{\rho.g}. \qquad [\because\ 1+\cos\alpha = 2\cos^2(\alpha/2).$$

Or, $\quad H = 2\cos(\alpha/2)\sqrt{T/\rho g}$,

whence, it is clear that

(*a*) *all flat drops of a liquid, resting on a horizontal solid plate, which they do not wet, must have the same height;*

(*b*) *all liquids that wet solid surface spread out indefinitely on it for, in their case,* $\theta = 0°$ *and, therefore,* $\alpha = \pi$; so that, $\cos\alpha/2 = 0$ and , consequently, $H = 0$.

5. Rayleigh's Jet Method. A liquid jet, issuing horizontally out of an orifice, shows a strange recurrence of forms in its surface– a phenomenon, in which surface tension plays its own part. This has been utilized by *Rayleigh, Pedersen, Bohr* and *Stocker* for the determination of the surface tension of a liquid. We shall consider here only the simpler *Rayleigh's method*, by way of an illustration as to how this may be done.

We know that, under ordinary conditions, a liquid jet is quite *'steady,'* in the hydro-dynamical sense of the term; for, although made up of drops *in motion*, *(i)* its surface is *fixed in space* and *(ii)* the velocity at any point remains constant. But, on making a closer and careful examination of it, we find that, in its initial formative stage, *its various parts are in motion with respect ot each other, its form* oscillating this way and that about a mean spherical one, with its lateral dimensions, measured along a particular direction, exhibiting a cyclic change. Thus, for example, points *P* and *Q* in the jet, (Fig. 14.36), include one such cyclic change in its cross-section.

The *time-period*, *t*, of the oscillation of such a drop, may be easily determined by the method of dimensions. For, obviously, it will depend upon *(i) r*, the *radius* of the orifice *(O)* of the horizontal tube whence the jet emerges, *(ii) T*, the *surface tension* and *(iii)*, the *density* of the liquid. Thus, let

$$t = K\, r^a.\rho^b.T^c, \quad \text{where } K \text{ is a dimensionless constant.}$$

Now, the dimensions of *t*, (*time*) = $[T]$; of *r* (*radius*) = $[L]$; *of density* or *mass/volume* = $[ML^{-3}]$:

and, of *surface tension T or force/length* = $[MLT^{-2}].[L^{-1}] = MT^{-2}$.

Hence $\qquad [T] = K[L]^a\,[ML^{-3}]^b\,[MT^{-2}]^c = K.L^{a-3b}M^{b+c}T^{-2c}$.

So that, $\qquad a - 3b = 0,\ b + c = 0$ and $-2c = 1$.

whence, $\qquad c = -\frac{1}{2},\quad b = \frac{1}{2}$ and $a = \frac{3}{2}$.

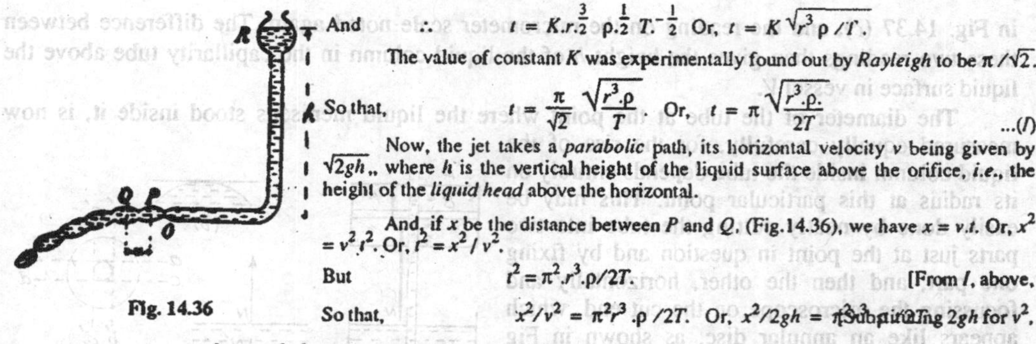

Fig. 14.36

And $\therefore$ $t = K.r.^{\frac{3}{2}}\rho.^{\frac{1}{2}}T.^{-\frac{1}{2}}$ Or, $t = K\sqrt{r^3\rho/T}$.

The value of constant K was experimentally found out by *Rayleigh* to be $\pi/\sqrt{2}$.

So that, $t = \frac{\pi}{\sqrt{2}}\sqrt{\frac{r^3.\rho}{T}}$. Or, $t = \pi\sqrt{\frac{r^3.\rho.}{2T}}$ ...(*I*)

Now, the jet takes a *parabolic* path, its horizontal velocity v being given by $\sqrt{2gh}$,, where h is the vertical height of the liquid surface above the orifice, *i.e.*, the height of the *liquid head* above the horizontal.

And, if x be the distance between P and Q, (Fig.14.36), we have $x = v.t$. Or, $x^2 = v^2t^2$. Or, $t^2 = x^2/v^2$.

But $t^2 = \pi^2.r^3.\rho/2T$. [From *I*, above,

So that, $x^2/v^2 = \pi^2r^3.\rho/2T$. Or, $x^2/2gh = \pi^2r^3.\rho/2T$. [Substituting $2gh$ for v^2,

Or, $x^2/gh = \pi^2r^3.\rho/T$ whence, $T = \pi^2r^3.\rho.g.h./x^2$.

So that, measuring x with the help of a travelling microscope (by properly illuminating the jet), we can easily calculate the value of T, the surface tension of the given liquid.

6. The Capillary Rise Method. A direct application of the relation, $T = r(h + r/3)\rho g./2$ (§ 14.17 page 455) connecting the rise of a liquid in capillary tube with its surface tension, it is a simple *laboratory method* for the determination of surface tension of liquids which *wet* the walls of the tube, *i.e.*, for which the angle of contact may be taken to be zero.

A capillary tube is taken and the uniformity of its bore carefully tested by introducing a thread of mercury inside it and measuring its length, by jerking it into different positions along the tube. If the thread measures the same everywhere, the tube has a uniform bore. When a proper tube has thus been selected, it is thoroughly cleaned by rinsing it first with caustic soda, then, with nitric acid and and finally with distilled water. If the liquid, whose surface tension is to be determined, be water, the tube may be used straightway, but if it be some other liquid than water, the tube must be properly dried by passing a current of warm air through it, and then fixed *vertically*, alongside a plumb line, with its lower end immersed in the experimental liquid contained in a thin glass beaker V, as shown, (Fig. 14.37(*a*)). A bent pin N or bent glass rod, (with one end drawn to a point), is attached to the tube by means of rubber bands B and its position adjusted, until the lower pointed end of the pin or the glass rod *just touches* the liquid surface in the beaker.

Due to capillary action, the liquid rises up into the tube to a certain height h, say, above the liquid surface in the beaker. Usually, there comes about a slight, unavoidable, greasy contamination of the tube and the liquid, particularly in the case of water, with the result that, due to its lowered surface tension, and altered angle of contact, the water column *creeps down a little* below its initial position in the tube*. To eliminate this error, we either lower the capillary tube further down into the liquid or attach a piece of rubber tubing to its upper end and pinch it, so that, in either case, the liquid rises in the tube to a much higher level. If, on now raising the tube up to its initial position, or on releasing the rubber tubing, the liquid falls readily to a lower fixed position and stays there, the tube may be assumed to be free of any greasy contamination and may safely be used.

Now, when the liquid column in the tube has become quite steady, we focus a travelling microscope, placed right in front of the tube, on the liquid meniscus at the top and arrange the horizontal cross-wire of the micrometer eye-piece to be tangential to its lower curved surface. Since, however, the meniscus appears inverted, and therefore convex upwards, through the microscope, the cross-wire is actually arranged to be tangential to this convex surface, as shown in Fig.14.37 (*b*) and the reading noted on the micrometer scale of the eye-piece. The position of the needle is next so adjusted that its lower tip just makes contact with the liquid surface in the vessel, the microscope focused on it and the horizontal cross-wire of the eye-piece so arranged that it just touches the tip of needle N as also that of its reflected image N' in the liquid surface, as indicated

* No such thine occurs in the case of liquids like ether and benzene etc. This is probably because they dissolve out any greasy contamination present.

in Fig. 14.37 (*c*), and the reading on the micrometer scale noted again. The difference between these two readings then gives the height *h* of the liquid column in the capillarity tube above the liquid surface in vessel *V*.

The diameter of the tube at the point where the liquid meniscus stood inside it, is now measured equally carefully, for the rise of the liquid column inside the tube depends entirely on its radius at this particular point. This may be easily done by neatly cutting the tube into two parts just at the point in question and by fixing one part, and then the other, horizontally and focussing the microscope on the cut end, which appears like an annular disc, as shown in Fig 14.37 (*d*). The horizontal crosswire of the eye-piece is adjusted to be tangential to the inner circle first at its upper and then at its lower edge, as indicated by the dotted lines *ab* and *cd* in the figure. The difference between the two readings of the micrometer scale then gives the inner vertical diameter of the tube. Its horizontal diameter is similarly measured by making the vertical crosswire of the eye-piece tangential to the inner circle first on one side and then on the other, as shown by the dotted lines *a′ b′* the *c′ d′*. In fact, several pairs of perpendicular diameters are thus measured for either piece of the tube and their mean is then taken as its correct inner diameter.

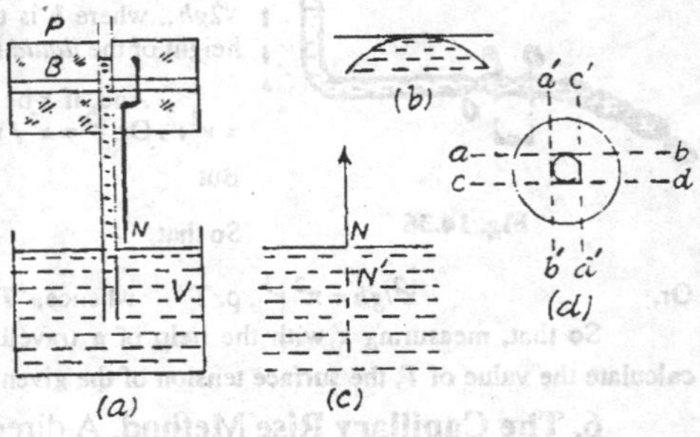

Fig. 14.37

A more accurate method of determining the diameter of the tube is to measure the length of a mercury thread introduced into it, and then to carefully take the mass of the mercury by withdrawing it in a weighed and clean watch glass. Thus, if *l* be the length of the mercury thread in the tube *m*, its mass ; ρ, its density and *r*, the radius of the tube, we have $m = \pi.r^2.l.\rho$, whence, $r = \sqrt{m/\pi .l. \rho}$.. *This would however be so, only if the ends of the mercury thread were flat or plane, which, as we know, is seldom the case, and the length of the thread must, therefore, be* corrected for the curvature of its ends, – usually semi-ellipsoidal in shape, (Fig. 14.38). The two curved *ends* of the thread are, therefore, measured separately, and if *x* / 2 be the length of each curved part, so that the *total length of the curved end is* x, then, if *l* be the length of the cylindrical part, we shall have

$$m = \pi r^2 . \left(l + \frac{2}{3}x \right).\rho, \quad \text{whence, } r = \sqrt{m/\pi\,\rho\,(l + \tfrac{2}{3}x)}.$$

Substituting the values of *x* and *r*, thus obtained, in the expression above, the value of *T* can be easily calculated.

The experiment is repeated with another capillary tube of a different bore and the mean of the two values of *T* obtained is then taken as the correct value of the surface tension of the given liquid.

Comparison with Jaeger's method. Drawbacks of the method. Simple though it is, the capillarity tube method does not compare favourably with *Jaeger's* method discussed earlier [§ 14. 24 (2)]. For one thing, it can not be used for any of the purposes for which Jaeger's method is so eminently suitable and due to which it derives its deserved importance (see page 464) and, for another, it suffers from some obvious drawbacks. Thus (*i*) *the bore of the capillary tube, must be uniform all along its length, which need not be so in Jaeger's method, where we require its value only at the lower tip.* As can be well imagined, it is extremely difficult to obtain a capillary of a really uniform bore throughout and equally difficult to maintain it clean.

(*ii*) *the air-liquid surface, being clearly static here, is highly prone to contamination, thereby greatly vitiating the results obtained;*

(iii) it is not quite so easy to correctly measure the height up to which the liquid column rises in the capillary tube; and finally,

(iv) the angle of contact (θ) for the given liquid and glass, not involved in Jaeger's method, is very much in the picture here. Although we assume it to be zero for the sake of simplicity and convenience, there is really hardly any justification for it. Indeed, as pointed out by *Ferguson* and *Dowson*, despite all possible care taken in various measurements, the value obtained is that of $T \cos\theta$ rather than that of T. So that, the method, though widely used, is good enough only as a junior laboratory exercise.

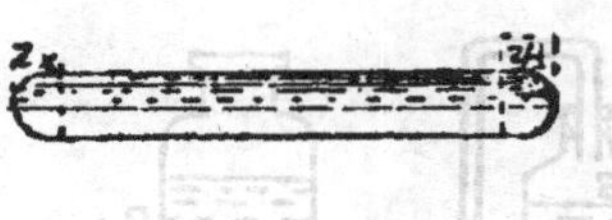

Fig. 14.38

Perhaps, the one and only point in its favour is that a suitable modification of it, suggested by *Ferguson* (see next article) enables the quantity of the liquid used to be reduced to the very minimum.

7. Ferguson and Kennedy's Method. This is an ingenious modification of the capillary-tube method, discussed above, and is *most suitable for liquids available in small quantities*, –the more so, because *the density of the liquid is not required for the purpose.*

A clean glass tube, with a small capillary portion PQ, (Fig. 14.39) is mounted horizontally and connected to a manometer M and a pressure-varying arrangement, consisting of a Woulff's bottle and a water reservoir R, which can be raised or lowered at will (or a tap funnel containing water, is fitted into the Woulff's bottle itself). The small quantity of the liquid, the surface tension of which is to be determined, (and which may be as small in quantity as 2 *cubic millimeters*), is introduced into the capillary at the end P, *which must be carefully ground plane.*

Now, so long as the pressure at either end of the liquid is the same, *i.e.*, there is no pressure-difference across it, the two ends remain equally concave outwards, as shown in inset (*a*). If the pressure be increased on the inner side, by raising the reservoir (or opening the tap funnel) to run more water into the Woulff's bottle, the liquid column not only moves outwards but the curvature of its outer end (towards P) also goes on progressively decreasing (to maintain its equilibrium), tending first to become plane and then convex outwards. The increase of pressure is stopped the moment the outer end of the liquid column becomes *perfectly accurately plane*, as shown in inset (*b*).

This is easily tested by illuminating the liquid surface at P by means of an electric lamp L, placed obliquely some distance (about 30 to 40 *cm*) away, and observing the reflected light through a microscope E. So long as the liquid surface at P remains curved, –either concave or convex, –a regular image of the lamp would be visible. As soon, however, as it becomes perfectly plane, the image vanishes and the surface at P appears to be *uniformly illuminated*, instead.

Now, with the liquid surface *plane* at P, there is no difference of pressure due to surface tension on its two sides. At the end Q, however, the excess pressure due to the curved meniscus is given by the manometer. If the difference of level of the liquid columns in the manometer be h and the density of the liquid in it be ρ, this excess pressure = $h.\rho.\gamma$. Hence, if r be the radius of the capillary tube, the total thrust on the meniscus, acting *towards the left* = $\pi.r^2.h.\ \rho.g$. And, clearly, the force due to surface tension, acting on it, *towards the right*, is equal to $2\pi r.T \cos\theta$, where θ is the angle of contact for the liquid and glass. Since the liquid column is in equilibrium, we have

$$2\pi.r.T\cos\theta = \pi r^2.h.\rho.g. \quad \text{whence } T = r.h.\rho.g/2\ \cos\theta$$

And, if the liquid be one which *wets* the tube, *i.e.*, for which. $\theta = 0°$; so that, $\cos\theta = 1$, the expression for T reduces to

$$T = r.h.\theta.g/2,$$

from which the surface tension T of the given liquid can be easily calculated.

There is only one slight error possible in this method, *viz.*, that the shape of the liquid meniscus at Q may not be really perfectly spherical (as assumed), due to the distortion effects on account of gravitational forces. But *Ferguson* and *Kennedy* have shown that that can be safely neglected, if the bore of the tube be of a really small diameter, about 1 *cm*, or less.

8. Ripples method. Before proceeding with the method proper, we must try to understand the difference between *waves* and *ripples*.

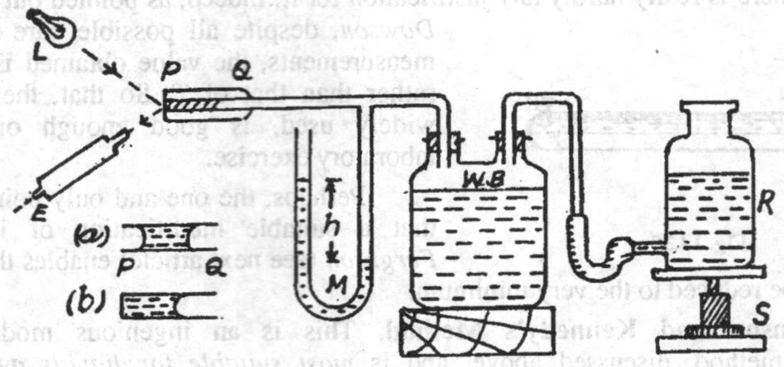

Fig. 14.39.

Although we are familiar, in a general way, with transverse waves on the surface of a liquid, these are actually of different well-defined types according as the dominant factor controlling their propagation is the *force of gravity*, *surface tension* or the *depth of the liquid*, when they are referred to us *gravity waves* or simply, *waves*, *ripples* or *cappillary waves* and *long or canal waves* respectively. The *viscosity* of the liquid comes into the picture only if we wish to estimate the rate of decay of the waves with distance or time.

We shall concern ourselves herewith only the first two types, *viz.*, the *gravity waves* and *ripples*.

(*i*) **Gravity waves.** Since liquids are incompressible, it follows that if the surface of a deep liquid be disturbed, a transverse motion of the particles on the surface must necessarily be accompanied by their longitudinal motion, or else, whereas a large pressure would be built up under a trough, a cavity would be created under a crest, thereby disrupting the continuity of the liquid. To preserve its continuity, therefore, the liquid must move from a trough to an adjoining crest. In other words, a particle on the liquid surface must simultaneously execute vertical and horizontal oscillations which are passed on from particle to particle, producing a *transverse wave*.

Further, if the pattern of the transverse wave is to be maintained unaltered as it moves along the liquid surface, tne two types of oscillations must have a *small amplitude* as also an *identical time-period*, in the absence of which the wave pattern gets progressively distorted and ultimately breaks down.

Now, the vertical and horizontal velocities of a particle on the liquid surface can only increase (or decrease), one at the expense of the other. So that, at a crest or a trough, where the vertical velocity of a particle is *zero*, its horizontal velocity is the *maximum* (in the opposite direction to that of the wave velocity at a crest and in the same direction as the wave velocity at a trough, as we shall presently see). And, on the undisturbed part of the liquid surface, the horizontal velocity of a particle is *zero* and its vertical velocity, the maximum (and at right angles to the direction of the wave velocity). The vertical and horizontal oscillations thus differ in phase by $\pi/2$.

Remembering that at every point on the liquid surface, the wave velocity is superimposed on the particle velocity there, let us consider a wave *OACBD* (Fig. 14.40) travelling from left to right with velocity V and let OX represent the undisturbed surface of the liquid or the datum plane.

If dx/dt be the maximum horizontal velocity of the particle at the crest A and at the trough B, directed opposite to, and along, the direction of the wave velocity respectively, as shown, (the vertical velocity here being zero), we have *resultant velocity of the particle at crest A*, say, $v_1 = V - dx/dt$ and *resultant velocity of the particle at trough B*, say, $v_2 = V + (dx/dt)^2$

And, if the vertical velocity of the particle at C on the undisturbed part of the liquid surface by dy/dt (its horizontal velocity being zero), we have *resultant velocity of the particle at C*, say

$$v_3 = \sqrt{V^2 + (dy/dt)^2}$$

since, as we know, dy/dt is perpendicular to V.

Now, applying Bernoulli's equation to points A and C of the liquid surface, we have

$$p/\rho + ag + \frac{1}{2}v_1^2 = p/\rho + 0 + \frac{1}{2}v_3^2$$

Or, $$ag + \frac{1}{2}v_1^2 = \frac{1}{2}v_3^2, \qquad ...(i)$$

where a is the amplitude of the wave (or the vertical oscillation of the particle), p, the atmospheric pressure on the liquid surface (the same at A and C, ρ, the density of the liquid, which is constant, in view of the incompressibility of the liquid.

Similarly, for points B and C, we have

$$p/\rho - ag + \frac{1}{2}v_2^2 = p/\rho + 0 + \frac{1}{2}v_3^2$$

Or, $$-ag + \frac{1}{2}v_2^2 = \frac{1}{2}v_3^2, \qquad ...(ii)$$

Adding relations (*i*) and (*ii*), therefore, we have

$$\frac{1}{2}(v_1^2 + v_2^2) = v_3^2,$$

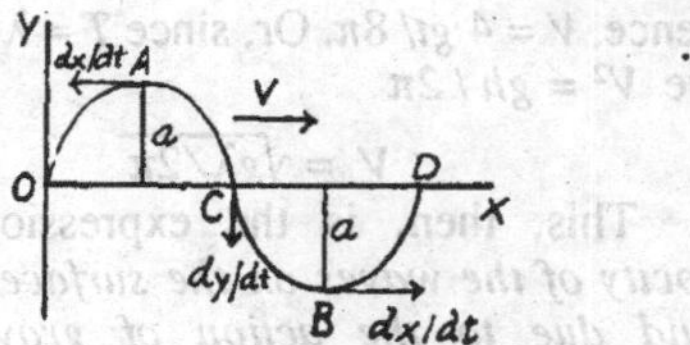

Fig. 14.40

or, substituting the values of v_1, v_2 and v_3 from above, we have

$$\frac{1}{2}[(V - dx/dt)^2 + (V + dx/dt)^2] = V^2 + (dy/dt)^2$$

or, $$V^2 + (dx/dt)^2 = V^2 + (dy/dt)^2,$$

whence, $$dx/dt = dy/dt$$

i.e., the maximum velocity of the horizontal oscillation of the particle is the same as that of its vertical oscillation. Obviously therefore, the amplitude of the horizontal oscillation is also equal to the amplitude (a) of the vertical oscillation. If the amplitude be small, the two oscillations are simple harmonic in nature.

We thus see that each particle on the liquid surface is simultaneously subjected to two equal simple harmonic oscillations, *i.e.*, of the same amplitude a and time-period T, differing in phase by $\pi/2$. Since two such simple harmonic motions combine into a uniform circular motion of a radius equal to the amplitude of either oscillation, *the resultant motion of each particle on the liquid surface is along a circle of radius a in the vertical plane,* – in the anticlockwise direction if the wave be travelling from left to right, as shown in Fig. 14.41, which represents a vertical section of the liquid by a plane parallel to the direction of propagation of the wave.

This means, in other words, that there is a *local circulation* of the liquid on its surface which makes it flow upwards to form a crest and downwards to form a trough. Below the liquid surface too, the liquid particles move in circeles but due to viscosity, the radii of the circles go on decreasing until at a certain depth this local motion ceases.

The form of the wave is similar to the one given by the successive positions of a particle fixed to a circle of radius a, rolling along a straight line. So that, during the time T that the particle takes to make one full round of the circle, the wave advances through a distance λ, its *wave length*, equal to $2\pi a$, the circumference of the circles, since the velocity of particle along its circular path is

$2\pi a/T$ and is clearly directed opposite to the wave velocity V at the crest A and along the wave velocity at the trough B, we have

resultant velocity of the particle at crest A, i.e., $v_1 = V - 2\pi a/T$

and *resultant velocity of the particle at trough B, i.e.,* $v_2 = V + 2\pi a/T$

Applying Bernoulli's equation again to points A and B of the liquid surface, therefore, we have

$$p/\rho + ag + \frac{1}{2}v_1^2 = p/\rho - ag + \frac{1}{2}v_2^2.$$

Or, $$ag + \frac{1}{2}v_1^2 = -ag + \frac{1}{2}v_2^2. \text{ Or, } \frac{1}{2}(v_2^2 - v_1^2) = 2ag.$$

Or, $$v_2^2 - v_1^2 = 4ag$$

(We could also obtain the same result directly from the fact that the increase in the particle velocity from v_1 at A to v_2 at B is due solely to its having fallen through a vertical distance $h = 2a$ under the action of gravity. So that, $v_2^2 - v_1^2 = 2gh = 2g(2a) = 4ag$)

Substituting the values of v_1 and v_2, therefore, we have

$$(V + 2\pi a/T)^2 - (V - 2\pi a/T)^2 = 4ag \text{ or, } 8\pi vT = 4ag$$

whence, $V = 4gt/8\pi$. Or, since $T = \lambda/V$ we have $V^2 = gh/2\pi$

Or, $$V = \sqrt{g\lambda/2\pi} \qquad ...I$$

This, then, is the expression for the *velocity of the waves on the surface of a deep liquid due to the action of gravity alone, provided their amplitude is very much less than their wavelength – hence the name 'gravity waves'* given to them.

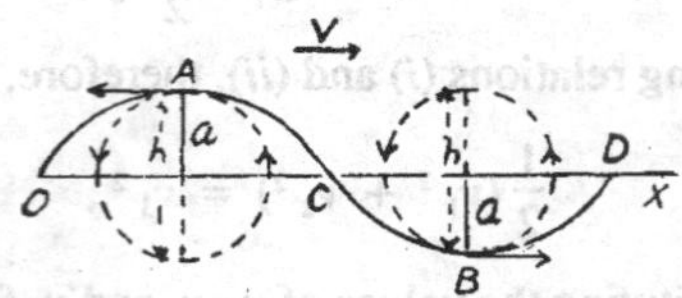

Fig. 14.41

As will be seen at once, *V is quite independent of the density of the liquid.*

(*iii*) **Effect of surface tension - Ripples.** Let us now see how the surface tension of the liquid affects the velocity of the waves discussed above.

Let Fig. 14.42 represent a vertical section of the liquid, with *OACBD* as the *harmonic wave* travelling over its surface and *CX*, its undisturbed surface.

Let $PQ = y$ be the displacement of a particle Q of the liquid at a distance x from an arbitrary origin O, at any given instant. Then, clearly,

$$y = a \sin(2\pi x/\lambda + \phi), \qquad ...(i)$$

where a is the *amplitude* of the particle and ϕ, the *phase constant.*

Now, due to the action of gravity alone, the pressure at Q will increase by $y\,\rho.g$, where ρ *is the density of the liquid.*

Owing to surface tension T of the liquid, however, there will also be an excess pressure equal to $T(1/R + 1/R')$, *acting along the normal PN* at P (*i.e.* from the concave towards the convex side), where R and R' are the respective radii of curvature of the liquid surface *in the plane,* and *perpendicular to the plane* of the paper. But, since the wave system is a cylindrical one, one surface of which is plane, *R′ is infinite and hence* $1/R' = 0$. *The pressure at P* due to surface tension thus reduces to $T(1/R) = T/R$. And, therefore, *net increase of pressure at* $Q = y\,\rho.g - (T/R)\cos\theta$, where θ is the angle that the normal PN at P makes with the vertical through P. Since, however, the amplitude of the wave is small compared with its wavelength, θ is extremely small ad the normal at P (i.e., PN) almost coincides with the vertical through P. We, therefore, have

net increase of pressure at $Q = y\,\rho.g - T/R$

Now, $1/R = (d^2y/dx^2)/[1 + (dy/dx)^2]^{3/2}$ which reduces to d^2y/dx^2 if $a << 1$ and, therefore $dy/dx << 1$ so that,

net increase of pressure at R, say, $p = y\,\rho.g - T\,(d^2y/dx^2)$ ... (*ii*)

To obtain the value of d^2y/dx^2, we differentiate relation (*i*) twice, when we get

$$d^2y/dx^2 = -4\pi^2y/\lambda^2$$

Putting this value of d^2y/dx^2 in relation (*ii*) above, we have

$$p = y\,\rho g + 4\pi^2yT/\lambda^2 = y\,\rho(g + 4\pi^2T/\lambda^2\rho). \quad ...(iii)$$

It is thus clear that *the effect of surface tension is tantamount to increasing the value of g by* $4\pi^2T/\lambda^2.\rho$.

Clearly, therefore, if V be the velocity of the wave under the action of both gravity and surface tension, we obtain its value by substituting $(g + 4\pi^2T/\lambda^2\rho)$ for g in relation (*iii*) *above; i.e.*,

$$V = \sqrt{\frac{\lambda}{2\pi}\left(g + \frac{4\pi^2T}{\lambda^2.\rho}\right)} = \sqrt{\frac{\lambda g}{2\pi} + \frac{2\pi T}{\lambda.\rho}}. \quad ...(iv)$$

Fig. 14.42

A mere glance at this relation tells us that $V = \infty$, both when $\lambda = 0$ and when $\lambda = \infty$. In between these two extreme values of λ, there must be a certain value of it for which V has the *minimum* value. Clearly, the product of the two terms $\lambda g/2\pi$ and $2\pi T/\lambda.\rho$, *viz.*, gT/ρ is a constant, and it follows, therefore, that *their sum will be a minimum when they are equal; i.e.*, when

$$\lambda g/2\pi = 2\pi T/\lambda.\rho. \quad \text{Or, when } \lambda^2.g.\rho = 4\pi^2T.$$

or, when $\lambda^2 = 4\pi^2T/g.\rho$. Or, when $\lambda = 2\pi\sqrt{T/g.\rho}$.

This value of λ, for which the velocity of the wave is the *minimum*, is called the ***critical wave-length***, and may be denoted by the symbol λ_c. Thus, $\lambda_c = 2\pi\sqrt{T/g.\rho}$. ...(*v*)

Substituting this value of λ, therefore, in expression (*iv*) for V above, we get the ***minimum velocity*** (V_m) of the wave given by the relation,

$$V_m = \sqrt{2}\,(T.g./\rho)^{\frac{1}{4}}.*$$

Now, examining relation (*iv*) again, we find that

(*i*) If $\lambda > \lambda_c$ the first term $\lambda g/2\pi$ becomes more important as λ increases and, therefore, neglecting the second term in comparison with it, we have $V = \sqrt{\lambda g/2\pi}$.

Disturbance of this type, *whose wave-length is greater than the critical value*, are, as pointed out earlier, known as **gravity waves** or waves, their propagation being mainly due to the force of gravity and, as can be readily seen, *their velocity increases as* λ *increases.*

(*ii*) If $\lambda < \lambda_c$ the second term becomes more predominant, and the first term may, therefore, be neglected. So that, in this case, $V = \sqrt{2\pi T/\lambda\rho}$.

Waves of this type *for which the wave-length is less than the critical wave-length*, are called **ripples or capillarity waves.** *Their propagation is, in main, due to surface tension*, and, as can be easily seen, *their velocity decreases as* λ *increases.*

N.B. Taking the case of water for which $T = 75$ *dynes/cm* and $\rho = 1$ *gm/cc* we get $\lambda_c = 1.7$ *cm* and $V_m = 23$ *cm/sec.*

The Method Proper. *Lord Rayleigh* (1890) was the first to have used ripples, excited on the surface of a liquid, to measure its surface tension, by the direct application of relation (*iv*) above, although the method has subsequently been greatly improved by a host of other workers, including *Dorsey, Ghosh, Banerji* and *Datta*. And, still more recent improvements have been effected by *Tyler* and *Brown*, enabling the ripples to be photographed and their images thrown on a screen.

We shall deal only in brief outline, here, with the essentials of the method employed. The experimental liquid is taken in a shallow rectangular porcelain trough, about 10" long, 6" wide and 1.5" deep, (Fig. 14.43), and above it is arranged an electrically maintained tuning fork F, of a large frequency, with its prongs horizontal and one above the other in their position of rest, so as to vibrate in the vertical plane. A light style or *'dipper' P*, of polished silver or aluminium, about 3" long, is attached to the lower prong of the fork, with its plane also vertical but perpendicular to the plane of vibration of the fork, such that it just touches the liquid surface in the trough.

Now, as the fork is set vibrating, the style or the dipper alternately moves in and out of the liquid, thus exciting trains of ripples on its surface. These, on reflection from the walls of the trough, give rise to stationary ripples.

To enable these ripples to be observed and their wave-length measured, the liquid surface must be properly illuminated. This is done by completing the tuning fork circuit through the primary P' of a small *induction coil*, to the secondary S of which is connected a *neon-discahrge-tube* ; so that, every time the tuning fork circuit is made, an electric discharge passes through the *neon-tube*, the light from which then brilliantly illuminates the liquid surface in the trough, thus enabling the wave-form of the ripples to be easily seen and photographed.

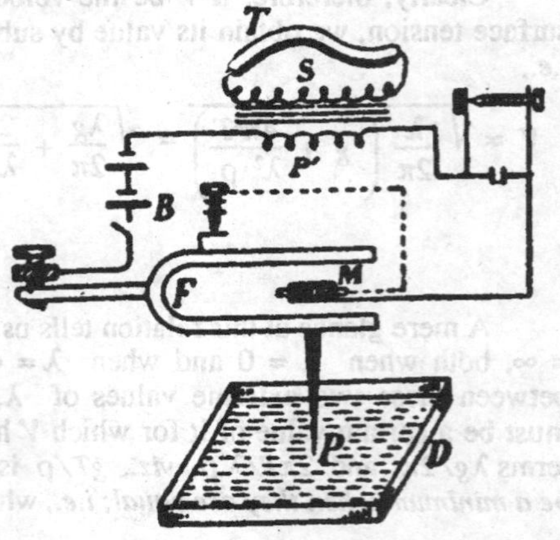

Fig. 14.43

Now, *during one vibration of the fork, there is one alternate make and break of the tuning fork circuit, giving us one view of the ripples on the liquid surface.* The same *recurs* when the fork has made one vibration, and when, therefore, the ripples have advanced through a distance equal to their one full wavelength; so that, as the liquid surface is lit up again, as before, we get *a second* view of it, *identical with the first*, the ripples *appearing to be stationary* in their earlier position. This is so, because we do not see the liquid surface continuously but only at intervals equal to the time-period of the ripples, the frequency with which the liquid surface is lit up being the *same* as that of the fork and, therefore, of the ripples themselves. The successive views we have of this liquid surface are thus only those in which one ripple has just replaced another *similar* ripple and hence, *due to persistence of vision*, (the frequency of the fork being large), the surface of the liquid *appears* to us to be exactly the same as before, with the ripples in their original position.

The distance between as large a number of ripples as possible is measured with the help of travelling microscope and their mean wave-length λ thus determined. Then, if n be the frequency

---

* For, substituting the value of $T/\rho = \lambda^2 g/4\pi^2$ from (v) in relation (iv), above, we have

$$V_m = \sqrt{\frac{\lambda g}{2\pi} + \frac{2\pi}{\lambda} \cdot \frac{\lambda^2 g}{4\pi}} = \sqrt{\frac{\lambda g}{2\pi} + \frac{\lambda g}{2\pi}} = \sqrt{\frac{2\lambda g}{2\pi}} = \sqrt{\frac{\lambda g}{\pi}}.$$

Now, substituting $2\pi\sqrt{T/g.\rho}$ for λ, we have

$$V_m = \sqrt{\frac{2\pi\sqrt{\frac{T}{g.\rho}}.g}{\pi}} = \sqrt{2}\left(\frac{Tg^2}{g\rho}\right)^{\frac{1}{4}} = \sqrt{2}\left(\frac{T.g}{\rho}\right)^{\frac{1}{4}}$$

of the fork (which is known to us), we have $V = n\lambda$, where V is the velocity of the ripples. Substituting this value of V in relation (*iv*) above, we have

$$n\lambda = \left(\frac{\lambda g}{2\pi} + \frac{2\pi T}{\lambda.\rho}\right)^{\frac{1}{2}}. \quad \text{Or,}\ n^2\lambda^2 = \frac{\lambda g}{2\pi} + \frac{2\pi T}{\lambda.\rho}.$$

or $$n^2\lambda^2 - \frac{\lambda g}{2\pi} = \frac{2\pi T}{\lambda.\rho}, \quad \text{whence, } T = \frac{\lambda.\rho}{2\pi}.n^2\lambda^2 - \frac{\lambda.\rho}{2\pi}.\frac{\lambda g}{2\pi}.$$

or $$T = \frac{n^2.\lambda^3.\rho}{2\pi} - \frac{\lambda^2.\rho.g}{4\pi^2},$$

from which, λ being known, the value of T, *the surface tension of the given liquid*, may be easily calculated.

9. The Ring Method. This method derives its importance from the fact that it enables us to study *the changes that come about in the surface tensions of different liquids with the passages of time*. It has been used with great accuracy by *Harkins, Young* and *Cheng*, and is being increasingly employed in Applied Physics.

A metal ring of a wire circular cross-section is suspended in the experimental liquid, with its plane horizontal. It is then raised gradually out of the liquid, when, in addition to its own weight, the extra downward pull on it due to surface tension passes through a *maximum value*. If the wire, constituting the ring, be a thin one, this maximum pull is approximately given by $P = 4\pi RT$ where R is the radius of the ring and T, the surface tension of the liquid. More correctly, however, P is only proportional to $4\pi RT$, so that $\alpha P = 4\pi RT$ where α is a non-dimensional factor. To avoid calculating the value of α, *Harkins* and his co-workers assumed that

$$\frac{4\pi RT}{P} = \phi_1\left(\frac{R}{r}\right) \quad ...(i) \qquad \text{Or,}\ \frac{4\pi RT}{P} = \phi_2\left(\frac{R^3}{V}\right) \quad ...(ii)$$

where r is the *radius of the wire V*, the *volume of the liquid held up by the maximum pull of the ring*, equal to $P/\rho.g$, (ρ being the density of the liquid).

They used three liquids of known surface tensions, *viz., water, benzene* and *bromobenzene* and measured the maximum pull on three different rings, with different values of R and r, but the *same constant ratio R/r*, and then plotted a graph between R^3/V, (along the x-axis,) and $4\pi RT/P$ (along the y-axis), when the points for all the three liquids were found to lie on the same smooth curve. From this, they came to the legitimate conclusion that, if the *same* three rings were to be used with other liquids, the points corresponding to the values of R^3/V and $4\pi RT/P$ for them too would lie on the same curve.

Thus, the procedure to determine the surface tension of a given liquid by this method reduces to the following :

We take a ring, identical with one of the rings used by *Harkins, Youngs* and *Cheng*, (*i.e.*, of the same material and the same value of R/r), measure, with its help, the maximum pull as the ring is raised out of the given liquid and then calculate the value of R^3/V for it, which is equal to $R^3.g.\rho/P$. Next, we locate the point corresponding to *this* value of R^3/V as abscissa, on the graph plotted by them for the ring of the same value of R/r. The ordinates y, corresponding to it, then gives the value of $4\pi RT/P$ for the liquid and is carefully noted. Thus

$$y = 4\pi RT/P, \text{ and, therefore, } T = y.P/4\pi R,$$

whence the value of T for the liquid can be easily calculated.

The method has the merit of *quickness, ease* and *accuracy*, but the following precautions are necessary for its success :

(*i*) *The liquid surface must be perfectly clean*. This is ensured by sweeping the surface of the liquid by means of what are called 'barriers' before performing the actual experiment.

(*ii*) *The liquid surface must be kept properly covered by an inverted glass funnel, to avoid evaporation.*

(*iii*) *The dish containing the liquid must be wide enough*, so that errors due to curvature of the mensicus etc. are almost altogether eliminated.

(*iv*) *The whole apparatus must be arranged inside a thermostat*, to ensure that the temperature of the liquid remains constant throughout, and the thermostat must be supported independently of the ring and other apparatus, to avoid disturbance or agitation of the experimental liquid.

14.25. Surface Tension of Liquid Interfaces. If we consider a system of immiscible liquids in contact, we naturally expect a new phase to develop at their *interface*, with a definite energy of its own, depending upon the nature of the two liquids.

Antonow gave a rule that the interfacial tension between two liquids, in equilibrium, is equal to the difference between their individual surface tensions. Thus, if T_1 and T_2 be the surface tensions of two liquids separately, their interfacial tension T_{12}, when they are in equilibrium, is given by $T_{12} = T_1 - T_2$.

This rule would, however, apply only to mutually saturated solutions, for then alone would they be in equilibrium, or else the addition of one to the other would reduce its surface tension. Being the difference of the individual surface tensions of the two liquids, the surface tension of their interface is obviously a small quantity, in general, increasing with the decrease in the solubility of one liquid into the other, and decreasing with a rise in temperature.

Since a knowledge of interfacial tension is of importance while considering problems like spreading of one liquid over another, or those relating to chemical constitution, we shall now proceed to see how it may be determined for a given liquid-liquid interface.

Most of the methods discussed above, the measurement of surface tension of pure liquids would, in principle, be equally applicable in this case also. Thus, for example, the *drop-weight* method can be easily used for the purpose and probably gives the best results. We shall discuss this method in proper detail.

(*i*) The Drop-weight Method. Here, we proceed in the same manner as in determining the surface tension of a pure liquid, (see page 464), with the difference that the tube, with a fine orifice, of about 3 to 5 *mm* radius, containing one liquid, dips inside the other, *so that the drop of the first liquid is formed inside the second, instead of in air*, the weight of the drop being obtained by collecting and weighing a known number of them.

Then, if ρ_1 and ρ_2 be the densities of the two liquids respectively, we have *volume of the drop* = m/ρ_1. This is obviously also the *volume of the second liquid displaced by it*. So that,

weight of the second liquid displaced or, upthrust on the drop

$$= \frac{m}{\rho_1} \cdot \rho_2 . g. \qquad \left[m \text{ being the mass of the drop} \right.$$

And, therefore, *apparent or effective weight of the drop*

$$= mg - \frac{m}{\rho_1} \cdot \rho_2 . g = mg \left(1 - \frac{\rho_2}{\rho_1} \right) = mg \left(\frac{\rho_1 - \rho_2}{\rho_1} \right).$$

Now, assuming, as before, that, just at the time of being detached from the orifice, the drop is *cylindrical* in shape and has the same cross-sectional area as that of the orifice, the excess pressure inside it is given by

$$p = T_{12}/r, \text{*}$$

where T_{12} is the surface tension of the interface.

The downward force on the drop due to it is thus

$$= p \times \pi r^2 = \frac{T_{12}}{r} \cdot \pi r^2 = \pi . T_{12} . r.$$

And, therefore, *total downward force on the drop*
= *effective weight of the drop* + $\pi . T_{12} . r.$ = $mg \dfrac{\rho_1 - \rho_2}{\rho_1} + \pi . T_{12} . r$.

And, the upward force on it due to surface tension = $T_{12} . 2\pi r = 2\pi T_{12} . v$, where $2\pi v$ is the circumference of the cross-section of the drop.

In the equilibrium position of the drop, therefore, we have

*** The radius of the other face of the cylinder being infinite.**

$$\frac{mg(\rho_1 - \rho_2)}{\rho_1} + \pi.T_{12}.r = 2\pi.T_{12}.r. \quad \text{Or,} \quad \frac{mg(\rho_1 - \rho_2)}{\rho_1} = \pi.T_{12}.r.$$

or
$$T_{12} = \frac{mg(\rho_1 - \rho_2)}{\rho_1.\pi.r}.$$

In the above treatment, we have assumed that the drop attains static equilibrium when, having attained its cylindrical form, it is just about to be detached. This is, however, not so, the detachment of the drop being essentially a dynamical process, for which we obtain the following relation, by dimensional analysis :

$$\frac{mg(\rho_1 - \rho_2)}{\rho_1} = T_{12}.r.\phi\left[\frac{T_{12}}{gr^2(\rho_1 - \rho_2)}\right].$$

where ϕ is an arbitrary function of the non-dimensional variable $T_{12}/gr^2(\rho_1 - \rho_2)$.

Lord Rayleigh has, however, shown from his work on water drops in air that the value of $\phi\left[\frac{T_{12}}{gr^2(\rho_1 - \rho_2)}\right]$ comes to 3.8 for a large variation in the value of r.

So that, we have $\frac{mg(\rho_1 - \rho_2)}{\rho_1} = T_{12} \times r \times 3.8$ Or, $T_{12} = \frac{mg(\rho_1 - \rho_2)}{3.8\,\rho_1.r}$

whence, the surface tension T_{12} for the given liquid interface can be easily calculated.

(*ii*) **Wilhelmy Method.** *Wilhelmy* had suggested a straight forward method of determining the surface tension of a liquid by measuring with the help of a balance, the additional force necessary to counteract the vertical pull of surface tension on a vertical metal plate, suspended form the balance arm, when it is dragged away from the surface. Thus, if the lower edge of the metal plate be in level with the undisturbed liquid surface and if l be the length of the its line of contact with it, the vertical pull on it due to surface tension is equal to $T.l \cos\theta$, where θ is the angle of contact of the liquid-plate surface. And, therefore, if mg be the additional weight required in the scale-pan to balance this additional force, we have

$$mg = T.l \cos\theta.$$

If we use a *torsion balance*, in place of the ordinary balance, the method can be easily used for interfacial tension also, as will be clear from the following :

The two liquids are taken in a beaker, one above the other, a vedritcal plate suspended from the arm of the torsion balance and adjustment made for equilibrium, with the plate wholly immersed but well above the interface of the two liquids. The beaker is now gradually raised and the equilibrium continues to be maintained, until the plate *just* approaches the *interface*, when the additional downward pull on it, due to the interfacial tension T_{12} of the two liquids, disturbs the balance. The torsion head has thus to be turned through an additional angle α, say, to resore, the plate back to its equilibrium position. This angle of twist (α) measures the force $l.T_{12}$ on the plate, where l is the *horizontal perimeter*, the angle of contact (θ) being assumed to be *zero* (or $\cos\theta$ very nearly equal to 1). The value of T_{12} for the given interface can thus be easily determined.

(*iii*) **Mack and Bartell Method.** *Mack* and *Bartell* used the following simple method for the measurement of interfacial tension of water and organic liquids. Besides its simplicity, it has the great merit of precision and of requiring only a very small quantity of the liquid, just 2 *cc* of it or so.

The apparatus used by them is as shown in Fig. 14.44, where A and B are two wide glass cups, communicating with a central wide tube C through two *capillary tubes*, P and Q, of slightly differing radii R_1 and R_2, sealed on to them, where R_2, the larger of the two, is less than 1 *mm*,

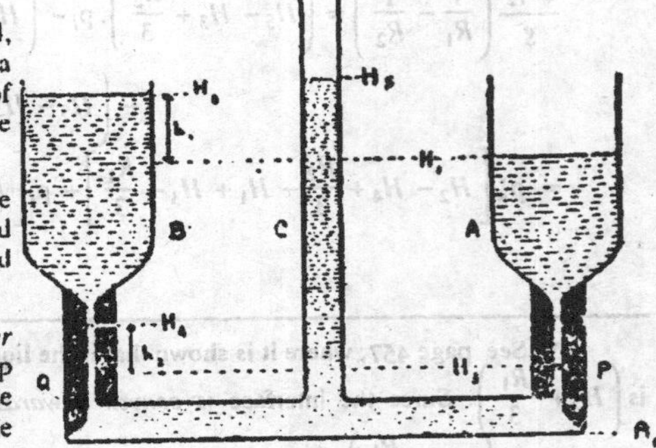

Fig. 14.44

Water is first poured into the cup A, connected to the narrow capillary P, to ensure that no air bubbles get entrapped, and a larger quantity of it is then poured into cup B, a little over and above that required to fill the two capillaries.

The organic liquid, *which here is supposed to be denser than water*, is poured into the central tube C, when it rises up inside the two capillaries, where its levels are adjusted to coincide with two etched marks, at heights H_3 and H_4 respectively from the horizontal tube at level H_2, the interface at H_3 and H_4 being *convex upwards.* The whole system is allowed sufficient time to attain equilibrium and the following levels measured accurately.

(i) levels H_1 and H_2 of the columns in A and B,

(ii) level H_5 of the liquid column in the central tube C, and

(iii) levels H_3 and H_4 up to the horizontal tangent planes to the liquid-water interfaces in P and Q respectively.

Since the interfaces are assumed to be *hemispherical* (due to the narrow bores of the capillaries), the *effective heights* of the liquid columns are

$$\left(H_3 - \frac{R_1}{3}\right) \text{ and } \left(H_4 - \frac{R_2}{3}\right) \text{ respectively.}$$

Now, clearly, *upward pressure in P due to the liquid column of the interface*

$$= \left[H_5 - \left(H_3 - \frac{R_1}{3}\right)\right].\rho_l.g = \left(H_5 - H_2 + \frac{R_1}{3}\right).\rho_l.g$$

where ρ_l is the *density* of the organic liquid.

And, since *thrust = pressure × area*, we have

upthrust in P due to the liquid column $= \left(H_5 - H_3 + \dfrac{R_1}{3}\right)\rho_l . g.\pi R_1^2$,

where πR_1^2 is the area of cross-section of the tube.

And, *downwards force acting at the interface = downward thrust due to water + force due to intefacial tension* T_{12}.

$$= \left(H_1 - H_3 + \frac{R_1}{3}\right).\rho_w.g.\pi R_1^2 + 2\pi R_1.T_{12}.$$

where ρ_w is the density of water and $2\pi R_1$ the length of the line of contact of the interface with the capillary.

So that, for equilibrium, we have

$$\left(H_5 - H_3 + \frac{R_1}{3}\right).\rho_l.g.\pi R_1^2 = \left(H_1 - H_3 + \frac{R_1}{3}\right)\rho_w.g.\pi.R_1^2 + 2\pi R_1.T_{12}.$$

Or, $$\pi R_1^2 g\left[\left(H_5 - H_3 + \frac{R_1}{3}\right)\rho_l - \left(H_1 - H_3 + \frac{R_1}{3}\right)\rho_w\right] = 2\pi R_1 . T_{12} \quad ...(i)$$

Similarly, for equilibrium in capillary Q, we have

$$\pi R_2^2 g\left[\left(H_5 - H_4 + \frac{R_2}{3}\right)\rho_l - \left(H_2 - H_4 + \frac{R_2}{3}\right)\rho_w\right] = 2\pi R_1 . T_{12} \quad ...(ii)$$

Or, $$\left(H_5 - H_3 + \frac{R_1}{3}\right)\rho_l - \left(H_1 - H_3 + \frac{R_1}{3}\right)\rho_w = \frac{2\pi R_1.T_{12}}{\pi R_1^2 g} = \frac{2T_{12}}{R_1.g} \quad ...(iii)$$

and $$\left(H_5 - H_4 + \frac{R_2}{3}\right)\rho_l - \left(H_2 - H_4 + \frac{R_2}{3}\right)\rho_w = \frac{2\pi R_2.T_{12}}{\pi R_2^2 g} = \frac{2T_{12}}{R_2.g} \quad ...(iv)$$

Subtracting relation *(iv)* from *(iii)*, we have

$$\frac{2T_{12}}{g}\left(\frac{1}{R_1} - \frac{1}{R_2}\right) = \left(H_5 - H_3 + \frac{R_2}{3}\right).\rho_l - \left(H_1 - H_3 + \frac{R_1}{3}\right).\rho_w$$
$$- \left(H_5 - H_4 + \frac{R_2}{3}\right)\rho_l + \left(H_2 - H_4 + \frac{R_2}{3}\right)\rho_w$$
$$= \rho_w\left[H_2 - H_4 + \frac{R_2}{3} - H_1 + H_3 - \frac{R_1}{3}\right] + \rho_l\left[H_5 - H_3 + \frac{R_1}{3} - H_5 + H_4 - \frac{R_2}{3}\right]$$

---

* See page 457, where it is shown that if the liquid meniscus be *concave*, the effective height of the liquid column is $\left(H_3 + \dfrac{R_1}{3}\right)$. Since the interface is *convex upwards*, the effective height of the liqwuid column can similarly be shown to be $\left(H_3 - \dfrac{R_1}{3}\right)$

$$= \rho_w \left[(H_2 - H_1) - (H_4 - H_3) \right] + \left[\frac{R_2 - R_1}{3} \right] \rho_w + \rho_l \left[(H_4 - H_3) \right] - \left[\frac{R_2 - R_1}{3} \right] \rho_l$$

So that, putting $(H_2 - H_1) = h_1$, $(H_4 - H_3) = h_2$ and $(R_2 - R_1) = r$, we have

$$\frac{2T_{12}}{g} \cdot \left(\frac{1}{R_1} - \frac{1}{R_2} \right) = \rho_w (h_1 - h_2) = \rho_l h_2 - \frac{1}{3} r (\rho_l - \rho_w).$$

Or,
$$T_{12} = \frac{g}{2\left(\frac{1}{R_1} - \frac{1}{R_2} \right)} \cdot \left[\rho_w (h_1 - h_2) + \rho_l h_2 - \frac{1}{3} r (\rho_l - \rho_w) \right]$$

And, thus the value of the surface tension of the interface of the given organic liquid and water can be easily determined.

N.B. It will be noted that ρ_l, the density of the liquid, occurs only in the small term $\rho l.h_2 - \frac{1}{3} r (\rho_l - \rho_w)$ of the relation for T_{12}, and need not, therefore, be known with any high degree of accuracy, – up to about two decimal places being enough for our purpose. This may be taken to be an additional advantage of the method.

14.26 Factors affecting Surface Tension. We shall touch only briefly here on the various factors which affect the surface tension of a liquid.

(i) Contamination of the liquid surface. The surface tension of a liquid reacts immediately, and more or less sharply, to any contamination of the liquid surface. The presence of impurities of any knid on the surface of a liquid may, therefore, bring about a marked change in the value of its surface tesion, depending, of course, on the degree of contamination thus produced.

(ii) Presence of a dissolved substance in the liquid. Here, two distinct cases arise, *viz.*, (1) *when the substance is highly soluble in the liquid*, and (2) *when it is only feebly so.*

In case (1), obviously, the solute molecules are attracted more strongly by the liquid molecules. If this attraction exceeds that between the liquid molecules among themselves, it follows that the presence of the solute molecules in the surface layer of the solution will result in an increase in its surface energy and hence in its *surface tension*, as is in general, the case with aqueous solutions of inorganic salts like $NaCl$, $ZnSO_4$ etc.*

Since, like any other physical system, the potential energy of a solution also tends to assume its minimum value, its surface tenison (which is a measure of the energy asssociated with its surface layer) tends to be a minimum. The molecules of the solute, therefore, get pulled into the interior of the liquid, resulting in an increased cocentration there, depleting that in the surface layers.

In case (2), on the other hand, the molecules of the liquid itself move inwards, *i.e.*, into the interior of the liquid, their place in the outer layers being taken up by the solute molecules, resulting in a decrease in the surface tension of the solution, the concentration being higher in the outer layers than in the interior of the solution. This happens in a solution of *phenol* in water, and, to a much more marked degree, in the case of *soap solution*. Indeed it has been clearly shown by *Lord Rayleigh* that although the surface tension of a *freshly prepared* soap solution is about the same as that of water, it falls by as much as 40 per cent when equilibrium is attained. This explains the great stability of soap films- and bubbles. For, the concentration is much greater in the film-surface than in the interior and fresh liquid is brought on to the surface** at any place where the film gets unduly stretched and is likely to snap. *The film thus gets automatically reinforced at just those points where a possible rupture threatens to occur.*

(iii) Change of temperature. Liquids, for our present purpose, may be divided into two general categories, *viz.*, *(i) unassociated* and *(ii) associated* ones. In the former class of liquids, there is nothing but *individual molecules* present, as, for example, in the case of *benzene* and *carbon tetrachloride*, at the ordinary temperature. But, in the latter class, we have groups of molecules of quite another type. These groups, however, tend to break up into single mlecules with a rise in temperature.

Thus, water, at the ordinary temperatures, is known to consist of groups, consisting of two H_2O molecules, in addition to ordinary single H_2O molecules, and is thus an exmaple of an associated liquid at these temperatures.

Now, it is found that the surface tension of unassociated liquids (with the notable exceptions of molten cadmium and copper) *decreases with rise of temperature*. For small temperature ranges, the relation is a *linear* one, and is given by

$$T = T_0(1 - \alpha t),$$

where T is the surface tension at $t^{\circ}C$; T_0 at $0^{\circ}C$ and α, the *temperature coefficient of surface tension* for the liquid.

So that, $\quad dT/dt = -T_0 \alpha = -k \qquad$ | putting $T_0 . \alpha = k$.

---

* Thus, for example, a solution of $NaCl$, containing 5 *gram-molecules* of it *per litre*, has surface tension of 82 *dynes/cm*, as against 75 *dynes/cm* for water alone.

** The amount of solute, transferred in this manner from the interior of the liquid to the surface, is called the **'adsorption'** *per unit area of the surface*. And, as we have seen, it may be positive or negative.

Vander waals suggested, in the year 1893, the relation $T = A(1 - \theta/\theta_c)^{3/2}$, modified later by *Ferguson* in 1915 to $T = A(1 - \theta/\theta_c)^n$, where T is the *surface tension* of the liquid at $\theta°$ *abc* ; $\theta_c°$ *abs.*, its *critical temperature*, *A*, a *constant* for the liquid and *n*, different for different liquids, but equal to 1.21 for most of them.

But, perhaps one of the best relations connecting surface tension and temperature for both associated and unassociated liquids is the one due to Eotvos, called the Eotvos law, *viz.*, $T\,(M/\rho)^{2/3} = K\,(\theta_c - \theta)$, where $(M/\rho)^{2/3}$, *i.e.*, *(mol. wt/density)* $^{2/3}$ is called the *molar surface* of the liquid and $T\,(M/\rho)^{2/3}$, its *molar energy*, and K is a constant.

The law is, obviously, analogous to the well known gas law $P.v = R.\theta$, where $P.v$ corresponds to the free molar energy of the gas at absolute temperature θ.

Ramsay and *Shields* later modified the law to $T\,(M/\rho)^{2/3} = K\,(\theta_c - \theta - d)$, which may also be put in the form

$$T(M\,v\,x)^{3/2} = K(\theta_c\ \theta - d),$$

where x is called the *coefficient of association** of the liquid at $\theta°$ *abs.*, *d*, a constant, introduced by *Ramsay* and *Shields* (its value varying from 6 to 8 for most of the liquids), *K*, a constant, varying from 2.12 for associated liquids to 2.22 for un-associated ones (for which $x = 1$); *M*, the *molecular weight* of the un-associated liquid and *v*, its *specific volume*.

This shows that the surface tension is zero not at the critical temperature θ_c, but at a temperature $\theta = (\theta_c - d)$, a little below the critical temperature. This is amply borne out by *Callendar's* experiment on water, when he found that the surface tension of water vanishes at 647° *absolute*, although its critical temperature is actually 653° *absolute*.

14.28. Experimental study of the variation of surface tension with temperature. This may be easily done with the help of the following simple experiment due to Jaeger, applicable only for a small range of temperature.

A capillary glass tube *AB* is connected to another glass tube *CD*, of a comparatively wider bore, so as to form a complete *U*-tube, (Fig. 14.45)

The experimental liquid is then poured into this *U*-tube, with its *two limbs of different bores*. Obviously, if the angle of contact for the liquid and glass be less than 90°, the liquid stands much higher in the narrower limb *AB* of the tube than its wider limb *CD*.

If r_1 and r_2 be the radii of the bores of the two limbs respectively, T, the surface tension of the liquid and θ, its *angle of contact* with glass, we have

pressure just below the liquid meniscus in AB

$$= P - 2T\cos\theta/r_1,$$

and pressure just below the liquid meniscus in CD

$$= P - 2T\cos\theta/r_2,$$

where P is the *atmospheric pressure*.

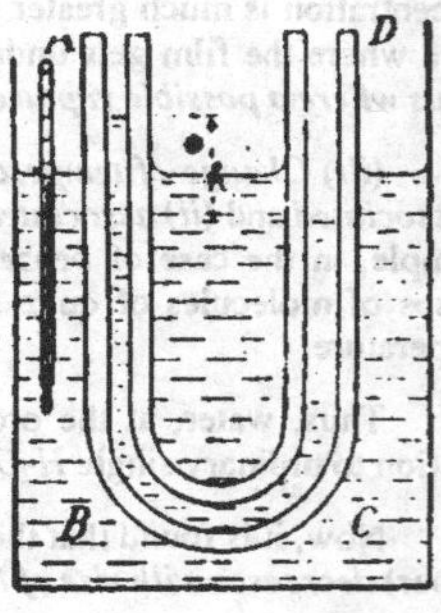

Fig. 14.45

So that, *difference of pressure in the two limbs*

$$= 2T.\cos\theta\,(1/r_1 - 1/r_2).$$

This is obviously balanced by the *hydrostatic pressure* $h.\rho.g.$, due to the liquid column in the *U*-tube, where h is the difference of liquid-levels in the two limbs, and ρ, its density.

Thus, we have $$2T\cos\theta\,(1/r_1 - 1/r_2) = h\,\rho\,g,$$

whence, $$T = h\,\rho\,g/2(1/r_1 - 1/r_2)\cos\theta,$$

and can be easily calculated.

The experiment is repeated with the liquid maintained at different temperatures by means of suitable temperture baths and a graph is then plotted between

---

* This is equal to the ratio between the *effective* molecular weight of the associated liquid and the molecular weight of an un-associated liquid, with the same molecules.

the temperatures (t^0) and the corresponding values of the surface tension (T), which gives at a glance, the variation of the surface tension of the given liquid with a change in its temperature.

N.B. 1. In case $\theta > 90°$, the liquid column in AB will be depressed into the tube to a position h below the liquid column in CD. This will be so, when the liquid does not wet the walls of the tube, as, for example, in the case of mercury and glass, or that of a wax-coated tube and water.

2. For higher ranges of temperature, we substitute a platinum capillary tube, in place of the glass tube, in the above experiment.

14.29. Surface tension and vapour pressure over a liquid surface. Let a capillary tube be dipped vertically in a liquid, of surface tenison T and density ρ and let the whole arrangement be enclosed in an exhausted bell jar or chamber, as shwon in Fig. 14.46, so that the effect of the atmosphere may be neglected.

Obviously, then, there is nothing but its own vapour over the liquid and, when equilibrium is attained, the space above the liquid becomes saturated with its vapour.

Now, *two cases* arise: (*i*) *when the liquid is like water, alcohal or ether etc. which wet the tube* and (*ii*) *when the liquid is like mercury (or water in a tube whose inside is coated with wax) which does not wet the tube.* Let us deal with each separately.

Case (i)–(*When the liquid wets the tube*). In this case, the liquid rises up in the tube above the liqid surface outside the tube, as shown in Fig. 14.46 (*a*), and the meniscus at the top is *concave upwards.*

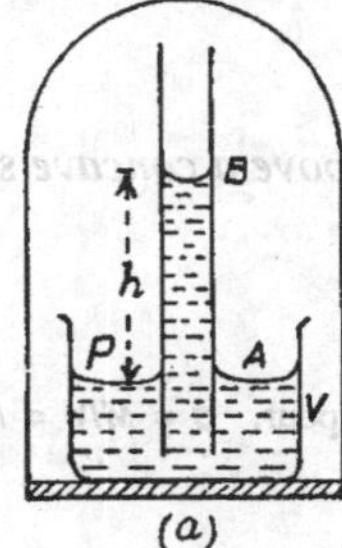

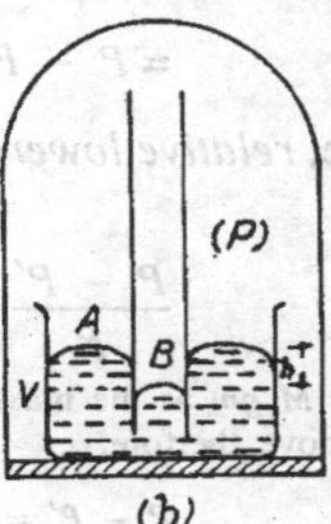

Fig. 14.46

Suppose the liquid rises into the tube up to a height h above the horizontal liquid surface A in the vessel.

Then, if P be the vapour pressure at the horizontal surface A, and σ, the density of the vapour at this pressure, the vapour pressure above the concave surface B of the liquid, in the tube, will clearly be less than P by an amount equal to that of a column h of the vapour, *i.e.*, by an amount $h.\sigma.g$; so that, vapour pressure *just above the meniscus* at B say, $P' = P - h.\sigma.g$. And, clearly, the pressure in the liquid just below the meniscus is equal to $(P - h.\sigma.g)$, where ρ is the density of the liquid.

If the tube be narrow, the meniscus may be regarded as spherical, of radius r, *nearly the same as that of the tube*, and so the excess pressure, just above the meniscus over that just below it, will be equal to $2T/r$.

Clearly, *this pressure just above the meniscus is greater than the presure just below it by* $(P - h\sigma g) - (P - h\rho g)$

$$= P' - (P - h\rho g) = (P' - P) + h\rho g \qquad [\because P - h\sigma g = P'$$

$$\therefore \qquad (P' - P) + h\rho g = 2T/r. \text{ Or, } (P - P') = h\rho g - 2T/r.$$

Since $\qquad h\sigma g = (P - P')$, we have $hg = (P - P')/\sigma$.

And, therfore, $\quad (P - P') = \left(\dfrac{P - P'}{\sigma}\right)\rho - \dfrac{2T}{r}$. or, $\dfrac{2T}{r} = \left(\dfrac{P - P'}{\sigma}\right)\rho - (P - P')$

$$= (P - P')\left(\frac{\rho}{\sigma} - 1\right)$$

whence,
$$P - P' = \frac{2T}{r}\left(\frac{\sigma}{\rho - \sigma}\right) \quad ...(I)$$

Since ρ is greater than σ, the expression for $(P-P')$ is *positive*, and , therefore, *the vapour pressure above the concave surface of a liquid is less than the vapour pressure at the horizontal surface of the same liquid by an amount* $2T\sigma/r\,(\rho - \sigma)$.

Or, neglecting σ compared with ρ, we have, as a near approximation,

$$P - P' = \frac{2T}{r.\rho}\sigma. \quad ...(II)$$

Now if we treat the vapour as a *gas*, for which the velation $PV = R.\theta$, is applicable, (denoting the *absolute temperature* by θ, instead of by *T*, which we have used for surface tension here), we have $V = R\theta / P$ and ∴ $\sigma = 1/V = P/R\theta$,

where *V* is the volume of 1 *gm. molecule* (or 1 *mole*) of the vapour.

Hence, substituting this value of σ in the expression for $(P-P')$ above, we have

excess vapour pressure on the horizontal surface of a liquid over that on its concave surface

$$= P - P' = \frac{2T}{r\rho}\sigma = \frac{2T}{r\rho}\cdot\frac{P}{R\theta}. \quad ...(III)$$

And, therfore, *relative lowering of vapour pressure above a concave* surface is given by the expression

$$\frac{P - P'}{P} = \frac{2T}{r\rho}\cdot\frac{1}{R\theta} \quad ...(IV)$$

N.B. It we take *M gm* as the mass of 1 *gm. molecule* of the vapour, $\sigma = M/V = MP/R\theta$. Substituting in the expression for $(P-P')$ above, therfore,

we have
$$P - P' = \frac{2T}{r\rho}\sigma = \frac{2T}{r\rho}\frac{MP}{R\theta}.$$

And ∴
$$\frac{P - P'}{P} = \frac{2T}{r\rho}\cdot\frac{M}{R\theta}$$

Case (ii) – (*When the liquid does not wet the tube*), In this case, the liquid inside the tube gets depressed below the level of the liquid surface outside the tube, as shown in Fig.14.46 (*b*), and the meniscus is *convex upwards.*

Suppose the liquid gets depressed through a distance *h* below the liquid surface in the vessel. Then if the pressure due to the vapour on the liquid surface in the vessel be *P*, that above the convex meniscus inside the tube is $P' = P + h\sigma g$. In this case, therefore, $P' > P$ and we have $P' - P = h\sigma g$.

Now, pressure in the liquid just below the meniscus is $(P + h\rho g)$ which is clearly greater than the pressure $(P + h\sigma g)$ above it, since $\rho > \sigma$.

∴ *excess pressure below the meniscus over that above it*

$$= (P + h\rho g) - (P + h\sigma g) = P + h\rho g - P' = (P - P') + h\rho g.$$

Again, the convex meniscus at ρ may be taken to be spherical, of radius *r* (the same as that of the tube). So that, excess pressure below it over that above it is equal to $2T/r$. We, therefore, have $(P - P') + h\rho g = 2T/r$. Or, $(P' - P) = h\rho g - 2T/r$,

which, on proceeding exactly as in case (*i*), gives

$$P' - P = \frac{2T}{r}\left(\frac{\sigma}{\rho - \sigma}\right) \quad ...(V)$$

Again since $\rho > \delta$, the expression for $(P'-P)$ is *positive*, and, therefore, *the vapour pressure above the convex surface of a liquid is greater than the vapour pressure at the horizontal surface of the same liquid by an amount* $2T\sigma/r(\rho - \sigma)$.

Or, neglecting σ, compared with ρ, we have, as a near approximation,

$$P' - P = \frac{2T}{r\rho}\sigma, \qquad ...(VI)$$

which, on a treatment, similar to that in case (*i*), then gives

$$P' - P = \frac{2T}{r\rho} \cdot \frac{P}{R\theta} \quad ...(vii) \quad \text{And, therefore,} \quad \frac{P' - P}{P} = \frac{2T}{r\rho} \cdot \frac{1}{R\theta} \quad ...(VIII)$$

N.B. In terms of M, the mass of 1 *gm. molecule* of the vapour, we similarly have

$$P' - P = \frac{2T}{r\rho} \cdot \frac{MP}{R\theta} \quad \text{And therefore,} \quad \frac{P' - P}{P} = \frac{2T}{\rho R} \cdot \frac{M}{R\theta}$$

It should be noted that the *excess pressure is inversely proportional to r, the radius of curvature of the surface*; so that, the smaller the value of r, or the greater the curvature, the higher the value of the saturation vapour pressure at the curved surface. *It thus follows that the saturation vapour pressure over a small drop of a liquid will be greater than over a large drop of it.*

Let us consider a *large drop of water of diameter* 1 *mm*, or $r = .5$ *mm* $= .05$ *cm*.

Putting $\theta = 273°$ *Abs.*, $\rho = 1$ *gm/cc*, $\sigma = .61 \times 10^{-3}$ *gm/cc*, and $T = 75$ *dynes/cm*, we have

$$P' - P = \frac{2.T.\sigma}{r.\rho} = \frac{2 \times 75 \times .61 \times 10^{-3}}{.05 \times 1} = 1.83 \text{ dynes/cm}^2.$$

For a drop of water, of diamete, 1/1000 *mm*

$$P' - P = 1.83 \times 1000 = 1.83 \times 10^3 \text{ dynes/cm}^2.$$

And, for a drop of water, of diameter equal to one-millionth of a *mm*, as is the case when it just starts forming, we have $P' - P = 1.83 \times 10^3 \times 10^3 = 1.83 \times 10^6$ *dynes/cm*^{2} which is *greater than a pressure of one atmosphere.*

Thus, we see how the saturation vapour pressure rises with diminition in the size of the drop.

Note. In our discussion above, we have tacitly assumed the density of the vapour (σ) to remain constant over the height h. This assumption is more or less valid if h be small. If, howerer, h be large enough, there is hardly any justification for it and we should not, therefore, take $(P-P')$ in *case* (*i*) or $(P'-P)$ in case (*ii*) to be equal to $h\sigma g$, but put it more correctly as equal to $\int g\sigma dh$..

Thus, considering case (*ii*) above, we have $P' - P = \int g\sigma dh$., where P is the pressure due to the vapour on the liquid surface A outside the tube, and P'_1 the pressure due to it above the liquid meniscus B inside the tube.

Since the *pressure in the liquid just below the meniscus B* is $P + h\rho g$ and that above the meniscus, P', we have

excess pressure below the meniscus over that above it $=(P + h\rho g) - P'$.

This must obviously be equal to $2T/r$, so that,

$$P + h\rho g - P' = 2T/r, \quad \text{or,} \quad P' - P = h\rho g - 2T/r.$$

Or, *substituting the value of* $(P'-P)$, we have $\int g\sigma dh = h\rho g - 2T/r$.

Or. $\quad 2T/r = h\rho g - \int g\sigma dh = \int g\,(\rho - \sigma)\,dh$

Since $dP = g\sigma dh$, we have $dh = dP/g\sigma$.

Substituting for dh, therefore, we have $\dfrac{2T}{r} = \displaystyle\int_P^{P'} \frac{\rho - \sigma}{\sigma}\, dP = \int_P^{P'} \frac{\rho}{\sigma}\, dP$

neglecting σ, compared with ρ.

And, since, as we have seen, $\sigma = P/R\theta$, we have

$$\frac{2T}{r} = \int_P^{P'} \frac{\rho R\theta}{P} dP = \rho R\theta \int_P^{P'} \frac{1}{P} dP.$$

Or, $$\frac{2T}{r} = \rho R\theta \ log \frac{P'}{P}, \quad \text{whence, } log \frac{P'}{P} = \frac{2T}{r\rho} \cdot \frac{1}{R\theta} \qquad ...(IX)$$

This, then gives the relation between P and P' for large values of h and is equally applicable to case (i).

14.30. Effect on Evaporation and Condensation. We have seen in the previous article how the saturation vapour pressure is less for a concave, and greater for a convex, liquid surface than that for a plane or a horizonatal surface. If, therefore, we place a drop of the liquid in a space in which its vapour is at the saturation value for a plane surface, it will begin to evaporate for the simple reason that the vapour pressure in the space being less than the saturation vapour pressure for it, it converts itself into vapour in order to increase the vapour pressure to its own saturation value.

This will naturally result in a decrease in the radius of the drop or an increase in its curvature (*i.e.*, its convexity) and a consequent rise in the saturation value of its vapour presure, and it will, therefore, evaporate more and more rapidly. As to whether the whole of the drop will thus evaporate, will depend upon its size. For, obviously, the drop will need its latent heat of vaporisation to be able to evaporate at constant temperature and this is supplied by the surface energy set free due to the reduction in its size and hence its surface area. Thus, if r be the radius of the drop, the surface energy associated with it is clearly equal to *its surface area × its surface tension* = $4\pi r^2 T$. And if L be the *latent heat of vaporisation of the liquid per unit volume*, the quantity of heat required for vaporisation of the entire drop will obviously be $\frac{4}{3}\pi r^3 L$. If, therefore, $\frac{4}{3}\pi . r^3 L = 4\pi . r^2 T$, or $r = 3T/L$, the whole of the drop will evaporate. Thus, r gives the *critical radius* of the drop and all drops of this *critical size*, or less, will automatically evaporate. For water, the *cirtical radius* of the drop that will completely evaporate, works out to about 1.2×10^{-8} *cm* and this gives a general idea as to the critical size of the liquid drops that can wholely evaporate. The radii of such drops are thus comparable to an atomic diameter, which is physically quite understandable.

It will be easily seen, therefore, why a saturated vapour does not automatically start condensing into drops. For, as soon as a tiny drop is formed, the saturation vapour pressure for it being much higher that of the saturated vapour, it at once starts evaporating. Indeed, condensation may not occur even if the vapour be supersaturated, *i.e.*, even if the vapour pressure be higher than its normal saturation vapour presure.

It was first shown by *Wilson* that condensation can be induced by the introduction of suitable nuclei into the supersaturated vapour, in the form of *dust particles* or *charged ions.*

A dust particle provides a coparatively large and flat surface to the vapour and condensation starts around it; for, the radius of the drop formed is not small even at the initial stage of its formation and it has, therefore, little tendency to evaporate. And, as its radius increases due to condensation on it, its curvature or convexity, decreases further and hence also its tendency to evaporate. On the other hand, the saturation value of its vapour pressure thus progressively decreasing, its goes on increasing in size.

A charged ion, though neither large nor flat like a dust particle, induces condensation because it first attracts the neutral molecules of the vapour to itself and a tiny drop of the liquid is thus formed around it. The drop then, naturally shares the charge with the ion and thus experiences an outward force equal to $2\pi q^2$, where q is the *surface density of the charge* on it. This reduces the inward pull on it due to surface tension from $2T/r$ to $(2T/r - 2\pi q^2)$, so that the saturation vapour pressure above the drop over that on a plane or a horizontal surface, *i.e.*, $(P'-P)$ is thus reduced from $\frac{T}{\rho.R\theta}\left(\frac{2T}{r}\right)$ as given by relation *(VIII)* above, to $\frac{T}{\rho R\theta}\left(\frac{2T}{r} - 2\pi^2 q\right)$. In consequence of this reduction in its saturation vapour pressure, the vapour continues to condense on the drop.

As is well known, this is made use of in *Wilson's Cloud Chamber* for tracking the paths of fast moving charged particles. Thus, for example, when an α-particle passes through the saturated water vapour in the chamber, it ionises the vapour and a tiny drop of water is formed around each ion thus produced. The path of the α-particle is thus made *'visible'* by the string of water droplets it leaves behind in its wake.

It is obvious that a particle of the same size as an ion, but with no charge on it, will not be able to induce any condensation if introduced into a supersaturated vapour.

Again, because the saturation vapour pressure over a concave surface is less than on a plane or horizontal surface, *the vapour will condense more readily on a concave surface than on a plane surface*. This might be clearly seen by closing the bottom of the tube in the last experiment and removing some liquid from it, when condensation would set in on the liquid inside the tube and will go on, until the liquid column in the tube attains its previous height.

WORKED EXAMPLES

1. A sphere of water, of radius 1 mm, is sprayed into a million drops of equal size. Find the work expended in doing so. *(Allahabad, Vikram)*

Breaking the liquid drop means an increase in surface area and, therefore, work is required to be done for the purpose, which is equal to the product of the surface tension and the *new* surface area formed. This work done becomes the *surface energy* of the new surface area. We shall, therefore, first calculate *the initial surface energy of the sphere of water*, when it is intact, and then *the surface energy of the million drops*, into which it is broken up when sparayed. The difference of the two will give the work expended in the process.

Let the surface tension of water be taken to be 72 *dynes/cm*. Then, this is equal to the *free surface energy* of the sphere of water,

Or free surface energy = 72 *ergs/cm*^{2} and radius of the sphere = 1 *mm* = .1 *cm*

∴ surface area of the sphere = $4\pi r^2 = 4\pi \times (.1)^2$ *sq.cm*

And, ∴ *initial surface energy of the sphere* = $72 \times 4\pi \times (.1)^2 = 9.048$ *ergs*.

Now, volume of the sphere $= \frac{4}{3}\pi r^3 = \frac{4}{3} \times \pi \times (.1)^3 cc.$

And, number of drops into which it is broken $= 10^6$.

∴ Volume of 10^6 drops $= \frac{4}{3} \times \pi \times (.1)^3\ cc.$

And ∴ volume of *one* drop $= \frac{4}{3} \times \pi \times (.1)^3 / 10^6 = \frac{4}{3} \times \pi \times (.1/10^2)^3\ cc.$

Hence, radius of *one* drop $= .1/10^2\ cm = .001\ cm$.

∴ surface area of *one* drop $= 4\pi(.001)^2$ *sqcm*.

And surface area of *all* drops $= 4\pi(.001)^2 \times 10^6 = 4\pi \times 1 = 4\pi$ *sqcm*.

∴ surface energy of *all* drops, or, *final surface energy* $= 4\pi \times 72 = 904.8$ *ergs*.

Hence, increase in *surface energy* = (904.8 — 9.048) = 895.752 *ergs*.

Or, work expended in the process = 895.752 *ergs*.

2. A glass plate, of length 10 cm, breadth 1.54 cm and thickness .20 cm weights 8.2 gm in air. If it is held vertically, with its long side horizontal, and its lower half immersed in water, what will be its apparent weight ? (Surface tension of water = 73 dynes per cm) *(Cambridge Higher School Certificate; Gorakhpur)*

Here, apart from its weight, there will be two other forces acting on the plate, *viz.*, (*i*) the upthrust of water, and (*ii*) downward force due to surface tension along its edges inside the water.

Now, upthrust of water = weight of water displaced by it

= *half the volume of the plate* × *density of water* × *g*,

because half the plate is immersed in water.

[∵ volume of the plate = 10 × 1.54 × .20 *cc*.

Or, $= \frac{10 \times 1.54 \times .20}{2} \times 1 \times 981$ *dynes*

$= 1.54 \times 1 \times 981$ *dynes* $= 1.54$ *gm wt.*

And, the force due to surface tension acts downwards along a distance equal to twice the sum of the length and the thickness of the plate, *i.e.*, along total length = 2(10 + .2) = 20.4 *cm*

∴ **total downward force on the plate due to surface tension**

$= 20.4 \times 73$ *dynes* $= \frac{0.4 \times 73}{981}$ *gm.wt* $= 1.518$ *gm.wt.*

∴ *net upward thrust* on the plate = 1.54 − 1.518 = 0.022 *gm.wt.*

Hence, *apparent weight of the plate in water = weight in air — upthrust on it.*

= 8.2 − .022 = 8.178 *gm.wt.*

3. The pressure of air in a soap bubble of 0.7 cm diameter is 8 mm of water above the atmospheric pressure. Calculate the surface tension of the soap solution. ***(Delhi)***

We know that *excess pressure* inside a soap bubble over that outside it is given by $p = 4T/r$, where T is the surface tension of the soap solution and r, the radius of the bubble.

Here, the excess pressure inside a soap bubble over the outside atmosphere is given to be equal to 8 *mm* or .8 *cm* of water column.

Taking density of water = 1 *gm/cc*. and value of $g = 981$ *cm/sec*^{2}.

$p = .8$ *cm* of water column

$= .8 \times 1 \times 981$ *dynes/cm*^{2}

$= 784.8$ *dynes / cm*^{2}

and, $r = .7/2 = .35$ *cm.*

∴ substituting the value of p and r in the above relation, we have

$784.8 = 4T/.35$, whence, $4T = 784.8 \times .35$.

Or, $T = 784.8 \times .35/4 = 68.66$ *dynes / cm.*

Therefore, surface tension of the soap solution is 68.66 *dynes/cm*.

4. Why is the upper surface of mercury in a glass capillary tube convex upward, while for water it is concave? Assuming the surface tension of rain water to be 72 dynes/cm, find the difference of pressure inside and outside a rain drop of diameter .02 cm. What would this difference of pressure amount to, if the drop were to be decresed by evaporation to a diameter of 0.00002 cm ? ***(Punjab)***

For answer to part one of the question, see § 14.14,

We know that the excess pressure inside a liquid drop over that outside it is given by $p = 2T/r$, where T is the surface tension of the liquid drop, and r, its radius.

Here, $T = 72$ *dynes/cm.* and $r = .02/2 = .01$ *cm.*

∴ $p = 2T/r = 2 \times 72/.01 = 144/.01 = 14400 = 1.44 \times 10^4$ *dynes/cm*^{2}.

If the diameter of the drop be reduced to .00002 *cm* its radius becomes .00002/2 = .00001 *cm* and we, therefore, have $p = 2 \times 72/.00001 = 14400000$ *dynes/cm*^{2} $= 1.44 \times 10^7$ *dynes/cm*^{2}.

Thus, the excess pressure inside the drop in the two cases will be 1.44×10^4 *dynes/cm*^{2}, and 1.44×10^7 *dynes cm*^{2}, respectively.

5. What will be the pressure in a spherical cavity within a mass of paraffin oil ? The cavity is at a depth of 20 cm below the surface and has diameter of 0.0026 cm. The specific gravity of the oil is 0.85 and its surface tension is 26 c.g.s. units. The pressure of the air over the liquid surface is equal to 76 cm of mercury. ***(Bombay)***

Here, pressure on the surface of the oil = 76 *cm* of mercury column.

$= 76 \times 13.6 \times 981 = 1.014 \times 10^6$ *dynes/cm*^{2},

pressure due to the oil column, 20 *cm* long

$= 20 \times .85 \times 981 = 1.667 \times 10^4$ *dynes/cm*^{2}, because $P = h.\rho.g.$

∴ *total pressure on the spherical cavity, i.e., on the bubble*

$= 1.014 \times 10^6 + 1.667 \times 10^4 = 10^4 (1.014 \times 10^2 + 1.667)$.

$= 10^4 (101.4 + 1.667) = 103.067 \times 10^4 = 1030670$ *dynes/cm*^{2}.

Now, the pressure inside the spherical bubble is in excess of the pressure outside it by $p = 2T/r$.

Here, $T = 26$ *dynes / cm*^{2} and $r = .0026/2 = .0013$ *cm.*

$\therefore \quad p = 2 \times 26/.0013 = 2 \times 2/.0001 = 40000 \; dynes/cm^2$.

Hence *pressure inside the cavity* = 1030670 + 40000 = 107067

$= 10.7067 \times 10^5 \; dynes/cm^2$.

6. A minute spherical air bubble is rising slowly through a column of mercury contained in a deep jar. If the radius of the bubble at a depth of 100 cm is 0.1 mm, calculate its depth when its radius is 0.126 mm; given that the surface tension of mercury is 567. Assume that the atmospheric pressure is 760 mm of mercury.

Here, pressure on mercury surface in the jar = 760 *mm*.

$= 76 \; cm$ of mercury $= 76 \times 13.6 \times 981 \; dynes/cm^2$.

and pressure due to 100 *cm* of mercury column $= 100 \times 13.6 \times 981 \; dynes/cm^2$.

$\therefore$ total pressure on air bubble $= 76 \times 13.6 \times 981 + 100 \times 13.6 \times 981$

$= (76 + 100) \times 13.6 \times 981 = 176 \times 13.6 \times 981 = 2348000 \; dynes/cm^2$.

And, excess pressure inside the air bubble

$2T/r = 2 \times 567/.01 = 113400 \; dynes/cm^2$,

because $T = 567 \; dynes/cm$ and $r = .1 \; mm = .01 \; cm$.

$\therefore$ total pressure inside the bubble at depth 100 *cm*

$= 2348000 + 113400 = 2461400 \; dynes/cm^2$.

And, volume of the bubble $= \frac{4}{3}\pi r^3 = \frac{4}{3}\pi (.01)^3 \; cc$.

Let depth at which its radius becomes .126 *mm* be *h cm*..

Then, total pressure on the bubble at this depth

$= 76 \times 13.6 \times 981 + h \times 13.6 \times 981$

$= (76 + h) \times 981 \times 13.6 = 13350 \; (76 + h) \; dynes/cm^2$.

And, excess pressure inside the bubble over *this* pressure on its outside

$= \frac{2T}{r} = \frac{2 \times 567}{.0126} = 89990 \; dynes/cm^2$. [$\because$ here, $r = .126 \; mm = .0126 \; cm$.]

$\therefore$ total pressure inside the bubble at depth *h cm*

$= 13350\,(76 + h) + 89990 \; dynes/cm^2$

$= 1014000 + 13350\,h + 89990 = 1103990 + 13350\,h$

And, volume of the bubble at this depth $= \frac{4}{3}\pi . (.0126)^3 \; cc$.

Now, in accordance with Boyle's law, the product of pressure and volume of the bubble must be the same in the two cases so that,

$$2461400 \times \frac{4}{3}\pi . (.01)^3 = (1103990 + 13350\,h) \times \frac{4}{3}\pi . (.0126)^3.$$

Or, $$2461400 \times (.01)^3 = (1103990 + 13350\,h) \times (.0126)^3$$

$$= 1103990 \times (.0126)^3 + 13350 \times (.0126)^3 \times h.$$

Or, $$13350 \times (.0126)^3 \times h = 2461400 \times (.01)^3 - 1103990 \times (.0126)^3.$$

Or, $$.02669\,h = 2.461 - 2.209 = .252,$$

whence, $$h = .252 / .02669 = 9.441 \; cm$$

Thus, the required depth of the bubble = 9.441 *cm*.

7. If a number of little droplets of water, all of the same radius, coalesce to form a single drop of radius R, show that the rise in temperature of water will be given by

$$\theta = \frac{3T}{J}\left(\frac{1}{r} - \frac{1}{R}\right),$$

where T is the surface tension of water and J, the mechanical equivalent of heat.

[Poona (supplementary), Sagar]

Let the number of little droplets be *n* and the radius of each droplet *r*. Then, ***surface area of all the droplets*** $= n\,4\pi r^2$ and surface area of the single drop formed by their coalescing together $= 4\pi R^2$. And, therefore, *decrese in surface area* $= n \cdot 4\pi r^2 - 4\pi R^2$

and $\therefore$ *decrease in surface energy* $= (n.4\pi r^2 - 4\pi R^2) \cdot T$ *ergs.*

Hence, *heat produced* $= (n \cdot 4\pi r^2 - 4\pi R^2)\, T/J$ *cal.*

This heat is obviously taken up by the single drop formed, of volume $\frac{4}{3}\pi R^3$ and hence of mass $\frac{4}{3}\pi R^3 \times 1$, taking density of water to be 1 *gm / cc*

If, therefore, $\theta^o C$ be the rise in temperatur, we have

$$\left(\frac{4}{3}\pi R^3 \times 1\right) \times 1 \times \theta = (n \cdot 4\pi r^2 - 4.\pi R^2)\, T/J,$$

Or, $$\theta = \frac{3T}{J}\left(\frac{n r^2}{R^3} - \frac{1}{R}\right)$$

Now, *volume of the single drop* = *volume of n droplets,*

i.e. $$\frac{4}{3}\pi R^3 = n \cdot \frac{4}{3}\pi r^3, \text{ whence } R^3 = n r^3$$

So that, $$\theta = \frac{3T}{J}\left(\frac{1}{r} - \frac{1}{R}\right).$$ [substituting the value of R^3

8. A capillary tube of 0.5 mm bore stands vertically in a wide vessel containing a liquid of surface tension 0.03 N/m. The liquid wets the tube and has a specific gravity 0.8. Calculate the rise of the liquid in the tube.

Here, *radius of the tube* $r = 0.5 / 2 \times 1000 = 0.00025\ m,$

surface tension $T = 0.03\ N/m,$

density of the liquid $\rho = 0.8 \times 10^3\ kg/c.m$ ($\because$ density of water $= 10^3\ kg/c.m$)

and *angle of contact* $(\theta) = 0$ (because the liquid wets the tube)

Now $T = rh\rho g / 2\cos\theta = rh\rho g/2.$ [$\because \theta = 0$ and $\therefore \cos\theta = 1$

So that, $h = 2T/r\rho g = 2 \times .03/.00025 \times .8 \times 10^3 \times 9.81 = .03058\ m.$

Thus, *the liquid will rise in the tube to a height of* 0.03058 *metres.*

9. A capillary tube of internal diameter 1 mm and external diameter 5 mm hangs vertically from the arm of a balance, the lower end of the tube being in a liquid of surface tension 40 dynes / cm. Assuming that the liquid wets the tube, what is the change in the apparent weight of the tube due to surface tension ? (g = 980 cm sec^{-2}).

(London Higher School Certificate)

Here, clearly, the force of surface tension will act downward on the capillary tube along the inner as well as the outer circumference of its lower end dipping into the liquid, *i.e.*, along a length $2\pi \times .25 + 2\pi \times .05 = 2\pi(.25 + .05)$.

$= 2\pi \times .30\ cm = .60\,\pi\ cm.$

$\therefore$ force acting downward on the tube due to surface tension

$= .60\,\pi \times T\ dynes = .60\,\pi \times 40\ dynes.$

$= \frac{.60\,\pi \times 40}{980} = .07696\ gm\ wt.$

$\because$ the external radius $= 5/2 = 2.5\ mm = .25\ cm$ and the internal radius $= 1/2 = .5\ mm = .05\ cm.$

Or, increase in apparent weight of the tube is equal to .07696 *gm.wt.*

10. The stem of a common hydrometer is a circular cylinder of diameter 2 mm. It floats, with its stem wetted, in alcohol, whose specific gravity is .796, and surface tension, 25.5 dynes / cm. Calculate how much deeper it floats than if alcohol had had zero surface tension. *(Cambridge Higher School Certificate)*

Because of the downward force due to surface tension, the hydrometer goes deeper down into alcohol than it would otherwise do, such that the weight of the alcohol displaced by this *additional* immersion of the stem, or *the upthrust due to this displaced alcohol*, equal to the force of surface tension acting on it..

Now, diameter of the stem $= 2\ mm$ and $\therefore$ radius of the stem $(r) = 1\ mm = .1\ cm.$

And $\therefore$ circumference of the stem $= 2\pi r = 2\pi \times .1 = .2\pi\ cm.$

So that, the surface tension acts along this lenght; and, since its value is 25.5 *dynes / cm* the force on the stem due to surface tension

$$= 25.5 \times .2\pi = 16.03\ dynes.$$

Let the stem go further down through a distance *x cm* than it would do if alcohol had had *zero* surface tension. Then, additional volume of stem immersed, or the additional volume of alcohol displaced, is equal to $\pi.r^2.x.c.c.$

∴ *mass* of this alcohol displaced, $= \pi . r^2 . x . \rho gm$,

where ρ is the density of alcohol..

And ∴ *weight* of alcohol displaced, or *upward thrust due to alcohol displaced*,

$$= \pi r^2 \times x \times \rho \; gmwt = \pi r^2 \times x \times \rho \times g \; dynes$$

$$= \pi \times (.1)^2 \times x \times .796 \times 981 \; dynes = 24.53 \; x \; dynes.$$

Since upward thrust due to displaced alcohol is equal to downward force due to surface tension, we have

$$24.53 \, x = 16.03,$$

whence, $x = 16.03 / 24.53 = .6530 \, cm = 6.53 \, mm.$

Thus, the hydrometer floats deeper by 6.53 *mm* than it would, if alcohol had had zero surface tension.

11. Water rises to a height of 5.0 cm in a certain capillary tube. In the same tube the level of mercury surface is depressed by 1.54 cm. Compare the surface tensions of water and mercury, (the specific gravity of mercury is 13.6, the angle of contact for water is 0° and for mercury 130°). ***(London Inter-Science)***

We know that surface tension of a liquid is related to the rise or depression of a liquid in a capillary tube, density and radius of the tube, by the relation, $T = \dfrac{r h \rho g}{2 \cos \theta}$, where r is the radius of the tube, h the rise or depression of the liquid, ρ, the density of the liquid and θ, the angle of contact for the liquid and the tube..

Let T_w be the surface tension of water and T_m that of mercury. Then, we have

$$T_m = \frac{r \times (-1.54) \times 13.6 \times g}{2 \cos 130°},$$

$\because \; h = 1.54 \, cm$ (depression), $\rho = 13.6 \, gm / cc$ and $\theta = 130°$

Or, $$T_m = \frac{-1.54 \times 13.6 \times r \times g}{2 \times (-.6428)} = \frac{1.54 \times 13.6 \times r \times g}{2 \times .6428}$$

And $T_w = 5 \times r \times 1 \times g/2 \times 1 = 5 \times r \times g/2$,

$\because \; h = 5 \, cm$, $\rho = 1 \, gm / cc$ and $\theta = 0°$; so that, $\cos \theta = 1$.

$$\therefore \; \frac{T_m}{T_w} = \frac{1.54 \times 13.6 \times r \times g}{2 \times .6428} \times \frac{2}{5 \times r \times g}$$

$$= \frac{1.54 \times 13.6}{.6428 \times 5} = 6.516. \quad \text{Or,} \quad \frac{T_w}{T_m} = \frac{1}{6.516}$$

∴ the surface tensions of water and mercury are in the ratio, 1 : 6.516.

12. A U-tube is made up of two capillaries of bore 1 mm and 2 mm respectively. The tube is held vertically and partially filled with a liquid of surface tension 49 dynes cm^{-1}. and zero contact angle. Calculate the density of the liquid, if the difference in the levels of the menisci is 1.25 cm. ***(London Inter-Science)***

We have the relation $T = r \,.\, h \,.\, \rho \,.\, g/2$ for a liquid for which the angle of contact is *zero*, ($\cos \theta$ being = 1).

Let h_1 and h_2 be the heights of the liquid columns in two limbs, and r_1 and r_2, their respective radii. Then, we have

$$T = \frac{r_1 \,.\, h_1 \,.\, \rho \,.\, g}{2} \; \therefore \; \text{Or,} \; h_1 = \frac{2T}{r_1 \rho \,.\, g}$$ [where ρ = density of the liquid.

And, similarly, $$T = \frac{r_2 \,.\, h_2 \,.\, \rho \,.\, g}{2} \quad \text{Or,} \quad h_2 = \frac{2T}{r_2 \,.\rho \,.\, g}.$$

$$\therefore \; (h_1 - h_2) = \left(\frac{2T}{r_1 \,.\, \rho \,.\, g} - \frac{2T}{r_2 \,.\, \rho \,.\, g} \right) = \frac{2T}{\rho \,.\, g} \times \left(\frac{1}{r_1} - \frac{1}{r_2} \right).$$

Or, $$1.25 = \frac{4 \times 49}{\rho \times 981} \times \left(\frac{1}{.05} - \frac{1}{0.1} \right)$$

$\because \; h_1 - h_2 = 1.25 cm$; $r_1 = .5 \; mm = .05 \; cm$ and $r_2 = 1 \; mm = .1 \; cm$.

Or, $$1.25 = \frac{98}{\rho \times 981} \times (20 - 10). \quad \text{Or,} \; 1.25 = 980/\rho \times 981,$$

whence, $\rho = 980/1.25 \times 981 = 1.7991 \; gm/cc$

Or, the density of the given liquid is .7991 *gm / cc.*

13. Find the difference in the levels of mercury in the two limbs of a U-tube, if the diameter of the bore of one limb is 1 mm and of the other 8 mm. The surface tension of mercury is 440 c.g.s. units, its density 13.6 gm / cc and the angle of contact with the walls of the tube 140°. ***(Joint Matriculation Board Higher School Certificate)***

We have the relation, $T = \frac{r.h.\rho.g}{2\cos\theta}$, whence, $h = \frac{2T\cos\theta}{r.\rho..g.}$

Let the depression of mercury in one limb be h_1 and that in the other h_2.

Then, since $r_1 = .5\ mm = .05\ cm$ and $r_2 = 4\ mm = .4\ cm$, we have

$$h_1 = \frac{2 \times 440 \times .7660}{.05 \times 13.6 \times 981} \text{ and } h_2 = \frac{2 \times 440 \times .7660}{.4 \times 13.6 \times 981} \quad \left[\because \cos 440° = -.7660\right.$$

$$\therefore \quad (h_1 - h_2) = \frac{2 \times 440 \times .7660}{13.6 \times 981} \cdot \left(\frac{1}{.05} - \frac{1}{.4}\right) = \frac{2 \times 440 \times .7660 \times 35}{13.6 \times 981 \times 2} = 0.8843\ cm$$

∴ the difference in the levels of the two columns is equal to 0.8843 *cm*.

14. What is the work done in blowing a soap bubble of radius 100 mm ? (T = 0.03 j / sq. m). What additional work will be performed in further blowing it, so that its radius becomes 150 mm ?

(i) We know that work done in blowing a soap bubble is equal to its surface area (*inner* and *outer*) × (its free surface energy), *i.e.* equal to *surface area* × surface tension (which is the free surface energy or the work done in creating unit area of the film).

Here *surface area of the bubble* $= 2 \times 4\pi (r^2) = 8\pi (0.1)^2 = 8\pi \times .01$ *sqm.*

∴ *Work done in blowing the bubble* $= 8\pi \times .01 \times .03 = 7.558 \times 10^{-3}$ *Joule*

(ii) Surface area of the bubble of 150 *mm* radius $= 2 \times 4\pi\ (.15)^2 = 8\pi \times .0225$ *sq.m*.

So that, *increase in surface area of the bubble or the surface area of the soap film* $= 8\pi\ (.0225 - .01)$
$= 8\pi \times 0.0125$ *sq m.*

∴ *additional work done* $= 8\pi \times 0.0125 \times .03 = 9.426 \times 10^{-3}$ *joule*

15. There is a pin hole, 0.1 mm in diameter, at the bottom of a vessel containing mercury. What depth of mercury may be put in the vessel without any leak occuring ? (Surface tension of mercury = 550 dynes / cm^{-1} and density = 13.6 gm / cc. Neglect the angle of contact.)

We have to calculate, here, the length of mercury column that would be supported by the surface tension acting along the circumference of the hole.

Clearly, $$T = \frac{r.h.\rho.g}{2\cos\theta},$$

where h is the height of mercury column, without any leak occuring.

Or, since $\theta = 0°$, and $\therefore \cos\theta = 1$,

we have $T = r.h.\rho.g/2$, whence, $h = 2T/r.\rho.g$.

Here, $T = 550$ *dynes/cm*, $r = .1/2 = 0.5$ *mm* $= .005$ *cm* and $\rho = 13.6$ *gm/cc*.

$$\therefore \quad h = \frac{2 \times 550}{.005 \times 13.6 \times 981} = 16.49\ cm$$

∴ mercury can be poured into the vessel to a depth of 16.49 *cm, without any leak occurring.*

16. The pressure inside a soap bubble of radius 1 cm. balances a 1.4 mm column of oil, of specific gravity 0.80. Calculate the surface tension of the soap solution.

We know that the excess pressure inside a soap bubble is equal to $4T/r$, where T is the surface tension of the soap solution and r, the radius of the bubble. Since this is balanced by the column of oil, we have

$$4\,T/r = h.\rho.g,$$

where h is the length of the oil column, ρ, the density of oil, and g the, value of acceleration due to gravity.

$$\therefore \quad 4\,T/1 = .14 \times .80 \times 981. \quad \left[\because r = 1\ cm,\ h = 1.4\ mm = 0.14\ cm \text{ and } \rho = 0.8\ gm/cc\right.$$

$$\text{Or,} \quad T = \frac{.14 \times .80 \times 981}{4}$$

$$= .14 \times .2 \times 981 = 27.47\ dynes/cm.$$

Thus, the surface tension of the soap solution is 27.47 *dynes / cm*

17. Two spherical soap bubbles unite to form one spherical soap bubble without any leakage of air. If V be the consequent change in volume of the contained air and S, the change in surface area, show that 3 pV + 4 ST = 0, where T is the surface tension and p, the atmosphere pressure.

(Gorakhpur, 1964, Allahabad, 1968)

Let r_1 and r_2 be the radii of the two spherical soap bubbles respectively and R, that of the bubble formed by their coalescing together. Then, pressure inside them will respectively. be $(p + 4T/r_1)$, $(p + 4T/r_2)$ and $(p + 4T/R)$ and the volumes $\frac{4}{3}\pi r_1^3$, $\frac{4}{3}\pi r_2^3$ and $\frac{4}{3}\pi R^3$.

Since there is no leakage of air, the temperature must remain constant and Boyls' law must, therefore, hold good. We, therefore have

$$\left(p + \frac{4T}{r_1}\right) \cdot \frac{4}{3}\pi r_1^3 + \left(p + \frac{4T}{r_2}\right) \cdot \frac{4}{3}\pi r_2^3 = \left(p + \frac{4T}{R}\right) \cdot \frac{4}{3}\pi R^3.$$

Or, $$p\left(\frac{4}{3}\pi r_1^3 + \frac{4}{3}\pi r_2^3 - \frac{4}{3}\pi R^3\right) + \frac{4T}{3}\left(4\pi r_1^2 + 4\pi r_2^2 - 4\pi R^2\right) = 0$$

Clearly, $\frac{4}{3}\pi r_1^3 + \frac{4}{3}\pi r_2^3 - \frac{4}{3}\pi R^3 = V$; *the total change in volume*

and $4\pi r_1^2 + 4\pi r_2^2 - 4\pi R^2 = S$, *the total change in surface area.*

So that $$pV + \frac{4T}{3} . S = 0,$$

whence $$3pV + 4ST = 0$$

EXERCISE XIV

1. Show that the surface tension of a liquid is equal to the mechanical part of its surface energy. Calculate the work done on the film in blowing a soap bubble from a diameter of 4 *cm* to one of 30 *cm* if its surface tension be 45 in *c.g.s. units*.
Ans. 2.5×10^5 *ergs*.

2. Show that the excess pressure inside a soap bubble of radius *r* over the atmospheric pressure outside it is equal to $4T/r$, where *T* is the surface tension of the soap solution.

How may the surface tension of a bubble be determined?

3. Define surface tension, and show how it can be regarded as involving potential energy. From consideration of energy, show that *(i)* oil will spread on the surface of water, *(ii)* mercury will collect on a clean glass plate in drops of different shapes according to their size, and *(iii)* water will rise in a capillary tube. *(Bombay)*

4. What would be the pressure inside a small air bubble of 1 *mm* radius, situated just below the surface of water? Surface tension of water may be taken to be 70 *dynes / cm* and the atmosphere pressure to be 1.012×10^6 *dynes/cm*^{2}.
Ans. 1.027×10^6 *dynes/cm*^{2}

5. Calculate the loss of energy if 1000 drops of water, each of diameter 2 *mm* coalesce to form one large drop. The surface tension of water is equal to 72 *dynes / cm*.
Ans. 8143 *ergs*.

6. Calculate the work done in breaking up a drop of pertrol of volume 1 *cc* into a thousand million drops. The surface tension of petrol is 26 *dynes / cm*.
Ans. 125574 *ergs*.

7. Calculate the amount of energy needed to break a drop of water, 2 *mm* in diameter, into 10^9 droplets of equal size, taking surface tension of water as 73 *dynes / cm*. *(Madras)*
Ans. 9160 *ergs*.

8. Define surface tension. Show that the excess pressure acting on the curved surface of a curved membrane is given by

$$P = 2T\left[1/r_1 + 1/r_2\right],$$

where $r1$ and $r2$ are the radii of curvature and T, the surface tension of the membrane.
[Bombay (Principal paper), 1962]

9. Calculate the amount of energy evolved when eight droplets of water (surface tension 0.072 *N/m*), of radius 0.5 *mm* each, combine into one.
Ans. 9.048×10^{-7} *joules*.

10. Explain Rayleigh's jet method for determining the surface tension of a liquid. Derive the necessary formula.
[Poona (Principal paper), 1962]

11. Define surface teunsion and describe how you would determine its value for water, using a capiluary tube.

A spherical soap bubble of radius 1.0 *cm* is formed inside another of radius 2.0 *cm*. Find the radius of a single soap bubble whose internal pressure is the same as that of the smaller bubble. Prove any formula used in the calculation. *(Delhi, 1965)*
Ans. 0.666 *cm*.

12. Calculate the force required to separate two plates of glass, of area 10 *sq cm* each with a layer of water .001 *mm* thick in between them. The surface tension of water = 72 *dynes / cm*.
Ans. 1.44×10^7 *dynes*

13. Describe a method of determining the surface tension of a soap bubble. Deduce the formula used.

Two equal spherical soap bubbles coalesce ; if V is the consequent change in volume of the contained air and S, the change in the total surface area, show that $3PV = 4ST$, where T is the surface tension of the soap bubble and P, the atmospheric pressure. *(B.Sc. (Hons), Delhi)*

14. If a number of little droplets of water, all of the same radius r *cm* coalesce to form a single drop of radius R *cm*, show that the rise of temperature of water will be given by $\frac{3S}{J}\left(\frac{1}{r} - \frac{1}{R}\right)$, where S is the surface tension of water and J, the mechanical equivalent of heat. *(Sagar)*

15. Find the relation between the radius of (*a*) a sphericla drop, (*b*) a spherical bubble of a liquid, the surface tension and pressure.

Two soap bubbles of radii 2 and 3 *cm* coalesce into a single bubble of radius R *cm*. If the surface tension of the soap solution is 25 *dynes per cm* and the atmospheric pressure is 76 *cm* of mercury (whose density = 13.6 *gm / c.c*), find the equation to determine R. *(Madras)*

Ans. $1.014 \times 10^4 . (R^3 - 35) + (R^2 - 13) = 0$.

16. A soap film, .001 *mm* thick and at 0°C, is stretched *adiabatically*, until its area is doubled. What is the resulting fall of temperature, if we assume that the specific heat and density of the film are both unity, that $dT/d\theta$ at 0°C. = – .15 *dyne per cm per* 1°C, and that $J = 4.2 \times 107$ *ergs per calorie ?* **Ans.** .0195°C.

17. Explain the terms surface tension and angle of contact. Show that the pressure inside a spherical bubble of radius r exceeds that outside it by $4T/r$. If this excess pressure is balanced by that due to a coloumn of oil, (sp. gr. 0.8) 2*mm* high, when r = 10 *mm*, find the surface tension of the soap bubble. **Ans.** .3918 *N/m*

18. Calculate the work done in blowing a soap bubble of radius 10 *cm*. (T = 30 *dynes/cm*). What additional work will be performed in further blowing it so that its radius is doubled ? *[Bombay (Principal paper), 1962]*

Ans. 7.54×10^4 *ergs*; 2.26×10^5 *ergs*.

19. A soap film is formed on a rectangular framework of length .7 *cm* dipping into a soap solution. This framework hangs from the arm of a balance. An extra weight of 0.38 *gm* must be placed on the opposite pan to balance the pull of the film. What is the surface tension of the soap solution? **Ans.** 26.6 *dynes/cm*.

20. Calculate the density of a liquid, a column 2.09 *cm* of which balances the excess pressure inside a soap bubble of radius 1 *mm*. Surface tension for the soap solution may be taken to be 40 *dynes/cm*. **Ans.** .78 *gm /cc*.

21. Describe and explain Jaeger's method of measuring the surface tension of a liquid.

(Agra, B.Sc. Part II, 1963 ; Pubjab, 1963; Vikram, 1961)

Discuss whether the result obtained in this way should be the same as that given by the capillary tube method.

What would be the pressure inside a small air bubble of 0.1 *mm* radius, situated just below the surface ? Surface tension of water = 72 *dynes per cm* and atmospheric pressure = 1.013×10^6 *dynes per sq cm*.

(Agra, B.Sc. Part II, 1962)

Ans. 1.0274×10^6 *dynes/cm*^{2}.

22. Show that the excess pressure inside a drop of water, of diameter 1/100th of a *mm* is .C 37 *cm* of mercury column at 0°C. Density of water at 0°C = 1 *gm /cc*. density of water vapour at 0°C and 760 *mm* pressure = $.61 \times 10^{-3}$ *gm / cc* and surface tension for water = 75 *dynes per cm*..

23. Explain the method of finding the surface tension of a liquid by the method of drops.

A drop of water, 0.5 *cm*. radius, is split into 1000 tiny drops. Find the mechanical work expended. Calculate the pressure inside one of these small drops (surface tension of water = 75 *dynes per cm*). *(Madras)*

Ans. .675 π *ergs* ;(P + 3000) *dynes /cm*^{2}

(where P is the atmospheric pressure).

24. In a drop-weight determination of the surface tension between water and chloroform, a glass tubes 4 *mm* external diameter was used and 50 drops of chloroform, density 1.5 *gm per cc*, were allowed to fall in the water. The weight of these drops was 3.43 *gm*. Find the interfacial surface tension. **Ans.** 29.5 *dynes per cm*.

25. What should be the radius of a capillary tube so that a liquid (surface tension 70 *dynes/cm*) will riso to a height of a 8 *cm* in it? *(Delhi, 1961)*

Ans. 0.018 *cm*

26. (*a*) Define surface energy. Give the theory and practice of the method of determining the surface tension of a liquid by weighing drops. *(Punjab)*

(*b*) If a globe of water of diameter 2 *cm* suddenly splits into 100 equal globules under isothermal conditions, determine the gain in surface energy which occurs, given that the surface tension of water is 75 *dynes per cm*.

(A.M.I.E., 1950)

Ans. 3432 *ergs*.

27. Explain clearly from where the energy comes when a liquid rises against gravity in a capillary tube. Derive an expression for the height h through which the liquid of surface tension T will rise in a capillary tube of radius r. What will happen if the length of the tube is smaller than h ? *(Punjab)*

28. Describe *Jaeger's* maximum bubble pressure method of detemining surface tension.

A large soap bubble of radius R contains inside it a smaller soap bubble of radius r. If the smaller bubble now bursts *isothermally*, with no leakage of air from the system as a whole, so that a new bubble of radius R' is formed, show that the radii of the three bubbles are connected by the relation $P(R'^3 - R^3) + 4T(R'^2 - r^2 - R^2) = 0$, where P stands for the atmospheric pressure and T, for surface tension of the soap solution.

29. Define surface tension. Describe Quincke's methods of masuring the surface tension and the angle of contact for mercury in the form of a large flat drop. Derive the necessary formula. *[Baroda (Principal paper), 1962]*

30. Define surface tension and angle of contact. Describe Quincke's drop method of measuring these quantities for mercury in contact with glass. Give the theory of the method.

Calculate the work done in spraying a spherical drop of mercury of one millimetre radius into a million identical droplets, the surface tension of mercury being 550 *dynes/cm*. *[Bombay (Subsidiary), 1962]*

Ans. 6839 *ergs*.

31. What forces determine the shape of a liquid drop on a horizontal plate ? show that all large drops of mercury resting on clean glass plate will have the same height.

32. Describe the laboratory method of finding the surface tension of a liquid by the rise of the liquid in a capillary tube.

A tube of conical bore is dipped into water with apex upwards. The length of the tube is 20 *cm* and the radii at the upper and lower ends are 0.1 and 0.3 *cm*. Find the height to which the liquid rises in the tube. (Surface tension of water = 80 *dynes/cm*.) *(Agra ; Allahabad)*

Ans. 0.55 *cm*.

33. Show how the existence of an acute angle of contact and of a pressure difference due to curvature accounts for the rise of liquid in a capillary tube.

A capillary tube of internal diameter 1 *mm* and external diameter 5 *mm* hangs vertically from the arm of a balanced, the lower end of the tube being in water (angle of contact = 0°). Calculate the change in the apparent weight of the tube due to surface tension (Surface tension of water = 78 *dynes/cm* and g = 980 *cm/sec*^{2}). *(Allahabad)*

Ans. 0.15 *gm*.

34. Explain the capillary tube method of determining the surface tension of a liquid. Why is this method not suitable for temperatures other than that of the surroundings ? Suggest and explain other methods for doing so, in which the surface tension may be calculated by measuring the pressure necessary to force the meniscus back in a level with the surrounding liquid.

35. A capillary tube is dipped in water. Water rises to a height of 4 *cm* above the surrounding liquid. If the angle of contact is zero and the radius of the tube is 0.1 *mm* what is the surface tension of the liquid.? *(A.M.I.E., 1961)*

Ans. 19.63 dynes/cm.

36. A vertical U-tube containing mercury has one limb of diameter 5 *cm* and the other, of diameter. 1 *cm*. Calculate the difference in level of the mercury columns in the two limbs. (T for mercury = 550 *dynes/cm* density of mercury = 13.6 *gm per cc* and its angle of contact with the walls of the tube = 140°). **Ans. 1.04 *cm*.**

37. A liquid of density 1.05 *gm/cc* and angle of contact 20° has a vertical capillary tube of 2 *mm* diameter dipping into it. If the surface tension of the liquid be 0.0235 *N/m*, find the rise of the liquid in the capillary tube. **Ans. 4.28 × 10^{-3} *m*.**

38. Define the terms surface tensions and surface energy. How is surface tension determined, using the ripple method ? *[Punjab, 1963 and Poona (Principal paper), 1962]*

39. A U-tube, whose ends are open and whose limbs are vertical, contains oil of sp. gravity 0.85 and surface tension 28 *dynes/cm*. If one limb has a diameter of 2.2 *mm* and the other, a diameter of 0.8 *mm* what is the difference in level of the oil in the two limbs ? Assume the angle of contact between the oil and the glass to be zero.

(A.M.I. Mech. Engineering)

Ans. 1.07 *cm*.

40. Define surface tension. Explain how you can determine the surface tension of a solution with the help of a bubble blown out of it. *(Bombay)*

41. A U-tube made up of two capillary tubes, of diameters 0.5 *mm* and 1.0 *mm* respectively, contains water (*S.T.* 72 *dynes/cm*). What would be the difference of levels between the two arms? *(Poona (Subsidiary). 1962]*

Ans. 2.944 *cm*.

42. Deduce the relation between surface tension and vapour pressure at a curved surface and discuss its effects on evaporation and condensation. *(Bombay)*

43. Calculate the difference in vapour pressure of water for a plane surface and that for a drop of radius 0.1 *mm*. *Density of water vapour* = 6.1 × 10^{-4} *gm/cc* and *surface tension* of water = 75 *dynes/cm*. *(Bombay)*

Ans. 9.15 *dynes/cm*^{2}.

44. Show that the maximum vapour pressure over a curved liquid surface differs from that over a plane surface. Deduce the value of this difference in terms of the densities of the liquid and the vapour, the radius of curvature of the surface and the surface tension of the liquid. Discuss the bearing of the above fact on the formation of rain drops. *(Madras)*

45. Show that the vapour pressure over the curved surface of a liquid drop of radius *r* exceeds that over a flat surface by an amount equal to $\frac{2T}{r} \cdot \frac{\sigma}{\rho - \sigma}$ where *T* is the surface tension of the liquid, ρ, its density and σ, the density of its vapour. Discuss the application of this result to the condensation of super-saturated vapour on dust particles and other nuclei. *(Bombay)*

46. If the aqueous vapour tension at 0°*C* be 4.6 *mm* of mercury, calculate the radius of a water drop, at 0°*C*, which would be in equilibrium with its vapour at twice this pressure. (Surface tension of water = 75 *dynes per cm*. 18 *gm* of water at *N.T.P.* measure 22.4 *litres*). **Ans.** 2.42×10^{-6} ***cm.***

47. A capillary tube is immersed in water and, by exerting a pressure of 15.65 *cm* of water, the meniscus in the tube was kept 1.25 *cm* below the surface of the water outside. The radius of the tube was 0.104 *cm*. Calculate the surface tension of water. **Ans. 73.5 *dynes/cm.***

48. Explain concisely, but clearly, the following :

(*i*) pieces of camphor scurry about on the surface of water but their motion slows down if we immerse our finger in the water ;

(*ii*) it is difficult to introduce mercury in a fine glass tube ;

(*iii*) small pieces of cork and straw etc., floating on water, appear to attract each other ;

(*iv*) small drops of mercury on a plane glass sheet are spherical in shape, but large drops of it are flat at the top ;

(*v*) ploughing of fields helps retain moisture in them.

49. Distinguish between a wave and a ripple. How may the surface tension of a liquid be determined by the method of ripples ?

50. What is the effect of temperature on the surface tension of a liquid? Describe how it may be studied experimentally.

51. Two soap bubbles of radii *a* and *b* coalesce to form a single bubble of radius *r*. If the external pressure is *p*, prove that the surface tension of the solution, from which the bubble is formed, is given by $T = \frac{1}{4} p(r^3 - a^3 - b^3)/(a^2 + b^2 - r^2)$.

(Delhi, B.Sc. (Hons.) 1961)

Hint. See worked example 17.

52. Discuss the stability of the cylindrical film. Give the verification of the result. ***[Delhi, B.Sc. (Hons.)]***

CHAPTER XV

GASES—KINETIC THEORY

15.1. The Kinetic Theory–Introduction. In order to connect and co-relate the many facts accumulated by experience, some hypothesis or theory became necessary, and the *Kinetic theory of matter* is the one that has proved most helpful for the purpose. The distinction of being the founder of the modern Kinetic Theory goes, by common consent, to **Daniel Bernoulli** (1730), as he was the first to explain Boyle's law on its basis, though the theory may be said to have had its beginnings in the speculations and ideas of *Thales of Miletus* (640 to 547 B.C.) about the possible structure of matter, and a host of other early workers. And, the credit for having established it on a firm *mathematical* basis is due to **Clausius** and **Maxwell**, in whose hands it attained its present form. The triumph of the theory lies in its success in explaining known results and predicting new ones.

The theory is based on two assumptions :

(1) That *matter is not continuous*, but consists of small aggregations or lumps, called *molecules*, very much like a handful of sand, composed of fine granules; so that, even when the molecules are in contact with each other, there are *inter-spaces* in between them.

The molecule of a substance is the smallest part of it that possesses the characteristic properties of that substance, and can have an independent existence of its own. It can be broken up into smaller bits, called *atoms*, by various methods, but then, these no longer exhibit the distinctive properties of that substance. A molecule may consist of one or more atoms, *e.g.*, the molecule of Helium is monatomic, *i.e.*, consists of only one atom, that of Hydrogen or Oxygen is diatomic, *i.e.*, consists of two atoms, and that of Carbon dioxide is triatomic, *i.e.*, consists of three atoms, and so on.

It should be noted with care that although there may be any number of different kinds of molecules, there can be only a hundred and odd different kinds of atoms (including isotopes), corresponding to the different elements. These atoms are the smallest particles that can take part in chemical reactions.

(2) That the molecules are generally not in contact with each other but are in a *continuous state of agitation*, moving about with great speed, haphazardly, in all directions, —their freedom of movement, however, being different in the three states of matter, *viz.*, the solid, the liquid and the gaseous.

Now, there is a huge mass of evidence in favour of both these assumptions, *e.g.*, (*a*) the phenomena of *diffusion* and *solution*, which clearly suggest the molecular structure of matter and agitation of molecules; for, we find heavier gases, like Carbon dioxide, diffusing into lighter ones, like Hydrogen, in defiance of gravity, as it were, and even a gas diffusing into lead, (*b*) the familiar phenomenon of *expansibility* of gases, which is clearly due to the tendency of the molecules to fly away from each other, (*c*) the phenomena of *vapour pressure* and *evaporation*, and (*d*) phenomenon of **Brownian movement**, which is perhaps the one single experimental fact, which enables us to actually see, so vividly, the molecules of a substance moving about hither and thither before our very eyes.

This last phenomenon derives its name from **Robert Brown**, an English botanist, who first discovered it in 1827 while observing, under a high-power microscope, the suspension of plant pollen in water, when he found the pollen grains dancing about in the wildest manner, and thought that they were perhaps very tiny living creatures. The phenomenon has been fully investigated by **Parrin** (in 1908) and may be readily observed by examining a colloidal suspension under an ultra-microscope, when a strong beam of light is passed through the liquid. The particles, as small in size as 6×10^{-7} *cm* which just appear as mere points of light surrounded by diffraction rings, are visible by the light they scatter at right angles to the beam and are seen in a spontaneous and eternal dance of a most irregular and haphazard manner,—now spinning, now resting, and now rising again, and so on and on.

Now, this mad movement of the molecules is the clearest proof of *molecular agitation.* For, the movement of the tiny particles is due to the large number of molecular impacts they receive simultaneously on all sides. Since these impacts are not necessarily uniformly distributed, there is a resultant unbalanced force on the tiny particules which caues then to move. And, because the force varies most haphazardly and irregularly, the motion of the particle also exhibits the same haphazardness and irregularity. Quite obviously, the smaller the particles, the more readily are they subject to these irregular motions; for, on a larger or a heavier particle, the impacts will almost balance and no resultant motion will ensue, *e.g., when* a large body like a glass bead, or a marble piece, is immersed in water or any other liquid, it is not tossed about in this manner.

15.2. Kinetic Theory of Gases. Whereas the kinetic theory of solids and liquids is still in a formative stage, the kinetic theory of gases has made rapid strides and can fully explain the various properties of gases. The reason is not far to seek and will be clear if we try to picture to ourselves the structure of a solid, a liquid and a gas.

(*i*) In the case of a *solid,* due to the great force of cohesion, its molecules are all compactly or closely packed and every molecule is more and less *fired in its position, having only the freedom to vibrate about this position.* It is not free to move over the whole volume of the solid, much less to escape away from it.

(*ii*) In the case of a *liquid,* the cohesive force is still there, but is not so strong, with the result that although it is sufficient to hold the liquid together and to give it a *free surface,* it cannot prevent a molecule from roaming over the whole volume of the liquid, with the result that while *a liquid has a definite volume and a well defined free surface, it has no shape of its own.* The average distance between two molecules of a liquid is estimated to be more or less of the same order as the size of its molecule.

(*iii*) In the case of a *gas, on* the other hand, these *cohesive forces are almost negligible* under ordinary conditions of temperature and pressure, so that the molecules lie far apart from each other, the average distance between two molecules being about a hundred times the size of a molecule. A molecule is, therefore, free to wander over the entire space available to it, *with the result that a gas has neither a volume nor a free surface of its own.*

It will thus be clear from the above why the problem becomes simpler in the case of gas, compared with that in the case of a solid or a liquid. It is obviously so, because in the case of a gas, (*i*) *the molecules lie far apart and we can, therefore, neglect, under ordinary conditions, the actual volume occupied by the molecules themselves* (if compactly arranged) *compared with that occupied by the gas, as a whole.* Clearly, this cannot be done in the case of a liquid or a solid, (*ii*) *the molecular forces* can be neglected because of the large distance between the molecules.*

Now, in order to further simplify our problem, we shall first consider, here, a gas whose molecules have negligible size, *i.e.,* whose molecules are mere *mass points,* and in which the molecular forces are also negligible. Such a gas, with *zero molecular size and zero molecular forces, is called an ideal or* a **perfect** gas. No such gas, however, exists in reality, and the properties of an actual or a real gas only approximate to those of *this* ideal gas.

15.3. Pressure exerted by a perfect gas. Due to the constant random motion of the molecules of a gas and their high speed, they *'bombard'* the walls of the containing vessel, and thus exert pressure. To calculate this pressure exerted by a gas, we make the following further assumptions, to simplify matters.

(1) That the molecules of gas are *all alike,* (though different from those of another), and are *perfectly elastic spheres,* and that there is *no force of attraction or repulsion between them, or*

* These molecular forces are entirely different from Newton's gravitational forces. They are electrical in nature and *do not obey the ordinary gravitational inverse square law,* as has been indicated already.

between them and the walls of the containing vessel. In other words, that *all their energy is kinetic and that they do not suffer any loss of momentum, or kinetic energy, on a direct or 'head on' impact with the walls of the vessel, only their direction of motion being reversed.*

(2) That *large numbers of molecules exist in the smallest volume of a gas* with which we can deal. Under ordinary conditions of temperature and pressure, the number of molecules present in 1 *cc* of a gas is estimated to be of the order of 10^{19}, *their size or diameter being very small compared with the inter-molecular space,*— the actual volume occupied by the molecules being only about 3 *cc* in 10 *litres* of the gas.

(3) That due to their large number and ceaseless, haphazard motion in all directions, with velocities, ranging from *zero to infinite, there are frequent collisions against one another and consequent changes in the direction and magnitude of their velocities at each collision; and that this does not affect the molecular density of the gas, i.e.*, the molecules do not, in the steady state, collect more at one place than another.

(4) That the molecules, being material bodies, subject to the laws of motion, *move in straight lines, with uniform velocity, between any two collisions.*

(5) *That the time for which a collision lasts is small compared with the time-interval between two collisions or the time taken to cover the distance traversed between two collisions.* This distance is called the *mean free path of the molecule* and depends upon the temperature and pressure of the gas.

We are now in a position to calculate the pressure exerted by an *ideal or a perfect gas* on the walls of the containing vessel.

Let there be a gas, enclosed in a cubical vessel *of unit edge, with its walls perfectly elastic,* (Fig. 15.1), and let the number of molecules present in it be n, the mass of each molecule being m.

Since the molecules are constantly moving about with different velocities in different directions, they are bombarding the walls of the cube. Consider any one molecule m, having a velocity c_1 at a given instant. This velocity of the molecule may be resolved into three *rectangular components,* u_1, v_1 *and* w_1 *along the axes of x, y, and z, parallel to the three edges of the cubical vessel respectively.* Then, clearly,

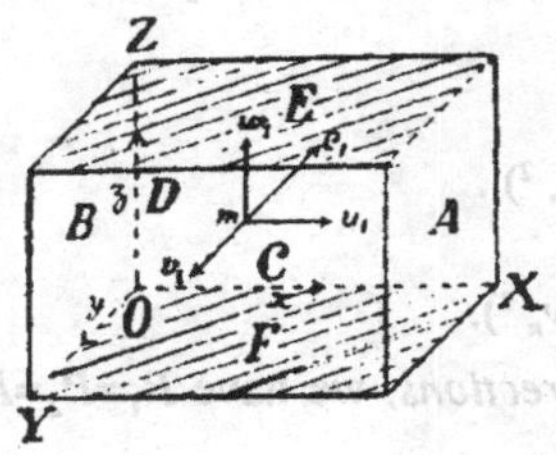

Fig. 15.1

$$c_1^2 = u_1^2 + v_1^2 + w_1^2.$$

Now, consider the motion of this molecule along the axis of x, *i.e.*, perpendicular to the walls A and B of the vessel ; so that, striking the wall A with velocity u_1, it rebounds with the *same* velocity in the opposite direction (the molecules as well as the wall being perfectly elastic), strikes wall B, rebounds back to A and so on.

Clearly, the momentum of the molecule as it approaches A is mu_1 and is obviously directed towards A. When, however, it collides against A, it rebounds *with only its velocity reversed in direction, its magnitude remaining unchanged* ; so that, *its momentum still remains the same in magnitude but is now directed oppositely, i.e.*, is now $-m_1u_1$.

Therefore, *change in its momentum* $= mu_1 - (-mu_1) = 2\,m\,u_1$.

This, then, is the change in momentum of the molecule at *each* collision with wall A and, clearly, as it collides against the wall, it imparts *this* momentum $(2mu_1)$ to the wall.

Hence, *momentum imparted by the molecule to wall A per second*

$= 2mu_1 \times$ *number of collisions it makes with the wall per second.*

Now, for each successive collision or impact with wall A, the molecule must traverse the distance from A to B and back, *i.e.*, a distance equal to twice the length of the *unit* cube, or a distance of 2*cm*. Since it covers a distance u_1, along this direction, in 1 *second,* it will cover the distance 2 *cm* (from *A* to *B* and back) in $2/u_1$ seconds ; and, therefore, *the number of collisions it will have with the wall in one second is equal to* $u_1/2$.

So that, *momentum imparted to wall A by the molecule per second*

$$= 2mu_1 \times u_1/2 = mu_1^2 .$$

In other words, this is the force f_1, say, exerted by the molecule on wall *A* ; for, in accordance with Newton's second law of motion, *force is just rate of change of momentum.* Thus,

$$f_1 = mu_1^2$$

Similarly, forces $f_2, f_3, \ldots, f_n$, exerted on wall *A* by other molecules, having velocities $c_2, c_3, \ldots, c_n$ and rectangular components $(u_1, v_1, w_1), (u_2, v_2, w_2), \ldots, (u_n, v_n, w_n)$, along the three axes respectively, are given by

$$f_2 = mu_2^2,\ f_3 = mu_3^2 \ldots,\ f_n = mu_n^2,$$

the mass of each molecule being the same *m*, in each case.

Thus, *total force on wall A due to all the n molecules* is given by

$$F_A = f_1 + f_2 + f_3 + \ldots\ldots + f_n$$

$$= m u_1^2 + m u_2^2 + m u_3^2 + \ldots\ldots + m u_n^2$$

$$= m(u_1^2 + u_2^2 + u_3^2 + \ldots\ldots + u_n^2).$$

Now, area of wall *A* $= 1 \times 1 = 1$ *sq cm* [$\because$ each edge of the cube = 1 *cm*]. So that, F_A is the force exerted by the molecule *on unit area* and is thus equal to pressure P_1 exerted by them on wall *A*, because, as we know, *pressure is equal to force per unit area.* Thus, pressure P_1 on wall *A* is equal to F_A.

Or, $$P_1 = m(u_1^2 + u_2^2 + u_3^3 + \ldots\ldots + u_n^2).$$

Exactly in the same manner, considering the motion of the molecule along the axes of *y* and *z*, – *i.e.*, perpendicular to walls *C* and *L* and walls *E* and *F* of the cubical vessel respectively, – we have

pressure exerted by the molecules on wall C given by

$$P_2 = m(v_1^2 + v_2^2 + v_3^2 + \ldots\ldots + v_n^2).$$

and *pressure exerted by the molecules on wall E* given by

$$P_3 = m(w_1^2 + w_2^2 + w_3^2 + \ldots\ldots + w_n^2).$$

Since the pressure exerted by the gas is the *same in all directions*, we have $P_1 = P_2 = P_3 = P$, say.

And, therefore, $$P_1 + P_2 + P_3 = 3P.$$

So that,

$$3P = m(u_1^2 + u_2^2 + \ldots u_n^2) + m(v_1^2 + v_2^2 + \ldots v_n^2) + m(w_1^2 + w_2^2 + \ldots w_n^2).$$

Or, $$P = \frac{1}{3} m(u_1^2 + u_2^2 + \ldots u_n^2) + (v_1^2 + v_2^2 + \ldots v_n^2) + (w_1^2 + w_2^2 + \ldots w_n^2)$$

Or, $$P = \frac{1}{3} m(c_1^2 + c_2^2 + \ldots c_n^2).$$

Now, $$\frac{(c_1^2 + c_2^2 + \ldots c_n^2)}{n} = \bar{c}^2 .$$

where the symbol $\bar{c}$ stands for the root mean square velocity of the molecules*, usually written, for brevity, as *r.m.s. velocity*.

* i.e. the root mean square of the mean

So that, $(\bar{c}_1{}^2 + \bar{c}_2{}^2 + \ldots + \bar{c}_n{}^2) = n\,\bar{c}^2$,

And therefore, $$P = \frac{1}{3} m\, n \bar{c}^2. \qquad \ldots (i)$$

Thus, the pressure exerted by the gas enclosed in the vessel is equal to $\frac{1}{3} m\, n\, \bar{c}^2$.

15.4. The value of $\bar{c}$. In relation (*i*) above, we have

$m.n$ = mass of the molecules in *unit* volume of the gas.

Or, = *mass per unit volume of the gas = density of the gas, ρ.*

Substituting this value of *mn* in relation (*i*) for *P*, we have

$$P = \frac{1}{3} \rho \bar{c}^2, \text{ whence, } \bar{c}^2 = 3\,P/\rho.$$

Or, $$\bar{c} = \sqrt{3\,P/\rho}\,.$$

This enables us to determine the value of $\bar{c}$ for any particular gas, *at a given temperature*, if *P* and ρ for it be known.

Thus, for example, taking *P* for air, at 0°*C* to be $= 76 \times 13.6 \times 981$ *dynes/cm*^{2}.

$= 10^6$ *dynes/cm*^{2}. (approx.)

and ρ for air, at 0°C = 1/1000 *th* of that of water $= 1/1000 = 10^{-3}$ *gm/cc*

we have $$\bar{c} = \sqrt{\frac{3 \times 10^6}{10^{-3}}}\ cm/sec = \sqrt{3 \times 10^6 \times 10^3}$$

$$= \sqrt{3 \times 10^9} = \sqrt{30 \times 10^8} = 5.5 \times 10^4\ cm/sec$$

$$= 5.5 \times 10^4/100 = 5.5 \times 10^2 = 550\ metres/sec$$

$$= \frac{1}{2}\ mile/sec = \text{the velocity of a rifle bullet at } 0°\,C.$$

We thus see how tremendous is the speed with which the molecules of the gas move about haphazardly in any given volume of it, and, obviously, *the lighter the molecule, the faster it moves.*

The velocity of the molecule of any other gas may also be calculated in the same manner, *at any desired temperature.*

15.5. Relation between $\bar{c}$ and *T*. We have the experimental *gas equation*, connecting *Boyle's law* and *Charles' law, viz.*,

$$PV = RT,$$

where *R* is the *gas constant, T, the absolute temperature of the gas* and *P* and *V*, its *pressure* and *volume*, respectively.

Now, $$P = \frac{1}{3} m\, n \bar{c}^2 \ \therefore\ PV = \frac{1}{3} m\, nV . \bar{c}^2.$$

Or, putting $nV = N$*, the total number of molecules in volume *V* of the gas, we have

$$PV = \frac{1}{3} m . N . \bar{c}^2.$$

Or, $$\frac{1}{3} m.N\, \bar{c}^2 = R.T, \quad \text{whence, } \bar{c}^2 = 3\,R.T/m.N.$$

Or, $$\bar{c}^2 \propto T/m. \qquad [\because R \text{ and } N \text{ are constants.}$$

---

* For the sake of simplicity, *N* is taken to be the number of molcules in a *gram-molecule* of the gas; for, at the same temperature and pressure, its value will be the same for *all* gases. It is called the Avogadro Number, and its value is founds to be 6.0×10^{23}.

Thus, *the absolute temperature of a gas is directly proportional to the square of the r.m.s. velocity of its molecules,* – the greater the velocity (and, therefore, the greater the kinetic energy) of the molecules, the greater the temperature of the gas.

15.6. Deduction of Gas Laws on the basis of the Kinetic Theory :-

1. Boyle's Law. We have the relation, $= \frac{1}{3}\, m.n.\, \bar{c}^2$.

Multiplying it by V, the volume of the gas, we have

$$PV = \frac{1}{3} m.n.V.\bar{c}^2 = \frac{1}{3} mN.\bar{c}^2.$$

where $nV=N$, the total number of molecules in the given volume V of the gas.

Hence, m and N being constants, it follows that $PV \propto \bar{c}^2$.

Or, $P.V$ = a constant, if $\bar{c}$ be constant.

Now, $\bar{c}$ is constant, if T, (the temperature of the gas) be constant. So that, *at constant, temperature, PV = a constant* which is our *Boyle's law*.

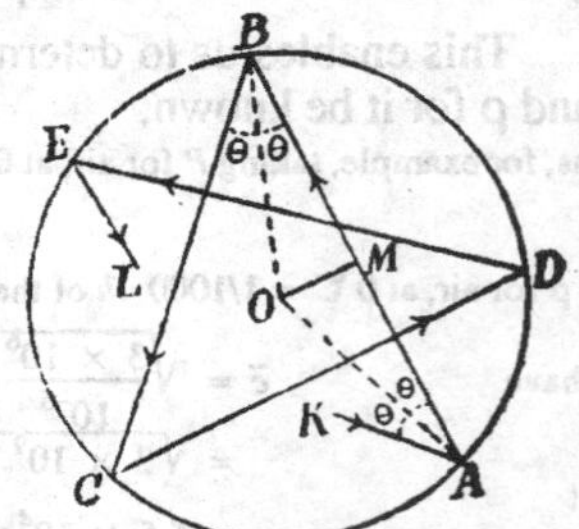

Fig. 15. 2

Alternative deduction of Boyle's Law. A more rigid derivation of the law from the kinetic theory is the following :

Let N molecules of a gas, each of mass m, be enclosed in a spherical vessel of radius r and volume V; so that, $V = \frac{4}{3}\pi r^3$.

Consider one molecule striking the wall of the vessel at A with a velocity, (*i.e., the root mean square velocity*), $\bar{c}$ at an angle θ with the normal at A, (Fig. 15.2). Let the molecule then follow the path *ABCDE*. The component of the velocity $\bar{c}$ perpendicular to the wall is clearly $\bar{c}\cos\theta$, and gets *reversed*, the other component $\bar{c}\sin\theta$, at right angles to it, remaining unchanged. The change of momentum at A is, therefore, equal to $2m\bar{c}\cos\theta$, *inwards*.

Now , the distance travelle d by th e molecule between collisions at A and B is equal to

$$AB = 2AM = 2AO.\cos\theta = 2r.\cos\theta.$$

And, therefore, the time taken by the molecule to travel *this* distance is equal to $2r.\cos\theta/\bar{c}$ *sec*; so that, the number of collisions in *one* second $= \bar{c}/2r.\cos\theta$.

∴ change of momentum is *one second*; or the *rate of change of momentum* due to *one* molecule is

$$= \frac{2m.\bar{c}\cos\theta \times \bar{c}}{2r\cos\theta} = \frac{m\bar{c}^2}{r} \text{ per second.}$$

Since there are N molecules in all, the total rate of change of momentum$=m.N.\bar{c}^2/r$, *inwards*.

Again, since the rate of change of momentum is equal to force, the force exerted by the molecules on the walls of the vessel, *outwards* $= m.N.\bar{c}^2/r$.

Now, the area of the vessel $= 4\pi r^2$, [Because it is spherical].

∴ force exerted by the molecules per unit area of the vessel

$$= \frac{mN.\bar{c}^2}{r.4\pi r^2} = \frac{mN.\bar{c}^2}{4\pi r^3}.$$

Or, *pressure P exerted by the molecules on the wall of the vessel is given by* $P = mN\bar{c}^2/4\pi r^3$.

[∵ force per unit area= pressure.]

Now, volume of the vessel, $V = 4\pi r^3/3$, whence, $4\pi r^3 = 3V$.

Substituting this value of $4\pi r^3$ in the expression for P above, we have

$$P = mN.\bar{c}^2/3V. \quad \text{or,} \quad PV = mN.\bar{c}^2/3. \qquad ...(ii)$$

And since m and N, as also $\bar{c}$ (at a given temperature), are all constants, PV =*a constant*, at a given temperature, which is Boyle's law.

N.B. *Note how the above treatment may be used to show that* $P = \frac{1}{3} m.n.\bar{c}^2$ .. *For,* $N = nV$, *where n is the number of molecules per unit volume. Substituting this value of N in relation (ii) above, we have* $P = \frac{1}{3} m.\ n\ .\bar{c}^2$.

2. Avogadro's Hypothesis. According to this *hypothesis, equal volumes of all gases and vapours, under the same conditions of temperature and pressure, contain the same number of molecules.* Let us see how we can deduce it from the kinetic theory.

Let N_1 and N_2 be the numbers of molecules in two equal volumes of gases respectively, at the same temperature and pressure.

Then, clearly, $$PV = \frac{1}{3} m_1 N_1 \bar{c}_1^{\ 2} = \frac{1}{3} m_2 N_2 \bar{c}_2^{\ 2}, \qquad ...(a)$$

where m_1 and m_2, and $\bar{c}_1$ and $\bar{c}_2$ are the masses and the root mean square velocities of the molecules of the two gases in the two cases respectively.

Now, *kinetic energy* = $\frac{1}{2}$ × *mass* × (*velocity*)^2, and, therefore, the kinetic energy of each molecule, in the *first case* = $\frac{1}{2} m_1 \bar{c}_1^2$, and the kinetic energy of each molecule, in the *second case* = $\frac{1}{2} m_2 \bar{c}_2^2$.

As shown by **Maxwell**, the average kinetic energy of *any* gas is the *same* at a given temperature ; so that, $\frac{1}{2} m_1 \bar{c}_1^{\ 2} = \frac{1}{2} m_2 \bar{c}_2^{\ 2}$.

Multiplying both sides by 2/3, we have $\frac{1}{3} m_1 \bar{c}_1^{\ 2} = \frac{1}{3} m_2 \bar{c}_2^{\ 2}$. ...(b)

Now, dividing (*a*) by (*b*), we have $N_1 = N_2$,

i.e., the number of molecules in the case of the two gases is the *same*, which is *Avogadro's hypothesis.*

Thus, *since a gram-molecule of every gas contains the same number of molecules under the same conditions of temperature and pressure*, it follows that the greater the density of the gas, the greater its molecular weight ; or, the molecular weights of two gases are proportional to their densities at the same temperature and pressure.

3. Graham's Law Diffusion. This law states that the *rates of diffusion of two gases* (which depend upon the velocities of the molecules of the gases) *are inversely proportional to the square roots of their densities.* Let us see how we can arrive at this result from the kinetic theory.

Let M_1 and M_2 be the molecular weights, (*i.e.*, weights of *one gram-molecules*), of the two gases respectively.

Then, since $PV = \frac{1}{3} m_1 N. \bar{c}_1^{\ 2} = \frac{1}{3} m_2 . N \bar{c}_2^{\ 2}$ at the same temperature and pressure, V being the volume of a *gram-molecule* of the gas, N, the *Avogadro number*, and m_1 and m_2 and $\bar{c}_1$ and $\bar{c}_2$, the respective masses and *r.m.s.* velocities of the molecules in the two cases, we have .

$$m_1 N = M_1 \quad \text{and} \quad m_2\ N = M_2.$$

And, therefore, $$PV = \frac{1}{3} M_1 . \bar{c}_1^{\ 2} = \frac{1}{3} M_2 . \bar{c}_2^{\ 2}.$$

Or, $$M_1 \bar{c}_1^{\ 2} = M_2 c_2^{\ 2}. \quad \text{Or,} \quad \bar{c}_1^{\ 2} / \bar{c}_2^{\ 2} = M_2/M_1 ,$$

whence, $$\bar{c}_1 / \bar{c}_2 = \sqrt{M_2 / M_1} .$$

But $M_1 \propto \rho_1$ and $M_2 \propto \rho_2$, where ρ_1 and ρ_2 are the densities of the two gases at the *same* temperature and pressure.

$$\therefore \quad \bar{c}_1 / \bar{c}_2 = \sqrt{\rho_2 / \rho_1} .$$

Since $\overline{c}_1 \propto r_1$ and $\overline{c}_2 \propto r_2$ where r_1 and r_2 are the *rates of diffusion* of the two gases respectively, we have

$$r_1/r_2 = \sqrt{\rho_2/\rho_1},$$

which is the mathematical expression of *Graham's law*.

For deduction of Charles's law and Dalton's law, see solved example 3, at the end of the chapter.

15.7. Kinetic energy of a Molecule. We have seen that the pressure of a gas is given by the relation, $P = \frac{1}{3} m.n. \overline{c}^2$.

Multiplying by V, where V is the *volume of a gram-molecule of the gas*, we have

$$\frac{1}{3} m.N. \overline{c}^2 = R.T.$$

Now, $nV = N$, the *Avogadro number*. $\therefore \; PV = \frac{1}{3} m.N. \overline{c}^2$.

Also, $PV = R.T.$, where R is the *gas constant*. $\therefore \; m.N. \overline{c}^2 = R.T.$

Multiplying both sides by 3/2, we have $\frac{1}{2} m.N. \overline{c}^2 = \frac{3}{2} R.T.$

Or, $$\frac{1}{2} m. \overline{c}^2 = \frac{3}{2} .(R/N).T.$$

But $\frac{1}{2} m\overline{c}^2$ is, clearly, the *kinetic energy* of a molecule.

Therefore, kinetic energy of a molecule $= \frac{3}{2}. (R/N).T.$

Now, R/N is a constant K, called *Boltzmann's constant*.

$\therefore$ *K.E. of the molecule of a gas* $= \frac{3}{2} .K.T.$

Obviously, therefore, the kinetic energy of the *gram-molecule* of a gas is equal to $\frac{3}{2} K.N.T. = \frac{3}{2} . (R/N).N.T.$

Or, *K.E. of the gram-molecule of a gas* $= \frac{3}{2} . R.T.$

This tells us at once that the kinetic energy of a molecule (*i*) *depends upon T, the absolute temperature of the gas* and (*ii*) *is quite independent of its mass*. Hence, a tiny molecule will be more active than a bulky one.

This act is often referred to as the *kinetic interpretation of temperature*.

Again, examining the above expressions for kinetic energy a little more carefully, we see that the factor $\frac{1}{2}$ appears there because of its presence in the expression for kinetic energy, and the factor 3 appears there because the molecule has *three degrees of freedom* of straight line motion, (*i.e.*, along the three axes). Therefore, the kinetic energy per degree of freedom is equal to *one-third* of the above.

Thus, *K.E. of one molecule of a gas, per degree of freedom* $= \frac{1}{2}$ *R.T.* And, therefore, *K.E.* of the gram-molecule of a gas, per degree of freedom $= \frac{1}{2} RT$.

15.8. Value of the Gas Constant (R). We know that a *gram-molecule* of every gas occupies the same volume at the same temperature and pressure. This volume is 22.4 *litres* at the normal temperature and pressure, (*N.T.P.*), *i.e.*, at 0°*C* and 76 *cm* pressure. And, since *PV=RT*, it follows that the value of *R* must be the same for all gases at the same temperature and pressure.

Now, $$PV/T = R.$$

So that, putting the value of *P*, *V* and *T* in the above relation, we have

$$R = 76 \times 13.6 \times 981 \times 22400/273$$
$$= 8.29 \times 10^7 \text{ ergs per degree C}$$

$\because \; P = 76 \; cm$
$= 76 \times 13.6 \times 981 \; dynes/cm^2$.
$V = 22.4$ litres

$= 8.29$ *joules per degree C.*

$= 22.4 \times 1000$
$= 22400\ cc$
and $0°C = 273°\ Abs.$

Now, 1 *calorie* $= 4.18 \times 10^7$ *ergs*, (value of J)

$\therefore \quad R = 8.29 \times 10^7 / 4.18 \times 10^7$

$= 1.98$ *calories/degree C*

$= 2$ *calories / °C* very nearly.

This, then, is the value of R for *one gram-molecule* of a gas. If we consider only *one gram* of a gas, however, the value of R will, naturally, be 1.98/*molecular weight of the gas,* and will, therefore, be different for different gases, depending upon their molecular weights.

Note. In the Constant Tables, (*Kaye and Laby's*), the relation $PV=RT$ is used with a different meaning, –*viz.*, P is taken in *atmospheres*, and V, as the ratio between the volume of the gas at pressure P and temperature T and its volume at normal atmospheric pressure and 0°C ; so that, in this case,

$$P = 1\ atm.,\quad V = 1 \quad \text{and} \quad T = 0°C = 273°\ Abs.$$

$$R = PV/T = 1 \times 1/273 = 1/273 \text{ for all gases.}$$

15.9. Van der Waals' Equation. In our discussion so far, we have concerned ourselves only with *ideal* gases, *i.e.*, gases *in which the molecular size is zero, and the molecular force*, (*i.e.*, force of cohesion between the molecules), is *zero*. These assumptions are far from valid in the case of any *real* gas, even at ordinary pressures, and they become absolutely inadmissible at higher pressures.

This fact was recognised as early as 1827, when *Desperetz* discovered that the resulting volumes of originally equal volumes of air and carbon dioxide differed from each other, when subjected to a pressure of 15 *atmospheres*. Surely, then, he argued, they could not both be obeying Boyle's law. The problem of obtaining a relationship between pressure, volume and temperature of a gas, more in conformity with actual facts, therefore, engaged the attention of many investigators and many a different equation were proposed. The one, however, most satisfactory from both the theoretical and the practical stand-points, is the one due to **Van der Waals**, known after him, as *Van der Waals' equation*. He takes into consideration both the factors mentioned above, *viz.*, *(i) the size of the gas-molecules, and (ii) the force of cohesion between them.*

(*i*) **Correction for molecular size.** A gas molecule has a *finite size* and, therefore, occupies some volume ; so that, the actual space in which it is free to roam about, when enclosed in a vessel, is a little less than the volume of the containing vessel, (or the gas) and this is much more so when the gas is compressed. A correction must, therefore, be applied for this decrease in the free space available to the gas.

Now, it may be recalled that we have imagined a molecule to be a perfectly hard or elastic sphere, so that when the total volume occupied by a gas is decreased, the volume of the free space available for the movement of its molecules is decreased to a greater extent due to the incompressibility of the latter, which still continue to occupy the *same volume* as before. For precisely the same reason, when the total volume occupied by a gas is increased, the free space available to the molecules increases in a greater ratio. Thus, the volume of the free space that undergoes change with pressure is not V but something less than V, and, at first sight, it might appear that this volume should be V *minus* the volume actually occupied by the molecules, when compactly arranged together. This would be quite a valid assumption, only if the motion of the molecules were orderly, and if only some of them moved and others remained at rest. Actually, however, the molecules fly about in a chaotic manner, in all possible directions, and, therefore, they interfere with, and obstruct, each other to a very much greater extent, thus greatly reducing the space for free movement. For this reason, the volume of the free space is taken to be $(V - b)$, where b is near about four times the total volume of the molecules. We must, therefore, substitute this value of volume $(V - b)$ for V in the relation $PV=RT$ for a perfect gas.

(*ii*) **Correction for the force of cohesion.** In an actual gas, the force of cohesion is not zero and, therefore, every molecule in a gas is attracting, and is being attracted by, every other molecule near about it. Now, a molecule, well inside the containing vessel, is being attracted equally in all directions, with the result that the resultant force on it is *zero*, and its speed remains undiminished. But a molecule close to a wall of the vessel is only being attracted backwards by molecules in the body of the gas and hence its speed is somewhat diminished, with the result that the force with which it strikes the wall of the vessel is lessened. In other words, the pressure exerted by the gas on the wall is now smaller.

Now, it is obvious that if we double the number of molecules *per cc* of the gas, this reduction of the observed pressure will become *four times as great*, for the simple reason that

(*i*) *there will now be twice as many molecules striking the walls of the containing vessel, and*

(*ii*) *each molecule, near to a wall, will now be attracted inwards or backwards by twice as many molecules.*

Now, both these depend upon the number of molecules per *cc* of the gas ; so that,

reduction of observed pressure $\propto$ (*number of molecules of the gas per cc*)^2.

Or, *reduction of observed pressure* $\propto \dfrac{1}{(\text{volume of the gas})^2}$, *i.e., reduction of observed pressure* $\propto 1/V^2$.

Or, *reduction of observed pressure* $= a/V^2$, where a is a constant.

It follows, therefore, that if the observed pressure be P for an *actual* gas, it would be $(P+a/V^2)$ for an *ideal* gas, and, therefore, we must substitute $(P+a/V^2)$ for P in the relation, $PV=RT$.

Thus, substituting the values $(V-b)$ for V, and $(P+a/V^2)$ for P in the perfect gas equation $PV=RT$, we have

$$(P+a/V^2)(V-b)=RT.$$

This is known as **Van der Waals' equation**, and as stated above, is more in conformity with experimental results, but is still far from representing the behaviour of gases accurately, and has its own defects.

15.10. Mean Free Path of a Molecule (λ). *The mean free path of a molecule is the average distance covered by it between two successive collisions.** The following is a simple way of calculating its approximate value.

Let us assume that only the particular molecule that we are considering is in motion, all the others being at rest. This moving molecule will, then, naturally collide with all those molecules whose centres happen to lie within its *sphere of influence* (see *f* 14.4). If the radius of the sphere of influence of the molecule be r, then, all these molecules lie in a sphere of radius r, described about this molecule as centre. If, therefore, $\bar{c}$ be the *r.m.s.* velocity of the molecule through the gas, it will, in one second, collide with all the molecules lying in the region traversed by its sphere of influence as it covers a distance $\bar{c}$. Now, the region thus traversed in one second is, obviously a cylinder of length $\bar{c}$, and an area of cross-section πr^2, *i.e.*, of a volume $\pi r^2\,\bar{c}$. Therefore, if n be the number of molecules *per cc*, there will be $\pi r^2\,.\,\bar{c}n$ number of molecules enclosed within this cylinder, and hence the number of collisions in *one second* will be $\pi r^2\,.\,\bar{c}n$. The average time between collisions is thus equal to $1/\pi r^2\,.\,\bar{c}n$ and, therefore, average distance covered by a molecule between successive collisions is equal to *velocity* $\times$ *this time, i.e.*,

$$= \bar{c} \times 1/\pi r^2\,.\,\bar{c}n = 1/\pi r^2\,.\,n.$$

Or, *the mean free path of a molecule*, $\lambda = 1/\pi r^2.n.$

Further, if m be the mass of one molecule, the mass per unit volume, of *density* ρ, of the gas, is equal to mass of one molecule *into* number of molecules per unit volume $= m \times n$.

$\therefore \quad \lambda = m/m.n.\pi r^2 n = m/\pi r^2 \rho, \qquad [\because m.n = \rho$

*The exact nature of these collisions is not yet known, and it is not quite clear whether molecules come into actual contact or whether they recede away from each other when at a distance, close to each other.

i.e., λ is inversely proportional to ρ, the *density* of the gas ; and, since density varies directly as the pressure of the gas, λ varies inversely as the pressure of the gas, and directly as its absolute temperature.

To have an idea as to the magnitude of λ, it might be mentioned that, at ordinary pressure, its value is of the order of 10^{-5} *cm*, but at low pressure, such as 10^{-4} *cm* of *Hg column*, (as in electric glow lamps etc.), its value ranges between 5 to 10 *cm*.

In the above discussion, we have made the simplifying assumption that all molecules, but the one under consideration, are at rest. *Maxwell* has taken into consideration the motion of all the molecules and has shown, by a more rigorous treatment that $\lambda = 1\sqrt{2}\,\pi\, r^2.n.$, which gives a smaller value of λ than the one obtained above.

15.11. Viscosity of Gases. The viscosity of a gas is just a mechanical property of it, and its viscosity coefficient may be defined in precisely the same manner as that of a liquid (see § 12.9, page 439).

Suppose we have a gas, flowing from left to right, over a solid horizontal surface with a velocity which is very small compared with the velocity of its molecules. This velocity will be the same everywhere in a plane *XOY*, (Fig. 15.3) parallel to the horizontal surface, but will increase upward in the direction of the *z-axis*, being the least, (*i.e., zero*) for the layers in contact with the solid surface, and increasing with the distance of the layer from it, *i.e.*, there will be a *velocity-gradient* dv/dz along the *z-axis*.

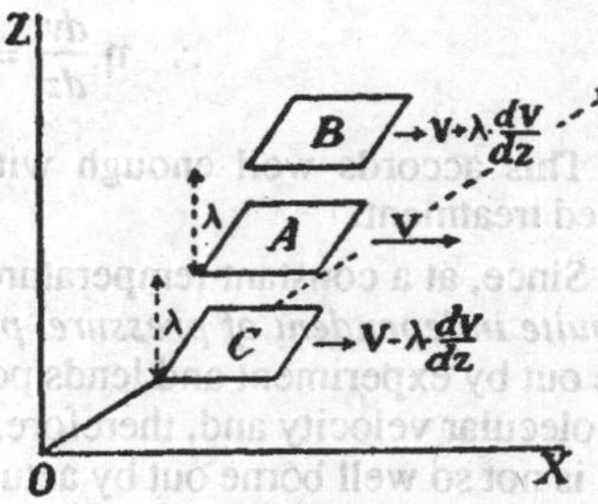

Fig. 15.3

Now, we may imagine the molecules of the gas to be divided up into three distinct parts, moving parallel to the three mutually perpendicular axes, *x, y*, and *z*, in either direction ; so that, the average number of molecules moving in *one* direction along any *one* axis will be *one-sixth* of the total number of molecules in the gas.

Consider an area *A*, parallel to the plane *XOY*. Let *V* be the velocity of the gas in *this* plane. Let there be two other layers *B* and *C*, parallel to *A*, above and below it respectively, *each at a distance from it equal to the mean free path of the molecules*, so that the molecules from it, moving normally upwards to *B* or downwards to *C*, do so *without any collision*.

Then, clearly, velocity of the molecules in layer *B* is $V + \lambda.dv/dz$, and, that of the molecules in layer *C* is equal to $V - \lambda.dv/dz$,

Now, due to molecules of the gas moving about indiscriminately in all directions, there is a continual interchange of molecules, between the two layers *B* and *C*. Since the molecules from *B*, crossing *A* downwards, have a velocity $(V + \lambda.dv/dz)$, the *forward momentum carried by them per unit area of A* is equal to *mass of molecules* $\times$ $(V + \lambda.dv/dz)$.

If *n* be the number of molecules per unit volume of the gas, and $\bar{c}$ and *m*, the velocity and mass of each molecule respectively, the number of molecules crossing unit area of *A* downward, *in unit time*, is equal to $n\bar{c}/6$, and their *mass* $= n.\bar{c}.m/6$.

So that, the forward momentum carried by them *downward, per unit area of A, per unit time* is

$$= \frac{n.\bar{c}}{6} m. \left(V + \lambda. \frac{dv}{dz} \right) = \frac{\rho.\bar{c}}{6} . \left(V + \lambda. \frac{dv}{dz} \right),$$

because $n.m = \rho$, the density of the gas in *grams per cc*.

This, then is the momentum lost per unit area, per unit time, in the direction of flow, by the faster moving layer *B*.

Similarly, the momentum carried *upward* per unit area of *A*, per unit time, in the direction of flow, from the slower moving layer $C = \frac{n.\bar{c}}{6} .m \left(V - \lambda. \frac{dv}{dz} \right) = \frac{\rho.\bar{c}}{6} \left(V - \lambda. \frac{dv}{dz} \right)$, and this is, therefore, the momentum gained per unit time by the layer *B*.

Thus, *the net momentum lost by layer B above A, per second*

$$= \frac{\rho\bar{c}}{6}\left(V + \lambda.\frac{dv}{dz}\right) - \frac{\rho.\bar{c}}{6}\left(V - \lambda.\frac{dv}{dz}\right)$$

$$= \frac{\rho.\bar{c}}{6}.2\lambda.\frac{dv}{dz} = \frac{1}{3}\rho.\bar{c}.\lambda.\frac{dv}{dz}.$$

And, clearly, the same amount of momentum is being gained by the layer *below A*. Thus, the layer above A tends to accelerate its motion, and that below A tends to retard it ; so that, the *backward dragging force acting per unit area on it is equal to* $\frac{1}{3}\rho.\bar{c}.\lambda.dv/dz$.

This must be equal to the tangential force acting per unit area of layer A, i.e., $\eta.dv/dz$, *where* η *is the coefficient of viscosity of the gas.*

$$\therefore\ \eta.\frac{dv}{dz} = \frac{1}{3}\eta.\bar{c}.\lambda.\frac{dv}{dz}.\quad \text{Or,}\quad \eta = \frac{1}{3}\rho.\bar{c}.\lambda.$$

This accords well enough with the value $\rho.\bar{c}\,\lambda/2.64$, obtained by *Chapman* by a more detailed treatment.

Since, at a constant temperature ρ increases with pressure and λ decreases in the same ratio, η *is quite independent of pressure, provided the temperature remains constant.* This fact is amply borne out by experiment and lends powerful support to the kinetic theory of gases. Further, $\eta \propto \bar{c}$, the molecular velocity and, therefore, to the square root of the absolute temperature of the gas. This result is not so well borne out by actual experiment and is only approximately true.

15.12. Production and measurement of low pressure – Vacuum technique. As is probably fairly well known, the ancient philosophers, including the celebrated **Aristotle** and even **Descartes**, who came much later, preached down the centuries that empty space or vacuum could never be created. They actually gave it out as a maxim that '*Nature abhorred vacuum*' and, indeed, had a horror for it – '*horror vacui*', as they termed it. Needless to add, with the weight of their great authority behind it, it came to be accepted as a gospel truth.

When, therefore, towards the middle of the seventeenth century, in the year 1744, **Toricelli's** *famous experiment* in connection with the mercury barometer, later repeated by **Pascal** (using red wine instead of mercury), showed, for the first time, the possibility of an empty space, it created a sensational stir in the then scientific world and spurred many a worker, the world over, to take up the challenge and devise to devise ways and means to create a perfect vacuum. These dedicated efforts soon bore fruit when, in the year 1750, **Otto von Guericke**, in Germany, and, in 1760, **Robert Boyle**, in England, succeeded in producing their exhaust pumps. Crude devices, these, relatively speaking, they, nevertheless, exploded the myth as to the utter impossibility of creating vacuum in defiance of Nature's dictum, as it were. No wonder, the barometer and the exhaust pump came to be hailed with the same sense of owe and wonder as the telescope, invented by Galileo in the same century.

From then on, it is all a story of renewed and relentless effort to improve upon those earlier devices, develop newer and better techniques and to continually refine them until they have attained today their high level of sophistication, enabling us to produce a pressure as low as 10^{-9} *mm of the* mercury, – a near perfect vacuum.

Among those who valiantly carried the battle forward, are a host of workers whose names will figure at appropriate places, along with their respective contributions, as we proceed along with our study of the subject but, perhaps, a special mention may be made here of just two of them, **Gaede and Langmuir**, as much for the specific devices introduced by them as for their over-all contribution to vacuum technique, in general. Indeed, but for their pioneering efforts, the present day staggering development – a virtual explosion – of scientific knowledge would have been well nigh impossible.

Apart from the well known use of vacuum technique in radio and X-ray equipment, we entirely owe to it our initiation into the new and fascinating realm of atomic physics in which we come across the ever-increasing tribe of fundamental particles and other interesting phenomena. And this, in turn, has had the reciprocal effect of enabling us to make further improvement in our high vacuum technique.

Simultaneously with the development of this high vacuum technique has, naturally, gone on the development of delicate and sensitive gauges to be able to measure accurately the low pressures produced. We shall now proceed, therefore, to study these twin devices.

15.13. Exhaust Pumps – Their Characteristics. Before dealing with any specific types of pumps, it will be worthwhile to understand the characteristics of a good vacuum pump. These are *(i) the exhaust pressure* ; *(ii) the degree of attainable vacuum* ; and *(iii) the speed of the pump.* Let us take each, in turn.

(i) **Exhaust pressure.** In any vacuum pump, there is an *inlet, fine or intake* side, where the gas or vapour from the vessel to be exhausted is drawn into the pump, and an *outlet* or *exhaust* side, where it is expelled out. As its very name indicates, *the exhaust pressure is the pressure on the exhaust side of the pump,* and may be just atmospheric or much lower than it, varying from pump to pump. But, generally speaking, *the higher the vacuum desired to be produced on the fine or the intake side of the pump, (i.e., in the vessel connected to it), the smaller must be the exhaust pressure, or the 'rough vacuum', as it is alternatively called, on its outlet side.* For, the recognised procedure in high-vacuum technique today is to first reduce the pressure from atmospheric to a small fraction of it, say to 0.1 *mm* or so, *i.e.*, to create a *'fore-vacuum'*, by means of an ordinary pump, here called *'auxiliary'*, the *'rough'* or the *'backing' pump.* This *fore-vacuum* or *backing pressure,* as it is also termed, is then reduced further from 10^{-4} to 10^{-7} *mm* by means of a suitable *fine* or *high vacuum pump.* For this purpose, the backing and the high-vacuum pumps are arranged *in series* or *tandem,* so that the gas or vapour from the vessel to be exhausted is drawn in at the inlet of the latter and expelled at its outlet into the fore-vacuum of the former, which, then finally expels it out into the atmosphere.

(i) **Degree of attainable vacuum.** By this we understand the lower limit of the pressure that it is possible to obtain in the vessel, connected to the pump. This depends to a very large extent on the exhaust pressure. For, if it be very low, it may result in the passage of the gas or vapour in the reverse direction, *i.e.*, in its leakage from the exhaust to the intake side of the pump.

Now, theoretically speaking, there is no lower limit to the attainable pressure in the case of a *diffusion-condensation pump,* but, in a *molecular pump,* a definite limit is set by the constant ratio it bears to the exhaust pressure. The limit may, however, in general, be considerably extended by using connecting tubes of wide bores in between the vessel to be exhausted and the pump, as it greatly minimizes the resistance to the flow of the gas or the vapour from the former to the latter.

(iii) **Speed of the pump.** The *speed of a pump* has sometimes been defined either *(i)* as the *rate of decrease of mass of the gas or vapour in a given volume, i.e., as* $-dm/dt$ or *(ii)* as the *rate of change of volume of the gas or vapour in the vessel at any given instant, the volume being measured at the pressure obtaining at that very instant, i.e., as* $-dv/dt$.

The two definitions are hardly compatible with each other and create unnecessary confusion. In fact, the second definition is called the *speed of exhaust of the pump* or its *pumping speed* and is seldom referred to as the speed of the pump. In any case, since it is ultimately the change of pressure that is involved in either definition (this being the primary function of the pump), it is perhaps desirable to define the speed of a pump in a more general and satisfying manner in terms of the change of pressure brought about by it.

The speed of a pump may, therefore, be defined as the **relative rate of reduction of pressure in a given volume.**

Thus, if p_0 be the limiting value of the attainable pressure, with the help of a given pump, p, the pressure at an instant t in the vessel of volume V, connected to it, and S, the *speed* of the pump *at this pressure,* the rate of reduction of pressure in the vessel, *i.e., dp/dt*, is given by the relation.

$-\frac{dp}{dt} = \frac{S(p - p_0)}{V}$ ...(i) whence, we have $-\frac{dp}{p - P_0} = \frac{S.dt}{v}$ ...(ii)

So that, if p_1 and p_2 be the values of the pressure in the vessel at instants t_1 and t_2 respectively, we have, by integrating expression (ii) for the limits $p = p_1$, $t = t_1$ and $p = p_2, t = t_2$,

$$\log_e\left(\frac{p_1 - p_2}{p_2 - p_0}\right) = \frac{S}{V}.(t_2 - t_1), \text{ which gives } S = \frac{V}{t_2 - t_1}.\log_e\left[\frac{p_1 - p_0}{p_2 - p_0}\right]^* \quad ...(iii)$$

This is referred to as *Langmuir's equation for the speed of a pump*, since it was he who drew attention to the fact that the limiting value of the attainable pressure (p_0) is *never zero* or that perfect vacuum is not attainable by means of an exhaust pump.

If, however, p_0 be negligibly small the eqation reduces to

$$S = \frac{V}{t_2 - t_1}\log_e\frac{p_1}{p_2}.$$

This is known as *Gaede' equation* and gives us a definition of *S*, the *intrinsic speed* of the pump. For, if $P_1/P_2 = e$ and $(t_2 - t_1) = 1$ *sec* we have $S = V$.

Thus, the *intrinsic speed of a pump may be defined as the volume in which it can reduce the pressure to 1/e th (which comes to about 36.79%) of its instantaneous value in 1 second*, ($\because e = 2.17828$).

Putting $p_0 = 0$, *i.e.*, taking the attainable vacuum to be perfect, we have, from (i) above, $-dp/dt = S.p/V$, whence, $S = -\frac{V}{p}.\frac{dp}{dt}$. This gives the *speed of exhaust* of the pump, or the *pumping speed E*, as defined by *Langmuir*.

So that, $$E = -\frac{V}{p}\frac{dp}{dt} \quad ...(iv)$$

Now if *dV* be the volume of the gas or vapour, measured at pressure *p*, extracted in the time-interval *dt*, from the volume *V*, we have

$$pV = (p - dp)(V + dV), \quad [\because \text{pressure decreases.}$$

whence, neglecting the product *dp.dV*, compared with the other terms, we have

$$dV = -\frac{V}{p}dp. \text{ And } \therefore \frac{dV}{dt} = -\frac{V}{p}.\frac{dp}{dt} \quad ...(v)$$

Or, from relations (iv) and (v), we have $E = \frac{dV}{dt} = -\frac{V}{p}.\frac{dp}{dt}$.

This is called *Langmuir's equation for exhaust speed of a pump*.

Thus, *the speed of exhaust of a pump may be defined as the rate of change of volume of the gas or vapour in the vessel at any given instant, the measurement of volume being effected at the pressure attained by the pump at that very instant*.

Substituting the value of *dp/dt* from relation (i) above in expression, (iv) for *E*, we have,

$$E = -\frac{V}{p}.\left[-\frac{S}{V}.(p - p_0)\right] = \frac{S.(p - P_0)}{p}$$

* As will be readily seen, this equation enables us to determine the interval ($t_2 - t_1$) taken by the vacuum system to recover from the rush or '*surge*' of the gas, which raises the pressure to p_1, —the working pressure required by the system being p_2.

$$E = S(1 - p_0/p).$$

This relation tells us that, in the beginning, when the pump starts working, p is very much greater the p_0, so that p_0/p is practically *zero*, and, therefore, E is almost equal to S, *i.e.*, *in the beginning, the pumping speed of the pump is practically equal to its intrinsic speed.* But as p is progressively reduced and approaches p_0, E gradually decreases, and finally becomes *zero* when $p = p_0$. In other words, a *pump loses all its pumping speed at the lowest attainable pressure.* It is, therefore, important to design pumps not only with a view to producing high vacuum, but also with a view to having as high a pumping speed as possible, at all pressures.

And, since the pumping speed (E) is found to depend not only upon its intrinsic speed (S) and the lower limit of attainable pressure (p_0) but also upon the resistance to its flow, it follows that the wide bores of the connecting tubes, referred to under (*ii*) above, also help achieve this end.

N.B. It will be seen from the following how definitions (*i*) and (*ii*) of the speed of a pump also lead to ***Langmuir's and Gaede's equations.***

Let a mass δm of a gas be removed by the pump at pressure p and temperature $T°$ *abs* and let its volume be δV. Then, if r be the *gas constant per unit mass*, we have $p\,\delta V = \delta m\; r.T$.

Or, $$p\frac{dV}{dt} = \frac{dm}{dt}.r.T$$

Since the speed of the pump (according to the second definition) is given by $S = -dV/dt$, we have

$$S = -\frac{dV}{dt} = -\frac{dm}{dt}.\frac{r.T}{p}$$

Now, if m be the mass of the gas in the vessel at the same pressure and temperature (p and T respectively) and V, the *volume of the vessel*, we have

$$pV = m.r.T.$$

Or, $$\frac{dp}{dt} = \frac{dm}{dt}.r\frac{T}{V}, \quad \text{whence,} \quad \frac{dm}{dt} = \frac{V}{r.T}\frac{dp}{dt}$$

Substituting this value of dm/dt in the relation for S above, we have

$$S = -\frac{V}{r.T}.\frac{dp}{dt}.\frac{r.T}{p}. \quad \text{Or,} \quad S = -\frac{V}{p}.\frac{dp}{dt}, \quad \text{or,}\; S\,dt = -\frac{V}{p}dp.$$

If at time t_1, the pressure be p_0 and time t_2 it is reduced to p_2, we have

$$S\int_{t_1}^{t_2} dt = -V\int_{p_1}^{p_2}\frac{1}{p}dp. \quad \text{or,} \quad S(t_2 - t_1) = V\log_e\frac{p_1}{p_2}.$$

whence, $$S = \frac{V}{(t_2 - t_1)}\log_e\frac{p_1}{p_2},$$

which is *Gaede's equation for the speed of the pump.*

If, on the other hand, the attainable pressure limit be p_0, we shall have

$$S = -\frac{V}{(p - p_0)}.\frac{d(p - p_0)}{dt}. \quad \text{or,}\; S\,dt = -\frac{V}{(p - p_0)}d(p - p_0).$$

$$\therefore \quad S\int_{t_1}^{t_2} dt = -V\int_{p_1}^{p_2}\frac{1}{(p - p_0)}d(p - p_o). \quad \text{or,}\; S(t_2 - t_1) = V\log_e\frac{p_1 - p_0}{p_2 - p_o}$$

whence, $$S = \frac{V}{(t_2 - t_1)}\log_e\frac{p_1 - p_o}{p_2 - p_o}.$$

which is Langmuir's equation for the speed of the pump.

15.14. Different types of pumps. The following shows at a glance the classification of the different types of exhaust or vacuum pumps.

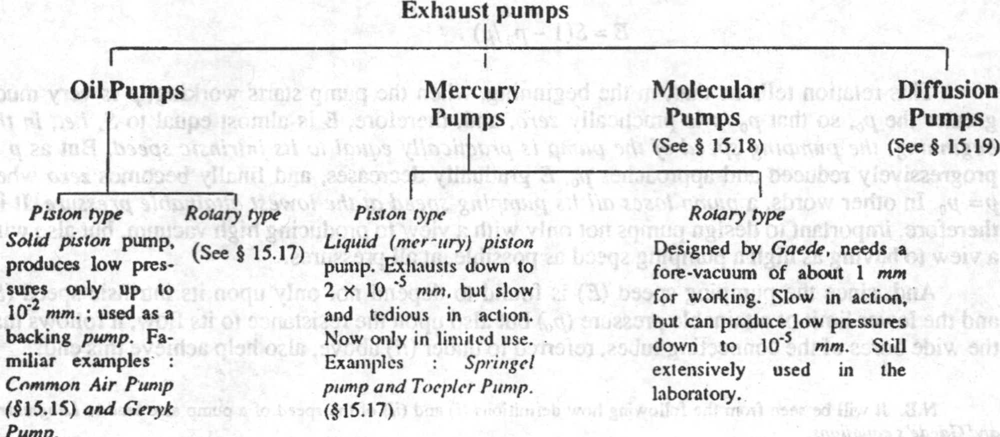

We shall now deal in detail with some of the simple and important types of pumps.

15.15. The common airpump. The student is no doubt already familiar with this pump from his junior classes and knows full well its limitations, *viz.*, that, for various reasons, it can hardly reduce the pressure in a vessel beyond 1 *cm* of mercury column.

For obtaining low pressures, therefore, we use other types of pumps, the Rotary oil pumps being the more suitable for the purpose. We shall now proceed to discuss these.

15.16. Rotary Oil Pumps. Originally devised by **Gaede**, these are of two types, *viz.*, *(i) the rotary vane oil pumps, and (ii) the stationary-vane oil pump*. The principle underlying both is, however, the same, – a massive cylindrical shaft or '*rotor*' revolving eccentrically inside a hollow stout steel cylinder, or, '*stator*', compressing the gas or vapour entering it, and finally ejecting it through a 'non-return' outlet valve, the whole pump being kept immersed in oil, which serves a three-fold purpose, (*i*) *providing automatic lubrication*, (*ii*) *preventing leakage of gas or vapour into the high vacuum created*, and (*iii*) *making for efficient cooling of the pump*. Let us study each type in a little more detail.

1. The Rotary-vane Oil Pump, or the Gaede Rotary Oil pump. The main parts of the pump are shown diagrammatically in Fig. 15.4, where *S* is the hollow, cylindrical steel chamber or '*stator*' and *R*, the stout and massive cylindrical shaft or '*rotor*', rotating *eccentrically* (by means of an electric motor) about its own axis, parallel to (but eccentric from) the axis of the chamber, such that it is always in contact with the chamber at a fixed peripheral point, such as *P*.

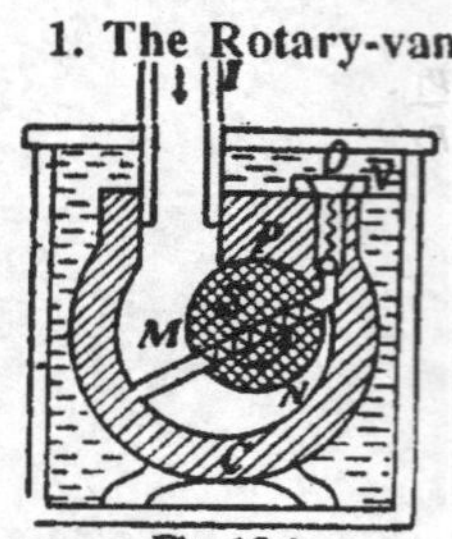

Fig. 15.4

A slot, cut *diametrically* right across the rotor, carries two *vanes*, (or discs), V_1 and V_2, slid into it, which are not only kept apart from each other, but also pressed against the walls of the stator by means of one or more springs in between them, thus dividing the space between the stator and the rotor into two separate compartments, *M* and *N*: Obviously, the slot, along with the vanes inside it, also rotated with the rotates — hence the name *rotary vane* oil pump.

On sides of the point *P*, where the rotor and the stator remain in contact, the stator is provided with an *inlet* and an *outlet port, I and O* respectively, the latter being fitted with spring-operated valve *V*. The whole pump is kept immersed in oil for the reasons explained above.

As the rotor rotates in the direction shown, the space between the rotor and the stator, on the inlet side, goes on increasing, while that between the rotor and the outlet side of the stator goes on decreasing, so that the gas or vapour from the vessel connected to *I* is *continually* drawn into the former, and that, in the latter, gets progressively compressed, until when its pressure becomes sufficiently high, it forces open valve V and escapes out of the outlet O. The process goes on repeating itself until a pressure as low as 10^{-3} *mm* is produced in the vessel connected to the pump. A special *self-sealing oil-valve* prevents the gas or vapour from being sucked back into the exhausted vessel, even when the pump stops working. *No fore-vacuum is required for the working of this pump*, and it can, therefore, be used directly from the atmospheric pressure.

2. The stationary vane or the Cenco-Hyvac Rotary Oil Pump. This is the more commonly used oil pump, in view of its better performance. It consists of a stout outer cylinder C,

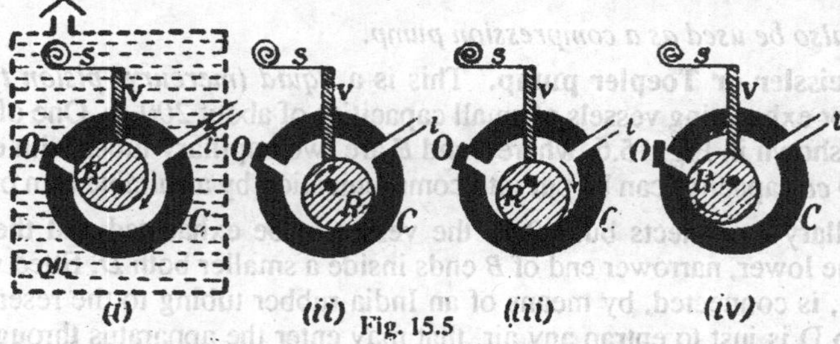

Fig. 15.5

inside which is mounted *eccentrically*, a cylinder R, called the *rotor*, (Fig. 15.5). Kept pressed against the rotor, with the help of a spring S, is a partition, called the *vane V*, which keeps the gas or air, already inside the cylinder, apart from the fresh incoming gas or air and remains fixed or stationary in its position,—hence the name *stationary vane oil pump*. The outer cylinder is provided with an inlet tube i, which is connected to the vessel to be exhausted, and an outlet tube O, which is provided with a valve, *opening outwards*. To prevent any leakage, the whole pump is immersed in oil, as shown in Fig. 15.5 (*i*), but a special type of valve prevents the oil from getting into the vessel being exhausted, when the pump is stopped. The rotor is driven at a very high speed by means of a separate electric motor in the direction shown by the arrow heads.

Fig. (*i*) shows the condition to start with, when the inlet tube is connected to the vessel to be exhausted, and when the gas or air from the vessel has just been admitted into the space in between the cylinder and the rotor R.

Fig. (*ii*) shows the condition when the rotor has started rotating eccentrically, and the gas or air is being compressed. Fresh gas or air comes into the cylinder through the inlet tube i, behind the rotor, and is kept apart from that already present by the vane V, as explained above.

Fig. (*iii*) shows the process of compression, taken a step further.

Fig. (*iv*) shows the final stage of compression, when due to increased pressure of the gas in C, the valve at the mouth of the outlet tube O is forced open, and the gas is expelled out.

The gas or air behind the rotor is similarly compressed and forced out and the cycle is repeated, until a high vacuum is produced in the vessel, connected to it.

In practice, the pumping system consists of two such units, in series with each other and mounted side by side, worked by the same motor. The first unit works directly from the atmosphere, as explained above, and the second then works from the fore-vacuum created by it. The maximum speed of working attainable is about 6 *litres per minute*, and the vacuum obtainable, about 10^{-3} *mm*, as mentioned already. It is thus particularly suited for exhausting vessel of large capacities.

It will be readily seen that if a vessel be connected to the outlet tube O, the gas or air will be compressed into it, and, therefore, this pump can also be used as a compression pump.

Chief advantages of the pump. This pump obviously possesses the following advantages:

(*i*) *It is run by an electric motor; so that the whole operation is quick and automatic, with no manual labour involved.*

(*ii*) *Being much faster than an ordinary exhaust pump, it is capable of creating a high degree of vacuum* (10^{-3}*mm*) *(iii)* Kept immersed in oil, *it has all the three advantages mentioned under § 15.16 above viz; (a) automatic lubrication, (b) prevention of leakage of gas or vapour into the high vacuum created and (c) efficient cooling.*

(iv) Can be used directly from the atmospheric pressure and no fore-vaccum is thus necessary.

(v) Occupies considerably much less space compared with the other plunger type exhaust pumps.

(vi) Can also be used as a compression pump.

15.17. Geissler or Toepler pump. This is a *liquid (mercury) piston type of pump* and is quite suitable for exhausting vessels of small capacities of about *200 cc*. One of the common forms of the pump is shown in Fig. 15.6, where *I* and *E* are two capillary tubes with either of which both *B* (of about 200 *cc* capacity) can be put into communication by a suitable turn of tap *T*.

The capillary *I* connects bulb *B* to the vessel to be exhausted and the capillary *E* to the atmosphere. The lower, narrower end of *B* ends inside a smaller both *D*, fitted with *h* one side, and thus, in its turn, is connected, by means of an India rubber tubing to the reservoir of mercury, *R*. The use of bulb D is just to entrap any air, that may enter the apparatus through the rubber tubing and then to remove it through tap *T*′.

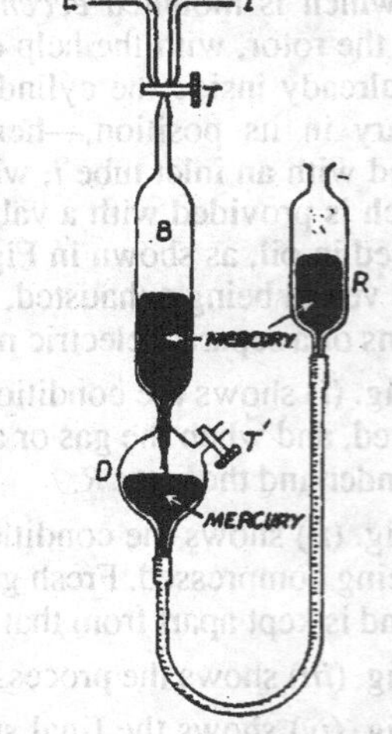

Fig. 15.6

To start with both B is connected to capillary *E* and the mercury reservoir (*R*) raised up, so that, as mercury rises up in *bulb B*, it forces out any gas or air there out into the atmosphere through *E*. When the whole of bulb *B* is full of mercury, tap *T* is turned, cutting off communication between *B* and *E*. The reservoir is now lowered and capillary I connected to *B*, so that the gas in the vessel connected to I rushes into *B*, forcing the mercury down in it. The reservoir is gradully lowered further until almost the whole of both B is filled with the gas. Tap T is then turned to cut off communication between *B* and *I*, reservoir *R* raised up to increase the pressure of the gas in both *B* to a valve higher than that of the atmosphere and B put into communication with *E* to enable the gas in it to escape into the atmosphere.

The process repeated a number of times until no more of the gas from the vessel can be exhausted, this stage being indicated by the mercury rising in *B* (as the reservoir is raised) and, in the near absence of any gas there, striking against tap T with a characteristic metallic sound.

Although comparatively much slower and rather tenders to work, the pump can create a vacuum as high as 10^{-4} *mm* of mercury an important feature of it being that, with simple modification, it can be used to transfer gas into a previously exhausted vessel and, later, to recover it, if so desired.

It will be readily seen that, as in the case of a solid-piston or the common air pump, so also, here, the pressure P_n in the vessel after *n* repetitions of the process, will be given by

$$P_n = P\left(\frac{V}{V+v}\right)^n$$

where *P* is the original pressure in the vessel, *v*, the volume of bulb *B* and *V*, that of the vessel.

15.18. Molecular Pumps. These too are the result of the labours of *Gaede*, together with **Langmuir**, and are based on the principle that if there be a rapidly rotating surface (called the *rotor*) very close or adjacent to a stationary one (called the *stator*), the space or *clearance* between the two being as small as .03 *mm*, it exerts, due to viscosity, a dragging force, in the direction of its own motion, on the molecules of the gas or vapour in that space. This is so because, at low pressures, the mean free path (λ) of the molecules of the gas or vapour is greater than the linear dimensions of this small annular gap between the rotor and the stator, so that there are too few collisions between molecules and molecules, compared with those between molecules and the walls of this gap, and **Knudsen** has shown that when such is the case, the molecules acquire the '*drift velocity*' of the rotating surface they impinge upon and rebound from, and that this velocity is not altered by any subsequent collisions amongst themselves.

Unlike the oil pumps which can start working straightaway from the atmospheric pressure, down to a pressure of 10^{-1} to 10^{-2} *mm*, *the molecular pumps (as also the diffusion pumps) operate only from a reduced pressure or a fore-vacuum.* They are, therefore, always used in series with a *backing pump*, connected to the exhaust port, which creates the necessary fore-vacuum or '*rough*' vacuum to receive the gas or vapour driven into it by the rotor, *there being a continuous unbroken communication* between the receiver and the rough vacuum, in contrast with the other types of pumps*, where there is only an intermittent or interrupted communication between the two, — a solid or a liquid piston first separating out a part of the gas or vapour in the receiver and then putting it into communication with the atmosphere or a rough vacuum, into which it si expelled.

Fig. 15.7. shows a diagrammatic representation of the pump, where *A* represents the *rotor*, revolving round** about an axis through its centre, inside and closely adjacent to the walls of the *hermetically* sealed shell or *stator C*. There are an *inlet* and an *outlet* in the stator at *P* and *Q* respectively, with a *slot* in between the two, where the annular gap is consequently wider than at all other points.

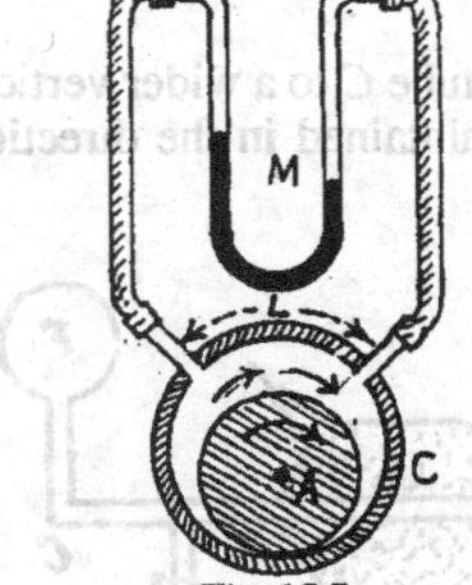

Fig. 15.7

As the rotor revolves in the direction shown, it drags with it the gas or vapour from the inlet port *P* to the outlet port *Q*, thus creating a pressure difference between the two, as indicated by the manometer *M*. This pressure difference cannot possibly be due to the viscosity of the gas alone ; for, at ordinary pressure, the viscosity of a gas is found to be quite independent of the pressure. Indeed, *Gaede* has shown that the pressure difference $(p_1 - p_2)$ between *P* and *Q* is given by the relation, $p_1 - p_2 = \eta.6lu/h^2$.
where η is the *viscosity* of the gas in question, *u*, the *speed* of the rotor and *l* and *h*, the *length* and *radial depth* of the slot between *P* and *Q*. Thus, at ordinary pressures, since η undergoes no change, the pressure difference $(p_1 - p_2)$ remains constant for a constant speed (*u*) of the rotor.

At low pressure, on the other hand, it has been shown by *Gaede* that it is the *pressure ratio* p_1/p_2 (and *not the pressure difference*) that remains constant for a constant speed of the rotor, and which is also quite independent of the fore-vacuum. This pressure ratio is given by the relation $p_1/p_2 = e^{cu}$, where *c* is a *constant*, depending upon both the nature of the gas or vapour and the dimensions of the slot.

In *Gaede's* own form of the pump, *there are a set of twelve slots or grooves along the circumference of the rotor*, their depths decreasing progressively from about .6 *cm* in the inner to about .15 *cm* in the outer section, the sections being all connected in series. Into these rotor-grooves fit projections from the stator, the clearance between the two being .03 *mm* and the gas or vapour is swept along this small clearance. The arrangement of the slots is such that the pressure has its lowest value at the centre and goes on gradually increasing as we proceed outwards to the ends, where we have the backing pump connected.

* Except the diffusion pumps.

** The speed of its rotation should never be less than 5000 revolutions per minute.

In the *Holweck* type of pump, the working is on similar lines, but *the slots are made in the stator*, with no corresponding projections on the rotor, and the clearance between the two can here be reduced to a figure even lower than .03 *mm*. The low pressure obtained by these pumps is conditioned by (*i*) the speed of the rotor and (*ii*) the fore-vacuum at which they are worked. Thus, for example, with a fore-vacuum of about 2 *mm* and with a rotor speed of 10,000 *revolutions per minute*, the pressure may be reduced by Gaede's pump down to the figure of 10^{-6} *mm* within a matter of minutes. *Holweck* even succeeded in evacuating nitrogen gas down to a pressure of 10^{-7} *mm* with the help of his pump, with a rough vacuum of only 15 *mm* and a rotor-speed of 4500 *rev. per. mnt.* Such low pressures are, however, well-nigh impossible to attain when there are vapours present.

The one serious drawback of these pumps is the recurrence of mechanical trouble, due to the small clearance between the rotor and the stator. And then, while they can easily deal with gases and vapours, slowly vaporising substances, like mercury, and traces of grease etc., give a lot of trouble.

15.19. Diffusion-Condensation Pump. The inter-diffusion of one gas into another has been used to create rapid vacuum in vessels. The method was first used by *Gaede* in the year 1915, and *depends upon the principle that, in a mixture of gases, the diffusion of a gas takes place from a region, where its concentration is greater to the one where its concentration is smaller, or, in other words, from a region where its partial pressure is higher, to the one, where it is lower, irrespective of the total pressure in the two regions.* The action of the pump will be clear from the diagrammatic representation of it in Fig. 15.8.

The bulb *F* to be exhausted of air or gas is connected by means of a tube *C* to a wider vertical tube *AB*, through which a regular stream of air-free liquid vapour is maintained in the direction shown, the tube *C* being kept at a low temperature, by circulating cold water round it.

Now, clearly, the concentration or the pressure of air is greater at *C* than at the other end, because the liquid vapour being air-free, there is little or no air here and, therefore, diffusion of air takes place from *C* into *AB*, where it is swept out by the stream of the liquid vapour; so that, again, the concentration and partial pressure of air are greater at *C* than at the end opposite and hence more air diffuses from *C* into *AB*. This goes on so long as the pressure of air at *C* is greater than its partial pressure in *AB*.

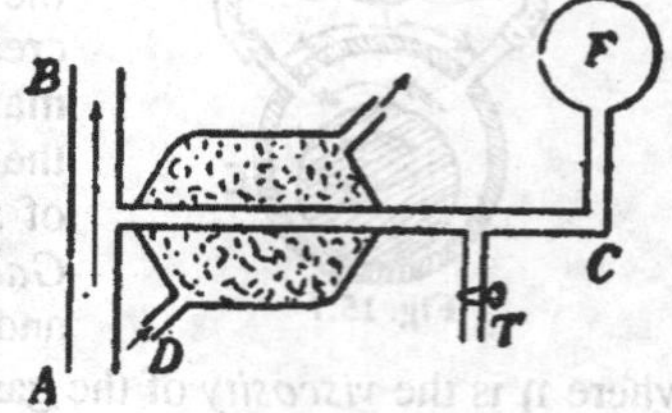

Fig. 15.8

The liquid vapour also tends to diffuse from *AB*, where its concentration and, therefore, its pressure is greater towards *C*, where its concentration and pressure are lower, thus driving the air diffusing from *C* into *AB backwards*, but this is prevented by (*i*) making the aperture of *C*, opening into *AB*, narrow, *its dimensions being smaller than the mean free path of the molecules of the gas or air in AB*, so that there are much fewer vapour-gas or vapour-air molecular collisions and the diffusion velocity of mercury vapour towards *C* is smaller than the diffusion velocity of air from *C* into *B*, and (*ii*) condensing the mercury vapour by cooling the tube *C*, thus not only preventing it from proceeding further and entering the vessel *F*, but also reducing the residual vapour pressure there is less than 10^{-3} *mm*, thereby ensuring an uninterrupted diffusion of the gas from *C* into *AB*. The condensed mercury is then conveyed back to the boiler, (not shown), to be used over again.

This pump can, however, be used successfully only when the pressure in the receiver, or vessel *F*, is about or less than 1 *cm* of mercury. The pressure in the vessel is, therefore, first reduced to this value, or this *fore-vacuum* created, by means of rotary vacuum pump, which can be connected to the vessel through the inside tube *T*, *the maximum effect being obtained when the pressure of the mercury vapour is just above the fore-vacuum, thus produced.*

Such pumps may be used *in series*, each pump, in turn, carrying the evacuation a step further than the last, and thus a pressure as low as a *millionth of a centimetre* of mercury column can be rapidly attained, although, theoretically speaking, there can be no limit to the vacuum produced by them. For, under the ideal conditions, mentioned above, there would be a continuous diffusion of the air or gas from *C* into *AB*, until the air or gas pressure there is reduced to *zero*.

Since diffusion as well as condensation both play their part in the working of these pumps, they are referred to variously as *diffusion-condensation pumps, diffusion pumps* or even *condensation pumps*.

The vacuum obtained with the help of these pumps of is a very high order,— 10^{-6} *mm* or less, with a fore-vacuum of 10^{-1} *mm* and they are, therefore, being increasingly used for the evacuation of *X*-ray tubes and wireless valves and for such other industrial purposes.

There are two main *drawbacks* of these pumps, *viz.* ;

(i) they have a comparatively slow exhaust speed, and

(ii) they need a rather vigilant regulation of the vapour-temperature.

Langmuir, in the year 1916, removed both these drawbacks and improved greatly the working of the pump by (*a*) *replacing the narrow slit of Gaede's form of the pump by a wider one, thereby greatly increasing the speed of the pump* ; (*b*) *so directing the stream of vapour as to more efficiently sweep out the air molecules* and (*c*) making the speed of the pump almost independent of the liquid vapour pressure by increasing the speed of the vapour stream, thereby obviating the necessity of a vigilant and tiresome regulation of the vapour-temperature.

A simple and yet an efficient form of diffusion pump was, however, designed by **Waran** in the year 1923, and is shown in Fig. 15.9. It consists of a conical glass vessel *A*, containing mercury, which is heated, so that the mercury vapour passes up to the tube *B*, which ends in a nozzle or a '*parallel jet*', about 8 *mm* in diameter, into a wider tube *C*, surrounded with a cold water jacket. An inlet tube *i* connects the tube *C* to the vessel to be exhausted, through a *liquid-air-trap*, (not shown). The tube *B* is lagged with asbestos to prevent any condensation of mercury vapour in it.

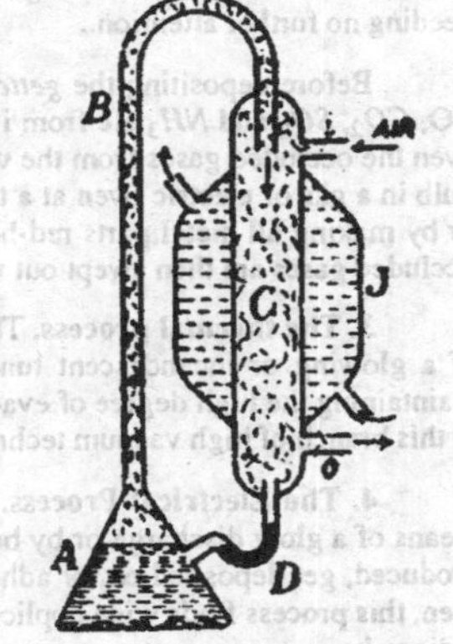

Fig. 15.9

As the mercury vapour issues out of the nozzle, it carries along with it the air, diffusing into *C* through *i* from the vessel to be exhausted, —the diffusion of the mercury vapour into the vessel being prevented by its almost immediate condensation on entering *C*. The air is then discharged through the outlet tube and the mercury vapour collected in the bend *D* below *C* and returned back to *A*. The process goes on, until a high vacuum, of the order of about 10^{-5} *mm* of mercury is created in the vessel, connected to *i*. A still higher vacuum, of the order of about 10^{-7} *mm* of the mercury can be produced with the help of *the process of absorption*, (see § 15.20), *i.e.*, by placing in the vessel some loose coconut charcoal, recently heated under reduced pressure which, on being cooled externally with liquid air, absorbs the gas and the vapour.

Oil diffusion pumps. The trouble with mercury as the working substance in a diffusion pump is that if its vapour once gets into the vacuum system, it can be a real source of trouble.

Burch, therefore, suggested in the year 1928, the use of some *high-boiling-point oils*, obtained as distillates of petroleum, in place of mercury. On account of their much lower vapour pressures, no liquid-air trap need be used (unless the operation happens to be a prolonged one) and the maximum speed of the pump can thus be easily utilised. In fact, size for size, an oil diffusion pump has invariably a much higher speed than a mercury one. The one serious drawback of this type of pump, however, is that since, unlike mercury, an organic oil, like the one used here, dissolves gases and vapours, it has to be subjected to periodical cleansing.

Some two years later, in 1930, **Hickman** synthetically prepared pure organic compounds like *n-dibutyle phthalate*, (*mol. wt*; 278) and *n-dibutyl sebacate* (*mol. wt.* 314) etc., the former having a vapour pressure *one-hundredth* of that of mercury at the room temperature and far less at lower temperatures.

With suitable modification in their design from that of a mercury diffusion pump, these oil diffusion pumps can, with a *fore-vacuum* of 0.1 *mm*, produce as high a vacuum as 10^{-6} *mm*. Other liquids that have been used successfully in place of mercury are what are called *'Apiezon oils'*, mostly in diffusion pumps with an all-metal construction.

15.20. Other Methods of producing Vacua. As we have seen above, the lowest pressure attainable with the help of a pumping device is 10^{-6} *mm*. To produce still lower pressures, other devices have to be used. The following is a very brief description of these :

1. The Sorption Process. This consists in connecting the system to be evacuated to a tube containing some freshly heated coconut charcoal (preferably under reduced pressure) and surrounded by liquid air, when gases, like *carbon dioxide*, *nitrogen* and *ammonia*, are absorbed by the charcoal. To remove *hydrogen*, palladium black may be used in place of coconut charcoal. The pressure, thus obtained, is as low as 10^{-7} *mm* which may be even further reduced if the method be used in conjunction with a backing pump.

2. The Chemical Process. Known as the *chemical process of 'flashing'*, and first suggested by **Soddy**, in the year 1907, it consists in suddenly vaporizing an electropositive metal, like *calcium*, (first used by Soddy), *barium*, *maganesium* of their alloys in a vessel in communication with the system desired to be evacuated, when most of the vaporized metal condenses back on the walls of the vessel, the rest forming with the gases present, a compound of negligibly small vapour pressure, thus, reducing the pressure in the system to a very large extent. Usually, the vessel used is a glass bulb, properly fitted with a tungsten filament, on which the chosen metal, called the getter, is placed. A momentary large current is then passed through the filament, when the metal burns out with a flash and gets vaporized, the condensing part of it forming a bright, mirror-like deposit on the inner side of the bulb. This is the method largely employed in evacuating radio valves, X-ray tubes, electric lamps etc. and maintaining the vacuum produced.

This deposit of the metal layer (or the *getter*) is itself *gas-hungry*, so to speak, and can absorb large amounts of gas, It thus serves as a *perpetual high speed pump* incorporated inside the sealed bulb, maintaining the high vacuum there and needing no further attention..

Before depositing the *getter* on the inside of the bulb, however, it is necessary not only to remove *water vapour*, CO, CO_2, SO_2 and NH_3 etc from it with the help of calcium (as *Soddy* did) but also to **'degas'** it thorughly, *i.e.*, to remove even the occluded gases from the walls of the bulb and the metallic parts inside it. This is done either by suitably baking the bulb in a gas or electric oven at a temperature some 50^0C. below the lowest softening point of any of the materials present, or by making all metal parts red-hot be means of eddy currents at radio frequencies in the coils surrounding the bulb. The occluded gases are then swept out with the help of a diffusion pump and the bulb subjected to *gettering* and sealed.

3. The thermal process. This consists in the removal of gases like *nitrogen*, which disappear slowly in the presence of a glowing or incandescent tungsten filament. As can be easily understood, this process plays an important part in maintaining the high degree of evacuation in the ordinary incandescent electric lamps. *Langmuir* is indisputably the pioneer in this branch of high vacuum technique.

4. The Electrical Process. This consists in ionizing the atoms or molecules of the gas to be removed, either by means of a glow discharge or by bombarding them by means of electrons, obtained by thermoinic emission. The ions, thus produced, get deposited on, or adhere to, the walls of the containing vessel, if it be kept suitably cooled. As can be readily seen, this process finds wide application in the production and maintenance of high order vacuum in incandescent lamps and radio valves.

In actual practice, in most cases, the chemical, the thermal, the electrical processes, although each a completely independent process in itself, operate conjointly and simultaneously, as, for example, in the case of the evacuation of radio valves, where the first one produces the initial vacuum and the latter two augment and maintain it.

And, finally, however effective the method or the means employed, it is imperative for a satisfactory maintenance of a high degree vacuum that the whole evacuated system should be perfectly leak-proof. Hence it is that it should preferably consist wholly of glass or wholly of metal, but, if both needs must be used, they must be directly and carefully sealed together air-tight.

15.21. Measurement of Low Pressures-Manometers and Gauges. As mentioned earlier, the production of low pressures, of necessity, led to the development of the proper means to measure them. Such measuring instruments or devices fall under the heading *'manometers and gauges'*. Since we are concerned with the detailed working of only the more important or useful ones, here, we shall content ourselves with only a brief and rapid survey of the rest, not so much with a view to studying their working in detail as to acquainting ourselves with the different principles on which such devices may possibly be based. Here, then, are the different measuring devices at a glance.

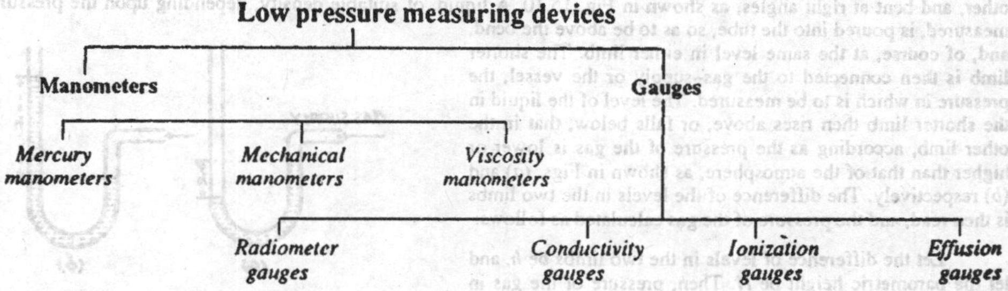

1. Mercury Manometers. These are of two important types, *viz.*,

(*i*) *Differential and Optical Lever Manometers*, which are improved modifications of the ordinary mercury manometers, with more sensitive methods of observation and measurement, —suitable for measurement of pressures upto 10^{-3} *mm.*

(*ii*) *McLeod Gauge*, which is a standard device of its type, based on the validity of *Boyle's Law* at low pressures, suitable for accurate measurement of pressures, down to 10^{-5} *mm*. (See pages 509-511).

2. Mechanical Manometers. These are based on the principle of *mechanical deformation* produced in a thin wall or diaphragm, due to pressure. They are calibrated against the McLeod Gauge and their range too does not go below 10^{-3} *mm*. The two well known manometers of this class are:

(i) the Bourdon Spiral Gauge, (page 508), and

(*ii*) the Aneroid Barometer type.

3. Viscosity Manometers. The principle underlying these is that, at low pressures, the viscous drag between two surfaces in relative motion is proportional to P/M, where P is the *pressure of the gas*, and M, *its molecular weight*.

They are of two types, *viz.*,

(*i*) *The Damping or Decrement type*, a good example of which is *Coolidge's Quartz fibre gauge*, which is suitable for the measurement of pressures, ranging between 10^{-2} *mm* and 10^{-5} *mm*.

(*ii*) *The Molecular type*, a well known example of which is *Langmuir and Dushman's Molecular Gauge*, suitable for the pressure range 10^{-3} *mm* to 10^{-7} *mm*.

4. Radiometer gauges. These are based on the measurement of the rate of transference of momentum from a hot to a cold surface due to molecular bombardment. Among gauges of this type may be mentioned

(*i*) *Crooke's Radiometer*, which is suitable only for qualitative work. (See Foot note-page 517).

(*ii*) *Knudsen's absolute gauge*, which is a standard gauge of its type and a close rival of McLoed gauge in its sensitiveness and accuracy, having a wide range, from 10^{-2} *mm* to 10^{-7} *mm*. (See page 517).

5. Conductivity gauges. The underlying principle of these is the effect of pressure on the rate of transfer of heat by the process of conduction, their range being comparatively small, from 10^{-1} *mm* to 10^{-4} *mm*. Among gauges of this type may be mentioned

(*i*) *The Thermocouple gauges* (page 513),

(*ii*) *Pirani-Hall Resistance Gauge* (page 511), and

(*iii*) *Gauges, based on linear expansion of metallic wires or strips.*

6. Ionization gauges. These depend for their action on the variation of electrical conductivity of a gas with pressure. Mention may be made here of two gauges of this type, *viz.*, (*i*) *Buckley's type*, and (*ii*) *Dushman and Found type*, this latter one being suitable for measurement of pressures from 10^{-2} *mm* down to the lowest attainable pressure. (*iii*) *x-ray ionization gauge* (page 516).

Effusion gauges. Used only for the measurement of vapour pressures of metals.

Let us now proceed on with a detailed consideration of some of these.

15.22. Common Mercury Manometers. These are of two types, *viz.*, (*i*) *Open* and (*ii*) *Closed.*

1. The Open Manometer. It consists of a U-tube, with both limbs open, one limb being a little shorter than the other, and bent at right angles, as shown in Fig. 15.10. A liquid, of suitable density, depending upon the pressure* to be measured, is poured into the tube, so as to be above the bend, and, of course, at the same level in either limb. The shorter limb is then connected to the gas–supply or the vessel, the pressure in which is to be measured. The level of the liquid in the shorter limb then rises above, or falls below, that in the other limb, according as the pressure of the gas is lower or higher than that of the atmosphere, as shown in Figs. (*a*) and (*b*) respectively. The difference of the levels in the two limbs is then read, and the pressure of the gas calculated as follows:

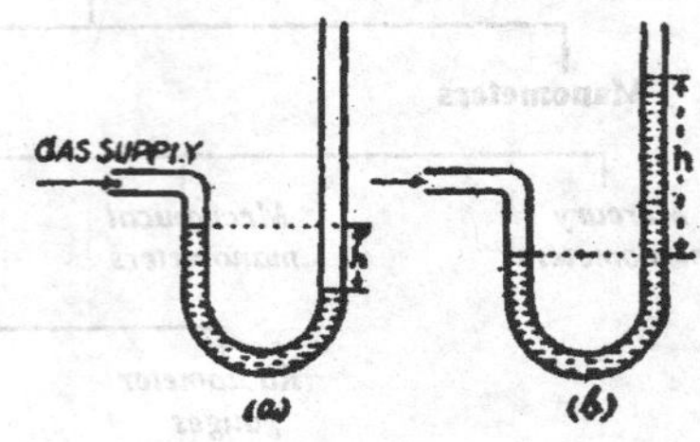

Fig. 15.10

Let the difference of levels in the two limbs be h, and let the barometric height be H. Then, pressure of the gas in case (*a*) is $(H\text{-}h)$, and in case (*b*), it is $(H+h)$ *cm* of mercury column, if the liquid used is to be mercury. In case, however, the liquid be oil or water, of density ρ, the pressure in the two cases will be $(H - h.\rho/13.6)$ and $(H+h.\rho/13.6)$ respectively, (where 13.6 *gm / cc* is the density of mercury).

2. The Closed Manometer. It is used for the measurement of high pressures. It is just like the open manometer in construction, but with the longer limb closed at the top, as shown, (Fig. 15.11), and containing some air, at *atmospheric pressure*, in the closed space above the liquid, with the level of the liquid columns in the two limbs *the same, to start with.*

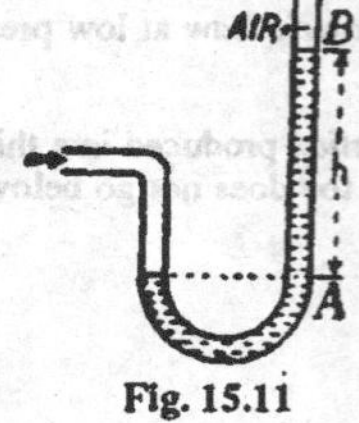

Fig. 15.11

When the shorter limb is connected to the gas supply, the level of the liquid column in the shorter limb is pushed down, and that in the other pushed up, so that the air in it gets compressed. The pressure of this enclosed air being inversely proportional to its volume, it can be determined by noting its new volume. And, from this pressure and the difference in the levels of the liquid columns in the two limbs, the pressure of the gas-supply, in communication with the shorter limb, can be easily calculated as follows:

Let original volume of the enclosed air be V *cc*, its pressure being *one atmosphere* or 76 *cm* of the mercury column. Then, if v be its volume after the shorter limb is connected to the gas-supply, we have, by Boyle's law,

$$76 \times V = H \times v.$$

[where H is now the pressure of the enclosed air, in terms of the length of mercury column.]

whence, $$H = 76 \times V/v.$$

Thus, knowing the original volume V and the new volume v of the enclosed air, H, the pressure of the enclosed air can be known. This, then, is the pressure at B in the longer closed limb. If therefore, the difference of the levels in the two limbs be h, and the liquid used in the manometer be mercury, the pressure at A, *i.e.*, the pressure of the gas-supply $= (H + h)$ *cm* of mercury column. In case the liquid used be oil or water, of density ρ, we have

$$\textit{pressure of the gas supply} = \left(H + \frac{h.\rho}{13.6}\right) \textit{cm of mercury.}$$

15.23. The Bourdon Gauge. For the measurement of both high and low pressures, what is called a *Bourdon Gauge* is used. The principle underlying it is the same as that in the case of the *Aneroid barometer*, which is in fact a modification of the gauge.

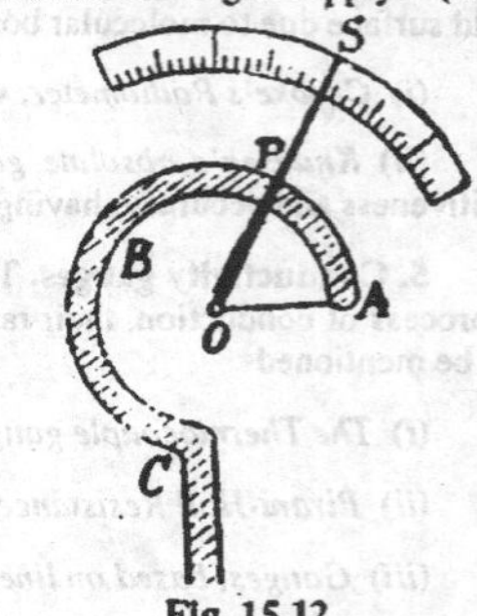

Fig. 15.12

It consists of a *glass or* metal tube *ABC*, (Fig. 15.12), *elliptical in section*, with the end A closed and the end C open, so that it can be put into communication with the gas-supply, the pressure of which is desired to be determined.

Due to the high pressure of the gas entering the tube, it tends to become more circular in section and this results in the end A of the tube being forced away from C. This movement of A, in its turn, moves by means of a suitable mechanical arrangement, the pointer P over a scale, graduated directly *in atmospheres*, by comparing its indications with a standard gauge, as in the Aneroid barometer. The instrument is thus a direct reading one.

---

* If the pressure to be measured differs from that of the atomosphere by only a small amount, the difference of level in the two limbs will be very small, if a heavy liquid, like mercury, be used. But, with a lighter liquid, like oil, the difference is quite considerable. With mercury as the liquid used, pressure ranging between $\frac{1}{2}$ to $1\frac{1}{2}$ atmospheres only can be measured correctly.

It is obvious that the deformation of the spiral tube must not be allowed to go beyond the elastic limit or else, due to a '*permanent set*' in it, its readings will no longer be reliable. Even so, it can be used to measure pressures from 4000 *to* 5000 *kgm wt /sq.cm.*

For measuring *low pressures, Ladenburg* has used the gauge in the form of a spiral of thin glass tubing, the movement of end *A* of the spiral being noted by an optical mirror arrangement, for which purpose a mirror is attached to it. Actually, by applying an equal external pressure, the mirror is brought back to the zero-position, so that the method becomes a *null method.* The external pressure is read off on a *Mecleod gauge* some distance away. Thus, even the pressures of gases, which may react with mercury, can be easily measured.

15.24 McLoed Vacuum Gauge. An ordinary manometer is not quite suitable for the measurement of very low pressures, like those obtaining in incandescent electric lamps and *X*-ray tubes etc., where the pressue is as low as 10^{-5} *mm.* For such purpose, the *Mcleod Vacuum Gauge,* designed by McLeoad in the year 1874, is used. The form of the instrument, shown in Fig 15.13 is a slight modification of the original, in main due to *Gaede*, and consists of a cylindrical or spherical bulb *B*, *of known volume*, ending above in a *graduated capillary tube CA*, and connected to a reservoir of mercury *R* and a side tube *EF*, which can be put into communication with the vessel or the pump in which the pressure is desired to be determined. A side capillary tube *G* is attached to it, as shown, whose diameter is the *same* as that of *CA*. Its use is *to counteract the depression of the mercury column in CA due to capillarity;* for, being of the *same diameter as CA*, the depression of the mercury column in it is also the *same* as that in *CA*.

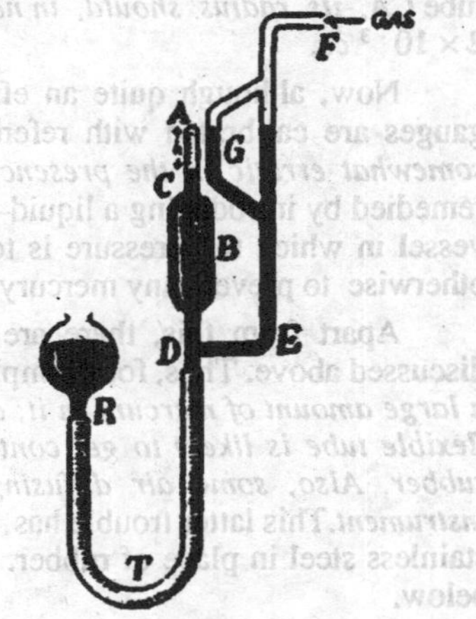

Fig. 15.13

When the reservoir *R* is lowered until the mercury falls below the **bend** *D*, the bulb *B* and the vessel in which the pressure is to be measured are put into communication with each other and the bulb is filled with the gas, whose pressure *P* is to be determined. On raising the reservoir, mercury rises into the bulb as well as the side tube, thus cutting off *EF* from *B*, and the gas enclosed in the bulb is compressed as mercury rises further and further up into it, until the whole of it is forced into the capillary tube *CA*. The reservoir is raised still further, until the whole of the bulb *B* and a part of the capillary tube *CA* are filled with mercury, and the mercury in the capillary tube *G*, attached to *EF*, rises up in a level with the top-end *A* of *CA*. The depression of the column *CA* having been compensated for, as explained above the difference of level between the columns of mercury in the two capillary tubes *CA* and *G*, or what is the same thing, between the end *A* of *CA* and the top of mercury column in it (because *A* is at the same height as the mercury column in *G)* gives straightaway the pressure of the gas in *CA*. Let it be *h cm.*

Then, if *V* be the volume of the capillary tube *CA* and the bulb *B* up to the bend *D* and *v*, the volume of the gas after the mercury has risen into it, we have $p \times V = h \times v$, in accordance with *Boyle's law*

Or $$p = h.v / V,$$

whence *p*, the pressure of the gas, can be easily calculated.

It will be readily seen that the greater the value of *V* and the smaller that of *v*, the smaller the value of *p* that can be measured. Thus, *the sensitiveness of the gauge depends upon the ratio V/v.*

If *a* be the area of cross section of the capillary tube *CA*, we have $v = a.h$ and therefore, $p = h^2 . a / V$ or $\propto p \propto h^2$ since a / V is a constant for a given instrument. This obviously gives a scale of pressure which is a *non-linear* one, –infact, a parabolic one. The great redeeming feature, however, is that it spreads out more and more as the pressure decreases, giving us *a higher accuracy at lower pressures.* For measurement of low pressures (in the range 10^{-3} to 10^{-5} *mm,* say) the gauge must necessarily be used in the above manner and to make it direct reading, the non-linear scale may be arranged along side the capillary tube *CA*.

For comparatively higher pressures, up to 10^{-3} *mm*, the level of mercury in capillary *CA* may be raised up to only a fixed point marked on it. If h_o be now the difference of mercury levels in the two capillaries (*AC* and *G*), and *l*, the length of the air column in *CA*, we have $p = \left(\frac{l.a}{V}\right).h_0$. Or, $p \propto h_0$. , thus making the pressure scale a *linear one*. Such a scale may, therefore be placed alongside the capillary *CA* to make the instrument a direct reading one.

It may as well be pointed out here that, although the actual range of *McLeod gauge* will depend upon its dimensions, *viz., volume of the bulb (B)* and the radius and length of the capillary tube *CA*, *its radius should, in no case, be less than* 1 *mm. or the volume of the gas in it less than* 2×10^{-3} *cc.*

Now, although quite an efficient gauge, – so much so that practically all other types of gauges are calibrated with reference to it, – *the performance of the McLeod gauge becomes somewhat erratic in the presence of easily condensible vapours.* This can, however, be easily remedied by introducing a liquid–air trap in between the gauge and the high vacuum side (*i.e.*, the vessel in which the pressure is to be determined). In fact, the liquid-air trap must be used even otherwise to prevent any mercury vapours entering the evacuated vessel.

Apart from this, there are quite a few other drawbacks in the form of the instrument, discussed above. Thus, for example, *it is rather inconvenient to manipulate the reservoir with such a large amount of mercury in it; and the mercury which remains in contact with the rubber of the flexible tube is likely to get contaminated due to the presence of sulphur in the composition of rubber. Also, some air diffusing through the walls of the tube may find its way into the instrument.* This latter trouble has, in recent years, been sought to be got over by the use of a tube of stainless steel in place of rubber. Even so, a better modification of the gauge is the one described below.

Improved modification of McLeod Gauge. In this improved version of the gauge, the reservoir, with its attached flexible rubber tube is dispensed with altogether. Instead, tube *D* is made longer and fitted into a rubber bung in one mouth of a *Woulff's bottle W*, so as to dip inside mercury contained therein, as shown in Fig. 15.14 Into the other mouth of the bottle is fitted a side-tube *N*, connected, through a stop-cock *S*, to (*i*) a small *soda-lime tower T*, and (*ii*) *a* tube *L*, leading to *some simple form* of a backing pump. The tower *T* has a packing of glass wool at either end to prevent any particles of soda-lime getting into the gauge, and is connected at the top to a long capillary tube *J*, through a small rubber tubing, provided with a spring-clip, so as to enable it to be put into communication with, or cut off from, the outside air, as desired.

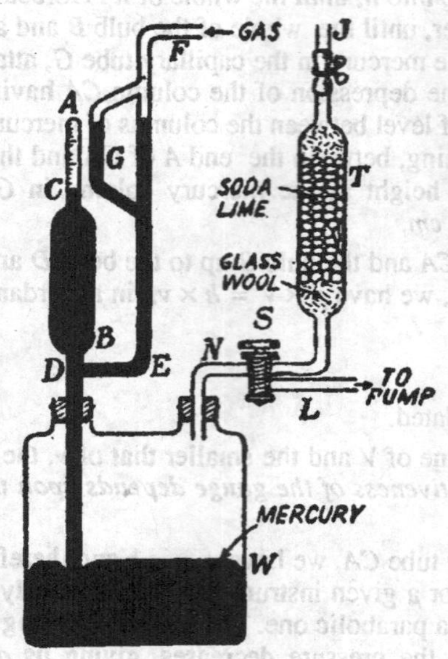

Fig. 15.14

The proceduse consists in first putting the Woulff's bottle in communication with the pump, through the stop-cock, so that the whole of the mercury in the gauge comes down into the bottle, the pressure throughout being the same as produced by the backing pump, and, of course, very much lower than that of the atmosphere. The communication between the bottle and the pump is now cut off and that between the former and the soda-lime tower partially established by a slight rotation of S, so that the air from outside gradually enters the bottle, losing its moisture during its passage through soda-lime in the tower. This results in an increase of pressure on the surface of mercury in the bottle and it being forced up into D. The gauge is then used in the manner already explained. It will easily be seen that the labour involved in moving the reservoir up and down for necessary adjustment of the mercury columns in C and G, and the possibility of contamination of mercury by the rubber tubing and the possibility of air diffusing through its walls are all thus neatly obviated.

When all is said and done, however, the McLeod gauge still continues to have its inherent defects of (*a*) *being rather unwieldy in size* and (*b*) *its inability to give a continuous record of pressure changes in the vessel.* Then, again,(*c*) *the use of the liquid-air trap adversely affects the rate of pumping and the readings obtained on the gauge may not be truly indicative of the actual pressure inside the vessel at any given instant.* In any case, *its readings in the last lap of its range, from* 10^{-4} *to* 10^{-5} *mm are really never quite so reliable.*

15.25. The Pirani Resistance Gauge. It is fairly well known that whereas at high pressures, the thermal conductivity (K) of a gas is quite independent of pressure; at pressures below 10^{-2} *mm* of mercury column, when the mean free path of gas molecules is of the same order of magnitude as the diameter of the containing vessel, it is directly proportional to the pressure (p) *i.e.*, K *is a linear function* of p. Or, $K = \alpha.p$, where α is a constant.

This fact, first made use of by **Warburg**, in the year 1907, for the measurement of low pressures, is really the basis of the *Pirani Gauge, P. G.*, in Fig. 15.15 (*a*) which consists of a tungsten or platinum filament (F), enclosed in a small detachable glass bulb (B),* very much similar in construction to that of the '*cage-type*' incandescent lamp, and maintained at a temperature, higher than that of the surroundings. The bulb is open at the lower end which can be connected to the vessel in which the pressure is to be determined.

With change in the pressure of the gas in between the filament and the walls of the bulb, the rate of heat conduction across the gas also changes, resulting in a change in the temperature of the filament and hence in its resistance. We measure this change in the resistance of the filament which gives the change in the thermal conductivity and hence, indirectly, the pressure of the gas. A *calibration curve* for the gauge, is, therefore, plotted by measuring simultaneously the resistance of the filament and the pressure of the gas around it,– the former by means of a Wheastone's bridge, to which a constant potential difference is applied, sufficient to heat the filament to a temperature of about 120^0C (but never to red heat), and the latter, by means of a McLeod gauge. The pressure, corresponding to any value of the resistance of the filament, can then be read directly on this curve.

Now, for the reason already stated, this calibration curve is a straight line, so long as the pressure of the gas is below 10^{-2} *mm* but at higher pressures, the relation between K and p no longer remains linear and K varies in a somewhat complicated manner with both the pressure of the gas and the temperature of the surroundings. In fact, the relation between K and p can be represented by no simple formula, with the result that the calibration curve now ceases to be straight. To tide over this difficulty, **Campbell** suggested that instead of keeping the voltage across the bridge constant and measuring the resistance of the filament, the temperature, and hence the resistance of the filament, must be kept constant and the potential difference, required to be applied to the bridge for the purpose, must be measured.

Accordingly, the gauge (*P.G.*) is connected in one arm of the bridge, [Fig.15.15 (*a*)], together with fixed resistances R_1 and R_2 and the variable resistance R_3 in the other three arms, as indicated, –all these resistances being made of an alloy like '*manganin*' and '*minalpha*', having an almost zero coefficient of temperature. The potential difference is applied to the bridge at A and C, by means of a *potential-divider, i.e*, through a rheostat included in the battery circuit, so as to be varied at will and its value read on the voltmeter V connected across the bridge terminals A and C,

* There are also other forms of the gauge in which the bulb is not detachable.

as shown. To ensure constancy of the temperature of the surroundings of the filament, the bulb of the gauge is placed in a thermostat at 0°C.

Procedure. (*i*). With a known potential difference across *A* and *C*, the bridge is balanced in the usual manner by adjusting the variable resistance R_3, so that the deflection in the galvanometer is *zero* and there is sufficient current through the filament of the gauge to raise its temperature to about 100°C. Now, every time the pressure changes, the voltage across *A* and *C* is adjusted to restore the balance of the bridge and make the galvanometer deflection *zero*. Then, if θ be the excess temperature of the filament over that of the surroundings, the

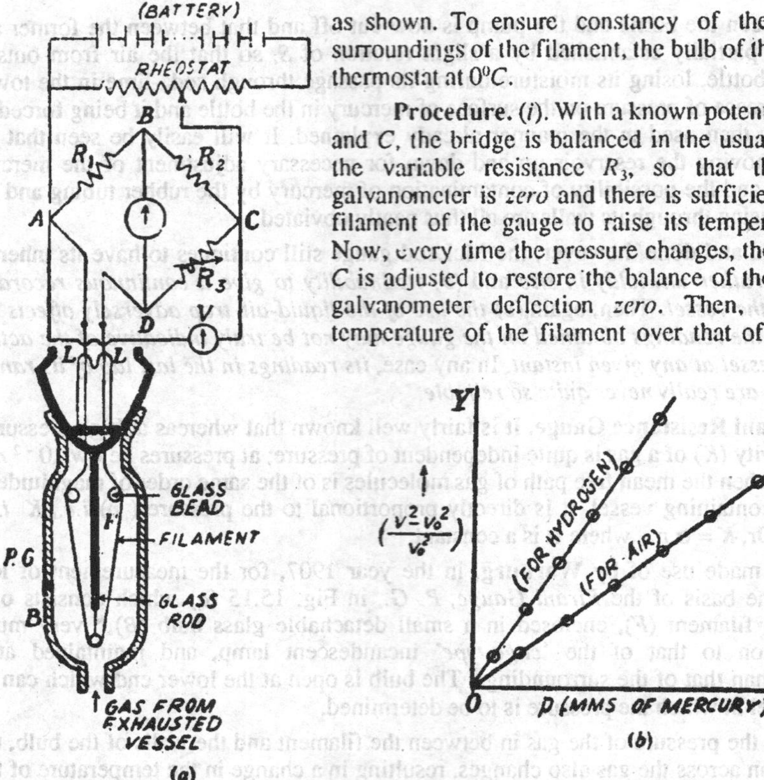

Fig. 15.15

heat-loss along the leads *L* and *L*, *assumed small*, will be proportional to θ, say equal to βθ, where β is a constant. And, if *V* be the voltage applied across the bridge, the heat dissipated per second in the filament is equal to αV^2, where α is another constant. And, further, if *p* be the pressure of the gas around the filament, the heat lost per second by conduction across it is $f(p)$, where $f(p)$ is some function of pressure. We, therefore, have

$$\alpha V^2 = \beta.\theta + f(p) \qquad ...(i)$$

Now, for the same value of θ, but with $p = 0$,(*i.e.*, with vacuum around the filament), if the voltage required to be applied to the bridge be V_0, clearly, $\alpha V_0^2 = \beta\,\theta$. ...(*ii*)

Dividing relation (*i*) by (*ii*), therefore, we have

$$\frac{V^2}{V_0^2} = 1 + \frac{f(p)}{\beta.\theta}. \quad \text{Or,} \quad \frac{V^2 - V_0^2}{\beta\theta} = kf(p),$$

where $1/\beta\theta = k$, a constant, which is almost quite independent of the material and the length of the filament and varies only with the nature of the gas.

But, as we have seen, measurement of the pressure of the gas by means of a McLeod (or any other absolute) gauge shows that $f(p) = \gamma p$, where γ is also constant. So that

$$(V^2 - V_0^2)/V_0^2 = k\gamma p,$$

i.e., $(V^2 - V_0^2)/V_0^2$ is directly proportional to pressure.

If, therefore, we plot a graph between *p* and $(V^2 - V_0^2)/V_0^2$ for the gas, we get a straight line, its inclination with the axis of *p* depending upon the nature of the gas in question, as is clear from the two curves drawn by *Campbell* for hydrogen and air, [Fig. 15.15, (*b*)], where *p* was measured by a McLeod gauge.

(*ii*) In actual practice, it becomes rather tedious to use Campbell's method and, therefore, the following *simplified procedure* is adopted.

The bridge is first balanced with only vacuum about the filament and then, keeping the voltage across the bridge constant at this very value, the gas or air is allowed into the gauge. The balance of the bridge is thus naturally upset, and a current, corresponding to this upset, or the '*out of the balance current*', as it is aptly called, passes through the galvanometer, the deflections of which, in terms of scale divisions are noted for various values of pressure of the gas (as indicated by a McLeod gauge).

Then, if N be the number of scale divisions through which the galvanometer needle is deflected for a pressure p of the gas surrounding the filament, we have $N \propto f(p)$. A graph between the two is therefore, a straight line and gives the required calibration curve for the gauge, from which the pressure of the gas for any deflection in the galvanometer can be read off straightaway.

Essential points of the gauge. For the success of the gauge as a low-pressure-measurer, the essential points are that (*i*) *the material of the filament must have a high coefficient of temperature*, so that the change in its resistance must be appreciable for a small change in its temperature. It is, therefore, made of a tungsten or a platinum wire, of a diameter of about .06 *mm* ; (*ii*) *the filament must throughout be kept taut*, so that the distance between it and the walls of the enclosing bulb remains unaltered ; (*iii*) *the heat losses along the filament- support must be as small as possible*.

To ensure both conditions (*ii*) and (*iii*), *a poor conductor of heat like a glass rod is used by way of support for the filament and it is taken round glass breds*, as shown, (Fig.15.15), *with its longer portions equidistant from the walls of the bulb on either side*, and *the galvanometer should have a high current sensitivity*.

Range, Utility and Drawbacks of the Gauge. The range of the gauge is rather small, *viz.*, between 10^{-2} to 10^{-4} *mm* of mercury, but because of its almost instantaneous action, *it is extremely useful for the measurement of pressure fluctuations*.

Its chief **drawbacks** are that (*i*) *it is wholly unsuitable for use with organic vapours, as its filament gets 'poisoned' by them* ; (*ii*) *it is rather much too sensitive to sudden or accidental thermal or mechanical shocks and vibrations*, which must, therefore, be avoided as far as possible. In fact, as a safeguard against the latter disturbance, it is usual to provide a shock-absorbing mounting for it; (*iii*) *it is not an absolute gauge and has to be calibrated against a McLeod or some other abslute gauge; (iv) at pressures below 10^{-3} mm of mercury column, the heat loss occurs more by radiation than conduction; (v) pressures below 10^{-4} mm of mercury cannot be measured with its help with any reasonable amount of certainty*.

Finally, it may as well be mentioned that, like most other gauges in the pressure range 10^{-3} to 10^{-5} *mm* of mercury, this too requires some manual adjustments to be made, which are never quite as reliable as mechanical or automatic ones. This has, however, been remedied, comparatively recently, (in the year 1939) by **Scott** by his clever introduction of a Triode valve in the gauge circuit, thereby marking the working of the gauge at once *quick*, *smooth* and *automatic*.

15.26. Thermocouple Gauge. It is just a variant of the Pirani Hot Wire or Resistance Gauge, in which, instead of measuring the resistance of the filament, we measure the temperature of the hot junction of a thermocouple, attached to the filament, from the thermo-electric *e.m.f.* developed in it. The temperature of the hot junction will obviously depend upon the thermal conductvity K of the gas in between itself (or the filament) and the walls of the containing glass bulb, the outside of which is maintained at 0^0C and which is connected at its upper open end (Fig. 15.16) to the vessel in

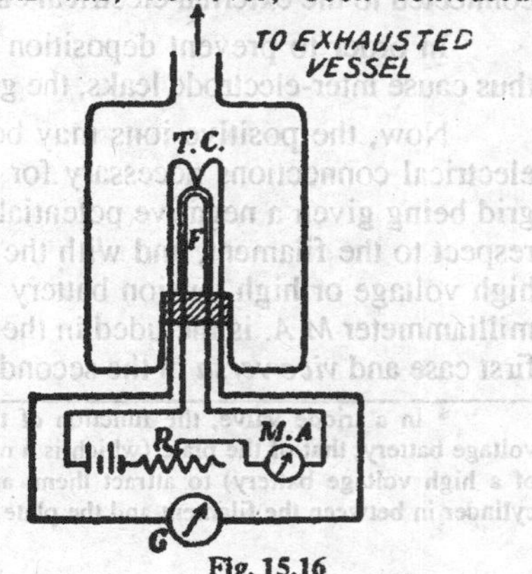

Fig. 15.16

which the (low) pressure is to be measured. The value of K for the gas, at such low pressures, is as we have seen, directly proportional to its pressure.

The filament F, here, is a short ribbon of constantan and is heated by a current of upto 50 *milliamperes*, and the hot junction of a suitable thermocouple *T.C.*, – *iron-constantan, chromelalumel, antimony* – bismuth, or any other--is attached to the midpoint of the filament, which is connected through a rheostat R to a low-voltage battery and a milliammeter *M.A.* The thermocouple circuit includes a sensitive galvanometer, the deflections of which give an indication of the thermo-electric *e.m.f.* developed and hence that of the pressure of the gas.

The gauge is, as usual, calibrated against a McLeod gauge, with the *same* gas in both, if strict accuracy is aimed at.

15.27. Ionisation gauges. Ionisation, as we know, is the process of knocking out an electron from the outer shell of a gas atom. Among other ways, this may also be done by means of a fast moving electron colliding against the gas atom, the process then being spoken of as *ionisation by electron collision.* Before an atom can thus ionise a gas, however, it must possess a certain minimum amount of energy, depending upon the gas in question, and must, therefore, be accelerated through a certain minimum potential difference, called the *ionisation potential*,. V_i, say, for that particular gas. The energy thus acquired by the electron is measured in terms of *electron volts,* where one electron volt is the energy acquired by an electron on being accelerated through a potential difference of 1 volt.

Now, when as electron is thus knocked off from a gas atom, naturally what remains behind is a positive ion, since the atom, as a whole, is neutral. Thus, ionisation produces positive ions and electrons (negative). If these positive ions be collected on another auxiliary electrode (*i.e.*, an electrode in addition to the positive electrode or the anode and the negative electrode or the cathode), we get a positive ion current, or an ionisation current, for a given value V of the accelerating voltage above the ionisation potential V_i. *This ionisation, or the positive ion current, at low pressures, below* 10^{-3} *mm of mercury, varies linearly with the pressure of the gas,* because at such pressures, an electron is hardly likely to collide with more than one atom on its way from the cathode to the auxiliary electrode. Thus, in the ordinary triode valve, the grid* may very well act as the auxiliary electrode, if it be given a negative potential with respect to the filament ; so that, any triode valve may be used as an ionisation gauge. In order to avoid the possibility of electrical leaks between the electrodes, however, the triodes, meant to be used as ionisation gauges, are specially constructed with this end in view.

Since the electrons are emitted on heating the filament or the cathode, such a gauge is called a *hot cathode ionisation gauge* and the first satisfactory form of it was due to ***Dushman*** and ***Found.*** A later modification, now in common use is the one shown in Fig. 15.17. The usual *tungsten filament F* is here supported on a *glass rod R* in the manner indicated, with a co-axial *grid G*, (of tungsten or molybdenum) around it, and a nickel or silver coating on the interior of the glass bulb, enclosing the two, acts as the plate P, with a platinum wire w sealed on to it, to enable it to be connected to the external electrical circuit.

In order to prevent deposition of any metal film on it, (from the filament or the plate) and thus cause inter-electrode leaks, the glass rod is provided with loose glass collars C, C, as shown.

Now, the positive ions may be collected either on the grid or on the plate, as desired, the electrical connections necessary for the purpose being shown in Fig. 15.17 (*a*) and (*b*), with the grid being given a negative potential in the former and a positive potential in the latter case, with respect to the filament, and with the plate connected to the positive and the negative poles of the high voltage or high tension battery (*H.T.B.*) of about 120 volts, in the two cases respectively. A milliammeter *M.A.* is included in the plate circuit and a galvanometer G' in the grid-circuit, in the first case and *vice versa* in the second.

*** In a triode valve, the function of the filament is to emit electrons, when heated by the current from a low voltage battery, that of the plate (which is a metal cylinder around the filament and is ordinarily connected to the positive of a high voltage battery) to attract them, and that of the grid, (which is a spiral, a wre-guage or a perforated metal cylinder in between the filament and the plate) to control their flow.**

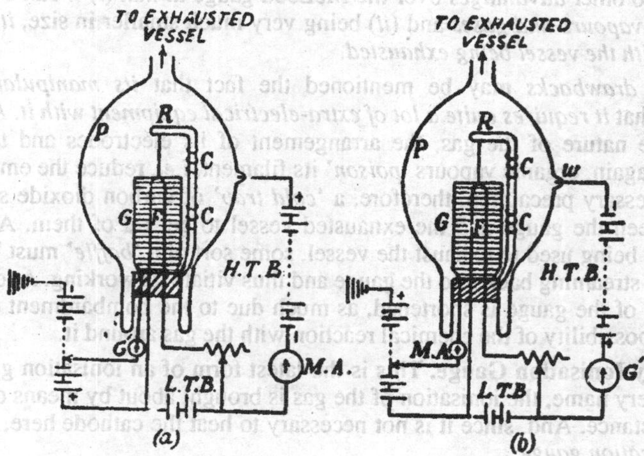

Fig. 15.1

Working.(i) **First Case.** In the case, the plate being at a positive potential with respect to the filament, electrons emitted by the latter (when heated by the current from the low voltage battery *L.T.B.*) are attracted towards the plate and pass through the meshes or holes in the grid. Thus, on their way to the plate, they bring about ionisation of the gas between the grid and the plate. The positive ions formed are collected by the grid which is at a negative potential with respect to the filament. A small ionisation current, or a positive ion current, thus flows through the grid-filament circuit and can be easily read on the galvanometer G' included in the circuit, the usual electron current being given by the milliammeter *M.A.*

(ii) **Second Case.** Here, *since the grid is at a positive and the plate at a negative potential with respect to the filament,* the electrons emitted by the filament or the hot cathode, are strongly attracted by the grid (it being nearer to the filament), but quite a number of them get through it on account of their momentum and thus ionise the gas in the space between the grid and the plate (in case, of course, the accelerating voltage applied is higher than the ionisation potential for the gas). The positive ions thus released are collected at the plate, any electrons straying into the region being repelled back by it. The positive ion, or the ionisation current, is then read on galvanometer G' included in the plate circuit and the electron current on the milliammeter, as before.

Of the two arrangements discussed above, the latter, *i.e.*, the —ve plate one, is by far the more sensitive, but the first one is simpler to work with.

As already indicated, the relation between the ionisation current and pressure is a linear one only at pressures below 10^{-3} *mm* of mercury. If, however, the emission of electrons from the filament be relatively small, the relation also becomes linear at higher pressures, but, then, the sensitivity of the gauge is considerably impaired.

This gauge too is not an absolute one and has to be calibrated against a McLeod gauge, with the *same* gas in it as the one, the pressure of which is to be determined by the ionisation gauge.

Once calibrated, the galvanometer G' is replaced by a microammeter, graduated in pressure units.

The gauge can measure much lower pressures, in the range 10^{-3} *mm* to 10^{-7} of mercury column. It has also other *advantages* over the McLeod gauge in that (*i*) *it can be used to measure pressures of both vapours and gases* and (*ii*) being very much smaller in size, *it can be located in close proximity with the vessel being exhausted.*

Among its *drawbacks* may be mentioned the fact that *its manipulation is somewhat complicated* and that *it requires quite a lot of extra-electrical equipment with it.* Also, *its sensitivity* depends upon the nature of the gas, the arrangement of its electrodes and the electric circuit employed. Then, again, organic vapours '*poison*' its filament, *i.e.*, reduce the emission of electrons from it. As a necessary precaution, therefore, a '*cold trap*' of carbon dioxide snow or acetone is arranged in between the gauge and the exnausted vessel to get rid of them. And, in case an oil diffusion pump is being used to exhaust the vessel, some sort of a '*baffle*' must be used to prevent any oil molecules streaming back into the gauge and thus vitiate its working. And, finally, at higher pressures, the life of the gauge is shortened, as much due to the bombardment of the filament by the ions as to the possibility of the chemical reaction with the gas around it.

15.28. α-ray Ionisation Gauge. This is the latest form of an ionisation gauge, in which, as indicated by its very name, the ionisation of the gas is brought about by means of α particles from a radioactive substance. And, since it is not necessary to heat the cathode here, it may be called a *cold cathode ionisation gauge.*

Designed by **Downing** and **Mellen** in the year 1946, it consists of a closed ionisation chamber *C* (Fig. 15.18.), inside an outer protective shell, and perforated at its top and bottom to allow free access to the gas inside it. At the bottom of the chamber is securely held in position a small saucer-shaped plaque *P*, 1 cm^2 in area, and with its upper surface made of an alloy of gold and radium, which is in equilibrium with its products of decay, *viz., radon, radium A* and *radium B*, of which the first one is a gas. To prevent any of this gas escaping out, the upper surface of the plaque is electrolytically coated with a layer of nickel which also serves the additional purpose of preventing contamination by mercury vapour. The losses from the plaque (due to radio-active emission) are so small that the instrument needs to be tested only once in a number of years. This plaque thus forms a highly efficient α-ray emitter, though with a slow emanating power.

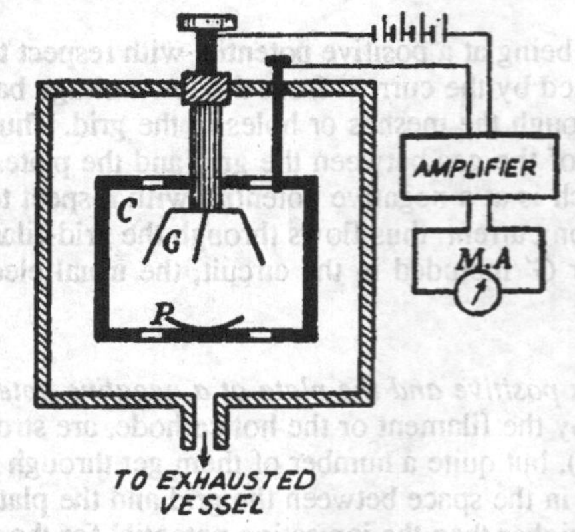

Fig.15.18

The inner electrode, or the grid *G*, consists of four wires spread or stretched out, as shown. This limits the distance to be covered by the positive ions produced by the ionisation of the gas (by the α -particles), thus facilitating their 'capture' before they have time to cover longer distances. For, in the latter case, the ions may re-unite and thus the linear relation between the ionisation current and pressure may no longer remain valid, with the whole basis of the gauge knocked out. The small ionisation current produced is first amplified and then read on the microammeter *M.A.*, the gauge otherwise functioning just like the hot-cathode one, discussed earlier, and having, likewise, to be calibrated against a McLeod gauge.

Now, although the relationship between ionisation current and pressure of the gas no longer remains valid beyond a pressure of 10 *mm* of mercury, *the gauge may be used* (in three successive stages) *to measure pressures within a wide range from* 10^{-3} *mm to* 1000 *mm.* Further, the gauge is *a continuous reading one.*

The one serious drawback of the gauge is that extra precautions are necessary to work with it, if one is to save oneself from the hazards of exposure to the radio-active substance used.

15.29. The Knudsen Gauge. This simple and efficient gauge, which has been used to measure the lowest pressures yet produced*, depends upon what is called the *radiometric effect*. It, therefore, becomes necessary to first understand clearly what this '*effect*' is.

Radiometric Effect. At high and moderately high pressures, a gas behaves like a viscous liquid and its flow through narrow tubes is governed by *Poiseuille's law*, the rate of flow being proportional to the *fourth power* of the radius of the tube and being limited by the frequency of intra-molecular collisions. At very low pressures, however, the mean free path of the molecules becomes greater than the radius of the tube, the frequency of collisions with the containing walls preponderates over the frequency of intra-molecular collisions and considerations of conductivity and viscosity etc. cease to be of any consequence. The flow of a gas, under these conditions, is very suggestively called '*molecular flow*' by *Knudsen*, whose extensive work on the behaviour of gases under *vacuum* conditions has earned its deserved recognition in a gas, at such pressures, being known as a '*knudsen gas*'.

He assumes that a typical gas-molecule is accelerated by the pressure gradient along the tube and that all its drift velocity is lost when it collides against a wall ; for, according to him, the smoothest walls are '*molecularly rough*', having, here and there, minute projections of molecular proportions, due to the piling up of one or two atoms above the surrounding ones, so that the directions of motion of the molecules, before and after striking the wall, bear no relation to each other.

Assuming these '*Knudsen conditions*', if a molecule acquires extra energy on rebounding from a heated surface, it will collide with the wall a number of times before it reaches the surface again. It follows, therefore, from the *law of conservation of momentum*, that there is a net force acting normally to the heated surface which, for moderate differences of temperature between the surface and the wall, is directly proportional to these differences of temperature.

This mechanical force exerted between two surfaces, very close to each other and maintained at a difference of temperature, is called **radiometric effect.**** Since it becomes manifest only when the mean free path is longer than the distance between the heated surface and the wall, it is a typically *Knudsen effect* and must not be confused with that physical effect of radiation we call *radiation pressure* – the latter being (*i*) quite independent of the presence or absence of any gas, (*ii*) about twice as great on a polished as on a blackened surface, and (*iii*) of a much smaller order of magnitude.

The Gauge Proper. Knudsen very ingeniously exploited and harnessed these radiometric forces at low pressures in devising his *absolute gauge* +, which enables us to measure pressure in an evacuated vessel, from 10^{-3} *mm* down to 10^{-7} *mm*, by simply noting the deflection of a cold plate suspended in the vessel, due to its bombardment by the molecules rebounding from a nearby hot plate. If the dimensions of the plates be very large, compared with the distance between them, (as though we were considering only a very small portion of an infinite area), so that all edge-effects could be neglected, and, further, *if this distance be smaller than the mean free path of the molecules*, so that the molecules arrive at one in exactly the same condition in which they leave the other, the deflecting repulsive force on the cold plate is found to be proportional to the gas pressure in the vessel, right up to a pressure of 10^{-7} *mm*.

Fig. 15.19. shows, diagrammatically the essentials of the gauge, where P_1 and P_2 are two fixed plates, kept electrically heated and arranged on opposite sides of a cold plate A, in the form of

* Up to 3×10^{-3} *mm*., by *Woodraw*, with a temperature difference of 100° *C* between the two plates. And, up to 5×12^{-9} *mm*., by *Shrader* and *Sherwood* who used a slight modification of it.

** A familiar example of this effect is the *Crook's radiometer*, so often used to detect thermal radiations. It consists of a freely pivoted 'spider', fitted with mica vanes with their plante parallel to the vertical axis, and coated on one side with lamp black, the other being polished, the whole instrument being arranged inside an evacuated glass bulb. On exposure to thermal radiation, the blackened sides of the vanes become hotter than the polished ones and radiometric forces come into play, making the spider rotate, the blackened side moving away from the incident radiation.

+ It is termed '*absolute*' because it ives the pressure directly in absolute units, *i.e.*, in *dynes*/*cm*^{2}.

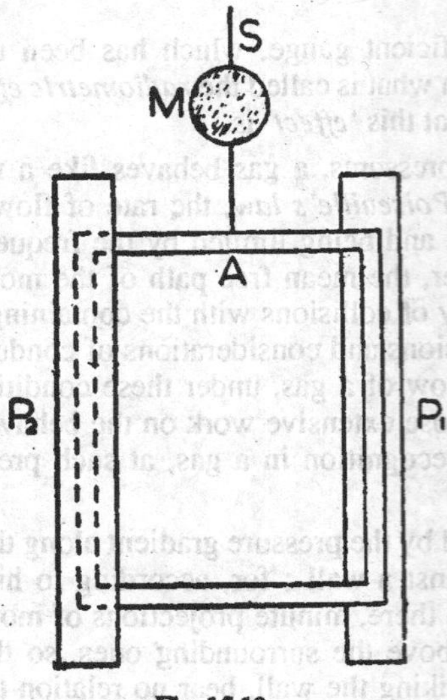

Fig. 15.19

a rectangular picture-frame, suspended (in the evacuated vessel) by means of a quartz fibre, carrying a mirror M to enable its deflection to be measured by the usual *'lamp and scale'* method.

Let the temperature of P_1 and P_2 be T_1 and that of A (and hence also that of the evacuated vessel, in which it is suspended) be T_2 and let n_1 and n_2 be the *number of molecules per cubic centimeter* travelling from P_1 to A and from A to P_1, with *root mean square velocities* $\bar{c}_1$ and $\bar{c}_2$ respectively. Then, since in the equilibrium state, the rate of molecular collisions per square centimeter must be the same, we have

$$n_1 \bar{c}_1 = n_2 \bar{c}_2. \quad ...(i)$$

And, if a be *the number of molecules per cc* in the vessel, outside the space between P_1 and A, then, since the number of molecules flowing out of the latter into the former and *vice versa*, must be the same, we have

$$n\bar{c}_2 = n_1\bar{c}_1 + n_2\bar{c}_2 = 2n_1\bar{c}_1 = 2n_2\bar{c}_2 \quad ...(ii)$$ [From relation (*i*) above.

So that, if m be the mass of each molecule, we have

total pressure between the plates $= \dfrac{m\,n_1\,\bar{c}_2{}^2}{3} + \dfrac{m\,n_2\,\bar{c}_2{}^2}{3}$, $\left[\because P = \frac{1}{3}mn\bar{c}^2.\right.$

whereas *pressure on the sides, remote from* $A = m\,n\,\bar{c}_2{}^2/3$.

Obviously, therefore, *excess pressure on the side of A nearer to* P_1 *over that away from it, is given by* $\dfrac{m\,n_1\,\bar{c}_1{}^2}{3} + \dfrac{m\,n_2\,\bar{c}_2{}^2}{3} - \dfrac{m\,n\,\bar{c}_2{}^2}{3}$.

Or, *substituting* $n\bar{c}_2/2$ *for* $n_1\bar{c}_1$ *and for* $n_2\bar{c}_2$, [From relation (*ii*) above.

we have

$$\textit{excess pressure on } A = \frac{m\,n\bar{c}_1\,\bar{c}_2}{6} + \frac{m\,n\,\bar{c}_2{}^2}{6} - \frac{m\,n\,\bar{c}_2{}^2}{3} = \frac{m\,n\,\bar{c}_1\,\bar{c}_2}{6} - \frac{m\,n\,\bar{c}_2{}^2}{6}.$$

Or, $$\textit{excess pressure on } A = \frac{m\,n\,\bar{c}_2{}^2}{6}\left(\frac{\bar{c}_1}{\bar{c}_2} - 1\right).$$

Assuming the molecules striking P_1 and A to take up their respective temperatures T_1 and T_2, we have

$$\bar{c}_1/\bar{c}_2 = \sqrt{T_1/T_2},$$ [See page 489]

And, thus, $$\textit{excess pressure on } A = \frac{m\,n\,\bar{c}_2{}^2}{6}\left(\sqrt{\frac{T_1}{T_2}} - 1\right).$$

$$= \frac{1}{2}\cdot\frac{m\,n\,\bar{c}_2{}^2}{3}\left(\sqrt{\frac{T_1}{T_2}} - 1\right).$$

Now, $m\,n\,\bar{c}_2{}^2/3 = p$, the pressure inside the vessel.

So that, we have *excess pressure on A*, or, *force per sq cm* on A, say,

$$F = \frac{p}{2}\left(\sqrt{\frac{T_1}{T_2}} - 1\right). \qquad ...(iii)$$

If the temperature difference $(T_1 - T_2)$ be not very large (*i.e.*, not more than 250°C), we may consider the flow of molecules in between the space P_1A and the rest of the vessel as a *thermal transpiration** effect between the temperature $\left(\frac{T_1 + T_2}{2}\right)$ inside $P_1 A$ and T_2, outside ; so that, if p' be the pressure between the plates, we have

$$\frac{p'}{p} = \sqrt{\frac{(T_1 + T_2)/2}{T_2}} = \sqrt{\frac{T_1 + T_2}{2T_2}} = \sqrt{\frac{2T_2 + T_1 - T_2}{2T_2}}$$

$$= \left(1 + \frac{T_1 - T_2}{2T_2}\right)^{\frac{1}{2}} = 1 + \frac{1}{4}\cdot\frac{(T_1 - T_2)}{T_2}.$$

So that,

$$p' = p + \frac{p}{4}\cdot\frac{T_1 - T_2}{T_2}.$$

And, therefore,

$$F = \frac{p}{4}\left(\frac{T_1 - T_2}{T_2}\right). \qquad ...(iv) \qquad \left[\begin{array}{l}\text{Since } F = (p' - p).\\ \text{Also from (iii) above,}\end{array}\right.$$

showing that the force exerted on plate A *is quite independent of the nature of the gas in the vessel.*

Now, if a be the area of each vertical strip of A, we have

force acting on each strip $= F.a.$

Since these forces act in *opposite directions* on the two strips, they constitute a couple, equal to $F.a.2r$, where r is the *mean* distance of each strip from the suspension wire.

The frame is thus deflected, giving rise to a *restoring torsional couple* in the suspension wire and, therefore, comes to rest, when the two couples just balance each other, say, when it has deflected through an angle θ.

Then if c be the torsional couple, set up in the suspension wire per unit twist in it (or per unit deflection of the frame) the total torsional couple tending to restore the frame back to its original position is $c\,\theta$; and we, therefore, have

$$c\,\theta = 2F.a.r. \qquad \text{Or,} \qquad F = c\,\theta/2a\,r.$$

Substituting this value of F in relation (iv) above, we have

$$\frac{c\,\theta}{2a\,r} = \frac{p}{4}\cdot\left(\frac{T_1 - T_2}{T_2}\right)$$

Or,

$$p = \frac{4T_2.c\,\theta}{2a\,r\,(T_1 - T_2)} = \frac{2c\,\theta}{a\,r}\cdot\frac{T_2}{T_1 - T_2}. \qquad ...(v)$$

Further, if I be the *moment of inertia* of the frame about the *suspension wire* as axis and t, its *time-period* of vibration, we have

$$t = 2\pi\sqrt{I/c}, \quad \text{whence,} \quad c = 4\pi^2 I/t^2,$$

Hence

$$p = \frac{8\,\pi^2 I.\theta}{t^2.a\,r}\cdot\frac{T_2}{T_1 - T_2}, \qquad \text{[From (v) above.}$$

* It is the phenomenon of the flow of a gas from a colder to a hotter chamber, both containing the gas at a low pressure and connected to each other by means of a capillary tubing, the flow continuing until a pressure difference, depending upon the temperature difference between two chambers, is established. The ratio between the final pressures attained can be shown to be the same as that between the square roots of the absolute temperatures of the two chambers

whence p, the pressure of the gas in the vessel, can be easily evaluated.

It will be readily seen that apart from the value of p being quite independent of the molecular weight of the gas, the gauge possesses the following advantages :

(i) It gives a continuous indication of pressure in the vessel.

(ii) It is unaffected by any outside influences.

(iii) It is stable and yet very sensitive at low pressures, down to 10^{-7} *mm.*

(iv) It can be used to measure the pressure of all kinds of gases and vapours, irrespective of their mass or condensibility.

(v) It does not require the use of objectionable liquids, like mercury.

(vi) It is simple, cheap and easy to construct and work with, though obviously, it cannot be used above a pressure of 10^{-3} *mm. ; for then, the mean free path of the molecules becomes comparable with the distance between the plates.*

(vii) It is an absolute gauge, giving the pressure directly in dynes /cm², and has not to be calibrated against a McLeod gauge etc.

WORKED EXAMPLES

1. Given that the gas constant R is 8.3 × 10⁷ joules per 1°C, and the atomic weight of chlorine is 35.5, find the root mean square (r.m.s.) velocity of the chlorine molecules at 0°C.

We have the relation, $P = \frac{1}{3} m n \bar{c}^2$. [See page 489

Multiplying by V, the volume of 1 *gram-molecule* of the gas, we have

$$PV = \tfrac{1}{3} mn. V. \bar{c}^2.$$

Where N is the *Avogadro number* or number of molecules in 1 *gram–molecule* of the gas.

Now, $nV = N$.

$\therefore$ $PV = \frac{1}{3} m. N. \bar{c}^2$

And, clearly, $m N = M$, the *molecular weight* of the gas.

$\therefore$ $PV = \frac{1}{3} M \bar{c}^2$. Also, $PV = RT$.

So that, $\frac{1}{3} M \bar{c}^2 = R.T.$ Or, $\bar{c}^2 = 3RT/M$,

whence, $\bar{c} = \sqrt{3RT/M}$.

Here, $R = 8.3$ *joules/°C*, $T = 0 + 273 = 273^0$ *Abs.* and $M = 35.5 \times 2 = 71$ *gm* = 0.071 *kg*.

because a chlorine molecule consists of 2 atoms.

$\therefore$ substituting the values of R, T and M, we have

$$\bar{c} = \sqrt{\frac{3 \times 8.3 \times 273}{0.071}} = 3.094 \times 10^2 \ m/sec.$$

Thus, the r.m.s. velocity of the chlorine molecule is 3.094×10^2 *m /sec.*

2. If the density of nitrogen is 1.25 gm. / litre at normal temperature and pressure, calculate the root mean square velocity of its molecule.

We have the relation $P = \frac{1}{3} mn\bar{c}^2$, for a gas.

But $m \cdot n = \rho$, the *density* of the gas. therefore, $P = \frac{1}{3} \rho \bar{c}^2$.

Or, $\rho \bar{c}^2 = 3P$, whence, $\bar{c} = \sqrt{3P/\rho}$.

Here $P = 76\ cm = 76 \times 13.6 \times 981$ *dynes / cm²*,

and $\rho = 1.25$ *gm/litre* $= 1.25/1000 = .00125$ *gm/cc*.

$$\therefore \quad \bar{c} = \sqrt{\frac{3 \times 76 \times 13.6 \times 981}{.00125}} = 4.933 \times 10^4 \ cm/sec.$$

The velocity of the nitrogen molecule is, therefore, equal to 4.933×10^4 *cm /sec*.

3. Derive Charles' law and Dalton's law on the basis of the kinetic theory of gases, as also the Standard Gas Equation, PV = RT, and write a brief note on Absolute Zero from the standpoint of the theory.

(i) We have $P = \frac{1}{3} m n \bar{c}^2 = \frac{1}{3} \rho \bar{c}^2$ [$\because$ $m.n = \rho$, the density of the gas.

Now, $\rho = 1/V$, and $\bar{c}^2 \propto T$, [where V is the volume of the gas.

where T is the *absolute temperature* of the gas.

$\therefore$ $P \propto \frac{1}{3V} . T$. ... (i) If $\therefore$ V be constant, $P \propto T$.

Or, at constant volume, the pressure of a given mass of gas varies directly as its absolute temperature, which is *Charles' law*.

Or, we can put relation (i) above, as $V \propto \frac{1}{3P} . T$; so that, if P be constant, $V \propto T$.

Or, *at constant pressure, the volume of a given mass of gas varies directly as its absolute temperature*, which is also *Charles' law*.

(ii) Let a number of gases or vapours, having no chemical reaction with each other, and having densities and mean square velocities, ρ_1, ρ_2, ρ_3 .... and $\bar{c}_1$, $\bar{c}_2$ $\bar{c}_3$ .... respectively, be mixed together in the *same* volume. Then, the resultant pressure P will clearly be given by

$P = \frac{1}{3}\rho_1\bar{c}_1^{\,2} + \frac{1}{3}\rho_2\bar{c}_2^{\,2} + \frac{1}{3}\rho_3\bar{c}_3^{\,2} + \ldots$, considering each set of molecules of the different gases or vapours.

Now, $\frac{1}{3}\rho_1\bar{c}_1^{\,2}, \frac{1}{3}\rho_2\bar{c}_2^{\,2}, \frac{1}{3}\rho_3\bar{c}_3^{\,2}, \ldots$, are the individual pressures exerted by the different gases or vapours. Putting these as $p_1 + p_2 + p_3 \ldots$, we have

$$P = p_1 + p_2 + p_3 .$$

i.e., the resultant or total pressure exerted by the mixture of gases is equal to the sum of their individual or p[artial] pressures, which is Dalton's law of partial pressures.

(iii) We have, $P = \frac{1}{3}\rho\bar{c}^2$.

Multiplying both sides by V, the volume of the gas, we have

$$PV = \frac{1}{3}\rho.V.\bar{c}^2.$$

Now, $\rho V = M$, *mass* of the gas. [$\because$ *mass = volume × density*.

$\therefore$ $PV = \frac{1}{3}M\bar{c}^2 = \frac{2}{3}.\frac{1}{2}M\bar{c}^2$.

Now, $T \propto K.E$ *of the gas*, Or, $T \propto \frac{1}{2}M\bar{c}$.

Or, $PV \propto T$, Or, PV = a constant × T. Or, $PV = RT$,

where R is a constant, called the *gas constant*, whose value, for a given mass of gas, is quite independent of the values of P, V and T.

Thus, the *standard gas equation*, ($PV = RT$), can be deduced from the kinetic theory of gases.

(iv) According to the kinetic interpretation of temperature, the temperature of a gas is, as we have seen, proportional to the mean $K.E.$ of the molecules, because the $K.E.$ of a molecule $= \frac{1}{2}m\bar{c}^2$ and $\bar{c}^2$ is proportional to T, the absolute temperature of the gas.

Clearly, $T = 0$, when $\bar{c} = 0$, *i.e.*, when the molecules have zero velocity, or are devoid of all motion.

Hence, *on the basis of the kinetic theory, the absolute zero of temperature is the temperature at which the gas molecules are devoid of all motion*. This, obviously, is not quite correct, for the above deduction is made on the assumption of the gas being an ideal or a perfect one, and no actual gas approaches this ideal, so that, even if at the ordinary temperatures and pressures, the actual gases may be taken to approximate, more or less, to the perfect gas condition, these ideal gas conditions do not hold down to the absolute zero.

A more satisfactory interpretation of temperature is afforded by **thermodynamics**, which does not require the cessation of all molecular motion at the absolute zero.

4. Deduce the relation $\lambda = 1/\pi r^2 n$, for the mean free path of a molecule, and use it to calculate the diameter of a molecule of benzene if $n = 2.79 \times 10^{19}$ mol. per cc and λ for benzene $= 2.2 \times 10^{-6}$ cm.

For answer to first part, see page 494.

Substituting the values of λ and n in the above relation, we have

$$2.2 \times 10^{-6} = \frac{1}{\pi r^2 \times 2.79 \times 10^{19}}.$$

Or, $\pi r^2 \times 2.79 \times 10^{19} \times 2.2 \times 10^{-6} = 1$. Or, $\pi r^2 \times 2.79 \times 10^{13} \times 2.2 = 1$.

Or, $r^2 = \dfrac{1}{\pi \times 2.79 \times 2.2 \times 10^{13}}$ whence, $r = \sqrt{\dfrac{1}{\pi \times 2.79 \times 2.2 \times 10^{13}}}$

$= 7.201 \times 10^{-8}$ *cm*

Or, the radius of the sphere of influence of the benzene molecule is equal to 7.201×10^{-8} *cm*.

Now, *the radius of the sphere of influence of a molecule is equal to its diameter**. Hence, the diameter of the benzene molecule is equal to 7.201×10^{-8} *cm*.

5. Find the mean free molecular path in air, taken as a uniform gas; given that the density of air $=1.2 \times 10^{-3}$ gm /cm^3 at 0°C and a pressure of 10^6 dynes/cm^2, and that its coefficient of viscosity=1.7×10^{-4} dynes per cm^2 per unit velocity-gradient.

We have the relation, $\eta = \frac{1}{3}\rho \bar{c}\lambda$, for the coefficient of Viscosity of a gas. Also $P = \frac{1}{3}\rho \bar{c}^2$,

whence $\bar{c}^2 = 3P/\rho$, and $\therefore \bar{c} = \sqrt{3P/\rho}$.

So that, $\eta = \frac{1}{3}\rho.\lambda.\sqrt{3P/\rho} = \lambda\sqrt{3P.\rho^2/9\rho} = \lambda\sqrt{P.\rho/3}$

And $\therefore$ $\lambda = \eta\sqrt{\dfrac{3}{P.\rho}}$.

Substituting the value of η, P and ρ, therefore, we have

$$\lambda = 1.7 \times 10^{-4}\sqrt{\frac{3}{10^6 \times 1.2 \times 10^{-3}}} = 8.498 \times 10^{-6}\ cm.$$

Therefore, the mean free molecular path in air = 8.498×10^{-6} *cm*.

6. The mean velocity of a molecule of nitrogen gas is 4.5×10^4 cm per sec. Its density at N.T.P. is 1.25×10^{-3} gm /cc, and its coefficient of viscosity= 166×10^{-6} gm/cm per sec. Calculate (*i*) its molecular mean free path (*ii*) the number of collisions made per second, and (*iii*) its molecular diameter. (Assume n = 2.7×10^{19} per cc)

(*i*) We have $\eta = \frac{1}{3}\rho\bar{c}\lambda$, whence, $\lambda = 3\eta/\rho\bar{c}$,

$$= \frac{3 \times 166 \times 10^{-6}}{1.25 \times 10^{-3} \times 4.5 \times 10^4} = \frac{3 \times 166 \times 10^{-6}}{1.25 \times 10 \times 4.5} = 8.853 \times 10^{-5}\ cm.$$

Or, *molecular mean free path of nitrogen, i.e.*, $\lambda = 8.853 \times 10^{-5}$ *cm*.

(*ii*) Number of collisions made *per second*

$$= \frac{\bar{c}}{\lambda} = \frac{4.5 \times 10^4}{8.853 \times 10^{-5}} = 5.083 \times 10^9.$$

(*iii*) Now, $\lambda = \dfrac{1}{\sqrt{2}\,n\pi r^2}$, whence, $r^2 = \dfrac{1}{\sqrt{2}\,n\pi\lambda}$.

Or, $$r = \sqrt{\frac{1}{\sqrt{2n}\,\pi\lambda}} = \sqrt{\frac{1}{\sqrt{2 \times 2.7 \times 10^{19}} \times \pi \times 8.853 \times 10^{-5}}}$$

$$= \sqrt{\frac{1}{\sqrt{2 \times 2.7 \times 10^{10}} \times \pi \times 8.853}} = 3.068 \times 10^{-8}\ cm.$$

$\therefore$ the *molecular diameter* of nitrogen = 3.068×10^{-8} *cm*.

7. Calculate the molecular K.E. of 1 gm of Hydrogen (*i*) at 0°C and (*ii*) at 100° C. (R = 8.3×10^7 ergs).

We know that the *K.E.* of 1 *gram-molecule* of a gas = $\frac{3}{2}$.*R.T.*

Now, *mol. wt. of hydrogen* = 2 *gm*.

$\therefore$ *K.E.* of 1 *gm* of Hydrogen at 0°C or 273° *Abs.* = $\frac{3}{2}$ *R.T./2*

* Atoms and molecules are really not the simple bodies they were once imagined to be. To speak of the diameter of a molecule in the geometrical sense has, therefore, no meaning. *The diameter of a molecule is taken to be the distance between the centres of two molecules beyond which they do not exert any influence over each other, which is clearly equal to the radius of the sphere of influence of the molecule.*

$= 3 \times 8.3 \times 10^7 \times 273/4 = 16.99 \times 10^9$ *ergs.*

And, *K.E.* at 100°*C* or 373° *Abs.*

$$= 16.99 \times 10^9 \times \frac{373}{273} = 23.21 \times 10^9 \text{ ergs.}$$

Thus, the *K.E.* of 1 *gm* of Hydrogen is 16.99×10^9 *ergs* at 0 °*C*, and 23.21×10^9 *ergs* at 100° *C*.

8. The lowest attainable pressure with a given exhaust pump is 10^{-3} mm. How long would it take to reduce the pressure in a vessel of capacity 2 litres from 60 mm to 1 mm if the intrinsic speed of the pump be 500 cc/sec ?

The intrinsic speed of a pump, as we know is given by $S = \frac{V}{t_2 - t_1} \log_e \left(\frac{p_1 - p_0}{p_2 - p_0}\right)$. [See page 498

Here, obviously, $S = 500$ *cc /sec*, $V = 2$ *litres* $= 2 \times 10^3$ *cc*, $p_1 = 60$ *mm*, $p_2 = 1$ *mm* and $p_0 = 10^{-3}$ *mm*, almost negligible compared with p_1 and p_2.

So that, $500 = \frac{2 \times 10^3}{(t_2 - t_1)} \cdot \log_e \frac{60}{1} = \frac{2 \times 10^3}{(t_2 - t_1)} \times 2.3026 \log_{10} 60,$

whence $(t_2 - t_1)$ or $t = \frac{2 \times 10^3}{500} \times 2.3026 \log_{10} 60 = 4 \times 2.3026 \times 1.7782 = 16.39$ *sec.*

Thus, the pressure in the vessel will be reduced from 60 *mm* to 1 *mm* in 16.39 *sec.*

9 (a). The pressure of air in a given vessel is atmospheric to start with. An air pump (or a Toepler pump) reduces it to one fourth of an atmosphere in 4 successive operations. What will be the pressure in the vessel after the sixth operation?

***(b)* The volume of the barrel of an air pump (or the bulb of a Toepler pump) is one-fifth of that of the vessel to be exhausted. How many operations of the pump will reduce the density of the air (or the gas) in the vessel to 125/216 th of its original value?**

(a) Let *P* be the initial pressure of the air or gas in the vessel and P_4 and P_6, the pressures after the fourth and the sixth operation of the pump.

Then, $P_4/P = [V/(V+v)]^4$ and $P_6/P = [V/(V+v)]^6$.

Since $P_4/P = 1/4$, we have $P_6 = P\,(1/4).\sqrt{1/4} = 1.(1/4).(1/2) = 1/8$ *atmopsphere*.

Thus, the pressure in the vessel, after the sixth operation, will be 1/ 8 *atmosphere*.

(b) For a given volume of air or gas, the density is directly proportional to pressure, we have $\rho_n = \rho\,(V/v + v)^n$ where ρ and ρ_n are the densities of the air or gas in the beginning and after *n* operations respectively.

So that $\frac{\rho_n}{\rho} = \left(\frac{V}{V+v}\right)^n$. Or, $\frac{125}{216} = \left(\frac{5}{6}\right)^n$, whence, $n = 3$.

Thus, the air or gas in the vessel will have he desired density after three successive operations of the pump.

10. The volume of the bulb of a McLeod Gauge is 250 cc and the length and diameter of the capillary tube 10 cm and 2 mm respectively. Estimate the range of the gauge.

We know that the pressure in the exhausted vessel is given by the relation $P = \frac{v}{V} h$, , where the symbols have their usual meanings (See § 15.23 page 509).

Now, *for the upper limit of P*, the volume *v* of the gas in the capillary tube is equal to the volume of the tube itself, *i.e.*, $v = (\pi \times .1)^2 \times 10$ *cc.*, the volume *V* of the bulb = 250 *cc* and *h* = 10 *cm*. We, therefore, have

$$P = \frac{\pi (.1)^2 \times 10}{250} \times 10 = \frac{\pi}{250} = 1.257 \times 10^{-2} \text{ cm or } 1.257 \times 10^{-1} \text{ mm.}$$

For the lower limit of P, the volume *v* of the gas in the capillary tube $= 2 \times 10^{-3}$. (See page 510).

And $\therefore$ $h = v/\pi r^2 = 2 \times 10^{-3}/\pi \times (.1)^2$. So that, here,

$$P = \frac{2 \times 10^{-3}}{250} \times \frac{2 \times 10^{-3}}{\pi \times (.1)^2} = 3.183 \times 10^{-6} \text{ cm or } 3.183 \times 10^{-5} \text{ mm.)}$$

Thus, the range of the McLeod gauge is from 1.257×10^{-1} *mm* to 3.183×10^{-5} *mm.*

EXERCISE XV

1. Obtain the expression $\eta = \frac{1}{3}\rho\bar{c}.\lambda$, for the viscosity of a gas, and use it to obtain the mean free path for molecules of benzene vapour, (C_6H_6), at 0°C ; given that η at 0°C = 0.69 × 10⁴ *c.g.s. units* ; *r.m.s.* velocity of benzene molecules at 0°C = 2.95 × 10⁴ *cm/sec* ; *atomic wt.* of carbon = 12 ; *cc* in 1 *gm - molecule* = 22,400.

Ans. 2.015 × 10⁻⁶ *cm*.

2. Find the root mean square velocity of the hydrogen molecule at *N.T.P.*, the density of hydrogen being .00009 *gm/cc*. Ans.1.839 × 10⁶ *cm/sec*.

3. Calculate the molecular velocity (*i.e.*, *r.m.s.* velocity) in the case of a gas whose density is 1.4 *gm per litre* at a pressure of 7.6 *cm* of mercury. Density of mercury = 13.6 *gm/cc.*; *g* = 981 *cm/sec*². (*Manchester, B.Sc.*)

Ans. 4.66 × 10⁴ *cm/sec*.

4. Describe the working of a rotary oil pump for producing low pressures. How are these pressures measured? (*Allahabad*)

5. Show that the pressure of a gas is equal to two-thirds of the kinetic energy of translation per unit volume. Calculate the kinetic energy of hydrogen *per gm molecule* at 0°C. (*Allahabad*)

Ans. 5.67 × 10¹⁰ *ergs*.

6. Discuss the assumptions which form the basis of the simple kinetic theory of gases. Distinguish between the average velocity and the root-mean-square velocity. Deduce an expression for the pressure of a gas from the considerations of kinetic theory. (*Delhi*)

7. Give an account, with a neat sketch, of some form of a gas pump, suitable for the attainment of low pressures. (*Punjab*)

8. An imperfect barometer tube contained some air above the mercury. When the mercury stood at 18 *inches* above the outside level, the air space was 6 *inches*. On pushing the tube down, so that the air space was only 3 *inches*, the mercury stood at 26 *inches*. Find the barometric height. Ans. 30 *inches*.

9. One hundred litres of a gas at atmospheric pressure are compressed into a cylinder, 100 *cm* long and 12 *cm* in diameter. What is the resulting pressure ? Ans. 672 *cm of mercury*.

10. Describe, with a neat diagram, the construction and working of (*a*) any modem high vacuum pump and (*b*) a gauge which can measure the low pressure this pump produces. (*Madras*)

11. The reading of a mercury barometer, the brass scale of which was correct when at 0°C, was 76.69 *cm*, the room temperature being 17.8°C. If the coefficient of cubical expansion of mercury be .009180 and the coefficient of linear expansion of brass be .000019, calculate the reading of the barometer reduced to 0°C. Ans. 73.54 *cm*.

12. Calculate the number of molecules *per c.c.* of a gas, taking the mean free path as 1.83 × 10⁻⁵ *cm* and the molecular diameter equal to 2.3 × 10⁻⁸ *cm*.

[*Hint*. Use the more accurate relation, $\eta = 1/\sqrt{2}\,\pi r^2 . n$] Ans. 2.3 × 10¹⁹.

13. Describe in detail, with a neat sketch, the construction and working of a *Mcleod Gauge*.. What is the recent modification effected in its working? Enumerate the advantages and disadvantages of the gauge.

14. Describe a diffusion pump and explain its working principle. How is the vacuum measured ? (*Delhi*)

15. Give an account of the construction and working of a good diffusion pump. Explain with the help of a diagram how such a pump is utilized to obtain a high vacuum.

Describe a method of measuring accurately low pressures. [*Punjab (Suppl.)*]

16. Describe, with a neat diagram, the construction and working of (*a*) any modem high vacuum pump and (*b*) a gauge which can measure the low pressure this pump produces. [*Delhi* ; *Agra (B.Sc. Part I)*]

17. What are ionisation gauges ? Describe the construction and working of one such gauge, pointing out its advantages and disadvantages over a McLeod Gauge. [*Lucknow* ; *Banaras* ; *Vikram*]

18. Describe some form of a *thermocouple gauge*.

19. Describe in detail the *Pirani resistance gauge*, illustrating your answer with a neat diagram and explaining clearly the essential points of the gauge, its range, utility and drawbacks.

20. Explain the principle and working of the *Knudsen gauge*. What are its particular advantages over the other forms of gauges ?

21. Write an essay on the production and measurement of low pressures. (*Gorakhpur*)

22. Explain the principle and working of a modem high vacuum pump ? Why do these pumps require an auxiliary pump ? How can you measure the pressure so obtained ? (*Raipur*).

CHAPTER XVI

SPECIAL THEORY OF RELATIVITY

16.1. Some common terms. Before proceeding with our subject proper, we had better acquaint ourselves with the precise meanings of the following terms we shall have occasion to use all too frequently.

(i) **Particle.** A particle is, ideally speaking, a *point-body*, just a piece or quantity of matter, *having no linear dimensions but only a position*, the measure of this quantity of matter being the *mass* of the particle.

Such point-bodies are, of course, only convenient mathematical fictions and do not actually exist. In practice, therefore, a body, very much smaller in dimensions than the other quantities involved, may be regarded as a particle.

There are, however, situations in which even the tiny electron may not be treated as a particle. So that, it ultimately depends upon the circumstances obtaining at the time whether or not a body may be considered to be a particle. Nevertheless, for most of our purposes, the above definition of a particle is satisfactory enough.

(ii) **Event.** Unlike its commonly understood meaning as 'something out of the ordinary' or 'something very special', an event, in our mathematical model, stands for *anything that occurs suddenly or instantaneously at a point in space and thus involves both a position and a time of occurrence.*

(iii) **Frame of reference.** A frame of reference is just *a system of coordinate axes which clearly specifies the position of a particle or an event in two or three-dimensional space.*

The one fundamental requirements of a frame of reference is that it must be *rigid.* We, therefore, locate our coordinate axes on a suitably selected rigid body.

The simplest reference frame is the familiar *Cartesian* system of coordinates, in which the position of a particle or an event is exactly defined by its co-ordinates x, y and z along the three perpendicular axes.

It is not necessary that the origin of the frame of reference should exactly coincide with the position of the observer but it must certainly be at rest with respect to him.

Now, for a complete information about an event, we must not only know its true location or position, but also its exact time of occurrence. We must, therefore, have yet another co-ordinate axis, perpendicular to the three spatial coordinate axes of x, y and z, namely, that of time, t. Such a reference frame, with four co-ordinate axes of x, y, z and t, is spoken of as a **space-time frame,** its four axes defining a **four-dimensional continuum,** called *space-time.*

Since, however, all rest and motion are relative, the position or the state of motion of the same particle may appear to be different from different frames of reference. Thus, for example, a ball dropped from the window of a running railway carriage appears to fall vertically to the person at the window, but to take a parabolic path to a person outside the train; or, a point on the rim of a rotating wheel appears to us, (from our frame of reference on the earth) to be taking a cycloidal path, but to a person situated at the centre of the wheel, the point appears to be moving along a circular path.

To solve our problems easily, we naturally prefer to so locate our frame of reference that the motion in question appears to be the simplest from it, - for example, in the window of the railway carriage in the first case and on the wheel, with the origin coinciding with its centre, in the second and imagine ourselves to be stationed at the origin.

16.2. Limitations of Newton's laws of motion - Inertial frames of reference. Let us now examine a little more closely our fundamental laws of mechanics, *viz*; Newton's laws of motion, as to any possible limitations they may have and to see whether they hold good in all frames of reference.

The three laws of motion, with which we are familiar from our junior classes, are the following:

First Law. *A body must continue in its state of rest or of uniform motion along a straight line, unless acted upon by an external force.*

Second Law. *The rate of change of momentum is proportional to the impressed force and takes place in the direction of the force.*

Third Law. *To every action there is an equal and opposite reaction.*

Clearly, the first law gives us a definition of force as the '*casual agent of motion change*', meaning thereby that if no force be applied, there is no change in velocity and, therefore, no acceleration., *i.e.*, if F=0, a=0.

The second law gives a measure of force as the rate of change of momentum, *i.e.*,

$$F = \frac{dp}{dt} = \frac{d}{dt}(mv),$$

where m is the *mass* of the particle, v, its *velocity and* p, its momentum, equal to mv.

So that, if m be constant and independent of velocity, we have

$$F = m\frac{dv}{dt} = ma.$$

Again, therefore, if $a = 0$, $F = 0$, which is the first law, *i.e.*, the first law is only a special case of the second law when $F = 0$ and is thus contained in it. That is why the second law is spoken of as the *real law of motion*.

Now, the obvious limitation of the second law is that it would not hold good if m does not remain constant as, for example, in the case of (i) a falling rain drop which increases in mass as it falls, due to more water vapour condensing around it, or (ii) a rocket hurtling into space, ejecting burnt-up fuel and thus shedding mass continuously or (iii) as we shall shortly see, a body moving with a velocity, which is an appreciable fraction of the velocity of light in free space (c).

The third law too has its own limitation, in that the forces on two bodies interacting with each other are equal and opposite *only if they be measured simultaneously* and it is clearly impossible to do so. If, however, the time of interaction of the two bodies be sufficiently large compared with that taken by light to travel from one to the other, this condition of simultaneity may be said to be more or less satisfied for ordinary purposes. But this is simply out of question in the case of particles of atomic dimensions and the law, therefore, ceases to hold good for them.

Then, again, the first two laws do not hold good in any and every reference frame. For, a body at rest in one reference frame may not appear to be so in another reference frame. For example, if the second reference frame be rotating with respect to the first, the body appears to an observer in it to be moving in a circle.

In fact, it is only in a very special frame of reference that the two laws hold good. Such a frame of reference is referred to as a *Newtonian, Galilian* or, usually, an **inertial frame of reference.**

The first law defines and the second law identifies an inertial frame of reference to be *non-accelerating* and hence also *non-rotating*. All other frames of reference moving with a constant velocity with respect to it are also referred to as *inertial frames of reference*.

Now, an inertial frame of reference being an unaccelerated one, the question arises as to whether it can be located on the earth which, as we know, not only goes round the Sun but also spins about its own axis and would thus impart a centrapetal acceleration ($\omega^2 R$=3.4 cm/sec^2 at the equator) to a body in the reference frame based on it. Even the Sun is not at rest but is moving about the centre of our galaxy, which is itself moving through space.

The nearest to being stationary or at rest, and hence ideal for locating an inertial frame of reference, are the *'fixed stars'*, so called because they seem to preserve their own arrangement or their relative distance from each other. This is however so, not because we know them to be actually fixed in their positions but because from the huge distance from them, here on the earth, their velocity appears to be almost nil. It follows, therefore, that *any rigid body, situated in remote space, far from attracting matter and fixed with respect to the 'fixed stars', may be taken to be an inertial frame of reference and, naturally, so also any other reference frame moving with constant velocity relative to it.*

For our ordinary purposes, however, *the earth serves well enough as an inertial frame of reference and hence also the interior of a vehicle* (train, car etc) *moving smoothly and with a constant velocity on its surface.*

It may as well be pointed out here that the *importance of an inertial frame of reference* lies in that (*i*) *all our fundamental laws of physics have been formulated with respect to such a frame*, (*ii*) it is a basic assumption of the special theory of relativity that *all fundamental physical laws can be so expressed as to assume the same form in all inertial frames of reference* and (*iii*) *such a frame of reference is supposed to be isotropic* with respect to mechanical and optical experiments.*

16.3. Galilian transformation. If we observe an event from two different frames of reference, it will naturally have a set of coordinates in one, different from that in the other. Equations relating the two sets of coordinates of the event in the two frames of reference are called *transformation equations*, obviously because they enable us to transform the observations made in one reference frame into those made in the other. If the two reference frames be inertial or Galilian ones and if mass and time be assumed to be independent of any frame of reference, in accordance with Newtonian mechanics, the transformation is referred to as *Galilian transformation*. And if a quantity is characterised by the same number in the two frames of reference, it is said to be *invariant to the transformation.*

Let us now consider some typical cases of Galilian transformation.

(*i*) **Position.** Let S and S' be two inertial frames of reference, with S at rest and S' moving with a constant velocity v with respect to it in the positive direction of x, where v is very much less than the velocity of light in free space (c). (Fig. 16.1). Let the two observers be stationed at the origins of the two frames, O and O', with their clocks set to zero when O and O' coincide with, or lie opposite, each other, so that time is measured from the instant that the observers just cross each other.

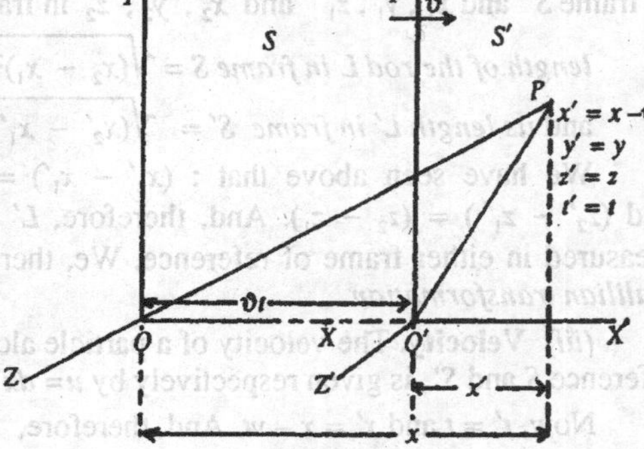

Fig. 16.1

Let the position of an event at P, as determined by the observer in frame S, after time t, be given by the coordinates x, y and z and its position, as determined by the observer in frame S', by the coordinates x', y' and z' and its time of occurrence by t and t', as noted on their respective clocks.

Clearly, since S' is moving relative to S with a constant velocity v in the $+x$ direction, the x coordinate of P in frame S will be greater than that in frame S' by a distance vt covered by S' in time t. So that, $x' = x - v\,t$. And, as there is no relative motion between the two frames of reference along the axes of y and z, we have $y' = y$ and $z' = z$. And, since time is supposed to be independent

* *i.e.* having the same properties in all directions.

of any frame of reference, we $t' = t$. Thus, the four *transformation equations* for the position of the event as observed in frame S' are

$$x' = x - vt, \; y' = y, \; z' = z \text{ and } t' = t$$

The *inverse transformation equations* (*i.e*, if S' be supposed to be at rest and S moving relative to it with velocity v in the negative direction of x) are easily obtained by replacing v by $-v$, when we have

$$x = x' + vt, \; y = y', \; z = z' \text{ and } t = t'$$

Since the position of the event is not the same as measured in the two frames of reference, we say that *position is not invariant to Galilian transformation.*

(*ii*) **Distance or length.** Suppose two events occur at two points P_1 and P_2, some distance apart along the axis of x, such that their coordinates in the stationary frame S are x_1, y_1, z_1 and x_2, y_2, z_2 and in the moving frame S', x_1', y_1', z_1' and x_2', y_2', z_2' respectively. The distance between the two events is thus $(x_2 - x_1) = \Delta x$, as measured in frame S and $(x_2' - x_1') = \Delta x'$, as measured in frame S'.

Now, the transformation equations for the event at P_1, in accordance with (*i*) above, are $x_1' = x_1 - vt$, $y_1' = y_1$ and $z_1' = z_1$. Similarly, the transformation equations for the event at P_2 are $x_2' = x_2 - vt$, $y_2' = y_2$ and $z_2' = z_2$.

Clearly, therefore, $x_2' - x_1' = \Delta x' = (x_2 - vt) - (x_1 - vt) = x_2 - x_1 = \Delta x$ *i.e., the distance between the two events, as measured in either frame of reference, is the same, irrespective of the values of v and t, provided the observations are made simultaneously in the two frames.*

It can be easily seen that if Δx be the length L of a rod, as measured in frame S, with its end-points at x_1 and x_2, and $\Delta x'$, its length L', *as measured in frame* S', with its end-points at x_1' and x_2' we have $\Delta x' = L' = \Delta x = L$, *i.e.*, the length of the rod is the same as measured in either frame of reference.

Alternatively, taking the coordinates of the end-points of the rod to be x_1, y_1, z_1 and x_2, y_2, z_2 in frame S' and x_1', y_1', z_1' and x_2', y_2', z_2' in frame S', we have

length of the rod L in frame S $= \sqrt{(x_2 - x_1)^2 + (y_2 - y_1)^2 + (z_2 - z_1)^2}$

and *its length L' in frame* $S' = \sqrt{(x_2' - x_1')^2 + (y_2' - y_1')^2 + (z_2' - z_1')^2}$

We have seen above that : $(x_2' - x_1') = (x_2 - x_1)$. Similarly, $(y_2' - y_1') = (y_2 - y_1)$ and $(z_2' - z_1') = (z_2 - z_1)$. And, therefore, $L' = L$, *i.e.*, the length of the rod is the same, as measured in either frame of reference. We, therefore, say that *distance or length is invariant to Galilian transformation.*

(*iii*) **Velocity.** The velocity of a particle along the axis of x, as observed in the two frames of reference S and S', is given respectively by $u = dx/dt$ and $u' = dx'/dt'$

Now, $t' = t$ and $x' = x - vt$. And, therefore,

$$u' = \frac{dx'}{dt'} = \frac{d}{dt}(x - vt) = \frac{dx}{dt} - v = u - v.$$

The inverse transformation obviously gives $u = u' + v$, showing that the velocities of the particle, as observed in the two frames, are not the same. In other words, *velocity is not invariant to Galilian transformation.*

Either of the relations for u and u', it will be noted, is the same as obtained by the Newtonian law of composition of velocities and is referred to here as the *Galilian law of addition of velocities.*

(*iv*) **Acceleration.** Let the acceleration of a particle along the axis of x, as measured in the two frames of reference S and S' be a and a' *respectively. Then, clearly,* $a = du/dt$ and $a' = du'/dt$. Since $u' = u - v$, as we have seen under (*iii*) above, we have

$$a' = \frac{d}{dt}(u - v) = \frac{du}{dt} - 0 = a,$$

i.e., the acceleration of the particle is the same, as measured in either frame of reference. And thus, *acceleration is invariant to Galilian transformation.*

16.4. Classical or Newtonian relativity – Galilian invariance. That Newton himself had his own conception of relativity is borne out by the statement in one of the corollaries to his laws of motion that '*the motions of bodies included in a given space are the same among themselves whether that space is at rest or moves uniformly forward in a straight line*'. This means, in other words, that even if we are drifting along at a uniform speed as, for example, in a closed space-ship or even a smoothly running railway carriage, all the experiments performed and all the phenomena observed inside the ship or the carriage will appear to be the same as though the ship or the carriage were at rest and we cannot, therefore, determine whether we are at rest or in uniform motion.

This is the concept of Newtonian or Classical relativity and clearly implies that the fundamental laws of physics are identical in all inertial or Galilian frames of reference, *i.e.*, they are invariant to Galilian transformation – hence the term '*Galilian invariance*' applied to them.

Let us now make a rapid survey of some of these fundamental laws and see whether they are really characterised by Galilian invariance.

1. Newton's law of motion. Considering the second law as the real law of motion, let F and F' be the forces acting on a particle and a and a', its accelerations respectively as observed in two inertial frames of reference S and S' such as those shown in Fig. 16.1. Then, if m be the mass of the particle, supposed to be quite independent of velocity, we have $F = ma$ and $F' = ma'$. Since $a = a'$, as we have seen under § 16.3, (iv) above, we have $F = F'$, indicating that the force acting on the particle is the same in either frame of reference or *that Newton's law of motion is invariant to Galilian transformation.*

2. Law of conservation of momentum. Let two particles of masses m_1 and m_2 moving initially with velocities u_1 and u_2 respectively, along the axis of x, as observed in a reference frame S, collide head on with each other (elastically or inelastically*) and let their respective velocities, after the collision be v_1 and v_2, also along the axis of x. Then, in accordance with the law of conservation of momentum, we have

Total momentum before collision = total momentum after collision,

i.e., $$m_1u_1 + m_2u_2 = m_1v_1 + m_2v_2 \qquad ...(i)$$

If the velocities of the particles before and after the collision, as observed in the reference frame S' moving with constant velocity v relative to S along the x-axis, be u_1', u_2' and v_1', v_2' respectively, we should have, in accordance with the law of conservation of momentum,

$$m_1u_1' + m_2u_2' = m_1v_1' + m_2v_2' \qquad ...(ii)$$

Let us see if this is so.

We have, in accordance with Galilian transformation,

$$u_1 = u_1' + v,\ u_2 = u_2' + v,\ v_1 = v_1' + v \text{ and } v_2 = v_2' + v.$$

Substituting these values of u_1, u_2, v_1 and v_2 in relation (*i*) above, we have

$$m_1(u_1' + v) + m_2(u_2' + v) = m_1(v_1' + v) + m_2(v_2' + v).$$

or, $$m_1u_1' + m_2u_2' + v(m_1 + m_2) = m_1v_1' + m_2v_2' + v(m_1 + m_2),$$

* The law of conservation of momentum holds good in either case, but whereas in an elastic collision, the entire kinetic energy of the particles before collision appears as kinetic energy even after the collision, in the case of an inelastic collision, part of it may appear in the form of heat or some other form of excitation energy after the collision and only the rest is kinetic energy.

whence, $m_1u_1' + m_2u_2' = m_1v_1' + m_2v_2'$

showing clearly that the law of conservation of momentum holds good in either frame of reference and is thus *invariant to Galilian transformation.*

It may be carefully noted, however, that whereas the momenta of the individual particles are not the same in the two reference frames, their total momentum remains the same, *i.e.*, the momentum of the system, as a whole, remains conserved, thus, *not momentum by itself, but the law of conservation of momentum, is invariant to Galilian transformation.*

3. Law of conservation of energy. In case 2 above, the total energy of the particles before and after the collision, as observed in frame S is respectively $\left(\frac{1}{2}m_1u_1^2 + \frac{1}{2}m_2u_2^2\right)$ and $(\frac{1}{2}m_1v^2 + \frac{1}{2}m_1v_2^2 + \Delta E)$, where, in the event of the collision being inelastic, (ΔE) is part of the energy of the system, before collision, appearing in the form of heat or some other form of excitation energy after the collision. So that, in accordance with the law of conservation of energy, we have

$$\frac{1}{2}m_1u_1^2 + \frac{1}{2}m_2u_2^2 = \frac{1}{2}m_1v_2^2 + \frac{1}{2}m_2v_2^2 + \Delta E.$$

Now, as we know, $u_1 = u_1' + v$, $u_2 = u_2' + v$, $v_1 = v_1' + v$ and $v_2 = v_2' + v$, where u_1', u_2' and v_1', v_2' are the velocities of the two particles before and after the collision respectively, as observed in frame S' and v, its velocity relative to S.

Since m_1 and m_2 are supposed to remain unchanged and, (as shown by actual experiment), (ΔE) remains the same, we have from (i) above

$$\frac{1}{2}m_1(u_1' + v)^2 + \frac{1}{2}m_2(u_2' + v)^2 = \frac{1}{2}m_1(v_1' + v)^2 + \frac{1}{2}m_2(v_2' + v)^2 + \Delta E.$$

Or, $$\frac{1}{2}m_1(u_1'^2 + v^2 + 2u_1'v) + \frac{1}{2}m_2(u_2'^2 + v^2 + 2u_2'v) = \frac{1}{2}m_1(v_1'^2 + v^2 + 2v_1'v) + \frac{1}{2}m_2(v_2'^2 + v^2 + 2v_2'v) + \Delta E.$$

Or, $$\frac{1}{2}m_1u_1'^2 + \frac{1}{2}m_2u_2'^2 + v(m_1u_1' + m_2u_2') = \frac{1}{2}m_1v_1'^2 + \frac{1}{2}m_2v_2'^2 + v(m_1v_1' + m_2v_2') + \Delta E.$$

As we have seen under 2 above, the momentum of the system remains conserved in either frame of reference, so that

$$m_1u_1' + m_2u_2' = m_1v_1' + m_2v_2'.$$

And, therefore, $$\frac{1}{2}m_1u_1'^2 + \frac{1}{2}m_2u_2'^2 = \frac{1}{2}m_1v_1'^2 + \frac{1}{2}m_2v_2'^2 + \Delta E,$$

showing clearly that the law of conservation of energy holds good in either frame of reference or that it is invariant to Galilian transformation.

Again, however, it will be seen that, although the energies of the individual particles are different in the two frames of reference, their total energy remains the same. *It is thus the law of conservation of energy which is invariant to Galilian transformation and not energy by itself.*

Now, although all these fundamental laws of physics are thus found to be invariant to Galilian transformation, *i.e.*, to be the same in all inertial frames of reference, this *is so under the severe limitation of mass and time being arbitrarily assumed to be invariant in all frames of reference, there being really hardly any justification for it.* As we shall shortly see, when we deal with Einstein's theory of relativity, they do change with velocity, though the change manifests itself to any appreciable extent only at very high velocities, comparable with that of light in free space, *i.e.*, c, the constancy of which alone in all frames of reference, it may be pointed out, forms a basic postulate of Eeinstein's theory. Our fundamental laws of physics are, therefore, invariant, strictly, not to Galilian but to what is called *Lorentz transformation* (see § 16.10). At velocities very much lower than c, however, all Lorentz transformations too get reduced to Galilian ones.

16.5 Non-inertial frames of reference - Fictitious forces. An *inertial* frame of reference, as we know, is either at rest or moves with a uniform velocity and is thus *unaccelerated.* It follows, therefore, that *a frame of reference which has acceleration, either in its linear motion or because it moves in a circle, is a non-inertial one.* It is found that even if no force be acting on a particle in an inertial frame, a force seems to be acting on it, as observed in an accelerated or a non-inertial frame, whether in linear motion with respect to the inertial frame or in circular motion about it, as will be seen from the following.

(*i*) A reference frame with linear acceleration. Let a frame of reference S' be moving with a linear acceleration a_0 with respect to an inertial frame S. Then, clearly, a particle at rest in frame S will appear to have an acceleration $-a_0$ in frame S'. And, therefore, a particle having an acceleration a in the former frame will appear to have an acceleration $a = (a - a_0)$ in the latter. So that, if m be the mass of the particle (Supposed to remain the same in either frame), we have

force acting on the particle in frame S, say, $F = ma$

and *force acting on the particle in frame S'*, say, $F' = m a'$.

Since $a' = (a - a_0)$, we have $F' = m(a - a_0) = ma - ma_0 = F - ma_0$.

or, putting $-m a_0 = F_0$, we have $F' = F + F_0$.

If therefore, $F = 0$, we have $F' = F_0$,

i.e., even if no force be acting on the particle in frame S, a force $F_0 = -m a_0$ seems to be acting on it in frame S'.

Newton's law of motion, therefore, does not hold good in frame S', which again shows that it is a *non-inertial* frame.

Now, this force $F_0 = -m a_0$ does not actually exist, but appears to come into being *purely in consequence of the acceleration (a_0) of frame S' with respect to S.* It is, therefore, called a **fictitious** or a **pseudo** (*i.e.*,false) **force** and is clearly equal to (*mass of particle*)× (*acceleration of the non-inertial frame, with sign reversed*) $= m(-a_0)$ or $-ma_0$.

As will be readily seen, this *fictitious force* F_0, when added to the *true force (F)* acting on the particle in the inertial frame S, gives the *observed force* F' acting on it in the accelerated frame S'. It follows at once, therefore, that Newton's law of motion will also be obeyed in frame S', *i.e.* it will also behave as an inertial frame, if we add to the true force F, the fictitious force $F_0 = -ma_0$.

A common example of fictitious force is the force that appears to act on a particle (or a body) at rest in a lift which is moving down with an acceleration equal to g. The fictitious force acting on the particle is clearly given by F_0 =(*mass of the particle*) × (*acceleration of the lift, or the moving frame, with sign reversed*) $= m(-g) = -mg$. So that, the resultant force on it, as observed by an observer in the moving frame or the lift is $F' = mg - mg = 0$. The particle thus appears to be *weightless* in the lift or to be just suspended in the air – a condition we hear so much about in artificial satellites.

An important aspect of a fictitious force is that it enables us to determine whether or not a given frame of reference is accelerated or non-inertial. For if two reference frames be in uniform relative motion (*i.e* with zero relative acceleration), they are obviously inertial frames and it is impossible to determine which one of them is at rest and which one in motion. If , however , one of them has a linear acceleration with respect to the other, a fictitious force comes into play in the accelerated frame.

Now, whereas a true force diminishes rapidly with distance and is almost negligible at a large distance from the earth or any other national body, the fictitious force, arising purely due to the acceleration of the frame, goes on increasing with the acceleration. So that, at a large distance from a material body, the real force is practically nil but the fictitious force is quite appreciable

Hence, if a particle or a body, far away from other material particles or bodies, experiences an appreciable force, its frame of reference is clearly an accelerated or a non-inertial one.

***(ii)* A reference frame in uniform circular motion.** If a frame of reference S' be rotating about an inertial frame S with an angular velocity ω, it can be shown that even if no force be acting on a moving particle in the inertial frame, a fictitious force appears to be acting on it in the rotating frame S'. given by $F_0 = -2\,m\,\omega\,v' - m\,\omega^2\,r$, where v' is the linear velocity of the particle, as observed in frame S'.

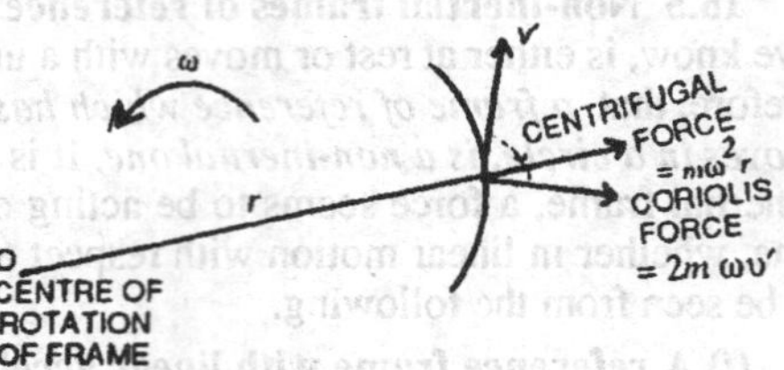

Fig. 16.2

Here, $m\,\omega^2\,r$ is the familiar **centrifugal force** and $-2\,m\omega\,v'$, what is called **Coriolis force (and, therefore, $-2\,\omega\,v'$, *(Coriolis acceleration)*, after *G.Coriolis,* who first discovered it. So that, we have fictitious force F_0 = Coriolis force + Centrifugal force.**

It follows, therefore, that *centrifugal force is actually a fictitious force* acting on a particle at rest with respect to the rotating frame. And, since it keeps the particle at rest in the rotating frame, it is equal in magnitude but opposite in direction to the centripetal force acting on the particle on account of rotation of the frame.

The *Coriolis force too is a fictitious force* which appears to act on the particle in the rotating frame, when it is in linear motion with respect to the frame, as, for example, a fly or an ant walking on a rotating piece of card board.

As will be readly seen, *it is proportional to the angular velocity* (ω) *of the rotating frame and the linear velocity of the particle* (v') *with respect to it.* Its direction, being perpendicular to that of v', is easily obtained by a 90° rotation from v' in the opposite sense to that of ω, as shown by the dotted arrow head in Fig. 16.2.

The Coriolis force is obviously *zero* when either v' is *zero* or ω is *zero, i.e.*, when either the particle is at rest with respect to the rotating frame (in which case, the centrifugal force alone acts on the particle) or when the frame is a non-rotating or an inertial one.

It will also be seen that Newton's law of motion would be obeyed in the rotating frame S' and it would, therefore, behave as an inertial frame if we add to the observed force on the particle in it the two fictitious forces. *viz.*, the *Coriolis* and the centrifugal forces.

16.6 Search for the absolute frame of reference. It was undoubtedly the genius of *Newton* that, being physically located in an earth-bound (and, therefore, accelerated) frame of reference, he could mentally detach himself from it and imagine and interpret events as though he were in an unaccelerated or inertial frame of reference and enunciate his laws of motion accordingly.

Unfortunately, however, knowing fully well that '*translatory motion can be perceived only in the form of motion relative to other material bodies*', he still insisted that there must exist a *fundamental* or an *absolute frame of reference*, which he called *absolute space* and which, according to him, '*in its own nature, without regard to anything external, remains always similar and immovable*'. With the weight of his great authority behind the idea, the search for this hypothetical absolute space began and continued for a long time in right earnest. Here is a brief account of this infructuous quest :

In the year 1864, *Maxwell* had clearly shown the inter-relationship of electricity, magnetism and light when, from the known properties of electricity and magnetism, he formulated his justly celebrated theory of electromagnetic radiation and gave his famous equations of electromagnetic field, representing a wave phenomenon. He thus established the existence of electromagnetic waves in space, travelling with the speed of light. In other words, ***he showed light to be an electromagnetic wave.***

Now all the waves known hitherto *(e.g.* water waves, sound waves etc) required a material medium (water, air etc) for their propagation. It was thought imperative, therefore, that there should also be a suitable medium for the propagation of these newfound electromagnetic waves, which travelled even through the empty space between the earth and the stars. Although no one had the slightest inkling as to what this medium could possibly be, it came to be called '*luminiferous ether*' (or, usually, simply *ether*) and, though quite invisible, it was supposed to pervade all space, empty or otherwise.

Then, in order to be able to transmit light – a transverse wave motion – it had to be a perfectly transparent, rigid solid; and, in view of the high velocity of light, it had to have a large shear modulus and yet the earth and the other planets could continue in their regular courses without encountering the slightest resistance from it.

All this may seem quite incredible to us today but no one, at the time, seemed to object to the existence of a medium with such contradictory and mutually exclusive properties. On the other hand, it was felt that this was perhaps the very absolute space or fundamental frame of reference Newton was looking for and in which (and in other reference frames fixed relative to it) his laws of motion will hold good with perfection.

Quite a few workers engaged themselves in this challenging task of establishing the existence of *ether* and devised many a brilliant experiments, all based naturally on light waves, for ether was little else but a vehicle for them.

The high sensitivity demanded of the apparatus may be seen from the fact that whereas the velocity of light in free space is 3×10^{-5} *km/sec*, the orbital velocity of the earth around the Sun and, therefore, through the ether, is only 30 *km/sec*, which must also be the velocity of the expected '*ether drift*' resulting from it, and, as can be easily seen, it is a mere $30/3 \times 10^{-5}$ or 10^{-4} of the velocity of light. In view of this, *Maxwell* had actually given up all hope that any terrestrial experiment would be sensitive enough to detect this small ether-drift and started looking for any astronomical evidence that might be of help.

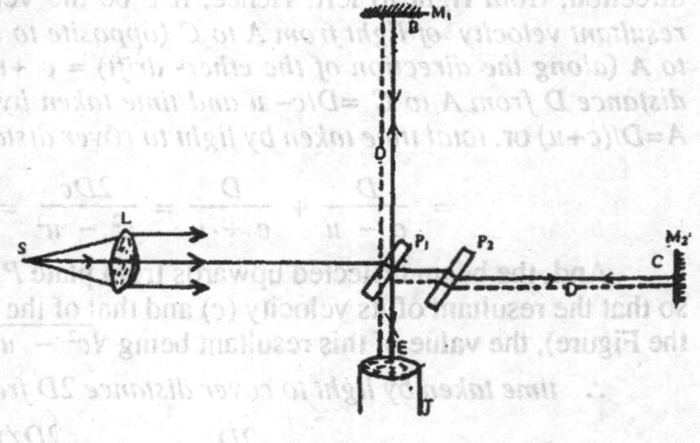

Fig. 16.3

Luckily, *A.A.Michelson* of the USA, who learnt about the matter in the 1879, accepted the challenge and in collaboration with his colleague, *E.W.Morley*, designed and performed the historic experiment in the year 1887, which dealt a fatal flow to the laboriously built up ether - hypothesis, (see § 16.7)

16.7 Michelson-Morley experiment. The apparatus, based on interference of light and hence called '*interferometer*', is shown in Fig.16.3

A parallel beam of monochromatic light from a source S is allowed to fall on an optically plane, thin and parallel-sided glass plate P_1, thinly silvered on its back face, such that it splits up the incident beam into two separate beams of equal intensity and at right angles to each other at A, a reflected beam upwards and a refracted or transmitted beam in its original direction.

The two beams suffer normal reflections at B and C respectively as they fall on plane mirrors M_1 and M_2, silvered heavily on their front faces to avoid multiple reflections and arranged at right angles to each other at the same distance from A. Travelling back along their own paths, the two beams again fall on plate P_1, with the former passing through it and the latter getting reflected at its back surface at A, so as to travel along the same path AE on to the telescope T.

Since the beam AB, suffering reflection at M_1, has to traverse the thickness of plate P_1 thrice and the beam AC, reflected at M_2, only once, their paths through glass are obviously not equal. To make them exactly equal, therefore, an unsilvered compensating plate P_2, identical with plate P_1 in all other respects, is arranged parallel to P_1 as shown, so that beam AC has to traverse it *twice* on its way to, and back from, M_2 and is thus made to cover the same total thickness of glass as beam AB.

Now, if the incident beam falling on plate P_1 be an exactly parallel one and distances AB and AC be exactly equal, the two beams travelling along AB and AC and getting reflected at M_1 and M_2 respectively, arrive at the telescope in the same phase and thus reinforce each other, making the field of view appear bright. But if their paths differ by an old number of half wavelengths, they arrive at the telescope in opposite phases and thus nullify each other and the field of view appears dark. Since, however, the incident beam falling on plate P_1 is almost in variably slightly convergent, some of the rays constituting the two beams arrive at the telescope in phase and some out of phase, so that what we actually see through it is an *interference pattern of alternate bright and dark fringes of equal width* which, if the mirrors M_1 *and* M_2 be not exactly perpendicular to each other, take the form of *straight and parallel fringes*. The apparatus is sensitive enough to be able to detect and measure a shift as small as one-hundredth of a fringe.

Clearly, if u be the orbital velocity of the earth, and hence that of the apparatus relative to the ether, say, from left to right, the drift (velocity of ether relative to it will be $-u$, *i.e.*, in the opposite direction, from right to left. Hence, if c be the velocity of light, *the same in all directions, the resultant velocity of light from A to C (opposite to that of the ether) drift)* $=c-u$ and *that from C to A (along the direction of the ether- drift)* $= c + u$. And, therefore, *time taken by light to cover distance D from A to C* $=D/c-u$ *and time taken by light to cover the same distance D from C to A*$=D/(c+u)$ or, *total time taken by light to cover distance 2 D from A to C and back*—

$$= \frac{D}{c-u} + \frac{D}{c+u} = \frac{2Dc}{c^2-u^2} = \frac{2Dc}{c^2(1-u^2/c^2)} = \frac{2D/c}{(1-u^2/c^2)} = t, \text{ say.}$$

And, the beam reflected upwards from plate P_1 at A, must clearly be made to travel along AG so that the resultant of its velocity (c) and that of the ether- drift ($-u$) may be along AB (see inset in the Figure), the value of this resultant being $\sqrt{c^2-u^2}$ both along AB and back.

$\therefore$ *time taken by light to cover distance 2D from A to B and back*

$$= \frac{2D}{\sqrt{c^2-u^2}} - \frac{2D/c}{\sqrt{1-u^2/c^2}} = t'$$

It may be noted that the numerators of the expressions for t and t are the same ($2D/c$), representing the time that light would take to travel from A to either mirror and back *if the apparatus were at rest*.

Since u^2/c^2 is always a very small quantity, we have $t > t'$,even though the distance covered is the same ($2D$) in either case, and we have

$$t - t' = \Delta t = \frac{2D/c}{1-u^2/c^2} - \frac{2D/c}{\sqrt{1-u^2/c^2}} = \frac{2D}{c}\left(1-\frac{u^2}{c^2}\right)^{-1} - \frac{2D}{c}\left(1-\frac{u^2}{c^2}\right)^{-\frac{1}{2}}$$

Expanding by Binomial theorem and remembering that $u << c$, we have

$$\Delta t = \frac{2D}{c}\left[\left(1+\frac{u^2}{c^2}\right) - \left(1+\frac{1}{2}\frac{u^2}{c^2}\right)\right] = \frac{2D}{c}\left(\frac{u^2}{2c^2}\right) = \frac{Du^2}{c^3}$$

Clearly, *distance covered by light in this interval of time* $= \Delta t \times c = \frac{D u^2}{c^3} \times c = \frac{D u^2}{c^3}$, indicating that the optical path along AC is longer by Du^2/c^2 than that along AB. This, then, is the *path - difference introduced between the two parts of the incident beam (suffering reflection at mirror* M_1 *and* M_2 *respectively), due to motion of the apparatus.*

The entire apparatus is rotated through 90°, so that AB now lies along the direction of motion of the apparatus and AC at right angles to it, with the optical path along AB now longer by Du^2/c^2 than that along AC. *The total path difference* introduced between the two interfering beams is thus $2Du^2/c^2$ and this naturally brings about a shift in the interference pattern, say, through n fringes.

Since a path difference of one full wavelength λ results in a shift of the interference pattern through one fringe, a shift through n fringes means a path difference of $n\ \lambda$. We, therefore, have

$$n\lambda = 2\,Du^2/c^2\,, \quad \text{whence,}\ n = 2\,Du^2/c^2\,\lambda$$

Michelson and *Morley* effectively increased the distance D to nearly 11 *metres* (by repeated reflections back and forth) and, in order that the apparatus may rotate without any strain, they mounted it on a stone slab, floated in mercury and kept rotating slowly at the rate of 10 rotation per hour. So that, with the velocity of light $c = 3 \times 10^{10}$ *cm/sec*, the orbital velocity of the earth, *i.e.*, the velocity of the apparatus, $u = 30$ *km/sec* or 3×10^{6} *m /sec* and wavelength of the light used, ($\lambda = 6 \times 10^{-5}$ *cm*), the value of n was expected to be $2 \times 1100 \times 9 \times 10^{12} / 9/\times 10^{20} \times 6 \times 10^{-5} = 0.37 \approx 0.4$or two-fifth of a fringe and could thus be easily measured with the given apparatus which could, as mentioned earlier, measure even a shift equal to one-hundredth of a fringe-width.

The actual fringe-shift observed was negligibly small, *showing that there was no relative motion between the earth and the ether.*

The experiment has since been repeated both below the earth's surface and at high attitudes, at different times of the year and even with highly monochromatic light of a laser beam. The result, however, has been a negative one all along. Other experiments too to discover any ether- drift have similarly failed miserably.

Still, obsessed with the idea that there must exist this elusive medium ether, representing the absolute reference frame conceived by *Newton*, vigorous attempts were made to explain away the negative results of all these experiments. *Michelson* himself came forward with the suggestion that since the earth dragged along with it the ether in its immediate neighbourhood, no relative drift between the two could be observed. *Lodge* showed this explanation to be quite untenable when, in the year 1892, he measured the velocity of light near rapidly rotating bodies and came to the conclusion that hardly half a percent of the velocity of light could thus be communicated to the ether.

Fitzgerald and *Lorentz*, in that very year, hazarded the suggestion that due to an interaction between ether and a material body moving relative to it, there was a shrinkage in the dimensions of the body along the direction of its relative velocity, such that if L_0 be the length of the body when at rest and u, its velocity along the direction of its length, it will shrink to a smaller length $L = L_0\sqrt{1 - u^2/c^2}$. According to this view, although distance D along AB, at right angles to the direction of motion of the apparatus, will remain unaltered, the same distance along AC, along the direction of motion of the apparatus, will get shortened to $D\sqrt{1 - u^2/c^2}$. . So that, putting this in place of D in the expression for t above, we shall have

$$t = \frac{(2D/c)\,(\sqrt{1 - u^2/c^2}\,)}{1 - u^2/c^2} = \frac{2D/c}{\sqrt{1 - u^2/c^2}} = t'.\ \text{Or,}\ t - t' = \Delta t = 0.$$

Although this contraction hypothesis thus readily explained the negative result of *Michelson-Morley* experiment, it could not gain acceptance for the simple reason that it was put forward with the specific purpose of explaining the observed fact, with hardly any logic behind it.

The true explanation of the failure of Michelson-Morley and other experiments to discover any ether- drift was furnished by *Einstein* in the year 1905, when, based on the velocity of light (c) in free space being a universal constant, he announced to the world his *special theory of relativity*, with which we shall deal very shortly.

16.8. Einstein's concept of relativity. With the failure of the experiments searching for the ether- drift, it began to be increasingly accepted that there was no such thing as an absolute, fixed or a privileged frame of reference and that the basic laws of physics assumed the same form in all inertial frames.

Poincare came nearest to suggesting the theory of relativity when, emphasising the implications of Galilian invariance, he declared: '*the laws of physical phenomena [are] the same whether for an observer fixed or for an observer carried along in a uniform motion of translation, so that we have not and could not have any means of discerning whether or not we are carried along in such a motion*'. This means, in plain and simple language, that, drifting along with uniform speed in a space- ship, with the, windows closed, we cannot find out by means of any experiments we might choose to perform inside the ship, whether we are at rest or in motion. Even if we work out of an open window, we can at best say that we are in relative motion with respect to the 'fixed stars' but not whether we or the 'fixed stars' are in actual motion.

Unfortunately, instead of following up the idea and building it into a theory, he fell into the same error as the others and tried to adjust and modify his conclusions with the aim of bringing them in a line with the classical or the Newtonian view of an ether- drift.

It was only *Einstein* who understood the correct meaning of the failure of the ether-drift experiments and used his own great powers of objective reasoning to come to the *conclusion that the introduction of luminiferous ether was superfluous since no absolutely stationary frame of reference having special properties, was really necessary.* thus, boldly breaking away from the ideas witherto held about *space* and *time*, he formulated and announced his *special theory of relativity* in the year 1905 and his more complex *General* theory of relativity in the year 1915, the former being concerned with problems associated with unaccelerated frames of reference and the latter with those associated with accelerated ones.

We shall restrict ourselves, here, to the special theory of relativity which we shall now proceed to study in necessary detail.

16.9. The special theory of relativity. The following are the two basic postulates of the theory :

1. **All the laws of physics have the same identical form in all inertial frames.**
2. **The velocity of light in free space is the same (c) in all inertial frame, i.e.,** ***is invariant to transformation from one inertial frame to another. It has thus the same value*** $c = 3 \times 10^{10}$ ***cm/sec for all observers, irrespective of their state of motion.***

As will be easily seen, the first postulate is just a generalised statement of the observed fact that there is no distinction between the state of rest and of uniform motion— *a direct consequence of the absence of an absolute frame of reference.*

And the second postulate is obviously a statement of the result obtained by *Michelson* and *Morley* in their historic experiment. It is clearly at variance with Galilian transformation ; for, in accordance with the latter, if the velocity of light be c in an inertial frame S its value in another inertial frame S', *moving with a relative velocity* v with respect to it, should be $c-v$, whereas, in accordance with this postulate, it must be the same c whatever the state of motion of the frame of reference.

It is in consequence of this constancy of the value of c in all frames of reference that if a flash of light he produced at the origin O of an inertial frame S just at the moment that the origin O' of another inertial frame S' passes it with a relative velocity v along the $+x$ direction, the observers at O and O', even when they have drifted apart, will each claim to be at the centre of the expanding spherical wavefront, as shown in Fig. 16.4. For, if the wavefront does not appear to be spherical to the observer at O', with himself at its centre, it will enable him to find out his state of relative motion with respect to

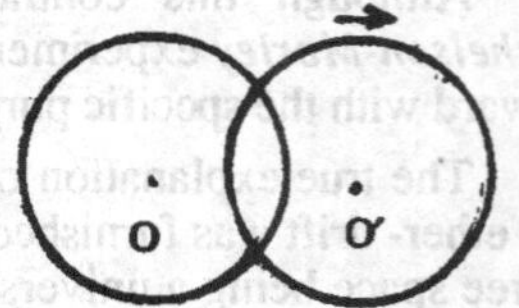

Fig. 16.4

the observer at O and this the first postulate clearly forbids.

The case will, of course, be entirely different if the two observers, instead of being in free space, were stationed in two boats S and S', *and if instead of a flash being produced, a stone were dropped into the water from boat* S just at the moment that boat S', crosses it with relative velocity v. For, the ripple pattern formed will not appear to be the same to the two observers and they will thus be able to decide whether or not they are in motion or whether one is at rest and the other in motion.

The difference in the two cases is obviously due to the fact that whereas water, by itself, functions as a frame of reference, free space does not, and whereas the speed of the water waves varies with the motion of the observer, that of the waves in free space (*i.e.*, light waves or electrodynamic waves) is quite independent of it.

16.10. Lorentz transformation. As we have just seen in the preceding article, Galilian transformation is at variance with the constancy of the velocity of light in free space and thus being quite independent of the motion of its own source or that of the observer. We, therefore, look for a transformation consistent with this new concept.

Referring back again to the last article, we have seen how if a flash of light is produced at the origin O of a frame S just when the origin O' of frame S' crosses it with relative velocity v along the $+x$ direction, each of the two observers, at O and O' respectively, considers himself to be at the centre of the expanding spherical wavefront. So that, the *equation of the wavefront for the observer at O in frame S is* $x^2 + y^2 + z^2 = c^2 t^2$*...(i) and equation of the wavefront for the observer at O′ in frame S′ is* $x'^2 + y'^2 + z'^2 = c^2 t^2$*,...(ii) the velocity of light being the same, c*, in either frame.

In order, therefore, that both the observers may seem to lie at the centre of the same expanding spherical wavefront, we must clearly have equation (*i*) equal to equation (*ii*).

Now, in accordance with Galilian transformation, $x' = x-vt$, $y' = y$, $z' = z$ and $t' = t$. Substituting these values in equation (*ii*), we have $x^2 - 2xvt + v^2t^2 + y^2 + z^2 = c^2 t^2$, ..(*iii*) which is clearly not the same as equation (*i*), showing once again the failure of Galilian transformation in the face of the constancy of the value of c, the velocity of light in free space.

In our new transformation, therefore, the relation $x' = x-vt$ will no longer remain valid. Since, however, it is consistent with the classical or Newtonian mechanics, the x-coordinate must, in the new transformation also, acquire this value when the relative velocity v is negligibly small compared with c.

The relation for the x-coordinate in the new transformation must thus be a *linear* one to provide for a uniformly expanding wavefront but must admit of one and only one interpretation of any observation made from frame S in frame S', *i.e.*, must give only one value of x' for any given value of x, without imposing any further conditions. A quadratic equation will give two and a higher order equation more than two values of x' for a given value of x and these are, therefore, out of the question. The simplest relation that suggests itself is thus

$$x' = \gamma (x - v.t) \quad ...(iv)$$

where γ is a constant of proportionably, independent of both x and t, though possibly dependent on v.

The inverse transformation connecting measurements in frame S with these made in frame S' will be obtained by simply reversing the sign of v, when we shall have $x = \gamma (x' + vt')$....(*v*) the constant of proportionality (γ) remaining the same, because the first postulate denies the existence of any preferred or absolute frame. and, the motion of frame S' being along the $+x$ direction, we have $y' = y$ and $z' = z$

But, clearly, $t' \neq t$ or else relation (*iv*) and (*v*) cannot both be true.

The value of t' may be obtained by substitating the value of x' from equation (*iv*) in equation (*v*), when we have

$$x = \gamma\, [\gamma(x - v.t) + v.t'] = \gamma^2 (x - v.t) + \gamma v.t$$

Or, $\quad \gamma v.t' = x - \gamma^2 (x - vt) = x - \gamma^2 x + \gamma^2 .vt = \gamma^2 v.t' + x(1 - \gamma^2)$

which gives
$$t' = \gamma.t + x\left(\frac{1-\gamma^2}{\gamma.v}\right) \qquad ...(vi)$$

And the value of γ may also be easily obtained, if we remember that frame S' moves with relative velocity v with respect to S in the $+x$ direction and that the flash of light is emitted when the observers at O and O' are just opposite each other, so that at the time the flash is emitted, $t = t' = 0$ and $x = x' = 0$.

Since the velocity of light is the same, c, in either frame, we have

x-coordinate of the light pulse in frame S given by $x = ct$ ......*(vii)*

and *x′-coordinate of the light pulse in frame S′ given by* $x' = ct'$ ......*(viii)*

Substituting in relation (*viii*) the values of x' and t' from relations (*iv*) and (*vi*) respectively, we have

$$\gamma(x - v.t) = c.\gamma t + cx\left(\frac{1-\gamma^2}{\gamma.v}\right)$$

which, on solving for x gives
$$x = c.t\left[\frac{1+v/c}{1-(c/v)(1/\gamma^2-1)}\right]$$

or, substituting $c.t$ for x from relation *(vii)*, we have

$$\frac{1+v/c}{1-(c/v)(1/\gamma^2-1)} = 1 \quad \text{Or,} \quad 1+\frac{v}{c} = 1-\left(\frac{c}{v}\right)\left(\frac{1}{\gamma^2}-1\right) \qquad ...(ix)$$

whence,
$$\gamma = \frac{1}{\sqrt{1-v^2/c^2}}$$

Since v/c is usually represented by β, we have $\gamma = 1/\sqrt{1-\beta^2}$.

Now, if we substitute this value of γ in relation (*iv*), we have

$$x' = \frac{x - v.t}{\sqrt{1-v^2/c^2}} = \frac{x - v.t}{\sqrt{1-\beta^2}}$$

Since frame S' moves along the $+x$ direction relative to frame S, we have $y' = y$ and $z' = z$

Again,
$$t' = \gamma.t + x\left(\frac{1-\gamma^2}{\gamma.v}\right) = \gamma\left[t + x\left(\frac{1-\gamma^2}{\gamma^2.v}\right)\right], \text{ from relation (vi) above.}$$

And, from relation (*ix*), we have $(1/\gamma^2 - 1) = (1-\gamma^2)/\gamma^2 = -v^2/c^2$. So that,

$$t' = \gamma\left[t - x\frac{v^2}{c^2.v}\right] = \gamma\left[t - \frac{x.v}{c^2}\right]$$

or, substituting the value of γ, we have

$$t' = \frac{t - v.x/c^2}{\sqrt{1-v^2/c^2}}.$$

On substituting these values of x', y', z' and t' in relation (*ii*) above, we obtain relation (*i*), indicating that the two observers find themselves at the centre of the same expanding spherical wavefront.

This, then, is the *relativistic transformation* we have been looking for, in conformity with the new concept of the constancy of the velocity of light in free space, ir-respective of the motion of the source of light or the observer. It is referred to as *Lorentz transformations*, after the Dutch physicist *H.A. Lorentz*, because he was the first to suggest that this was the only possible transformation under which the laws of electricity and magnetism assume the same form in all inertial frames, though it was *Einstein* who later showed it to be *the only proper transformation for all physical laws in general*.

The relations for x', y', z' and t', deduced above, are thus *Lorentz coordinate transformation equations* or, simply, *Lorentz transformation equations.*

For ready reference, the Galilian and Lorentz transformation equations for observations made from frame S in frame S', moving with a uniform relative velocity with respect to it, in the positive direction of x, are set out side by side in the table below :

Galilian transformation	**Lorentz transformation**
$x' = x - v.t$	$x' = \gamma(x - v.t) = \dfrac{x - v.t}{\sqrt{1 - v^2/c^2}}$
$y' = y$	$y' = y$
$z' = z$	$z' = z$
$t' = t$	$t' = \gamma\left(t - \dfrac{vx}{c^2}\right) = \dfrac{t - v.x/c^2}{\sqrt{1 - v^2/c^2}}$

The *inverse transformation equations* are, in both cases, obtained, as we know, by substituting $-v$ for v.

N.B. Since v/c is usually small, only the first two terms of the Binomial expansion of $1/\sqrt{1 - v^2/c^2}$ or $(1 - v^2/c^2)^{-1/2}$ are retained and γ, therefore, taken to be equal to $\left(1 + \frac{1}{2}v^2/c^2\right)$ — this approximation being quite valid upto a value of $v/c = 0.5$.

The above discussion will make it amply clear that *measurements in space and time are no longer absolute but depend upon the relative motion between the observer and the phenomenon observed.*

It must be carefully noted, however, that *if $v \ll c$, i.e.. if $v/c \rightarrow 0$, we have $\gamma = 1/\sqrt{1 - v^2/c^2} \approx 1$ and all Lorentz transformations reduce to corresponding Galilian transformations*, showing that the classical or Newtonian physics is just a particular case of relativistic physics. And, since the orbital velocity of the earth (v) is only about 10^{-4} of the value of c, $v/c \approx 0$, so that *the classical physics is perfectly valid for most of our every day purposes except, of course, when we are dealing with particles like electrons, protons, mesons etc which move with velocities comparable with c.*

Let us now consider a few important cases of this relativistic or Lorentz transformation.

1. Transformation of Length–Length contraction. Let L_0 be the length of a rod laid along the axis of x in a reference frame S at rest with respect to the observer, this length being referred to as its **proper length.** Then, if the coordinates of the two end-points of the rod be x_1 and x_2, clearly, *proper length of the rod,* $L_0 = x_2 - x_1$.

If the coordinates of the end-points of the rod in reference frame S', moving with a uniform relative velocity v with respect to frame S, be x_1', and x_2', *as noted simultaneously* at the same instant t', *we have*

length of the rod in the moving frame S'. say, $L = x_2' - x_1'$

Now, in accordance with inverse Lorentz transformation,

$$x_1 = \frac{x_1' + v.t'}{\sqrt{1 - v^2/c^2}} \quad \text{and} \quad x_2 = \frac{x_2' + v.t'}{\sqrt{1 - v^2/c^2}}.$$

so that,
$$L_0 = x_2 - x_1 = \frac{x_2' + v.t'}{\sqrt{1 - v^2/c^2}} - \frac{x_1' + v.t'}{\sqrt{1 - v^2/c^2}} = \frac{x_2' - x_1'}{\sqrt{1 - v^2/c^2}} = \frac{L}{\sqrt{1 - v^2/c^2}},$$

whence, $L = L_0\sqrt{1 - v^2/c^2}$,

showing that *the length of the rod in the moving frame S' is shorter than its proper length L_0.*

As will be easily seen, the faster the motion of the reference frame (S'), carrying the rod, *i.e.*, of the rod itself, the shorter its length, compared with its proper length L_0 and if the velocity of the rod, $v = c$, *i.e.*, if the rod be moving with the velocity of light in free space, its length L would be *zero.* It is thus clear that *its proper length (when its velocity $v = 0$) is the largest and its length when it moves with the velocity of light (i.e., when $v = c$), the shortest.*

There will obviously be no shortening of the rod if it lies along the axis of y or z, perpendicular to the axis of x or the direction of motion of the reference frame S'. If, however, the rod be lying inclined to the axis of x, its length will have a component along this axis or the direction of motion of the reference frame and there will thus be some slight shortening of its length.

This shortening or contraction in the length of a material object, along its direction of motion, is known as **Lorentz-Filzgerald contraction,** for, as will be recalled (§ 16.7), both *Lorentz* and *G.F. Filzgerald* (an Irish physicist) had independently tried to explain, in the 1890's, the null result of Michelson and Morley's experiment, by making the *ad hoc* suggestion that the length of a material object contracted by a fraction $\sqrt{1 - v^2/c^2}$ if it moves with velocity v through ether, in the direction of its length. The contraction in length is certainly this fraction, as we have just seen above, but for reasons entirely different from that adduced by them (*viz.* interaction with ether).

It may as well be mentioned here that *this contraction in length is a reciprocal effect and works both ways, i.e., is independent of the direction of velocity v, but depends only on its magnitude.* Thus, a length along the axis of x in frame S appears contracted by a fraction $\sqrt{1 - v^2/c^2}$ to an observer in frame S' and a length along the axis of x' in frame S' appears equally contracted to an observer in frame S.

We may, therefore, generalise and say that *with relative motion between a material object and an observer, the length of the object, as measured by the observer, is always shorter than its proper length.*

It follows at once, therefore, that whereas a straight line along the axis of x in a stationary frame S will appear to be shorter in length to an observer in a frame S', in relative motion with respect to S, along the axis of x, a straight line along the axis of y or z will suffer no contraction. As a result, a square and a circle in frame S will appear to be a rectangle and an ellipse respectively to an observer in frame S'.

Now, whereas, strictly, the correct relation for L is the one given above, in actual practice, we hardly come across speeds (v) comparable with that of light in free space (c), so that $v/c \approx 0$ and $\sqrt{1 - v^2/c^2} \approx 1$ and, therefore, $L = L_0$, *i.e.*, the Lorentz-Fitzgeralld contraction is nil and the length of a moving object is the same as its proper length or its length, when at rest with respect to the observer. In other words, the length in all such cases remains an absolute quantity in accordance with classical physics. The relativistic effect of length contraction manifests itself only when v is comparable with c. [See worked example 8].

2. Transformation of time – Time dilation. Suppose we have a reference frame S at rest and a reference frame S' moving with respect to it along the positive direction of x with velocity v. Suppose further that the two reference frames carry *identical* clocks at their origins O and O', such that when frame S' just crosses S, *i.e.*, when O' lies just opposite O, the two clocks show zero time.

Let two events occur at a point x' *in frame* S' at times t_1' and t_2', as noted on the clock carried by it, and at times t_1 and t_2, as noted on the clock carried by frame S. Then, clearly, *time-interval between the two events, as noted on the clock in* S', say $\Delta t' = t_2' - t_1'$ and *time-interval between the two events as noted on the clock in S, say* $\Delta t = t_2 - t_1$.

As we know, in accordance with *Inverse Lorentz transformation,*

$$t_1 = \frac{t_1' + vx'/c^2}{\sqrt{1 - v^2/c^2}} \quad \text{and} \quad t_2 = \frac{t_2' + vx'/c^2}{\sqrt{1 - v^2/c^2}}.$$

$$\therefore \quad \Delta t = t_2 - t_1 = \frac{t_2' + vx'/c^2}{\sqrt{1 - v^2/c^2}} - \frac{t_1' + vx'/c^2}{\sqrt{1 - v^2/c^2}} = \frac{t_2' - t_1'}{\sqrt{1 - v^2/c^2}} = \frac{\Delta t'}{\sqrt{1 - v^2/c^2}} = \gamma \Delta T'.$$

Since $1/\sqrt{1 - v^2/c^2}$ or γ is greater than 1, we have $\Delta t > \Delta t'$.

Thus, *the time interval* ($\Delta t'$) at a given point in the moving frame (S') *appears to be longer, or dilated, by a factor* $\left(\frac{1}{\sqrt{1 - v^2/c^2}}\right)$ *or* γ *to the observer in the stationary frame* (S).

Exactly the same will happen if frame S' were at rest and frame S in motion, showing that *time-dilation too, like length contraction, is a reciprocal effect and works both ways, i.e., is independent of the direction of velocity, depending only on its magnitude.*

Again, therefore, we may generalise and say that *the interval between two events occurring at a point in a moving frame, appears to be longer by a factor* γ *to an observer in a stationary frame.*

The time-interval $\Delta t'$ between the two events occurring at the same point in the moving frame, as measured on the clock carried by it, and hence at rest with respect to it, is called the **proper time-interval** or, simply, the **proper time** and is usually denoted by the symbol τ. We may thus state that *the proper time-interval in a moving reference frame is always smaller than the corresponding time-interval, measured in a stationary one.*

It follows, as a natural consequence, therefore, that the passage of time and hence also all physiological processes that go with it, like pulse and heart-beats, in short, *the process of ageing itself gets slowed down in a fast moving reference frame and if the latter be moving with the velocity of light, the passage of time and hence also the process of ageing will stop altogether* ! This explains what is known as the **Twin paradox**, namely, that if *one of two identical twins stays back on the earth, while the other goes about in space in a high speed rocket, the latter, on returning to the earth, will be found to be younger than the former,* obviously because his ageing process got slowed down during the time that he spent in the high speed rocket.

This also shows, in a way, that *no material object can travel faster than light*, an established fact, now regarded as a *fundamental law.*

Again, however, in our every day life, we seldom come across speeds comparable with that of light (c), so that ($v/c \to 0$) and $\left(\frac{1}{\sqrt{1 - v^2/c^2}} \approx 1\right)$ and, therefore, ($\Delta t = \Delta t'$). This clearly implies that, for most of our purposes, the time-interval between the two events, as measured in both the stationary and the moving frames, is the same, *i.e.*, time remains an absolute quantity, which is a result quite in accord with the Classical or Newtonian physics, the time-dilation effect manifesting itself only at speeds comparable with that of light (c). (See worked example 11).

3. Simultaneity. The simultaneity of two events means their occurrence *at exactly the same time*. Let us see whether two events occurring simultaneously in a stationary frame S also appear to be so in a reference frame S', moving with respect to it with velocity v along the positive direction of x.

Let A and B be two points along the axis of x in the stationary frame S, distant x_1 and x_2 from the origin, with P as their carefully determined mid-point. Then, a flash of light emitted at P will arrive at A and B at exactly the same time t, because of speed of light being the same in all directions. The two events (*viz.* the arrival of the light signals) thus occur simultaneously at A and B in frame S.

To an observer in the moving frame S', the corresponding values of t for the arrival of the time signals at A and B will, in accordance with Lorentz transformation, appear to be

$$t_1' = \gamma(t - vx_1/c^2) \quad \text{and} \quad t_2' = \gamma(t - vx_2/c^2)$$

Clearly, $x_1 \neq x_2$ and, therefore, $t_1' \neq t_2'$

i.e., the two events will *not* appear to be simultaneous to the observer in S'. In fact, the time-interval between the two events will appear to be $\Delta t' = t_2' - t_1' = \gamma.v(x_1 - x_2)/c^2$, which clearly depends upon the distance between the two events and the velocity of the moving frame with respect to the stationary one.

In exactly the same manner, the two events occurring simultaneously in the moving frame S' *will not appear to be so to an observer in the stationary frame S.*

Thus, *the simultaneity of event, separated in space, is not absolute in frames of reference in relative motion with respect to each other.* This is known as **relativity of simultaneity** and is clearly a consequence of the invariance of the velocity of light (c) in all frames of reference and in all directions.

Simultaneity is, in fact, a local affair of each individual reference frame, whether at rest or in motion, *i.e.*, it is there in each individual frame, taken by itself, but the simultaneity in one frame is not the same as in another in relative motion with respect to it.

4. Transformation of velocity. Suppose a frame of reference S' is moving with velocity v relative to a frame of reference S at rest, along the $+x$ direction and that a particle is moving with velocity u in frame S, with components $u_x = dx/dt$, $u_y = dy/dt$ and $u_z = dz/dt$ along the three coordinate axes. If the velocity of the particle, as measured in frame S', be u' and its components along the three coordinate axes, u_x', u_y' and u_z', *we have, in accordance with Lorentz transformation,*

$$x' = \gamma(x - v.t), \quad y' = y, \quad z' = z \text{ and } t' = (t - v.x/c^2).$$

So that,
$$u'_x = \frac{dx'}{dt'} = \frac{\gamma.(dx - v.dt)}{\gamma(dt - v.dx/c^2)} = \frac{dx - v.dt}{dt - v.dx/c^2}.$$

or,
$$u'_x = \frac{dx/dt - v}{1 - (v/c^2)(dx/dt)} = \frac{u_x - v}{1 - v.u_x/c^2}$$

And
$$u'_y = \frac{dy'}{dt'} = \frac{dy}{\gamma(dt - v.dx/c^2)} = \frac{dy/dt}{\gamma[1 - (v/c^2)(dx/dt)]} = \frac{u_y}{\gamma(1 - v.u_x/c^2)}$$

Similarly,
$$u'_z = \frac{u_z}{\gamma(1 - v.u_x/c^2)}$$

If, however, velocity u of the particle in frame S be itself along the axis of x, the components u_y and u_z in frame S, and so also components u'_y and u'_z in frame S', are each equal to zero, with $u_x = u$ and $u'_x = u'$ and we have

velocity of the particle in frame S' i.e., $u' = \dfrac{u - v}{1 - u.v/c^2}$.

It is thus clear, that as *in the case of Galilian transformation; so also under Lorentz transformation, velocity is not invariant in the two reference frames.*

The Inverse Lorentz transformation (*i.e.*, replacing v by $-v$) will obviously give

$$u_x = \frac{u'_x + v}{1 + v.u'_x/c^2}, \quad u_y = \frac{u'_y}{\gamma(1 + v.u'_x/c^2)} \text{ and } u_z = \frac{u'_z}{\gamma(1 + v.u'_x/c^2)}$$

And, if the velocity u' be along the axis of x', we shall have

$$u = \frac{u' + v}{1 + u'.v/c^2}$$

the components u'_y and u'_z in frame S', and hence also components u_y and u_z in frame S, being zero.

Now, velocity u of the particle in frame S may be taken to be the resultant of its velocity u' in frame S' and the velocity v of frame S' itself with respect to frame S. So that, the expressions above give us a method of adding up relativistic velocities (*i.e.*, velocities comparable with c) and this

constitutes the law for the addition of relativistic velocities, which may (from the expression for *u* above) be stated thus :

If a particle has a velocity u′ in a frame of reference S′ which is moving with velocity v with respect to another frame of reference S, the velocity of the particle in the latter frame (S) is given by $u=(u'+v)/(1+u.v/c^2)$.

It may be pointed out that

(i) If $u'x << c$, ***so that*** $(u'_x/c) \rightarrow 0$ ***, all the above expressions reduce to those obtained by Galilian transformation, viz.,*** $u_x=u'_x+v$, $u_y=u'_y$ ***and*** $u_z=u'_z$.

(ii) If velocity u'_x ***be high enough to be comparable to c, the denominator in the expression for*** u_x ***is greater than 1 and hence the resultant velocity*** u_x ***is smaller that*** (u'_x+v)***, as given by Galilian transformation.***

(iii) If $u'_x = c$ ***or, the velocity of the particle along the axis of x′ in frame S′ be the same as the velocity of light, i.e., if the particle be a photon,*** **its velocity in frame *S* is**

$$u_x = \frac{c+v}{1+v.c/c^2} = \frac{c+v}{(c+v)/c} = c$$

the same as in frame *S′*.

Thus, *only this velocity c is invariant in the two frames of reference.* This is only to be expected, since the *Lorentz transformation itself is based on the invariance of c in any frame of reference.*

Another *important result* that emerges from the above discussion is that *under Lorentzs transformation, no two velocities can add up to more than c.* This is obviously not the case under Galilian transformation, as the following example will clearly show :

Let the velocity of frame *S′* relative to frame *S* be$v = 3c/4$ and that of the particle moving along the *x′* -axis in this frame be $u'_x = c/2$. . Then, Lorentz transformation gives the resultant velocity of the particle in frame *S* to be

$$u_x = \frac{c/2 + 3\,c/4}{1-(3\,c/4)\,(c/2) + c^2} = \frac{5\,c/4}{1+(3/8)} = \frac{5\,c/4}{11/8} = \frac{10}{11}\,c = 0.909\,c \text{ or less than } c.$$

Galilian transformation, on the other hand, would give its value to be $u_x = u'_x + v = c/2 + 3c/4 = 5c/4 = 1.25\,c$ *or greater than c*.

16.11 Relativity of mass. **We have already seen how *length* and *time* are no longer absolute quantities and that, in fact, there is *length contraction* and *time dilation* in a moving frame of reference. A direct consequence of time dilation is that the mass of a particle too no longer remains an absolute quantity, but varies with its velocity, as will be seen from the following discussion :**

Consider two inertial frames *S* and *S′*, the former at rest and the latter moving with uniform velocity *v* with respect to it along the *+x* direction. Then, as we know, the displacements of a particle along the axes of *y* and *z*, as measured in either frame of reference, remain unaffected, whatever the value if *v*, so that $\Delta y = \Delta.y'$ and $\Delta z = \Delta z'$. Since, however, time is not invariant to Lorentz transformation, the time taken by particle to cover these displacements in the two frames will not be the same and hence the velocity of the particle too will not be the same in both the frames, *i.e.*, velocity too is not invariant to Lorentz transformation, as we have seen already under § 16.10, (4).

But again, the *proper time*, $\Delta\tau$, as noted on the clock carried by the frame itself to cover the displacement Δy (or $\Delta y'$) will be the *same in either frame* and so will, therefore, be the quantity $\Delta y/\Delta\tau$.

Hence, taking $\Delta y/\Delta\tau$ as the *y*-component of the velocity of the particle in the moving frame *S′* and m_0, its *rest mass* or its mass as noted by an observer at rest with respect to it (in this case, by the observer in frame *S′*), we have

y-component of the momention of the particle, as noted in frame S′ given by

$$p_y = m_0 \Delta y/\Delta \tau .$$

And if Δt be the time taken to cover the displacement Δy of the particle (in frame S'), *as noted, by an observer in the stationary frame S*, on a clock identical with that carried in frame S', we have *y-component of the velocity of the particle, as noted in frame S*, is given by $v_y = \Delta y/\Delta t$

In accordance with Lorentz ttransformation, $\Delta \tau = \Delta t \sqrt{1 - v^2/c^2}$ and, therefore, $\Delta t = \Delta \tau / \sqrt{1 - v^2/c^2} = \gamma \Delta \tau$, so that $v_y = \Delta y / \gamma \Delta \tau$.

If, therefore, m be the mass of the partide, as measured by an observer in the stationary frame S, we have

y-component of the momentum of the particle, as observed in the stationary frame is given by

$$m v_y = m \Delta y/\gamma \Delta \tau$$

Since, despite relativistic effects, the fundamental laws of physics must assume the same

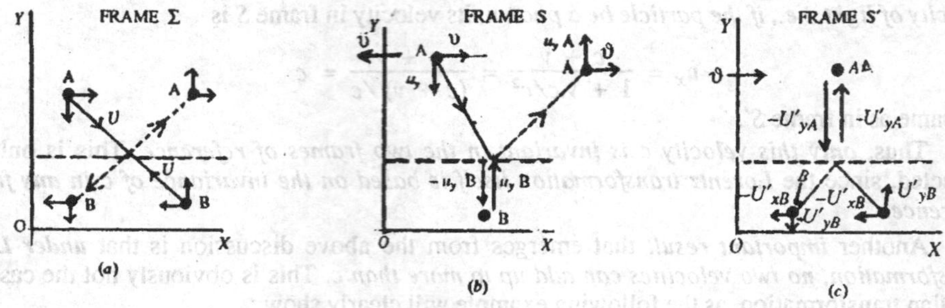

Fig. 16.5

mathematical form in all inertial reference frames, the law of conservation of momentum must hold good in either frame and we must, therefore, have

$$p_y = m.v_y = m_0 \Delta y/\Delta \tau = m_0 \Delta y \ \Delta t / \Delta t \ \Delta \tau \qquad [\because \Delta y/\Delta t = v_y$$

or, $$m.v_y = m_0 v_y \Delta t/\Delta \tau = m_0 \Delta v_y \gamma \Delta \tau / \Delta \tau = \gamma m_0 v_y , \qquad \text{and } \Delta t = \gamma \Delta \tau]$$

whence, $$m = \gamma m_0 = m_0 / \sqrt{1 - v^2/c^2},$$

i.e., the mass of a particle in a moving frame of reference, as noted by an observer in a stationary frame, is γ times its rest mass.

Since $\gamma = 1/\sqrt{1 - v^2/c^2}$ is always greater than 1, *the mass of the particle in the moving frame of reference is greater than its rest mass.*

Here, again, there is the same reciprocity effect as in the case of length and time, so that whereas for an observer moving with the particle itself, its mass will be m_0, for an observer in the stationary frame, it will be $m = m_0 / \sqrt{1 - v^2/c^2}$.

Alternative method.

Let Σ be a reference frame at rest, in which two identical particles A and B, like two *protons*, for instance, undergo a symmetrical glancing collision, moving with equal and opposite velocities after the collision, as shown in Fig. 16.5(*a*), where the dotted line, parallel to the x-axis, bisects the angle between their trajectories.

If the collision be a perfectly elastic one, the x-components of the velocities of the two particles remain the same in direction as well as magnitude, both before and after the collision, so that, their masses being equal, *there is no change in momentum along the x-direction.*

The y-components of the velocities of the particles, which are initially oppositely directed, get reversed in direction, their magnitude remaining unchanged. The change in momentum of particle A *is* thus equal and opposite to the change in momentum of particle B and the change in momentum along the y-direction too is, therefore, zero. So that, the momentum of system remains conserved, in accordance with the Classical or Newtonian physics, in which the mass of a particle is quite independent of its velocity.

If we observe the collision from a reference frame S, moving relative to the stationary frame Σ with a velocity equal to the x-component of the velocity of particle B (so that this latter component is neutralised), particle B appears to move along the axis of y with a velocity U_{yB}, say, and to bounce back after collision with particle A, with its velocity reversed [Fig. 16.5 (*b*)]. Particle A, on the other hand, has an x-component v and a y-component U_{yA}. The former remains unchanged in magnitude as well as direction on collision with particle B and the latter gets reversed in direction.

Similarly, if we observe the collision from a reference frame S', moving relative to frame S with velocity v, equal to the x-component of the velocity of particle A (so that this latter component is neutralised), particle A appears to move downwards with velocity $-U'y_A$, say, and to bounce back after collision with particle B, with its velocity reversed. And, particle B retains its x-component of velocity, $-U'x_B$, unchanged in magnitude and direction, with its y-component, $-U'y_B$, reversed in direction, [Fig. 16.5(*c*)].

In view of the initial symmetry of the glancing collision, the situations in frames S and S' also appear to be symmetrical, with particles A and B having just interchanged their roles. So that, we have $U'_{yA} = U_{yB}$.

Since in accordance with Lorentz transformation, $U_{yA} = \dfrac{U'_{yA}}{\gamma(1 + v.U'_{xA}/c^2)}$, where U'_{xA} is the x-component of the velocity of particle A in frame S', equal to *zero* in this case, we have

$$U_{yA} = (U'_{yA}/\gamma = U_{yB}/\gamma) = U_{yB}\sqrt{1 - v^2/c^2},$$

showing that the y-components of the velocities of the two particles, which are equal to each other in the stationary frame Σ, are no longer so in the moving frame S.

It will also be seen that in frame S, the x-component of the velocity of particle A remains unchanged in magnitude and direction after the collision and the x-component of the velocity of particle B is zero. There is, therefore, *no change in momentum along the x-direction. The entire change in momentum thus occurs along the y-direction.*

Clearly, change in the y-component of the velocity of particle A is $U_{yA} - (-U_{yA}) = 2U_{yA}$ and change in the y-component of the velocity of particle B is $-U_{yB} - U_{yB} = -2U_{yB}$.

If, therefore, m_0 be the mass of each particle, supposed to be invariant in accordance with the Classical or Newtonian physics, we have
change in the momentum of the system $= 2m_0 U_{yA} - 2m_0 U_{yB}$ which is clearly *not zero*, because $U_{yA} \neq U_{yB}$.

Obviously, therefore, for the law of conservation of momentum to hold good, as we know it must, the masses of particles A and B in the moving frame S must necessarily be different from their Newtonian or rest masses in the stationary frame Σ. Let these be m_A and m_B respectively. Hence, for the law of conservation of momentum to hold good, we must have

$$2m_A U_{yA} - 2m_B U_{yB} = 0 \text{ Or, } m_A U_{yA} = m_B U_{yB}.$$

But $U_{yA} = U_{yB}/\gamma$ and, therefore, $U_{yB} = \gamma U_{yA}$. so that, $m_A U_{yA} = \gamma m_B U$

Or, $m_A = m_B = m_B/\sqrt{1 - v^2/c^2}$.

Now, in frame S, the y-components, U_{yA} and U_{yB} of the velocities of particles A and B respectively will be extremely small or negligible if the x-component v of the velocity of particle A be sufficiently large and particle B will be practically at rest, with particle A moving relative to it with velocity v. If, therefore, m_{0B} be the rest mass of particle B, we have

$$m_A = m_{0B}/\sqrt{1 - v^2/c^2}.$$

Since the particles are identical in all respects, we have $m_{0B} = m_{0A}$, and, therefore,

$$m_A = m_{0A}/\sqrt{1 - v^2/c^2}\ .$$

or, *in general, if m_0 be the rest mass (or inertial mass) of a particle or body and m, its mass when moving with velocity v (i.e., its relativistic mass), we have*. $m = \left|\dfrac{m_0}{\sqrt{1 - v^2/c^2}}\right| = \gamma m_0$

This is the important *mass-velocity relation of the Special theory of relativity*, showing clearly that the mass of a particle or a body is no longer an absolute, invariant quantity, as assumed in the Classical or Newtonian physics, but depends upon its speed or velocity.

This relativistic effect of change in mass with velocity, just like the change in length and time, becomes appreciable, however, only when the mass is moving with a velocity comparable with the velocity of light in free space (c) as, for example, in the case of atomic particles. At ordinary speeds,(v/c) $\rightarrow 0$ and, therefore, $m = m_0$, *i.e.*, the mass of a particle or body remains the same as its rest mass, a result in accord with the Classical or Newtonian physics.

The relativistic increase in mass with velocity was first confirmed experimentally by *Bucherer*, in the year 1908, when he showed that the charge of mass ratio (e/m) for fast moving particles was smaller than for slow moving ones due, obviously, to the higher value of m for the former, the value of c being invariant.

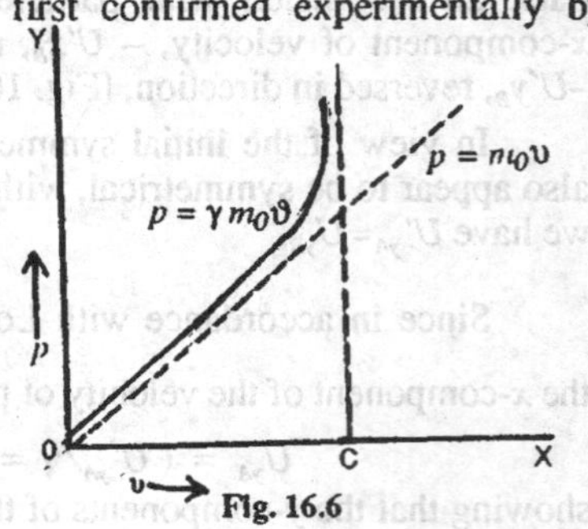

Fig. 16.6

Another spectacular confirmation of this relativistic effect is provided by the fact that to deflect high speed electrons in a synchroton, the magnetic field must be 2000 times as strong as would be required on the basis of Newton's laws, showing very clearly that the mass of an electron increases, at its high speed, to 2000 times its normal mass, *i.e.*, it acquires practically the same mass as a proton.

The relativity of mass, along with that of length (or distance) and time; now stands fully established and has to be accepted as a basic formulation in atomic physics.

16.12 Relativistic law of conservation of momentum-Force. We have just seen in our discussion above that the Newtonian law of conservation of momentum, $p = m_o v$ a constant, where m_o is the rest mass of the body (supposed to be the same in all frames of reference) is not invariant to Lorentz transformation and cannot, therefore, be accepted as the correct law at relativistic velocities. What remains constant in different inertial frames of reference at such speeds and is invariant to Lorentz transformation, is *not* $m_0 v$ but $\gamma m_0 v$ *or* $m_0 v / \sqrt{1 - v^2/c^2}$. So that, the *relativistic law of conservation of momentum* is

$$p + \gamma m_0 v = m_0 v / \sqrt{1 - v^2/c^2}\ , \text{ a constant.}$$

As in all other cases, so also here, at non-relativistic velocities $v/c \rightarrow 0$ or $\gamma \approx 1$ and, therefore, $p = m_0 v$, *i.e.*, at such velocities, the relativistic law of conservation of momentum reduces to the Newtonian law, which is thus a particular case of the former general law. So that, unless we are dealing with high speed particles, like electrons, protons, mesons etc, the Newtonian law is good enough for most of our daily problems.

A good insight into the distinction between Newtonian momentum ($p=m_0 v$) and relativistic momentum ($p = m_0 v$) is provided by plotting the momentum (p) of a particle against its velocity (Fig. 16.6). The graph in the former case would be a straight line passing through the origin (shown dotted), indicating that the momentum goes on increasing with velocity, even if the velocity attains the value c or exceeds it. The graph in the latter case, on the other hand, would take the form shown by the full line curve, where, it may be noted, the velocity does not increase beyond c, although the momentum goes on increasing. (For explanation, see § 16.13).

Force. As we know, force in classical mechanics is defined as the rate of change of momentum, *i.e.*, $F = dp/dt$. This definition is equally valid in relativistic mechanics. Since, however, mass, in the latter case, means relativistic mass, the expression for force in terms of mass and acceleration, *in the case of linear motion,* turns out to be different from the classical expression $F = ma = m\,dv/dt$.

For, under relativistic mechanics,

$$F = \frac{dp}{dt} = \frac{d}{dt}(m.v) = \frac{d}{dt}\left(\frac{m_0.v}{\sqrt{1 - v^2/c^2}}\right) \qquad ...(i)$$

Since in the case of linear motion, only the magnitude of the velocity changes and not its direction, we may put the above relation as

$$F = \frac{d}{dt}\left(\frac{m_0.v}{(1 - v^2/c^2)^{1/2}}\right) = \frac{m_0\,(dv/dt)}{(1 - v^2/c^2)^{3/2}}.$$

But $\frac{m_0}{(1 - v^2/c^2)^{1/2}} = m$. And, therefore, $F = \frac{m}{(1 - v^2/c^2)} \cdot \frac{dv}{dt}$, ...(ii)

where dv/dt is the *rate of change of velocity* or *acceleration.*

This is clearly different from the classical expression, $F = ma = m.\,dv/dt$.

Or, we may also look at it in a different way. For.

$$F = \frac{d}{dt}(m.v) = m\frac{dv}{dt} + v\frac{dm}{dt} = m.a + v\frac{dm}{dt}.$$

Clearly, $v\,(dm/dt) \neq 0$ if v varies with time. So that, $F \neq m.a$

Thus, *the correct definition of force, under the special theory of relativity, is 'time-rate of change of momentum'* and not *'mass × acceleration'*.

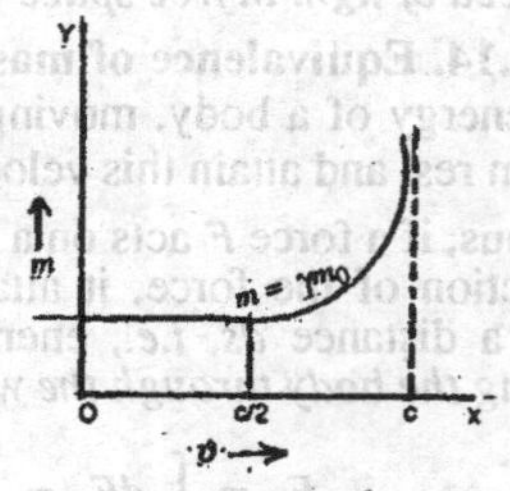

Fig. 16.7

In the case of **circular motion,** since only the direction of velocity changes continuously, its magnitude remaining the same, expression *(i)* takes the form

$$F = \frac{m_0}{\sqrt{1 - v^2/c^2}} \cdot \frac{dv}{dt} = m\frac{dv}{dt},$$

where dv/dt, in this case, is the *centripetal acceleration* v^2/r, normal to the circular path of radius r. So that,

$$F = \frac{m_0}{\sqrt{1 - v^2/c^2}} \cdot \frac{v^2}{r} = m.a;$$

the same as the classical expression for force, except that mass m, here, is the ***relativistic mass.***

16.13. Ultimate speed of a material particle or body. As we know, there is no upper limit for the speed or velocity of a particle in classical or Newtonian physics. It simply goes on increasing with continuous acceleration and may attain the value c or even exceed it. This is, however, firmly denied by the theory of relativity.

The question, therefore, arises as to what happens if a particle be continually accelerated. The answer is that although the speed of the particle certainly goes on increasing, but so does its mass, in accordance with the mass-velocity relation we have derived above, *viz.*, $m = m_0/\sqrt{1 - v^2/c^2} = \gamma m_0$. So that, as v approaches c, $v^2/c^2 \rightarrow 1$ and $\sqrt{1 - v^2/c^2} \rightarrow 0$ and, with m approaching infinity. This will be amply clear from the graph shown in Fig. 16.7, which shows that for values of v up to $c/2$, the increase in mass is not quite so appreciable, but as v exceeds $c/2$, the increase becomes quite rapid and as $v \rightarrow c$, $m \rightarrow \infty$. This is also clear from the earlier graph between velocity and momentum (Fig. 16.6), where although v remains constant at c, the momentum goes on increasing, implying obviously that the accelerating force applied only increases the mass of the particle but not its velocity beyond c.

Now, if the mass of a particle or a body be infinite, *(i)* it would require an infinite force to accelerate it to a speed equal to c, *(ii)* its length, and hence also its volume, would be reduced to zero in accordance with Lorentz transformation and *(iii)* it will exert an indifinite gravitational pull on all other bodies in the universe, etc., etc.

We, therefore, discard the idea of infinite mass as making no sense and interpret the above result to mean that a material body can never attain a speed as high as c, the speed of light in free space, equal to 3×10^{10} *cm* or 3×10^{8} *m per sec.*

An objection may perhaps be raised that a *photon* travels in free space with a speed c, but, then, a photon is *not* a material particle in the accepted sense of the term, since *it possesses no rest mass.* Again, it may perhaps be said that phase velocities do actually attain, and even exceed, the value c. This too is easily countered by pointing out that *phase* is not material particle but an abstract mathematical function.

Finally, though some of the atomic particles may actually travel in media like air, water, glass etc with velocities higher than that of light *in these media* but *by no means higher than the speed of light in free space, viz., c.*

The conclusion is thus inescapable that *the ultimately speed of any material particle or body is the speed of light in free space (c) which it can never actually attain, much less exceed.*

16.14. Equivalence of mass and energy. In classical as well as Relativistic mechanics, the kinetic energy of a body, moving with velocity v, is equal to the work done in making the body start from rest and attain this velocity.

Thus, if a force F acts on a body such that when the body is displaced through a distance s in the direction of the force, it attains the velocity v, we have work done in displacing the body through a distance *ds, i.e.,* energy imparted to it, say, $dE_k = F.ds$ and, therefore, *work done in displacing the body through the whole distance s or energy imparted to it is given by*

$$E_k = \int dE_k = \int_0^s \frac{d}{dt}(m.v)\ ds \qquad \left[\because F = \frac{d}{dt}(mv)\right.$$

$$\text{Now,} \quad \frac{d}{dt}(m.v)\,ds = \frac{ds}{dt}\,d\,(m.v) = vd\,(m.v) \qquad \left[\because \frac{ds}{dt} = v\right.$$

And, since the body starts from rest and finally attains velocity v, we have

$$E_k = \int_0^v v.d\,(m.v) = \int_0^v v.d\,\frac{m_0.v}{\sqrt{1 - v^2/c^2}}.$$

Integrating by parts* , we have

$$E_k = \frac{m_0.v^2}{\sqrt{1 - v^2/c^2}} - m_0 \int_0^v \frac{v}{1 - v^2/c^2}\,dv = \frac{m_0.v^2}{\sqrt{1 - v^2/c^2}} + m_0.c^2\left[\sqrt{1 - v^2/c^2}\,\right]_0^v$$

$$\text{or,} \quad E_k = \frac{m_0.c^2}{\sqrt{1 - v^2/c^2}} - m_0.c^2 = m.c^2 - m_0.c^2 = (m - m_0)\,c^2 = \Delta\,m.c^2,$$

where $\Delta m = (m - m_0)$ is the relativistic increase in the mass of the body with increase in its velocity.

$\therefore$ *Kinetic energy* $= E_k =$ *(increase in mass of the body)* $\times\, c^2$

The above equation may be put in the form

$$m.c^2 = m_0\,c^2 + E_k = m_0\,c^2 + \frac{m_0\,c^2}{\sqrt{1 - v^2/c^2}} - m_0\,c^2 = \frac{m_0\,c^2}{\sqrt{1 - v^2/c^2}} = \gamma m_0.c^2 \qquad ...(i)$$

Here, m_0c^2 measures the energy of the body when it is at rest with respect to the observer and is called its **rest energy or proper energy**, denoted by the symbol E_0 and mc^2 measures the **total energy** E possessed by the body. So that, we have

* *i.e.,* $\int x.dy = x - \int y.dy.$

total energy of the body, E = its rest energy, E_0 + its kinetic energy, E_k.

The relativistic expression for kinetic energy or the *mass-energy* equation [$E_k + (m - m_0)c^2 = \Delta m.c^2$], is called **Einstein's mass-energy relation. It is perhaps the most famous of all the equations deduced by him and stands fully vindicated by experimental results.**

As will be readily seen, on account of the high value of c^2, (equal to 9×10^{20}), even a small amount of mass is equivalent to a large amount of energy. In our daily lives, however, we hardly come by any appreciable changes in mass for the energy-effects to be discernable. But a good example is provided by radioactive substances, where the conversion of mass results in such a large amount of energy that α and β particles get ejected with high velocities and over long periods of time. Another strikingly good example is that of the reduction in the mass of the Sun at the rate of 4×10^{12} *gm/sec*, resulting in the radiation of an energy equivalent of 36×10^{32} *erg/sec*.

It may be noted that although the classical expression for kinetics energy, *viz.*, $\frac{1}{2}mv^2$ does not give the true kinetic energy of a body, even if we substitude the relativistic values of m and v in it, but the relativistie expression $E_k = m.c^2 - m_0.c^2$ reduces to the classical expression ($E_k = \frac{1}{2}mv^2$) for non-relativistic values of v, *i.e.* for values of $v << c$. For, then,

$$E_k = \frac{m_0 c^2}{\sqrt{1 - v^2/c^2}} - m_0.c^2 = m_0 c^2 (1 - v^2/c^2)^{-1/2} - m_0.c^2$$

$$= m_0.c^2 (1 + \frac{1}{2}v^2/c^2 \ldots\ldots) - m_0c^2 = \frac{1}{2}m_0v^2 \qquad \ldots(ii)$$

It will thus be seen that the relativistic expression for kinetic energy too, like those for length, time and mass, reduces to the classical form values of $v << c$, *i.e.*, when $v/c \to 0$.

Importance of mass-energy relation. It will be seen from relation *(ii)* above that

$$E_k = \frac{1}{2}m_0 v^2 = \frac{m_0 c^2}{\sqrt{1 - v^2/c^2}} - m_0.c^2 = m.c^2 - m_0.c^2, \text{ or } m.c^2 = m_0.c^2 + \frac{1}{2}m_0 c^2$$

whence, $$m = m_0 + \frac{1}{2}m_0.v^2/c^2 = m_0 + E_k/c^2.$$

i.e., the total inertial mass of a particle, moving with velocity v relative to the observer, is the sum of its rest mass m_0 and an additional mass (or inertia) E_k/c^2.

This clearly implies that the addition of energy to a system increases its inertial mass. In other words, *mass may appear as energy and energy as mass.*

The scope of the law of conservation of energy is thus broadened into what may be called the *law of conservation of mass-energy, viz., that in any interaction, neither mass alone nor energy alone, is conserved. What is conserved is mass inclusive of energy, in terms of mass or energy, inclusive of mass in terms of energy.*

16.15 Transformation of relativistic momentum and energy. We know that an interval of time dt, as noted on a clock in a reference frame S is related to the *proper interval* of time $d\tau$, as noted on an identical clock carried by the particle itself (*i.e.*, with respect to which the particle is at rest), as $dt = d\tau / \sqrt{1 - v^2/c^2}$, whence, $dt/d\tau = 1/\sqrt{1 - v^2/c^2} = \gamma$.

Now, consider a particle moving in the frame of reference S, such that the components of its *momentum* along the three coordinate axes are p_x, p_y and p_z respectively.

Clearly, $p_x = m.\ dx/dt$, where dx/dt is the component of the velocity of the particle along the axis of x. If, therefore, m_0 be the *rest mass* of the particle, we have

$$p_x = \frac{m_0}{\sqrt{1 - v^2/c^2}} \cdot \frac{dx}{dt} = m_0 \cdot \frac{dx}{dt} \cdot \frac{dt}{d\tau} = m_0 \frac{dx}{d\tau}.$$

Similarly, $p_y = m_0.dy/d\tau$ and $p_z = m_0.dz/d\tau$.

And, the *energy of the particle, i.e.*, $E = \gamma m_0 c^2 = m_0 c^2 . dt/d\tau$. or, $E/c^2 = m_0 . dt/d\tau$.

Since m_0, τ and c are all Lorentz invariants, the quantities p_x, p_y and p_z get transformed in exactly the same manner as x, y, z, and t in a reference frame S' moving with velocity v relative to the reference frame S along the axis of x. And we, threfore, have $p'_x = \gamma(p_x - v\,E/c^2)$, $p'_y = p_y$ and $p'_z = p_z$.

$$\text{And, } E'/c^2 = \gamma(E/c^2 - \frac{v}{c^2}p_x) . \text{ or } E' = \gamma(E - v\,p_x).$$

The Inverse Lorent transformation will thus obviously give

$$p_x = \gamma(p'_x + VE'/c^2),\ p_y = p'_y \text{ and } p_z = p'_z$$

$$\text{And, } \quad E = \gamma(E' + v\,p'_x).$$

16.16. Conservation of relativistic energy. In an indirect manner, the law of conservation of relativistic energy stands established from the very fact that in the relativistic expression for it, *viz.*, $E = m.c^2 = \gamma m_0 c^2$, the value of $m = \gamma m_0$ has been obtained on the basis of the validity of the law of conservation of relativistic momentum. Clearly, therefore, the relativistic energy too must be governed by a law of conservation similar to that for relativistic momentum.

The law may, however, be deduced *directly* as follows :

Consider a collision between two particles of initial momenta p_1 and p_2 and energies E_1 and E_2 in a reference frame S, such that their momenta and energies after the collision are p_3, p_4 and E_3 and E_4 respectively. Then, in accordance with the law of conservation of momentum, we have

$$p_1 + p_2 = p_3 + p_4.$$

Or, considering the x-components of the momenta alone, we have

$$p_{1x} + p_{2x} = p_{3x} + p_{4x}.$$

These components of momenta, in another frame of reference S' moving with velocity v along the x-direction relative to frame S will be p'_{1x}, p'_{2x}, p'_{3x} and p'_{4x} and the corresponding energies of the particles will be E_1', E_2' and E_3', E_4'.

The law of conservation of momentum, being a Lorentz invariant, is equally valid in the reference frame S' and we, therefore, have

$$p_{1x}' + p_{2x}' = p_{3x}' + p_{4x}' .$$

Or, by Lorentz transformation of momentum § (16.15), we have

$$\gamma(p_{1x} - v.E_1/c^2) + \gamma(p_{2x} - v.E_2/c^2) = \gamma(p_{3x} - v.E_3/c^2) + \gamma(p_{4x} - v.E_4/c^2)$$

$$\text{Or, } (p_{1x} + p_{2x}) - \frac{v}{c^2}(E_1 + E_2) = (p_{3x} + p_{4x}) - \frac{v}{c^2}(E_3 + E_4)$$

Since $(p_{1x} + p_{2x}) = (p_{3x} + p_{4x})$, we have

$$E_1 + E_2 = E_3 + E_4,$$

i.e., the energy of the two particles too, like momentum, remains conserved in frame S.

Similarly, it can be shown that $E_1' + E_2' = E_3' + E_4'$, *i.e.*, the energy of the two particles also remains conserved in frame S'.

Obviously, what is true for two particles is equally true for any number of them. We thus conclude that, *like the law of conservation of relativistic momentum, the law of conservation of relativistic energy too is a Lorentz invariant and is valid in all inertial frames of reference.*

16.17. Relation between relativistic momentum and energy. As we know, the relativistic momentum of a particle of rest mass m_0, moving with velocity v, is given by $p = \gamma m_0.v$. We therefore, have $p^2 = \gamma^2 m_0^2 v^2$.

Now, $\gamma = 1/\sqrt{1 - v^2/c^2}$ or $\gamma^2 = 1/(1 - v^2/c^2) = c^2/(c^2 - v^2)$.

And $\therefore \gamma^2.v^2 = c^2.v^2/(c^2 - v^2) = c^2 [c^2/(c^2 - v^2) - 1] = c^2 (\gamma^2 - 1) = \gamma^2c^2 - c^2$.

Substituting this value of $\gamma^2.v^2$ in the expression for p^2 above, we have

$p^2 = (\gamma^2 c^2 - c^2)m_0^2$ which, when multiplied by c^2, gives

$$p^2.c^2 = m_0^2 \gamma^2.c^4 - m_0^2.c^4 \text{. or, } m_0^2 \gamma^2.c^4 = p^2c^2 + m_0^2.c^4$$

Since $\gamma m_0 c^2 = E$, the *total energy* of the particle, we clearly have

$$E^2 = p^2c^2 + m_0^2c^4 \text{ or, } E^2 - p^2c^2 = m_0^2.c^4,$$

a neat expression connecting relativistic energy and momentum.

Since m_0 and c are both Lorentz invariants, $m_0^2c^4$ is a *constant*. It follows, therefore, that the quantity $E^2 - p^2 c^2$ too is a Lorentz invariant, its value remaining the same in all inertial frames of reference.

Also, $p = \gamma m_0.v$ and $E = \gamma m_0 c^2$ or $\gamma m_0 = E/c^2$

And, $\therefore p = E.v/c^2$, whence, $v = p\, c^2/E$,

a relation that enables us to obtain the velocity of a particle from its momentum and energy.

16.18. Mass, velocity, momentum and energy of a particle of zero rest mass. We have just seen in § 16.17 above how the energy of a particle of rest mass m_0 and momentum p is given by the relation $E^2 = (p^2c^2 + m_0^2c^4)$. If , therefore , we have a particle of *zer* o rest mass , *i.e.* , with $m_0 = 0$, as for example, a *photon* or a *nutrino*, clearly, $E = pc$, whence, $p = E/c$. And, since p is also *equal to* Ev/c^2, we have $Ev/c^2 = E/c$, or $v = c$, showing that the velocity of the particle is the same as that of light in free space.

Thus, both the rest mass of the particle (zero) and its speed (c) are Lorentz invariants and hence the same in all inertial reference frames.

The energy of the particle, $E = pc = m.c^2$, where m is the *mass equivalent of energy E, i.e.*, equal to E/c^2.

We thus have *(i) mass of a particle of zero rest mass, i.e.,* $m = E / c^2$, *(ii) its speed or velocity* $= c$, *(iii) its momentum* $p = mc = E/c$ and *(iv) its energy* $E = m.c^2 = p.c$.

If the particle happens to be *photon*, of frequency υ, we have $E = h.\upsilon$, where h is *Planck's constant* (equal to 6.626×10^{-27} *erg-sec*). And, therefore, its *mass* $m = E/c^2 = h.\upsilon / c^2$ and its *momentum* $p = E/c = h.\upsilon / c$.

16.19. Atomic mass unit. The basic unit of mass in atomic physics is the *atomic mass unit* (*amu*, for short) and is equal to 1/12*th* of the rest mass of Carbon-12, (C^{12}) $\approx$ mass of a proton, *i.e.*, 1.66×10^{-27} *kg*.

Now, energy associated with a rest mass m_0 is $m_0 c^2$ and, therefore, 1 *amu, in terms of energy* $= 1.66 \times 10^{-27} \times (3 \times 10^8)^2 = 1.49 \times 10^{-10}$ *joules* or $1.49 \times 10^{-10} \times 10^7 = 1.49 \times 10^3$ *ergs*.

And, since 1.60×10^{-19} *joule* = 1 *eV (electron volt)*, we have.

$$1\ amu = \frac{1.49 \times 10^{-10}}{1.60 \times 10^{-19}} = 9.31 \times 10^8\ eV$$

$$= 9.31 \times 10^8 / 10^6 = 931\ MeV \text{ (mega electron volts).}$$

WORKED EXAMPLES

1. To an observer at rest on the ground, a body thrown vertically upwards in a uniformly movingt frame S appears to describe a parabolic path. What is the path of the body as it would appear to an observer in another reference frame S' in a uniform motion, parallel to S, when S' has *(i)* an identical velocity with S, *(ii)* a velocity equal and opposite to that of S and *(iii)* a velocity twice that of S ?

Let the parabolic path described by the body in frame S be in the x-y plane, with the frame moving along the $+x$ direction. If v_x and v_y be the components of the velocity of the body along the axes of x and y respectively in this frame, clearly, v_x is also the velocity of the frame itself along the axis of x.

Now, let us consider the three given cases :

Case (i). Here, frame S' moves with a velocity identical with that of frame, S, *viz.* v_x . *So that,* x-component of the velocity of the body, as seen from frame S', *i.e.,* $v_x' = v_x - v_x = 0$. And, since neither of the two frames has a vertical velocity, the y-component of the velocity of the body remains unchanged.

Clearly, therefore, the body will appear to an observer in frame S' to be moving along the axis of y, *i.e., its path will appear to be a vertical straight line.*

Case (ii). The velocity of frame S' , in this case, being equal and opposite to that of frame S is, clearly $-v_x$. So that, x–component of the velocity of the body in frame S, as observed from frame S', is $v_x-(-v_x) = 2v_x$, the y-component of its velocity remaining unchanged, as before, *i.e.,* $v'_y = v_y$.

Thus, *distance covered by the body in time t along the axis of x, i.e.,* $x = 2\,v_x\,t$, whence, $t = x/2v_x$, and the *distance covered by it in the same time along the axis of y, i.e.*

$$y = v_y\,t - \frac{1}{2}g\,t^2 = v_y\,(x/2v_x) - \frac{1}{2}g\,(x/2v_x)^2.$$

$$\text{Or,} \qquad y = \frac{v_y}{2\,v_x}x - \frac{g}{8\,v_x^2}x^2,$$

which, being an equation of the form $y = ax + bx^2$, represents a *parabola.*

The path of the body in frame S will thus appear to the observer in frame S' *to be a parabola, described in the same sense as seen by the observer on the ground.* Since, however, the velocity of the body along the x-axis is now twice that observed by the ground-based observer, the *latus rectum of the parabola will appear to be twice that observed by the latter.*

Case (iii). The frame S', here, moving with twice the velocity of frame S, the x-component of the velocity of the body in frame S will appear to the observer in frame S' to be $v_x' = v_x - 2\,v_x$, the y-component of its velocity again remaining unchanged, *i.e.,* $v_y' = v_y$.

This is, clearly, a case similar to that of the ground-based observer, with only the x-component of the velocity of the body in S now reversed.

To the observer in S', *therefore, the path of the body in S will again appear to be a parabola, similar to that seen by ground-based observer but described in the opposite sense to it, i.e., the parabola seen by the observer in S' will be the mirror-image of that seen by the observer on the ground.*

2. The earth, revolving around its axis, is not, strictly speaking, an inertial frame of reference. If, however, we take it to be so, what would be the error involved in 1 sec in the position of a particle close to its surface. (Radius of the earth = 6.4×10^8 cm)

Clearly, the particle will experience a centripetal acceleration $a = \omega^2 R$ where ω is the angular velocity of the earth $= 2\,\pi/24 \times 60 \times 60$ *radians per sec* and R, the *radius of the earth* $= 6.4 \times 10^8$ *cm., i.e.,*

$$a = \frac{2\pi}{24 \times 60 \times 60} \times 6.4 \times 10^8 = 3.4\,cm/sec^2.$$

The acceleration of the particle, as measured in the earth's frame of reference, will thus differ from that measured in a truly inertial frame of reference by 3.4 cm/sec^2. The position of the particle, as measured in the latter frame will thus differ from that measured in the former by a distance $\frac{1}{2}at^2$ in time t and hence by a distance $\frac{1}{2}a = \frac{1}{2} \times 3.4 = 1.7\,cm$ in 1 *sec.* This will, therefore, be the error involved in the position of the particle in 1 sec.

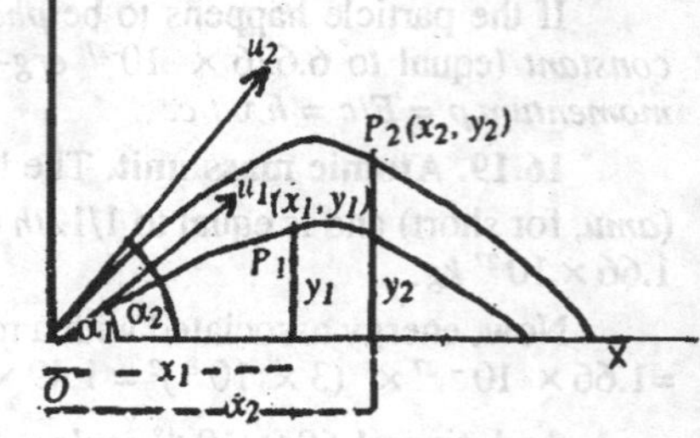

Fig. 16.8

3. Show that the motion of one projectile, as seen from another projectile, will always be a straight line motion. *(Agra)*

Let two projectiles be projected from the origin O in the x-y plane, with their trajectories as shown in Fig. 16.8, and let their velocities and angles of projection be u_1, u_2 and α_1, α_2 respectively.

Consider two points P_1 and P_2 on the two trajectories, with x_1, y_1 and x_2, y_2 as their respective positon coordinates at a given instant t. Then, clearly, $x_1 = u_1 \cos\alpha_1\,t$, $x_2 = u_2\cos\alpha_2\,t$, $y_1 = u_1 \sin\alpha_1\,t - \frac{1}{2}gt^2$ and $y_2 = u_2 \sin\alpha_2\;t - \frac{1}{2}gt^2$.

We, therefore, have $\dfrac{y_2 - y_1}{x_2 - x_1} = \dfrac{y}{x} = \dfrac{u_2 \sin\alpha_2 - u_1 \sin\alpha_1}{u_2\cos\alpha_2 - u_1\cos\alpha_1} = m$, say, So that, $y = mx$, which is the equation to a straight line.

The motion of one projectile, as seen from the other, will therefore appear to be a staight line motion.

4. Calculate the fictitious force and the observed (or total) force on a body of mass 5 kg in a frame of reference moving (i) vertically upwards, (ii) vertically downwards, with an acceleration of 4 m/sec². (Take g = 9.80 m/sec²)

Taking the earth to be a reference frame at rest and the upward direction as positive, so that g is negative, equal to -9.8 *m/sec*², let us consider the two given cases which are obviously the same as those of a lift moving *(i)* upwards and *(ii)* downwards.

Case (i). Here, *true wight of the body or the true force acting on it, i.e.,*

$$F = mg = 5(-9.8) = -49.0 \text{ newton.}$$

And, *fictitious force acting on the body = (mass of the body) (reversed acceleration of the moving reference frame), i.e.,*

$$F_0 = 5(-4) = -20.0 \text{ newton.}$$

$\therefore$ *total or observed force on the body,* $F' = F + F_0 = -49.0 - 20.0 = -69.0\,N$, *or* 69.0 *newton downwards.*

Case (ii). Here, again, *true weight of the body or true force acting on it, i.e.,*

$$F = mg = 5(-9.8) = -49.0 \text{ newton}$$

And, *fictitious force acting on the body i.e.,*

$$F_0 = 5 - [(-4)] = 20.0 \text{ newton.}$$

$\therefore$ *total or observed force on the body,* $F' = F + F_0 = -49.0 + 20.0 = -29.0\,N$, *or* 29.0 *N downwards.*

5. A mass of 1 kg is hurled horizontally due north with a velocity of half a kilometer per sec in latitude 30° North. Obtain the magnitude and direction of the Coriolis force actng on the mass and its deflection in consequence thereof in a time interval of $\frac{1}{2}$ sec.

Clearly, *the effective angular velocity* of the earth, here, is $\omega \cos(90 - \Phi) = \omega \sin \Phi$, where $\omega = 2\pi/24 \times 60 \times 60 = 0.727 \times 10^{-5}$ *radian/sec*. and velocity of the mass, $v' = \frac{1}{2}$ *km/sec* = 500 *m/sec*. So that,

Coriolis acceleration, $a_c = 2(\omega \sin \Phi)\, v' = 2(\omega \sin 30^0)\, v'$

$$= 2\left(0.727 \times 10^{-5} \times \frac{1}{2}\right) \times 500 = 36.35 \times 10^{-3}\ m/sec.^2$$

$\therefore$ *Coriolis force acting on the mass of* 1 *kg* $= m \times a_c = 1 \times 36.35 \times 10^{-3}$

$$= 36.35 \times 10^{-3} \text{ newton, towards the east.}$$

And, *deflection of the mass due to Coriolis force, in* $\frac{1}{2}$ *sec.*

$$= \frac{1}{2} a_c t^2 = \frac{1}{2} \times 36.35 \times 10^{-3} \times \left(\frac{1}{2}\right)^{-3} = 4.544 \times 10^{-3}\ m$$

6. In the Michelson-Morley experiment, the wavelength of the monochromatic light used is 5000 A.U. What will be the expected fringe shift on the basis of the stationary either hypothesis if the effective lenth of each path is 5 m? (Velocity of the earth = 3×10^4 m/sec and c = 3×10^8 m/sec.)

As we know the fringe shift is given by the relation $n = 2Du^2/c^2\lambda$. Here, $D = 5m$, $u = 3 \times 10^4$ *m/sec*. $c = 3 \times 10^8$ *m/sec* and $\lambda = 5000 \times 10^{-8}$ *cm* $= 5000 \times 10^{-8}/100 = 5 \times 10^{-7}$ *m*. So that,

$$n = \frac{2 \times 5\,(3 \times 10^4)^2}{(3 \times 10^8)^2 \times 5 \times 10^{-7}} = 0.2 \text{ or } 1/5.$$

Thus, *the expected shift is* $\frac{1}{5}$ *th of a fringe.*

7. Use Lorentz transformation equations to prove the relation $x'^2 - c^2 t'^2 = x^2 - c^2 t^2$.

In accordance with Lorentz transfomation equations,

$$x' = \gamma(x - vt) \text{ and } t' = \gamma(t - vx/c^2),\ c \text{ being invariant.}$$

$$\therefore \quad x'^2 - c^2 t'^2 = \gamma^2 (x - vt)^2 - c^2 \gamma^2 (t - v.x/c^2)^2.$$

$$= \gamma^2 (x^2 - 2xvt + v^2 t^2) - c^2 \gamma^2 (t^2 - 2v.x\, t/c^2 + v^2 x^2/c^4)$$

$$= \gamma^2 x^2 (1 - v^2/c^2) - \gamma^2 t^2 c^2 (1 - v^2/c^2) = \gamma^2 (x^2 - c^2 t^2)(1 - v^2/c^2).$$

Since $\gamma = 1/\sqrt{1 - v^2/c^2}$ and, therefore, $\gamma^2 (1 - v^2/c^2) = 1$, we have

$$x'^2 - c^2 t'^2 = x^2 - c^2.t^2.$$

8. A rod 1 metre long is moving along its lenth with a velocity 0.6 c. Calculate its length as it appears to an observer (a) on the earth, (b) moving with the rod itself.

(a) Let L be the length of the rod as it appears to an observer on the earth (a stationary frame of reference).

Then, clearly, $L = L_0\sqrt{1 - v^2/c^2}$, where L_0 is the *proper length* of the rod, equal to 1 metre.

So that, $L = 1\sqrt{1 - (0.6c)^2/c^2} = \sqrt{1 - 0.36} = 0.8m$ or 80 *cm*.

the lenth of the rod thus appears to be 80 cm to the observer on the earth.

(*b*) Since observer, here, is moving with the rod itself, there is *no relative motion* between the two, *i.e.*, the rod is always at rest with respect to him. *The length of the rod, as measured by him, thus remains the same as its proper length, viz.*, 1 *metre.*

9. How fast should a rocket ship move relative to an observer in order that its length may appear to him to be 99% of its proper length?

Let v be the required speed of the rocket ship when its *length L* would appear to be 99% of its *proper length L_0* to an observer on the earth.

Now, $L = L_0\sqrt{1 - v^2/c^2}$, or, $\frac{L}{L_0} = (99/100) = \sqrt{1 - v^2/c^2}$

or, $1 - v^2/c^2 = (99/100)^2$, Or, $v^2/c^2 = 1 - (99/100)^2 = 1 - 0.98 = 0.02$.

Thus, $v^2 = 0.02\,c^2$. Or, $v = \sqrt{0.02}\,c = 0.1415 \times 3 \times 10^{10} = 4.245 \times 10^9$ *cm/sec*.

The speed of the rocket should, therefore, be 4.245×10^9 *cm/sec with respect to the observer.*

10. A beam of particles of half life 2 x 10^{-6} sec travels in the laboratory with 0.96 times the speed of light. How much distance does the beam travel before the flux falls to $\left(\frac{1}{2}\right)$ times the initial flux? ***(Agra)***

Here, clearly, the *proper half life* of the particles, $(\tau/2) = 2 \times 10^{-6}$ *sec.*). If, therefore, ($\tau'/2$ be their observed half life when travelling with velocity $v = 0.96\,c$, we have

$$\tau'/2 = \gamma\tau/2 = (2 \times 10^{-6})/\sqrt{1 - v^2/c^2} = (2 \times 10^{-6})/\sqrt{1 - (0.96c/c)^2} = 2 \times 10^{-6}/0.28 \text{ sec.}$$

The flux thus falls to half its initial value in $2 \times 10^{-6}/0.28$ *sec and the distance covered by the beam in this time* $= (0.96\,c \times \tau'/2) = 0.96 \times 3 \times 10^{10} \times 2 \times 10^{-6}/0.28 = 2.057 \times 10^5$ *cm.*

11. The life time of μ-mensons is 2.2 x 10^{-6} sec and their speed 0.998 c, so that they can cover only a distance of 0.998c × 2.2 × 10^{-6} or 658.6 metres in their entire life-time and yet they are found in profusion at sea level, i.e., at a depth of 10 km from the upper atmosphere where they are produced. How may this be explained on the basis of (i) Lorentz-Fitzgerald contraction, (ii) time dilation?

(*i*) Here, *the proper life-time of* μ-mesons (*i.e.*, their life-time, as measured in their own frame of reference) is $(\tau = 2.2 \times 10^{-6}$ *sec*) in which they can cover only a distance h_0, say, equal to 658.6 *m*.

Therefore, *the observed distance covered by them, as seen from the reference frame of the earth is given by*

$$h = \gamma h_0 = h_0/\sqrt{1 - v^2/c^2} = 658.6/\sqrt{1 - (0.998\,c/c)^2} = 658.6/0.064$$

$$= 10290\,m = 10.29\,km.$$

(*ii*) Here, again, *the proper life time of μ-mesons is* $(\tau = 2.2 \times 10^{-6}$ *sec.*)

If, therefore, *their observed life time*, as seen from the reference frame of the earth be τ' we have

$$\tau' = \gamma\tau' = 2.2 \times 10^{-6}/\sqrt{1 - v^2/c^2} = 2.2 \times 10^{-6}/\sqrt{1 - (0.998.c/c)^2}$$

$$= 2.2 \times 10^{-6}/0.064 = 34.38 \times 10^{-6} \text{ sec.}$$

i.e., about 16 times longer than their proper life time and in this much time, they can cover a distance $0.998 \times 3 \times 10^8 \times 34.38 \times 10^{-6} = 10290\,m = 10.29\,km$.

Thus, on the bais of both length (or distance) contraction and time dilation, the μ -mesons cover a distance 10.29 *km. That is why they are found in profusion at sea level,* 10 *km below the upper atmosphere where they are produced.*

12. Show from Lorentz transformation that two events simultaneous ($t_1 = t_2$) at different position $x_1 \neq x_2$ in a reference frame S are not, in general, simultaneous in another reference frame S′.

As we know, Inverse Lorentz transformation gives

$$t_1 = \gamma(t_1' + vx_1'/c^2) = t_2 = \gamma(t_2' + vx_2'/c^2) \quad \text{....}(i)$$

where $t1'$ and $t2'$ are the corresponding values to $t1$ and $t2$ in reference frame S', moving with velocity v relative to S. Obviously, for the two events to appear simultaneous in frame S', we must have $t'1 = t'2$. Let us see if this is so.

From relation (*i*), we have $t_1' = t_2' + \frac{v}{c^2}(x_2' - x_1')$ ....(*ii*)

Now, $x_2' = \gamma(x_2 - vt_2)$ and $x_1' = \gamma(x_1 - v.t)$. so that,

$(x_2' - x_1') = \gamma(x_2 - v.t_2) - \gamma(x_1 - v.t_1)$

Since $t_1 = t_2$, we *have* $(x_2' - x_1') = \gamma(x_2 - x_1)$

Substituting this value of $(x'_2 - x'_1)$ in relation *(ii)* above

we have $$t'_1 = t'_2 + \frac{\gamma.v}{c^2}(x_2 - x_1).$$

Thus, *if* $x_2 \neq x_1$, t_2' *can not be equal to* t', *and the two events, simultaneous in frame S, can not appear to be so in frame S' unless* $v/c = 0$, *in which case, there is no relativistic effect at all.*

13. In the laboratory, two particles are observed to travel in opposite directions with speed 2.8 × 10^10 cm/sec. Deduce the relative speed of the particles. Agra (Supp.)

Let the two particles be *A* and *B*, moving along the axis of *x* with velocities 2.8×10^{10} *cm/sec* and -28×10^{10} *cm/sec* respectively.

Let us assume particle *B* to be in a reference frame *S* at rest. Then the laboratory consitutes a frame *S'*, moving along the axis of *x* with velocity $v = -v_x = -(-2.8 \times 10^{10})$ or 2.8×10^{10} *cm/sec.*

We have thus to determine the velocity of particle *A* in the moving frame *S'* as it would be appear to an observer in the stationary frame *S*.

The Inverse Lorentz transformation gives $u_x = \dfrac{u_x' + v}{1 + \dfrac{v}{c^2}u_x'}$

where ux is the velocity of particle *A*, as seen from the moving frame S' to be determined.

Clearly, u_{x_0}' (the velocity of the partide *A* in frame *S'*) $= 2.8 \times 10^{10}$ *cm/sec; v*, (the velocity of the moving frame *S'*) $= -v_x = 2.8 \times 10^{10}$ *cm/sec* and $c = 3 \times 10^{10}$ *cm/sec*. We, therefore, have

$$u_x = \frac{2.8 \times 10^{10} + 2.8 \times 10^{10}}{1 + \dfrac{2.8 \times 10^{10}\ (2.8 \times 10^{10})}{(3 \times 10^{10})^2}} = \frac{5.6 \times 10^{10}}{1 + \left(\dfrac{2.8 \times 10^{10}}{3 \times 10^{10}}\right)^2} = \frac{5.6 \times 10^{10}}{1.8714} = 2.993 \times 10^{10} \text{ cm/sec.}$$

The relative velocity of the two particles is thus 2.993×10^{10} *cm/sec.*

14. A particle in a stationary frame *S* lies in the X-Y plane and has a velocity 0.8 c, inclined at an angle of 60° to the axis of x. What will be the velocity of the particle, as observed by a person in a frame *S'* moving relative to frame *S* with a velocity 0.4 cm/sec along the axis of x ?

Here, *velocity of the particle*, $u = 0.8c$ and *hence its component along the axis of x, i.e.*, $u_x = 0.8c \cos 60° = 0.8c \times 1/2 = 0.4\,c$ and *its component along the axis of y, i.e.*, $u_y = 0.8c \sin 60° = 0.8c \times \left(\sqrt{\frac{3}{2}}\right) = (0.4\sqrt{3}\,c)$. *And, the velocity of the moving frame S' relative to the stationary frame S, i.e.*, $v = 0.4\,c$ *along the axis of x.*

The values of the components u'_x and u'_y as observed by the person in frame *S'* are, in accordance with Lorentz transfomation, given by

$$u'_x = \frac{u_x - v}{1 - u_x.v/c^2} = \frac{0.4\,c - 0.4\,c}{1 - 0.4\,c\,(0.4\,c)/c^2} = 0,$$

and $$u'_y = \frac{u_y}{\gamma(1 - u_x.v/c^2)} = \frac{0.4\sqrt{3}\,c\sqrt{1 - v^2/c^2}}{1 - 0.4\,c(0.4c)/c^2} = \frac{0.4\sqrt{3}\,c\sqrt{1 - (0.4\,c/c)^2}}{1 - (0.4c/c)^2}$$

$$= \frac{0.4\sqrt{3}\sqrt{1 - 0.16}}{1 - 0.16}c = \frac{0.4\sqrt{3}\sqrt{0.84}}{0.84}c = \frac{0.4\sqrt{3}}{\sqrt{0.84}}c = 0.756\,c$$

∴ *resultant velocity of the particle* $= (\sqrt{u'^2_x + u'^2_y})$

Since $u'_x = 0$, *the resultant velocityis* $u'y = 0.756\,c$.

Thus , *the velocity of the particle, as observed by the person in frame* $S' = 0.756\,c$ *along the axis of y.*

15. Find the velocity at which the mass of a particle is double its rest mass. *(Punjab)*

Let the required velocity be *v* and let the mass of the particle at this veloicty be *m*, equal to twice its rest mass m_0, *i.e.*, $m = 2m_0$. Now, from the relation for the mass of a body at velocity *v*, *viz*, $m = m_0/\sqrt{1 - v^2/c^2}$, we have $m_0/m = m_0/2m_0 = 1/2 = \sqrt{1 - v^2/c^2}$. Putting $(v/c = \sin\theta)$, we have $\sqrt{1 - v^2/c^2} = \sqrt{1 - \sin^2\theta} = \cos\theta = \frac{1}{2}$, whence, $\theta = 60°$. And, therefore, $v/c = \sin\theta = 60° = \sqrt{3}/2$ or $v = (\sqrt{3}/2)\,c$ *i.e.*, $v = (1.732/2) \times 3 \times 10^{10} = 0.866 \times 3 \times 10^{10} = 2.598 \times 10^{10}$ *cm/sec.*

Thus, *the velocity at which a particle will have doube its rest mass is* 2.598×10^{10} *cm/sec.*

16. Given a proton for which $\beta = 0.995$, measured in the laboratory. What are its relativistic energy and momentum?

As we know, the rest mass of a proton, $m_0 = 1.67 \times 10^{-24}$ *gm*, and since $\beta = v/c = 0.995$ (given), we have $v = .995\,c$, where v is, of course, *velocity of the proton.* If, therefore, m be the mass of the proton at this velocity, we have,

$$m = m_0/\sqrt{1 - v^2/c^2} = 1.67 \times 10^{-24}/\sqrt{1 - (0.995)^2}$$

Hence, *relativistic energy of the proton, i.e.,*

$$E = m.c^2 = \frac{1.67 \times 10^{-24}}{\sqrt{1-(0.995)^2}} \times (3 \times 10)^2 = 1.503 \times 10^{-2}\,erg.$$

or, since 1.60×10^{-12} *ergs* $= 1$ *eV*, we have

$$\text{energy of the proton, } E = \frac{1.503 \times 10^{-2}}{1.60 \times 10^{-12}} = 9.39 \times 10^9\,eV = 9.39\,BeV.$$

Now, we have the relation $E^2 - p^2c^2 = m_0^2\,c^4$,

whence, $p.c = \sqrt{E^2 - m_0^2\,c^4}$ Or, $p = \sqrt{E^2 - m_0^2\,c^4/c}$.

$$\text{or,}\quad p = \frac{\sqrt{(1.503 \times 10^{-2})^2 - (1.67 \times 10^{-24})^2\,(3 \times 10^{10})^2}}{3 \times 10^{10}} = \frac{\sqrt{2.225 \times 10^{-4}}}{3 \times 10^{10}} = 4.983 \times 10^{-10}\ gm\text{—}cm/sec.$$

Thus, the *momentum of the proton* $= 4.903 \times 10^{-10}$ *gm - cm/sec*

17. (*a*) Calculate the speed of an electron which has kinetic energy equal to 1.02 *MeV*. ***(Agra)***

(*b*) Calculate the amount of energy when a neutron decays into a proton and an electron, (rest mass of neutron = 1.6747×10^{-24} gm, rest mass of proton = 1.6724×10^{-24} gm and rest mass of electron (or positron) = 9.11×10^{-28} gm) ***(Agra)***

(*a*) The mass of the electron will certainly increase by say, Δm, when it acquires a speed v and kinetic energy E_k=1.02 *MeV* ; so that $\Delta mc^2 = E_k$

whence, $\Delta m = E_k/c^2$.

Here $E_k = 1.02\,MeV = 1.02 \times 10^6\,eV = 1.02 \times 10^6 \times 1.60 \times 10^{-19} = 1.632 \times 10^{-13}$ *joules.*

$$\therefore\ \Delta m = \frac{1.632 \times 10^{-13}}{(3 \times 10^8)^2} = 18.13 \times 10^{-31}\ kg$$

Hence *mass of the electron, on acquiring velocity v i.e.,* $m = m_0 + \Delta m$

$= (9.11 \times 18.13)\,10^{-31} = 27.24 \times 10^{-31}$ *kg.* $[\because m_0 = 9.11 \times 10^{-31}$ *kg*

Now $m = m_0/\sqrt{1 - v^2/c^2}$, or putting $v/c = sin\theta$, $(m_0/m) = \sqrt{1 - sin^2\theta} = cos\theta$

whence, $cos\,\theta = (9.11 \times 10^{-31})/(27.24 \times 10^{-31}) = 0.3344$; $\therefore\ \theta = cos^{-1}(.3344) = 70^\circ\,28'$,

So that, $sin\theta = sin\,70^\circ\,28' = 0.9425$.

And therefore, $v/c = 0.9425$, whence $v = 0.9425\,c = 0.9425 \times 3 \times 10^8 = 2.83 \times 10^8$ *m/sec.*

The speed of the electron, with kinetic energy 1.02 *MeV is thus* 2.83×10^8 *m/Sec.*

(*b*) Clearly, *decrease in mass when a neutron decays into a proton and an electron* is given by

Δm = mass of neutron – (*mass of proton = mass of electron*)

$= 1.647 \times 10^{-24} - (1.6724 \times 10^{-24} + 9.11 \times 10^{-28})\ gm$

$= (1.674 - 1.673) \times 10^{-24} = 0.0014 \times 10^{-24}$ *gm*

$\therefore$ *energy released* $= \Delta\,m.c^2 = (0.0014 \times 10^{-24}) \times (3 \times 10^{10})^2 = 0.0126 \times 10^{-4}$ *ergs.*

$$= \frac{0.0126 \times 10^{-4}}{1.60 \times 10^{-12}} = 0.007875 \times 10^{8}$$

$$= 7.875 \times 10^{5}\ eV = 0.079\ MeV.$$

Exercise XVI

1. What is a *frame of reference* ? Illustrate with an example or two how the motion of a body in one frame of reference appears, in general, to be different when observed from another frame of reference.

2. Discuss the limitations of Newton's laws of motion and explain what is meant by an *inertial frame of reference*, mentioning clearly its characteristic properties and its importance. Can the earth be regarded as such a frame of reference?

3. Explain what you understand by *Galilian transformation* and *Galilian invariance*. Show that length (or distance) and acceleration are both invariant to Galilian transformation but not velocity.

4. Enunciate the laws of conservation of momentum and energy and show that they are both invariant to Galilian transformation.

5. In a head - on collision, a particle with an initial speed v_0 strikes a stationary particle of the same mass. Find the velocities v_1 and v_2 of the two particles respectively after the collision if (*a*) half the original kinetic energy is lost, (*b*) the final kinetic energy is 50% greater than the original kinetic energy

Ans: (*a*) $v_1 = v_2 = \frac{1}{2}v_0$, (*b*) $v_l = \frac{1}{2}v_0(1+\sqrt{2})$, $v_2 = \frac{1}{2}v_0(1-\sqrt{2})$.

6. A particle of mass m_1 moving with velocity v collides head-on with a particle of mass m_2 at rest. (*i*) if the collision be an elastic one, obtain the velocities of the two particles after the collision and the fraction of the total energy transferred to the second particle. How will this fraction change if $m_1 = m_2$? (*ii*) If the collision be an inelastic one, obtain the common velocity of the coalescing particles after the collision and the loss of kinetic energy suffered.

Ans : (*i*) $v_1 = \frac{m_1 v - m_2 v_2}{m_1}$, $v_2 = \frac{2v}{1 + m_1/m_2}$; $\frac{4 m_1 m_2}{(m_1+m_2)^2}$; the fraction will change to 1, the entire energy of m_1 being transferred to m_2.

(*ii*) *Common velocity* $= \frac{m_1 v}{m_1 + m_2}$; *loss of energy suffered* $= \frac{1}{2}m_1 v^2 \frac{m_2}{m_1+m_2}$

7. What is meant by (*a*) *a non-intertial frame of reference*, (*b*) *a fictitious force*? Why is this force so called ? Can a non-inertial frame of reference serve as an inertial frame of reference ? If so, under what condition ?

8. A person weighs 50 *kg* on the earth. Calculate the fictitious force acting on him and his effective weight in a lift moving (i) down, (ii) up, with an acceleration of $4 m/sec^2$.

Ans. (i) 200*N upwards*, 29.6 *kg*; (ii) 200 *N downwards*, 70.4*kg*.

9. What is *Coriolis force* ? Under what conditions does it come into play ? Calculate the centrigugal and Coriolis forces on a mass of 20 *gm* placed at a distance of 10 *cm* from the axis of a frame of reference rotating at the rate of 10 radians per second.

Ans. 2×10^4 *dynes* ; 4×10^4 *dynes*.

10. Describe the *Michelson-Morley experiment* and explain the physical significance of the negative result.

(*Bombay*, 1970; *B.Sc.* (*Hons*), *Delhi*, 1970; *M.Sc.*, *Kanpur*, 1968)

11. Write a brief accound of the stationary ether hypothesis, explaining clearly why the medium of ether was invented in the first place and discarded later.

12. If the effective length of each path in Michelson-Morley experiment be 6 *m*, velocity of the earth, 3×10^4 *m/sec.* and the wavelength of the monochromatic light used, 5000 *A.U.*, calculate the expected fringe shift.

Ans. 0.24*fringe-width*.

13. Explain the basic postulates of *Einstein's special theory of relativity*. Derive the Lorentz space -time transformation formulae.

[Punjab (Supp.); Agra]

14. Write down the transformation equations for x', y', z' and t' in a reference frame S', in terms of x, y, z and t in a reference frame S, if S' be moving with velocity v along the $+x$ direction with respect to S and if the origins of the two frames coincide at $t = 0$. Show that all Lorentz transformation equations reduce to Galilian ones if $v \ll c$.

15. Show that the relativistic equations to transform space and time to a system S from observations in a system S' are respectively $x = \gamma(x' + v.t')$ and $t = \gamma(t' + vx'/c^2)$, where S' is moving relative to S with velocity v along the positive direction of x.

16. What is meant by relativistic *length contraction* ? Show that it is a reciprocal effect and is the same when observed from one frame of reference in the other, if one of them be moving with velocity v relative to the other. What do you understand by the *proper length* of a body ?

17. The length of the side of a square, as measured by an observer in a stationary frame S is L. What is the area of the square as observed in a reference frame S', moving relative to S with velocity v along one of the sides of the square?

Ans. $L^2 \sqrt{1 - v^2/c^2}$

18. The *proper length* of a rod is 5 *metres*. What would its legth appear to be to an obnserver if it be moving relative to him in a direction parallel to its length with velocity (*i*) 0.8 *c*, (*ii*) *c* ? What would be its length as observed by an observer moving with the rod itself? **Ans.**(*i*) 3 *metres*, (*ii*) *zero* ; 5 *metres*.

19. How fast should a body move relative to an observer in a direction parallel to its length, so that its length appears to be contracted to 99% of its proper length?. **Ans** 4.245×10^9 *cm / sec.*

20. What do you understand by time dilation? What is proper interval of time? Briefly discuss one experiment in support of time dilation in special theory of relativity. (*Agra*)

21. State the basic postulates of the special theory of relativity and hence obtain Lorentz transformation. Discuss length contraction and time dilation.

22. The proper means life time of *pions* is 2.5×10^{-8} *sec*. At what velocity will they appear to have a mean life time 2.5×10^7 *sec* ?

23. State and deduce the law of addition of relativistic velocities and show that in no case can the resultant velocity of a material particle be greater than *c*.

24. (*a*) Two electrons move towards each other, the speed of each being 0.9 *c* in a Galilian frame of reference. What is their speed relative to each other?

(*b*) In the laboratory, electrons from two accelerators are projected with the same speed of 2×10^8 *m/sec* in opposite directions. What is the relative velocity of the two sets of electrons? **Ans.** (*a*) 0.994 *c* ; (*b*) 2.993×10^8 *m/sec.*

25. Obtain the resultant of two velociteis of 0.6 *c* and 0.8 *c* respectively inclined to one another at an angle of 60°. **Ans.** 0.992 *c at an angle of* 29° 15′ *with velocity* 0.6 *c*.

26. Deduce the formula for relativistic variation of mass with velocity. Briefly explain its significance. (*Punjab*)

Also show that the ultimate speed of a material particle is *c*.

27. (*a*) What speed must a body have for its mass to be (*i*) three times, (*ii*) five times its rest mass?

(*b*) Show that the rest mass of an electron (9.11×10^{-28} *gm*) is equivalent to 0.511 *MeV*.

(*c*) The electron beam in a television picture tube can move across the screen faster than the speed of light. Why is it not a violation of the special theory or relativity?

Ans. (*a*) (*i*) 2.828×10^8 *m/sec*; (*ii*) (*b*) 2.94×10^8 *m/sec.*

28. What is the meaning of mass-energy equivalence? Obtain Einstein's mass-energy relation. Show that 1 *amu* = 931 *MeV*. (*B.Sc. (Hons), Delhi*)

29. Write down the relativistic expression for kinetic energy of a body and show that for smaller speeds, it reduces to the classical expression. (*Meerut*)

30. A proton has velocity equal to 0.999 *c* in the laboratory. Find its energy and momentum as observed in a frame travelling in the same direction with velocity 0.990 *c* with respect the laboratory. **Ans.** 1.6 *BeV*; 1.3 *BeV/c*.

[**Hint.** First obtain the velocity of the proton in the moving frame by Lorentz transformation ; then obtain its energy from the relation $E = m_p c^2/\sqrt{1 - v^2/c^2}$ where m_p, the mass of the proton = 1.6725×10^{-24} *gm*, and finally, its momentum from the relation $p.c. = (E^2 - m_p c^4)^{\frac{1}{2}}$].

TABLES OF CONSTANTS

I–DENSITIES OF COMMON SUBSTANCES

Solids				**Liquids**		**Gases**	
Substance	***Density (gm. per c.c.)***	***Substance***	***Density (gm. per c.c.)***	***Substance***	***Density (gm. per c.c.)***	***Substance***	***Density (gm. per c.c.) at N.T.P.***
Aluminium	2.70	Iron (*Wrought*)	7.8–7.9	Alcohol	.79	Air	1.2928
Brass	8.4–8.7	Ivory	1.8–1.9	Benzene	.88	Carbon dioxide	1.977
Bronze	8.7–8.9	Lead	11.37	Carbon disulphide	1.27	Hydrogen	.084882
Constantan	8.8	Leather	.85–1	Ether	.72	Helium	1.7848
Common salt	2.15	Nickel	8.9	Glycerine	1.26	Neon	0.8999
Copper	8.93	Paper	.71–1.1	Mercury (0°*C*)	13.596	Nitrogen	1.25050
Cork	.22–.26	Plantinum	21.50	Methylated Spirit	.83	Oxygen	1.42895
German Silver	8.5–8.9	Porcelain	2.2–2.4	Milk	1.03	Steam (*Saturated at* 100°*C*)	.6059
Glass (*Crown*)	2.4–2.6	Quartz	2.66	Naphtha	.85		
Glass (*Flint*)	2.9-4.5	Sand	2.63	Oil, *Castor*	.97		
Glass (*Pyrex*)	2.25	Silver	10.5	,, *Olive*	.91–.93		
Gold	19.32	Sugar	1.59	Paraffin	.80		
Gun metal	8.0–8.4	Tar	1.02	Petrol	.68–.72		
Ice (0°*C*)	.9168	Tin	7.29	Tupentine	.87		
India-rubber	.91–.93	Wax (*Bee's*)	.95–.96	Vinegar	1.02		
Iron (*Cast*)	7.1–7.7	Zinc	7.1	Water (4°*C*)	1.00		
Iron (*Pure*)	7.87						

II-ELASTIC CONSTANTS

Substance	*Young's modulus (dynes/cm^2.)*	*Bulk modulus (dynes/cm^2.)*	*Rigidity modulus (dynes/cm^2.)*	*Poisso's Ratio τ*	*Breaking stress (for wires) (dynes/cm^2.)*
Aluminium	7.05×10^{11}	7.46×10^{11}	2.63×10^{11}	.310	$1.7–2.0\times10^{9}$
Brass	$9.7–10.2\times10^{11}$	10.65×10^{11}	3.5×10^{11}	.34–40	$3.1–3.9\times10^{9}$
Copper	$12.4–12.9\times10^{11}$	14.3×10^{11}	$3.9–4.0\times10^{11}$	.26	$4.0–4.6\times10^{9}$
Glass (*Flint*)	$5.0–6.8\times10^{11}$	$3.6–3.8\times10^{11}$	$2.0–2.5\times10^{11}$	.24	....
Gold	8.00×10^{11}	16.6×10^{11}	2.80×10^{11}	.495	2.6×10^{9}
Iron (C)	$10.0–13\text{-}0\times10^{11}$	9.6×10^{11}	$3.5–5.3\times10^{11}$	.27	...
Iron (W)	$19.0–20.0\times10^{11}$	14.3×10^{11}	$3.9–4.0\times10^{11}$	.27	4.6×10^{9}
Mercury	...	2.6×10^{11}	...	...	...
Silver	7.90×10^{11}	10.90×10^{11}	2.86×10^{11}	.369	2.9×10^{8}
Steel	$19.0–20.6\times10^{11}$	18.1×10^{11}	$7.9–8.9\times10^{11}$	$.25–.33\times10^{11}$	$18.6–23.3\times10^{9}$
Water	...	$.205\times10^{11}$	...	...	...

III—COEFFICIENTS OF RESTITUTION

1.	Clay	0.20
2.	Cork	0.70
3.	Iron (balls)	0.70
4.	Glass (Brass)	0.90
5.	" (Iron)	0.10
6.	Ivory	0.80

IV—COEFFICIENTS OF VISCOSITY

Liquids (20°*C*)

1.	Alcohol	0.0119
2.	Benzene	0.00649
3.	Carbon disulphide	0.00367
4.	Carbon tetrachloride	0.00969
5.	Chlorofonn	0.00564
6.	Ether	0.00234
7.	Glycerine	8.500
8.	Mercurv	0.016
9.	Turoentine	0.0149
10.	Water	0.01006

Gases (15°*C*)

1.	Air	.000181
2.	Carbon dioxide	.000144
3.	Hydrogen	.000089
4	Nitrogen	.000174
5.	Oxygen	.000198

Vapours

1.	Alcohol	(190°*C*)	.000109
2.	Benzene	(100°*C*)	.000094
3.	Ether	(100°*C*)	.000097
4.	Mercury	(300°*C*)	.000532
5.	Water	(300°*C*)	.000120

V—MOLECULAR ELEVATION OF BOILING POINTS OF SOLVENTS

(Elevation per 1 gm. -molecule of salt per 100 c.c. of solvent)

Solvent	*Normal Boiling Point (under 1 atm. pressure)*	*Molecular Elevation*
1. Acetone	56.0°C	17.2°C
2. Alcohol	78.4°C	11.9°C
3. Benzene	80.4°C	26.1°C
4. Chloroform	61.0°C	38.0°C
5. Ether	35.5°C	20.9°C
6. Water	100.0°C	5.15°C

VI—MOLECULAR DEPRESSION OF FREEZING POINTS OF SOLVENTS

(Depression per 1 gm.-molecule of salt per 100 c.c. of solvent)

Solvent	*Freezing Points*	*Molecular Depression*
1. Acetic Acid	17.0°C	38.5°C
2. Benzene	5.5°C	51.6°C
3. Phenol	40.0°C	78.6°C
4. Water	0°C	18.6°C

VII—SURFACE TENSIONS OF IMPORTANT LIQUIDS

(In contact with air)

Liquid	*Temp.*	*Surface-tension (dynes per cm.)*	*Angle of contact with glass & air*	*Liquid*	*Temp.*	*Surface-tension (dynes per cm.)*	*Angle of contact with glass & air*
Alcohol	20.0°C	22.0	0°.0	Paraffin Oil	25.0°C	26.4	26°.0
Benzene	17.5°C	29.2	0°.0	Turpentine	15.0°C	27.3	17°.0
Carbon disulphide	19.4°C	33.6	...	Water	15.0°C	74.0	8°.9
Chloroform	15.0°C	27.2	0°.0				
Ether	20.0°C	16.5	16 .0				
Mercury	20.0°C	46.5	127°.20				
Olive oil	20.0°C	32.0					

VIII—MOLECULAR CONSTANTS
(At N.T.P.)

1. *Number of molecule per c.c. of a gas* = 2.75×10^{19}
2. *Avogadro Number, or number of molecules per gm.-molecule of a gas* = 6.022×10^{24}
3. *Mass of Hydrogen atom* = 1.67×10^{-21}

Gas	*Molecular Diameter*	*Molecular Velocity (r.m.s.)*	*Mean Free Path*	*Collision Frequency*
Carbon dioxide	4.32×10^{-8} *cm.*	3.92×10^{4} *cm./sec.*	6.29×10^{-6} *cm.*	5.74×10^{9} *per sec.*
Hydrogen	2.40×10^{-8} *cm.*	18.39×10^{4} *cm./sec.*	18.3×10^{-6} *cm.*	9.255×10^{9} *per sec.*
Nitrogen	3.31×10^{-8} *cm.*	4.93×10^{4} *cm./sec.*	9.44×10^{-6} *cm.*	4.899×10^{9} *per sec.*
Oxygen	3.11×10^{-8} *cm.*	4.61×10^{4} *cm./sec.*	9.95×10^{-6} *cm.*	$5.00 \times 10^{9} n$

Mean molecular velocity = .921 r.m.s. velocity.

Collision Frequency = Mean molecular velocity/Mean free path.

Logarithmic and Antilogarithmic Tables

This page is intentionally left blank

Logarithms of Numbers

	0	1	2	3	4	5	6	7	8	9	Mean Differences. 1	2	3	4	5	6	7	8	9
10	00000	00432	00860	01284	01703						42	85	127	170	212	254	297	339	381
						02119	02531	02938	03342	03743	40	81	121	162	202	242	283	323	364
11	04139	04532	04922	05308	05690						37	77	116	154	193	232	270	309	348
						06070	06446	06819	07188	07555	37	74	111	148	185	222	259	296	333
12	07918	08279	08636	08991	09342						36	71	106	142	177	213	248	284	319
						09691	10037	10380	10721	11059	34	68	102	136	170	204	238	272	307
13	11394	11727	12057	12385	12710						33	66	98	131	164	197	229	262	295
						13033	13354	13672	13988	14301	32	63	95	126	158	190	221	253	284
14	14613	14922	15229	15534	15836						30	61	91	122	152	183	213	244	274
						16137	16435	16732	17026	17319	29	59	88	118	147	177	206	236	265
15	17609	17898	18184	18469	18752						28	57	85	114	142	171	199	228	256
						19033	19312	19590	19866	20140	28	55	83	110	138	165	193	221	248
16	20412	20683	20951	21219	21484						27	53	80	107	134	160	187	214	240
						21748	22011	22272	22531	22789	26	52	78	104	130	156	182	208	233
17	23045	23300	23553	23805	24055						26	50	76	101	126	151	176	201	227
						24304	24551	24797	25042	25285	25	49	73	98	122	147	171	196	220
18	25527	25768	26007	26245	26482						24	48	71	95	119	143	167	190	214
						26717	26951	27184	27416	27646	23	46	69	93	116	139	162	185	208
19	27875	28103	28330	28556	28780						23	45	68	90	113	135	158	180	203
						29003	29226	29447	29667	29885	22	44	66	88	110	132	154	176	198
20	30103	30320	30535	30750	30963	31175	31387	31597	31806	32015	21	43	64	85	106	127	148	170	190
21	32222	32428	32634	32838	33041	33244	33445	33646	33846	34044	20	41	61	81	101	121	141	162	182
22	34242	34439	34635	34830	35025	35218	35411	35603	35793	35984	20	39	58	77	97	116	135	154	174
23	36173	36361	36549	36736	36922	37107	37291	37475	37658	37840	19	37	56	74	93	111	130	148	167
24	38021	38202	38382	38561	38739	38917	39094	39270	39445	39620	18	35	53	71	89	106	124	142	159
25	39794	39967	40140	40312	40483	40654	40824	40993	41162	40330	17	34	51	68	85	102	119	136	153
26	41497	41664	41830	41996	42160	42325	42488	42651	42813	42975	16	33	49	66	82	98	115	131	148
27	43136	43297	43457	43616	43775	43933	44091	44248	44404	44560	16	32	47	63	79	95	111	126	142
28	44716	44871	45025	45179	45332	45484	45637	45788	45939	46090	15	30	46	61	76	91	107	122	137
29	46240	46389	46538	46687	46835	46982	47129	47276	47422	47567	15	29	44	59	74	88	103	118	132
30	47712	47857	48001	48144	48287	48430	48572	48714	48855	48996	14	29	43	57	72	86	100	114	129
31	49136	49276	49415	49554	49693	49831	49969	50106	50243	50379	14	28	41	55	69	83	97	110	124
32	50515	50650	50786	50920	51054	51188	51322	51455	51587	51720	13	27	40	54	67	80	94	107	121
33	51851	51983	52114	52244	52375	52504	52634	52763	52892	53020	13	26	39	52	65	78	91	104	117
34	53148	53275	53403	53529	53656	53782	53908	54033	54158	54283	13	25	38	50	63	76	88	101	113
35	54407	54531	54654	54777	54900	55023	55145	55267	55388	55509	12	24	37	49	61	73	85	98	110
36	55630	55751	55871	55991	56110	56229	56348	56467	56585	56703	12	24	36	48	60	71	83	95	107
37	56820	56937	57054	57171	57287	57403	57519	57634	57749	57864	12	23	35	46	58	70	81	93	104
38	57978	58092	58206	58320	58433	58546	58659	58771	58883	58995	11	23	34	45	57	68	79	90	102
39	59106	59218	59329	59439	59550	59660	59770	59879	59988	60097	11	22	33	44	55	66	77	88	99
40	60206	60314	60423	60531	60638	60746	60853	60959	61066	61172	11	21	32	43	54	64	75	86	97
41	61278	61384	61490	61595	61700	61805	61909	62014	62118	62221	10	21	31	42	53	63	74	84	95
42	62325	62428	62531	62634	62737	62839	62941	63043	63144	63246	10	20	31	41	51	61	71	82	92
43	63347	63448	63548	63649	63749	63849	63949	64048	64147	64246	10	20	30	40	50	60	70	80	90
44	64345	64444	64542	64640	64738	64836	64933	65031	65128	65225	10	20	29	39	49	59	68	78	88
45	65321	65418	65514	65610	65706	65801	65896	65992	66087	66181	10	19	29	38	48	57	67	76	86
46	66276	66370	66464	66558	66652	66745	66839	66932	67025	67117	9	19	28	37	47	56	65	74	84
47	67210	67302	67394	67486	67578	67669	67761	67852	67943	68034	9	18	27	36	46	55	64	73	82
48	68124	68215	68305	68395	68485	68574	68664	68753	68842	68931	9	18	27	36	45	53	63	72	81
49	69020	69108	69197	69285	69373	69461	69548	69636	69723	69810	9	18	26	35	44	53	62	70	79

Logarithms. (Contd.)

	0	1	2	3	4	5	6	7	8	9	Mean Differences. 1	2	3	4	5	6	7	8	9
50	69897	69984	70070	70157	70243	70329	70415	70501	70586	70672	9	17	26	34	43	52	60	69	77
51	70757	70842	70927	71012	71096	71181	71265	71349	71433	71517	8	17	25	34	42	50	59	67	76
52	71600	71684	71767	71850	71933	72016	72099	72181	72263	72346	8	17	25	33	42	50	58	66	75
53	72428	72509	72591	72673	72754	72835	72916	72997	73078	73159	8	16	24	32	41	49	57	65	73
54	73239	73320	73400	73480	73560	73640	73719	73799	73878	73957	8	16	24	32	40	48	56	64	72
55	74036	74115	74194	74273	74351	74429	74507	74586	74663	74741	8	16	23	31	39	47	55	63	70
56	74819	74896	74974	75051	75128	75205	75282	75358	75435	75511	8	15	23	31	39	46	54	62	69
57	75587	75664	75740	75815	75891	75967	76042	76118	76193	76268	8	15	23	30	38	45	53	60	68
58	76343	76418	76492	76567	76641	76716	76790	76864	76938	77012	7	15	22	30	37	44	52	59	67
59	77085	77159	77232	77305	77379	77452	77525	77597	77670	77743	7	15	22	29	37	44	51	58	66
60	77815	77887	77960	78032	78104	78176	78247	78319	78390	78462	7	14	22	29	36	43	50	58	65
61	78533	78604	78675	78746	78817	78888	78958	79029	79099	79169	7	14	21	28	36	43	50	57	64
62	79239	79309	79379	79449	79518	79588	79657	79727	79796	79565	7	14	21	28	35	41	48	55	62
63	79934	80003	80072	80140	80209	80277	80346	80414	80482	80550	7	14	20	27	34	41	48	54	61
64	80618	80686	80754	80821	80889	80956	81023	81090	81158	81224	7	13	20	27	34	40	47	54	60
65	81291	81358	81425	81491	81558	81624	81690	81757	81823	81889	7	13	20	26	33	40	46	53	59
66	81954	82020	82086	82151	82217	82282	82347	82413	82478	82543	7	13	20	26	33	39	46	52	59
67	82607	82672	82737	82802	82866	82930	82995	83059	83123	83187	6	13	19	26	32	38	45	51	58
68	83251	83315	83378	83442	83506	83569	83632	83696	83759	83822	6	13	19	25	32	38	44	50	57
69	83885	83948	84011	84073	84136	84198	84261	84323	84386	84448	6	12	19	25	31	37	43	50	56
70	84510	84572	84634	84696	84757	84819	84880	84942	85003	85065	6	12	19	25	31	37	43	50	56
71	85126	85187	85248	85309	85370	85431	85491	85552	85612	85673	6	12	18	24	31	37	43	49	55
72	85733	85794	85854	85914	85974	86034	86094	86153	86213	86273	6	12	18	24	30	36	42	48	54
73	86332	86392	86451	86510	86570	86629	86688	86747	86806	86864	6	12	18	24	30	35	41	47	53
74	86923	86982	87040	87099	87157	87216	87274	87332	87390	87448	6	12	17	23	29	35	41	46	52
75	87506	87564	87622	87679	87737	87795	87852	87910	87967	88024	6	12	17	23	29	35	41	46	52
76	88081	88138	88195	88252	88309	88366	88423	88480	88536	88593	6	11	17	23	29	34	40	46	51
77	88649	88705	88762	88818	88874	88930	88986	89042	89098	89154	6	11	17	22	28	34	39	45	50
78	89209	89265	89321	89376	89432	89487	89542	89597	89653	89708	6	11	17	22	28	33	39	44	50
79	89763	89818	89873	89927	89982	90037	90091	90146	90200	90255	6	11	17	22	28	33	39	44	50
80	90309	90363	90417	90472	90526	90580	90634	90687	90741	90795	5	11	16	22	27	32	38	43	49
81	90848	90902	90956	91009	91062	91116	91169	91222	91275	91328	5	11	16	21	27	32	37	42	48
82	91381	91434	91487	91540	91593	91645	91698	91751	91803	91855	5	11	16	21	27	32	37	42	48
83	91908	91960	92012	92064	92117	92169	92221	92273	92324	92376	5	10	16	21	26	31	36	42	47
84	92428	92480	92531	92583	92634	92686	92737	92788	92840	92891	5	10	15	20	26	31	36	41	46
85	92942	92993	93044	93095	93146	93197	93247	93298	93349	93399	5	10	15	20	26	31	36	41	46
86	93450	93500	93551	93601	93651	93702	93752	93802	93852	93902	5	10	15	20	25	30	35	40	45
87	93952	94002	94052	94101	94151	94201	94250	94300	94349	94399	5	10	15	20	25	30	35	40	45
88	94448	94498	94547	94596	94645	94694	94743	94792	94841	94890	5	10	15	20	25	29	34	39	44
89	94939	94988	95036	95085	95134	95182	95231	95279	95328	95376	5	10	15	19	24	29	34	39	44
90	95424	95472	95521	95569	95617	95665	95713	95761	95809	95856	5	10	14	19	24	29	34	38	43
91	95904	95952	95999	96047	96095	96142	96190	96237	96284	96332	5	9	14	19	24	28	33	38	42
92	96379	96426	96473	96520	96567	96614	96661	96708	96755	96802	5	9	14	19	24	28	33	38	42
93	96848	96895	96942	96988	97035	97081	97128	97174	97220	97267	5	9	14	18	23	28	32	38	42
94	97313	97359	97405	97451	97497	97543	97589	97635	97681	97727	5	9	14	18	23	28	32	37	42
95	97772	97818	97864	97909	97955	98000	98046	98091	98137	98182	5	9	14	18	23	27	32	36	41
96	98227	98272	98318	98363	98408	98453	98498	98543	98588	98632	5	9	14	18	23	27	32	36	41
97	98677	98722	98767	98811	98856	98900	98945	98989	99034	99078	4	9	13	18	22	27	31	36	40
98	99123	99167	99211	99255	99300	99344	99388	99432	99476	99520	4	9	13	18	22	26	31	35	40
99	99564	99607	99651	99695	99739	99782	99826	99870	99913	99957	4	9	13	17	22	26	31	35	39

Antilogrithms of Numbers

	0	1	2	3	4	5	6	7	8	9	Mean Differences.								
											1	2	3	4	5	6	7	8	9
.00	10000	10023	10046	10069	10093	10116	10139	10162	10186	10209	2	5	7	9	12	14	16	19	21
.01	10233	10257	10280	10304	10328	10351	10375	10399	10423	10447	2	5	7	10	12	14	17	19	21
.02	10471	10495	10520	10544	10568	10593	10617	10641	10666	10691	2	5	7	10	12	15	17	20	22
.03	10715	10740	10765	10789	10814	10839	10864	10889	10914	10940	3	5	8	10	13	15	18	20	23
.04	10965	10990	11015	11041	11066	11092	11117	11143	11169	11194	3	5	8	10	13	15	18	20	23
.05	11220	11246	11272	11298	11324	11350	11376	11402	11429	11455	3	5	8	11	13	16	18	21	24
.06	11482	11508	11535	11561	11588	11614	11641	11668	11695	11722	3	5	8	11	13	16	19	21	24
.07	11749	11776	11803	11830	11858	11885	11912	11940	11967	11995	3	5	8	11	14	16	19	22	25
.08	12023	12050	12078	12106	12134	12162	12190	12218	12246	12274	3	6	8	11	14	17	20	22	25
.09	12303	12331	12359	12388	12417	12445	12474	12503	12531	12560	3	6	9	11	14	17	20	23	26
.10	12589	12618	12647	12677	12706	12735	12764	12794	12823	12853	3	6	9	12	15	18	21	24	26
.11	12882	12912	12942	12972	13002	13032	13062	13092	13122	13152	3	6	9	12	15	18	21	24	27
.12	13183	13213	13243	13274	13305	13335	13366	13397	13428	13459	3	6	9	12	15	18	21	25	28
.13	13490	13521	13552	13583	13614	13646	13677	13709	13740	13772	3	6	9	13	16	19	22	25	28
.14	13804	13836	13868	13900	13932	13964	13996	14028	14060	14093	3	6	10	13	16	19	22	26	29
.15	14125	14158	14191	14223	14256	14289	14322	14355	14388	14421	3	7	10	13	16	20	23	26	30
.16	14454	14488	14521	14555	14588	14622	14655	14689	14723	14757	3	7	10	13	17	20	24	27	30
.17	14791	14825	14859	14894	14928	14962	14997	15031	15066	15101	3	7	10	14	17	21	24	28	31
.18	15136	15171	15205	15241	15276	15311	15346	15382	15417	15453	4	7	11	14	18	21	25	28	32
.19	15488	15524	15560	15596	15631	15668	15704	15740	15776	15812	4	7	11	14	18	22	25	29	32
.20	15849	15885	15922	15959	15996	16032	16069	16106	16144	16181	4	7	11	15	18	22	26	30	33
.21	16218	16255	16293	16331	16368	16406	16444	16482	16520	16558	4	8	11	15	19	23	26	30	34
.22	16596	16634	16672	16711	16749	16788	16827	16866	16904	16943	4	8	12	15	19	23	27	31	35
.23	16982	17022	17061	17100	17140	17179	17219	17258	17298	17338	4	8	12	16	20	24	28	32	36
.24	17378	17418	17458	17498	17539	17579	17620	17660	17701	17742	4	8	12	16	20	24	28	32	36
.25	17783	17824	17865	17906	17947	17989	18030	18072	18113	18155	4	8	12	17	21	25	29	33	37
.26	18197	18239	18281	18323	18365	18408	18450	18493	18535	18578	4	8	13	17	21	25	30	34	38
.27	18621	18664	18707	18750	18793	18836	18880	18923	18967	19011	4	9	13	17	22	26	30	35	39
.28	19055	19099	19143	19187	19231	19275	19320	19364	19409	19454	4	9	13	18	22	26	31	35	40
.29	19498	19543	19588	19634	19679	19724	19770	19815	19861	19907	5	9	14	18	23	27	32	36	41
.30	19953	19999	20045	20091	20137	20184	20230	20277	20324	20370	5	9	14	19	23	28	32	37	42
.31	20417	20464	20512	20559	20606	20654	20701	20749	20797	20845	5	10	14	19	24	29	33	38	43
.32	20893	20941	20989	21038	21086	21135	21184	21232	21281	21330	5	10	15	19	24	29	34	39	44
.33	21380	21429	21478	21528	21577	21627	21677	21727	21777	21827	5	10	15	20	25	30	35	40	45
.34	21878	21928	21979	22029	22080	22131	22182	22233	22284	22336	5	10	15	20	25	31	36	41	46
.35	22387	22439	22491	22542	22594	22646	22699	22751	22803	22856	5	10	16	21	26	31	37	42	47
.36	22909	22961	23014	23067	23121	23174	23227	23281	23336	23388	5	11	16	21	27	32	37	43	48
.37	23442	23496	23550	23605	23659	23714	23768	23823	23878	23933	5	11	16	22	27	33	38	44	49
.38	23988	24044	24099	24155	24210	24266	24322	24378	24434	24491	6	11	17	22	28	34	39	45	50
.39	24547	24604	24660	24717	24774	24831	24889	24946	25003	25061	6	11	17	23	29	34	40	46	51
.40	25119	25177	25236	25293	25351	25410	25468	25527	25586	25645	6	12	18	23	29	35	41	47	53
.41	25704	25763	25823	25882	25942	26002	26062	26122	26182	26242	6	12	18	24	30	36	42	48	54
.42	26303	26363	26424	26485	26546	26607	26669	26730	26792	26853	6	12	18	24	31	37	43	49	55
.43	26915	26977	27040	27102	27164	27227	27290	27353	27416	27479	6	13	19	25	31	38	44	50	56
.44	27542	27606	27669	27733	27797	27861	27925	27990	28054	28119	6	13	19	26	32	39	45	51	58
.45	28184	28249	28314	28379	28445	28510	28576	28642	28708	28774	7	13	20	26	33	39	46	52	59
.46	28840	28907	28973	29040	29107	29174	29242	29309	29376	29444	7	13	20	27	34	40	47	54	60
.47	29512	29580	29648	29717	29785	29854	29923	29992	30061	30130	7	14	21	28	34	41	48	55	62
.48	30200	30269	30339	30409	30479	30549	30620	30690	30761	30832	7	14	21	28	35	42	49	56	63
.49	30903	30974	31046	31117	31189	31261	31333	31405	31477	31550	7	14	22	29	36	43	50	58	65

Antilogrithms. (Contd.)

	0	1	2	3	4	5	6	7	8	9	Mean Differences.								
											1	2	3	4	5	6	7	8	9
.50	31623	31696	31769	31842	31916	31989	32063	32137	32211	32285	7	15	22	29	37	44	52	59	66
.51	32359	32434	32509	32584	32659	32735	32809	32885	32961	33037	8	15	23	30	38	45	53	60	68
.52	33113	33189	33266	33343	33420	33497	33574	33651	33729	33806	8	15	23	31	39	46	54	62	69
.53	33884	33963	34041	34119	34198	34277	34356	34435	34514	34594	8	16	24	32	40	47	55	63	71
.54	34674	34754	34834	34914	34995	35075	35156	35237	35318	35400	8	16	24	32	40	48	56	65	73
.55	35481	35563	35645	35727	35810	35892	35975	36058	36141	36224	8	16	25	33	41	50	58	66	74
.56	36308	36392	36475	36559	36644	36728	36813	36898	36983	37068	8	17	25	34	42	51	59	68	76
.57	37154	37239	37325	37411	37497	37584	37670	37757	37844	37931	9	17	26	35	43	52	61	69	78
.58	38019	38107	38194	38282	38371	38459	38548	38637	38726	38815	9	18	27	35	44	53	62	71	80
.59	38905	38994	39084	39174	39264	39355	39446	39537	39628	39719	9	18	27	36	45	54	63	72	82
.60	39811	39902	39994	40087	40179	40272	40365	40458	40551	40644	9	19	28	37	46	56	65	74	83
.61	40738	40832	40926	41020	41115	41210	41305	41400	41495	41591	9	19	28	38	47	57	66	76	85
.62	41687	41783	41879	41976	42073	42170	42267	42364	42462	42560	10	19	29	39	49	58	68	78	87
.63	42658	42756	42855	42954	43053	43152	43251	43351	43451	43551	10	20	30	40	50	60	70	80	89
.64	43652	43752	43853	43954	44055	44157	44259	44361	44463	44566	10	20	30	41	51	61	71	81	91
.65	44668	44771	44875	44978	45082	45186	45290	45394	45499	45604	10	21	31	42	52	62	73	83	94
.66	45709	45814	45920	46026	46132	46238	46345	46452	46559	46666	11	21	32	43	53	64	75	85	96
.67	46774	46881	46989	47098	47206	47315	47424	47534	47643	47753	11	22	33	44	54	65	76	87	98
.68	47863	47973	48084	48195	48306	48417	48529	48641	48753	48865	11	22	33	45	56	67	78	89	100
.69	48978	49091	49204	49317	49431	49545	49659	49774	49888	50003	11	23	34	46	57	68	80	91	103
.70	50119	50234	50350	50466	50582	50699	50816	50933	51050	51168	12	23	35	47	58	70	82	93	105
.71	51286	51404	51523	51642	51761	51880	52000	52119	52240	52360	12	24	36	48	60	72	84	96	108
.72	52481	52602	52723	52845	52966	53088	53211	53333	53456	53580	12	24	37	49	61	73	85	98	110
.73	53703	53827	53951	54075	54200	54325	54450	54576	54702	54828	13	25	38	50	63	75	88	100	113
.74	54954	55081	55208	55336	55463	55590	55719	55847	55976	56105	13	26	38	51	64	77	90	102	115
.75	56234	56364	56494	56624	56754	56885	57016	57148	57280	57412	13	26	39	52	66	79	92	105	118
.76	57544	57677	57810	57943	58076	58210	58345	58479	58614	58749	13	27	40	54	67	80	94	107	121
.77	58884	59020	59156	59293	59429	59566	59704	59841	59979	60117	14	27	41	55	69	82	96	110	123
.78	60256	60395	60534	60674	60814	60954	61094	61235	61376	61518	14	28	42	56	70	84	98	112	126
.79	61659	61802	61944	62087	62230	62373	62517	62661	62806	62951	14	29	43	58	72	86	101	115	130
.80	63096	63241	63387	63533	63680	63826	63973	64121	64269	64417	15	29	44	59	74	88	103	118	132
.81	64565	64714	64863	65013	65163	65313	65464	65615	65766	65917	15	30	45	60	75	90	105	120	135
.82	66069	66222	66374	66527	66681	66834	66988	67143	67298	67453	15	31	46	62	77	92	108	123	139
.83	67608	67764	67920	68077	68234	68391	68549	68707	68865	69024	16	32	47	63	79	95	110	126	142
.84	69183	69343	69503	69663	69823	69984	70146	70307	70469	70632	16	32	48	64	81	97	113	129	145
.85	70795	70958	71121	71285	71450	71614	71779	71945	72111	72277	17	33	50	66	83	99	116	132	149
.86	72444	72611	72778	72946	73114	73282	73451	73621	73790	73961	17	34	51	68	85	101	118	135	152
.87	74131	74302	74473	74645	74817	74989	75162	75336	75509	75683	17	35	52	69	87	104	121	138	156
.88	75858	76033	76208	76384	76560	76736	76913	77090	77268	77446	18	35	53	71	89	107	125	142	159
.89	77625	77804	77983	78163	78343	78524	78705	78886	79068	79250	18	36	54	72	91	109	127	145	163
.90	79433	79616	79799	79983	80168	80353	80538	80724	80910	81096	19	37	56	74	93	111	130	148	167
.91	81283	81470	81658	81846	82035	82224	82414	82604	82794	82985	19	38	57	76	95	113	132	151	170
.92	83176	83368	83560	83753	83946	84140	84333	84528	84723	84918	19	39	58	78	97	116	136	155	175
.93	85114	85310	85507	85704	85901	86099	86298	86497	86696	86896	20	40	60	79	99	119	139	158	178
.94	87096	87297	87498	87700	87902	88105	88308	88512	88716	88920	20	41	61	81	102	122	142	162	183
.95	89125	89331	89536	89743	89950	90157	90365	90573	90782	90991	21	42	62	83	104	125	146	166	187
.96	91201	91411	91622	91833	92045	92257	92470	92683	92897	32111	21	42	64	85	106	127	149	170	191
.97	93325	93541	93756	93972	94189	94406	94624	94842	95060	95280	22	43	65	87	109	130	152	174	195
.98	95499	95719	95940	96161	96383	96605	96828	97051	97275	97499	22	44	67	89	111	133	155	178	200
.99	97724	97949	98175	98401	98628	98855	99083	99312	99541	99770	23	46	68	91	114	137	160	182	205